Biology

EIGHTH EDITION
Volume 1

Eldra P. Solomon
University of South Florida

Linda R. Berg
St. Petersburg College

Diana W. Martin
Rutgers University

THOMSON
BROOKS/COLE

Australia • Canada • Mexico • Singapore • Spain • United Kingdom • United States

THOMSON
★
BROOKS/COLE

Biology, Eighth Edition
Eldra P. Solomon, Linda R. Berg, and Diana W. Martin

Publisher: Peter Adams
Development Editor: Suzannah Alexander
Assistant Editor: Lauren Oliveira
Editorial Assistant: Kate Franco
Technology Project Manager: Keli Amann
Marketing Manager: Kara Kindstrom
Marketing Assistant: Catie Ronquillo
Marketing Communications Manager: Bryan Vann
Project Manager, Editorial Production: Cheryll Linthicum,
 Jennifer Risden
Creative Director: Rob Hugel
Art Director: Lee Friedman

Print Buyer: Judy Inouye
Permissions Editor: Bob Kauser
Production Service: Jamie Armstrong, Newgen–Austin
Text Designer: John Walker
Photo Researcher: Kathleen Olson
Copy Editor: Cynthia Lindlof
Illustrator: Precision Graphics, Newgen–Austin
Cover Designer: Robin Terra
Cover Image: Gail Shumway/Taxi/Getty Images
Cover Printer: Courier/Kendallville
Compositor: Newgen–Austin
Printer: Courier/Kendallville

Thomson Higher Education
10 Davis Drive
Belmont, CA 94002-3098
USA

Library of Congress Control Number: 2006924995

Student Edition: ISBN 13: 978-0-495-10705-7

Student Edition: ISBN 10: 0-495-10705-0

Volume 1: ISBN 13: 978-0-495-39266-8

Volume 1: ISBN 10: 0-495-39266-9

For more information about our products, contact us at:
Thomson Learning Academic Resource Center
1-800-423-0563
For permission to use material from this text or product, submit a request online at **http://www.thomsonrights.com**.
Any additional questions about permissions can be submitted by e-mail to **thomsonrights@thomson.com**.

DEDICATION

To our families, friends, and colleagues who gave freely of their love, support, knowledge, and time as we prepared this eighth edition of *Biology*

ESPECIALLY TO
Rabbi Theodore and Freda Brod
Alan and Jennifer
Chuck and Margaret

ABOUT THE AUTHORS

ELDRA P. SOLOMON has written several leading college-level textbooks in biology and in human anatomy and physiology. Her books have been translated into more than 10 languages. Dr. Solomon earned an M.S. from the University of Florida and an M.A. and Ph.D. from the University of South Florida. Dr. Solomon taught biology and nursing students for more than 20 years. She is adjunct professor and member of the Graduate Faculty of the University of South Florida.

In addition to being a biologist and science author, Dr. Solomon is a biopsychologist with a special interest in the neurophysiology of traumatic experience. Her research has focused on the relationships among stress, emotions, and health and on post-traumatic stress disorder.

Dr. Solomon has presented her work in plenary sessions and scientific meetings at many national and international conferences. She has been profiled more than 20 times in leading publications, including *Who's Who in America, Who's Who in Science and Engineering, Who's Who in Medicine and Healthcare, Who's Who in American Education, Who's Who of American Women,* and *Who's Who in the World.*

LINDA R. BERG is an award-winning teacher and textbook author. She received a B.S. in science education, an M.S. in botany, and a Ph.D. in plant physiology from the University of Maryland. Her research focused on the evolutionary implications of steroid biosynthetic pathways in various organisms.

Dr. Berg taught at the University of Maryland at College Park for 17 years and at St. Petersburg College in Florida for 8 years. During her career, she taught introductory courses in biology, botany, and environmental science to thousands of students. At the University of Maryland, she received numerous teaching and service awards. Dr. Berg is also the recipient of many national and regional awards, including the National Science Teachers Association Award for Innovations in College Science Teaching, the Nation's Capital Area Disabled Student Services Award, and the Washington Academy of Sciences Award in University Science Teaching.

During her career as a professional science writer, Dr. Berg has authored or co-authored several leading college science textbooks. Her writing reflects her teaching style and love of science.

DIANA W. MARTIN is the Director of General Biology, Division of Life Sciences, at Rutgers University, New Brunswick Campus. She received an M.S. at Florida State University, where she studied the chromosomes of related plant species to understand their evolutionary relationships. She earned a Ph.D. at the University of Texas at Austin, where she studied the genetics of the fruit fly, *Drosophila melanogaster,* and then conducted postdoctoral research at Princeton University. She has taught general biology and other courses at Rutgers for more than 20 years and has been involved in writing textbooks since 1988. She is immensely grateful that her decision to study biology in college has led to a career that allows her many ways to share her excitement about all aspects of biology.

Brief Contents

Contents

Contents   **xiii**

Preface

This eighth edition of Solomon, Berg, Martin *Biology* continues to convey our vision of the dynamic science of biology and how it affects every aspect of our lives, from our health and behavior to the challenging environmental issues that confront us. Recent discoveries in the biological sciences have increased our understanding of both the unity and diversity of life's processes and adaptations. With this understanding, we have become more aware of our interdependence with the vast diversity of organisms with which we share planet Earth.

BIOLOGY IS A BOOK FOR STUDENTS

The study of biology can be an exciting journey of discovery for beginning biology students. To that end, we seek to help students better appreciate Earth's diverse organisms, their remarkable adaptations to the environment, and their evolutionary and ecological relationships. We also want students to appreciate the workings of science and the contributions of scientists whose discoveries not only expand our knowledge of biology but also help shape and protect the future of our planet. *Biology* provides insight into what science is, how scientists work, what the roles are of the many scientists who have contributed to our current understanding of biology, and how scientific knowledge affects daily life.

Since the first edition of *Biology,* we have tried to present the principles of biology in an integrated way that is accurate, interesting, and conceptually accessible to students. In this eighth edition of *Biology,* we continue this tradition. We also continue to present biology in an inquiry-based framework. Many professors interpret inquiry as a learning method that takes place in the laboratory as students perform experiments. Laboratory research is certainly an integral part of inquiry-based learning. But inquiry is also a way of learning in which the student actively pursues some line of knowledge outside the laboratory. In *Biology,* we have always presented the history of scientific advances, including scientific debates, to help students understand that science is a process (that is, a field of investigative inquiry) as well as a body of knowledge (the product of inquiry). **NEW** In *Biology,* Eighth Edition, we make a concerted effort to further integrate inquiry-based learning into the textbook with the inclusion of two new features: *Key Experiment* figures and *Analyzing Data* questions. *Key Experiment* figures encourage students to evaluate investigative approaches that real scientists have taken.

Analyzing Data **questions** require students to actively interpret experimental data presented in the text.

Throughout the text, we stimulate interest by relating concepts to experiences within the student's frame of reference. By helping students make such connections, we facilitate their mastery of general concepts. We hope the combined effect of an engaging writing style and interesting features will fascinate students and encourage them to continue their study of biology.

THE SOLOMON/BERG/MARTIN LEARNING SYSTEM

In the eighth edition we have refined our highly successful Learning System, which concentrates on learning objectives and outcomes. Mastering biology is challenging, particularly because the subject of biology is filled with so many facts that must be integrated into the framework of general biological principles. To help students, we provide Learning Objectives both for the course and for each major section of every chapter. At the end of each section, we provide Review questions based on the learning objectives so students can assess their mastery of the material presented in the section.

Throughout the book, students are directed to Thomson-NOW, a powerful online diagnostic tool that helps students assess their study needs and master the chapter objectives. After taking a pretest on ThomsonNOW, students receive feedback based on their answers as well as a Personalized Study plan with links to animations and other resources keyed to their specific learning needs. Selected illustrations in the text are also keyed to *Animated* figures in ThomsonNOW.

Course Learning Objectives

At the end of a successful study of introductory biology, the student can demonstrate mastery of the subject by responding accurately to the following Course Learning Objectives, or Outcomes:

- Design an experiment to test a given hypothesis, using the procedure and terminology of the scientific method.

- Cite the cell theory, and relate structure to function in both prokaryotic and eukaryotic cells.

- Describe the theory of evolution, explain why it is the principal unifying concept in biology, and discuss natural selection as the primary agent of evolutionary change.

- Explain the role of genetic information in all species, and discuss applications of genetics that affect society.

- Describe several mechanisms by which cells and organisms transfer information, including the use of nucleic acids, chemical signals (such as hormones and pheromones), electrical signals (for example, neural transmission), signal transduction, sounds, and visual displays.

- Argue for or against the classification of organisms in three domains and six kingdoms, characterizing each of these clades; based on your knowledge of genetics and evolution, give specific examples of the unity and diversity of these organisms.

- Compare the structural adaptations, life processes, and life cycles of a prokaryote, protist, fungus, plant, and animal.

- Define *homeostasis,* and give examples of regulatory mechanisms, including feedback systems.

- Trace the flow of matter and energy through a photosynthetic cell and a nonphotosynthetic cell, and through the biosphere, comparing the roles of producers, consumers, and decomposers.

- Describe the study of ecology at the levels of an individual organism, a population, a community, and an ecosystem.

Pedagogical Features

We use numerous learning strategies to increase the student's success:

- **NEW** A list of *Key Concepts* at the beginning of each chapter provides a chapter overview.

- *Learning Objectives* at the beginning of each major section in the chapter indicate, in behavioral terms, what the student must do to demonstrate mastery of the material in that section.

- Each major section of the chapter is followed by a series of *Review* **questions** that assess comprehension by asking the student to describe, explain, compare, contrast, or illustrate important concepts. The Review questions are based on the section Learning Objectives.

- *Concept Statement Subheads* introduce sections, previewing and summarizing the key idea or ideas to be discussed in that section.

- *Sequence Summaries* within the text simplify and summarize information presented in paragraph form. For example, paragraphs describing blood circulation through the body or the steps by which cells take in certain materials are followed by a Sequence Summary listing the structures or steps.

- *Focus On* **boxes** explore issues of special relevance to students, such as the effects of smoking or alcohol abuse. These boxes also provide a forum for discussing certain topics of current interest in more detail, such as the smallest ancient humans (*Homo floresiensis*), worldwide declining amphibian populations, and Alzheimer's disease.

- Numerous *Tables,* many illustrated, help the student organize and summarize material presented in the text.

- A *Summary with Key Terms* at the end of each chapter is organized around the chapter Learning Objectives. This summary provides a review of the material, and because selected key terms are boldface in the summary, students are provided the opportunity to study vocabulary words within the context of related concepts.

- End-of-chapter questions provide students with the opportunity to evaluate their understanding of the material in the chapter. *Test Your Understanding* consists of multiple-choice questions, some of which are based on the recall of important terms, whereas others challenge students to integrate their knowledge. Answers to the Test Your Understanding questions are provided in Appendix E. A series of thought-provoking *Critical Thinking* questions encourages the student to apply the concepts just learned to new situations or to make connections among important concepts. **NEW** Every chapter has one or more *Evolution Link* **questions** in the Critical Thinking section. **NEW** Many chapters contain one or more *Analyzing Data* **questions** based on data presented in the chapter.

- The *Glossary* at the end of the book, the most comprehensive glossary found in any biology text, provides precise definitions of terms. The Glossary is especially useful because it is extensively cross-referenced and includes pronunciations. The vertical blue bar along the margin facilitates rapid access to the Glossary. The companion website also includes glossary flash cards with audio pronunciations.

- **NEW** An updated and expanded *art program* brings to life, reinforces, and expands concepts discussed in the text. This edition includes *Key Experiment* **figures,** which emphasize the scientific process in both classic and modern research; examples include Figures 9-7, 17-2, 36-7, and 50-8. Also new to this edition are *Key Points* stated in process diagrams of complex topics; examples include Figures 28-19 and 54-2. Many of these figures have numbered parts that show sequences of events in biological processes or life cycles. Numerous photographs, both alone and combined with line art, help students grasp concepts by connecting the "real" to the "ideal." The line art uses devices such as *orientation icons* to help the student put the detailed figures into the broad context. We use symbols and colors consistently throughout the book to help students connect concepts. For example, the same four colors and shapes are used throughout the book to identify guanine, cytosine, adenine, and thymine.

WHAT'S NEW: AN OVERVIEW OF *BIOLOGY*, EIGHTH EDITION

Three themes are interwoven throughout *Biology:* the evolution of life, the transmission of biological information, and the flow of energy through living systems. As we introduce the concepts of modern biology, we explain how these themes are connected and how life depends on them.

Educators present the major topics of an introductory biology course in a variety of orders. For this reason, we carefully designed the eight parts of this book so that they do not depend heavily on preceding chapters and parts. This flexible organization means that an instructor can present the 56 chapters in any number of sequences with pedagogical success. Chapter 1, which introduces the student to the major principles of biology, provides a good springboard for future discussions, whether the professor prefers a "top-down" or "bottom-up" approach.

In this edition, as in previous editions, we examined every line of every chapter for accuracy and currency, and we made a serious attempt to update every topic and verify all new material. The following brief survey provides a general overview of the organization of *Biology* and some changes made to the eighth edition.

Part 1: The Organization of Life

The six chapters that make up Part 1 provide basic background knowledge. We begin Chapter 1 with a new discussion of research on heart repair and then introduce the main themes of the book—evolution, energy transfer, and information transfer. Chapter 1 then examines several fundamental concepts in biology and the nature of the scientific process, including a discussion of systems biology. Chapters 2 and 3, which focus on the molecular level of organization, establish the foundations in chemistry necessary for understanding biological processes. Chapters 4 and 5 focus on the cell level of organization. Chapter 4 contains additional information on microtubules, microfilaments, and motor proteins; Chapter 5 has expanded coverage of transport proteins and a new section on uniporters, symporters, and antiporters.

NEW In the eighth edition we have added a new chapter, Chapter 6, on cell communication, because recent studies of cell signaling, including receptor function and signal transduction, are providing new understanding of many life processes, particularly at the cell level.

Part 2: Energy Transfer through Living Systems

Because all living cells need energy for life processes, the flow of energy through living systems—that is, capturing energy and converting it to usable forms—is a basic theme of *Biology*. Chapter 7 examines how cells capture, transfer, store, and use energy. Chapters 8 and 9 discuss the metabolic adaptations by which organisms obtain and use energy through cellular respiration and photosynthesis. New to Chapter 9 is a summary section on the importance of photosynthesis to plants and other organisms.

Part 3: The Continuity of Life: Genetics

We have updated and expanded the eight chapters of Part 3 for the eighth edition. We begin this unit by discussing mitosis and meiosis in Chapter 10, which includes new sections on binary fission and prometaphase in mitosis. Chapter 11, which considers Mendelian genetics and related patterns of inheritance, has a new section on recognition of Mendel's work, including discussion of the chromosome theory of inheritance. We then turn our attention to the structure and replication of DNA in Chapter 12, including a new section on proofreading and repair of errors in DNA. Chapter 13 has a discussion of RNA and protein synthesis, including a new section and summary table on the different kinds of eukaryotic RNA. Gene regulation is discussed in Chapter 14, which includes new paragraphs on genomic imprinting and epigenetic inheritance. In Chapter 15, we focus on DNA technology and genomics, including new material on DNA analysis, genomics, expressed sequence tags, RNAi, and the importance of GM crops to U.S. agriculture. These chapters build the necessary foundation for exploring the human genome in Chapter 16, which has expanded discussion of abnormalities in chromosome structure and on X-linked genes affecting intelligence. In Chapter 17, we introduce the role of genes in development, emphasizing studies on specific model organisms that have led to spectacular advances in this field; changes include new material on RNAi as a powerful tool in developmental genetics and discussion of cancer as a stem cell disease.

NEW The art program in the genetics section includes many new pieces, such as Figures 10-8 (cohesins), 10-11 (binary fission), 12-14 (DNA repair), 13-11 (linkage of amino acid to its specific RNA), 13-18 (photo of plants with altered miRNA), 14-1 (lean pig with *IGF2* mutation), 14-13 (protein degradation by ubiquitin-proteasomes), 15-9 (Southern blotting technique), 15-15 (DNA fingerprinting), 16-5 (common abnormalities in chromosome structure), 16-10 (gene therapy in mouse bone marrow cells), and 17-6 (six model organisms).

Part 4: The Continuity of Life: Evolution

Although we explore evolution as the cornerstone of biology throughout the book, Part 4 delves into the subject in depth. We provide the history behind the discovery of the theory of evolution, the mechanisms by which it occurs, and the methods by which it is studied and tested. Chapter 18 introduces the Darwinian concept of evolution and presents several kinds of evidence that support the theory of evolution. In Chapter 19, we examine evolution at the population level. Chapter 20 describes the evolution of new species and discusses aspects of macroevolution. Chapter 21 summarizes the evolutionary history of life on Earth. In Chapter 22, we recount the evolution of the primates, including humans. Many topics and examples have been added to the eighth edition, such as new material on developmental biology as evidence for evolution in Galápagos finches, the founder effect and genetic drift with respect to Finns and Icelanders, ways of defining a species, the Archaeon and Proterozoic eons, and a new *Focus On* box on *Homo floresiensis*.

Part 5: The Diversity of Life

In this edition of *Biology*, we continue to emphasize the cladistic approach. We use an evolutionary framework to discuss each group of organisms, presenting current hypotheses of how groups of organisms are related. Chapter 23 discusses *why* organisms are classified and provides insight into the scientific process of deciding *how* they are classified. New advances have enabled us to further clarify the connection between evolutionary history and systematics in the eighth edition. Chapter 24, which focuses on the viruses and prokaryotes, contains a major revision of prokaryote classification and new sections on evolution in bacteria and on biofilms. Chapter 25 reflects the developing consensus on protist diversity and summarizes evolutionary relationships among the eight major eukaryote groups. Chapter 26 describes the fungi and includes an updated phylogeny with a discussion of members of phylum Glomeromycota and their role as mycorrhizal fungi. Chapters 27 and 28 present the members of the plant kingdom; Chapter 28 includes new material that updates the evolution of the angiosperms. In Chapters 29 through 31, which cover the diversity of animals, we place more emphasis on molecular systematics. We have also added new sections on animal origins and the evolution of development.

Part 6: Structure and Life Processes in Plants

Part 6 introduces students to the fascinating plant world. It stresses relationships between structure and function in plant cells, tissues, organs, and individual organisms. In Chapter 32, we introduce plant structure, growth, and differentiation. Chapters 33 through 35 discuss the structural and physiological adaptations of leaves, stems, and roots. Chapter 36 describes reproduction in flowering plants, including asexual reproduction, flowers, fruits, and seeds. Chapter 37 focuses on growth responses and regulation of growth. In the eighth edition, we present the latest findings generated by the continuing explosion of knowledge in plant biology, particularly at the molecular level. New topics include the molecular basis of mycorrhizal and rhizobial symbioses, molecular aspects of floral initiation at apical meristems, self-incompatibility, and the way that auxin acts by signal transduction. Chapter 36 also discusses a new experiment figure on the evolutionary implications of a single mutation in a gene for flower color that resulted in a shift in animal pollinators.

Part 7: Structure and Life Processes in Animals

In Part 7, we provide a strong emphasis on comparative animal physiology, showing the structural, functional, and behavioral adaptations that help animals meet environmental challenges. We use a comparative approach to examine how various animal groups have solved similar and diverse problems. In Chapter 38, we discuss the basic tissues and organ systems of the animal body, homeostasis, and the ways that animals regulate their body temperature. Chapter 39 focuses on body coverings, skeletons, and muscles. In Chapters 40 through 42, we discuss neural signaling, neural regulation, and sensory reception. In Chapters 43 through 50, we compare how different animal groups carry on specific life processes, such as internal transport, internal defense, gas exchange, digestion, reproduction, and development. Each chapter in this part considers the human adaptations for the life processes being discussed. Part 7 ends with a discussion of

behavioral adaptations in Chapter 51. Reflecting recent research findings, we have added new material on homeostasis and the process of contraction in whole muscle. We have also updated and added new material on neurotransmitters, information processing, thermoreceptors, electroreceptors and magnetoreception, endocrine regulation of blood pressure, Toll-like receptors, cytokines, cancer treatment, HIV and autoimmune diseases, new nutrition guidelines, obesity, maintenance of fluid and electrolyte balance, steroid action that allows rapid signaling, melanocyte-stimulating hormones (MSH), diabetes, anabolic steroids, menopause, the estrous cycle, contraception, and STDs. In Chapter 51, we have added new information on cognition, communication, interspecific and intraspecific selection, parental care, and helping behavior.

Part 8: The Interactions of Life: Ecology

Part 8 focuses on the dynamics of populations, communities, and ecosystems and on the application of ecological principles to disciplines such as conservation biology. Chapters 52 through 55 give the student an understanding of the ecology of populations, communities, ecosystems, and the biosphere, whereas Chapter 56 focuses on global environmental issues. Among the many changes in this unit are new material on secondary productivity, top-down and bottom-up processes, characteristics of endangered species, deforestation, and a greatly expanded section on conservation biology.

A COMPREHENSIVE PACKAGE FOR LEARNING AND TEACHING

A carefully designed supplement package is available to further facilitate learning. In addition to the usual print resources, we are pleased to present student multimedia tools that have been developed in conjunction with the text.

Resources for Students

Study Guide to Accompany **Biology, Eighth Edition,** by Ronald S. Daniel of California State Polytechnic University, Pomona; Sharon C. Daniel of Orange Coast College; and Ronald L. Taylor. Extensively updated for this edition, the study guide provides the student with many opportunities to review chapter concepts. Multiple-choice study questions, coloring-book exercises, vocabulary-building exercises, and many other types of active-learning tools are provided to suit different cognitive learning styles.

A Problem-Based Guide to Basic Genetics by Donald Cronkite of Hope College. This brief guide provides students with a systematic approach to solving genetics problems, along with numerous solved problems and practice problems.

Spanish Glossary. **NEW** This Spanish glossary of biology terms is available to Spanish-speaking students.

Website. The content-rich companion website that accompanies *Biology,* Eighth Edition, gives students access to high-quality resources, including focused quizzing, a Glossary complete with pronunciations, *InfoTrac® College Edition* readings and exercises, Internet activities, and annotated web links. For these and other resources, visit www.thomsonedu.com/biology/solomon.

ThomsonNOW. NEW This updated and expanded online learning tool helps assess students' personal study needs and focus their time. By taking a pretest, they are provided with a Personalized Study plan that directs them to text sections and narrated animations—many of which are new to this edition—that they need to review. If they need to brush up on basic skills, the ***How Do I Prepare?*** feature walks them through tutorials on basic math, chemistry, study skills, and word roots.

Audio. NEW This edition of *Biology* is accompanied by a range of "mobile content" resources, including downloadable study skill tips and concept reviews in MP3 format for use on a portable MP3 player.

Virtual Biology Laboratory 3.0 by Beneski and Waber. **NEW** These 14 online laboratory experiments, designed within a simulation format, allow students to "do" science by acquiring data, performing experiments, and using data to explain biological concepts. Assigned activities automatically flow to the instructor's grade book. New self-designed activities ask students to plan their procedures around an experimental question and write up their results.

Additional Resources for Instructors

The instructors' Examination Copy for this edition lists a comprehensive package of print and multimedia supplements, including online resources, available to qualified adopters. Please ask your local sales representative for details.

ACKNOWLEDGMENTS

The development and production of the eighth edition of *Biology* required extensive interaction and cooperation among the authors and many individuals in our family, social, and professional environments. We thank our editors, colleagues, students, family, and friends for their help and support. Preparing a book of this complexity is challenging and requires a cohesive, talented, and hardworking professional team. We appreciate the contributions of everyone on the editorial and production staff at Brooks/Cole/Thomson Learning who worked on this eighth edition of *Biology*. We thank Michelle Julet, Vice President and Editor-in-Chief, and Peter Adams, Executive Editor, for their commitment to this book and for their support in making the eighth edition happen. We appreciate Stacy Best and Kara Kindstrom, our Marketing Managers, whose expertise ensured that you would know about our new edition.

We appreciate the hard work of our dedicated Developmental Editor, Suzannah Alexander, who provided us with valuable input as she guided the eighth edition through its many phases. We especially thank Suzannah for sharing her artistic talent and for her great ideas for visual presentations. We appreciate the help of Senior Content Project Manager Cheryll Linthicum, Content Project Manager Jennifer Risden, and Project Editor Jamie Armstrong, who expertly shepherded the project. We thank Editorial Assistants Kristin Marrs and Kate Franco for quickly providing us with resources whenever we needed them. We appreciate the efforts of photo editor Don Murie. We thank Creative Director Rob Hugel, Art Director Lee Friedman, Text Designer John Walker, and Cover Designer Robin Terra. We also thank Joy Westberg for developing the Instructor's Preface.

We are grateful to Keli Amann, Technology Project Manager, who coordinated the many high-tech components of the computerized aspects of our Learning System. We thank Lauren Oliveira, Assistant Editor, for coordinating the print supplements. These dedicated professionals and many others at Brooks/Cole provided the skill, attention, and good humor needed to produce *Biology,* Eighth Edition. We thank them for their help and support throughout this project.

A ***Biology Advisory Board*** greatly enhanced our preparation of the eighth edition of *Biology*. We thank these professionals for their insight and suggestions:

Susan R. Barnum, Miami University, Oxford, OH, Molecular biology of cyanobacteria

Virginia McDonough, Hope College, Molecular biology of lipid metabolism

David K. Bruck, San Jose State University, Plant cell biology

Lee F. Johnson, The Ohio State University, Molecular genetics, biochemistry

Karen A. Curto, University of Pittsburgh, Cell signaling

Robert J. Kosinski, Clemson University, Introductory biology laboratory development

Tim Schuh, St. Cloud State University, Developmental biology

Robert W. Yost, Indiana University–Purdue University Indianapolis, Physiology, animal form and function

We are grateful to Bruce Mohn of Rutgers University for his expert advice on dinosaur and bird evolution and to Bill Norris of Western New Mexico University for his help with the book's cladograms. We thank Dr. Susan Pross, University of South Florida, College of Medicine, for her helpful suggestions for updating the immunology chapter. We greatly appreciate the expert assistance of Mary Kay Hartung of Florida Gulf Coast University, who came to our rescue whenever we had difficulty finding needed research studies from the Internet. We thank doctoral student Lois Ball of the University of South Florida, Department of Biology, who reviewed several chapters and offered helpful suggestions. We thank obstetrician, gynecologist Dr. Amy Solomon for her input regarding pregnancy, childbirth, conception, and sexually transmitted diseases.

We thank our families and friends for their understanding, support, and encouragement as we struggled through many revisions and deadlines. We especially thank Mical Solomon, Dr. Amy Solomon, Dr. Kathleen M. Heide, Alan Berg, Jennifer Berg, Dr. Charles Martin, and Margaret Martin for their support and input. We greatly appreciate the many hours Alan Berg devoted to preparing the manuscript.

Our colleagues and students who have used our book have provided valuable input by sharing their responses to past editions of *Biology*. We thank them and ask again for their comments and suggestions as they use this new edition. We can be reached through the Internet at our website www.thomsonedu.com/biology/solomon or through our editors at Brooks/Cole, a division of Thomson Learning.

We express our thanks to the many biologists who have read the manuscript during various stages of its development and provided us with valuable suggestions for improving it. Eighth edition reviewers include the following:

Joseph J. Arruda, Pittsburg State University

Amir M. Assadi-Rad, San Joaquin Delta College

Douglas J. Birks, Wilmington College

William L. Bischoff, University of Toledo

Catherine S. Black, Idaho State University

Andrew R. Blaustein, Oregon State University

Scott Bowling, Auburn University

W. Randy Brooks, Florida Atlantic University

Mark Browning, Purdue University

Arthur L. Buikema Jr., Virginia Tech

Anne Bullerjahn, Owens Community College

Carolyn J. W. Bunde, Idaho State University

Scott Burt, Truman State University

David Byres, Florida Community College–Jacksonville

Jeff Carmichael, University of North Dakota

Domenic Castignetti, Loyola University of Chicago

Geoffrey A. Church, Fairfield University

Barbara Collins, California Lutheran University

Linda W. Crow, Montgomery College

Karen J. Dalton, Community College of Baltimore County

Mark Decker, University of Minnesota

Jonathan J. Dennis, University of Alberta

Philippa M. Drennan, Loyola Marymount University

David W. Eldridge, Baylor University

H. W. Elmore, Marshall University

Cheryld L. Emmons, Alfred University

Robert C. Evans, Rutgers University, Camden

John Geiser, Western Michigan University

William J. Higgins, University of Maryland

Jeffrey P. Hill, Idaho State University

Walter S. Judd, University of Florida

Mary Jane Keleher, Salt Lake Community College

Scott L. Kight, Montclair State University

Joanne Kivela Tillotson, Purchase College SUNY

Will Kleinelp, Middlesex County College

Kenneth M. Klemow, Wilkes University

Jonathan Lyon, Merrimack College

Blasé Maffia, University of Miami

Kathleen R. Malueg, University of Colorado, Colorado Springs

Patricia Matthews, Grand Valley State University

David Morgan, Western Washington University

Darrel L. Murray, University of Illinois at Chicago

William R. Norris, Western New Mexico University

James G. Patton, Vanderbilt University

Mitch Price, Penn State

Susan Pross, University of South Florida

Jerry Purcell, San Antonio College

Kenneth R. Robinson, Purdue University

Darrin Rubino, Hanover College

Julie C. Rutherford, Concordia College

Andrew M. Scala, Dutchess Community College

Pramila Sen, Houston Community College

Mark A. Sheridan, North Dakota State University

Marcia Shofner, University of Maryland

Phillip Snider, Gadsden State Community College

David Stanton, Saginaw Valley State University

William Terzaghi, Wilkes University

Keti Venovski, Lake Sumter Community College

Steven D. Wilt, Bellarmine University

James R. Yount, Brevard Community College

We would also like to thank the hundreds of previous edition reviewers, both professors and students, who are too numerous to mention. Without their contributions, *Biology,* Eighth Edition, would not have been the same. They asked thoughtful questions, provided new perspectives, offered alternative wordings to clarify difficult passages, and informed us of possible errors. We are truly indebted to their excellent feedback.

To the Student

Biology is a challenging subject. The thousands of students we have taught have differed in their life goals and learning styles. Some have had excellent backgrounds in science; others, poor ones. Regardless of their backgrounds, it is common for students taking their first college biology course to find that they must work harder than they expected. You can make the task easier by using approaches to learning that have been successful for a broad range of our students over the years. Be sure to use the Learning System we use in this book. It is described in the Preface.

Make a Study Schedule

Many college professors suggest that students study 3 hours for every hour spent in class. This major investment in study time is one of the main differences between high school and college. To succeed academically, college students must learn to manage their time effectively. The actual number of hours you spend studying biology will vary depending on how quickly you learn the material, as well as on your course load and personal responsibilities, such as work schedules and family commitments.

The most successful students are often those who are best organized. At the beginning of the semester, make a detailed daily calendar. Mark off the hours you are in each class, along with travel time to and from class if you are a commuter. After you get your course syllabi, add to your calendar the dates of all exams, quizzes, papers, and reports. As a reminder, it also helps to add an entry for each major exam or assignment 1 week before the test or due date. Now add your work schedule and other personal commitments to your calendar. Using a calendar helps you find convenient study times.

Many of our successful biology students set aside 2 hours a day to study biology rather than depend on a weekly marathon session for 8 or 10 hours during the weekend (when that kind of session rarely happens). Put your study hours into your daily calendar, and stick to your schedule.

Determine Whether the Professor Emphasizes Text Material or Lecture Notes

Some professors test almost exclusively on material covered in lecture. Others rely on their students' learning most, or even all, of the content in assigned chapters. Find out what your professor's requirements are, because the way you study will vary accordingly.

How to study when professors test lecture material

If lectures are the main source of examination questions, make your lecture notes as complete and organized as possible. Before going to class, skim over the chapter, identifying key terms and examining the main figures, so that you can take effective lecture notes. Spend no more than 1 hour on this.

Within 24 hours after class, rewrite (or type) your notes. Before rewriting, however, read the notes and make marginal notes about anything that is not clear. Then read the corresponding material in your text. Highlight or underline any sections that clarify questions you had in your notes. Read the entire chapter, including parts that are not covered in lecture. This extra information will give you breadth of understanding and will help you grasp key concepts.

After reading the text, you are ready to rewrite your notes, incorporating relevant material from the text. It also helps to use the Glossary to find definitions for unfamiliar terms. Many students develop a set of flash cards of key terms and concepts as a way to study. Flash cards are a useful tool to help you learn scientific terminology. They are portable and can be used at times when other studying is not possible, for example, when riding a bus.

Flash cards are not effective when the student tries to second-guess the professor. ("She won't ask this, so I won't make a flash card of it.") Flash cards are also a hindrance when students rely on them exclusively. Studying flash cards instead of reading the text is a bit like reading the first page of each chapter in a mystery novel: It's hard to fill in the missing parts, because you are learning the facts in a disconnected way.

How to study when professors test material in the book

If the assigned readings in the text are going to be tested, you must use your text intensively. After reading the chapter introduction, read the list of Learning Objectives for the first section. These objectives are written in behavioral terms; that is, they ask you to "do" something in order to demonstrate mastery. The objectives give you a concrete set of goals for each section the chapter. At the end of each section, you will find Review questions keyed to the Learning Objectives. Test yourself, going back over the material to check your responses.

Read each chapter section actively. Many students read and study passively. An active learner always has questions in mind and is constantly making connections. For example, there are many processes that must be understood in biology. Don't try to blindly memorize these; instead, think about causes and effects so that every process becomes a story. Eventually, you'll see that many processes are connected by common elements.

You will probably have to read each chapter two or three times before mastering the material. The second and third times through will be much easier than the first, because you'll be reinforcing concepts that you have already partially learned.

After reading the chapter, write a four- to six-page chapter outline by using the subheads as the body of the outline (first-level heads are boldface, in color, and all caps; second-level heads are in color and not all caps). Flesh out your outline by adding important concepts and boldface terms with definitions. Use this outline when preparing for the exam.

Now it is time to test yourself. Answer the Test Your Understanding questions, and check your answers. Write answers to each of the Critical Thinking questions. Finally, review the Learning Objectives in the Chapter Summary, and try to answer them before reading the summary provided. If your professor has told you that some or all of the exam will be short-answer or essay format, write out the answer for each Learning Objective. Remember that this is a self-test. If you do not know an answer to a question, find it in the text. If you can't find the answer, use the Index.

Learn the Vocabulary

One stumbling block for many students is learning the many terms that make up the language of biology. In fact, it would be much more difficult to learn and communicate if we did not have this terminology, because words are really tools for thinking. Learning terminology generally becomes easier if you realize that most biological terms are modular. They consist of mostly Latin and Greek roots; once you learn many of these, you will have a good idea of the meaning of a new word even before it is defined. For this reason, we have included an Appendix on Understanding Biological Terms. To be sure you understand the precise definition of a term, use the Index and Glossary. The more you use biological terms in speech and writing, the more comfortable you will be.

Form a Study Group

Active learning is facilitated if you do some of your studying in a small group. In a study group, the roles of teacher and learner can be interchanged: a good way to learn material is to teach. A study group lets you meet challenges in a nonthreatening environment and can provide some emotional support. Study groups are effective learning tools when combined with individual study of text and lecture notes. If, however, you and other members of your study group have not prepared for your meetings by studying individually in advance, the study session can be a waste of time.

Prepare for the Exam

Your calendar tells you it is now 1 week before your first biology exam. If you have been following these suggestions, you are well prepared and will need only some last-minute reviewing. No all-nighters will be required.

During the week prior to the exam, spend 2 hours each day actively studying your lecture notes or chapter outlines. It helps many students to read these notes out loud (most people listen to what they say!). Begin with the first lecture/chapter covered on the exam, and continue in the order on the lecture syllabus. Stop when you have reached the end of your 2-hour study period. The following day, begin where you stopped the previous day. When you reach the end of your notes, start at the beginning and study them a second time. The material should be very familiar to you by the second or third time around. At this stage, use your textbook only to answer questions or clarify important points.

The night before the exam, do a little light studying, eat a nutritious dinner, and get a full night's sleep. That way, you'll arrive in class on exam day with a well-rested body (and brain) and the self-confidence that goes with being well prepared.

Eldra P. Solomon
Linda R. Berg
Diana W. Martin

A View of Life

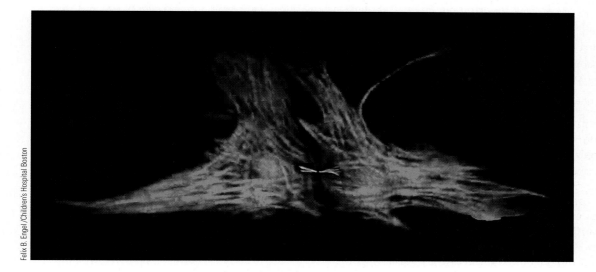

A cardiac muscle cell dividing into two new cells.

Felix B. Engel/Children's Hospital Boston

KEY CONCEPTS

Basic themes of biology include evolution, information transfer, and energy for life.

Characteristics of life include growth and development, self-regulated metabolism, response to stimuli, and reproduction.

Biological organization is hierarchical and includes chemical, cell, tissue, organ, organ system, and organism levels; ecological organization includes population, community, ecosystem, and biosphere levels.

The tree of life includes three major branches, or domains.

Species evolve by natural selection, adapting to changes in their environment.

Biologists ask questions, develop hypotheses, make predictions, and collect data by careful observation and experiment.

Heart disease is the leading cause of death in the United States and other developed countries. When cardiac muscle is damaged, for example, by a myocardial infarction (a "heart attack"), the injured muscle does not repair itself. Instead, the muscle is replaced by scar tissue that impairs the heart's ability to pump blood. Researcher Mark T. Keating and his team at the Howard Hughes Medical Institute at Children's Hospital Boston discovered that the small, tropical zebrafish can regenerate up to 20% of its heart muscle within two months after the tissue is injured or removed. The challenge is to find the mechanisms that stimulate regeneration and learn how to manipulate them.

Felix Engel, an investigator working with Keating, identified that a growth factor (known as fibroblast growth factor) can stimulate cardiac muscle cells to replicate their DNA. The research team also discovered that the protein known as p38 inhibits cardiac muscle growth in fetal rats. They wondered how they could inhibit the inhibitor so that cardiac muscle could repair itself. In 2005, these investigators reported that when they treated cardiac muscle cells with the growth factor in combination with a compound that inhibits p38, cardiac muscle cells activate a group of genes involved in cell division and finally divide (see photograph).

A research group in Germany is pioneering another approach to treating heart damage. This group, led by Gustav Steinhoff at the University of Rostock, injected stem cells (collected from the patient's own bone marrow) into the cardiac muscle of patients who had heart damage as a result of myocardial infarction. One year after receiving the injected stem cells, patients who had this treatment in addition to bypass surgery showed significantly better heart function compared with patients who had only surgery. Searching for ways to repair damaged hearts is one example of the thousands of new research studies in progress.

This is an exciting time to begin studying **biology,** the science of life. The remarkable new discoveries biologists are making almost daily affect every aspect of our lives, including our health, food, safety, relationships with humans and other organisms, and our ability to enjoy the life that surrounds us. New knowledge provides new insights into the human species and the millions of other organisms with which we share this planet.

Biology affects our personal, governmental, and societal decisions. For example, the U.S. Supreme Court abolished the death penalty for juvenile offenders in 2005, based on research findings that the brain does not complete its development until after ado-lescence. (At the time of this ruling, 20 states still permitted capital punishment for offenders younger than age 18.) Adolescents have not reached neurophysiological maturity and cannot be held accountable for their crimes to the same extent as adults.

Whatever your college major or career goals, knowledge of biological concepts is a vital tool for understanding this world and for meeting many of the personal, societal, and global challenges that confront us. Among these challenges are decreasing biological diversity, diminishing natural resources, the expanding human population, and prevention and cure of diseases, such as cancer, Alzheimer's disease, malaria, acquired immunodeficiency syndrome (AIDS), and avian flu. Meeting these challenges will require the combined efforts of biologists and other scientists, politicians, and biologically informed citizens.

This book is a starting point for an exploration of biology. It provides you with the basic knowledge and the tools to become a part of this fascinating science and a more informed member of society. ■

THREE BASIC THEMES

Learning Objective
1   Describe three basic themes of biology.

In this first chapter we introduce three basic themes of biology:

1. **Evolution.** Populations of organisms have evolved through time from earlier forms of life. Scientists have accumulated a wealth of evidence showing that the diverse life-forms on this planet are related and that populations have *evolved,* that is, have changed over time, from earlier forms of life. The process of *evolution* is the framework for the science of biology and is a major theme of this book.

2. **Information transfer.** Information must be transmitted within organisms and among organisms. The survival and function of every cell and every organism depend on the orderly transmission of information. Evolution depends on the transmission of genetic information from one generation to another.

3. **Energy for life.** Energy from the sun flows through living systems from producers to consumers. All life processes, including the thousands of chemical transactions that maintain life's organization, require a continuous input of energy.

Evolution, information transmission, and energy flow are forces that give life its unique characteristics. We begin our study of biology by developing a more precise understanding of the fundamental characteristics of living systems.

Review
■   Why are evolution, information transfer, and energy considered basic to life?

■   What does the term *evolution* mean as applied to populations of organisms?

CHARACTERISTICS OF LIFE

Learning Objective
2   Distinguish between living and nonliving things by describing the features that characterize living organisms.

We easily recognize that a pine tree, a butterfly, and a horse are living things, whereas a rock is not. Despite their diversity, the organisms that inhabit our planet share a common set of characteristics that distinguish them from nonliving things. These features include a precise kind of organization, growth and development, self-regulated metabolism, the ability to respond to stimuli, reproduction, and adaptation to environmental change.

Organisms are composed of cells

Although they vary greatly in size and appearance, all organisms consist of basic units called **cells.** New cells are formed only by the division of previously existing cells. These concepts are expressed in the **cell theory** (discussed in Chapter 4), a fundamental unifying concept of biology.

Some of the simplest life-forms, such as protozoa, are **unicellular** organisms, meaning that each consists of a single cell (■ Fig. 1-1). In contrast, the body of a cat or a maple tree is made of billions of cells. In such complex **multicellular** organisms, life processes depend on the coordinated functions of component

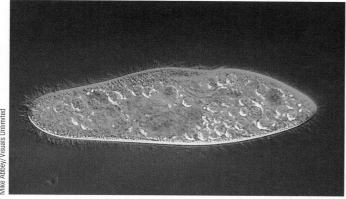

(a) Unicellular organisms consist of one intricate cell that performs all the functions essential to life. Ciliates, such as this *Paramecium*, move about by beating their hairlike cilia.

250 µm

Figure 1-1   Unicellular and multicellular life-forms

(b) Multicellular organisms, such as this African buffalo (*Syncerus caffer*) and the plants on which it grazes, may consist of billions of cells specialized to perform specific functions.

cells that may be organized to form tissues, organs, and organ systems.

Every cell is enveloped by a protective **plasma membrane** that separates it from the surrounding external environment. The plasma membrane regulates passage of materials between cell and environment. Cells have specialized molecules that contain genetic instructions and transmit genetic information. In most cells, the genetic instructions are encoded in deoxyribonucleic acid, more simply known as **DNA.** Cells typically have internal structures called **organelles** that are specialized to perform specific functions.

There are two fundamentally different types of cells: prokaryotic and eukaryotic. **Prokaryotic cells** are exclusive to bacteria and to microscopic organisms called *archaea.* All other organisms are characterized by their **eukaryotic cells.** These cells typically contain a variety of organelles enclosed by membranes, including a **nucleus,** which houses DNA. Prokaryotic cells are structurally simpler; they do not have a nucleus or other membrane-enclosed organelles.

Organisms grow and develop

Biological growth involves an increase in the size of individual cells of an organism, in the number of cells, or in both. Growth may be uniform in the various parts of an organism, or it may be greater in some parts than in others, causing the body proportions to change as growth occurs. Some organisms—most trees, for example—continue to grow throughout their lives. Many animals have a defined growth period that terminates when a characteristic adult size is reached. An intriguing aspect of the growth process is that each part of the organism typically continues to function as it grows.

Living organisms develop as well as grow. **Development** includes all the changes that take place during an organism's life. Like many other organisms, every human begins life as a fertilized egg that then grows and develops. The structures and body form that develop are exquisitely adapted to the functions the organism must perform.

Organisms regulate their metabolic processes

Within all organisms, chemical reactions and energy transformations occur that are essential to nutrition, the growth and repair of cells, and the conversion of energy into usable forms. The sum of all the chemical activities of the organism is its **metabolism.**

Metabolic processes occur continuously in every organism, and they must be carefully regulated to maintain **homeostasis,** an appropriate, balanced internal environment. When enough of a cell product has been made, its manufacture must be decreased or turned off. When a particular substance is required, cell processes that produce it must be turned on. These *homeostatic mechanisms* are self-regulating control systems that are remarkably sensitive and efficient.

The regulation of glucose (a simple sugar) concentration in the blood of complex animals is a good example of a homeostatic mechanism. Your cells require a constant supply of glucose, which they break down to obtain energy. The circulatory system delivers glucose and other nutrients to all the cells. When the concentration of glucose in the blood rises above normal limits, glucose is stored in the liver and in muscle cells. When you have not eaten for a few hours, the glucose concentration begins to fall. Your body converts stored nutrients to glucose, bringing the glucose concentration in the blood back to normal levels. When the glucose concentration decreases, you also feel hungry and restore nutrients by eating.

Organisms respond to stimuli

All forms of life respond to **stimuli,** physical or chemical changes in their internal or external environment. Stimuli that evoke a response in most organisms are changes in the color, intensity, or direction of light; changes in temperature, pressure, or sound; and changes in the chemical composition of the surrounding soil, air, or water. Responding to stimuli involves movement, though not always locomotion (moving from one place to another).

In simple organisms, the entire individual may be sensitive to stimuli. Certain unicellular organisms, for example, respond to bright light by retreating. In some organisms, locomotion is achieved by the slow oozing of the cell, the process of *amoeboid movement.* Other organisms move by beating tiny, hairlike extensions of the cell called **cilia** or longer structures known as **flagella** (❚ Fig. 1-2). Some bacteria move by rotating their flagella.

Most animals move very obviously. They wiggle, crawl, swim, run, or fly by contracting muscles. Sponges, corals, and oysters have free-swimming larval stages, but most are **sessile** as adults, meaning that they do not move from place to place. In fact, they may remain firmly attached to a surface, such as the sea bottom or a rock. Many sessile organisms have cilia or flagella that beat rhythmically, moving the surrounding water, which contains the food and oxygen the organisms require. Complex animals, such as grasshoppers, lizards, and humans, have highly specialized cells that respond to specific types of stimuli. For example, cells in the retina of the human eye respond to light.

Although their responses may not be as obvious as those of animals, plants do respond to light, gravity, water, touch, and other stimuli. For example, plants orient their leaves to the sun and grow toward light. Many plant responses involve different growth rates of various parts of the plant body.

A few plants, such as the Venus flytrap of the Carolina swamps, are very sensitive to touch and catch insects (❚ Fig. 1-3). Their leaves are hinged along the midrib, and they have a scent that attracts insects. Trigger hairs on the leaf surface detect the arrival of an insect and stimulate the leaf to fold. When the edges come together, they interlock, preventing the insect's escape. The leaf then secretes enzymes that kill and digest the insect. The Venus flytrap usually grows in soil deficient in nitrogen. The plant obtains part of the nitrogen required for its growth from the insects it "eats."

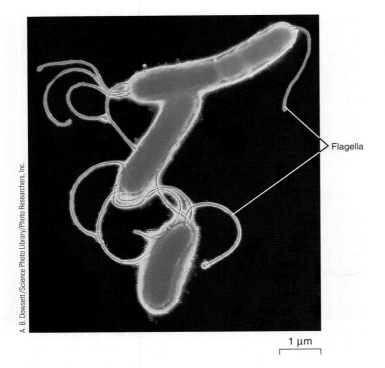

A. B. Dowsett/Science Photo Library/Photo Researchers, Inc.

1 µm

Figure 1-2 Biological movement

These bacteria (*Helicobacter pylori*), equipped with flagella for locomotion, have been linked to stomach ulcers. The photograph was taken using a scanning electron microscope. The bacteria are not really red and blue. Their color has been artificially enhanced.

David M. Dennis/Tom Stack & Associates

(a) Hairs on the leaf surface of the Venus flytrap (*Dionaea muscipula*) detect the touch of an insect, and the leaf responds by folding.

David M. Dennis/Tom Stack & Associates

(b) The edges of the leaf come together and interlock, preventing the fly's escape. The leaf then secretes enzymes that kill and digest the insect.

Figure 1-3 Plants respond to stimuli

Organisms reproduce

At one time, people thought worms arose spontaneously from horsehair in a water trough, maggots from decaying meat, and frogs from the mud of the Nile. Thanks to the work of several scientists, including the Italian physician Francesco Redi in the 17th century and French chemist Louis Pasteur in the 19th century, we now know that organisms arise only from previously existing organisms.

Simple organisms, such as amoebas, perpetuate themselves by **asexual reproduction** (▌Fig. 1-4a). When an amoeba has grown to a certain size, it reproduces by splitting in half to form two new amoebas. Before an amoeba divides, its hereditary material (set of genes) duplicates, and one complete set is distributed to each new cell. Except for size, each new amoeba is similar to the parent cell. The only way that variation occurs among asexu-

ally reproducing organisms is by genetic *mutation*, a permanent change in the genes.

In most plants and animals, **sexual reproduction** is carried out by the fusion of an egg and a sperm cell to form a fertilized egg (▌Fig. 1-4b). The new organism develops from the fertilized egg. Offspring produced by sexual reproduction are the product of the interaction of various genes contributed by the mother and the father. This genetic variation is important in the vital processes of evolution and adaptation.

Populations evolve and become adapted to the environment

The ability of a population to evolve over many generations and adapt to its environment equips it to survive in a changing world. **Adaptations** are inherited characteristics that enhance an organism's ability to survive in a particular environment. The long, flexible tongue of the frog is an adaptation for catching insects, the feathers and lightweight bones of birds are adaptations for flying, and the thick fur coat of the polar bear is an adaptation for surviving frigid temperatures. Adaptations may be structural, physiological, biochemical, behavioral, or a combination of all four (▌Fig. 1-5). Every biologically successful organism is a complex collection of coordinated adaptations produced through evolutionary processes.

Review

- ▌ What characteristics distinguish a living organism from a nonliving object?
- ▌ What would be the consequences to an organism if its homeostatic mechanisms failed? Explain your answer.
- ▌ What do we mean by adaptations?

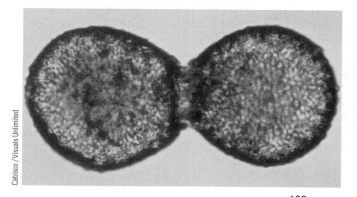

100 μm

(a) Asexual reproduction. One individual gives rise to two or more offspring that are similar to the parent. *Difflugia*, a unicellular amoeba, is shown dividing to form two amoebas.

(b) Sexual reproduction. Typically, two parents each contribute a gamete (sperm or egg). Gametes fuse to produce the offspring, which has a combination of the traits of both parents. A pair of tropical flies are shown mating.

▌ **Figure 1-4** Asexual and sexual reproduction

▌ **Figure 1-5** Adaptations

These Burchell's zebras (*Equus burchelli*), photographed at Ngorongoro Crater in Tanzania, are behaviorally adapted to position themselves to watch for lions and other predators. Stripes are thought to be an adaptation for visual protection against predators. They serve as camouflage or to break up form when spotted from a distance. The zebra stomach is adapted for feeding on coarse grass passed over by other grazers, an adaptation that helps the animal survive when food is scarce.

LEVELS OF BIOLOGICAL ORGANIZATION

Learning Objective

3 Construct a hierarchy of biological organization, including levels characteristic of individual organisms and ecological levels.

Whether we study a single organism or the world of life as a whole, we can identify a hierarchy of biological organization (❚ Fig. 1-6). At every level, structure and function are precisely coordinated. One way to study a particular level is by looking at its components. Biologists can gain insights about cells by studying atoms and molecules. Learning about a structure by studying its parts is called **reductionism.** However, the whole is more than the sum of its parts. Each level has **emergent properties,** characteristics not found at lower levels. Populations of organisms have emergent properties such as population density, age structure, and birth and death rates. The *individuals* that make up a population do not have these characteristics. Consider also the human brain. The brain is composed of millions of neurons (nerve cells). We could study millions of individual neurons and have no clue about the functional capacities of the brain. Only when the neurons are wired together in precise fashion are the emergent properties, such as the capacity for thought, judgment, and motor coordination, evident.

Organisms have several levels of organization

The chemical level, the most basic level of organization, includes atoms and molecules. An **atom** is the smallest unit of a chemical element that retains the characteristic properties of that element. For example, an atom of iron is the smallest possible amount of iron. Atoms combine chemically to form **molecules.** Two atoms of hydrogen combine with one atom of oxygen to form a single molecule of water. Although composed of two types of atoms that are gases, water is a liquid with very different properties, an example of emergent properties.

At the cell level, many types of atoms and molecules associate with one another to form *cells*. However, a cell is much more than a heap of atoms and molecules. Its emergent properties make it the basic structural and functional unit of life, the simplest component of living matter that can carry on all the activities necessary for life.

During the evolution of multicellular organisms, cells associated to form **tissues.** For example, most animals have muscle tissue and nervous tissue. Plants have epidermis, a tissue that serves as a protective covering, and vascular tissues that move materials throughout the plant body. In most complex organisms, tissues organize into functional structures called **organs,** such as the heart and stomach in animals and roots and leaves in plants. In animals, each major group of biological functions is performed by a coordinated group of tissues and organs called an **organ system.** The circulatory and digestive systems are ex-

amples of organ systems. Functioning together with great precision, organ systems make up a complex, multicellular **organism.** Again, emergent properties are evident. An organism is much more than its component organ systems.

Several levels of ecological organization can be identified

Organisms interact to form still more complex levels of biological organization. All the members of one species living in the same geographic area at the same time make up a **population.** The populations of various types of organisms that inhabit a particular area and interact with one another form a **community.** A community can consist of hundreds of different types of organisms. As populations within a community evolve, the community changes.

A community together with its nonliving environment is an **ecosystem.** An ecosystem can be as small as a pond (or even a puddle) or as vast as the Great Plains of North America or the Arctic tundra. All of Earth's ecosystems together are known as the **biosphere.** The biosphere includes all of Earth that is inhabited by living organisms—the atmosphere, the hydrosphere (water in any form), and the lithosphere (Earth's crust). The study of how organisms relate to one another and to their physical environment is called **ecology** (derived from the Greek *oikos,* meaning "house").

Review

❚ What are the levels of organization within an organism?
❚ What are the levels of ecological organization?

INFORMATION TRANSFER

Learning Objective

4 Summarize the importance of information transfer to living systems, giving specific examples.

For an organism to grow, develop, carry on self-regulated metabolism, respond to stimuli, and reproduce, it must have precise instructions and its cells must be able to communicate. The information an organism requires to carry on these life processes is coded and delivered in the form of chemical substances and electrical impulses. Organisms must also communicate information to one another.

DNA transmits information from one generation to the next

Humans give birth only to human babies, not to giraffes or rosebushes. In organisms that reproduce sexually, each offspring is a combination of the traits of its parents. In 1953, James Watson

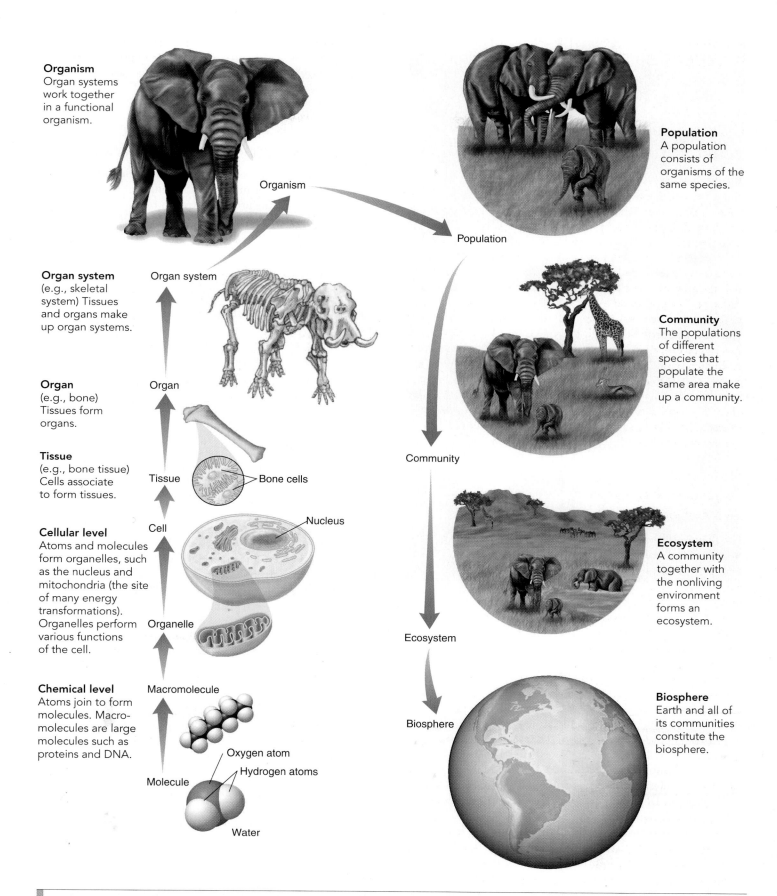

Organism
Organ systems work together in a functional organism.

Organism

Population
A population consists of organisms of the same species.

Population

Organ system
(e.g., skeletal system) Tissues and organs make up organ systems.

Organ system

Community
The populations of different species that populate the same area make up a community.

Organ
(e.g., bone) Tissues form organs.

Organ

Community

Tissue
(e.g., bone tissue) Cells associate to form tissues.

Tissue

Bone cells

Cellular level
Atoms and molecules form organelles, such as the nucleus and mitochondria (the site of many energy transformations). Organelles perform various functions of the cell.

Cell

Nucleus

Ecosystem
A community together with the nonliving environment forms an ecosystem.

Organelle

Ecosystem

Chemical level
Atoms join to form molecules. Macromolecules are large molecules such as proteins and DNA.

Macromolecule

Biosphere

Biosphere
Earth and all of its communities constitute the biosphere.

Molecule

Oxygen atom

Hydrogen atoms

Water

Figure 1-6 *Animated* The hierarchy of biological organization

isms as diverse as bacteria, frogs, and redwood trees. The genetic code is universal, that is, virtually identical in all organisms—a dramatic example of the unity of life.

Information is transmitted by chemical and electrical signals

Genes control the development and functioning of every organism. The DNA that makes up the genes contains the "recipes" for making all the proteins required by the organism. **Proteins** are large molecules important in determining the structure and function of cells and tissues. For example, brain cells differ from muscle cells in large part because they have different types of proteins. Some proteins are important in communication within and among cells. Certain proteins on the surface of a cell serve as markers so that other cells "recognize" them. Some cell-surface proteins serve as receptors that combine with chemical messengers.

Cells use proteins and many other types of molecules to communicate with one another. In a multicellular organism, cells produce chemical compounds, such as **hormones**, that signal other cells. Hormones and other chemical messengers can signal cells in distant organs to secrete a particular required substance. In this way chemical signals help regulate growth, development, and metabolic processes. The mechanisms involved in **cell signaling** often involve complex biochemical processes.

Cell signaling is currently an area of intense research. A major focus has been the transfer of information among cells of the immune system. A better understanding of how cells communicate promises new insights into how the body protects itself against disease organisms. Learning to manipulate cell signaling may lead to new methods of delivering drugs into cells and new treatments for cancer and other diseases.

Some organisms use electrical signals to transmit information. Most animals have nervous systems that transmit information by way of both electrical impulses and chemical compounds known as **neurotransmitters.** Information transmitted from one part of the body to another is important in regulating life processes. In complex animals, the nervous system gives the animal information about its outside environment by transmitting signals from sensory receptors such as the eyes and ears to the brain.

Information must also be transmitted from one organism to another. Mechanisms for this type of communication include the release of chemicals, visual displays, and sounds. Typically, organisms use a combination of several types of communication signals. A dog may signal aggression by growling, using a particular facial expression, and laying its ears back. Many animals perform complex courtship rituals in which they display parts of their bodies, often elaborately decorated, to attract a mate.

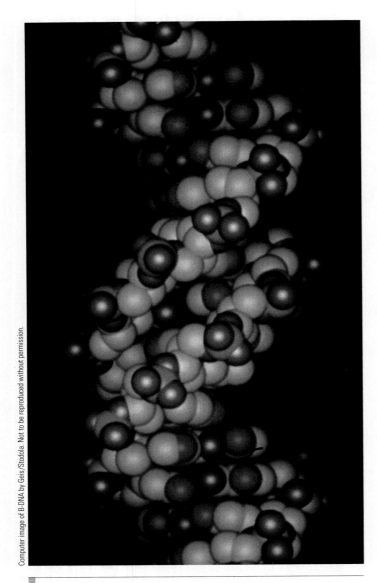

Figure 1-7 DNA

Organisms transmit information from one generation to the next by DNA, the hereditary material. As shown in this model, DNA consists of two chains of atoms twisted into a helix. Each chain consists of subunits called nucleotides. The sequence of nucleotides makes up the genetic code.

Computer image of B-DNA by Geis/Stodola. Not to be reproduced without permission.

and Francis Crick worked out the structure of DNA, the large molecule that makes up the **genes,** the units of hereditary material (▌Fig. 1-7). A DNA molecule consists of two chains of atoms twisted into a helix. Each chain is made up of a sequence of chemical subunits called **nucleotides.** There are four types of nucleotides in DNA; and each sequence of three nucleotides is part of the genetic code.

Watson and Crick's work led to the understanding of this genetic code. The information coded in sequences of nucleotides in DNA transmits genetic information from generation to generation. The code works somewhat like an alphabet. The nucleotides can "spell" an amazing variety of instructions for making organ-

Review
▌ What is the function of DNA?
▌ What are two examples of cell signaling?

8   Chapter 1

www.thomsonedu.com/biology/solomon

EVOLUTION: THE BASIC UNIFYING CONCEPT OF BIOLOGY

Learning Objectives

5 Demonstrate the binomial system of nomenclature by using specific examples, and classify an organism (such as a human) in its domain, kingdom, phylum, class, order, family, genus, and species.

6 Identify the three domains and six kingdoms of living organisms, and give examples of organisms assigned to each group.

7 Give a brief overview of the theory of evolution, and explain why it is the principal unifying concept in biology.

8 Apply the theory of natural selection to any given adaptation, and suggest a logical explanation of how the adaptation may have evolved.

The **theory of evolution** explains how populations of organisms have changed over time. This theory has become the most important unifying concept of biology. We can define the term **evolution** as the process by which populations of organisms change over time. As we will discuss, evolution involves passing genes for new traits from one generation to another, leading to differences in populations.

The evolutionary perspective is important in every specialized field within biology. Biologists try to understand the structure, function, and behavior of organisms and their interactions with one another by considering them in light of the long, continuing process of evolution. Although we discuss evolution in depth in Chapters 18 through 22, we present a brief overview here to give you the background necessary to understand other aspects of biology. First we examine how biologists organize the millions of organisms that have evolved, and then we summarize the mechanisms that drive evolution.

Biologists use a binomial system for naming organisms

Biologists have identified about 1.8 million species of extant (currently living) organisms and estimate that several million more remain to be discovered. To study life, we need a system for organizing, naming, and classifying its myriad forms. **Systematics** is the field of biology that studies the diversity of organisms and their evolutionary relationships. **Taxonomy,** a sub-specialty of systematics, is the science of naming and classifying organisms.

In the 18th century Carolus Linnaeus, a Swedish botanist, developed a hierarchical system of naming and classifying organisms that, with some modification, is still used today. The most narrow category of classification is the **species,** a group of organisms with similar structure, function, and behavior. A species consists of one or more populations whose members are capable of breeding with one another; in nature, they do not breed with members of other species. Members of a population

TABLE 1-1			
Category	Cat	Human	White Oak
Domain	Eukarya	Eukarya	Eukarya
Kingdom	Animalia	Animalia	Plantae
Phylum	Chordata	Chordata	Anthophyta
Subphylum	Vertebrata	Vertebrata	None
Class	Mammalia	Mammalia	Eudicotyledones
Order	Carnivora	Primates	Fagales
Family	Felidae	Hominidae	Fagaceae
Genus	*Felis*	*Homo*	*Quercus*
Species	*Felis catus*	*Homo sapiens*	*Quercus alba*

have a common *gene pool* (all the genes present in the population) and share a common ancestry. Closely related species are grouped in the next broader category of classification, the **genus** (pl., *genera*).

The Linnaean system of naming species is known as the **binomial system of nomenclature** because each species is assigned a two-part name. The first part of the name is the genus, and the second part, the **specific epithet,** designates a particular species belonging to that genus. The specific epithet is often a descriptive word expressing some quality of the organism. It is always used together with the full or abbreviated generic name preceding it. The generic name is always capitalized; the specific epithet is generally not capitalized. Both names are always italicized or underlined. For example, the domestic dog, *Canis familiaris* (abbreviated *C. familiaris*), and the timber wolf, *Canis lupus* (*C. lupus*), belong to the same genus. The domestic cat, *Felis catus*, belongs to a different genus. The scientific name of the American white oak is *Quercus alba*, whereas the name of the European white oak is *Quercus robur*. Another tree, the white willow, *Salix alba*, belongs to a different genus. The scientific name for our own species is *Homo sapiens* ("wise man").

Taxonomic classification is hierarchical

Just as closely related species may be grouped in a common genus, related genera can be grouped in a more inclusive group, a **family.** Families are grouped into **orders,** orders into **classes,** and classes into **phyla** (sing., *phylum*). Biologists group phyla into **kingdoms,** and kingdoms are assigned to **domains.** Each formal grouping at any given level is a **taxon** (pl., *taxa*). Note that each taxon is more inclusive than the taxon below it. Together they form a hierarchy ranging from species to domain (❚ Table 1-1 and ❚ Fig. 1-8).

Consider a specific example. The family Canidae, which includes all doglike carnivores (animals that eat mainly meat), consists of 12 genera and about 34 living species. Family Canidae, along with family Ursidae (bears), family Felidae (catlike animals), and several other families that eat mainly meat, are all placed in order Carnivora. Order Carnivora, order Primates (to

Key Point

Biologists use a hierarchical classification scheme with a series of taxonomic categories from species to domain; each category is more general and more inclusive than the one below it.

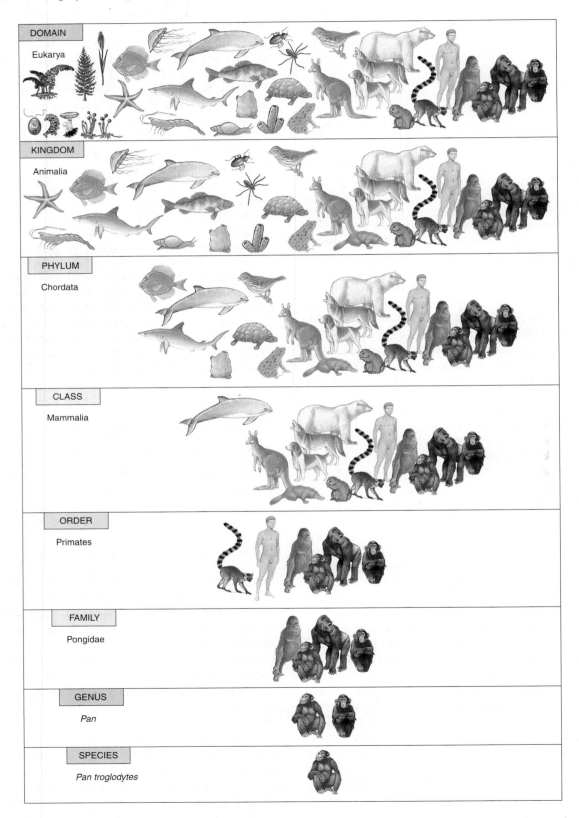

DOMAIN
Eukarya

KINGDOM
Animalia

PHYLUM
Chordata

CLASS
Mammalia

ORDER
Primates

FAMILY
Pongidae

GENUS
Pan

SPECIES
Pan troglodytes

Figure 1-8 Classification of the chimpanzee (*Pan troglodytes*)

which chimpanzees and humans belong), and several other orders belong to class Mammalia (mammals). Class Mammalia is grouped with several other classes that include fishes, amphibians, reptiles, and birds in subphylum Vertebrata. The vertebrates belong to phylum Chordata, which is part of kingdom Animalia. Animals are assigned to domain Eukarya.

The tree of life includes three domains and six kingdoms

Systematics has itself evolved as scientists have developed new molecular techniques. As researchers report new data, the classification of organisms changes. Systematists have developed a **tree of life**, a family tree showing the relationships among organisms, that is based on similarities in the organisms' molecules. Although the tree of life is a work in progress, most biologists now assign organisms to three domains and six kingdoms (❙ Fig. 1-9).

Bacteria have long been recognized as unicellular prokaryotic cells; they differ from all other organisms (except archaea) in that they are **prokaryotes.** Microbiologist Carl Woese (pronounced "woes") has been a pioneer in developing molecular approaches to systematics. Woese and his colleagues selected a molecule known as small subunit ribosomal RNA (rRNA) that functions in the process of manufacturing proteins in all organisms. Because its molecular structure differs somewhat in various organisms, Woese hypothesized that the molecular composition of rRNA in closely related organisms would be more similar than in distantly related organisms.

Woese's findings showed that there are two distinct groups of prokaryotes. He established the domain level of taxonomy and assigned these prokaryotes to two domains: **Bacteria** and **Archaea.** The **eukaryotes,** organisms with eukaryotic cells, are classified in domain **Eukarya.** Woese's work became widely accepted in the mid-1990s.

In the classification system used in this book, every organism is assigned to a domain and to one of six kingdoms. Two kingdoms correspond to the prokaryotic domains: kingdom Archaea corresponds to domain Archaea, and kingdom Bacteria corresponds to domain Bacteria. The remaining four kingdoms are assigned to domain Eukarya.

Kingdom **Protista** consists of protozoa, algae, water molds, and slime molds. These are unicellular or simple multicellular organisms. Some protists are adapted to carry out photosynthesis, the process in which light energy is converted to the chemical energy of food molecules. Members of kingdom **Plantae** are complex multicellular organisms adapted to carry out photosynthesis. Among characteristic plant features are the *cuticle* (a waxy covering over aerial parts that reduces water loss) and *stomata* (tiny openings in stems and leaves for gas exchange); many plants have multicellular *gametangia* (organs that protect developing reproductive cells). Kingdom Plantae includes both nonvascular plants (mosses) and vascular plants (ferns, conifers, and flowering plants), those that have tissues specialized for transporting materials throughout the plant body. Most plants are adapted to terrestrial environments.

Kingdom **Fungi** is composed of the yeasts, mildews, molds, and mushrooms. These organisms do not photosynthesize. They obtain their nutrients by secreting digestive enzymes into food and then absorbing the predigested food. Kingdom **Animalia** is made up of multicellular organisms that eat other organisms for nutrition. Most animals exhibit considerable tissue specialization and body organization. These characters have evolved along with complex sense organs, nervous systems, and muscular systems.

We have provided an introduction here to the groups of organisms that make up the tree of life. We will refer to them throughout this book, as we consider the many kinds of challenges organisms face and the various adaptations that have evolved in response to them. We discuss the diversity of life in more detail in Chapters 23 through 31, and we summarize classification in Appendix B.

Species adapt in response to changes in their environment

Every organism is the product of numerous interactions between environmental conditions and the genes inherited from its ancestors. If all individuals of a species were exactly alike, any change in the environment might be disastrous to all, and the species would become extinct. Adaptations to changes in the environment occur as a result of evolutionary processes that take place over time and involve many generations.

Natural selection is an important mechanism by which evolution proceeds

Although philosophers and naturalists discussed the concept of evolution for centuries, Charles Darwin and Alfred Wallace first brought a theory of evolution to general attention and suggested a plausible mechanism, **natural selection,** to explain it. In his book *On the Origin of Species by Natural Selection,* published in 1859, Darwin synthesized many new findings in geology and biology. He presented a wealth of evidence supporting his hypothesis that present forms of life descended, with modifications, from previously existing forms. Darwin's book raised a storm of controversy in both religion and science, some of which still lingers.

Darwin's theory of evolution has helped shape the biological sciences to the present day. His work generated a great wave of scientific observation and research that has provided much additional evidence that evolution is responsible for the great diversity of organisms on our planet. Even today, the details of evolutionary processes are a major focus of investigation and discussion.

Darwin based his theory of natural selection on the following four observations: (1) Individual members of a species show some variation from one another. (2) Organisms produce many more offspring than will survive to reproduce (❙ Fig. 1-10). (3) Organisms compete for necessary resources such as food, sunlight, and space. Individuals with characteristics that enable them to obtain

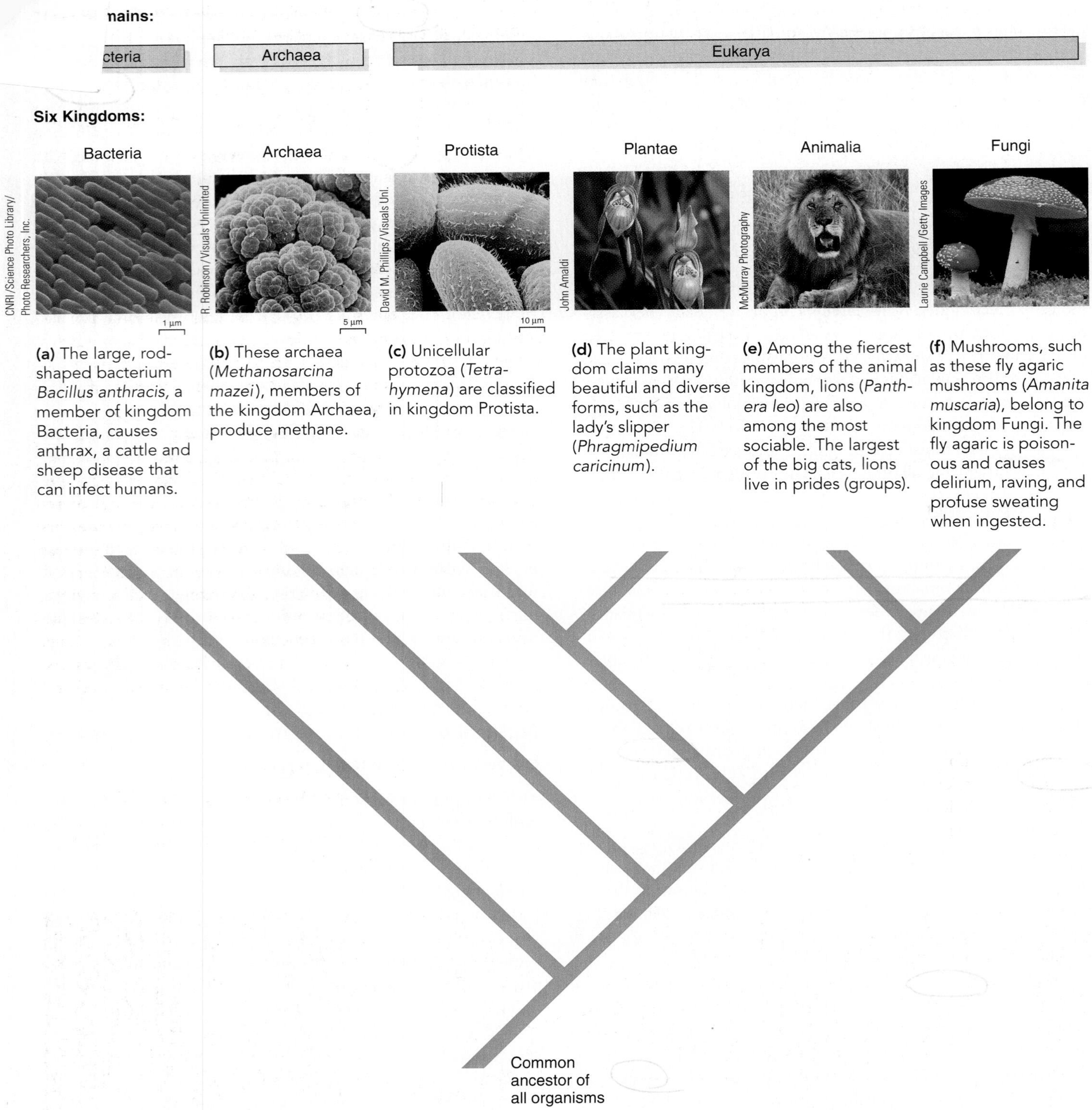

| cteria | Archaea | Eukarya |

Six Kingdoms:

| Bacteria | Archaea | Protista | Plantae | Animalia | Fungi |

(a) The large, rod-shaped bacterium *Bacillus anthracis,* a member of kingdom Bacteria, causes anthrax, a cattle and sheep disease that can infect humans.

(b) These archaea (*Methanosarcina mazei*), members of the kingdom Archaea, produce methane.

(c) Unicellular protozoa (*Tetrahymena*) are classified in kingdom Protista.

(d) The plant kingdom claims many beautiful and diverse forms, such as the lady's slipper (*Phragmipedium caricinum*).

(e) Among the fiercest members of the animal kingdom, lions (*Panthera leo*) are also among the most sociable. The largest of the big cats, lions live in prides (groups).

(f) Mushrooms, such as these fly agaric mushrooms (*Amanita muscaria*), belong to kingdom Fungi. The fly agaric is poisonous and causes delirium, raving, and profuse sweating when ingested.

Common ancestor of all organisms

Figure 1-9 *Animated* A survey of the kingdoms of life

In this book, organisms are assigned to three domains and six kingdoms.

and use resources are more likely to survive to reproductive maturity and produce offspring. (4) The survivors that reproduce pass their adaptations for survival on to their offspring. Thus, the best-adapted individuals of a population leave, on average, more offspring than do other individuals. Because of this differential reproduction, a greater proportion of the population becomes adapted to the prevailing environmental conditions. The environment *selects* the best-adapted organisms for survival. Note that *adaptation involves changes in populations rather than in individual organisms.*

Figure 1-10 Egg masses of the wood frog (*Rana sylvatica*)

Many more eggs are produced than can possibly develop into adult frogs. Random events are largely responsible for determining which of these developing frogs will hatch, reach adulthood, and reproduce. However, certain traits of each organism also contribute to its probability for success in its environment. Not all organisms are as prolific as the frog, but the generalization that more organisms are produced than survive is true throughout the living world.

Darwin did not know about DNA or understand the mechanisms of inheritance. Scientists now understand that most variations among individuals are a result of different varieties of genes that code for each characteristic. The ultimate source of these variations is random **mutations,** chemical or physical changes in DNA that persist and can be inherited. Mutations modify genes and by this process provide the raw material for evolution.

Populations evolve as a result of selective pressures from changes in their environment

All the genes present in a population make up its **gene pool.** By virtue of its gene pool, a population is a reservoir of variation. Natural selection acts on individuals within a population. Selection favors individuals with genes specifying traits that allow them to respond effectively to pressures exerted by the environment. These organisms are most likely to survive and produce offspring. As successful organisms pass on their genetic recipe for survival, their traits become more widely distributed in the population. Over time, as populations continue to change (and as the environment itself changes, bringing different selective pressures), the members of the population become better adapted to their environment and less like their ancestors.

As a population adapts to environmental pressures and exploits new opportunities for finding food, maintaining safety, and avoiding predators, the population diversifies and new species may evolve. The Hawaiian honeycreepers, a group of related birds, are a good example. When honeycreeper ancestors first reached Hawaii, few other birds were present, so there was little competition. Genetic variation among honeycreepers allowed some to move into different food zones, and over time, species with various types of bills evolved (❚ Fig. 1-11; see also Chapter 20 and Fig. 20-16). Some honeycreepers now have long, curved bills, adapted for feeding on nectar from tubular flowers. Others have short, thick bills for foraging for insects, and still others have adapted for eating seeds.

Review

- ❚ What is the binomial system of nomenclature?
- ❚ What are the three domains and six kingdoms of living organisms?
- ❚ How might you explain the sharp claws and teeth of tigers in terms of natural selection?

(a) The bill of this 'Akiapola'au male (*Hemignathus munroi*) is adapted for extracting insect larvae from bark. The lower mandible (jaw) is used to peck at and pull off bark, whereas the maxilla (upper jaw) and tongue remove the prey.

(b) 'I'iwi (*Vestiaria coccinea*) in 'ohi'a blossoms. The bill is adapted for feeding on nectar in tubular flowers.

(c) Palila (*Loxiodes bailleui*) in mamane tree. This finch-billed honeycreeper feeds on immature seeds in pods of the mamane tree. It also eats insects, berries, and young leaves.

Figure 1-11 Adaptation and diversification in Hawaiian honeycreepers

All three species shown here are endangered, mainly because their habitats have been destroyed by humans.

THE ENERGY FOR LIFE

Learning Objective

9   Summarize the flow of energy through ecosystems, and contrast the roles of producers, consumers, and decomposers.

Life depends on a continuous input of energy from the sun, because every activity of a living cell or organism requires energy. Whenever energy is used to perform biological work, some is converted to heat and dispersed into the environment.

Recall that all the energy transformations and chemical processes that occur within an organism are referred to as its *metabolism.* Energy is necessary to carry on the metabolic activities essential for growth, repair, and maintenance. Each cell of an organism requires nutrients that contain energy. Certain nutrients are used as fuel for **cellular respiration,** a process that releases some of the energy stored in the nutrient molecules (Fig. 1-12). The cell can use this energy to do work, including the synthesis of required materials, such as new cell components. Virtually all cells carry on cellular respiration.

Like individual organisms, ecosystems depend on a continuous input of energy. A self-sufficient ecosystem contains three types of organisms—producers, consumers, and decomposers—and includes a physical environment in which they can survive. These organisms depend on one another and on the environment for nutrients, energy, oxygen, and carbon dioxide. However, there is a one-way flow of energy through ecosystems. Organisms can neither create energy nor use it with complete efficiency.

During every energy transaction, some energy disperses into the environment as heat and is no longer available to the organism (Fig. 1-13).

Producers, or **autotrophs,** are plants, algae, and certain bacteria that produce their own food from simple raw materials. Most of these organisms use sunlight as an energy source and carry out **photosynthesis,** the process in which producers synthesize complex molecules from carbon dioxide and water. The light energy is transformed into chemical energy, which is stored within the chemical bonds of the food molecules produced. Oxy-

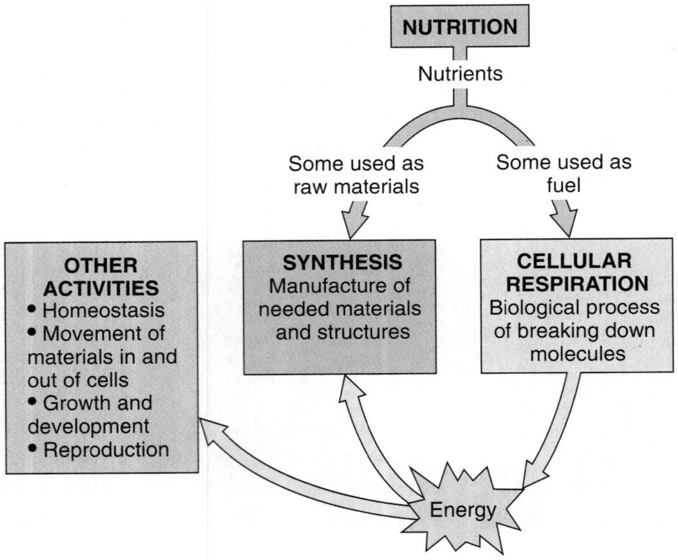

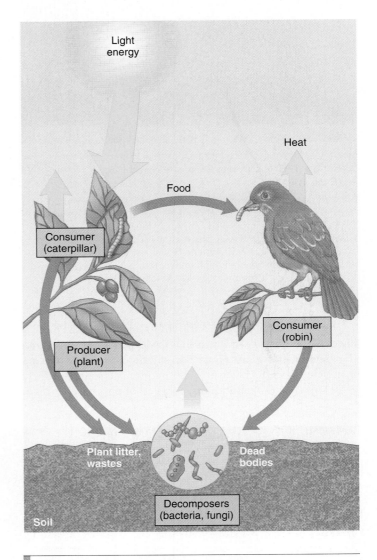

Figure 1-13 *Animated* Energy flow

Continuous energy input from the sun operates the biosphere. During photosynthesis, producers use the energy from sunlight to make complex molecules from carbon dioxide and water. Consumers, such as the caterpillar and robin shown here, obtain energy, carbon, and other required materials when they eat producers or consumers that have eaten producers. Wastes and dead organic material supply decomposers with energy and carbon. During every energy transaction, some energy is lost to biological systems, dispersing into the environment as heat.

Figure 1-12 Relationships among metabolic processes

These processes occur continuously in the cells of living organisms. Cells use some of the nutrients in food to synthesize required materials and cell parts. Cells use other nutrients as fuel for cellular respiration, a process that releases energy stored in food. This energy is required for synthesis and for other forms of cell work.

gen, which is required by plant cells and the cells of most other organisms, is produced as a by-product of photosynthesis:

Carbon dioxide + water + light energy →
$\qquad\qquad$ sugars (food) + oxygen

Animals are **consumers,** or **heterotrophs**—that is, organisms that depend on producers for food, energy, and oxygen. Consumers obtain energy by breaking down sugars and other food molecules originally produced during photosynthesis. When chemical bonds are broken during this process of cellular respiration, their stored energy is made available for life processes:

Sugars (and other food molecules) + oxygen →
$\qquad\qquad$ carbon dioxide + water + energy

Consumers contribute to the balance of the ecosystem. For example, consumers produce carbon dioxide required by producers. (Note that producers also carry on cellular respiration.) The metabolism of consumers and producers helps maintain the life-sustaining mixture of gases in the atmosphere.

Most bacteria and fungi are **decomposers,** heterotrophs that obtain nutrients by breaking down nonliving organic material such as wastes, dead leaves and branches, and the bodies of dead organisms. In their process of obtaining energy, decomposers make the components of these materials available for reuse. If decomposers did not exist, nutrients would remain locked up in wastes and dead bodies, and the supply of elements required by living systems would soon be exhausted.

Review

▪ What components do you think a balanced forest ecosystem might include?

▪ In what ways do consumers depend on producers? On decomposers? Include energy considerations in your answer.

THE PROCESS OF SCIENCE

Learning Objectives

10 Design a study to test a given hypothesis, using the procedure and terminology of the scientific method.

11 Compare the reductionist and systems approaches to biological research.

Biology is a science. The word *science* comes from a Latin word meaning "to know." Science is a way of thinking and a method of investigating the natural world in a systematic manner. We test ideas, and based on our findings, we modify or reject these ideas. The **process of science** is investigative, dynamic, and often controversial. The observations made, the range of questions asked, and the design of experiments depend on the creativity of the individual scientist. However, science is influenced by cultural, social, historical, and technological contexts, so the process changes over time.

The **scientific method** involves a series of ordered steps. Using the scientific method, scientists make careful observa-

Figure 1-14 Biologist at work

This biologist studying the rainforest canopy in Costa Rica is part of an international effort to study and preserve tropical rain forests. Researchers study the interactions of organisms and the effects of human activities on the rain forests.

Mark Moffett / Minden Pictures

tions, ask critical questions, and develop *hypotheses*, which are tentative explanations. Using their hypotheses, scientists make predictions that can be tested by making further observations or by performing experiments. They gather **data,** information that they can analyze, often using computers and sophisticated statistical methods. They interpret the results of their experiments and draw conclusions from them. As we will discuss, scientists pose many hypotheses that cannot be tested by using all of the steps of the scientific method in a rigid way. Scientists use the scientific method as a generalized framework or guide.

Biologists explore every imaginable aspect of life from the structure of viruses and bacteria to the interactions of the communities of our biosphere. Some biologists work mainly in laboratories, and others do their work in the field (▮ Fig. 1-14). Perhaps you will decide to become a research biologist and help unravel the complexities of the human brain, discover new hormones that cause plants to flower, identify new species of animals or bacteria, or develop new stem cell strategies to treat cancer, AIDS, or heart disease. Applications of basic biological research have provided the technology to transplant kidneys, livers, and hearts; manipulate genes; treat many diseases; and increase world food production. Biology has been a powerful force in providing the quality of life that most of us enjoy. You may choose to enter an applied field of biology, such as environmental science, dentistry, medicine, pharmacology, or veterinary medicine. Many interesting careers in the biological sciences are discussed in the Careers section on our website.

C. Dani /Peter Arnold

Figure 1-15   Is this animal a bird?

The kiwi bird of New Zealand is about the size of a chicken. Its tiny 2-inch wings cannot be used for flight. The survivor of an ancient order of birds, the kiwi has bristly, hairlike feathers and other characteristics that qualify it as a bird.

Science requires systematic thought processes

Science is systematic. Scientists organize, and often quantify, knowledge, making it readily accessible to all who wish to build on its foundation. In this way, science is both a personal and a social endeavor. Science is not mysterious. Anyone who understands its rules and procedures can take on its challenges. What distinguishes science is its insistence on rigorous methods to examine a problem. Science seeks to give precise knowledge about the natural world; the supernatural is not accessible to scientific methods of inquiry. Science is not a replacement for philosophy, religion, or art. Being a scientist does not prevent one from participating in other fields of human endeavor, just as being an artist does not prevent one from practicing science.

Deductive reasoning begins with general principles

Scientists use two types of systematic thought processes: deduction and induction. With **deductive reasoning,** we begin with supplied information, called *premises,* and draw conclusions on the basis of that information. Deduction proceeds from general principles to specific conclusions. For example, if you accept the premise that all birds have wings and the second premise that sparrows are birds, you can conclude deductively that sparrows have wings. Deduction helps us discover relationships among known facts. A **fact** is information, or knowledge, based on evidence.

Inductive reasoning begins with specific observations

Inductive reasoning is the opposite of deduction. We begin with specific observations and draw a conclusion or discover a general principle. For example, you know that sparrows have wings, can fly, and are birds. You also know that robins, eagles, pigeons, and hawks have wings, can fly, and are birds. You might induce that

all birds have wings and fly. In this way, you can use the inductive method to organize raw data into manageable categories by answering this question: What do all these facts have in common?

A weakness of inductive reasoning is that conclusions generalize the facts to all possible examples. When we formulate the general principle, we go from many observed examples to all possible examples. This is known as an *inductive leap.* Without it, we could not arrive at generalizations. However, we must be sensitive to exceptions and to the possibility that the conclusion is not valid. For example, the kiwi bird of New Zealand does *not* have functional wings (▌ Fig. 1-15). We can never conclusively prove a universal generalization. The generalizations in inductive conclusions come from the creative insight of the human mind, and creativity, however admirable, is not infallible.

Scientists make careful observations and ask critical questions

In 1928, British bacteriologist Alexander Fleming observed that a blue mold had invaded one of his bacterial cultures. He almost discarded it, but then he noticed that the area contaminated by the mold was surrounded by a zone where bacterial colonies did not grow well. The bacteria were disease organisms of the genus *Staphylococcus,* which can cause boils and skin infections. Anything that could kill them was interesting! Fleming saved the mold, a variety of *Penicillium* (blue bread mold). He isolated the antibiotic penicillin from the mold. However, he had difficulty culturing it.

Even though Fleming recognized the potential practical benefit of penicillin, he did not develop the chemical techniques needed to purify it, and more than 10 years passed before the drug was put to significant use. In 1939, Sir Howard Florey and Ernst Boris Chain developed chemical procedures to extract and produce the active agent penicillin from the mold. Florey took the process to laboratories in the United States, and penicillin was first produced to treat wounded soldiers in World War II. In recognition of their work, Fleming, Florey, and Chain shared the 1945 Nobel Prize in Physiology or Medicine.

Chance often plays a role in scientific discovery

Fleming did not set out to discover penicillin. He benefited from the chance growth of a mold in one of his culture dishes. However, we may wonder how many times the same type of mold grew on the cultures of other bacteriologists who failed to make the connection and simply threw away their contaminated cultures. Fleming benefited from chance, but his mind was prepared to make observations and formulate critical questions, and his pen was prepared to publish them. Significant discoveries are usually made by those who are in the habit of looking critically at nature and recognizing a phenomenon or problem. Of course, the technology necessary for investigating the problem must also be available.

A hypothesis is a testable statement

Scientists make careful observations, ask critical questions, and develop hypotheses. A **hypothesis** is a tentative explanation for observations or phenomena. Hypotheses can be posed as "if . . . then . . ." statements. For example, *if* students taking introductory biology attend classes, *then* they will make a higher grade on the exam than students who do not attend classes.

In the early stages of an investigation, a scientist typically thinks of many possible hypotheses. A good hypothesis exhibits the following characteristics: (1) It is reasonably consistent with well-established facts. (2) It is capable of being tested; that is, it should generate definite predictions, whether the results are positive or negative. Test results should also be repeatable by independent observers. (3) It is falsifiable, which means it can be proven false, as we will discuss in the next section.

After generating hypotheses, the scientist decides which, if any, could and should be subjected to experimental test. Why not test them all? Time and money are important considerations in conducting research. Scientists must establish priority among the hypotheses to decide which to test first.

A falsifiable hypothesis can be tested

In science, a well-stated hypothesis can be tested. If no evidence is found to support it, the hypothesis is rejected. The hypothesis can be shown to be false. Even results that do not support the hypothesis may be valuable and may lead to new hypotheses. If the results do support a hypothesis, a scientist may use them to generate related hypotheses.

Let us consider a hypothesis that we can test by careful observation: Female mammals (animals that have hair and produce milk for their young) bear live young. The hypothesis is based on the observations that dogs, cats, cows, lions, and humans all are mammals and all bear live young. Consider further that a new species, species X, is identified as a mammal. Biologists predict that females of species X will bear live young. (Is this inductive or deductive reasoning?) If a female of the new species gives birth to live offspring, the hypothesis is supported.

Before the Southern Hemisphere was explored, most people would probably have accepted the hypothesis without question, because all known furry, milk-producing animals did, in fact, bear live young. But biologists discovered that two Australian animals (the duck-billed platypus and the spiny anteater) had fur and produced milk for their young but laid eggs (❙ Fig. 1-16). The hypothesis, as stated, was false no matter how many times it had previously been supported. As a result, biologists either had to consider the platypus and the spiny anteater as nonmammals or had to broaden their definition of mammals to include them. (They chose to do the latter.)

A hypothesis is not true just because some of its predictions (the ones people happen to have thought of or have thus far been able to test) have been shown to be true. After all, they could be true by coincidence. In fact, a hypothesis can be supported by data, but it cannot really be *proven* true.

An **unfalsifiable hypothesis** cannot be proven false; in fact, it cannot be scientifically investigated. Belief in an unfalsifiable hypothesis, such as the existence of invisible and undetectable elves, must be rationalized on grounds other than scientific ones.

Figure 1-16 Is this animal a mammal?

The duck-billed platypus (*Ornithorhynchus anatinus*) is classified as a mammal because it has fur and produces milk for its young. However, unlike most mammals, it lays eggs.

Models are important in developing and testing hypotheses

Hypotheses have many potential sources, including direct observations or even computer simulations. Increasingly in biology, hypotheses may be derived from **models** that scientists have developed to provide a comprehensive explanation for a large number of observations. Examples of such testable models include the model of the structure of DNA and the model of the structure of the plasma membrane (discussed in Chapter 5).

The best design for an experiment can sometimes be established by performing computer simulations. Virtual testing and evaluation are undertaken before the experiment is performed in the laboratory or field. Modeling and computer simulation save time and money.

Many predictions can be tested by experiment

A hypothesis is an abstract idea, but based on their hypotheses, scientists can make predictions that can be tested. For example, we might predict that biology students who study for 10 hours will do better on an exam than students who do not study. As used here, a **prediction** is a deductive, logical consequence of a hypothesis. It does not have to be a future event.

Some predictions can be tested by controlled experiments. Early biologists observed that the nucleus was the most prominent part of the cell, and they hypothesized that cells would be adversely affected if they lost their nuclei. Biologists predicted that if the nucleus were removed from the cell, then the cell would die. They then experimented by surgically removing the nucleus of a unicellular amoeba. The amoeba continued to live and move, but it did not grow, and after a few days it died. These results suggested that the nucleus is necessary for the metabolic processes that provide for growth and cell reproduction.

But, the investigators asked, what if the operation itself, not the loss of the nucleus, caused the amoeba to die? They performed a controlled experiment, subjecting two groups of amoebas to the same operative trauma (■ Fig. 1-17). In the experimental group, the nucleus was removed; in the control group, it was not. Ideally, an **experimental group** differs from a control group only with respect to the variable being studied. In the **control group,** the researcher inserted a microloop into each amoeba and pushed it around inside the cell to simulate removal of the nucleus; then the instrument was withdrawn, leaving the nucleus inside. Amoebas treated with such a sham operation recovered and subsequently grew and divided, but the amoebas without nuclei died. This experiment showed that the removal of the nucleus, not simply the operation, caused the death of the amoebas. The conclusion is that amoebas cannot live without their nuclei. The results support the hypothesis that if cells lose their nuclei, they are adversely affected. The nucleus is essential for the survival of the cell.

Researchers must avoid bias

In scientific studies, researchers must avoid bias or preconceived notions of what should happen. For example, to prevent bias, most medical experiments today are carried out in a double-blind fashion. When a drug is tested, one group of patients receives the new medication, and a control group of matched patients receives a placebo (a harmless starch pill similar in size, shape, color, and taste to the pill being tested). This is a *double-blind study,* because neither the patient nor the physician knows who is getting the experimental drug and who is getting the placebo. The pills or treatments are coded in some way, and the code is broken only after the experiment is over and the results are recorded. Not all experiments can be so neatly designed; for example, it is often difficult to establish appropriate controls.

Scientists interpret the results of experiments and make conclusions

Scientists gather data in an experiment, interpret their results, and then draw conclusions from them. In the amoeba experiment described earlier, investigators concluded that the data supported the hypothesis that the nucleus is essential for the survival of the cell. Even results that do not support the hypothesis may be valuable and may lead to new hypotheses. If the results do support a hypothesis, a scientist may use them to generate related hypotheses.

OBSERVATION: The nucleus is the most prominent part of the cell.
ASK CRITICAL QUESTIONS: Why is the nucleus so large? What is its importance?
DEVELOP HYPOTHESIS: Cells will be adversely affected if they lose their nuclei.
MAKE A PREDICTION THAT CAN BE TESTED: If the nucleus is removed from an amoeba, the amoeba will die.
PERFORM EXPERIMENTS TO TEST THE PREDICTION: Using a microloop, researchers removed the nucleus from each amoeba in the experimental group. Amoebas in the control group were subjected to the same surgical procedure, but their nuclei were not removed.

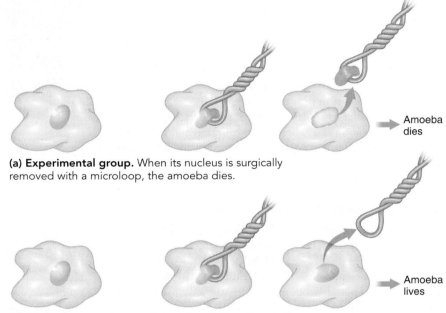

Amoeba dies

(a) Experimental group. When its nucleus is surgically removed with a microloop, the amoeba dies.

Amoeba lives

(b) Control group. A control amoeba subjected to similar surgical procedures (including insertion of a microloop), but without actual removal of the nucleus, does not die.

RESULTS: Amoebas without nuclei died. Amoebas in the control group lived.
CONCLUSION: Amoebas cannot live without their nuclei. The hypothesis is supported.

■ **Figure 1-17** Testing a prediction regarding the importance of the nucleus

Let us discuss another experiment. Research teams studying chimpanzee populations in Africa have observed that chimpanzees can learn specific ways to use tools from one another. Behavior that is learned from others in a population and passed to future generations is what we call culture. In the past, most biologists have thought that only humans had culture. It has been difficult to test this type of learning in the field, and the idea has been controversial. Investigators at Yerkes National Primate Research Center in Atlanta recently performed an experiment to test their hypothesis that chimpanzees can learn particular ways to use tools by observing other chimps (■ Fig. 1-18).

The investigators predicted that if they taught one chimp to use a stick to obtain food from a dispenser, other chimps would learn the technique from the educated one. They divided chimpanzees into two experimental groups with 16 in each group. Then they taught a high-ranking female in each group to use a stick to obtain food from an apparatus. The two chimps were taught different methods. One chimp was taught to poke the stick inside the device to free the food. The other was taught to use the

OBSERVATION: African chimpanzees appear to mimic the use of tools.

ASK CRITICAL QUESTIONS: Do chimpanzees learn how to use tools from one another?

DEVELOP HYPOTHESIS: Chimpanzees can learn particular ways to use tools by observing other chimps.

MAKE A PREDICTION THAT CAN BE TESTED: If one chimp is taught to use a stick to obtain food from a dispenser, other chimps will learn the technique from the educated chimp.

PERFORM EXPERIMENTS TO TEST THE PREDICTION: One female in each of two groups of 16 chimps was educated in a specific way to use a stick to obtain food. The two educated chimps were then returned to their respective groups. In a control group, chimpanzees were not taught how to use a stick.

David Bygott

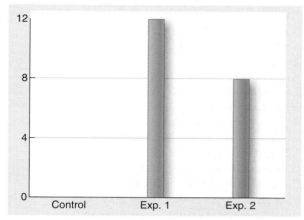

(a) Number of chimpanzees successfully employing specific method of tool use

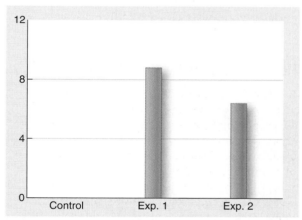

(b) Number of chimpanzees successfully employing learned method of tool use two months later

RESULTS: Chimps in each experimental group observed the use of the stick by the educated chimp, and a majority began to use the stick in the same way. When tested two months later, many of the chimps in each group continued to use the stick. The results presented here have been simplified and are based on the number of chimps observed to use the learned method at least 10 times. All but one chimp in each group learned the technology, but a few used it only a few times. Some chimps taught themselves the alternative method and used that alternative. However, most conformed to the group's use of the method that the investigator taught to the educated chimp.

CONCLUSION: Chimpanzees learn specific ways to use tools by observing other chimps. The hypothesis is supported.

Figure 1-18 Testing a prediction about learning in chimpanzee populations

Wild chimpanzees are shown observing a member of their group using a tool.

stick to lift a hook that removed a blockage, allowing the food to roll forward out of the device.

A third group served as a control group. The chimps in the control group were given access to the sticks and the apparatus with the food inside but were not taught how to use the sticks. All of the control-group chimps manipulated the apparatus with the stick, but none succeeded in releasing food.

When the chimps were returned to their groups, other chimps observed how the educated chimps used the stick, and a large majority began to use sticks in the same way. The chimps in each group learned the specific style of using the stick that their educated chimp had been taught. All but two of the chimps learned to use the stick. Two months later, the apparatus was reintroduced to the chimps. Again, most of the chimps used the learned technique for obtaining food. The results of the experiment supported the hypothesis. The researchers concluded that chimpanzees are capable of culturally transmitting learned technology.

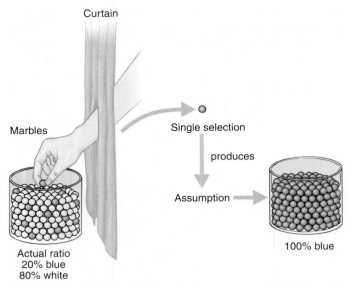

(a) Taking a single selection can result in sampling error. If the only marble selected is blue, we might assume all the marbles are blue.

Curtain

Marbles

Single selection

produces

Assumption

100% blue

Actual ratio
20% blue
80% white

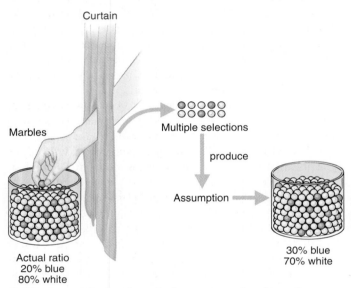

Curtain

Marbles

Multiple selections

produce

Assumption

30% blue
70% white

Actual ratio
20% blue
80% white

(b) The greater the number of selections we take of an unknown, the more likely we can make valid assumptions about it.

Figure 1-19 *Animated* Statistical probability

Sampling error can lead to inaccurate conclusions

One reason for inaccurate conclusions is *sampling error*. Because not *all* cases of what is being studied can be observed or tested (scientists cannot study every amoeba or every chimpanzee population), scientists must be content with a sample. How can scientists know whether that sample is truly representative of whatever they are studying? If the sample is too small, it may not be representative because of random factors. A study with only two, or even nine, amoebas may not yield reliable data that can be generalized to other amoebas. If researchers test a large number of subjects, they are more likely to draw accurate scientific conclusions (▌Fig. 1-19). The scientist seeks to state with some level of confidence that any specific conclusion has a certain statistical probability of being correct.

Experiments must be repeatable

When researchers publish their findings in a scientific journal, they typically describe their methods and procedures in sufficient detail so that other scientists can repeat the experiments. When the findings are replicated, the conclusions are, of course, strengthened.

A theory is supported by tested hypotheses

Nonscientists often use the word *theory* incorrectly to refer to a hypothesis or even to some untestable idea they wish to promote. A **theory** is actually an integrated explanation of some aspect of the natural world that is based on a number of hypotheses, each supported by consistent results from many observations or experiments. A theory relates data that previously appeared unrelated. A good theory grows, building on additional facts as they become known. It predicts new facts and suggests new relationships among phenomena. It may even suggest practical applications.

A good theory, by showing the relationships among classes of facts, simplifies and clarifies our understanding of the natural world. As Einstein wrote, "In the whole history of science from Greek philosophy to modern physics, there have been constant attempts to reduce the apparent complexity of natural phenomena to simple, fundamental ideas and relations." Developing theories is indeed a major goal of science.

Many hypotheses cannot be tested by direct experiment

Some well-accepted theories do not lend themselves to hypothesis testing by ordinary experiments. Often, these theories describe events that occurred in the distant past. We cannot directly observe the origin of the universe from a very hot, dense state about 13.7 billion years ago (the Big Bang theory). However, physicists and cosmologists have been able to formulate many hypotheses related to the Big Bang and to test many of the predictions derived from these hypotheses.

Similarly, humans did not observe the evolution of major groups of organisms because that process took place over millions of years and occurred before humans had evolved. However, many hypotheses about evolution have been posed, and predictions based on them have been tested. For example, if complex organisms evolved from simple life-forms, we would find the fossils of the simplest organisms in the oldest strata (rock layers). As we explore more recent strata, we would expect to find increasingly complex organisms. Indeed, scientists have found this progression of simple to complex fossils. In addition to fossils, evidence for evolution comes from many sources, including physical and molecular similarities between organisms. Evidence also comes from recent and current studies of evolution in action. Many aspects of ongoing evolution can, in fact, be studied in the laboratory or in the field. The evidence for evolution is so compelling that almost all scientists today accept evolution as a well-established theory.

Paradigm shifts allow new discoveries

A **paradigm** is a set of assumptions or concepts that constitute a way of thinking about reality. For example, from the time of Aristotle to the mid-19th century, biologists thought that organisms were either plants (kingdom Plantae) or animals (kingdom Animalia). This concept was deeply entrenched. However, with the development of microscopes, investigators discovered tiny life-forms—bacteria and protists—that were neither plant nor animal. Some biologists were able to make a **paradigm shift,** that is, they changed their view of reality, to accommodate this new knowledge. They assigned these newly discovered organisms to new kingdoms.

Systems biology integrates different levels of information

In the *reductionist* approach to biology, researchers study the simplest components of biological processes. Their goal is to synthesize their knowledge of many small parts to understand the whole. Reductionism has been (and continues to be) important in biological research. However, as biologists and their tools have become increasingly sophisticated, huge amounts of data have been generated, bringing the science of biology to a different level.

Systems biology is a field of biology that builds on information provided by the reductionist approach, adding large data sets generated by computers. Systems biology is also referred to as *integrative biology* or *integrative systems biology.* Reductionism and systems biology are complementary approaches. Using reductionism, biologists have discovered basic information about components, such as molecules, genes, cells, and organs. Systems biologists, who focus on systems as a whole rather than on individual components, need this basic knowledge to study, for example, the *interactions* among various parts and levels of an organism.

Systems biologists integrate data from various levels of complexity with the goal of understanding the big picture—how biological systems function. For example, systems biologists are developing models of different aspects of cell function. One group of researchers has developed a model consisting of nearly 8000 chemical signals involved in a molecular network that leads to programmed cell death. By understanding cell communication, the interactions of genes and proteins in metabolic pathways, and physiological processes, systems biologists hope to eventually develop a model of the whole organism.

The development of systems biology has been fueled by the huge amount of data generated by the Human Genome Project.

Researchers working in this project identified the DNA sequences that make up the estimated 25,000 genes of the *human genome,* the complete set of genes that make up the human genetic material. Computer software developed for the Human Genome Project can analyze large data sets. These programs are being used to integrate data about protein interactions and many other aspects of molecular biology. Systems biologists view biology in terms of information systems. Increasingly, they depend on mathematics, statistics, and engineering principles.

Science has ethical dimensions

Scientific investigation depends on a commitment to practical ideals, such as truthfulness and the obligation to communicate results. Honesty is particularly important in science. Consider the great (though temporary) damage done whenever an unprincipled or even desperate researcher, whose career may depend on the publication of a research study, knowingly disseminates false data. Until the deception is uncovered, researchers may devote thousands of dollars and hours of precious professional labor to futile lines of research inspired by erroneous reports. Deception can also be dangerous, especially in medical research. Fortunately, science tends to correct itself through consistent use of the scientific process. Sooner or later, someone's experimental results are sure to cast doubt on false data.

In addition to being ethical about their own work, scientists face many societal and political issues surrounding areas such as genetic research, stem cell research, cloning, and human and animal experimentation. For example, some stem cells that show great potential for treating human disease come from early embryos. The cells can be taken from 5- or 6-day-old human embryos and then cultured in laboratory glassware. (At this stage, the embryo is a group of cells about 0.15 mm long [0.006 in].) Such cells could be engineered to treat failing hearts or brains harmed by stroke, injury, Parkinson's disease, or Alzheimer's disease. They could save the lives of burn victims and perhaps be engineered to treat specific cancers. Scientists, and the larger society, will need to determine whether the potential benefits of any type of research outweigh its ethical risks.

The era of the genome brings with it ethical concerns and responsibilities. How do people safeguard the privacy of genetic information? How can we be certain that knowledge of our individual genetic codes would not be used against us when we seek employment or health insurance? Scientists must be ethically responsible and must help educate people about their work, including its benefits relative to its risks. Interestingly, at the very beginning of the Human Genome Project, part of its budget was allocated for research on the ethical, legal, and social implications of its findings.

Review

- What are the characteristics of a good hypothesis?
- What is meant by a "controlled" experiment?
- What is systems biology?

Learning Objectives

1 Describe three basic themes of biology (page 2).
   - Three basic themes of biology are evolution, transfer of information, and energy for life. The process of **evolution** results in populations changing over time and explains how the ancestry of organisms can be traced back to earlier forms of life. Information must be transmitted within cells, among cells, among organisms, and from one generation to the next. Life requires continuous energy from the sun.

2 Distinguish between living and nonliving things by describing the features that characterize living organisms (page 2).
   - Every living organism is composed of one or more **cells.** Living things grow by increasing the size and/or number of their cells.
   - **Metabolism** includes all the chemical activities that take place in the organism, including the chemical reactions essential to nutrition, growth and repair, and conversion of energy to usable forms. **Homeostasis** refers to the appropriate, balanced internal environment.
   - Organisms respond to **stimuli,** physical or chemical changes in their external or internal environment. Responses typically involve movement. Some organisms use tiny extensions of the cell, called **cilia,** or longer **flagella** to move from place to place. Other organisms are **sessile** and remain rooted to some surface.
   - In **asexual reproduction,** offspring are typically identical to the single parent; in **sexual reproduction,** offspring are the product of the fusion of gametes, and genes are typically contributed by two parents.
   - Populations evolve and become adapted to their environment. **Adaptations** are traits that increase an organism's ability to survive in its environment.

3 Construct a hierarchy of biological organization, including levels characteristic of individual organisms and ecological levels (page 6).
   - Biological organization is hierarchical. A complex organism is organized at the chemical, cell, **tissue, organ, organ system,** and organism levels. Cells associate to form tissues that carry out specific functions. In most multicellular organisms, tissues organize to form functional structures called organs, and an organized group of tissues and organs form an organ system. Functioning together, organ systems make up a complex, multicellular organism.
   - The basic unit of ecological organization is the **population.** Various populations form **communities;** a community and its physical environment are an **ecosystem;** all of Earth's ecosystems together make up the **biosphere.**

ThomsonNOW **Learn more about biological organization by clicking on the figure in ThomsonNOW.**

4 Summarize the importance of information transfer to living systems, giving specific examples (page 6).
   - Organisms transmit information chemically, electrically, and behaviorally.
   - **DNA,** which makes up the **genes,** is the hereditary material. Information encoded in DNA is transmitted from one generation to the next. DNA contains the instructions for the development of an organism and for carrying out life processes. DNA codes for **proteins,** which are impor-

tant in determining the structure and function of cells and tissues.
   - **Hormones,** chemical messengers that transmit messages from one part of an organism to another, are an important type of **cell signaling.**
   - Many organisms use electrical signals to transmit information; most animals have nervous systems that transmit electrical impulses and release **neurotransmitters.**

5 Demonstrate the binomial system of nomenclature by using several specific examples, and classify an organism (such as a human) in its domain, kingdom, phylum, class, order, family, genus, and species (page 9).
   - Millions of species have evolved. A **species** is a group of organisms with similar structure, function, and behavior that, in nature, breed only with one another. Members of a species have a common **gene pool** and share a common ancestry.
   - Biologists use a **binomial system of nomenclature** in which the name of each species includes a **genus** name and a **specific epithet.**
   - Taxonomic classification is hierarchical; it includes species, genus, **family, order, class, phylum, kingdom,** and **domain.** Each grouping is referred to as a **taxon.**

6 Identify the three domains and six kingdoms of living organisms, and give examples of organisms assigned to each group (page 9).
   - Bacteria and archaea have **prokaryotic cells;** all other organisms have **eukaryotic cells.** Prokaryotes make up two of the three domains.
   - Organisms are classified in three domains: **Archaea, Bacteria,** and **Eukarya;** and six kingdoms: **Archaea, Bacteria, Protista** (protozoa, algae, water molds, and slime molds), **Fungi** (molds and yeasts), **Plantae,** and **Animalia.**

ThomsonNOW **Learn more about life's diversity by clicking on the figure in ThomsonNOW.**

7 Give a brief overview of the theory of evolution, and explain why it is the principal unifying concept in biology (page 9).
   - **Evolution** is the process by which populations change over time in response to changes in the environment. The theory of evolution explains how millions of species came to be and helps us understand the structure, function, behavior, and interactions of organisms.
   - **Natural selection,** the major mechanism by which evolution proceeds, favors individuals with traits that enable them to cope with environmental changes. These individuals are most likely to survive and to produce offspring.
   - Charles Darwin based his theory of natural selection on his observations that individuals of a species vary; organisms produce more offspring than survive to reproduce; individuals that are best adapted to their environment are more likely to survive and reproduce; and as successful organisms pass on their hereditary information, their traits become more widely distributed in the population.
   - The source of variation in a population is random **mutation.**

8 Apply the theory of natural selection to any given adaptation, and suggest a logical explanation of how the adaptation may have evolved (page 9).

When the ancestors of Hawaiian honeycreepers first reached Hawaii, few other birds were present, so there was little competition for food. Through many generations, honeycreepers with longer, more curved bills became adapted for feeding on nectar from tubular flowers. Perhaps those with the longest, most curved bills were best able to survive in this food zone and lived to transmit their genes to their offspring. Those with shorter, thicker bills were more successful foraging for insects and passed their genes to new generations of offspring. Eventually, different species evolved that were adapted to specific food zones.

9 Summarize the flow of energy through ecosystems, and contrast the roles of producers, consumers, and decomposers (page 14).

   ■ Activities of living cells require energy; life depends on continuous energy input from the sun. During **photosynthesis,** plants, algae, and certain bacteria use the energy of sunlight to synthesize complex molecules from carbon dioxide and water.

   ■ Virtually all cells carry on **cellular respiration,** a biochemical process in which they capture the energy stored in nutrients by producers. Some of that energy is then used to synthesize required materials or to carry out other cell activities.

   ■ A self-sufficient ecosystem includes **producers,** or **autotrophs,** which make their own food; **consumers,** which eat producers or organisms that have eaten producers; and **decomposers,** which obtain energy by breaking down wastes and dead organisms. Consumers and decomposers are **heterotrophs,** organisms that depend on producers as an energy source and for food and oxygen.

ThomsonNOW **Learn more about energy flow by clicking on the figure in ThomsonNOW.**

10 Design a study to test a given hypothesis, using the procedure and terminology of the scientific method (page 15).

   ■ The **process of science** is a dynamic approach to investigation. The **scientific method** is a general framework that scientists use in their work; it includes observing, recognizing a problem or stating a critical question, developing a hypothesis, making a prediction that can be tested, making further observations and/or performing experiments, interpreting results, and drawing conclusions that support or falsify the hypothesis.

   ■ Deductive reasoning and inductive reasoning are two categories of systematic thought used in the scientific method. **Deductive reasoning** proceeds from general principles to specific conclusions and helps people discover relationships among known facts. **Inductive reasoning** begins with specific observations and draws conclusions from them. Inductive reasoning helps people discover general principles.

   ■ A **hypothesis** is a tentative explanation for observations or phenomena. A hypothesis can be tested. If no evidence is found to support it, the hypothesis is rejected.

   ■ A well-designed scientific experiment typically includes both a **control group** and an **experimental group** and must be as free as possible from bias. The control group should be as closely matched to the experimental group as possible. Ideally, the experimental group differs from the control group only with respect to the variable being studied.

ThomsonNOW **Do your own random sampling by clicking on the figure in ThomsonNOW.**

   ■ A **theory** is an integrated explanation of some aspect of the natural world that is based on a number of hypotheses, each supported by consistent results from many observations or experiments.

11 Compare the reductionist and systems approaches to biological research (page 15).

   ■ Using reductionism, researchers study the simplest components of biological processes, for example, molecules or cells.

   ■ **Systems biology** uses knowledge provided by reductionism. Systems biologists integrate data from various levels of complexity with the goal of understanding how biological systems function.

TEST YOUR UNDERSTANDING

1. Metabolism (a) is the sum of all the chemical activities of an organism (b) results from an increase in the number of cells (c) is characteristic of plant and animal kingdoms only (d) refers to chemical changes in an organism's environment (e) does not take place in producers

2. Homeostasis (a) is the appropriate, balanced internal environment (b) generally depends on the action of cilia (c) is the long-term response of organisms to changes in their environment (d) occurs at the ecosystem level, not in cells or organisms (e) may be sexual or asexual

3. Structures used by some organisms for locomotion are (a) cilia and nuclei (b) flagella and DNA (c) nuclei and membranes (d) cilia and sessiles (e) cilia and flagella

4. An amoeba splits into two smaller amoebas. This is an example of (a) locomotion (b) neurotransmission (c) asexual reproduction (d) sexual reproduction (e) metabolism

5. Cells (a) are the building blocks of living organisms (b) always have nuclei (c) are not found among the bacteria (d) a, b, and c (e) a and b

6. An increase in the size or number of cells best describes (a) homeostasis (b) biological growth (c) chemical level of organization (d) asexual reproduction (e) adaptation

7. DNA (a) makes up the genes (b) functions mainly to transmit information from one species to another (c) cannot be changed (d) is a neurotransmitter (e) is produced during cellular respiration

8. Cellular respiration (a) is a process whereby sunlight is used to synthesize cell components with the release of energy (b) occurs in heterotrophs only (c) is carried on by both autotrophs and heterotrophs (d) causes chemical changes in DNA (e) occurs in response to environmental changes

9. Which of the following is a correct sequence of levels of biological organization? (a) cell, organ, tissue, organ system (b) chemical, cell, organ, tissue (c) chemical, cell, tissue, organ (d) tissue, organ, cell, organ system (e) chemical, cell, ecosystem, population

10. Which of the following is a correct sequence of levels of biological organization? (a) organism, population, ecosystem, community (b) organism, population, community, ecosystem (c) population, biosphere, ecosystem, community (d) species, population, ecosystem, community (e) ecosystem, population, community, biosphere

11. Protozoa are assigned to kingdom (a) Protista (b) Fungi (c) Bacteria (d) Animalia (e) Plantae

12. Yeasts and molds are assigned to kingdom (a) Protista (b) Fungi (c) Bacteria (d) Archaea (e) Plantae

13. In the binomial system of nomenclature, the first part of an organism's name designates the (a) specific epithet (b) genus (c) class (d) kingdom (e) phylum

14. Which of the following is a correct sequence of levels of classification? (a) genus, species, family, order, class, phylum, kingdom (b) genus, species, order, phylum, class, kingdom (c) genus, species, order, family, class, phylum, kingdom (d) species, genus, family, order, class, phylum, kingdom (e) species, genus, order, family, class, kingdom, phylum

15. Darwin suggested that evolution takes place by (a) mutation (b) changes in the individuals of a species (c) natural selection (d) interaction of hormones (e) homeostatic responses to each change in the environment

16. A testable statement is a(an) (a) theory (b) hypothesis (c) principle (d) inductive leap (e) critical question

17. Ideally, an experimental group differs from a control group (a) only with respect to the hypothesis being tested (b) only with respect to the variable being studied (c) in that it is less subject to bias (d) in that it is less vulnerable to sampling error (e) in that its subjects are more reliable

18. A systems biologist would most likely work on (a) better understanding the components of cells (b) developing a better system of classification of organisms (c) devising a new series of steps for the scientific method (d) researching a series of reactions that communicate information in the cell (e) identifying the connections and interactions of neurons in order to learn about brain function

CRITICAL THINKING

1. What would happen if a homeostatic mechanism failed? Describe a homeostatic mechanism at work in your body (other than the regulation of glucose cited in the chapter).

2. Contrast the reductionist approach with systems biology. How are the two approaches complementary? Which approach is more likely to consider emergent properties?

3. What are some characteristics of a good hypothesis?

4. Make a prediction and devise a suitably controlled experiment to test each of the following hypotheses: (a) A type of mold found in your garden does not produce an effective antibiotic. (b) The growth rate of a bean seedling is affected by temperature. (c) Estrogen alleviates symptoms of Alzheimer's disease in elderly women.

5. **Evolution Link.** In what ways does evolution depend on transfer of information? In what ways does transfer of information depend on evolution?

6. **Evolution Link.** How might an understanding of evolutionary processes help a biologist doing research in (a) animal behavior, for example, the hunting behavior of lions? (b) the development of a new antibiotic to replace one to which bacteria have become resistant? (c) conservation of a specific plant in a rain forest?

Additional questions are available in ThomsonNOW at www.thomsonedu.com/login

2

Atoms and Molecules: The Chemical Basis of Life

Frans Lanting/Minden Pictures

Water is a basic requirement for all life. A jaguar (*Panthera onca*), the largest cat in the Western Hemisphere, pauses to drink water from a rainforest stream.

KEY CONCEPTS

Carbon, hydrogen, oxygen, and nitrogen are the most abundant elements in living things.

The chemical properties of an atom are determined by its highest-energy electrons, known as valence electrons.

A molecule consists of atoms joined by covalent bonds. Other important chemical bonds include ionic bonds. Hydrogen bonds and van der Waals interactions are weak attractions.

The energy of an electron is transferred in a redox reaction.

Water molecules are polar, having regions of partial positive and partial negative charge that permit them to form hydrogen bonds with one another and with other charged substances.

Acids are hydrogen ion donors; bases are hydrogen ion acceptors. The pH scale is a convenient measure of the hydrogen ion concentration of a solution.

A knowledge of chemistry is essential for understanding organisms and how they function. This jaguar and the plants of the tropical rain forest, as well as abundant unseen insects and microorganisms, share fundamental similarities in their chemical composition and basic metabolic processes. These chemical similarities provide strong evidence for the evolution of all organisms from a common ancestor and explain why much of what biologists learn from studying bacteria or rats in laboratories can be applied to other organisms, including humans. Furthermore, the basic chemical and physical principles governing organisms are not unique to living things, for they apply to nonliving systems as well.

The success of the Human Genome Project and related studies has relied heavily on biochemistry and **molecular biology,** the chemistry and physics of the molecules that constitute living things. A biochemist may investigate the precise interactions among a cell's atoms and molecules that maintain the energy flow essential to life, and a molecular biologist may study how proteins interact with deoxyribonucleic acid (DNA) in ways that control the expression of certain genes. However, an understanding of chemistry is essential to *all* biologists. An evolutionary biologist may study evolutionary relationships by comparing the DNA of different types of organisms. An ecologist may study how energy is transferred among the organisms living in an estuary or monitor the biological effects of changes in the salinity of the water. A botanist may study unique compounds produced by plants and may even be a "chemical prospector," seeking new sources of medicinal agents.

In this chapter we lay a foundation for understanding how the structure of atoms determines the way they form chemical bonds to produce complex compounds. Most of our discussion focuses on small, simple substances known as **inorganic compounds.** Among the biologically important groups of inorganic compounds are water, many simple acids and bases, and simple salts. We pay particular attention to water, the most abundant substance in organisms and on Earth's surface, and we examine how its unique properties affect living things as well as their nonliving environment. In Chapter 3 we extend our discussion to **organic compounds,** carbon-containing compounds that are generally large and complex. In all but the simplest organic compounds, two or more carbon atoms are bonded to each other to form the backbone, or skeleton, of the molecule. ■

ELEMENTS AND ATOMS

Learning Objectives

1 Name the principal chemical elements in living things, and give an important function of each.
2 Compare the physical properties (mass and charge) and locations of electrons, protons, and neutrons. Distinguish between the atomic number and the mass number of an atom.
3 Define the terms *orbital* and *electron shell*. Relate electron shells to principal energy levels.

Elements are substances that cannot be broken down into simpler substances by ordinary chemical reactions. Each element has a **chemical symbol:** usually the first letter or first and second letters of the English or Latin name of the element. For example, O is the symbol for oxygen, C for carbon, H for hydrogen, N for nitrogen, and Na for sodium (from the Latin word *natrium*). Just four elements—oxygen, carbon, hydrogen, and nitrogen—are responsible for more than 96% of the mass of most organisms. Others, such as calcium, phosphorus, potassium, and magnesium, are also consistently present but in smaller quantities. Some elements, such as iodine and copper, are known as *trace elements*, because they are required only in minute amounts. ■ Table 2-1 lists the elements that make up organisms and briefly explains the importance of each in typical plants and animals.

An **atom** is defined as the smallest portion of an element that retains its chemical properties. Atoms are much too small to be visible under a light microscope. However, by sophisticated techniques (such as scanning tunneling microscopy, with magnifications as great as 5 million times) researchers have been able to photograph the positions of some large atoms in molecules.

The components of atoms are tiny particles of **matter** (anything that has mass and takes up space) known as subatomic particles. Physicists have discovered a number of subatomic particles, but for our purposes we need consider only three: electrons, protons, and neutrons. An **electron** is a particle that carries a unit of negative electric charge; a **proton** carries a unit of positive charge; and a **neutron** is an uncharged particle. In an electrically neutral atom, the number of electrons is equal to the number of protons.

Clustered together, protons and neutrons compose the **atomic nucleus.** Electrons, however, have no fixed locations and move rapidly through the mostly empty space surrounding the atomic nucleus.

TABLE 2-1

Functions of Elements in Organisms

Element (chemical symbol)	Functions
O Oxygen	Required for cellular respiration; present in most organic compounds; component of water
C Carbon	Forms backbone of organic molecules; each carbon atom can form four bonds with other atoms
H Hydrogen	Present in most organic compounds; component of water; hydrogen ion (H^+) is involved in some energy transfers
N Nitrogen	Component of proteins and nucleic acids; component of chlorophyll in plants
Ca Calcium	Structural component of bones and teeth; calcium ion (Ca^{2+}) is important in muscle contraction, conduction of nerve impulses, and blood clotting; associated with plant cell wall
P Phosphorus	Component of nucleic acids and of phospholipids in membranes; important in energy transfer reactions; structural component of bone
K Potassium	Potassium ion (K^+) is a principal positive ion (cation) in interstitial (tissue) fluid of animals; important in nerve function; affects muscle contraction; controls opening of stomata in plants
S Sulfur	Component of most proteins
Na Sodium	Sodium ion (Na^+) is a principal positive ion (cation) in interstitial (tissue) fluid of animals; important in fluid balance; essential for conduction of nerve impulses; important in photosynthesis in plants
Mg Magnesium	Needed in blood and other tissues of animals; activates many enzymes; component of chlorophyll in plants
Cl Chlorine	Chloride ion (Cl^-) is principal negative ion (anion) in interstitial (tissue) fluid of animals; important in water balance; essential for photosynthesis
Fe Iron	Component of hemoglobin in animals; activates certain enzymes

*Other elements found in very small (trace) amounts in animals, plants, or both include iodine (I), manganese (Mn), copper (Cu), zinc (Zn), cobalt (Co), fluorine (F), molybdenum (Mo), selenium (Se), boron (B), silicon (Si), and a few others.

The periodic table provides information about the elements: their compositions, structures, and chemical behavior.

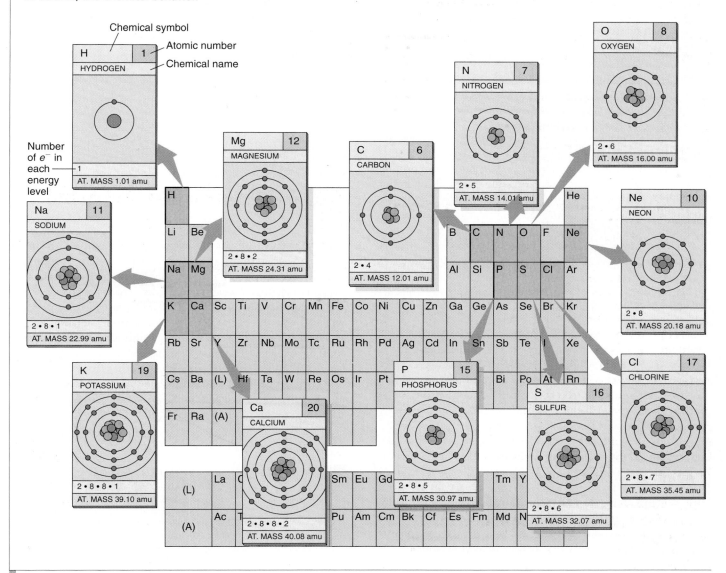

Figure 2-1 The periodic table

Note the Bohr models depicting the electron configuration of atoms of some biologically important elements. Although the Bohr model does not depict electron configurations accurately, it is commonly used because of its simplicity and convenience. A complete periodic table is given in Appendix A.

An atom is uniquely identified by its number of protons

Every element has a fixed number of protons in the atomic nucleus, known as the **atomic number.** It is written as a subscript to the left of the chemical symbol. Thus, $_1H$ indicates that the hydrogen nucleus contains 1 proton, and $_8O$ means that the oxygen nucleus contains 8 protons. The atomic number determines an atom's identity and defines the element.

The **periodic table** is a chart of the elements arranged in order by atomic number (Fig. 2-1 and Appendix A). The periodic table is useful because it lets us simultaneously correlate many of the relationships among the various elements.

Figure 2-1 includes representations of the **electron configurations** of several elements important in organisms. These *Bohr models*, which show the electrons arranged in a series of concentric circles around the nucleus, are convenient to use, but inaccurate. The space outside the nucleus is actually extremely large compared to the nucleus, and as you will see, electrons do not actually circle the nucleus in fixed concentric pathways.

Protons plus neutrons determine atomic mass

The mass of a subatomic particle is exceedingly small, much too small to be conveniently expressed in grams or even micrograms.[1] Such masses are expressed in terms of the **atomic mass unit (amu),** also called the **dalton** in honor of John Dalton, the English chemist who formulated an atomic theory in the early 1800s. One amu is equal to the approximate mass of a single proton or a single neutron. Protons and neutrons make up almost all the mass of an atom. The mass of a single electron is only about 1/1800 the mass of a proton or neutron.

The **atomic mass** of an atom is a number that indicates approximately how much matter it contains compared with another atom. This value is determined by adding the number of protons to the number of neutrons and expressing the result in atomic mass units or daltons.[2] The mass of the electrons is ignored because it is so small. The atomic mass number is indicated by a superscript to the left of the chemical symbol. The common form of the oxygen atom, with 8 protons and 8 neutrons in its nucleus, has an atomic number of 8 and a mass of 16 amu. It is indicated by the symbol $^{16}_{8}O$.

The characteristics of protons, electrons, and neutrons are summarized in the following table:

Particle	Charge	Approximate Mass	Location
Proton	Positive	1 amu	Nucleus
Neutron	Neutral	1 amu	Nucleus
Electron	Negative	Approx. 1/1800 amu	Outside nucleus

Isotopes of an element differ in number of neutrons

Most elements consist of a mixture of atoms with different numbers of neutrons and thus different masses. Such atoms are called **isotopes.** Isotopes of the same element have the same number of protons and electrons; only the number of neutrons varies. The three isotopes of hydrogen, $^{1}_{1}H$ (ordinary hydrogen), $^{2}_{1}H$ (deuterium), and $^{3}_{1}H$ (tritium), contain 0, 1, and 2 neutrons, respectively. Figure 2-2 shows Bohr models of two isotopes of carbon, $^{12}_{6}C$ and $^{14}_{6}C$. The mass of an element is expressed as an average of the masses of its isotopes (weighted by their relative abundance in nature). For example, the atomic mass of hydrogen is not 1.0 amu, but 1.0079 amu, reflecting the natural occurrence of small amounts of deuterium and tritium in addition to the more abundant ordinary hydrogen.

Because they have the same number of electrons, all isotopes of a given element have essentially the same chemical character-

---

[1]Tables of commonly used units of scientific measurement are printed inside the back cover of this text.

[2]Unlike weight, mass is independent of the force of gravity. For convenience, however, we consider mass and weight equivalent. Atomic weight has the same numerical value as atomic mass, but it has no units.

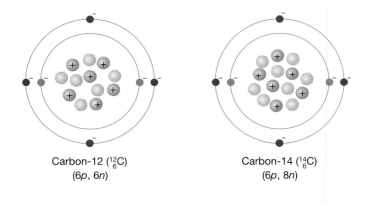

Carbon-12 ($^{12}_{6}C$)
(6*p*, 6*n*)

Carbon-14 ($^{14}_{6}C$)
(6*p*, 8*n*)

Figure 2-2 Isotopes

Carbon-12 ($^{12}_{6}C$) is the most common isotope of carbon. Its nucleus contains 6 protons and 6 neutrons, so its atomic mass is 12. Carbon-14 ($^{14}_{6}C$) is a rare radioactive carbon isotope. It contains 8 neutrons, so its atomic mass is 14.

istics. However, some isotopes are unstable and tend to break down, or decay, to a more stable isotope (usually becoming a different element); such **radioisotopes** emit radiation when they decay. For example, the radioactive decay of $^{14}_{6}C$ occurs as a neutron decomposes to form a proton and a fast-moving electron, which is emitted from the atom as a form of radiation known as a beta (β) particle. The resulting stable atom is the common form of nitrogen, $^{14}_{7}N$. Using sophisticated instruments, scientists can detect and measure β particles and other types of radiation. Radioactive decay can also be detected by a method known as **autoradiography,** in which radiation causes the appearance of dark silver grains in photographic film (Fig. 2-3).

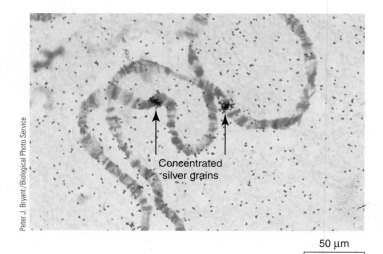

Concentrated silver grains

50 μm

Figure 2-3 Autoradiography

The chromosomes of the fruit fly, *Drosophila melanogaster,* shown in this light micrograph, have been covered with photographic film in which silver grains (*dark spots*) are produced when tritium (^{3}H) that has been incorporated into DNA undergoes radioactive decay. The concentrations of silver grains (*arrows*) mark the locations of specific DNA molecules.

Because the different isotopes of a given element have the same chemical characteristics, they are essentially interchangeable in molecules. Molecules containing radioisotopes are usually metabolized and/or localized in the organism in a similar way to their nonradioactive counterparts, and they can be substituted. For this reason, radioisotopes such as ^3H (tritium), ^{14}C, and ^{32}P are extremely valuable research tools used, for example, in dating fossils (see Fig. 18-9), tracing biochemical pathways, determining the sequence of genetic information in DNA (see Fig. 15-11), and understanding sugar transport in plants.

In medicine, radioisotopes are used for both diagnosis and treatment. The location and/or metabolism of a substance such as a hormone or drug can be followed in the body by labeling the substance with a radioisotope such as carbon-14 or tritium. Radioisotopes are used to test thyroid gland function, to provide images of blood flow in the arteries supplying the cardiac muscle, and to study many other aspects of body function and chemistry. Because radiation can interfere with cell division, radioisotopes have been used therapeutically in treating cancer, a disease often characterized by rapidly dividing cells.

Electrons move in orbitals corresponding to energy levels

Greater energy, farther from nucleus.

Electrons move through characteristic regions of 3-D space, or **orbitals.** Each orbital contains a maximum of 2 electrons. Because it is impossible to know an electron's position at any given time, orbitals are most accurately depicted as "electron clouds," shaded areas whose density is proportional to the probability that an electron is present there at any given instant. The energy of an electron depends on the orbital it occupies. Electrons in orbitals with similar energies, said to be at the same **principal energy level**, make up an **electron shell** (▌Fig. 2-4).

In general, electrons in a shell with a greater average distance from the nucleus have greater energy than those in a shell close to the nucleus. The reason is that energy is required to move a negatively charged electron farther away from the positively charged nucleus. The most energetic electrons, known as **valence electrons,** are said to occupy the **valence shell**. The valence shell is represented as the outermost concentric ring in a Bohr model. As we will see in the next sections, it is these valence electrons that play a key role in chemical reactions.

An electron can move to an orbital farther from the nucleus by receiving more energy, or it can give up energy and sink to a lower energy level in an orbital nearer the nucleus. Changes in electron energy levels are important in energy conversions in organisms. For example, during photosynthesis, light energy absorbed by chlorophyll molecules causes electrons to move to a higher energy level (see Fig. 9-3).

Review

- Do all atoms of an element have the same atomic number? The same atomic mass?
- What is a radioisotope? What are some ways radioisotopes are used in biological research?

- How do electrons in different orbitals of the same electron shell compare with respect to their energy?

CHEMICAL REACTIONS

Learning Objectives

4 Explain how the number of valence electrons of an atom is related to its chemical properties.
5 Distinguish among simplest, molecular, and structural chemical formulas.
6 Explain why the mole concept is so useful to chemists.

The chemical behavior of an atom is determined primarily by the number and arrangement of its valence electrons. The valence shell of hydrogen or helium is full (stable) when it contains 2 electrons. The valence shell of any other atom is full when it contains 8 electrons. When the valence shell is not full, the atom tends to lose, gain, or share electrons to achieve a full outer shell. The valence shells of all isotopes of an element are identical; for this reason, they have similar chemical properties and can substitute for each other in chemical reactions (for example, tritium can substitute for ordinary hydrogen).

Elements in the same vertical column (belonging to the same *group*) of the periodic table have similar chemical properties because their valence shells have similar tendencies to lose, gain, or share electrons. For example, chlorine and bromine, included in a group commonly known as the *halogens*, are highly reactive. Because their valence shells have 7 electrons, they tend to gain an electron in chemical reactions. By contrast, hydrogen, sodium, and potassium each have a single valence electron, which they tend to give up or share with another atom. Helium (He) and neon (Ne) belong to a group referred to as the *noble gases*. They are quite unreactive, because their valence shells are full. Notice in Figure 2-1 the incomplete valence shells of some of the elements important in organisms, including carbon, hydrogen, oxygen, and nitrogen, and compare them with the full valence shell of neon in Figure 2-4d.

Atoms form compounds and molecules

Two or more atoms may combine chemically. When atoms of *different* elements combine, the result is a chemical compound. A **chemical compound** consists of atoms of two or more different elements combined in a fixed ratio. For example, water is a chemical compound composed of hydrogen and oxygen in a ratio of 2:1. Common table salt, sodium chloride, is a chemical compound made up of sodium and chlorine in a 1:1 ratio.

Two or more atoms may become joined very strongly to form a stable particle called a **molecule.** For example, when two atoms of oxygen combine chemically, a molecule of oxygen is formed. Water is a molecular compound, with each molecule consisting of two atoms of hydrogen and one of oxygen. However, as you will see, not all compounds are made up of molecules. Sodium chloride (common table salt) is an example of a compound that is not molecular.

Electrons occupy orbitals corresponding to energy levels.

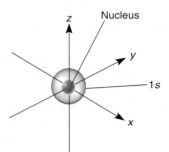

(a) The first principal energy level contains a maximum of 2 electrons, occupying a single spherical orbital (designated 1s). The electrons depicted in the diagram could be present anywhere in the blue area.

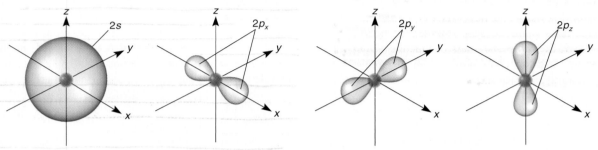

(b) The second principal energy level includes four orbitals, each with a maximum of 2 electrons: one spherical (2s) and three dumbbell-shaped (2p) orbitals at right angles to one another.

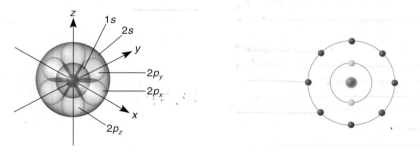

(c) Orbitals of the first and second principal energy levels of a neon atom are shown superimposed. Note that the single 2s orbital plus three 2p orbitals make up neon's full valence shell of 8 electrons. Compare this more realistic view of the atomic orbitals with the Bohr model of a neon atom at right.

(d) Neon atom (Bohr model)

Figure 2-4 *Animated* Atomic orbitals

Each orbital is represented as an "electron cloud." The arrows labeled x, y, and z establish the imaginary axes of the atom.

Simplest, molecular, and structural chemical formulas give different information

A **chemical formula** is a shorthand expression that describes the chemical composition of a substance. Chemical symbols indicate the types of atoms present, and subscript numbers indicate the ratios among the atoms. There are several types of chemical formulas, each providing specific kinds of information.

In a **simplest formula** (also known as an *empirical formula*), the subscripts give the smallest whole-number ratios for the at-

oms present in a compound. For example, the simplest formula for hydrazine is NH_2, indicating a 1:2 ratio of nitrogen to hydrogen. (Note that when a single atom of a type is present, the subscript number 1 is never written.)

In a **molecular formula,** the subscripts indicate the actual numbers of each type of atom per molecule. The molecular formula for hydrazine is N_2H_4, which indicates that each molecule of hydrazine consists of two atoms of nitrogen and four atoms of hydrogen. The molecular formula for water, H_2O, indicates that each molecule consists of two atoms of hydrogen and one atom of oxygen.

A **structural formula** shows not only the types and numbers of atoms in a molecule but also their arrangement. For example, the structural formula for water is H—O—H. As you will learn in Chapter 3, it is common for complex organic molecules with different structural formulas to share the same molecular formula.

One mole of any substance contains the same number of units

The **molecular mass** of a compound is the sum of the atomic masses of the component atoms of a single molecule; thus, the molecular mass of water, H_2O, is (hydrogen: 2×1 amu) + (oxygen: 1×16 amu), or 18 amu. (Because of the presence of isotopes, atomic mass values are not whole numbers, but for easy calculation each atomic mass value has been rounded off to a whole number.) Similarly, the molecular mass of glucose ($C_6H_{12}O_6$), a simple sugar that is a key compound in cell metabolism, is (carbon: 6×12 amu) + (hydrogen: 12×1 amu) + (oxygen: 6×16 amu), or 180 amu.

The amount of an element or compound whose mass in grams is equivalent to its atomic or molecular mass is 1 **mole (mol).** Thus, 1 mol of water is 18 grams (g), and 1 mol of glucose has a mass of 180 g. The mole is an extremely useful concept, because it lets us make meaningful comparisons between atoms and molecules of very different mass. The reason is that *1 mol of any substance always has exactly the same number of units,* whether those units are small atoms or large molecules. The very large number of units in a mole, 6.02×10^{23}, is known as **Avogadro's number,** named for the Italian physicist Amadeo Avogadro, who first calculated it. Thus, 1 mol (180 g) of glucose contains 6.02×10^{23} molecules, as does 1 mol (2 g) of molecular hydrogen (H_2). Although it is impossible to count atoms and molecules individually, a scientist can calculate them simply by weighing a sample. Molecular biologists usually deal with smaller values, either millimoles (mmol, one thousandth of a mole) or micromoles (μmol, one millionth of a mole).

The mole concept also lets us make useful comparisons among solutions. A 1-molar solution, represented by 1 *M*, contains 1 mol of that substance dissolved in a total volume of 1 liter (L). For example, we can compare 1 L of a 1-*M* solution of glucose with 1 L of a 1-*M* solution of sucrose (table sugar, a larger molecule). They differ in the mass of the dissolved sugar (180 g and 340 g, respectively), but they each contain 6.02×10^{23} sugar molecules.

Chemical equations describe chemical reactions

During any moment in the life of an organism—a bacterial cell, a mushroom, or a butterfly—many complex chemical reactions are taking place. Chemical reactions, such as the reaction between glucose and oxygen, can be described by means of chemical equations:

$$C_6H_{12}O_6 + 6\,O_2 \longrightarrow 6\,CO_2 + 6\,H_2O + energy$$

$$\phantom{C_6H_{12}O_6}\text{Glucose}\quad\text{Oxygen}\quad\text{Carbon dioxide}\quad\text{Water}$$

In a chemical equation, the **reactants,** the substances that participate in the reaction, are generally written on the left side, and the **products,** the substances formed by the reaction, are written on the right side. The arrow means "yields" and indicates the direction in which the reaction proceeds.

Chemical compounds react with each other in quantitatively precise ways. The numbers preceding the chemical symbols or formulas (known as *coefficients*) indicate the relative number of atoms or molecules reacting. For example, 1 mol of glucose burned in a fire or metabolized in a cell reacts with 6 mol of oxygen to form 6 mol of carbon dioxide and 6 mol of water.

Many reactions can proceed simultaneously in the reverse direction (to the left) and the forward direction (to the right). At **dynamic equilibrium,** the rates of the forward and reverse reactions are equal (see Chapter 7). Reversible reactions are indicated by double arrows:

$$CO_2 + H_2O \rightleftharpoons H_2CO_3$$

$$\text{Carbon dioxide}\quad\text{Water}\quad\text{Carbonic acid}$$

In this example, the arrows are drawn in different lengths to indicate that when the reaction reaches equilibrium, there will be more reactants (CO_2 and H_2O) than product (H_2CO_3).

Review

- Why is a radioisotope able to substitute for an ordinary (nonradioactive) atom of the same element in a molecule?
- Which kind of chemical formula provides the most information?
- How many atoms would be included in 1 g of hydrogen atoms? In 2 g of hydrogen molecules?

11:32 pm

CHEMICAL BONDS

Learning Objective

7 Distinguish among covalent bonds, ionic bonds, hydrogen bonds, and van der Waals interactions. Compare them in terms of the mechanisms by which they form and their relative strengths.

Atoms can be held together by forces of attraction called **chemical bonds.** Each bond represents a certain amount of chemical energy. **Bond energy** is the energy necessary to break a chemical bond. The valence electrons dictate how many bonds an atom can form. The two principal types of strong chemical bonds are covalent bonds and ionic bonds.

In covalent bonds electrons are shared

Covalent bonds involve the sharing of electrons between atoms in a way that results in each atom having a filled valence shell. A molecule consists of atoms joined by covalent bonds. A simple example of a covalent bond is the joining of two hydrogen atoms in a molecule of hydrogen gas, H_2. Each atom of hydrogen has 1 electron, but 2 electrons are required to complete its valence

Covalent bonds form when atoms share electrons.

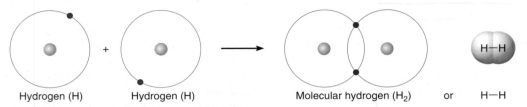

(a) Single covalent bond formation. Two hydrogen atoms achieve stability by sharing a pair of electrons, thereby forming a molecule of hydrogen. In the structural formula on the right, the straight line between the hydrogen atoms represents a single covalent bond.

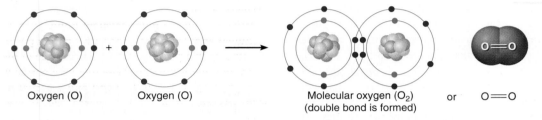

(b) Double covalent bond formation. In molecular oxygen, two oxygen atoms share two pairs of electrons, forming a double covalent bond. The parallel straight lines in the structural formula represent a double covalent bond.

Figure 2-5 Electron sharing in covalent compounds

shell. The hydrogen atoms have equal capacities to attract electrons, so neither donates an electron to the other. Instead, the two hydrogen atoms share their single electrons so that the 2 electrons are attracted simultaneously to the 2 protons in the two hydrogen nuclei. The 2 electrons whirl around both atomic nuclei, thus forming the covalent bond that joins the two atoms. Similarly, unlike atoms can also be linked by covalent bonds to form molecules; the resulting compound is a **covalent compound**.

A simple way of representing the electrons in the valence shell of an atom is to use dots placed around the chemical symbol of the element. Such a representation is called the *Lewis structure* of the atom, named for G. N. Lewis, an American chemist who developed this type of notation. In a water molecule, two hydrogen atoms are covalently bonded to an oxygen atom:

$$\text{H} \cdot + \text{H} \cdot + \cdot \overset{..}{\underset{..}{\text{O}}} \cdot \longrightarrow \text{H} \overset{..}{\underset{..}{\text{O}}} \text{H}$$

Oxygen has 6 valence electrons; by sharing electrons with two hydrogen atoms, it completes its valence shell of 8. At the same time, each hydrogen atom obtains a complete valence shell of 2. (Note that in the structural formula H—O—H, each pair of shared electrons constitutes a covalent bond, represented by a solid line. Unshared electrons are usually omitted in a structural formula.)

The carbon atom has 4 electrons in its valence shell, all of which are available for covalent bonding:

$$\cdot \overset{.}{\text{C}} \cdot$$

When one carbon and four hydrogen atoms share electrons, a molecule of the covalent compound methane, CH_4, is formed:

$$
\begin{array}{ccc}
\text{H} & & \text{H} \\
\text{H}:\overset{..}{\underset{..}{\text{C}}}:\text{H} & \text{or} & \text{H}-\overset{|}{\underset{|}{\text{C}}}-\text{H} \\
\text{H} & & \text{H}
\end{array}
$$

Lewis structure   Structural formula

The nitrogen atom has 5 electrons in its valence shell. Recall that each orbital can hold a maximum of 2 electrons. Usually 2 electrons occupy one orbital, leaving 3 available for sharing with other atoms:

$$\cdot \overset{..}{\text{N}} \cdot$$

When a nitrogen atom shares electrons with three hydrogen atoms, a molecule of the covalent compound ammonia, NH_3, is formed:

$$
\begin{array}{ccc}
& \overset{..}{} & \\
\text{H}:\text{N}:\text{H} & \text{or} & \text{H}-\text{N}-\text{H} \\
\text{H} & & \overset{|}{\text{H}}
\end{array}
$$

Lewis structure   Structural formula

When one pair of electrons is shared between two atoms, the covalent bond is called a **single covalent bond** (❙ Fig. 2-5a). Two hydrogen atoms share a single pair of electrons. Two oxygen atoms may achieve stability by forming covalent bonds with each other. Each oxygen atom has 6 electrons in its outer shell. To become stable, the two atoms share two pairs of electrons, forming molecular oxygen (▌Fig. 2-5b). When two pairs of electrons are shared in this way, the covalent bond is called a **double covalent bond,** which is represented by two parallel solid lines. Similarly, a **triple covalent bond** is formed when three pairs of electrons are shared between two atoms (represented by three parallel solid lines).

The number of covalent bonds usually formed by the atoms in biologically important molecules is summarized as follows:

Atom	Symbol	Covalent Bonds
Hydrogen	H	1
Oxygen	O	2
Carbon	C	4
Nitrogen	N	3
Phosphorus	P	5
Sulfur	S	2

The function of a molecule is related to its shape

In addition to being composed of atoms with certain properties, each kind of molecule has a characteristic size and a general overall shape. Although the shape of a molecule may change (within certain limits), the functions of molecules in living cells are dictated largely by their geometric shapes. A molecule that consists of two atoms is linear. Molecules composed of more than two atoms may have more complicated shapes. The geometric shape of a molecule provides the optimal distance between the atoms to counteract the repulsion of electron pairs.

When an atom forms covalent bonds with other atoms, the orbitals in the valence shell may become rearranged in a process known as **orbital hybridization,** thereby affecting the shape of the resulting molecule. For example, when four hydrogen atoms combine with a carbon atom to form a molecule of methane (CH_4), the hybridized valence shell orbitals of the carbon form a geometric structure known as a *tetrahedron,* with one hydrogen atom present at each of its four corners (▌Fig. 2-6; see Fig. 3-2a).

We will explore the importance of molecular shape in more detail in Chapter 3 and in our discussion of the properties of water in this chapter.

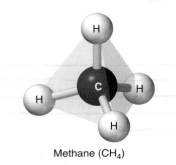

Methane (CH_4)

Figure 2-6 Orbital hybridization in methane

The four hydrogens are located at the corners of a tetrahedron because of hybridization of the valence shell orbitals of carbon.

Covalent bonds can be nonpolar or polar

Atoms of different elements vary in their affinity for electrons. **Electronegativity** is a measure of an atom's attraction for shared electrons in chemical bonds. Very electronegative atoms such as oxygen, nitrogen, fluorine, and chlorine are sometimes called "electron greedy." When covalently bonded atoms have similar electronegativities, the electrons are shared equally and the covalent bond is described as **nonpolar.** The covalent bond of the hydrogen molecule is nonpolar, as are the covalent bonds of molecular oxygen and methane.

In a covalent bond between two different elements, such as oxygen and hydrogen, the electronegativities of the atoms may be different. If so, electrons are pulled closer to the atomic nucleus of the element with the greater electron affinity (in this case, oxygen). A covalent bond between atoms that differ in electronegativity is called a **polar covalent bond.** Such a bond has two dissimilar ends (or poles), one with a partial positive charge and the other with a partial negative charge. Each of the two covalent bonds in water is polar, because there is a partial positive charge at the hydrogen end of the bond and a partial negative charge at the oxygen end, where the "shared" electrons are more likely to be.

Covalent bonds differ in their degree of polarity, ranging from those in which the electrons are equally shared (as in the nonpolar hydrogen molecule) to those in which the electrons are much closer to one atom than to the other (as in water). Oxygen is quite electronegative and forms polar covalent bonds with carbon, hydrogen, and many other atoms. Nitrogen is also strongly electronegative, although less so than oxygen.

A molecule with one or more polar covalent bonds can be polar even though it is electrically neutral as a whole. The reason is that a **polar molecule** has one end with a partial positive charge and another end with a partial negative charge. One example is water (▌Fig. 2-7). The polar bonds between the hydrogens and the oxygen are arranged in a V shape, rather than linearly. The oxygen end constitutes the negative pole of the molecule, and the end with the two hydrogens is the positive pole.

Ionic bonds form between cations and anions

Some atoms or groups of atoms are not electrically neutral. A particle with 1 or more units of electric charge is called an **ion.** An atom becomes an ion if it gains or loses 1 or more electrons. An atom with 1, 2, or 3 electrons in its valence shell tends to lose electrons to other atoms. Such an atom then becomes positively charged, because its nucleus contains more protons than the number of electrons orbiting around the nucleus. These positively charged ions are termed **cations.** Atoms with 5, 6, or 7 valence electrons tend to gain electrons from other atoms and become negatively charged **anions.**

The properties of ions are quite different from those of the electrically neutral atoms from which they were derived. For example, although chlorine gas is a poison, chloride ions (Cl^-) are essential to life (see Table 2-1). Because their electric charges provide a basis for many interactions, cations and anions are involved in energy transformations within the cell, the transmis-

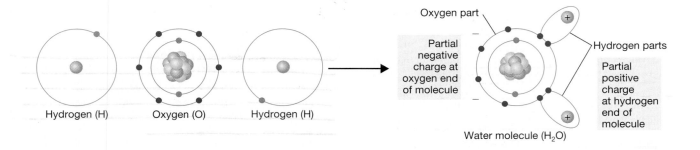

Oxygen part

Partial
negative
charge at
oxygen end
of molecule

Hydrogen parts

Partial
positive
charge
at hydrogen
end of
molecule

Water molecule (H$_2$O)

Figure 2-7 Water, a polar molecule

Note that the electrons tend to stay closer to the nucleus of the oxygen atom than to the hydrogen nuclei. This results in a partial negative charge on the oxygen portion of the molecule and a partial positive charge at the hydrogen end. Although the water molecule as a whole is electrically neutral, it is a polar covalent compound.

sion of nerve impulses, muscle contraction, and many other biological processes (■ Fig. 2-8).

A group of covalently bonded atoms can also become an ion (*polyatomic ion*). Unlike a single atom, a group of atoms can lose or gain protons (derived from hydrogen atoms) as well as electrons. Therefore, a group of atoms can become a cation if it loses 1 or more electrons or gains 1 or more protons. A group of atoms becomes an anion if it gains 1 or more electrons or loses 1 or more protons.

An **ionic bond** forms as a consequence of the attraction between the positive charge of a cation and the negative charge of an anion. An **ionic compound** is a substance consisting of anions and cations bonded by their opposite charges.

A good example of how ionic bonds are formed is the attraction between sodium ions and chloride ions. A sodium atom has 1 electron in its valence shell. It cannot fill its valence shell by ob-

taining 7 electrons from other atoms, because it would then have a large unbalanced negative charge. Instead, it gives up its single valence electron to a very electronegative atom, such as chlorine, which acts as an electron acceptor (■ Fig. 2-9). Chlorine cannot give up the 7 electrons in its valence shell, because it would then have a large positive charge. Instead, it strips an electron from an electron donor (sodium, in this example) to complete its valence shell.

When sodium reacts with chlorine, sodium's valence electron is transferred completely to chlorine. Sodium becomes a cation, with 1 unit of positive charge (Na$^+$). Chlorine becomes an anion, a chloride ion with 1 unit of negative charge (Cl$^-$). These ions attract each other as a result of their opposite charges. This electrical attraction in ionic bonds holds them together to form NaCl, sodium chloride, or common table salt.

The term *molecule* does not adequately explain the properties of ionic compounds such as NaCl. When NaCl is in its solid crystal state, each ion is actually surrounded by six ions of opposite charge. The simplest formula, NaCl, indicates that sodium ions and chloride ions are present in a 1:1 ratio, but the actual crystal has no discrete molecules composed of one Na$^+$ and one Cl$^-$ ion.

Compounds joined by ionic bonds, such as sodium chloride, have a tendency to *dissociate* (separate) into their individual ions when placed in water:

$$NaCl \xrightarrow{\text{in H}_2\text{O}} Na^+ + Cl^-$$

Sodium chloride   Sodium ion   Chloride ion

In the solid form of an ionic compound (that is, in the absence of water), ionic bonds are very strong. Water, however, is an excellent **solvent;** as a liquid it is capable of dissolving many substances, particularly those that are polar or ionic, because of the polarity of water molecules. The localized partial positive charge (on the hydrogen atoms) and partial negative charge (on the oxygen atom) on each water molecule attract and surround the anions and cations, respectively, on the surface of an ionic solid. As a result, the solid dissolves. A dissolved substance is referred to as a **solute.** In solution, each cation and anion of the ionic compound is surrounded by oppositely charged ends of the wa-

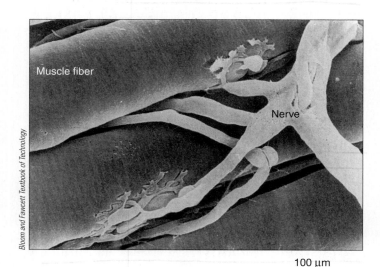

Bloom and Fawcett Textbook of Technology

Muscle fiber

Nerve

100 μm

Figure 2-8 Ions and biological processes

Sodium, potassium, and chloride ions are essential for this nerve cell to stimulate these muscle fibers, initiating a muscle contraction. Calcium ions in the muscle cell are required for muscle contraction.

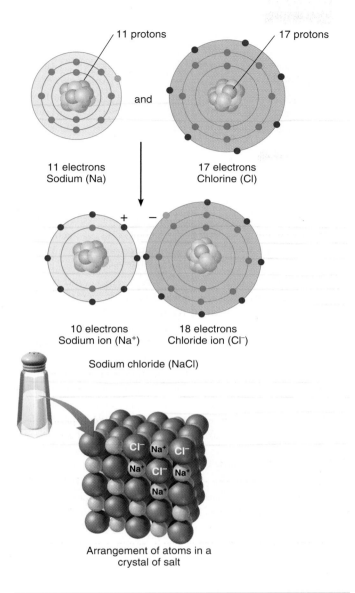

Figure 2-9 *Animated* Ionic bonding

Sodium becomes a positively charged ion when it donates its single valence electron to chlorine, which has 7 valence electrons. With this additional electron, chlorine completes its valence shell and becomes a negatively charged chloride ion. These sodium and chloride ions are attracted to one another by their unlike electric charges, forming the ionic compound sodium chloride.

ter molecules. This process is known as **hydration** (█ Fig. 2-10). Hydrated ions still interact with one another to some extent, but the transient ionic bonds formed are much weaker than those in a solid crystal.

Hydrogen bonds are weak attractions

Another type of bond important in organisms is the **hydrogen bond.** When hydrogen combines with oxygen (or with another relatively electronegative atom such as nitrogen), it acquires a

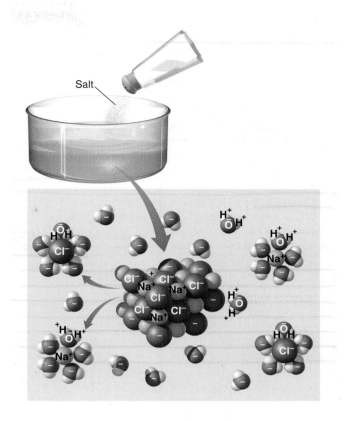

Figure 2-10 *Animated* Hydration of an ionic compound

When the crystal of NaCl is added to water, the sodium and chloride ions are pulled apart. When the NaCl is dissolved, each Na$^+$ and Cl$^-$ is surrounded by water molecules electrically attracted to it.

partial positive charge because its electron spends more time closer to the electronegative atom. Hydrogen bonds tend to form between an atom with a partial negative charge and a hydrogen atom that is covalently bonded to oxygen or nitrogen (█ Fig. 2-11). The atoms involved may be in two parts of the same large molecule or in two different molecules. Water molecules interact with one another extensively through hydrogen bond formation.

Hydrogen bonds are readily formed and broken. Although individually relatively weak, hydrogen bonds are collectively

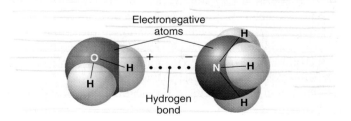

Figure 2-11 *Animated* Hydrogen bonding

A hydrogen bond (*dotted line*) can form between two molecules with regions of unlike partial charge. Here, the nitrogen atom of an ammonia molecule is joined by a hydrogen bond to a hydrogen atom of a water molecule.

strong when present in large numbers. Furthermore, they have a specific length and orientation. As you will see in Chapter 3, these features are very important in determining the 3-D structure of large molecules such as DNA and proteins.

van der Waals interactions are weak forces

Even electrically neutral, nonpolar molecules can develop transient regions of weak positive and negative charge. These slight charges develop as a consequence of the fact that electrons are in constant motion. A region with a temporary excess of electrons will have a weak negative charge, whereas one with an electron deficit will have a weak positive charge. Adjacent molecules may interact in regions of slight opposite charge. These attractive forces, called **van der Waals interactions,** operate over very short distances and are weaker and less specific than the other types of interactions we have considered. They are most important when they occur in large numbers and when the shapes of the molecules permit close contact between the atoms. Although a single interaction is very weak, the binding force of a large number of these interactions working together can be significant.

Review

- Are all compounds composed of molecules? Explain.
- What are the ways an atom or molecule can become an anion or a cation?
- How do ionic and covalent bonds differ?
- Under what circumstances can weak forces such as hydrogen bonds and van der Waals interactions play significant roles in biological systems?

REDOX REACTIONS

Learning Objective

8 Distinguish between the terms *oxidation* and *reduction,* and relate these processes to the transfer of energy.

Many energy conversions that go on in a cell involve reactions in which an electron transfers from one substance to another. The reason is that the transfer of an electron also involves the transfer of the energy of that electron. Such an electron transfer is known as an oxidation–reduction, or **redox reaction.** Oxidation and reduction always occur together. **Oxidation** is a chemical process in which an atom, ion, or molecule *loses* electrons. **Reduction** is a chemical process in which an atom, ion, or molecule *gains* electrons. (The term refers to the fact that the gain of an electron results in the reduction of any positive charge that might be present.)

Rusting—the combining of iron (symbol Fe) with oxygen—is a simple illustration of oxidation and reduction:

$$4\,Fe + 3\,O_2 \longrightarrow 2\,Fe_2O_3$$
Iron (III) oxide

In rusting, each iron atom becomes oxidized as it loses 3 electrons.

$$4\,Fe \longrightarrow 4\,Fe^{3+} + 12e^-$$

The e^- represents an electron; the $+$ superscript in Fe^{3+} represents an electron deficit. (When an atom loses an electron, it acquires 1 unit of positive charge from the excess of 1 proton. In our example, each iron atom loses 3 electrons and acquires 3 units of positive charge.) Recall that the oxygen atom is very electronegative, able to remove electrons from other atoms. In this reaction, oxygen becomes reduced when it accepts electrons from the iron.

$$3\,O_2{}^+ 12e^- \longrightarrow 6\,O^{2-}$$

Redox reactions occur simultaneously because one substance must accept the electrons that are removed from the other. In a redox reaction, one component, the *oxidizing agent,* accepts 1 or more electrons and becomes reduced. Oxidizing agents other than oxygen are known, but oxygen is such a common one that its name was given to the process. Another reaction component, the *reducing agent,* gives up 1 or more electrons and becomes oxidized. In our example, there was a complete transfer of electrons from iron (the reducing agent) to oxygen (the oxidizing agent). Similarly, Figure 2-9 shows that an electron was transferred from sodium (the reducing agent) to chlorine (the oxidizing agent).

Electrons are not easily removed from covalent compounds unless an entire atom is removed. In cells, oxidation often involves the removal of a hydrogen atom (an electron plus a proton that "goes along for the ride") from a covalent compound; reduction often involves the addition of the equivalent of a hydrogen atom (see Chapter 7).

Review

- Why must oxidation and reduction occur simultaneously?
- Why are redox reactions important in some energy transfers?

WATER

Learning Objective

9 Explain how hydrogen bonds between adjacent water molecules govern many of the properties of water.

A large part of the mass of most organisms is water. In human tissues the percentage of water ranges from 20% in bones to 85% in brain cells; about 70% of our total body weight is water. As much as 95% of a jellyfish and certain plants is water. Water is the source, through photosynthesis, of the oxygen in the air we breathe, and its hydrogen atoms become incorporated into many organic compounds. Water is also the solvent for most biological reactions and a reactant or product in many chemical reactions.

Water is important not only as an internal constituent of organisms but also as one of the principal environmental factors affecting them (Fig. 2-12). Many organisms live in the ocean or in freshwater rivers, lakes, or puddles. Water's unique combination of physical and chemical properties is considered to have been essential to the origin of life, as well as to the continued survival and evolution of life on Earth.

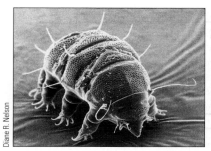

(a) Commonly known as "water bears," tardigrades, such as these members of the genus *Echiniscus*, are small animals (less than 1.2 mm long) that normally live in moist habitats, such as thin films of water on mosses.

100 μm

(b) When subjected to desiccation (dried out), tardigrades assume a barrel-shaped form known as a *tun*, remaining in this state, motionless but alive, for as long as 100 years. When rehydrated, they assume their normal appearance and activities.

10 μm

Figure 2-12 The effects of water on an organism

Hydrogen bonds form between water molecules

As discussed, water molecules are polar; that is, one end of each molecule bears a partial positive charge and the other a partial negative charge (see Fig. 2-7). The water molecules in liquid water and in ice associate by hydrogen bonds. The hydrogen atom of one water molecule, with its partial positive charge, is attracted to the oxygen atom of a neighboring water molecule, with its partial negative charge, forming a hydrogen bond. An oxygen atom in a water molecule has two regions of partial negative charge, and each of the two hydrogen atoms has a partial positive charge. Each water molecule can therefore form hydrogen bonds with a maximum of four neighboring water molecules (Fig. 2-13).

Water molecules have a strong tendency to stick to one another, a property known as cohesion. This is due to the hydrogen bonds among the molecules. Because of the cohesive nature

Figure 2-13 Hydrogen bonding of water molecules

Each water molecule can form hydrogen bonds (*dotted lines*) with as many as four neighboring water molecules.

of water molecules, any force exerted on part of a column of water is transmitted to the column as a whole. The major mechanism of water movement in plants (see Chapter 34) depends on the cohesive nature of water. Water molecules also display adhesion, the ability to stick to many other kinds of substances, most notably those with charged groups of atoms or molecules on their surfaces. These adhesive forces explain how water makes things wet.

A combination of adhesive and cohesive forces accounts for **capillary action,** which is the tendency of water to move in narrow tubes, even against the force of gravity (Fig. 2-14). For example, water moves through the microscopic spaces between soil particles to the roots of plants by capillary action.

Water has a high degree of surface tension because of the cohesion of its molecules, which have a much greater attraction for one another than for molecules in the air. Thus, water molecules at the surface crowd together, producing a strong layer as they are pulled downward by the attraction of other water molecules beneath them (Fig. 2-15).

Water molecules interact with hydrophilic substances by hydrogen bonding

Because its molecules are polar, water is an excellent solvent, a liquid capable of dissolving many kinds of substances, especially polar and ionic compounds. Earlier we discussed how polar water molecules pull the ions of ionic compounds apart so that they dissociate (see Fig. 2-10). Because of its solvent properties and the tendency of the atoms in certain compounds to form ions in solution, water plays an important role in facilitating chemical

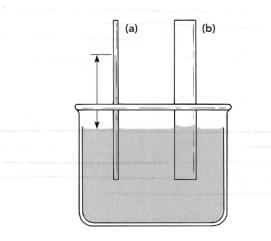

Figure 2-14 Capillary action

(a) In a narrow tube, there is adhesion between the water molecules and the glass wall of the tube. Other water molecules inside the tube are then "pulled along" because of cohesion, which is due to hydrogen bonds between the water molecules. (b) In the wider tube, a smaller percentage of the water molecules line the glass wall. As a result, the adhesion is not strong enough to overcome the cohesion of the water molecules beneath the surface level of the container, and water in the tube rises only slightly.

Figure 2-15 Surface tension of water

Hydrogen bonding between water molecules is responsible for the surface tension of water, which causes a dimpled appearance of the surface as these water striders (*Gerris*) walk across it. Fine hairs at the ends of the legs of these insects create highly water-repellent "cushions" of air.

reactions. Substances that interact readily with water are **hydrophilic** ("water-loving"). Examples include table sugar (sucrose, a polar compound) and table salt (NaCl, an ionic compound), which dissolve readily in water. Not all substances in organisms are hydrophilic, however. Many **hydrophobic** ("water-fearing") substances found in living things are especially important because of their ability to form associations or structures that are not disrupted. **Hydrophobic interactions** occur between groups of nonpolar molecules. Such molecules are insoluble in water and tend to cluster together. This is not due to formation of bonds between the nonpolar molecules but rather to the fact that the hydrogen-bonded water molecules exclude them and in a sense "drive them together." Hydrophobic interactions explain why oil tends to form globules when added to water. Examples of hydrophobic substances include fatty acids and cholesterol, discussed in Chapter 3.

Water helps maintain a stable temperature

Hydrogen bonding explains the way water responds to changes in temperature. Water exists in three forms, which differ in their degree of hydrogen bonding: gas (vapor), liquid, and ice, a crystalline solid (Fig. 2-16). Hydrogen bonds are formed or broken as water changes from one state to another.

Raising the temperature of a substance involves adding heat energy to make its molecules move faster, that is, to increase the energy of motion—**kinetic energy**—of the molecules (see Chapter 7). The term **heat** refers to the *total* amount of kinetic energy in a sample of a substance; **temperature** is a measure of the *average* kinetic energy of the particles. For the molecules to move more freely, some of the hydrogen bonds of water must be broken. Much of the energy added to the system is used up in breaking the hydrogen bonds, and only a portion of the heat energy is available to speed the movement of the water molecules, thereby increasing the temperature of the water. Conversely, when liquid water changes to ice, additional hydrogen bonds must be formed,

making the molecules less free to move and liberating a great deal of heat into the environment.

Heat of vaporization, the amount of heat energy required to change 1 g of a substance from the liquid phase to the vapor phase, is expressed in units called *calories*. A **calorie** (**cal**) is the amount of heat energy (equivalent to 4.184 joules [J]) required to raise the temperature of 1 g of water 1 degree Celsius (C). Water has a high heat of vaporization—540 cal—because its molecules are held together by hydrogen bonds. The heat of vaporization of most other common liquid substances is much less. As a sample of water is heated, some molecules are moving much faster than others (they have more heat). These faster-moving molecules are more likely to escape the liquid phase and enter the vapor phase (see Fig. 2-16a). When they do, they take their heat with them, lowering the temperature of the sample, a process called **evaporative cooling.** For this reason, the human body can dissipate excess heat as sweat evaporates from the skin, and a leaf can keep cool in the bright sunlight as water evaporates from its surface.

Hydrogen bonding is also responsible for water's high **specific heat**; that is, the amount of energy required to raise the temperature of water is quite large. The specific heat of water is 1 cal/g of water per degree Celsius. Most other common substances, such as metals, glass, and ethyl alcohol, have much lower specific heat values. The specific heat of ethyl alcohol, for example, is 0.59 cal/g/1°C (2.46 J/g/1°C).

Because so much heat input is required to raise the temperature of water (and so much heat is lost when the temperature is lowered), the ocean and other large bodies of water have relatively constant temperatures. Thus, many organisms living in the ocean are provided with a relatively constant environmental temperature. The properties of water are crucial in stabilizing temperatures on Earth's surface. Although surface water is only a thin film relative to Earth's volume, the quantity is enormous compared to the exposed landmass. This relatively large mass of water resists both the warming effect of heat and the cooling effect of low temperatures.

Hydrogen bonding causes ice to have unique properties with important environmental consequences. Liquid water expands as it freezes because the hydrogen bonds joining the water molecules in the crystalline lattice keep the molecules far enough apart to give ice a density about 10% less than the density of liquid water (see Fig. 2-16c). When ice has been heated enough to raise its temperature above 0°C (32°F), the hydrogen bonds are broken, freeing the molecules to slip closer together. The density of water is greatest at 4°C. Above that temperature water begins to expand again as the speed of its molecules increases. As a result, ice floats on the denser cold water.

This unusual property of water has been important to the evolution of life. If ice had a greater density than water, the ice would sink; eventually all ponds, lakes, and even the ocean would freeze solid from the bottom to the surface, making life impossible. When a deep body of water cools, it becomes covered with floating ice. The ice insulates the liquid water below it, retarding freezing and permitting organisms to survive below the icy surface.

The high water content of organisms helps them maintain relatively constant internal temperatures. Such minimizing of

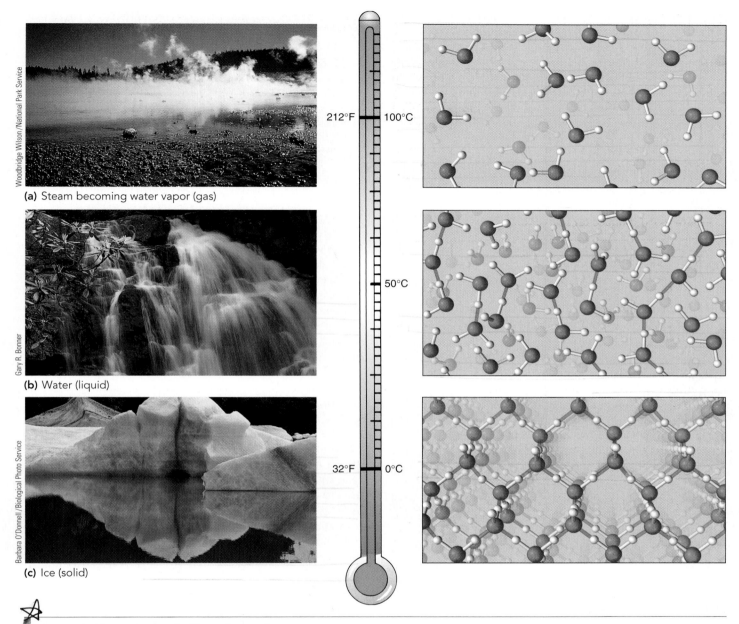

(a) Steam becoming water vapor (gas)

(b) Water (liquid)

(c) Ice (solid)

212°F 100°C

50°C

32°F 0°C

Figure 2-16 *Animated* Three forms of water

(a) When water boils, as in this hot spring at Yellowstone National Park, many hydrogen bonds are broken, causing steam, consisting of minuscule water droplets, to form. If most of the remaining hydrogen bonds break, the molecules move more freely as water vapor (a gas).

(b) Water molecules in a liquid state continually form, break, and re-form hydrogen bonds with one another. (c) In ice, each water molecule participates in four hydrogen bonds with adjacent molecules, resulting in a regular, evenly distanced crystalline lattice structure.

temperature fluctuations is important because biological reactions can take place only within a relatively narrow temperature range.

Review

▌ Why does water form hydrogen bonds?

▌ What are some properties of water that result from hydrogen bonding? How do these properties contribute to the role of water as an essential component of organisms?

▌ How can weak forces, such as hydrogen bonds, have significant effects in organisms?

ACIDS, BASES, AND SALTS

Learning Objectives

10 Contrast acids and bases, and discuss their properties.

11 Convert the hydrogen ion concentration (moles per liter) of a solution to a pH value, and describe how buffers help minimize changes in pH.

12 Describe the composition of a salt, and explain why salts are important in organisms.

Water molecules have a slight tendency to **ionize,** that is, to dissociate into hydrogen ions (H^+) and hydroxide ions (OH^-). The H^+ immediately combines with a negatively charged region of a water molecule, forming a hydronium ion (H_3O^+). However, by convention, H^+, rather than the more accurate H_3O^+, is used. In pure water, a small number of water molecules ionize. This slight tendency of water to dissociate is reversible as hydrogen ions and hydroxide ions reunite to form water.

$$HOH \rightleftharpoons H^+ + OH^-$$

Because each water molecule splits into one hydrogen ion and one hydroxide ion, the concentrations of hydrogen ions and hydroxide ions in pure water are exactly equal (0.0000001 or 10^{-7} mol/L for each ion). Such a solution is said to be neutral, that is, neither acidic nor basic (alkaline).

An **acid** is a substance that dissociates in solution to yield hydrogen ions (H^+) and anions.

$$Acid \longrightarrow H^+ + Anion$$

An acid is a proton *donor*. (Recall that a hydrogen ion, or H^+, is nothing more than a proton.) Hydrochloric acid (HCl) is a common inorganic acid.

A **base** is defined as a proton *acceptor*. Most bases are substances that dissociate to yield a hydroxide ion (OH^-) and a cation when dissolved in water.

$$NaOH \longrightarrow Na^+ + OH^-$$

$$OH^- + H^+ \longrightarrow H_2O$$

A hydroxide ion can act as a base by accepting a proton (H^+) to form water. Sodium hydroxide (NaOH) is a common inorganic base. Some bases do not dissociate to yield hydroxide ions directly. For example, ammonia (NH_3) acts as a base by accepting a proton from water, producing an ammonium ion (NH_4^+) and releasing a hydroxide ion.

$$NH_3 + H_2O \longrightarrow NH_4^+ + OH^-$$

pH is a convenient measure of acidity

The degree of a solution's acidity is generally expressed in terms of **pH,** defined as the negative logarithm (base 10) of the hydrogen ion concentration (expressed in moles per liter):

$$pH = -\log_{10}[H^+]$$

The brackets refer to concentration; therefore, the term $[H^+]$ means "the concentration of hydrogen ions," which is expressed in moles per liter because we are interested in the *number* of hydrogen ions per liter. Because the range of possible pH values is broad, a logarithmic scale (with a 10-fold difference between successive units) is more convenient than a linear scale.

Hydrogen ion concentrations are nearly always less than 1 mol/L. One gram of hydrogen ions dissolved in 1 L of water (a 1-*M* solution) may not sound impressive, but such a solution would be extremely acidic. The logarithm of a number less than 1 is a negative number; thus, the *negative* logarithm corresponds to

TABLE 2-2

Calculating pH Values and Hydroxide Ion Concentrations from Hydrogen Ion Concentrations

Substance	[H^+]*	log [H^+]	pH	[OH^-]†
Gastric juice	0.01, 10^{-2}	−2	2	10^{-12}
Pure water, neutral solution	0.0000001, 10^{-7}	−7	7	10^{-7}
Household ammonia	0.00000000001, 10^{-11}	−11	11	10^{-3}

*[H^+] = hydrogen ion concentration (mol/L)

†[OH^-] = hydroxide ion concentration (mol/L)

a *positive* pH value. (Solutions with pH values less than zero can be produced but do not occur under biological conditions.)

Whole-number pH values are easy to calculate. For instance, consider our example of pure water, which has a hydrogen ion concentration of 0.0000001 (10^{-7}) mol/L. The logarithm is −7. The negative logarithm is 7; therefore, the pH is 7. ■ Table 2-2 shows how to calculate pH values from hydrogen ion concentrations, and the reverse. For comparison, the table also includes the hydroxide ion concentrations, which can be calculated because the product of the hydrogen ion concentration and the hydroxide ion concentration is 1×10^{-14}:

$$[H^+][OH^-] = 1 \times 10^{-14}$$

Pure water is an example of a **neutral solution;** with a pH of 7, it has equal concentrations of hydrogen ions and hydroxide ions (the concentration of each is 10^{-7} mol/L). An **acidic solution** has a hydrogen ion concentration that is higher than its hydroxide ion concentration and has a pH value of less than 7. For example, the hydrogen ion concentration of a solution with pH 1 is 10 times that of a solution with pH 2. A **basic solution** has a hydrogen ion concentration that is lower than its hydroxide ion concentration and has a pH greater than 7.

The pH values of some common substances are shown in ▌ Figure 2-17. Although some very acidic compartments exist within cells (see Chapter 4), most of the interior of an animal or plant cell is neither strongly acidic nor strongly basic but an essentially neutral mixture of acidic and basic substances. Although certain bacteria are adapted to life in extremely acidic environments (discussed in Chapter 24), a substantial change in pH is incompatible with life for most cells. The pH of most types of plant and animal cells (and their environment) ordinarily ranges from around 7.2 to 7.4.

Buffers minimize pH change

Many homeostatic mechanisms operate to maintain appropriate pH values. For example, the pH of human blood is about 7.4 and must be maintained within very narrow limits. If the blood becomes too acidic (for example, as a result of respiratory disease), coma and death may result. Excessive alkalinity can result

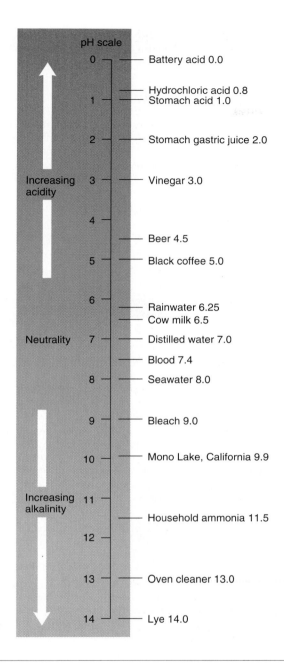

pH scale

- Battery acid 0.0
- Hydrochloric acid 0.8
- Stomach acid 1.0
- Stomach gastric juice 2.0
- Vinegar 3.0
- Beer 4.5
- Black coffee 5.0
- Rainwater 6.25
- Cow milk 6.5
- Distilled water 7.0
- Blood 7.4
- Seawater 8.0
- Bleach 9.0
- Mono Lake, California 9.9
- Household ammonia 11.5
- Oven cleaner 13.0
- Lye 14.0

Increasing acidity

Neutrality

Increasing alkalinity

Figure 2-17 *Animated* pH values of some common solutions

A neutral solution (pH 7) has equal concentrations of H^+ and OH^-. Acidic solutions, which have a higher concentration of H^+ than OH^-, have pH values less than 7; pH values greater than 7 characterize basic solutions, which have an excess of OH^-.

in overexcitability of the nervous system and even convulsions. Organisms contain many natural buffers. A **buffer** is a substance or combination of substances that resists changes in pH when an acid or base is added. A buffering system includes a weak acid or a weak base. A weak acid or weak base does not ionize completely. At any given instant, only a fraction of the molecules are ionized; most are not dissociated.

One of the most common buffering systems functions in the blood of vertebrates (see Chapter 45). Carbon dioxide, produced as a waste product of cell metabolism, enters the blood, the main constituent of which is water. The carbon dioxide reacts with the water to form carbonic acid, a weak acid that dissociates to yield a hydrogen ion and a bicarbonate ion. The following expression describes the buffering system:

$$CO_2 + H_2O \rightleftharpoons H_2CO_3 \rightleftharpoons H^+ + HCO_3^-$$

Carbon dioxide · Water · Carbonic acid · Bicarbonate ion

As the double arrows indicate, all the reactions are reversible. Because carbonic acid is a weak acid, undissociated molecules are always present, as are all the other components of the system. The expression describes the system when it is at *dynamic equilibrium,* that is, when the rates of the forward and reverse reactions are equal and the relative concentrations of the components are not changing. A system at dynamic equilibrium tends to stay at equilibrium unless a stress is placed on it, which causes it to shift to reduce the stress until it attains a new dynamic equilibrium. A change in the concentration of any component is one such stress. Therefore, the system can be "shifted to the right" by adding reactants or removing products. Conversely, it can be "shifted to the left" by adding products or removing reactants.

Hydrogen ions are the important products to consider in this system. The addition of excess hydrogen ions temporarily shifts the system to the left, as they combine with the bicarbonate ions to form carbonic acid. Eventually a new dynamic equilibrium is established. At this point the hydrogen ion concentration is similar to the original concentration, and the product of the hydrogen ion and hydroxide ion concentrations is restored to the equilibrium value of 1×10^{-14}.

If hydroxide ions are added, they combine with the hydrogen ions to form water, effectively removing a product and thus shifting the system to the right. As this occurs, more carbonic acid ionizes, effectively replacing the hydrogen ions that were removed.

Organisms contain many weak acids and weak bases, which allows them to maintain an essential reserve of buffering capacity and helps them avoid pH extremes.

An acid and a base react to form a salt

When an acid and a base are mixed in water, the H^+ of the acid unites with the OH^- of the base to form a molecule of water. The remainder of the acid (an anion) combines with the remainder of the base (a cation) to form a salt. For example, hydrochloric acid reacts with sodium hydroxide to form water and sodium chloride:

$$HCl + NaOH \longrightarrow H_2O + NaCl$$

A **salt** is a compound in which the hydrogen ion of an acid is replaced by some other cation. Sodium chloride, NaCl, is a salt in which the hydrogen ion of HCl has been replaced by the cation Na^+.

When a salt, an acid, or a base is dissolved in water, its dissociated ions can conduct an electric current; these substances are called **electrolytes.** Sugars, alcohols, and many other substances do not form ions when dissolved in water; they do not conduct an electric current and are referred to as **nonelectrolytes.**

Cells and extracellular fluids (such as blood) of animals and plants contain a variety of dissolved salts that are the source of the many important mineral ions essential for fluid balance and acid–base balance. Nitrate and ammonium ions from the soil are the important nitrogen sources for plants. In animals, nerve and muscle function, blood clotting, bone formation, and many other aspects of body function depend on ions. Sodium, potassium, calcium, and magnesium are the chief cations present; chloride, bicarbonate, phosphate, and sulfate are important anions. The concentrations and relative amounts of the various cations and anions are kept remarkably constant. Any marked change results in impaired cell functions and may lead to death.

Review

- A solution has a hydrogen ion concentration of 0.01 mol/L. What is its pH? What is its hydroxide ion concentration? Is it acidic, basic, or neutral? How does this solution differ from one with a pH of 1?
- What would be the consequences of adding or removing a reactant or a product from a reversible reaction that is at dynamic equilibrium?
- Why are buffers important in organisms? Why can't strong acids or bases work as buffers?
- Why are acids, bases, and salts referred to as *electrolytes*?

SUMMARY WITH KEY TERMS

Learning Objectives

1. Name the principal chemical elements in living things, and give an important function of each (page 26).
    - An **element** is a substance that cannot be decomposed into simpler substances by normal chemical reactions. About 96% of an organism's mass consists of carbon, the backbone of organic molecules; hydrogen and oxygen, the components of water; and nitrogen, a component of proteins and nucleic acids.

2. Compare the physical properties (mass and charge) and locations of electrons, protons, and neutrons. Distinguish between the atomic number and the mass number of an atom (page 26).
    - Each **atom** is composed of a **nucleus** containing positively charged **protons** and uncharged **neutrons.** Negatively charged **electrons** encircle the nucleus.
    - An atom is identified as belonging to a particular element by its number of protons (**atomic number**). The **atomic mass** of an atom is equal to the sum of its protons and neutrons.
    - A single proton or a single neutron each has a mass equivalent to one **atomic mass unit.** The mass of a single electron is only about 1/1800 amu.

3. Define the terms *orbital* and *electron shell*. Relate electron shells to principal energy levels (page 26).
    - In the space outside the nucleus, electrons move rapidly in electron **orbitals.** An **electron shell** consists of electrons in orbitals at the same **principal energy level.** Electrons in a shell distant from the nucleus have greater energy than those in a shell closer to the nucleus.

ThomsonNOW **Learn more about atomic orbitals by clicking on the figure in ThomsonNOW.**

4. Explain how the number of valence electrons of an atom is related to its chemical properties (page 29).
    - The chemical properties of an atom are determined chiefly by the number and arrangement of its most energetic electrons, known as **valence electrons.** The valence shell of most atoms is full when it contains 8 electrons; that of hydrogen or helium is full when it contains 2. An atom tends to lose, gain, or share electrons to fill its valence shell.

5. Distinguish among simplest, molecular, and structural chemical formulas (page 29).
    - Different atoms are joined by chemical bonds to form **compounds.** A **chemical formula** gives the types and relative numbers of atoms in a substance.
    - A **simplest formula** gives the smallest whole-number ratio of the component atoms. A **molecular formula** gives the actual numbers of each type of atom in a molecule. A **structural formula** shows the arrangement of the atoms in a molecule.

6. Explain why the mole concept is so useful to chemists (page 29).
    - One **mole** (the atomic or molecular mass in grams) of any substance contains 6.02×10^{23} atoms, molecules, or ions, enabling scientists to "count" particles by weighing a sample. This number is known as **Avogadro's number.**

7. Distinguish among covalent bonds, ionic bonds, hydrogen bonds, and van der Waals interactions. Compare them in terms of the mechanisms by which they form and their relative strengths (page 31).
    - **Covalent bonds** are strong, stable bonds formed when atoms share valence electrons, forming molecules. When covalent bonds are formed, the orbitals of the valence electrons may become rearranged in a process known as **orbital hybridization.** Covalent bonds are **nonpolar** if the electrons are shared equally between the two atoms. Covalent bonds are **polar** if one atom is more **electronegative** (has a greater affinity for electrons) than the other.
    - An **ionic bond** is formed between a positively charged **cation** and a negatively charged **anion.** Ionic bonds are strong in the absence of water but relatively weak in aqueous solution.

ThomsonNOW **Learn more about ionic bonding by clicking on the figure in ThomsonNOW.**

- **Hydrogen bonds** are relatively weak bonds formed when a hydrogen atom with a partial positive charge is attracted to an atom (usually oxygen or nitrogen) with a partial negative charge already bonded to another molecule or in another part of the same molecule.

ThomsonNOW™ **Learn more about hydrogen bonding by clicking on the figure in ThomsonNOW.**

- **van der Waals interactions** are weak forces based on fluctuating electric charges.

8 Distinguish between the terms *oxidation* and *reduction*, and relate these processes to the transfer of energy (page 36).

- **Oxidation** and **reduction** reactions (**redox reactions**) are chemical processes in which electrons (and their energy) are transferred from a reducing agent to an oxidizing agent. In oxidation, an atom, ion, or molecule loses electrons (and their energy). In reduction, an atom, ion, or molecule gains electrons (and their energy).

9 Explain how hydrogen bonds between adjacent water molecules govern many of the properties of water (page 36).

- Water is a **polar molecule** because one end has a partial positive charge and the other has a partial negative charge. Because its molecules are polar, water is an excellent **solvent** for ionic or polar **solutes.**
- Water molecules exhibit the property of **cohesion** because they form hydrogen bonds with one another; they also exhibit **adhesion** through hydrogen bonding to substances with ionic or polar regions.
- Because hydrogen bonds must be broken to raise its temperature, water has a high **specific heat,** which helps organisms maintain a relatively constant internal temperature; this property also helps keep the ocean and other large bodies of water at a constant temperature.
- Water has a high **heat of vaporization.** Hydrogen bonds must be broken for molecules to enter the vapor phase. These molecules carry a great deal of heat, which accounts for **evaporative cooling.**

- The hydrogen bonds between water molecules in ice cause it to be less dense than liquid water. The fact that ice floats makes the aquatic environment less extreme than it would be if ice sank to the bottom.

ThomsonNOW™ **Learn more about the three forms of water by clicking on the figure in ThomsonNOW.**

10 Contrast acids and bases, and discuss their properties (page 39).

- **Acids** are proton (hydrogen ion, H^+) donors; **bases** are proton acceptors. An acid dissociates in solution to yield H^+ and an anion. Many bases dissociate in solution to yield hydroxide ions (OH^-), which then accept protons to form water.

11 Convert the hydrogen ion concentration (moles per liter) of a solution to a pH value, and describe how buffers help minimize changes in pH (page 39).

- **pH** is the negative log of the hydrogen ion concentration of a solution (expressed in moles per liter). A **neutral solution** with equal concentrations of H^+ and OH^- (10^{-7} mol/L) has a pH of 7, an **acidic solution** has a pH less than 7, and a **basic solution** has a pH greater than 7.

ThomsonNOW™ **Learn more about the pH of common solutions by clicking on the figure in ThomsonNOW.**

- A buffering system is based on a weak acid or a weak base. A **buffer** resists changes in the pH of a solution when acids or bases are added.

12 Describe the composition of a salt, and explain why salts are important in organisms (page 39).

- A **salt** is a compound in which the hydrogen atom of an acid is replaced by some other cation. Salts provide the many mineral ions essential for life functions.

TEST YOUR UNDERSTANDING

1. Which of the following elements is *mismatched* with its properties or function? (a) carbon—forms the backbone of organic compounds (b) nitrogen—component of proteins (c) hydrogen—very electronegative (d) oxygen—can participate in hydrogen bonding (e) all of the above are correctly matched

2. Which of the following applies to a neutron? (a) positive charge and located in an orbital (b) negligible mass and located in the nucleus (c) positive charge and located in the nucleus (d) uncharged and located in the nucleus (e) uncharged and located in an orbital

3. $^{32}_{15}P$, a radioactive form of phosphorus, has (a) an atomic number of 32 (b) an atomic mass of 15 (c) an atomic mass of 47 (d) 32 electrons (e) 17 neutrons

4. Which of the following facts allows you to determine that atom A and atom B are isotopes of the same element? (a) they each have 6 protons (b) they each have 4 neutrons (c) the sum of the electrons and neutrons in each is 14 (d) they each have 4 valence electrons (e) they each have an atomic mass of 14

5. $^{1}_{1}H$ and $^{3}_{1}H$ have (a) different chemical properties, because they have different atomic numbers (b) the same chemical properties, because they have the same number of valence electrons

(c) different chemical properties, because they differ in their number of protons and electrons (d) the same chemical properties, because they have the same atomic mass (e) the same chemical properties, because they have the same number of protons, electrons, and neutrons

6. Sodium and potassium atoms behave similarly in chemical reactions because (a) they have the same number of neutrons (b) each has a single valence electron (c) they have the same atomic mass (d) they have the same number of electrons (e) they have the same number of protons

7. The orbitals composing an atom's valence electron shell (a) are arranged as concentric spheres (b) contain the atom's least energetic electrons (c) may change shape when covalent bonds are formed (d) never contain more than 1 electron each (e) more than one of the preceding is correct

8. Which of the following bonds and properties are correctly matched? (a) ionic bonds—are strong only if the participating ions are hydrated (b) hydrogen bonds—are responsible for bonding oxygen and hydrogen to form a single water molecule (c) polar covalent bonds—can occur between two atoms of the same element (d) covalent bonds—may be single, double, or triple (e) hydrogen bonds—are stronger than covalent bonds

9. In a redox reaction (a) energy is transferred from a reducing agent to an oxidizing agent (b) a reducing agent becomes oxidized as it accepts an electron (c) an oxidizing agent accepts a proton (d) a reducing agent donates a proton (e) the electrons in an atom move from its valence shell to a shell closer to its nucleus

10. Water has the property of adhesion because (a) hydrogen bonds form between adjacent water molecules (b) hydrogen bonds form between water molecules and hydrophilic substances (c) it has a high specific heat (d) covalent bonds hold an individual water molecule together (e) it has a great deal of kinetic energy

11. The high heat of vaporization of water accounts for (a) evaporative cooling (b) the fact that ice floats (c) the fact that heat is liberated when ice forms (d) the cohesive properties of water (e) capillary action

12. Water has a high specific heat because (a) hydrogen bonds must be broken to raise its temperature (b) hydrogen bonds must be formed to raise its temperature (c) it is a poor insulator (d) it has low density considering the size of the molecule (e) it can ionize

13. A solution at pH 7 is considered neutral because (a) its hydrogen ion concentration is 0 mol/L (b) its hydroxide ion concentration is 0 mol/L (c) the product of its hydrogen ion concentration and its hydroxide ion concentration is 0 mol/L (d) its hydrogen ion concentration is equal to its hydroxide ion concentration (e) it is nonpolar

14. A solution with a pH of 2 has a hydrogen ion concentration that is _____ the hydrogen ion concentration of a solution with a pH of 4. (a) 1/2 (b) 1/100 (c) 2 times (d) 10 times (e) 100 times

15. Which of the following cannot function as a buffer? (a) phosphoric acid, a weak acid (b) sodium hydroxide, a strong base (c) sodium chloride, a salt that ionizes completely (d) a and c (e) b and c

16. NaOH and HCl react to form Na^+, Cl^-, and water. Which of the following statements is *true*? (a) Na^+ is an anion, and Cl^- is a cation (b) Na^+ and Cl^- are both anions (c) a hydrogen bond can form between Na^+ and Cl^- (d) Na^+ and Cl^- are electrolytes (e) Na^+ is an acid, and Cl^- is a base

17. Which of the following statements is *true*? (a) the number of individual particles (atoms, ions, or molecules) contained in one mole varies depending on the substance (b) Avogadro's number is the number of particles contained in one mole of a substance (c) Avogadro's number is 10^{23} particles (d) one mole of ^{12}C has a mass of 12 g (e) b and d

CRITICAL THINKING

1. Element A has 2 electrons in its valence shell (which is complete when it contains 8 electrons). Would you expect element A to share, donate, or accept electrons? What would you expect of element B, which has 4 valence electrons, and element C, which has 7?

2. A hydrogen bond formed between two water molecules is only about 1/20 as strong as a covalent bond between hydrogen and oxygen. In what ways would the physical properties of water be different if these hydrogen bonds were stronger (for example, 1/10 the strength of covalent bonds)?

3. Consider the following reaction (in water).

$$HCl \longrightarrow H^+ + Cl^-$$

Name the reactant(s) and product(s). Does the expression indicate that the reaction is reversible? Could HCl be used as a buffer?

4. **Evolution Link.** Initiatives designed to discover evidence for life (biosignatures or biomarkers) in the atmospheres of distant planets have been proposed by the U.S. National Aeronautics and Space Administration (NASA) and the European Space Agency (ESA). If implemented, both the Terrestrial Planet Finder project (NASA) and the Darwin project (ESA) will use special space telescopes to detect atmospheric water vapor, as well as oxygen and carbon dioxide. Why is water considered the most fundamental indicator that life could have evolved on these planets?

Additional questions are available in ThomsonNOW at www.thomsonedu.com/login

The Chemistry of Life: Organic Compounds

Momatiuk Eastcott / The Image Works

This young girl is using a leaf to feed her baby brother.

KEY CONCEPTS

Carbon atoms join with one another or other atoms to form large molecules with a wide variety of shapes.

Hydrocarbons are nonpolar and hydrophobic, but their properties can be altered by adding functional groups: hydroxyl and carbonyl groups (polar), carboxyl and phosphate groups (acidic), and amino groups (basic).

Carbohydrates are composed of sugar subunits (monosaccharides), which can be joined to form disaccharides, storage polysaccharides, and structural polysaccharides.

Lipids store energy (triacylglycerol) and are the main structural components of cell membranes (phospholipids).

Proteins have multiple levels of structure and are composed of amino acid subunits joined by peptide bonds.

Nucleic acids (DNA and RNA) are informational molecules composed of long chains of nucleotide subunits. ATP and some other nucleotides have a central role in energy metabolism.

Both inorganic and organic forms of carbon occur widely in nature. Many types of organic compounds will become incorporated into the body of the baby in the photograph as he grows. **Organic compounds** are those in which carbon atoms are covalently bonded to one another to form the backbone of the molecule. Some very simple carbon compounds are considered inorganic if the carbon is not bonded to another carbon or to hydrogen. The carbon dioxide we exhale as a waste product from the breakdown of organic molecules to obtain energy is an example of an inorganic carbon compound. Organic compounds are so named because at one time it was thought that they could be produced only by living (organic) organisms. In 1828, the German chemist Friedrich Wühler synthesized urea, a metabolic waste product. Since that time, scientists have learned to synthesize many organic molecules and have discovered organic compounds not found in any organism.

Organic compounds are extraordinarily diverse; in fact, more than 5 million have been identified. There are many reasons for this diversity. Organic compounds can be produced in a wide variety of three-dimensional (3-D) shapes. Furthermore, the carbon atom forms bonds with a greater number of different elements than does any other type of atom. The addition of chemical groups containing atoms of other elements—especially oxygen, nitrogen, phosphorus, and sulfur—can profoundly change the properties of an organic molecule. Diversity also results from the fact that many organic compounds found in organisms are extremely large *macro-molecules,* which cells construct from simpler modular subunits. For example, protein molecules are built from smaller compounds called *amino acids.*

As you study this chapter, you will develop an understanding of the major groups of organic compounds found in organisms, including carbohydrates, lipids, proteins, and nucleic acids (DNA and RNA). Why are these compounds of central importance to all living things? The most obvious answer is that they constitute the structures of cells and tissues. However, they also participate in and regulate metabolic reactions, transmit information, and provide energy for life processes. ■

CARBON ATOMS AND MOLECULES

Learning Objectives

1   Describe the properties of carbon that make it the central component of organic compounds.
2   Define the term *isomer,* and distinguish among the three principal isomer types.
3   Identify the major functional groups present in organic compounds, and describe their properties.
4   Explain the relationship between polymers and macro-molecules.

Carbon has unique properties that allow the formation of the carbon backbones of the large, complex molecules essential to life (▌Fig. 3-1). Because a carbon atom has 4 valence electrons, it can complete its valence shell by forming a total of four covalent bonds (see Fig. 2-2). Each bond can link it to another carbon atom or to an atom of a different element. Carbon is particularly well suited to serve as the backbone of a large molecule because carbon-to-carbon bonds are strong and not easily broken. However, they are not so strong that it would be impossible for cells to break them. Carbon-to-carbon bonds are not limited to single bonds (based on sharing one electron pair). Two carbon atoms can share two electron pairs with each other, forming double bonds:

$$\text{>C=C<}$$

In some compounds, triple carbon-to-carbon bonds are formed:

$$\text{—C≡C—}$$

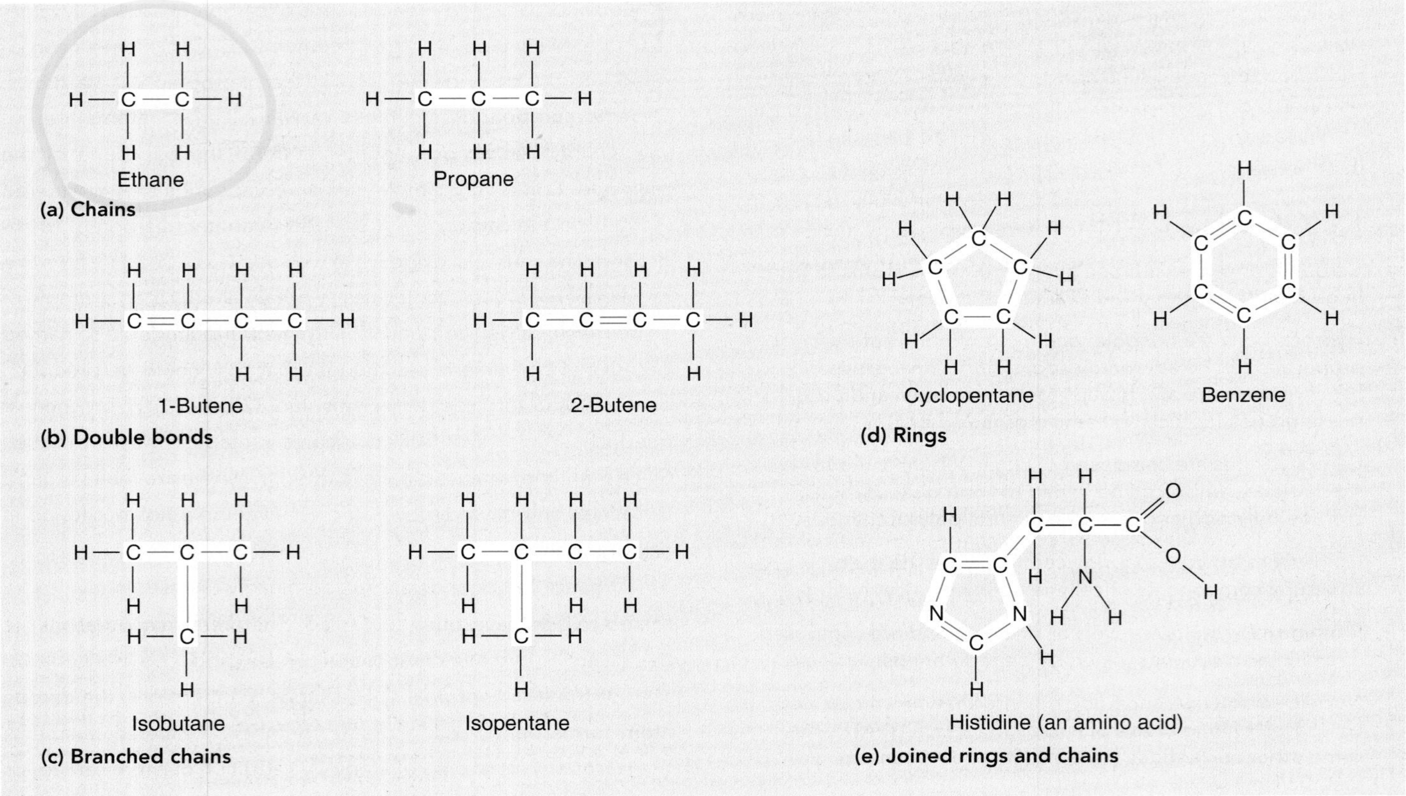

Figure 3-1   Organic molecules

Note that each carbon atom forms four covalent bonds, producing a wide variety of shapes.

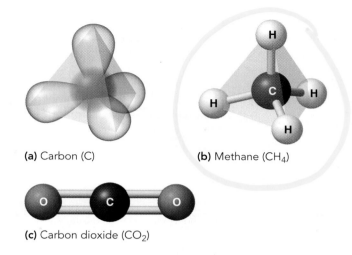

(a) Carbon (C)   **(b)** Methane (CH_4)

(c) Carbon dioxide (CO_2)

Figure 3-2   Carbon bonding

(a) The 3-D arrangement of the bonds of a carbon atom is responsible for **(b)** the tetrahedral architecture of methane. **(c)** In carbon dioxide, oxygen atoms are joined linearly to a central carbon by polar double bonds.

As shown in Figure 3-1, **hydrocarbons,** organic compounds consisting only of carbon and hydrogen, can exist as unbranched or branched chains, or as rings. Rings and chains are joined in some compounds.

The molecules in the cell are analogous to the components of a machine. Each component has a shape that allows it to fill certain roles and to interact with other components (often with a complementary shape). Similarly, the shape of a molecule is important in determining its biological properties and function. Carbon atoms can link to one another and to other atoms to produce a wide variety of 3-D molecular shapes, because the four covalent bonds of carbon do not form in a single plane. Instead, as discussed in Chapter 2, the valence electron orbitals become elongated and project from the carbon atom toward the corners of a tetrahedron (❙ Fig. 3-2). The structure is highly symmetrical, with an angle of about 109.5 degrees between any two of these bonds. Keep in mind that for simplicity, many of the figures in this book are drawn as two-dimensional (2-D) graphic representations of 3-D molecules. Even the simplest hydrocarbon chains, such as those in Figure 3-1, are not actually straight but have a 3-D zigzag structure.

Generally, there is freedom of rotation around each carbon-to-carbon single bond. This property permits organic molecules to be flexible and to assume a variety of shapes, depending on the extent to which each single bond is rotated. Double and triple bonds do not allow rotation, so regions of a molecule with such bonds tend to be inflexible.

Isomers have the same molecular formula but different structures

One reason for the great number of possible carbon-containing compounds is the fact that the same components usually can link in more than one pattern, generating an even wider variety of molecular shapes. Compounds with the same molecular for-

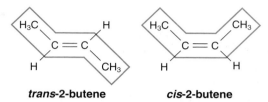

Ethanol (C_2H_6O)   Dimethyl ether (C_2H_6O)

(a) Structural isomers. Structural isomers differ in the covalent arrangement of their atoms.

trans*-2-butene   ***cis*-2-butene***

(b) Geometric isomers. Geometric, or *cis–trans*, isomers have identical covalent bonds but differ in the order in which groups are arranged in space.

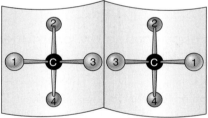

(c) Enantiomers. Enantiomers are isomers that are mirror images of one another. The central carbon is asymmetrical because it is bonded to four different groups. Because of their 3-D structure, the two figures cannot be superimposed no matter how they are rotated.

Figure 3-3   Isomers

Isomers have the same molecular formula, but their atoms are arranged differently.

mulas but different structures and thus different properties are called **isomers.** Isomers do not have identical physical or chemical properties and may have different common names. Cells can distinguish between isomers. Usually, one isomer is biologically active and the other is not. Three types of isomers are structural isomers, geometric isomers, and enantiomers.

Structural isomers are compounds that differ in the covalent arrangements of their atoms. For example, ❙ Figure 3-3a illustrates two structural isomers with the molecular formula C_2H_6O. Similarly, there are two structural isomers of the four-carbon hydrocarbon butane (C_4H_{10}), one with a straight chain and the other with a branched chain (isobutane). Large compounds have more possible structural isomers. There are only two structural isomers of butane, but there can be up to 366,319 isomers of $C_{20}H_{42}$.

Geometric isomers are compounds that are identical in the arrangement of their covalent bonds but different in the spatial arrangement of atoms or groups of atoms. Geometric isomers are present in some compounds with carbon-to-carbon double bonds. Because double bonds are not flexible, as single bonds are, atoms joined to the carbons of a double bond cannot rotate freely about the axis of the bonds. These *cis–trans* isomers may be represented as shown in ▌Figure 3-3b. The designation *cis* (Latin, "on this side") indicates that the two larger components are on the same side of the double bond. If they are on opposite sides of the double bond, the compound is designated a *trans* (Latin, "across") isomer.

Enantiomers are isomers that are mirror images of each other (▌Fig. 3-3c). Recall that the four groups bonded to a single carbon atom are arranged at the vertices of a tetrahedron. If the four bonded groups are all different, the central carbon is described as asymmetrical. Figure 3-3c illustrates that the four groups can be arranged about the asymmetrical carbon in two different ways that are mirror images of each other. The two molecules are enantiomers if they cannot be superimposed on each other no matter how they are rotated in space. Although enantiomers have similar chemical properties and most of their physical properties are identical, cells recognize the difference in shape, and usually only one form is found in organisms.

Functional groups change the properties of organic molecules

The existence of isomers is not the only source of variety among organic molecules. The addition of various combinations of atoms generates a vast array of molecules with different properties.

Because covalent bonds between hydrogen and carbon are nonpolar, hydrocarbons lack distinct charged regions. For this reason, hydrocarbons are insoluble in water and tend to cluster together, through **hydrophobic** interactions. "Water fearing," the literal meaning of the term *hydrophobic,* is somewhat misleading. Hydrocarbons interact with water, but much more weakly than the water molecules cohere to one another through hydrogen bonding. Hydrocarbons interact weakly with one another, but the main reason for hydrophobic interactions is that they are driven together in a sense, having been excluded by the hydrogen-bonded water molecules.

However, the characteristics of an organic molecule can be changed dramatically by replacing one of the hydrogens with one or more **functional groups,** groups of atoms that determine the types of chemical reactions and associations in which the compound participates. Most functional groups readily form associations, such as ionic and hydrogen bonds, with other molecules. Polar and ionic functional groups are **hydrophilic** because they associate strongly with polar water molecules.

The properties of the major classes of biologically important organic compounds—carbohydrates, lipids, proteins, and nucleic acids—are largely a consequence of the types and arrangement of functional groups they contain. When we know what kinds of functional groups are present in an organic compound, we can predict its chemical behavior. Note that the symbol R is used to represent the *remainder* of the molecule of which each functional group is a part. For example, the **methyl group,** a common nonpolar hydrocarbon group, is abbreviated R—CH_3. As you read the rest of this section, refer to ▌Table 3-1 for the structural formulas of other important functional groups, as well as for additional information.

The **hydroxyl group** (abbreviated R—OH) is polar because of the presence of a strongly electronegative oxygen atom. (Do not confuse it with the hydroxide ion, OH^-, discussed in Chapter 2.) If a hydroxyl group replaces one hydrogen of a hydrocarbon, the resulting molecule can have significantly altered properties. For example, ethane (see Fig. 3-1a) is a hydrocarbon that is a gas at room temperature. If a hydroxyl group replaces a hydrogen atom, the resulting molecule is ethyl alcohol, or ethanol, which is found in alcoholic beverages (see Fig. 3-3a). Ethanol is somewhat cohesive, because the polar hydroxyl groups of adjacent molecules interact; it is therefore liquid at room temperature. Unlike ethane, ethyl alcohol dissolves in water because the polar hydroxyl groups interact with the polar water molecules.

The **carbonyl group** consists of a carbon atom that has a double covalent bond with an oxygen atom. This double bond is polar because of the electronegativity of the oxygen; thus, the carbonyl group is hydrophilic. The position of the carbonyl group in the molecule determines the class to which the molecule belongs. An **aldehyde** has a carbonyl group positioned at the end of the carbon skeleton (abbreviated R—CHO); a **ketone** has an internal carbonyl group (abbreviated R—CO—R).

The **carboxyl group** (abbreviated R—COOH) in its non-ionized form consists of a carbon atom joined by a double covalent bond to an oxygen atom, and by a single covalent bond to another oxygen, which is in turn bonded to a hydrogen atom. Two electronegative oxygen atoms in such close proximity establish an extremely polarized condition, which can cause the hydrogen atom to be stripped of its electron and released as a hydrogen ion (H^+). The resulting ionized carboxyl group has 1 unit of negative charge (R—COO^-):

Carboxyl groups are weakly acidic; only a fraction of the molecules ionize in this way. This group therefore exists in one of two hydrophilic states: ionic or polar. Carboxyl groups are essential constituents of amino acids.

An **amino group** (abbreviated R—NH_2) in its non-ionized form includes a nitrogen atom covalently bonded to two hydrogen atoms. Amino groups are weakly basic because they are able to accept a hydrogen ion (proton). The resulting ionized amino group has 1 unit of positive charge (R—NH_3^+). Amino groups are components of amino acids and of nucleic acids.

TABLE 3-1

Some Biologically Important Functional Groups

Functional Group and Description	Structural Formula		Class of Compound Characterized by Group
Hydroxyl Polar because electronegative oxygen attracts covalent electrons	$R{-}OH$		Alcohols $$H{-}\underset{\underset{H}{\vert}}{\overset{\overset{H}{\vert}}{C}}{-}\underset{\underset{H}{\vert}}{\overset{\overset{H}{\vert}}{C}}{-}OH$$ Example, ethanol
Carbonyl **Aldehydes:** Carbonyl group carbon is bonded to at least one H atom; polar because electronegative oxygen attracts covalent electrons	$$R{-}\overset{\overset{O}{\|}}{C}{-}H$$		Aldehydes $$H{-}\overset{\overset{O}{\|}}{C}{-}H$$ Example, formaldehyde
Ketones: Carbonyl group carbon is bonded to two other carbons; polar because electronegative oxygen attracts covalent electrons	$$R{-}\overset{\overset{O}{\|}}{C}{-}R$$		Ketones $$H{-}\underset{\underset{H}{\vert}}{\overset{\overset{H}{\vert}}{C}}{-}\overset{\overset{O}{\|}}{C}{-}\underset{\underset{H}{\vert}}{\overset{\overset{H}{\vert}}{C}}{-}H$$ Example, acetone
Carboxyl Weakly acidic; can release an H^+	$$R{-}\overset{\overset{O}{\|}}{C}{-}OH$$ Non-ionized	$$R{-}\overset{\overset{O}{\|}}{C}{-}O^- + H^+$$ Ionized	Carboxylic acids (organic acids) $$R{-}\overset{\overset{O}{\|}}{C}{-}OH$$ Example, amino acid
Amino Weakly basic; can accept an H^+	$$R{-}\overset{\overset{H}{\vert}}{\underset{\underset{H}{\vert}}{N}}$$ Non-ionized	$$R{-}N^+{\overset{\overset{H}{\vert}}{\underset{\underset{H}{\vert}}{{-}H}}}$$ Ionized	Amines $$R{-}\underset{\underset{H}{\vert}}{\overset{\overset{NH_2}{\vert}}{C}}{-}\overset{\overset{O}{\|}}{C}{-}OH$$ Example, amino acid
Phosphate Weakly acidic; one or two H^+ can be released	$$R{-}O{-}\underset{\underset{OH}{\vert}}{\overset{\overset{O}{\|}}{P}}{-}OH$$ Non-ionized	$$R{-}O{-}\underset{\underset{O^-}{\vert}}{\overset{\overset{O}{\|}}{P}}{-}O^-$$ Ionized	Organic Phosphates $$HO{-}\underset{\underset{OH}{\vert}}{\overset{\overset{O}{\|}}{P}}{-}O{-}R$$ Example, phosphate ester (as found in ATP)
Sulfhydryl Helps stabilize internal structure of proteins	$R{-}SH$		Thiols $$H{-}\underset{\underset{SH}{\vert}}{\overset{\overset{H}{\vert}}{C}}{-}\underset{\underset{NH_2}{\vert}}{\overset{\overset{H}{\vert}}{C}}{-}\overset{\overset{O}{\|}}{C}{-}OH$$ Example, cysteine

A **phosphate group** (abbreviated $R{-}PO_4H_2$) is weakly acidic. The attraction of electrons by the oxygen atoms can result in the release of one or two hydrogen ions, producing ionized forms with 1 or 2 units of negative charge. Phosphates are constituents of nucleic acids and certain lipids.

The **sulfhydryl group** (abbreviated $R{-}SH$), consisting of an atom of sulfur covalently bonded to a hydrogen atom, is found in molecules called *thiols*. As you will see, amino acids that contain a sulfhydryl group can make important contributions to the structure of proteins.

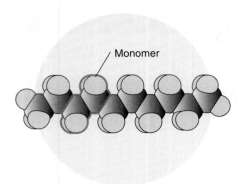

Figure 3-4 A simple polymer

This small polymer of polyethylene is formed by linking two-carbon ethylene (C_2H_4) monomers. One such monomer is outlined in red. The structure is represented by a space-filling model, which accurately depicts the actual 3-D shape of the molecule.

Many biological molecules are polymers

Many biological molecules such as proteins and nucleic acids are very large, consisting of thousands of atoms. Such giant molecules are known as **macromolecules.** Most macromolecules are **polymers,** produced by linking small organic compounds called **monomers** (Fig. 3-4). Just as all the words in this book have been written by arranging the 26 letters of the alphabet in various combinations, monomers can be grouped to form an almost infinite variety of larger molecules. The thousands of different complex organic compounds present in organisms are constructed from about 40 small, simple monomers. For example, the 20 monomers called *amino acids* can be linked end to end in countless ways to form the polymers known as *proteins.*

Polymers can be degraded to their component monomers by **hydrolysis** reactions ("to break with water"). In a reaction regulated by a specific enzyme (biological catalyst), a hydrogen from a water molecule attaches to one monomer, and a hydroxyl from water attaches to the adjacent monomer (Fig. 3-5).

Monomers are covalently linked by **condensation reactions.** Because the *equivalent* of a molecule of water is removed during the reactions that combine monomers, the term *dehydration synthesis* is sometimes used to describe condensation (see Fig. 3-5).

However, in biological systems the synthesis of a polymer is not simply the reverse of hydrolysis, even though the net effect is the opposite of hydrolysis. Synthetic processes such as condensation require energy and are regulated by different enzymes.

In the following sections we examine carbohydrates, lipids, proteins, and nucleic acids. Our discussion begins with the smaller, simpler forms of these compounds and extends to the linking of these monomers to form macromolecules.

Review

- What are some of the ways that the features of carbon-to-carbon bonds influence the stability and 3-D structure of organic molecules?
- Draw pairs of simple sketches comparing two (1) structural isomers, (2) geometric isomers, and (3) enantiomers. Why are these differences biologically important?
- Sketch the following functional groups: methyl, amino, carbonyl, hydroxyl, carboxyl, and phosphate. Include both non-ionized and ionized forms for acidic and basic groups.
- How is the fact that a group is nonpolar, polar, acidic, or basic related to its hydrophilic or hydrophobic properties?
- Why is the equivalent of a water molecule important to both condensation reactions and hydrolysis reactions?

CARBOHYDRATES

Learning Objective

5 Distinguish among monosaccharides, disaccharides, and polysaccharides. Compare storage polysaccharides with structural polysaccharides.

Sugars, starches, and cellulose are **carbohydrates.** Sugars and starches serve as energy sources for cells; cellulose is the main structural component of the walls that surround plant cells. Carbohydrates contain carbon, hydrogen, and oxygen atoms in a ratio of approximately one carbon to two hydrogens to one oxygen $(CH_2O)_n$. The term *carbohydrate,* meaning "hydrate (water) of carbon," reflects the 2:1 ratio of hydrogen to oxygen, the same ratio found in water (H_2O). Carbohydrates contain one sugar unit (monosaccharides), two sugar units (disaccharides), or many sugar units (polysaccharides).

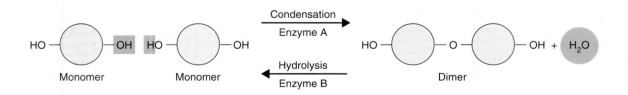

Figure 3-5 *Animated* Condensation and hydrolysis reactions

Joining two monomers yields a dimer; incorporating additional monomers produces a polymer. Note that condensation and hydrolysis reactions are catalyzed by different enzymes.

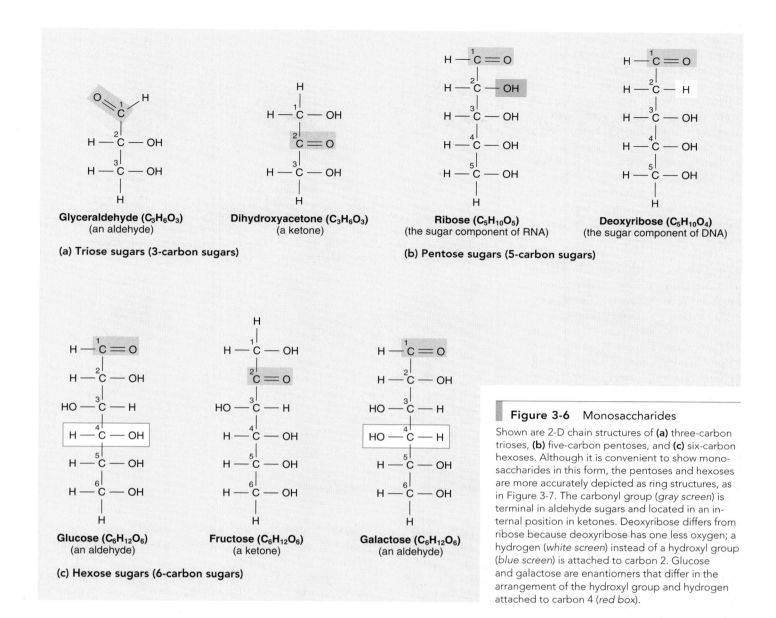

Glyceraldehyde (C₃H₆O₃)
(an aldehyde)

Dihydroxyacetone (C₃H₆O₃)
(a ketone)

(a) Triose sugars (3-carbon sugars)

Ribose (C₅H₁₀O₅)
(the sugar component of RNA)

Deoxyribose (C₅H₁₀O₄)
(the sugar component of DNA)

(b) Pentose sugars (5-carbon sugars)

Glucose (C₆H₁₂O₆)
(an aldehyde)

Fructose (C₆H₁₂O₆)
(a ketone)

Galactose (C₆H₁₂O₆)
(an aldehyde)

(c) Hexose sugars (6-carbon sugars)

Figure 3-6 Monosaccharides

Shown are 2-D chain structures of **(a)** three-carbon trioses, **(b)** five-carbon pentoses, and **(c)** six-carbon hexoses. Although it is convenient to show monosaccharides in this form, the pentoses and hexoses are more accurately depicted as ring structures, as in Figure 3-7. The carbonyl group (*gray screen*) is terminal in aldehyde sugars and located in an internal position in ketones. Deoxyribose differs from ribose because deoxyribose has one less oxygen; a hydrogen (*white screen*) instead of a hydroxyl group (*blue screen*) is attached to carbon 2. Glucose and galactose are enantiomers that differ in the arrangement of the hydroxyl group and hydrogen attached to carbon 4 (*red box*).

Monosaccharides are simple sugars

Monosaccharides typically contain from three to seven carbon atoms. In a monosaccharide, a hydroxyl group is bonded to each carbon except one; that carbon is double-bonded to an oxygen atom, forming a carbonyl group. If the carbonyl group is at the end of the chain, the monosaccharide is an aldehyde; if the carbonyl group is at any other position, the monosaccharide is a ketone. (By convention, the numbering of the carbon skeleton of a sugar begins with the carbon at or nearest the carbonyl end of the open chain.) The large number of polar hydroxyl groups, plus the carbonyl group, gives a monosaccharide hydrophilic properties.

Figure 3-6 shows simplified, 2-D representations of some common monosaccharides. The simplest carbohydrates are the three-carbon sugars (trioses): glyceraldehyde and dihydroxyacetone. Ribose and deoxyribose are common pentoses, sugars that

contain five carbons; they are components of nucleic acids (DNA, RNA, and related compounds). Glucose, fructose, galactose, and other six-carbon sugars are called **hexoses.** (Note that the names of carbohydrates typically end in -*ose*.)

Glucose (C₆H₁₂O₆), the most abundant monosaccharide, is used as an energy source in most organisms. During cellular respiration (see Chapter 8), cells oxidize glucose molecules, converting the stored energy to a form that can be readily used for cell work. Glucose is also used in the synthesis of other types of compounds such as amino acids and fatty acids. Glucose is so important in metabolism that mechanisms have evolved to maintain its concentration at relatively constant levels in the blood of humans and other complex animals (see Chapter 48).

Glucose and fructose are structural isomers: they have identical molecular formulas, but their atoms are arranged differently.

In fructose (a ketone) the double-bonded oxygen is linked to a carbon within the chain rather than to a terminal carbon as in glucose (an aldehyde). Because of these differences, the two sugars have different properties. For example, fructose, found in honey and some fruits, tastes sweeter than glucose.

Glucose and galactose are both hexoses and aldehydes. However, they are mirror images (enantiomers) because they differ in the arrangement of the atoms attached to asymmetrical carbon atom 4.

The linear formulas in Figure 3-6 give a clear but somewhat unrealistic picture of the structures of some common monosaccharides. As we have mentioned, molecules are not 2-D; in fact, the properties of each compound depend largely on its 3-D structure. Thus, 3-D formulas are helpful in understanding the relationship between molecular structure and biological function.

Molecules of glucose and other pentoses and hexoses in solution are actually rings rather than extended straight carbon chains.

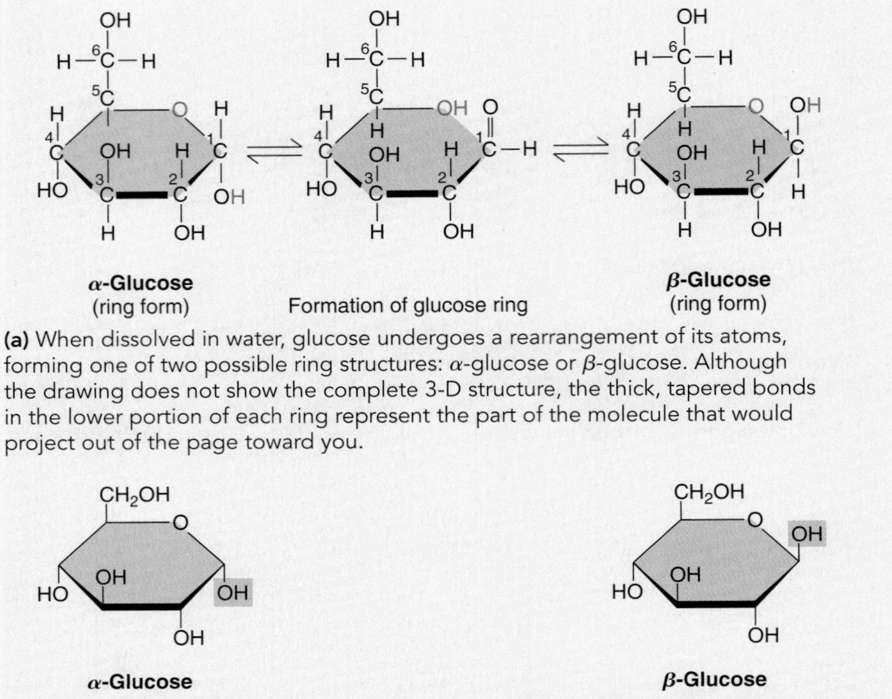

Glucose in solution (as in the cell) typically exists as a ring of five carbons and one oxygen. It assumes this configuration when its atoms undergo a rearrangement, permitting a covalent bond to connect carbon 1 to the oxygen attached to carbon 5 (Fig. 3-7). When glucose forms a ring, two isomeric forms are possible, differing only in orientation of the hydroxyl (—OH) group attached to carbon 1. When this hydroxyl group is on the same side of the plane of the ring as the —CH₂OH side group, the glucose is designated beta glucose (β-glucose). When it is on the side (with respect to the plane of the ring) opposite the —CH₂OH side group, the compound is designated alpha glucose (α-glucose). Although the differences between these isomers may seem small, they have important consequences when the rings join to form polymers.

(a) When dissolved in water, glucose undergoes a rearrangement of its atoms, forming one of two possible ring structures: α-glucose or β-glucose. Although the drawing does not show the complete 3-D structure, the thick, tapered bonds in the lower portion of each ring represent the part of the molecule that would project out of the page toward you.

(b) The essential differences between α-glucose and β-glucose are more readily apparent in these simplified structures. By convention, a carbon atom is assumed to be present at each angle in the ring unless another atom is shown. Most hydrogen atoms have been omitted.

Figure 3-7 α and β forms of glucose

Disaccharides consist of two monosaccharide units

A **disaccharide** (two sugars) contains two monosaccharide rings joined by a **glycosidic linkage,** consisting of a central oxygen covalently bonded to two carbons, one in each ring (Fig. 3-8). The glycosidic linkage of a disaccharide generally forms between carbon 1 of one molecule and carbon 4 of the other molecule. The disaccharide maltose (malt sugar) consists of two covalently linked α-glucose units. Sucrose, common table sugar, consists of

a glucose unit combined with a fructose unit. Lactose (the sugar present in milk) consists of one molecule of glucose and one of galactose.

As shown in Figure 3-8, a disaccharide can be hydrolyzed, that is, split by the addition of water, into two monosaccharide units. During digestion, maltose is hydrolyzed to form two molecules of glucose:

$$\text{Maltose} + \text{water} \rightarrow \text{glucose} + \text{glucose}$$

Similarly, sucrose is hydrolyzed to form glucose and fructose:

$$\text{Sucrose} + \text{water} \rightarrow \text{glucose} + \text{fructose}$$

Polysaccharides can store energy or provide structure

A **polysaccharide** is a macromolecule consisting of repeating units of simple sugars, usually glucose. The polysaccharides are the most abundant carbohydrates and include starches, glycogen, and cellulose. Although the precise number of sugar units varies, thousands of units are typically present in a single molecule. A polysaccharide may be a single long chain or a branched chain.

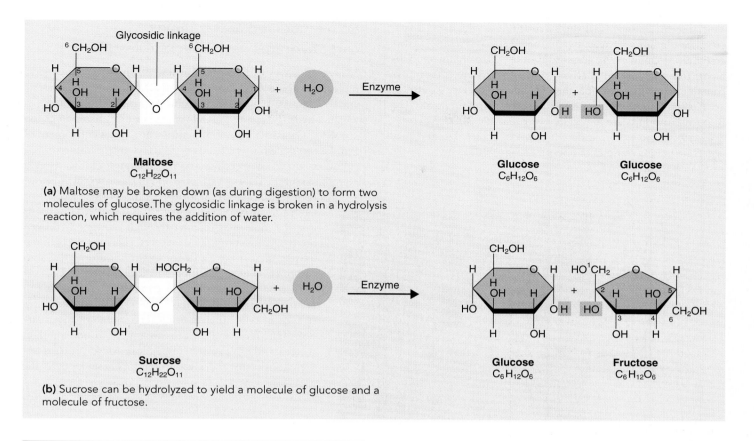

(a) Maltose may be broken down (as during digestion) to form two molecules of glucose. The glycosidic linkage is broken in a hydrolysis reaction, which requires the addition of water.

(b) Sucrose can be hydrolyzed to yield a molecule of glucose and a molecule of fructose.

Figure 3-8 Hydrolysis of disaccharides

Note that an enzyme is needed to promote these reactions.

Because they are composed of different isomers and because the units may be arranged differently, polysaccharides vary in their properties. Those that can be easily broken down to their subunits are well suited for energy storage, whereas the macromolecular 3-D architecture of others makes them particularly well suited to form stable structures.

Starch, the typical form of carbohydrate used for energy storage in plants, is a polymer consisting of α-glucose subunits. These monomers are joined by α 1—4 linkages, which means that carbon 1 of one glucose is linked to carbon 4 of the next glucose in the chain (❚ Fig. 3-9). Starch occurs in two forms: amylose and amylopectin. Amylose, the simpler form, is unbranched. Amylopectin, the more common form, usually consists of about 1000 glucose units in a branched chain.

Plant cells store starch mainly as granules within specialized organelles called **amyloplasts** (see Fig. 3-9a); some cells, such as those of potatoes, are very rich in amyloplasts. When energy is needed for cell work, the plant hydrolyzes the starch, releasing the glucose subunits. Virtually all organisms, including humans and other animals, have enzymes that can break α 1—4 linkages.

Glycogen (sometimes referred to as *animal starch*) is the form in which glucose subunits, joined by α 1—4 linkages, are stored as an energy source in animal tissues. Glycogen is similar in structure to plant starch but more extensively branched and more water soluble. In vertebrates, glycogen is stored mainly in liver and muscle cells.

Carbohydrates are the most abundant group of organic compounds on Earth, and **cellulose** is the most abundant carbohydrate; it accounts for 50% or more of all the carbon in plants (❚ Fig. 3-10). Cellulose is a structural carbohydrate. Wood is about half cellulose, and cotton is at least 90% cellulose. Plant cells are surrounded by strong supporting cell walls consisting mainly of cellulose.

Cellulose is an insoluble polysaccharide composed of many joined glucose molecules. The bonds joining these sugar units are different from those in starch. Recall that starch is composed of α-glucose subunits, joined by α 1—4 glycosidic linkages. Cellulose contains β-glucose monomers joined by β 1—4 linkages. These bonds cannot be split by the enzymes that hydrolyze the α linkages in starch. Because humans, like other animals, lack enzymes that digest cellulose, we cannot use it as a nutrient. The cellulose found in whole grains and vegetables remains fibrous and provides bulk that helps keep our digestive tract functioning properly.

Some microorganisms digest cellulose to glucose. In fact, cellulose-digesting bacteria live in the digestive systems of cows and sheep, enabling these grass-eating animals to obtain nourishment from cellulose. Similarly, the digestive systems of termites contain microorganisms that digest cellulose (see Fig. 25-5b).

Cellulose molecules are well suited for a structural role. The β-glucose subunits are joined in a way that allows extensive hydrogen bonding among different cellulose molecules, and they aggregate in long bundles of fibers (see Fig. 3-10a).

Some modified and complex carbohydrates have special roles

Many derivatives of monosaccharides are important biological molecules. Some form important structural components. The amino sugars galactosamine and glucosamine are compounds in which a hydroxyl group (—OH) is replaced by an amino group (—NH$_2$). Galactosamine is present in cartilage, a constituent of the skeletal system of vertebrates. *N*-acetyl glucosamine (NAG) subunits, joined by glycosidic bonds, compose **chitin,** a main component of the cell walls of fungi and of the external skeletons of insects, crayfish, and other arthropods (Fig. 3-11). Chitin forms very tough structures because, as in cellulose, its molecules interact through multiple hydrogen bonds. Some chitinous structures, such as the shell of a lobster, are further hardened by the addition of calcium carbonate (CaCO$_3$, an inorganic form of carbon).

Carbohydrates may also combine with proteins to form **glycoproteins,** compounds present on the outer surface of cells other than bacteria. Some of these carbohydrate chains allow cells to adhere to one another, whereas others provide protection. Most proteins secreted by cells are glycoproteins. These include the

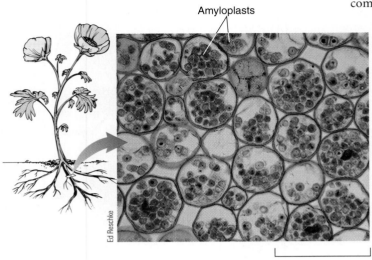

Amyloplasts

100 μm

Ed Reschke

(a) Starch (*stained purple*) is stored in specialized organelles, called *amyloplasts,* in these cells of a buttercup root.

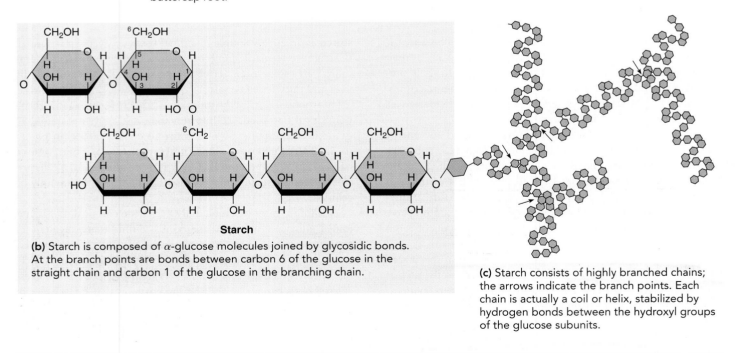

Starch

(b) Starch is composed of α-glucose molecules joined by glycosidic bonds. At the branch points are bonds between carbon 6 of the glucose in the straight chain and carbon 1 of the glucose in the branching chain.

(c) Starch consists of highly branched chains; the arrows indicate the branch points. Each chain is actually a coil or helix, stabilized by hydrogen bonds between the hydroxyl groups of the glucose subunits.

Figure 3-9 *Animated* Starch, a storage polysaccharide

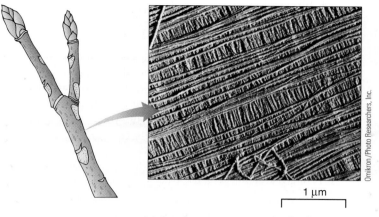

major components of mucus, a protective material secreted by the mucous membranes of the respiratory and digestive systems. Carbohydrates combine with lipids to form **glycolipids,** compounds on the surfaces of animal cells that allow cells to recognize and interact with one another.

Review

■ What features related to hydrogen bonding give storage polysaccharides, such as starch and glycogen, different properties from structural polysaccharides, such as cellulose and chitin?

■ Why can't humans digest cellulose?

1 μm

(a) An electron micrograph of cellulose fibers from a cell wall. The fibers consist of bundles of cellulose molecules that interact through hydrogen bonds.

Cellulose

(b) The cellulose molecule is an unbranched polysaccharide consisting of about 10,000 β-glucose units joined by glycosidic bonds.

Figure 3-10 *Animated* Cellulose, a structural polysaccharide

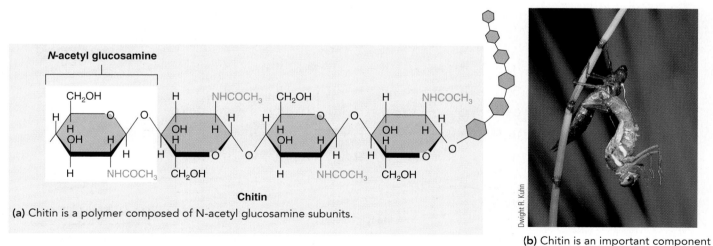

N-acetyl glucosamine

Chitin

(a) Chitin is a polymer composed of N-acetyl glucosamine subunits.

(b) Chitin is an important component of the exoskeleton (outer covering) this dragonfly is shedding.

Figure 3-11 Chitin, a structural polysaccharide

LIPIDS

Learning Objective

6 Distinguish among fats, phospholipids, and steroids, and describe the composition, characteristics, and biological functions of each.

Unlike carbohydrates, which are defined by their structure, **lipids** are a heterogeneous group of compounds that are categorized by the fact that they are soluble in nonpolar solvents (such as ether and chloroform) and are relatively insoluble in water. Lipid molecules have these properties because they consist mainly of carbon and hydrogen, with few oxygen-containing functional groups. Hydrophilic functional groups typically contain oxygen atoms; therefore, lipids, which have little oxygen, tend to be hydrophobic. Among the biologically important groups of lipids are fats, phospholipids, carotenoids (orange and yellow plant pigments), steroids, and waxes. Some lipids are used for energy storage, others serve as structural components of cell membranes, and some are important hormones.

Triacylglycerol is formed from glycerol and three fatty acids

The most abundant lipids in living organisms are triacylglycerols. These compounds, commonly known as *fats,* are an economical form of reserve fuel storage because, when metabolized, they yield more than twice as much energy per gram as do carbohydrates. Carbohydrates and proteins can be transformed by enzymes into fats and stored within the cells of adipose (fat) tissue of animals and in some seeds and fruits of plants.

A **triacylglycerol** molecule (also known as a *triglyceride*) consists of glycerol joined to three fatty acids (⏸ Fig. 3-12). **Glycerol** is a three-carbon alcohol that contains three hydroxyl (—OH) groups, and a **fatty acid** is a long, unbranched hydrocarbon chain with a carboxyl group (—COOH) at one end. A triacylglycerol molecule is formed by a series of three condensation reactions. In each reaction, the equivalent of a water molecule is removed as one of the glycerol's hydroxyl groups reacts with the carboxyl group of a fatty acid, resulting in the formation of a covalent linkage known as an **ester linkage** (see Fig. 3-12b). The first reaction yields a **monoacylglycerol** (*monoglyceride*); the second, a **diacylglycerol** (*diglyceride*); and the third, a triacylglycerol. During digestion, triacylglycerols are hydrolyzed to produce fatty acids and glycerol (see Chapter 46). Diacylglycerol is an important molecule for sending signals within the cell (see Chapters 6 and 48).

Saturated and unsaturated fatty acids differ in physical properties

About 30 different fatty acids are commonly found in lipids, and they typically have an even number of carbon atoms. For example, butyric acid, present in rancid butter, has 4 carbon atoms.

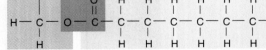

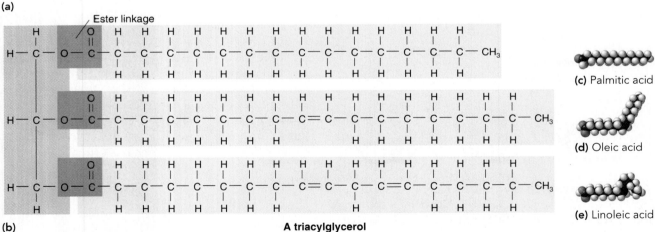

(b) **A triacylglycerol**

(c) Palmitic acid

(d) Oleic acid

(e) Linoleic acid

Figure 3-12 *Animated* Triacylglycerol, the main storage lipid

(a) Glycerol and fatty acids are the components of fats. **(b)** Glycerol is attached to fatty acids by ester linkages (*in gray*). The space-filling models show the actual shapes of the fatty acids. **(c)** Palmitic acid, a saturated fatty acid, is a straight chain. **(d)** Oleic acid (monounsaturated) and **(e)** linoleic acid (polyunsaturated) are bent or kinked wherever a carbon-to-carbon double bond appears.

Oleic acid, with 18 carbons, is the most widely distributed fatty acid in nature and is found in most animal and plant fats.

Saturated fatty acids contain the maximum possible number of hydrogen atoms. Palmitic acid, a 16-carbon fatty acid, is a common saturated fatty acid (see Fig. 3-12c). Fats high in saturated fatty acids, such as animal fat and solid vegetable shortening, tend to be solid at room temperature. The reason is that even electrically neutral, nonpolar molecules can develop transient regions of weak positive charge and weak negative charge. This occurs as the constant motion of their electrons causes some regions to have a temporary excess of electrons, whereas others have a temporary electron deficit. These slight opposite charges result in **van der Waals interactions** between adjacent molecules (see Chapter 2). Although van der Waals interactions are weak attractions, they are strong when many occur among long hydrocarbon chains. These van der Waals interactions tend to make a substance more solid by limiting the motion of its molecules.

Unsaturated fatty acids include one or more adjacent pairs of carbon atoms joined by a double bond. Therefore, they are not fully saturated with hydrogen. Fatty acids with one double bond are **monounsaturated fatty acids,** whereas those with more than one double bond are **polyunsaturated fatty acids.** Oleic acid is a monounsaturated fatty acid, and linoleic acid is a common polyunsaturated fatty acid (see Fig. 3-12d and e). Fats containing a high proportion of monounsaturated or polyunsaturated fatty acids tend to be liquid at room temperature. The reason is that each double bond produces a bend in the hydrocarbon chain that prevents it from aligning closely with an adjacent chain, thereby limiting van der Waals interactions and permitting freer molecular motion.

Food manufacturers commonly hydrogenate or partially hydrogenate cooking oils to make margarine and other foodstuffs, converting unsaturated fatty acids to saturated fatty acids and making the fat more solid at room temperature. This process makes the fat less healthful because saturated fatty acids in the diet are known to increase the risk of cardiovascular disease (see Chapter 43). The hydrogenation process has yet another effect. Note that in the naturally occurring unsaturated fatty acids oleic acid and linoleic acid shown in Figure 3-12, the two hydrogens flanking each double bond are on the same side of the hydrocarbon chain (the *cis* configuration). When fatty acids are artificially hydrogenated, the double bonds can become rearranged, resulting in a *trans* configuration, analogous to the arrangement shown in Figure 3-3b. *Trans* fatty acids are technically unsaturated, but they mimic many of the properties of saturated fatty acids. Because the *trans* configuration does not produce a bend at the site of the double bond, *trans* fatty acids are more solid at room temperature; like saturated fatty acids, they increase the risk of cardiovascular disease.

At least two unsaturated fatty acids (linoleic acid and arachidonic acid) are essential nutrients that must be obtained from food because the human body cannot synthesize them. However, the amounts required are small, and deficiencies are rarely seen. There is no dietary requirement for saturated fatty acids.

Phospholipids are components of cell membranes

Phospholipids belong to a group of lipids called **amphipathic lipids,** in which one end of each molecule is hydrophilic and the other end is hydrophobic (Fig. 3-13). The two ends of a phospholipid differ both physically and chemically. A **phospholipid** consists of a glycerol molecule attached at one end to two fatty acids and at the other end to a phosphate group linked to an organic compound such as choline. The organic compound usually contains nitrogen. (Note that phosphorus and nitrogen are absent in triacylglycerols, as shown in Figure 3-12b.) The fatty acid portion of the molecule (containing the two hydrocarbon "tails") is hydrophobic and not soluble in water. However, the portion composed of glycerol, phosphate, and the organic base (the "head" of the molecule) is ionized and readily water soluble. The amphipathic properties of phospholipids cause them to form lipid bilayers in aqueous (watery) solution. Thus, they are uniquely suited as the fundamental components of cell membranes (discussed in Chapter 5).

Carotenoids and many other pigments are derived from isoprene units

The orange and yellow plant pigments called **carotenoids** are classified with the lipids because they are insoluble in water and have an oily consistency. These pigments, found in the cells of plants, play a role in photosynthesis. Carotenoid molecules, such as β-carotene, and many other important pigments, consist of five-carbon hydrocarbon monomers known as **isoprene units** (Fig. 3-14).

Most animals convert carotenoids to vitamin A, which can then be converted to the visual pigment **retinal.** Three groups of animals—the mollusks, insects, and vertebrates—have eyes that use retinal in the process of light reception.

Notice that carotenoids, vitamin A, and retinal all have a pattern of double bonds alternating with single bonds. The electrons that make up these bonds can move about relatively easily when light strikes the molecule. Such molecules are *pigments;* they tend to be highly colored because the mobile electrons cause them to strongly absorb light of certain wavelengths and reflect light of other wavelengths.

Steroids contain four rings of carbon atoms

A **steroid** consists of carbon atoms arranged in four attached rings; three of the rings contain six carbon atoms, and the fourth contains five (Fig. 3-15). The length and structure of the side chains that extend from these rings distinguish one steroid from

A lipid bilayer forms when phospholipids interact with water.

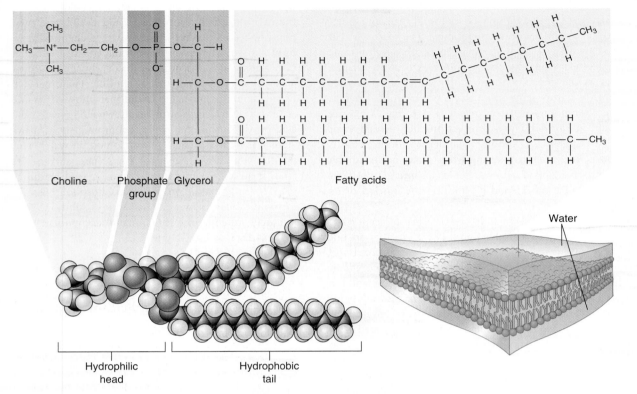

(a) Phospholipid (lecithin). A phospholipid consists of a hydrophobic tail, made up of two fatty acids, and a hydrophilic head, which includes a glycerol bonded to a phosphate group, which is in turn bonded to an organic group that can vary. Choline is the organic group in lecithin (or phosphatidylcholine), the molecule shown. The fatty acid at the top of the figure is monounsaturated; it contains one double bond that produces a characteristic bend in the chain.

(b) Phospholipid bilayer. Phospholipids form lipid bilayers in which the hydrophilic heads interact with water and the hydrophobic tails are in the bilayer interior.

Figure 3-13 *Animated*   A phospholipid and a phospholipid bilayer

---

another. Like carotenoids, steroids are synthesized from isoprene units.

Among the steroids of biological importance are cholesterol, bile salts, reproductive hormones, and cortisol and other hormones secreted by the adrenal cortex. Cholesterol is an essential structural component of animal cell membranes, but when excess cholesterol in blood forms plaques on artery walls, the risk of cardiovascular disease increases (see Chapter 43).

Plant cell membranes contain molecules similar to cholesterol. Interestingly, some of these plant steroids block the intestine's absorption of cholesterol. Bile salts emulsify fats in the intestine so they can be enzymatically hydrolyzed. Steroid hor-

mones regulate certain aspects of metabolism in a variety of animals, including vertebrates, insects, and crabs.

Some chemical mediators are lipids

Animal cells secrete chemicals to communicate with one another or to regulate their own activities. Some chemical mediators are produced by the modification of fatty acids that have been removed from membrane phospholipids. These include *prostaglandins*, which have varied roles, including promoting inflammation and smooth muscle contraction. Certain hormones, such

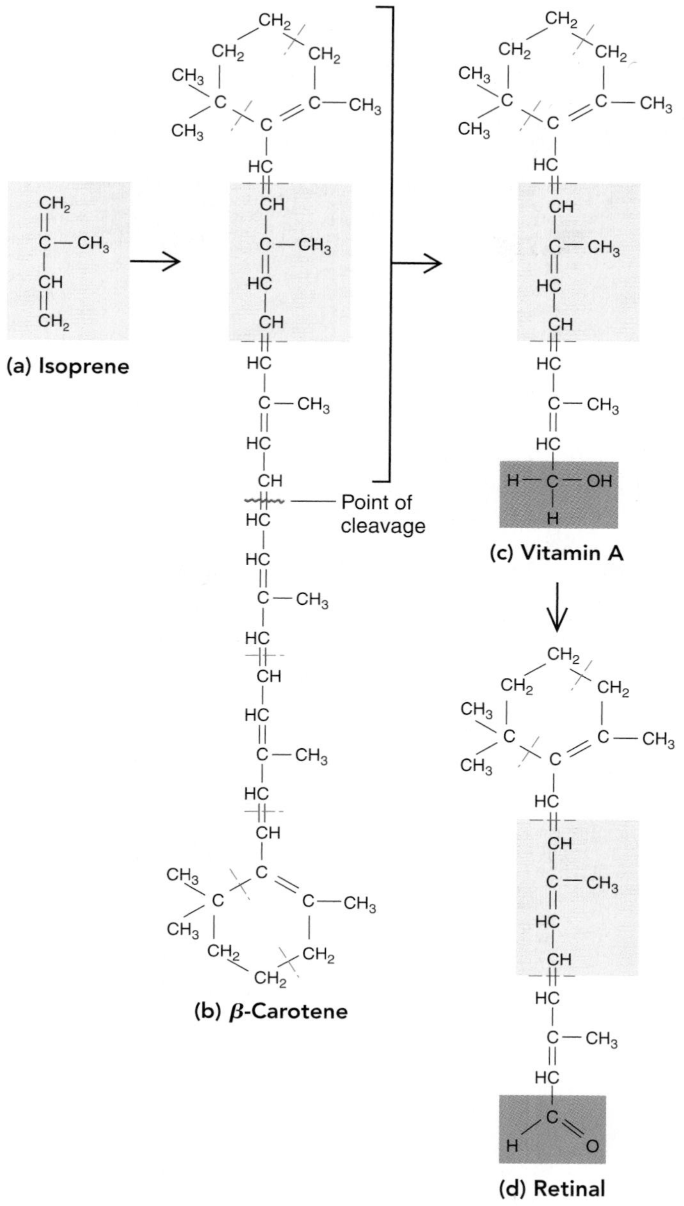

(a) Isoprene

(b) β-Carotene

Point of cleavage

(c) Vitamin A

(d) Retinal

Figure 3-14 Isoprene-derived compounds

(a) An isoprene subunit. **(b)** β-carotene, with dashed lines indicating the boundaries of the individual isoprene units within. The wavy line is the point at which most animals cleave the molecule to yield two molecules of **(c)** vitamin A. Vitamin A is converted to the visual pigment **(d)** retinal.

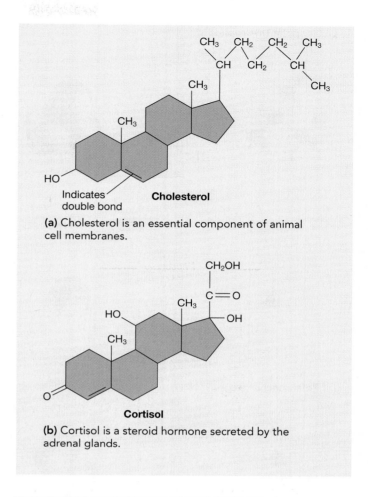

HO — Indicates double bond — **Cholesterol**

(a) Cholesterol is an essential component of animal cell membranes.

Cortisol

(b) Cortisol is a steroid hormone secreted by the adrenal glands.

Figure 3-15 Steroids

Four attached rings—three six-carbon rings and one with five carbons—make up the fundamental structure of a steroid (*shown in green*). Note that some carbons are shared by two rings. In these simplified structures, a carbon atom is present at each angle of a ring; the hydrogen atoms attached directly to the carbon atoms have not been drawn. Steroids are mainly distinguished by their attached functional groups.

as the juvenile hormone of insects, are also fatty acid derivatives (discussed in Chapter 48).

Review

- Why do saturated, unsaturated, and *trans* fatty acids differ in their properties?
- Why do phospholipids form lipid bilayers in aqueous conditions?

PROTEINS

Learning Objectives

7   Give an overall description of the structure and functions of proteins.

8   Describe the features that are shared by all amino acids, and explain how amino acids are grouped into classes based on the characteristics of their side chains.

9   Distinguish among the four levels of organization of protein molecules.

Proteins, macromolecules composed of amino acids, are the most versatile cell components. As will be discussed in Chapter 16, scientists have succeeded in sequencing virtually all the

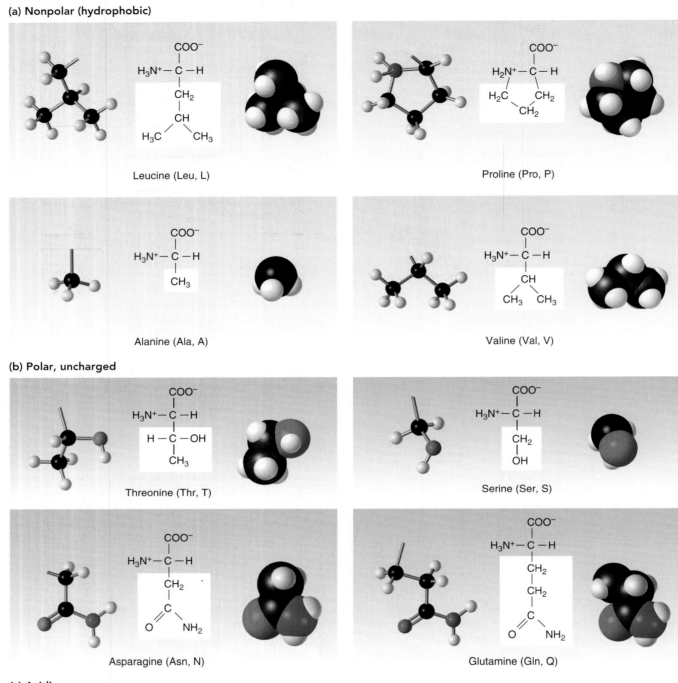

(a) Nonpolar (hydrophobic)

Leucine (Leu, L)

Proline (Pro, P)

Alanine (Ala, A)

Valine (Val, V)

(b) Polar, uncharged

Threonine (Thr, T)

Serine (Ser, S)

Asparagine (Asn, N)

Glutamine (Gln, Q)

(c) Acidic

Aspartic acid (Asp, D)

Glutamic acid (Glu, E)

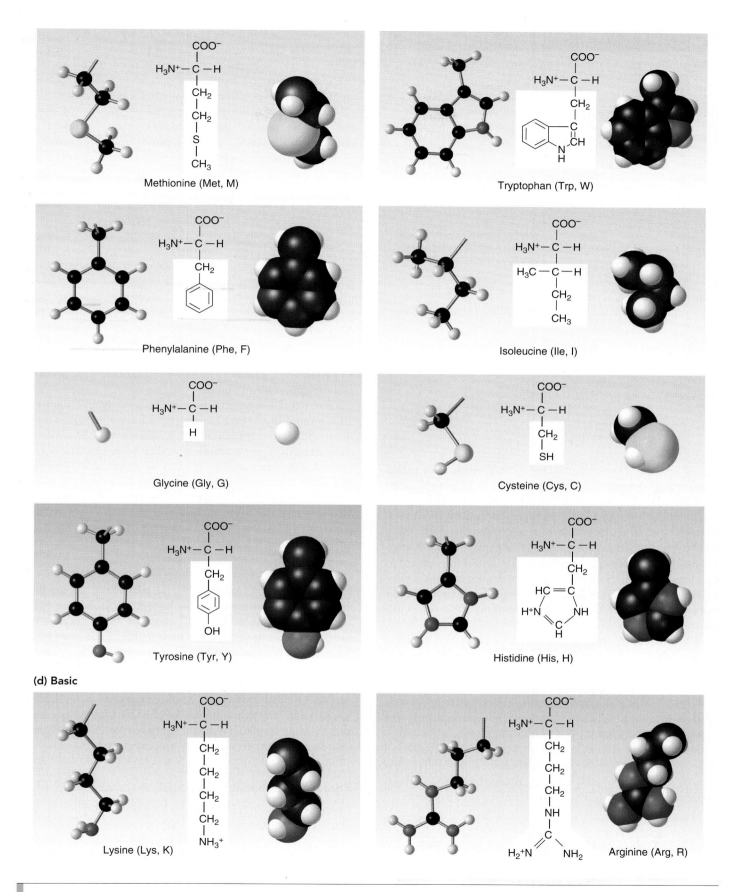

Methionine (Met, M)

Tryptophan (Trp, W)

Phenylalanine (Phe, F)

Isoleucine (Ile, I)

Glycine (Gly, G)

Cysteine (Cys, C)

Tyrosine (Tyr, Y)

Histidine (His, H)

(d) Basic

Lysine (Lys, K)

Arginine (Arg, R)

Figure 3-16 *Animated* The 20 common amino acids

(a) Nonpolar amino acids have side chains that are relatively hydrophobic, whereas **(b)** polar amino acids have relatively hydrophilic side chains. Carboxyl groups and amino groups are electrically charged at cell pH; therefore, **(c)** acidic and **(d)** basic amino acids are hydrophilic. The standard three-letter and one-letter abbreviations appear beside the amino acid names.

TABLE 3-2

Major Classes of Proteins and Their Functions

Protein Class	Functions and Examples
Enzymes	Catalyze specific chemical reactions
Structural proteins	Strengthen and protect cells and tissues (e.g., collagen strengthens animal tissues)
Storage proteins	Store nutrients; particularly abundant in eggs (e.g., ovalbumin in egg white) and seeds (e.g., zein in corn kernels)
Transport proteins	Transport specific substances between cells (e.g., hemoglobin transports oxygen in red blood cells); move specific substances (e.g., ions, glucose, amino acids) across cell membranes
Regulatory proteins	Some are protein hormones (e.g., insulin); some control the expression of specific genes
Motile proteins	Participate in cellular movements (e.g., actin and myosin are essential for muscle contraction)
Protective proteins	Defend against foreign invaders (e.g., antibodies play a role in the immune system)

genetic information in a human cell, and the genetic information of many other organisms is being studied. Some people might think that the sequencing of genes is the end of the story, but it is actually only the beginning. Most genetic information is used to specify the structure of proteins, and biologists may devote most of the 21st century to understanding these extraordinarily multifaceted macromolecules that are of central importance in the chemistry of life. In a real sense, proteins are involved in virtually all aspects of metabolism because most **enzymes,** molecules that accelerate the thousands of different chemical reactions that take place in an organism, are proteins.

Proteins are assembled into a variety of shapes, allowing them to serve as major structural components of cells and tissues. For this reason, growth and repair, as well as maintenance of the organism, depend on proteins. As shown in ▌ Table 3-2, proteins perform many other functions. Each cell type contains characteristic forms, distributions, and amounts of protein that largely determine what the cell looks like and how it functions. A muscle cell contains large amounts of the proteins myosin and actin, which are responsible for its appearance as well as its ability to contract. The protein hemoglobin, found in red blood cells, is responsible for the specialized function of oxygen transport.

Amino acids are the subunits of proteins

Amino acids, the constituents of proteins, have an amino group ($—NH_2$) and a carboxyl group ($—COOH$) bonded to the same asymmetrical carbon atom, known as the **alpha carbon.** Twenty amino acids are commonly found in proteins, each identified by the variable side chain (R group) bonded to the α carbon (▌ Fig. 3-16). Glycine, the simplest amino acid, has a hydrogen atom as its R group; alanine has a methyl ($—CH_3$) group.

Amino acids in solution at neutral pH are mainly dipolar ions. This is generally how amino acids exist at cell pH. Each carboxyl group ($—COOH$) donates a proton and becomes ionized ($—COO^-$), whereas each amino group ($—NH_2$) accepts a proton and becomes $—NH_3^+$ (▌ Fig. 3-17). Because of the ability of their amino and carboxyl groups to accept and release protons, amino acids in solution resist changes in acidity and alkalinity and therefore are important biological buffers.

The amino acids are grouped in Figure 3-16 by the properties of their side chains. These broad groupings actually include amino acids with a fairly wide range of properties. Amino acids classified as having *nonpolar* side chains tend to have hydrophobic properties, whereas those classified as *polar* are more hydrophilic. An acidic amino acid has a side chain that contains a carboxyl group. At cell pH the carboxyl group is dissociated, giving the R group a negative charge. A basic amino acid becomes positively charged when the amino group in its side chain accepts a hydrogen ion. Acidic and basic side chains are ionic at cell pH and therefore hydrophilic.

Some proteins have unusual amino acids in addition to the 20 common ones. These rare amino acids are produced by the modification of common amino acids after they have become part of a protein. For example, after they have been incorporated into collagen, lysine and proline may be converted to hydroxylysine and hydroxyproline. These amino acids can form cross links between the peptide chains that make up collagen. Such cross links produce the firmness and great strength of the collagen molecule, which is a major component of cartilage, bone, and other connective tissues.

With some exceptions, prokaryotes and plants synthesize all their needed amino acids from simpler substances. If the proper raw materials are available, the cells of animals can manufacture some, but not all, of the biologically significant amino acids. **Essential amino acids** are those an animal cannot synthesize in amounts sufficient to meet its needs and must obtain from the diet. Animals differ in their biosynthetic capacities; what is an essential amino acid for one species may not be for another.

The essential amino acids for humans are isoleucine, leucine, lysine, methionine, phenylalanine, threonine, tryptophan, valine, and histidine. Arginine is added to the list for children because they do not synthesize enough to support growth.

Peptide bonds join amino acids

Amino acids combine chemically with one another by a condensation reaction that bonds the carboxyl carbon of one molecule to the amino nitrogen of another (▌ Fig. 3-18). The covalent carbon-to-nitrogen bond linking two amino acids is a **peptide bond.** When two amino acids combine, a **dipeptide** is formed; a longer chain of amino acids is a **polypeptide.** A protein consists

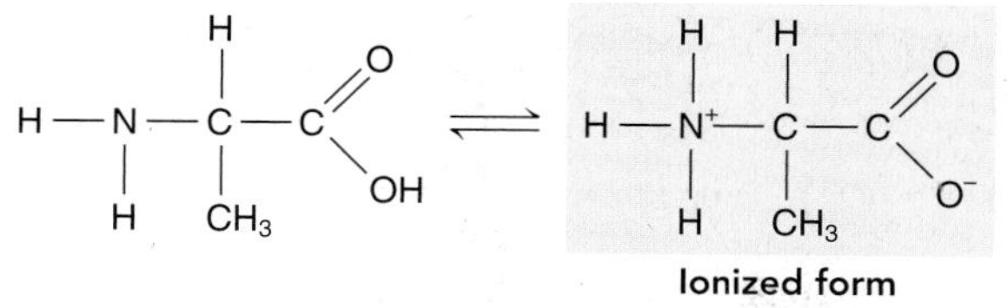

Figure 3-17 An amino acid at pH 7

In living cells, amino acids exist mainly in their ionized form, as dipolar ions.

of one or more polypeptide chains. Each polypeptide has a free amino group at one end and a free carboxyl group (belonging to the last amino acid added to the chain) at the opposite end. The other amino and carboxyl groups of the amino acid monomers (except those in side chains) are part of the peptide bonds. The process by which polypeptides are synthesized is discussed in Chapter 13.

A polypeptide may contain hundreds of amino acids joined in a specific linear order. The backbone of the polypeptide chain includes the repeating sequence

$$N-C-C-N-C-C-N-C-C$$

plus all other atoms *except those in the R groups*. The R groups of the amino acids extend from this backbone.

An almost infinite variety of protein molecules is possible, differing from one another in the number, types, and sequences of amino acids they contain. The 20 types of amino acids found in proteins may be thought of as letters of a protein alphabet; each protein is a very long sentence made up of amino acid letters.

Proteins have four levels of organization

The polypeptide chains making up a protein are twisted or folded to form a macromolecule with a specific *conformation*, or 3-D shape. Some polypeptide chains form long fibers. **Globular pro-**

teins are tightly folded into compact, roughly spherical shapes. There is a close relationship between a protein's conformation and its function. For example, a typical enzyme is a globular protein with a unique shape that allows it to catalyze a specific chemical reaction. Similarly, the shape of a protein hormone enables it to combine with receptors on its target cell (the cell on which the hormone acts). Scientists recognize four main levels of protein organization: primary, secondary, tertiary, and quaternary.

Primary structure is the amino acid sequence

The sequence of amino acids, joined by peptide bonds, is the **primary structure** of a polypeptide chain. As discussed in Chapter 13, this sequence is specified by the instructions in a gene. Using analytical methods, investigators can determine the exact sequence of amino acids in a protein molecule. The primary structures of thousands of proteins are known. For example, glucagon, a hormone secreted by the pancreas, is a small polypeptide, consisting of only 29 amino acid units (**Fig. 3-19**).

Primary structure is always represented in a simple, linear, "beads-on-a-string" form. However, the overall conformation of a protein is far more complex, involving interactions among the various amino acids that make up the primary structure of the molecule. Therefore, the higher orders of structure—secondary, tertiary, and quaternary—ultimately derive from the specific amino acid sequence (the primary structure).

Secondary structure results from hydrogen bonding involving the backbone

Some regions of a polypeptide exhibit **secondary structure,** which is highly regular. The two most common types of secondary structure are the α-helix and the β-pleated sheet; the designations α and β refer simply to the order in which these two types of secondary structure were discovered. An α-helix is a region where a polypeptide chain forms a uniform helical coil (**Fig. 3-20a**). The helical structure is determined and maintained by the formation

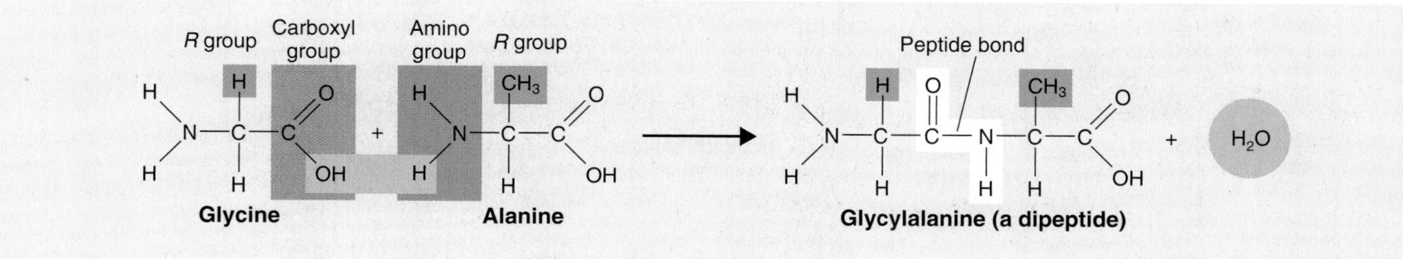

Figure 3-18 *Animated* Peptide bonds

A dipeptide is formed by a condensation reaction, that is, by the removal of the equivalent of a water molecule from the carboxyl group of one amino acid and the amino group of another amino acid. The resulting peptide bond is a covalent, carbon-to-nitrogen bond. Note

that the carbon is also part of a carbonyl group, and that the nitrogen is also covalently bonded to a hydrogen. Additional amino acids can be added to form a long polypeptide chain with a free amino group at one end and a free carboxyl group at the other.

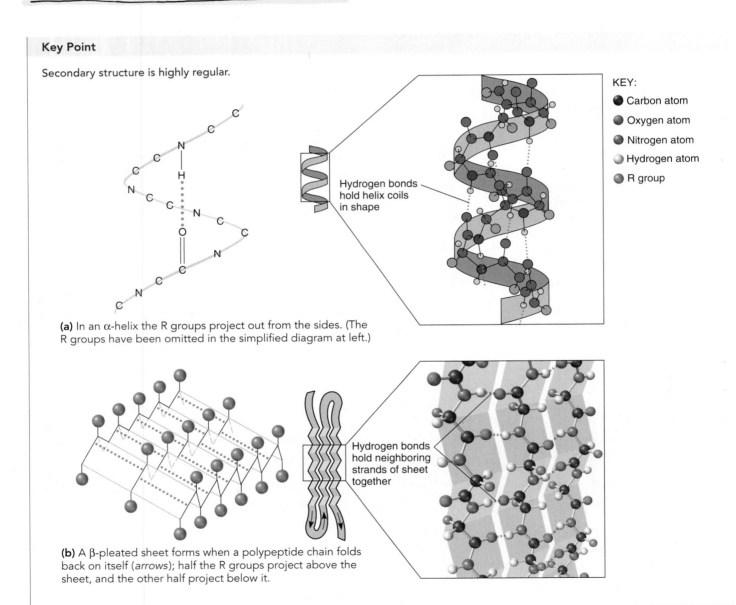

^+H_3N—His Ser Gln Gly Thr Phe Thr Ser Asp Tyr Ser Lys Tyr Leu Asp Ser Arg Arg Ala Gln Asp Phe Val Gln Trp Leu Met Asn Thr —COO$^-$
     1   2   3   4   5   6   7   8   9 10 11 12 13 14 15 16 17 18 19 20 21 22 23 24 25 26 27 28 29

Figure 3-19 Primary structure of a polypeptide

Glucagon is a very small polypeptide made up of 29 amino acids. The linear sequence of amino acids is indicated by ovals containing their abbreviated names (see Fig. 3-16).

of hydrogen bonds between the backbones of the amino acids in successive turns of the spiral coil.

Each hydrogen bond forms between an oxygen with a partial negative charge and a hydrogen with a partial positive charge. The oxygen is part of the remnant of the carboxyl group of one amino acid; the hydrogen is part of the remnant of the amino group of the fourth amino acid down the chain. Thus, 3.6 amino acids are included in each complete turn of the helix. Every amino acid in an α-helix is hydrogen bonded in this way.

The α-helix is the basic structural unit of some fibrous proteins that make up wool, hair, skin, and nails. The elasticity of these fibers is due to a combination of physical factors (the heli-

Key Point

Secondary structure is highly regular.

KEY:
- Carbon atom
- Oxygen atom
- Nitrogen atom
- Hydrogen atom
- R group

Hydrogen bonds hold helix coils in shape

(a) In an α-helix the R groups project out from the sides. (The R groups have been omitted in the simplified diagram at left.)

Hydrogen bonds hold neighboring strands of sheet together

(b) A β-pleated sheet forms when a polypeptide chain folds back on itself (*arrows*); half the R groups project above the sheet, and the other half project below it.

Figure 3-20 *Animated* Secondary structure of a protein

Tertiary structure depends on side chain interactions.

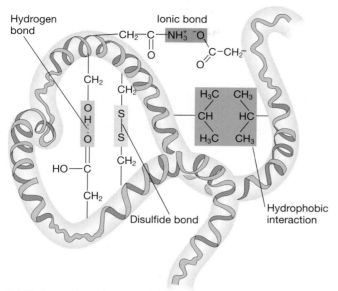

(a) Hydrogen bonds, ionic bonds, hydrophobic interactions, and disulfide bridges between *R* groups hold the parts of the molecule in the designated shape.

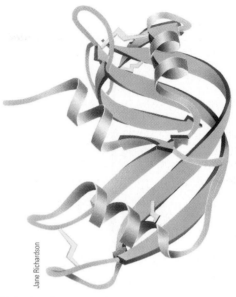

Jane Richardson

(b) In this drawing, α-helical regions are represented by purple coils, β-pleated sheets by broad green ribbons, and connecting regions by narrow tan ribbons. The interactions among *R* groups that stabilize the bends and foldbacks that give the molecule its overall conformation (tertiary structure) are represented in yellow. This protein is bovine ribonuclease a.

Figure 3-21 *Animated* Tertiary structure of a protein

cal shape) and chemical factors (hydrogen bonding). Although hydrogen bonds maintain the helical structure, these bonds can be broken, allowing the fibers to stretch under tension (like a telephone cord). When the tension is released, the fibers recoil and hydrogen bonds re-form. This explains why you can stretch the hairs on your head to some extent and they will snap back to their original length.

The hydrogen bonding in a β-pleated sheet takes place between different polypeptide chains, or different regions of a polypeptide chain that has turned back on itself (■ Fig. 3-20b). Each chain is fully extended; however, because each has a zigzag structure, the resulting "sheet" has an overall pleated conformation (much like a sheet of paper that has been folded to make a fan). Although the pleated sheet is strong and flexible, it is not elastic, because the distance between the pleats is fixed, determined by the strong covalent bonds of the polypeptide backbones. Fibroin, the protein of silk, is characterized by a β-pleated sheet structure, as are the cores of many globular proteins.

It is not uncommon for a single polypeptide chain to include both α-helical regions and regions with β-pleated sheet conformations. The properties of some biological materials result from such combinations. A spider's web is composed of a material that is extremely strong, flexible, and elastic. Once again we see

function and structure working together, as these properties derive from the fact that spider silk is a composite of proteins with α-helical conformations (providing elasticity) and others with β-pleated sheet conformations (providing strength).

Tertiary structure depends on interactions among side chains

The **tertiary structure** of a protein molecule is the overall shape assumed by each individual polypeptide chain (■ Fig. 3-21). This 3-D structure is determined by four main factors that involve interactions among *R* groups (side chains) belonging to the same polypeptide chain. These include both weak interactions (hydrogen bonds, ionic bonds, and hydrophobic interactions) and strong covalent bonds.

1. Hydrogen bonds form between *R* groups of certain amino acid subunits.

2. An ionic bond can occur between an *R* group with a unit of positive charge and one with a unit of negative charge.

3. Hydrophobic interactions result from the tendency of nonpolar *R* groups to be excluded by the surrounding water and therefore to associate in the interior of the globular structure.

Proteins with two or more polypeptide chains have quaternary structure.

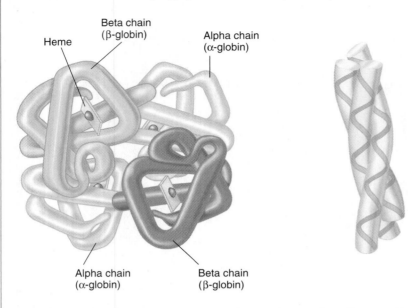

(a) Hemoglobin, a globular protein, consists of four polypeptide chains, each joined to an iron-containing molecule, a heme.

(b) Collagen, a fibrous protein, is a triple helix consisting of three long polypeptide chains.

Figure 3-22 Quaternary structure of a protein

4. Covalent bonds known as *disulfide bonds* or *disulfide bridges* (—S—S—) may link the sulfur atoms of two cysteine subunits belonging to the same chain. A disulfide bridge forms when the sulfhydryl groups of two cysteines react; the two hydrogens are removed, and the two sulfur atoms that remain become covalently linked.

Quaternary structure results from interactions among polypeptides

Many functional proteins are composed of two or more polypeptide chains, interacting in specific ways to form the biologically active molecule. **Quaternary structure** is the resulting 3-D structure. The same types of interactions that produce secondary and tertiary structure also contribute to quaternary structure; these include hydrogen bonding, ionic bonding, hydrophobic interactions, and disulfide bridges.

A functional antibody molecule, for example, consists of four polypeptide chains joined by disulfide bridges (see Chapter 44). Disulfide bridges are a common feature of antibodies and other proteins secreted from cells. These strong bonds stabilize the molecules in the extracellular environment.

Hemoglobin, the protein in red blood cells responsible for oxygen transport, is an example of a globular protein with a quaternary structure (⫿ Fig. 3-22a). Hemoglobin consists of 574 amino acids arranged in four polypeptide chains: two identical chains called *alpha chains* and two identical chains called *beta chains.*

Collagen, mentioned previously, has a fibrous type of quaternary structure that allows it to function as the major strengthener of animal tissues. It consists of three polypeptide chains wound about one another and bound by cross links between their amino acids (⫿ Fig. 3-22b).

The amino acid sequence of a protein determines its conformation

In 1996, researchers at the University of Illinois at Champaign–Urbana devised a sophisticated test of the hypothesis that the conformation of a protein is dictated by its amino acid sequence. They conducted an experiment in which they completely unfolded myoglobin, a polypeptide that stores oxygen in muscle cells, and then tracked the refolding process. They found that within a few fractions of a microsecond the molecule had coiled up to form α-helices, and within 4 microseconds formation of the tertiary structure was completed. Thus, these researchers demonstrated that, at least under defined experimental conditions in vitro (outside a living cell), a polypeptide can spontaneously undergo folding processes that yield its normal, functional conformation. Since this pioneering work, these and other researchers studying various proteins and using a variety of approaches have amassed evidence supporting the widely held conclusion that amino acid sequence is the ultimate determinant of protein conformation.

However, because conditions in vivo (in the cell) are quite different from defined laboratory conditions, proteins do not always fold spontaneously. On the contrary, scientists have learned that proteins known as **molecular chaperones** mediate the folding of other protein molecules. Molecular chaperones are thought to make the folding process more orderly and efficient and to prevent partially folded proteins from becoming inappropriately aggregated. However, there is no evidence that molecular chaperones actually dictate the folding pattern. For this reason, the existence of chaperones is not an argument against the idea that amino acid sequence determines conformation.

Protein conformation determines function

The overall structure of a protein helps determine its biological activity. A single protein may have more than one distinct structural region, called a **domain,** each with its own function. Many proteins are modular, consisting of two or more globular domains, connected by less compact regions of the polypeptide chain.

Each domain may have a different function. For example, a protein may have one domain that attaches to a membrane and another that acts as an enzyme. The biological activity of a protein can be disrupted by a change in amino acid sequence that results in a change in conformation. For example, the genetic disease known as *sickle cell anemia* is due to a mutation that causes the substitution of the amino acid valine for glutamic acid at position 6 (the sixth amino acid from the amino end) in the beta chain of hemoglobin. The substitution of valine (which has a nonpolar side chain) for glutamic acid (which has a charged side chain) makes the hemoglobin less soluble and more likely to form crystal-like structures. This alteration of the hemoglobin affects the red blood cells, changing them to the crescent or sickle shapes that characterize this disease (see Fig. 16-7).

The biological activity of a protein may be affected by changes in its 3-D structure. When a protein is heated, subjected to significant pH changes, or treated with certain chemicals, its structure becomes disordered and the coiled peptide chains unfold, yielding a more random conformation. This unfolding, which is mainly due to the disruption of hydrogen bonds and ionic bonds, is typically accompanied by a loss of normal function. Such changes in shape and the accompanying loss of biological activity are termed **denaturation** of the protein.

For example, a denatured enzyme would lose its ability to catalyze a chemical reaction. An everyday example of denaturation occurs when we fry an egg. The consistency of the egg white protein, known as *albumin,* changes to a solid. Denaturation generally cannot be reversed (you cannot "unfry" an egg). However, under certain conditions, some proteins have been denatured and have returned to their original shape and biological activity when normal environmental conditions were restored.

Protein conformation is studied through a variety of methods

The architecture of a protein can be ascertained directly through various types of analysis, such as the X-ray diffraction studies discussed in Chapter 12. Because these studies are tedious and costly, researchers are developing alternative approaches, which rely heavily on the enormous databases generated by the Human Genome Project and related initiatives.

Today a protein's primary structure can be determined rapidly through the application of genetic engineering techniques (discussed in Chapter 15) or by the use of sophisticated technology such as mass spectrometry. Researchers use these amino acid sequence data to predict a protein's higher levels of structure. As you have seen, side chains interact in relatively predictable ways, such as through ionic and hydrogen bonds. In addition, regions with certain types of side chains are more likely to form α-helices or β-pleated sheets. Computer programs make such predictions, but these are often imprecise because of the many possible combinations of folding patterns.

Computers are an essential part of yet another strategy. Once the amino acid sequence of a polypeptide has been determined, researchers use computers to search databases to find polypeptides with similar sequences. If the conformations of any of those polypeptides or portions have already been determined directly by X-ray diffraction or other techniques, this information can be extrapolated to make similar correlations between amino acid sequence and 3-D structure for the protein under investigation. These predictions are increasingly reliable, as more information is added to the databases every day.

Review

- Draw the structural formula of a simple amino acid. What is the importance of the carboxyl group, amino group, and *R* group?
- How does the primary structure of a polypeptide influence its secondary and tertiary structures?

NUCLEIC ACIDS

Learning Objective

10 Describe the components of a nucleotide. Name some nucleic acids and nucleotides, and discuss the importance of these compounds in living organisms.

Nucleic acids transmit hereditary information and determine what proteins a cell manufactures. Two classes of nucleic acids are found in cells: deoxyribonucleic acid and ribonucleic acid. **Deoxyribonucleic acid (DNA)** composes the genes, the hereditary material of the cell, and contains instructions for making all the proteins, as well as all the RNA the organism needs. **Ribonucleic acid (RNA)** participates in the process in which amino acids are linked to form polypeptides. Some types of RNA, known as **ribozymes,** can even act as specific biological catalysts. Like proteins, nucleic acids are large, complex molecules. The name *nucleic acid* reflects the fact that they are acidic and were first identified, by Swiss biochemist Friedrich Miescher in 1870, in the nuclei of pus cells.

Nucleic acids are polymers of **nucleotides,** molecular units that consist of (1) a five-carbon sugar, either **deoxyribose** (in DNA) or **ribose** (in RNA); (2) one or more phosphate groups, which make the molecule acidic; and (3) a nitrogenous base, a ring compound that contains nitrogen. The nitrogenous base may be either a double-ring **purine** or a single-ring **pyrimidine** (▌Fig. 3-23). DNA commonly contains the purines adenine (A) and guanine (G), the pyrimidines cytosine (C) and thymine (T), the sugar deoxyribose, and phosphate. RNA contains the purines adenine and guanine, and the pyrimidines cytosine and uracil (U), together with the sugar ribose, and phosphate.

The molecules of nucleic acids are made of linear chains of nucleotides, which are joined by **phosphodiester linkages,** each consisting of a phosphate group and the covalent bonds that attach it to the sugars of adjacent nucleotides (▌Fig. 3-24). Note that each nucleotide is defined by its particular base and that nucleotides can be joined in any sequence. A nucleic acid molecule is uniquely defined by its specific sequence of nucleotides, which constitutes a kind of code (see Chapter 13). Whereas RNA is usually composed of one nucleotide chain, DNA consists of two nucleotide chains held together by hydrogen bonds and entwined around each other in a double helix (see Fig. 1-7).

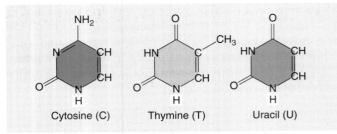

(a) **Pyrimidines.** The three major pyrimidine bases found in nucleotides are cytosine, thymine (in DNA only), and uracil (in RNA only).

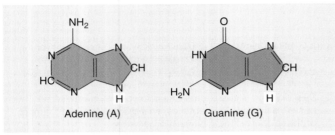

(b) **Purines.** The two major purine bases found in nucleotides are adenine and guanine.

Figure 3-23 Components of nucleotides

The hydrogens indicated by the white boxes are removed when the base is attached to a sugar.

Some nucleotides are important in energy transfers and other cell functions

In addition to their importance as subunits of DNA and RNA, nucleotides perform other vital functions in living cells. **Adenosine triphosphate (ATP),** composed of adenine, ribose, and three phosphates (see Fig. 7-5), is of major importance as the primary energy currency of all cells (see Chapter 7). The two terminal phosphate groups are joined to the nucleotide by covalent bonds. These are traditionally indicated by wavy lines, which indicate that ATP can transfer a phosphate group to another molecule, making that molecule more reactive. In this way ATP is able to donate some of its chemical energy. Most of the readily available chemical energy of the cell is associated with the phosphate groups of ATP. Like ATP, **guanosine triphosphate (GTP),** a nucleotide that contains the base guanine, can transfer energy by transferring a phosphate group and also has a role in cell signaling (see Chapter 6).

A nucleotide may be converted to an alternative form with specific cell functions. ATP, for example, is converted to **cyclic adenosine monophosphate (cyclic AMP)** by the enzyme adenylyl cyclase (Fig. 3-25). Cyclic AMP regulates certain cell functions, such as cell signaling, and is important in the mechanism by which some hormones act. A related molecule, **cyclic guanosine monophosphate (cGMP),** also plays a role in certain cell signaling processes.

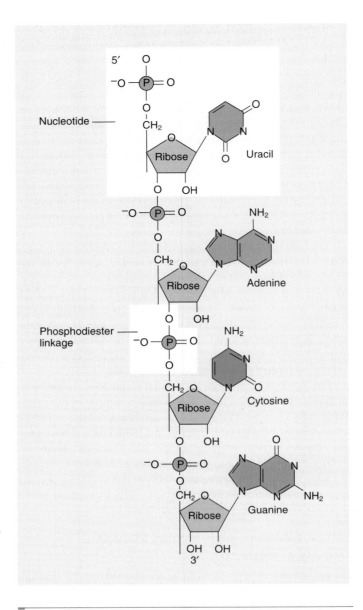

Figure 3-24 RNA, a nucleic acid

Nucleotides, each with a specific base, are joined by phosphodiester linkages.

Cells contain several dinucleotides, which are of great importance in metabolic processes. For example, as discussed in Chapter 7, **nicotinamide adenine dinucleotide** has a primary role in oxidation and reduction reactions in cells. It can exist in an oxidized form (NAD^+) that is converted to a reduced form (**NADH**) when it accepts electrons (in association with hydrogen; see Fig. 7-7). These electrons, along with their energy, are transferred to other molecules.

Review

■ How does the structure of a nucleotide relate to its function?

TABLE 3-3

Classes of Biologically Important Organic Compounds

Class and Component Elements	Description	How to Recognize	Principal Function in Living Systems
Carbohydrates C, H, O	Contain approximately 1 C:2 H:1 O (but make allowance for loss of oxygen when sugar units as nucleic acids and glycoproteins are linked)	Count the carbons, hydrogens, and oxygens.	Cell fuel; energy storage; structural component of plant cell walls; component of other compounds such as nucleic acids and glycoproteins
	1. *Monosaccharides* (simple sugars). Mainly five-carbon (pentose) molecules such as ribose or six-carbon (hexose) molecules such as glucose and fructose	Look for the ring shapes:	Cell fuel; components of other compounds
	2. *Disaccharides.* Two sugar units linked by a glycosidic bond, e.g., maltose, sucrose	Count sugar units.	Components of other compounds; form of sugar transported in plants
	3. *Polysaccharides.* Many sugar units linked by glycosidic bonds, e.g., glycogen, cellulose	Count sugar units.	Energy storage; structural components of plant cell walls
Lipids C, H, O (sometimes N, P)	Contain much less oxygen relative to carbon and hydrogen than do carbohydrates		Energy storage; cellular fuel, components of cells; thermal insulation
	1. *Fats.* Combination of glycerol with one to three fatty acids. Monoacylglycerol contains one fatty acid; diacylglycerol contains two fatty acids; triacylglycerol contains three fatty acids. If fatty acids contain double carbon-to-carbon linkages (C=C), they are unsaturated; otherwise, they are saturated.	Look for glycerol at one end of molecule:	Cell fuel; energy storage
	2. *Phospholipids.* Composed of glycerol attached to one or two fatty acids and to an organic base containing phosphorus	Look for glycerol and side chain containing phosphorus and nitrogen.	Components of cell membranes
	3. *Steroids.* Complex molecules containing carbon atoms arranged in four attached rings (Three rings contain six carbon atoms each, and the fourth ring contains five.)	Look for four attached rings:	Some are hormones, others include cholesterol, bile salts, vitamin D, components of cell membranes
	4. *Carotenoids.* Orange and yellow pigments; consist of isoprene units	Look for isoprene units.	Converted to retinal (important in photoreception) and vitamin A
Proteins C, H, O, N (usually S)	One or more polypeptides (chains of amino acids) coiled or folded in characteristic shapes	Look for amino acid units joined by C—N bonds.	Serve as enzymes; structural components; muscle proteins; hemoglobin
Nucleic Acids C, H, O, N, P	Backbone composed of alternating pentose and phosphate groups, from which nitrogenous bases project. DNA contains the sugar deoxyribose and the bases guanine, cytosine, adenine, and thymine. RNA contains the sugar ribose and the bases guanine, cytosine, adenine, and uracil. Each molecular subunit, called a nucleotide, consists of a pentose, a phosphate, and a nitrogenous base.	Look for a pentose-phosphate backbone. DNA forms a double helix.	Storage, transmission, and expression of genetic information

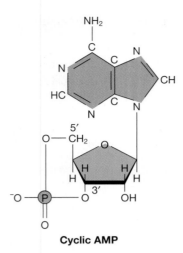

Cyclic AMP

Figure 3-25 Cyclic adenosine monophosphate (cAMP)

The single phosphate is part of a ring connecting two regions of the ribose.

IDENTIFYING BIOLOGICAL MOLECULES

Learning Objective

11 Compare the functions and chemical compositions of the major groups of organic compounds: carbohydrates, lipids, proteins, and nucleic acids.

Although the classes of biological molecules may seem overwhelming at first, you will learn to distinguish them readily by understanding their chief attributes. These are summarized in ▌Table 3-3.

Review

▌ How can you distinguish a pentose sugar from a hexose sugar? A disaccharide from a sterol? An amino acid from a monosaccharide? A phospholipid from a triacylglycerol? A protein from a polysaccharide? A nucleic acid from a protein?

SUMMARY WITH KEY TERMS

Learning Objectives

1 Describe the properties of carbon that make it the central component of organic compounds (page 46).
   - ▌ Each carbon atom forms four covalent bonds with up to four other atoms; these bonds are single, double, or triple bonds. Carbon atoms form straight or branched chains or join into rings. Carbon forms covalent bonds with a greater number of different elements than does any other type of atom.

2 Define the term *isomer,* and distinguish among the three principal isomer types (page 46).
   - ▌ **Isomers** are compounds with the same molecular formula but different structures.
   - ▌ **Structural isomers** differ in the covalent arrangements of their atoms. **Geometric isomers,** or *cis–trans* isomers, differ in the spatial arrangements of their atoms. **Enantiomers** are isomers that are mirror images of each other. Cells can distinguish between these configurations.

3 Identify the major functional groups present in organic compounds, and describe their properties (page 46).
   - ▌ Hydrocarbons, organic compounds consisting of only carbon and hydrogen, are nonpolar and hydrophobic. The **methyl group** is a hydrocarbon group.
   - ▌ Polar and ionic functional groups interact with one another and are hydrophilic. Partial charges on atoms at opposite ends of a bond are responsible for the polar property of a functional group. **Hydroxyl** and **carbonyl groups** are polar.
   - ▌ **Carboxyl** and **phosphate groups** are acidic, becoming negatively charged when they release hydrogen ions. The **amino group** is basic, becoming positively charged when it accepts a hydrogen ion.

4 Explain the relationship between polymers and macromolecules (page 46).
   - ▌ Long chains of **monomers** (similar organic compounds) linked through **condensation reactions** are called poly-

mers. Large polymers such as polysaccharides, proteins, and DNA are referred to as **macromolecules.** They can be broken down by **hydrolysis reactions.**

ThomsonNOW™ **Learn more about condensation and hydrolysis reactions by clicking on the figure in ThomsonNOW.**

5 Distinguish among monosaccharides, disaccharides, and polysaccharides. Compare storage polysaccharides with structural polysaccharides (page 50).
   - ▌ **Carbohydrates** contain carbon, hydrogen, and oxygen in a ratio of approximately one carbon to two hydrogens to one oxygen. **Monosaccharides** are simple sugars such as glucose, fructose, and ribose. Two monosaccharides join by a **glycosidic linkage** to form a **disaccharide** such as maltose or sucrose.
   - ▌ Most carbohydrates are **polysaccharides,** long chains of repeating units of a simple sugar. Carbohydrates are typically stored in plants as the polysaccharide **starch** and in animals as the polysaccharide **glycogen.** The cell walls of plants are composed mainly of the structural polysaccharide **cellulose.**

ThomsonNOW™ **Learn more about starch and cellulose by clicking on the figure in ThomsonNOW.**

6 Distinguish among fats, phospholipids, and steroids, and describe the composition, characteristics, and biological functions of each (page 56).
   - ▌ **Lipids** are composed mainly of hydrocarbon-containing regions, with few oxygen-containing (polar or ionic) functional groups. Lipids have a greasy or oily consistency and are relatively insoluble in water.
   - ▌ **Triacylglycerol,** the main storage form of fat in organisms, consists of a molecule of **glycerol** combined with three **fatty acids. Monoacylglycerols** and **diacylglycerols** contain one and two fatty acids, respectively. A fatty acid can be either **saturated** with hydrogen or **unsaturated.**

- **Phospholipids** are structural components of cell membranes. A phospholipid consists of a glycerol molecule attached at one end to two fatty acids and at the other end to a phosphate group linked to an organic compound such as choline.
- **Steroid** molecules contain carbon atoms arranged in four attached rings. Cholesterol, bile salts, and certain hormones are important steroids.

ThomsonNOW Learn more about triacylglycerol and other lipids by clicking on the figures in ThomsonNOW.

7 Give an overall description of the structure and functions of proteins (page 59).
- **Proteins** are large, complex molecules made of simpler subunits, called **amino acids,** joined by **peptide bonds.** Two amino acids combine to form a **dipeptide.** A longer chain of amino acids is a **polypeptide.** Proteins are the most versatile class of biological molecules, serving a variety of functions, such as **enzymes,** structural components, and cell regulators.
- Proteins are composed of various linear sequences of 20 different amino acids.

ThomsonNOW Learn more about amino acids and peptide bonds by clicking on the figures in ThomsonNOW.

8 Describe the features that are shared by all amino acids, and explain how amino acids are grouped into classes based on the characteristics of their side chains (page 59).
- All amino acids contain an amino group and a carboxyl group. Amino acids vary in their side chains, which dictate their chemical properties—nonpolar, polar, acidic, or basic. Amino acids generally exist as dipolar ions at cell pH and serve as important biological buffers.

9 Distinguish among the four levels of organization of protein molecules (page 59).

- **Primary structure** is the linear sequence of amino acids in the polypeptide chain.
- **Secondary structure** is a regular conformation, such as an α-helix or a β-pleated sheet; it is due to hydrogen bonding between elements of the backbones of the amino acids.
- **Tertiary structure** is the overall shape of the polypeptide chains, as dictated by chemical properties and interactions of the side chains of specific amino acids. Hydrogen bonds, ionic bonds, hydrophobic interactions, and disulfide bridges contribute to tertiary structure.
- **Quaternary structure** is determined by the association of two or more polypeptide chains.

ThomsonNOW Learn more about the structure of a protein by clicking on the figure in ThomsonNOW.

10 Describe the components of a nucleotide. Name some nucleic acids and nucleotides, and discuss the importance of these compounds in living organisms (page 67).
- **Nucleotides** are composed of a two-ring **purine** or one-ring **pyrimidine** nitrogenous base, a five-carbon sugar (**ribose** or **deoxyribose**), and one or more phosphate groups.
- The **nucleic acids DNA** and **RNA,** composed of long chains of nucleotide subunits, store and transfer information that specifies the sequence of amino acids in proteins and ultimately the structure and function of the organism.
- **ATP (adenosine triphosphate)** is a nucleotide of special significance in energy metabolism. **NAD$^+$** is also involved in energy metabolism through its role as an electron (hydrogen) acceptor in biological oxidation and reduction reactions.

11 Compare the functions and chemical compositions of the major groups of organic compounds: carbohydrates, lipids, proteins, and nucleic acids (page 70).
- Review Table 3-3.

TEST YOUR UNDERSTANDING

1. Which of the following is generally considered an inorganic form of carbon? (a) CO_2 (b) C_2H_4 (c) CH_3COOH (d) b and c (e) all of the preceding

2. Carbon is particularly well suited to be the backbone of organic molecules because (a) it can form both covalent bonds and ionic bonds (b) its covalent bonds are very irregularly arranged in three-dimensional space (c) its covalent bonds are the strongest chemical bonds known (d) it can bond to atoms of a large number of other elements (e) all the bonds it forms are polar

3. The structures depicted are (a) enantiomers (b) different views of the same molecule (c) geometric (*cis–trans*) isomers (d) both geometric isomers and enantiomers (e) structural isomers

$$
\begin{array}{cc}
\begin{array}{c}
CH_3 \; CH_3 \\
\vert \;\;\; \vert \\
H{-}C{-}C{-}H \\
\vert \;\;\; \vert \\
H \;\;\; H
\end{array}
&
\begin{array}{c}
H \;\;\; CH_3 \\
\vert \;\;\; \vert \\
H{-}C{-}C{-}H \\
\vert \;\;\; \vert \\
CH_3 \; H
\end{array}
\end{array}
$$

4. Which of the following are generally hydrophobic? (a) polar molecules and hydrocarbons (b) ions and hydrocarbons (c) nonpolar molecules and ions (d) polar molecules and ions (e) none of the preceding

5. Which of the following is a nonpolar molecule? (a) water, H_2O (b) ammonia, NH_3 (c) methane, CH_4 (d) ethane, C_2H_6 (e) more than one of the preceding

6. Which of the following functional groups normally acts as an acid? (a) hydroxyl (b) carbonyl (c) sulfhydryl (d) phosphate (e) amino

7. The synthetic process by which monomers are covalently linked is (a) hydrolysis (b) isomerization (c) condensation (d) glycosidic linkage (e) ester linkage

8. A monosaccharide designated as an aldehyde sugar contains (a) a terminal carboxyl group (b) an internal carboxyl group (c) a terminal carbonyl group (d) an internal carbonyl group (e) a terminal carboxyl group and an internal carbonyl group

9. Structural polysaccharides typically (a) have extensive hydrogen bonding between adjacent molecules (b) are much more hydrophilic than storage polysaccharides (c) have much stronger covalent bonds than do storage polysaccharides (d) consist of alternating α-glucose and β-glucose subunits (e) form helical structures in the cell

10. A carboxyl group is always found in (a) organic acids and sugars (b) sugars and fatty acids (c) fatty acids and amino acids (d) alcohols (e) glycerol

11. Fatty acids are components of (a) phospholipids and carotenoids (b) carotenoids and triacylglycerol (c) steroids and triacylglycerol (d) phospholipids and triacylglycerol (e) carotenoids and steroids

12. Saturated fatty acids are so named because they are saturated with (a) hydrogen (b) water (c) hydroxyl groups (d) glycerol (e) double bonds

13. Fatty acids in phospholipids and triacylglycerols interact with one another by (a) disulfide bridges (b) van der Waals interactions (c) covalent bonds (d) hydrogen bonds (e) fatty acids do not interact with one another

14. Which pair of amino acid side groups would be most likely to associate with each other by an ionic bond?

    1. —CH₃

    2. —CH₂—COO⁻

    3. —CH₂—CH₂—NH₃⁺

    4. —CH₂—CH₂—COO⁻

    5. —CH₂—OH

    (a) 1 and 2 (b) 2 and 4 (c) 1 and 5 (d) 2 and 5 (e) 3 and 4

15. Which of the following levels of protein structure may be affected by hydrogen bonding? (a) primary and secondary (b) primary and tertiary (c) secondary, tertiary, and quaternary (d) primary, secondary, and tertiary (e) primary, secondary, tertiary, and quaternary

16. Which of the following associations between R groups are the strongest? (a) hydrophobic interactions (b) hydrogen bonds (c) ionic bonds (d) peptide bonds (e) disulfide bridges

17. Each phosphodiester linkage in DNA or RNA includes a phosphate joined by covalent bonds to (a) two bases (b) two sugars (c) two additional phosphates (d) a sugar, a base, and a phosphate (e) a sugar and a base

CRITICAL THINKING

1. Like oxygen, sulfur forms two covalent bonds. However, sulfur is far less electronegative. In fact, it is approximately as electronegative as carbon. How would the properties of the various classes of biological molecules be altered if you were to replace all the oxygen atoms with sulfur atoms?

2. Hydrogen bonds and van der Waals interactions are much weaker than covalent bonds, yet they are vital to organisms. Why?

3. **Evolution Link.** In what ways are all species alike biochemically? Why? Identify some ways in which species may differ from one another biochemically? Why?

4. **Evolution Link.** The total number of possible amino acid sequences in a polypeptide chain is staggering. Given that there are 20 amino acids, potentially there could be 20^{100} different amino acid sequences (an impossibly large number), just for polypeptides only 100 amino acids in length. However, the actual number of different polypeptides occurring in organisms is only a tiny fraction of this. Why?

5. **Evolution Link.** Each amino acid could potentially exist as one of two possible enantiomers, known as the D-form and the L-form (based on the arrangement of the groups attached to the asymmetric α carbon). However, in all organisms, only L-amino acids are found in proteins. What does this suggest about the evolution of proteins?

Additional questions are available in ThomsonNOW at www.thomsonedu.com/login

4

Organization of the Cell

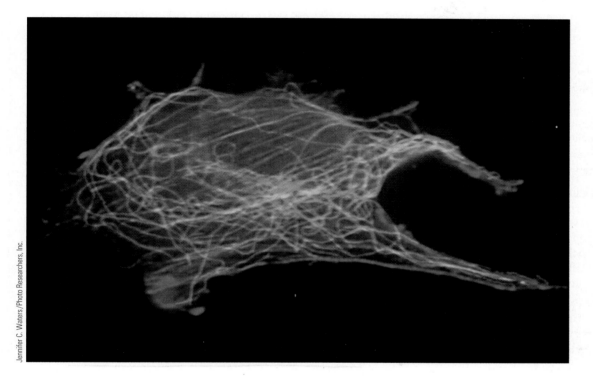

Jennifer C. Waters/Photo Researchers, Inc.

The cytoskeleton. The cell shown here was stained with fluorescent antibodies (specific proteins) that bind to proteins associated with DNA (*purple*) and to a protein (tubulin) in microtubules (*green*). Microfilaments (*red*) are also visible. This type of microscopy, known as confocal fluorescence microscopy, shows the extensive distribution of microtubules in this cell.

KEY CONCEPTS

The organization and size of the cell are critical in maintaining homeostasis.

Unlike prokaryotic cells, eukaryotic cells have internal membranes that divide the cell into compartments, allowing cells to conduct specialized activities within separate, small areas.

In eukaryotic cells, genetic information coded in DNA is located in the nucleus.

Proteins synthesized on ribosomes and processed in the endoplasmic reticulum are further processed by the Golgi complex and then transported to specific destinations.

Mitochondria and chloroplasts convert energy from one form to another.

The cytoskeleton is a dynamic internal framework that functions in various types of cell movement.

The cell is the smallest unit that can carry out all activities we associate with life. When provided with essential nutrients and an appropriate environment, some cells can be kept alive and growing in the laboratory for many years. By contrast, no isolated part of a cell is capable of sustained survival. Composed of a vast array of inorganic and organic ions and molecules, including water, salts, carbohydrates, lipids, proteins, and nucleic acids, most cells have all the physical and chemical components needed for their own maintenance, growth, and division. Genetic information is stored in DNA molecules and is faithfully replicated and passed to each new generation of cells during cell division. Information in DNA codes for specific proteins that, in turn, determine cell structure and function.

Most prokaryotes and many protists and fungi consist of a single cell. In contrast, most plants and animals are composed of millions of cells. Cells are the building blocks of complex multicellular organisms. Although they are basically similar, cells are also extraordinarily diverse and versatile. They are modified in a variety of ways to carry out specialized functions.

Cells exchange materials and energy with the environment. All living cells need one or more sources of energy, but a cell rarely obtains energy in a form that is immediately usable. Cells convert energy from one form to another, and that energy is used to carry out various activities, ranging from mechanical work to chemical synthesis. Cells convert energy to a convenient form, usually chemical energy stored in adenosine triphosphate, or ATP (see Chapter 3). The chemical reactions that convert energy from one form to another are essentially the same in all cells, from those in bacteria to those of complex plants and animals.

Thanks to advances in technology, cell biologists use increasingly sophisticated tools in their search to better understand the structure and function of cells. For example, investigation of the cytoskeleton (cell skeleton), currently an active and exciting area of research, has been greatly enhanced by advances in microscopy. In the photomicrograph, we see the extensive distribution of microtubules in cells. Microtubules are key components of the cytoskeleton. They help maintain cell shape, function in cell movement, and facilitate transport of materials within the cell. ∎

THE CELL THEORY

Learning Objective

1   Describe the cell theory, and relate it to the evolution of life.

Cells are dramatic examples of the underlying unity of all living things. This idea was first expressed by two German scientists, botanist Matthias Schleiden in 1838 and zoologist Theodor Schwann in 1839. Using their own observations and those of other scientists, these early investigators used inductive reasoning to conclude that all plants and animals consist of cells. Later, Rudolf Virchow, another German scientist, observed cells dividing and giving rise to daughter cells. In 1855, Virchow proposed that new cells form only by the division of previously existing cells.

The work of Schleiden, Schwann, and Virchow contributed greatly to the development of the **cell theory**, the unifying concept that (1) cells are the basic living units of organization and function in all organisms and (2) that all cells come from other cells. About 1880, another German biologist, August Weismann, added an important corollary to Virchow's concept by pointing out that the ancestry of all the cells alive today can be traced back to ancient times. Evidence that all living cells have a common origin is provided by the basic similarities in their structures and in the molecules of which they are made. When we examine a variety of diverse organisms, ranging from simple bacteria to the most complex plants and animals, we find striking similarities at the cell level. Careful studies of shared cell characteristics help us trace the evolutionary history of various organisms and furnish powerful evidence that all organisms alive today had a common origin.

Review

∎   How does the cell theory contribute to our understanding of the evolution of life? ✓

CELL ORGANIZATION AND SIZE

Learning Objectives

2   Summarize the relationship between cell organization and homeostasis.
3   Explain the relationship between cell size and homeostasis.

The organization of cells and their small size allow them to maintain **homeostasis,** an appropriate internal environment. Cells experience constant changes in their environments, such as deviations in salt concentration, pH, and temperature. They must work continuously to restore and maintain the internal conditions that enable their biochemical mechanisms to function.

The organization of all cells is basically similar

In order for the cell to maintain homeostasis, its contents must be separated from the external environment. The **plasma membrane** is a structurally distinctive surface membrane that surrounds all cells. By making the interior of the cell an enclosed compartment, the plasma membrane allows the chemical composition of the cell to be different from that outside the cell. The plasma membrane serves as a selective barrier between the cell contents and the outer environment. Cells exchange materials with the environment and can accumulate needed substances and energy stores.

Most cells have internal structures, called **organelles**, that are specialized to carry out metabolic activities, such as converting energy to usable forms, synthesizing needed compounds, and manufacturing structures necessary for functioning and reproduction. Each cell has genetic instructions coded in its DNA, which is concentrated in a limited region of the cell.

Cell size is limited

Although their sizes vary over a wide range (∎ Fig. 4-1), most cells are microscopic and must be measured by very small units. The basic unit of linear measurement in the metric system (see inside back cover) is the meter (m), which is just a little longer than a yard. A millimeter (mm) is 1/1000 of a meter and is about as long as the bar enclosed in parentheses (-). The micrometer (μm) is the most convenient unit for measuring cells. A bar 1 μm long is 1/1,000,000 (one millionth) of a meter, or 1/1000 of a millimeter—far too short to be seen with the unaided eye.

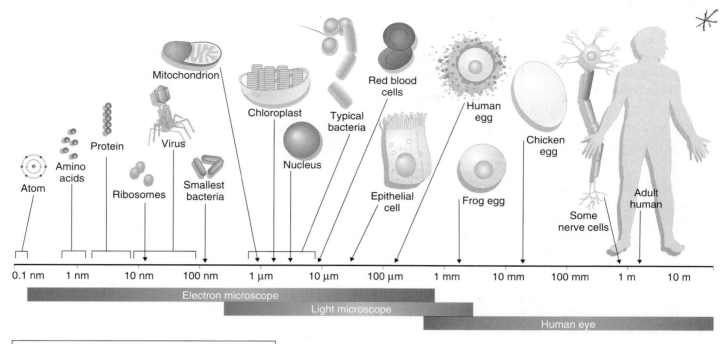

Measurements		
1 meter	=	1000 millimeters (mm)
1 millimeter	=	1000 micrometers (μm)
1 micrometer	=	1000 nanometers (nm)

Figure 4-1 Biological size and cell diversity

We can compare relative size from the chemical level to the organismic level by using a logarithmic scale (multiples of 10). The prokaryotic cells of bacteria typically range in size from 1 to 10 μm long. Most eukaryotic cells are between 10 and 30 μm in diameter. The nuclei of animal and plant cells range from about 3 to 10 μm in diameter. Mitochondria are about the size of small bacteria, whereas chloroplasts are usually larger, about 5 μm long. Ova (egg cells) are among the largest cells. Although microscopic, some nerve cells are very long. The cells shown here are not drawn to scale.

Most of us have difficulty thinking about units that are too small to see, but it is helpful to remember that a micrometer has the same relationship to a millimeter that a millimeter has to a meter (1/1000).

As small as it is, the micrometer is actually too large to measure most cell components. For this purpose biologists use the nanometer (nm), which is 1/1,000,000,000 (one billionth) of a meter, or 1/1000 of a micrometer. To mentally move down to the world of the nanometer, recall that a millimeter is 1/1000 of a meter, a micrometer is 1/1000 of a millimeter, and a nanometer is 1/1000 of a micrometer.

A few specialized algae and animal cells are large enough to be seen with the naked eye. A human egg cell, for example, is about 130 μm in diameter, or approximately the size of the period at the end of this sentence. The largest cells are birds' eggs, but they are atypical because both the yolk and the egg white consist of food reserves. The functioning part of the cell is a small mass on the surface of the yolk.

Why are most cells so small? If you consider what a cell must do to maintain homeostasis and to grow, it may be easier to understand the reasons for its small size. A cell must take in food and other materials and must rid itself of waste products generated by metabolic reactions. Everything that enters or leaves a cell must pass through its plasma membrane. The plasma membrane contains specialized "pumps" and channels with "gates" that selectively regulate the passage of materials into and out of the cell. The plasma membrane must be large enough relative to the cell volume to keep up with the demands of regulating the passage

of materials. Thus, a critical factor in determining cell size is the ratio of its surface area (the plasma membrane) to its volume (Fig. 4-2).

As a cell becomes larger, its volume increases at a greater rate than its surface area (its plasma membrane), which effectively places an upper limit on cell size. Above some critical size, the number of molecules required by the cell could not be transported into the cell fast enough to sustain its needs. In addition, the cell would not be able to regulate its concentration of various ions or efficiently export its wastes.

Of course, not all cells are spherical or cuboid. Because of their shapes, some very large cells have relatively favorable ratios of surface area to volume. In fact, some variations in cell shape represent a strategy for increasing the ratio of surface area to volume. For example, many large plant cells are long and thin, which increases their surface area–to-volume ratio. Some cells, such as epithelial cells lining the small intestine, have fingerlike projections of the plasma membrane, called **microvilli**, that significantly increase the surface area for absorbing nutrients and other materials (see Fig. 46-10).

Another reason for the small size of cells is that, once inside, molecules must be transported to the locations where they are converted into other forms. Because cells are small, the distances molecules travel within them are relatively short, which speeds up many cell activities.

Cell size and shape are related to function

The sizes and shapes of cells are related to the functions they perform. Some cells, such as the amoeba and the white blood cell, change their shape as they move about. Sperm cells have long, whiplike tails, called *flagella,* for locomotion. Nerve cells have long, thin extensions that enable them to transmit messages over great distances. The extensions of some nerve cells in the human body may be as long as 1 m! Certain epithelial cells are almost rectangular and are stacked much like building blocks to form sheetlike structures.

Review

▮ How does the plasma membrane help maintain homeostasis?

▮ Why is the relationship between surface area and volume of a cell important in determining cell-size limits?

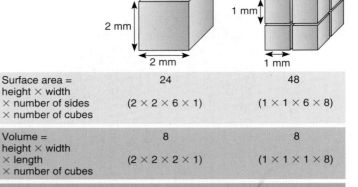

		2 mm cube	1 mm cubes
Surface Area (mm²)	Surface area = height × width × number of sides × number of cubes	24 (2 × 2 × 6 × 1)	48 (1 × 1 × 6 × 8)
Volume (mm³)	Volume = height × width × length × number of cubes	8 (2 × 2 × 2 × 1)	8 (1 × 1 × 1 × 8)
Surface Area/ Volume Ratio	Surface area/ volume	3 (24:8)	6 (48:8)

Figure 4-2 Surface area–to-volume ratio

The surface area of a cell must be large enough relative to its volume to allow adequate exchange of materials with the environment. Although their volumes are the same, eight small cells have a much greater surface area (plasma membrane) in relation to their total volume than one large cell does. In the example shown, the ratio of the total surface area to total volume of eight 1-mm cubes is double the surface area–to-volume ratio of the single large cube.

METHODS FOR STUDYING CELLS

Learning Objective

4   Describe methods that biologists use to study cells, including microscopy and cell fractionation.

One of the most important tools biologists use for studying cell structures is the microscope. Cells were first described in 1665 by the English scientist Robert Hooke in his book *Micrographia.* Using a microscope he had made, Hooke examined a piece of cork and drew and described what he saw. Hooke chose the term *cell* because the tissue reminded him of the small rooms monks lived in. Interestingly, what Hooke saw were not actually living cells but the walls of dead cork cells (▮ Fig. 4-3a). Much later, scientists recognized that the interior enclosed by the walls is the important part of living cells.

A few years later, inspired by Hooke's work, the Dutch naturalist Anton van Leeuwenhoek viewed living cells with small lenses that he made. Leeuwenhoek was highly skilled at grinding lenses and was able to magnify images more than 200 times. Among his important discoveries were bacteria, protists, blood cells, and sperm cells. Leeuwenhoek was among the first scientists to report cells in animals. Leeuwenhoek was a merchant and not formally trained as a scientist. However, his skill, curiosity, and diligence in sharing his discoveries with scientists at the Royal Society of London brought an awareness of microscopic life to the scientific world. Unfortunately, Leeuwenhoek did not share his techniques, and not until more than 100 years later, in the late 19th century, were microscopes sufficiently developed for biologists to seriously focus their attention on the study of cells.

Light microscopes are used to study stained or living cells

The **light microscope (LM),** the type used by most students, consists of a tube with glass lenses at each end. Because it contains several lenses, the modern light microscope is referred to as a *compound microscope.* Visible light passes through the specimen being observed and through the lenses. Light is refracted (bent) by the lenses, magnifying the image.

Two features of a microscope determine how clearly a small object can be viewed: magnification and resolving power. **Magnification** is the ratio of the size of the image seen with the microscope to the actual size of the object. The best light microscopes usually magnify an object no more than 1000 times. **Resolution,** or **resolving power,** is the capacity to distinguish fine detail in an image; it is defined as the minimum distance between two points at which they can both be seen separately rather than as a single, blurred point. Resolving power depends on the quality of the lenses and the wavelength of the illuminating light. As the wavelength decreases, the resolution increases.

The visible light used by light microscopes has wavelengths ranging from about 400 nm (violet) to 700 nm (red); this limits the resolution of the light microscope to details no smaller than the diameter of a small bacterial cell (about 0.2 μm). By the early

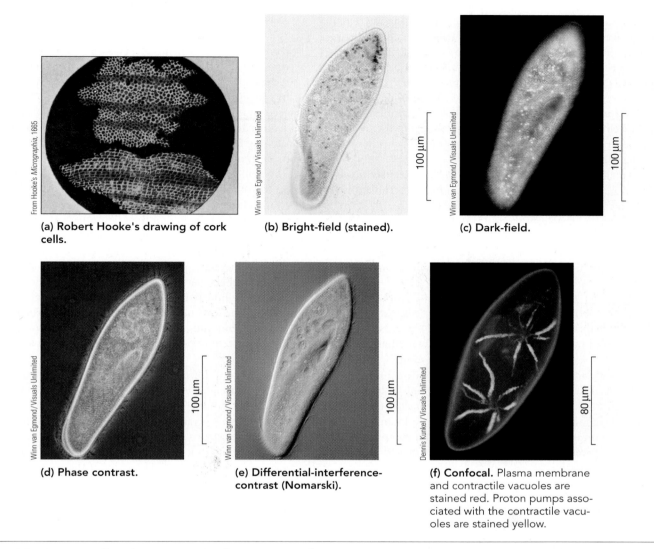

(a) Robert Hooke's drawing of cork cells.

From Hooke's *Micrographia*, 1665

(b) Bright-field (stained).

Winn van Egmond/Visuals Unlimited

100 μm

(c) Dark-field.

Winn van Egmond/Visuals Unlimited

100 μm

(d) Phase contrast.

Winn van Egmond/Visuals Unlimited

100 μm

(e) Differential-interference-contrast (Nomarski).

Winn van Egmond/Visuals Unlimited

100 μm

(f) Confocal. Plasma membrane and contractile vacuoles are stained red. Proton pumps associated with the contractile vacuoles are stained yellow.

Dennis Kunkel/Visuals Unlimited

80 μm

Figure 4-3 Viewing cells with various types of microscopes. The specimen visualized in (b) through (f) is a *Paramecium* (a protist).

(a) Using a crude microscope that he constructed, Robert Hooke looked at a thin slice of cork and drew what he saw. More sophisticated microscopes and techniques enable biologists to view cells in more detail. Bright-field microscopy can be enhanced by staining. The phase contrast and differential-interference-contrast microscopes enhance detail by increasing the differences in optical density in different regions of the cells.

20th century, refined versions of the light microscope, as well as compounds that stain different cell structures, became available. Using these tools, biologists discovered that cells contain many different internal structures, the organelles. The contribution of organic chemists in developing biological stains was essential to this understanding, because the interior of many cells is transparent. Most methods used to prepare and stain cells for observation, however, also kill them in the process.

Living cells can now be studied using light microscopes with special optical systems. In *bright-field microscopy,* an image is formed by transmitting light through a cell (Fig. 4-3b). Because there is little contrast, the details of cell structure are not visible. In *dark-field microscopy,* rays of light are directed from the side and only scattered light enters the lenses. The cell is visible as a bright object against a dark background (Fig. 4-3c).

Phase contrast microscopy and *Nomarski differential-interference-contrast microscopy* take advantage of variations in density within the cell (Fig. 4-3d and e). These variations in density cause differences in the way various regions of the cytoplasm refract (bend) light. Using these microscopes, scientists can observe living cells in action, with numerous internal structures that are constantly changing shape and location.

Cell biologists use the *fluorescence microscope* to detect the locations of specific molecules in cells. In the fluorescence microscope, filters transmit light that is emitted by fluorescently stained molecules. Fluorescent stains (like paints that glow under black light) are molecules that absorb light energy of one wavelength and then release some of that energy as light of a longer wavelength. These stains bind specifically to DNA or to specific protein molecules. The molecules absorb ultraviolet light and emit light of a different color. Cells can be stained, and the location of the labeled molecules can be determined, by observing the source of the fluorescent light within the cell. Biologists commonly use a fluorescent molecule known as green fluorescent protein (GFP) that occurs naturally in jellyfish. GFP has been valuable in observing specific proteins in living cells.

Some fluorescent stains are chemically bonded to *antibodies*, protein molecules important in internal defense. The antibody binds to a highly specific region of a molecule in the cell. A single type of antibody molecule binds to only one type of structure, such as a part of a specific protein or some of the sugars in a specific polysaccharide. Purified fluorescent antibodies known to bind to a specific protein are used to determine where that protein is located within the cell.

Confocal microscopy produces a sharper image than standard fluorescence microscopy (Fig. 4-3f). Living cells that have been labeled with fluorescent dye are mounted on a microscope slide. A beam of ultraviolet light from a laser is focused at a specific depth within the cell. The fluorescent label emits visible light, and a single plane through the cell is visible. The microscope produces optical sections (see the photomicrograph in the chapter introduction). A computer reassembles the images so that a series of optical sections from different planes of the cell can be used to construct a three-dimensional image. Powerful computer-imaging methods greatly improve the resolution of structures labeled by fluorescent dyes.

Electron microscopes provide a high-resolution image that can be greatly magnified

Even with improved microscopes and techniques for staining cells, ordinary light microscopes can distinguish only the gross details of many cell parts (Fig. 4-4a). In most cases, you can clearly see only the outline of an organelle and its ability to be stained by some dyes and not by others. With the development of the **electron microscope (EM),** which came into wide use in the 1950s, researchers began to study the fine details, or **ultrastructure,** of cells.

Whereas the best light microscopes have about 500 times the resolution that the human eye does, the electron microscope multiplies the resolving power by more than 10,000. The reason is that electrons have very short wavelengths, on the order of about 0.1 to 0.2 nm. Although such resolution is difficult to achieve with biological material, researchers can approach that resolution when examining isolated molecules such as proteins and DNA. This high degree of resolution permits very high magnifications of 250,000 times or more as compared to typical magnifications of no more than 1000 times in light microscopy.

The image formed by the electron microscope is not directly visible. The electron beam itself consists of energized electrons, which, because of their negative charge, can be focused by electromagnets just as images are focused by glass lenses in a light microscope (Fig. 4-4b). Two types of electron microscopes are the **transmission electron microscope (TEM)** and the **scanning electron microscope (SEM).** The acronyms TEM and SEM also identify that a micrograph was prepared using a transmission or scanning EM.

In transmission electron microscopy, the specimen is embedded in plastic and then cut into extraordinarily thin sections (50 to 100 nm thick) with a glass or diamond knife. A section is then placed on a small metal grid. The electron beam passes through the specimen and then falls onto a photographic plate or a fluorescent screen. When you look at TEMs in this chapter (and elsewhere), keep in mind that each represents only a thin cross section of a cell.

Researchers detect certain specific molecules in electron microscope images by using antibody molecules to which very tiny gold particles are bound. The dense gold particles block the electron beam and identify the location of the proteins recognized by the antibodies as precise black spots on the electron micrograph.

In the scanning electron microscope, the electron beam does not pass through the specimen. Instead, the specimen is coated with a thin film of gold or some other metal. When the electron beam strikes various points on the surface of the specimen, secondary electrons are emitted whose intensity varies with the contour of the surface. The recorded emission patterns of the secondary electrons give a 3-D picture of the surface (Fig. 4-4c). The SEM provides information about the shape and external features of the specimen that cannot be obtained with the TEM.

Note that the LM, TEM, and SEM are focused by similar principles. A beam of light or an electron beam is directed by the condenser lens onto the specimen and is magnified by the objective lens and the eyepiece in the light microscope or by the objective lens and the projector lens in the TEM. The TEM image is focused onto a fluorescent screen, and the SEM image is viewed on a type of television screen. Lenses in electron microscopes are actually magnets that bend the beam of electrons.

Biologists use biochemical techniques to study cell components

The EM is a powerful tool for studying cell structure, but it has limitations. The methods used to prepare cells for electron microscopy kill them and may alter their structure. Furthermore, electron microscopy provides few clues about the functions of organelles and other cell components. To determine what organelles actually do, researchers use a variety of biochemical techniques.

Cell fractionation is a technique for purifying different parts of cells so that they can be studied by physical and chemi-

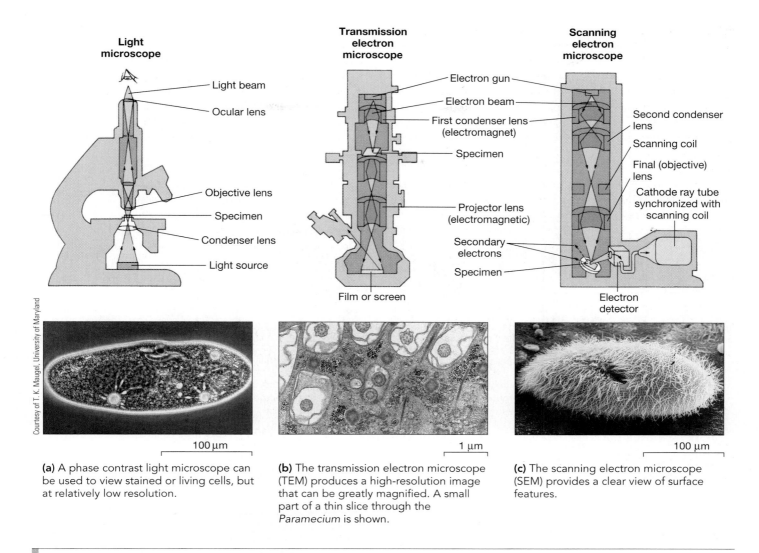

Light microscope

- Light beam
- Ocular lens
- Objective lens
- Specimen
- Condenser lens
- Light source

Courtesy of T. K. Maugel, University of Maryland

Transmission electron microscope

- Electron gun
- Electron beam
- First condenser lens (electromagnet)
- Specimen
- Projector lens (electromagnetic)
- Film or screen

Scanning electron microscope

- Second condenser lens
- Scanning coil
- Final (objective) lens
- Cathode ray tube synchronized with scanning coil
- Secondary electrons
- Specimen
- Electron detector

100 μm

1 μm

100 μm

(a) A phase contrast light microscope can be used to view stained or living cells, but at relatively low resolution.

(b) The transmission electron microscope (TEM) produces a high-resolution image that can be greatly magnified. A small part of a thin slice through the *Paramecium* is shown.

(c) The scanning electron microscope (SEM) provides a clear view of surface features.

Figure 4-4 Comparing light and electron microscopy

A phase contrast light microscope and two types of electron microscopes produce distinctive images of cells, such as the protist *Paramecium* shown in these photomicrographs.

cal methods. Generally, cells are broken apart as gently as possible; and the mixture, referred to as the *cell extract,* is subjected to centrifugal force by spinning in a device called a **centrifuge** (❙ Fig. 4-5a). The powerful ultracentrifuge can spin at speeds exceeding 100,000 revolutions per minute (rpm), generating a centrifugal force of 500,000 × G (a G is equal to the force of gravity). Centrifugal force separates the extract into two fractions: a pellet and a supernatant. The *pellet* that forms at the bottom of the tube contains heavier materials, such as nuclei, packed together. The *supernatant,* the liquid above the pellet, contains lighter particles, dissolved molecules, and ions.

After the pellet is removed, the supernatant can be centrifuged again at a higher speed to obtain a pellet that contains the next-heaviest cell components, for example, mitochondria and chloroplasts. In **differential centrifugation,** the supernatant is spun at successively higher speeds, permitting various cell components to be separated on the basis of their different sizes and densities (❙ Fig. 4-5b).

Cell components in the resuspended pellets can be further purified by **density gradient centrifugation.** In this procedure, the centrifuge tube is filled with a series of solutions of decreasing density. For example, sucrose solutions can be used. The concentration of sucrose is highest at the bottom of the tube and decreases gradually so that it is lowest at the top. The resuspended pellet is placed in a layer on top of the density gradient. Because the densities of organelles differ, each will migrate during centrifugation and form a band at the position in the gradient where its own density equals that of the sucrose solution. Purified or-

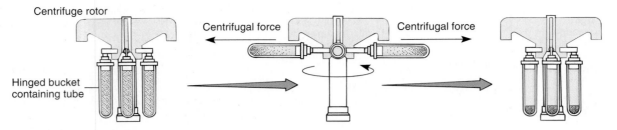

(a) **Centrifugation.** Due to centrifugal force, large or very dense particles move toward the bottom of a tube and form a pellet.

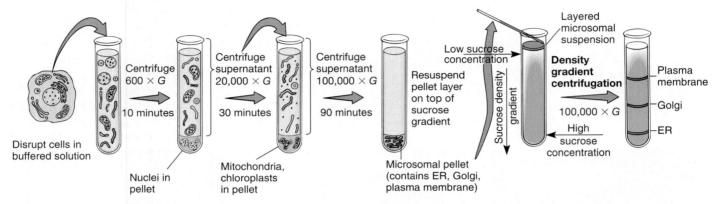

(b) **Differential centrifugation.** Cell structures can be separated into various fractions by spinning the suspension at increasing revolutions per minute. Membranes and organelles from the resuspended pellets can then be further purified by density gradient centrifugation (shown as last step). *G* is the force of gravity. ER is the endoplasmic reticulum.

Figure 4-5   Cell fractionation

ganelles are examined to determine what kinds of proteins and other molecules they might contain, as well as the nature of the chemical reactions that take place within them.

Review

▪ What is the main advantage of the electron microscope? Explain.

▪ What is cell fractionation? Describe the process.

PROKARYOTIC AND EUKARYOTIC CELLS

See chart done in class

Learning Objective

5   Compare and contrast the general characteristics of prokaryotic and eukaryotic cells, and contrast plant and animal cells.

Recall from Chapter 1 that two basic types of cells are known: **prokaryotic cells** and **eukaryotic cells.** Bacteria and archaea are prokaryotic cells. All other known organisms consist of eukaryotic cells. Prokaryotic cells are typically smaller than eukaryotic cells. In fact, the average prokaryotic cell is only about 1/10 the

diameter of the average eukaryotic cell. In prokaryotic cells, the DNA is not enclosed in a nucleus. Instead, the DNA is located in a limited region of the cell called a **nuclear area,** or **nucleoid,** which is not enclosed by a membrane (Fig. 4-6). The term *prokaryotic,* meaning "before the nucleus," refers to this major difference between prokaryotic and eukaryotic cells. Other types of internal membrane–enclosed organelles are also absent in prokaryotic cells.

Like eukaryotic cells, prokaryotic cells have a plasma membrane that confines the contents of the cell to an internal compartment. In some prokaryotic cells, the plasma membrane may be folded inward to form a complex of membranes along which many of the cell's metabolic reactions take place. Most prokaryotic cells have **cell walls,** which are extracellular structures that enclose the entire cell, including the plasma membrane.

Many prokaryotes have **flagella** (sing., *flagellum*), long fibers that project from the surface of the cell. Prokaryotic flagella, which operate like propellers, are important in locomotion. Their structure is different from that of flagella found in eukaryotic cells.

The dense internal material of the bacterial cell contains **ribosomes,** small complexes of ribonucleic acid (RNA) and protein that synthesize polypeptides. The ribosomes of prokaryotic cells are smaller than those found in eukaryotic cells. Prokaryotic cells also contain storage granules that hold glycogen,

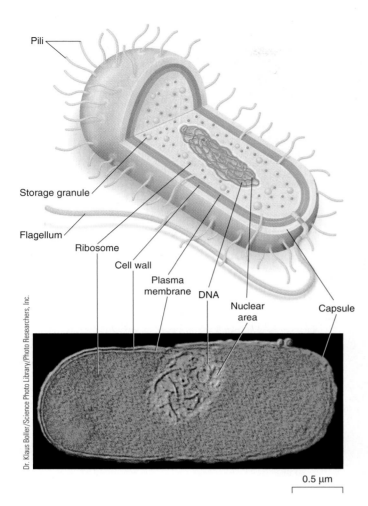

Pili

Storage granule

Flagellum

Ribosome

Cell wall

Plasma
membrane   DNA

Nuclear
area

Capsule

Dr. Klaus Boller/Science Photo Library/Photo Researchers, Inc.

0.5 µm

Figure 4-6 *Animated* Structure of a prokaryotic cell

This colorized TEM shows a thin lengthwise slice through an *Escherichia coli* bacterium. Note the prominent nuclear area containing the genetic material (DNA). *E. coli* is a normal inhabitant of the human intestine, but under certain conditions some strains can cause infections.

lipid, or phosphate compounds. This chapter focuses primarily on eukaryotic cells. Prokaryotes are discussed in more detail in Chapter 24.

Eukaryotic cells are characterized by highly organized membrane-enclosed organelles, including a prominent *nucleus,* which contains the hereditary material, DNA. The term *eukaryotic* means "true nucleus." Early biologists thought cells consisted of a homogeneous jelly, which they called *protoplasm*. With the electron microscope and other modern research tools, perception of the environment within the cell has been greatly expanded. We now know that the cell is highly organized and complex (❚ Figs. 4-7 and 4-8). The eukaryotic cell has its own control center, internal transportation system, power plants, factories for making needed materials, packaging plants, and even a "self-destruct" system.

Biologists refer to the part of the cell outside the nucleus as **cytoplasm** and the part of the cell within the nucleus as **nucleoplasm.** Various organelles are suspended within the fluid component of the cytoplasm, which is called the **cytosol**. Therefore, the term *cytoplasm* includes both the cytosol and all the organelles other than the nucleus.

The many specialized organelles of eukaryotic cells solve some of the problems associated with large size, so eukaryotic cells can be larger than prokaryotic cells. Eukaryotic cells also differ from prokaryotic cells in having a supporting framework, or cytoskeleton, important in maintaining shape and transporting materials within the cell.

Some organelles are present only in specific cells. For example, *chloroplasts,* structures that trap sunlight for energy conversion, are only in cells that carry on photosynthesis, such as certain plant or algal cells. Most bacteria, fungi, and plant cells are surrounded by a *cell wall* external to the plasma membrane. Plant cells also contain a large, membrane-enclosed *vacuole*. We discuss these and other differences among major types of cells throughout this chapter. Plant and animal cells are compared in Figures 4-7 and 4-8 and also in Figures 4-9 and 4-10.

Review

❚ What are two important differences between prokaryotic and eukaryotic cells?

❚ What are three ways that a plant cell might differ from an animal cell?

CELL MEMBRANES

Learning Objective

6   Describe three functions of cell membranes.

Membranes divide the eukaryotic cell into compartments, and their unique properties enable membranous organelles to carry out a wide variety of functions. For example, cell membranes never have free ends; therefore, a membranous organelle always contains at least one enclosed internal space or compartment. These membrane-enclosed compartments allow certain cell activities to be localized within specific regions of the cell. Reactants located in only a small part of the total cell volume are far more likely to come in contact, dramatically increasing the rate of the reaction. On the other hand, membrane-enclosed compartments keep certain reactive compounds away from other parts of the cell that they might adversely affect. Compartmentalizing also allows many different activities to go on simultaneously.

Membranes serve as important work surfaces. For example, many chemical reactions in cells are carried out by enzymes that are bound to membranes. Because the enzymes that carry out successive steps of a series of reactions are organized close together on a membrane surface, certain series of chemical reactions occur more rapidly.

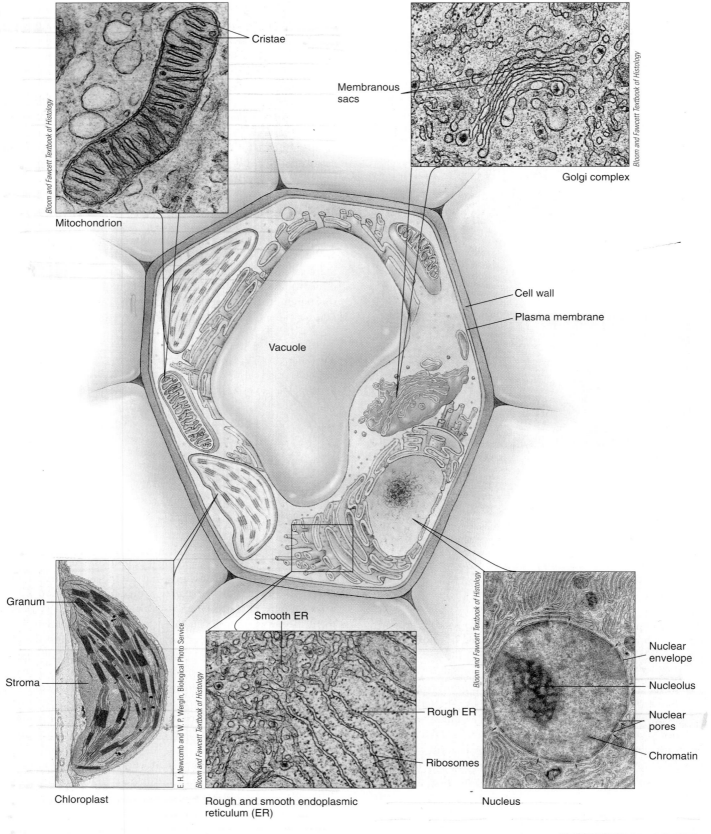

Cristae

Membranous sacs

Bloom and Fawcett Textbook of Histology

Golgi complex

Mitochondrion

Bloom and Fawcett Textbook of Histology

Cell wall

Plasma membrane

Vacuole

Granum

Stroma

E. H. Newcomb and W. P. Wergin, Biological Photo Service

Smooth ER

Bloom and Fawcett Textbook of Histology

Rough ER

Ribosomes

Chloroplast

Rough and smooth endoplasmic reticulum (ER)

Bloom and Fawcett Textbook of Histology

Nuclear envelope

Nucleolus

Nuclear pores

Chromatin

Nucleus

Figure 4-7 *Animated* Composite diagram of a plant cell

Chloroplasts, a cell wall, and prominent vacuoles are characteristic of plant cells. The TEMs show specific structures or areas of the cell. Some plant cells do not have all the organelles shown here. For example, leaf and stem cells that carry on photosynthesis contain chloroplasts, whereas root cells do not. Many of the organelles, such as the nucleus, mitochondria, and endoplasmic reticulum (ER), are characteristic of all eukaryotic cells.

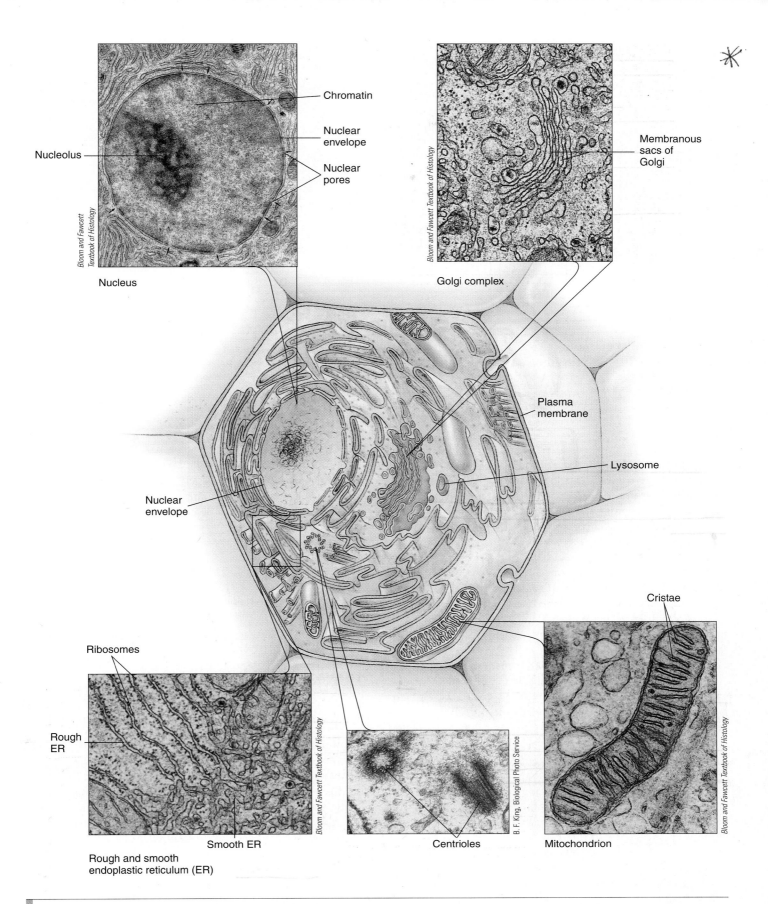

Chromatin

Nuclear envelope

Nuclear pores

Nucleolus

Nucleus

Membranous sacs of Golgi

Golgi complex

Plasma membrane

Nuclear envelope

Lysosome

Cristae

Ribosomes

Rough ER

Smooth ER

Rough and smooth endoplastic reticulum (ER)

Centrioles

Mitochondrion

Bloom and Fawcett Textbook of Histology

B. F. King, Biological Photo Service

Figure 4-8 *Animated* Composite diagram of an animal cell

This generalized animal cell is shown in a realistic context surrounded by adjacent cells, which cause it to be slightly compressed. The TEMs show the structure of various organelles. Depending on the cell type, certain organelles may be more or less prominent.

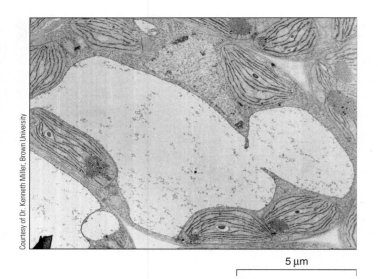
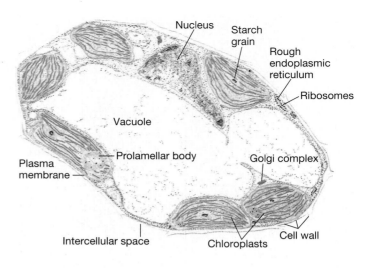

Nucleus
Starch grain
Rough endoplasmic reticulum
Ribosomes
Vacuole
Prolamellar body
Golgi complex
Plasma membrane
Intercellular space
Chloroplasts
Cell wall

Courtesy of Dr. Kenneth Miller, Brown University

5 µm

Figure 4-9 TEM of a plant cell and an interpretive drawing

Most of this cross section of a cell from the leaf of a young bean plant (*Phaseolus vulgaris*) is dominated by a vacuole. Prolamellar bodies are membranous regions typically seen in developing chloroplasts.

Membranes allow cells to store energy. The membrane serves as a barrier that is somewhat analogous to a dam on a river. As we will discuss later in this chapter, there is both an electric charge difference and a concentration difference on the two sides of the membrane. These differences constitute an *electrochemical gradient*. Such gradients store energy and so have *potential energy* (see Chapter 7). As particles of a substance move across the membrane from the side of higher concentration to the side of lower concentration, the cell can convert some of this potential energy to the chemical energy of ATP molecules. This process of energy conversion (discussed in Chapters 8 and 9) is a basic mechanism that cells use to capture and convert the energy necessary to sustain life.

In a eukaryotic cell, several types of membranes are generally considered part of the internal membrane system, or **endomembrane system.** In ▌Figures 4-9 and 4-10 (also see Figs. 4-7 and 4-8), notice how membranes divide the cell into many compartments: the nucleus, endoplasmic reticulum (ER), Golgi complex, lysosomes, vesicles, and vacuoles. Although it is not internal, the plasma membrane is also included because it participates in the activities of the endomembrane system. (Mitochondria and chloroplasts are also separate compartments but are not generally considered part of the endomembrane system, because they function somewhat independently of other membranous organelles.)

Some organelles have direct connections between their membranes and compartments. Others transport materials in **vesicles,** small, membrane-enclosed sacs formed by "budding" from the membrane of another organelle. Vesicles also carry cargo from one organelle to another. A vesicle can form as a "bud" from the membrane of one organelle and then move to another organelle to which it fuses, thus delivering its contents into another compartment.

Review

▌ How do membrane-enclosed organelles facilitate cell metabolism?

▌ What organelles belong to the endomembrane system?

THE CELL NUCLEUS

Learning Objective

7   Describe the structure and functions of the nucleus.

Typically, the **nucleus** is the most prominent organelle in the cell. It is usually spherical or oval in shape and averages 5 µm in diameter. Because of its size and the fact that it often occupies a relatively fixed position near the center of the cell, some early investigators guessed long before experimental evidence was available that the nucleus served as the control center of the cell (see *Focus On: Acetabularia and the Control of Cell Activities*). Most cells have one nucleus, although there are exceptions.

The **nuclear envelope** consists of two concentric membranes that separate the nuclear contents from the surrounding cyto-

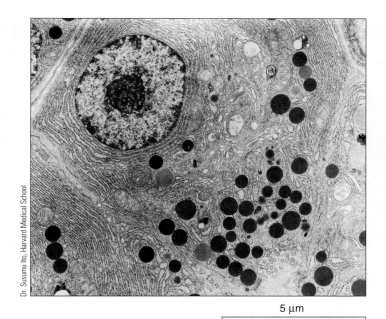

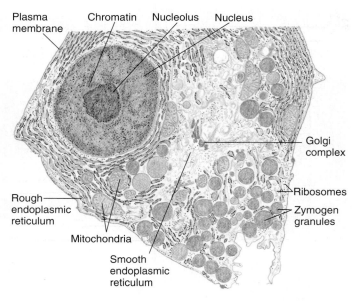

Plasma membrane — Chromatin — Nucleolus — Nucleus

Golgi complex

Rough endoplasmic reticulum

Ribosomes

Zymogen granules

Mitochondria

Smooth endoplasmic reticulum

5 μm

Dr. Susumu Ito, Harvard Medical School

Figure 4-10 TEM of a human pancreas cell and an interpretive drawing

Most of the structures of a typical animal cell are present. However, like most cells, this one has certain structures associated with its specialized functions. Pancreas cells such as the one shown here secrete large amounts of digestive enzymes. The large, dark, circular bodies in the TEM and the corresponding structures in the drawing are zymogen granules, which contain inactive enzymes. When released from the cell, the enzymes catalyze chemical reactions such as the breakdown of peptide bonds of ingested proteins in the intestine. Most of the membranes visible in this section are part of the rough endoplasmic reticulum, an organelle specialized to manufacture protein.

plasm (█ Fig. 4-11). These membranes are separated by about 20 to 40 nm. At intervals the membranes come together to form **nuclear pores,** which consist of protein complexes. Nuclear pores regulate the passage of materials between nucleoplasm and cytoplasm. A fibrous network of protein filaments, called the **nuclear lamina,** forms an inner lining for the nuclear envelope. The nuclear lamina supports the inner nuclear membrane and may have other important functions, such as organizing the nuclear contents. Mutations in proteins associated with the nuclear lamina and inner nuclear membrane are associated with several human genetic diseases, including muscular dystrophies, as well as premature aging.

The cell stores information in the form of **DNA,** and most of the cell's DNA is located inside the nucleus. When a cell divides, the information stored in DNA must be reproduced and passed intact to the two daughter cells. DNA has the unique ability to make an exact duplicate of itself through a process called replication. DNA molecules consist of sequences of nucleotides called **genes,** which contain the chemically coded instructions for producing the proteins needed by the cell. The nucleus controls protein synthesis by transcribing its information in **messenger RNA (mRNA) molecules.** Messenger RNA moves into the cytoplasm, where proteins are manufactured.

DNA is associated with proteins, forming a complex known as **chromatin,** which appears as a network of granules and strands in cells that are not dividing. Although chromatin appears disorganized, it is not. Because DNA molecules are extremely long and thin, they must be packed inside the nucleus in a very regular fashion as part of structures called **chromosomes.** In dividing cells, the chromosomes become visible as distinct threadlike structures. If the DNA in the 46 chromosomes of one human cell could be stretched end to end, it would extend for 2 m!

Most nuclei have one or more compact structures called **nucleoli** (sing., *nucleolus*). A nucleolus, which is *not* enclosed by a membrane, usually stains differently than the surrounding chromatin. Each nucleolus contains a **nucleolar organizer,** made up of chromosomal regions containing instructions for making the type of RNA in ribosomes. This **ribosomal RNA is synthesized in the nucleolus.** The proteins needed to make ribosomes are synthesized in the cytoplasm and imported into the nucleolus. Ribosomal RNA and proteins are then assembled into ribosomal subunits that leave the nucleus through the nuclear pores.

Review

- How does the nucleus store information?
- What is the function of the nuclear envelope?

ACETABULARIA AND THE CONTROL OF CELL ACTIVITIES

Figure A   Light micrograph of *Acetabularia*

To the romantically inclined, the little seaweed *Acetabularia* resembles a mermaid's wineglass, although the literal translation of its name, "vinegar cup," is somewhat less elegant (**Fig. A**). In the 19th century, biologists discovered that this marine eukaryotic alga consists of a single cell. At up to 5 cm (2 in) in length, *Acetabularia* is small for a seaweed but gigantic for a cell. It consists of a rootlike holdfast; a long, cylindrical stalk; and a cuplike cap. The nucleus is in the holdfast, about as far away from the cap as it can be. Because it is a single giant cell, *Acetabularia* is easy for researchers to manipulate.

If the cap of *Acetabularia* is removed experimentally, another one grows after a few weeks. Such a response, common among simple organisms, is called **regeneration**. This fact attracted the attention of investigators, especially Danish biologist J. Hämmerling and Belgian biologist J. Brachet, who became interested in whether a relationship exists between the nucleus and the physical characteristics of the alga. Because of its large size, *Acetabularia* could be subjected to surgery that would be difficult with smaller cells. During the 1930s and 1940s, these researchers performed brilliant experiments that in many ways laid the foundation for much of our modern knowledge of the nucleus. Two species were used for most experiments:

A. mediterranea, which has a smooth cap, and *A. crenulata*, which has a cap divided into a series of fingerlike projections. The kind of cap that is regenerated depends on the species of *Acetabularia* used in the experiment. As you might expect, *A. crenulata* regenerates a "cren" cap, and *A. mediterranea* regenerates a "med" cap.

Experiment 1

Question: What controls the shape of the cap in *Acetabularia*?

Hypothesis: Something in the stalk or holdfast of *Acetabularia* controls the shape of the cap.

Experiment: Hämmerling and Brachet removed the caps from *A. mediterranea* and *A. crenulata*. They then grafted together the two capless algae (**Fig. B**).

Results and Conclusion: The algae regenerated a common cap with characteristics intermediate between those of the two species involved. This experiment demonstrated that something in the stalk or holdfast controls cap shape.

Experiment 2

Question: Does the stalk or the holdfast control the shape of the cap in *Acetabularia*?

Hypothesis: Something in the holdfast of *Acetabularia* controls the shape of the cap.

Experiment: Hämmerling and Brachet removed the caps from *A. mediterranea* and *A. crenulata*. Then they severed the stalks from the holdfasts. By telescoping the cell walls of the two species into one another, Hämmerling and Brachet were able to attach a section of the stalk of one species to a holdfast of the other species (**Fig. C**).

Results and Conclusion: The results were surprising. The caps that regenerated were characteristic not of the species donating the holdfasts but of those donating the stalks! However, when the caps were removed once again, this time the caps that regenerated were characteristic of the species that donated the holdfasts. This continued to be the case no matter how many more times the regenerated caps were removed.

From these results Hämmerling and Brachet deduced that the ultimate control of the *Acetabularia* cell is associated with the holdfast. Because there is a time lag before the holdfast appears to take over, they hypothesized it produces some temporary cytoplasmic messenger substance whereby it exerts its control. They further hypothe-

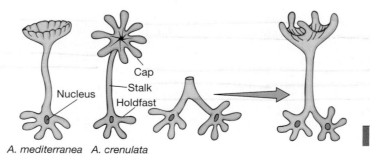

A. mediterranea   *A. crenulata*

Figure B

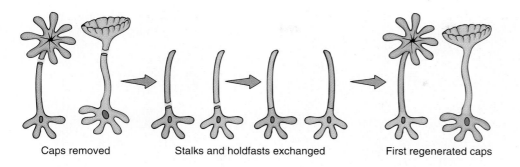

Caps removed   Stalks and holdfasts exchanged   First regenerated caps

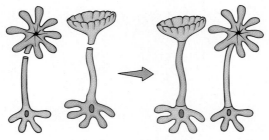

Caps removed again   Second regenerated caps

Results and Conclusion: A new cap regenerated that was characteristic of the species of the nucleus. When two kinds of nuclei were inserted, the regenerated cap was intermediate in shape between those of the species that donated the nuclei.

As a result of these and other experiments, biologists began to develop some basic ideas about the control of cell activities. The holdfast controls the cell because the nucleus is located there. Further, the nucleus is the apparent source of some "messenger substance" that temporarily exerts control but is limited in quantity and cannot be produced without the nucleus (▮ **Fig. F**). This information helped provide a starting point for research on the role of nucleic acids in the control of all cells.

▮ **Figure C**

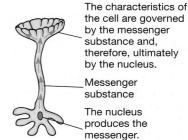

The characteristics of the cell are governed by the messenger substance and, therefore, ultimately by the nucleus.

Messenger substance

The nucleus produces the messenger.

▮ **Figure F**

sized that the grafted stalks initially contain enough of the substance from their former holdfasts to regenerate a cap of the former shape. But this still left the question of how the holdfast exerts its apparent control. An obvious suspect was the nucleus.

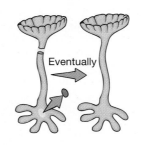

Eventually

▮ **Figure D**

Experiment 3

Question: Does the nucleus in the holdfast control the shape of the cap in *Acetabularia*?

Hypothesis: The nucleus in the holdfast of *Acetabularia* controls the shape of the cap.

Experiment: The investigators removed the nucleus and cut off the cap of *Acetabularia* (▮ **Fig. D**). A new cap typical of the species regenerated. *Acetabularia*, however, can usually regenerate only once without a nucleus. The researchers inserted a nucleus of another species and cut the cap off once again (▮ **Fig. E**).

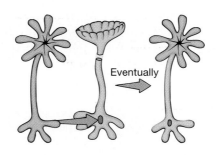

Eventually

▮ **Figure E**

Cell biologists extended these early findings as they developed our modern view of information flow and control in the cell. We now know that the nucleus of eukaryotes controls the cell's activities because it contains DNA, the ultimate source of biological information. DNA precisely replicates itself and passes its information to successive generations. The information in DNA specifies the sequence of amino acids in all the proteins of the cell. To carry out its function, DNA uses a type of ribonucleic acid (RNA) called messenger RNA (mRNA) to transmit information from the nucleus into the cytoplasm, where proteins are manufactured.

The nucleus contains DNA and is the control center of the cell.

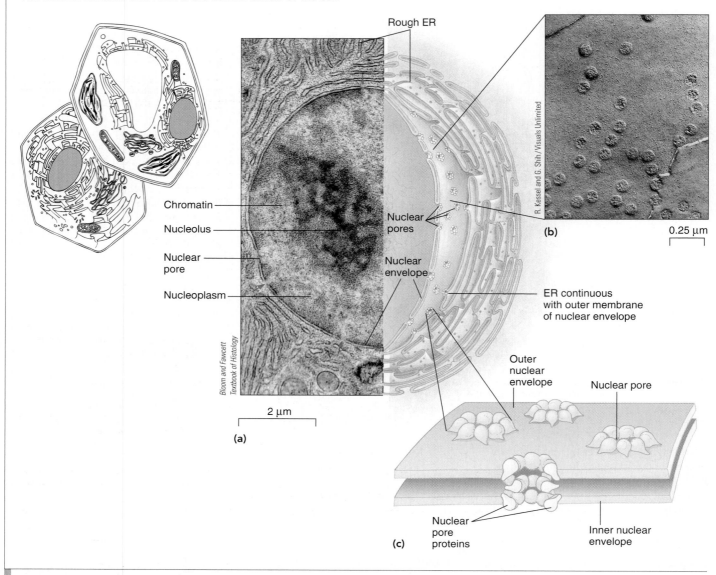

Rough ER

R. Kessel and G. Shih/Visuals Unlimited

(b)   0.25 μm

Chromatin

Nucleolus

Nuclear
pore

Nucleoplasm

Nuclear
pores

Nuclear
envelope

ER continuous
with outer membrane
of nuclear envelope

Bloom and Fawcett
Textbook of Histology

2 μm

(a)

Outer
nuclear
envelope

Nuclear pore

Nuclear
pore
proteins

Inner nuclear
envelope

(c)

Figure 4-11   *Animated*   The cell nucleus

(a) The TEM and interpretive drawing show that the nuclear envelope, composed of two concentric membranes, is perforated by nuclear pores (*red arrows*). The outer membrane of the nuclear envelope is continuous with the membrane of the ER (endoplasmic reticulum). The nucleolus is not bounded by a membrane. (b) TEM of nuclear pores. A technique known as freeze-fracture was used to split the membrane. (c) The nuclear pores, which are made up of proteins, form channels between the nucleoplasm and cytoplasm.

ORGANELLES IN THE CYTOPLASM

Learning Objectives

8 Distinguish between smooth and rough endoplasmic reticulum in terms of both structure and function.

9 Trace the path of proteins synthesized in the rough endoplasmic reticulum as they are processed, modified, and sorted by the Golgi complex and then transported to specific destinations.

10 Describe the functions of lysosomes, vacuoles, and peroxisomes.

11 Compare the functions of mitochondria and chloroplasts, and discuss ATP synthesis by each of these organelles.

Cell biologists have identified many types of organelles in the cytoplasm of eukaryotic cells. Among them are the ribosomes, endoplasmic reticulum, Golgi complex, lysosomes, peroxisomes, vacuoles, mitochondria, and chloroplasts. ❚ Table 4-1 summarizes eukaryotic cell structures and functions.

TABLE 4-1

Eukaryotic Cell Structures and Their Functions

Structure	Description	Function
Cell Nucleus		
Nucleus	Large structure surrounded by double membrane; contains nucleolus and chromosomes	Information in DNA is transcribed in RNA synthesis; specifies cell proteins
Nucleolus	Granular body within nucleus; consists of RNA and protein	Site of ribosomal RNA synthesis; ribosome subunit assembly
Chromosomes	Composed of a complex of DNA and protein known as chromatin; condense during cell division, becoming visible as rodlike structures	Contain genes (units of hereditary information) that govern structure and activity of cell
Cytoplasmic Organelles		
Plasma membrane	Membrane boundary of cell	Encloses cell contents; regulates movement of materials in and out of cell; helps maintain cell shape; communicates with other cells (also present in prokaryotes)
Ribosomes	Granules composed of RNA and protein; some attached to ER, some free in cytosol	Synthesize polypeptides in both prokaryotes and eukaryotes
Endoplasmic reticulum (ER)	Network of internal membranes extending through cytoplasm	Synthesizes lipids and modifies many proteins; origin of intracellular transport vesicles that carry proteins
Smooth	Lacks ribosomes on outer surface	Lipid synthesis; drug detoxification; calcium ion storage
Rough	Ribosomes stud outer surface	Manufactures proteins
Golgi complex	Stacks of flattened membrane sacs	Modifies proteins; packages secreted proteins; sorts other proteins to vacuoles and other organelles
Lysosomes	Membranous sacs (in animals)	Contain enzymes that break down ingested materials, secretions, wastes
Vacuoles	Membranous sacs (mostly in plants, fungi, algae)	Store materials, wastes, water; maintain hydrostatic pressure
Peroxisomes	Membranous sacs containing a variety of enzymes	Site of many diverse metabolic reactions; e.g., break down fatty acids
Mitochondria	Sacs consisting of two membranes; inner membrane is folded to form cristae and encloses matrix	Site of most reactions of cellular respiration; transformation of energy originating from glucose or lipids into ATP energy
Plastids (e.g., chloroplasts)	Double-membrane structure enclosing internal thylakoid membranes; chloroplasts contain chlorophyll in thylakoid membranes	Chloroplasts are site of photosynthesis; chlorophyll captures light energy; ATP and other energy-rich compounds are formed and then used to convert CO_2 to carbohydrate
Cytoskeleton		
Microtubules	Hollow tubes made of subunits of tubulin protein	Provide structural support; have role in cell and organelle movement and cell division; components of cilia, flagella, centrioles, basal bodies
Microfilaments	Solid, rodlike structures consisting of actin protein	Provide structural support; play role in cell and organelle movement and cell division
Intermediate filaments	Tough fibers made of protein	Help strengthen cytoskeleton; stabilize cell shape
Centrioles	Pair of hollow cylinders located near nucleus; each centriole consists of nine microtubule triplets (9×3 structure)	Mitotic spindle forms between centrioles during animal cell division; may anchor and organize microtubule formation in animal cells; absent in most plant cells
Cilia	Relatively short projections extending from surface of cell; covered by plasma membrane; made of two central and nine pairs of peripheral microtubules ($9 + 2$ structure)	Movement of some unicellular organisms; used to move materials on surface of some tissues
Flagella	Long projections made of two central and nine pairs of peripheral microtubules ($9 + 2$ structure); extend from surface of cell; covered by plasma membrane	Cell locomotion by sperm cells and some unicellular eukaryotes

Ribosomes manufacture proteins

Ribosomes are tiny particles found free in the cytoplasm or attached to certain membranes. They consist of RNA and proteins and are synthesized by the nucleolus. Each ribosome has two main components: a large subunit and a small subunit. Ribosomes contain the enzyme necessary to form peptide bonds (see Chapter 3), and when their two subunits join, they function as manufacturing plants that assemble polypeptides. Cells that actively produce a lot of proteins may have millions of ribosomes, and the cell can change the number of ribosomes present to meet its metabolic needs. We will discuss much more about ribosomes in Chapter 13.

The endoplasmic reticulum is a network of internal membranes

One of the most prominent features in the electron micrographs in Figures 4-7 and 4-8 is a maze of parallel internal membranes that encircle the nucleus and extend into many regions of the cytoplasm. This complex of membranes, the **endoplasmic reticulum (ER),** forms a network that makes up a significant part of the total volume of the cytoplasm in many cells. A higher-magnification TEM of the ER is shown in ▌Figure 4-12. Remember that a TEM represents only a thin cross section of the cell, so there is a tendency to interpret the ER as a series of tubes. In fact, many ER membranes consist of a series of tightly packed and flattened, saclike structures that form interconnected compartments within the cytoplasm.

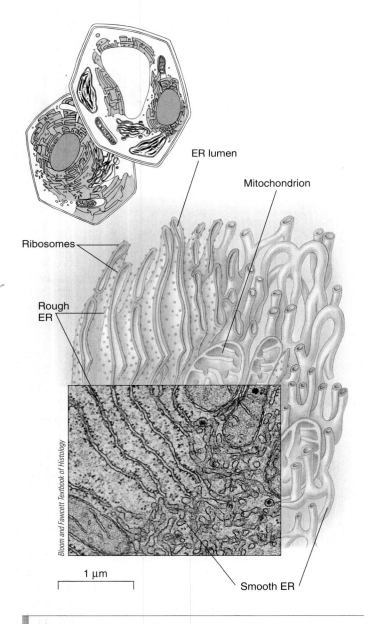

ER lumen

Mitochondrion

Ribosomes

Rough ER

Bloom and Fawcett Textbook of Histology

1 μm

Smooth ER

The internal space the membranes enclose is called the **ER lumen**. In most cells the ER lumen forms a single internal compartment that is continuous with the compartment formed between the outer and inner membranes of the nuclear envelope (see Fig. 4-11). The membranes of other organelles are not directly connected to the ER; they form distinct and separate compartments within the cytoplasm.

The ER membranes and lumen contain enzymes that catalyze many types of chemical reactions. In some cases the membranes serve as a framework for enzymes that carry out sequential biochemical reactions. The two surfaces of the membrane contain different sets of enzymes and represent regions of the cell with different synthetic capabilities, just as different regions of a factory make different parts of a particular product. Still other enzymes are located within the ER lumen.

Two distinct regions of the ER can be distinguished in TEMs: rough ER and smooth ER. Although these regions have different functions, their membranes are connected and their internal spaces are continuous.

Smooth ER synthesizes lipids

Smooth ER has a tubular appearance, and its outer membrane surfaces appear smooth. Enzymes in the membranes of the smooth ER catalyze the synthesis of many lipids and carbohydrates. The smooth ER is the primary site for the synthesis of phospholipids and cholesterol needed to make cell membranes. Smooth ER synthesizes steroid hormones, including reproductive hormones, from cholesterol. In liver cells, smooth ER is important in enzymatically breaking down stored glycogen. (The liver helps regulate the concentration of glucose in the blood.) The smooth ER also stores calcium ions.

Whereas the smooth ER may be a minor membrane component in some cells, extensive amounts of smooth ER are present in others. For example, extensive smooth ER is present in human liver cells, where it synthesizes and processes cholesterol and other lipids and serves as a major detoxification site. Enzymes located along the smooth ER of liver cells break down toxic chemicals such as carcinogens (cancer-causing agents) and many drugs, including alcohol, amphetamines, and barbiturates. The cell then converts these compounds to water-soluble products that it excretes. Interestingly, alcohol and many other drugs stimulate liver cells to produce more smooth ER, increasing the rate that these cells can detoxify the drugs. Alcohol abuse causes liver inflammation that can lead to cirrhosis and eventual liver failure.

The rough ER is important in protein synthesis

The outer surface of the **rough ER** is studded with ribosomes that appear as dark granules. Notice in Figure 4-12 that the lumen side of the rough ER appears bare, whereas the outer surface (the cytosolic side) looks rough. The ribosomes attached to the rough ER are known as *bound ribosomes; free ribosomes* are suspended in the cytosol.

The rough ER plays a central role in the synthesis and assembly of proteins. Many proteins that are exported from the cell (such as digestive enzymes), and those destined for other organelles, are synthesized on ribosomes bound to the ER membrane.

Figure 4-12 Endoplasmic reticulum (ER)

The TEM shows both rough and smooth ER in a liver cell.

The ribosome forms a tight seal with the ER membrane. A tunnel within the ribosome connects to an ER pore. Polypeptides are transported through the tunnel and the pore in the ER membrane into the ER lumen.

In the ER lumen, proteins are assembled and may be modified by enzymes that add carbohydrates or lipids to them. Other enzymes, called **molecular chaperones,** in the ER lumen catalyze the efficient folding of proteins into proper conformations. Proteins that are not processed correctly, for example, misfolded, are transported to the cytosol. There, they are degraded by **proteasomes,** protein complexes in the cytosol that direct the destruction of defective proteins. Properly processed proteins are transferred to other compartments within the cell by small **transport vesicles,** which bud off the ER membrane and then fuse with the membrane of some target organelle.

The Golgi complex processes, sorts, and modifies proteins

The **Golgi complex** (also known as the *Golgi body* or *Golgi apparatus*) was first described in 1898 by the Italian microscopist Camillo Golgi, who found a way to specifically stain this organelle. However, many investigators thought the Golgi complex was an artifact, and its legitimacy as a cell organelle was not confirmed until cells were studied with the electron microscope in the 1950s.

In many cells, the Golgi complex consists of stacks of flattened membranous sacs called **cisternae** (sing., *cisterna*). In certain regions, cisternae may be distended because they are filled with cell products (▌Fig. 4-13). Each of the flattened sacs has an internal space, or lumen. The Golgi complex contains a number of separate compartments, as well as some that are interconnected.

Each Golgi stack has three areas, referred to as the *cis face*, the *trans face*, and a *medial region* in between. Typically, the *cis* face (the entry surface) is located nearest the nucleus and receives materials from transport vesicles bringing molecules from the ER. The *trans* face (the exit surface) is closest to the plasma membrane. It packages molecules in vesicles and transports them out of the Golgi.

In a cross-sectional view like that in the TEM in Figure 4-13, many ends of the sheetlike layers of Golgi membranes are distended, an arrangement characteristic of well-developed Golgi complexes in many cells. In some animal cells, the Golgi complex lies at one side of the nucleus; other animal cells and plant cells have many Golgi complexes dispersed throughout the cell. Cells that secrete large amounts of glycoproteins have many Golgi stacks. (Recall from Chapter 3 that a glycoprotein is a protein with a covalently attached carbohydrate.) Golgi complexes of plant cells produce extracellular polysaccharides that are used as components of the cell wall. In animal cells, the Golgi complex manufactures lysosomes.

The Golgi complex processes, sorts, and modifies proteins. Researchers have studied the function of the ER, Golgi complex, and other organelles by radioactively labeling newly manufactured molecules and observing their movement through the cell. The general pathway is from ribosomes to the lumen of the

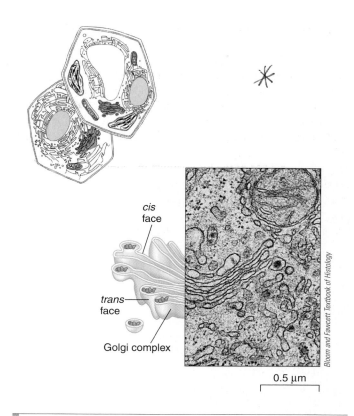

cis face

trans face

Golgi complex

0.5 μm

Bloom and Fawcett Textbook of Histology

Figure 4-13 TEM and an interpretive drawing of the Golgi complex

rough ER to the Golgi complex and then to some final destination (▌Fig. 4-14).

Proteins that are accurately assembled are transported from the rough ER to the *cis* face of the Golgi complex in small transport vesicles formed from the ER membrane. Transport vesicles fuse with one another to form clusters that move along microtubules (part of the cytoskeleton) to the Golgi complex. What happens to glycoprotein molecules released into the Golgi complex? One hypothesis holds that the glycoproteins are enclosed in new vesicles that shuttle them from one compartment to another within the Golgi complex. A competing hypothesis postulates that the cisternae themselves may move from *cis* to *trans* positions. Both hypotheses may prove true; glycoproteins may be transported by both methods.

Regardless of how proteins are moved through the Golgi complex, while there they are modified in different ways, resulting in the formation of complex biological molecules. For example, the carbohydrate part of a glycoprotein (first added to proteins in the rough ER) may be modified. In some cases the carbohydrate component may be a "sorting signal," a cellular zip code that tags the protein, routing it to a specific organelle.

Glycoproteins are packaged in transport vesicles in the *trans* face. These vesicles pinch off from the Golgi membrane and transport their contents to a specific destination. Vesicles transporting products for export from the cell fuse with the plasma membrane. The vesicle membrane becomes part of the plasma membrane, and the glycoproteins are secreted from the cell. Other vesicles may store glycoproteins for secretion at a later

Key Point

After proteins are synthesized, they are transported through a series of compartments where they are successively modified.

1. Polypeptides synthesized on ribosomes are inserted into ER lumen.

2. Sugars are added, forming glycoproteins.

3. Transport vesicles deliver glycoproteins to *cis* face of Golgi.

4. Glycoproteins modified further in Golgi.

5. Glycoproteins move to *trans* face where they are packaged in transport vesicles.

6. Glycoproteins transported to plasma membrane (or other organelle).

7. Contents of transport vesicle released from cell.

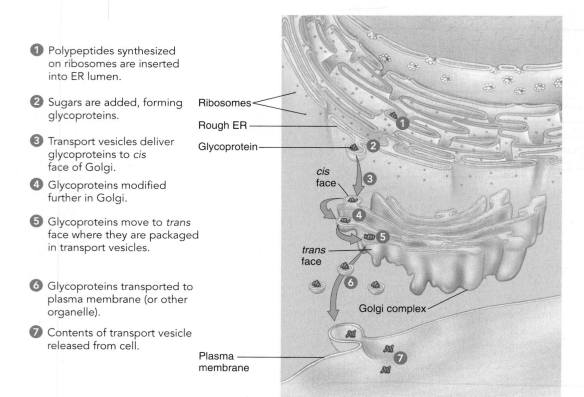

Figure 4-14 *Animated* Protein transport within the cell

Glycoproteins are transported from ribosomes into the ER. They are then transported to the Golgi complex, where they are modified. This diagram shows the passage of glycoproteins through compartments of the endomembrane system of a mucus-secreting goblet cell that lines the intestine. Mucus consists of a complex mixture of covalently linked proteins and carbohydrates.

time, and still others are routed to various organelles of the endomembrane system. In summary, here is a typical sequence followed by a glycoprotein destined for secretion from the cell:

Polypeptides synthesized on ribosomes ⟶ protein assembled and carbohydrate component added in lumen of ER ⟶ transport vesicles move glycoprotein to Golgi (*cis* face) ⟶ glycoprotein further modified in Golgi ⟶ in *trans* face, glycoproteins packaged in transport vesicles ⟶ glycoproteins transported to plasma membrane ⟶ contents released from cell

Lysosomes are compartments for digestion

Lysosomes are small sacs of digestive enzymes dispersed in the cytoplasm of most animal cells (❚ Fig. 4-15). Researchers have identified about 40 different digestive enzymes in lysosomes.

Because lysosomal enzymes are active under rather acidic conditions, the lysosome maintains a pH of about 5 in its interior. Lysosomal enzymes break down complex molecules in bacteria and debris ingested by scavenger cells. The powerful enzymes and low pH that the lysosome maintains provide an excellent example of the importance of separating functions within the cell into different compartments. Under most normal conditions, the lysosome membrane confines its enzymes and their actions. However, some forms of tissue damage are related to "leaky" lysosomes.

Primary lysosomes are formed by budding from the Golgi complex. Their hydrolytic enzymes are synthesized in the rough ER. As these enzymes pass through the lumen of the ER, sugars attach to each molecule, identifying it as bound for a lysosome. This signal permits the Golgi complex to sort the enzyme to the lysosomes rather than to export it from the cell.

Bacteria (or debris) engulfed by scavenger cells are enclosed in a vesicle formed from part of the plasma membrane. One or

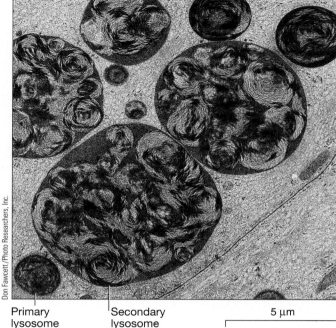

Primary   Secondary   5 µm
lysosome   lysosome

Don Fawcett/Photo Researchers, Inc.

Figure 4-15 Lysosomes

The dark vesicles in this TEM are lysosomes, compartments that separate powerful digestive enzymes from the rest of the cell. Primary lysosomes bud off from the Golgi complex. After a lysosome encounters and takes in material to be digested, it is known as a secondary lysosome. The large vesicles shown here are secondary lysosomes containing various materials being digested.

more primary lysosomes fuse with the vesicle containing the ingested material, forming a larger vesicle called a *secondary lysosome*. Powerful enzymes in the secondary lysosome come in contact with the ingested molecules and degrade them into their components. Under some conditions, lysosomes break down organelles and allow their components to be recycled or used as an energy source.

In certain genetic diseases of humans, known as *lysosomal storage diseases*, one of the digestive enzymes normally present in lysosomes is absent. Its substrate (a substance the enzyme would normally break down) accumulates in the lysosomes, ultimately interfering with cell activities. An example is Tay–Sachs disease (see Chapter 16), in which a normal lipid cannot be broken down in brain cells. The lipid accumulates in the cells and causes mental retardation, blindness, and death before age 5.

Vacuoles are large, fluid-filled sacs with a variety of functions

Although lysosomes have been identified in almost all kinds of animal cells, their occurrence in plant and fungal cells is open to debate. Many functions carried out in animal cells by lysosomes are performed in plant and fungal cells by a large, single, membrane-enclosed sac called a **vacuole**. The membrane of the vacuole, part of the endomembrane system, is called the **tono-**

plast. The term *vacuole*, which means "empty," refers to the fact that these organelles have no internal structure. Although some biologists use the terms *vacuole* and *vesicle* interchangeably, vacuoles are usually larger structures, sometimes produced by the merging of many vesicles.

Vacuoles play a significant role in plant growth and development. Immature plant cells are generally small and contain numerous small vacuoles. As water accumulates in these vacuoles, they tend to coalesce, forming a large central vacuole. A plant cell increases in size mainly by adding water to this central vacuole. As much as 90% of the volume of a plant cell may be occupied by a large central vacuole containing water, as well as stored food, salts, pigments, and metabolic wastes (see Figs. 4-7 and 4-9). The vacuole may serve as a storage compartment for inorganic compounds. In seeds, vacuoles store molecules such as proteins.

Because the vacuole contains a high concentration of solutes (dissolved materials), it takes in water and pushes outward on the cell wall. This hydrostatic pressure, called *turgor pressure*, provides much of the mechanical strength of plant cells. The vacuole is important in maintaining homeostasis. For example, it helps maintain appropriate pH by taking in excess hydrogen ions.

Plants lack organ systems for disposing of toxic metabolic waste products. Plant vacuoles are like lysosomes in that they contain hydrolytic enzymes and break down wastes, as well as unneeded organelles and other cell components. Wastes may be recycled in the vacuole, or they may aggregate and form small crystals inside the vacuole. Compounds that are noxious to herbivores (animals that eat plants) may also be stored in some plant vacuoles as a means of defense.

Vacuoles are also present in many types of animal cells and in unicellular protists such as protozoa. Most protozoa have **food vacuoles,** which fuse with lysosomes that digest the food (Fig. 4-16). Some protozoa also have **contractile vacuoles,** which remove excess water from the cell (see Chapter 25).

Food vacuoles containing diatoms

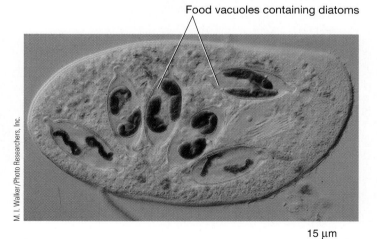

M. I. Walker/Photo Researchers, Inc.

15 µm

Figure 4-16 LM of food vacuoles

This protist, *Chilodonella*, has ingested many small, photosynthetic protists called diatoms (*dark areas*) that have been enclosed in food vacuoles. From the number of diatoms scattered about its cell, one might conclude that *Chilodonella* has a rather voracious appetite.

Peroxisomes metabolize small organic compounds

Peroxisomes are membrane-enclosed organelles containing enzymes that catalyze an assortment of metabolic reactions in which hydrogen is transferred from various compounds to oxygen (∥ Fig. 4-17). Peroxisomes get their name from the fact that during these oxidation reactions, they produce hydrogen peroxide (H_2O_2). Hydrogen peroxide detoxifies certain compounds; but if it were to escape from the peroxisome, it would damage other membranes in the cell. Peroxisomes contain the enzyme *catalase*, which rapidly splits excess hydrogen peroxide to water and oxygen, rendering it harmless.

Peroxisomes are found in large numbers in cells that synthesize, store, or degrade lipids. One of their main functions is to break down fatty acid molecules. Peroxisomes also synthesize certain phospholipids that are components of the insulating covering of nerve cells. In fact, certain neurological disorders occur when peroxisomes do not perform this function. Mutations that cause abnormal peroxisome-membrane synthesis are linked to some forms of mental retardation.

When yeast cells are grown in an alcohol-rich medium, they manufacture large peroxisomes containing an enzyme that degrades the alcohol. Peroxisomes in human liver and kidney cells detoxify certain toxic compounds, including ethanol, the alcohol in alcoholic beverages.

In plant seeds, specialized peroxisomes, called *glyoxysomes,* contain enzymes that convert stored fats to sugars. The sugars are used by the young plant as an energy source and as a component for synthesizing other compounds. Animal cells lack glyoxysomes and cannot convert fatty acids into sugars.

Mitochondria and chloroplasts are energy-converting organelles

The energy a cell obtains from its environment is usually in the form of chemical energy in food molecules (such as glucose) or in the form of light energy. These types of energy must be converted to forms that cells can use more conveniently. Some energy conversions occur in the cytosol, but other types take place in mitochondria and chloroplasts, organelles specialized to facilitate the conversion of energy from one form to another.

Chemical energy is most commonly stored in ATP. Recall from Chapter 3 that the chemical energy of ATP can be used to drive a variety of chemical reactions in the cell. ∥ Figure 4-18 summarizes the main activities that take place in mitochondria, found in almost all eukaryotic cells (including algae and plants), and in chloroplasts, found only in algae and certain plant cells.

Mitochondria and chloroplasts grow and reproduce themselves. They contain small amounts of DNA that code for a small number of the proteins found in these organelles. These proteins are synthesized by mitochondrial or chloroplast ribosomes, which are similar to the ribosomes of prokaryotes. The existence of a separate set of ribosomes and DNA molecules in mitochondria and chloroplasts and their similarity in size to many bacteria provide support for **serial endosymbiosis** (discussed in Chapters 21 and 25; see Figs. 21-7 and 25-2). According to this hypothesis, mitochondria and chloroplasts evolved from prokaryotes that took up residence inside larger cells and eventually lost the ability to function as autonomous organisms.

Mitochondria make ATP through cellular respiration

Virtually all eukaryotic cells (plant, animal, fungal, and protist) contain complex organelles called **mitochondria** (sing., *mitochondrion*). These organelles are the site of **aerobic respiration,** an oxygen-requiring process that includes most of the reactions that convert the chemical energy present in certain foods to ATP (discussed in Chapter 8). During aerobic respiration, carbon and oxygen atoms are removed from food molecules, such as glucose, and converted to carbon dioxide and water.

Mitochondria are most numerous in cells that are very active and therefore have high energy requirements. More than 1000 mitochondria have been counted in a single liver cell! These organelles vary in size, ranging from 2 to 8 μm in length, and change size and shape rapidly. Mitochondria usually give rise to other mitochondria by growth and subsequent division.

Each mitochondrion is enclosed by a double membrane, which forms two *different* compartments within the organelle: the intermembrane space and the matrix (∥ Fig. 4-19; we will provide more detailed descriptions of mitochondrial structure in Chap-

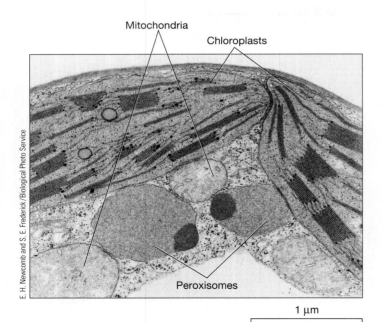

Mitochondria

Chloroplasts

E. H. Newcomb and S. E. Frederick/Biological Photo Service

Peroxisomes

1 μm

Figure 4-17 Peroxisomes

In this TEM of a tobacco (*Nicotiana tabacum*) leaf cell, peroxisomes are in close association with chloroplasts and mitochondria. These organelles may cooperate in carrying out some metabolic processes.

Mitochondria and chloroplasts convert energy into forms that can be used by cells.

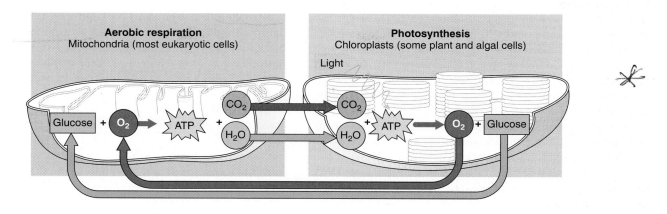

Figure 4-18 Cellular respiration and photosynthesis

Cellular respiration takes place in the mitochondria of virtually all eukaryotic cells. In this process, some of the chemical energy in glucose is transferred to ATP. Photosynthesis, which is carried out in chloroplasts in some plant and algal cells, converts light energy to ATP and to other forms of chemical energy. This energy is used to synthesize glucose from carbon dioxide and water.

ter 8). The **intermembrane space** is the compartment formed between the outer and inner mitochondrial membranes. The **matrix,** the compartment enclosed by the inner mitochondrial membrane, contains enzymes that break down food molecules and convert their energy to other forms of chemical energy.

The **outer mitochondrial membrane** is smooth and allows many small molecules to pass through it. By contrast, the **inner mitochondrial membrane** has numerous folds and strictly regulates the types of molecules that can move across it. The folds, called **cristae** (sing., *crista*), extend into the matrix. Cristae greatly increase the surface area of the inner mitochondrial membrane, providing a surface for the chemical reactions that transform the chemical energy in food molecules into the energy of ATP. The membrane contains the enzymes and other proteins needed for these reactions.

Mitochondria affect health and aging by leaking electrons. These electrons form **free radicals,** which are toxic, highly reactive compounds with unpaired electrons. The electrons bond with other compounds in the cell, interfering with normal function.

Mitochondria play an important role in programmed cell death, or **apoptosis.** Unlike **necrosis,** which is uncontrolled cell death that causes inflammation and damages other cells, apoptosis is a normal part of development and maintenance. For example, during the metamorphosis of a tadpole to a frog, the cells of the tadpole tail must die. The hand of a human embryo is webbed until apoptosis destroys the tissue between the fingers. Cell death also occurs in the adult. For example, cells in the upper layer of human skin and in the intestinal wall are continuously destroyed and replaced by new cells.

Mitochondria initiate cell death in several different ways. For example, they can interfere with energy metabolism or activate enzymes that mediate cell destruction. When a mitochondrion is

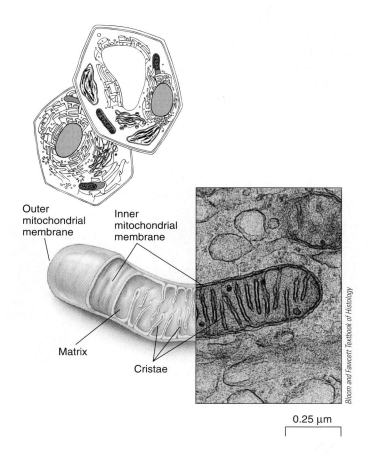

Outer mitochondrial membrane

Inner mitochondrial membrane

Matrix

Cristae

Bloom and Fawcett Textbook of Histology

0.25 µm

Figure 4-19 Mitochondria

Aerobic respiration takes place within mitochondria. Cristae are evident in the TEM as well as in the drawing. The drawing shows the relationship between the inner and outer mitochondrial membranes.

injured, large pores open in its membrane, and cytochrome *c*, a protein important in energy production, is released into the cytoplasm. Cytochrome *c* triggers apoptosis by activating enzymes known as **caspases,** which cut up vital compounds in the cell. Inappropriate inhibition of apoptosis may contribute to a variety of diseases, including cancer. Too much apoptosis may deplete needed cells and lead to acquired immunodeficiency syndrome (AIDS). Pharmaceutical companies are developing drugs that block apoptosis. However, cell dynamics are extremely complex, and blocking apoptosis could lead to a worse fate, including necrosis.

Each mitochondrion in a mammalian cell has 5 to 10 identical, circular molecules of DNA, accounting for up to 1% of the total DNA in the cell. Mitochondrial DNA mutates far more frequently than nuclear DNA. Mutations in mitochondrial DNA have been associated with certain genetic diseases, including a form of young adult blindness, and certain types of progressive muscle degeneration. Mutations that promote apoptosis may be an important mechanism in mammalian aging.

Chloroplasts convert light energy to chemical energy through photosynthesis

Certain plant and algal cells carry out **photosynthesis,** a set of reactions during which light energy is transformed into the chemical energy of glucose and other carbohydrates. **Chloroplasts** are organelles that contain **chlorophyll,** a green pigment that traps

light energy for photosynthesis. Chloroplasts also contain a variety of light-absorbing yellow and orange pigments known as **carotenoids** (see Chapter 3). A unicellular alga may have only a single large chloroplast, whereas a leaf cell may have 20 to 100. Chloroplasts tend to be somewhat larger than mitochondria, with lengths usually ranging from about 5 to 10 μm or longer.

Chloroplasts are typically disc-shaped structures and, like mitochondria, have a system of folded membranes (Fig. 4-20; we will provide a more detailed description of chloroplast structure in Chapter 9). Two membranes enclose the chloroplast and separate it from the cytosol. The inner membrane encloses a fluid-filled space called the **stroma,** which contains enzymes. These enzymes produce carbohydrates from carbon dioxide and water, using energy trapped from sunlight. A system of internal membranes, which consist of an interconnected set of flat, disclike sacs called **thylakoids,** is suspended in the stroma. The thylakoids are arranged in stacks called **grana** (sing., *granum*).

The thylakoid membranes enclose the innermost compartments within the chloroplast, the **thylakoid lumens.** Chlorophyll is present in the thylakoid membranes, which like the inner mitochondrial membranes, are involved in the formation of ATP. Energy absorbed from sunlight by the chlorophyll molecules excites electrons; the energy in these excited electrons is then used to produce ATP and other molecules that transfer chemical energy.

Chloroplasts belong to a group of organelles, known as **plastids,** that produce and store food materials in cells of plants and algae. All plastids develop from **proplastids,** precursor organelles found in less specialized plant cells, particularly in growing, un-

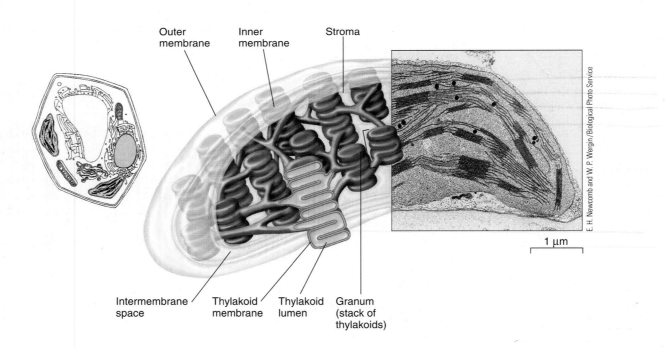

Figure 4-20 A chloroplast, the organelle of photosynthesis

The TEM shows part of a chloroplast from a corn leaf cell. Chlorophyll and other photosynthetic pigments are in the thylakoid membranes. One granum is cut open to show the thylakoid lumen. The inner chlo-

roplast membrane may or may not be continuous with the thylakoid membrane (*as shown*).

developed tissues. Depending on the specific functions a cell will eventually have, its proplastids can develop into a variety of specialized mature plastids. These are extremely versatile organelles; in fact, under certain conditions even mature plastids can convert from one form to another.

Chloroplasts are produced when proplastids are stimulated by exposure to light. **Chromoplasts** contain pigments that give certain flowers and fruits their characteristic colors; these colors attract animals that serve as pollinators or as seed dispersers. **Leukoplasts** are unpigmented plastids; they include **amyloplasts** (see Fig. 3-9), which store starch in the cells of many seeds, roots, and tubers (such as white potatoes).

Review

- How do the structure and function of rough ER differ from those of smooth ER?
- What are the functions of the Golgi complex?
- What sequence of events must take place for a protein to be manufactured and then secreted from the cell?
- How are chloroplasts like mitochondria? How are they different?
- Draw and label a chloroplast and a mitochondrion.

THE CYTOSKELETON

Learning Objectives

12 Describe the structure and functions of the cytoskeleton.
13 Compare cilia and flagella, and describe their functions.

Scientists watching cells growing in the laboratory see that they frequently change shape and that many types of cells move about. The **cytoskeleton,** a dense network of protein fibers, gives cells mechanical strength, shape, and their ability to move (Fig. 4-21). The cytoskeleton also functions in cell division and in the transport of materials within the cell.

The cytoskeleton is highly dynamic and constantly changing. Its framework is made of three types of protein filaments: microtubules, microfilaments, and intermediate filaments. Both microfilaments and microtubules are formed from beadlike, globular protein subunits, which can be rapidly assembled and disassembled. Intermediate filaments are made from fibrous protein subunits and are more stable than microtubules and microfilaments.

Microtubules are hollow cylinders

Microtubules, the thickest filaments of the cytoskeleton, are rigid, hollow rods about 25 nm in outside diameter and up to several micrometers in length. In addition to playing a structural role in the cytoskeleton, these extremely adaptable structures are involved in the movement of chromosomes during cell division. They serve as tracks for several other kinds of intracellular movement and are the major structural components of cilia and flagella—specialized structures used in some cell movements.

The cytoskeleton consists of networks of several types of fibers that support the cell and are important in cell movement.

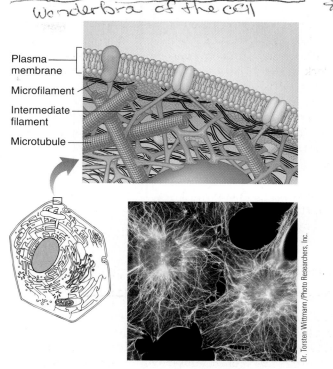

Plasma membrane
Microfilament
Intermediate filament
Microtubule

Dr. Torsten Wittmann /Photo Researchers, Inc.

Figure 4-21 *Animated* The cytoskeleton

Eukaryotic cells have a cytoskeleton consisting of networks of several types of fibers, including microtubules, microfilaments, and intermediate filaments. The cytoskeleton contributes to the shape of the cell, anchors organelles, and sometimes rapidly changes shape during cell locomotion. The cytoskeleton of two fibroblast cells is visible in the fluorescent light micrograph (microtubules, *yellow*; microfilaments, *blue*; nuclei, *green*).

Microtubules consist of two similar proteins: **α-tubulin** and **β-tubulin.** These proteins combine to form a dimer. (Recall from Chapter 3 that a dimer forms from the association of two simpler units, referred to as monomers.) A microtubule elongates by the addition of tubulin dimers (Fig. 4-22). Microtubules are disassembled by the removal of dimers, which are recycled to form new microtubules. Each microtubule has polarity, and its two ends are referred to as *plus* and *minus*. The plus end elongates more rapidly.

Other proteins are important in microtubule function. **Microtubule-associated proteins (MAPs)** are classified into two groups: structural MAPs and motor MAPs. *Structural MAPs* may help regulate microtubule assembly, and they cross-link microtubules to other cytoskeletal polymers. *Motor MAPs* use ATP energy to produce movement.

What are the mechanisms by which organelles and other materials move within the cell? Nerve cells typically have long extensions called axons that transmit signals to other nerve cells, muscle cells, or cells that produce hormones. Because of the

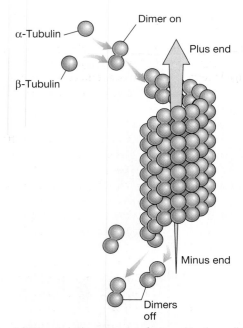

(a) Microtubules are manufactured in the cell by adding dimers of α-tubulin and β-tubulin to an end of the hollow cylinder. Notice that the cylinder has polarity. The end shown at the top of the figure is the fast-growing, or plus, end; the opposite end is the minus end. Each turn of the spiral requires 13 dimers.

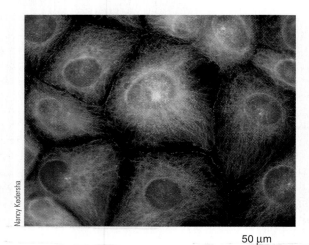

(b) Fluorescent LM showing microtubules in green. A microtubule-organizing center (*pink dot*) is visible beside or over most of the cell nuclei (*blue*).

Figure 4-22 Organization of microtubules

axon's length and accessibility and because other cells use similar transport mechanisms, researchers have used the axon as a model for studying the transport of organelles within the cell. They have found that many organelles, including mitochondria and transport and secretory vesicles, attach to microtubules. The

microtubules serve as tracks along which organelles move to different cell locations.

One motor protein, **kinesin,** moves organelles toward the plus end of a microtubule (❙ Fig. 4-23). **Dynein,** another motor protein, transports organelles in the opposite direction, toward the minus end. This dynein movement is referred to as *retrograde transport*. A protein complex called *dynactin* is also required for retrograde transport. Dynactin is an adapter protein that links dynein to the microtubule and the organelle. At times, for example, during peroxisome transport, kinesins and dyneins may work together.

Centrosomes and centrioles function in cell division

For microtubules to act as a structural framework or participate in cell movement, they must be anchored to other parts of the cell. In nondividing cells, the minus ends of microtubules appear to be anchored in regions called **microtubule-organizing centers (MTOCs).** In animal cells, the main MTOC is the cell center or **centrosome,** a structure that is important in cell division.

In many cells, including almost all animal cells, the centrosome contains two structures called **centrioles** (❙ Fig. 4-24). These structures are oriented within the centrosome at right angles to each other. They are known as *9 × 3 structures* because they consist of nine sets of three attached microtubules arranged to form a hollow cylinder. The centrioles are duplicated before cell division and may play a role in some types of microtubule assembly. Most plant cells and fungal cells have an MTOC but lack centrioles. This suggests that centrioles are not essential to most microtubule assembly processes and that alternative assembly mechanisms are present.

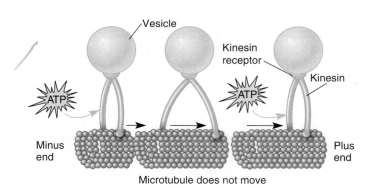

Figure 4-23 *Animated* A hypothetical model of a kinesin motor

A kinesin molecule attaches to a specific receptor on the vesicle. Energy from ATP allows the kinesin molecule to change its conformation and "walk" along the microtubule, carrying the vesicle along. (Size relationships are exaggerated for clarity.)

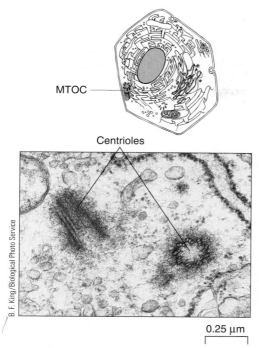

MTOC

Centrioles

B. F. King/Biological Photo Service

0.25 μm

(a) In the TEM, the centrioles are positioned at right angles to each other, near the nucleus of a nondividing animal cell.

(b) Note the 9 × 3 arrangement of microtubules. The centriole on the right has been cut transversely.

Figure 4-24 Centrioles

The ability of microtubules to assemble and disassemble rapidly is seen during cell division, when much of the cytoskeleton disassembles (see Chapter 10). At that time tubulin subunits organize into a structure called the **mitotic spindle,** which serves as a framework for the orderly distribution of chromosomes during cell division.

Cilia and flagella are composed of microtubules

Thin, movable structures, important in cell movement, project from surfaces of many cells. If a cell has one, or only a few, of these appendages and if they are long (typically about 200 μm) relative to the size of the cell, they are called **flagella.** If the cell

has many short (typically 2 to 10 μm long) appendages, they are called **cilia** (sing., *cilium*).

Cilia and flagella help unicellular and small multicellular organisms move through a watery environment. In animals and certain plants, flagella serve as the tails of sperm cells. In animals, cilia occur on the surfaces of cells that line internal ducts of the body (such as respiratory passageways). Cells use cilia to move liquids and particles across the cell surface.

Eukaryotic cilia and flagella are structurally alike (but different from bacterial flagella). Each consists of a slender, cylindrical stalk covered by an extension of the plasma membrane. The core of the stalk contains a group of microtubules arranged so there are nine attached pairs of microtubules around the circumference and two unpaired microtubules in the center (Fig. 4-25). This *9 + 2 arrangement* of microtubules is characteristic of virtually all eukaryotic cilia and flagella.

The microtubules in cilia and flagella move by sliding in pairs past each other. The sliding force is generated by dynein proteins, which are attached to the microtubules like small arms. These proteins use the energy from ATP to power the cilia or flagella. The dynein proteins (arms) on one pair of tubules change their shape and "walk" along the adjacent microtubule pair. Thus, the microtubules on one side of a cilium or a flagellum extend farther toward the tip than those on the other side. This sliding of microtubules translates into a bending motion (see Fig. 4-25e). Cilia typically move like oars, alternating power and recovery strokes and exerting a force that is parallel to the cell surface. A flagellum moves like a whip, exerting a force perpendicular to the cell surface.

Each cilium or flagellum is anchored in the cell by a **basal body,** which has nine sets of three attached microtubules in a cylindrical array (*9 × 3 structure*). The basal body appears to be the organizing structure for the cilium or flagellum when it first begins to form. However, experiments have shown that as growth proceeds, the tubulin subunits are added much faster to the tips of the microtubules than to the base. Basal bodies and centrioles may be functionally related as well as structurally similar. In fact, centrioles are typically found in the cells of eukaryotic organisms that produce flagellated or ciliated cells; these include animals, certain protists, a few fungi, and a few plants. Both basal bodies and centrioles replicate themselves.

Microfilaments consist of intertwined strings of actin

Microfilaments, also called *actin filaments,* are flexible, solid fibers about 7 nm in diameter. Each microfilament consists of two intertwined polymer chains of beadlike **actin** molecules (Fig. 4-26). Microfilaments are linked with one another and with other proteins by linker proteins. They form bundles of fibers that provide mechanical support for various cell structures.

In many cells, a network of microfilaments is visible just inside the plasma membrane, a region called the **cell cortex.** Micro-

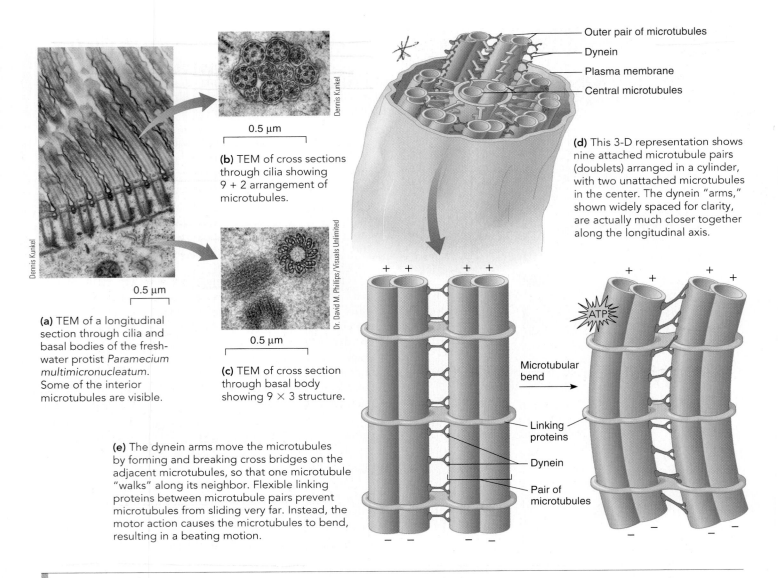

(a) TEM of a longitudinal section through cilia and basal bodies of the freshwater protist *Paramecium multimicronucleatum*. Some of the interior microtubules are visible.

(b) TEM of cross sections through cilia showing 9 + 2 arrangement of microtubules.

(c) TEM of cross section through basal body showing 9 × 3 structure.

(d) This 3-D representation shows nine attached microtubule pairs (doublets) arranged in a cylinder, with two unattached microtubules in the center. The dynein "arms," shown widely spaced for clarity, are actually much closer together along the longitudinal axis.

Outer pair of microtubules
Dynein
Plasma membrane
Central microtubules

Microtubular bend
ATP
Linking proteins
Dynein
Pair of microtubules

(e) The dynein arms move the microtubules by forming and breaking cross bridges on the adjacent microtubules, so that one microtubule "walks" along its neighbor. Flexible linking proteins between microtubule pairs prevent microtubules from sliding very far. Instead, the motor action causes the microtubules to bend, resulting in a beating motion.

Figure 4-25 *Animated* Structure of cilia

filaments give the cell cortex a gel-like consistency compared to the more fluid state of the cytosol deeper inside the cell. The microfilaments in the cell cortex help determine the shape of the cell and are important in its movement.

Microfilaments themselves cannot contract, but they can generate movement by rapidly assembling and disassembling. In muscle cells, microfilaments are associated with filaments composed of another protein, **myosin.** ATP bound to myosin provides energy for muscle contraction. When ATP is hydrolyzed to ADP, myosin binds to actin and causes the microfilament to slide. When thousands of filaments slide in this way, the muscle cell shortens. Thus, ATP, actin, and myosin generate the forces that contract muscles (discussed in Chapter 39). In nonmuscle cells, actin also associates with myosin, forming contractile structures involved in various cell movements. For example, in animal cell division, contraction of a ring of actin associated with myo-sin constricts the cell, forming two daughter cells (discussed in Chapter 10).

Some cells change their shape quickly in response to changes in the outside environment. Amoebas, human white blood cells, and cancer cells are among the many cell types that can crawl along a surface, a process that includes changes in shape. Such responses depend on external signals that affect microfilament, as well as microtubule, assembly. Actin filaments push the plasma membrane outward, forming cytoplasm-filled bulges called **pseudopodia** ("false feet"). The pseudopodia adhere to the surface. Contractions of microfilaments at the opposite end of the cell force the cytoplasm forward in the direction of locomotion. Microtubules, myosin, and other proteins also appear necessary for cell crawling.

As mentioned earlier in the chapter, some types of cells have microvilli, projections of the plasma membrane that increase

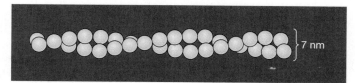

(a) A microfilament consists of two intertwined strings of beadlike actin molecules.

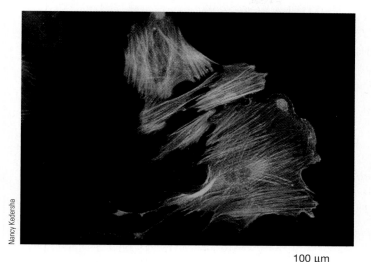

Nancy Kedersha

100 μm

(b) Many bundles of microfilaments (*green*) are evident in this fluorescent LM of fibroblasts, cells found in connective tissue.

Figure 4-26 Microfilaments

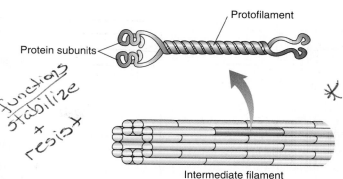

Protofilament

Protein subunits

functions: stabilize + resist

Intermediate filament

(a) Intermediate filaments are flexible rods about 10 nm in diameter. Each intermediate filament consists of components, called protofilaments, composed of coiled protein subunits.

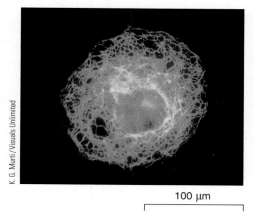

K. G. Murti/Visuals Unlimited

100 μm

(b) Intermediate filaments are stained green in this human cell isolated from a tissue culture.

Figure 4-27 Intermediate filaments

the surface area of the cell for transporting materials across the plasma membrane. Composed of bundles of microfilaments, microvilli extend and retract as the microfilaments assemble and disassemble.

Intermediate filaments help stabilize cell shape

Intermediate filaments are tough, flexible fibers about 10 nm in diameter (▌ Fig. 4-27). They provide mechanical strength and help stabilize cell shape. These filaments are abundant in regions of a cell that may be subject to mechanical stress applied from outside the cell. Intermediate filaments prevent the cell from stretching excessively in response to outside forces. Certain proteins cross-link intermediate filaments with other types of filaments and mediate interactions between them.

All eukaryotic cells have microtubules and microfilaments, but only some animal groups, including vertebrates, are known to have intermediate filaments. Even when present, intermediate filaments vary widely in protein composition and size among different cell types and different organisms. Examples of inter-

forming rope-like assemblages in the cells

mediate filaments are the keratins found in the epithelial cells of vertebrate skin and neurofilaments found in vertebrate nerve cells.

Certain mutations in genes coding for intermediate filaments weaken the cell and are associated with several diseases. For example, in the neurodegenerative disease amyotrophic lateral sclerosis (ALS, or Lou Gehrig's disease), abnormal neurofilaments have been identified in nerve cells that control muscles. This condition interferes with normal transport of materials in the nerve cells and leads to degeneration of the cells. The resulting loss of muscle function is typically fatal.

Review

▌ What are the main functions of the cytoskeleton?

▌ How are microfilaments and microtubules similar? How are they different?

▌ How are cilia and flagella similar? How are they different?

CELL COVERINGS

Learning Objective

14 Describe the glycocalyx, extracellular matrix, and cell wall.

Most eukaryotic cells are surrounded by a **glycocalyx,** or **cell coat,** formed by polysaccharide side chains of proteins and lipids that are part of the plasma membrane. The glycocalyx protects the cell and may help keep other cells at a distance. Certain molecules of the glycocalyx enable cells to recognize one another, to make contact, and in some cases to form adhesive or communicating associations (discussed in Chapter 5). Other molecules of the cell coat contribute to the mechanical strength of multicellular tissues.

Many animal cells are also surrounded by an **extracellular matrix (ECM),** which they secrete. It consists of a gel of carbohydrates and fibrous proteins (Fig. 4-28). The main structural protein in the ECM is collagen, which forms very tough fibers. Certain glycoproteins of the ECM, called **fibronectins,** help organize the matrix and help cells attach to it. Fibronectins bind to protein receptors that extend from the plasma membrane.

Integrins are proteins that serve as membrane receptors for the ECM. These proteins activate many **cell signaling** pathways that communicate information from the ECM. Integrins may be important in cell movement and in organizing the cytoskeleton so that cells assume a definite shape. In many types of cells, integrins anchor the external ECM to the intermediate filaments and microfilaments of the internal cytoskeleton. When these cells are not appropriately anchored, apoptosis results.

Most bacteria, fungi, and plant cells are surrounded by a cell wall. Plant cells have thick cell walls that contain tiny fibers composed of the polysaccharide **cellulose** (see Fig. 3-10). Other polysaccharides in the plant cell wall form cross links between the bundles of cellulose fibers. Cell walls provide structural support, protect plant cells from disease-causing organisms, and help keep excess water out of cells so they do not burst.

A growing plant cell secretes a thin, flexible *primary cell wall*. As the cell grows, the primary cell wall increases in size (Fig. 4-29). After the cell stops growing, either new wall material is secreted that thickens and solidifies the primary wall or multiple layers of a *secondary cell wall* with a different chemical composition are formed between the primary wall and the plasma membrane. Wood is made mainly of secondary cell walls. The **middle lamella,** a layer of gluelike polysaccharides called *pectins,* lies between the primary cell walls of adjacent cells. The middle lamella causes the cells to adhere tightly to one another. (For more information on plant cell walls, see the discussion of the ground tissue system in Chapter 32.)

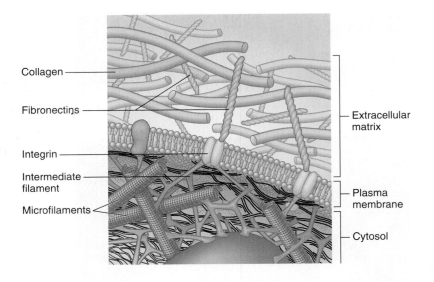

Figure 4-28 The extracellular matrix (ECM)

Fibronectins, glycoproteins of the ECM, bind to integrins and other receptors in the plasma membrane.

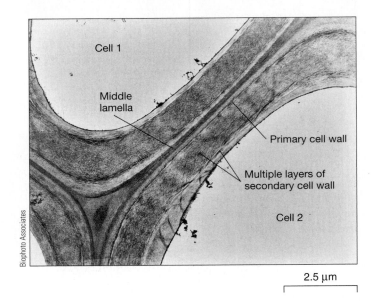

Figure 4-29 *Animated* Plant cell walls

The cell walls of two adjacent plant cells are labeled in this TEM. The cells are cemented together by the middle lamella, a layer of gluelike polysaccharides called pectins. A growing plant cell first secretes a thin primary wall that is flexible and can stretch as the cell grows. The thicker layers of the secondary wall are secreted inside the primary wall after the cell stops elongating.

Review

■ What are the functions of the glycocalyx?
■ How do the functions of fibronectins and integrins differ?
■ What is the main component of plant cell walls?
■ How are plant cell walls formed?

Learning Objectives

1 Describe the cell theory, and relate it to the evolution of life (page 74).

- The **cell theory** holds that (1) cells are the basic living units of organization and function in all organisms and (2) all cells come from other cells. It explains that the ancestry of all the cells alive today can be traced back to ancient times. Evidence that all living cells have evolved from a common ancestor is supported by the basic similarities in their structures and in their molecular composition.

2 Summarize the relationship between cell organization and homeostasis (page 74).

- The organization of cells is important in maintaining **homeostasis,** an appropriate internal environment. Cells have many **organelles,** internal structures that carry out specific functions, that help maintain homeostasis.

- Every cell is surrounded by a **plasma membrane** that separates it from its external environment. The plasma membrane allows the cell to maintain internal conditions that may be very different from those of the outer environment. The plasma membrane also allows the cell to exchange materials with its outer environment.

3 Explain the relationship between cell size and homeostasis (page 74).

- A critical factor in determining cell size is the ratio of the plasma membrane (surface area) to the cell's volume. The plasma membrane must be large enough relative to the cell volume to regulate the passage of materials into and out of the cell. For this reason, most cells are microscopic.

- Cell size and shape are related to function and are limited by the need to maintain homeostasis.

4 Describe methods that biologists use to study cells, including microscopy and cell fractionation (page 76).

- Biologists use **light microscopes, electron microscopes,** and a variety of chemical methods to study cells and learn about cell structure. The electron microscope has superior **resolving power,** enabling investigators to see details of cell structures not observable with conventional microscopes.

- Cell biologists use the technique of **cell fractionation** for purifying organelles to gain information about the function of cell structures.

5 Compare and contrast the general characteristics of prokaryotic and eukaryotic cells, and contrast plant and animal cells (page 80).

- **Prokaryotic cells** are bounded by a plasma membrane but have little or no internal membrane organization. They have a **nuclear area** rather than a membrane-enclosed nucleus. Prokaryotic cells typically have a **cell wall** and **ribosomes** and may have propeller-like **flagella.**

- **Eukaryotic cells** have a membrane-enclosed nucleus, and their **cytoplasm** contains a variety of organelles; the fluid component of the cytoplasm is the **cytosol.**

- Plant cells differ from animal cells in that plant cells have rigid cell walls, plastids, and large vacuoles; cells of most plants lack centrioles. Vacuoles are important in plant growth and development.

ThomsonNOW Test yourself on the structure of prokaryotic and eukaryotic cells by clicking on the interaction in ThomsonNOW.

6 Describe three functions of cell membranes (page 81).

- Membranes divide the cell into compartments, allowing it to conduct specialized activities within small areas of the cytoplasm, concentrate reactants, and organize metabolic reactions. Small membrane-enclosed sacs, called **vesicles,** transport materials between compartments.

- Membranes are important in energy storage and conversion.

- A system of interacting membranes forms the **endomembrane system.**

ThomsonNOW Learn more about the endomembrane system by clicking on the figure in ThomsonNOW.

7 Describe the structure and functions of the nucleus (page 84).

- The **nucleus,** the control center of the cell, contains genetic information coded in DNA. The nucleus is bounded by a **nuclear envelope** consisting of a double membrane perforated with **nuclear pores** that communicate with the cytoplasm.

- DNA in the nucleus associates with protein to form **chromatin,** which is organized into **chromosomes.** During cell division, the chromosomes condense and become visible as threadlike structures.

- The **nucleolus** is a region in the nucleus that is the site of ribosomal RNA synthesis and ribosome assembly.

8 Distinguish between smooth and rough endoplasmic reticulum in terms of both structure and function (page 88).

- The **endoplasmic reticulum (ER)** is a network of folded internal membranes in the cytosol. **Smooth ER** is the site of lipid synthesis, calcium ion storage, and detoxifying enzymes.

- **Rough ER** is studded along its outer surface with **ribosomes** that manufacture polypeptides. Polypeptides synthesized on rough ER may be moved into the ER lumen, where they are assembled into proteins and modified by the addition of a carbohydrate or lipid.

9 Trace the path of proteins synthesized in the rough endoplasmic reticulum as they are processed, modified, and sorted by the Golgi complex and then transported to specific destinations (page 88).

- The **Golgi complex** consists of stacks of flattened membranous sacs called **cisternae** that process, sort, and modify proteins synthesized on the ER. The Golgi complex also manufactures lysosomes.

- Glycoproteins are transported from the ER to the *cis* face of the Golgi complex by **transport vesicles,** which are formed by membrane budding. The Golgi complex modifies carbohydrates and lipids that were added to proteins by the ER and packages them in vesicles.

- Glycoproteins exit the Golgi at its *trans* face. The Golgi routes some proteins to the plasma membrane for export from the cell. Others are transported to lysosomes or other organelles within the cytoplasm.

10 Describe the functions of lysosomes, vacuoles, and peroxisomes (page 88).

■ **Lysosomes** contain enzymes that break down worn-out cell structures, bacteria, and other substances taken into cells.

■ **Vacuoles** store materials, water, and wastes. They maintain hydrostatic pressure in plant cells.

■ **Peroxisomes** are important in lipid metabolism and detoxify harmful compounds such as ethanol. They produce hydrogen peroxide but contain the enzyme catalase, which degrades this toxic compound.

11 Compare the functions of mitochondria and chloroplasts, and discuss ATP synthesis by each of these organelles (page 88).

■ **Mitochondria,** organelles enclosed by a double membrane, are the sites of aerobic respiration. The inner membrane is folded, forming **cristae** that increase its surface area.

■ The cristae and the compartment enclosed by the inner membrane, the **matrix,** contain enzymes for the reactions of **aerobic respiration.** During aerobic respiration, nutrients are broken down in the presence of oxygen. Energy captured from nutrients is packaged in ATP, and carbon dioxide and water are produced as by-products.

■ Mitochondria play an important role in **apoptosis,** which is programmed cell death.

■ **Plastids** are organelles that produce and store food in the cells of plants and algae. **Chloroplasts** are plastids that carry out **photosynthesis.**

■ The inner membrane of the chloroplast encloses a fluid-filled space, the **stroma. Grana,** stacks of disclike membranous sacs called **thylakoids,** are suspended in the stroma.

■ During photosynthesis, **chlorophyll,** the green pigment found in the thylakoid membranes, traps light energy. This energy is converted to chemical energy in ATP and used to synthesize carbohydrates from carbon dioxide and water.

12 Describe the structure and functions of the cytoskeleton (page 97).

■ The **cytoskeleton** is a dynamic internal framework made of microtubules, microfilaments, and intermediate filaments. The cytoskeleton provides structural support and functions in various types of cell movement, including transport of materials in the cell.

■ **Microtubules** are hollow cylinders assembled from subunits of the protein **tubulin.** In cells that are not dividing, the minus ends of microtubules are anchored in **microtubule-organizing centers (MTOCs).**

■ **Microtubule-associated proteins (MAPs)** include structural MAPs and motor MAPs.

■ The main MTOC of animal cells is the **centrosome,** which usually contains two centrioles. Each centriole has a 9×3 arrangement of microtubules.

■ **Microfilaments,** or actin filaments, formed from subunits of the protein **actin,** are important in cell movement.

■ **Intermediate filaments** strengthen the cytoskeleton and stabilize cell shape.

ThomsonNOW Learn more about the cytoskeleton by clicking on the figure in ThomsonNOW.

13 Compare cilia and flagella, and describe their functions (page 97).

■ **Cilia** and **flagella** are thin, movable structures that project from the cell surface and function in movement. Each consists of a $9 + 2$ arrangement of microtubules, and each is anchored in the cell by a basal body that has a 9×3 organization of microtubules. Cilia are short, and flagella are long.

14 Describe the glycocalyx, extracellular matrix, and cell wall (page 102).

■ Most cells are surrounded by a **glycocalyx,** or **cell coat,** formed by polysaccharides extending from the plasma membrane.

■ Many animal cells are also surrounded by an **extracellular matrix (ECM)** consisting of carbohydrates and protein. **Fibronectins** are glycoproteins of the ECM that bind to **integrins,** receptor proteins in the plasma membrane.

■ Most bacteria, fungi, and plant cells are surrounded by a cell wall made of carbohydrates. Plant cells secrete **cellulose** and other polysaccharides that form rigid cell walls.

TEST YOUR UNDERSTANDING

1. The ability of a microscope to reveal fine detail is known as (a) magnification (b) resolving power (c) cell fractionation (d) scanning electron microscopy (e) phase contrast

2. A plasma membrane is characteristic of (a) all cells (b) prokaryotic cells only (c) eukaryotic cells only (d) animal cells only (e) eukaryotic cells except plant cells

3. Detailed information about the shape and external features of a specimen can best be obtained by using a (a) differential centrifuge (b) fluorescence microscope (c) transmission electron microscope (d) scanning electron microscope (e) light microscope

4. In eukaryotic cells, DNA is found in (a) chromosomes (b) chromatin (c) mitochondria (d) a, b, and c (e) a and b

5. Which of the following structures would *not* be found in prokaryotic cells? (a) cell wall (b) ribosomes (c) nuclear area (d) nucleus (e) propeller-like flagellum

6. Which of the following is/are most closely associated with protein synthesis? (a) ribosomes (b) smooth ER (c) mitochondria (d) microfilaments (e) lysosomes

7. Which of the following is/are most closely associated with the breakdown of ingested material? (a) ribosomes (b) smooth ER (c) mitochondria (d) microfilaments (e) lysosomes

8. Which of the following are most closely associated with photosynthesis? (a) basal bodies (b) smooth ER (c) cristae (d) thylakoids (e) MTOCs

9. A $9 + 2$ arrangement of microtubules best describes (a) cilia (b) centrosomes (c) basal bodies (d) microfilaments (e) microvilli

10. Use the numbered choices to select the sequence that most accurately describes information flow in the eukaryotic cell.
    1. DNA in nucleus 2. RNA 3. mitochondria 4. protein synthesis 5. ribosomes
    (a) 1, 2, 5, 4 (b) 3, 2, 5, 1 (c) 5, 2, 3, 1 (d) 4, 3, 2, 1 (e) 1, 2, 3, 4

11. Use the numbered choices to select the sequence that most accurately describes glycoprotein processing in the eukaryotic cell.

1. ER 2. ribosomes 3. *cis* face of Golgi 4. *trans* face of Golgi 5. transport vesicle 6. grana 7. plasma membrane

(a) 1, 6, 2, 3, 4, 7 (b) 2, 1, 3, 4, 5, 7 (c) 7, 1, 2, 4, 3, 5 (d) 2, 1, 4, 3, 5, 7 (e) 2, 1, 7, 4, 5, 1

12. Which of the following is/are *not* associated with mitochondria? (a) cristae (b) aerobic respiration (c) apoptosis (d) free radicals (e) thylakoids

13. Which of the following organelles contain small amounts of DNA and convert energy? (a) microfilaments and microtubules (b) lysosomes and peroxisomes (c) ER and ribosomes (d) nucleus and ribosomes (e) mitochondria and chloroplasts

14. Which of the following function(s) in cell movement? (a) microtubules (b) nucleolus (c) grana (d) smooth ER (e) rough ER

15. The extracellular matrix (a) consists mainly of myosin and RNA (b) projects to form microvilli (c) houses the centrioles (d) contains fibronectins that bind to integrins (e) has an elaborate system of cristae

16. Label the diagrams of the animal and plant cells. How is the structure of each organelle related to its function? Use Figures 4-7 and 4-8 to check your answers.

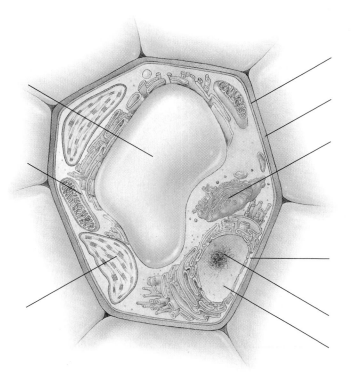

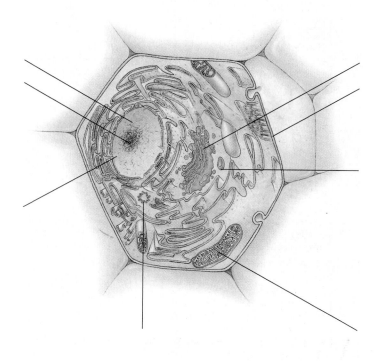

CRITICAL THINKING

1. Explain why the cell is considered the basic unit of life, and discuss some of the implications of the cell theory.

2. Why does a eukaryotic cell need both membranous organelles and fibrous cytoskeletal components?

3. Describe a specific example of the correlation between cell structure and function. (*Hint:* Think of mitochondrial structure.)

4. The *Acetabularia* experiments described in this chapter suggest that DNA is much more stable in the cell than is messenger RNA. Is this advantageous or disadvantageous to the cell? Why? *Acetabularia* continues to live for a few days after its nucleus is removed. How can you explain this?

5. **Evolution Link.** Biologists use similarities in cells to trace the evolutionary history of various groups of organisms. Explain their rationale. What do such similarities in cell structure and function tell biologists about the common origin of organisms? Explain.

Additional questions are available in ThomsonNOW at www.thomsonedu.com/login

5

Biological Membranes

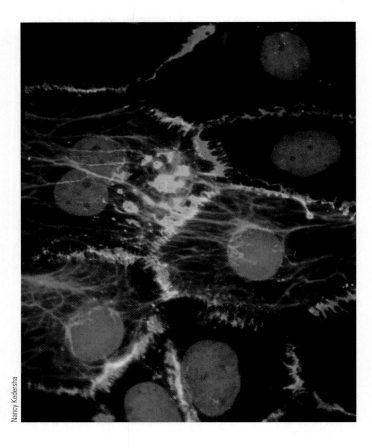

Nancy Kedersha

Cadherins. The human skin cells shown in this LM were grown in culture and stained with fluorescent antibodies. Cadherins, a group of membrane proteins, are seen as green belts around each cell in this sheet of cells. The nuclei appear as blue spheres; myosin in the cells appears red.

KEY CONCEPTS

Biological membranes are selectively permeable membranes that help maintain homeostasis in the cell.

According to the fluid mosaic model, cell membranes consist of a fluid bilayer composed of phospholipids in which a variety of proteins are embedded.

Membrane proteins, and thus membranes, have many functions, including transport of materials, enzymatic activity, transmission of information, and recognition of other cells.

Diffusion is a passive process that does not require the cell to expend metabolic energy, whereas active transport requires a direct expenditure of energy.

Cells in close contact form specialized junctions between one another.

The evolution of biological membranes that separate the cell from its external environment was an essential step in the origin of life. Later, membranes made the evolution of complex cells possible. The extensive internal membranes of eukaryotic cells form multiple compartments with unique environments for highly specialized activities.

An exciting area of cell membrane research focuses on membrane proteins. Some proteins associated with the plasma membrane transport materials, whereas others transmit information or serve as enzymes. Still others, known as *cell adhesion molecules,* are important in connecting cells to one another to form tissues.

The principal cell adhesion molecules in vertebrates and in many invertebrates are **cadherins.** These molecules are responsible for calcium-dependent adhesion between cells that form multicellular sheets. For example, cadherins form cell junctions important in maintaining the structure of the epithelium that makes up human skin (see photograph). An absence of these membrane proteins is associated with the invasiveness of some malignant tumors. Certain cadherins mediate the way cells adhere in the early embryo, and thus they are important in development.

In Chapter 4, we discussed a variety of cell organelles. In this chapter, we will focus on the structure and functions of the biological membranes that surround many organelles as well as the plasma membrane that surrounds the cell. We first consider what is known about the composition and structure of biological membranes. Then we discuss how cells transport various materials, from ions to complex molecules and even bacteria, across membranes. Finally, we examine specialized structures that allow membranes of different cells to interact. Although much of our discussion centers on the structure and functions of plasma membranes, many of the concepts apply to other cell membranes. ■

THE STRUCTURE OF BIOLOGICAL MEMBRANES

Learning Objectives

1 Evaluate the importance of membranes to the homeostasis of the cell, emphasizing their various functions.
2 Describe the fluid mosaic model of cell membrane structure.
3 Relate properties of the lipid bilayer to properties and functions of cell membranes.
4 Describe the ways that membrane proteins associate with the lipid bilayer, and discuss the functions of membrane proteins.

To carry out the many chemical reactions necessary to sustain life, the cell must maintain an appropriate internal environment. Every cell is surrounded by a **plasma membrane** that physically separates it from the outside world and defines it as a distinct entity. By regulating passage of materials into and out of the cell, the plasma membrane helps maintain a life-supporting internal environment. As we discussed in Chapter 4, eukaryotic cells are characterized by numerous organelles that are surrounded by membranes. Some of these organelles—including the nuclear envelope, endoplasmic reticulum, Golgi complex, lysosomes, vesicles, and vacuoles—form the endomembrane system, which extends throughout the cell.

Biological membranes are complex, dynamic structures made of lipid and protein molecules that are in constant motion. The properties of membranes allow them to perform vital functions in the cell. They regulate the passage of materials, divide the cell into compartments, serve as surfaces for chemical reactions, adhere to and communicate with other cells, and transmit signals between the environment and the interior of the cell. Membranes are also an essential part of energy transfer and storage systems. How do the properties of cell membranes enable the cell to carry on such varied functions?

Long before the development of the electron microscope, scientists knew that membranes consist of both lipids and proteins. Work by researchers in the 1920s and 1930s had provided clues that the core of cell membranes consists of lipids, mostly phospholipids (see Chapter 3).

Phospholipids form bilayers in water

Phospholipids are primarily responsible for the physical properties of biological membranes, because certain phospholipids have unique attributes, including features that allow them to form bilayered structures. A phospholipid contains two fatty acid chains linked to two of the three carbons of a glycerol molecule (see Fig. 3-13). The fatty acid chains make up the nonpolar, *hydrophobic* ("water-fearing") portion of the phospholipid. Bonded to the third carbon of the glycerol is a negatively charged, *hydrophilic* ("water-loving") phosphate group, which in turn is linked to a polar, hydrophilic organic group. Molecules of this type, which have distinct hydrophobic and hydrophilic regions, are called **amphipathic molecules.** All lipids that make up the core of biological membranes have amphipathic characteristics.

Because one end of each phospholipid associates freely with water and the opposite end does not, the most stable orientation for them to assume in water results in the formation of a bilayer structure (Fig. 5-1a). This arrangement allows the hydrophilic heads of the phospholipids to be in contact with the aqueous medium while their oily tails, the hydrophobic fatty acid chains, are buried in the interior of the structure away from the water molecules.

Amphipathic properties alone do not predict the ability of lipids to associate as a bilayer. Shape is also important. Phospholipids tend to have uniform widths; their roughly cylindrical shapes, together with their amphipathic properties, are responsible for bilayer formation. In summary, phospholipids form bilayers because the molecules have (1) two distinct regions, one strongly hydrophobic and the other strongly hydrophilic (making them strongly amphipathic); and (2) cylindrical shapes that allow them to associate with water most easily as a bilayer.

Do you know why detergents remove grease from your hands or from dirty dishes? Many common detergents are amphipathic molecules, each containing a single hydrocarbon chain (like a fatty acid) at one end and a hydrophilic region at the other. These

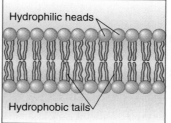
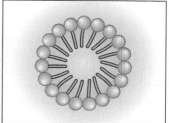

(a) Phospholipids in water. Phospholipids associate as bilayers in water because they are roughly cylindrical amphipathic molecules. The hydrophobic fatty acid chains are not exposed to water, whereas the hydrophilic phospholipid heads are in contact with water.

(b) Detergent in water. Detergent molecules are roughly cone-shaped amphipathic molecules that associate in water as spherical structures.

Figure 5-1 *Animated* Properties of lipids in water

The Davson–Danielli model was the accepted view until about 1970 when advances in biology and chemistry led to new findings about biological membranes that were incompatible with this model. The fluid mosaic model fits the new data.

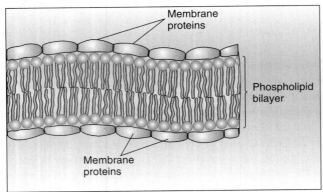

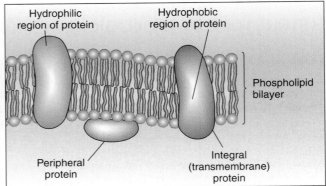

(a) The Davson–Danielli model. According to this model, the membrane is a sandwich of phospholipids spread between two layers of protein. Although accepted for many years, this model was shown to be incorrect.

(b) Fluid mosaic model. According to this model, a cell membrane is a fluid lipid bilayer with a constantly changing "mosaic pattern" of associated proteins.

Figure 5-2 Two models of membrane structure

molecules are roughly cone shaped, with the hydrophilic end forming the broad base and the hydrocarbon tail leading to the point. Because of their shapes, these molecules do not associate as bilayers but instead tend to form spherical structures in water (▌ Fig. 5-1b). Detergents can "solubilize" oil because the oil molecules associate with the hydrophobic interiors of the spheres.

Current data support a fluid mosaic model of membrane structure

By examining the plasma membrane of the mammalian red blood cell and comparing its surface area with the total number of lipid molecules per cell, early investigators calculated that the membrane is no more than two phospholipid molecules thick. In 1935, these findings, together with other data, led Hugh Davson and James Danielli, working at London's University College, to propose a model in which they envisioned a membrane as a kind of "sandwich" consisting of a *lipid bilayer* (a double layer of lipid) between two protein layers (▌ Fig. 5-2a). This useful model had a great influence on the direction of membrane research for more than 20 years. Models are important in the scientific process; good ones not only explain the available data but are testable. Scientists use models to help them develop hypotheses that can be tested experimentally (see Chapter 1).

With the development of the electron microscope in the 1950s, cell biologists were able to see the plasma membrane for the first time. One of their most striking observations was how uniform and thin the membranes are. The plasma membrane is

no more than 10 nm thick. The electron microscope revealed a three-layered structure, something like a railroad track, with two dark layers separated by a lighter layer (▌ Fig. 5-3). Their findings seemed to support the protein–lipid–protein sandwich model.

During the 1960s, a paradox emerged regarding the arrangement of the proteins. Biologists assumed membrane proteins were uniform and had shapes that would allow them to lie like

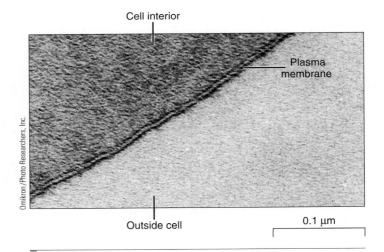

Figure 5-3 TEM of the plasma membrane of a mammalian red blood cell

The plasma membrane separates the cytosol (*darker region*) from the external environment (*lighter region*). The hydrophilic heads of the phospholipids are the parallel dark lines, and the hydrophobic tails are the light zone between them.

thin sheets on the membrane surface. But when purified by cell fractionation, the proteins were far from uniform; in fact, they varied widely in composition and size. Some proteins are quite large. How could they fit within a surface layer of a membrane less than 10 nm thick? At first, some researchers tried to answer this question by modifying the model with the hypothesis that the proteins on the membrane surfaces were a flattened, extended form, perhaps a β-pleated sheet (see Figure 3-20b).

Other cell biologists found that instead of having sheetlike structures, many membrane proteins are rounded, or globular. Studies of many membrane proteins showed that one region (or domain) of the molecule could always be found on one side of the bilayer, whereas another part of the protein might be located on the opposite side. Rather than form a thin surface layer, many membrane proteins extended completely through the lipid bilayer. Thus, the evidence suggested that membranes contain different types of proteins of different shapes and sizes that are associated with the bilayer in a mosaic pattern.

In 1972, S. Jonathan Singer and Garth Nicolson of the University of California at San Diego proposed a model of membrane structure that represented a synthesis of the known properties of biological membranes. According to their **fluid mosaic model,** a cell membrane consists of a fluid bilayer of phospholipid molecules in which the proteins are embedded or otherwise associated, much like the tiles in a mosaic picture. This mosaic pattern is not static, however, because the positions of the proteins are constantly changing as they move about like icebergs in a fluid sea of phospholipids. This model has provided great impetus to research; it has been repeatedly tested and has been shown to accurately predict the properties of many kinds of cell membranes. ▌ Figure 5-2b depicts the plasma membrane of a eukaryotic cell according to the fluid mosaic model; prokaryotic plasma membranes are discussed in Chapter 24.

Biological membranes are two-dimensional fluids

An important physical property of phospholipid bilayers is that they behave like *liquid crystals*. The bilayers are crystal-like in that the lipid molecules form an ordered array with the heads on the outside and fatty acid chains on the inside; they are liquidlike in that despite the orderly arrangement of the molecules, their hydrocarbon chains are in constant motion. Thus, molecules are free to rotate and can move laterally within their single layer (▌ Fig. 5-4). Such movement gives the bilayer the property of a *two-dimensional fluid*. This means that under normal conditions a single phospholipid molecule can travel laterally across the surface of a eukaryotic cell in seconds.

The fluid qualities of lipid bilayers also allow molecules embedded in them to move along the plane of the membrane (as long as they are not anchored in some way). David Frye and Michael Edidin elegantly demonstrated this in 1970. They conducted experiments in which they followed the movement of membrane proteins on the surface of two cells that had been joined (▌ Fig. 5-5). When the plasma membranes of a mouse cell

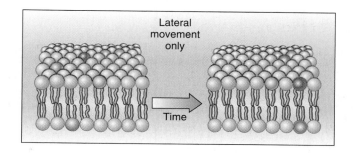

Figure 5-4 Membrane fluidity

The ordered arrangement of phospholipid molecules makes the cell membrane a liquid crystal. The hydrocarbon chains are in constant motion, allowing each molecule to move laterally on the same side of the bilayer.

and a human cell are fused, within minutes at least some of the membrane proteins from each cell migrate and become randomly distributed over the single, continuous plasma membrane that surrounds the joined cells. Frye and Edidin showed that the fluidity of the lipids in the membrane allows many of the proteins to move, producing an ever-changing configuration. Occasionally, with the help of enzymes in the cell membrane, phospholipid molecules flip-flop from one layer to the other.

For a membrane to function properly, its lipids must be in a state of optimal fluidity. The membrane's structure is weakened if its lipids are too fluid. However, many membrane functions, such as the transport of certain substances, are inhibited or cease if the lipid bilayer is too rigid. At normal temperatures, cell membranes are fluid, but at low temperatures the motion of the fatty acid chains is slowed. If the temperature decreases to a critical point, the membrane is converted to a more solid gel state.

Certain properties of membrane lipids have significant effects on the fluidity of the bilayer. Recall from Chapter 3 that molecules are free to rotate around single carbon-to-carbon covalent bonds. Because most of the bonds in hydrocarbon chains are single bonds, the chains themselves twist more and more rapidly as the temperature rises.

The fluid state of the membrane depends on its component lipids. You have probably noticed that when melted butter is left at room temperature, it solidifies. Vegetable oils, however, remain liquid at room temperature. Recall from our discussion of fats in Chapter 3 that animal fats such as butter are high in saturated fatty acids that lack double bonds. In contrast, a vegetable oil may be polyunsaturated, with most of its fatty acid chains having two or more double bonds. At each double bond there is a bend in the molecule that prevents the hydrocarbon chains from coming close together and interacting through van der Waals interactions. In this way, unsaturated fats lower the temperature at which oil or membrane lipids solidify.

Many organisms have regulatory mechanisms for maintaining cell membranes in an optimally fluid state. Some organisms compensate for temperature changes by altering the fatty acid content of their membrane lipids. When the outside temperature

QUESTION: Can molecules embedded in a biological membrane move about?

HYPOTHESIS: Proteins are able to move laterally through the plasma membrane.

EXPERIMENT: Larry Frye and Michael Edidin labeled membrane proteins of mouse and human cells using different-colored fluorescent dyes to distinguish between them. Then they fused mouse and human cells to produce hybrid cells.

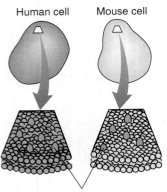

Human cell   Mouse cell

1 Labeled membrane proteins. Membrane proteins of mouse cells and human cells were labeled with fluorescent dye markers in two different colors.

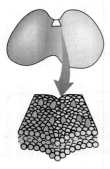 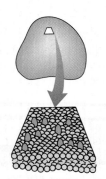

2 Human-mouse hybrid cell forming. When plasma membranes of mouse cell and human cell were fused, mouse proteins migrated to human side and human proteins to mouse side.

3 Proteins randomly distributed. After short time, mouse and human proteins became randomly distributed through membrane.

RESULTS AND CONCLUSION: After a brief period of incubation, mouse and human cells intermixed over the surface of the hybrid cells. After about 40 minutes, the proteins of each species had become randomly distributed through the entire hybrid plasma membrane. This experiment demonstrated that proteins in the plasma membrane do move.

Figure 5-5   Frye and Edidin's experiment

---

is cold, the membrane lipids contain relatively high proportions of unsaturated fatty acids.

Some membrane lipids stabilize membrane fluidity within certain limits. One such "fluidity buffer" is cholesterol, a steroid found in animal cell membranes. A cholesterol molecule is largely hydrophobic but is slightly amphipathic because of the presence of a single hydroxyl group (see Fig. 3-15a). This hydroxyl group associates with the hydrophilic heads of the phospholipids; the hydrophobic remainder of the cholesterol molecule fits between the fatty acid hydrocarbon chains (∎ Fig. 5-6).

At low temperatures cholesterol molecules act as "spacers" between the hydrocarbon chains, restricting van der Waals interactions that would promote solidifying. Cholesterol also helps prevent the membrane from becoming weakened or unstable at higher temperatures. The reason is that the cholesterol molecules interact strongly with the portions of the hydrocarbon chains closest to the phospholipid head. This interaction restricts motion in these regions. Plant cells have steroids other than cholesterol that carry out similar functions.

Biological membranes fuse and form closed vesicles

Lipid bilayers, particularly those in the liquid-crystalline state, have additional important physical properties. Bilayers tend to resist forming free ends; as a result, they are self-sealing and under most conditions spontaneously round up to form closed vesicles. Lipid bilayers are also flexible, allowing cell membranes to change shape without breaking. Under appropriate conditions lipid bilayers fuse with other bilayers.

Membrane fusion is an important cell process. When a vesicle fuses with another membrane, both membrane bilayers and their compartments become continuous. Various transport vesicles form from, and also merge with, membranes of the ER and Golgi complex, facilitating the transfer of materials from one compartment to another. A vesicle fuses with the plasma membrane when a product is secreted from the cell.

Membrane proteins include integral and peripheral proteins

The two major classes of membrane proteins, integral proteins and peripheral proteins, are defined by how tightly they are associated with the lipid bilayer (∎ Fig. 5-7). **Integral membrane proteins** are firmly bound to the membrane. Cell biologists usually can release them only by disrupting the bilayer with detergents. These proteins are amphipathic. Their hydrophilic regions extend out of the cell or into the cytoplasm, whereas their hydrophobic regions interact with the fatty acid tails of the membrane phospholipids.

Some integral proteins do not extend all the way through the membrane. Many others, called **transmembrane proteins,** extend completely through the membrane. Some span the mem-

According to the fluid mosaic model, a cell membrane is composed of a fluid bilayer of phospholipids in which proteins move about like icebergs in a sea.

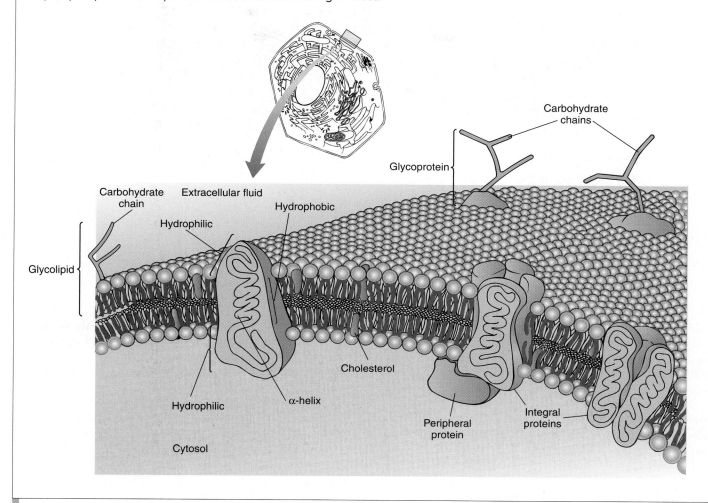

Figure 5-6 Detailed structure of the plasma membrane

Although the lipid bilayer consists mainly of phospholipids, other lipids, such as cholesterol and glycolipids, are present. Peripheral proteins are loosely associated with the bilayer, whereas integral proteins are tightly bound. The integral proteins shown here are transmembrane proteins that extend through the bilayer. They have hydrophilic regions on both sides of the bilayer, connected by a membrane-spanning α-helix. Glycolipids (carbohydrates attached to lipids) and glycoproteins (carbohydrates attached to proteins) are exposed on the extracellular surface; they play roles in cell recognition and adhesion to other cells.

brane only once, whereas others wind back and forth as many as 24 times. The most common kind of transmembrane protein is an α-helix (see Chapter 3) with hydrophobic amino acid side chains projecting out from the helix into the hydrophobic region of the lipid bilayer. Some proteins span the membrane in the form of rolled-up β-pleated sheets. These protein formations are barrel shaped and form pores through which water and other substances can pass.

Peripheral membrane proteins are not embedded in the lipid bilayer. They are located on the inner or outer surface of the plasma membrane, usually bound to exposed regions of integral proteins by noncovalent interactions. Peripheral proteins

can be easily removed from the membrane without disrupting the structure of the bilayer.

Proteins are oriented asymmetrically across the bilayer

One of the most remarkable demonstrations that proteins are actually embedded in the lipid bilayer comes from *freeze–fracture* electron microscopy, a technique that splits the membrane into halves. The researcher can literally see the two halves of the membrane from "inside out." When cell biologists examine membranes

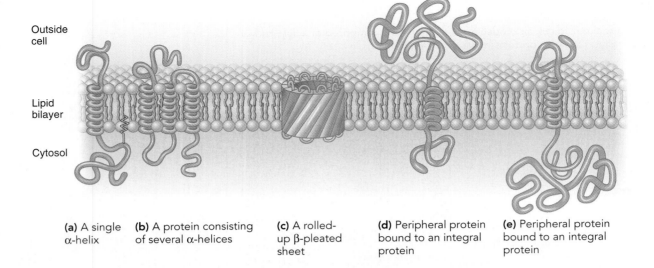

(a) A single α-helix

(b) A protein consisting of several α-helices

(c) A rolled-up β-pleated sheet

(d) Peripheral protein bound to an integral protein

(e) Peripheral protein bound to an integral protein

Figure 5-7 *Animated* Membrane proteins

Three transmembrane proteins are shown in **(a)**, **(b)**, and **(c)**. The rolled-up β-pleated sheet shown in **(c)** forms a pore through the membrane. Water molecules and ions pass through specific types of pores.

(d, e) Peripheral proteins are bound to integral proteins by non-covalent interactions.

in this way, they observe numerous particles on the fracture faces (■ Fig. 5-8). The particles are clearly integral membrane proteins, because researchers never see them in freeze-fractured artificial lipid bilayers. These findings profoundly influenced Singer and Nicolson in developing the fluid mosaic model.

Membrane protein molecules are *asymmetrically oriented*. The asymmetry is produced by the highly specific way in which each protein is inserted into the bilayer. Membrane proteins that

will become part of the inner surface of the plasma membrane are manufactured by free ribosomes and move to the membrane through the cytoplasm. Membrane proteins that will be associated with the cell's outer surface are manufactured like proteins destined to be exported from the cell.

As discussed in Chapter 4, proteins destined for the cell's outer surface are initially manufactured by ribosomes on the rough ER. They pass through the ER membrane into the ER lumen, where

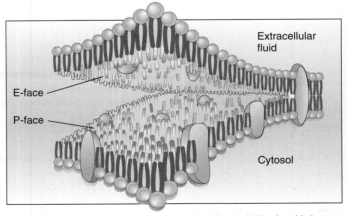

(a) Transmembrane proteins are inserted through the lipid bilayer.

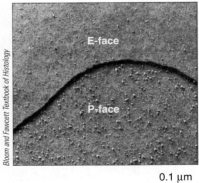

Bloom and Fawcett Textbook of Histology

0.1 μm

(b) A freeze–fracture TEM. The particles (which appear as bumps) represent large transmembrane proteins.

Figure 5-8 A view inside the plasma membrane

In the freeze–fracture method, the membrane splits along the hydrophobic interior of the lipid bilayer. Two complementary fracture faces result. The inner half-membrane presents the P-face (or protoplasmic face), and the outer half-membrane presents the E-face (or external face).

The orientation of a protein in the plasma membrane is determined by the pathway of its synthesis and transport.

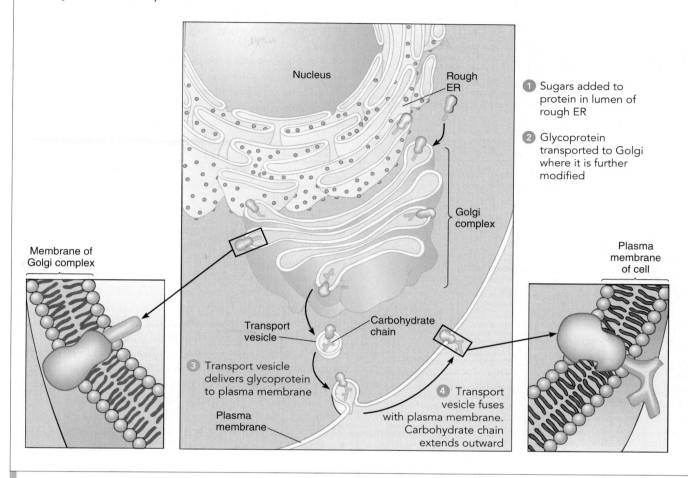

① Sugars added to protein in lumen of rough ER

② Glycoprotein transported to Golgi where it is further modified

③ Transport vesicle delivers glycoprotein to plasma membrane

④ Transport vesicle fuses with plasma membrane. Carbohydrate chain extends outward

Figure 5-9 Synthesis and orientation of a membrane protein

The surface of the rough ER membrane that faces the lumen of the rough ER also faces the lumen of the Golgi complex and vesicles. However, when a vesicle fuses with the plasma membrane, the inner surface of the vesicle becomes the extracellular surface of the plasma membrane.

sugars are added, making them **glycoproteins.** However, only a part of each protein passes through the ER membrane, so each completed protein has some regions that are located in the ER lumen and other regions that remain in the cytosol. Enzymes that attach the sugars to certain amino acids on the protein are found only in the lumen of the ER. Thus, carbohydrates can be added only to the parts of proteins located in that compartment.

In ▌ Figure 5-9, follow from top to bottom the vesicle budding and membrane fusion events that are part of the transport process. You can see that the same region of the protein that protruded into the ER lumen is also transported to the lumen of the Golgi complex. There additional enzymes further modify the carbohydrate chains. Within the Golgi complex, the glycoprotein is sorted and directed to the plasma membrane. The modified region of the protein remains inside the membrane compartment of a transport vesicle as it buds from the Golgi complex. Note that when the transport vesicle fuses with the plasma membrane, the inside layer of the transport vesicle becomes the outside layer of

the plasma membrane. The carbohydrate chain extends to the exterior of the cell surface.

In summary, this is the sequence:

Sugar added to protein in ER lumen ⟶ glycoprotein transported to Golgi complex, where it is further modified ⟶ glycoprotein transported to plasma membrane ⟶ transport vesicle fuses with plasma membrane (inside layer of transport vesicle becomes outer layer of plasma membrane)

Membrane proteins function in transport, in information transfer, and as enzymes

Why does the plasma membrane require so many different proteins? This diversity reflects the multitude of activities that take place in or on the membrane. Proteins associated with the membrane are essential for many of these activities. Generally, plasma

Key Point

Cell proteins perform many functions, including transporting materials, serving as enzymes for chemical reactions, and transmitting cell signals.

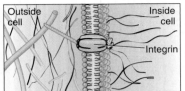

(a) Anchoring. Some membrane proteins, such as integrins, anchor the cell to the extracellular matrix; they also connect to microfilaments within the cell.

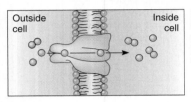

(b) Passive transport. Certain proteins form channels for selective passage of ions or molecules.

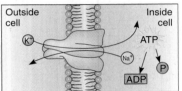

(c) Active transport. Some transport proteins pump solutes across the membrane, which requires a direct input of energy.

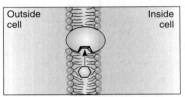

(d) Enzymatic activity. Many membrane-bound enzymes catalyze reactions that take place within or along the membrane surface.

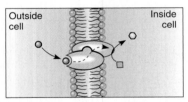

(e) Signal transduction. Some receptors bind with signal molecules such as hormones and transmit information into the cell.

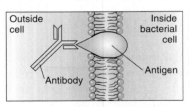

(f) Cell recognition. Some glycoproteins function as identification tags. For example, bacterial cells have surface proteins, or antigens, that human cells recognize as foreign.

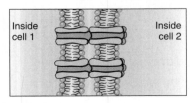

(g) Intercellular junction. Cell adhesion proteins attach membranes of adjacent cells.

Figure 5-10   Some functions of membrane proteins

membrane proteins fall into several broad functional categories, as shown in ▌Figure 5-10. Some membrane proteins anchor the cell to its substrate. For example, *integrins*, described in Chapter 4, attach to the extracellular matrix while simultaneously binding to microfilaments inside the cell (Fig. 5-10a). They also serve as receptors, or docking sites, for proteins of the extracellular matrix.

Many membrane proteins are involved in the transport of molecules across the membrane. Some form channels that selectively allow the passage of specific ions or molecules (Fig. 5-10b). Other proteins form pumps that use ATP, or other energy sources, to actively transport solutes across the membrane (Fig. 5-10c).

Certain membrane proteins are enzymes that catalyze reactions near the cell surface (Fig. 5-10d). In mitochondrial or chloroplast membranes, enzymes may be organized in a sequence to regulate a series of reactions, as in cellular respiration or photosynthesis.

Some membrane proteins are receptors that receive information from other cells in the form of chemical or electrical signals. Endocrine glands release hormones that signal other cells. Information may be transmitted to the cell interior by *signal transduction*, discussed in Chapter 6 (Fig. 5-10e).

Some membrane proteins serve as identification tags that other cells recognize. For example, certain cells recognize the surface proteins, or *antigens*, of bacterial cells as foreign. Antigens stimulate immune defenses that destroy the bacteria (Fig. 5-10f). Certain cells that recognize one another connect to form tissues. Some membrane proteins form junctions between adjacent cells (Fig. 5-10g). These proteins may also serve as anchoring points for networks of cytoskeletal elements.

Review

▌ What molecules are responsible for the physical properties of a cell membrane?

▌ How might a transmembrane protein be positioned in a lipid bilayer? How do the hydrophilic and hydrophobic regions of the protein affect its orientation?

▌ What is the pathway used by cells to place carbohydrates on plasma membrane proteins?

▌ Describe at least five functions of the plasma membrane.

PASSAGE OF MATERIALS THROUGH CELL MEMBRANES

Learning Objective

5   Describe the importance of selectively permeable membranes, and compare the functions of carrier proteins and channel proteins.

A membrane is *permeable* to a given substance if it allows that substance to pass through, and impermeable if it does not. Biological membranes are **selectively permeable membranes**—they allow some but not all substances to pass through them. In response to varying environmental conditions or cell needs, a membrane may be a barrier to a particular substance at one time and actively

Biological membranes present a barrier to polar molecules

In general, biological membranes are most permeable to small nonpolar (hydrophobic) molecules. Such molecules can pass through the hydrophobic lipid bilayer. Gases such as oxygen and carbon dioxide are small, nonpolar molecules that cross the lipid bilayer rapidly. Although they are polar, water molecules are small enough to pass through gaps that occur as a fatty acid chain momentarily moves out of the way, and they cross the lipid bilayer slowly.

The lipid bilayer of the plasma membrane is relatively impermeable to ions and to most large polar molecules. Ions are important in cell signaling and many other physiological processes. For example, many cell processes, such as muscle contraction, depend on changes in the cytoplasmic concentration of calcium ions. However, because of their electric charges, ions cannot easily cross a lipid bilayer, which is relatively impermeable to charged ions of any size, so ions pass through the bilayer slowly.

Glucose, amino acids, and most other compounds required in metabolism are polar molecules that also pass through the lipid bilayer slowly. This is advantageous to cells because the impermeability of the plasma membrane prevents them from diffusing out. How then do cells obtain the ions and polar molecules they require?

Transport proteins transfer molecules across membranes

Systems of **transport proteins** that move ions, amino acids, sugars, and other needed polar molecules through membranes apparently evolved very early in the origin of cells. These transmembrane proteins have been found in all biological membranes. Two main types of membrane transport proteins are carrier proteins and channel proteins. Each type of transport protein transports a specific type of ion or molecule or a group of related substances.

Carrier proteins, also called **transporters,** bind the ion or molecule and undergo changes in shape, resulting in movement of the molecule across the membrane. Transfer of solutes by carrier proteins located within the membrane is called **carrier-mediated transport.** As we will discuss, the two forms of carrier-mediated transport—facilitated diffusion and carrier-mediated active transport—differ in their capabilities and energy sources.

ABC transporters make up a large, important group of carrier proteins. The acronym *ABC* stands for ATP-binding cassette. Found in the cell membranes of all species, ABC transporters use the energy of ATP hydrolysis to transport certain ions, sugars, and polypeptides across cell membranes. Scientists have identi-fied about 48 ABC transporters in human cells. Mutations in the genes encoding these proteins cause or contribute to many human disorders, including cystic fibrosis and certain neurological diseases. ABC transporters transport hydrophobic drugs out of the cell. This response can be a problem clinically because certain transporters remove antibiotics, antifungal drugs, and anticancer drugs.

Channel proteins form tunnels, called pores, through the membrane. Many of these channels are *gated,* which means that they can be opened and closed. Cells regulate the passage of materials through the channels by opening and closing the gates in response to electrical changes, chemical stimuli, or mechanical stimuli. Water and specific types of ions are transported through channels. There are numerous ion channels in every membrane of every cell.

Porins are transmembrane proteins that allow various solutes or water to pass through membranes. These proteins are rolled-up, barrel-shaped β-pleated sheets that form pores. Researchers Peter Agre of Johns Hopkins School of Medicine in Baltimore, Maryland, and Roderick MacKinnon of the Howard Hughes Medical Institute at Rockefeller University in New York shared the 2003 Nobel Prize in Chemistry for their work on transport proteins. Agre identified transmembrane proteins called **aquaporins** that function as gated water channels. These channel proteins facilitate the rapid transport of water through the plasma membrane in response to osmotic gradients. About a billion water molecules per second can pass through an aquaporin! These channels are very selective and do not permit passage of ions and other small molecules. Aquaporins have been identified in all cells studied—from bacterial cells to human cells. In some cells, such as those lining the kidney tubules of mammals, aquaporins respond to specific signals from hormones. Aquaporins help prevent dehydration by returning water from the kidney tubules into the blood.

Review

- What types of molecules pass easily through the plasma membrane?
- What are the two main types of transport proteins? What are their functions?
- What are aquaporins?

PASSIVE TRANSPORT

Learning Objectives

6 Contrast simple diffusion with facilitated diffusion.
7 Define *osmosis,* and solve simple problems involving osmosis; for example, predict whether cells will swell or shrink under various osmotic conditions.

Passive transport does not require the cell to expend metabolic energy. Many ions and small molecules move through membranes by *diffusion.* Two types of diffusion are simple diffusion and facilitated diffusion.

(Text continued from previous page:) promote its passage at another time. By regulating chemical traffic across its plasma membrane, a cell controls its volume and its internal ionic and molecular composition, which can be quite different from its external environment.

Diffusion occurs down a concentration gradient

Some substances pass into or out of cells and move about within cells by **diffusion,** a physical process based on random motion. All atoms and molecules possess kinetic energy, or energy of motion, at temperatures above absolute zero (0 K, −273°C, or −459.4°F). Matter may exist as a solid, liquid, or gas, depending on the freedom of movement of its constituent particles. The particles of a solid are closely packed, and the forces of attraction between them let them vibrate but not move around. In a liquid the particles are farther apart; the intermolecular attractions are weaker, and the particles move about with considerable freedom. In a gas the particles are so far apart that intermolecular forces are negligible; molecular movement is restricted only by the walls of the container that encloses the gas. Atoms and molecules in liquids and gases move in a kind of "random walk," changing directions as they collide.

Although the movement of individual particles is undirected and unpredictable, we can nevertheless make predictions about the behavior of groups of particles. If the particles (atoms, ions, or molecules) are not evenly distributed, then at least two regions exist: one with a higher concentration of particles and the other with a lower concentration. Such a difference in the concentration of a substance from one place to another establishes a **concentration gradient.**

In diffusion, the random motion of particles results in their net movement "down" their own concentration gradient, from the region of higher concentration to the one of lower concentration. This does not mean individual particles are prohibited from moving "against" the gradient. However, because there are initially more particles in the region of high concentration, it logically follows that more particles move randomly from there into the low-concentration region than the reverse (∎ Fig. 5-11).

Thus, if a membrane is permeable to a substance, there is net movement from the side of the membrane where it is more highly concentrated to the side where it is less concentrated. Such a gradient across the membrane is a form of stored energy. Stored energy is *potential energy,* which is the capacity to do work as a result of position or state. The stored energy of the concentration gradient is released when ions or molecules move from a region of high concentration to one of low concentration. For this reason, movement down a concentration gradient is spontaneous. (Forms of energy and spontaneous processes are discussed in greater detail in Chapter 7.)

Diffusion occurs rapidly over very short distances. The rate of diffusion is determined by the movement of the particles, which in turn is a function of their size and shape, their electric charges, and the temperature. As the temperature rises, particles move faster and the rate of diffusion increases.

Particles of different substances in a mixture diffuse independently of one another. Diffusion moves solutes toward a state of equilibrium. If particles are not added to or removed from the system, a state of **dynamic equilibrium** is reached. In this condition, the particles are uniformly distributed and there is no net change in the system. Particles continue to move back and forth across the membrane, but they move at equal rates and in both directions.

In organisms, equilibrium is rarely attained. For example, human cells continually produce carbon dioxide as sugars and other molecules are metabolized during aerobic respiration. Carbon dioxide readily diffuses across the plasma membrane but then is rapidly removed by the blood. This limits the opportunity for the molecules to re-enter the cell, so a sharp concentration gradient of carbon dioxide molecules always exists across the plasma membrane.

In **simple diffusion** through a biological membrane, small, nonpolar (uncharged) solute molecules move directly through the membrane down their concentration gradient. Oxygen and carbon dioxide can rapidly diffuse through the membrane. The rate of simple diffusion is directly related to the concentration of the solute; the more concentrated the solute, the more rapid the diffusion.

Osmosis is diffusion of water across a selectively permeable membrane

Osmosis is a special kind of diffusion that involves the net movement of water (the principal *solvent* in biological systems) through a selectively permeable membrane from a region of higher concentration to a region of lower concentration. Water molecules pass freely in both directions, but as in all types of diffusion, *net* movement is from the region where the water molecules are more concentrated to the region where they are less concentrated. Most *solute molecules* (such as sugar and salt) cannot diffuse freely through the selectively permeable membranes of the cell.

The principles involved in osmosis can be illustrated using an apparatus called a U-tube (∎ Fig. 5-12). The U-tube is divided into two sections by a selectively permeable membrane that al-

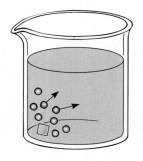 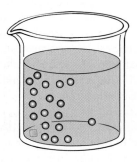 

① When lump of sugar is dropped into beaker of pure water, sugar molecules begin to dissolve and diffuse through water.

② Sugar molecules continue to dissolve and diffuse through water.

③ Eventually sugar molecules become distributed randomly throughout water.

Figure 5-11 *Animated* Diffusion

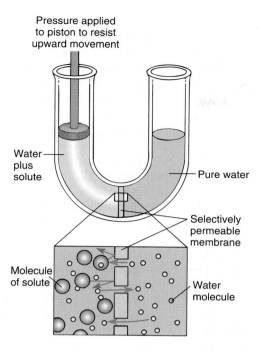

Pressure applied
to piston to resist
upward movement

Water
plus
solute

Pure water

Selectively
permeable
membrane

Molecule
of solute

Water
molecule

Figure 5-12 *Animated* Osmosis

The U-tube contains pure water on the right and water plus a solute on the left, separated by a selectively permeable membrane. Water molecules cross the membrane in both directions (*red arrows*). Solute molecules cannot cross (*green arrows*). The fluid level would normally rise on the left and fall on the right because net movement of water would be to the left. However, the piston prevents the water from rising. The force that must be exerted by the piston to prevent the rise in fluid level is equal to the osmotic pressure of the solution.

lows solvent (water) molecules to pass freely but excludes solute molecules. A water/solute solution is placed on one side, and pure water is placed on the other. The side containing the solute has a lower effective concentration of water than the pure water side does. The reason is that the solute particles, which are charged (ionic) or polar, interact with the partial electric charges on the polar water molecules. Many of the water molecules are thus "bound up" and no longer free to diffuse across the membrane.

Because of the difference in effective water concentration, there is net movement of water molecules from the pure water side (with a high effective concentration of water) to the water/solute side (with a lower effective concentration of water). As a result, the fluid level drops on the pure water side and rises on the water/solute side. Because the solute molecules do not diffuse across the membrane, equilibrium is never attained. Net movement of water continues, and the fluid level rises on the side containing the solute. The weight of the rising column of fluid eventually exerts enough pressure to stop further changes in fluid levels, although water molecules continue

to pass through the selectively permeable membrane in both directions.

We define the **osmotic pressure** of a solution as the pressure that must be exerted on the side of a selectively permeable membrane containing the higher concentration of solute to prevent the diffusion of water (by osmosis) from the side containing the lower solute concentration. In the U-tube example, you could measure the osmotic pressure by inserting a piston on the water/solute side of the tube and measuring how much pressure must be exerted by the piston to prevent the rise of fluid on that side of the tube. A solution with a high solute concentration has a low effective water concentration and a high osmotic pressure; conversely, a solution with a low solute concentration has a high effective water concentration and a low osmotic pressure.

Two solutions may be isotonic

Salts, sugars, and other substances are dissolved in the fluid compartment of every cell. These solutes give the cytosol a specific osmotic pressure. ▌ Table 5-1 summarizes the movement of water into and out of a solution (or cell) depending on relative solute concentrations.

When a cell is placed in a fluid with exactly the same osmotic pressure, no net movement of water molecules occurs, either into or out of the cell. The cell neither swells nor shrinks. Such a fluid is of equal solute concentration, or **isotonic,** to the fluid within the cell. Normally, your blood plasma (the fluid component of blood) and all your other body fluids are isotonic to your cells; they contain a concentration of water equal to that in the cells. A solution of 0.9% sodium chloride (sometimes called *physiological saline*) is isotonic to the cells of humans and other mammals. Human red blood cells placed in 0.9% sodium chloride neither shrink nor swell (▌ Fig. 5-13a).

One solution may be hypertonic and the other hypotonic

If the surrounding fluid has a concentration of dissolved substances greater than the concentration within the cell, the fluid has a higher osmotic pressure than the cell and is said to be **hypertonic** to the cell. Because a hypertonic solution has a lower effective water concentration, a cell placed in such a solution shrinks as it loses water by osmosis. Human red blood cells placed in a solution of 1.3% sodium chloride shrivel (▌ Fig. 5-13b).

If the surrounding fluid contains a lower concentration of dissolved materials than does the cell, the fluid has a lower

TABLE 5-1			
Osmotic Terminology			
Solute Concentration in Solution A	Solute Concentration in Solution B	Tonicity	Direction of Net Movement of Water
Greater	Less	A hypertonic to B; B hypotonic to A	B to A
Less	Greater	B hypertonic to A; A hypotonic to B	A to B
Equal	Equal	A and B are isotonic to each other	No net movement

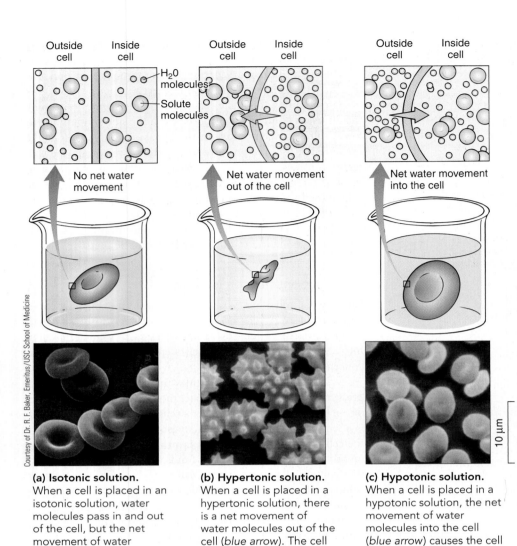

(a) Isotonic solution.
When a cell is placed in an isotonic solution, water molecules pass in and out of the cell, but the net movement of water molecules is zero.

(b) Hypertonic solution.
When a cell is placed in a hypertonic solution, there is a net movement of water molecules out of the cell (*blue arrow*). The cell becomes dehydrated and shrunken.

(c) Hypotonic solution.
When a cell is placed in a hypotonic solution, the net movement of water molecules into the cell (*blue arrow*) causes the cell to swell or even burst.

Figure 5-13 *Animated* The responses of animal cells to osmotic pressure differences

Water moves into the cells by osmosis, filling their central vacuoles and distending the cells. The cells swell, building up **turgor pressure** against the rigid cell walls (Fig. 5-14a). The cell walls stretch only slightly, and a steady state is reached when their resistance to stretching prevents any further increase in cell size and thereby halts the *net movement* of water molecules into the cells. (Of course, molecules continue to move back and forth across the plasma membrane.) Turgor pressure in the cells is an important factor in supporting the body of a nonwoody plant.

If a cell that has a cell wall is placed in a hypertonic medium, the cell loses water to its surroundings. Its contents shrink, and the plasma membrane separates from the cell wall, a process known as **plasmolysis** (Fig. 5-14b and c). Plasmolysis occurs in plants when the soil or water around them contains high concentrations of salts or fertilizers. It also explains why lettuce becomes limp in a salty salad dressing and why a picked flower wilts from lack of water.

Facilitated diffusion occurs down a concentration gradient

In all processes in which substances move across membranes by diffusion, the net transfer of those molecules from one side to the other occurs as a result of a concentration gradient. We have seen that small, uncharged (nonpolar) solute molecules, such as oxygen and carbon dioxide, move directly through the membrane down their concentration gradient by simple diffusion. In **facilitated diffusion,** a specific transport protein makes the membrane permeable to a particular solute, such as a specific ion or polar molecule. A specific solute can be transported from inside the cell to the outside or from the outside to the inside, but net movement is always from a region of higher solute concentration to a region of lower concentration. Channel proteins and carrier proteins facilitate diffusion by different mechanisms.

Channel proteins form hydrophilic channels through membranes

Some channel proteins are porins that form rather large tunnels through which water and other solutes pass. However, most channel proteins form narrow channels that transport specific ions down their gradients (Fig. 5-15). (As we will discuss, be-

osmotic pressure and is said to be **hypotonic** to the cell; water then enters the cell and causes it to swell. Red blood cells placed in a solution of 0.6% sodium chloride gain water, swell (Fig. 5-13c), and may eventually burst. Many cells that normally live in hypotonic environments have adaptations to prevent excessive water accumulation. For example, certain protists such as *Paramecium* have contractile vacuoles that expel excess water (see Fig. 25-7).

Turgor pressure is the internal hydrostatic pressure usually present in walled cells

Plant cells, algae, bacteria, and fungi have relatively rigid cell walls. These cells can withstand, without bursting, an external medium that is very dilute, containing only a very low concentration of solutes. Because of the substances dissolved in the cytoplasm, the cells are hypertonic to the outside medium (conversely, the outside medium is hypotonic to the cytoplasm).

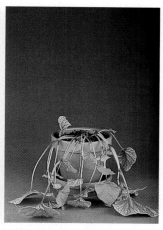

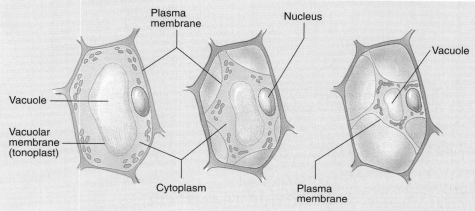

Plasma membrane

Nucleus

Vacuole

Vacuole

Vacuolar membrane (tonoplast)

Cytoplasm

Plasma membrane

(a) In hypotonic surroundings, the vacuole of a plant cell fills with water, but the rigid cell walls prevent the cell from expanding. The cells of this healthy begonia plant are turgid.

(b) When the begonia plant is exposed to a hypertonic solution, its cells become plasmolyzed as they lose water.

(c) The plant wilts and eventually dies.

Figure 5-14 *Animated* Turgor pressure and plasmolysis

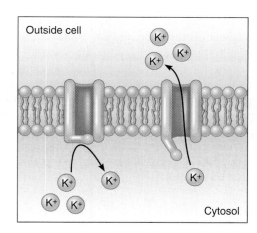

Outside cell

Cytosol

Figure 5-15 Facilitated diffusion of potassium ions

In response to an electrical stimulus, the gate of the potassium ion channel opens, allowing potassium to diffuse out of the cell.

cause ions are charged particles, these gradients are electrochemical gradients.) These *ion channels* are referred to as *gated channels* because they can open and close. As many as 100 million ions per second can pass through an open ion channel! Channels can only facilitate transport down a gradient. They cannot actively transport ions from a region of lower concentration to a region of higher concentration.

Carrier proteins undergo a change in shape

Transport of solutes through carrier proteins is slower than through channel proteins. The carrier protein binds with one or more solute molecules on one side of the membrane. The protein then undergoes a conformational change (change in shape) that moves the solute to the other side of the membrane.

As an example of facilitated diffusion by a carrier protein, let us consider glucose transport. A carrier protein known as *glucose transporter 1*, or *GLUT 1*, transports glucose into red blood cells (Fig. 5-16). The concentration of glucose is higher in the blood plasma than in red blood cells, so glucose diffuses down its concentration gradient into these blood cells. The GLUT 1 transporter facilitates glucose diffusion, allowing glucose to enter the cell about 50,000 times as rapidly as it could by simple diffusion.

Red blood cells keep the internal concentration of glucose low by immediately adding a phosphate group to entering glucose molecules, converting them to highly charged glucose phosphates that cannot pass back through the membrane. Because glucose phosphate is a different molecule, it does not contribute to the glucose concentration gradient. Thus, a steep concentration gradient for glucose is continually maintained, and glucose rapidly diffuses into the cell, only to be immediately changed to the phosphorylated form. Facilitated diffusion is powered by the concentration gradient.

Researchers have studied facilitated diffusion of glucose using **liposomes,** artificial vesicles surrounded by phospholipid bilayers. The phospholipid membrane of a liposome does not allow the passage of glucose unless a glucose transporter is present in the liposome membrane. Glucose transporters and similar carrier proteins temporarily bind to the molecules they transport. This mechanism appears to be similar to the way an enzyme binds with its substrate, the molecule on which it acts (discussed in Chapter 7). In addition, as in enzyme action, binding apparently changes the shape of the carrier protein. This change allows the glucose molecule to be released on the inside of the cell. Accord-

Facilitated diffusion requires the potential energy of a concentration gradient.

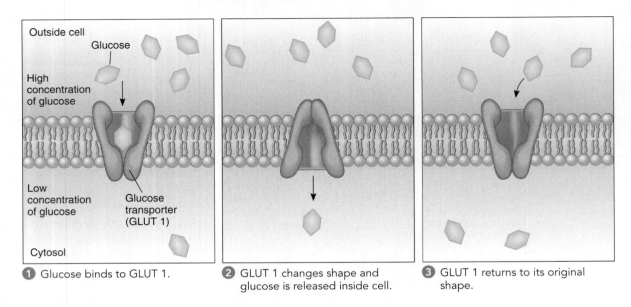

① Glucose binds to GLUT 1.

② GLUT 1 changes shape and glucose is released inside cell.

③ GLUT 1 returns to its original shape.

Figure 5-16 *Animated* Facilitated diffusion of glucose molecules

ing to this model, when the glucose is released into the cytoplasm, the carrier protein reverts to its original shape and is available to bind another glucose molecule on the outside of the cell.

Another similarity to enzyme action is that carrier proteins become saturated when there is a high concentration of the molecule being transported. This saturation may occur because a finite number of carrier proteins are available and they operate at a defined maximum rate. When the concentration of solute molecules to be transported reaches a certain level, all the carrier proteins are working at their maximum rate.

It is a common misconception that diffusion, whether simple or facilitated, is somehow "free of cost" and that only active transport mechanisms require energy. Because diffusion always involves the net movement of a substance down its concentration gradient, we say that the concentration gradient "powers" the process. However, energy is required to do the work of establishing and maintaining the gradient. In our example of facilitated diffusion of glucose, the cell maintains a steep concentration gradient (high outside, low inside) by phosphorylating the glucose molecules once they enter the cell. One ATP molecule is spent for every glucose molecule phosphorylated, and there are additional costs, such as the energy required to make the enzymes that carry out the reaction.

Review

- What is the immediate source of energy for simple diffusion? For facilitated diffusion?

- In what direction do particles move along their concentration gradient? Would your answers be different for facilitated diffusion compared with simple diffusion?

- What would happen if a plant cell were placed in an isotonic solution? A hypertonic environment? A hypotonic environment? How would you modify your predictions for an animal cell?

ACTIVE TRANSPORT

Learning Objective

8   Describe active transport, and compare direct and indirect active transport.

Although adequate amounts of a few substances move across cell membranes by diffusion, cells must actively transport many solutes *against* a concentration gradient. The reason is that cells require many substances in concentrations higher than their concentration outside the cell.

Both diffusion and active transport require energy. The energy for diffusion is provided by a concentration gradient for the substance being transported. Active transport requires the cell to expend metabolic energy directly to power the process.

An **active transport** system can pump materials from a region of low concentration to a region of high concentration. The energy stored in the concentration gradient not only is unavailable to the system but actually works against it. For this reason, the cell needs some other source of energy. In many cases, cells use ATP energy directly. However, active transport may be coupled to ATP indirectly. In indirect active transport, a concentration gradient provides energy for the cotransport of some other substance, such as an ion.

Active transport systems "pump" substances against their concentration gradients

One of the most striking examples of an active transport mechanism is the **sodium–potassium pump** found in all animal cells (Fig. 5-17). The pump is an ABC transporter, a specific carrier

Key Point

The sodium–potassium pump is a carrier protein that maintains an electrochemical gradient across the plasma membrane.

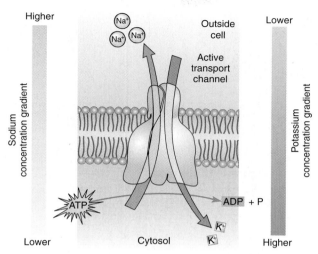

(a) The sodium–potassium pump is a carrier protein that requires energy from ATP. In each complete pumping cycle, the energy of one molecule of ATP is used to export three sodium ions (Na^+) and import two potassium ions (K^+).

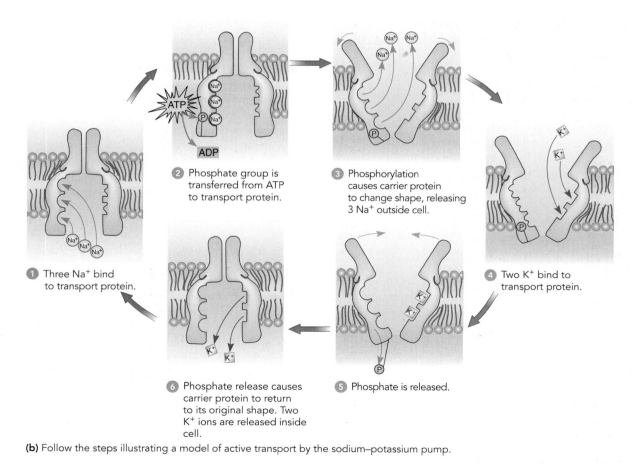

1 Three Na^+ bind to transport protein.

2 Phosphate group is transferred from ATP to transport protein.

3 Phosphorylation causes carrier protein to change shape, releasing 3 Na^+ outside cell.

4 Two K^+ bind to transport protein.

5 Phosphate is released.

6 Phosphate release causes carrier protein to return to its original shape. Two K^+ ions are released inside cell.

(b) Follow the steps illustrating a model of active transport by the sodium–potassium pump.

Figure 5-17 *Animated*   A model for the pumping cycle of the sodium–potassium pump

protein in the plasma membrane. This transporter uses the energy from ATP hydrolysis to pump sodium ions out of the cell and potassium ions into the cell. The exchange is unequal: usually only two potassium ions are imported for every three sodium ions exported. Because these particular concentration gradients involve ions, an electrical potential (separation of electric charges) is generated across the membrane; that is, the membrane is **polarized.**

Both sodium and potassium ions are positively charged, but because there are fewer potassium ions inside relative to the sodium ions outside, the inside of the cell is negatively charged relative to the outside. The unequal distribution of ions establishes an **electrical gradient** that drives ions across the plasma membrane. Sodium–potassium pumps help maintain a separation of charges across the plasma membrane. This separation is called a **membrane potential.** Because there is both an electric charge difference and a concentration difference on the two sides of the membrane, the gradient is called an **electrochemical gradient.** Such gradients store energy (somewhat like the energy of water stored behind a dam) that is used to drive other transport systems. So important is the electrochemical gradient produced by these pumps that some cells (such as nerve cells) expend 70% of their total energy just to power this one transport system.

Sodium–potassium pumps (as well as all other ATP-driven pumps) are transmembrane proteins that extend entirely through the membrane. By undergoing a series of conformational changes, the pumps exchange sodium for potassium across the plasma membrane. Unlike what occurs in facilitated diffusion, at least one of the conformational changes in the pump cycle requires energy, which is provided by ATP. The shape of the pump protein changes as a phosphate group from ATP first binds to it and is subsequently removed later in the pump cycle.

The use of electrochemical potentials for energy storage is not confined to the plasma membranes of animal cells. Bacterial, fungal, and plant cells use carrier proteins, known as proton pumps, to actively transport hydrogen ions (which are protons) out of the cell. These ATP-driven membrane pumps transfer protons from the cytosol to the outside (Fig. 5-18). Removal of positively charged protons from the cytoplasm of these cells results in a large difference in the concentration of protons between the outside and inside of the cell. The outside of the cells is positively charged relative to the inside of the plasma membrane. The energy stored in these electrochemical gradients can be used to do many kinds of cell work.

Other proton pumps are used in "reverse" to synthesize ATP. Bacteria, mitochondria, and chloroplasts use energy from food or sunlight to establish proton concentration gradients (discussed in Chapters 8 and 9). When the protons diffuse through the proton carriers from a region of high proton concentration to one of low concentration, ATP is synthesized. These electrochemical gradients form the basis for the major energy conversion systems in virtually all cells.

Ion pumps have other important roles. For example, they are instrumental in the ability of an animal cell to equalize the osmotic pressures of its cytoplasm and its external environment. If an animal cell does not control its internal osmotic pressure, its contents become hypertonic relative to the exterior. Water enters by osmosis, causing the cell to swell and possibly burst (see

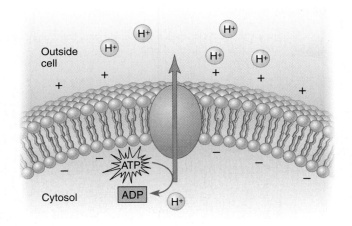

Figure 5-18 A model of a proton pump

Proton pumps use the energy of ATP to transport protons (hydrogen ions) across membranes. The energy of the electrochemical gradient established can then be used for other processes.

Fig. 5-13c). By controlling the ion distribution across the membrane, the cell indirectly controls the movement of water, because when ions are pumped out of the cell, water leaves by osmosis.

Carrier proteins can transport one or two solutes

You may have noticed that some carrier proteins, such as proton pumps, transport one type of substance in one direction. These carrier proteins are called **uniporters.** Other carrier proteins are **symporters** that move two types of substances in one direction. For example, a specific carrier protein transports both sodium and glucose into the cell. Still other carrier proteins are **antiporters** that move two substances in opposite directions. Sodium–potassium pumps transport sodium ions out of the cell and potassium ions into the cell. Both symporters and antiporters cotransport solutes.

Cotransport systems indirectly provide energy for active transport

A **cotransport system** moves solutes across a membrane by **indirect active transport.** Two solutes are transported at the same time. The movement of one solute down its concentration gradient provides energy for transport of some other solute up its concentration gradient. However, an energy source such as ATP is required to power the pump that produces the concentration gradient.

Sodium–potassium pumps (and other pumps) generate electrochemical concentration gradients. Sodium is pumped out of the cell and then diffuses back in by moving down its concentration gradient. This process generates sufficient energy to power the active transport of other essential substances. In these systems, a carrier protein *cotransports* a solute *against* its concentra-

A carrier protein transports sodium ions down their concentration gradient and uses that energy to cotransport glucose molecules against their concentration gradient.

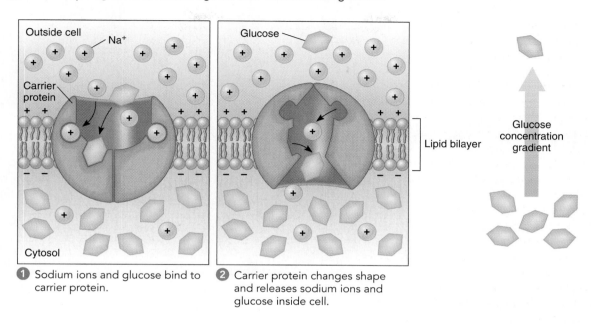

① Sodium ions and glucose bind to carrier protein.

② Carrier protein changes shape and releases sodium ions and glucose inside cell.

Figure 5-19   A model for the cotransport of glucose and sodium ions
Note that this carrier protein is a symporter.

tion gradient, while sodium, potassium, or hydrogen ions move *down* their gradient. Energy from ATP produces the ion gradient. Then the energy of this gradient drives the active transport of a required substance, such as glucose, against its gradient.

We have seen how glucose can be moved into the cell by facilitated diffusion. Glucose can also be cotransported into the cell. The sodium concentration inside the cell is kept low by the ATP-requiring sodium–potassium pumps that actively transport sodium ions out of the cell. In glucose cotransport, a carrier protein transports both sodium and glucose (▌ Fig. 5-19). As sodium moves into the cell along its concentration gradient, the carrier protein captures the energy released and uses it to transport glucose into the cell. Thus, this indirect active transport system for glucose is "driven" by the cotransport of sodium.

Review

▌   What is the energy source for active transport?
▌   What is the energy source for linked cotransport?

EXOCYTOSIS AND ENDOCYTOSIS

Learning Objective

9   Compare exocytotic and endocytotic transport mechanisms.

In both simple and facilitated diffusion and in carrier-mediated active transport, individual molecules and ions pass through the plasma membrane. Some larger materials, such as large mole-

cules, particles of food, and even small cells, are also moved into or out of cells. They can be moved by exocytosis and endocytosis, active transport mechanisms that require cells to expend energy directly.

In exocytosis, vesicles export large molecules

In **exocytosis,** a cell ejects waste products, or specific products of secretion such as hormones, by the fusion of a vesicle with the plasma membrane (▌ Fig. 5-20). Exocytosis results in the incorporation of the membrane of the secretory vesicle into the plasma membrane as the contents of the vesicle are released from the cell. This is also the primary mechanism by which plasma membranes grow larger.

In endocytosis, the cell imports materials

In **endocytosis,** materials are taken into the cell. Several types of endocytotic mechanisms operate in biological systems, including phagocytosis, pinocytosis, and receptor-mediated endocytosis.

In **phagocytosis** (literally, "cell eating"), the cell ingests large solid particles such as bacteria and food (▌ Fig. 5-21). Phagocytosis is used by certain protists to ingest food and by some types of vertebrate cells, including certain white blood cells, to ingest bacteria and other particles. During ingestion, folds of the plasma membrane enclose the particle, which has bound to the surface of the cell, and form a large membranous sac, or vacuole. When

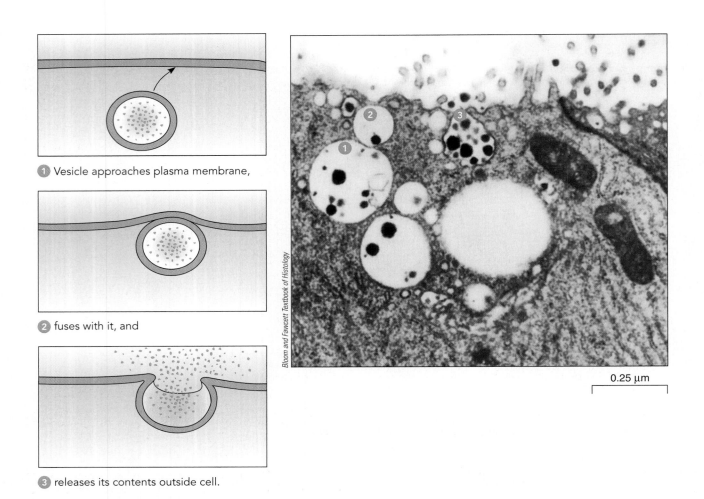

① Vesicle approaches plasma membrane,

② fuses with it, and

③ releases its contents outside cell.

Bloom and Fawcett Textbook of Histology

0.25 μm

Figure 5-20 Exocytosis

The TEM shows exocytosis of the protein components of milk by a mammary gland cell.

the membrane has encircled the particle, the membrane fuses at the point of contact. The vacuole then fuses with lysosomes, and the ingested material is degraded.

In **pinocytosis** ("cell drinking"), the cell takes in dissolved materials (⫿ Fig. 5-22). Tiny droplets of fluid are trapped by folds in the plasma membrane, which pinch off into the cytosol as tiny vesicles. The liquid contents of these vesicles are then slowly transferred into the cytosol; the vesicles become progressively smaller.

In a third type of endocytosis, **receptor-mediated endocytosis,** specific molecules combine with receptor proteins in the plasma membrane. Receptor-mediated endocytosis is the main mechanism by which eukaryotic cells take in macromolecules. As an example, let us look at how cells take up cholesterol from the blood. Cholesterol is transported in the blood as part of particles called *low-density lipoproteins* (*LDLs;* popularly known as "bad cholesterol"). Cells use cholesterol as a component of cell membranes and as a precursor of steroid hormones.

Much of the receptor-mediated endocytosis pathway was detailed through studies at the University of Texas Health Science Center by Michael Brown and Joseph Goldstein on the LDL re-

ceptor. In 1985, these researchers were awarded the Nobel Prize in Physiology or Medicine for their pioneering work. Their findings have important medical implications because cholesterol that remains in the blood instead of entering the cells can be deposited in the artery walls, which increases the risk of cardiovascular disease.

When it needs cholesterol, the cell makes LDL receptors. The receptors are concentrated in *coated pits,* depressed regions on the cytoplasmic surface of the plasma membrane. Each pit is coated by a layer of a protein, called *clathrin,* found just below the plasma membrane. A molecule that binds specifically to a receptor is called a **ligand.** In this case, LDL is the ligand. After the LDL binds with a receptor, the coated pit forms a *coated vesicle* by endocytosis.

⫿ **Figure 5-23** shows the uptake of an LDL particle. Seconds after the vesicle moves into the cytoplasm, the coating dissociates from it, leaving an uncoated vesicle. The vesicles fuse with small compartments called *endosomes.* The LDL and LDL receptors separate, and the receptors are transported to the plasma membrane, where they are recycled. LDL is transferred to a lysosome, where it is broken down. Cholesterol is released into the cytosol

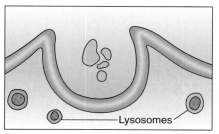

1 Folds of plasma membrane surround particle to be ingested, forming small vacuole around it.

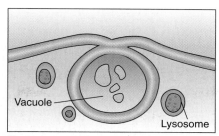

2 Vacuole then pinches off inside cell.

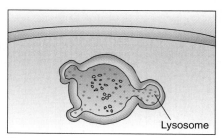

3 Lysosomes fuse with vacuole and pour potent hydrolytic enzymes onto ingested material.

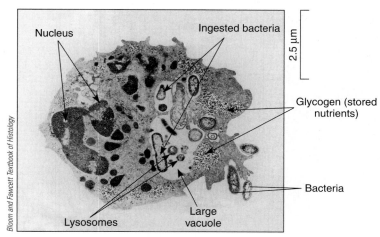

Figure 5-21 *Animated* Phagocytosis

In this type of endocytosis, a cell ingests relatively large solid particles. The white blood cell (a neutrophil) shown in the TEM is phagocytizing bacteria. The vacuoles contain bacteria that have already been ingested. Lysosomes contain digestive enzymes that break down the ingested material. Other bacteria are visible outside the cell.

for use by the cell. A simplified summary of receptor-mediated endocytosis follows:

> Ligand binds to receptors in coated pits of plasma membrane ⟶ coated vesicle forms by endocytosis ⟶ coating detaches from vesicle ⟶ uncoated vesicle fuses with endosome ⟶ ligands separate from receptors and are recycled; endosome fuses with primary lysosome, forming secondary lysosome ⟶ contents of secondary lysosome are digested and released into the cytosol

The recycling of LDL receptors to the plasma membrane through vesicles causes a problem common to all cells that use endocytotic and exocytotic mechanisms: the plasma membrane changes size as the vesicles bud off from it or fuse with it. A type of phagocytic cell known as a *macrophage*, for example, ingests the equivalent of its entire plasma membrane in about 30 minutes, requiring an equivalent amount of recycling or new membrane synthesis for the cell to maintain its surface area. On the other hand, cells that are constantly involved in secretion must return an equivalent amount of membrane to the interior of the cell for each vesicle that fuses with the plasma membrane; if not, the cell

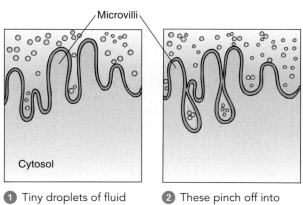
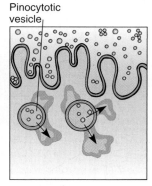

1 Tiny droplets of fluid are trapped by folds of plasma membrane.

2 These pinch off into cytosol as small fluid-filled vesicles.

3 Contents of these vesicles are then slowly transferred to cytosol.

Figure 5-22 Pinocytosis, or "cell drinking"

Key Point

In receptor-mediated endocytosis specific macromolecules bind to receptor proteins, accumulate in coated pits, and enter the cell in clathrin-coated vesicles.

1. LDL attaches to specific receptors in coated pits on plasma membrane.

2. Endocytosis results in formation of a coated vesicle in cytosol. Seconds later coat is removed.

3. Uncoated vesicle fuses with endosome.

4. Receptors are returned to plasma membrane and recycled.

5. Vesicle containing LDL particles fuses with a lysosome, forming a secondary lysosome. Hydrolytic enzymes then digest cholesterol from LDL particles for use by cell.

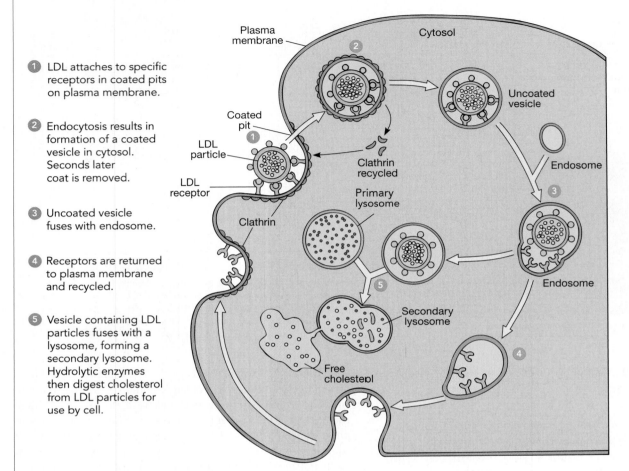

(a) Uptake of low-density lipoprotein (LDL) particles, which transport cholesterol in the blood.

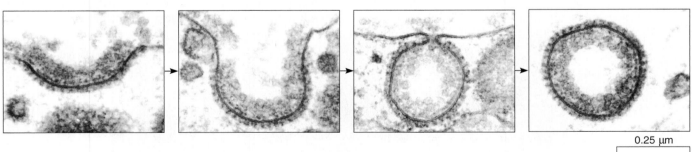

0.25 μm

(b) This series of TEMs shows the formation of a coated vesicle from a coated pit.

Figure 5-23 *Animated* Receptor-mediated endocytosis

From M. M. Perry and A. B. Gilbert, *Journal of Cell Science* 39:257–272, 1979. © 1979 The Company of Biologists Ltd.

surface would continue to expand even though the growth of the cell itself may be arrested.

Review

- How are exocytosis and endocytosis similar?
- How are the processes of phagocytosis and pinocytosis different?
- What is the sequence of events in receptor-mediated endocytosis?

CELL JUNCTIONS

Learning Objective

10 Compare the structures and functions of anchoring junctions, tight junctions, gap junctions, and plasmodesmata.

Cells in close contact with one another typically develop specialized intercellular junctions. These structures may allow neighboring cells to form strong connections with one another, prevent the passage of materials, or establish rapid communication between adjacent cells. Several types of junctions connect animal cells, including anchoring junctions, tight junctions, and gap junctions. Plant cells are connected by plasmodesmata.

Anchoring junctions connect cells of an epithelial sheet

Adjacent epithelial cells, such as those found in the outer layer of the skin, are so tightly bound to each other by **anchoring junctions** that strong mechanical forces are required to separate them. Cadherins, transmembrane proteins shown in the chapter opening photograph, are important components of anchoring junctions. These junctions do not affect the passage of materials between adjacent cells. Two common types of anchoring junctions are desmosomes and adhering junctions.

Desmosomes are points of attachment between cells (Fig. 5-24). They hold cells together at one point as a rivet or a spot weld does. As a result, cells form strong sheets, and substances still pass freely through the spaces between the plasma membranes. Each desmosome is made up of regions of dense material associated with the cytosolic sides of the two plasma membranes, plus protein filaments that cross the narrow intercellular space between them. Desmosomes are anchored to systems of intermediate filaments inside the cells. Thus, the intermediate filament networks of adjacent cells are connected. As a result, mechanical stresses are distributed throughout the tissue.

Adhering junctions cement cells together. Cadherins form a continuous adhesion belt around each cell, binding the cell to neighboring cells. These junctions connect to microfilaments of the cytoskeleton. The cadherins of adhering junctions are a potential path for signals from the outside environment to be transmitted to the cytoplasm.

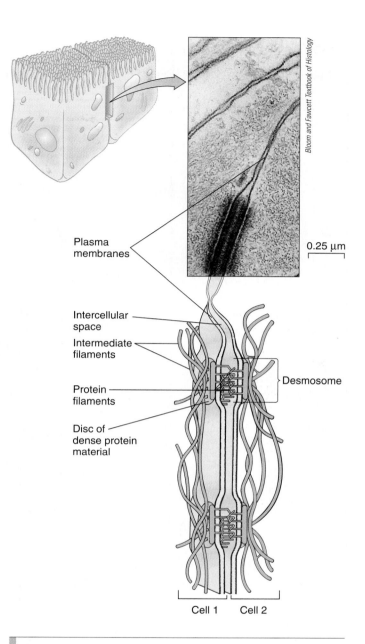

Figure 5-24 Desmosomes

The dense structure in the TEM is a desmosome. Each desmosome consists of a pair of buttonlike discs associated with the plasma membranes of adjacent cells, plus the intercellular protein filaments that connect them. Intermediate filaments in the cells are attached to the discs and are connected to other desmosomes.

Tight junctions seal off intercellular spaces between some animal cells

Tight junctions are literally areas of tight connections between the membranes of adjacent cells. These connections are so tight that no space remains between the cells and substances cannot leak between them. TEMs of tight junctions show that in the region of the junction the plasma membranes of the two cells are in actual contact with each other, held together by proteins linking

Tight junctions prevent the passage of materials through spaces between cells.

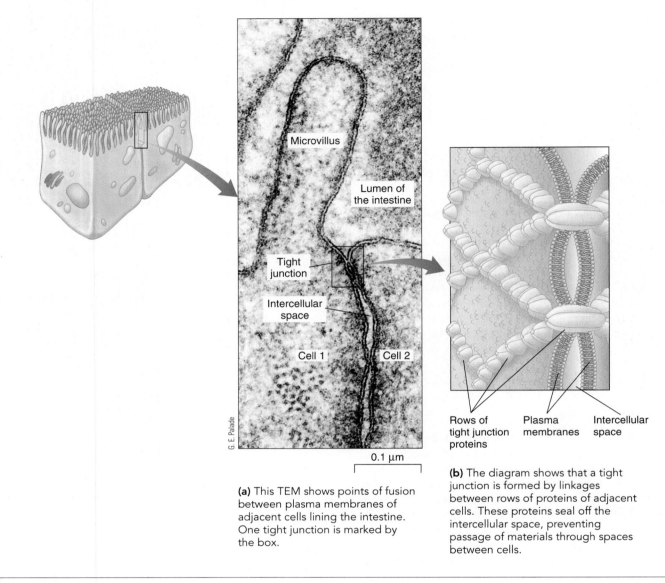

Microvillus

Lumen of the intestine

Tight junction

Intercellular space

Cell 1　　Cell 2

G. E. Palade

0.1 μm

(a) This TEM shows points of fusion between plasma membranes of adjacent cells lining the intestine. One tight junction is marked by the box.

Rows of tight junction proteins

Plasma membranes

Intercellular space

(b) The diagram shows that a tight junction is formed by linkages between rows of proteins of adjacent cells. These proteins seal off the intercellular space, preventing passage of materials through spaces between cells.

Figure 5-25　Tight junctions

the two cells. However, as shown in ▌Figure 5-25, tight junctions are located intermittently. The plasma membranes of the two cells are not fused over their entire surface.

Cells connected by tight junctions seal off body cavities. For example, tight junctions between cells lining the intestine prevent substances in the intestine from passing between the cells and directly entering the body or the blood. The sheet of cells thus acts as a selective barrier. Food substances must be transported across the plasma membranes and *through* the intestinal cells before they enter the blood. This arrangement helps prevent toxins and other unwanted materials from entering the blood and also prevents nutrients from leaking out of the intestine. Tight junc-

tions are also present between the cells that line capillaries in the brain. They form the *blood–brain barrier,* which prevents many substances in the blood from passing into the brain.

Gap junctions allow the transfer of small molecules and ions

A **gap junction** is like a desmosome in that it bridges the space between cells; however, the space it spans is somewhat narrower (▌Fig. 5-26). Gap junctions also differ in that they are communicating junctions. They not only connect the plasma membranes

Gap junctions allow the transfer of small molecules and ions between adjacent cells.

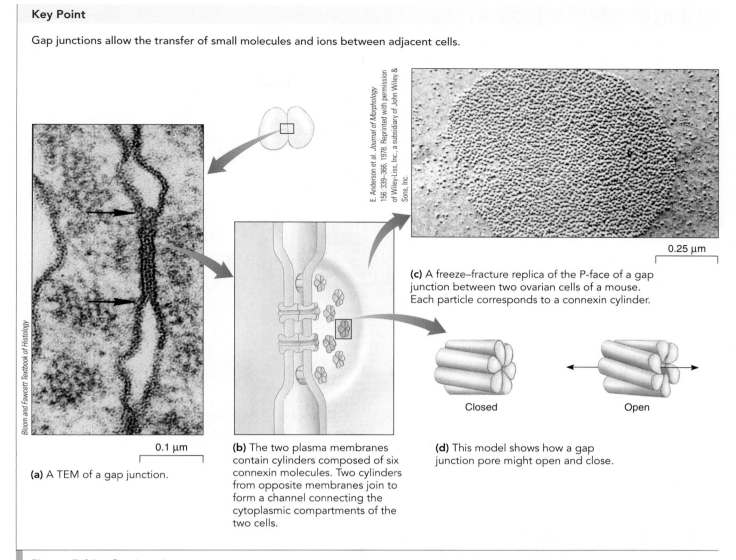

Bloom and Fawcett Textbook of Histology

E. Anderson et al. Journal of Morphology 156: 339–366, 1978. Reprinted with permission of Wiley-Liss, Inc., a subsidiary of John Wiley & Sons, Inc.

0.1 µm

(a) A TEM of a gap junction.

(b) The two plasma membranes contain cylinders composed of six connexin molecules. Two cylinders from opposite membranes join to form a channel connecting the cytoplasmic compartments of the two cells.

0.25 µm

(c) A freeze–fracture replica of the P-face of a gap junction between two ovarian cells of a mouse. Each particle corresponds to a connexin cylinder.

Closed   Open

(d) This model shows how a gap junction pore might open and close.

Figure 5-26   Gap junctions

The model of a gap junction shown in **(b)** is based on electron microscopic and X-ray diffraction data.

but also contain channels connecting the cytoplasm of adjacent cells. Gap junctions are composed of *connexin,* an integral membrane protein. Groups of six connexin molecules cluster to form a cylinder that spans the plasma membrane. The connexin cylinders on adjacent cells become tightly joined. The two cylinders form a channel, about 1.5 nm in diameter. Small inorganic molecules (such as ions) and some regulatory molecules (such as cyclic AMP) pass through the channels, but larger molecules are excluded. When a marker substance is injected into one of a group of cells connected by gap junctions, the marker passes rapidly into the adjacent cells but does not enter the space between the cells.

Gap junctions provide for rapid chemical and electrical communication between cells. Cells control the passage of materials through gap junctions by opening and closing the channels (see

Fig. 5-26d). Cells in the pancreas, for example, are linked by gap junctions in such a way that if one of a group of cells is stimulated to secrete insulin, the signal is passed through the junctions to the other cells in the cluster, ensuring a coordinated response to the initial signal. Gap junctions allow some nerve cells to be electrically coupled. Cardiac muscle cells are linked by gap junctions that permit the flow of ions necessary to synchronize contractions of the heart.

Plasmodesmata allow certain molecules and ions to move between plant cells

Because plant cells have walls, they do not need desmosomes for strength. However, these same walls could isolate the cells, preventing them from communicating. For this reason, plant cells require connections that are functionally equivalent to the gap junctions of some animal cells. **Plasmodesmata** (sing., *plasmodesma*), channels 20–40 nm wide through adjacent cell walls, connect the cytoplasm of neighboring cells (❚ Fig. 5-27). The plasma membranes of adjacent cells are continuous with each other through the plasmodesmata. Most plasmodesmata contain a narrow cylindrical structure, called the **desmotubule,** which runs through the channel and connects the smooth ER of the two adjacent cells.

Plasmodesmata generally allow molecules and ions, but not organelles, to pass through the openings from cell to cell. The movement of ions through the plasmodesmata allows for a very slow type of electrical signaling in plants. Whereas the channels of gap junctions have a fixed diameter, plants cells can dilate the plasmodesmata channels.

Review

❚ How are desmosomes and tight junctions functionally similar? How do they differ?

❚ What is the justification for considering gap junctions and plasmodesmata to be functionally similar? How do they differ structurally?

Key Point

Most cells of complex plants are connected by plasmodesmata that connect the cytoplasm of adjacent cells.

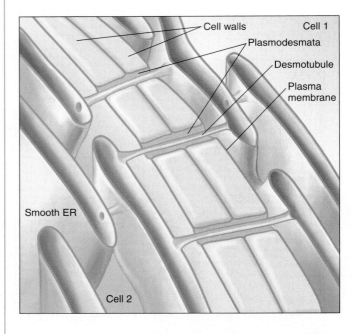

Figure 5-27 Plasmodesmata

Cytoplasmic channels through the cell walls of adjacent plant cells allow passage of water, ions, and small molecules. The channels are lined with the fused plasma membranes of the two adjacent cells.

SUMMARY WITH KEY TERMS

Learning Objectives

1 Evaluate the importance of membranes to the homeostasis of the cell, emphasizing their various functions (page 107).

   ❚ The **plasma membrane** physically separates the interior of the cell from the extracellular environment, receives information about changes in the environment, regulates the passage of materials into and out of the cell, and communicates with other cells.

   ❚ Biological membranes form compartments within eukaryotic cells that allow a variety of separate functions. Membranes participate in and serve as surfaces for biochemical reactions.

2 Describe the fluid mosaic model of cell membrane structure (page 107).

   ❚ According to the **fluid mosaic model,** membranes consist of a fluid phospholipid bilayer in which a variety of proteins are embedded.

   ❚ The phospholipid molecules are **amphipathic:** they have hydrophobic and hydrophilic regions. The hydrophilic heads of the phospholipids are at the two surfaces of the

bilayer, and their hydrophobic fatty acid chains are in the interior.

3 Relate properties of the lipid bilayer to properties and functions of cell membranes (page 107).

   ❚ In almost all biological membranes, the lipids of the bilayer are in a fluid or liquid-crystalline state, which allows the lipid molecules to move rapidly in the plane of the membrane. Proteins also move within the membrane.

   ❚ Lipid bilayers are flexible and self-sealing and can fuse with other membranes. These properties allow the cell to transport materials from one region of the cell to another; materials are transported in vesicles that bud from one cell membrane and then fuse with some other membrane.

4 Describe the ways that membrane proteins associate with the lipid bilayer, and discuss the functions of membrane proteins (page 107).

   ❚ **Integral membrane proteins** are embedded in the bilayer with their hydrophilic surfaces exposed to the aqueous environment and their hydrophobic surfaces in contact with the hydrophobic interior of the bilayer. **Transmembrane proteins** are integral proteins that extend completely through the membrane.

- **Peripheral membrane proteins** are associated with the surface of the bilayer, usually bound to exposed regions of integral proteins, and are easily removed without disrupting the structure of the membrane.
- Membrane proteins have many functions, including transporting materials, acting as enzymes or receptors, recognizing cells, and structurally linking cells.

5 Describe the importance of selectively permeable membranes, and compare the functions of carrier proteins and channel proteins (page 114).

- Biological membranes are selectively permeable membranes: they allow the passage of some substances but not others. By regulating passage of molecules that enter and leave the cell and its compartments, the cell controls its volume and the internal composition of ions and molecules.
- Membrane **transport proteins** facilitate the passage of certain ions and molecules through biological membranes. **Carrier proteins** are transport proteins that undergo a series of conformational changes as they bind and transport a specific solute. **ABC transporters** are carrier proteins that use energy from ATP to transport solutes.
- **Channel proteins** are transport proteins that form passageways through which water and certain ions travel through the membrane. **Porins** are channel proteins that form relatively large pores through the membrane for passage of water and certain solutes.

ThomsonNOW™ **Learn more about the plasma membrane and membrane proteins by clicking on the figures in ThomsonNOW.**

6 Contrast simple diffusion with facilitated diffusion (page 115).

- **Diffusion** is the net movement of a substance down its **concentration gradient** from a region of greater concentration to one of lower concentration. Diffusion and osmosis are physical processes that do not require the cell to directly expend metabolic energy.
- In **simple diffusion** through a biological membrane, solute molecules or ions move directly through the membrane down their concentration gradient. **Facilitated diffusion** uses specific transport proteins to move solutes across a membrane. As in simple diffusion, net movement is always from a region of higher concentration to a region of lower solute concentration. Facilitated diffusion cannot work against a gradient.

7 Define osmosis, and solve simple problems involving osmosis; for example, predict whether cells will swell or shrink under various osmotic conditions (page 115).

- **Osmosis** is a kind of diffusion in which molecules of water pass through a selectively permeable membrane from a region where water has a higher effective concentration to a region where its effective concentration is lower.
- The concentration of dissolved substances (solutes) in a solution determines its **osmotic pressure**. Cells regulate their internal osmotic pressures to prevent shrinking or bursting.
- An **isotonic** solution has an equal solute concentration compared to that of another fluid, for example, the fluid within the cell.

- When placed in a **hypertonic** solution, one that has a greater solute concentration than the cell's, a cell loses water to its surroundings; plant cells undergo **plasmolysis,** a process in which the plasma membrane separates from the cell wall.
- When cells are placed in a **hypotonic** solution, one that has a lower solute concentration than the cell's, water enters the cells and causes them to swell.
- Plant cells withstand high internal hydrostatic pressure because their cell walls prevent them from expanding and bursting. Water moves into plant cells by osmosis and fills the central vacuoles. The cells swell, building up **turgor pressure** against the rigid cell walls.

8 Describe active transport, and compare direct and indirect active transport (page 120).

- In **active transport,** the cell expends metabolic energy to move ions or molecules across a membrane against a concentration gradient. For example, the **sodium–potassium pump** uses ATP to pump sodium ions out of the cell and potassium ions into the cell.
- In **indirect active transport,** two solutes are transported at the same time. An ATP-powered pump such as the sodium–potassium pump maintains a concentration gradient. Then a carrier protein **cotransports** two solutes. It transports one solute (such as sodium) down its concentration gradient and uses the energy released to move another solute against its concentration gradient.

9 Compare exocytotic and endocytotic transport mechanisms (page 123).

- The cell expends metabolic energy to carry on exocytosis and endocytosis.
- In **exocytosis,** the cell ejects waste products or secretes substances such as hormones or mucus by fusion of vesicles with the plasma membrane. In this process, the surface area of the plasma membrane increases.
- In **endocytosis,** materials such as food particles are moved into the cell. A portion of the plasma membrane envelops the material, enclosing it in a vesicle or vacuole that is then released inside the cell. This process decreases the surface area of the plasma membrane. Three types of endocytosis are phagocytosis, pinocytosis, and receptor-mediated endocytosis.
- In **phagocytosis,** the plasma membrane encloses a large particle such as a bacterium or protist, forms a vacuole around it, and moves it into the cell. In **pinocytosis,** the cell takes in dissolved materials by forming tiny vesicles around droplets of fluid trapped by folds of the plasma membrane.
- In **receptor-mediated endocytosis,** specific receptors in coated pits along the plasma membrane bind **ligands.** These pits, coated by the protein clathrin, form coated vesicles by endocytosis. The vesicles eventually fuse with lysosomes, and their contents are digested and released into the cytosol.

ThomsonNOW™ **Learn more about membrane transport by clicking on the figures in ThomsonNOW.**

10 Compare the structures and functions of anchoring junctions, tight junctions, gap junctions, and plasmodesmata (page 127).

- Cells in close contact with one another may develop intercellular junctions. **Anchoring junctions** include desmosomes and adhering junctions; they are found between cells that form a sheet of tissue. **Desmosomes** spot-weld adjacent animal cells together. **Adhering junctions** are formed by cadherins that cement cells together.

▌ **Tight junctions** seal membranes of adjacent animal cells together, preventing substances from moving through the spaces between the cells.

▌ **Gap junctions,** composed of the protein connexin, form channels that allow communication between the cytoplasm of adjacent animal cells.

▌ **Plasmodesmata** are channels connecting adjacent plant cells. Openings in the cell walls allow the plasma membranes and cytosol to be continuous; certain molecules and ions pass from cell to cell.

TEST YOUR UNDERSTANDING

1. Which of the following statements is *not* true? Biological membranes (a) are composed partly of amphipathic lipids (b) have hydrophobic and hydrophilic regions (c) are typically in a fluid state (d) are made mainly of lipids and of proteins that lie like thin sheets on the membrane surface (e) function in signal transduction

2. According to the fluid mosaic model, membranes consist of (a) a lipid–protein sandwich (b) mainly phospholipids with scattered nucleic acids (c) a fluid phospholipid bilayer in which proteins are embedded (d) a fluid phospholipid bilayer in which carbohydrates are embedded (e) a protein bilayer that behaves as a liquid crystal

3. Transmembrane proteins (a) are peripheral proteins (b) are receptor proteins (c) extend completely through the membrane (d) extend along the surface of the membrane (e) are secreted from the cell

4. Which of the following is *not* a function of the plasma membrane? (a) transports materials (b) helps structurally link cells (c) manufactures proteins (d) anchors the cell to the extracellular matrix (e) has receptors that relay signals

5. ABC transporters (a) use the energy of ATP hydrolysis to transport certain ions and sugars (b) are important in facilitated diffusion of certain ions (c) are a small group of channel proteins (d) are found mainly in plant cell membranes (e) permit passive diffusion through their channels

6. When plant cells are in a hypotonic medium, they (a) undergo plasmolysis (b) build up turgor pressure (c) wilt (d) decrease pinocytosis (e) lose water to the environment

7. A laboratory technician accidentally places red blood cells in a hypertonic solution. What happens? (a) they undergo plasmolysis (b) they build up turgor pressure (c) they swell (d) they pump solutes out (e) they become dehydrated and shrunken

8. Which of the following processes requires the cell to expend metabolic energy directly (for example, from ATP)? (a) active transport (b) facilitated diffusion (c) all forms of carrier-mediated transport (d) osmosis (e) simple diffusion

9. Which of the following requires a transport protein? (a) simple diffusion (b) facilitated diffusion (c) establishment of turgor pressure (d) osmosis (e) osmosis when a cell is in a hypertonic solution

10. The action of sodium–potassium pumps is an example of (a) direct active transport (b) pinocytosis (c) aquaporin transport (d) exocytosis (e) facilitated diffusion

11. Electrochemical gradients (a) power simple diffusion (b) are established by pinocytosis (c) are necessary for transport by aquaporins (d) are established by concentration gradients (e) are a result of both an electric charge difference and a concentration difference between the two sides of the membrane

12. Which carrier protein is a symporter? (a) one that cotransports glucose and sodium (b) one that carries on facilitated diffusion (c) one that moves two substances in opposite directions (d) an ion channel (e) the sodium–potassium pump

13. In indirect active transport (a) a uniporter moves a solute across a membrane against its concentration gradient (b) a channel protein moves ions by facilitated diffusion (c) the movement of one solute down its concentration gradient provides energy for transport of some other solute up its concentration gradient (d) osmosis powers the movement of ions against their concentration gradient (e) sodium is directly transported in one direction, and potassium is indirectly transported in the same direction

14. A cell takes in dissolved materials by forming tiny vesicles around fluid droplets trapped by folds of the plasma membrane. This process is (a) indirect active transport (b) pinocytosis (c) receptor-mediated endocytosis (d) exocytosis (e) facilitated diffusion

15. Anchoring junctions that hold cells together at one point as a spot weld does are (a) tight junctions (b) adhering junctions (c) desmosomes (d) gap junctions (e) plasmodesmata

16. Junctions that permit the transfer of water, ions, and molecules between adjacent plant cells are (a) tight junctions (b) adhering junctions (c) desmosomes (d) gap junctions (e) plasmodesmata

17. Junctions that help form the blood–brain barrier are (a) tight junctions (b) adhering junctions (c) desmosomes (d) gap junctions (e) plasmodesmata

1. Why can't larger polar molecules and ions diffuse through the plasma membrane? Would it be advantageous to the cell if they could? Explain.

2. Describe one way that an ion gradient can be established and maintained.

3. Most adjacent plant cells are connected by plasmo-desmata, whereas only certain adjacent animal cells are associated through gap junctions. What might account for this difference?

4. Evaluate the importance of membranes to the cell, discussing their various functions.

5. **Evolution Link.** Hypothesis: The evolution of biological membranes was an essential step in the origin of life. Give arguments supporting (or challenging) this hypothesis.

6. **Evolution Link.** Transport proteins have been found in all biological membranes. What hypothesis could you make regarding whether these molecules evolved early or later in the history of cells? Argue in support of your hypothesis.

Additional questions are available in ThomsonNOW at www.thomsonedu.com/login

6

Cell Communication

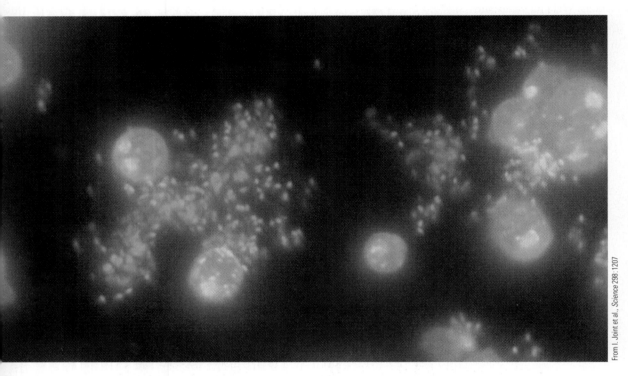

From I. Joint et al., *Science* 298:1207

Cell-to-cell communication across the prokaryote–eukaryote boundary. Spores of the green seaweed *Enteromorpha* (Domain Eukarya, Kingdom Protista) attach to biofilm-forming bacteria (Domain Bacteria, Kingdom Bacteria) in response to chemical compounds released by the bacteria. The bacteria (*blue*) were stained and visualized with blue light. The spores appear red because of the fluorescence of chlorophyll within them.

KEY CONCEPTS

Cells communicate by signaling one another using chemical compounds such as neurotransmitters, hormones, and other regulatory molecules.

A signaling molecule binds to a receptor molecule on the cell surface or within the target cell.

In signal transduction, a receptor molecule converts an extracellular signal into an intracellular signal that causes some change in the cell.

The cell responds to signals by opening or closing ion channels, activating or inhibiting enzymes, or altering the activity of specific genes.

In Chapter 1, we introduced three basic themes of biology. One of these, transmission of information, is the focus of this chapter. To maintain homeostasis, cells must continuously receive and respond to information from the outside environment. Some of that information is sent by other cells. The cells of a multicellular organism must also communicate with one another. In Chapter 5, we discussed how gap junctions provide a conduit for rapid electrical and chemical communication between neighboring cells in many animal tissues. We also described plasmodesmata, which connect adjacent plant cells. Many types of cells that are not directly connected, such as cells of the immune system, also communicate with one another.

Prokaryotes, protists, fungi, plants, and animals all communicate with other members of their species by secreting chemical signals. For example, bacteria release chemical signals that diffuse among nearby bacteria. As the population of bacteria increases, the concentration of the chemical signal increases. Through a process known as *quorum sensing*, the bacteria sense when a certain critical concentration of a signal molecule is reached. The bacteria respond by activating a specific biological process. They may form a **biofilm,** a community of microorganisms attached to a solid surface. Forming a biofilm requires the coordinated activity of numerous bacteria.

Over billions of years, elaborate systems of cell signaling have evolved. Organisms of different species, and even different kingdoms and domains, communicate with one another. As an example, the chemical signals released by bacteria can be intercepted by other organisms. During its life cycle, the green seaweed *Enteromorpha* produces spores that move about and temporarily attach to a surface. The spores sense a chemical signal released by bacteria that form biofilms. The spores move toward and attach to the individual bacteria that are part of the biofilm surface.

Cell biologists are greatly expanding their understanding of communication among cells. They have discovered that faulty signaling can cause or contribute to a variety of diseases, including cancer and diabetes. Increased understanding of cell communication mechanisms suggests new strategies for preventing or treating diseases. In this chapter, we discuss how cells send and receive signals. We then consider how information crosses the plasma membrane and is transmitted through a signal relay system. Finally, we describe some of the responses that cells make. ■

CELL SIGNALING: AN OVERVIEW

Learning Objective

1   Describe the sequence of events that takes place in cell signaling.

When food is scarce, the amoeba-like cellular slime mold *Dictyostelium* secretes the compound cyclic adenosine monophosphate (cAMP). This chemical compound diffuses through the cell's environment and binds to receptors on the surfaces of nearby cells. The activated receptors send signals into the cells that result in movement toward the cAMP. Hundreds of slime molds come together and form a multicellular slug-shaped colony (Fig. 6-1; also see Fig. 25-24).

Plants also communicate with one another. For example, diseased maple trees send airborne chemical signals that are received by uninfected trees nearby. Cells of the uninfected trees increase their chemical defenses so that they are more resistant to the disease-causing organisms. Plants also send signals to insects. When tobacco plants are attacked by herbivorous insects, the plants release volatile chemicals. These signals are received by the insects, which respond by laying fewer eggs, thus reducing the number of insects feeding on the plants. Predator insects that eat the eggs of the herbivorous insects respond to the plant signals by eating more of the herbivorous insect eggs. Thus, natural selection has resulted in plant signals that herbivorous insects detect and avoid and that carnivorous insects detect and approach. This system of information transfer helps protect plants from herbivorous insects.

The term **cell signaling** refers to the mechanisms by which cells communicate with one another. If the cells are physically close to one another, a signaling molecule on one cell may combine with a receptor on another cell. Most commonly, cells communicate by sending chemical signals over some distance.

The development and functioning of an organism require precise internal communication as well as effective responses to the outside environment. In plants and animals, hormones serve as important chemical signals between various cells and organs. In most animals, **neurons** (nerve cells) transmit information electrically and chemically.

Cell signaling involves a series of processes (Fig. 6-2). First, a cell must *send a signal.* In chemical signaling, a cell must synthesize and release signaling molecules. For example, specialized cells in the vertebrate pancreas secrete the signaling molecule insulin, a hormone. If the **target cells,** the cells that can respond to the signal, are not in close proximity, the signal must be transported to them. Blood transports insulin to target cells throughout the body. Next, target cells must receive the information being signaled. This process is called **reception.** In many cases, the

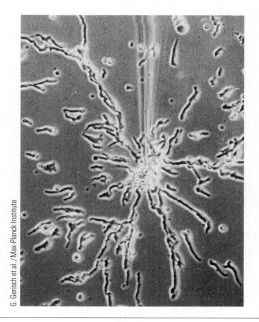

G. Gerisch et al./Max-Planck Institute

Figure 6-1   Cell signaling in cellular slime molds

When food is scarce, the amoeba-like cellular slime mold *Dictyostelium* secretes cyclic AMP. The slime molds respond to this chemical signal by aggregating. Converging streams of hundreds of individuals come together and form a multicellular colony.

signal molecule binds to a receptor on the surface of the target cell. Insulin receptors are found on the surfaces of target cells.

Many signals do not actually enter the target cell. **Signal transduction** is the process by which a cell converts an extracellular signal into an intracellular signal that results in some **response.** When insulin binds to an insulin receptor, a series of chemical reactions take place in the cell. Enzyme activity is altered, leading to metabolic changes. (Recall that *enzymes* catalyze specific chemical reactions; most enzymes are proteins.) Insulin stimulates several metabolic responses. One critical response is that the cell takes up glucose, thus lowering the concentration of glucose in the blood. Other signaling molecules may stimulate ion channels in the plasma membrane to open or close. Some signaling molecules activate or inhibit specific genes in the nucleus. These responses can result in changes in cell shape, cell growth, cell division, and cell differentiation (specialization).

The signal is *amplified* as it is relayed through the cell, so the response is much greater than would be possible if each signaling molecule acted alone. As we will discuss, the signaling process is carefully regulated. After a signal has done its job, its action must be stopped, so during the signaling process, certain mechanisms operate to *terminate the signal.*

Review

▌ What is the sequence of events that takes place in cell signaling?

SENDING SIGNALS

Learning Objective

2 Describe three types of signaling molecules: local regulators, neurotransmitters, and hormones.

Cells synthesize many different types of signals and deliver them in various ways. Some types of cells signal one another by making direct contact. Others are local regulators that affect only cells close to the signaling cell. In animals, neurons signal one another with chemical compounds called neurotransmitters. Specialized cells in plants and animals produce signaling molecules called hormones. Signaling molecules may be synthesized by neighboring cells or by specialized tissues some distance from the target cells. In animals, these distant molecules reach target cells by diffusion or via the circulatory system.

In animals, many cells in the immune system signal one another by making direct contact (▌ Fig. 6-3a). A local regulator is a signaling molecule that diffuses through the interstitial fluid (the fluid surrounding the cells) and acts on nearby cells. This is called **paracrine regulation** (▌ Fig. 6-3b). Local regulators include local chemical mediators such as growth factors, histamine, substances called prostaglandins, and nitric oxide. More than 50 **growth fac-**

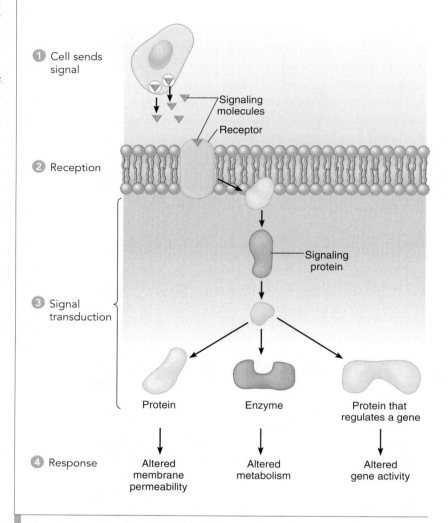

Key Point

When a signaling molecule binds to a receptor molecule on a target cell, signal transduction occurs, leading to some response in the cell.

① Cell sends signal

Signaling molecules

Receptor

② Reception

Signaling protein

③ Signal transduction

Protein   Enzyme   Protein that regulates a gene

④ Response

Altered membrane permeability   Altered metabolism   Altered gene activity

Figure 6-2 Overview of cell signaling

tors, typically peptides, stimulate cell division and normal development in specific types of cells. **Histamine** is stored in certain cells of the immune system and is released in response to allergic reactions, injury, or infection. Histamine causes blood vessels to dilate and capillaries to become more permeable.

Prostaglandins are paracrine regulators that modify cAMP levels and interact with other signaling molecules to regulate metabolic activities. Some prostaglandins target smooth muscle. Certain prostaglandins stimulate smooth muscle to contract, whereas others cause relaxation. When smooth muscle in blood vessel walls contracts, blood vessels constrict and blood pressure rises. In contrast, when smooth muscle in blood vessel walls relaxes, blood vessels dilate and blood pressure decreases.

Nitric oxide (NO), another local regulator, is a gas produced by many types of cells, including plant and animal cells. Nitric

Cell signal one another via neurotransmitters, hormones,

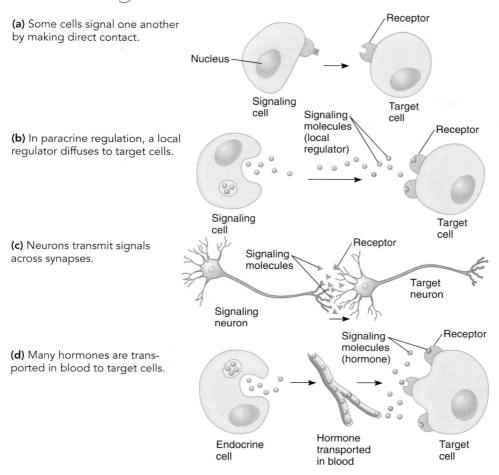

(a) Some cells signal one another by making direct contact.

Nucleus

Signaling cell

Receptor

Target cell

(b) In paracrine regulation, a local regulator diffuses to target cells.

Signaling molecules (local regulator)

Signaling cell

Receptor

Target cell

(c) Neurons transmit signals across synapses.

Signaling molecules

Receptor

Target neuron

Signaling neuron

(d) Many hormones are transported in blood to target cells.

Signaling molecules (hormone)

Receptor

Endocrine cell

Hormone transported in blood

Target cell

Figure 6-3 Some types of cell signaling

Various cells communicate in different ways.

oxide released by cells lining blood vessels relaxes smooth muscle in the blood vessel walls. As a result, the blood vessels dilate, decreasing blood pressure.

Some neurons communicate with electrical signals. However, most neurons signal one another by releasing chemical compounds called **neurotransmitters** (Fig. 6-3c). Neurotransmitter molecules diffuse across **synapses,** the tiny gaps between neurons. More than 60 different neurotransmitters have been identified, including acetylcholine, norepinephrine, dopamine, serotonin, and several amino acids and peptides.

In endocrine signaling in animals, chemical messengers called **hormones** are secreted by endocrine glands. **Endocrine glands** have no ducts, so they secrete their hormones into the surrounding interstitial fluid. Typically, hormones diffuse into capillaries and are transported by the blood to target cells (Fig. 6-3d).

Review

- What are local regulators?
- What are neurotransmitters?
- How are hormones typically transported to target cells?

RECEPTION

Learning Objectives

3 Identify mechanisms that make reception a highly specific process.
4 Compare ion channel–linked receptors, G protein–linked receptors, enzyme-linked receptors, and intracellular receptors.

Hundreds of different types of signaling molecules surround the cells of a multicellular organism. How do cells know which messages are for them? The answer is that each type of cell is genetically programmed to receive and respond to specific types of signals. Which signals a cell responds to depends on the specific receptors it is programmed to synthesize.

Receptors are large proteins or glycoproteins that bind with signaling molecules. A signaling molecule that binds to a specific receptor is called a **ligand**. Most ligands are hydrophilic molecules that bind to protein receptors on the surface of target cells (Fig. 6-4a). Some signaling molecules are small enough or sufficiently hydrophobic to move through the plasma membrane and enter the cell (Fig. 6-4b). These signaling molecules bind with intracellular receptors. Reception occurs when a ligand binds to a specific receptor protein on the surface of or inside a target cell. The signaling molecule activates the receptor.

EXP: Dove trans. msg.

A receptor on the cell surface generally has at least three domains. (In biochemistry, the term *domain* refers to a structural and functional region of a protein.) The external domain is a docking site for a signaling molecule. A second domain extends through the plasma membrane, and a third domain is a "tail" that extends into the cytoplasm. The tail transmits the signal to the inside of the cell.

Reception is highly selective. Each type of receptor has a specific shape. The receptor binding site is somewhat like a lock, and signaling molecules are like different keys. Only the signaling molecule that fits the specific receptor can influence the metabolic machinery of the cell. Receptors are important in determining the specificity of cell communication.

Different types of cells can have different types of receptors. Furthermore, the same cell can make different receptors at different stages in its life cycle or in response to different conditions. Another consideration is that the same signal can have different meanings for various target cells.

Some receptors are specialized to respond to signals other than chemical signals. For example, in the vertebrate eye, a receptor called **rhodopsin** is activated by light. Rhodopsin is part of a signal transduction pathway that leads to vision in dim light. Plants and some algae have **phytochromes,** a family of blue-green pigment proteins that are activated by red light. Activation can lead to changes, such as flowering. Plants, some algae, and at least some animals have **cryptochromes,** pigments that absorb blue light. Cryptochromes play a role in biological rhythms.

Cells regulate reception

An important mechanism that cells use to regulate reception is increasing or decreasing the number of each type of receptor. Depending on the needs of the cell, receptors are synthesized or degraded. As an example, when the concentration of the hormone insulin is too high for an extended period, cells decrease the number of their insulin receptors. This process is called **receptor down-regulation.** In the case of insulin, receptor down-regulation suppresses the sensitivity of target cells to the hormone. Insulin stimulates cells to take in glucose by facilitated diffusion, so receptor down-regulation decreases the ability of cells to take in glucose. Receptor down-regulation often involves transporting receptors to lysosomes, where they are destroyed.

Receptor up-regulation occurs in response to low hormone concentrations. A greater number of receptors are synthesized, and their increased numbers on the plasma membrane make it more likely that the signal will be received by a receptor on the cell. Receptor up-regulation thus amplifies the signaling molecule's effect on the cell. Receptor up-regulation and down-regulation are regulated in part by signals to genes that code for the receptors.

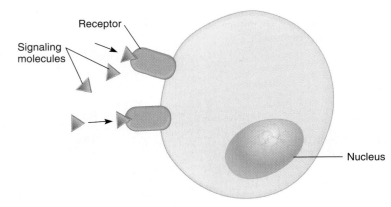

(a) Some signaling molecules bind to receptors in the plasma membrane.

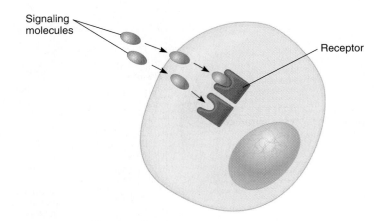

(b) Some signaling molecules bind to receptors inside the cell.

Figure 6-4 Cell-surface and intracellular receptors

Three types of receptors occur on the cell surface

Three main types of receptors on the cell surface are ion channel–linked receptors, G protein–linked receptors, and enzyme-linked receptors. **Ion channel–linked receptors** are found in the plasma membranes of many cells and have been extensively studied in neurons and muscle cells (∥ Fig. 6-5). These receptors convert chemical signals into electrical signals. In many cases the receptor itself serves as a channel.

Ion channel–linked receptors are also called **ligand-gated channels,** which means that the ion channel opens or closes in response to the binding of the signaling molecule (ligand). Part of the receptor (protein) that makes up the channel forms the gate. The receptor responds to certain signals by changing its shape, opening or closing the gate. Typically, the gate of an ion channel remains closed until a ligand binds to the receptor.

The signaling molecule can be a neurotransmitter, or it can be an ion inside the cell. For example, **acetylcholine** is a neurotransmitter that binds to an acetylcholine receptor. This receptor is a ligand-gated sodium ion channel, important in muscle contraction. When acetylcholine binds to the receptor, the channel opens, allowing sodium ions to enter the cell. The influx of sodium ions decreases the electric charge difference across the membrane (depolarization), which can lead to muscle contraction.

G protein–linked receptors (also called G protein–coupled receptors) are transmembrane proteins that loop back and forth through the plasma membrane seven times (see Fig. 6-5b). The receptor consists of seven transmembrane alpha helices connected by loops that extend into the cytosol or outside the cell. G protein–linked receptors couple certain signaling molecules to various signal transduction pathways inside the cell. The outer part of the receptor has a binding site for a signaling molecule, and the part of the receptor that extends into the cytosol has a binding site for a specific **G protein.** The *G* stands for *guanosine triphosphate* (*GTP*), the compound to which the G protein binds

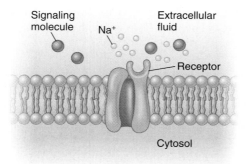

① Ion channel is closed.

(a) Ion channel–linked receptor

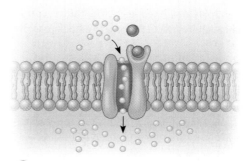

② When signaling molecule binds to receptor, the channel opens. Sodium ions enter cell.

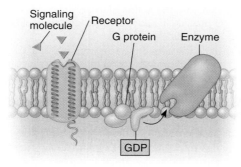

① In inactive state, the three subunits of G protein are joined.

(b) G protein–linked receptor

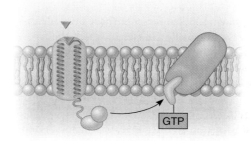

② When signaling molecule (ligand) binds to receptor, the ligand-receptor complex associates with G protein. This causes GDP to be replaced with GTP. One subunit of G protein then separates from the other two parts.

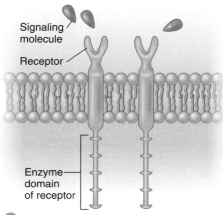

① Receptors are in inactive state.

(c) Enzyme-linked receptors

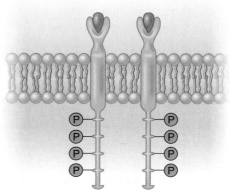

② When signaling molecule (ligand) binds to receptors, receptors are enzymatically phosphorylated. Phosphate comes from ATP.

Figure 6-5   Three types of cell-surface receptors

when activated. GTP will be discussed in a later section. When a ligand binds with a G protein–linked receptor, the receptor changes shape. This change allows the G protein to associate with the receptor.

Most receptors on animal cell surfaces are G protein–linked receptors, and understanding how they work is medically important. About 60% of prescription medications currently in use act on these receptors. About 900 G protein–linked receptors have been identified by studying the human genome, and more than 400 of these receptors are potential targets for pharmaceutical interventions.

Enzyme-linked receptors function directly as enzymes or are linked to enzymes. These receptors are transmembrane proteins with a binding site for a signaling molecule outside the cell and an enzyme component inside the cell. Some enzyme-linked receptors do not have an enzyme component but have a binding site for an enzyme.

Several groups of enzyme-linked receptors have been identified. One group consists of **tyrosine kinases** in which the tyrosine kinase enzyme is the domain of the receptor that extends into the cytosol. (Recall that tyrosine is an amino acid.) Tyrosine kinase phosphorylates specific tyrosines in certain signaling proteins inside the cell.

Tyrosine kinase receptors bind signal molecules known as *growth factors,* including insulin and nerve growth factor. When the ligand binds to the receptor, two receptor molecules come together. Then the tyrosine kinase part of each receptor phosphorylates the neighboring receptor (see Fig. 6-5c).

Many plant cell-surface receptors are enzyme-linked receptors. One example is **brassinolide,** the most active member of a group of plant steroid hormones (brassinosteroids), which binds with a protein kinase receptor in the plasma membrane. Brassinolide regulates many plant processes, including cell division, cell elongation, and flower development (discussed in Chapter 37).

The gas **ethylene** is a plant hormone that regulates a variety of processes, including seed germination and ripening of fruit. Ethylene is also important in plant responses to stressors. The ethylene receptor has two components. Each component has a domain that is an enzyme (a histidine kinase) that extends into the cell. Receptors with histidine kinase domains are also present in bacterial and yeast cells.

Some receptors are intracellular

Some receptors are located inside the cell (see Fig. 6-4). Most of these **intracellular receptors** are **transcription factors,** proteins that regulate the expression of specific genes. The ligands that bind with intracellular receptors are small, hydrophobic signaling molecules that can diffuse across the membranes of target cells.

In animal cells, steroid hormones, such as the molting hormone *ecdysone* in insects and *cortisol* and *sex hormones* in vertebrates, enter target cells and combine with receptor molecules in the cytosol. Vitamins A and D and nitric oxide also bind with intracellular receptors. After binding, the ligand–receptor com-

plex moves into the nucleus. *Thyroid hormones* bind to receptors already bound to DNA inside the nucleus.

Review

- How does a receptor "know" which signaling molecules to bind?
- Under what conditions might receptor up-regulation occur? Receptor down-regulation?
- What are the three main types of cell-surface receptors?

SIGNAL TRANSDUCTION

Learning Objectives

5   Compare the action of ion channel–linked receptors, G protein–linked receptors, enzyme-linked receptors, and intracellular receptors.

6   Trace the sequence of events in signal transduction for each of the following second messengers: cyclic AMP, inositol trisphosphate, diacylglycerol, and calcium ions.

As we have discussed, many regulatory molecules transmit information to the cell's interior without physically crossing the plasma membrane. These signal molecules rely on membrane proteins to transduce the signal. During signal transduction, the original signal is amplified. Each component of a signal transduction system acts as a relay "switch," which can be in an active ("on") state or an inactive ("off") state.

The first component in a signal transduction system is typically the receptor, which may be either a transmembrane protein with a domain exposed on the extracellular surface or an intracellular receptor. In a typical signaling pathway, the ligand binds with a cell-surface receptor and activates it by changing the shape of the receptor tail that extends into the cytoplasm. The signal may then be relayed through a sequence of proteins. Different receptors activate different signal transduction pathways.

Ion channel–linked receptors open or close channels

The gates of many ion channels remain closed until a ligand binds to the receptor. For example, when the neurotransmitter acetylcholine binds to an acetylcholine receptor (which is an ion channel–linked receptor), the channel opens, allowing sodium ions to enter the cell (see Fig. 6-5a). Depending on the type of cell, the influx of sodium ions can result in transmission of a neural impulse or in muscle contraction.

Gamma-aminobutyric acid (GABA) is a neurotransmitter that binds to ligand-gated chloride ion channels in neurons. When GABA binds to the receptor, the channel opens. Chloride ions, which are negatively charged, rush into the neuron, changing the electrical potential; this inhibits transmission of neural impulses. Thus, GABA *inhibits* neural signaling. Barbiturates and benzodiazepine drugs such as Valium bind to GABA receptors. When this occurs, lower amounts of GABA are required to open the chloride channels and inhibit neural impulses. This action results in a tranquilizing effect. (Not all GABA receptors are *themselves* ion channels; some act *indirectly* on other proteins that serve as ion channels.)

G protein–linked receptors initiate signal transduction

As discussed earlier, G protein–linked receptors activate G proteins, a group of regulatory proteins important in many signal transduction pathways. These proteins are found in yeasts, protists, plants, and animals. They are involved in the action of some plant and many animal hormones. Some G proteins regulate channels in the plasma membrane, allowing ions to enter the cell. Other G proteins are involved in the perception of sight and smell. In 1994, Alfred G. Gilman of the University of Texas and Martin Rodbell of the National Institute of Environmental Health Sciences were awarded the Nobel Prize in Physiology or Medicine for their groundbreaking research on G proteins.

In its inactive state, the G protein consists of three subunits that are joined (see Fig. 6-5b). One subunit is linked to a molecule of **guanosine diphosphate (GDP)**, a molecule similar to ADP but containing the base guanine instead of adenine. When a ligand binds with the receptor, the GDP is released and is replaced by **guanosine triphosphate (GTP)**. The subunit of the G protein that is linked to the GTP separates from the other two subunits. The G protein subunit is a GTPase, an enzyme that catalyzes the hydrolysis of GTP to GDP. This action, a process that releases energy, deactivates the G protein. Now in its inactive state, the G protein subunit rejoins the other two subunits.

When activated, G protein initiates signal transduction by binding with a specific protein in the cell. In some cases, G proteins *directly* activate enzymes that catalyze changes in certain proteins. These changes lead to alterations in cell function. More commonly, the signaling molecule (ligand) serves as the **first messenger** and information is relayed by way of the G protein to a second messenger, an intracellular signaling agent.

Second messengers are intracellular signaling agents

Second messengers are ions or small molecules that relay signals inside the cell. When receptors are activated, second messengers may be produced in large quantities, thus amplifying the signal. Second messengers rapidly diffuse through the cell (or membrane), relaying the signal. They are not enzymes, but some regulate specific enzymes, such as protein kinases. Others bind to ion channels, opening or closing them.

Some second messengers signal other molecules that pass the signal along through a pathway of proteins and other molecules. The last molecule in the sequence stimulates the final response. A chain of molecules in the cell that relays a signal is called a **signaling cascade.**

Cyclic AMP is a second messenger

Most G proteins shuttle a signal between the receptor and a second messenger. In many signaling cascades in prokaryotic and animal cells, the second messenger is **cyclic AMP (cAMP)** (▮ Fig. 6-6). Researcher Earl Sutherland identified cAMP as a second messenger in the 1960s and was awarded the 1971 Nobel Prize in Physiology or Medicine for his pioneering work.

When the G protein undergoes a conformational change (change in shape), it binds with and activates

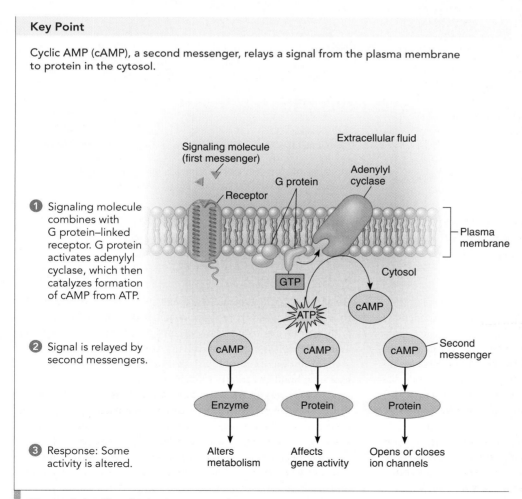

Key Point

Cyclic AMP (cAMP), a second messenger, relays a signal from the plasma membrane to protein in the cytosol.

1 Signaling molecule combines with G protein–linked receptor. G protein activates adenylyl cyclase, which then catalyzes formation of cAMP from ATP.

2 Signal is relayed by second messengers.

3 Response: Some activity is altered.

Signaling molecule (first messenger)
Extracellular fluid
G protein
Receptor
Adenylyl cyclase
Plasma membrane
GTP
Cytosol
ATP
cAMP
cAMP
cAMP
cAMP
Second messenger
Enzyme
Protein
Protein
Alters metabolism
Affects gene activity
Opens or closes ion channels

Figure 6-6 Signal relay by a second messenger

Key Point

The binding of a signaling molecule to a G protein–linked receptor activates a signal transduction pathway in which the signal is relayed through a cascade of molecules, leading to some response.

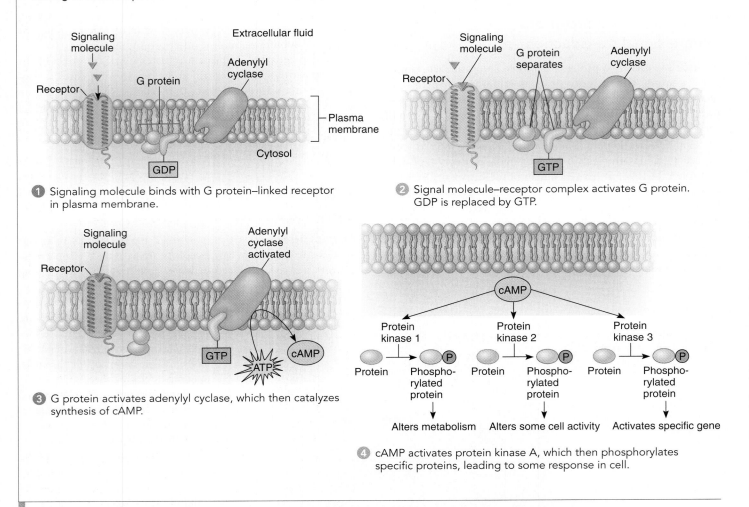

1 Signaling molecule binds with G protein–linked receptor in plasma membrane.

2 Signal molecule–receptor complex activates G protein. GDP is replaced by GTP.

3 G protein activates adenylyl cyclase, which then catalyzes synthesis of cAMP.

4 cAMP activates protein kinase A, which then phosphorylates specific proteins, leading to some response in cell.

Figure 6-7   Signal transduction involving a G protein and cyclic AMP

---

adenylyl cyclase, an enzyme on the cytoplasmic side of the plasma membrane. The type of G protein that activates adenylyl cyclase is known as a stimulatory G protein, or G_s. (Some G proteins, denoted as G_i, inhibit enzymes.) Note that the G protein couples the ligand–receptor complex to adenylyl cyclase action (Fig. 6-7).

When activated, adenylyl cyclase catalyzes the formation of cAMP from ATP (Fig. 6-8). By coupling the signaling molecule–receptor complex to an enzyme that generates a signal, G proteins amplify the effects of signaling molecules and many second-messenger molecules are rapidly produced.

Cyclic AMP activates **protein kinase A,** an enzyme that belongs to a large group of protein kinase enzymes. A protein ki-

nase transfers the terminal phosphate group from ATP molecules to a specific hydroxyl group in a target protein. The process of adding a phosphate to a molecule is called *phosphorylation*. When a protein is phosphorylated, its function is altered and it triggers a chain of reactions leading to some response in the cell, such as a metabolic change. The substrates (the substances on which an enzyme acts) for protein kinase A are different in various cell types. Consequently, the effect of the enzyme varies depending on the substrate. In skeletal muscle cells, protein kinase A activates enzymes that break down glycogen to glucose, providing the muscle cells with energy; whereas in cells of the hypothalamus (a part of the brain), the same enzyme activates the gene that encodes a growth-inhibiting hormone.

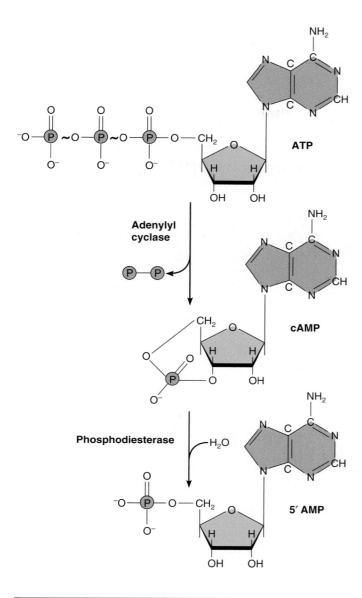

Figure 6-8 Synthesis and inactivation of cyclic AMP

Cyclic AMP (cAMP) is a second messenger produced from ATP. The enzyme adenylyl cyclase catalyzes the reaction. Cyclic AMP is inactivated by the enzyme phosphodiesterase, which converts it to adenosine monophosphate (AMP).

We can summarize the sequence of events beginning with the binding of the signaling molecule to the receptor and leading to a change in some cell function:

> Signaling molecule binds to G protein–linked receptor $\longrightarrow$ activates G protein $\longrightarrow$ activates adenylyl cyclase $\longrightarrow$ catalyzes the formation of cAMP $\longrightarrow$ activates protein kinase A $\longrightarrow$ phosphorylates proteins $\longrightarrow$ leads to some response in the cell

Some G proteins use phospholipids as second messengers

Certain signaling molecule–receptor complexes activate a G protein that then activates the membrane-bound enzyme phospholipase C (❙ Fig. 6-9). This enzyme splits a membrane phospholipid, PIP_2 (phosphotidylinositol-4,5-bisphosphate), into two phospholipid products, **inositol trisphosphate (IP$_3$)** and **diacylglycerol (DAG)**. Both act as second messengers.

DAG remains in the plasma membrane, where in combination with calcium ions, it activates **protein kinase C** enzymes. Members of this enzyme family phosphorylate a variety of target proteins. Depending on the type of cell and the specific protein kinase C, the response of the cell can include growth, a change in cell pH, or regulation of certain ion channels.

IP_3 is a member of a family of inositol phosphate messengers, some of which can donate phosphate groups to proteins. IP_3 binds to calcium channels in the endoplasmic reticulum (ER), causing them to open and release calcium ions into the cytosol. We can summarize this sequence of events as follows:

> Signaling molecule binds to G protein–linked receptor $\longrightarrow$ activates G protein $\longrightarrow$ activates phospholipase $\longrightarrow$ splits PIP_2 $\longrightarrow$ inositol trisphosphate (IP$_3$) + diacylglycerol (DAG)

> DAG $\longrightarrow$ activates protein kinase enzymes $\longrightarrow$ phosphorylates proteins $\longrightarrow$ some response in cell

> IP$_3$ $\longrightarrow$ binds to calcium channels in ER $\longrightarrow$ calcium ions released into the cytosol $\longrightarrow$ some response in the cell

Calcium ions are important messengers

Calcium ions (Ca^{2+}) have important functions in many cell processes, including microtubule disassembly, muscle contraction, blood clotting, and activation of certain cells in the immune system. These ions are critical in neural signaling, including the pathways involved in learning. Calcium ions are also essential in the fertilization of an egg and in the initiation of development.

Ion pumps in the plasma membrane normally maintain a low calcium ion concentration in the cytosol compared to its concentration in the extracellular fluid. Calcium ions are also stored in the endoplasmic reticulum. When Ca^{2+} gates open in the plasma membrane or endoplasmic reticulum, the Ca^{2+} concentration rises in the cytosol.

Calcium ions can act alone, but typically they exert their effects by binding to certain proteins. **Calmodulin,** found in all eukaryotic cells, is an important Ca^{2+} binding protein. When four Ca^{2+} bind to a calmodulin molecule, it changes shape and can then activate certain enzymes. Calmodulin combines with a number of different enzymes, including protein kinases and protein phosphatases, and alters their activity. (A phosphatase is an enzyme that catalyzes the removal of a phosphate group by hydrolysis.) Various types of cells have different calmodulin-binding proteins. How a target cell responds depends on which proteins are present.

Enzyme-linked receptors function directly

Recall that most enzyme-linked receptors are tyrosine kinases, enzymes that phosphorylate the amino acid tyrosine in proteins (see Fig. 6-5c). When a growth factor binds to an enzyme-linked receptor, the receptor is activated. Tyrosine kinase receptors activate several different signal transduction pathways, including the IP_3 and Ras pathways.

Ras proteins are a group of small G proteins that, like other G proteins, are active when bound to GTP. When activated, Ras triggers a cascade of reactions called the Ras pathway. In this pathway, the amino acid tyrosine is phosphorylated in specific cell proteins, leading to critical cell responses.

One reaction in the Ras pathway activates a series of molecules known as **MAP kinases** (MAP is an acronym for mitogen-activated protein). These kinases phosphorylate various proteins, including a nuclear protein that combines with other proteins to form a transcription factor, a protein that activates or inhibits the expression of specific genes. When certain genes are activated, proteins needed for cell growth, cell division, and cell differentiation (specialization) are synthesized. The MAP kinase cascade is the main signaling cascade for cell division and differentiation. *Ras* genes code for Ras proteins. Certain mutations in *Ras* genes result in unregulated cell division, which can lead to cancer. In fact, mutations in *Ras* genes have been identified in about 30% of all human cancers.

Ras proteins may work with growth factors in signaling pathways that regulate gene expression, cell division, movement of cells, and embryonic development. For example, to initiate DNA synthesis, fibroblasts (a type of connective tissue cell) require the presence of two growth factors, epidermal growth factor and platelet-derived growth factor. In one study, investigators injected fibroblasts with antibodies that inactivate Ras proteins by binding to them. The fibroblasts no longer synthesized DNA in response to growth factors. Data from this and similar experiments led to the conclusion that Ras proteins are important in signal transduction involving growth factors.

Phospholipase splits the phospholipid PIP_2, producing the second messengers DAG and IP_3.

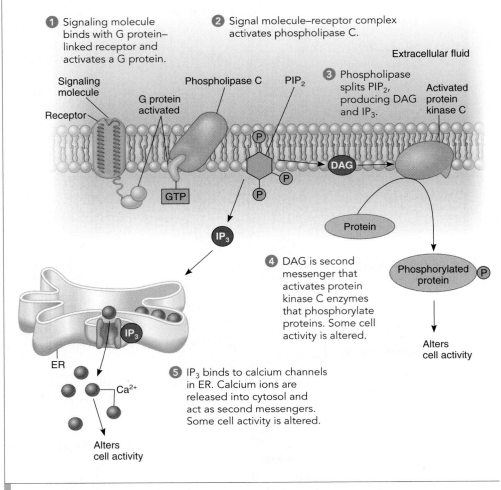

① Signaling molecule binds with G protein–linked receptor and activates a G protein.

② Signal molecule–receptor complex activates phospholipase C.

Extracellular fluid

Signaling molecule

Receptor

G protein activated

Phospholipase C

PIP_2

③ Phospholipase splits PIP_2, producing DAG and IP_3.

Activated protein kinase C

GTP

DAG

IP_3

Protein

Phosphorylated protein

IP_3

ER

④ DAG is second messenger that activates protein kinase C enzymes that phosphorylate proteins. Some cell activity is altered.

Alters cell activity

Ca^{2+}

⑤ IP_3 binds to calcium channels in ER. Calcium ions are released into cytosol and act as second messengers. Some cell activity is altered.

Alters cell activity

Figure 6-9 Phospholipid products as second messengers

Many activated intracellular receptors are transcription factors

As stated earlier, some hydrophobic signaling molecules diffuse across the membranes of target cells and bind with intracellular receptors in the cytosol or in the nucleus. Many of these receptors are transcription factors that regulate the expression of specific genes. When a ligand binds to a receptor, the receptor is activated. The ligand–receptor complex binds to a specific region of DNA and activates or represses specific genes (‖ Fig. 6-10).

Gene activation can take place quickly—within about 30 minutes. Messenger RNA is produced and carries the code for synthesis of a particular protein into the cytoplasm. In combination with ribosomes, messenger RNA manufactures specific proteins that can alter cell activity.

Nitric oxide is a gaseous signaling molecule that passes into target cells. In many target cells, nitric oxide binds directly with

Key Point

Many intracellular receptors are transcription factors. When activated, they activate or repress specific genes.

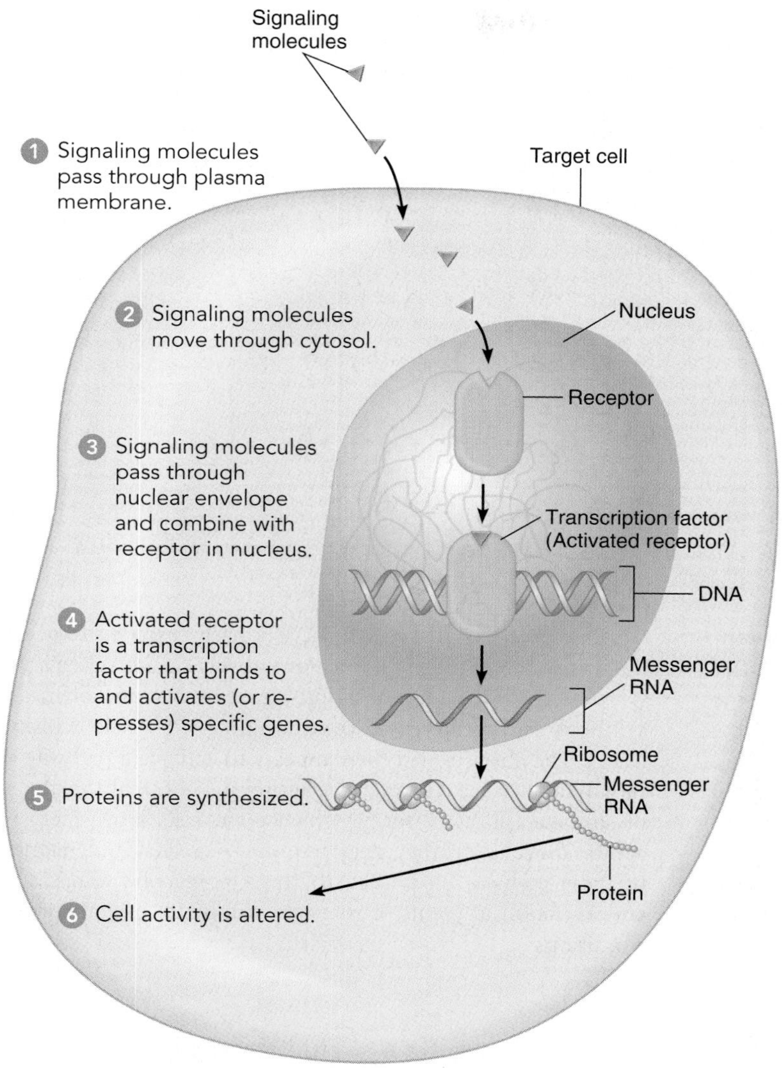

1. Signaling molecules pass through plasma membrane.

2. Signaling molecules move through cytosol.

3. Signaling molecules pass through nuclear envelope and combine with receptor in nucleus.

4. Activated receptor is a transcription factor that binds to and activates (or represses) specific genes.

5. Proteins are synthesized.

6. Cell activity is altered.

Labels: Signaling molecules; Target cell; Nucleus; Receptor; Transcription factor (Activated receptor); DNA; Messenger RNA; Ribosome; Messenger RNA; Protein

Figure 6-10 Intracellular receptors

the enzyme guanylyl cyclase. This enzyme catalyzes the production of cyclic GMP, which causes relaxation of the smooth muscle in blood vessel walls. When their smooth muscle relaxes, blood vessels dilate and blood pressure decreases.

During sexual arousal, neurons in the penis produce nitric oxide, which causes relaxation of smooth muscle in the lining of penile blood vessels. The blood vessels dilate, and erectile tissue in the penis becomes engorged with blood so that the penis becomes erect. The drug sildenafil (Viagra) inhibits the enzyme that catalyzes cyclic GMP breakdown. As a result, cyclic GMP increases in the erectile tissue of the penis and leads to sustained erection.

Scaffolding proteins increase efficiency

Signal transduction is a rapid, precise process. Enzymes must be organized so that they are available as needed for signaling pathways. **Scaffolding proteins** organize groups of intracellular signaling molecules into signaling complexes (Fig. 6-11). Scaffolding proteins position enzymes close to their substrates and prevent the enzymes from being co-opted by other pathways. For example, cells have many different kinases and phosphatases, enzymes that regulate various cell activities. Scaffolding proteins hold the needed kinases and phosphatases near the proteins they regulate and so decrease the chance that they might act on the wrong proteins. Thus, scaffolding proteins ensure that signals are relayed accurately, rapidly, and more efficiently.

Scaffolding proteins have been identified in many pathways, and similar scaffolding proteins have been found in diverse organisms. Both yeasts and mammals have scaffolding proteins that bind kinases in MAP kinase pathways.

Signals can be transmitted in more than one direction

Cell biologists have demonstrated that when certain ligands bind to **integrins** (transmembrane proteins that connect the cell to the extracellular matrix) in the plasma membrane, specific signal transduction pathways are activated. Interestingly, growth factors and certain molecules of the extracellular matrix may modulate each other's messages. Integrins also respond to information received from inside the cell. This *inside-out signaling* affects how selective integrins are with respect to the molecules to which they bind and how strongly they bind to them.

Review

- How is an extracellular signal converted to an intracellular signal in signal transduction? Give a specific example.
- What is the action of cyclic AMP? Of DAG?
- What are the functions of scaffolding proteins?

RESPONSES TO SIGNALS

Learning Objectives

7 Describe three types of responses that cells make to signals.
8 Contrast signal amplification with signal termination.

We have mentioned many responses that cells make to signaling molecules. Most of these responses fall into three categories: ion channels open or close; enzyme activity is altered, leading to metabolic changes and other effects; and specific gene activity may be turned on or off. These responses can result in changes in cell shape, cell growth, cell division, and cell differentiation.

In animals, neurons release neurotransmitters that excite or inhibit other neurons or muscle cells by affecting ion channels. For example, when acetylcholine binds with a receptor on a target neuron, an ion channel opens and allows passage of sodium and potassium ions. The resulting change in ion permeability can activate the neuron so that it transmits a neural impulse. Serotonin and some other neurotransmitters work indirectly through G proteins and cyclic AMP. In this chain of events, cAMP activates a kinase that phosphorylates a protein that then closes potassium ion channels. This action leads to transmission of a neural impulse. Some G proteins directly open or close ion channels.

Some receptors directly affect enzyme activity, whereas others initiate signal transduction pathways in which enzymes are altered by components of the pathway. When bacteria infect the body, they release certain peptides. Neutrophils, a type of white blood cell, have cell-surface receptors that detect these peptides. When the peptide binds to this receptor, enzymes are activated that lead to assembly of microfilaments (actin filaments) and microtubules. Contractions of microfilaments at the far end of the cell force the cytoplasm forward, which allows the neutrophils to move toward the invading bacteria. Microtubules and a variety of proteins, including the contractile protein myosin, appear necessary for this movement.

Some signaling molecules affect gene activity. For example, the Ras pathway appears to activate genes that lead to the manufacture of proteins needed for growth and cell division. In both plants and animals, steroid hormones regulate development by causing changes in the expression of specific genes. In animal cells, steroid hormones bind to nuclear receptors and directly regulate expression of specific target genes. In plant cells, steroid hormones bind to receptors on the cell surface. The signal is then transmitted through a chain of events that eventually lead to changes in gene expression. Plant hormones will be discussed in greater detail in Chapter 37. Animal hormones are the focus of Chapter 48.

Researchers have investigated a specific example of altered gene expression in plant cells. In *Arabidopsis thaliana,* a member of the mustard family, drought conditions cause osmotic stress, which acts as a signal (❚ Fig. 6-12). The receptor is thought to be a G protein–linked receptor. At least four signal transduction pathways have been identified leading to altered gene expression and resulting in increased tolerance to osmotic stress.

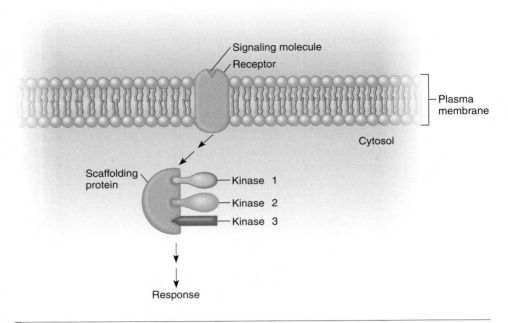

Figure 6-11 A scaffolding protein

Scaffolding proteins organize groups of signaling molecules into a signaling complex, making signal transduction more efficient.

One pathway involves a MAP kinase cascade. Two other pathways involve the plant hormone **abscisic acid**. When a plant is exposed to drought conditions, gene activation and protein synthesis lead to increased concentration of abscisic acid in its leaves. The abscisic acid then appears to activate a pathway that rapidly affects ion channels, resulting in a decrease in the size of stomata (pores) through which plant cells lose water. This action prevents further dehydration. A greater understanding of these pathways may lead to ways to increase abscisic acid concentration in plants so that they can better tolerate drought conditions.

The response to a signal is amplified

Signaling molecules are typically present in very low concentration, yet their effects on the cell are often profound. The binding of a single signaling molecule can lead to changes in millions of molecules at the end of a signaling cascade. This process of magnifying the strength of a signaling molecule is called **signal amplification.** As a result of signal amplification, just a few signaling molecules can lead to major responses in the cell (❚ Fig. 6-13).

As an example of signal amplification, let us examine how the action of a signaling molecule such as the hormone epinephrine is magnified as a signal passes through a series of proteins inside the cell. Epinephrine is released by the adrenal glands in response to stress. This hormone prepares the body for fight or flight. Among its many actions, it increases heart rate, blood flow to skeletal muscle, and glucose concentration in the blood.

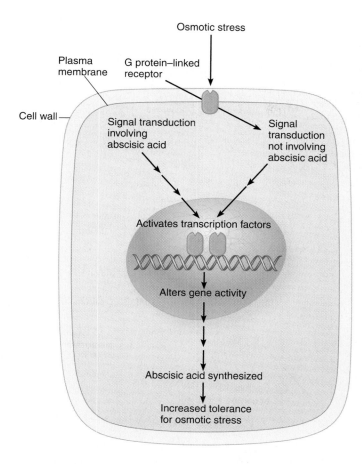

Osmotic stress

Plasma membrane

G protein–linked receptor

Cell wall

Signal transduction involving abscisic acid

Signal transduction not involving abscisic acid

Activates transcription factors

Alters gene activity

Abscisic acid synthesized

Increased tolerance for osmotic stress

Figure 6-12 Response to osmotic stress in plant cells

Biologists are just beginning to understand the pathways shown here. Just how osmotic stress activates receptors is not known. Several signal transduction pathways transmit signals that alter gene activity. The response increases the cell's tolerance to osmotic stress. Note that abscisic acid is shown in two places in this pathway. Apparently, after it is synthesized, it becomes part of the signal transduction pathway.

Epinephrine binds to a G protein–linked receptor, causing the receptor to change shape and activate a G protein. A single molecule of a hormone such as epinephrine can activate many G proteins. Each activated G protein does its job and then returns to its inactive state. The hormone–receptor complex is then free to activate another G protein.

Each G protein activates an adenylyl cyclase molecule. Then, each adenylyl cyclase molecule produces many cAMP molecules. In turn, each cAMP activates a protein kinase that phosphorylates many protein molecules. Some activated enzymes activate other intracellular enzymes or catalyze reactions between other proteins. Thus, the amplification process proceeds through a cascade of intracellular signaling molecules, enzymes, and other proteins. Note that amplification occurs at several steps along the signal cascade. Many of the molecules in a signaling cascade can activate numerous additional molecules and responses.

Signals must be terminated

Once a signal has done its job, it must be terminated. **Signal termination** returns the receptor and each of the components of the signal transduction pathway to their inactive states. This action allows the magnitude of the response to reflect the strength of the signal. Molecules in the system must also be ready to respond to new signals. We have seen that after a G protein is activated, a subunit of the G protein, a GTPase, catalyzes the hydrolysis of GTP to GDP. This action inactivates the G protein.

In the cyclic AMP pathway, any increase in cAMP concentration is temporary. Cyclic AMP is rapidly inactivated by a **phosphodiesterase,** which converts it to adenosine monophosphate (AMP). Thus, the concentration of cAMP depends on the activity of both adenylyl cyclase, which produces it, and of phosphodiesterase, which breaks it down (see Fig. 6-8). Similarly, a kinase activates a protein by phosphorylating it, and then a phosphatase inactivates it by removing the phosphate group.

Failure to terminate signals can lead to dire consequences. For example, the bacterium that causes cholera is ingested when people drink contaminated water. Cholera is prevalent in areas where water is contaminated with human feces. The cholera bacterium releases a toxin that activates G proteins in the epithelial cells lining the intestine. The toxin chemically changes the G protein so that it can no longer switch itself off. As a result, the G protein continues to stimulate adenylyl cyclase to make cAMP. The cells lining the intestine malfunction, allowing a large flow of chloride ions into the intestine. Water and other ions follow, leading to the severe watery diarrhea that characterizes cholera. The disease is treated by replacing the lost fluid. If untreated, this G protein malfunction can cause death.

Review

- What are some cell responses to signals?
- How do cells amplify signals?
- What is signal termination?

EVOLUTION OF CELL COMMUNICATION

Learning Objective

9   Cite evidence supporting a long evolutionary history for cell signaling molecules.

In this chapter, we have discussed how cells transmit information to one another. We have examined how the cells of a multicellular organism signal one another and described some of the signal transduction pathways within cells. We have described quorum sensing and other examples of communication between members of a species. We have also discussed communication among members of different species, such as signaling between plants and insects. In our discussion, we have noted many similarities in the types of signals used and in the molecules that cells use to relay signals from the cell surface to the molecules that carry out a specific response. Such similarities suggest evolutionary relationships.

Similar cell signaling molecules, receptors, and intracellular components of signal transduction pathways have been identified in protists and fungi, as well as in plants and animals. For example, G protein–linked receptors and some of their signal transduction pathways have been identified in yeast, some protists, plants, and animals. Certain disease-causing bacteria have signal transduction pathways similar to those found in eukaryotes. Recent findings suggest that bacteria can use some of these signal mechanisms to interfere with normal function in the cells they infect.

Similarities in cell signaling suggest that the molecules and mechanisms used in cell communication are very old. The evidence suggests that cell communication first evolved in prokaryotes and continued to change over time as new types of organisms evolved. However, some cell signaling molecules have not changed very much over time. Some signal transduction pathways found in organisms as diverse as yeasts and animals are quite similar. These similarities suggest that the importance of these pathways to cell survival has restricted any evolutionary changes that might have made them less effective. Thus, these pathways have weathered the demands of natural selection through millions of years of evolution.

Choanoflagellates, unicellular protists that are closely related to animals, are used as models for studying the early evolution of animals. Choanoflagellates have many of the same proteins found in animals. These tiny organisms have protein kinases similar to those of animals, and recently a G protein–linked receptor was identified. These findings indicate that important proteins necessary for cell communication in animals had evolved long before the evolution of animals. As we will discuss in later chapters, similarities and differences in basic molecules, such as G proteins, can be used to trace evolutionary pathways.

Cell biologists are only beginning to identify the many ways that cells communicate and the similarities of signaling mechanisms among kingdoms and domains. New receptors are being discovered at a rapid pace, and researchers are working to un-

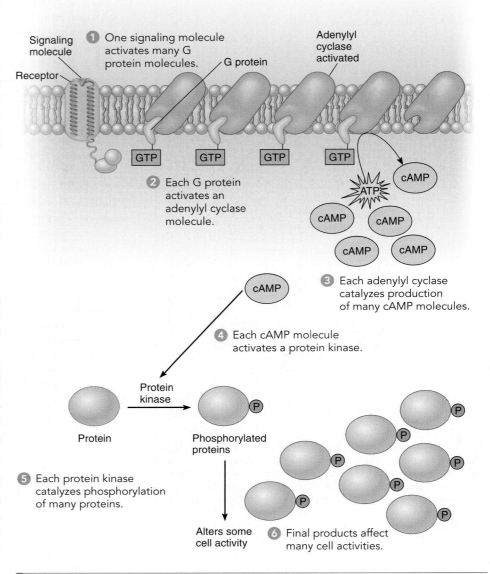

Figure 6-13 Signal amplification

The original signal is amplified many times so that the response is far greater than what you might expect.

derstand how proteins interact in cell signaling pathways. The relay sequences we have described here are oversimplified. Some proteins in signaling pathways come together briefly, producing large molecular complexes that are shared among several pathways. We have much to learn about how cells "talk" to one another.

Review

▪ Choanoflagellates and animals have similar protein kinases. What does that suggest about their cell signaling mechanisms?

Learning Objectives

1 Describe the sequence of events that takes place in cell signaling (page 135).

- Cells communicate by **cell signaling,** which involves (1) synthesis, release, and transport of signaling molecules, which can be neurotransmitters, hormones, and other regulatory molecules (a signaling molecule that binds to a specific receptor is called a **ligand**); (2) **reception** of information by **target cells;** (3) **signal transduction,** the process in which a receptor converts an extracellular signal into an intracellular signal that causes some change in the cell (signal transduction typically involves a series of molecules that relay information); and (4) **response** by the cell.

2 Describe three types of signaling molecules: local regulators, neurotransmitters, and hormones (page 136).

- Local regulators act by **paracrine regulation,** that is, they diffuse through the interstitial fluid and act on nearby cells. Local regulators include histamine, growth factors, substances called **prostaglandins,** and **nitric oxide,** a gaseous signaling molecule that passes into target cells.
- Most **neurons** (nerve cells) signal one another by releasing chemical compounds called **neurotransmitters.**
- **Hormones** are chemical messengers in plants and animals. In animals, they are secreted by **endocrine glands,** glands that have no ducts. Most hormones diffuse into capillaries and are transported by the blood to target cells.

3 Identify mechanisms that make reception a highly specific process (page 137).

- Each type of receptor has a specific shape. Only the signaling molecule that fits the specific receptor can affect the cell.
- Different types of cells can have different types of receptors.
- The same cell can make different receptors at different stages in its life cycle or in response to different conditions.

4 Compare ion channel–linked receptors, G protein–linked receptors, enzyme-linked receptors, and intracellular receptors (page 137).

- When a signaling molecule binds to an **ion channel–linked receptor,** the ion channel opens, or in some cases, closes.
- **G protein–linked receptors** are transmembrane proteins composed of seven alpha helices connected by loops that extend into the cytosol or outside the cell. These receptors couple certain signaling molecules to signal transduction pathways inside the cell. The tail of the receptor that extends into the cytosol has a binding site for a specific **G protein,** a regulatory protein that binds to GTP.
- **Enzyme-linked receptors** are transmembrane proteins with a binding site for a signaling molecule outside the cell and a binding site for an enzyme inside the cell. Some of these receptors function directly as enzymes. Many enzyme-linked receptors are **tyrosine kinases** in which the enzyme is part of the receptor.
- **Intracellular receptors** are located in the cytosol or in the nucleus. Their ligands are small, hydrophobic molecules that diffuse across the plasma membrane.

5 Compare the action of ion channel–linked receptors, G protein–linked receptors, enzyme-linked receptors, and intracellular receptors (page 140).

- Ion channel–linked receptors convert chemical signals into electrical signals. The gates of many ion channels remain closed until ligands bind to them.
- G protein–linked receptors activate G proteins. A G protein consists of three subunits. It is linked to a molecule of **guanosine diphosphate (GDP),** a molecule similar to ADP but containing the base guanine instead of adenine. When a ligand binds with the receptor, the GDP is released and is replaced by **guanosine triphosphate (GTP).** Then one subunit of the G protein separates from the other two subunits. When activated, G protein initiates signal transduction by binding with a specific protein in the cell. Some G proteins directly activate enzymes that catalyze changes in certain proteins, leading to changes in cell function.
- Most enzyme-linked receptors are tyrosine kinases, enzymes that phosphorylate proteins. Tyrosine kinase receptors activate several different signal transduction pathways, including the IP_3 and Ras pathways. **Ras proteins** are a group of small G proteins that, when activated, trigger a cascade of reactions. In this pathway, the amino acid tyrosine is phosphorylated in specific cell proteins, leading to critical cell responses. In some of these pathways, **scaffolding proteins** organize enzymes such as kinases into signaling complexes, increasing speed and efficiency.
- Intracellular receptors are located in the cytosol or nucleus. Steroid hormones are examples of signaling molecules that combine with intracellular receptors in animal cells. These receptors are **transcription factors** that activate or repress the expression of specific genes.

6 Trace the sequence of events in signal transduction for each of the following second messengers: cyclic AMP, inositol trisphosphate, diacylglycerol, and calcium ions (page 140).

- In many cases, the signaling molecule serves as the **first messenger.** Information is relayed by the G protein to a **second messenger,** an intracellular signaling agent.
- When certain G proteins undergo a conformational change, they bind with and activate **adenylyl cyclase,** an enzyme on the cytoplasmic side of the plasma membrane. Adenylyl cyclase catalyzes the formation of **cyclic AMP (cAMP)** from ATP. Cyclic AMP activates **protein kinase A,** an enzyme that phosphorylates certain proteins. The phosphorylated protein triggers a chain of reactions that lead to some response in the cell.
- Certain G proteins activate the membrane-bound enzyme phospholipase C. This enzyme splits a phospholipid, PIP_2 (phosphotidylinositol- 4,5-bisphosphate), into two products, **inositol trisphosphate (IP_3)** and **diacylglycerol (DAG).** IP_3 is a second messenger that can donate phosphate groups to proteins. IP_3 binds to calcium channels in the endoplasmic reticulum, which causes the channels to open and release calcium ions into the cytosol. DAG is a second messenger that activates **protein kinase C** enzymes. These enzymes phosphorylate a variety of target proteins.
- Calcium ions also act as second messengers. They typically combine with the protein **calmodulin,** which then affects the activity of protein kinases and protein phosphatases.

7 Describe three types of responses that cells make to signals (page 145).

- There are three types of responses: ion channels open or close; enzyme activity changes lead to metabolic changes and other effects; and specific genes are activated or repressed. These responses can affect many aspects of cell development and activity, including cell shape, cell growth, cell division, cell differentiation, and metabolism.

8 Contrast signal amplification with signal termination (page 145).
- **Signal amplification** is the process of enhancing signal strength as a signal is relayed through a signal transduction pathway.
- **Signal termination** is the process of inactivating the receptor and each component of the signal transduction pathway once they have done their jobs. Signal termination allows molecules in the system to respond to new signals.

9 Cite evidence supporting a long evolutionary history for cell signaling molecules (page 147).
- Molecules important in cell signaling first evolved in prokaryotes. Similar, but not identical, receptors and intracellular signal molecules are found in organisms in all kingdoms.
- Important proteins necessary for cell communication in animals have been found in choanoflagellates, indicating that these molecules were present before animals evolved.

TEST YOUR UNDERSTANDING

1. Which of the following statements is *not* true? During signal transduction (a) the cell converts an extracellular signal into an intracellular signal (b) nitric oxide is typically produced (c) the signal is amplified (d) relay of the signal is regulated (e) the signal is terminated

2. Which of the following statements is *not* true? Local regulators (a) are signaling molecules (b) typically carry on paracrine regulation (c) include growth factors (d) transmit electrical signals (e) signal target cells

3. In animals, hormones (a) are secreted by endocrine glands (b) are electrical signals (c) include neurotransmitters (d) terminate signals during signal transduction (e) are primarily produced by glands with ducts

4. When a ligand binds with a receptor, (a) G proteins are inactivated (b) a third messenger is activated (c) cell signaling is stopped (d) the signaling cell is activated (e) the receptor is activated

5. A G protein–linked receptor (a) inactivates G proteins (b) activates first messengers (c) consists of 18 transmembrane alpha helices (d) is a member of a family of about 10 different receptors (e) has a tail that extends into the cytosol and has a binding site for a G protein

6. An enzyme-linked receptor (a) is an integral membrane protein (b) would not be found on plant cell surfaces (c) aggregates with another enzyme-linked receptor when a ligand binds to it (d) is typically an adenylyl cyclase molecule (e) typically activates ion channels

7. G proteins (a) relay a message from the activated receptor to an enzyme that activates a second messenger (b) are GTP molecules (c) terminate cell signaling (d) directly activate protein kinases (e) function as first messengers

8. When activated, a G protein (a) consists of four subunits (b) releases GDP and binds with GTP (c) detaches from adenylyl cyclase (d) acts as a second messenger (e) typically terminates a signaling cascade

9. Which is the correct sequence?

   1. protein kinase activated 2. adenylyl cyclase activated 3. cAMP produced 4. proteins phosphorylated 5. G protein activated

   (a) 1, 2, 3, 5, 4 (b) 5, 3, 2, 1, 4 (c) 5, 2, 3, 4, 1 (d) 5, 2, 3, 1, 4 (e) 2, 3, 1, 4, 5

10. Which is the correct sequence?

    1. phospholipase activated 2. G protein activated 3. PIP_2 split 4. proteins phosphorylated 5. DAG produced

    (a) 1, 2, 5, 3, 4 (b) 4, 2, 3, 1, 5 (c) 2, 1, 3, 5, 4 (d) 5, 2, 3, 1, 4 (e) 2, 3, 5, 4, 1

11. Calcium ions (a) act as second messengers (b) split calmodulin (c) are kept at higher concentration in the cytosol than in the extracellular fluid (d) are produced in the ER by protein kinases and protein phosphatases (e) typically terminate signaling cascades

12. Ras proteins (a) act as second messengers (b) are growth factors (c) are transcription factors (d) are a group of small R proteins (e) trigger a cascade of reactions that can activate genes

13. Nitric oxide (a) binds directly with GTP (b) stimulates the blood vessel dilation necessary for penile erection (c) is released by many target cells (d) acts as a second messenger (e) is a signaling molecule that combines with enzyme-linked receptors

14. Scaffolding proteins (a) release kinases and phosphatases into the extracellular fluid (b) increase the probability that an enzyme can be used by several different pathways (c) increase accuracy but slow signaling cascades (d) organize groups of intracellular signaling molecules into signaling complexes (e) are found mainly in plant cells

15. Each adenylyl cyclase molecule produces many cAMP molecules. This is an example of (a) receptor up-regulation (b) receptor down-regulation (c) signal amplification (d) scaffolding (e) paracrine regulation

1. In many instances, the pathway from signaling molecule to final response in the cell involves complex signaling cascades. Give an example of a complex signaling system, and explain why such complexity may have advantages.

2. More than 500 genes have been identified in the human genome that code for protein kinases. What does this imply regarding protein kinases? Explain your answer.

3. **Evolution Link.** Cell signaling in plant and animal cells is similar in some ways and different in others. Offer a hypothesis for these similarities and differences, and cite some specific examples.

4. **Evolution Link.** Some of the same G protein–linked receptors and signal transduction pathways found in plants and animals have been identified in fungi and protists. What does this suggest about the evolution of these molecules? What does it suggest about these molecules and pathways?

Additional questions are available in ThomsonNOW at www.thomsonedu.com/login

7

Energy and Metabolism

Giant panda (*Ailuropoda mela-noleuca*). The chemical energy produced by photosynthesis and stored in bamboo leaves transfers to the panda as it eats.

KEY CONCEPTS

Energy, the capacity to do work, can be kinetic energy (energy of motion) or potential energy (energy due to position or state).

Energy cannot be created or destroyed (the first law of thermodynamics), but the total amount of energy available to do work in a closed system decreases over time (the second law of thermodynamics). Organisms do not violate the laws of thermodynamics because, as open systems, they use energy obtained from their surroundings to do work.

In cells, energy-releasing (exergonic) processes drive energy-requiring (endergonic) processes.

ATP plays a central role in cell energy metabolism by linking exergonic and endergonic reactions. ATP transfers energy by transferring a phosphate group.

The transfer of electrons in redox reactions is another way that cells transfer energy.

As biological catalysts, enzymes increase the rate of specific chemical reactions. The activity of an enzyme is influenced by temperature, pH, the presence of cofactors, and inhibitors and/or activators.

All living things require energy to carry out life processes. It may seem obvious that cells need energy to grow and reproduce, but even nongrowing cells need energy simply to maintain themselves. Cells obtain energy in many forms, but that energy can seldom be used directly to power cell processes. For this reason, cells have mechanisms that convert energy from one form to another. Because most components of these energy conversion systems evolved very early in the history of life, many aspects of energy metabolism tend to be similar in a wide range of organisms.

The sun is the ultimate source of almost all the energy that powers life; this *radiant energy* flows from the sun as electromagnetic waves. Plants and other photosynthetic organisms capture about 0.02% of the sun's energy that reaches Earth. As discussed in Chapter 9, photosynthetic organisms convert radiant energy to *chemical energy* in the bonds of organic molecules. This chemical energy becomes available to plants, animals such as the giant panda shown in the photograph, and other organisms through the process of cellular respiration. In cellular respiration, discussed in Chapter 8, organic molecules are broken apart and their energy converted to more immediately usable forms.

This chapter focuses on some of the basic principles that govern how cells capture, transfer, store, and use energy. We discuss the functions of adenosine triphosphate (ATP) and other molecules used in energy conversions, including those that transfer electrons in oxidation–reduction (redox) reactions. We also pay particular attention to the essential role of enzymes in cell energy dynamics. The flow of energy in ecosystems is discussed in Chapter 54. ■

BIOLOGICAL WORK

Learning Objectives

1 Define *energy,* emphasizing how it is related to work and to heat.
2 Use examples to contrast potential energy and kinetic energy.

Energy, one of the most important concepts in biology, can be understood in the context of **matter,** which is anything that has mass and takes up space. **Energy** is defined as the capacity to do work, which is any change in the state or motion of matter. Technically, mass is a form of energy, which is the basis behind the energy generated by the sun and other stars. More than 4 billion kilograms of matter per second are converted into energy in the sun.

Biologists generally express energy in units of work—**kilojoules (kJ).** It can also be expressed in units of *heat energy*—**kilocalories (kcal)**—thermal energy that flows from an object with a higher temperature to an object with a lower temperature. One kilocalorie is equal to 4.184 kJ. Heat energy cannot do cell work, because a cell is too small to have regions that differ in temperature. For that reason, the unit most biologists prefer today is the kilojoule. However, we will use both units because references to the kilocalorie are common in the scientific literature.

Organisms carry out conversions between potential energy and kinetic energy

When an archer draws a bow, **kinetic energy,** the energy of motion, is used and work is performed (▌Fig. 7-1). The resulting tension in the bow and string represents stored, or potential, energy. **Potential energy** is the capacity to do work as a result of position or state. When the string is released, this potential energy is converted to kinetic energy in the motion of the bow, which propels the arrow.

Most actions of an organism involve a series of energy transformations that occur as kinetic energy is converted to potential energy or as potential energy is converted to kinetic energy. **Chemical energy,** potential energy stored in chemical bonds, is of particular importance to organisms. In our example, the chemical energy of food molecules is converted to kinetic energy in the muscle cells of the archer. The contraction of the archer's muscles, like many of the activities performed by an organism, is an example of *mechanical energy,* which performs work by moving matter.

Review

▌ You exert tension on a spring and then release it. How do these actions relate to work, potential energy, and kinetic energy?

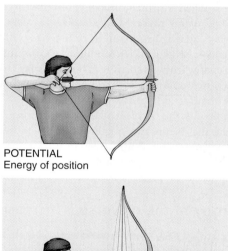
POTENTIAL
Energy of position

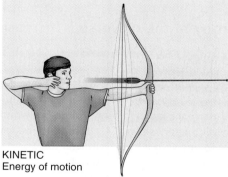
KINETIC
Energy of motion

▌ **Figure 7-1 Potential versus kinetic energy**

The potential chemical energy released by cellular respiration is converted to kinetic energy in the muscles, which do the work of drawing the bow. The potential energy stored in the drawn bow is transformed into kinetic energy as the bowstring pushes the arrow toward its target.

THE LAWS OF THERMODYNAMICS

Learning Objective

3 State the first and second laws of thermodynamics, and discuss the implications of these laws as they relate to organisms.

Thermodynamics, the study of energy and its transformations, governs all activities of the universe, from the life and death of cells to the life and death of stars. When considering thermodynamics, scientists use the term *system* to refer to an object that they are studying, whether a cell, an organism, or planet Earth. The rest of the universe other than the system being studied constitutes the *surroundings*. A **closed system** does not exchange energy with its surroundings, whereas an **open system** can exchange energy with its surroundings (Fig. 7-2). Biological systems are open systems. Two laws about energy apply to all things in the universe: the first and second laws of thermodynamics.

The total energy in the universe does not change

According to the **first law of thermodynamics,** energy cannot be created or destroyed, although it can be transferred or converted from one form to another, including conversions between matter and energy. As far as we know, the total mass-energy present in the universe when it formed, almost 14 billion years ago, equals the amount of energy present in the universe today. This is all the energy that can ever be present in the universe. Similarly, the energy of any system plus its surroundings is constant. A system may absorb energy from its surroundings, or it may give up some energy to its surroundings, but the total energy content of that system plus its surroundings is always the same.

As specified by the first law of thermodynamics, organisms cannot create the energy they require in order to live. Instead, they must capture energy from the environment and transform it to a form that can be used for biological work.

The entropy of the universe is increasing

The **second law of thermodynamics** states that when energy is converted from one form to another, some usable energy—that is, energy available to do work—is converted into heat that disperses into the surroundings (see Fig. 54-1). As you learned in Chapter 2, **heat** is the kinetic energy of randomly moving particles. Unlike *heat energy,* which flows from an object with a higher temperature to one with a lower temperature, this random motion cannot perform work. As a result, the amount of usable energy available to do work in the universe decreases over time.

It is important to understand that the second law of thermodynamics is consistent with the first law; that is, the total amount of energy in the universe is *not* decreasing with time. However, the total amount of energy in the universe that is available to do work is decreasing over time.

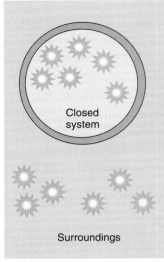
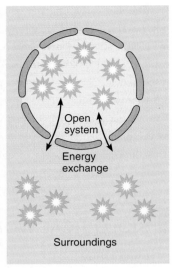

(a) A closed system does not exchange energy with its surroundings.

(b) An open system exchanges energy with its surroundings.

Figure 7-2 Closed and open systems

Less-usable energy is more diffuse, or disorganized. **Entropy** (**S**) is a measure of this disorder, or randomness; organized, usable energy has a low entropy, whereas disorganized energy, such as heat, has a high entropy.

Entropy is continuously increasing in the universe in all natural processes. Maybe at some time, billions of years from now, all energy will exist as heat uniformly distributed throughout the universe. If that happens, the universe will cease to operate, because no work will be possible. Everything will be at the same temperature, so there will be no way to convert the thermal energy of the universe into usable mechanical energy.

As a consequence of the second law of thermodynamics, no process requiring an energy conversion is ever 100% efficient, because much of the energy is dispersed as heat, increasing entropy. For example, an automobile engine, which converts the chemical energy of gasoline to mechanical energy, is between 20% and 30% efficient. Thus, only 20% to 30% of the original energy stored in the chemical bonds of the gasoline molecules is actually transformed into mechanical energy; the other 70% to 80% dissipates as waste heat. Energy use in your cells is about 40% efficient, with the remaining energy given to the surroundings as heat.

Organisms have a high degree of organization, and at first glance they may appear to refute the second law of thermodynamics. As organisms grow and develop, they maintain a high level of order and do not appear to become more disorganized. However, organisms are open systems; they maintain their degree of order over time only with the constant input of energy from their surroundings. This is why plants must photosynthesize and animals must eat. Although the order within organisms may tend to increase temporarily, the total entropy of the universe (organisms plus surroundings) always increases over time.

■ What is the first law of thermodynamics? The second law?

■ Life is sometimes described as a constant struggle against the second law of thermodynamics. How do organisms succeed in this struggle without violating the second law?

ENERGY AND METABOLISM

Learning Objectives

4 Discuss how changes in free energy in a reaction are related to changes in entropy and enthalpy.

5 Distinguish between exergonic and endergonic reactions, and give examples of how they may be coupled.

6 Compare the energy dynamics of a reaction at equilibrium with the dynamics of a reaction not at equilibrium.

The chemical reactions that enable an organism to carry on its activities—to grow, move, maintain and repair itself; reproduce; and respond to stimuli—together make up its metabolism. Recall from Chapter 1 that **metabolism** is the sum of all the chemical activities taking place in an organism. An organism's metabolism consists of many intersecting series of chemical reactions, or pathways, which are of two main types: anabolism and catabolism. **Anabolism** includes the various pathways in which complex molecules are synthesized from simpler substances, such as in the linking of amino acids to form proteins. **Catabolism** includes the pathways in which larger molecules are broken down into smaller ones, such as in the degradation of starch to form monosaccharides.

As you will see, these changes involve not only alterations in the arrangement of atoms but also various energy transformations. Catabolism and anabolism are complementary processes; catabolic pathways involve an overall release of energy, some of which powers anabolic pathways, which have an overall energy requirement. In the following sections we discuss how to predict whether a particular chemical reaction requires energy or releases it.

Enthalpy is the total potential energy of a system

In the course of any chemical reaction, including the metabolic reactions of a cell, chemical bonds break and new and different bonds may form. Every specific type of chemical bond has a certain amount of *bond energy,* defined as the energy required to break that bond. The total bond energy is essentially equivalent to the total potential energy of the system, a quantity known as **enthalpy (H).**

Free energy is available to do cell work

Entropy and enthalpy are related by a third type of energy, termed **free energy (G),** which is the amount of energy available to do work under the conditions of a biochemical reaction.

(G, also known as "Gibbs free energy," is named for J. W. Gibbs, a Yale professor who was one of the founders of the science of thermodynamics.) Free energy, the only kind of energy that can do cell work, is the aspect of thermodynamics of greatest interest to a biologist. Enthalpy, free energy, and entropy are related by the following equation:

$$enthalpy = f\mathcal{E} + Entropy$$
$$H = G + TS$$

in which H is enthalpy; G is free energy; T is the absolute temperature of the system, expressed in Kelvin units; and S is entropy. Disregarding temperature for the moment, enthalpy (the total energy of a system) is equal to free energy (the usable energy) plus entropy (the unusable energy).

A rearrangement of the equation shows that as entropy increases, the amount of free energy decreases:

$$G = H - TS$$

If we assume that entropy is zero, the free energy is simply equal to the total potential energy (enthalpy); an increase in entropy reduces the amount of free energy.

What is the significance of the temperature (T)? Remember that as the temperature increases, there is an increase in random molecular motion, which contributes to disorder and multiplies the effect of the entropy term.

Chemical reactions involve changes in free energy

Biologists analyze the role of energy in the many biochemical reactions of metabolism. Although the total free energy of a system (G) cannot be effectively measured, the equation $G = H - TS$ can be extended to predict whether a particular chemical reaction will release energy or require an input of energy. The reason is that *changes* in free energy can be measured. Scientists use the Greek capital letter delta (Δ) to denote any change that occurs in the system between its initial state before the reaction and its final state after the reaction. To express what happens with respect to energy in a chemical reaction, the equation becomes

$$\Delta G = \Delta H - T\Delta S$$

Notice that the temperature does not change; it is held constant during the reaction. Thus, the change in free energy (ΔG) during the reaction is equal to the change in enthalpy (ΔH) minus the product of the absolute temperature (T) in Kelvin units multiplied by the change in entropy (ΔS). Scientists express ΔG and ΔH in kilojoules or kilocalories per mole; they express ΔS in kilojoules or kilocalories per Kelvin unit.

Free energy decreases during an exergonic reaction

An **exergonic reaction** releases energy and is said to be a spontaneous or a "downhill" reaction, from higher to lower free energy (■ Fig. 7-3a). Because the total free energy in its final state is less

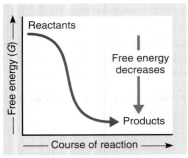

(a) In an exergonic reaction, there is a net loss of free energy. The products have less free energy than was present in the reactants, and the reaction proceeds spontaneously.

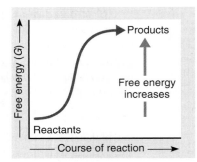

(b) In an endergonic reaction, there is a net gain of free energy. The products have more free energy than was present in the reactants.

Figure 7-3 *Animated* Exergonic and endergonic reactions

than the total free energy in its initial state, ΔG is a negative number for exergonic reactions.

The term *spontaneous* may give the false impression that such reactions are always instantaneous. In fact, spontaneous reactions do not necessarily occur readily; some are extremely slow. The reason is that energy, known as *activation energy,* is required to initiate every reaction, even a spontaneous one. We discuss activation energy later in the chapter.

Free energy increases during an endergonic reaction

An **endergonic reaction** is a reaction in which there is a gain of free energy (⊩ Fig. 7-3b). Because the free energy of the products is greater than the free energy of the reactants, ΔG has a positive value. Such a reaction cannot take place in isolation. Instead, it must occur in such a way that energy can be supplied from the surroundings. Of course, many energy-requiring reactions take place in cells, and as you will see, metabolic mechanisms have evolved that supply the energy to "drive" these nonspontaneous cell reactions in a particular direction.

Diffusion is an exergonic process

In Chapter 5, you saw that randomly moving particles diffuse down their own concentration gradient (⊩ Fig. 7-4). Although the movements of the individual particles are random, net movement of the group of particles seems to be directional. What provides energy for this apparently directed process? A **concentration gradient,** with a region of higher concentration and another region of lower concentration, is an orderly state. A cell must expend energy to produce a concentration gradient. Because work is done to produce this order, a concentration gradient is a form of potential energy. As the particles move about randomly, the gradient becomes degraded. Thus, free energy decreases as entropy increases.

In cellular respiration and photosynthesis, the potential energy stored in a concentration gradient of hydrogen ions (H^+) is transformed into chemical energy in adenosine triphosphate (ATP) as the hydrogen ions pass through a membrane down their concentration gradient. This important concept, known as *chemiosmosis,* is discussed in detail in Chapters 8 and 9.

Free-energy changes depend on the concentrations of reactants and products

According to the second law of thermodynamics, any process that increases entropy can do work. As we have discussed, differences in the concentration of a substance, such as between two different parts of a cell, represent a more orderly state than that when the substance is diffused homogeneously throughout the cell. Free-energy changes in any chemical reaction depend mainly on the difference in bond energies (enthalpy, H) between reactants and products. Free energy also depends on *concentrations* of both reactants and products.

In most biochemical reactions there is little intrinsic free-energy difference between reactants and products. Such reactions are reversible, indicated by drawing double arrows (⇌):

$$A \rightleftharpoons B$$

At the beginning of a reaction, only the reactant molecules (A) may be present. As the reaction proceeds, the concentration of the reactant molecules decreases and the concentration of the product molecules (B) increases. As the concentration of the product molecules increases, they may have enough free energy to initiate the reverse reaction. The reaction thus proceeds in both directions simultaneously; if undisturbed, it eventually reaches a state of **dynamic equilibrium,** in which the rate of the reverse reaction equals the rate of the forward reaction. At equilibrium there is no net change in the system; a reverse reaction balances every forward reaction.

Concentration gradient

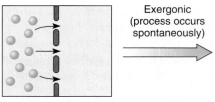 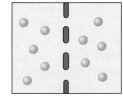

(a) A concentration gradient is a form of potential energy.

Exergonic (process occurs spontaneously)

(b) When molecules are evenly distributed, they have high entropy.

Figure 7-4 Entropy and diffusion

The tendency of entropy to increase can be used to produce work, in this case, diffusion.

At a given temperature and pressure, each reaction has its own characteristic equilibrium. For any given reaction, chemists can perform experiments and calculations to determine the relative concentrations of reactants and products present at equilibrium. If the reactants have much greater intrinsic free energy than the products, the reaction goes almost to completion; that is, it reaches equilibrium at a point at which most of the reactants have been converted to products. Reactions in which the reactants have much less intrinsic free energy than the products reach equilibrium at a point where very few of the reactant molecules have been converted to products.

If you increase the initial concentration of A, then the reaction will "shift to the right," and more A will be converted to B. A similar effect can be obtained if B is removed from the reaction mixture. The reaction always shifts in the direction that reestablishes equilibrium so that the proportions of reactants and products characteristic of that reaction at equilibrium are restored. The opposite effect occurs if the concentration of B increases or if A is removed; here the system "shifts to the left." The actual free-energy change that occurs during a reaction is defined mathematically to include these effects, which stem from the relative initial concentrations of reactants and products. Cells use energy to manipulate the relative concentrations of reactants and products of almost every reaction. Cell reactions are virtually never at equilibrium. By displacing their reactions far from equilibrium, cells supply energy to endergonic reactions and direct their metabolism according to their needs.

Cells drive endergonic reactions by coupling them to exergonic reactions

Many metabolic reactions, such as protein synthesis, are anabolic and endergonic. Because an endergonic reaction cannot take place without an input of energy, endergonic reactions are coupled to exergonic reactions. In **coupled reactions,** the thermodynamically favorable exergonic reaction provides the energy required to drive the thermodynamically unfavorable endergonic reaction. The endergonic reaction proceeds only if it absorbs free energy released by the exergonic reaction to which it is coupled.

Consider the free-energy change, ΔG, in the following reaction:

$$(1)\ A \longrightarrow B \qquad \Delta G = +20.9\ \text{kJ/mol}\ (+5\ \text{kcal/mol})$$

Because ΔG has a positive value, you know that the product of this reaction has more free energy than the reactant. This is an endergonic reaction. It is not spontaneous and does not take place without an energy source.

By contrast, consider the following reaction:

$$(2)\ C \longrightarrow D \qquad \Delta G = -33.5\ \text{kJ/mol}\ (-8\ \text{kcal/mol})$$

The negative value of ΔG tells you that the free energy of the reactant is greater than the free energy of the product. This exergonic reaction proceeds spontaneously.

You can sum up reactions 1 and 2 as follows:

$$(1)\ A \longrightarrow B \qquad \Delta G = +20.9\ \text{kJ/mol}\ (+5\ \text{kcal/mol})$$
$$\underline{(2)\ C \longrightarrow D \qquad \Delta G = -33.5\ \text{kJ/mol}\ (-8\ \text{kcal/mol})}$$
$$\text{Overall} \qquad \Delta G = -12.6\ \text{kJ/mol}\ (-3\ \text{kcal/mol})$$

Because thermodynamics considers the overall changes in these two reactions, which show a net negative value of ΔG, the two reactions taken together are exergonic.

The fact that scientists can write reactions this way is a useful bookkeeping device, but it does not mean that an exergonic reaction mysteriously transfers energy to an endergonic "bystander" reaction. However, these reactions are coupled if their pathways are altered so a common intermediate links them. Reactions 1 and 2 might be coupled by an intermediate (I) in the following way:

$$(3)\ A + C \longrightarrow I \qquad \Delta G = -8.4\ \text{kJ/mol}\ (-2\ \text{kcal/mol})$$
$$\underline{(4)\ I \longrightarrow B + D \qquad \Delta G = -4.2\ \text{kJ/mol}\ (-1\ \text{kcal/mol})}$$
$$\text{Overall} \qquad \Delta G = -12.6\ \text{kJ/mol}\ (-3\ \text{kcal/mol})$$

Note that reactions 3 and 4 are sequential. Thus, the reaction pathways have changed, but overall the reactants (A and C) and products (B and D) are the same, and the free-energy change is the same.

Generally, for each endergonic reaction occurring in a living cell there is a coupled exergonic reaction to drive it. Often the exergonic reaction involves the breakdown of ATP. Now let's examine specific examples of the role of ATP in energy coupling.

Review

- Consider the free-energy change in a reaction in which enthalpy decreases and entropy increases. Is ΔG zero, or does it have a positive value or a negative value? Is the reaction endergonic or exergonic?
- Why can't a reaction at equilibrium do work?

ATP, THE ENERGY CURRENCY OF THE CELL

Learning Objective

7 Explain how the chemical structure of ATP allows it to transfer a phosphate group, and discuss the central role of ATP in the overall energy metabolism of the cell.

In all living cells, energy is temporarily packaged within a remarkable chemical compound called **adenosine triphosphate (ATP),** which holds readily available energy for very short periods. We may think of ATP as the energy currency of the cell. When you work to earn money, you might say your energy is symbolically stored in the money you earn. The energy the cell requires for immediate use is temporarily stored in ATP, which is like cash. When you earn extra money, you may deposit some in the bank; similarly, a cell may deposit energy in the chemical bonds of lipids, starch, or glycogen. Moreover, just as you dare not make less money than you spend, the cell must avoid energy bankruptcy,

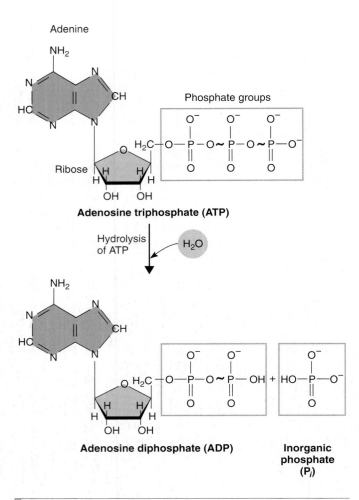

Adenine

Phosphate groups

Adenosine triphosphate (ATP)

Ribose

Hydrolysis
of ATP

H_2O

Adenosine diphosphate (ADP)

Inorganic phosphate (P_i)

Figure 7-5 *Animated* ATP and ADP

ATP, the energy currency of all living things, consists of adenine, ribose, and three phosphate groups. The hydrolysis of ATP, an exergonic reaction, yields ADP and inorganic phosphate. (The black wavy lines indicate unstable bonds. These bonds allow the phosphates to be transferred to other molecules, making them more reactive.)

which would mean its death. Finally, just as you probably do not keep money you make very long, the cell continuously spends its ATP, which must be replaced immediately.

ATP is a nucleotide consisting of three main parts: adenine, a nitrogen-containing organic base; ribose, a five-carbon sugar; and three phosphate groups, identifiable as phosphorus atoms surrounded by oxygen atoms (❙ Fig. 7-5). Notice that the phosphate groups are bonded to the end of the molecule in a series, rather like three cars behind a locomotive, and, like the cars of a train, they can be attached and detached.

ATP donates energy through the transfer of a phosphate group

When the terminal phosphate is removed from ATP, the remaining molecule is **adenosine diphosphate (ADP)** (see Fig. 7-5). If the phosphate group is not transferred to another molecule, it

is released as inorganic phosphate (P_i). This is an exergonic reaction. ATP is sometimes called a "high-energy" compound because the hydrolysis reaction that releases a phosphate has a relatively large negative value of ΔG. (Calculations of the free energy of ATP hydrolysis vary somewhat, but range between about -28 and -37 kJ/mol, or -6.8 to -8.7 kcal/mol.)

(5) $ATP + H_2O \longrightarrow ADP + P_i$

$$\Delta G = -32 \text{ kJ/mol (or } -7.6 \text{ kcal/mol)}$$

Reaction 5 can be coupled to endergonic reactions in cells. Consider the following endergonic reaction, in which two monosaccharides, glucose and fructose, form the disaccharide sucrose.

(6) Glucose + fructose $\longrightarrow$ sucrose + H_2O

$$\Delta G = +27 \text{ kJ/mol (or } +6.5 \text{ kcal/mol)}$$

With a free-energy change of -32 kJ/mol (-7.6 kcal/mol), the hydrolysis of ATP in reaction 5 can drive reaction 6, but only if the reactions are coupled through a common intermediate.

The following series of reactions is a simplified version of an alternative pathway that some bacteria use:

(7) Glucose + ATP $\longrightarrow$ glucose-P + ADP

(8) Glucose-P + fructose $\longrightarrow$ sucrose + P_i

Reaction 7 is a **phosphorylation reaction,** one in which a phosphate group is transferred to some other compound. Glucose becomes phosphorylated to form glucose phosphate (glucose-P), the intermediate that links the two reactions. Glucose-P, which corresponds to I in reactions 3 and 4, reacts exergonically with fructose to form sucrose. For energy coupling to work in this way, reactions 7 and 8 must occur in sequence. It is convenient to summarize the reactions thus:

(9) Glucose + fructose + ATP $\longrightarrow$ sucrose + ADP + P_i

$$\Delta G = -5 \text{ kJ/mol } (-1.2 \text{ kcal/mol})$$

When you encounter an equation written in this way, remember that it is actually a summary of a series of reactions and that transitory intermediate products (in this case, glucose-P) are sometimes not shown.

ATP links exergonic and endergonic reactions

We have just discussed how the transfer of a phosphate group from ATP to some other compound is coupled to endergonic reactions in the cell. Conversely, adding a phosphate group to adenosine monophosphate, or AMP (forming ADP) or to ADP (forming ATP) requires coupling to exergonic reactions in the cell.

$$AMP + P_i + energy \longrightarrow ADP$$

$$ADP + P_i + energy \longrightarrow ATP$$

Thus, ATP occupies an intermediate position in the metabolism of the cell and is an important link between exergonic reactions, which are generally components of *catabolic pathways,*

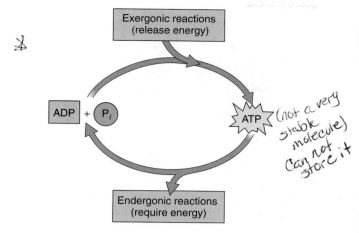

(handwritten note near ATP: "(not a very stable molecule) Can not store it")

Figure 7-6 ATP links exergonic and endergonic reactions

Exergonic reactions in catabolic pathways (*top*) supply energy to drive the endergonic formation of ATP from ADP. Conversely, the exergonic hydrolysis of ATP supplies energy to endergonic reactions in anabolic pathways (*bottom*).

and endergonic reactions, which are generally part of *anabolic pathways* (▌Fig. 7-6).

The cell maintains a very high ratio of ATP to ADP

The cell maintains a ratio of ATP to ADP far from the equilibrium point. ATP constantly forms from ADP and inorganic phosphate as nutrients break down in cellular respiration or as photosynthesis traps the radiant energy of sunlight. At any time, a typical cell contains more than 10 ATP molecules for every ADP molecule. The fact that the cell maintains the ATP concentration at such a high level (relative to the concentration of ADP) makes its hydrolysis reaction even more strongly exergonic and more able to drive the endergonic reactions to which it is coupled.

Although the cell maintains a high ratio of ATP to ADP, the cell cannot store large quantities of ATP. The concentration of ATP is always very low, less than 1 mmol/L. In fact, studies suggest that a bacterial cell has no more than a 1-second supply of ATP. Thus, it uses ATP molecules almost as quickly as they are produced. A healthy adult human at rest uses about 45 kg (100 lb) of ATP each day, but the amount present in the body at any given moment is less than 1 g (0.035 oz). Every second in every cell, an estimated 10 million molecules of ATP are made from ADP and phosphate, and an equal number of ATPs transfer their phosphate groups, along with their energy, to whatever chemical reactions need them.

Review

- Why do coupled reactions typically have common intermediates?
- Give a generalized example involving ATP, distinguishing between the exergonic and endergonic reactions.
- Why is the ATP concentration in a cell about 10 times the concentration of ADP?

ENERGY TRANSFER IN REDOX REACTIONS

Learning Objective

8   Relate the transfer of electrons (or hydrogen atoms) to the transfer of energy.

You have seen that cells transfer energy through the transfer of a phosphate group from ATP. Energy is also transferred through the transfer of electrons. As discussed in Chapter 2, **oxidation** is the chemical process in which a substance loses electrons, whereas **reduction** is the complementary process in which a substance gains electrons. Because electrons released during an oxidation reaction cannot exist in the free state in living cells, every oxidation reaction must be accompanied by a reduction reaction, in which the electrons are accepted by another atom, ion, or molecule. Oxidation and reduction reactions are often called **redox reactions,** because they occur simultaneously. The substance that becomes oxidized gives up energy as it releases electrons, and the substance that becomes reduced receives energy as it gains electrons.

(handwritten margin note: "oxidation and reduction occur at same time")

Redox reactions often occur in a series, as electrons are transferred from one molecule to another. These electron transfers, which are equivalent to energy transfers, are an essential part of cellular respiration, photosynthesis, and many other chemical reactions. Redox reactions, for example, release the energy stored in food molecules so that ATP can be synthesized using that energy.

Most electron carriers transfer hydrogen atoms

Generally it is not easy to remove one or more electrons from a covalent compound; it is much easier to remove a whole atom. For this reason, redox reactions in cells usually involve the transfer of a hydrogen atom rather than just an electron. A hydrogen atom contains an electron, plus a proton that does not participate in the oxidation–reduction reaction.

When an electron, either singly or as part of a hydrogen atom, is removed from an organic compound, it takes with it some of the energy stored in the chemical bond of which it was a part. That electron, along with its energy, is transferred to an acceptor molecule. An electron progressively loses free energy as it is transferred from one acceptor to another.

One of the most frequently encountered acceptor molecules is **nicotinamide adenine dinucleotide (NAD⁺).** When NAD⁺ becomes reduced, it temporarily stores large amounts of free energy. Here is a generalized equation showing the transfer of hydrogen from a compound, which we call X, to NAD^+:

$$XH_2 + NAD^+ \longrightarrow X + NADH + H^+$$

Oxidized   Reduced

Notice that the NAD^+ becomes reduced when it combines with hydrogen. NAD^+ is an ion with a net charge of +1. When 2 electrons and 1 proton are added, the charge is neutralized and the

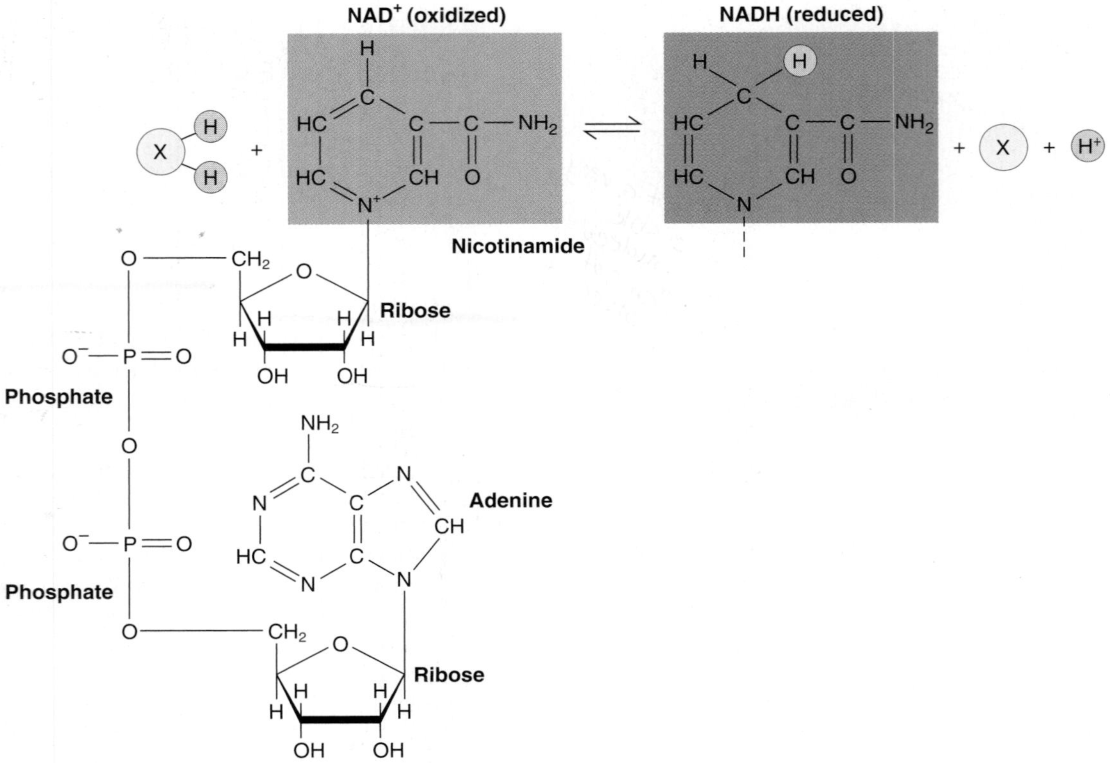

Figure 7-7 NAD⁺ and NADH

NAD^+ consists of two nucleotides, one with adenine and one with nicotinamide, that are joined at their phosphate groups. The oxidized form of the nicotinamide ring in NAD^+ (*left*) becomes the reduced form in NADH (*right*) by the transfer of 2 electrons and 1 proton from another organic compound (XH_2), which becomes oxidized (to X) in the process.

reduced form of the compound, **NADH,** is produced (Fig. 7-7). (Although the correct way to write the reduced form of NAD^+ is NADH + H^+, for simplicity we present the reduced form as NADH in this book.) Some energy stored in the bonds holding the hydrogen atoms to molecule X has been transferred by this redox reaction and is temporarily held by NADH. When NADH transfers the electrons to some other molecule, some of their energy is transferred. This energy is usually then transferred through a series of reactions that ultimately result in the formation of ATP (discussed in Chapter 8).

Nicotinamide adenine dinucleotide phosphate ($NADP^+$) is a hydrogen acceptor that is chemically similar to NAD^+ but has an extra phosphate group. Unlike NADH, the reduced form of $NADP^+$, abbreviated **NADPH,** is not involved in ATP synthesis. Instead, the electrons of NADPH are used more directly to provide energy for certain reactions, including certain essential reactions of photosynthesis (discussed in Chapter 9).

Other important hydrogen acceptors or electron acceptors are FAD and the cytochromes. **Flavin adenine dinucleotide** (**FAD**) is a nucleotide that accepts hydrogen atoms and their electrons; its reduced form is **FADH₂.** The **cytochromes** are proteins that contain iron; the iron component accepts electrons from hydrogen atoms and then transfers these electrons to some other compound. Like NAD^+ and $NADP^+$, FAD and the cytochromes are electron transfer agents. Each exists in a *reduced state,* in

which it has more free energy, or in an *oxidized state,* in which it has less. Each is an essential component of many redox reaction sequences in cells.

Review

■ Which has the most energy, the oxidized form of a substance or its reduced form? Why?

ENZYMES

Learning Objectives

9 Explain how an enzyme lowers the required energy of activation for a reaction.

10 Describe specific ways enzymes are regulated.

The principles of thermodynamics help us predict whether a reaction can occur, but they tell us nothing about the speed of the reaction. The breakdown of glucose, for example, is an exergonic reaction, yet a glucose solution stays unchanged virtually indefinitely in a bottle if it is kept free of bacteria and molds and not subjected to high temperatures or strong acids or bases. Cells cannot wait for centuries for glucose to break down, nor can they use extreme conditions to cleave glucose molecules. Cells regulate the rates of chemical reactions with **enzymes,** which are biological

Figure 7-8 Catalase as a defense mechanism

When threatened, a bombardier beetle (*Stenaptinus insignis*) uses the enzyme catalase to decompose hydrogen peroxide. The oxygen gas formed in the decomposition ejects water and other chemicals with explosive force. Because the reaction releases a great deal of heat, the water comes out as steam. (A wire attached by a drop of adhesive to the beetle's back immobilizes it. The researcher prodded its leg with the dissecting needle on the left to trigger the ejection.)

catalysts that increase the speed of a chemical reaction without being consumed by the reaction. Although most enzymes are proteins, scientists have learned that some types of RNA molecules have catalytic activity as well (discussed in Chapter 13).

Cells require a steady release of energy, and they must regulate that release to meet metabolic energy requirements. Metabolic processes generally proceed by a series of steps such that a molecule may go through as many as 20 or 30 chemical transformations before it reaches some final state. Even then, the seemingly completed molecule may enter yet another chemical pathway and become totally transformed or consumed to release energy. The changing needs of the cell require a system of flexible metabolic control. The key directors of this control system are enzymes.

The catalytic ability of some enzymes is truly impressive. For example, hydrogen peroxide (H_2O_2) breaks down extremely slowly if the reaction is uncatalyzed, but a single molecule of the enzyme **catalase** brings about the decomposition of 40 million molecules of hydrogen peroxide per second! Catalase has the highest catalytic rate known for any enzyme. It protects cells by destroying hydrogen peroxide, a poisonous substance produced as a by-product of some cell reactions. The bombardier beetle uses the enzyme catalase as a defense mechanism (Fig. 7-8).

All reactions have a required energy of activation

All reactions, whether exergonic or endergonic, have an energy barrier known as the **energy of activation (E_A), or activation energy**, which is the energy required to break the existing bonds

Figure 7-9 The space shuttle *Challenger* explosion

This disaster resulted from an explosive exergonic reaction between hydrogen and oxygen. All seven crew members died in the accident on January 28, 1986.

and begin the reaction. In a population of molecules of any kind, some have a relatively high kinetic energy, whereas others have a lower energy content. Only molecules with a relatively high kinetic energy are likely to react to form the product.

Even a strongly exergonic reaction, one that releases a substantial quantity of energy as it proceeds, may be prevented from proceeding by the activation energy required to begin the reaction. For example, molecular hydrogen and molecular oxygen can react explosively to form water:

$$2\ H_2 + O_2 \longrightarrow 2\ H_2O$$

This reaction is spontaneous, yet hydrogen and oxygen can be safely mixed as long as all sparks are kept away, because the required activation energy for this particular reaction is relatively high. A tiny spark provides the activation energy that allows a few molecules to react. Their reaction liberates so much heat that the rest react, producing an explosion. Such an explosion occurred on the space shuttle *Challenger* on January 28, 1986 (Fig. 7-9). The failure of a rubber O-ring to properly seal caused the liquid hydrogen in the tank attached to the shuttle to leak and start burning. When the hydrogen tank ruptured a few seconds later, the resulting force burst the nearby oxygen tank as well, mixing hydrogen and oxygen and igniting a huge explosion.

An enzyme lowers a reaction's activation energy

Like all catalysts, enzymes affect the rate of a reaction by lowering the activation energy (E_A) necessary to initiate a chemical reaction (Fig. 7-10). If molecules need less energy to react because the activation barrier is lowered, a larger fraction of the reactant molecules reacts at any one time. As a result, the reaction proceeds more quickly.

Although an enzyme lowers the activation energy for a reaction, it has no effect on the overall free-energy change; that is, an enzyme can promote only a chemical reaction that could proceed without it. If the reaction goes to equilibrium, no catalyst can cause a reaction to proceed in a thermodynamically unfavorable direction or can influence the final concentrations of reactants and products. Enzymes simply speed up reaction rates.

An enzyme works by forming an enzyme–substrate complex

An uncatalyzed reaction depends on random collisions among reactants. Because of its ordered structure, an enzyme reduces this reliance on random events and thereby controls the reaction. The enzyme accomplishes this by forming an unstable intermediate complex with the **substrate,** the substance on which it acts. When the **enzyme–substrate complex,** or **ES complex,** breaks up, the product is released; the original enzyme molecule is regenerated and is free to form a new ES complex:

$$\text{Enzyme} + \text{substrate(s)} \longrightarrow \text{ES complex}$$

$$\text{ES complex} \longrightarrow \text{enzyme} + \text{product(s)}$$

The enzyme itself is not permanently altered or consumed by the reaction and can be reused.

As shown in Figure 7-11a, every enzyme contains one or more **active sites,** regions to which the substrate binds, to form the ES complex. The active sites of some enzymes are grooves or cavities in the enzyme molecule, formed by amino acid side chains. The active sites of most enzymes are located close to the surface. During the course of a reaction, substrate molecules occupying these sites are brought close together and react with one another.

The shape of the enzyme does not seem exactly complementary to that of the substrate. The binding of the substrate to the enzyme molecule causes a change, known as **induced fit,** in the shape of the enzyme (Fig. 7-11b). Usually the shape of the substrate also changes slightly, in a way that may distort its chemical bonds. The proximity and orientation of the reactants, together with

strains in their chemical bonds, facilitate the breakage of old bonds and the formation of new ones. Thus, the substrate is changed into a product, which diffuses away from the enzyme. The enzyme is then free to catalyze the reaction of more substrate molecules to form more product molecules.

Scientists usually name enzymes by adding the suffix *-ase* to the name of the substrate. The enzyme sucrase, for example, splits sucrose into glucose and fructose. A few enzymes retain tradi-

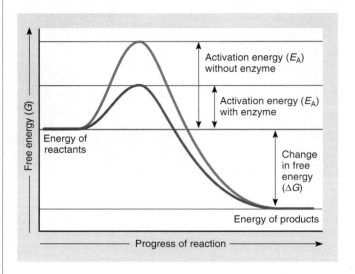

Key Point

An enzyme lowers the activation energy of a reaction, but does not alter the free energy change.

Figure 7-10 *Animated* Activation energy and enzymes

An enzyme speeds up a reaction by lowering its activation energy (E_A). In the presence of an enzyme, reacting molecules require less kinetic energy to complete a reaction.

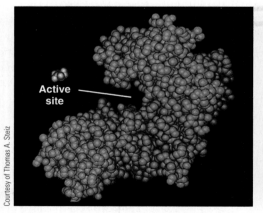

Courtesy of Thomas A. Steiz

(a) Prior to forming an ES complex, the enzyme's active site is the furrow where the substrate will bind.

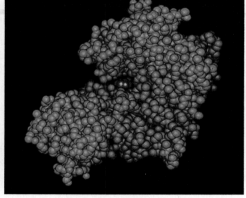

(b) The binding of the substrate to the active site induces a change in the conformation of the active site.

Figure 7-11 An enzyme–substrate complex

This computer graphic model shows the enzyme hexokinase (*blue*) and its substrate, glucose (*red*).

TABLE 7-1

Important Classes of Enzymes

Enzyme Class	Function
Oxidoreductases	Catalyze oxidation–reduction reactions
Transferases	Catalyze the transfer of a functional group from a donor molecule to an acceptor molecule
Hydrolases	Catalyze hydrolysis reactions
Isomerases	Catalyze conversion of a molecule from one isomeric form to another
Ligases	Catalyze certain reactions in which two molecules join in a process coupled to the hydrolysis of ATP
Lyases	Catalyze certain reactions in which double bonds form or break

tional names that do not end in -ase; some of these end in -zyme. For example, lysozyme (from the Greek lysis, "a loosening") is an enzyme found in tears and saliva; it breaks down bacterial cell walls. Other examples of enzymes with traditional names are pepsin and trypsin, which break peptide bonds in proteins.

Enzymes are specific

Enzymes catalyze virtually every chemical reaction that takes place in an organism. Because the shape of the active site is closely related to the shape of the substrate, most enzymes are highly specific. Most catalyze only a few closely related chemical reactions or, in many cases, only one particular reaction. The enzyme urease, which decomposes urea to ammonia and carbon dioxide, attacks no other substrate. The enzyme sucrase splits only sucrose; it does not act on other disaccharides, such as maltose or lactose. A few enzymes are specific only to the extent that they require the substrate to have a certain kind of chemical bond. For example, lipase, secreted by the pancreas, splits the ester linkages connecting the glycerol and fatty acids of a wide variety of fats.

Scientists classify into groups enzymes that catalyze similar reactions, although each particular enzyme in the group may catalyze only one specific reaction. Table 7-1 describes the six classes of enzymes that biologists recognize. Each class is divided into many subclasses. For example, sucrase, mentioned earlier, is called a glycosidase, because it cleaves a glycosidic linkage (see Chapter 3). Glycosidases are a subclass of the hydrolases. Phosphatases, enzymes that remove phosphate groups by hydrolysis, are also hydrolases. Kinases, enzymes that transfer phosphate groups to substrates, are transferases.

Many enzymes require cofactors

Some enzymes consist only of protein. The enzyme pepsin, which is secreted by the animal stomach and digests dietary protein by breaking certain peptide bonds, is exclusively a protein molecule. Other enzymes have two components: a protein called the apoenzyme and an additional chemical component called a cofactor. Neither the apoenzyme nor the cofactor alone has catalytic activity; only when the two are combined does the enzyme function. A cofactor may be inorganic, or it may be an organic molecule.

Some enzymes require a specific metal ion as a cofactor. Two very common inorganic cofactors are magnesium ions and calcium ions. Most of the trace elements, such as iron, copper, zinc, and manganese, all of which organisms require in very small amounts, function as cofactors.

An organic, nonpolypeptide compound that binds to the apoenzyme and serves as a cofactor is called a coenzyme. Most coenzymes are carrier molecules that transfer electrons or part of a substrate from one molecule to another. We have already introduced some examples of coenzymes in this chapter. NADH, NADPH, and FADH$_2$ are coenzymes; they transfer electrons.

ATP functions as a coenzyme; it is responsible for transferring phosphate groups. Yet another coenzyme, coenzyme A, is involved in the transfer of groups derived from organic acids. Most vitamins, which are organic compounds that an organism requires in small amounts but cannot synthesize itself, are coenzymes or components of coenzymes (see Table 46-3).

Enzymes are most effective at optimal conditions

Enzymes generally work best under certain narrowly defined conditions, such as appropriate temperature, pH (Fig. 7-12), and ion concentration. Any departure from optimal conditions adversely affects enzyme activity.

Each enzyme has an optimal temperature

Most enzymes have an optimal temperature, at which the rate of reaction is fastest. For human enzymes, the temperature optima are near the human body temperature (35°C to 40°C). Enzymatic reactions occur slowly or not at all at low temperatures. As the temperature increases, molecular motion increases, resulting in more molecular collisions. The rates of most enzyme-controlled reactions therefore increase as the temperature increases, within limits (see Fig. 7-12a). High temperatures rapidly denature most enzymes. The molecular conformation (3-D shape) of the protein becomes altered as the hydrogen bonds responsible for its secondary, tertiary, and quaternary structures are broken. Because this inactivation is usually not reversible, activity is not regained when the enzyme is cooled.

Most organisms are killed by even a short exposure to high temperature; their enzymes are denatured, and they are unable to continue metabolism. There are a few stunning exceptions to this rule. Certain species of archaea (see Chapter 1) can survive in the waters of hot springs, such as those in Yellowstone Park, where the temperature is almost 100°C; these organisms are responsible for the brilliant colors in the terraces of the hot springs (Fig. 7-13). Still other archaea live at temperatures much above that of boiling water, near deep-sea vents, where the extreme pressure keeps water in its liquid state (discussed in Chapter 24; see also Chapter 54, Focus On: Life without the Sun).

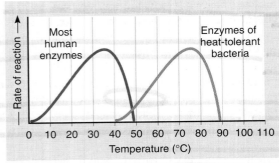

(a) Generalized curves for the effect of temperature on enzyme activity. As temperature increases, enzyme activity increases until it reaches an optimal temperature. Enzyme activity abruptly falls after it exceeds the optimal temperature because the enzyme, being a protein, denatures.

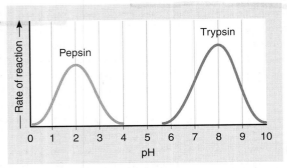

(b) Enzyme activity is very sensitive to pH. Pepsin is a protein-digesting enzyme in the very acidic stomach juice. Trypsin, secreted by the pancreas into the slightly basic small intestine, digests polypeptides.

Figure 7-12 *Animated* The effects of temperature and pH on enzyme activity

Substrate and enzyme concentrations are held constant in the reactions illustrated.

Each enzyme has an optimal pH

Most enzymes are active only over a narrow pH range and have an optimal pH, at which the rate of reaction is fastest. The optimal pH for most human enzymes is between 6 and 8. (Recall from Chapter 2 that *buffers* minimize pH changes in cells so that the pH is maintained within a narrow limit.) Pepsin, a protein-digesting enzyme secreted by cells lining the stomach, is an exception; it works only in a very acidic medium, optimally at pH 2 (see Fig. 7-12b). In contrast, trypsin, a protein-splitting enzyme secreted by the pancreas, functions best under the slightly basic conditions found in the small intestine.

The activity of an enzyme may be markedly changed by any alteration in pH, which in turn alters electric charges on the enzyme. Changes in charge affect the ionic bonds that contribute to tertiary and quaternary structure, thereby changing the protein's conformation and activity. Many enzymes become inactive, and usually irreversibly denatured, when the medium is made very acidic or very basic.

Figure 7-13 Grand Prismatic Spring in Yellowstone National Park

The world's third-largest spring, about 61 m (200 ft) in diameter, the Grand Prismatic Spring teems with heat-tolerant archaea. The rings around the perimeter, where the water is slightly cooler, get their distinctive colors from the various kinds of archaea living there.

Enzymes are organized into teams in metabolic pathways

Enzymes play an essential role in reaction coupling because they usually work in sequence, with the product of one enzyme-controlled reaction serving as the substrate for the next. You can picture the inside of a cell as a factory with many different assembly (and disassembly) lines operating simultaneously. An assembly line consists of a number of enzymes. Each enzyme carries out one step, such as changing molecule A into molecule B. Then molecule B is passed along to the next enzyme, which converts it into molecule C, and so on. Such a series of reactions is called a **metabolic pathway.**

$$A \xrightarrow{\text{Enzyme 1}} B \xrightarrow{\text{Enzyme 2}} C$$

Each of these reactions is reversible, despite the fact that an enzyme catalyzes it. An enzyme does not itself determine the direction of the reaction it catalyzes. However, the overall reaction sequence is portrayed as proceeding from left to right. Recall that if there is little intrinsic free-energy difference between the reactants and products for a particular reaction, the direction of the reaction is determined mainly by the relative concentrations of reactants and products.

In metabolic pathways, both intermediate and final products are often removed and converted to other chemical compounds. Such removal drives the sequence of reactions in a particular direction. Let us assume that reactant A is continually supplied and that its concentration remains constant. Enzyme 1 converts reactant A to product B. The concentration of B is always lower than the concentration of A, because B is removed as it is converted to C in the reaction catalyzed by enzyme 2. If C is removed as quickly as it is formed (perhaps by leaving the cell), the entire reaction pathway is "pulled" toward C.

In some cases the enzymes of a pathway bind to one another to form a multienzyme complex that efficiently transfers intermediates in the pathway from one active site to another. An example of one such multienzyme complex, pyruvate dehydrogenase, is discussed in Chapter 8.

The cell regulates enzymatic activity

Enzymes regulate the chemistry of the cell, but what controls the enzymes? One regulatory mechanism involves controlling the amount of enzyme produced. A specific gene directs the synthesis of each type of enzyme. The gene, in turn, may be switched on by a signal from a hormone or by some other signal molecule. When the gene is switched on, the enzyme is synthesized. The total amount of enzyme present then influences the overall cell reaction rate.

If the pH and temperature are kept constant (as they are in most cells), the rate of the reaction can be affected by the substrate concentration or by the enzyme concentration. If an excess of substrate is present, the enzyme concentration is the rate-limiting factor. The initial rate of the reaction is then directly proportional to the enzyme concentration (❙ Fig. 7-14a).

If the enzyme concentration is kept constant, the rate of an enzymatic reaction is proportional to the concentration of substrate present. Substrate concentration is the rate-limiting factor at lower concentrations; the rate of the reaction is therefore directly proportional to the substrate concentration. However, at higher substrate concentrations, the enzyme molecules become saturated with substrate; that is, substrate molecules are bound to all available active sites of enzyme molecules. In this situation, increasing the substrate concentration does not increase the net reaction rate (❙ Fig. 7-14b).

The product of one enzymatic reaction may control the activity of another enzyme, especially in a sequence of enzymatic reactions. For example, consider the following metabolic pathway:

$$A \xrightarrow{\text{Enzyme 1}} B \xrightarrow{\text{Enzyme 2}} C \xrightarrow{\text{Enzyme 3}} D \xrightarrow{\text{Enzyme 4}} E$$

A different enzyme catalyzes each step, and the final product E may inhibit the activity of enzyme 1. When the concentration of E is low, the sequence of reactions proceeds rapidly. However, an increasing concentration of E serves as a signal for enzyme 1 to slow down and eventually to stop functioning. Inhibition of enzyme 1 stops the entire reaction sequence. This type of enzyme regulation, in which the formation of a product inhibits an earlier reaction in the sequence, is called **feedback inhibition** (❙ Fig. 7-15).

Another method of enzymatic control focuses on the activation of enzyme molecules. In their inactive form, the active sites of the enzyme are inappropriately shaped, so the substrates do not fit. Among the factors that influence the shape of the enzyme are pH, the concentration of certain ions, and the addition of phosphate groups to certain amino acids in the enzyme.

Some enzymes have a receptor site, called an **allosteric site,** on some region of the enzyme molecule other than the active site. (The word *allosteric* means "another space.") When a substance binds to an enzyme's allosteric site, the conformation of

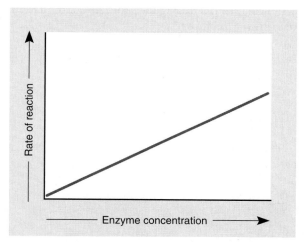

(a) In this example, the rate of reaction is measured at different enzyme concentrations, with an excess of substrate present. (Temperature and pH are constant.) The rate of the reaction is directly proportional to the enzyme concentration.

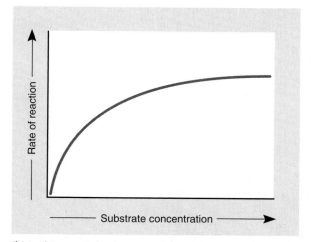

(b) In this example, the rate of the reaction is measured at different substrate concentrations, and enzyme concentration, temperature, and pH are constant. If the substrate concentration is relatively low, the reaction rate is directly proportional to substrate concentration. However, higher substrate concentrations do not increase the reaction rate, because the enzymes become saturated with substrate.

Figure 7-14 The effects of enzyme concentration and substrate concentration on the rate of a reaction

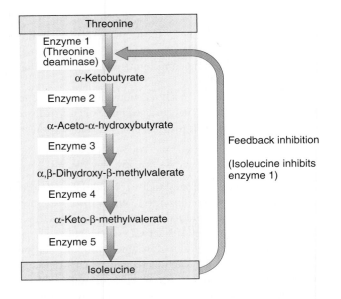

Figure 7-15 *Animated* Feedback inhibition

Bacteria synthesize the amino acid isoleucine from the amino acid threonine. The isoleucine pathway involves five steps, each catalyzed by a different enzyme. When enough isoleucine accumulates in the cell, the isoleucine inhibits threonine deaminase, the enzyme that catalyzes the first step in this pathway.

the enzyme's active site changes, thereby modifying the enzyme's activity. Substances that affect enzyme activity by binding to allosteric sites are called **allosteric regulators.** Some allosteric regulators are allosteric inhibitors that keep the enzyme in its inactive shape. Conversely, the activities of allosteric activators result in an enzyme with a functional active site.

The enzyme *cyclic AMP-dependent protein kinase* is an allosteric enzyme regulated by a protein that binds reversibly to the allosteric site and inactivates the enzyme. Protein kinase is in this inactive form most of the time (Fig. 7-16). When protein kinase activity is needed, the compound cyclic AMP (cAMP; see Fig. 3-25) contacts the enzyme-inhibitor complex and removes the inhibitory protein, thereby activating the protein kinase. Activation of protein kinases by cAMP is an important aspect of the mechanism of cell signaling, including the action of certain hormones (see Chapters 6 and 48).

Enzymes are inhibited by certain chemical agents

Most enzymes are inhibited or even destroyed by certain chemical agents. Enzyme inhibition may be reversible or irreversible. **Reversible inhibition** occurs when an inhibitor forms weak chemical bonds with the enzyme. Reversible inhibition can be competitive or noncompetitive.

In **competitive inhibition,** the inhibitor competes with the normal substrate for binding to the active site of the enzyme (Fig. 7-17a). Usually a competitive inhibitor is structurally similar to the normal substrate and fits into the active site and combines with the enzyme. However, it is not similar enough to substitute fully for the normal substrate in the chemical reaction, and the enzyme cannot convert it to product molecules. A competitive inhibitor occupies the active site only temporarily and does not permanently damage the enzyme. In competitive inhibition, an active site is occupied by the inhibitor part of the time and by the normal substrate part of the time. If the concentration of the substrate is increased relative to the concentration of the inhibitor, the active site is usually occupied by the substrate. Biochemists demonstrate competitive inhibition experimentally by showing that increasing the substrate concentration reverses competitive inhibition.

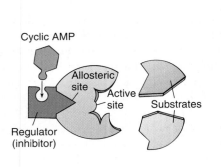
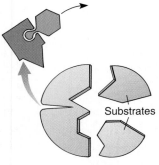
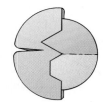

(a) Inactive form of the enzyme. The enzyme protein kinase is inhibited by a regulatory protein that binds reversibly to its allosteric site. When the enzyme is in this inactive form, the shape of the active site is modified so that the substrate cannot combine with it.

(b) Active form of the enzyme. Cyclic AMP removes the allosteric inhibitor and activates the enzyme.

(c) Enzyme–substrate complex. The substrate can then combine with the active site.

Figure 7-16 *Animated* An allosteric enzyme

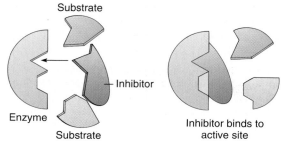

(a) Competitive inhibition. The inhibitor competes with the normal substrate for the active site of the enzyme. A competitive inhibitor occupies the active site only temporarily.

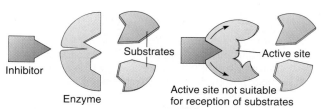

(b) Noncompetitive inhibition. The inhibitor binds with the enzyme at a site other than the active site, altering the shape of the enzyme and thereby inactivating it.

Figure 7-17 *Animated* Competitive and noncompetitive inhibition

In **noncompetitive inhibition,** the inhibitor binds with the enzyme at a site other than the active site (❚ Fig. 7-17b). Such an inhibitor inactivates the enzyme by altering its shape so that the active site cannot bind with the substrate. Many important noncompetitive inhibitors are metabolic substances that regulate enzyme activity by combining reversibly with the enzyme. Allosteric inhibition, discussed previously, is a type of noncompetitive inhibition in which the inhibitor binds to a special site, the allosteric site.

In **irreversible inhibition,** an inhibitor permanently inactivates or destroys an enzyme when the inhibitor combines with one of the enzyme's functional groups, either at the active site or elsewhere. Many poisons are irreversible enzyme inhibitors. For example, heavy metals such as mercury and lead bind irreversibly to and denature many proteins, including enzymes. Certain nerve gases poison the enzyme acetylcholinesterase, which is important for the functioning of nerves and muscles. Cytochrome oxidase, one of the enzymes that transports electrons in cellular respiration, is especially sensitive to cyanide. Death results from cyanide poisoning because cytochrome oxidase is irreversibly inhibited and no longer transfers electrons from its substrate to oxygen.

Some drugs are enzyme inhibitors

Physicians treat many bacterial infections with drugs that directly or indirectly inhibit bacterial enzyme activity. For example, sulfa drugs have a chemical structure similar to that of the nutrient

Para-aminobenzoic acid
(PABA)

Generic sulfonamide
(Sulfa drug)

Figure 7-18 Para-aminobenzoic acid and sulfonamides
Sulfa drugs inhibit an enzyme in bacteria necessary for the synthesis of folic acid, an important vitamin required for growth. (Note the unusual structure of the sulfonamide molecule; sulfur, which commonly forms two covalent bonds, forms six instead.)

para-aminobenzoic acid (PABA) (❚ Fig. 7-18). When PABA is available, microorganisms can synthesize the vitamin *folic acid,* which is necessary for growth. Humans do not synthesize folic acid from PABA. For this reason, sulfa drugs selectively affect bacteria. When a sulfa drug is present, the drug competes with PABA for the active site of the bacterial enzyme. When bacteria use the sulfa drug instead of PABA, they synthesize a compound that cannot be used to make folic acid. Therefore, the bacterial cells are unable to grow.

Penicillin and related antibiotics irreversibly inhibit a bacterial enzyme called *transpeptidase.* This enzyme establishes some of the chemical linkages in the bacterial cell wall. Bacteria susceptible to these antibiotics cannot produce properly constructed cell walls and are prevented from multiplying effectively. Human cells do not have cell walls and therefore do not use this enzyme. Thus, except for individuals allergic to it, penicillin is harmless to humans. Unfortunately, during the years since it was introduced, many bacterial strains have evolved resistance to penicillin. The resistant bacteria fight back with an enzyme of their own, penicillinase, which breaks down the penicillin and renders it ineffective. Because bacteria evolve at such a rapid rate, drug resistance is a growing problem in medical practice.

Review

❚ What effect does an enzyme have on the required activation energy of a reaction?

❚ How does the function of the active site of an enzyme differ from that of an allosteric site?

❚ How are temperature and pH optima of an enzyme related to its structure and function?

❚ Is allosteric inhibition competitive or noncompetitive?

Learning Objectives

1 Define *energy*, emphasizing how it is related to work and to heat (page 153).

   ■ **Energy** is the capacity to do work (expressed in **kilojoules, kJ**). Energy can be conveniently measured as **heat energy,** thermal energy that flows from an object with a higher temperature to an object with a lower temperature; the unit of heat energy is the **kilocalorie (kcal),** which is equal to 4.184 kJ. Heat energy cannot do cell work.

2 Use examples to contrast potential energy and kinetic energy (page 153).

   ■ **Potential energy** is stored energy; **kinetic energy** is energy of motion.

   ■ All forms of energy are interconvertible. For example, photosynthetic organisms capture radiant energy and convert some of it to **chemical energy,** a form of potential energy that powers many life processes, such as muscle contraction.

3 State the first and second laws of thermodynamics, and discuss the implications of these laws as they relate to organisms (page 154).

   ■ A **closed system** does not exchange energy with its surroundings. Organisms are **open systems** that do exchange energy with their surroundings.

   ■ The **first law of thermodynamics** states that energy cannot be created or destroyed but can be transferred and changed in form. The first law explains why organisms cannot produce energy; but as open systems, they continuously capture it from the surroundings.

   ■ The **second law of thermodynamics** states that disorder (entropy) in the universe, a closed system, is continuously increasing. No energy transfer is 100% efficient; some energy is dissipated as **heat,** random motion that contributes to **entropy** (*S*), or disorder. As open systems, organisms maintain their ordered states at the expense of their surroundings.

4 Discuss how changes in free energy in a reaction are related to changes in entropy and enthalpy (page 155).

   ■ As entropy increases, the amount of **free energy** decreases, as shown in the equation $G = H - TS$, in which G is the free energy, H is the **enthalpy** (total potential energy of the system), T is the absolute temperature (expressed in Kelvin units), and S is entropy.

   ■ The equation $\Delta G = \Delta H - T\Delta S$ indicates that the change in free energy (ΔG) during a chemical reaction is equal to the change in enthalpy (ΔH) minus the product of the absolute temperature (T) multiplied by the change in entropy (ΔS).

5 Distinguish between exergonic and endergonic reactions, and give examples of how they may be coupled (page 155).

   ■ An **exergonic reaction** has a negative value of ΔG; that is, free energy decreases. Such a reaction is spontaneous; it releases free energy that can perform work.

   ■ Free energy increases in an **endergonic reaction.** Such a reaction has a positive value of ΔG and is nonspontaneous. In a **coupled reaction,** the input of free energy required to drive an endergonic reaction is supplied by an exergonic reaction.

ThomsonNOW™ **Learn more about exergonic and endergonic reactions by clicking on the figure in ThomsonNOW.**

6 Compare the energy dynamics of a reaction at equilibrium with the dynamics of a reaction not at equilibrium (page 155).

   ■ When a chemical reaction is in a state of **dynamic equilibrium,** the rate of change in one direction is exactly the same as the rate of change in the opposite direction; the system can do no work because the free-energy difference between the reactants and products is zero.

   ■ When the concentration of reactant molecules is increased, the reaction shifts to the right and more product molecules are formed until equilibrium is re-established.

7 Explain how the chemical structure of ATP allows it to transfer a phosphate group, and discuss the central role of ATP in the overall energy metabolism of the cell (page 157).

   ■ **Adenosine triphosphate (ATP)** is the immediate energy currency of the cell. It donates energy by means of its terminal phosphate group, which is easily transferred to an acceptor molecule. ATP is formed by the **phosphorylation** of **adenosine diphosphate (ADP),** an endergonic process that requires an input of energy.

   ■ ATP is the common link between exergonic and endergonic reactions and between **catabolism** (degradation of large complex molecules into smaller, simpler molecules) and **anabolism** (synthesis of complex molecules from simpler molecules).

8 Relate the transfer of electrons (or hydrogen atoms) to the transfer of energy (page 159).

   ■ Energy is transferred in **oxidation–reduction (redox) reactions.** A substance becomes oxidized as it gives up one or more electrons to another substance, which becomes reduced in the process. Electrons are commonly transferred as part of hydrogen atoms.

   ■ NAD^+ and $NADP^+$ accept electrons as part of hydrogen atoms and become reduced to form **NADH** and **NADPH,** respectively. These electrons (along with some of their energy) can be transferred to other acceptors.

9 Explain how an enzyme lowers the required energy of activation for a reaction (page 160).

   ■ An **enzyme** is a biological **catalyst;** it greatly increases the speed of a chemical reaction without being consumed.

   ■ An enzyme works by lowering the **activation energy (E_A),** the energy necessary to get a reaction going. The **active site** of an enzyme is a 3-D region where **substrates** come into close contact and thereby react more readily. When a substrate binds to an active site, an **enzyme–substrate complex** forms in which the shapes of the enzyme and substrate change slightly. This **induced fit** facilitates the breaking of bonds and formation of new ones.

ThomsonNOW™ **Learn more about activation energy by clicking on the figure in ThomsonNOW.**

10 Describe specific ways enzymes are regulated (page 160).

   ■ Enzymes work best at specific temperature and pH conditions.

   ■ A cell can regulate enzymatic activity by controlling the amount of enzyme produced and by regulating metabolic conditions that influence the shape of the enzyme.

   ■ Some enzymes have **allosteric sites,** noncatalytic sites to which an **allosteric regulator** binds, changing the enzyme's activity. Some allosteric enzymes are subject to **feedback inhibition,** in which the formation of an end product inhibits an earlier reaction in the **metabolic pathway.**

■ **Reversible inhibition** occurs when an inhibitor forms weak chemical bonds with the enzyme. Reversible inhibition may be **competitive,** in which the inhibitor competes with the substrate for the active site, or **noncompetitive,** in

which the inhibitor binds with the enzyme at a site other than the active site. **Irreversible inhibition** occurs when an inhibitor combines with an enzyme and permanently inactivates it.

ThomsonNOW™   **Learn more about enzymes by clicking on the figures in ThomsonNOW.**

TEST YOUR UNDERSTANDING

1. According to the first law of thermodynamics, (a) energy may be changed from one form to another but is neither created nor destroyed (b) much of the work an organism does is mechanical work (c) the disorder of the universe is increasing (d) free energy is available to do cell work (e) a cell is in a state of dynamic equilibrium

2. According to the second law of thermodynamics, (a) energy may be changed from one form to another but is neither created nor destroyed (b) much of the work an organism does is mechanical work (c) the disorder of the universe is increasing (d) free energy is available to do cell work (e) a cell is in a state of dynamic equilibrium

3. In thermodynamics, _____ is a measure of the amount of disorder in the system. (a) bond energy (b) catabolism (c) entropy (d) enthalpy (e) work

4. The _____ energy of a system is that part of the total energy available to do cell work. (a) activation (b) bond (c) kinetic (d) free (e) heat

5. A reaction that requires a net input of free energy is described as (a) exergonic (b) endergonic (c) spontaneous (d) both a and c (e) both b and c

6. A reaction that releases energy is described as (a) exergonic (b) endergonic (c) spontaneous (d) both a and c (e) both b and c

7. A spontaneous reaction is one in which the change in free energy (ΔG) has a _____ value. (a) positive (b) negative (c) positive or negative (d) none of these (ΔG has no measurable value)

8. To drive a reaction that requires an input of energy, (a) an enzyme–substrate complex must form (b) the concentration of ATP must be decreased (c) the activation energy must be increased (d) some reaction that releases energy must be coupled to it (e) some reaction that requires energy must be coupled to it

9. Which of the following reactions could be coupled to an endergonic reaction with $\Delta G = +3.56$ kJ/mol? (a) A $\longrightarrow$ B, $\Delta G = +6.08$ kJ/mol (b) C $\longrightarrow$ D, $\Delta G = +3.56$ kJ/mol (c) E $\longrightarrow$ F, $\Delta G = 0$ kJ/mol (d) G $\longrightarrow$ H, $\Delta G = -1.22$ kJ/mol (e) I $\longrightarrow$ J, $\Delta G = -5.91$ kJ/mol

10. Consider this reaction: Glucose + 6 O_2 $\longrightarrow$ 6 CO_2 + 6 H_2O ($\Delta G = -2880$ kJ/mol). Which of the following statements about this reaction is *not* true? (a) the reaction is spontaneous in a thermodynamic sense (b) a small amount of energy (activation energy) must be supplied to start the reaction, which then proceeds with a release of energy (c) the reaction is exergonic (d) the reaction can be coupled to an endergonic reaction (e) the reaction must be coupled to an exergonic reaction

11. The energy required to initiate a reaction is called (a) activation energy (b) bond energy (c) potential energy (d) free energy (e) heat energy

12. A biological catalyst that affects the rate of a chemical reaction without being consumed by the reaction is a(an) (a) product (b) cofactor (c) coenzyme (d) substrate (e) enzyme

13. The region of an enzyme molecule that combines with the substrate is the (a) allosteric site (b) reactant (c) active site (d) coenzyme (e) product

14. Which inhibitor binds to the active site of an enzyme? (a) noncompetitive inhibitor (b) competitive inhibitor (c) irreversible inhibitor (d) allosteric regulator (e) PABA

15. In the following reaction series, which enzyme(s) is/are most likely to have an allosteric site to which the end product E binds? (a) enzyme 1 (b) enzyme 2 (c) enzyme 3 (d) enzyme 4 (e) enzymes 3 and 4

  Enzyme 1   Enzyme 2   Enzyme 3   Enzyme 4
A $\longrightarrow$ B $\longrightarrow$ C $\longrightarrow$ D $\longrightarrow$ E

CRITICAL THINKING

1. Given what you have learned in this chapter, explain why an extremely high fever (body temperature above 40°C, or 105°F) is often fatal.

2. **Evolution Link.** What does the fact that all organisms use ATP/ADP as central links between exergonic and endergonic reactions suggest about the evolution of energy metabolism?

3. **Evolution Link.** Some have argued that "evolution is impossible because the second law of thermodynamics states that entropy always increases; therefore natural processes cannot give rise to greater complexity." In what ways is this statement a misunderstanding of the laws of thermodynamics?

4. **Analyzing Data.** Reactions 1 and 2 happen to have the same standard free-energy change: $\Delta G - 41.8$ kJ/mol (-10 kcal/mol). Reaction 1 is at equilibrium, but reaction 2 is far from equilibrium. Is either reaction capable of performing work? If so, which one?

5. **Analyzing Data.** You are performing an experiment in which you are measuring the rate at which succinate is converted to fumarate by the enzyme succinic dehydrogenase. You decide to add a little malonate to make things interesting. You observe that the reaction rate slows markedly and hypothesize that malonate is inhibiting the reaction. Design an experiment that will help you decide whether malonate is acting as a competitive inhibitor or a noncompetitive inhibitor.

Additional questions are available in ThomsonNOW at www.thomsonedu.com/login

How Cells Make ATP: Energy-Releasing Pathways

Grizzly bear (*Ursus arctos*). This grizzly, shown attempting to eat a jumping salmon, may also eat fruit, nuts, roots, insects, and small vertebrates such as mice and ground squirrels.

Digital Vision/Getty Images

KEY CONCEPTS

Aerobic respiration is an exergonic redox process in which glucose becomes oxidized, oxygen becomes reduced, and energy is captured to make ATP.

Aerobic respiration consists of four stages: glycolysis, formation of acetyl coenzyme A, the citric acid cycle, and the electron transport chain and chemiosmosis.

Nutrients other than glucose, including many carbohydrates, lipids, and amino acids, can be oxidized by aerobic respiration.

In anaerobic respiration carried out by some bacteria, ATP is formed during a redox process in which glucose becomes oxidized and an inorganic substance becomes reduced.

Fermentation is an inefficient anaerobic redox process in which glucose becomes oxidized and an organic substance becomes reduced. Some fungi and bacteria, as well as muscle cells under conditions of low oxygen, obtain low yields of ATP through fermentation.

Cells are tiny factories that process materials on the molecular level, through thousands of metabolic reactions. Cells exist in a dynamic state and are continuously building up and breaking down the many different cell constituents. As you learned in Chapter 7, metabolism has two complementary components: **catabolism,** which releases energy by splitting complex molecules into smaller components, and **anabolism,** the synthesis of complex molecules from simpler building blocks. Anabolic reactions produce proteins, nucleic acids, lipids, polysaccharides, and other molecules that help maintain the cell or the organism. Most anabolic reactions are endergonic and require ATP or some other energy source to drive them.

Every organism must extract energy from food molecules that it either manufactures by photosynthesis or obtains from the environment. Grizzly bears, such as the one in the photograph, obtain organic molecules from their varied plant and animal diets. How do they obtain energy from these organic molecules? First, the complex food molecules are broken down by digestion into simpler components that are absorbed into the blood and transported to all the cells. The catabolic processes that convert the energy in the chemical bonds of nutrients to chemical energy stored in ATP then occur inside cells, usually through a process known as **cellular respiration.** (The term *cellular respiration* is used to distinguish it from *organismic respiration*, the exchange of oxygen and carbon dioxide with the environment by animals that have special organs, such as lungs or gills, for gas exchange.)

Cellular respiration may be either aerobic or anaerobic. *Aerobic respiration* requires oxygen, whereas *anaerobic pathways*, which include anaerobic respiration and fermentation, do not require oxygen. A steady supply of oxygen enables your cells to capture energy through aerobic respiration, which is by far the most common pathway and the main subject of this chapter. All three pathways—aerobic respiration, anaerobic respiration, and fermentation—are exergonic and release free energy that can be captured by the cell. ■

REDOX REACTIONS

Learning Objective

1   Write a summary reaction for aerobic respiration that shows which reactant becomes oxidized and which becomes reduced.

Most eukaryotes and prokaryotes carry out **aerobic respiration,** a form of cellular respiration requiring molecular oxygen (O_2). During aerobic respiration, nutrients are catabolized to carbon dioxide and water. Most cells use aerobic respiration to obtain energy from glucose, which enters the cell though a specific transport protein in the plasma membrane (see discussion of facilitated diffusion in Chapter 5). The overall reaction pathway for the aerobic respiration of glucose is summarized as follows:

$$C_6H_{12}O_6 + 6\,O_2 + 6\,H_2O \longrightarrow$$
$$6\,CO_2 + 12\,H_2O + \text{energy (in the chemical bonds of ATP)}$$

Note that water is shown on both sides of the equation, because it is a reactant in some reactions and a product in others. For purposes of discussion, the equation for aerobic respiration can be simplified to indicate that there is a net yield of water:

$$\underbrace{C_6H_{12}O_6 + 6\,O_2}_{\text{Reduction}} \overset{\text{Oxidation}}{\longrightarrow} 6\,CO_2 + 6\,H_2O + \text{energy} \quad \text{(in the chemical bonds of ATP)}$$

If we analyze this summary reaction, it appears that CO_2 is produced by the removal of hydrogen atoms from glucose. Conversely, water seems to be formed as oxygen accepts the hydrogen atoms. Because the transfer of hydrogen atoms is equivalent to the transfer of electrons, this is a **redox reaction** in which glucose becomes **oxidized** and oxygen becomes **reduced** (see Chapters 2 and 7).

The products of the reaction would be the same if the glucose were simply placed in a test tube and burned in the presence of oxygen. However, if a cell were to burn glucose, its energy would be released all at once as heat, which not only would be unavailable to the cell but also would actually destroy it. For this reason, cells do not transfer hydrogen atoms directly from glucose to oxygen. Aerobic respiration includes a series of redox reactions in which electrons associated with the hydrogen atoms in glucose are transferred to oxygen in a series of steps (■ Fig. 8-1). During this process, the free energy of the electrons is coupled to ATP synthesis.

Review

■   In the overall reaction of aerobic respiration, which reactant becomes oxidized and which becomes reduced?

■   What is the specific role of oxygen in most cells?

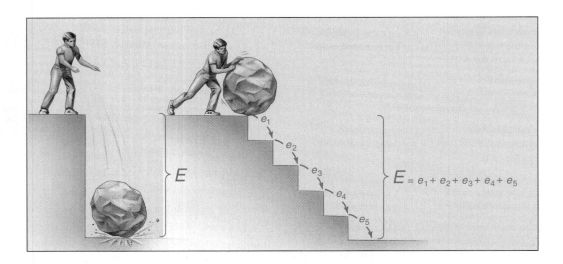

Figure 8-1   Changes in free energy
The release of energy from a glucose molecule is analogous to the liberation of energy by a falling object. The total energy released (E) is the same whether it occurs all at once or in a series of steps.

THE FOUR STAGES
OF AEROBIC RESPIRATION

Learning Objectives

2   List and give a brief overview of the four stages of aerobic respiration.

3   Indicate where each stage of aerobic respiration takes place in a eukaryotic cell.

4   Add up the energy captured (as ATP, NADH, and $FADH_2$) in each stage of aerobic respiration.

5   Define *chemiosmosis*, and explain how a gradient of protons is established across the inner mitochondrial membrane.

6   Describe the process by which the proton gradient drives ATP synthesis in chemiosmosis.

The chemical reactions of the aerobic respiration of glucose are grouped into four stages (▮ Fig. 8-2 and ▮ Table 8-1; see also the summary equations at the end of the chapter). In eukary-

---

Key Point

The stages of aerobic respiration occur in specific locations.

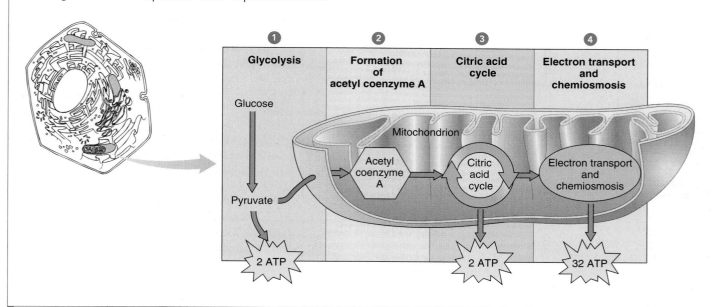

Figure 8-2 *Animated* The four stages of aerobic respiration

① Glycolysis, the first stage of aerobic respiration, occurs in the cytosol. ② Pyruvate, the product of glycolysis, enters a mitochondrion, where cellular respiration continues with the formation of acetyl CoA, ③ the citric acid cycle, and ④ electron transport and chemiosmosis. Most ATP is synthesized by chemiosmosis.

TABLE 8-1

Summary of Aerobic Respiration

Stage	Summary	Some Starting Materials	Some End Products
1. Glycolysis (in cytosol)	Series of reactions in which glucose is degraded to pyruvate; net profit of 2 ATPs; hydrogen atoms are transferred to carriers; can proceed anaerobically	Glucose, ATP, NAD^+, ADP, P_i	Pyruvate, ATP, NADH
2. Formation of acetyl CoA (in mitochondria)	Pyruvate is degraded and combined with coenzyme A to form acetyl CoA; hydrogen atoms are transferred to carriers; CO_2 is released	Pyruvate, coenzyme A, NAD^+	Acetyl CoA, CO_2, NADH
3. Citric acid cycle (in mitochondria)	Series of reactions in which the acetyl portion of acetyl CoA is degraded to CO_2; hydrogen atoms are transferred to carriers; ATP is synthesized	Acetyl CoA, H_2O, NAD^+, FAD, ADP, P_i	CO_2, NADH, $FADH_2$, ATP
4. Electron transport and chemiosmosis (in mitochondria)	Chain of several electron transport molecules; electrons are passed along chain; released energy is used to form a proton gradient; ATP is synthesized as protons diffuse down the gradient; oxygen is final electron acceptor	NADH, $FADH_2$, O_2, ADP, P_i	ATP, H_2O, NAD^+, FAD

otes, the first stage (glycolysis) takes place in the cytosol, and the remaining stages take place inside mitochondria. Most bacteria and archaea also carry out these processes, but because prokaryotic cells lack mitochondria, the reactions of aerobic respiration occur in the cytosol and in association with the plasma membrane.

1. *Glycolysis.* A six-carbon glucose molecule is converted to two three-carbon molecules of pyruvate.[1] Some of the energy of glucose is captured with the formation of two kinds of energy carriers, ATP and NADH.[2] See Chapter 7 to review how ATP transfers energy by transferring a phosphate group (see Figs. 7-5 and 7-6). NADH is a reduced molecule that transfers energy by transferring electrons as part of a hydrogen atom (see Fig. 7-7).

2. *Formation of acetyl coenzyme A.* Each pyruvate enters a mitochondrion and is oxidized to a two-carbon group (acetate) that combines with coenzyme A, forming acetyl coenzyme A. NADH is produced, and carbon dioxide is released as a waste product.

3. *The citric acid cycle.* The acetate group of acetyl coenzyme A combines with a four-carbon molecule (oxaloacetate) to form a six-carbon molecule (citrate). In the course of the cycle, citrate is recycled to oxaloacetate, and carbon dioxide is released as a waste product. Energy is captured as ATP and the reduced, high-energy compounds NADH and $FADH_2$ (see Chapter 7 to review $FADH_2$).

4. *Electron transport and chemiosmosis.* The electrons removed from glucose during the preceding stages are transferred from NADH and $FADH_2$ to a chain of electron acceptor compounds. As the electrons are passed from one electron acceptor to another, some of their energy is used to transport hydrogen ions (protons) across the inner mitochondrial membrane, forming a proton gradient. In a process known as *chemiosmosis* (described later), the energy of this proton gradient is used to produce ATP.

Most reactions involved in aerobic respiration are one of three types: dehydrogenations, decarboxylations, and those we informally categorize as preparation reactions. **Dehydrogenations** are reactions in which two hydrogen atoms (actually, 2 electrons plus 1 or 2 protons) are removed from the substrate and transferred to NAD^+ or FAD. **Decarboxylations** are reactions in which part of a carboxyl group (—COOH) is removed from the substrate as a molecule of CO_2. The carbon dioxide you exhale with each breath is derived from decarboxylations that occur in your cells. The rest of the reactions are preparation reactions in which molecules undergo rearrangements and other changes so that they

can undergo further dehydrogenations or decarboxylations. As you examine the individual reactions of aerobic respiration, you will encounter these three basic types.

In following the reactions of aerobic respiration, it helps to do some bookkeeping as you go along. Because glucose is the starting material, it is useful to express changes on a per glucose basis. We will pay particular attention to changes in the number of carbon atoms per molecule and to steps in which some type of energy transfer takes place.

In glycolysis, glucose yields two pyruvates

The word **glycolysis** comes from Greek words meaning "sugar splitting," which refers to the fact that the sugar glucose is metabolized. Glycolysis does not require oxygen and proceeds under aerobic or anaerobic conditions. ▌Figure 8-3 shows a simplified overview of glycolysis, in which a glucose molecule consisting of six carbons is converted to two molecules of **pyruvate,** a three-carbon molecule. Some of the energy in the glucose is captured; there is a net yield of two ATP molecules and two NADH molecules. The reactions of glycolysis take place in the cytosol, where the necessary reactants, such as ADP, NAD^+, and inorganic phosphate, float freely and are used as needed.

The glycolysis pathway consists of a series of reactions, each of which is catalyzed by a specific enzyme (▌Fig. 8-4). Glycolysis is divided into two major phases: the first includes endergonic reactions that require ATP, and the second includes exergonic reactions that yield ATP and NADH.

The first phase of glycolysis requires an investment of ATP

The first phase of glycolysis is sometimes called the *energy investment phase* (see Fig. 8-4, steps ❶ to ❺). Glucose is a relatively stable molecule and is not easily broken down. In two separate **phosphorylation reactions,** a phosphate group is transferred from ATP to the sugar. The resulting phosphorylated sugar (fructose-1,6-bisphosphate) is less stable and is broken enzymatically into two three-carbon molecules, dihydroxyacetone phosphate and glyceraldehyde-3-phosphate (G3P). The dihydroxyacetone phosphate is enzymatically converted to G3P, so the products at this point in glycolysis are two molecules of G3P per glucose. We can summarize this portion of glycolysis as follows:

$$\text{Glucose} \ + \ 2 \text{ ATP} \longrightarrow 2 \text{ G3P} \ + \ 2 \text{ ADP}$$

Six-carbon compound   Three-carbon compound

The second phase of glycolysis yields NADH and ATP

The second phase of glycolysis is sometimes called the *energy capture phase* (see Fig. 8-4, steps ❻ to ❿). Each G3P is converted to pyruvate. In the first step of this process, each G3P is oxidized

---

[1]Pyruvate and many other compounds in cellular respiration exist as anions at the pH found in the cell. They sometimes associate with H^+ to form acids. For example, pyruvate forms pyruvic acid. In some textbooks these compounds are presented in the acid form.

[2]Although the correct way to write the reduced form of NAD^+ is NADH + H^+, for simplicity we present the reduced form as NADH throughout the book.

Key Point

Glycolysis includes both energy investment and energy capture.

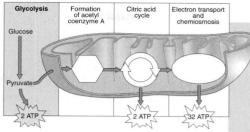

| Glycolysis | Formation of acetyl coenzyme A | Citric acid cycle | Electron transport and chemiosmosis |

Glucose

Pyruvate

2 ATP   2 ATP   32 ATP

Figure 8-3 *Animated* An overview of glycolysis

The black spheres represent carbon atoms. The energy investment phase of glycolysis leads to the splitting of sugar; ATP and NADH are produced during the energy capture phase. During glycolysis, each glucose molecule is converted to two pyruvates, with a net yield of two ATP molecules and two NADH molecules.

by the removal of 2 electrons (as part of two hydrogen atoms). These immediately combine with the hydrogen carrier molecule, NAD^+:

$$NAD^+ \ + \ 2\,H \longrightarrow NADH + H^+$$

Oxidized   (From G3P)   Reduced

Because there are two G3P molecules for every glucose, two NADH are formed. The energy of the electrons carried by NADH is used to form ATP later. This process is discussed in conjunction with the electron transport chain.

In two of the reactions leading to the formation of pyruvate, ATP forms when a phosphate group is transferred to ADP from a phosphorylated intermediate (see Fig. 8-4, steps ⑦ and ⑩). This process is called **substrate-level phosphorylation.** Note that in the energy investment phase of glycolysis two molecules of ATP are consumed, but in the energy capture phase four molecules of ATP are produced. Thus, glycolysis yields a net energy profit of *two* ATPs per glucose.

We can summarize the energy capture phase of glycolysis as follows:

$$2\ G3P + 2\ NAD^+ + 4\ ADP \longrightarrow$$
$$2\ \text{pyruvate} + 2\ NADH + 4\ ATP$$

Pyruvate is converted to acetyl CoA

In eukaryotes, the pyruvate molecules formed in glycolysis enter the mitochondria, where they are converted to **acetyl coenzyme A (acetyl CoA).** These reactions occur in the cytosol of aerobic prokaryotes. In this series of reactions, pyruvate undergoes a process known as **oxidative decarboxylation.** First, a carboxyl group is removed as carbon dioxide, which diffuses out of the cell (∎ Fig. 8-5). Then the remaining two-carbon fragment becomes oxidized, and NAD^+ accepts the electrons removed during the oxidation. Finally, the oxidized two-carbon fragment, an acetyl group, becomes attached to **coenzyme A,** yielding acetyl

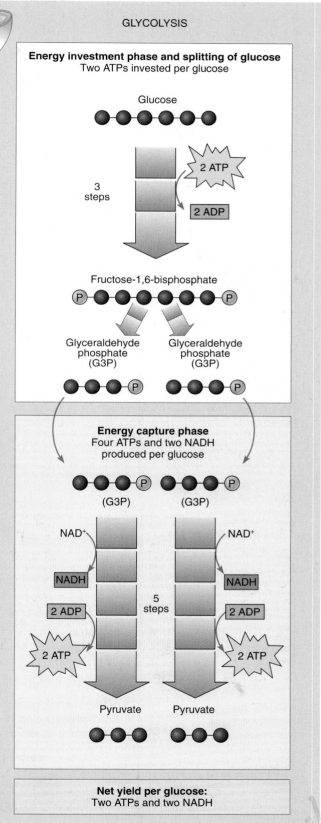

GLYCOLYSIS

Energy investment phase and splitting of glucose
Two ATPs invested per glucose

Glucose

3 steps

2 ATP

2 ADP

Fructose-1,6-bisphosphate

Glyceraldehyde phosphate (G3P)

Glyceraldehyde phosphate (G3P)

Energy capture phase
Four ATPs and two NADH produced per glucose

(G3P)   (G3P)

NAD^+ NAD^+

NADH   NADH

2 ADP   2 ADP

5 steps

2 ATP   2 ATP

Pyruvate   Pyruvate

Net yield per glucose:
Two ATPs and two NADH

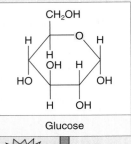

Glucose

Hexokinase

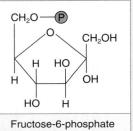

Glucose-6-phosphate

1. Glycolysis begins with preparation reaction in which glucose receives phosphate group from ATP molecule. ATP serves as source of both phosphate and energy needed to attach phosphate to glucose molecule. (Once ATP is spent, it becomes ADP and joins ADP pool of cell until turned into ATP again.) Phosphorylated glucose is known as glucose-6-phosphate. (Note phosphate attached to its carbon atom 6.) Phosphorylation of glucose makes it more chemically reactive.

Phosphoglucoisomerase

Fructose-6-phosphate

2. Glucose-6-phosphate undergoes another preparation reaction, rearrangement of its hydrogen and oxygen atoms. In this reaction glucose-6-phosphate is converted to its isomer, fructose-6-phosphate.

Phosphofructokinase

Fructose-1,6-bisphosphate

3. Next, another ATP donates phosphate to molecule, forming fructose-1,6-bisphosphate. So far, two ATP molecules have been invested in process without any being produced. Phosphate groups are now bound at carbons 1 and 6, and molecule is ready to be split.

Aldolase

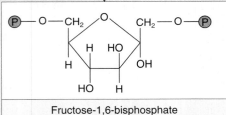

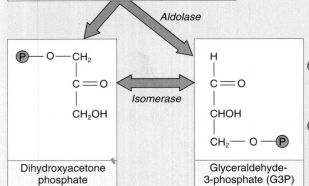

Isomerase

Dihydroxyacetone phosphate

Glyceraldehyde-3-phosphate (G3P)

4. Fructose-1,6-bisphosphate is then split into two 3-carbon sugars, glyceraldehyde-3-phosphate (G3P) and dihydroxyacetone phosphate.

5. Dihydroxyacetone phosphate is enzymatically converted to its isomer, glyceraldehyde-3-phosphate, for further metabolism in glycolysis.

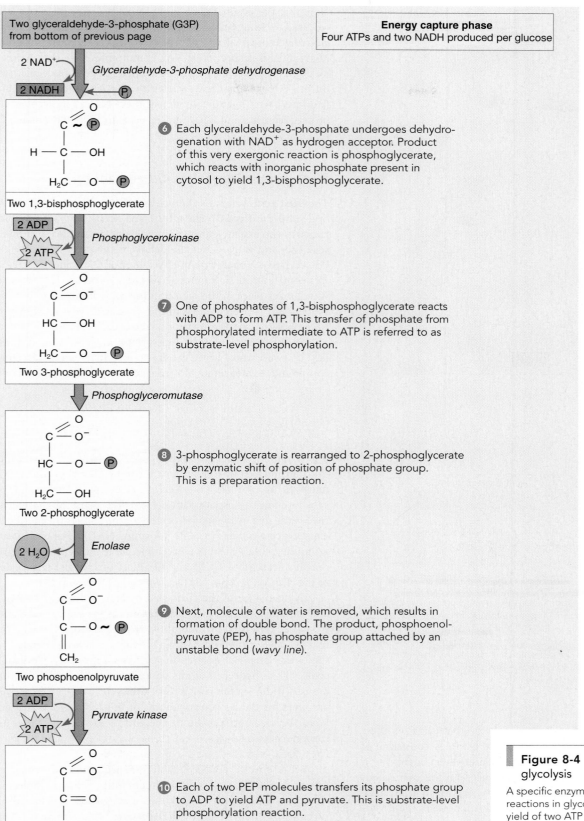

Two glyceraldehyde-3-phosphate (G3P) from bottom of previous page

2 NAD$^+$

Glyceraldehyde-3-phosphate dehydrogenase

2 NADH ← P

Two 1,3-bisphosphoglycerate

6 Each glyceraldehyde-3-phosphate undergoes dehydrogenation with NAD$^+$ as hydrogen acceptor. Product of this very exergonic reaction is phosphoglycerate, which reacts with inorganic phosphate present in cytosol to yield 1,3-bisphosphoglycerate.

2 ADP

Phosphoglycerokinase

2 ATP

Two 3-phosphoglycerate

7 One of phosphates of 1,3-bisphosphoglycerate reacts with ADP to form ATP. This transfer of phosphate from phosphorylated intermediate to ATP is referred to as substrate-level phosphorylation.

Phosphoglyceromutase

Two 2-phosphoglycerate

8 3-phosphoglycerate is rearranged to 2-phosphoglycerate by enzymatic shift of position of phosphate group. This is a preparation reaction.

2 H$_2$O

Enolase

Two phosphoenolpyruvate

9 Next, molecule of water is removed, which results in formation of double bond. The product, phosphoenolpyruvate (PEP), has phosphate group attached by an unstable bond (*wavy line*).

2 ADP

Pyruvate kinase

2 ATP

Two pyruvate

10 Each of two PEP molecules transfers its phosphate group to ADP to yield ATP and pyruvate. This is substrate-level phosphorylation reaction.

Figure 8-4 A detailed look at glycolysis

A specific enzyme catalyzes each of the reactions in glycolysis. Note the net yield of two ATP molecules and two NADH molecules. (The black wavy lines indicate bonds that permit the phosphates to be readily transferred to other molecules; in this case, ADP.)

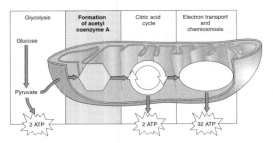

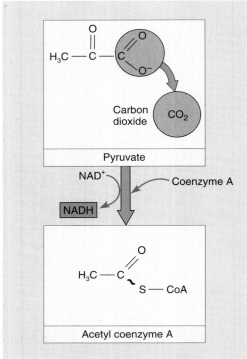

Note that the original glucose molecule has now been partially oxidized, yielding two acetyl groups and two CO_2 molecules. The electrons removed have reduced NAD^+ to NADH. At this point in aerobic respiration, four NADH molecules have been formed as a result of the catabolism of a single glucose molecule: two during glycolysis and two during the formation of acetyl CoA from pyruvate. Keep in mind that these NADH molecules will be used later (during electron transport) to form additional ATP molecules.

The citric acid cycle oxidizes acetyl CoA

The **citric acid cycle** is also known as the **tricarboxylic acid (TCA) cycle** and the **Krebs cycle,** after Hans Krebs, the German biochemist who assembled the accumulated contributions of many scientists and worked out the details of the cycle in the 1930s. He received a Nobel Prize in Physiology or Medicine in 1953 for this contribution. A simplified overview of the citric acid cycle, which takes place in the matrix of the mitochondria, is given in ▌ Figure 8-6. The eight steps of the citric acid cycle are shown in ▌ Figure 8-7. A specific enzyme catalyzes each reaction.

The first reaction of the cycle occurs when acetyl CoA transfers its two-carbon acetyl group to the four-carbon acceptor compound **oxaloacetate,** forming **citrate,** a six-carbon compound (step ❶):

$$\text{Oxaloacetate } + \text{ acetyl CoA} \longrightarrow \text{ citrate } + \text{ CoA}$$

Four-carbon compound   Two-carbon compound   Six-carbon compound

The citrate then goes through a series of chemical transformations, losing first one and then a second carboxyl group as CO_2 (steps ❷, ❸, and ❹). One ATP is formed (per acetyl group) by substrate-level phosphorylation (step ❺). Most of the energy made available by the oxidative steps of the cycle is transferred as energy-rich electrons to NAD^+, forming NADH. For each acetyl group that enters the citric acid cycle, three molecules of NADH are produced (steps ❸, ❹, and ❽). Electrons are also transferred to the electron acceptor FAD, forming $FADH_2$ (step ❻).

In the course of the citric acid cycle, two molecules of CO_2 and the equivalent of eight hydrogen atoms (8 protons and 8 electrons) are removed, forming three NADH and one $FADH_2$. You may wonder why more hydrogen equivalents are generated by these reactions than entered the cycle with the acetyl CoA molecule. These hydrogen atoms come from water molecules that are added during the reactions of the cycle. The CO_2 produced accounts for the two carbon atoms of the acetyl group that entered the citric acid cycle. At the end of each cycle, the four-carbon oxaloacetate has been regenerated (step ❽), and the cycle continues.

Because two acetyl CoA molecules are produced from each glucose molecule, two cycles are required per glucose molecule. After two turns of the cycle, the original glucose has lost all its carbons and may be regarded as having been completely consumed. To summarize, the citric acid cycle yields four CO_2, six NADH, two $FADH_2$, and two ATPs per glucose molecule.

At this point in aerobic respiration, only four molecules of ATP have been formed per glucose by substrate-level phosphorylation: two during glycolysis and two during the citric acid cycle

Figure 8-5 The formation of acetyl CoA

This series of reactions is catalyzed by the multienzyme complex pyruvate dehydrogenase. Pyruvate, a three-carbon molecule that is the end product of glycolysis, enters the mitochondrion and undergoes oxidative decarboxylation. First, the carboxyl group is split off as carbon dioxide. Then, the remaining two-carbon fragment is oxidized, and its electrons are transferred to NAD^+. Finally, the oxidized two-carbon group, an acetyl group, is attached to coenzyme A. CoA has a sulfur atom that forms a bond, shown as a black wavy line, with the acetyl group. When this bond is broken, the acetyl group can be readily transferred to another molecule.

CoA. *Pyruvate dehydrogenase,* the enzyme that catalyzes these reactions, is an enormous multienzyme complex consisting of 72 polypeptide chains! Recall from Chapter 7 that coenzyme A transfers groups derived from organic acids. In this case, coenzyme A transfers an acetyl group, which is related to acetic acid. Coenzyme A is manufactured in the cell from one of the B vitamins, pantothenic acid.

The overall reaction for the formation of acetyl coenzyme A is

$$2 \text{ Pyruvate} + 2 \text{ NAD}^+ + 2 \text{ CoA} \longrightarrow$$
$$2 \text{ Acetyl CoA} + 2 \text{ NADH} + 2 \text{ CO}_2$$

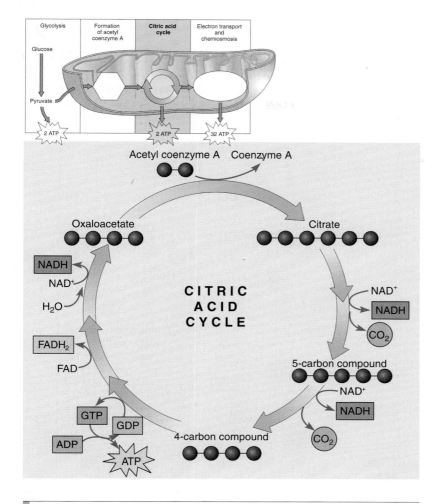

Figure 8-6 *Animated* Overview of the citric acid cycle

For every glucose, two acetyl groups enter the citric acid cycle (*top*). Each two-carbon acetyl group combines with a four-carbon compound, oxaloacetate, to form the six-carbon compound citrate. Two CO_2 molecules are removed, and energy is captured as one ATP, three NADH, and one $FADH_2$ per acetyl group (or two ATPs, six NADH, and two $FADH_2$ per glucose molecule).

(step ⑤). Most of the energy of the original glucose molecule is in the form of high-energy electrons in NADH and $FADH_2$. Their energy will be used to synthesize additional ATP through the electron transport chain and chemiosmosis.

The electron transport chain is coupled to ATP synthesis

Let us consider the fate of all the electrons removed from a molecule of glucose during glycolysis, acetyl CoA formation, and the citric acid cycle. Recall that these electrons were transferred as part of hydrogen atoms to the acceptors NAD^+ and FAD, forming NADH and $FADH_2$. These reduced compounds now enter the **electron transport chain,** where the high-energy electrons of their hydrogen atoms are shuttled from one acceptor to another. As the electrons are passed along in a series of exergonic redox reactions, some of their energy is used to drive the synthesis of ATP, which is an endergonic process. Because ATP synthesis (by

phosphorylation of ADP) is coupled to the redox reactions in the electron transport chain, the entire process is known as **oxidative phosphorylation.**

The electron transport chain transfers electrons from NADH and $FADH_2$ to oxygen

The electron transport chain is a series of electron carriers embedded in the inner mitochondrial membrane of eukaryotes and in the plasma membrane of aerobic prokaryotes. Like NADH and $FADH_2$, each carrier exists in an oxidized form or a reduced form. Electrons pass down the electron transport chain in a series of redox reactions that works much like a bucket brigade, the old-time chain of people who passed buckets of water from a stream to one another, to a building that was on fire. In the electron transport chain, each acceptor molecule becomes alternately reduced as it accepts electrons and oxidized as it gives them up. The electrons entering the electron transport chain have a relatively high energy content. They lose some of their energy at each step as they pass along the chain of electron carriers (just as some of the water spills out of the bucket as it is passed from one person to another).

Members of the electron transport chain include the flavoprotein *flavin mononucleotide (FMN),* the lipid *ubiquinone* (also called *coenzyme Q* or *CoQ*), several *iron–sulfur proteins,* and a group of closely related iron-containing proteins called *cytochromes* (❚ Fig. 8-8). Each electron carrier has a different mechanism for accepting and passing electrons. As cytochromes accept and donate electrons, for example, the charge on the iron atom, which is the electron carrier portion of the cytochromes, alternates between Fe^{2+} (reduced) and Fe^{3+} (oxidized).

Scientists have extracted and purified the electron transport chain from the inner mitochondrial membrane as four large, distinct protein complexes, or groups, of acceptors. *Complex I (NADH–ubiquinone oxidoreductase)* accepts electrons from NADH molecules that were produced during glycolysis, the formation of acetyl CoA, and the citric acid cycle. *Complex II (succinate–ubiquinone reductase)* accepts electrons from $FADH_2$ molecules that were produced during the citric acid cycle. Complexes I and II both produce the same product, reduced ubiquinone, which is the substrate of *complex III (ubiquinone–cytochrome c oxidoreductase)*. That is, complex III accepts electrons from reduced ubiquinone and passes them on to cytochrome *c*. *Complex IV (cytochrome c oxidase)* accepts electrons from cytochrome *c* and uses these electrons to reduce molecular oxygen, forming water in the process. The electrons simultaneously unite with protons from the surrounding medium to form hydrogen, and the chemical reaction between hydrogen and oxygen produces water.

Because oxygen is the final electron acceptor in the electron transport chain, organisms that respire aerobically require oxygen. What happens when cells that are strict aerobes are deprived of oxygen? The last cytochrome in the chain retains its electrons when no oxygen is available to accept them. When that occurs, each acceptor molecule in the chain retains its electrons (each

Figure 8-7 A detailed look at the citric acid cycle

Begin with step **1**, in the upper right corner, where acetyl coenzyme A attaches to oxaloacetate. Follow the steps in the citric acid cycle to see that the entry of a two-carbon acetyl group is balanced by the release of two molecules of CO_2. Electrons are transferred to NAD^+ or FAD, yielding NADH and $FADH_2$, respectively, and ATP is formed by substrate-level phosphorylation.

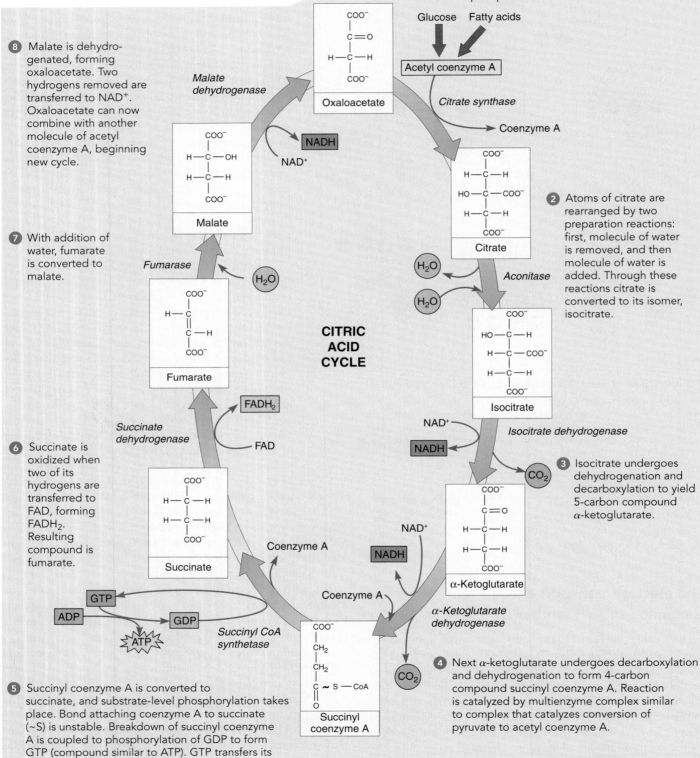

1 Unstable bond attaching acetyl group to coenzyme A breaks. 2-carbon acetyl group becomes attached to 4-carbon oxaloacetate molecule, forming citrate, 6-carbon molecule with three carboxyl groups. Coenzyme A is free to combine with another 2-carbon group and repeat process.

8 Malate is dehydrogenated, forming oxaloacetate. Two hydrogens removed are transferred to NAD^+. Oxaloacetate can now combine with another molecule of acetyl coenzyme A, beginning new cycle.

7 With addition of water, fumarate is converted to malate.

6 Succinate is oxidized when two of its hydrogens are transferred to FAD, forming $FADH_2$. Resulting compound is fumarate.

5 Succinyl coenzyme A is converted to succinate, and substrate-level phosphorylation takes place. Bond attaching coenzyme A to succinate (~S) is unstable. Breakdown of succinyl coenzyme A is coupled to phosphorylation of GDP to form GTP (compound similar to ATP). GTP transfers its phosphate to ADP, yielding ATP.

2 Atoms of citrate are rearranged by two preparation reactions: first, molecule of water is removed, and then molecule of water is added. Through these reactions citrate is converted to its isomer, isocitrate.

3 Isocitrate undergoes dehydrogenation and decarboxylation to yield 5-carbon compound α-ketoglutarate.

4 Next α-ketoglutarate undergoes decarboxylation and dehydrogenation to form 4-carbon compound succinyl coenzyme A. Reaction is catalyzed by multienzyme complex similar to complex that catalyzes conversion of pyruvate to acetyl coenzyme A.

CITRIC ACID CYCLE

Malate dehydrogenase
Oxaloacetate
Glucose Fatty acids
Acetyl coenzyme A
Citrate synthase
Coenzyme A
Malate
NADH
NAD^+
Citrate
Fumarase
H_2O
H_2O
H_2O
Aconitase
Fumarate
$FADH_2$
Succinate dehydrogenase
FAD
Isocitrate
NAD^+
NADH
Isocitrate dehydrogenase
CO_2
Succinate
Coenzyme A
NAD^+
NADH
α-Ketoglutarate
GTP
ADP
GDP
ATP
Succinyl CoA synthetase
Coenzyme A
α-Ketoglutarate dehydrogenase
Succinyl coenzyme A
CO_2

Electron carriers in the mitochondrial inner membrane transfer electrons from NADH and FADH$_2$ to oxygen.

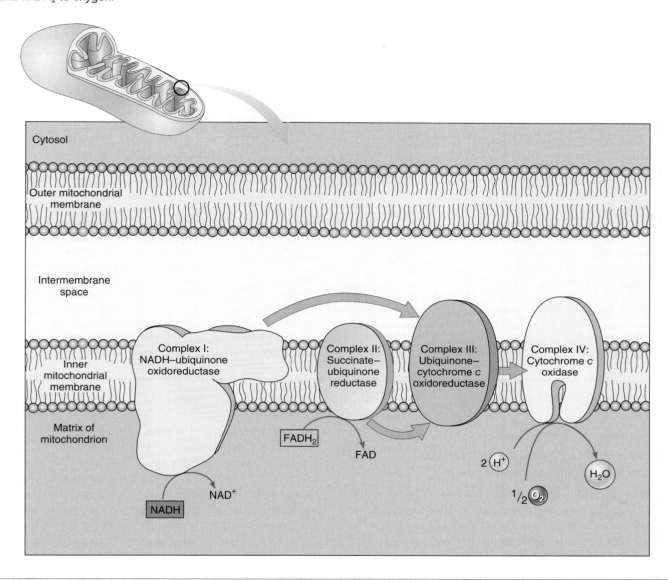

Figure 8-8 *Animated* An overview of the electron transport chain

Electrons fall to successively lower energy levels as they are passed along the four complexes of the electron transport chain located in the inner mitochondrial membrane. (The orange arrows indicate the pathway of electrons.) The carriers within each complex become alternately reduced and oxidized as they accept and donate electrons. The terminal acceptor is oxygen; one of the two atoms of an oxygen molecule (written as $\frac{1}{2}$ O$_2$) accepts 2 electrons, which are added to 2 protons from the surrounding medium to produce water.

remains in its reduced state), and the entire chain is blocked all the way back to NADH. Because oxidative phosphorylation is coupled to electron transport, no additional ATP is produced by way of the electron transport chain. Most cells of multicellular organisms cannot live long without oxygen, because the small amount of ATP they produce by glycolysis alone is insufficient to sustain life processes.

Lack of oxygen is not the only factor that interferes with the electron transport chain. Some poisons, including cyanide,

inhibit the normal activity of the cytochromes. Cyanide binds tightly to the iron in the last cytochrome in the electron transport chain (cytochrome a_3), making it unable to transport electrons to oxygen. This blocks the further passage of electrons through the chain, and ATP production ceases.

Although the flow of electrons in electron transport is usually tightly coupled to the production of ATP, some organisms uncouple the two processes to produce heat (see *Focus On: Electron Transport and Heat*).

ELECTRON TRANSPORT AND HEAT

What is the source of our body heat? Essentially, it is a by-product of various exergonic reactions, especially those involving the electron transport chains in our mitochondria. Some cold-adapted animals, hibernating animals, and newborn animals produce unusually large amounts of heat by uncoupling electron transport from ATP production. These animals have adipose tissue (tissue in which fat is stored) that is brown. The brown color comes from the large number of mitochondria found in the brown adipose tissue cells. The inner mitochondrial membranes of these mitochondria contain an uncoupling protein that produces a passive proton channel through which protons flow into the mitochondrial matrix. As a consequence, most of the energy of glucose is converted to heat rather than to chemical energy in ATP.

Certain plants, which are not generally considered "warm" organisms, also have the ability to produce large amounts of heat. Skunk cabbage (*Symplocarpus foeti-*

Leonard Lee Rue III /Animals Animals

Skunk cabbage (*Symplocarpus foetidus*)
This plant not only produces a significant amount of heat when it flowers but also regulates its temperature within a specific range.

dus), for example, lives in North American swamps and wet woodlands and generally flowers during February and March when the ground is still covered with snow

(see figure). Its uncoupled mitochondria generate large amounts of heat, enabling the plant to melt the snow and attract insect pollinators by vaporizing certain odiferous molecules into the surrounding air. The flower temperature of skunk cabbage is 15° to 22°C (59° to 72°F) when the air surrounding it is −15° to 10°C (5° to 50°F). Skunk cabbage flowers maintain this temperature for two weeks or more. Other plants, such as splitleaf philodendron (*Philodendron selloum*) and sacred lotus (*Nelumbo nucifera*), also generate heat when they bloom and maintain their temperatures within precise limits.

Some plants generate as much or more heat per gram of tissue than animals in flight, which have long been considered the greatest heat producers in the living world. The European plant lords-and-ladies (*Arum maculatum*), for example, produces 0.4 J (0.1 cal) of heat per second per gram of tissue, whereas a hummingbird in flight produces 0.24 J (0.06 cal) per second per gram of tissue.

The chemiosmotic model explains the coupling of ATP synthesis to electron transport in aerobic respiration

For decades, scientists were aware that oxidative phosphorylation occurs in mitochondria, and many experiments had shown that the transfer of 2 electrons from each NADH to oxygen (via the electron transport chain) usually results in the production of up to three ATP molecules. However, for a long time, the connection between ATP synthesis and electron transport remained a mystery.

In 1961, Peter Mitchell, a British biochemist, proposed the *chemiosmotic model*, which was based on his experiments with bacteria. Because the respiratory electron transport chain is located in the plasma membrane of an aerobic bacterial cell, the bacterial plasma membrane can be considered comparable to the inner mitochondrial membrane. Mitchell demonstrated that if bacterial cells are placed in an acidic environment (that is, an environment with a high hydrogen ion, or proton, concentration), the cells synthesized ATP even if electron transport was not taking place. On the basis of these and other experiments, Mitchell proposed that electron transport and ATP synthesis are coupled by means of a proton gradient across the inner mitochondrial membrane in eukaryotes (or across the plasma membrane in bacteria). His model was so radical that it was not immediately accepted, but by 1978, so much evidence had accumulated in sup-

port of **chemiosmosis** that Peter Mitchell was awarded a Nobel Prize in Chemistry.

The electron transport chain establishes the proton gradient; some of the energy released as electrons pass down the electron transport chain is used to move protons (H^+) across a membrane. In eukaryotes the protons are moved across the inner mitochondrial membrane into the intermembrane space (❙ Fig. 8-9). Hence, the inner mitochondrial membrane separates a space with a higher concentration of protons (the intermembrane space) from a space with a lower concentration of protons (the mitochondrial matrix).

Protons are moved across the inner mitochondrial membrane by three of the four electron transport complexes (complexes I, III, and IV) (❙ Fig. 8-10a). Like water behind a dam, the resulting proton gradient is a form of potential energy that can be harnessed to provide the energy for ATP synthesis.

Diffusion of protons from the intermembrane space, where they are highly concentrated, through the inner mitochondrial membrane to the matrix of the mitochondrion is limited to specific channels formed by a fifth enzyme complex, **ATP synthase,** a transmembrane protein. Portions of these complexes project from the inner surface of the membrane (the surface that faces the matrix) and are visible by electron microscopy (❙ Fig. 8-10b). Diffusion of the protons down their gradient, through the ATP synthase complex, is exergonic because the entropy of the system increases. This exergonic process provides the energy for ATP

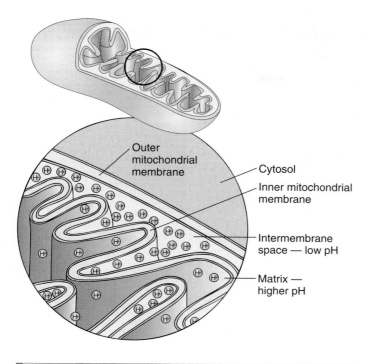

Outer mitochondrial membrane

Cytosol

Inner mitochondrial membrane

Intermembrane space — low pH

Matrix — higher pH

Figure 8-9 The accumulation of protons (H⁺) within the intermembrane space

As electrons move down the electron transport chain, the electron transport complexes move protons (H^+) from the matrix to the intermembrane space, creating a proton gradient. The high concentration of H^+ in the intermembrane space lowers the pH.

production, although the exact mechanism by which ATP synthase catalyzes the phosphorylation of ADP is still not completely understood. In 1997, Paul Boyer of the University of California at Los Angeles and John Walker of the Medical Research Council Laboratory of Molecular Biology, Cambridge, England, shared the Nobel Prize in Chemistry for the discovery that ATP synthase functions in an unusual way. Experimental evidence strongly suggests that ATP synthase acts like a highly efficient molecular motor: During the production of ATP from ADP and inorganic phosphate, a central structure of ATP synthase rotates, possibly in response to the force of protons moving through the enzyme complex. The rotation apparently alters the conformation of the catalytic subunits in a way that allows ATP synthesis.

Chemiosmosis is a fundamental mechanism of energy coupling in cells; it allows exergonic redox reactions to drive the endergonic reaction in which ATP is produced by phosphorylating ADP. In photosynthesis (discussed in Chapter 9), ATP is produced by a comparable process.

Aerobic respiration of one glucose yields a maximum of 36 to 38 ATPs

Let us now review where biologically useful energy is captured in aerobic respiration and calculate the total energy yield from the complete oxidation of glucose. ▮ Figure 8-11 summarizes the arithmetic involved.

1. In glycolysis, glucose is activated by the addition of phosphates from 2 ATP molecules and converted ultimately to 2 pyruvates + 2 NADH + 4 ATPs, yielding a net profit of 2 ATPs.

2. The 2 pyruvates are metabolized to 2 acetyl CoA + 2 CO_2 + 2 NADH.

3. In the citric acid cycle the 2 acetyl CoA molecules are metabolized to 4 CO_2 + 6 NADH + 2 $FADH_2$ + 2 ATPs.

   Because the oxidation of NADH in the electron transport chain yields up to 3 ATPs per molecule, the total of 10 NADH molecules can yield up to 30 ATPs. The 2 NADH molecules from glycolysis, however, yield either 2 or 3 ATPs each. The reason is that certain types of eukaryotic cells must expend energy to shuttle the NADH produced by glycolysis across the mitochondrial membrane (to be discussed shortly). Prokaryotic cells lack mitochondria; hence, they have no need to shuttle NADH molecules. For this reason, bacteria are able to generate 3 ATPs for every NADH, even those produced during glycolysis. Thus, the maximum number of ATPs formed using the energy from NADH is 28 to 30.

   The oxidation of $FADH_2$ yields 2 ATPs per molecule (recall that electrons from $FADH_2$ enter the electron transport chain at a different location than those from NADH), so the 2 $FADH_2$ molecules produced in the citric acid cycle yield 4 ATPs.

4. Summing all the ATPs (2 from glycolysis, 2 from the citric acid cycle, and 32 to 34 from electron transport and chemiosmosis), you can see that the complete aerobic metabolism of one molecule of glucose yields a maximum of 36 to 38 ATPs. Most ATP is generated by oxidative phosphorylation, which involves the electron transport chain and chemiosmosis. Only 4 ATPs are formed by substrate-level phosphorylation in glycolysis and the citric acid cycle.

We can analyze the efficiency of the overall process of aerobic respiration by comparing the free energy captured as ATP to the total free energy in a glucose molecule. Recall from Chapter 6 that although heat energy cannot power biological reactions, it is convenient to measure energy as heat. This is done through the use of a calorimeter, an instrument that measures the heat of a reaction. A sample is placed in a compartment surrounded by a chamber of water. As the sample burns (becomes oxidized), the temperature of the water rises, providing a measure of the heat released during the reaction.

When 1 mol of glucose is burned in a calorimeter, some 686 kcal (2870 kJ) are released as heat. The free energy temporarily held in the phosphate bonds of ATP is about 7.6 kcal (31.8 kJ) per mole. When 36 to 38 ATPs are generated during the aerobic respiration of glucose, the free energy trapped in ATP amounts to 7.6 kcal/mol × 36, or about 274 kcal (1146 kJ) per mole. Thus, the efficiency of aerobic respiration is 274/686, or about 40%. (By comparison, a steam power plant has an efficiency of 35% to 36% in converting its fuel energy into electricity.) The remainder of the energy in the glucose is released as heat.

Key Point

The electron transport chain forms a concentration gradient for H⁺, which diffuses through ATP synthase complexes, producing ATP.

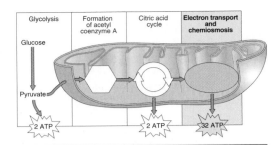

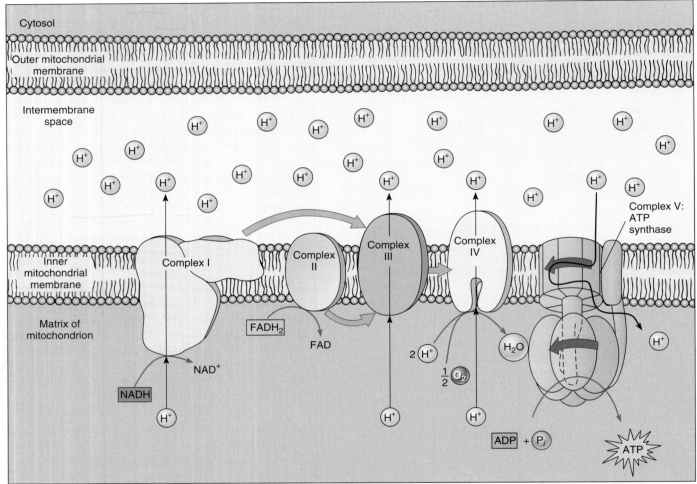

(a) The electron transport chain in the inner mitochondrial membrane includes three proton pumps that are located in three of the four electron transport complexes. (The orange arrows indicate the pathway of electrons; and the black arrows, the pathway of protons.) The energy released during electron transport is used to transport protons (H⁺) from the mitochondrial matrix to the intermembrane space, where a high concentration of protons accumulates. The protons cannot diffuse back into the matrix except through special channels in ATP synthase in the inner membrane. The flow of the protons through ATP synthase provides the energy for generating ATP from ADP and inorganic phosphate (P_i). In the process, the inner part of ATP synthase rotates (thick red arrows) like a motor.

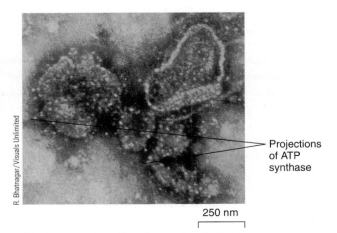

R. Bhatnagar/ Visuals Unlimited

250 nm

Projections of ATP synthase

(b) This TEM shows hundreds of projections of ATP synthase complexes along the surface of the inner mitochondrial membrane.

Figure 8-10 A detailed look at electron transport and chemiosmosis

Key Point

Most ATP is produced by electron transport and chemiosmosis (oxidative phosphorylation).

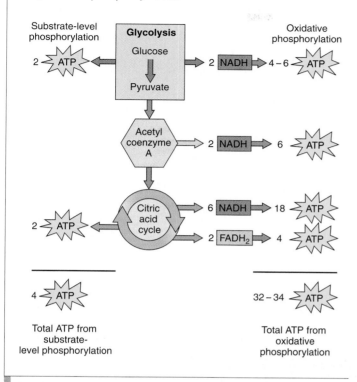

Figure 8-11 Energy yield from the complete oxidation of glucose by aerobic respiration

Mitochondrial shuttle systems harvest the electrons of NADH produced in the cytosol

The inner mitochondrial membrane is not permeable to NADH, which is a large molecule. Therefore, the NADH molecules produced in the cytosol during glycolysis cannot diffuse into the mitochondria to transfer their electrons to the electron transport chain. Unlike ATP and ADP, NADH does not have a carrier protein to transport it across the membrane. Instead, several systems have evolved to transfer just the *electrons* of NADH, not the NADH molecules themselves, into the mitochondria.

In liver, kidney, and heart cells, a special shuttle system transfers the electrons from NADH through the inner mitochondrial membrane to an NAD^+ molecule in the matrix. These electrons are transferred to the electron transport chain in the inner mitochondrial membrane, and up to three molecules of ATP are produced per pair of electrons.

In skeletal muscle, brain, and some other types of cells, another type of shuttle operates. Because this shuttle requires more energy than the shuttle in liver, kidney, and heart cells, the electrons are at a lower energy level when they enter the electron transport chain. They are accepted by ubiquinone rather than by NAD^+ and so generate a maximum of 2 ATP molecules per pair of electrons. For this reason, the number of ATPs produced

by aerobic respiration of one molecule of glucose in skeletal muscle cells is 36 rather than 38.

Cells regulate aerobic respiration

Aerobic respiration requires a steady input of fuel molecules and oxygen. Under normal conditions these materials are adequately provided and do not affect the rate of respiration. Instead, the rate of aerobic respiration is regulated by how much ADP and phosphate are available, with ATP synthesis continuing until most of the ADP has been converted to ATP. At this point oxidative phosphorylation slows considerably, which in turn slows down the citric acid cycle.

Glycolysis is partly controlled by feedback regulation (see Fig. 7-15) exerted on the enzyme phosphofructokinase, which catalyzes an early reaction of glycolysis (see Fig. 8-4). The active site of phosphofructokinase binds ATP and fructose-6-phosphate. However, the enzyme has two allosteric sites: an inhibitor site to which ATP binds when present at very high levels, and an activator site to which AMP (adenosine monophosphate, a molecule formed when two phosphates are removed from ATP) binds. Therefore, this enzyme is inactivated when ATP levels are high and activated when they are low. Respiration proceeds when the enzyme becomes activated, thus generating more ATP.

Review

- How much ATP is made available to the cell from a single glucose molecule by the operation of (1) glycolysis, (2) the formation of acetyl CoA, (3) the citric acid cycle, and (4) the electron transport chain and chemiosmosis?
- Why is each of the following essential to chemiosmotic ATP synthesis: (1) electron transport chain, (2) proton gradient, and (3) ATP synthase complex?
- What are the roles of NAD^+, FAD, and oxygen in aerobic respiration?
- What are some of the ways aerobic respiration is controlled?

ENERGY YIELD OF NUTRIENTS OTHER THAN GLUCOSE

Learning Objective

7   Summarize how the products of protein and lipid catabolism enter the same metabolic pathway that oxidizes glucose.

Many organisms, including humans, depend on nutrients other than glucose as a source of energy. In fact, you usually obtain more of your energy by oxidizing fatty acids than by oxidizing glucose. Amino acids derived from protein digestion are also used as fuel molecules. Such nutrients are transformed into one of the metabolic intermediates that are fed into glycolysis or the citric acid cycle (Fig. 8-12).

Amino acids are metabolized by reactions in which the amino group ($—NH_2$) is first removed, a process called **deamination.** In mammals and some other animals, the amino group is converted to urea (see Fig. 47-1) and excreted, but the carbon chain is metabolized and eventually is used as a reactant in one of the

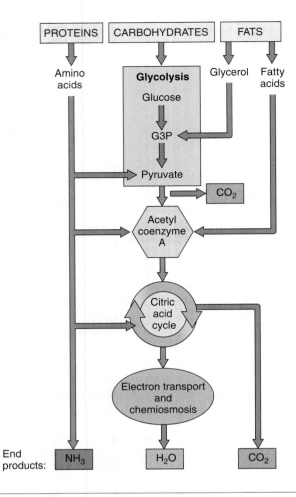

Figure 8-12 Energy from proteins, carbohydrates, and fats

Products of the catabolism of proteins, carbohydrates, and fats enter glycolysis or the citric acid cycle at various points. This diagram is greatly simplified and illustrates only a few of the principal catabolic pathways.

glycolysis. Fatty acids are oxidized and split enzymatically into two-carbon acetyl groups that are bound to coenzyme A; that is, fatty acids are converted to acetyl CoA. This process, which occurs in the mitochondrial matrix, is called **β-oxidation (beta-oxidation)**. Acetyl CoA molecules formed by β-oxidation enter the citric acid cycle.

Review

▌ How can a person obtain energy from a low-carbohydrate diet?

▌ What process must occur before amino acids enter the aerobic respiratory pathway?

▌ Where do fatty acids enter the aerobic respiratory pathway?

ANAEROBIC RESPIRATION AND FERMENTATION

Learning Objective

8   Compare and contrast anaerobic respiration and fermentation. Include the mechanism of ATP formation, the final electron acceptor, and the end products.

Anaerobic respiration, which does not use oxygen as the final electron acceptor, is performed by some prokaryotes that live in anaerobic environments, such as waterlogged soil, stagnant ponds, and animal intestines. As in aerobic respiration, electrons are transferred in anaerobic respiration from glucose to NADH; they then pass down an electron transport chain that is coupled to ATP synthesis by chemiosmosis. However, an inorganic substance such as nitrate (NO_3^-) or sulfate (SO_4^{2-}) replaces molecular oxygen as the terminal electron acceptor. The end products of this type of anaerobic respiration are carbon dioxide, one or more reduced inorganic substances, and ATP. One representative type of anaerobic respiration, which is part of the biogeochemical cycle known as the **nitrogen cycle** (discussed in Chapter 54), is summarized below.

$$C_6H_{12}O_6 + 12\ KNO_3 \longrightarrow$$

Potassium
nitrate

$$6\ CO_2 + 6\ H_2O + 12\ KNO_2 + \text{energy}$$

Potassium
nitrite   (in the chemical
bonds of ATP)

Certain other bacteria, as well as some fungi, regularly use **fermentation,** an anaerobic pathway that does not involve an electron transport chain. During fermentation only two ATPs are formed per glucose (by substrate-level phosphorylation during glycolysis). One might expect that a cell that obtains energy from glycolysis would produce pyruvate, the end product of glycolysis. However, this cannot happen because every cell has a limited supply of NAD^+, and NAD^+ is required for glycolysis to continue. If virtually all NAD^+ becomes reduced to NADH during glycolysis, then glycolysis stops and no more ATP is produced.

In fermentation, NADH molecules transfer their hydrogen atoms to organic molecules, thus regenerating the NAD^+ needed to keep glycolysis going. The resulting relatively reduced organic

steps of aerobic respiration. The amino acid alanine, for example, undergoes deamination to become pyruvate, the amino acid glutamate is converted to α-ketoglutarate, and the amino acid aspartate yields oxaloacetate. Pyruvate enters aerobic respiration as the end product of glycolysis, and α-ketoglutarate and oxaloacetate both enter aerobic respiration as intermediates in the citric acid cycle. Ultimately, the carbon chains of all the amino acids are metabolized in this way.

Each gram of lipid in the diet contains 9 kcal (38 kJ), more than twice as much energy as 1 g of glucose or amino acids, which have about 4 kcal (17 kJ) per gram. Lipids are rich in energy because they are highly reduced; that is, they have many hydrogen atoms and few oxygen atoms. When completely oxidized in aerobic respiration, a molecule of a six-carbon fatty acid generates up to 44 ATPs (compared with 36 to 38 ATPs for a molecule of glucose, which also has six carbons).

Both the glycerol and fatty acid components of a triacylglycerol (see Chapter 3) are used as fuel; phosphate is added to glycerol, converting it to G3P or another compound that enters

TABLE 8-2

A Comparison of Aerobic Respiration, Anaerobic Respiration, and Fermentation

	Aerobic Respiration	Anaerobic Respiration	Fermentation
Immediate fate of electrons in NADH	Transferred to electron transport chain	Transferred to electron transport chain	Transferred to organic molecule
Terminal electron acceptor of electron transport chain	O_2	Inorganic substances such as NO_3^- or SO_4^{2-}	No electron transport chain
Reduced product(s) formed	Water	Relatively reduced inorganic substances	Relatively reduced organic compounds (commonly, alcohol or lactate)
Mechanism of ATP synthesis	Oxidative phosphorylation/chemiosmosis; also substrate-level phosphorylation	Oxidative phosphorylation/chemiosmosis; also substrate-level phosphorylation	Substrate-level phosphorylation only (during glycolysis)

molecules (commonly, alcohol or lactate) tend to be toxic to the cells and are essentially waste products.

▌Table 8-2 compares aerobic respiration, anaerobic respiration, and fermentation.

Alcohol fermentation and lactate fermentation are inefficient

Yeasts are **facultative anaerobes** that carry out aerobic respiration when oxygen is available but switch to **alcohol fermentation** when deprived of oxygen (▌Fig. 8-13a). These eukaryotic, unicel-

lular fungi have enzymes that decarboxylate pyruvate, releasing carbon dioxide and forming a two-carbon compound called *acetaldehyde*. NADH produced during glycolysis transfers hydrogen atoms to acetaldehyde, reducing it to *ethyl alcohol* (▌Fig. 8-13b). Alcohol fermentation is the basis for the production of beer, wine, and other alcoholic beverages. Yeast cells are also used in baking to produce the carbon dioxide that causes dough to rise; the alcohol evaporates during baking.

Certain fungi and bacteria perform **lactate (lactic acid) fermentation.** In this alternative pathway, NADH produced during glycolysis transfers hydrogen atoms to pyruvate, reducing it to *lactate* (▌Fig. 8-13c). The ability of some bacteria to produce

Key Point

Fermentation regenerates NAD^+ needed for glycolysis.

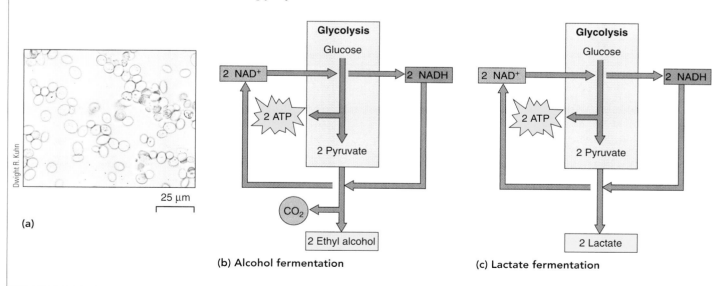

(b) **Alcohol fermentation**

(c) **Lactate fermentation**

▌ **Figure 8-13** *Animated* Fermentation

(a) Light micrograph of live brewer's yeast (*Saccharomyces cerevisiae*). Yeast cells have mitochondria and carry on aerobic respiration when O_2 is present. In the absence of O_2, yeasts carry on alcohol fermentation. **(b, c)** Glycolysis is the first part of fermentation pathways. In alcohol fermentation **(b)**, CO_2 is split off, and the two-carbon compound ethyl

alcohol is the end product. In lactate fermentation **(c)**, the final product is the three-carbon compound lactate. In both alcohol and lactate fermentation, there is a net gain of only two ATPs per molecule of glucose. Note that the NAD^+ used during glycolysis is regenerated during both alcohol fermentation and lactate fermentation.

lactate is exploited by humans, who use these bacteria to make yogurt and to ferment cabbage for sauerkraut.

Vertebrate muscle cells also produce lactate. Exercise can cause fatigue and muscle cramps possibly due to insufficient oxygen, the depletion of fuel molecules, and the accumulation of lactate during strenuous activity. This buildup of lactate occurs because muscle cells shift briefly to lactate fermentation if the amount of oxygen delivered to muscle cells is insufficient to support aerobic respiration. The shift is only temporary, however, and oxygen is required for sustained work. About 80% of the lactate is eventually exported to the liver, where it is used to regenerate more glucose for the muscle cells. The remaining 20% of the lactate is metabolized in muscle cells in the presence of oxygen. For this reason, you continue to breathe heavily after you have stopped exercising: the additional oxygen is needed to oxidize lactate, thereby restoring the muscle cells to their normal state.

Although humans use lactate fermentation to produce ATP for only a few minutes, a few animals can live without oxygen for much longer periods. The red-eared slider, a freshwater turtle, remains underwater for as long as 2 weeks. During this time, it is relatively inactive and therefore does not expend a great deal of energy. It relies on lactate fermentation for ATP production.

Both alcohol fermentation and lactate fermentation are highly inefficient, because the fuel is only partially oxidized. Alcohol, the end product of fermentation by yeast cells, can be burned and is even used as automobile fuel; obviously, it contains a great deal of energy that the yeast cells cannot extract using anaerobic methods. Lactate, a three-carbon compound, contains even more energy than the two-carbon alcohol. In contrast, all available energy is removed during aerobic respiration, because the fuel molecules become completely oxidized to CO_2. A net profit of only 2 ATPs is produced by the fermentation of one molecule of glucose, compared with up to 36 to 38 ATPs when oxygen is available.

The inefficiency of fermentation necessitates a large supply of fuel. To perform the same amount of work, a cell engaged in fermentation must consume up to 20 times as much glucose or other carbohydrate per second as a cell using aerobic respiration. For this reason, your skeletal muscle cells store large quantities of glucose in the form of glycogen, which enables them to metabolize anaerobically for short periods.

Review

- What is the fate of hydrogen atoms removed from glucose during glycolysis when oxygen is present in muscle cells? How does this compare to the fate of hydrogen atoms removed from glucose when the amount of available oxygen is insufficient to support aerobic respiration?
- Why is the ATP yield of fermentation only a tiny fraction of the yield from aerobic respiration?

SUMMARY WITH KEY TERMS

Learning Objectives

1   Write a summary reaction for aerobic respiration that shows which reactant becomes oxidized and which becomes reduced (page 172).

$$\overset{\overbrace{\qquad\qquad \text{oxidation} \qquad\qquad}}{C_6H_{12}O_6 + 6\ O_2 \longrightarrow 6\ CO_2 + 6\ H_2O + \text{energy}}\underset{\underbrace{\qquad\qquad \text{reduction} \qquad\qquad}}{}$$

- **Aerobic respiration** is a catabolic process in which a fuel molecule such as glucose is broken down to form carbon dioxide and water. It includes **redox reactions** that result in the transfer of electrons from glucose (which becomes **oxidized**) to oxygen (which becomes **reduced**).
- Energy released during aerobic respiration is used to produce up to 36 to 38 ATPs per molecule of glucose.

2   List and give a brief overview of the four stages of aerobic respiration (page 173).

- The chemical reactions of aerobic respiration occur in four stages: glycolysis, formation of acetyl CoA, the citric acid cycle, and the electron transport chain and chemiosmosis.
- During **glycolysis,** a molecule of glucose is degraded to two molecules of **pyruvate.** Two ATP molecules (net) are produced by **substrate-level phosphorylation** during glycolysis. Four hydrogen atoms are removed and used to produce two NADH.

ThomsonNOW   **See the process of glycolysis unfold by clicking on the figure in ThomsonNOW.**

- During the formation of **acetyl CoA,** the two pyruvate molecules each lose a molecule of carbon dioxide, and the remaining acetyl groups each combine with **coenzyme A,** producing two molecules of acetyl CoA; one NADH is produced per pyruvate.
- Each acetyl CoA enters the **citric acid cycle** by combining with a four-carbon compound, **oxaloacetate,** to form **citrate,** a six-carbon compound. Two acetyl CoA molecules enter the cycle for every glucose molecule. For every two carbons that enter the cycle as part of an acetyl CoA molecule, two leave as carbon dioxide. For every acetyl CoA, hydrogen atoms are transferred to three NAD$^+$ and one FAD; only one ATP is produced by substrate-level phosphorylation.

ThomsonNOW   **Interact with the citric acid cycle by clicking on the figure in ThomsonNOW.**

- Hydrogen atoms (or their electrons) removed from fuel molecules are transferred from one electron acceptor to another down an **electron transport chain** located in the mitochondrial inner membrane; ultimately, these electrons reduce molecular oxygen, forming water. In **oxidative phosphorylation,** the redox reactions in the electron transport chain are coupled to synthesis of ATP through the mechanism of **chemiosmosis.**

ThomsonNOW   **See the electron transport chain in action by clicking on the figure in ThomsonNOW.**

3   Indicate where each stage of aerobic respiration takes place in a eukaryotic cell (page 173).

- Glycolysis occurs in the cytosol, and the remaining stages of aerobic respiration take place in the mitochondria

4   Add up the energy captured (as ATP, NADH, and FADH$_2$) in each stage of aerobic respiration (page 173).

- In glycolysis, each glucose molecule produces 2 NADH and 2 ATPs (net). The conversion of 2 pyruvates to acetyl CoA results in the formation of 2 NADH. In the citric acid cycle, the 2 acetyl CoA molecules are metabolized to form 6 NADH, 2 $FADH_2$, and 2 ATPs. To summarize, we have 4 ATPs, 10 NADH, and 2 $FADH_2$.

- When the 10 NADH and 2 $FADH_2$ pass through the electron transport chain, 32 to 34 ATPs are produced by chemiosmosis. Therefore, each glucose molecule yields a total of up to 36 to 38 ATPs.

5 Define *chemiosmosis*, and explain how a gradient of protons is established across the inner mitochondrial membrane (page 173).

- In chemiosmosis, some of the energy of the electrons in the electron transport chain is used to pump protons across the inner mitochondrial membrane into the intermembrane space. This pumping establishes a proton gradient across the inner mitochondrial membrane. Protons (H^+) accumulate within the intermembrane space, lowering the pH.

6 Describe the process by which the proton gradient drives ATP synthesis in chemiosmosis (page 173).

- The diffusion of protons through channels formed by the enzyme **ATP synthase**, which extends through the inner mitochondrial membrane from the intermembrane space to the mitochondrial matrix, provides the energy to synthesize ATP.

7 Summarize how the products of protein and lipid catabolism enter the same metabolic pathway that oxidizes glucose (page 185).

- Amino acids undergo **deamination**, and their carbon skeletons are converted to metabolic intermediates of aerobic respiration.

- Both the glycerol and fatty acid components of lipids are oxidized as fuel. Fatty acids are converted to acetyl CoA molecules by the process of **β-oxidation.**

8 Compare and contrast anaerobic respiration and fermentation. Include the mechanism of ATP formation, the final electron acceptor, and the end products (page 186).

- In **anaerobic respiration**, electrons are transferred from fuel molecules to an electron transport chain that is coupled to ATP synthesis by chemiosmosis; the final electron acceptor is an inorganic substance such as nitrate or sulfate, not molecular oxygen.

- **Fermentation** is an anaerobic process that does not use an electron transport chain. There is a net gain of only two ATPs per glucose; these are produced by substrate-level phosphorylation during glycolysis. To maintain the supply of NAD^+ essential for glycolysis, hydrogen atoms are transferred from NADH to an organic compound derived from the initial nutrient.

- Yeast cells carry out **alcohol fermentation,** in which ethyl alcohol and carbon dioxide are the final waste products.

- Certain fungi, prokaryotes, and animal cells carry out **lactate (lactic acid) fermentation,** in which hydrogen atoms are added to pyruvate to form lactate, a waste product.

ThomsonNOW™ **Learn more about fermentation by clicking on the figure in ThomsonNOW.**

Summary Reactions for Aerobic Respiration

Summary reaction for the complete oxidation of glucose:

$$C_6H_{12}O_6 + 6\ O_2 + 6\ H_2O \longrightarrow 6\ CO_2 + 12\ H_2O + energy\ (36\ to\ 38\ ATP)$$

Summary reaction for glycolysis:

$$C_6H_{12}O_6 + 2\ ATP + 2\ ADP + 2\ P_i + 2\ NAD^+ \longrightarrow 2\ pyruvate + 4\ ATP + 2\ NADH + H_2O$$

Summary reaction for the conversion of pyruvate to acetyl CoA:

$$2\ Pyruvate + 2\ coenzyme\ A + 2\ NAD^+ \longrightarrow 2\ acetyl\ CoA + 2\ CO_2 + 2\ NADH$$

Summary reaction for the citric acid cycle:

$$2\ Acetyl\ CoA + 6\ NAD^+ + 2\ FAD + 2\ ADP + 2\ P_i + 2\ H_2O \longrightarrow 4\ CO_2 + 6\ NADH + 2\ FADH_2 + 2\ ATP + 2\ CoA$$

Summary reactions for the processing of the hydrogen atoms of NADH and $FADH_2$ in the electron transport chain:

$$NADH + 3\ ADP + 3\ P_i + \tfrac{1}{2}\ O_2 \longrightarrow NAD^+ + 3\ ATP + H_2O$$

$$FADH_2 + 2\ ADP + 2\ P_i + \tfrac{1}{2}\ O_2 \longrightarrow FAD^+ + 2\ ATP + H_2O$$

Summary Reactions for Fermentation

Summary reaction for lactate fermentation:

$$C_6H_{12}O_6 \longrightarrow 2\ lactate + energy\ (2\ ATP)$$

Summary reaction for alcohol fermentation:

$$C_6H_{12}O_6 \longrightarrow 2\ CO_2 + 2\ ethyl\ alcohol + energy\ (2\ ATP)$$

TEST YOUR UNDERSTANDING

1. The process of splitting larger molecules into smaller ones is an aspect of metabolism called (a) anabolism (b) fermentation (c) catabolism (d) oxidative phosphorylation (e) chemiosmosis

2. The synthetic aspect of metabolism is called (a) anabolism (b) fermentation (c) catabolism (d) oxidative phosphorylation (e) chemiosmosis

3. A chemical process during which a substance gains electrons is called (a) oxidation (b) oxidative phosphorylation (c) deamination (d) reduction (e) dehydrogenation

4. The pathway through which glucose is degraded to pyruvate is called (a) aerobic respiration (b) the citric acid cycle (c) the oxidation of pyruvate (d) alcohol fermentation (e) glycolysis

5. The reactions of _____ take place within the cytosol of eukaryotic cells. (a) glycolysis (b) oxidation of pyruvate (c) the citric acid cycle (d) chemiosmosis (e) the electron transport chain

6. Before pyruvate enters the citric acid cycle, it is decarboxylated, oxidized, and combined with coenzyme A, forming acetyl CoA, carbon dioxide, and one molecule of (a) NADH (b) $FADH_2$ (c) ATP (d) ADP (e) $C_6H_{12}O_6$

7. In the first step of the citric acid cycle, acetyl CoA reacts with oxaloacetate to form (a) pyruvate (b) citrate (c) NADH (d) ATP (e) CO_2

8. Dehydrogenase enzymes remove hydrogen atoms from fuel molecules and transfer them to acceptors such as (a) O_2 and H_2O (b) ATP and FAD (c) NAD^+ and FAD (d) CO_2 and H_2O (e) CoA and pyruvate

9. Which of the following is a major source of electrons for the electron transport chain? (a) H_2O (b) ATP (c) NADH (d) ATP synthase (e) coenzyme A

10. In the process of _____, electron transport and ATP synthesis are coupled by a proton gradient across the inner mitochondrial membrane. (a) chemiosmosis (b) deamination (c) anaerobic respiration (d) glycolysis (e) decarboxylation

11. Which of the following is a common energy flow sequence in aerobic respiration, starting with the energy stored in glucose? (a) glucose ⟶ NADH ⟶ pyruvate ⟶ ATP (b) glucose ⟶ ATP ⟶ NADH ⟶ electron transport chain (c) glucose ⟶ NADH ⟶ electron transport chain ⟶ ATP (d) glucose ⟶ oxygen ⟶ NADH ⟶ water (e) glucose ⟶ $FADH_2$ ⟶ NADH ⟶ coenzyme A

12. Which multiprotein complex in the electron transport chain is responsible for reducing molecular oxygen? (a) complex I (NADH–ubiquinone oxidoreductase) (b) complex II (succinate–ubiquinone reductase) (c) complex III (ubiquinone–cytochrome c oxidoreductase)

(d) complex IV (cytochrome c oxidase) (e) complex V (ATP synthase)

13. A net profit of only 2 ATPs can be produced anaerobically from the _____ of one molecule of glucose, compared with a maximum of 38 ATPs produced in _____. (a) fermentation; anaerobic respiration (b) aerobic respiration; fermentation (c) aerobic respiration; anaerobic respiration (d) dehydrogenation; decarboxylation (e) fermentation; aerobic respiration

14. When deprived of oxygen, yeast cells obtain energy by fermentation, producing carbon dioxide, ATP, and (a) acetyl CoA (b) ethyl alcohol (c) lactate (d) pyruvate (e) citrate

15. During strenuous muscle activity, the pyruvate in muscle cells may accept hydrogen from NADH to become _____. (a) acetyl CoA (b) ethyl alcohol (c) lactate (d) pyruvate (e) citrate

16. Label the 10 blank lines in the figure. Use Figure 8-2 to check your answers.

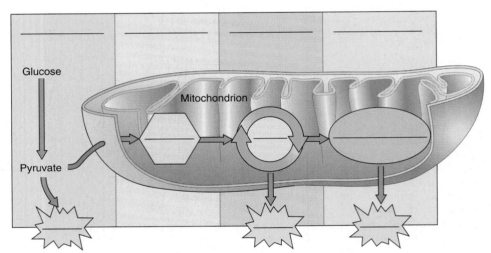

CRITICAL THINKING

1. How are the endergonic reactions of the first phase of glycolysis coupled to the hydrolysis of ATP, which is exergonic? How are the exergonic reactions of the second phase of glycolysis coupled to the endergonic synthesis of ATP and NADH?

2. In what ways is the inner mitochondrial membrane essential to the coupling of electron transport and ATP synthesis? Could the membrane carry out its function if its lipid bilayer were readily permeable to hydrogen ions (protons)?

3. Based on what you have learned in this chapter, explain why a schoolchild can run 17 miles per hour in a 100-yard dash, but a trained athlete can run only about 11.5 miles per hour in a 26-mile marathon.

4. **Evolution Link.** The reactions of glycolysis are identical in *all* organisms—prokaryotes, protists, fungi, plants,

and animals—that obtain energy from glucose catabolism. What does this universality suggest about the evolution of glycolysis?

5. **Evolution Link.** Molecular oxygen is so reactive that it would not exist in Earth's atmosphere today if it were not constantly replenished by organisms that release oxygen as a waste product of photosynthesis. What does that fact suggest about the evolution of aerobic respiration and oxygen-releasing photosynthetic processes?

Additional questions are available in ThomsonNOW at www.thomsonedu.com/login

Photosynthesis: Capturing Energy

Skip Moody/Dembinsky Photo Associates

Photosynthesis. These blue lupines (*Lupinus hirsutus*) and the trees behind them use light energy to power the processes that incorporate CO_2 into organic molecules. This photograph was taken in southern Michigan.

KEY CONCEPTS

Light energy powers photosynthesis, which is essential to plants and most life on Earth.

Photosynthesis, which occurs in chloroplasts, is a redox process.

Light-dependent reactions convert light energy to the chemical energy of NADPH and ATP.

Carbon fixation reactions incorporate CO_2 into organic molecules.

Most photosynthetic organisms are photoautotrophs.

Look at all the living things that surround you—the trees, your pet goldfish, your own body. Most of that biomass is made up of carbon-based biological molecules. What is the ultimate source of all that carbon? Surprising to some, the source is carbon dioxide from the air. Your cells cannot take carbon dioxide from the air and incorporate it into organic molecules—but some plant cells can. They do this through **photosynthesis,** the sequence of events by which light energy is converted into the stored chemical energy of organic molecules. Photosynthesis is the first step in the flow of energy through most of the living world, capturing the vast majority of the energy that living organisms use. Photosynthesis not only sustains plants (see photograph) and other photosynthetic organisms such as algae and photosynthetic bacteria but also indirectly supports most nonphotosynthetic organisms such as animals, fungi, protozoa, and most bacteria. Each year photosynthetic organisms convert CO_2 into billions of tons of organic molecules. These molecules have two important roles in both photosynthetic and nonphotosynthetic organisms: they are both the building blocks of cells and, as we saw in Chapter 8, a source of chemical energy that fuels the metabolic reactions that sustain almost all life.

In this chapter we first examine how light energy is used in the synthesis of ATP and other molecules that temporarily hold chemical energy but are unstable and cannot be stockpiled in the cell. We then see how their energy powers the anabolic pathway by which a photosynthetic cell synthesizes stable organic molecules from the simple inorganic compounds CO_2 and water. Finally, we explore the role of photosynthesis in plants and in Earth's environment. ■

LIGHT

Learning Objective

1   Describe the physical properties of light, and explain the relationship between a wavelength of light and its energy.

Because most life on this planet depends on light, either directly or indirectly, it is important to understand the nature of light and its essential role in photosynthesis. Visible light represents a very small portion of a vast, <u>continuous range of radiation</u> called the *electromagnetic spectrum* (█ Fig. 9-1). All radiation in this spectrum travels as waves. A **wavelength** is the distance from one wave peak to the next. At one end of the electromagnetic spectrum are gamma rays, which have very short wavelengths measured in fractions of nanometers, or nm (1 nanometer equals 10^{-9} m, one billionth of a meter). At the other end of the spectrum are radio waves, with wavelengths so long they can be measured in kilometers. The portion of the electromagnetic spectrum from 380 to 760 nm is called the *visible spectrum,* because we humans can see it. The visible spectrum includes all the colors of the rainbow (█ Fig. 9-2); violet has the shortest wavelength, and red has the longest.

Light is composed of small particles, or packets, of energy called **photons.** The energy of a photon is inversely proportional to its wavelength: Shorter-wavelength light has more energy per photon than longer-wavelength light.

Why does photosynthesis depend on light detectable by the human eye (visible light) rather than on some other wavelength of radiation? We can only speculate on the answer. Perhaps the reason is that radiation within the visible-light portion of the spectrum excites certain types of biological molecules, moving electrons into higher energy levels. Radiation with wavelengths longer than those of visible light does not have enough energy to excite these biological molecules. Radiation with wavelengths shorter than those of visible light is so energetic that it disrupts the bonds of many biological molecules. Thus, visible light has just the right amount of energy to cause the kinds of reversible changes in molecules that are useful in photosynthesis.

When a molecule absorbs a photon of light energy, one of its electrons becomes energized, which means that the electron shifts from a lower-energy atomic orbital to a high-energy orbital that is more distant from the atomic nucleus. One of two things then happens, depending on the atom and its surroundings (█ Fig. 9-3). The atom may return to its **ground state,** which is the condition in which all its electrons are in their normal, lowest-energy levels. When an electron returns to its ground state, its energy dissipates as heat or as an emission of light of a longer wavelength than the absorbed light; this emission of light is called **fluorescence.** Alternatively, the energized electron may leave the atom and be accepted by an electron acceptor molecule, which becomes reduced in the process; this is what occurs in photosynthesis.

Now that you understand some of the properties of light, let us consider the organelles that use light for photosynthesis.

Review

█   Why does photosynthesis require visible light?
█   Which color of light has the longer wavelength, violet or red? Which color of light has the higher energy per photon, violet or red?

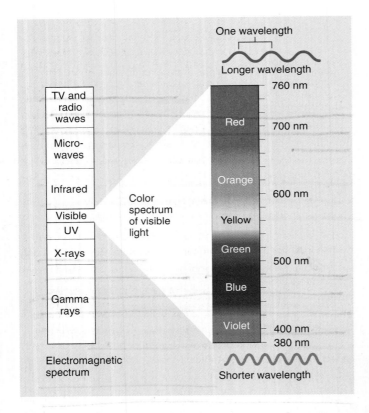

Figure 9-1   *Animated*   The electromagnetic spectrum

Waves in the electromagnetic spectrum have similar properties but different wavelengths. Radio waves are the longest (and least energetic) waves, with wavelengths as long as 20 km. Gamma rays are the shortest (and most energetic) waves. Visible light represents a small fraction of the electromagnetic spectrum and consists of a mixture of wavelengths ranging from about 380 to 760 nm. The energy from visible light is used in photosynthesis.

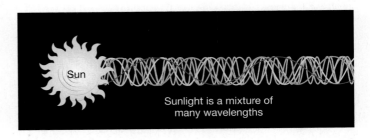

Figure 9-2   Visible radiation emitted from the sun

Electromagnetic radiation from the sun includes ultraviolet radiation and visible light of varying colors and wavelengths.

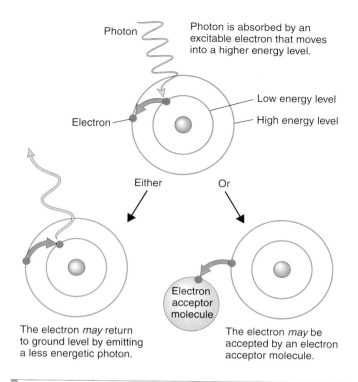

Photon

Photon is absorbed by an excitable electron that moves into a higher energy level.

Electron

Low energy level

High energy level

Either   Or

Electron acceptor molecule

The electron *may* return to ground level by emitting a less energetic photon.

The electron *may* be accepted by an electron acceptor molecule.

Figure 9-3 Interactions between light and atoms or molecules

(*Top*) When a photon of light energy strikes an atom or a molecule of which the atom is a part, the energy of the photon may push an electron to an orbital farther from the nucleus (that is, into a higher energy level). (*Lower left*) If the electron returns to the lower, more stable energy level, the energy may be released as a less energetic, longer-wavelength photon, known as fluorescence (*shown*), or as heat. (*Lower right*) If the appropriate electron acceptors are available, the electron may leave the atom. During photosynthesis, an electron acceptor captures the energetic electron and passes it to a chain of acceptors.

CHLOROPLASTS

Learning Objectives

2   Diagram the internal structure of a chloroplast, and explain how its components interact and facilitate the process of photosynthesis.

3   Describe what happens to an electron in a biological molecule such as chlorophyll when a photon of light energy is absorbed.

If you examine a section of leaf tissue in a microscope, you see that the green pigment, chlorophyll, is not uniformly distributed in the cell but is confined to organelles called **chloroplasts.** In plants, chloroplasts lie mainly inside the leaf in the cells of the **mesophyll,** a layer with many air spaces and a very high concentration of water vapor (❚ Fig. 9-4a). The interior of the leaf exchanges gases with the outside through microscopic pores, called **stomata** (sing., *stoma*). Each mesophyll cell has 20 to 100 chloroplasts (❚ Fig. 9-4b).

The chloroplast, like the mitochondrion, is enclosed by outer and inner membranes (❚ Fig. 9-4c). The inner membrane encloses a fluid-filled region called the **stroma,** which contains most of the enzymes required to produce carbohydrate molecules. Suspended in the stroma is a third system of membranes that forms an interconnected set of flat, disclike sacs called **thylakoids.**

The thylakoid membrane encloses a fluid-filled interior space, the **thylakoid lumen.** In some regions of the chloroplast, thylakoid sacs are arranged in stacks called **grana** (sing., *granum*). Each granum looks something like a stack of coins, with each "coin" being a thylakoid. Some thylakoid membranes extend from one granum to another. Thylakoid membranes, like the inner mitochondrial membrane (see Chapter 8), are involved in ATP synthesis. (Photosynthetic prokaryotes have no chloroplasts, but thylakoid membranes are often arranged around the periphery of the cell as infoldings of the plasma membrane.)

Chlorophyll is found in the thylakoid membrane

Thylakoid membranes contain several kinds of *pigments,* which are substances that absorb visible light. Different pigments absorb light of different wavelengths. **Chlorophyll,** the main pigment of photosynthesis, absorbs light primarily in the blue and red regions of the visible spectrum. Green light is not appreciably absorbed by chlorophyll. Plants usually appear green because some of the green light that strikes them is scattered or reflected.

A chlorophyll molecule has two main parts, a complex ring and a long side chain (❚ Fig. 9-5). The ring structure, called a *porphyrin ring,* is made up of joined smaller rings composed of carbon and nitrogen atoms; the porphyrin ring absorbs light energy. The porphyrin ring of chlorophyll is strikingly similar to the heme portion of the red pigment hemoglobin in red blood cells. However, unlike heme, which contains an atom of iron in the center of the ring, chlorophyll contains an atom of magnesium in that position. The chlorophyll molecule also contains a long, hydrocarbon side chain that makes the molecule extremely nonpolar and anchors the chlorophyll in the membrane.

All chlorophyll molecules in the thylakoid membrane are associated with specific *chlorophyll-binding proteins;* biologists have identified about 15 different kinds. Each thylakoid membrane is filled with precisely oriented chlorophyll molecules and chlorophyll-binding proteins that facilitate the transfer of energy from one molecule to another.

There are several kinds of chlorophyll. The most important is **chlorophyll *a,*** the pigment that initiates the light-dependent reactions of photosynthesis. **Chlorophyll *b*** is an accessory pigment that also participates in photosynthesis. It differs from chlorophyll *a* only in a functional group on the porphyrin ring: The methyl group ($-CH_3$) in chlorophyll *a* is replaced in chlorophyll *b* by a terminal carbonyl group ($-CHO$). This difference shifts the wavelengths of light absorbed and reflected by chlorophyll *b,* making it appear yellow-green, whereas chlorophyll *a* appears bright green.

(a) This leaf cross section reveals that the mesophyll is the photosynthetic tissue. CO_2 enters the leaf through tiny pores or stomata, and H_2O is carried to the mesophyll in veins.

(b) Notice the numerous chloroplasts in this LM of plant cells.

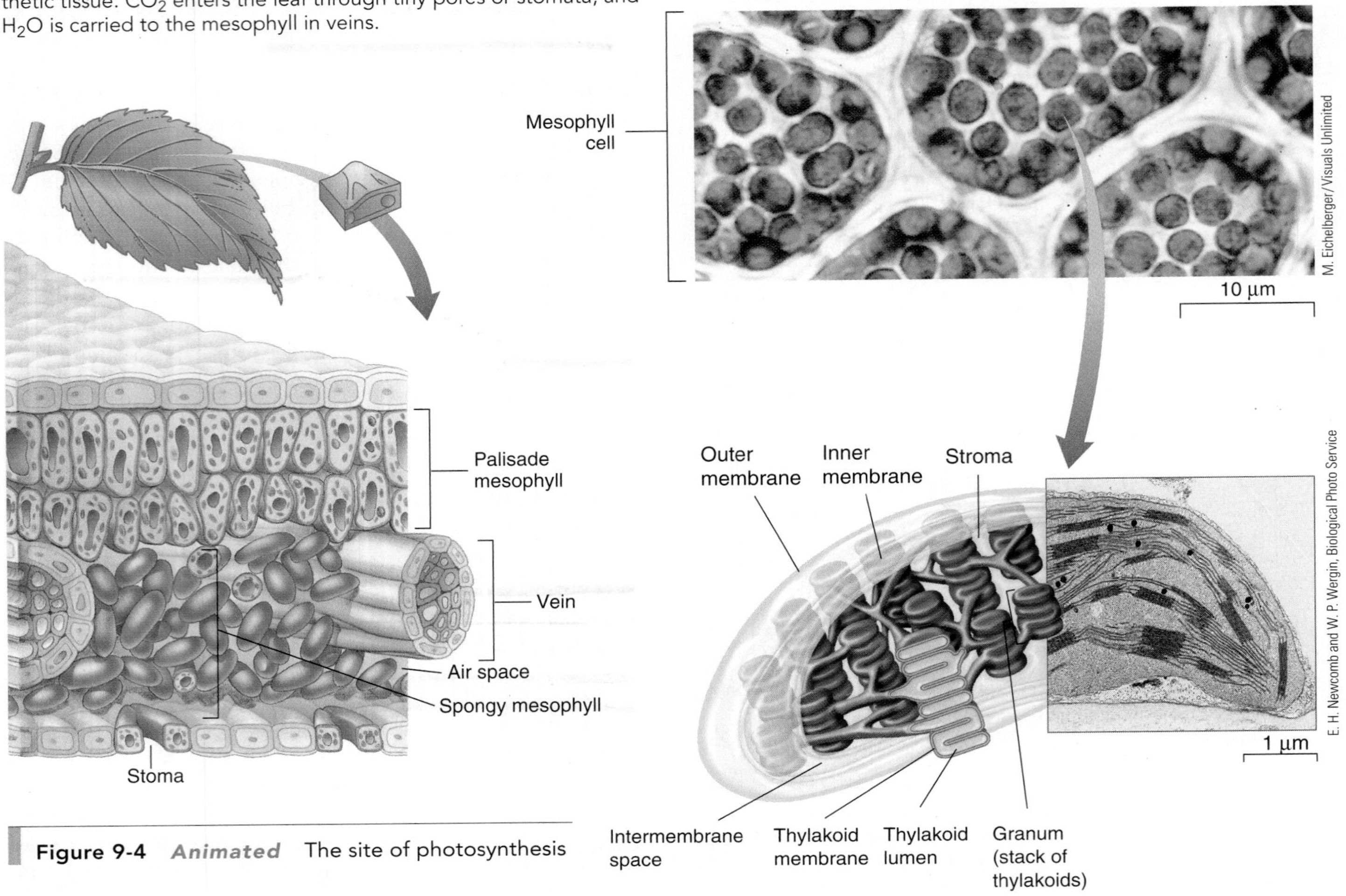

Mesophyll cell

10 μm

M. Eichelberger/Visuals Unlimited

Palisade mesophyll

Vein

Air space

Spongy mesophyll

Stoma

Outer membrane

Inner membrane

Stroma

Intermembrane space

Thylakoid membrane

Thylakoid lumen

Granum (stack of thylakoids)

1 μm

E. H. Newcomb and W. P. Wergin, Biological Photo Service

Figure 9-4 *Animated* The site of photosynthesis

Chloroplasts have other accessory photosynthetic pigments, such as **carotenoids,** which are yellow and orange (see Fig. 3-14). Carotenoids absorb different wavelengths of light than chlorophyll, thereby expanding the spectrum of light that provides energy for photosynthesis. Chlorophyll may be excited by light directly by energy passed to it from the light source, or indirectly by energy passed to it from accessory pigments that have become excited by light. When a carotenoid molecule is excited, its energy can be transferred to chlorophyll *a*. Carotenoids also protect chlorophyll and other parts of the thylakoid membrane from excess light energy that could easily damage the photosynthetic components. (High light intensities often occur in nature.)

Chlorophyll is the main photosynthetic pigment

As you have seen, the thylakoid membrane contains more than one kind of pigment. An instrument called a *spectrophotometer* measures the relative abilities of different pigments to absorb different wavelengths of light. The **absorption spectrum** of a pigment is a plot of its absorption of light of different wavelengths. Figure 9-6a shows the absorption spectra for chlorophylls *a* and *b*.

(c) In the chloroplast, pigments necessary for the light-capturing reactions of photosynthesis are part of thylakoid membranes, whereas the enzymes for the synthesis of carbohydrate molecules are in the stroma.

An **action spectrum** of photosynthesis is a graph of the relative effectiveness of different wavelengths of light. To obtain an action spectrum, scientists measure the rate of photosynthesis at each wavelength for leaf cells or tissues exposed to monochromatic light (light of one wavelength) (Fig. 9-6b).

In a classic biology experiment, the German biologist T. W. Engelmann obtained the first action spectrum in 1883. Engelmann's experiment, described in Figure 9-7, took advantage of the shape of the chloroplast in a species of the green alga *Spirogyra*. Its long, filamentous strands are found in freshwater habitats, especially slow-moving or still waters. *Spirogyra* cells each contain a long, spiral-shaped, emerald-green chloroplast embedded in the cytoplasm. Engelmann exposed these cells to a color spectrum produced by passing light through a prism. He hypothesized that if chlorophyll were indeed responsible for photosynthesis, the process would take place most rapidly in the areas where the chloroplast was illuminated by the colors most strongly absorbed by chlorophyll.

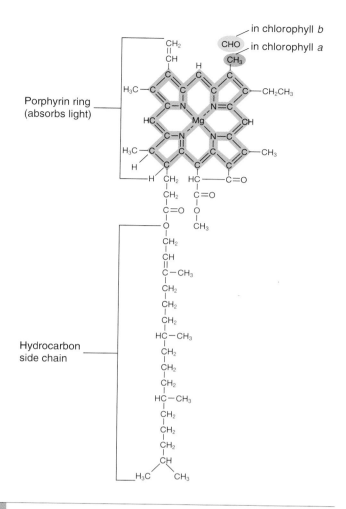

in chlorophyll *b*
in chlorophyll *a*

Porphyrin ring
(absorbs light)

Hydrocarbon
side chain

Figure 9-5 The structure of chlorophyll

Chlorophyll consists of a porphyrin ring and a hydrocarbon side chain. The porphyrin ring, with a magnesium atom in its center, contains a system of alternating double and single bonds; these are commonly found in molecules that strongly absorb certain wavelengths of visible light and reflect others (chlorophyll reflects green). Notice that at the top right corner of the diagram, the methyl group ($-CH_3$) distinguishes chlorophyll *a* from chlorophyll *b*, which has a carbonyl group ($-CHO$) in this position. The hydrophobic hydrocarbon side chain anchors chlorophyll to the thylakoid membrane.

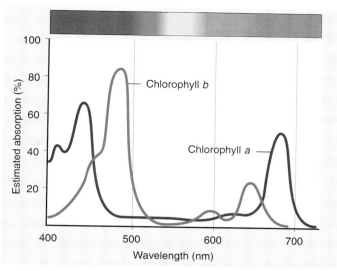

(a) Chlorophylls *a* and *b* absorb light mainly in the blue (422 to 492 nm) and red (647 to 760 nm) regions.

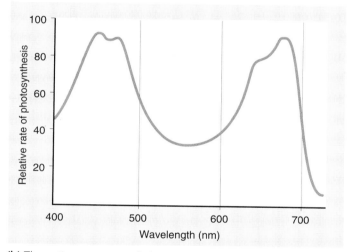

(b) The action spectrum of photosynthesis indicates the effectiveness of various wavelengths of light in powering photosynthesis. Many plant species have action spectra for photosynthesis that resemble the generalized action spectrum shown here.

Figure 9-6 A comparison of the absorption spectra for chlorophylls *a* and *b* with the action spectrum for photosynthesis

Yet how could photosynthesis be measured in those technologically unsophisticated days? Engelmann knew that photosynthesis produces oxygen and that certain motile bacteria are attracted to areas of high oxygen concentration. He determined the action spectrum of photosynthesis by observing that the bacteria swam toward the parts of the *Spirogyra* filaments in the blue and red regions of the spectrum. How did Engelmann know bacteria were not simply attracted to blue or red light? Engelmann exposed bacteria to the spectrum of visible light in the absence of *Spirogyra* as a control. The bacteria showed no preference for any particular wavelength of light. Because the action spectrum of photosynthesis closely matched the absorption spectrum of chlorophyll, Engelmann concluded that chlorophyll in the chloroplasts (and

not another compound in another organelle) is responsible for photosynthesis. Numerous studies using sophisticated instruments have since confirmed Engelmann's conclusions.

If you examine Figure 9-6 closely, you will observe that the action spectrum of photosynthesis does not parallel the absorption spectrum of chlorophyll exactly. This difference occurs because accessory pigments, such as carotenoids, transfer some of the energy of excitation produced by green light to chlorophyll molecules. The presence of these accessory photosynthetic pigments can be demonstrated by chemical analysis of almost any

Key Experiment

QUESTION: Is a pigment in the chloroplast responsible for photosynthesis?

HYPOTHESIS: Engelmann hypothesized that chlorophyll was the main photosynthetic pigment. Accordingly, he predicted that he would observe differences in the amount of photosynthesis, as measured by the amount of oxygen produced, depending on the wavelengths of light used, and that these wavelengths would be consistent with the known absorption spectrum of chlorophyll.

EXPERIMENT: The photograph (a) shows cells of the filamentous alga *Spirogyra*, which has a long spiral-shaped chloroplast. The drawing (b) shows how Engelmann used a prism to expose the cells to light that had been separated into various wavelengths. He estimated the formation of oxygen (which he knew was a product of photosynthesis) by exploiting the fact that certain aerobic bacteria would be attracted to the oxygen. As a control (*not shown*), he also exposed the bacteria to the spectrum of light in the absence of *Spirogyra* cells.

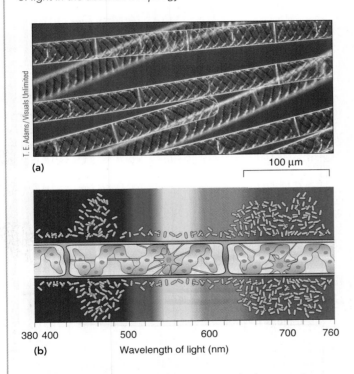

T. E. Adams/Visuals Unlimited

(a)   100 μm

(b)   Wavelength of light (nm)
380 400   500   600   700   760

RESULTS AND CONCLUSION: Although the bacteria alone (*control*) showed no preference for any particular wavelength, large numbers were attracted to the photosynthesizing cells in red or blue light, wavelengths that are strongly absorbed by chlorophyll (see Fig. 9-6). Thus, Engelmann concluded that chlorophyll is responsible for photosynthesis.

Figure 9-7 *Animated* The first action spectrum of photosynthesis

leaf, although it is obvious in temperate climates when leaves change color in the fall. Toward the end of the growing season, chlorophyll breaks down (and its magnesium is stored in the permanent tissues of the tree), leaving orange and yellow accessory pigments in the leaves.

Review

■ What chloroplast membrane is most important in photosynthesis? What two spaces does it separate?

■ What is the significance of the fact that the combined absorption spectra of chlorophylls *a* and *b* roughly match the action spectrum of photosynthesis? Why do they not coincide exactly?

■ Does fluorescence play a role in photosynthesis?

OVERVIEW OF PHOTOSYNTHESIS

Learning Objectives

4   Describe photosynthesis as a redox process.

5   Distinguish between the light-dependent reactions and carbon fixation reactions of photosynthesis.

During photosynthesis, a cell uses light energy captured by chlorophyll to power the synthesis of carbohydrates. The overall reaction of photosynthesis can be summarized as follows:

$$6\,CO_2 + 12\,H_2O \xrightarrow[\text{Chlorophyll}]{\text{Light energy}} C_6H_{12}O_6 + 6\,O_2 + 6\,H_2O$$

Carbon dioxide   Water   Glucose   Oxygen   Water

The equation is typically written in the form just given, with H_2O on both sides, because water is a reactant in some reactions and a product in others. Furthermore, all the oxygen produced comes from water, so 12 molecules of water are required to produce 12 oxygen atoms. However, because there is no net yield of H_2O, we can simplify the summary equation of photosynthesis for purposes of discussion:

$$6\,CO_2 + 6\,H_2O \xrightarrow[\text{Chlorophyll}]{\text{Light}} C_6H_{12}O_6 + 6\,O_2$$

When you analyze this process, it appears that hydrogen atoms are transferred from H_2O to CO_2 to form carbohydrate, so you can recognize it as a redox reaction. Recall from Chapter 7 that in a **redox reaction** one or more electrons, usually as part of one or more hydrogen atoms, are transferred from an electron donor (a reducing agent) to an electron acceptor (an oxidizing agent).

$$\underset{\text{Oxidation}}{\overset{\text{Reduction}}{6\,CO_2 + 6\,H_2O \xrightarrow[\text{Chlorophyll}]{\text{Light}} C_6H_{12}O_6 + 6\,O_2}}$$

When the electrons are transferred, some of their energy is transferred as well. However, the summary equation of photosynthesis is somewhat misleading, because no direct transfer of hydrogen atoms actually occurs. The summary equation describes *what* happens but not *how* it happens. The "how" is more complex and involves multiple steps, many of which are redox reactions.

The reactions of photosynthesis are divided into two phases: the light-dependent reactions (the *photo* part of photosynthesis) and the carbon fixation reactions (the *synthesis* part of photosynthesis). Each set of reactions occurs in a different part of the chloroplast: the light-dependent reactions in association with the thylakoids, and the carbon fixation reactions in the stroma (■ Fig. 9-8).

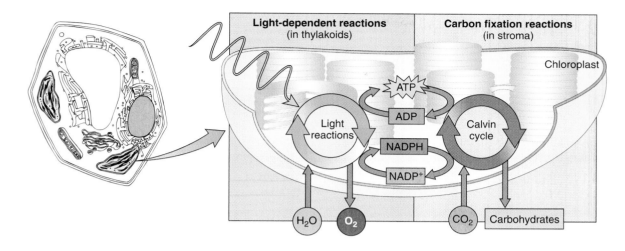

Figure 9-8 *Animated* An overview of photosynthesis

ATP and NADPH are the products of the light-dependent reactions: An overview

Light energy is converted to chemical energy in the **light-dependent reactions,** which are associated with the thylakoids. The light-dependent reactions begin as chlorophyll captures light energy, which causes one of its electrons to move to a higher energy state. The energized electron is transferred to an acceptor molecule and is replaced by an electron from H_2O. When this happens, H_2O is split and molecular oxygen is released (▌ Fig. 9-9). Some energy of the energized electrons is used to phosphorylate **adenosine diphosphate (ADP),** forming **adenosine triphosphate (ATP).** In addition, the coenzyme **nicotinamide adenine dinucleotide phosphate (NADP⁺)** becomes reduced, forming **NADPH.**[1] The products of the light-dependent reactions, ATP and NADPH, are both needed in the energy-requiring carbon fixation reactions.

Carbohydrates are produced during the carbon fixation reactions: An overview

The ATP and NADPH molecules produced during the light-dependent phase are suited for transferring chemical energy but not for long-term energy storage. For this reason, some of their energy is transferred to chemical bonds in carbohydrates, which can be produced in large quantities and stored for future use. Known as **carbon fixation,** these reactions "fix" carbon atoms from CO_2 to existing skeletons of organic molecules. Because the carbon fixation reactions have no direct requirement for light,

they were previously referred to as the "dark" reactions. However, they do not require darkness; in fact, many of the enzymes involved in carbon fixation are much more active in the light than in the dark. Furthermore, carbon fixation reactions depend on the products of the light-dependent reactions. Carbon fixation reactions take place in the stroma of the chloroplast.

Now that we have presented an overview of photosynthesis, let us examine the entire process more closely.

Review

▌ Which is more oxidized, oxygen that is part of a water molecule or molecular oxygen?

▌ In what ways do the carbon fixation reactions depend on the light-dependent reactions?

Figure 9-9 Oxygen produced by photosynthesis

On sunny days, the oxygen released by aquatic plants is sometimes visible as bubbles in the water. This plant (*Elodea*) is actively carrying on photosynthesis.

---

[1]Although the correct way to write the reduced form of NADP⁺ is NADPH + H⁺, for simplicity's sake we present the reduced form as NADPH throughout the book.

THE LIGHT-DEPENDENT REACTIONS

Learning Objectives

6 Describe the flow of electrons through photosystems I and II in the noncyclic electron transport pathway and the products produced. Contrast this with cyclic electron transport.

7 Explain how a proton (H^+) gradient is established across the thylakoid membrane and how this gradient functions in ATP synthesis.

In the light-dependent reactions, the radiant energy from sunlight phosphorylates ADP, producing ATP, and reduces $NADP^+$, forming NADPH. The light energy that chlorophyll captures is temporarily stored in these two compounds. The light-dependent reactions are summarized as follows:

$$12\ H_2O + 12\ NADP^+ + 18\ ADP + 18\ P_i \xrightarrow[\text{Chlorophyll}]{\text{Light}}$$
$$6\ O_2 + 12\ NADPH + 18\ ATP$$

Photosystems I and II each consist of a reaction center and multiple antenna complexes

The light-dependent reactions of photosynthesis begin when chlorophyll *a* and/or accessory pigments absorb light. According to the currently accepted model, chlorophylls *a* and *b* and accessory pigment molecules are organized with pigment-binding proteins in the thylakoid membrane into units called **antenna complexes.** The pigments and associated proteins are arranged as highly ordered groups of about 250 chlorophyll molecules associated with specific enzymes and other proteins. Each antenna complex absorbs light energy and transfers it to the **reaction center,** which consists of chlorophyll molecules and proteins, including electron transfer components, that participate directly in photosynthesis (Fig. 9-10). Light energy is converted to chemical energy in the reaction centers by a series of electron transfer reactions.

Two types of photosynthetic units, designated photosystem I and photosystem II, are involved in photosynthesis. Their reaction centers are distinguishable because they are associated with proteins in a way that causes a slight shift in their absorption spectra. Ordinary chlorophyll *a* has a strong absorption peak at about 660 nm. In contrast, the reaction center of **photosystem I** consists of a pair of chlorophyll *a* molecules with an absorption peak at 700 nm and is referred to as **P700.** The reaction center of **photosystem II** is made up of a pair of chlorophyll *a* molecules with an absorption peak of about 680 nm and is referred to as **P680.**

When a pigment molecule absorbs light energy, that energy is passed from one pigment molecule to another within the antenna complex until it reaches the reaction center. When the energy reaches a molecule of P700 (in a photosystem I reaction center) or P680 (in a photosystem II reaction center), an electron is then raised to a higher energy level. As we explain in the next section, this energized electron can be donated to an electron acceptor that becomes reduced in the process.

Noncyclic electron transport produces ATP and NADPH

Let us begin our discussion of **noncyclic electron transport** with the events associated with photosystem I (Fig. 9-11). A pigment molecule in an antenna complex associated with photosystem I absorbs a photon of light. The absorbed energy is transferred from one pigment molecule to another until it reaches the reaction center, where it excites an electron in a molecule of P700. This energized electron is transferred to a primary electron acceptor, which is the first of several electron acceptors in a series. (Uncertainty exists regarding the exact chemical nature of the primary electron acceptor for photosystem I.) The energized electron is passed along an **electron transport chain** from one electron acceptor to another, until it is passed to *ferredoxin,* an iron-containing protein. Ferredoxin transfers the electron to $NADP^+$ in the presence of the enzyme *ferredoxin–NADP$^+$ reductase.*

When $NADP^+$ accepts two electrons, they unite with a proton (H^+); thus, the reduced form of $NADP^+$ is NADPH, which is released into the stroma. P700 becomes positively charged when it gives up an electron to the primary electron acceptor; the missing electron is replaced by one donated by photosystem II.

As in photosystem I, photosystem II becomes activated when a pigment molecule in an antenna complex absorbs a photon of light energy. The energy is transferred to the reaction center, where it causes an electron in a molecule of P680 to move to a

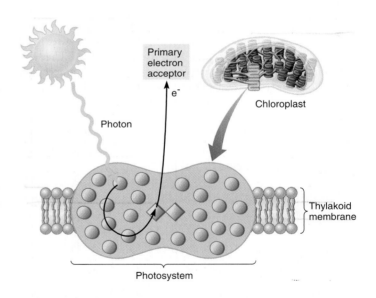

Figure 9-10 *Animated* Schematic view of a photosystem

Chlorophyll molecules (*green circles*) and accessory pigments (*not shown*) are arranged in light-harvesting arrays, or antenna complexes. A portion of one such complex within a photosystem is depicted. Each complex consists of several hundred pigment molecules, in association with special proteins (*not shown*). These proteins hold the pigments in a highly ordered spatial array, such that when a molecule in an antenna complex absorbs a photon, energy derived from that photon is readily passed from one pigment molecule to another. When this energy reaches one of the two chlorophyll molecules in the reaction center (*green diamonds*), an electron becomes energized and is accepted by a primary electron acceptor.

Key Point

Noncyclic electron transport converts light energy to chemical energy as ATP and NADPH.

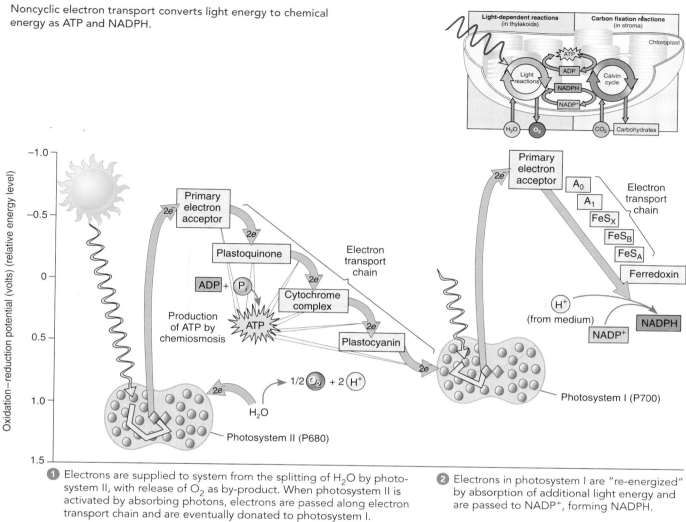

Figure 9-11 *Animated* Noncyclic electron transport

① Electrons are supplied to system from the splitting of H_2O by photosystem II, with release of O_2 as by-product. When photosystem II is activated by absorbing photons, electrons are passed along electron transport chain and are eventually donated to photosystem I.

② Electrons in photosystem I are "re-energized" by absorption of additional light energy and are passed to $NADP^+$, forming NADPH.

In noncyclic electron transport, the formation of ATP is coupled to one-way flow of energized electrons (*orange arrows*) from H_2O (*lower left*) to $NADP^+$ (*middle right*). Single electrons actually pass down the electron transport chain; two are shown in this figure because two electrons are required to form one molecule of NADPH.

higher energy level. This energized electron is accepted by a primary electron acceptor (a highly modified chlorophyll molecule known as *pheophytin*) and then passes along an electron transport chain until it is donated to P700 in photosystem I.

How is the electron that has been donated to the electron transport chain replaced? This occurs through **photolysis** (light splitting) of water, a process that not only yields electrons but also is the source of almost all the oxygen in Earth's atmosphere. A molecule of P680 that has given up an energized electron to the primary electron acceptor is positively charged ($P680^+$). $P680^+$ is an oxidizing agent so strong that it pulls electrons away from an oxygen atom that is part of a H_2O molecule. In a reaction catalyzed by a unique, manganese-containing enzyme, water is broken into its components: 2 electrons, 2 protons, and oxygen. Each electron is donated to a $P680^+$ molecule, which then loses

its positive charge; the protons are released into the thylakoid lumen. Because oxygen does not exist in atomic form, the oxygen produced by splitting one H_2O molecule is written O_2. Two water molecules must be split to yield one molecule of oxygen. The photolysis of water is a remarkable reaction, but its name is somewhat misleading because it implies that water is broken by light. Actually, light splits water indirectly by causing P680 to become oxidized.

Noncyclic electron transport is a continuous linear process

In the presence of light, there is a continuous, one-way flow of electrons from the ultimate electron source, H_2O, to the terminal electron acceptor, $NADP^+$. Water undergoes enzymatically catalyzed photolysis to replace energized electrons donated to the electron transport chain by molecules of P680 in photosystem II. These electrons travel down the electron transport chain that connects photosystem II with photosystem I. Thus, they provide a continuous supply of replacements for energized electrons that have been given up by P700.

As electrons are transferred along the electron transport chain that connects photosystem II with photosystem I, they lose energy. Some of the energy released is used to pump protons across the thylakoid membrane, from the stroma to the thylakoid lumen, producing a proton gradient. The energy of this proton gradient is harnessed to produce ATP from ADP by chemiosmosis, which we discuss later in the chapter. ATP and NADPH, the products of the light-dependent reactions, are released into the stroma, where both are required by the carbon fixation reactions.

Cyclic electron transport produces ATP but no NADPH

Only photosystem I is involved in **cyclic electron transport,** the simplest light-dependent reaction. The pathway is cyclic because energized electrons that originate from P700 at the reaction center eventually return to P700. In the presence of light, electrons flow continuously through an electron transport chain within the thylakoid membrane. As they pass from one acceptor to another, the electrons lose energy, some of which is used to pump protons across the thylakoid membrane. An enzyme (ATP synthase) in the thylakoid membrane uses the energy of the proton gradient to manufacture ATP. NADPH is not produced, H_2O is not split, and oxygen is not generated. By itself, cyclic electron transport could not serve as the basis of photosynthesis because, as we explain later in the chapter, NADPH is required to reduce CO_2 to carbohydrate.

The significance of cyclic electron transport to photosynthesis in plants is unclear. Cyclic electron transport may occur in plant cells when there is too little $NADP^+$ to accept electrons from ferredoxin. There is recent evidence that cyclic electron flow may help maintain the optimal ratio of ATP to NADPH required for carbon fixation, as well as provide extra ATP to power other ATP-requiring processes in chloroplasts. Biologists generally agree that ancient bacteria used this process to produce ATP from light energy. A reaction pathway analogous to cyclic electron transport in plants is present in some modern photosynthetic prokaryotes. Noncyclic and cyclic electron transport are compared in ▌Table 9-1.

ATP synthesis occurs by chemiosmosis

Each member of the electron transport chain that links photosystem II to photosystem I can exist in an oxidized (lower-energy) form and a reduced (higher-energy) form. The electron accepted from P680 by the primary electron acceptor is highly energized; it is passed from one carrier to the next in a series of exergonic redox reactions, losing some of its energy at each step. Some of the energy given up by the electron is not lost by the system, however; it is used to provide energy for ATP synthesis. Because the

TABLE 9-1

A Comparison of Noncyclic and Cyclic Electron Transport

	Noncyclic Electron Transport	Cyclic Electron Transport
Electron source	H_2O	None—electrons cycle through the system
Oxygen released?	Yes (from H_2O)	No
Terminal electron acceptor	$NADP^+$	None—electrons cycle through the system
Form in which energy is temporarily captured	ATP (by chemiosmosis); NADPH	ATP (by chemiosmosis)
Photosystem(s) required	PS I (P700) and PS II (P680)	PS I (P700) only

synthesis of ATP (that is, the phosphorylation of ADP) is coupled to the transport of electrons that have been energized by photons of light, the process is called **photophosphorylation.**

The chemiosmotic model explains the coupling of ATP synthesis and electron transport

As discussed earlier, the pigments and electron acceptors of the light-dependent reactions are embedded in the thylakoid membrane. Energy released from electrons traveling through the chain of acceptors is used to pump protons from the stroma, across the thylakoid membrane, and into the thylakoid lumen (▌Fig. 9-12). Thus, the pumping of protons results in the formation of a proton gradient across the thylakoid membrane. Protons also accumulate

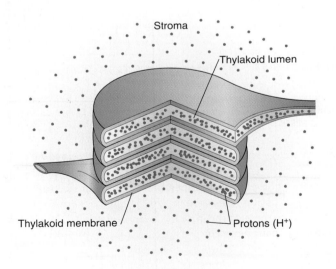

Figure 9-12 The accumulation of protons in the thylakoid lumen

As electrons move down the electron transport chain, protons (H^+) move from the stroma to the thylakoid lumen, creating a proton gradient. The greater concentration of H^+ in the thylakoid lumen lowers the pH.

in the thylakoid lumen as water is split during noncyclic electron transport. Because protons are actually hydrogen ions (H^+), the accumulation of protons causes the pH of the thylakoid interior to fall to a pH of about 5 in the thylakoid lumen, compared to a pH of about 8 in the stroma. This difference of about 3 pH units across the thylakoid membrane means there is an approximately 1000-fold difference in hydrogen ion concentration.

The proton gradient has a great deal of free energy because of its state of low entropy. How does the chloroplast convert that energy to a more useful form? According to the general principles of diffusion, the concentrated protons inside the thylakoid might be expected to diffuse out readily. However, they are prevented from doing so because the thylakoid membrane is impermeable to H^+ except through certain channels formed by the enzyme **ATP synthase.** This enzyme, a transmembrane protein also found in mitochondria, forms complexes so large they can be seen in electron micrographs (see Fig. 8-10b); these complexes project into the stroma. As the protons diffuse through an ATP synthase complex, free energy decreases as a consequence of an increase in entropy. Each ATP synthase complex couples this exergonic process of diffusion down a concentration gradient to the endergonic process of phosphorylation of ADP to form ATP, which is released into the stroma (‖ Fig. 9-13). The movement of protons through ATP synthase is thought to induce changes in the con-

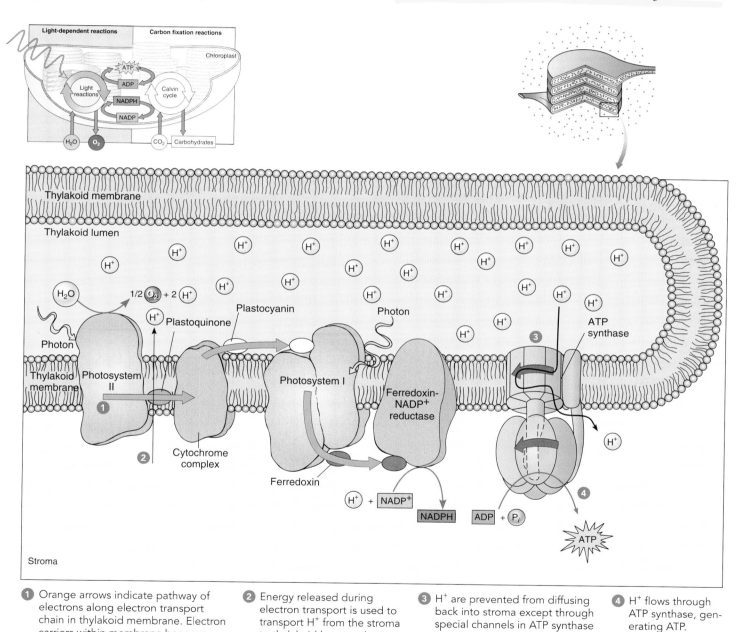

① Orange arrows indicate pathway of electrons along electron transport chain in thylakoid membrane. Electron carriers within membrane become alternately reduced and oxidized as they accept and donate electrons.

② Energy released during electron transport is used to transport H^+ from the stroma to thylakoid lumen, where high concentration of H^+ accumulates.

③ H^+ are prevented from diffusing back into stroma except through special channels in ATP synthase in thylakoid membrane.

④ H^+ flows through ATP synthase, generating ATP.

Figure 9-13　A detailed look at electron transport and chemiosmosis

TABLE 9-2

A Comparison of Photosynthesis and Aerobic Respiration

	Photosynthesis	Aerobic Respiration
Type of metabolic reaction	Anabolism	Catabolism
Raw materials	CO_2, H_2O	$C_6H_{12}O_6$, O_2
End products	$C_6H_{12}O_6$, O_2	CO_2, H_2O
Which cells have these processes?	Cells that contain chlorophyll (certain cells of plants, algae, and some bacteria)	Every actively metabolizing cell has aerobic respiration or some other energy-releasing pathway
Sites involved (in eukaryotic cells)	Chloroplasts	Cytosol (glycolysis); mitochondria
ATP production	By photophosphorylation (a chemiosmotic process)	By substrate-level phosphorylation and by oxidative phosphorylation (a chemiosmotic process)
Principal electron transfer compound	$NADP^+$ is reduced to form NADPH*	NAD^+ is reduced to form NADH*
Location of electron transport chain	Thylakoid membrane	Mitochondrial inner membrane (cristae)
Source of electrons for electron transport chain	In noncyclic electron transport: H_2O (undergoes photolysis to yield electrons, protons, and oxygen)	Immediate source: NADH, $FADH_2$ Ultimate source: glucose or other carbohydrate
Terminal electron acceptor for electron transport chain	In noncyclic electron transport: $NADP^+$ (becomes reduced to form NADPH)	O_2 (becomes reduced to form H_2O)

* NADPH and NADH are very similar hydrogen (i.e., electron) carriers, differing only in a single phosphate group. However, NADPH generally works with enzymes in anabolic pathways, such as photosynthesis. NADH is associated with catabolic pathways, such as cellular respiration.

formation of the enzyme that are necessary for the synthesis of ATP. It is estimated that for every 4 protons that move through ATP synthase, one ATP molecule is synthesized.

The mechanism by which the phosphorylation of ADP is coupled to diffusion down a proton gradient is called **chemiosmosis.** As the essential connection between the electron transport chain and the phosphorylation of ADP, chemiosmosis is a basic mechanism of energy coupling in cells. You may recall from Chapter 8 that chemiosmosis also occurs in aerobic respiration (see ▮ Table 9-2).

Review

- Why is molecular oxygen a necessary by-product of photosynthesis?
- What process is the actual mechanism of photophosphorylation?
- Why are both photosystems I and II required for photosynthesis? Can cyclic phosphorylation alone support photosynthesis?

THE CARBON FIXATION REACTIONS

Learning Objectives

8 Summarize the three phases of the Calvin cycle, and indicate the roles of ATP and NADPH in the process.

9 Discuss how photorespiration reduces photosynthetic efficiency.

10 Compare the C_4 and CAM pathways.

In carbon fixation, the energy of ATP and NADPH is used in the formation of organic molecules from CO_2. The carbon fixation reactions may be summarized as follows:

$$12 \text{ NADPH} + 18 \text{ ATP} + 6 \text{ CO}_2 \longrightarrow$$
$$C_6H_{12}O_6 + 12 \text{ NADP}^+ + 18 \text{ ADP} + 18 \text{ P}_i + 6 \text{ H}_2O$$

Most plants use the Calvin cycle to fix carbon

Carbon fixation occurs in the stroma through a sequence of 13 reactions known as the **Calvin cycle.** During the 1950s, University of California researchers Melvin Calvin, Andrew Benson, and others elucidated the details of this cycle. Calvin was awarded a Nobel Prize in Chemistry in 1961.

The 13 reactions of the Calvin cycle are divided into three phases: ❶ CO_2 uptake, ❷ carbon reduction, and ❸ RuBP regeneration (▮ Fig. 9-14). All 13 enzymes that catalyze steps in the Calvin cycle are located in the stroma of the chloroplast. Ten of the enzymes also participate in glycolysis (see Chapter 8). These enzymes catalyze reversible reactions, degrading carbohydrate molecules in cellular respiration and synthesizing carbohydrate molecules in photosynthesis.

❶ *CO_2 uptake.* The first phase of the Calvin cycle consists of a single reaction in which a molecule of CO_2 reacts with a phosphorylated five-carbon compound, **ribulose bisphosphate (RuBP).** This reaction is catalyzed by the enzyme *ribulose bisphosphate carboxylase/oxygenase,* also known as **rubisco.** More rubisco enzyme than any other protein is present in the chloroplast, and it may be one of the most abundant proteins in the biosphere. The product of this reaction is an unstable, six-carbon intermediate, which immediately breaks down into two molecules of **phosphoglycerate (PGA)** with three carbons each. The carbon that was originally part of a CO_2 molecule is now part of a carbon skeleton; the carbon has been "fixed." The Calvin cycle is also known as the C_3 **pathway** because the product of the initial carbon fixation reaction is a three-carbon compound. Plants that initially fix carbon in this way are called C_3 **plants.**

❷ *Carbon reduction.* The second phase of the Calvin cycle consists of two steps in which the energy and reducing

Key Point

ATP and NADPH provide the energy that drives carbon fixation in the Calvin cycle.

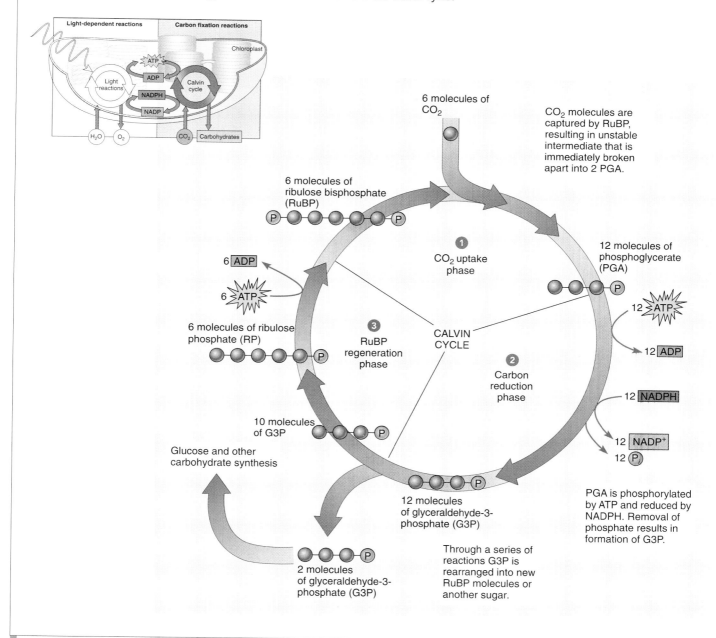

Figure 9-14 *Animated* A detailed look at the Calvin cycle

1 This diagram, in which carbon atoms are black balls, shows that six molecules of CO_2 must be "fixed" (incorporated into pre-existing carbon skeletons) in the CO_2 uptake phase to produce one molecule of a six-carbon sugar such as glucose. **2** Glyceraldehyde-3-phosphate (G3P) is formed in the carbon reduction phase. Two G3P molecules "leave" the cycle for every glucose formed. **3** Ribulose bisphosphate (RuBP) is regenerated, and a new cycle can begin. Although these reactions do not require light directly, the energy that drives the Calvin cycle comes from ATP and NADPH, which are the products of the light-dependent reactions.

power from ATP and NADPH (both produced in the light-dependent reactions) are used to convert the PGA molecules to **glyceraldehyde-3-phosphate (G3P).** As shown in Figure 9-14, for every six carbons that enter the cycle as CO_2, six carbons can leave the system as two molecules of G3P, to be used in carbohydrate synthesis. Each of these three-carbon

molecules of G3P is essentially half a hexose (six-carbon sugar) molecule. (In fact, you may recall that G3P is a key intermediate in the splitting of sugar in glycolysis; see Figs. 8-3 and 8-4.)

The reaction of two molecules of G3P is exergonic and leads to the formation of glucose or fructose. In some plants, glucose and fructose are then joined to produce sucrose (common table sugar). (Sucrose can be harvested from sugarcane, sugar beets, and maple sap.) The plant cell also uses glucose to produce starch or cellulose.

3. *RuBP regeneration.* Notice that although 2 G3P molecules are removed from the cycle, 10 G3P molecules remain; this represents a total of 30 carbon atoms. Through a series of 10 reactions that make up the third phase of the Calvin cycle, these 30 carbons and their associated atoms become rearranged into six molecules of ribulose phosphate, each of which becomes phosphorylated by ATP to produce RuBP, the five-carbon compound with which the cycle started. These RuBP molecules begin the process of CO_2 fixation and eventual G3P production once again.

In summary, the inputs required for the carbon fixation reactions are six molecules of CO_2, phosphates transferred from ATP, and electrons (as hydrogen) from NADPH. In the end, the six carbons from the CO_2 are accounted for by the harvest of a hexose molecule. The remaining G3P molecules are used to synthesize the RuBP molecules with which more CO_2 molecules may combine. ▌ Table 9-3 provides a summary of photosynthesis.

Photorespiration reduces photosynthetic efficiency

Many C_3 plants, including certain agriculturally important crops such as soybeans, wheat, and potatoes, do not yield as much carbohydrate from photosynthesis as might be expected, especially during periods of very hot temperature in summer. This phenomenon is a consequence of trade-offs between the plant's need for CO_2 and its need to prevent water loss. Recall that most photosynthesis occurs in mesophyll cells inside the leaf and that the entry and exit of gases from the interior of the leaf are regulated by stomata, tiny pores concentrated on the underside of the leaf (see Fig. 9-4a). On hot, dry days, plants close their stomata to conserve water. Once the stomata close, photosynthesis rapidly

uses up the CO_2 remaining in the leaf and produces O_2, which accumulates in the chloroplasts.

Recall that the enzyme RuBP carboxylase/oxygenase (rubisco) catalyzes CO_2 fixation in the Calvin cycle by attaching CO_2 to RuBP. As its full name implies, rubisco acts not only as a carboxylase but also as an oxygenase because high levels of O_2 compete with CO_2 for the active site of rubisco. Some of the intermediates involved in the Calvin cycle are degraded to CO_2 and H_2O in a process that is called **photorespiration,** because (1) it occurs in the presence of light; and as in aerobic respiration, (2) it requires oxygen and (3) produces CO_2 and H_2O. However, photorespiration does not produce ATP, and it reduces photosynthetic efficiency because it removes some of the intermediates used in the Calvin cycle.

The reasons for photorespiration are incompletely understood, although scientists hypothesize that it reflects the origin of rubisco at an ancient time when CO_2 levels were high and molecular oxygen levels were low. This view is supported by recent evidence that some amino acid sequences in rubisco are similar to sequences in certain bacterial proteins that apparently evolved prior to the evolution of the Calvin cycle. Genetic engineering to produce plants with rubisco that has a much lower affinity for oxygen is a promising area of research to improve yields of certain valuable crop plants.

The initial carbon fixation step differs in C_4 plants and in CAM plants

Photorespiration is not the only problem faced by plants engaged in photosynthesis. Because CO_2 is not a very abundant gas (composing only about 0.04% of the atmosphere), it is not easy for plants to obtain the CO_2 they need. As you have learned, when conditions are hot and dry, the stomata close to reduce the loss of water vapor, greatly diminishing the supply of CO_2. Ironically, CO_2 is potentially less available at the very times when maximum sunlight is available to power the light-dependent reactions.

Many plant species living in hot, dry environments have adaptations that facilitate carbon fixation. **C_4 plants** first fix CO_2

TABLE 9-3

Summary of Photosynthesis

Reaction Series	Summary of Process	Needed Materials	End Products
Light-dependent reactions (take place in thylakoid membranes)	Energy from sunlight used to split water, manufacture ATP, and reduce NADP⁺		
Photochemical reactions	Chlorophyll-activated; reaction center gives up photoexcited electron to electron acceptor	Light energy; pigments (chlorophyll)	Electrons
Electron transport	Electrons transported along chain of electron acceptors in thylakoid membranes; electrons reduce NADP⁺; splitting of water provides some H⁺ that accumulates inside thylakoid space	Electrons, NADP⁺, H_2O, electron acceptors	NADPH, O_2
Chemiosmosis	H⁺ permitted to diffuse across the thylakoid membrane down their gradient; they cross the membrane through special channels in ATP synthase complex; energy released is used to produce ATP	Proton gradient, ADP + P_i, ATP synthase	ATP
Carbon fixation reactions (take place in stroma)	Carbon fixation: Carbon dioxide used to make carbohydrate	Ribulose bisphosphate, CO_2, ATP, NADPH, necessary enzymes	Carbohydrates, ADP + P_i, NADP⁺

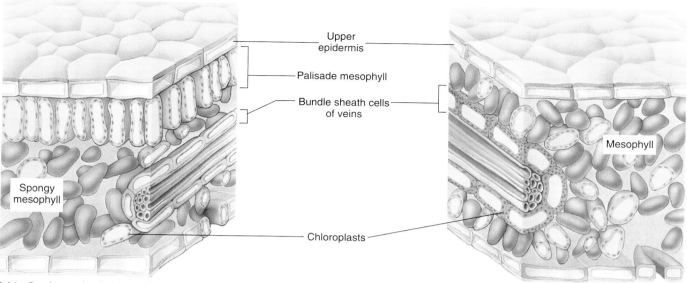

(a) In C_3 plants, the Calvin cycle takes place in the mesophyll cells and the bundle sheath cells are nonphotosynthetic.

(b) In C_4 plants, reactions that fix CO_2 into four-carbon compounds take place in the mesophyll cells. The four-carbon compounds are transferred from the mesophyll cells to the photosynthetic bundle sheath cells, where the Calvin cycle takes place.

Figure 9-15 C_3 and C_4 plant structure compared

---

into a four-carbon compound, **oxaloacetate. CAM plants** initially fix carbon at night through the formation of oxaloacetate. These special pathways found in C_4 and CAM plants precede the Calvin cycle (C_3 pathway); they do not replace it.

The C_4 pathway efficiently fixes CO_2 at low concentrations

The **C_4 pathway,** in which CO_2 is fixed through the formation of oxaloacetate, occurs not only before the C_3 pathway but also in different cells. Leaf anatomy is usually distinctive in C_4 plants. The photosynthetic mesophyll cells are closely associated with prominent, chloroplast-containing **bundle sheath cells,** which tightly encircle the veins of the leaf (❚ Fig. 9-15). The C_4 pathway occurs in the mesophyll cells, whereas the Calvin cycle takes place within the bundle sheath cells.

The key component of the C_4 pathway is a remarkable enzyme that has an extremely high affinity for CO_2, binding it effectively even at unusually low concentrations. This enzyme, **PEP carboxylase,** catalyzes the reaction by which CO_2 reacts with the three-carbon compound **phosphoenolpyruvate (PEP),** forming oxaloacetate (❚ Fig. 9-16).

In a step that requires NADPH, oxaloacetate is converted to some other four-carbon compound, usually malate. The malate then passes to chloroplasts within bundle sheath cells, where a different enzyme catalyzes the decarboxylation of malate to yield pyruvate (which has three carbons) and CO_2. NADPH is formed, replacing the one used earlier.

$$\text{Malate} + \text{NADP}^+ \longrightarrow \text{pyruvate} + CO_2 + \text{NADPH}$$

The CO_2 released in the bundle sheath cell combines with ribulose bisphosphate in a reaction catalyzed by rubisco and goes through the Calvin cycle in the usual manner. The pyruvate formed in the decarboxylation reaction returns to the mesophyll cell, where it reacts with ATP to regenerate phosphoenolpyruvate.

Because the C_4 pathway captures CO_2 and provides it to the bundle sheath cells so efficiently, CO_2 concentration within the bundle sheath cells is about 10 to 60 times as great as its concentration in the mesophyll cells of plants having only the C_3 pathway. Photorespiration is negligible in C_4 plants such as crabgrass, because the concentration of CO_2 in bundle sheath cells (where rubisco is present) is always high.

The combined C_3–C_4 pathway involves the expenditure of 30 ATPs per hexose, rather than the 18 ATPs used by the C_3 pathway alone. The extra energy expense required to regenerate PEP from pyruvate is worthwhile at high light intensities because it ensures a high concentration of CO_2 in the bundle sheath cells and permits them to carry on photosynthesis at a rapid rate. At lower light intensities and temperatures, C_3 plants are favored. For example, winter rye, a C_3 plant, grows lavishly in cool weather, when crabgrass cannot because it requires more energy to fix CO_2.

CAM plants fix CO_2 at night

Plants living in very dry, or *xeric*, conditions have a number of structural adaptations that enable them to survive. Many xeric plants have physiological adaptations as well, including a special carbon fixation pathway, the **crassulacean acid metabolism (CAM) pathway.** The name comes from the stonecrop plant family (the Crassulaceae), which possesses the CAM pathway, al-

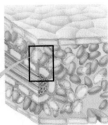

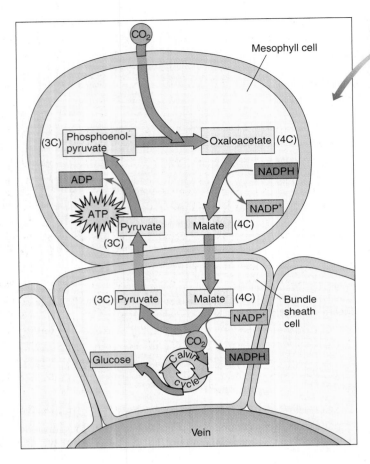

cur in *different locations* within the leaf of a C_4 plant. In CAM plants, the initial fixation of CO_2 occurs at night. Decarboxylation of malate and subsequent production of sugar from CO_2 by the normal C_3 photosynthetic pathway occur during the day. In other words, the CAM and C_3 pathways occur at *different times* within the same cell of a CAM plant.

Although it does not promote rapid growth the way that the C_4 pathway does, the CAM pathway is a very successful adaptation to xeric conditions. CAM plants can exchange gases for photosynthesis and reduce water loss significantly. Plants with CAM photosynthesis survive in deserts where neither C_3 nor C_4 plants can.

Review

▌ What are the three phases of the Calvin cycle?

▌ Which phase of the Calvin cycle requires both ATP and NADPH?

▌ In what ways does photorespiration differ from aerobic respiration?

▌ Do C_3, C_4, and CAM plants all have rubisco? PEP carboxylase?

Figure 9-16   *Animated*   Summary of the C_4 pathway

CO_2 combines with phosphoenolpyruvate (PEP) in the chloroplasts of mesophyll cells, forming a four-carbon compound that is converted to malate. Malate goes to the chloroplasts of bundle sheath cells, where it is decarboxylated. The CO_2 released in the bundle sheath cell is used to make carbohydrate by way of the Calvin cycle.

though the pathway has evolved independently in some members of more than 25 other plant families, including the cactus family (Cactaceae), the lily family (Liliaceae), and the orchid family (Orchidaceae) (▌Fig. 9-17).

Unlike most plants, CAM plants open their stomata at night, admitting CO_2 while minimizing water loss. They use the enzyme PEP carboxylase to fix CO_2, forming oxaloacetate, which is converted to malate and stored in cell vacuoles. During the day, when stomata are closed and gas exchange cannot occur between the plant and the atmosphere, CO_2 is removed from malate by a decarboxylation reaction. Now the CO_2 is available within the leaf tissue to be fixed into sugar by the Calvin cycle (C_3 pathway).

The CAM pathway is very similar to the C_4 pathway but with important differences. C_4 plants initially fix CO_2 into four-carbon organic acids in mesophyll cells. The acids are later decarboxylated to produce CO_2, which is fixed by the C_3 pathway in the bundle sheath cells. In other words, the C_4 and C_3 pathways oc-

METABOLIC DIVERSITY

Learning Objective

11 Contrast photoautotrophs and chemoheterotrophs with respect to their energy and carbon sources.

Land plants, algae, and certain prokaryotes are known as **photoautotrophs.** They are **phototrophs** because they use light energy to make ATP and NADPH, which temporarily hold chemical

Figure 9-17   A typical CAM plant

Prickly pear cactus (*Opuntia*) is a CAM plant. The more than 200 species of *Opuntia* living today originated in various xeric habitats in North and South America.

energy but are unstable and cannot be stockpiled in the cell. They are **autotrophs** (from the Greek *auto,* "self," and *trophos,* "nourishing") because the chemical energy of ATP and NADPH then drives carbon fixation, the anabolic pathway in which stable organic molecules are synthesized from CO_2 and water. These organic compounds are used not only as starting materials to synthesize all the other organic compounds the photosynthetic organism needs (such as complex carbohydrates, amino acids, and lipids) but also for energy storage. Glucose and other carbohydrates produced during photosynthesis are relatively reduced compounds that can be subsequently oxidized by aerobic respiration or by some other catabolic pathway (see Chapter 8).

In contrast, animals, fungi, and most bacteria are known as **chemoheterotrophs.** They are **chemotrophs** because they obtain energy from chemicals, typically by redox reactions (Chapters 7 and 8). They are **heterotrophs** (from the Greek *heter,* "other," and *trophos,* "nourishing") because they cannot fix carbon; they use organic molecules produced by other organisms as the building blocks from which they synthesize the carbon compounds they need.

We are so familiar with plants as photoautotrophs and animals such as ourselves as chemoheterotrophs that we tend to think that all organisms should fit in these two "mainstream" categories. Two other types of nutrition are found in certain bacteria. A few bacteria, known as nonsulfur purple bacteria, are **photoheterotrophs,** able to use light energy but unable to carry out carbon fixation and so must obtain carbon from organic compounds. Some other bacteria are **chemoautotrophs,** which obtain their energy from the oxidation of reduced inorganic molecules such as hydrogen sulfide (H_2S), nitrite (NO_2^-), or ammonia (NH_3). Some of this captured energy is subsequently used to carry out carbon fixation.

Review

▪ How does a green plant obtain energy? Carbon? How does your body obtain these things?

PHOTOSYNTHESIS IN PLANTS AND IN THE ENVIRONMENT

Learning Objective

12 State the importance of photosynthesis in a plant and to other organisms.

Although we characterize plants as photoautotrophs, not all plant cells carry out photosynthesis, and even cells with chloroplasts also possess mitochondria and carry out aerobic respiration. In fact, respiration utilizing the organic molecules the plant has made for itself is the direct source of ATP needed for most plant metabolism.

Several mechanisms regulate the relative activities of photosynthesis and aerobic respiration in plants. Although the enzymes of the Calvin cycle do not require light to function, they are actually regulated by light. As a consequence of the light-requiring reactions, the stroma becomes more basic (approximately pH 8), activating rubisco and other Calvin cycle enzymes. In contrast, light tends to inhibit the enzymes of glycolysis in the cytosol. Hence photosynthesis, not aerobic respiration, is favored in the light. When light is very dim, at a point known as the *light compensation point,* photosynthesis still occurs, but it is not evident because the rate of CO_2 fixation by photosynthesis is equal to the rate of CO_2 release through aerobic respiration. On the other hand, when light is very bright, photorespiration can significantly diminish photosynthetic yields.

As we have seen in this chapter, the reactions of the Calvin cycle provide a net yield of the three-carbon phosphorylated sugar, G3P. What are the various fates of G3P in the plant? Consider a leaf cell actively conducting photosynthesis. A series of enzymes may convert some of the G3P to glucose and then to starch. This starch is stored in starch granules that form inside chloroplasts. It has been recently shown that when this starch is broken down, the disaccharide maltose is typically formed (see Fig. 3-8a). Maltose is transported out of the chloroplast and then cleaved in the cytosol, providing glucose for aerobic respiration. Not all G3P ends up as carbohydrate; some is ultimately converted to amino acids, fatty acids, and other organic molecules needed by the photosynthetic cell.

Some of the G3P is exported to the cytosol, where enzymes convert it to the disaccharide sucrose (see Fig. 3-8b). Sucrose is then actively transported out of the cell, moves through the vascular system of the plant (see Chapter 34 for a discussion of plant transport), and is actively transported into the various cells. Sucrose can be broken down to glucose and fructose, which are used in aerobic respiration or as starting points for the synthesis of the various organic molecules the cells need, such as amino acids, lipids, and carbohydrates. Important carbohydrates include cellulose for cell walls (see Fig. 3-10) and starch, particularly in starch-storing structures such as roots (see Fig. 3-9a) and developing seeds and tubers (such as potatoes).

The benefits of photosynthesis in the environment are staggering. Of course, by fixing carbon, photoautotrophs are the ultimate source of virtually all organic molecules used as energy and carbon sources by chemoheterotrophs such as ourselves. Furthermore, in carrying out carbon fixation, they remove CO_2 from the atmosphere, thereby slowing global warming (see Chapter 56). Also of prime importance is the fact that photolysis of water by photosystem II releases the O_2 that all aerobic organisms require for aerobic respiration. Molecular oxygen is so reactive that it could not be maintained in the atmosphere if it were not constantly replenished in this way.

Review

▪ How does a root cell obtain energy? Organic molecules?

▪ What is the source of molecular oxygen in Earth's atmosphere?

Learning Objectives

1 Describe the physical properties of light, and explain the relationship between a wavelength of light and its energy (page 192).

▪ Light consists of particles called **photons** that move as waves.

▪ Photons with shorter **wavelengths** have more energy than those with longer wavelengths.

2 Diagram the internal structure of a chloroplast, and explain how its components interact and facilitate the process of photosynthesis (page 193).

▪ In plants, **photosynthesis** occurs in chloroplasts, which are located mainly within **mesophyll** cells inside the leaf.

▪ **Chloroplasts** are organelles enclosed by a double membrane; the inner membrane encloses the **stroma** in which membranous, saclike **thylakoids** are suspended. Each thylakoid encloses a **thylakoid lumen.** Thylakoids arranged in stacks are called **grana.**

▪ **Chlorophyll** *a*, **chlorophyll** *b*, **carotenoids,** and other photosynthetic pigments are components of the thylakoid membranes of chloroplasts.

3 Describe what happens to an electron in a biological molecule such as chlorophyll when a photon of light energy is absorbed (page 193).

▪ Photons excite biological molecules such as chlorophyll and other photosynthetic pigments, causing one or more electrons to become energized. These energized electrons may be accepted by electron acceptor compounds.

▪ The combined **absorption spectra** of chlorophylls *a* and *b* are similar to the **action spectrum** for photosynthesis.

4 Describe photosynthesis as a redox process (page 196).

▪ During photosynthesis, light energy is captured and converted to the chemical energy of carbohydrates; hydrogens from water are used to reduce carbon, and oxygen derived from water becomes oxidized, forming molecular oxygen.

5 Distinguish between the light-dependent reactions and carbon fixation reactions of photosynthesis (page 196).

▪ In the **light-dependent reactions,** electrons energized by light are used to generate **ATP** and **NADPH;** these compounds provide energy for the formation of carbohydrates during the **carbon fixation reactions.**

ThomsonNOW **Learn more about photosynthesis in plants by clicking on the figures in ThomsonNOW.**

6 Describe the flow of electrons through photosystems I and II in the noncyclic electron transport pathway and the products produced. Contrast this with cyclic electron transport (page 198).

▪ **Photosystems I** and **II** are the two types of photosynthetic units involved in photosynthesis. Each photosystem includes chlorophyll molecules and accessory pigments organized with pigment-binding proteins into **antenna complexes.**

▪ Only a special pair of chlorophyll *a* molecules in the **reaction center** of an antenna complex give up energized electrons to a nearby electron acceptor. **P700** is the reaction center for photosystem I; **P680** is the reaction center for photosystem II.

▪ During the noncyclic light-dependent reactions, known as **noncyclic electron transport,** ATP and NADPH are formed.

▪ Electrons in photosystem I are energized by the absorption of light and passed through an **electron transport chain** to $NADP^+$, forming NADPH. Electrons given up by P700 in photosystem I are replaced by electrons from P680 in photosystem II.

▪ A series of redox reactions takes place as energized electrons are passed along the electron transport chain from photosystem II to photosystem I. Electrons given up by P680 in photosystem II are replaced by electrons made available by the **photolysis** of H_2O; oxygen is released in the process.

▪ During **cyclic electron transport,** electrons from photosystem I are eventually returned to photosystem I. ATP is produced by chemiosmosis, but no NADPH or oxygen is generated.

ThomsonNOW **Experience the process of noncyclic electron transport by clicking on the figure in ThomsonNOW.**

7 Explain how a proton (H^+) gradient is established across the thylakoid membrane and how this gradient functions in ATP synthesis (page 198).

▪ **Photophosphorylation** is the synthesis of ATP coupled to the transport of electrons energized by photons of light. Some of the energy of the electrons is used to pump protons across the thylakoid membrane, providing the energy to generate ATP by **chemiosmosis.**

▪ As protons diffuse through **ATP synthase,** an enzyme complex in the thylakoid membrane, ADP is phosphorylated to form ATP.

8 Summarize the three phases of the Calvin cycle, and indicate the roles of ATP and NADPH in the process (page 202).

▪ The **carbon fixation reactions** proceed by way of the **Calvin cycle,** also known as the **C₃ pathway.**

▪ In the CO_2 uptake phase of the Calvin cycle, CO_2 is combined with **ribulose bisphosphate (RuBP),** a five-carbon sugar, by the enzyme ribulose bisphosphate carboxylase/oxygenase, commonly known as **rubisco,** forming the three-carbon molecule **phosphoglycerate (PGA).**

▪ In the carbon reduction phase of the Calvin cycle, the energy and reducing power of ATP and NADPH are used to convert PGA molecules to **glyceraldehyde-3-phosphate (G3P).** For every 6 CO_2 molecules fixed, 12 molecules of G3P are produced, and 2 molecules of G3P leave the cycle to produce the equivalent of 1 molecule of glucose.

▪ In the RuBP regeneration phase of the Calvin cycle, the remaining G3P molecules are modified to regenerate RuBP.

ThomsonNOW **See the Calvin cycle in action by clicking on the figure in ThomsonNOW.**

9 Discuss how photorespiration reduces photosynthetic efficiency (page 202).

▪ In **photorespiration,** C₃ plants consume oxygen and generate CO_2 by degrading Calvin cycle intermediates but do not produce ATP. Photorespiration is significant on bright, hot, dry days when plants close their stomata, conserving water but preventing the passage of CO_2 into the leaf.

10 Compare the C_4 and CAM pathways (page 202).

- In the **C_4 pathway,** the enzyme **PEP carboxylase** binds CO_2 effectively, even when CO_2 is at a low concentration. C_4 reactions take place within mesophyll cells. The CO_2 is fixed in **oxaloacetate,** which is then converted to malate. The malate moves into a **bundle sheath cell,** and CO_2 is removed from it. The released CO_2 then enters the Calvin cycle.
- The **crassulacean acid metabolism (CAM) pathway** is similar to the C_4 pathway. PEP carboxylase fixes carbon at night in the mesophyll cells, and the Calvin cycle occurs during the day in the same cells.

ThomsonNOW™ **See a comparison of the C_3 and C_4 pathways by clicking on the figure in ThomsonNOW.**

11 Contrast photoautotrophs and chemoheterotrophs with respect to their energy and carbon sources (page 206).

- **Photoautotrophs** use light as an energy source and are able to incorporate atmospheric CO_2 into pre-existing carbon skeletons. **Chemoheterotrophs** obtain energy by oxidizing chemicals and obtain carbon as organic molecules from other organisms.

12 State the importance of photosynthesis in a plant and to other organisms (page 207).

- Photosynthesis is the ultimate source of all chemical energy and organic molecules available to photoautotrophs, such as plants, and to virtually all other organisms as well. It also constantly replenishes the supply of oxygen in the atmosphere, vital to all aerobic organisms.

Summary Reactions for Photosynthesis

The light-dependent reactions (noncyclic electron transport):

$$12\ H_2O + 12\ NADP^+ + 18\ ADP + 18\ P_i \xrightarrow[\text{Chlorophyll}]{\text{Light}}$$
$$6\ O_2 + 12\ NADPH + 18\ ATP$$

The carbon fixation reactions (Calvin cycle):

$$12\ NADPH + 18\ ATP + 6\ CO_2 \longrightarrow$$
$$C_6H_{12}O_6 + 12\ NADP^+ + 18\ ADP + 18\ P_i + 6\ H_2O$$

By canceling out the common items on opposite sides of the arrows in these two coupled equations, we obtain the simplified overall equation for photosynthesis:

$$6\ CO_2 + 12\ H_2O \xrightarrow[\text{Chlorophyll}]{\text{Light energy}} C_6H_{12}O_6 + 6\ O_2 + 6\ H_2O$$

Carbon dioxide Water Glucose Oxygen Water

TEST YOUR UNDERSTANDING

1. Where is chlorophyll located in the chloroplast? (a) thylakoid membranes (b) stroma (c) matrix (d) thylakoid lumen (e) between the inner and outer membranes

2. In photolysis, some of the energy captured by chlorophyll is used to split (a) CO_2 (b) ATP (c) NADPH (d) H_2O (e) both b and c

3. Light is composed of particles of energy called (a) carotenoids (b) reaction centers (c) photons (d) antenna complexes (e) photosystems

4. The relative effectiveness of different wavelengths of light in photosynthesis is demonstrated by (a) an action spectrum (b) photolysis (c) carbon fixation reactions (d) photoheterotrophs (e) an absorption spectrum

5. In plants, the final electron acceptor in noncyclic electron flow is (a) $NADP^+$ (b) CO_2 (c) H_2O (d) O_2 (e) G3P

6. Most plants contain, in addition to chlorophyll, accessory photosynthetic pigments such as (a) PEP (b) G3P (c) carotenoids (d) PGA (e) $NADP^+$

7. The part of a photosystem that absorbs light energy is its (a) antenna complexes (b) reaction center (c) terminal quinone electron acceptor (d) pigment-binding protein (e) thylakoid lumen

8. In _____, electrons that have been energized by light contribute their energy to add phosphate to ADP, producing ATP. (a) crassulacean acid metabolism (b) the Calvin cycle (c) photorespiration (d) C_4 pathways (e) photophosphorylation

9. In _____, there is a one-way flow of electrons to $NADP^+$, forming NADPH. (a) crassulacean acid metabolism (b) the Calvin cycle (c) photorespiration (d) cyclic electron transport (e) noncyclic electron transport

10. The mechanism by which electron transport is coupled to ATP production by means of a proton gradient is called (a) chemiosmosis (b) crassulacean acid metabolism (c) fluorescence (d) the C_3 pathway (e) the C_4 pathway

11. In photosynthesis in eukaryotes, the transfer of electrons through a sequence of electron acceptors provides energy to pump protons across the (a) chloroplast outer membrane (b) chloroplast inner membrane (c) thylakoid membrane (d) inner mitochondrial membrane (e) plasma membrane

12. The inputs for _____ are CO_2, NADPH, and ATP. (a) cyclic electron transport (b) the carbon fixation reactions (c) noncyclic electron transport (d) photosystems I and II (e) chemiosmosis

13. The Calvin cycle begins when CO_2 reacts with (a) phosphoenolpyruvate (b) glyceraldehyde-3-phosphate (c) ribulose bisphosphate (d) oxaloacetate (e) phosphoglycerate

14. The enzyme directly responsible for almost all carbon fixation on Earth is (a) rubisco (b) PEP carboxylase (c) ATP synthase (d) phosphofructokinase (e) ligase

15. In C_4 plants, C_4 and C_3 pathways occur at different _____; whereas in CAM plants, CAM and C_3 pathways occur at different _____. (a) times of

day; locations within the leaf (b) seasons; locations (c) locations; times of day (d) locations; seasons (e) times of day; seasons

16. An organism characterized as a photoautotroph obtains energy from _____ and carbon from _____. (a) light; organic molecules (b) light; CO_2 (c) organic molecules; organic molecules (d) organic molecules; CO_2 (e) O_2; CO_2

17. Label the figure. Use Figure 9-8 to check your answers.

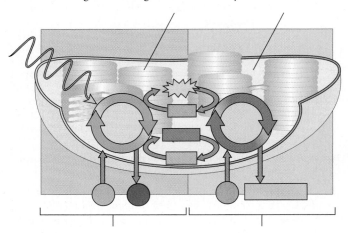

CRITICAL THINKING

1. Must all autotrophs use light energy? Explain.

2. Only some plant cells have chloroplasts, but all actively metabolizing plant cells have mitochondria. Why?

3. Explain why the proton gradient formed during chemiosmosis represents a state of low entropy. (You may wish to refer to the discussion of entropy in Chapter 7.)

4. The electrons in glucose have relatively high free energies. How did they become so energetic?

5. What strategies may be employed in the future to increase world food supply? Base your answer on your knowledge of photosynthesis and related processes.

6. What would life be like for photoautotrophs if there were no chemoheterotrophs? For chemoheterotrophs if there were no photoautotrophs?

7. What might you suspect if scientists learned that a distant planet has an atmosphere that is 15% molecular oxygen?

8. **Evolution Link.** Propose an explanation for the fact that bacteria, chloroplasts, and mitochondria all have ATP synthase complexes.

9. **Analyzing Data.** Examine Figure 9-6. Imagine a photosynthetic pigment that would be able to absorb the wavelengths of light under-utilized in plant photosynthesis. What possible colors might that pigment likely be (i.e., what colors/wavelengths might it reflect)?

Additional questions are available in ThomsonNOW at www.thomsonedu.com/login

Chromosomes, Mitosis, and Meiosis

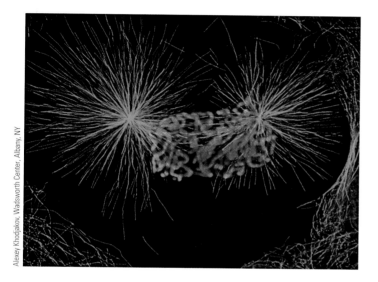

Alexey Khodjakov, Wadsworth Center, Albany, NY

Fluorescence LM of a cultured newt lung cell in mitosis (early prometaphase). Antibodies were used to attach fluorescent dyes to various structures. In this LM, the nuclear envelope has broken down, and the microtubules of the mitotic spindle (*green*) now interact with the chromosomes (*blue*).

KEY CONCEPTS

In eukaryotic cells, DNA is wound around specific proteins to form chromatin, which in turn is folded and packaged to make individual chromosomes.

Cell division is an important part of the cell cycle, which consists of the successive stages through which a cell passes.

An internal genetic program interacts with external signals to regulate the cell cycle.

In cell division by mitosis, duplicated chromosomes separate (split apart) and are evenly distributed into two daughter nuclei.

Meiosis, which reduces the number of chromosome sets from diploid to haploid, is necessary to maintain the normal chromosome number when two cells join during sexual reproduction.

Meiosis helps to increase genetic variation among offspring.

Pre-existing cells divide to form new cells. This remarkable process enables an organism to grow, repair damaged parts, and reproduce. Cells serve as the essential link between generations. Even the simplest cell contains a large amount of precisely coded genetic information in the form of deoxyribonucleic acid (DNA), collectively called the organism's *genome*. An individual's genome is organized into informational units called *genes*, which control the activities of the cell and are passed on to its descendants.

When a cell divides, the information contained in the DNA must be faithfully replicated and the copies then transmitted to each daughter cell through a precisely choreographed series of steps (see photograph). DNA is a very long, thin molecule that could easily become tangled and broken, and a eukaryotic cell's nucleus contains a huge amount of DNA. In this chapter we consider how eukaryotes accommodate the genetic material by packaging each DNA molecule with proteins to form a structure called a *chromosome*, each of which contains hundreds or thousands of genes.

We then consider *mitosis*, the highly regimented process that ensures a parent cell transmits one copy of every chromosome to each of its two daughter cells. In this way, the chromosome number is preserved through successive mitotic divisions. Most body cells of eukaryotes divide by mitosis. Mitosis is an active area of biological research, and for good reason: Errors in mitosis can result in a host of diseases and disorders, from Down syndrome to cancer, a disease condition in which cells divide at an uncontrolled rate and become invasive. Thus, a clearer understanding of mitosis has the potential to improve our treatment of many diseases.

Finally, we discuss *meiosis*, a process that reduces the chromosome number by half. Sexual life cycles in eukaryotes require meiosis. Sexual reproduction involves the fusion of two sex cells, or *gametes*, to form a fertilized egg called a *zygote*. Meiosis makes it possible for each gamete to contain only half the number of chromosomes in the parent cell, thereby preventing the zygotes from having twice as many chromosomes as the parents. ■

EUKARYOTIC CHROMOSOMES

Learning Objectives

1   Discuss the significance of chromosomes in terms of their information content.
2   Compare the organization of DNA in prokaryotic and eukaryotic cells.

The major carriers of genetic information in eukaryotes are the **chromosomes,** which lie within the cell nucleus. Although *chromosome* means "colored body," chromosomes are virtually colorless; the term refers to their ability to be stained by certain dyes. In the 1880s, light microscopes had been improved to the point that scientists such as the German biologist Walther Fleming began to observe chromosomes during cell division.

In 1903, American biologist Walter Sutton and German biologist Theodor Boveri noted independently that chromosomes were the physical carriers of genes, the genetic factors Gregor Mendel discovered in the 19th century (discussed in Chapter 11).

Chromosomes are made of **chromatin,** a material consisting of DNA and associated proteins. When a cell is not dividing, the chromosomes are present but in an extended, partially unraveled form. Chromatin consists of long, thin threads that are somewhat aggregated, giving them a granular appearance when viewed with the electron microscope (see Fig. 4-11). During cell division, the chromatin fibers condense and the chromosomes become visible as distinct structures (▮ Fig. 10-1).

DNA is organized into informational units called genes

An organism's **genome** may contain hundreds or even thousands of genes. For example, the **Human Genome Project** estimates that humans have about 25,000 genes that code for proteins (see Chapter 16). As you will see in later chapters, the concept of the gene has changed considerably since the science of genetics began, but our definitions have always centered on the gene as an informational unit. By providing information needed to carry out one or more specific cell functions, a **gene** affects some characteristic of the organism. For example, genes govern eye color in humans, wing length in flies, and seed color in peas.

DNA is packaged in a highly organized way in chromosomes

Prokaryotic and eukaryotic cells differ markedly in their DNA content as well as in the organization of DNA molecules. The bacterium *Escherichia coli* normally contains about 4×10^6 base pairs (almost 1.35 mm) of DNA in its single, circular DNA molecule. In fact, the total length of its DNA is about 1000 times as long as the length of the cell itself. Therefore, the DNA molecule is, with the help of proteins, twisted and folded compactly to fit inside the bacterial cell (see Fig. 24-9).

A typical eukaryotic cell contains much more DNA than a bacterium does, and it is organized in the nucleus as multiple chromosomes; these vary widely in size and number among different species. Although a human nucleus is about the size of a large bacterial cell, it contains more than 1000 times the amount of DNA found in *E. coli.* The DNA content of a human sperm cell is about 3×10^9 base pairs; stretched end to end, it would measure almost 1 m long. Remarkably, this DNA fits into a nucleus with a diameter of only 10 μm.

How does a eukaryotic cell pack its DNA into the chromosomes? Chromosome packaging is facilitated by certain proteins known as **histones.**[1] Histones have a positive charge because they

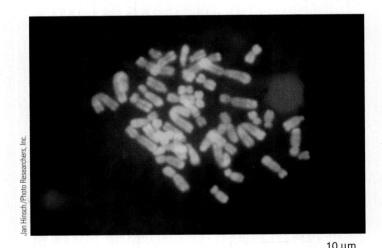

Figure 10-1 Chromosomes

Human chromosomes from an unidentified cell are shown in this fluorescence LM.

10 μm

Jan Hinsch/Photo Researchers, Inc.

---
[1] A few types of eukaryotic cells lack histones. Conversely, histones occur in one group of prokaryotes, the archaea (see Chapter 24).

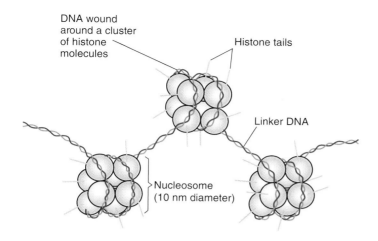

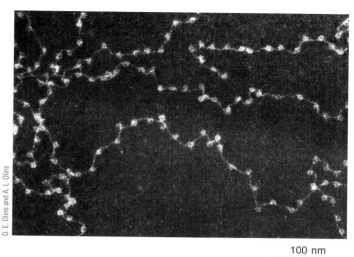

D. E. Olins and A. L. Olins

100 nm

(a) A model for the structure of a nucleosome. Each nucleosome bead contains a set of eight histone molecules, forming a protein core around which the double-stranded DNA winds. The DNA surrounding the histones consists of 146 nucleotide pairs; another segment of DNA, about 60 nucleotide pairs long, links nucleosome beads.

(b) TEM of nucleosomes from the nucleus of a chicken cell. Normally, nucleosomes are packed more closely together, but the preparation procedure has spread them apart, revealing the DNA linkers.

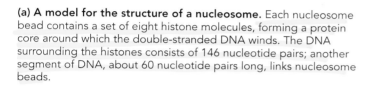

Figure 10-2   Nucleosomes

have a high proportion of amino acids with basic side chains (see Chapter 3). The positively charged histones associate with DNA, which has a negative charge because of its phosphate groups, to form structures called **nucleosomes.** The fundamental unit of each nucleosome consists of a beadlike structure with 146 base pairs of DNA wrapped around a disc-shaped core of eight histone molecules (two each of four different histone types) (❚ Fig. 10-2). Although the nucleosome was originally defined as a bead plus a DNA segment that links it to an adjacent bead, today the term more commonly refers only to the bead itself (that is, the eight histones and the DNA wrapped around them).

Nucleosomes function like tiny spools, preventing DNA strands from becoming tangled. You can see the importance of this role in ❚ Figure 10-3, which illustrates the enormous number of DNA fibers that unravel from a mouse chromosome after researchers have removed the histones. **Scaffolding proteins** are nonhistone proteins that help maintain chromosome structure. But the role of histones is more than simply structural, because their arrangement also affects the activity of the DNA with which they are associated. Histones are increasingly viewed as an important part of the regulation of gene expression—that is, whether genes are turned off or on. (We discuss gene regulation by histones in Chapter 14.)

The wrapping of DNA into nucleosomes represents the first level of chromosome structure. ❚ Figure 10-4 shows the higher-order structures of chromatin leading to the formation of a condensed chromosome. The nucleosomes themselves are 10 nm in

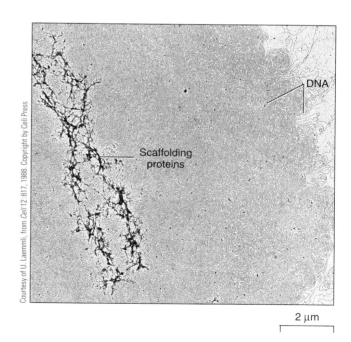

Courtesy of U. Laemmli, from *Cell* 12: 817, 1988. Copyright by Cell Press

2 μm

Figure 10-3   TEM of a mouse chromosome depleted of histones

Notice how densely packed the DNA strands are, even though they have been released from the histone proteins that organize them into tightly coiled structures.

As a cell prepares to divide, its chromosomes become thicker and shorter as their long chromatin fibers are compacted.

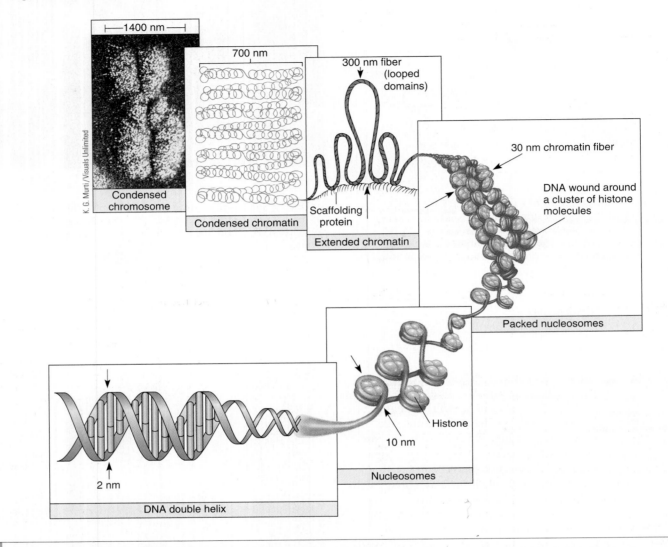

├── 1400 nm ──┤

K. G. Murti/Visuals Unlimited

Condensed chromosome

700 nm

Condensed chromatin

300 nm fiber (looped domains)

Scaffolding protein

Extended chromatin

30 nm chromatin fiber

DNA wound around a cluster of histone molecules

Packed nucleosomes

Histone

10 nm

Nucleosomes

2 nm

DNA double helix

Figure 10-4 *Animated* Organization of a eukaryotic chromosome

This diagram shows how DNA is packaged into highly condensed chromosomes. First, DNA is wrapped around histone proteins to form nucleosomes. Then, the nucleosomes are compacted into chromatin fibers, which are coiled into looped domains. The looped domains are compacted, ultimately forming chromosomes.

diameter. The packed nucleosome state occurs when a fifth type of histone, known as *histone H1*, associates with the linker DNA, packing adjacent nucleosomes together to form a compacted 30-nm chromatin fiber. In extended chromatin, these fibers form large, coiled loops held together by scaffolding proteins. The loops then interact to form the condensed chromatin found in a chromosome. Cell biologists have identified a group of proteins, collectively called **condensin,** required for chromosome compaction. Condensin binds to DNA and wraps it into coiled loops that are compacted into a mitotic or meiotic chromosome.

Chromosome number and informational content differ among species

Every individual of a given species has a characteristic number of chromosomes in most nuclei of its body cells. However, it is not the number of chromosomes that makes each species unique but the information the genes specify.

Most human body cells have exactly 46 chromosomes, but humans are not humans merely because we have 46 chromosomes. Some other species—the olive tree, for example—also have 46. Some humans have an abnormal chromosome composition with more or fewer than 46 (see Fig. 16-4).

Other species have different chromosome numbers. A certain species of roundworm has only 2 chromosomes in each cell, whereas some crabs have as many as 200, and some ferns have more than 1000. Most animal and plant species have between 8 and 50 chromosomes per body cell. Numbers above and below these are uncommon.

Review

- What are the informational units on chromosomes called? Of what do these informational units consist?
- How are bacterial DNA molecules and eukaryotic chromosomes similar? How do they differ?
- How is the large discrepancy between DNA length and nucleus size addressed in eukaryotic cells?

THE CELL CYCLE AND MITOSIS

Learning Objectives

3   Identify the stages in the eukaryotic cell cycle, and describe their principal events.

4   Describe the structure of a duplicated chromosome, including the sister chromatids, centromeres, and kinetochores.

5   Explain the significance of mitosis, and describe the process.

When cells reach a certain size, they usually either stop growing or divide. Not all cells divide; some, such as skeletal muscle and red blood cells, do not normally divide once they are mature. Other cells undergo a sequence of activities required for growth and cell division. The stages through which a cell passes from one cell division to the next are collectively referred to as the **cell cycle.** Timing of the cell cycle varies widely, but in actively growing plant and animal cells, it is about 8 to 20 hours. The cell cycle consists of two main phases, interphase and M phase, both of which can be distinguished under a light microscope (▮ Fig. 10-5).

M phase involves two main processes, mitosis and cytokinesis. *Mitosis,* a process involving the nucleus, ensures that each new nucleus receives the same number and types of chromosomes as were present in the original nucleus. *Cytokinesis,* which generally begins before mitosis is complete, is the division of the cell cytoplasm to form two cells.

Multinucleated cells form if mitosis is not followed by cytokinesis; this is a normal condition for certain cell types. For example, the body of plasmodial slime molds consists of a multinucleate mass of cytoplasm (see Fig. 25-23a).

Chromosomes duplicate during interphase

Most of a cell's life is spent in **interphase,** the time when no cell division is occurring. A cell that is capable of dividing is typically very active during interphase, synthesizing needed materials (proteins, lipids, and other biologically important molecules)

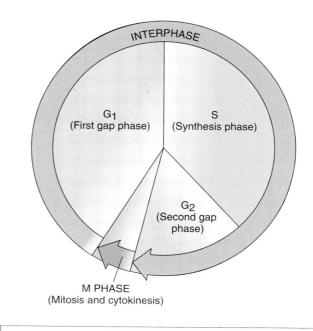

▮ **Figure 10-5**   *Animated*   The eukaryotic cell cycle

The cell cycle includes interphase (G_1, S, and G_2) and M phase (mitosis and cytokinesis). Proportionate amounts of time spent at each stage vary among species, cell types, and growth conditions. If the cell cycle were a period of 12 hours, G_1 would be about 5 hours, S would be 4.5 hours, G_2 would be 2 hours, and M phase would be 30 minutes.

and growing. Here is the sequence of interphase and M phase in the eukaryotic cell cycle:

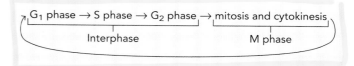

The time between the end of mitosis and the beginning of the S phase is termed the **G_1 phase** (*G* stands for *gap,* an interval during which no DNA synthesis occurs). Growth and normal metabolism take place during the G_1 phase, which is typically the longest phase. Cells that are not dividing usually become arrested in this part of the cell cycle and are said to be in a state called G_0. Toward the end of G_1, the enzymes required for DNA synthesis become more active. Synthesis of these enzymes, along with proteins needed to initiate cell division (discussed later in the chapter), enable the cell to enter the S phase.

During the **synthesis phase,** or **S phase,** DNA replicates and histone proteins are synthesized so that the cell can make duplicate copies of its chromosomes. How did researchers identify the S phase of the cell cycle? In the early 1950s, researchers demonstrated that cells preparing to divide duplicate their chromosomes

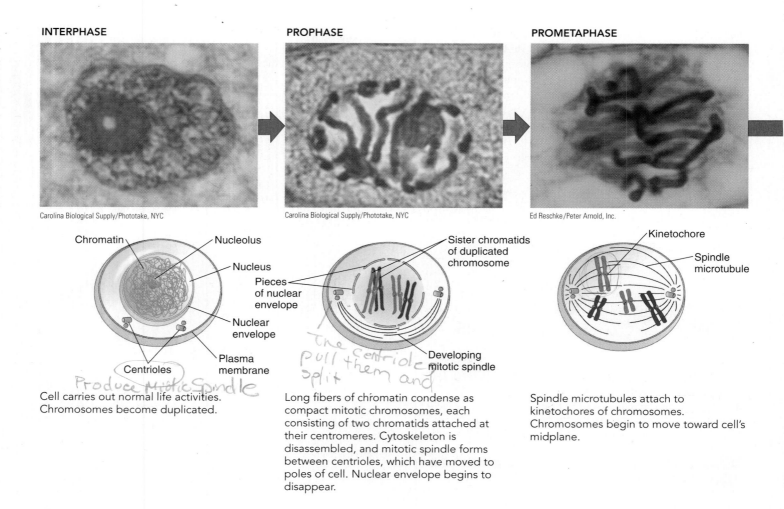

INTERPHASE

Carolina Biological Supply/Phototake, NYC

Chromatin
Nucleolus
Nucleus
Pieces of nuclear envelope
Nuclear envelope
Plasma membrane
Centrioles

Produce Mitotic Spindle

Cell carries out normal life activities. Chromosomes become duplicated.

PROPHASE

Carolina Biological Supply/Phototake, NYC

Sister chromatids of duplicated chromosome
Developing mitotic spindle

The centrioles pull them and split

Long fibers of chromatin condense as compact mitotic chromosomes, each consisting of two chromatids attached at their centromeres. Cytoskeleton is disassembled, and mitotic spindle forms between centrioles, which have moved to poles of cell. Nuclear envelope begins to disappear.

PROMETAPHASE

Ed Reschke/Peter Arnold, Inc.

Kinetochore
Spindle microtubule

Spindle microtubules attach to kinetochores of chromosomes. Chromosomes begin to move toward cell's midplane.

Figure 10-6 *Animated* Interphase and the stages of mitosis

The LMs show plant cells, which lack centrioles. The drawings depict generalized animal cells with a diploid chromosome number of 4; the sizes of the nuclei and chromosomes are exaggerated to show the structures more clearly.

at a relatively restricted interval during interphase and not during early mitosis, as previously hypothesized. These investigators used isotopes, such as ³H, to synthesize radioactive thymidine, a nucleotide that is incorporated specifically into DNA as it is synthesized. After radioactive thymidine was supplied for a brief period (such as 30 minutes) to actively growing cells, *autoradiography* (see Fig. 2-3) on exposed film showed that a fraction of the cells had silver grains over their chromosomes. The nuclei of these cells were radioactive, because during the experiment they had replicated DNA. DNA replication was not occurring in the cells that did not have radioactively labeled chromosomes. Researchers therefore inferred that the proportion of labeled cells out of the total number of cells provides a rough estimate of the length of the S phase relative to the rest of the cell cycle.

After it completes the S phase, the cell enters a second gap phase, the **G₂ phase.** At this time, increased protein synthesis occurs, as the final steps in the cell's preparation for division take place. For many cells, the G₂ phase is short relative to the G₁ and S phases.

Mitosis, the nuclear division that produces two nuclei containing chromosomes identical to the parental nucleus, begins at the end of the G₂ phase. Mitosis is a continuous process, but for descriptive purposes, it is divided into five stages:

Prophase ⟶ prometaphase ⟶ metaphase ⟶ anaphase ⟶ telophase

Look at ▌Figure 10-6 while you read the following descriptions of these stages as they occur in a typical plant or animal cell.

During prophase, duplicated chromosomes become visible with the microscope

The first stage of mitosis, **prophase,** begins with *chromosome compaction,* when the long chromatin fibers begin a coiling process that makes them shorter and thicker (see Fig. 10-4). The chromatin can then be distributed to the daughter cells with less likeli-

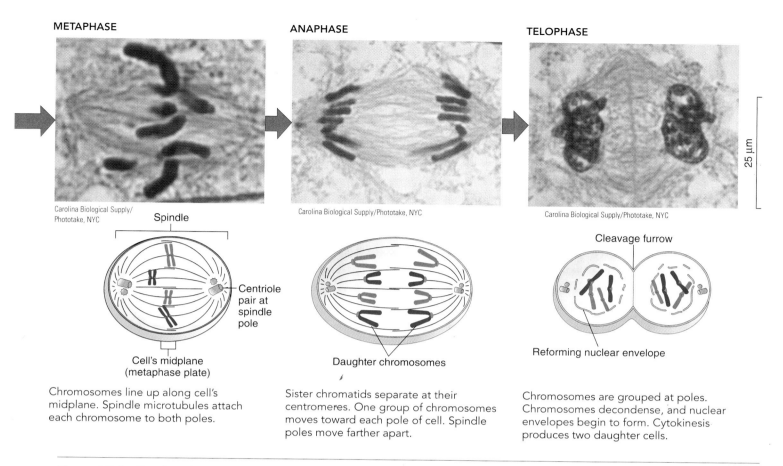

METAPHASE

Carolina Biological Supply/
Phototake, NYC

Spindle

Centriole pair at spindle pole

Cell's midplane (metaphase plate)

Chromosomes line up along cell's midplane. Spindle microtubules attach each chromosome to both poles.

ANAPHASE

Carolina Biological Supply/Phototake, NYC

Daughter chromosomes

Sister chromatids separate at their centromeres. One group of chromosomes moves toward each pole of cell. Spindle poles move farther apart.

TELOPHASE

Carolina Biological Supply/Phototake, NYC

25 μm

Cleavage furrow

Reforming nuclear envelope

Chromosomes are grouped at poles. Chromosomes decondense, and nuclear envelopes begin to form. Cytokinesis produces two daughter cells.

Figure 10-6 Continued

hood of tangling. After compaction, the chromatin is referred to as *chromosomes*.

When stained with certain dyes and viewed through the light microscope, chromosomes become visible as darkly staining bodies as prophase progresses. It is now apparent that each chromosome was duplicated during the preceding S phase and consists of a pair of **sister chromatids,** which contain identical, double-stranded DNA sequences. Each chromatid includes a constricted region called the **centromere.** Sister chromatids are tightly associated in the vicinity of their centromeres (▌ Fig. 10-7). Precise DNA sequences and proteins that bind to those DNA sequences are the chemical basis for this close association at the centromeres. For example, sister chromatids are physically linked by a ring-shaped protein complex called **cohesin.** Cohesins extend along the length of the sister chromatid arms and are particularly concentrated at the centromere (▌ Fig. 10-8).

Attached to each centromere is a **kinetochore,** a multiprotein complex to which **microtubules** can bind. These microtubules function in chromosome distribution during mitosis, in which one copy of each chromosome is delivered to each daughter cell.

A dividing cell can be described as a globe, with an equator that determines the midplane (equatorial plane) and two opposite poles. This terminology is used for all cells regardless of their actual shape. Microtubules radiate from each pole, and some of

these protein fibers elongate toward the chromosomes, forming the **mitotic spindle,** a structure that separates the duplicated chromosomes during anaphase (▌ Fig. 10-9). The organization and function of the spindle require the presence of motor proteins and a variety of signaling molecules.

Animal cells differ from plant cells in the details of mitotic spindle formation. In both types of dividing cells, each pole contains a region, the **microtubule-organizing center,** from which extend the microtubules that form the mitotic spindle. The electron microscope shows that microtubule-organizing centers in certain plant cells consist of fibrils with little or no discernible structure.

In contrast, animal cells have a pair of **centrioles** in the middle of each microtubule-organizing center (see Fig. 4-24). The centrioles are surrounded by fibrils that make up the **pericentriolar material.** The spindle microtubules terminate in the pericentriolar material, but they do not actually touch the centrioles. Although cell biologists once thought spindle formation in animal cells required centrioles, their involvement is probably coincidental. Current evidence suggests that centrioles organize the pericentriolar material and ensure its duplication when the centrioles duplicate.

Each of the two centrioles is duplicated during interphase, yielding two centriole pairs. Late in prophase, microtubules radi-

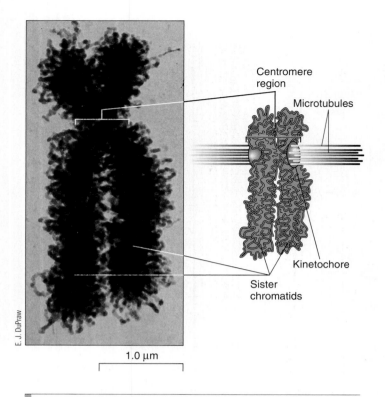

Centromere region

Microtubules

Kinetochore

Sister chromatids

1.0 µm

E. J. DuPraw

Figure 10-7 Sister chromatids and centromeres

The sister chromatids, each consisting of tightly coiled chromatin fibers, are tightly associated at their centromere regions, indicated by the brackets. Associated with each centromere is a kinetochore, which serves as a microtubule attachment site. Kinetochores and microtubules are not visible in this TEM of a metaphase chromosome.

ate from the pericentriolar material surrounding the centrioles; these clusters of microtubules are called **asters.** The two asters migrate to opposite sides of the nucleus, establishing the two poles of the mitotic spindle.

Prometaphase begins when the nuclear envelope breaks down

During **prometaphase** the nuclear envelope fragments and is sequestered in vesicles to be used later. The nucleolus shrinks and usually disappears, and the mitotic spindle is completely assembled. At the start of prometaphase, the duplicate chromosomes are scattered throughout the nuclear region (see the chapter introduction photograph, which shows early prometaphase). The spindle microtubules grow and shrink as they move toward the center of the cell in a "search and capture" process. Their random, dy-

namic movements give the appearance that they are "searching" for the chromosomes. If a microtubule comes near a centromere, one of the kinetochores of a duplicated chromosome "captures" it. As the now-tethered chromosome continues moving toward the cell's midplane, the unattached kinetochore of its sister chromatid becomes connected to a spindle microtubule from the cell's other pole.

During the chromosomes' movements toward the cell's midplane, long microtubules are shortened by the removal of tubulin subunits, and the short microtubules are lengthened by the addition of tubulin subunits (see Fig. 4-22). Evidence indicates that shortening and lengthening occurs at the kinetochore end (the plus end) of the microtubule, not at the spindle pole end (the minus end). This shortening or lengthening occurs while the spindle microtubule remains firmly tethered to the kinetochore. Motor proteins located at the kinetochores may be involved in this tethering of the spindle microtubule. These motor proteins may work in a similar fashion to the kinesin motor shown in Figure 4-23.

To summarize the events of prometaphase, sister chromatids of each duplicated chromosome become attached at their kinetochores to spindle microtubules extending from opposite poles of the cell, and the chromosomes begin to move toward the cell's midplane. As the cell transitions from prometaphase to metaphase, cohesins dissociate from the sister chromatid arms, freeing them from one another, although some cohesins remain in the vicinity of the centromere.

Duplicated chromosomes line up on the midplane during metaphase

During **metaphase,** all the cell's chromosomes align at the cell's midplane, or **metaphase plate.** As already mentioned, one of the two sister chromatids of each chromosome is attached by its

Key Point

When mitotic chromosomes replicate, sister chromatids are initially linked to one another by protein complexes called cohesins. Cohesin linkages are particularly concentrated in the vicinity of the centromere.

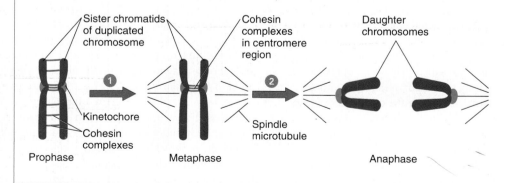

Sister chromatids of duplicated chromosome

Cohesin complexes in centromere region

Daughter chromosomes

Kinetochore
Cohesin complexes

Spindle microtubule

Prophase   Metaphase   Anaphase

Figure 10-8 Cohesins

❶ As mitosis progresses, cohesins dissociate from the duplicated chromosome arms. ❷ Cohesins then dissociate at the centromere to allow the daughter chromosomes to separate from one another during anaphase.

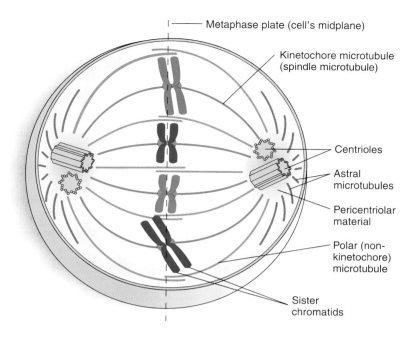

Metaphase plate (cell's midplane)

Kinetochore microtubule (spindle microtubule)

Centrioles

Astral microtubules

Pericentriolar material

Polar (non-kinetochore) microtubule

Sister chromatids

(a) One end of each microtubule of this animal cell is associated with one of the poles. Astral microtubules (*green*) radiate in all directions, forming the aster. Kinetochore microtubules (*red*) connect the kinetochores to the poles, and polar (non-kinetochore) microtubules (*blue*) overlap at the midplane.

10 µm

CNRI/Phototake, NYC

(b) This fluorescence LM of an animal cell at metaphase shows a well-defined spindle and asters (chromosomes, *orange*; microtubules, *green*).

Figure 10-9   The mitotic spindle

kinetochore to microtubules from one pole, and its sister chromatid is attached by its kinetochore to microtubules from the opposite pole. The mitotic spindle has two types of microtubules: polar microtubules and kinetochore microtubules (see Fig. 10-9). **Polar microtubules,** also known as *non-kinetochore microtubules,* extend from each pole to the equatorial region, where they generally overlap. **Kinetochore microtubules** extend from each pole and attach to the kinetochores.

Each chromatid is completely condensed and appears quite thick and distinct during metaphase. Because individual chromosomes are more distinct at metaphase than at any other time, they are usually photographed to check the **karyotype,** or chromosome makeup, at this stage if chromosome abnormalities are suspected (see Chapter 16). As the mitotic cell transitions from metaphase to anaphase, the remaining cohesin proteins joining the sister chromatids at their centromere dissociate.

During anaphase, chromosomes move toward the poles

Anaphase begins as the sister chromatids separate. Once the chromatids are no longer attached to their duplicates, each chromatid is called a *chromosome*. The now-separate chromosomes move to opposite poles, using the spindle microtubules as tracks. The kinetochores, still attached to kinetochore microtubules, lead the

way, with the chromosome arms trailing behind. Anaphase ends when all the chromosomes have reached the poles.

Cell biologists are making significant progress in understanding the overall mechanism of chromosome movement in anaphase. Chromosome movements are studied in several ways. The number of microtubules at a particular stage or after certain treatments is determined by carefully analyzing electron micrographs. Researchers also physically perturb living cells that are dividing, using laser beams or mechanical devices known as *micromanipulators.* A skilled researcher can move chromosomes, break their connections to microtubules, and even remove them from the cell entirely.

Microtubules lack elastic or contractile properties. Then how do the chromosomes move apart? Are they pushed or pulled, or do other forces operate? Microtubules are dynamic structures, with *tubulin* subunits constantly being removed from their ends and others being added. Evidence indicates that kinetochore microtubules shorten, or *depolymerize*, during anaphase. This shortening "pulls" the chromosomes toward the poles. One possible mechanism of this microtubule depolymerization is that chromosomes move poleward because they remain anchored to the kinetochore microtubules even as tubulin subunits are being removed at the kinetochore (that is, at the plus end of the kinetochore microtubule). Another possibility is the "reeling-in" mechanism, in which the minus end (the pole end) of the kinetochore microtubule is depolymerized. Motor proteins for both plus-end and minus-end depolymerizations have been isolated.

A second phenomenon also plays a role in chromosome separation. During anaphase the spindle as a whole elongates, at least partly because polar microtubules originating at opposite poles are associated with motors that let them slide past one another at the midplane. The sliding decreases the degree of overlap, thereby "pushing" the poles apart. This mechanism indirectly causes the chromosomes to move apart because they are attached to the poles by kinetochore microtubules.

During telophase, two separate nuclei form

The final stage of mitosis, **telophase,** is characterized by the arrival of the chromosomes at the poles and, in its final stage, by a return to interphase-like conditions. The chromosomes decondense by partially uncoiling. A new nuclear envelope forms around each set of chromosomes, made at least in part from small vesicles and other components derived from the old nuclear envelope. The spindle microtubules disappear, and the nucleoli reorganize.

Cytokinesis forms two separate daughter cells

Cytokinesis, the division of the cytoplasm to yield two daughter cells, usually overlaps mitosis, generally beginning during telophase. Cytokinesis of an animal or fungal cell begins as an *actomyosin contractile ring* attached to the plasma membrane encircles the cell in the equatorial region, at right angles to the spindle (Fig. 10-10a). The contractile ring consists of an association between actin and myosin filaments; it is thought that the motor activity of myosin moves actin filaments to cause the constriction, similar to the way actin and myosin cause muscle contraction (see Fig. 39-10). The ring contracts, producing a **cleavage furrow** that gradually deepens and eventually separates the cytoplasm into two daughter cells, each with a complete nucleus.

In plant cells, cytokinesis occurs by forming a **cell plate** (Fig. 10-10b), a partition constructed in the equatorial region of the spindle and growing laterally toward the cell wall. The cell plate forms as a line of vesicles originating in the *Golgi complex.* The vesicles contain materials to construct both a primary cell wall for each daughter cell and a middle lamella that cements the primary cell walls together. The vesicle membranes fuse to become the plasma membrane of each daughter cell.

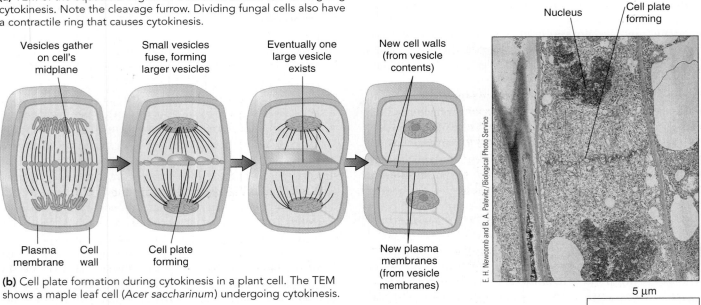

(a) TEM of the equatorial region of a cultured animal cell undergoing cytokinesis. Note the cleavage furrow. Dividing fungal cells also have a contractile ring that causes cytokinesis.

(b) Cell plate formation during cytokinesis in a plant cell. The TEM shows a maple leaf cell (*Acer saccharinum*) undergoing cytokinesis.

Figure 10-10   *Animated*   Cytokinesis in animal and plant cells

The nuclei in both TEMs are in telophase. The drawings show 3-D relationships.

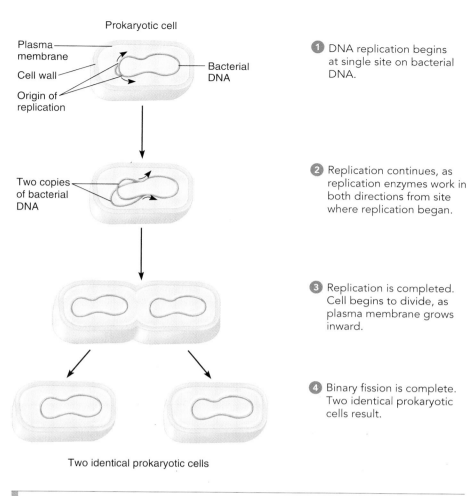

Prokaryotic cell

Plasma membrane

Cell wall

Origin of replication

Bacterial DNA

Two copies of bacterial DNA

Two identical prokaryotic cells

1 DNA replication begins at single site on bacterial DNA.

2 Replication continues, as replication enzymes work in both directions from site where replication began.

3 Replication is completed. Cell begins to divide, as plasma membrane grows inward.

4 Binary fission is complete. Two identical prokaryotic cells result.

Figure 10-11 Most prokaryotes reproduce by binary fission

The circular DNA molecule is much longer than depicted here.

Mitosis produces two cells genetically identical to the parent cell

The remarkable regularity of the process of cell division ensures that each daughter nucleus receives exactly the same number and kinds of chromosomes that the parent cell had. Thus, with a few exceptions, every cell of a multicellular organism has the same genetic makeup. If a cell receives more or fewer than the characteristic number of chromosomes through some malfunction of the cell division process, the resulting cell may show marked abnormalities and often cannot survive.

Mitosis provides for the orderly distribution of chromosomes (and of centrioles, if present), but what about the various cytoplasmic organelles? For example, all eukaryotic cells, including plant cells, require mitochondria. Likewise, photosynthetic plant cells cannot carry out photosynthesis without chloroplasts. These organelles contain their own DNA and appear to form by the division of previously existing mitochondria or plastids or their precursors. This nonmitotic division process is similar to prokaryotic cell division (discussed in the next section) and generally occurs during interphase, not when the cell divides. Because

many copies of each organelle are present in each cell, organelles are apportioned with the cytoplasm that each daughter cell receives during cytokinesis.

Lacking nuclei, prokaryotes divide by binary fission

Prokaryotic cells contain much less DNA than do most eukaryotic cells, but precise distribution of the genetic material into two daughter cells is still a formidable process. Prokaryotic DNA usually consists of a single, circular chromosome that is packaged with associated proteins. Although the distribution of genetic material in dividing prokaryotic cells is a simpler process than mitosis, it nevertheless is very precise, to ensure that the daughter cells are genetically identical to the parent cell.

Prokaryotes reproduce asexually, generally by **binary fission,** a process in which one cell divides into two offspring cells (❚ Fig. 10-11). First, the circular DNA replicates, resulting in two identical chromosomes that move to opposite ends of the elongating cell. Then the plasma membrane grows inward between the two DNA copies, and a new transverse cell wall forms between the two cells. (Bacterial reproduction is described further in Chapter 24.)

Review

■ What are the stages of the cell cycle? During which stage does DNA replicate?

■ What are sister chromatids?

■ Assume an animal has a chromosome number of 10. (1) How many chromosomes would it have in a typical body cell, such as a skin cell, during G_1? (2) How many chromosomes would be present in each daughter cell produced by mitosis? Assuming the daughter cells are in G_1, are these duplicated chromosomes?

REGULATION OF THE CELL CYCLE

Learning Objective

6 Explain some ways in which the cell cycle is controlled.

When conditions are optimal, some prokaryotic cells can divide every 20 minutes. The generation times of eukaryotic cells are generally much longer, although the frequency of cell division varies widely among different species and among different tissues of the same species. Some skeletal muscle cells usually stop dividing after the first few months of life, whereas blood-forming cells, digestive tract cells, and skin cells divide frequently throughout

Different cyclins associate with Cdks (cyclin-dependent kinases), triggering the onset of the different stages of the cell cycle.

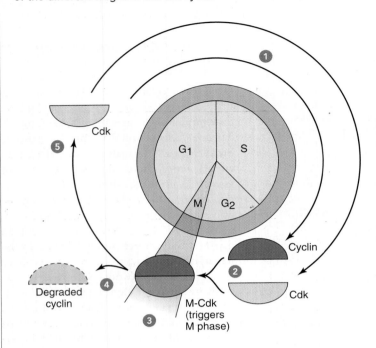

1. Cyclin is synthesized and accumulates.

2. Cdk associates with cyclin, forming a cyclin–Cdk complex, M-Cdk.

3. M-Cdk phosphorylates proteins, activating those that facilitate mitosis and inactivating those that inhibit mitosis.

4. An activated enzyme complex recognizes a specific amino acid sequence in cyclin and targets it for destruction. When cyclin is degraded, M-Cdk activity is terminated, and the cells formed by mitosis enter G_1.

5. Cdk is not degraded but is recycled and reused.

Figure 10-12　Molecular control of the cell cycle

This diagram is a simplified view of the control system that triggers the cell to move from G_2 to M phase.

the life of the organism. Under optimal conditions of nutrition, temperature, and pH, the length of the eukaryotic cell cycle is constant for any given cell type. Under less favorable conditions, however, the generation time may be longer.

Certain regulatory molecules that control the cell cycle are common to all eukaryotes. Genetically programmed in the cell's nucleus, these regulatory molecules are components of the *cell-cycle control system* found in organisms as diverse as yeast (a unicellular fungus), clams, frogs, humans, and plants. Molecular regulators trigger a specific sequence of events during the cell cycle. Because a failure to carefully control cell division can have disastrous consequences, signals in the genetic program, called **cell-cycle checkpoints,** ensure that all the events of a particular stage have been completed before the next stage begins. For example, if a cell produces damaged or unreplicated DNA, the cell cycle halts and the cell will not undergo mitosis.

❚ Figure 10-12 shows the key molecules involved in regulating the cell cycle. Among them are **protein kinases,** enzymes that activate or inactivate other proteins by *phosphorylating* (adding phosphate groups to) them. The protein kinases involved in controlling the cell cycle are **cyclin-dependent kinases (Cdks).** The activity of various Cdks increases and then decreases as the cell moves through the cell cycle. Cdks are active only when they bind tightly to regulatory proteins called **cyclins.** The cyclins are so

named because their levels fluctuate predictably during the cell cycle (that is, they "cycle," or are alternately synthesized and degraded as part of the cell cycle).

Three scientists who began their research in the 1970s and 1980s on the roles of protein kinases and cyclins in the cell cycle (Leland Hartwell from the United States, and Paul Nurse and Tim Hunt from Great Britain) were awarded the Nobel Prize in Physiology or Medicine in 2001. Their discoveries were cited as important not only in working out the details of the fundamental cell process of mitosis but also in understanding why cancer cells divide when they should not. For example, cyclin levels are often higher than normal in human cancer cells.

When a specific Cdk associates with a specific cyclin, it forms a **cyclin–Cdk complex.** Cyclin–Cdk complexes phosphorylate enzymes and other proteins. Some of these proteins become activated when they are phosphorylated, and others become inactivated. For example, phosphorylation of the protein **p27,** known to be a major inhibitor of cell division, is thought to initiate degradation of the protein. As various enzymes are activated or inactivated by phosphorylation, the activities of the cell change. Thus, a decrease in a cell's level of p27 causes a nondividing cell to resume division.

Eukaryotic cells form four major cyclin–Cdk complexes — G_1-Cdk, G_1/S-Cdk, S-Cdk, and M-Cdk. Each cyclin–Cdk complex phosphorylates a different group of proteins. G_1-Cdk prepares the cell to pass from the G_1 phase to the S phase, and then G_1/S-Cdk commits the cell to undergo DNA replication. S-Cdk initiates DNA replication. M-Cdk promotes the events of mitosis, including chromosome condensation, nuclear envelope breakdown, and mitotic spindle formation.

M-Cdk also activates another enzyme complex, the **anaphase-promoting complex (APC),** toward the end of metaphase. APC initiates anaphase by allowing degradation of the cohesins and other proteins that hold the sister chromatids together during metaphase. As a result, the sister chromatids separate as two daughter chromosomes. At this point, cyclin is degraded to negligible levels and M-Cdk activity drops, allowing the mitotic spindle to disassemble and the cell to exit mitosis.

Certain drugs can stop the cell cycle at a specific checkpoint. Some of these prevent DNA synthesis, whereas others inhibit the synthesis of proteins that control the cycle or inhibit the synthesis of structural proteins that contribute to the mitotic spindle. Because one of the distinguishing features of most cancer cells is their high rate of cell division relative to that of most normal body cells, cancer cells can be most affected by these drugs. Many side effects of certain anticancer drugs (such as nausea and hair loss) are due to the drugs' effects on normal cells that divide rapidly in the digestive system and hair follicles.

In plant cells, certain hormones stimulate mitosis. These include the **cytokinins,** a group of plant hormones that promote mitosis both in normal growth and in wound healing (see Chapter 37). Similarly, animal hormones, such as certain steroids, stimulate growth (see Chapter 48).

Protein **growth factors,** which are active at extremely low concentrations, stimulate mitosis in some animal cells. Of the approximately 50 protein growth factors known, some act only on specific types of cells, whereas others act on a broad range of cell types. For example, the effects of the growth factor *erythropoietin* are limited to cells that will develop into red blood cells, but *epidermal growth factor* stimulates many cell types to divide. Many types of cancer cells divide in the absence of growth factors.

Review

- What are cell-cycle checkpoints?
- What are two molecular controls that trigger the onset of different stages of the cell cycle?

SEXUAL REPRODUCTION AND MEIOSIS

human karyotype.
- *duplicated*
- *homologous: size, shape, genetic*
- *autosomes (22): not sex*
- *sex chrom (1)*

Learning Objectives

7 Differentiate between asexual and sexual reproduction.

8 Distinguish between haploid and diploid cells, and define *homologous chromosomes*.

9 Explain the significance of meiosis, and describe the process.

10 Contrast mitosis and meiosis, emphasizing the different outcomes.

11 Compare the roles of mitosis and meiosis in various generalized life cycles.

Although the details of the reproductive process vary greatly among different kinds of eukaryotes, biologists distinguish two basic types of reproduction: asexual and sexual. In **asexual reproduction** a single parent splits, buds, or fragments to produce two or more individuals. In most kinds of eukaryotic asexual reproduction, all the cells are the result of mitotic divisions, so their genes and inherited traits are like those of the parent. Such a group of genetically identical organisms is termed a **clone.** In asexual reproduction, organisms that are well adapted to their environment produce new generations of similarly adapted organisms. Asexual reproduction occurs rapidly and efficiently, partly because the organism does not need to expend time and energy finding a mate.

In contrast, **sexual reproduction** involves the union of two sex cells, or **gametes,** to form a single cell called a **zygote.** Usually two different parents contribute the gametes, but in some cases a single parent furnishes both gametes. In the case of animals and plants, the egg and sperm cells are the gametes, and the fertilized egg is the zygote.

Sexual reproduction results in genetic variation among the offspring. (*How* this genetic variation arises is discussed later in this chapter and in Chapter 11.) Because the offspring produced by sexual reproduction are not genetically identical to their parents or to each other, some offspring may be able to survive environmental changes better than either parent does. However, one disadvantage of sexual reproduction is that some offspring with a different combination of traits may be less likely to survive than their parents.

There is a potential problem in eukaryotic sexual reproduction: If each gamete had the same number of chromosomes as the parent cell that produced it, then the zygote would have twice as many chromosomes. This doubling would occur generation after generation. How do organisms avoid producing zygotes with ever-increasing chromosome numbers? To answer this question, we need more information about the types of chromosomes found in cells.

Each chromosome in a *somatic cell* (body cell) of a plant or animal normally has a partner chromosome. The two partners, known as **homologous chromosomes,** are similar in size, shape, and the position of their centromeres. Furthermore, special chromosome-staining procedures make a characteristic pattern of bands evident in the members of each chromosome pair. In most species, chromosomes vary enough in their structure that cell biologists can distinguish the different chromosomes and match up the homologous pairs. The 46 chromosomes in human cells constitute 23 homologous pairs.

The most important feature of homologous chromosomes is that they carry information about the same genetic traits, although this information is not necessarily identical. For example, each member of a homologous pair may carry a gene that specifies hemoglobin structure. But one member may have the information for the normal hemoglobin β chain (see Fig. 3-22a), whereas the other may specify the abnormal form of hemoglobin associated with sickle cell anemia (see Chapter 16). Homologous chromosomes can therefore be contrasted with the two members of a pair of sister chromatids, which are precisely identical to each other.

A *set* of chromosomes has one of each kind of chromosome; in other words, it contains one member of each homologous pair. If a cell or nucleus contains two sets of chromosomes, it is said to have a **diploid** chromosome number. If it has only a single set of chromosomes, it has the **haploid** number. In humans, the diploid chromosome number is 46 and the haploid number is 23. When a sperm and egg fuse at fertilization, each gamete is haploid, contributing one set of chromosomes; the diploid number is thereby restored in the fertilized egg (zygote). When the zygote divides by mitosis to form the first two cells of the embryo, each daughter cell receives the diploid number of chromosomes, and subsequent mitotic divisions repeat this. Thus, most human body cells are diploid.

If an individual's cells have three or more sets of chromosomes, we say that it is **polyploid.** Polyploidy is relatively rare among animals but quite common among plants (see Chapter 20). In fact, polyploidy has been important in plant evolution. As many as 80% of all flowering plants are polyploid. Polyploid plants are often larger and hardier than diploid members of the same group. Many commercially important plants, such as wheat and cotton, are polyploid.

The chromosome number found in the gametes of a particular species is represented as n, and the zygotic chromosome number is represented as $2n$. If the organism is not polyploid, the haploid chromosome number is equal to n, and the diploid number is equal to $2n$; thus, in humans, $n = 23$ and $2n = 46$. For simplicity, in the rest of this chapter the organisms used as examples are not polyploid. We use diploid and $2n$ interchangeably, and haploid and n interchangeably, although the terms are not strictly synonymous.

Meiosis produces haploid cells with unique gene combinations

We have examined the process of mitosis, which ensures that each daughter cell receives exactly the same number and kinds of chromosomes as the parent cell. A diploid cell that undergoes mitosis produces two diploid cells. Similarly, a haploid cell that undergoes mitosis produces two haploid cells. (Some eukaryotic organisms—certain yeasts, for example—are haploid, as are plants at certain stages of their life cycles.)

A cell division that reduces chromosome number is called **meiosis.** The term means "to make smaller," and the chromosome number is reduced by one half. In meiosis a diploid cell undergoes two cell divisions, potentially yielding four haploid cells. It is important to note that haploid cells do not contain just any combination of chromosomes, but one member of each homologous pair.

The events of meiosis are similar to the events of mitosis, with four important differences:

1. Meiosis involves two successive nuclear and cytoplasmic divisions, producing up to four cells.

2. Despite two successive nuclear divisions, the DNA and other chromosome components duplicate only once—during the interphase preceding the first meiotic division.

3. Each of the four cells produced by meiosis contains the haploid chromosome number, that is, only one chromosome set containing only one representative of each homologous pair.

4. During meiosis, each homologous chromosome pair is shuffled, so the resulting haploid cells each have a virtually unique combination of genes.

Meiosis typically consists of two nuclear and cytoplasmic divisions, designated the *first* and *second meiotic divisions,* or simply **meiosis I** and **meiosis II.** Each includes prophase, metaphase, anaphase, and telophase stages. During meiosis I, the members of each homologous chromosome pair first join and then separate and move into different nuclei. In meiosis II, the sister chromatids that make up each duplicated chromosome separate from each other and are distributed to two different nuclei. The following discussion describes meiosis in an organism with a diploid chromosome number of 4. Refer to ▌ Figure 10-13 as you read.

Prophase I includes synapsis and crossing-over

As occurs during mitosis, the chromosomes duplicate in the S phase of interphase, before the complex movements of meiosis actually begin. Each duplicated chromosome consists of two chromatids, which are linked by cohesins. During **prophase I,** while the chromatids are still elongated and thin, the homologous chromosomes come to lie lengthwise side by side. This process is called **synapsis,** which means "fastening together." In our example, because the diploid number is 4, synapsis results in two homologous pairs.

One member of each homologous pair is called the **maternal homologue,** because it was originally inherited from the female parent during the formation of the zygote; the other member of a homologous pair is the **paternal homologue,** because it was inherited from the male parent. Because each chromosome duplicated during interphase and now consists of two chromatids, synapsis results in the association of four chromatids. The resulting association is a **tetrad.** The number of tetrads per prophase I cell is equal to the haploid chromosome number. In our example of an animal cell with a diploid number of 4, there are 2 tetrads (and a total of 8 chromatids); in a human cell at prophase I, there are 23 tetrads (and a total of 92 chromatids).

Homologous chromosomes become closely associated during synapsis. Electron microscopic observations reveal that a characteristic structure, the **synaptonemal complex,** forms along the entire length of the synapsed homologues (▌ Fig. 10-14 on page 228). This proteinaceous structure holds the synapsed homologues together and is thought to play a role in chromo-

[handwritten top margin: genetic and physical]

some **crossing-over,** a process in which enzymes break and re-join DNA molecules, allowing paired homologous chromosomes to exchange genetic material (DNA). Crossing-over produces new combinations of genes. The **genetic recombination** from crossing-over greatly enhances the genetic variation—that is, new combinations of traits—among sexually produced offspring. Some biologists think that recombination is the main reason for sexual reproduction in eukaryotes.

[handwritten left margin, vertical: Similarity to mitosis]

In addition to the unique processes of synapsis and crossing-over, events similar to those in mitotic prophase also occur during prophase I. A spindle forms consisting of microtubules and other components. In animal cells, one pair of centrioles moves to each pole, and astral microtubules form. The nuclear envelope disappears in late prophase I, and in cells with large and distinct chromosomes, the structure of the tetrads can be seen clearly with the microscope.

The sister chromatids remain closely aligned along their lengths. However, the centromeres (and kinetochores) of the homologous chromosomes become separated from one another. In late prophase I, the homologous chromosomes are held together only at specific regions, termed **chiasmata** (sing., *chiasma*). Each chiasma originates at a crossing-over site, that is, a site at which homologous chromatids exchanged genetic material and rejoined, producing an X-shaped configuration (█ Fig. 10-15 on page 228). At the chiasmata, cohesins hold homologous chromosomes together after the synaptonemal complex has been disassembled. Later, the cohesins dissociate from the chiasmata, freeing the homologous chromosome arms from one another. The consequences of crossing-over and genetic recombination are discussed in Chapter 11 (for example, see Fig. 11-12).

During meiosis I, homologous chromosomes separate

Metaphase I occurs when the tetrads align on the midplane. Both sister kinetochores of one duplicated chromosome are attached by spindle fibers to the same pole, and both sister kinetochores of the duplicated homologous chromosome are attached to the opposite pole. (By contrast, sister kinetochores are attached to opposite poles in mitosis.)

During **anaphase I,** the paired homologous chromosomes separate, or disjoin, and move toward opposite poles. Each pole receives a random combination of maternal and paternal chromosomes, but only one member of each homologous pair is present at each pole. The sister chromatids remain united at their centromere regions. Again, this differs from mitotic anaphase, in which the sister chromatids separate and move to opposite poles.

During **telophase I,** the chromatids generally decondense somewhat, the nuclear envelope may reorganize, and cytokinesis may take place. Each telophase I nucleus contains the haploid number of chromosomes, but each chromosome is a duplicated chromosome (it consists of a pair of chromatids). In our example, 2 duplicated chromosomes lie at each pole, for a total of 4

chromatids; humans have 23 duplicated chromosomes (46 chromatids) at each pole.

An interphase-like stage usually follows. Because it is not a true interphase—there is no S phase and therefore no DNA replication—it is called **interkinesis.** Interkinesis is very brief in most organisms and absent in some.

Chromatids separate in meiosis II

Because the chromosomes usually remain partially condensed between divisions, the prophase of the second meiotic division is brief. **Prophase II** is similar to mitotic prophase in many respects. There is no pairing of homologous chromosomes (indeed, only one member of each pair is present in each nucleus) and no crossing-over.

During **metaphase II** the chromosomes line up on the midplanes of their cells. You can easily distinguish the first and second metaphases in diagrams; at metaphase I the chromatids are arranged in bundles of four (tetrads), and at metaphase II they are in groups of two (as in mitotic metaphase). This is not always so obvious in living cells.

During **anaphase II** the chromatids, attached to spindle fibers at their kinetochores, separate and move to opposite poles, just as they would at mitotic anaphase. As in mitosis, each former chromatid is now referred to as a *chromosome.* Thus, at **telophase II** there is one representative for each homologous pair at each pole. Each is an unduplicated (single) chromosome. Nuclear envelopes then re-form, the chromosomes gradually elongate to form chromatin fibers, and cytokinesis occurs.

The two successive divisions of meiosis yield four haploid nuclei, each containing *one* of each kind of chromosome. Each resulting haploid cell has a different combination of genes. This genetic variation has two sources: (1) DNA segments are exchanged between maternal and paternal homologues during crossing-over. (2) During meiosis, the maternal and paternal chromosomes of homologous pairs separate independently. The chromosomes are "shuffled" so that each member of a pair becomes randomly distributed to one of the poles at anaphase I.

Mitosis and meiosis lead to contrasting outcomes

Although mitosis and meiosis share many similar features, specific distinctions between these processes result in the formation of different types of cells (█ Fig. 10-16 on page 229). Mitosis is a single nuclear division in which sister chromatids separate from each other. If cytokinesis occurs, they are distributed to the two daughter cells, which are genetically identical to each other and to the original cell. Homologous chromosomes do not associate physically at any time in mitosis.

In meiosis, a diploid cell undergoes two successive nuclear divisions, meiosis I and meiosis II. In prophase I of meiosis, the homologous chromosomes undergo synapsis to form tetrads.

[handwritten bottom margin, left: ✳ Interkinesis · not interphase / NO S phase · chromosomes partially uncoil]

[handwritten bottom margin, right: → Not Needed; already have duplicated chromosomes.]

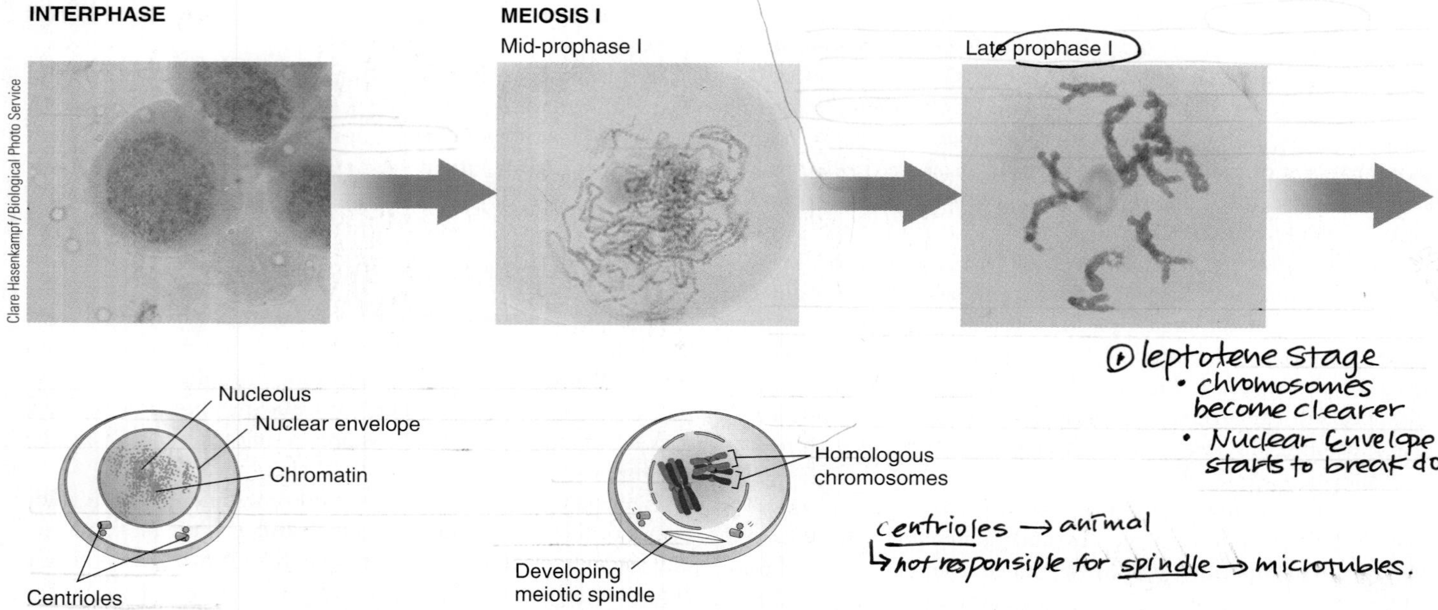
INTERPHASE
MEIOSIS I
Mid-prophase I
Late prophase I
Clare Hasenkampf/Biological Photo Service
Nucleolus
Nuclear envelope
Chromatin
Centrioles
Homologous chromosomes
Developing meiotic spindle

Interphase preceding meiosis; DNA replicates.

Homologous chromosomes synapse, forming tetrads; nuclear envelope breaks down.

(1) leptotene stage
• chromosomes become clearer
• Nuclear Envelope starts to break down

centrioles → animal
↳ not responsiple for spindle → microtubles.

flagellates ↓ centrioles.

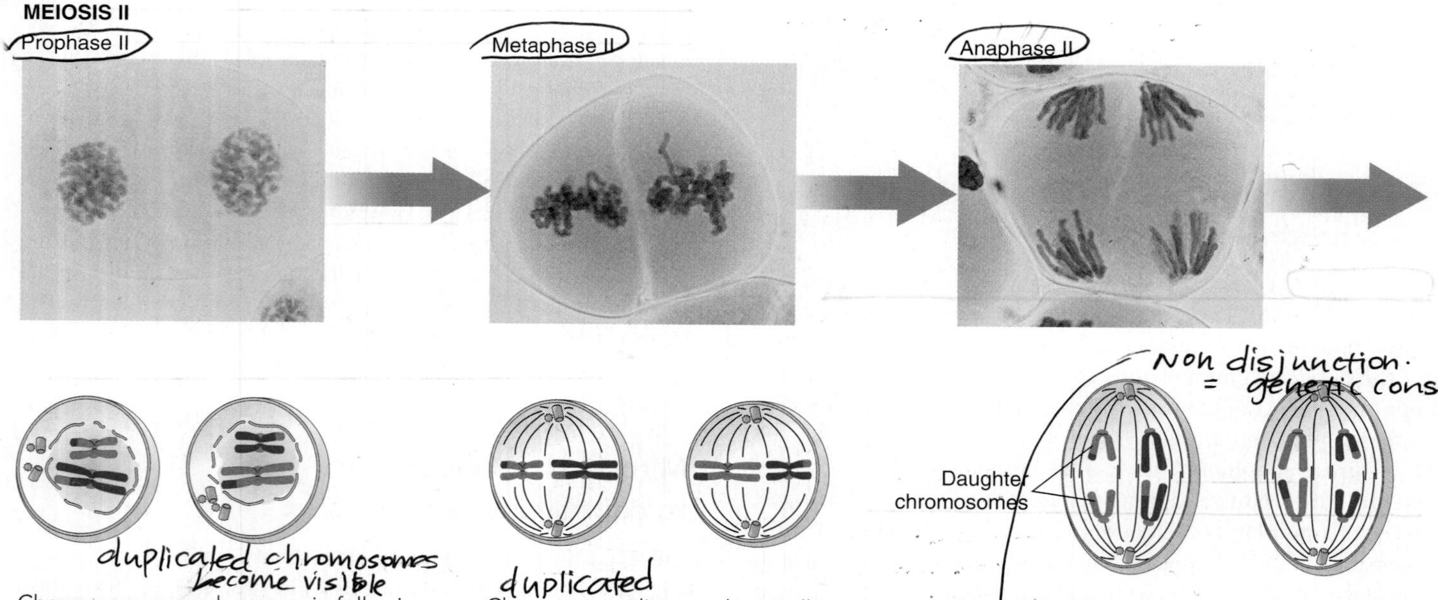
MEIOSIS II
Prophase II
Metaphase II
Anaphase II
Daughter chromosomes

Non disjunction. = genetic consequen

duplicated chromosomes become visible

Chromosomes condense again following brief period of interkinesis. DNA does *not* replicate again.

No more tetrads.
nucleus: 23 chromosomes.

duplicated

Chromosomes line up along cell's midplane.

• homologous kinetocores in different cells.

Sister chromatids separate, and chromosomes move to opposite poles.

↳• genetic consequences.

Figure 10-13 ***Animated*** Interphase and the stages of meiosis

Meiosis consists of two nuclear divisions, meiosis I (*top row*) and meiosis II (*bottom row*). The LMs show sectioned plant cells, which lack centrioles. The drawings depict generalized animal cells with a diploid chromosome number of 4; the sizes of the nuclei and chromosomes are exaggerated to show the structures more clearly.

Normal disjunction vs.

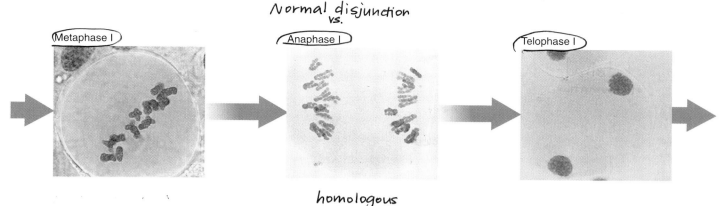

(Metaphase I)   (Anaphase I)   (Telophase I)

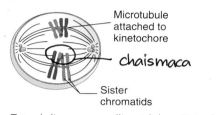

homologous

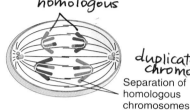

Microtubule attached to kinetochore

chaismaca

Sister chromatids

Tetrads line up on cell's midplane. Tetrads held together at chiasmata (sites of prior crossing-over).

- human: 23
- attach to same pole (sister)
  opposite pole (homologous)

duplicated chromosomes.

Separation of homologous chromosomes

Homologous chromosomes separate and move to opposite poles. Note that sister chromatids remain attached at their centromeres.

Cleavage furrow

duplicated chromosomes reach the poles.

One of each pair of homologous chromosomes is at each pole. Cytokinesis occurs.

chromosome # decreased.
phrophase : 4   nuclei : 2

---

unduplicated chromosomes reach the poles.

(Telophase II)   Four haploid cells

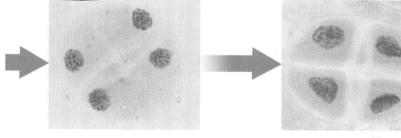

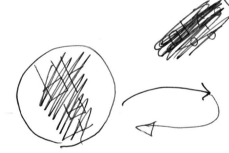

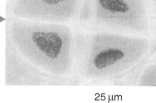

25 µm

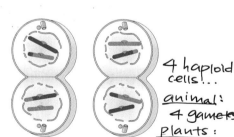

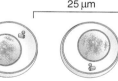

4 haploid cells...
animal:
4 gametes
plants:
4 spores

Nuclei form at opposite poles of each cell. Cytokinesis occurs.

Four gametes (animal) or four spores (plant) are produced.

Ready for meiosis

Premeetic Interphase

- S phase
  - chromosomes duplicated (just as before mitosis)

④ Diplotene stage
  - tetrads : looped out appearance
  - chiasmate (chiasma) sites where crossing over previously occured (during the parencyta stage)

Figure 10-13   Continued   No #'s

② Zygotene Cycle
  - Homo chromosomes begin to pair witheachother (syapsis) →
  - (Synaptonemal complexes) form in between
  - chromosome pair → (tetrad)

③ Pachytene Stage →
  - chromosomes are synapsed throughout length (synaptonemal complex is complete)
  - ✸ if human : 23 tetrads
    - synapsed pair : tetrad
  - Kinetic cores : 92

crossing over (can't visualize)
- biochemical test.
- chromoses are completey synapsed
- Special DNA enzymes break the DNA of homologous (non-sister) chromatids

- join cut ends /
chromatids exchange parts

tetrad:
of chromatids

Paternal sister chromatids

Maternal sister chromatids

sister chromaticles

Chromatin

Protein

Synaptonemal complex : protein ☆
little DNA.

Chromatin

Maternal sister chromatids

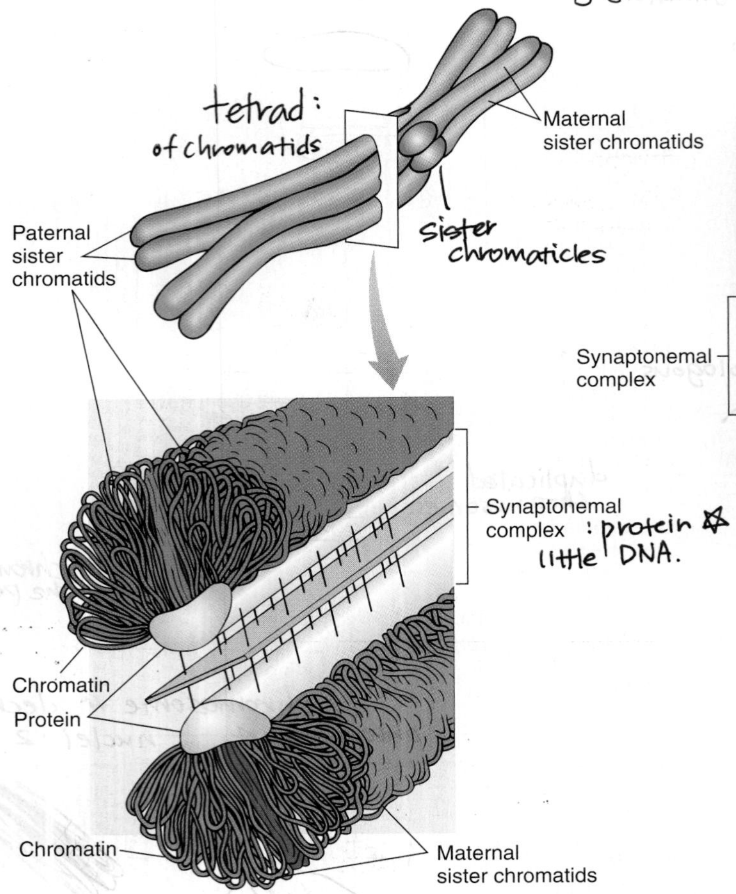

(a) A 3-D model of a tetrad with a complete synaptonemal complex.

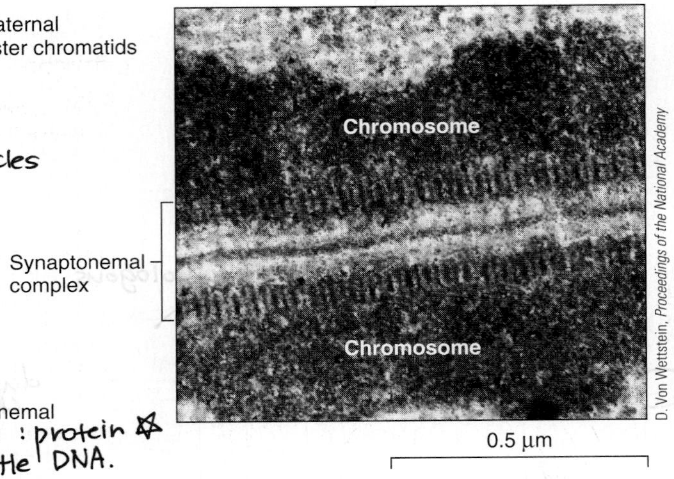

Homologous chromosomes separate during meiosis I, and sister chromatids separate during meiosis II. Meiosis ends with the formation of four genetically different, haploid daughter cells. The fates of these cells depend on the type of life cycle; in animals they differentiate as gametes, whereas in plants they become spores.

(b) TEM of a synaptonemal complex.
Not seen with light microscope.

Chromosome

Chromosome

0.5 μm

D. Von Wettstein, *Proceedings of the National Academy of Science* 68:851–855, 1971

Figure 10-14 A synaptonemal complex

Synapsing homologous chromosomes in meiotic prophase I are held together by a synaptonemal complex, composed mainly of protein.

The timing of meiosis in the life cycle varies among species

Because sexual reproduction is characterized by the fusion of two haploid sex cells to form a diploid zygote, it follows that in a sexual life cycle, meiosis must occur before gametes can form.

In animals and a few other organisms, meiosis leads directly to gamete production (∎ Fig. 10-17a). An organism's somatic (body) cells multiply by mitosis and are diploid; the only

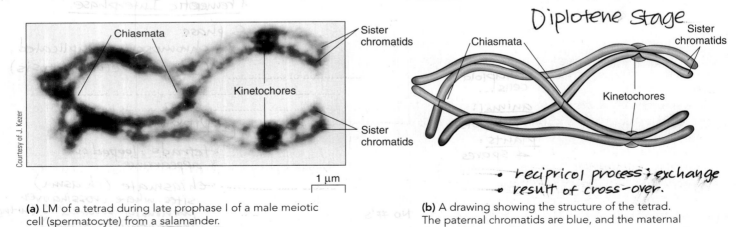

Chiasmata

Sister chromatids

Kinetochores

Sister chromatids

Courtesy of J. Kezer

1 μm

Diplotene Stage

Chiasmata

Sister chromatids

Kinetochores

• reciprical process : exchange
• result of cross-over.

(a) LM of a tetrad during late prophase I of a male meiotic cell (spermatocyte) from a salamander.

light microscope.

(b) A drawing showing the structure of the tetrad. The paternal chromatids are blue, and the maternal chromatids are red.

Figure 10-15 A meiotic tetrad with two chiasmata
The two chiasmata are the result of separate crossing-over events.

MITOSIS

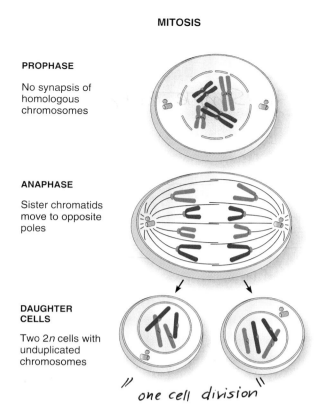

PROPHASE

No synapsis of homologous chromosomes

ANAPHASE

Sister chromatids move to opposite poles

DAUGHTER CELLS

Two *2n* cells with unduplicated chromosomes

"one cell division"

MEIOSIS

required in sex cycle

starts with a cell that can be diploid / haploid.

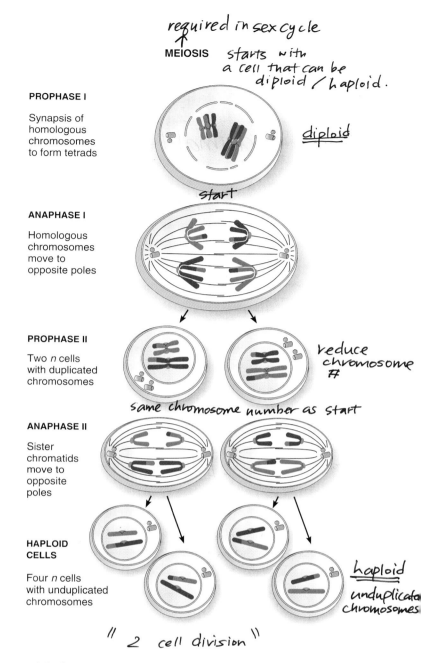

PROPHASE I

Synapsis of homologous chromosomes to form tetrads

diploid

start

ANAPHASE I

Homologous chromosomes move to opposite poles

PROPHASE II

Two *n* cells with duplicated chromosomes

reduce chromosome #

same chromosome number as start

ANAPHASE II

Sister chromatids move to opposite poles

HAPLOID CELLS

Four *n* cells with unduplicated chromosomes

haploid unduplicate chromosomes

"2 cell division"

Figure 10-16 *Animated* Mitosis and meiosis

This drawing compares the events and outcomes of mitosis and meiosis, in each case beginning with a diploid cell with four chromosomes (two pairs of homologous chromosomes). Because the chromosomes duplicated in the previous interphase, each chromosome consists of two sister chromatids. The chromosomes derived from one parent are shown in blue, and those from the other parent are red. Homologous pairs are similar in size and shape. Chiasmata are not shown, and some of the stages have been omitted for simplicity.

haploid cells produced are the gametes. Gametes develop when **germ line cells,** which give rise to the next generation, undergo meiosis.

The formation of gametes is known as **gametogenesis.** Male gametogenesis, termed **spermatogenesis,** forms four haploid sperm cells for each cell that enters meiosis. (See Chapter 49 and Fig. 49-5 for a detailed description of spermatogenesis.)

In contrast, female gametogenesis, termed **oogenesis,** forms a single egg cell, or *ovum,* for every cell that enters meiosis. In this process, most of the cytoplasm goes to only one of the two cells produced during each meiotic division. At the end of the first meiotic division, one nucleus is retained and the other, called the first *polar body,* often degenerates. Similarly, at the end of the second division, one nucleus becomes another polar body and the other nucleus survives. In this way, one haploid nucleus receives most of the accumulated cytoplasm and nutrients from the

original meiotic cell. (See Chapter 49 and Fig. 49-11 for a detailed description of oogenesis.)

Although meiosis occurs at some point in a sexual life cycle, it does not always *immediately* precede gamete formation. Many simple eukaryotes, including some fungi and algae, remain haploid (their cells dividing mitotically) throughout most of their life cycles, with individuals being unicellular or multicellular. Two haploid gametes (produced by mitosis) fuse to form a diploid zygote that undergoes meiosis to restore the haploid state (▌Fig. 10-17b). Examples of these types of life cycles are found in Figures 25-17 and 26-9.

Plants and some algae and fungi have some of the most complicated life cycles (▌Fig. 10-17c). These life cycles, characterized by an **alternation of generations,** consist of a multicellular diploid stage, the **sporophyte generation,** and a multicellular haploid stage, the **gametophyte generation.** Diploid sporophyte

Each species has a characteristic number of chromosomes that does not change.
In each life cycle, the doubling of chromosomes that occurs during fertilization is
compensated for by the reduction in chromosome number that occurs during meiosis.

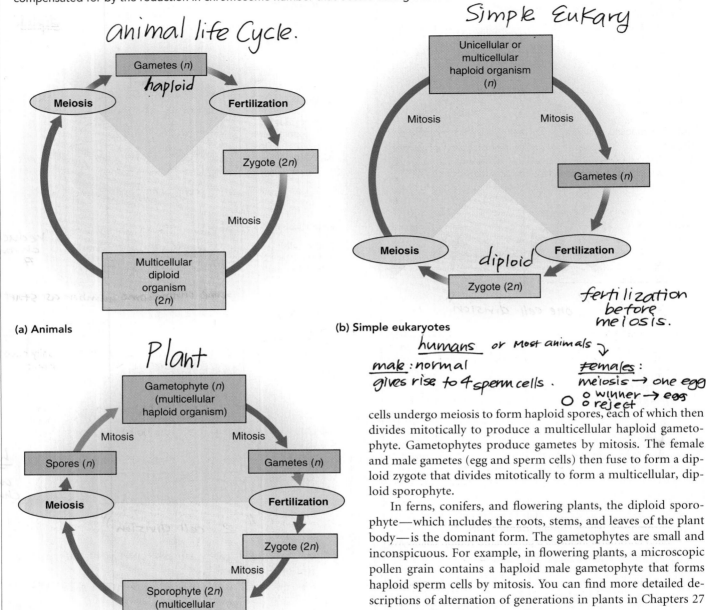

animal life Cycle. *haploid*

(a) Animals

Simple Eukary *diploid*

fertilization before meiosis.

(b) Simple eukaryotes

Plant *Spores = meiosis*

(c) Plants, some algae, and some fungi

Figure 10-17   Representative life cycles

The color code and design here is used throughout the rest of the
book. For example, in all life cycles the haploid (*n*) generation is
shown in purple, and the diploid (2*n*) generation is gold. The processes of meiosis and fertilization always link the haploid and diploid
generations.

humans   or Most animals
male: normal
gives rise to 4 sperm cells.
females:
meiosis → one egg cell.
o winner → egg
o reject

cells undergo meiosis to form haploid spores, each of which then
divides mitotically to produce a multicellular haploid gametophyte. Gametophytes produce gametes by mitosis. The female
and male gametes (egg and sperm cells) then fuse to form a diploid zygote that divides mitotically to form a multicellular, diploid sporophyte.

In ferns, conifers, and flowering plants, the diploid sporophyte—which includes the roots, stems, and leaves of the plant
body—is the dominant form. The gametophytes are small and
inconspicuous. For example, in flowering plants, a microscopic
pollen grain contains a haploid male gametophyte that forms
haploid sperm cells by mitosis. You can find more detailed descriptions of alternation of generations in plants in Chapters 27
and 28.

Review

- Are homologous chromosomes present in a diploid cell? Are
  they present in a haploid cell?
- Assume an animal cell has a diploid chromosome number of
  10. (1) How many tetrads would form in prophase I of meiosis? (2) How many chromosomes would be present in each
  gamete? Are these duplicated chromosomes?
- How does the outcome of meiosis differ from the outcome of
  mitosis?
- Can haploid cells divide by mitosis? By meiosis?

Learning Objectives

1 Discuss the significance of chromosomes in terms of their information content (page 212).

■ **Genes,** the cell's informational units, are made of DNA. In eukaryotes, DNA associates with protein to form the **chromatin** fibers that make up **chromosomes.** The organization of eukaryotic DNA into chromosomes allows the DNA, which is much longer than a cell's nucleus, to be accurately replicated and sorted into daughter cells without tangling.

2 Compare the organization of DNA in prokaryotic and eukaryotic cells (page 212).

■ Prokaryotic cells usually have circular DNA molecules.

■ Eukaryotic chromosomes have several levels of organization. The DNA is associated with **histones** (basic proteins) to form **nucleosomes,** each of which consists of a histone bead with DNA wrapped around it. The nucleosomes are organized into large, coiled loops held together by nonhistone **scaffolding proteins.**

3 Identify the stages in the eukaryotic cell cycle, and describe their principal events (page 215).

■ The eukaryotic **cell cycle** is the period from the beginning of one division to the beginning of the next. The cell cycle consists of interphase and M phase.

■ **Interphase** consists of the first gap phase (G_1), the synthesis phase (S), and the second gap phase (G_2). During the **G_1 phase,** the cell grows and prepares for the S phase. During the **S phase,** DNA and the chromosome proteins are synthesized, and chromosome duplication occurs. During the **G_2 phase,** protein synthesis increases in preparation for cell division.

■ **M phase** consists of **mitosis,** the nuclear division that produces two nuclei identical to the parental nucleus, and **cytokinesis,** the division of the cytoplasm to yield two daughter cells.

ThomsonNOW **Learn more about the cell cycle by clicking on the figure in ThomsonNOW.**

4 Describe the structure of a duplicated chromosome, including the sister chromatids, centromeres, and kinetochores (page 215).

■ A duplicated chromosome consists of a pair of **sister chromatids,** which contain identical DNA sequences. Each chromatid includes a constricted region called the **centromere.** Sister chromatids are tightly associated in the region of their centromeres.

■ Attached to each centromere is a **kinetochore,** a protein structure to which microtubules can bind.

5 Explain the significance of mitosis, and describe the process (page 215).

■ Mitosis assures that the chromosome number is preserved when one eukaryotic cell divides to form two. In mitosis, identical chromosomes are distributed to each pole of the cell, and a nuclear envelope forms around each set.

■ During **prophase,** chromatin condenses into duplicated chromosomes, each composed of a pair of sister chromatids. The nuclear envelope begins to disappear, and the **mitotic spindle** begins to form.

■ During **prometaphase,** spindle microtubules attach to kinetochores of chromosomes, and chromosomes begin to move toward the cell's midplane.

■ During **metaphase,** the chromosomes are aligned on the cell's midplane, or **metaphase plate;** the mitotic spindle is complete, and the kinetochores of the sister chromatids are attached by microtubules to opposite poles of the cell.

■ During **anaphase,** the sister chromatids separate and move to opposite poles. Each former chromatid is now a chromosome.

■ During **telophase,** a nuclear envelope re-forms around each set of chromosomes, nucleoli become apparent, the chromosomes uncoil, and the spindle disappears. Cytokinesis generally begins in telophase.

ThomsonNOW **Walk step-by-step through the stages of mitosis by clicking on the figure in ThomsonNOW.**

ThomsonNOW **Watch cytokinesis unfold by clicking on the figure in ThomsonNOW.**

6 Explain some ways in which the cell cycle is controlled (page 221).

■ **Cyclin-dependent kinases (Cdks)** are **protein kinases** involved in controlling the cell cycle. Cdks are active only when they bind tightly to regulatory proteins called **cyclins.** Cyclin levels fluctuate predictably during the cell cycle.

7 Differentiate between asexual and sexual reproduction (page 223).

■ Offspring produced by **asexual reproduction** usually have hereditary traits identical to those of the single parent. Mitosis is the basis for asexual reproduction in eukaryotic organisms.

■ In **sexual reproduction,** two haploid sex cells, or **gametes,** fuse to form a single diploid **zygote.** In a sexual life cycle, meiosis must occur before gametes can be produced.

8 Distinguish between haploid and diploid cells, and define *homologous chromosomes* (page 223).

■ A **diploid** cell has a characteristic number of chromosome pairs per cell. The members of each pair, called **homologous chromosomes,** are similar in length, shape, and other features and carry genes affecting the same kinds of attributes of the organism.

■ A **haploid** cell contains only one member of each homologous chromosome pair.

9 Explain the significance of meiosis, and describe the process (page 223).

■ A diploid cell undergoing **meiosis** completes two successive cell divisions, yielding four haploid cells. Sexual life cycles in eukaryotes require meiosis, which makes it possible for each gamete to contain only half the number of chromosomes in the parent cell.

■ **Meiosis I** begins with **prophase I,** in which the members of a homologous pair of chromosomes physically join by the process of **synapsis. Crossing-over** is a process of **genetic recombination** during which homologous (nonsister) chromatids exchange segments of DNA strands.

■ At **metaphase I,** tetrads—each consisting of a pair of homologous chromosomes held together by one or more **chiasmata**—line up on the metaphase plate. The mem-

[handwritten notes:]

✧ Diakinesis Stage

How Accurate are Meiosis? Meiosis → Chromosomal abnormalities.
- Chromosome abnormalities arise very frequently
- 17-20% of recognized pregnancies end in miscarriage
(about 60% of these have chromosomal abnormalities)

bers of each pair of homologous chromosomes separate during meiotic **anaphase I** and are distributed to different nuclei. Each nucleus contains the haploid number of chromosomes; each chromosome consists of two chromatids.

- During **meiosis II,** the two chromatids of each chromosome separate, and one is distributed to each daughter cell. Each former chromatid is now a chromosome.

ThomsonNOW Walk step-by-step through the stages of **meiosis by clicking on the figure in ThomsonNOW.**

10 Contrast mitosis and meiosis, emphasizing the different outcomes (page 223).

- Mitosis involves a single nuclear division in which the two daughter cells formed are genetically identical to each other and to the original cell. Synapsis of homologous chromosomes does not occur during mitosis.
- Meiosis involves two successive nuclear divisions and forms four haploid cells. Synapsis of homologous chromosomes occurs during prophase I of meiosis.

ThomsonNOW Watch a movie that features living cells undergoing mitosis and meiosis in ThomsonNOW.

11 Compare the roles of mitosis and meiosis in various generalized life cycles (page 223).

- The **somatic cells** of animals are diploid and are produced by mitosis. The only haploid cells are the gametes, produced by **gametogenesis,** which in animals occurs by meiosis.
- Simple eukaryotes may be haploid and are produced by mitosis. The only diploid stage is the zygote, which undergoes meiosis to restore the haploid state.
- The life cycle of plants and some algae includes an **alternation of generations.** The multicellular diploid **sporophyte generation** forms haploid spores by meiosis. Each spore divides mitotically to form a multicellular haploid **gametophyte generation,** which produces gametes by mitosis. Two haploid gametes then fuse to form a diploid zygote, which divides mitotically to produce a new diploid sporophyte generation.

TEST YOUR UNDERSTANDING

1. Chromatin fibers include (a) DNA and structural polysaccharides (b) RNA and phospholipids (c) protein and carbohydrate (d) DNA and protein (e) triacylglycerol and steroids

2. A nucleosome consists of (a) DNA and scaffolding proteins (b) scaffolding proteins and histones (c) DNA and histones (d) DNA, histones, and scaffolding proteins (e) histones only

3. The term *S phase* refers to (a) DNA synthesis during interphase (b) synthesis of chromosome proteins during prophase (c) gametogenesis in animal cells (d) synapsis of homologous chromosomes (e) fusion of gametes in sexual reproduction

4. At which of the following stages do human skin cell nuclei have the same DNA content? (a) early mitotic prophase and late mitotic telophase (b) G_1 and G_2 (c) G_1 and early mitotic prophase (d) G_1 and late mitotic telophase (e) G_2 and late mitotic telophase

5. In a cell at _____, each chromosome consists of a pair of attached chromatids. (a) mitotic prophase (b) meiotic prophase II (c) meiotic prophase I (d) meiotic anaphase I (e) all of the preceding

6. The molecular tether that links sister chromatids of a duplicated chromosome to each other is (a) condensin (b) actin (c) myosin (d) cohesin (e) actomyosin

7. In an animal cell at mitotic metaphase, you would expect to find (a) two pairs of centrioles located on the metaphase plate (b) a pair of centrioles inside the nucleus (c) a pair of centrioles within each microtubule-organizing center (d) a centriole within each centromere (e) no centrioles

8. In the spindle, the _____ ends of kinetochore microtubules are embedded in the kinetochore, and the _____ ends are at the spindle pole. (a) plus; minus (b) minus; plus (c) plus; plus (d) minus; minus (e) none of these is correct

9. Cell plate formation usually begins during (a) telophase in a plant cell (b) telophase in an animal cell (c) G_2 in a plant cell (d) G_2 in an animal cell (e) a and b

10. A particular plant species has a diploid chromosome number of 20. A haploid cell of that species at mitotic prophase contains a total of _____ chromosomes and _____ chromatids. (a) 20; 20 (b) 20; 40 (c) 10; 10 (d) 10; 20 (e) none of the preceding, because haploid cells cannot undergo mitosis

11. A diploid nucleus at early mitotic prophase has _____ set(s) of chromosomes; a diploid nucleus at mitotic telophase has _____ set(s) of chromosomes. (a) 1; 1 (b) 1; 2 (c) 2; 2 (d) 2; 1 (e) not enough information has been given

12. The life cycle of a sexually reproducing organism includes (a) mitosis (b) meiosis (c) fusion of sex cells (d) b and c (e) a, b, and c

13. Which of the following are genetically identical? (a) two cells resulting from meiosis I (b) two cells resulting from meiosis II (c) four cells resulting from meiosis I followed by meiosis II (d) two cells resulting from a mitotic division (e) all of the preceding

14. You would expect to find a synaptonemal complex in a cell at (a) mitotic prophase (b) meiotic prophase I (c) meiotic prophase II (d) meiotic anaphase I (e) meiotic anaphase II

15. A chiasma links a pair of (a) homologous chromosomes at prophase II (b) homologous chromosomes at late prophase I (c) sister chromatids at metaphase II (d) sister chromatids at mitotic metaphase (e) sister chromatids at metaphase I

CRITICAL THINKING

1. How can two species have the same chromosome number yet have very different attributes?

2. How does the DNA content of the cell change from the beginning of interphase to the end of interphase? Does the number of chromatids change? Explain. Does the number of chromosomes change? Explain.

In items 3–5, decide whether each is an example of sexual or asexual reproduction, and state why.

3. A diploid queen honeybee produces haploid eggs by meiosis. Some of these eggs are never fertilized and develop into haploid male honeybees (drones).

4. Seeds develop after a flower has been pollinated with pollen from the same plant.

5. After it has been placed in water, a cutting from a plant develops roots. After it is transplanted to soil, the plant survives and grows.

6. **Evolution Link.** Some organisms—for example, certain fungi—reproduce asexually when the environment is favorable and sexually when the environment becomes unfavorable. What might be the evolutionary advantage of sexual reproduction, with the associated process of meiosis, during unfavorable conditions?

Additional questions are available in ThomsonNOW at www.thomsonedu.com/login

The Basic Principles of Heredity

Bettmann/Corbis

Gregor Mendel. This painting shows Mendel with his pea plants in the monastery garden at Brünn, Austria (now Brüno, Czech Republic).

genetic factors do not blend together in hybrids.

Published 1866.

KEY CONCEPTS

The experiments of Gregor Mendel, a pioneer in the field of genetics, revealed the basic principles of inheritance.

Mendel's principle of segregation explains that members of a gene pair segregate (separate) from one another prior to gamete formation.

Mendel's principle of independent assortment explains that members of different gene pairs assort independently (randomly) into gametes.

Chromosome behavior during meiosis helps explain Mendel's principles of inheritance.

Distinctive inheritance patterns (that is, "exceptions" to Mendel's principles) characterize some traits.

Do you have your father's height and your mother's eye color and freckles? You have inherited these and a multitude of other characteristics, passed on from one generation to another. **Heredity,** the transmission of genetic information from parent to offspring, generally follows predictable patterns in organisms as diverse as humans, penguins, baker's yeast, and sunflowers. **Genetics,** the science of heredity, studies both genetic similarities and **genetic variation,** the differences, between parents and offspring or among individuals of a population.

The study of inheritance as a modern branch of science began in the mid-19th century with the work of Gregor Mendel (1822–1884), a monk who bred pea plants. Mendel was the first scientist to effectively apply quantitative methods to the study of inheritance. He did not merely describe his observations; he planned his experiments carefully, recorded the data, and analyzed the results mathematically. Although unappreciated during his lifetime, his work was rediscovered in 1900.

During the decades following the rediscovery of Mendel's findings, geneticists extended Mendel's principles by correlating the transmission of genetic information from generation to generation with the behavior of chromosomes during *meiosis*. By studying a variety of organisms, geneticists verified Mendel's findings and added to a growing list of so-called exceptions to his principles.

Geneticists study not only the transmission of genes but also the expression of genetic information. As you will see in this chapter and those that follow, understanding the relationships between an organism's genes and its characteristics has become increasingly sophisticated as biologists have learned more about the flow of information in cells. ■

MENDEL'S PRINCIPLES OF INHERITANCE

Learning Objectives

1   Define the terms *phenotype, genotype, locus, allele, dominant allele, recessive allele, homozygous,* and *heterozygous.*
2   Describe Mendel's principles of segregation and independent assortment.
3   Distinguish among monohybrid, dihybrid, and test crosses.
4   Explain Mendel's principles of segregation and independent assortment, given what scientists now know about genes and chromosomes.

Gregor Mendel was not the first plant breeder. At the time he began his work, breeders had long recognized the existence of **hybrid** plants and animals, the offspring of two genetically dissimilar parents. When Mendel began his breeding experiments in 1856, two main concepts about inheritance were widely accepted: (1) All hybrid plants that are the offspring of genetically pure, or **true-breeding,** parents are similar in appearance. (2) When these hybrids mate with each other, they do not breed true; their offspring show a mixture of traits. Some look like their parents, and some have features like those of their grandparents.

Mendel's genius lay in his ability to recognize a pattern in the way the parental traits reappear in the offspring of hybrids. Before Mendel, no one had categorized and counted the offspring and analyzed these regular patterns over several generations to the extent he did. Just as geneticists do today, Mendel chose the organism for his experiments very carefully. The garden pea, *Pisum sativum,* had several advantages. Pea plants are easy to grow, and many varieties were commercially available. Another advantage of pea plants is that controlled pollinations are relatively easy to conduct. Pea flowers have both male and female parts and naturally self-pollinate (❙ Fig. 11-1). However, the anthers (the male parts of the flower that produce pollen) can be removed to prevent self-fertilization. Pollen from a different source can then be applied to the stigma (the receptive surface of the female part). Pea flowers are easily protected from other sources of pollen because the petals completely enclose the reproductive structures.

Mendel obtained his original pea seeds from commercial sources and did some important preliminary work before starting his actual experiments. For two years he verified that the varieties were true-breeding lines for various inherited features. Today, scientists use the term **phenotype** to refer to the physical appearance of an organism. A true-breeding line produces only offspring expressing the same phenotype (for example, round seeds or tall plants) generation after generation. During this time Mendel apparently chose those traits of his pea strains that he could study most easily. He probably made the initial observations that later formed the basis of his hypotheses.

Mendel eventually chose strains representing seven **characters,** the attributes (such as seed color) for which heritable differences, or **traits,** are known (such as yellow seeds and green seeds). The characters Mendel selected had clearly contrasting phenotypes (❙ Fig. 11-2). Mendel's results were easy to analyze because he chose easily distinguishable phenotypes and limited the genetic variation studied in each experiment.

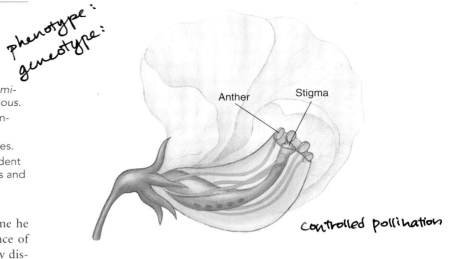

phenotype:
genotype:

controlled pollination

Figure 11-1 *Animated* Reproductive structures of a pea flower
This cutaway view shows the pollen-producing anthers and the stigma, the portion of the female part of the flower that receives the pollen.

Mendel began his experiments by crossing plants from two different true-breeding lines with contrasting phenotypes; these genetically pure individuals constituted the **parental generation,** or **P generation.** In every case, the members of the first generation of offspring all looked alike and resembled one of the two parents. For example, when he crossed tall plants with short plants, all the offspring were tall (❙ Fig. 11-3). These offspring were the first filial generation, or the **F₁ generation** (*filial* is from the Latin for "sons and daughters"). The second filial generation, or **F₂ generation,** resulted from a cross between F₁ individuals or by self-pollination of F₁ individuals. Mendel's F₂ generation in this experiment included 787 tall plants and 277 short plants.

Most breeders in Mendel's time thought that inheritance involved the blending of traits. In *blending inheritance,* male and female gametes supposedly contained fluids that blended together during reproduction to produce hybrid offspring with features intermediate between those of the mother and father. In fact, some plant breeders had obtained such hybrids. Although Mendel observed some intermediate types of hybrids, he chose for further study those F₁ hybrids in which "hereditary factors" (as he called them) from one of the parents apparently masked the expression of those factors from the other parent. Other breeders had also observed these types of hybrids, but they had not explained them. Using modern terms, we say that the factor expressed in the F₁ generation (tallness, in our example) is **dominant;** the one hidden in the F₁ (shortness) is **recessive.** Dominant traits mask recessive ones when both are present in the same individual. Although scientists know today that dominance is not always observed (we will explore exceptions later in this chapter), the fact that dominance can occur was not consistent with the notion of blending inheritance.

Mendel's results also argued against blending inheritance in a more compelling way. Once two fluids have blended, it is very difficult to imagine how they can separate. However, in the preceding example, in the F₁ generation the hereditary factor(s)

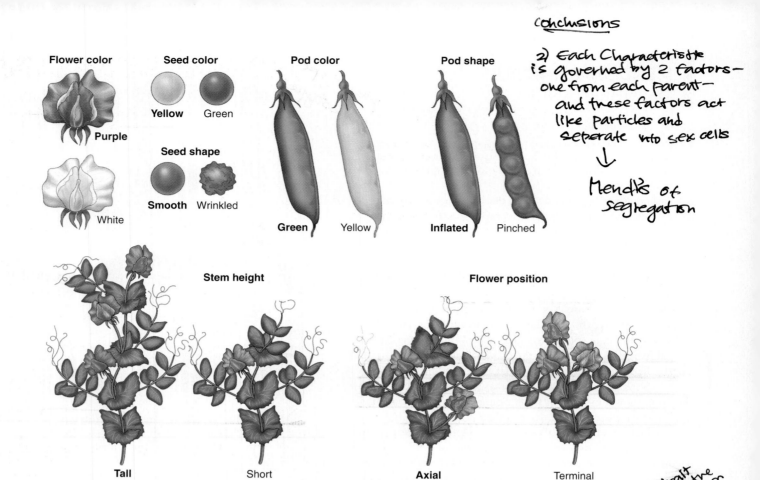

Figure 11-2   Seven characters in Mendel's study of pea plants

Each character had two clearly distinguishable phenotypes; the dominant phenotype is boldface.

that controlled shortness clearly was not lost or blended inseparably with the hereditary factor(s) that controlled tallness, because shortness reappeared in the F_2 generation. Mendel was very comfortable with the theoretical side of biology, because he was also a student of physics and mathematics. He therefore proposed that each kind of inherited feature of an organism is controlled by two factors that behave like discrete particles and are present in every individual. To Mendel these hereditary factors were abstractions—he knew nothing about chromosomes and DNA. These factors are essentially what scientists today call **genes.**

Mendel's experiments led to his discovery and explanation of the major principles of heredity, which we now know as the principles of segregation and independent assortment (see ▌Table 11-1 for a summary of Mendel's model of inheritance). We discuss the first principle next and the second later in the chapter.

Alleles separate before gametes are formed: the principle of segregation   *consequense of meiosis*

The term **alleles** refers to the alternative forms of a gene. In the example in Figure 11-3, each F_1 generation tall plant had two different alleles that control plant height: a **dominant allele** for tall

ness (which we designate *T*) and a **recessive allele** for shortness (designated *t*). Because the tall allele was dominant, these plants were tall. To explain his experimental results, Mendel proposed an idea now known as the principle of segregation. Using modern terminology, the **principle of segregation** states that before sexual reproduction occurs, the two alleles carried by an individual parent must become separated (that is, segregated).

Recall that during meiosis, homologous chromosomes (and therefore the alleles that reside on them) separate (see Fig. 10-13). As a result, each sex cell (egg or sperm) formed contains only one allele of each pair. (Later, at the time of fertilization, each haploid gamete contributes one chromosome from each homologous pair and therefore one allele for each gene pair.) An essential feature of meiosis is that the alleles remain intact (one does not mix with or eliminate the other); thus, recessive alleles are not lost and can reappear in the F_2 generation. In our example, before the F_1 plants formed gametes, the allele for tallness segregated from the allele for shortness; so half the gametes contained a *T* allele, and the other half, a *t* allele (▌Fig. 11-4).

The random process of fertilization led to three possible combinations of alleles in the F_2 offspring: one fourth with two tallness alleles (*TT*), one fourth with two shortness alleles (*tt*), and one half with one allele for tallness and one for shortness (*Tt*). Because both *TT* and *Tt* plants are tall, Mendel expected

Key Experiment

QUESTION: When the F_1 generation of tall pea plants is self-pollinated, what phenotypes appear in the F_2 generation?

HYPOTHESIS: Although only the "factor" (gene) for tall height is expressed in the F_1 generation, Mendel hypothesized that the factor for short height is not lost. He predicted that the short phenotype would reappear in the F_2 generation.

EXPERIMENT: Mendel crossed true-breeding tall pea plants with true-breeding short pea plants, yielding only tall offspring in the F_1 generation. He then allowed these F_1 individuals to self-pollinate to yield the F_2 generation.

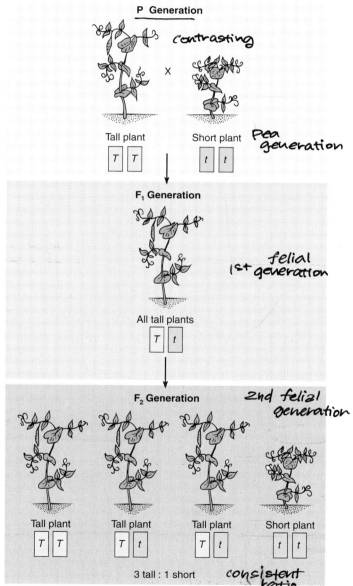

P Generation

contrasting

Tall plant   Short plant

Pea generation

F₁ Generation

1st felial generation

All tall plants

F₂ Generation

2nd felial generation

Tall plant   Tall plant   Tall plant   Short plant

3 tall : 1 short   *consistent ratio*

RESULTS AND CONCLUSION: The F_2 generation included 787 tall and 277 short plants, which resulted in a ratio of about 3:1. Thus, Mendelian traits pass to successive generations in fixed ratios.

Figure 11-3   One of Mendel's pea crosses

TABLE 11-1

Mendel's Model of Inheritance

1. Alternative forms of a "factor" (what we now call a *gene*) account for variations in inherited traits.

Although Mendel only observed two forms (what we now call *alleles*) for each factor he studied, we now know that many genes have more than two alleles.

2. Inherited traits pass from parents to offspring as unmodified factors.

Mendel did not observe offspring of intermediate appearance, as a hypothesis of blending inheritance would have predicted. Exceptions to this concept are known today.

3. Each individual has two sets of factors, one of each pair inherited from the mother and one from the father.

It does not matter which parent contributes which set of factors.

4. The paired factors separate during the formation of reproductive cells (the principle of segregation).

Because of *meiosis*, which was discovered after Mendel's time, each parent passes one set of factors to each offspring.

5. Factors may be expressed or hidden in a given generation, but they are never lost.

For example, factors not expressed in the F_1 generation reappear in some F_2 individuals.

6. Each factor is passed to the next generation independently from all other factors (the principle of independent assortment).

Research since Mendel's time has revealed that there are exceptions to this principle.

Key Point

Mendel's principle of segregation is related to the events of meiosis: The separation of homologous chromosomes during meiosis results in the segregation of alleles in a heterozygote.

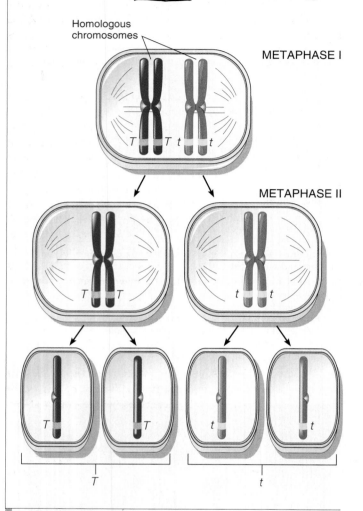

Figure 11-4 Chromosomes and segregation

Note that half of the gametes carry *T* and half carry *t*.

approximately three fourths (787 of the 1064 plants he obtained) to express the phenotype of the dominant allele (tall) and about one fourth (277/1064) to express the phenotype of the recessive allele (short). We will explain the mathematical reasoning behind these predictions later in the chapter.

Alleles occupy corresponding loci on homologous chromosomes

Today scientists know that each unduplicated chromosome consists of one long, linear DNA molecule and that each gene is actually a segment of that DNA molecule. We also know that homolo-

gous chromosomes not only are similar in size and shape but also usually have the same genes (often with different alleles) located in corresponding positions. The term **locus** (pl., *loci*) originally designated the location of a particular gene on the chromosome (▮ Fig. 11-5). We are actually referring to a segment of the DNA that has the information for controlling some aspect of the structure or function of the organism. One locus may govern seed color, another seed shape, still another the shape of the pods, and so on. Traditional genetic methods can infer the existence of a particular locus only if at least two allelic variants of that locus, producing contrasting phenotypes (for example, yellow peas versus green peas), are available for study. In the simplest cases an individual can express one (yellow) or the other (green) but not both. *today: characteristics.*

Alleles are, therefore, genes that govern variations of the same character (yellow versus green seed color) and occupy corresponding loci on homologous chromosomes. Geneticists assign each allele of a locus a single letter or group of letters as its symbol.[1] Although geneticists often use more complicated forms of notation, it is customary when working simple genetics problems to indicate a dominant allele with a capital letter and a recessive allele with the same letter in lowercase.

Remember that the term *locus* designates not only a position on a chromosome but also a type of gene controlling a particular character; thus, *Y* (yellow) and *y* (green) represent a specific pair of alleles of a locus involved in determining seed color in peas. Although you may initially be uncomfortable with the fact that geneticists sometimes use the term *gene* to specify a locus and at other times to specify one of the alleles of that locus, the meaning is usually clear from the context.

A monohybrid cross involves individuals with different alleles of a given locus

The basic principles of genetics and the use of genetics terms are best illustrated by examples. In the simplest case, a **monohybrid cross,** the inheritance of two different alleles of a single locus, is studied. ▮ Figure 11-6 illustrates a monohybrid cross featuring a locus that governs coat color in guinea pigs. The female comes from a true-breeding line of black guinea pigs. We say she is **homozygous** for black because the two alleles she carries for this locus are identical. The brown male is also from a true-breeding line and is homozygous for brown. What color would you expect the F₁ offspring to be? Actually, it is impossible to make such a prediction without more information.

In this particular case, the F₁ offspring are black, but they are **heterozygous,** meaning they carry two different alleles for this locus. The brown allele determines coat color only in a homozygous brown individual; it is a recessive allele. The black allele in-

---

[1] Early geneticists developed their own symbols to represent genes and alleles. Later, groups of scientists met and decided on specific symbols for a given research organism, such as the fruit fly, but each research group had its own rules for assigning symbols. Universally accepted rules for assigning symbols for genes and alleles still do not exist.

fluences coat color in both homozygous black and heterozygous individuals; it is a dominant allele. On the basis of this information, we can use symbols to designate the dominant black allele *B* and the recessive brown allele *b*.

During meiosis in the female parent (*BB*), the two *B* alleles separate, according to Mendel's principle of segregation, so each egg has only one *B* allele. In the male (*bb*) the two *b* alleles separate, so each sperm has only one *b* allele. The fertilization of each *B* egg by a *b* sperm results in heterozygous F₁ offspring, each with the alleles *Bb;* that is, each individual has one allele for brown coat and one for black coat. Because this is the only possible combination of alleles present in the eggs and sperm, all the F₁ offspring are *Bb*.

A Punnett square predicts the ratios of the various offspring of a cross

During meiosis in <u>heterozygous</u> black guinea pigs (*Bb*), the chromosome containing the *B* allele becomes separated from its homologue (the chromosome containing the *b* allele), so each normal sperm or egg contains *B* or *b* but never both. Heterozygous *Bb* individuals form gametes containing *B* alleles and gametes containing *b* alleles in equal numbers. Because no special attraction or repulsion occurs between an egg and a sperm containing the same allele, fertilization is a random process.

As you can see in Figure 11-6, the possible combinations of eggs and sperm at fertilization can be represented in the form of a grid known as a **Punnett square,** devised by the early English geneticist Sir Reginald Punnett. The types of gametes (and their expected frequencies) from one parent are listed across the top, and those from the other parent are listed along the left side. The squares are then filled in with the resulting F₂ combinations. Three fourths of all F₂ offspring have the genetic constitution *BB* or *Bb* and are phenotypically black; one fourth have the genetic constitution *bb* and are phenotypically brown. The genetic mechanism that governs the approximate 3:1 F₂ ratios obtained by Mendel in his pea-breeding experiments is again evident. These ratios are called *monohybrid F₂ phenotypic ratios.*

The phenotype of an individual does not always reveal its genotype

As mentioned earlier, an organism's phenotype is its appearance with respect to a certain inherited trait. However, because some alleles may be dominant and others recessive, we cannot always

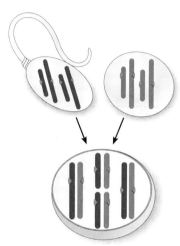

A gamete has one set of chromosomes, the *n* number. It carries *one* chromosome of *each* homologous pair. A given gamete can only have *one* gene of any particular pair of alleles.

When the gametes fuse, the resulting zygote is diploid (*2n*) and has homologous pairs of chromosomes. For purposes of illustration, these are shown physically paired.

(a) One member of each pair of homologous chromosomes is of maternal origin (*red*), and the other is paternal (*blue*).

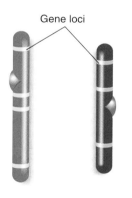

Gene loci

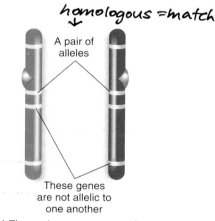

homologous =match

A pair of alleles

These genes are not allelic to one another

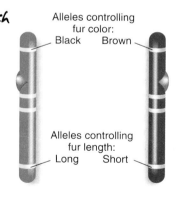

Alleles controlling fur color:
Black   Brown

Alleles controlling fur length:
Long   Short

Genotype:
homozygous dominant (BB)
Phenotype:
Black fur

Genotype (bb)
Phenotype:
Brown fur

Genotype: hetero (Bb)
Phenotyp: black fur

(b) These chromosomes are nonhomologous. Each chromosome is made up of hundreds or thousands of genes. A <u>locus is the specific place on a chromosome where a gene is located.</u>

(c) These chromosomes are <u>homologous</u>. Alleles are members of a gene pair that occupy corresponding loci on homologous chromosomes.

(d) Alleles govern the same character but do not necessarily contain the same information.

Figure 11-5   *Animated*   Gene loci and their alleles

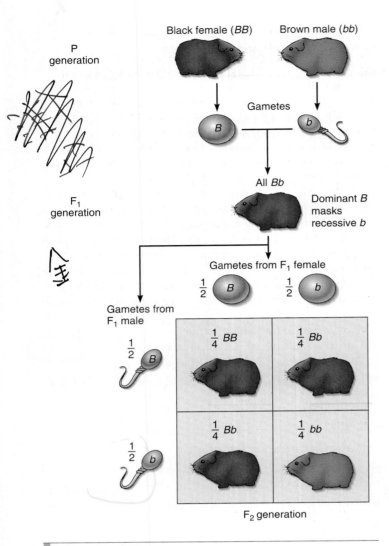

P generation

Black female (*BB*)   Brown male (*bb*)

Gametes

B b

All *Bb*

Dominant *B*
masks
recessive *b*

F₁ generation

Gametes from F₁ female

$\frac{1}{2}$ B $\frac{1}{2}$ b

Gametes from F₁ male

$\frac{1}{2}$ B

$\frac{1}{4}$ *BB* $\frac{1}{4}$ *Bb*

$\frac{1}{2}$ b

$\frac{1}{4}$ *Bb* $\frac{1}{4}$ *bb*

F₂ generation

Figure 11-6 *Animated*   A monohybrid cross in guinea pigs

In this example, a homozygous black guinea pig is mated with a homozygous brown guinea pig. The F₁ generation includes only black individuals. However, the mating of two of these F₁ offspring yields F₂ generation offspring in the expected ratio of 3 black to 1 brown, indicating that the F₁ individuals are heterozygous.

though the two parents contribute equally to their offspring's genetic constitution. Dominance is not predictable and can be determined only by experiment. In one species of animal, black coat may be dominant to brown; in another species, brown may be dominant to black. In a population, the dominant phenotype is not necessarily more common than the recessive phenotype.

A test cross can detect heterozygosity

Guinea pigs with the genotypes *BB* and *Bb* are alike phenotypically; they both have black coats. How, then, can we know the genotype of a black guinea pig? Geneticists accomplish this by performing a **test cross,** in which an individual of unknown genotype is crossed with a homozygous recessive individual (Fig. 11-7). In a test cross, the alleles carried by the gametes from the parent of unknown genotype are never "hidden" in the offspring by dominant alleles contributed by the other parent. Therefore, you can deduce the genotypes of all offspring directly from their phenotypes. If all the offspring were black, what inference would you make about the genotype of the black parent? If any of the offspring were brown, what conclusion would you draw regarding the genotype of the black parent? Would you be more certain about one of these inferences than about the other?[2]

Mendel conducted several test crosses; for example, he bred F₁ (tall) pea plants with homozygous recessive (*tt*) short ones. He reasoned that the F₁ individuals were heterozygous (*Tt*) and would be expected to produce equal numbers of *T* and *t* gametes. Because the homozygous short parents (*tt*) were expected to produce only *t* gametes, Mendel hypothesized that he would obtain equal numbers of tall (*Tt*) and short (*tt*) offspring. His results agreed with his hypothesis, providing additional evidence for the hypothesis that there is 1:1 segregation of the alleles of a heterozygous parent. Thus, Mendel's principle of segregation not only explained the known facts, such as the 3:1 monohybrid F₂ phenotypic ratio, but also let him successfully anticipate the results of other experiments—in this case, the 1:1 test-cross phenotypic ratio.

A dihybrid cross involves individuals that have different alleles at two loci

Monohybrid crosses involve a pair of alleles of a single locus. Mendel also analyzed crosses involving alleles of two or more loci. A mating between individuals with different alleles at two loci is called a **dihybrid cross.** Consider the case of two pairs of alleles carried on nonhomologous chromosomes (that is, one

determine, simply by examining its phenotype, which alleles are carried by an organism. The *genetic constitution* of that organism, most often expressed in symbols, is its **genotype.** In the cross we have been considering, the genotype of the female parent is homozygous dominant, *BB*, and her phenotype is black. The genotype of the male parent is homozygous recessive, *bb*, and his phenotype is brown. The genotype of all the F₁ offspring is heterozygous, *Bb*, and their phenotype is black. To prevent confusion, we always indicate the genotype of a heterozygous individual by writing the symbol for the dominant allele first and the recessive allele second (always *Bb*, never *bB*).

The phenomenon of dominance partly explains why an individual may resemble one parent more than the other, even

---

[2]If all the offspring were black, you could infer that the black parent is probably homozygous, *BB*. If any of the offspring were brown, you could infer that the black parent is heterozygous, *Bb*. You would be more certain that the second inference (about the *Bb* individual) is correct than the first inference (the *BB* individual).

pair of alleles is in one pair of homologous chromosomes, and the other pair of alleles is in a different pair of homologous chromosomes). Each pair of alleles is inherited independently; that is, each pair segregates during meiosis independently of the other.

An example of a dihybrid cross carried through the F_2 generation is shown in ▌Figure 11-8. In this example, black is dominant to brown, and short hair is dominant to long hair. When a homozygous, black, short-haired guinea pig (*BBSS*) and a homozygous, brown, long-haired guinea pig (*bbss*) are mated, the *BBSS* animal produces gametes that are all *BS*, and the *bbss* individual produces gametes that are all *bs*. Each gamete contains one allele for each of the two loci. The union of the *BS* and *bs* gametes yields only individuals with the genotype *BbSs*. All these F_1 offspring are heterozygous for hair color and for hair length, and all are phenotypically black and short-haired.

Each F_1 guinea pig produces four kinds of gametes with equal probability: *BS*, *Bs*, *bS*, and *bs*. Hence, the Punnett square has 16 (that is, 4^2) squares representing the F_2 offspring, some of which are genotypically or phenotypically alike. There are 9 chances in 16 of obtaining a black, short-haired individual; 3 chances in 16 of obtaining a black, long-haired individual; 3 chances in 16 of obtaining a brown, short-haired individual; and 1 chance in 16 of obtaining a brown, long-haired individual. This 9:3:3:1 phenotypic ratio is expected in a dihybrid F_2 if the hair color and hair length loci are on nonhomologous chromosomes.

Alleles on nonhomologous chromosomes are randomly distributed into gametes: the principle of independent assortment

On the basis of results similar to the guinea pig example, Mendel formulated the principle of inheritance, now known as Mendel's **principle of independent assortment,** which states that members of any gene pair segregate from one another independently of the members of the other gene pairs. This mechanism occurs in a regular way to ensure that each gamete contains one allele for each locus, but the alleles of different loci are assorted at random with respect to each other in the gametes.

Today we recognize that independent assortment is related to the events of meiosis. It occurs because two pairs of homologous chromosomes can be arranged in two different ways at metaphase I of meiosis (▌Fig. 11-9). These arrangements occur randomly, with approximately half the meiotic cells oriented one way and half oriented the opposite way. The orientation of the homolo-

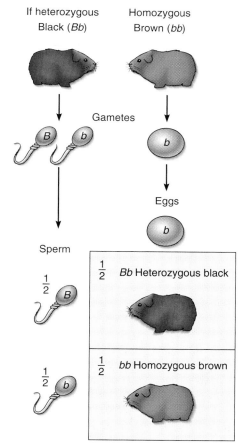

(a) If a black guinea pig is mated with a brown guinea pig and all the offspring are black, the black parent probably has a homozygous genotype.

(b) If any of the offspring is brown, the black guinea pig must be heterozygous. The expected phenotypic ratio is 1 black to 1 brown.

▌ **Figure 11-7** *Animated* A test cross in guinea pigs

In this illustration, a test cross is used to determine the genotype of a black guinea pig.

P1 = parental

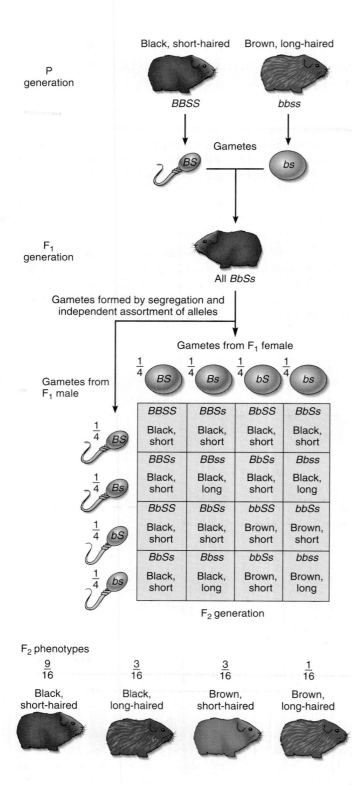

P generation

Black, short-haired   Brown, long-haired

BBSS   *bbss*

Gametes

BS   *bs*

F₁ generation

All *BbSs*

Gametes formed by segregation and independent assortment of alleles

Gametes from F₁ female

$\frac{1}{4}$ *BS* $\frac{1}{4}$ *Bs* $\frac{1}{4}$ *bS* $\frac{1}{4}$ *bs*

Gametes from F₁ male

	$\frac{1}{4}$ *BS*	$\frac{1}{4}$ *Bs*	$\frac{1}{4}$ *bS*	$\frac{1}{4}$ *bs*
$\frac{1}{4}$ *BS*	*BBSS* Black, short	*BBSs* Black, short	*BbSS* Black, short	*BbSs* Black, short
$\frac{1}{4}$ *Bs*	*BBSs* Black, short	*BBss* Black, long	*BbSs* Black, short	*Bbss* Black, long
$\frac{1}{4}$ *bS*	*BbSS* Black, short	*BbSs* Black, short	*bbSS* Brown, short	*bbSs* Brown, short
$\frac{1}{4}$ *bs*	*BbSs* Black, short	*Bbss* Black, long	*bbSs* Brown, short	*bbss* Brown, long

F₂ generation

F₂ phenotypes

$\frac{9}{16}$ $\frac{3}{16}$ $\frac{3}{16}$ $\frac{1}{16}$

Black, short-haired   Black, long-haired   Brown, short-haired   Brown, long-haired

Figure 11-8 *Animated* A dihybrid cross in guinea pigs

When a black, short-haired guinea pig is crossed with a brown, long-haired one, all the offspring are black and have short hair. However, when two members of the F₁ generation are crossed, the ratio of phenotypes is 9:3:3:1.

---

gous chromosomes on the metaphase plate then determines the way they subsequently separate and disperse into the haploid cells. (As you will soon see, however, independent assortment does not always occur.)

Recognition of Mendel's work came during the early 20th century

Mendel reported his findings at a meeting of the Brünn Society for the Study of Natural Science; he published his results in the society's report in 1866. At that time biology was largely a descriptive science, and biologists had little interest in applying quantitative and experimental methods such as Mendel had used. Other biologists of the time did not appreciate the importance of his results and his interpretations of those results. For 34 years his findings were largely neglected.

In 1900, Hugo DeVries in Holland, Carl Correns in Germany, and Erich von Tschermak in Austria recognized Mendel's principles in their own experiments; they later discovered Mendel's paper and found that it explained their own research observations. By this time biologists had a much greater appreciation of the value of quantitative experimental methods. Correns gave credit to Mendel by naming the basic laws of inheritance after him.

Although gametes and fertilization were known at the time Mendel carried out his research, mitosis and meiosis had not yet been discovered. It is truly remarkable that Mendel formulated his ideas mainly on the basis of mathematical abstractions. Today his principles are much easier to understand, because we relate the transmission of genes to the behavior of chromosomes.

The details of mitosis and meiosis were described during the late 19th century, and in 1902, American biologist Walter Sutton and German biologist Theodor Boveri independently pointed out the connection between Mendel's segregation of alleles and the separation of homologous chromosomes during meiosis. This connection developed into the **chromosome theory of inheritance,** also known as the *Sutton–Boveri theory,* which stated that inheritance can be explained by assuming that genes are linearly arranged in specific locations along the chromosomes.

The chromosome theory of inheritance was initially controversial, because at that time there was no direct evidence that genes are found on chromosomes. However, new research provided the findings necessary for wider acceptance and extension of these ideas and their implications. For example, the work of American geneticist Thomas Hunt Morgan in 1910 provided evidence for the location of a particular gene (white eye color) on a specific chromosome (the X chromosome) in fruit flies. Morgan and his graduate students also provided insight into the way genes are organized on chromosomes; we discuss some of Morgan's research contributions later in the chapter.

Review

■ What are the relationships among loci, genes, and alleles?
■ What is Mendel's principle of segregation?
■ What is Mendel's principle of independent assortment?

Key Point

Mendel's principle of independent assortment—that factors for different characteristics separate independently of one another prior to gamete formation—is a direct consequence of the events of meiosis.

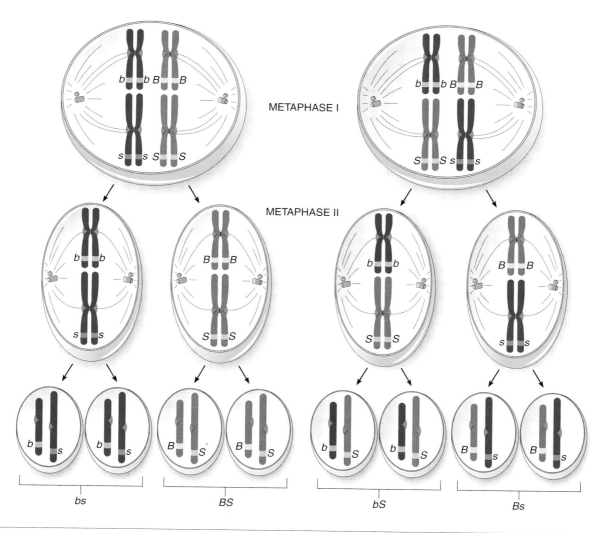

METAPHASE I

METAPHASE II

bs   BS   bS   Bs

Figure 11-9 *Animated* Meiosis and independent assortment

Two different pairs of homologous chromosomes can line up two different ways at metaphase I and be subsequently distributed. A cell with the orientation shown at the left produces half *BS* and half *bs* gametes.

Conversely, the cell at the right produces half *Bs* and half *bS* gametes. Because approximately half of the meiotic cells at metaphase I are of each type, the ratio of the four possible types of gametes is 1:1:1:1.

USING PROBABILITY TO PREDICT MENDELIAN INHERITANCE

Learning Objective

5 Apply the product rule and sum rule appropriately when predicting the outcomes of genetic crosses.

All genetic ratios are properly expressed in terms of probabilities. In monohybrid crosses, the expected ratio of the dominant and recessive phenotypes is 3:1. The probability of an event is its expected frequency. Therefore, we can say there are 3 chances in 4 (or $\frac{3}{4}$) that any particular individual offspring of two heterozygous individuals will express the dominant phenotype and 1 chance in 4 (or $\frac{1}{4}$) that it will express the recessive phenotype. Although we sometimes speak in terms of percentages, probabilities are calculated as fractions (such as $\frac{3}{4}$) or decimal fractions (such as 0.75). If an event is certain to occur, its probability is 1; if it is certain *not* to occur, its probability is 0. A probability can be 0, 1, or some number between 0 and 1.

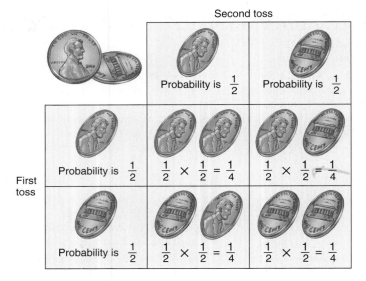

Second toss

First toss

Probability is $\frac{1}{2}$ Probability is $\frac{1}{2}$

Probability is $\frac{1}{2}$ $\frac{1}{2} \times \frac{1}{2} = \frac{1}{4}$ $\frac{1}{2} \times \frac{1}{2} = \frac{1}{4}$

Probability is $\frac{1}{2}$ $\frac{1}{2} \times \frac{1}{2} = \frac{1}{4}$ $\frac{1}{2} \times \frac{1}{2} = \frac{1}{4}$

Figure 11-10 The rules of probability

For each coin toss, the probability of heads is $\frac{1}{2}$, and the probability of tails is also $\frac{1}{2}$. Because the outcome of the first toss is independent of the outcome of the second, the combined probabilities of the outcomes of successive tosses are calculated by multiplying their individual probabilities (according to the product rule: $\frac{1}{2} \times \frac{1}{2} = \frac{1}{4}$). These same rules of probability predict genetic events.

The Punnett square lets you combine two or more probabilities. When you use a Punnett square, you are following two important statistical principles known as the product rule and the sum rule. The **product rule** predicts the combined probabilities of independent events. Events are *independent* if the occurrence of one does not affect the probability that the other will occur. For example, the probability of obtaining heads on the first toss of a coin is $\frac{1}{2}$; the probability of obtaining heads on the second toss (an independent event) is also $\frac{1}{2}$. If two or more events are independent of each other, the probability of both occurring is the product of their individual probabilities. If this seems strange, keep in mind that when we multiply two numbers that are less than 1, the product is a smaller number. Therefore, the probability of obtaining heads two times in a row is $\frac{1}{2} \times \frac{1}{2} = \frac{1}{4}$, or 1 chance in 4 (❙ Fig. 11-10).

Similarly, we can apply the product rule to genetic events. If both parents are *Bb*, what is the probability they will produce a child who is *bb*? For the child to be *bb*, he or she must receive a *b* gamete from each parent. The probability of a *b* egg is $\frac{1}{2}$, and the probability of a *b* sperm is also $\frac{1}{2}$. Like the outcomes of the coin tosses, these probabilities are independent, so we combine them by the product rule ($\frac{1}{2} \times \frac{1}{2} = \frac{1}{4}$). You might like to check this result using a Punnett square.

The **sum rule** predicts the combined probabilities of *mutually exclusive* events. In some cases, there is more than one way

to obtain a specific outcome. These different ways are mutually exclusive; if one occurs, the other(s) cannot. For example, if both parents are *Bb*, what is the probability that their first child will also have the *Bb* genotype? There are two different ways these parents can have a *Bb* child: Either a *B* egg combines with a *b* sperm (probability $\frac{1}{4}$), or a *b* egg combines with a *B* sperm (probability $\frac{1}{4}$).

Naturally, if there is more than one way to get a result, the chances of its being obtained improve; we therefore combine the probabilities of mutually exclusive events by summing (adding) their individual probabilities. The probability of obtaining a *Bb* child in our example is therefore $\frac{1}{4} + \frac{1}{4} = \frac{1}{2}$. (Because there is only one way these heterozygous parents can produce a homozygous recessive child, *bb*, that probability is only $\frac{1}{4}$. The probability of a homozygous dominant child, *BB*, is likewise $\frac{1}{4}$.)

The rules of probability can be applied to a variety of calculations

The rules of probability have wide applications. For example, what are the probabilities that a family with two (and only two) children will have two girls, two boys, or one girl and one boy? For purposes of discussion, let's assume male and female births are equally probable. The probability of having a girl first is $\frac{1}{2}$, and the probability of having a girl second is also $\frac{1}{2}$. These are independent events, so we combine their probabilities by multiplying: $\frac{1}{2} \times \frac{1}{2} = \frac{1}{4}$. Similarly, the probability of having two boys is $\frac{1}{4}$.

In families with both a girl and a boy, the girl can be born first or the boy can be born first. The probability that a girl will be born first is $\frac{1}{2}$, and the probability that a boy will be born second is also $\frac{1}{2}$. We use the product rule to combine the probabilities of these two independent events: $\frac{1}{2} \times \frac{1}{2} = \frac{1}{4}$. Similarly, the probability that a boy will be born first and a girl second is also $\frac{1}{4}$. These two kinds of families represent mutually exclusive outcomes, that is, two different ways of obtaining a family with one boy and one girl. Having two different ways of obtaining the desired result improves our chances, so we use the sum rule to combine the probabilities: $\frac{1}{4} + \frac{1}{4} = \frac{1}{2}$.

In working with probabilities, keep in mind a point that many gamblers forget: Chance has no memory. If events are truly independent, past events have no influence on the probability of the occurrence of future events. When working probability problems, common sense is more important than blindly memorizing rules. Examine your results to see whether they appear reasonable; if they don't, re-evaluate your assumptions. (See *Focus On: Solving Genetics Problems* for procedures to solve genetics problems, including when to use the rules of probability.)

Review

❙ How is probability used to predict the outcome of genetic crosses?

SOLVING GENETICS PROBLEMS

Simple Mendelian genetics problems are like puzzles. They can be fun and easy to work if you follow certain conventions and are methodical in your approach.

1. Always use standard designations for the generations. The generation in which a particular genetic experiment is begun is called the P, or parental, generation. Offspring of this generation are called the F_1, or first filial, generation. The offspring resulting when two F_1 individuals are bred constitute the F_2, or second filial, generation.

2. Write down a key for the symbols you are using for the allelic variants of each locus. Use an uppercase letter to designate a dominant allele and a lowercase letter to designate a recessive allele. Use the same letter of the alphabet to designate both alleles of a particular locus. If you are not told which allele is dominant and which is recessive, the phenotype of the F_1 generation is a good clue.

3. Determine the genotypes of the parents of each cross by using the following types of evidence:

   • Are they from true-breeding lines? If so, they should be homozygous.

   • Can their genotypes be reliably deduced from their phenotypes? This is usually true if they express the recessive phenotype.

   • Do the phenotypes of their offspring provide any information? Exactly how this is done is discussed shortly.

4. Indicate the possible kinds of gametes formed by each of the parents. It is helpful to draw a circle around the symbols for each kind of gamete.

   • If it is a monohybrid cross, we apply the principle of segregation; that is, a heterozygote *Aa* forms two kinds of gametes, *A* and *a*. A homozygote, such as *aa*, forms only one kind of gamete, *a*.

   • If it is a dihybrid cross, we apply the principles of segregation and independent assortment. For example, an individual heterozygous for two loci would have the genotype *AaBb*. Allele *A* segregates from *a*, and *B* segregates from *b*. The assortment of *A* and *a* into gametes is independent of the assortment of *B* and *b*. *A* is equally likely to end up in a gamete with *B* or *b*. The same is true for *a*. Thus, an individual with the genotype *AaBb* produces four kinds of gametes in equal amounts: *AB*, *Ab*, *aB*, and *ab*.

5. Set up a Punnett square, placing the possible types of gametes from one parent down the left side and the possible types from the other parent across the top.

6. Fill in the Punnett square. Avoid confusion by consistently placing the dominant allele first and the recessive allele second in heterozygotes (*Aa*, never *aA*). If it is a dihybrid cross, always write the two alleles of one locus first and the two alleles of the other locus second. It does not matter which locus you choose to write first, but once you have decided on the order, it is crucial to maintain it consistently. This means that if the individual is heterozygous for both loci, you will always use the form *AaBb*. Writing this particular genotype as *aBbA*, or even as *BbAa*, would cause confusion.

7. If you do not need to know the frequencies of all the expected genotypes and phenotypes, you can use the rules of probability as a shortcut instead of making a Punnett square. For example, if both parents are *AaBb*, what is the probability of an *AABB* offspring? To be *AA*, the offspring must receive an *A* gamete from each parent. The probability that a given gamete is *A* is $\frac{1}{2}$, and each gamete represents an independent event, so combine their probabilities by mul-

tiplying ($\frac{1}{2} \times \frac{1}{2} = \frac{1}{4}$). The probability of *BB* is calculated similarly and is also $\frac{1}{4}$. The probability of *AA* is independent of the probability of *BB*, so again, use the product rule to obtain their combined probabilities ($\frac{1}{4} \times \frac{1}{4} = \frac{1}{16}$).

8. Very often the genotypes of the parents can be deduced from the phenotypes of their offspring. In peas, for example, the allele for yellow seeds (*Y*) is dominant to the allele for green seeds (*y*). Suppose plant breeders cross two plants with yellow seeds, but you don't know whether the yellow-seeded plants are homozygous or heterozygous. You can designate the cross as: *Y_* × *Y_*. The offspring of the cross are allowed to germinate and grow, and their seeds are examined. Of the offspring, 74 produce yellow seeds, and 26 produce green seeds. Knowing that green-seeded plants are recessive, *yy*, the answer is obvious: Each parent contributed a *y* allele to the green-seeded offspring, so the parents' genotypes must have been *Yy* × *Yy*.

Sometimes it is impossible to determine from the data given whether an individual is homozygous dominant or heterozygous. For example, suppose a cross between a yellow-seeded plant and a green-seeded plant yields offspring that all produce yellow seeds. From the information given about the parents, you can designate the cross as: *Y_* × *yy*. Because all the offspring have yellow seeds, they have the genotype *Yy*. You can deduce this because each parent contributes one allele. (The offspring have yellow seeds, so they must have at least one *Y* allele; however, the green-seeded parent can contribute only a *y* allele to the offspring.) Therefore, the cross is most likely (*YY* × *yy*), but you cannot be absolutely certain about the genotype of the yellow-seeded parent. Further crosses would be necessary to determine this.

INHERITANCE AND CHROMOSOMES

Learning Objectives

6   Define *linkage*, and relate it to specific events in meiosis.

7   Show how data from a test cross involving alleles of two loci can be used to distinguish between independent assortment and linkage.

8   Discuss the genetic determination of sex and the inheritance of X-linked genes in mammals.

It is a measure of Mendel's genius that he worked out the principles of segregation and independent assortment without knowing anything about meiosis or the chromosome theory of inheritance. The chromosome theory of inheritance also helps explain certain apparent exceptions to Mendelian inheritance. One of these so-called exceptions involves linked genes.

Linked genes do not assort independently

Beginning around 1910, the research of geneticist Thomas Hunt Morgan and his graduate students extended the concept of the chromosome theory of inheritance. Morgan's research organism was the fruit fly (*Drosophila melanogaster*). Just as the garden pea was an excellent model research organism for Mendel's studies, the fruit fly was perfect for extending general knowledge about inheritance. Fruit flies have a short life cycle—just 14 days—and their small size means that thousands can be kept in a research lab (see Fig. 17-6b). The large number of individuals increases the chance of identifying mutants. Also, fruit flies have just four pairs of chromosomes, one of which is a pair of sex chromosomes.

By carefully analyzing the results of crosses involving fruit flies, Morgan and his students demonstrated that genes are arranged in a linear order on each chromosome. Morgan also showed that independent assortment does not apply if the two loci lie close together in the same pair of homologous chromosomes. In fruit flies there is a locus controlling wing shape (the dominant allele *V* for normal wings and the recessive allele *v* for abnormally short, or vestigial, wings) and another locus controlling body color (the dominant allele *B* for gray body and the recessive allele *b* for black body). If a homozygous *BBVV* fly is crossed with a homozygous *bbvv* fly, the F$_1$ flies all have gray bodies and normal wings, and their genotype is *BbVv*.

Because these loci happen to lie close to one another in the *same pair* of homologous chromosomes, their alleles do not assort independently; instead, they are **linked genes** that tend to be inherited together. **Linkage** is the tendency for a group of genes on the same chromosome to be inherited together in successive generations. You can readily observe linkage in the results of a test cross in which heterozygous F$_1$ flies (*BbVv*) are mated with homozygous recessive (*bbvv*) flies (Fig. 11-11). Because heterozygous individuals are mated to homozygous recessive individuals, this test cross is similar to the test cross described earlier. However, it is called a **two-point test cross** because alleles of two loci are involved.

Key Experiment

QUESTION: How can linkage be recognized in fruit flies?

HYPOTHESIS: Linkage can be recognized when an excess of parental-type offspring and a deficiency of recombinant-type offspring are produced in a two-point test cross.

EXPERIMENT: Fruit flies with gray, normal wings (*BbVv*) are crossed with flies that have black, vestigial wings (*bbvv*). If the alleles for color and wing shape are not linked (i.e., the alleles assort independently), the offspring will consist of an equal number of each of four phenotypes (*yellow row*).

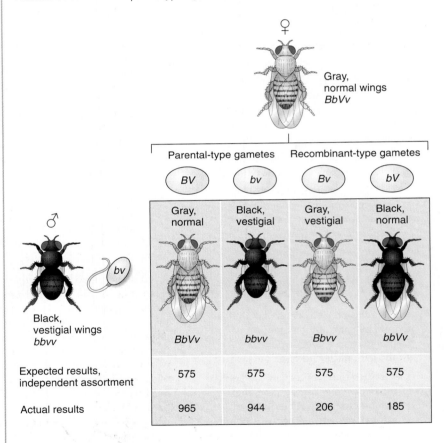

	Parental-type gametes		Recombinant-type gametes	
	BV	*bv*	*Bv*	*bV*
	Gray, normal	Black, vestigial	Gray, vestigial	Black, normal
	BbVv	*bbvv*	*Bbvv*	*bbVv*
Expected results, independent assortment	575	575	575	575
Actual results	965	944	206	185

RESULTS AND CONCLUSION: In the 2300 offspring (*bottom row*) of an actual cross, about 1909 of the offspring belong to each of the two parental classes (83% total), and 391 belong to each of the two recombinant classes (17% total). Thus, loci for wing length and body color are linked on a homologous chromosome pair.

Figure 11-11   A two-point test cross to detect linkage in fruit flies

If the loci governing these traits were unlinked—that is, on different chromosomes—the heterozygous parent in a test cross would produce four kinds of gametes (*BV*, *Bv*, *bV*, and *bv*) in equal numbers. This independent assortment would produce offspring with new gene combinations not present in the parental generation. Any process that leads to new gene combinations is called **recombination.** In our example, gametes *Bv* and *bV* are **recombinant types.** The other two kinds of gametes, *BV* and *bv*, are **parental types** because they are identical to the gametes produced by the P generation. Of course, the homozygous recessive parent produces only one kind of gamete, *bv*. Thus, if independent assortment were to occur in the F_1 flies, approximately 25% of the test-cross offspring would be gray-bodied and normal-winged (*BbVv*), 25% black-bodied and normal-winged (*bbVv*), 25% gray-bodied and vestigial-winged (*Bbvv*), and 25% black-bodied and vestigial-winged (*bbvv*). Notice that the two-point test cross lets us determine the genotypes of the offspring directly from their phenotypes.

By contrast, the alleles of the loci in our example do not undergo independent assortment, because they are linked. Alleles at different loci but close to one another on a given chromosome tend to be inherited together; because chromosomes pair and separate during meiosis as units, they therefore tend to be inherited as units. If linkage were complete, only parental-type flies would be produced, with approximately 50% having gray bodies and normal wings (*BbVv*) and 50% having black bodies and vestigial wings (*bbvv*). However, in our example, the offspring also include some gray-bodied, vestigial-winged flies and some black-bodied, normal-winged flies. These are recombinant flies, having received a recombinant-type gamete from the heterozygous F_1 parent. Each recombinant-type gamete arose by **crossing-over** between these loci in a meiotic cell of a heterozygous female fly. (Fruit flies are unusual in that crossing-over occurs only in females and not in males; it is far more common for crossing-over to occur in both sexes of a species.) Recall from Chapter 10 that when chromosomes pair and undergo synapsis, crossing-over occurs as homologous (nonsister) chromatids exchange segments of chromosome material by a process of breakage and rejoining catalyzed by enzymes (▌ Fig. 11-12; also see Fig. 10-15).

Calculating the frequency of crossing-over reveals the linear order of linked genes on a chromosome

In our fruit fly example (see Fig. 11-11), 391 of the offspring are recombinant types: gray flies with vestigial wings, *Bbvv* (206 of the total); and black flies with normal wings, *bbVv* (185 of the total). The remaining 1909 offspring are parental types. These data can be used to calculate the percentage of crossing-over between the loci (▌ Table 11-2). You do this by adding the number of individuals in the two recombinant classes of offspring, dividing by the *total number of offspring*, and multiplying by 100: $391 \div 2300 = 0.17; 0.17 \times 100 = 17\%$. Thus, the *V* locus and the *B* locus have 17% recombination between them.

During a single meiotic division, crossing-over may occur at several different points along the length of each homologous

Key Point

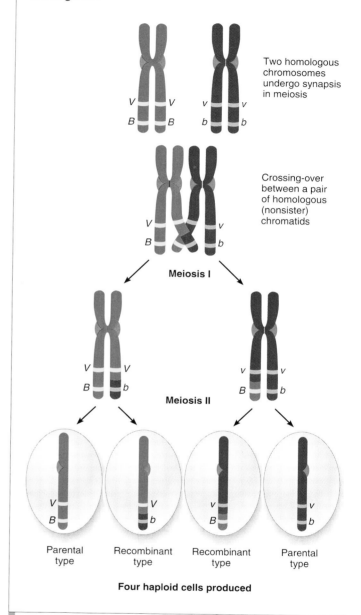

The exchange of segments between chromatids of homologous chromosomes is the mechanism of recombination of linked genes.

Two homologous chromosomes undergo synapsis in meiosis

Crossing-over between a pair of homologous (nonsister) chromatids

Meiosis I

Meiosis II

Parental type

Recombinant type

Recombinant type

Parental type

Four haploid cells produced

▌ **Figure 11-12** *Animated* Crossing-over

Genes located far apart on a chromosome have a greater probability of being separated by crossing-over than do genes that are closer together.

chromosome pair. In general, crossing-over is more likely to occur between two loci if they lie far apart on the chromosome and less likely to occur if they lie close together. Because of this rough correlation between the frequency of recombination of two loci and the linear distance between them, a genetic map of the chromosome can be generated by converting the percentage of re-

TABLE 11-2

How to Determine Recombination Frequency from a Two-Point Test Cross

Type of offspring	Test-Cross Results (from Fig. 11-11)			
	Parental Type		Recombinant Type	
Phenotype	Gray, normal	Black, vestigial	Gray, vestigial	Black, normal
Genotype	*BbVv*	*bbvv*	*Bbvv*	*bbVv*
Number of offspring	965	944	206	185
Calculations of recombination frequency	1. Number of parental-type offspring = 1909 2. Number of recombinant-type offspring = 391 3. Total number of offspring = 1909 + 391 = 2300 4. Recombination frequency = 391/2300 × 100 = 17%			

combination to **map units.** By convention, 1% recombination between two loci equals a distance of 1 map unit, so the loci in our example are 17 map units apart.

Scientists have determined the frequencies of recombination between specific linked loci in many species. All the experimental results are consistent with the hypothesis that genes are present in a linear order in the chromosomes. ▌ Figure 11-13 illustrates the traditional method for determining the linear order of genes in a chromosome.

More than one crossover between two loci in a single tetrad can occur in a given cell undergoing meiosis. (Recall from Chapter 10 that a *tetrad* is a group of four chromatids that make up a pair of synapsed homologous chromosomes.) We can observe only the frequency of offspring receiving recombinant-type gametes from the heterozygous parent, not the actual number of crossovers. In fact, the actual frequency of crossing-over is slightly more than the observed frequency of recombinant-type gametes. The reason is that the simultaneous occurrence of two crossovers involving the same two homologous chromatids reconstitutes the original combination of genes (▌ Fig. 11-14). When two loci are relatively close together, the effect of double crossing-over is minimized.

By putting together the results of numerous crosses, scientists have developed detailed linkage maps for many eukaryotes, including the fruit fly, the mouse, yeast, *Neurospora* (a fungus), and many plants, especially those that are important crops. In addition, researchers have used genetic methods to develop a detailed map for *Escherichia coli*, a bacterium with a single, circular DNA molecule, and many other prokaryotes and viruses. They have made much more sophisticated maps of chromosomes of various species by means of recombinant DNA technology (see Chapter 15). Using these techniques, the *Human Genome Project* has produced maps of human chromosomes (see Chapter 16).

Sex is generally determined by sex chromosomes

In some species, environmental factors exert a large control over an individual's sex. However, genes are the most important sex determinants in most organisms. The major sex-determining

genes of mammals, birds, and many insects are carried by **sex chromosomes.** Typically, members of one sex have a pair of similar sex chromosomes and produce gametes that are all identical in sex chromosome constitution. The members of the other sex have two different sex chromosomes and produce two kinds of gametes, each bearing a single kind of sex chromosome.

The cells of females of many animal species, including humans, contain two **X chromosomes.** In contrast, the males have a single X chromosome and a smaller **Y chromosome** that bears only a few active genes. For example, human males have 22 pairs of **autosomes,** which are chromosomes other than the sex chromosomes, plus one X chromosome and one Y chromosome; females have 22 pairs of autosomes plus a pair of X chromosomes. Domestic cats have 19 pairs of autosomes, to which are added a pair of X chromosomes in females, or an X plus a Y in males.

The Y chromosome determines male sex in most species of mammals

Do male humans have a male phenotype because they have only one X chromosome, or because they have a Y chromosome? Much of the traditional evidence bearing on this question comes

(a) If the recombination between *A* and *C* is 8% (8 map units), then *B* must be in the middle.

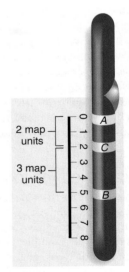

(b) If the recombination between *A* and *C* is 2%, then *C* must be in the middle.

▌ **Figure 11-13   Gene mapping**

Gene order (that is, which locus lies between the other two) is determined by the percentage of recombination between each of the possible pairs. In this hypothetical example, the percentage of recombination between locus *A* and locus *B* is 5% (corresponding to 5 map units) and that between *B* and *C* is 3% (3 map units). There are two alternatives for the linear order of these alleles.

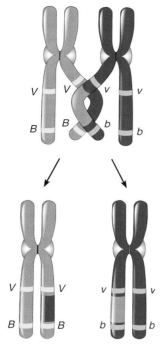

Figure 11-14   Double crossing-over

If the same homologous chromatids undergo double crossing-over between the genes of interest, the gametes formed are not recombinant for these genes.

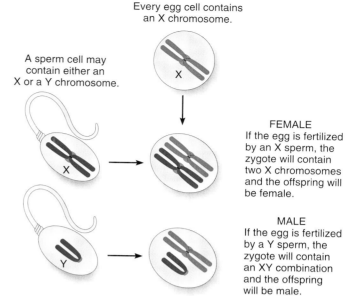

Every egg cell contains an X chromosome.

A sperm cell may contain either an X or a Y chromosome.

FEMALE
If the egg is fertilized by an X sperm, the zygote will contain two X chromosomes and the offspring will be female.

MALE
If the egg is fertilized by a Y sperm, the zygote will contain an XY combination and the offspring will be male.

Figure 11-15   Sex determination in mammals

The sperm determines sex in mammals. An X-bearing sperm produces a female, and a Y-bearing sperm produces a male.

from studies of people with abnormal sex chromosome constitutions (see Chapter 16). A person with an XXY constitution is a nearly normal male in external appearance, although his testes are underdeveloped (Klinefelter syndrome). A person with one X but no Y chromosome has the overall appearance of a female but has defects such as short stature and undeveloped ovaries (Turner syndrome). An embryo with a Y but no X does not survive. Thus, all individuals require at least one X, and the Y is the male-determining chromosome.

Geneticists have identified several genes on the Y chromosomes that are involved in determination of the male sex. The *sex reversal on Y (SRY) gene,* the major male-determining gene on the Y chromosome, acts as a "genetic switch" that causes testes to develop in the fetus. The developing testes then secrete the hormone **testosterone,** which causes other male characteristics to develop. A few other genes on the Y chromosome also play a role in sex determination, as do many genes on the X chromosome, which explains why an XXY individual does not have a completely normal male phenotype. Some genes on the autosomes also affect sex development.

Evidence suggests that the X and Y chromosomes of mammals originated as a homologous pair of autosomes. During the evolution of the X and Y chromosomes, almost all of the original functional genes were retained on the X chromosome and lost from the Y chromosome. At the same time, the Y chromosome retained certain genes that coded for proteins that determine maleness, or perhaps mutations occurred in existing genes on the Y chromosome that made it the male-determining chromosome.

Thus, the sex chromosomes are not truly homologous in their present forms, because they are not similar in size, shape, or genetic constitution. Nevertheless, they have retained a short homologous "pairing region" at the chromosome tips that lets them synapse and separate from one another during meiosis. Half the sperm contain an X chromosome, and half contain a Y chromosome (Fig. 11-15). All normal eggs bear a single X chromosome. Fertilization of an X-bearing egg by an X-bearing sperm results in an XX (female) zygote, or fertilized egg; fertilization by a Y-bearing sperm results in an XY (male) zygote.

You might expect to have equal numbers of X- and Y-bearing sperm and a 1:1 ratio of females to males. However, in humans more males are conceived than females, and more males die before birth. Even at birth the ratio is not 1:1; about 106 boys are born for every 100 girls. Although we do not know why this occurs, Y-bearing sperm seem to have some selective advantage.

An XX/XY sex chromosome mechanism, similar to that of humans, operates in many species of animals. However, it is not universal, and many of the details may vary. For example, the fruit fly, *Drosophila,* has XX females and XY males, but the Y does not determine maleness; a fruit fly with only one X chromosome and no Y chromosome has a male phenotype. In birds and butterflies the mechanism is reversed, with males having the equivalent of XX and females the equivalent of XY.

X-linked genes have unusual inheritance patterns

The human X chromosome contains many loci that are required in both sexes. Genes located in the X chromosome, such as those governing color perception and blood clotting, are sometimes called **sex-linked genes.** It is more appropriate, however, to re-

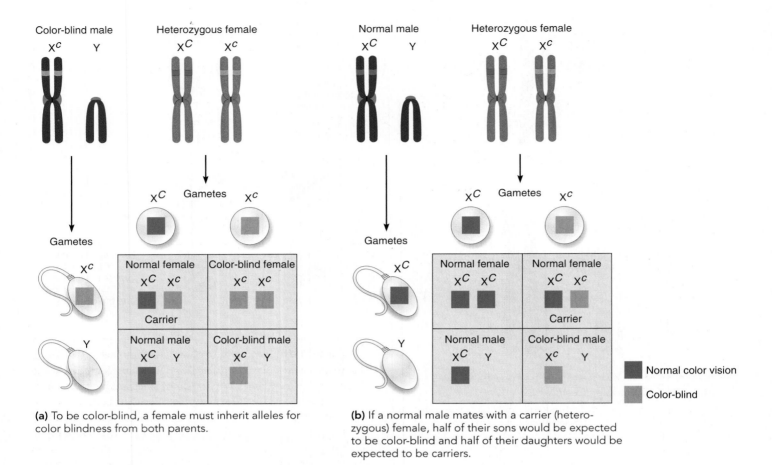

Color-blind male   Heterozygous female
X^c Y X^C X^c

Normal male   Heterozygous female
X^C Y X^C X^c

Gametes   Gametes
X^C X^c X^C X^c

Gametes
X^c

| Normal female X^C X^c Carrier | Color-blind female X^c X^c |
| Normal male X^C Y | Color-blind male X^c Y |

Y

Gametes
X^C

| Normal female X^C X^C | Normal female X^C X^c Carrier |
| Normal male X^C Y | Color-blind male X^c Y |

Y

■ Normal color vision
■ Color-blind

(a) To be color-blind, a female must inherit alleles for color blindness from both parents.

(b) If a normal male mates with a carrier (heterozygous) female, half of their sons would be expected to be color-blind and half of their daughters would be expected to be carriers.

Figure 11-16   X-linked red–green color blindness
Note that the Y chromosome does not carry a gene for color vision.

fer to them as **X-linked genes** because they follow the transmission pattern of the X chromosome and, strictly speaking, are not linked to the sex of the organism per se.

A female receives one X from her mother and one X from her father. A male receives his Y chromosome, which makes him male, from his father. From his mother he inherits a single X chromosome and therefore all his X-linked genes. In the male, every allele present on the X chromosome is expressed, whether that allele was dominant or recessive in the female parent. A male is neither homozygous nor heterozygous for his X-linked alleles; instead, he is always **hemizygous**—that is, he has only one copy of each X-linked gene. The significance of hemizygosity in males is that damaging recessive phenotypes are expressed, making males more likely to be affected by the many genetic diseases associated with the X chromosome (see Chapter 16).

We will use a simple system of notation for problems involving X linkage, indicating the X chromosome and incorporating specific alleles as superscripts. For example, the symbol X^c signifies a recessive X-linked allele for color blindness and X^C the dominant X-linked allele for normal color vision. The Y chromosome is written without superscripts because it does not carry the locus of interest. Two recessive X-linked alleles must be present in a female for the abnormal phenotype to be expressed (X^cX^c), whereas in the hemizygous male a single recessive allele (X^cY)

causes the abnormal phenotype. As a consequence, these abnormal alleles are much more frequently expressed in male offspring. A heterozygous female may be a *carrier*, an individual who possesses one copy of a mutant recessive allele but does not express it in the phenotype (X^CX^c).

To be expressed in a female, a recessive X-linked allele must be inherited from both parents. A color-blind female, for example, must have a color-blind father and a mother who is homozygous or heterozygous for a recessive color-blindness allele (■ Fig. 11-16). The homozygous combination is unusual, because the frequency of alleles for color blindness is relatively low. In contrast, a color-blind male need only have a mother who is heterozygous for color blindness; his father can have normal vision. Therefore, X-linked recessive traits are generally much more common in males than in females, a fact that may partially explain why human male embryos are more likely to die.

Dosage compensation equalizes the expression of X-linked genes in males and females

The X chromosome contains numerous genes required by both sexes, yet a normal female has two copies ("doses") for each locus, whereas a normal male has only one. **Dosage compensation** is a mechanism that makes equivalent the two doses in the female

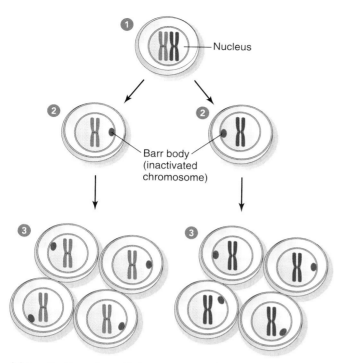

- **1** The zygote and early embryonic cells have two X chromosomes, one from each parent.

- **2** Random inactivation of one X chromosome occurs early in development. Roughly half the cells inactivate one X chromosome (*left cell*), and the other half inactivate the second X chromosome (*right cell*). The inactive X chromosome is visible as a Barr body near the nuclear envelope.

- **3** Chromosome inactivation persists through subsequent mitotic divisions, resulting in patches of cells in the adult body.

(a) Inactivation of one X chromosome in female cells.

(b) Persian calico cat. A calico cat has X-linked genes for both black and yellow (or orange) fur. Because of random X chromosome inactivation, black is expressed in some groups of cells and orange is expressed in others. (The patches of white fur are due to the presence of other genes that affect fur color.)

Figure 11-17 Dosage compensation in female mammals

and the single dose in the male. Because of dosage compensation, males and females produce the same amounts of X-linked proteins and other gene products. Male fruit flies accomplish this by making their single X chromosome more active. In most fly tissues, the metabolic activity of a single male X chromosome is equal to the combined metabolic activity of the two X chromosomes present in the female.

Dosage compensation in humans and other mammals generally involves the random inactivation of one of the two X chromosomes in the female (█ Fig. 11-17a). During interphase, a dark spot of chromatin is visible at the edge of the nucleus of each female mammalian cell when stained and observed with a microscope. This dark spot, known as a **Barr body,** is a dense, metabolically inactive X chromosome. The other X chromosome resembles the metabolically active autosomes; during interphase, it is a greatly extended thread that is not evident in light microscopy. From this and other evidence, the British geneticist Mary Lyon hypothesized in 1961 that in most cells of a female mammal, only one of the two X chromosomes is active; the other is inactive and is condensed as a Barr body. Actually, however, X chromosome inactivation is never complete; as many as 25% of the genes on the inactive chromosome are expressed to some degree.

X chromosome inactivation is a random event in each somatic (body) cell of the female embryo. A female mammal that is heterozygous at an X-linked locus expresses one of the alleles in about half her cells and the other allele in the other half. This is sometimes evident in the phenotype. Mice and cats have several X-linked genes for certain coat colors. Females that are heterozygous for such genes may show patches of one coat color in the middle of areas of the other coat color. This phenomenon, termed **variegation,** is evident in tortoiseshell and calico cats (█ Fig. 11-17b). Early in development, when relatively few cells are present, X chromosome inactivation occurs randomly in each cell. X inactivation is then maintained during all subsequent divisions of that cell line. When any one of these cells divides by mitosis, the cells of the resulting *clone* (a group of genetically identical cells) all have the same active X chromosome, and therefore a patch develops of cells that all express the same color.

Why, you might ask, is variegation not always apparent in females heterozygous at X-linked loci? The answer is that, although variegation usually occurs, we may need to use special techniques to observe it. For example, color blindness is caused by a defect involving the pigments in the cone cells in the retina of the eye (see Chapter 42). In at least one type of red–green color blindness, the retina of a heterozygous female actually contains patches of abnormal cones, but the patches of normal cones are enough to provide normal color vision. Variegation can be very hard to observe in cases where cell products become mixed in bodily fluids. For instance, in females heterozygous for the allele that causes hemophilia, only half the cells responsible for producing a specific blood-clotting factor do so, but they produce enough to ensure that the blood clots normally.

Review

- What are linked genes?
- How do geneticists use a two-point test cross to detect linkage?
- Which chromosome determines the male sex in humans and most species of mammals?
- What is dosage compensation, and what does it generally involve in mammals?

EXTENSIONS OF MENDELIAN GENETICS

Learning Objectives

9 Explain some of the ways genes may interact to affect the phenotype; discuss how a single gene can affect many features of the organism simultaneously.

10 Distinguish among incomplete dominance, codominance, multiple alleles, epistasis, and polygenes.

11 Describe *norm of reaction*, and give an example.

The relationship between a given locus and the trait it controls may or may not be simple. A single pair of alleles of a locus may regulate the appearance of a single trait (such as tall versus short in garden peas). Alternatively, a pair of alleles may participate in the control of several traits, or alleles of many loci may interact to affect the phenotypic expression of a single character. Not surprisingly, these more complex relationships are quite common.

You can assess the phenotype on one or many levels. It may be a morphological trait, such as shape, size, or color. It may be a physiological trait or even a biochemical trait, such as the presence or absence of a specific enzyme required for the metabolism of some specific molecule. In addition, changes in the environmental conditions under which the organism develops may alter the phenotypic expression of genes.

Dominance is not always complete

Studies of the inheritance of many traits in a wide variety of organisms have shown that one member of a pair of alleles may not be completely dominant over the other. In such instances, it is inaccurate to use the terms *dominant* and *recessive*.

The plants commonly known as four o'clocks (*Mirabilis jalapa*) may have red or white flowers. Each color breeds true when these plants are self-pollinated. What flower color might you expect in the offspring of a cross between a red-flowering plant and a white-flowering one? Without knowing which is dominant, you might predict that all would have red flowers or all would have white flowers. German botanist Carl Correns, one of the rediscoverers of Mendel's work, first performed this cross and found that all F₁ offspring have pink flowers!

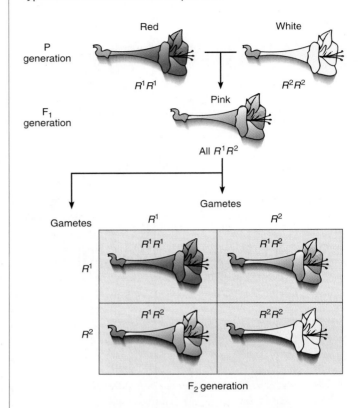

Figure 11-18 *Animated* Incomplete dominance in four o'clocks

Two incompletely dominant alleles, R^1 and R^2, are responsible for red, white, and pink flower colors. Red-flowered plants are R^1R^1, white-flowered plants are R^2R^2, and heterozygotes (R^1R^2) are pink. Note that uppercase letters are used for both alleles, because neither is recessive to the other.

Does this result in any way indicate that Mendel's assumptions about inheritance are wrong? Did the parental traits blend inseparably in the offspring? Quite the contrary, for when two of these pink-flowered plants are crossed, red-flowered, pink-flowered, and white-flowered offspring appear in a ratio of 1 : 2 : 1 (Fig. 11-18). In this instance, as in all other aspects of the scientific process, results that differ from those hypothesized prompt scientists to re-examine and modify their hypotheses to account for the exceptional results. The pink-flowered plants are clearly the heterozygous individuals, and neither the red allele nor the white allele is completely dominant. When the heterozygote has a phenotype intermediate between those of its two parents, the genes are said to show **incomplete dominance.** In these crosses, the genotypic and phenotypic ratios are identical.

Incomplete dominance is not unique to four o'clocks, and additional examples of incomplete dominance are known in both plants and animals. For example, true-breeding white chickens and true-breeding black chickens produce bluish gray offspring, known as Andalusian blues, when crossed.

TABLE 11-3

ABO Blood Types

Phenotype (blood type)	Genotypes	Antigen on RBC	Antibodies to A or B Antigens in Plasma
A	$I^A I^A$, $I^A i$	A	Anti-B
B	$I^B I^B$, $I^B i$	B	Anti-A
AB	$I^A I^B$	A, B	None
O	ii	None	Anti-A, anti-B

* This table and the discussion of the ABO system have been simplified somewhat. Note that the body produces antibodies against the antigens *lacking* on its own red blood cells (RBCs). Because of their specificity for the corresponding antigens, these antibodies are used in standard tests to determine blood types.

In both cattle and horses, reddish coat color is not completely dominant to white coat color. Heterozygous individuals have a mixture of reddish hairs and white hairs, which is called *roan*. If you saw a white mare nursing a roan foal, what would you guess was the coat color of the foal's father? Because the reddish and white colors are expressed independently (hair by hair) in the roan heterozygote, scientists sometimes refer to this as a case of **codominance.** Strictly speaking, the term *incomplete dominance* refers to instances in which the heterozygote is intermediate in phenotype, and *codominance* refers to instances in which the heterozygote simultaneously expresses the phenotypes of both types of homozygotes.

Humans have four blood types (A, B, AB, and O), collectively called the **ABO blood group.** The human ABO blood group is an excellent example of codominant alleles. Blood types A, B, AB, and O are controlled by three alleles representing a single locus (**Table 11-3**). Allele I^A codes for the synthesis of a specific glycoprotein, antigen A, which is expressed on the surface of red blood cells. (Immune responses are discussed in Chapter 44; for now, we define *antigens* as "substances capable of stimulating an immune response.") Allele I^B leads to the production of antigen *B*, a different but related glycoprotein. Allele *i* is allelic to I^A and I^B, but it does not code for an antigen. Individuals with the genotype $I^A I^A$ or $I^A i$ have blood type A. People with genotype $I^B I^B$ or $I^B i$ have blood type B. Those with genotype $I^A I^B$ have blood type AB, whereas those with genotype *ii* have blood type O. These results show that neither allele I^A nor allele I^B is dominant to the other. Both alleles are expressed phenotypically in the heterozygote and are therefore codominant to each other, although each is dominant to allele *i*.

At one time, determining blood types was used to settle cases of disputed parentage. Although blood-type tests can *exclude* someone as a possible parent of a particular child, they can never prove a certain person is the parent; they only determine he or she *could* be. Could a man with blood type AB be the father of a child with blood type O? Could a woman with blood type O be the mother of a child with blood type AB? Could a type B child with a type A mother have a type A father or a type O father?[3]

---

[3]The answer to all these questions is no.

Multiple alleles for a locus may exist in a population

Most of our examples so far have dealt with situations in which each locus was represented by a maximum of two allelic variants. It is true that a single diploid individual has a maximum of two different alleles for a particular locus and that a haploid gamete has only one allele for each locus. However, if you survey a population, you may find more than two alleles for a particular locus, as you saw with the ABO blood group. If three or more alleles for a given locus exist within the population, we say that locus has **multiple alleles.** Research has shown that many loci have multiple alleles. Some alleles can be identified by the activity of a certain enzyme or by some other biochemical feature but do not produce an obvious phenotype. Others produce a readily recognizable phenotype, and certain patterns of dominance can be discerned when the alleles are combined in various ways.

In rabbits, four alleles occur at the locus for coat color (**Fig. 11-19**). A *C* allele causes a fully colored dark gray coat. The homozygous recessive genotype, *cc*, causes albino (white) coat color. There are two additional allelic variants of the same locus, c^{ch} and c^h. An individual with the genotype $c^{ch} c^{ch}$ has the chinchilla pattern, in which the entire body has a light, silvery gray color. The genotype $c^h c^h$ causes the Himalayan pattern, in which the body is white but the tips of the ears, nose, tail, and legs are colored, like the color pattern of a Siamese cat. On the basis of the results of genetic crosses, these alleles can be arranged in the following series:

$$C > c^{ch} > c^h > c$$

Each allele is dominant to those following it and recessive to those preceding it. For example, a $c^{ch} c^{ch}$, $c^{ch} c^h$, or $c^{ch} c$ rabbit has the chinchilla pattern, whereas a $c^h c^h$ or $c^h c$ rabbit has the Himalayan pattern. (We revisit the Himalayan coat pattern later in the chapter when we discuss the influence of the environment on gene expression.)

In other series of multiple alleles, certain alleles may be codominant and others incompletely dominant. In such cases, the heterozygotes commonly have phenotypes intermediate between those of their parents.

A single gene may affect multiple aspects of the phenotype

In our examples, the relationship between a gene and its phenotype has been direct, precise, and exact, and the loci have controlled the appearance of single traits. However, the relationship of gene to trait may not have such a simple genetic basis. Most genes affect several different characters. The ability of a single gene to have multiple effects is known as **pleiotropy.** Most cases of pleiotropy can be traced to a single fundamental cause. For example, a defective enzyme may affect the functioning of many types of cells. Pleiotropy is evident in many genetic diseases in which a single pair of alleles causes multiple symptoms. For example, people who are homozygous for the recessive allele that causes cystic fibrosis produce abnormally thick mucus in many

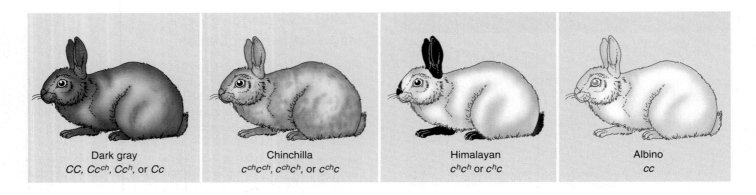

Dark gray	Chinchilla	Himalayan	Albino
CC, Cc^{ch}, Cc^h, or Cc	$c^{ch}c^{ch}$, $c^{ch}c^h$, or $c^{ch}c$	c^hc^h or c^hc	cc

Figure 11-19 *Animated* Multiple alleles in rabbits

In rabbits the locus for coat color has four alleles, designated C, c^{ch}, c^h, and c. An individual rabbit, being diploid, may have any two alleles of this series. The drawing shows phenotypes and their associated genotypes.

parts of the body, including the respiratory, digestive, and reproductive systems (see Chapter 16).

Alleles of different loci may interact to produce a phenotype

Several pairs of alleles may interact to affect a single phenotype, or one pair may inhibit or reverse the effect of another pair. One example of gene interaction is illustrated by the inheritance of combs in chickens, where two genes may interact to produce a novel phenotype (■ Fig. 11-20). The allele for a rose comb, R, is dominant to that for a single comb, r. A second, unlinked pair of alleles governs the inheritance of a pea comb, P, versus a single comb, p. A single-comb chicken is homozygous for the recessive allele at both loci ($pprr$). A rose comb chicken is either $ppRR$ or $ppRr$, and a pea comb chicken is either $PPrr$ or $Pprr$. When an R and P occur in the same individual, the phenotype is neither a pea nor a rose comb but a completely different type, a walnut comb. The walnut comb phenotype is produced whenever a chicken has one or two R alleles, plus one or two P alleles (that is, $PPRR$, $PpRR$, $PPRr$, or $PpRr$). What would you hypothesize about the types of combs among the offspring of two heterozygous walnut comb chickens, $PpRr$? How does this form of gene interaction affect the ratio of phenotypes in the F$_2$ generation? Is it the typical Mendelian 9:3:3:1 ratio?[4]

Epistasis is a common type of gene interaction in which the presence of certain alleles of one locus can prevent or mask the expression of alleles of a different locus and express their own phenotype instead. (The term *epistasis* means "standing on.") Unlike the chicken example, no novel phenotypes are produced in epistasis.

Coat color in Labrador retrievers is an example of epistasis that involves a gene for pigment and a gene for depositing color in the coat (■ Fig. 11-21). The two alleles for the pigment gene are B for black coat and its recessive counterpart, b, for brown coat. The gene for depositing color in the coat has two alleles, E for

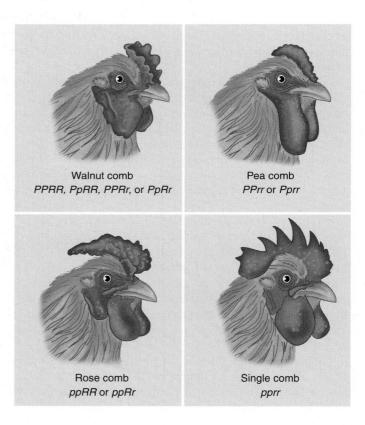

Walnut comb	Pea comb
$PPRR$, $PpRR$, $PPRr$, or $PpRr$	$PPrr$ or $Pprr$
Rose comb	**Single comb**
$ppRR$ or $ppRr$	$pprr$

Figure 11-20 *Animated* Gene interaction in chickens

Two gene pairs govern four chicken comb phenotypes. Chickens with walnut combs have the genotype $P\_R\_$. Chickens with pea combs have the genotype P_rr, and those with rose combs have the genotype $ppR\_$. Chickens that are homozygous recessive for both loci, $pprr$, have a single comb. (The blanks represent either dominant or recessive alleles.)

---

[4]The offspring of two heterozygous walnut comb chickens will have four genotypes in what appears to be a Mendelian 9:3:3:1 ratio: 9 walnut comb, 3 pea comb, 3 rose comb, and 1 single comb. This example is not a typical Mendelian 9:3:3:1 ratio because it involves a single character (comb shape) coded by alleles at two different loci. In Mendelian inheritance, the 9:3:3:1 ratio involves *two* characters (such as seed color and seed shape) coded by alleles at two loci.

Black
BBEE, BbEE, BBEe or *BbEe*

Yellow
BBee, Bbee or *bbee*

Chocolate
bbEE or *bbEe*

Figure 11-21 *Animated* Epistasis in Labrador retrievers

Two gene pairs interact to govern coat color in Labrador retrievers. Black Labs have the genotype *B_E_*; yellow Labs have the genotype *B_ee* or *bbee*; and chocolate Labs have the genotype *bbE_*. (The blanks represent either dominant or recessive alleles.)

the expression of black and brown coats and *e*, which is epistatic and blocks the expression of the *B/b* gene. The epistatic allele is recessive and therefore is only expressed as a yellow coat in the homozygous condition (*ee*), regardless of the combination of *B* and *b* alleles in the genotype.

Polygenes act additively to produce a phenotype

Many human characters—such as height, body form, and skin pigmentation—are not inherited through alleles at a single locus. The same holds true for many commercially important characters in domestic plants and animals, such as milk and egg production. Alleles at several, perhaps many, loci affect each character. The term **polygenic inheritance** is applied when multiple independent pairs of genes have similar and additive effects on the same character.

Polygenes—as many as 60 loci—account for the inheritance of skin pigmentation in humans. To keep things simple in this example, we illustrate the principle of polygenic inheritance in human skin pigmentation with pairs of alleles at three unlinked loci. These can be designated *A* and *a*, *B* and *b*, and *C* and *c*. The

capital letters represent incompletely dominant alleles producing dark skin. The more capital letters, the darker the skin, because the alleles affect skin pigmentation in an additive fashion. A person with the darkest skin possible would have the genotype *AABBCC*, and a person with the lightest skin possible would have the genotype *aabbcc*. The F_1 offspring of an *aabbcc* person and an *AABBCC* person are all *AaBbCc* and have an intermediate skin color. The F_2 offspring of two such triple heterozygotes (*AaBbCc* × *AaBbCc*) would have skin pigmentation ranging from very dark to very light (Fig. 11-22).

Polygenic inheritance is characterized by an F_1 generation that is intermediate between the two completely homozygous parents and by an F_2 generation that shows wide variation between the two parental types. When the number of individuals in a population is plotted against the amount of skin pigmentation and the points are connected, the result is a bell-shaped curve, a **normal distribution curve.** Most of the F_2 generation individuals have one of the intermediate phenotypes; only a few show the extreme phenotypes of the original P generation (that is, the grandparents). On average, only 1 of 64 is as dark as the very dark grandparent, and only 1 of 64 is as light as the very light grandparent. The alleles *A*, *B*, and *C* each produce about the same amount of darkening of the skin; hence, the genotypes *AaBbCc*,

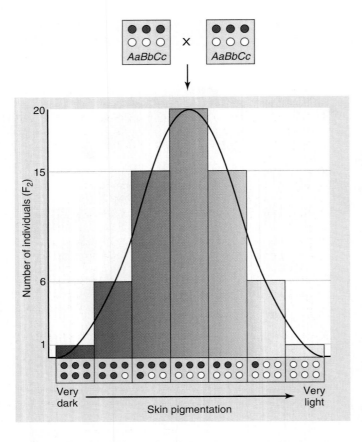

Figure 11-22 Polygenic inheritance in human skin pigmentation

This simplified example assumes that skin pigmentation in humans is governed by alleles of three unlinked loci. The alleles producing dark skin (A, B, and C) are represented by capital letters, but they are not dominant. Instead, their effects are additive. The number of dark dots, each signifying an allele producing dark skin, is counted to determine the phenotype. A wide range of phenotypes is possible when individuals of intermediate phenotype mate and have offspring (AaBbCc × AaBbCc). The expected distribution of phenotypes is consistent with the superimposed normal distribution curve.

AABbcc, AAbbCc, AaBBcc, aaBBCc, AabbCC, and *aaBbCC* all produce similar intermediate phenotypes.

Genes interact with the environment to shape phenotype

Imagine taking two cuttings from a coleus plant. You place one cutting in a growth chamber where it receives a lot of light and the other in a growth chamber programmed to deliver dim light. Although the cuttings are genetically identical (they came from the same plant), the appearance of the two plants would be dramatically different in a very short period. The plant with adequate light would thrive and grow vigorously, whereas the plant with inadequate light would be spindly and weak. During an organism's development, both genotype and environmental factors af-

fect the expression of phenotype, and genetically identical individuals may develop differently in different environments.

Consider another example of the effect of environment on gene expression, this one involving the Himalayan rabbits discussed earlier. Recall that the phenotype of these rabbits is white fur except for dark patches on the ears, nose, and paws. The local surface temperature of a rabbit's ears, nose, and paws is colder in the rabbit's natural environment, and this temperature difference causes the production of dark fur. If you raise rabbits with the Himalayan genotype at a warm temperature (30°C), the rabbits are completely white, with no dark patches on the ears, nose, or paws. If you raise Himalayan rabbits at a cooler temperature (25°C), however, they develop the characteristic dark patches of fur. Thus, genes can function differently in different environments.

Now let's examine a human example—height—in the context of genes and the environment. The inheritance of height in humans is polygenic and involves alleles representing 10 or more loci. Because many genes are involved and because height is modified by a variety of environmental conditions, such as diet and general health, the height of most adults ranges from 4 ft 2 in (1.25 m) to 7 ft 2 in (2.15 m). The genes that affect height set limits for the phenotype—no human is 12 ft tall, for example—but the environment shapes the phenotype within its genetic limits. The range of phenotypic possibilities that can develop from a single genotype under different environmental conditions is known as the **norm of reaction.** In certain genotypes the norm of reaction is quite limited. In other genotypes, such as those involved in human height, the norm of reaction is quite broad.

Although interactions between genes and the environment influence the phenotypes of many characters, it is difficult to determine the exact contributions of genes and the environment to a given phenotype. In some cases the environment regulates the activity of certain genes, turning them on in some environmental conditions and off in others. This subject, known as the *nature–nurture question,* is particularly controversial, because it relates to human characters ranging from mental illnesses, such as depression, to level of intelligence.

Intelligence is known to have a genetic component, but environment is also an important influence. To test for the relative effects of genes and environment on intelligence, however, we would have to separate and raise genetically identical children in different environments. Because it is ethically unacceptable to manipulate humans in this way, such tests cannot be performed. All we can say with certainty is that an individual's genes provide the potential for developing a particular phenotype but that environmental influences also shape the phenotype. Thus, phenotypic expression is much more complex than a simple nature-versus-nurture argument would imply.

Review

- What is the difference between incomplete dominance and codominance?
- What is the difference between multiple alleles and polygenes?
- What is the difference between pleiotropy and epistasis?
- What is a norm of reaction?

Learning Objectives

1 Define the terms *phenotype, genotype, locus, allele, dominant allele, recessive allele, homozygous,* and *heterozygous* (page 235).

▪ **Genes** are in chromosomes; the site a gene occupies in the chromosome is its **locus.** Different forms of a particular gene are **alleles;** they occupy corresponding loci on homologous chromosomes. Therefore, genes exist as pairs of alleles in diploid individuals.

▪ An individual that carries two identical alleles is said to be **homozygous** for that locus. If the two alleles are different, that individual is said to be **heterozygous** for that locus.

▪ One allele, the **dominant allele,** may mask the expression of the other allele, the **recessive allele,** in a heterozygous individual. For this reason two individuals with the same appearance, or **phenotype,** may differ from each other in genetic constitution, or **genotype.**

2 Describe Mendel's principles of segregation and independent assortment (page 235).

▪ According to Mendel's **principle of segregation,** during meiosis the alleles for each locus separate, or segregate, from each other. When haploid gametes are formed, each contains only one allele for each locus.

▪ According to Mendel's **principle of independent assortment,** alleles of different loci are distributed randomly into the gametes. This can result in **recombination,** the production of new gene combinations that were not present in the **parental (P) generation.**

ThomsonNOW **Learn more about independent assortment by clicking on the figure in ThomsonNOW.**

3 Distinguish among monohybrid, dihybrid, and test crosses (page 235).

▪ A cross between homozygous parents (P generation) that differ from each other with respect to their alleles at one locus is called a **monohybrid cross;** if they differ at two loci, it is called a **dihybrid cross.**

▪ The first generation of offspring is heterozygous and is called the F_1 **generation;** the generation produced by a cross of two F_1 individuals is the F_2 **generation.**

▪ A **test cross** is between an individual of unknown genotype and a homozygous recessive individual.

ThomsonNOW **Learn more about monohybrid, dihybrid, and test crosses by clicking on the figures in ThomsonNOW.**

4 Explain Mendel's principles of segregation and independent assortment, given what scientists now know about genes and chromosomes (page 235).

▪ Segregation of alleles is a direct result of homologous chromosomes separating during meiosis.

▪ Independent assortment occurs because there are two ways in which two pairs of homologous chromosomes can be arranged at metaphase I of meiosis. The orientation of homologous chromosomes on the metaphase plate determines the way chromosomes are distributed into haploid cells.

5 Apply the product rule and sum rule appropriately when predicting the outcomes of genetic crosses (page 243).

▪ Genetic ratios can be expressed in terms of probabilities. Any probability is expressed as a fraction or decimal fraction, calculated as the number of favorable events divided by the total number of events. Probabilities range from 0 (an impossible event) to 1 (a certain event).

▪ According to the **product rule,** the probability of two independent events occurring together can be calculated by multiplying the probabilities of each event occurring separately.

▪ According to the **sum rule,** the probability of an outcome that can be obtained in more than one way can be calculated by adding the separate probabilities.

6 Define *linkage,* and relate it to specific events in meiosis (page 246).

▪ Each chromosome behaves genetically as if it consisted of genes arranged in a linear order. **Linkage** is the tendency for a group of genes on the same chromosome to be inherited together. Groups of genes on the same chromosome are **linked genes.** Independent assortment does not apply if two loci are linked close together on the same pair of homologous chromosomes.

▪ Recombination of linked genes can result from **crossing-over** (breaking and rejoining of homologous chromatids) in meiotic prophase I. (Recombination can also result from independent assortment of unlinked genes.)

▪ By measuring the frequency of recombination between linked genes, it is possible to construct a linkage map of a chromosome.

7 Show how data from a test cross involving alleles of two loci can be used to distinguish between independent assortment and linkage (page 246).

▪ To distinguish between independent assortment and linked genes, perform a **two-point test cross** between an individual that is heterozygous at both loci and an individual that is homozygous recessive for both.

▪ Linkage is recognized when an excess of parental-type offspring and a deficiency of recombinant-type offspring are produced in a two-point test cross.

8 Discuss the genetic determination of sex and the inheritance of X-linked genes in mammals (page 246).

▪ The sex of humans and many other animals is determined by the **X** and **Y chromosomes** or their equivalents. Normal female mammals have two X chromosomes; normal males have one X and one Y. The fertilization of an X-bearing egg by an X-bearing sperm results in a female (XX) zygote. The fertilization of an X-bearing egg by a Y-bearing sperm results in a male (XY) zygote.

▪ The Y chromosome determines male sex in mammals. The X chromosome contains many important genes unrelated to sex determination that are required by both males and females. A male receives all his **X-linked genes** from his mother. A female receives X-linked genes from both parents.

▪ A female mammal shows **dosage compensation** of X-linked genes. Only one of the two X chromosomes is expressed in each cell; the other is inactive and is seen as a dark-staining **Barr body** at the edge of the interphase nucleus.

9 Explain some of the ways genes may interact to affect the phenotype; discuss how a single gene can affect many features of the organism simultaneously (page 252).

▪ The relationship between genotype and phenotype may be quite complex. **Pleiotropy** is the ability of one gene to have several effects on different characters. Most cases of pleiotropy can be traced to a single cause, such as a defective enzyme. Alternatively, alleles of many loci may interact to affect the phenotypic expression of a single character.

10 Distinguish among incomplete dominance, codominance, multiple alleles, epistasis, and polygenes (page 252).

■ Dominance does not always apply, and some alleles demonstrate **incomplete dominance,** in which the heterozygote is intermediate in phenotype, or **codominance,** in which the heterozygote simultaneously expresses the phenotypes of both homozygotes.

■ **Multiple alleles,** three or more alleles that can potentially occupy a particular locus, may exist in a population. A diploid individual has any two of the alleles; a haploid individual or gamete has only one.

■ In **epistasis,** an allele of one locus can mask the expression of alleles of a different locus.

■ In **polygenic inheritance,** multiple independent pairs of genes may have similar and additive effects on the phenotype.

11 Describe *norm of reaction,* and give an example (page 252).

■ The range of phenotypic possibilities that can develop from a single genotype under different environmental conditions is known as the **norm of reaction.**

■ Many genes are involved in the inheritance of height in humans. Also, height is modified by a variety of environmental conditions, such as diet and general health. The genes that affect height set the norm of reaction—that is, the limits—for the phenotype, and the environment molds the phenotype within its norm of reaction.

TEST YOUR UNDERSTANDING

1. One reason Mendel discovered the basic principles of inheritance is that he (a) understood the behavior of chromosomes in mitosis and meiosis (b) studied a wide variety of experimental organisms (c) began by establishing true-breeding lines (d) studied various types of linkage (e) studied hybrids between parents that differed in many, often not clearly defined, ways

2. One of the autosomal loci controlling eye color in fruit flies has two alleles, one for brown eyes and the other for red eyes. Fruit flies from a true-breeding line with brown eyes were crossed with flies from a true-breeding line with red eyes. The F_1 flies had red eyes. What conclusion can be drawn from this experiment? (a) these alleles underwent independent assortment (b) these alleles underwent segregation (c) these genes are X-linked (d) the allele for red eyes is dominant to the allele for brown eyes (e) all of the preceding are true

3. The F_1 flies described in question 2 were mated with brown-eyed flies from a true-breeding line. What phenotypes would you expect the offspring to have? (a) all red eyes (b) all brown eyes (c) half red eyes and half brown eyes (d) red-eyed females and brown-eyed males (e) brown-eyed females and red-eyed males

4. The type of cross described in question 3 is a(an) (a) F_2 cross (b) dihybrid cross (c) test cross (d) two-point test cross (e) none of the preceding

Use the following information to answer questions 5 through 8:

In peas, the allele for round seeds (R) is dominant to that for wrinkled seeds (r); the allele for yellow seeds (Y) is dominant to that for green seeds (y). These loci are unlinked. Plants from a true-breeding line with round, green seeds are crossed with plants from a true-breeding line with wrinkled, yellow seeds. These parents constitute the P generation.

5. The genotypes of the P generation are (a) *RRrr* and *Yyyy* (b) *RrYy* (c) *RRYY* and *rryy* (d) *RRyy* and *rrYY* (e) *RR* and *YY*

6. What are the expected genotypes of the F_1 hybrids produced by the described cross? (a) *RRrr* and *YYyy* (b) all *RrYy* (c) *RRYY* and *rryy* (d) *RRyy* and *rrYY* (e) *RR* and *YY*

7. What kinds of gametes can the F_1 individuals produce? (a) *RR* and *YY* (b) *Rr* and *Yy* (c) *RR, rr, YY,* and *yy* (d) *R, r, Y,* and *y* (e) *RY, Ry, rY,* and *ry*

8. What is the expected proportion of F_2 wrinkled, yellow seeds? (a) $\frac{9}{16}$ (b) $\frac{1}{16}$ (c) $\frac{3}{16}$ (d) $\frac{1}{4}$ (e) 0

9. Individuals of genotype *AaBb* were crossed with *aabb* individuals. Approximately equal numbers of the following classes of offspring were produced: *AaBb, Aabb, aaBb,* and *aabb.* These results illustrate Mendel's principle(s) of (a) linkage (b) independent assortment (c) segregation (d) a and c (e) b and c

10. Assume that the ratio of females to males is 1:1. A couple already has two daughters and no sons. If they plan to have a total of six children, what is the probability they will have four more girls? (a) $\frac{1}{4}$ (b) $\frac{1}{8}$ (c) $\frac{1}{16}$ (d) $\frac{1}{32}$ (e) $\frac{1}{64}$

11. Red−green color blindness is an X-linked recessive disorder in humans. Your friend is the daughter of a color-blind father. Her mother had normal color vision, but her maternal grandfather was color-blind. What is the probability your friend is color-blind? (a) 1 (b) $\frac{1}{2}$ (c) $\frac{1}{4}$ (d) $\frac{3}{4}$ (e) 0

12. When homozygous, a particular allele of a locus in rats causes abnormalities of the cartilage throughout the body, an enlarged heart, slow development, and death. This is an example of (a) pleiotropy (b) polygenic inheritance (c) epistasis (d) codominance (e) dosage compensation

13. In peas, yellow seed color is dominant to green. Determine the phenotypes (and their proportions) of the offspring of the following crosses: (a) homozygous yellow × green (b) heterozygous yellow × green (c) heterozygous yellow × homozygous yellow (d) heterozygous yellow × heterozygous yellow

14. If two animals heterozygous for a single pair of alleles are mated and have 200 offspring, about how many would be expected to have the phenotype of the dominant allele (that is, to look like the parents)?

15. When two long-winged flies were mated, the offspring included 77 with long wings and 24 with short wings. Is the

short-winged condition dominant or recessive? What are the genotypes of the parents?

16. Outline a breeding procedure whereby a true-breeding strain of red cattle could be established from a roan bull and a white cow.

17. What is the probability of rolling a 7 with a pair of dice? Which is a more likely outcome, rolling a 6 with a pair of dice or rolling an 8?

18. In rabbits, spotted coat (S) is dominant to solid color (s), and black (B) is dominant to brown (b). These loci are not linked. A brown, spotted rabbit from a pure line is mated to a solid black one, also from a pure line. What are the genotypes of the parents? What would be the genotype and phenotype of an F_1 rabbit? What would be the expected genotypes and phenotypes of the F_2 generation?

19. The long hair of Persian cats is recessive to the short hair of Siamese cats, but the black coat color of Persians is dominant to the brown-and-tan coat color of Siamese. Make up appropriate symbols for the alleles of these two unlinked loci. If a pure black, long-haired Persian is mated to a pure brown-and-tan, short-haired Siamese, what will be the appearance of the F_1 offspring? If two of these F_1 cats are mated, what is the chance that a long-haired, brown-and-tan cat will be produced in the F_2 generation? (Use the shortcut probability method to obtain your answer; then, check it with a Punnett square.)

20. Mr. and Mrs. Smith are concerned because their own blood types are A and B, respectively, but their new son, Richard, is blood type O. Could Richard be the child of these parents?

21. The expression of an allele called *frizzle* in fowl causes abnormalities of the feathers. As a consequence, the animal's body temperature is lowered, adversely affecting the functions of many internal organs. When one gene affects many traits of the organism in this way, we say that gene is _____.

22. A walnut comb rooster is mated to three hens. Hen A, which has a walnut comb, has offspring in the ratio of 3 walnut to 1 rose. Hen B, which has a pea comb, has offspring in the ratio of 3 walnut to 3 pea to 1 rose to 1 single. Hen C, which has a walnut comb, has only walnut comb offspring. What are the genotypes of the rooster and the three hens?

23. What kinds of matings result in the following phenotypic ratios? (a) 3:1 (b) 1:1 (c) 9:3:3:1 (d) 1:1:1:1

24. The weight of the fruit in a certain variety of squash is determined by alleles at two loci: *AABB* produces fruits that average 2 kg each, and *aabb* produces fruits that weigh 1 kg each. Each allele represented by a capital letter adds 0.25 kg. When a plant that produces 2-kg fruits is crossed with a plant that produces 1-kg fruits, all the offspring produce fruits that weigh 1.5 kg each. If two of these F_1 plants were crossed, how much would the fruits produced by the F_2 plants weigh?

25. The X-linked *barred* locus in chickens controls the pattern of the feathers, with the alleles B for barred pattern and b for no bars. If a barred female ($X^B Y$) is mated to a nonbarred male ($X^b X^b$), what will be the appearance of the male and female progeny? (Recall that in birds, males are XX and females are XY.) Do you see any commercial usefulness for this result? (*Hint:* It is notoriously difficult to determine the sex of newly hatched chicks.)

26. Individuals of genotype *AaBb* were mated to individuals of genotype *aabb*. One thousand offspring were counted, with the following results: 474 *Aabb*, 480 *aaBb*, 20 *AaBb*, and 26 *aabb*. What type of cross is this? Are these loci linked? What are the two parental classes and the two recombinant classes of offspring? What is the percentage of recombination between these two loci? How many map units apart are they?

27. Genes A and B are 6 map units apart, and A and C are 4 map units apart. Which gene is in the middle if B and C are 10 map units apart? Which is in the middle if B and C are 2 map units apart?

CRITICAL THINKING

1. Would the science of genetics in the 20th century have developed any differently if Gregor Mendel had never lived?

2. Sketch a series of diagrams showing each of the following, making sure to end each series with haploid gametes:

   a. How a pair of alleles for a single locus segregates in meiosis

   b. How the alleles of two unlinked loci assort independently in meiosis

   c. How the alleles of two linked loci undergo genetic recombination

3. Can you always ascertain an organism's genotype for a particular locus if you know its phenotype? Conversely, if you are given an organism's genotype for a locus, can you always reliably predict its phenotype? Explain.

4. **Evolution Link.** Darwin's theory of evolution by natural selection is based on four observations about the natural world. One of these is that each individual has a combination of traits that makes it uniquely different. Darwin recognized that much of this variation among individuals must be inherited, but he did not know about Mendel's mechanism of inheritance. Based on what you have learned in this chapter, briefly explain the variation among individuals that Darwin observed.

5. **Analyzing Data.** Using the graph in Figure 11-22, determine how many individuals were involved in the hypothetical cross studying skin color. What percentage had the lightest skin possible? The darkest skin?

Additional questions are available in ThomsonNOW at www.thomsonedu.com/login

DNA: The Carrier of Genetic Information

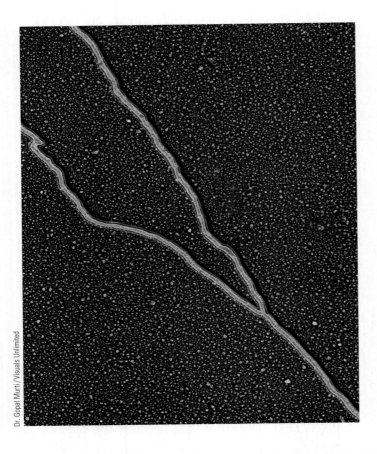

Dr. Gopal Murti / Visuals Unlimited

Electron micrograph of DNA replication. During replication, two DNA molecules are synthesized from the original parent molecule. Replication is occurring at the Y-shaped structure, which is called a *replication fork.*

KEY CONCEPTS

Beginning in the 1920s, evidence began to accumulate that DNA is the hereditary material.

The DNA molecule consists of two strands that wrap around each other to form a double helix; the order of its building blocks provides for the storage of genetic information.

The DNA building blocks consist of four different nucleotide subunits, designated T, C, A, and G.

The pairing of nucleotide subunits occurs based on precise pairing rules: T pairs with A, and C pairs with G. Hydrogen bonding between base pairs holds the two strands of DNA together.

DNA replication, which results in two identical double-stranded DNA molecules, provides the molecular mechanism for passing genetic information from one generation to the next.

After the rediscovery of Mendel's principles in 1900, geneticists conducted experiments to learn how genes are arranged in chromosomes and how they are transmitted from generation to generation. However, basic questions remained unanswered through most of the first half of the 20th century: What are genes made of? How do genes work? The studies of inheritance patterns described in Chapter 11 did not answer these questions. However, they provided a foundation of knowledge that enabled scientists to make predictions about the molecular (chemical) nature of genes and how genes function.

Scientists generally agreed that genes would have to store information in a form that the cell could retrieve and use. But scientists also had to account for other properties of genes. For example, experiments on a variety of organisms had demonstrated that genes are usually stable and passed unchanged from generation to generation. However, occasionally a gene converted to a different form; such genetic changes, called *mutations,* were then transmitted unchanged to future generations.

As the science of genetics was developing, biochemists were making a growing effort to correlate the known properties of genes with the nature of various biological molecules. What kind of molecule could store information? How could that information be retrieved and used to direct cell functions? What kind of molecule could be relatively stable but have the capacity to change, resulting in a mutation, under certain conditions?

As they learned more about the central role of proteins in virtually every aspect of cell structure and metabolism, some scientists considered proteins the prime candidates for the genetic material. However, protein did not turn out to be the molecule that governs inheritance. In this chapter we discuss how researchers discovered that **deoxyribonucleic acid (DNA),** a nucleic acid once thought unremarkable, is the molecular basis of inheritance. We explore the unique features of DNA, including its replication (see photograph), that enable it to carry out this role. ∎

EVIDENCE OF DNA AS THE HEREDITARY MATERIAL

Learning Objectives

1 Summarize the evidence that accumulated during the 1940s and early 1950s demonstrating that DNA is the genetic material.

2 State the questions that these classic experiments addressed: Griffith's transformation experiment, Avery's contribution to Griffith's work, and the Hershey–Chase experiments.

During the 1930s and early 1940s, most geneticists paid little attention to DNA, convinced that the genetic material must be protein. Given the accumulating evidence that genes control production of proteins (discussed in Chapter 13), it certainly seemed likely that genes themselves must also be proteins. Scientists knew proteins consisted of more than 20 different kinds of amino acids in many different combinations, which conferred unique properties on each type of protein. Given their complexity and diversity compared with other molecules, proteins seemed to be the "stuff" of which genes are made.

In contrast, scientists had established that DNA and other nucleic acids were made of only four nucleotides, and what was known about their arrangement made them relatively uninteresting to most researchers. For this reason, several early clues to the role of DNA were not widely noticed.

DNA is the transforming principle in bacteria

One of these clues had its origin in 1928, when Frederick Griffith, a British medical officer, made a curious observation concerning two strains of pneumococcus bacteria (▌ Fig. 12-1). A smooth (S) strain, named for its formation of smooth colonies on a solid growth medium, was known to exhibit **virulence,** the ability to cause disease, and often death, in its host. When living cells of this strain were injected into mice, the animals contracted pneumonia and died. Not surprisingly, the injected animals survived if the cells were first killed with heat. A related rough (R) strain of bacteria, which forms colonies with a rough surface, was known to exhibit **avirulence,** or inability to

Key Experiment

QUESTION: Can a genetic trait be transmitted from one bacterial strain to another?

HYPOTHESIS: The ability of pneumococcus bacteria to cause disease can be transmitted from the virulent strain (smooth, or S cells) to the avirulent strain (rough, or R cells).

EXPERIMENT: Griffith performed four experiments on mice, using the two strains of pneumococci: (1) injection of mice with live rough cells, (2) injection with live smooth cells, (3) injection with heat-killed smooth cells, and (4) injection with both live rough cells and heat-killed smooth cells.

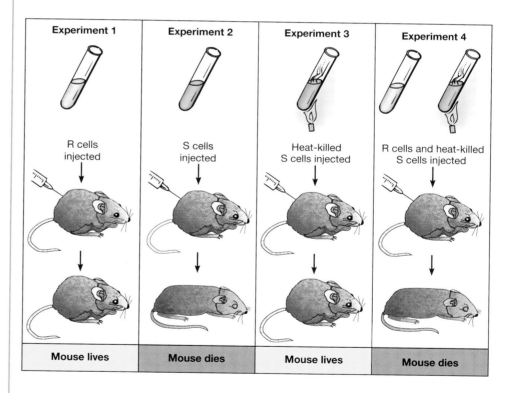

RESULTS AND CONCLUSION: Although neither the rough strain nor the heat-killed smooth strain could kill a mouse, a combination of the two did. Autopsy of the dead mouse showed the presence of living S-strain pneumococci. These results indicated that some substance in the heat-killed S cells had transformed the living R cells into a virulent form.

Figure 12-1 *Animated* Griffith's transformation experiments
Griffith was trying to develop a vaccine against pneumonia when he serendipitously discovered the phenomenon of transformation.

QUESTION: Is DNA or protein the genetic material in bacterial viruses (phages)?

HYPOTHESIS: DNA (or protein) is the genetic material in bacterial viruses.

EXPERIMENT: Hershey and Chase produced phage populations with either radioactively labeled DNA or radioactively labeled protein. In both cases, they infected bacteria with the phages and then determined whether DNA or protein was injected into bacterial cells to direct the formation of new viral particles.

RESULTS AND CONCLUSION: The researchers could separate phage protein coats labeled with the radioactive isotope ^{35}S (*left side*) from infected bacterial cells without affecting viral reproduction. However, they could not separate viral DNA labeled with the radioactive isotope ^{32}P (*right side*) from infected bacterial cells. This demonstrated that viral DNA enters the bacterial cells and is required for the synthesis of new viral particles. Thus, DNA is the genetic material in phages.

Figure 12-2 *Animated* The Hershey–Chase experiments

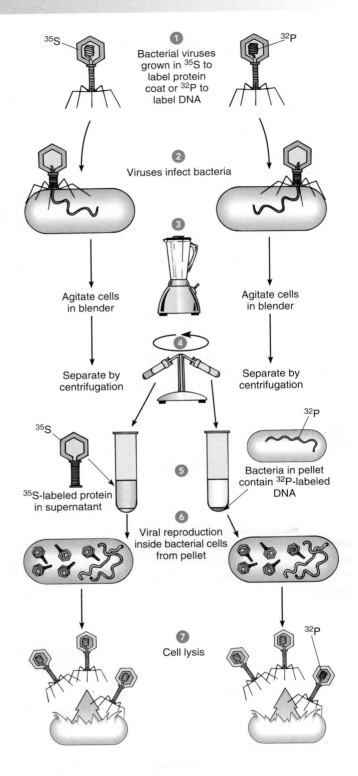

produce pathogenic effects; mice injected with either living or heat-killed cells of this strain survived. However, when Griffith injected mice with a mixture of *heat-killed* virulent S cells and live avirulent R cells, a high proportion of the mice died. Griffith then isolated living S cells from the dead mice.

Because neither the heat-killed S strain nor the living R strain could be converted to the living virulent form when injected by itself in the control experiments, it seemed that something in the heat-killed cells converted the avirulent cells to the lethal form. This type of permanent genetic change in which the properties of one strain of dead cells are conferred on a different strain of living cells became known as **transformation.** Scientists hypothesized that some chemical substance (a "transforming principle") was transferred from the dead bacteria to the living cells and caused transformation.

In 1944, American biologist Oswald T. Avery and his colleagues Colin M. MacLeod and Maclyn McCarty chemically identified Griffith's transforming principle as DNA. They did this through a series of careful experiments in which they lysed (split open) S cells and separated the cell contents into several fractions: lipids, proteins, polysaccharides, and nucleic acids (DNA and RNA). They tested each fraction to see if it could transform living R cells into S cells. The experiments using lipids, polysaccharides, and proteins did not cause transformation. However, when Avery treated living R cells with nucleic acids extracted from S cells, the R cells were transformed into S cells.

Although today scientists consider these results to be the first definitive demonstration that DNA is the genetic material, not all scientists of the time were convinced. Many thought the findings might apply only to bacteria and might not have any relevance for the genetics of eukaryotes. During the next few years, new evidence accumulated that the haploid nuclei of pollen grains and gametes such as sperm contain only half the amount of DNA found in diploid somatic cells of the same species. (*Somatic cells*

are body cells and never become gametes.) Because scientists generally accepted that genes are located on chromosomes, these findings correlating DNA content with chromosome number provided strong circumstantial evidence for DNA's importance in eukaryotic inheritance.

DNA is the genetic material in certain viruses

In 1952, American geneticists Alfred Hershey and Martha Chase performed a series of elegant experiments (❚ Fig. 12-2) on the reproduction of viruses that infect bacteria, known as **bacteriophages** or **phages** (discussed in Chapter 24). When they planned their experiments, they knew that bacteriophages reproduce inside a bacterial cell, eventually causing the cell to break open and release large numbers of new viruses. Because electron microscopic studies had shown that only part of an infecting bacteriophage actually enters the cell, they reasoned that the genetic material should be included in that portion.

As shown in Figure 12-2 ❶, they labeled the viral protein of one sample of bacteriophages with ^{35}S, a radioactive isotope of sulfur, and the viral DNA of a second sample with ^{32}P, a radioactive isotope of phosphorus. (Recall from Chapter 3 that proteins contain sulfur as part of the amino acids cysteine and methionine and that nucleic acids contain phosphate groups.) ❷ The bacteriophages in each sample attached to bacteria, and ❸ the researchers then shook them off by agitating them in a blender. ❹ Then they centrifuged the cells.

In the sample in which they had labeled the proteins with ^{35}S, they subsequently found the label in the supernatant, indicating that the protein had not entered the cells (left side of steps ❺, ❻, and ❼). In the sample in which they had labeled the DNA with ^{32}P, they found the label associated with the bacterial cells (in the pellet): DNA had actually entered the cells (right side of steps ❺, ❻, and ❼). Hershey and Chase concluded that bacteriophages inject their DNA into bacterial cells, leaving most of their protein on the outside. This finding emphasized the significance of DNA in viral reproduction, and many scientists saw it as an important demonstration of DNA's role as the hereditary material.

Review

- How did the experiments of Avery and his colleagues point to DNA as the essential genetic material?
- How did the Hershey–Chase experiment establish that DNA is the genetic material in bacteriophages?

THE STRUCTURE OF DNA

Learning Objectives

3   Explain how nucleotide subunits link to form a single DNA strand.
4   Describe how the two strands of DNA are oriented with respect to each other.
5   State the base-pairing rules for DNA, and describe how complementary bases bind to each other.

Scientists did not generally accept DNA as the genetic material until 1953, when American scientist James Watson and British scientist Francis Crick, both working in England, proposed a model for its structure that had extraordinary explanatory power. The story of how the structure of DNA was figured out is one of the most remarkable chapters in the history of modern biology (❚ Table 12-1). As you will see in the following discussion, scientists already knew a great deal about DNA's physical and chemical properties when Watson and Crick became interested in the

TABLE 12-1

A Time Line of Selected Historical DNA Discoveries

Date	Discovery
1871	**Friedrich Miescher** reports discovery of new substance, *nuclein*, from cell nuclei. Nuclein is now known to be a mixture of DNA, RNA, and proteins.
1928	**Frederick Griffith** finds a substance in heat-killed bacteria that "transforms" living bacteria.
1944	**Oswald Avery, Colin MacLeod,** and **Maclyn McCarty** chemically identify Griffith's transforming principle as DNA.
1949	**Erwin Chargaff** reports relationships among DNA bases that provide a clue to the structure of DNA.
1952	**Alfred Hershey** and **Martha Chase** demonstrate that DNA, not protein, is involved in viral reproduction.
1952	**Rosalind Franklin** produces X-ray diffraction images of DNA.
1953	**James Watson** and **Francis Crick** propose a model of the structure of DNA; this contribution is widely considered the start of a revolution in molecular biology that continues to the present.
1958	**Matthew Meselson** and **Franklin Stahl** demonstrate that DNA replication is semiconservative.
1962	**James Watson, Francis Crick,** and **Maurice Wilkins** are awarded the Nobel Prize in Medicine for discoveries about the molecular structure of nucleic acids.*
1969	**Alfred Hershey** is awarded the Nobel Prize in Medicine for discovering the replication mechanism and genetic structure of viruses.

*Rosalind Franklin could not have shared the prize because she was deceased.

problem; in fact, they did not conduct any experiments or gather any new data. Their all-important contribution was to integrate all the available information into a model that demonstrated how the molecule can both carry information for making proteins and serve as its own **template** (pattern or guide) for its duplication.

Nucleotides can be covalently linked in any order to form long polymers

As discussed in Chapter 3, each DNA building block is a **nucleotide** consisting of the pentose sugar **deoxyribose,** a phosphate, and one of four nitrogenous bases (❚ Fig. 12-3). It is conventional to number the atoms in a molecule using a system devised by organic chemists. Accordingly, in nucleic acid chemistry the individual carbons in each sugar and each base are numbered. The carbons in a base are designated by numerals, but the carbons in a sugar are distinguished from those in the base by prime symbols, such as 2'. The nitrogenous base is attached to the 1' carbon of the sugar, and the phosphate is attached to the 5' carbon. The bases include two **purines, adenine (A)** and **guanine (G),** and two **pyrimidines, thymine (T)** and **cytosine (C).**

The nucleotides are linked by covalent bonds to form an alternating sugar–phosphate backbone. The 3' carbon of one sugar is bonded to the 5' phosphate of the adjacent sugar to form a 3', 5' **phosphodiester linkage.** The result is a polymer of indefinite length, with the nucleotides linked in any order. Scientists now know that most DNA molecules found in cells are millions of bases long. Figure 12-3 also shows that a single polynucleotide

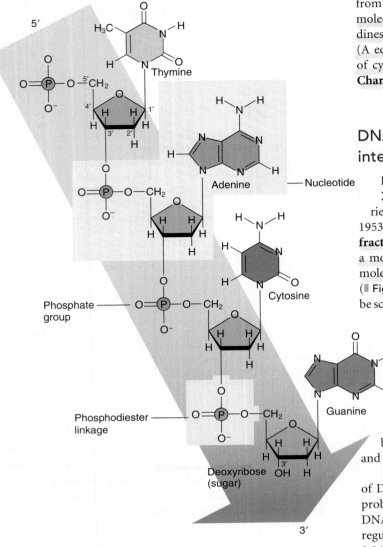

from 1" (▮ Table 12-2). In other words, in double-stranded DNA molecules, the number of purines equals the number of pyrimidines, the number of adenines equals the number of thymines (A equals T), and the number of guanines equals the number of cytosines (G equals C). These equalities became known as **Chargaff's rules.**

DNA is made of two polynucleotide chains intertwined to form a double helix

Key information about the structure of DNA came from X-ray diffraction studies on crystals of purified DNA, carried out by British scientist Rosalind Franklin from 1951 to 1953 in the laboratory of Maurice Wilkins in England. **X-ray diffraction,** a powerful method for elucidating the 3-D structure of a molecule, can determine the distances between the atoms of molecules arranged in a regular, repeating crystalline structure (▮ Fig. 12-4a). X-rays have such short wavelengths that they can be scattered by the electrons surrounding the atoms in a molecule. Atoms with dense electron clouds (such as phosphorus and oxygen) tend to deflect electrons more strongly than atoms with lower atomic numbers. Exposing a crystal to an intense beam of X-rays causes the regular arrangement of its atoms to diffract, or scatter, the X-rays in specific ways. The pattern of diffracted X-rays appears on photographic film as dark spots. Mathematical analysis of the pattern and distances between the spots yields the precise distances between atoms and their orientation within the molecules.

Franklin had already produced X-ray crystallographic films of DNA patterns when Watson and Crick began to pursue the problem of DNA structure. Her pictures clearly showed that DNA has a type of helical structure, and three major types of regular, repeating patterns in the molecule (with the dimensions 0.34 nm, 3.4 nm, and 2.0 nm) were evident (▮ Fig. 12-4b). Franklin and Wilkins had inferred from these patterns that the nucleotide bases (which are flat molecules) are stacked like the rungs of a ladder. Using this information, Watson and Crick began to build scale models of the DNA components and then fit them together to correlate with the experimental data.

▮ **Figure 12-3** *Animated* The nucleotide subunits of DNA

A single strand of DNA consists of a backbone (*superimposed on blue screen*) made of phosphate groups alternating with the sugar deoxyribose (*green*). Phosphodiester linkages (*pink*) join sugars of adjacent nucleotides. (*The nucleotide containing the base adenine is highlighted yellow.*) Linked to the 1′ carbon of each sugar is one of four nitrogenous bases (*top to bottom*): thymine, adenine, cytosine, and guanine. Note the polarity of the polynucleotide chain, with the 5′ end at the top of the figure and the 3′ end at the bottom.

chain is directional. No matter how long the chain may be, one end, the **5′ end,** has a 5′ carbon attached to a phosphate and the other, the **3′ end,** has a 3′ carbon attached to a hydroxyl group.

In 1949, Erwin Chargaff and his colleagues at Columbia University had determined the base composition of DNA from several organisms and tissues. They found a simple relationship among the bases that turned out to be an important clue to the structure of DNA. Regardless of the source of the DNA, in Chargaff's words the "ratios of purines to pyrimidines and also of adenine to thymine and of guanine to cytosine were not far

TABLE 12-2

Base Compositions in DNA from Selected Organisms

| Source of DNA | Percentage of DNA Bases | | | | Ratios | |
	A	T	G	C	A/T	G/C
E. coli	26.1	23.9	24.9	25.1	1.09	0.99
Yeast	31.3	32.9	18.7	17.1	0.95	1.09
Sea urchin sperm	32.5	31.8	17.5	18.2	1.02	0.96
Herring sperm	27.8	27.5	22.2	22.6	1.01	0.98
Human liver	30.3	30.3	19.5	19.9	1.00	0.98
Corn (*Zea mays*)	25.6	25.3	24.5	24.6	1.01	1.00

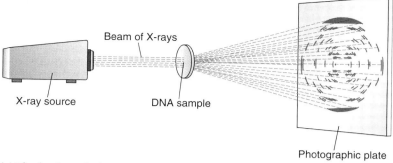

Beam of X-rays

X-ray source

DNA sample

Photographic plate

(a) The basic technique. Researchers direct a narrow beam of X-rays at a single crystal of DNA. Important clues about DNA structure are provided by detailed mathematical analysis of measurements of the spots, which are formed by X-rays hitting the photographic plate.

Figure 12-4 X-ray diffraction of DNA

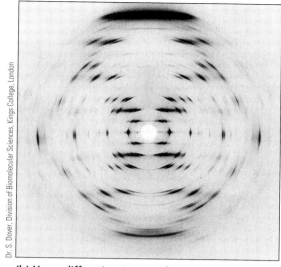

Dr. S. Dover, Division of Biomolecular Sciences, Kings College, London

(b) X-ray diffraction image of DNA. The diagonal pattern of spots stretching from 11 to 5 and from 1 to 7 (as on a clock face) provides evidence for the helical structure of DNA. The elongated horizontal patterns at the top and bottom indicate that the purine and pyrimidine bases are stacked 0.34 nm apart and are perpendicular to the axis of the DNA molecule.

After several trials, they worked out a model that fit the existing data (▌Fig. 12-5). The nucleotide chains conformed to the dimensions of the X-ray data only if each DNA molecule consisted of *two* polynucleotide chains arranged in a coiled **double helix.** In their model, the sugar–phosphate backbones of the two chains form the outside of the helix. The bases belonging to the two chains associate as pairs along the central axis of the helix. The reasons for the repeating patterns of 0.34-nm and 3.4-nm measurements are readily apparent from the model: Each pair of bases is exactly 0.34 nm from the adjacent pairs above and below. Because exactly 10 base pairs are present in each full turn of the helix, each turn constitutes 3.4 nm of length. To fit the data, the two chains must run in opposite directions; therefore, each end of the double helix must have an exposed 5′ phosphate on one strand and an exposed 3′ hydroxyl group (—OH) on the other. Because the two strands run in opposite directions, they are said to be **antiparallel** to each other (▌Fig. 12-6a).

In double-stranded DNA, hydrogen bonds form between A and T and between G and C

Other features of the Watson and Crick model integrated critical information about the chemical composition of DNA with the X-ray diffraction data. The X-ray diffraction studies indicated that the double helix has a precise and constant width, as shown by the 2.0-nm measurements. This finding is actually consistent with Chargaff's rules. As Figure 12-3 shows, each pyrimidine (cytosine or thymine) contains only one ring of atoms, whereas each purine (guanine or adenine) contains two rings. Study of the models made it clear to Watson and Crick that if each rung of the ladder contained one purine and one pyrimidine, the width of the helix at each base pair would be exactly 2.0 nm. By contrast, the combination of two purines (each of which is 1.2 nm wide) would be wider than 2.0 nm and that of two pyrimidines would be narrower, so the diameter would not be constant. Further examination of the model showed that adenine can pair with

thymine (and guanine with cytosine) in such a way that hydrogen bonds form between them; the opposite combinations, cytosine with adenine and guanine with thymine, do not lead to favorable hydrogen bonding.

The nature of the hydrogen bonding between adenine and thymine and between guanine and cytosine is shown in ▌Figure 12-6b. Two hydrogen bonds form between adenine and thymine, and three between guanine and cytosine. This concept of specific base pairing neatly explains Chargaff's rules. The amount of cytosine must equal the amount of guanine, because every cytosine in one chain must have a paired guanine in the other chain. Similarly, every adenine in the first chain must have a thymine in the second chain. The sequences of bases in the two chains are **complementary** to each other—that is, the sequence of nucleotides in one chain dictates the complementary sequence of nucleotides in the other. For example, if one strand has this sequence,

$$3'—AGCTAC—5'$$

then the other strand has the complementary sequence:

$$5'—TCGATG—3'$$

The double-helix model strongly suggested that the sequence of bases in DNA provides for the storage of genetic information and that this sequence ultimately relates to the sequences of amino acids in proteins. Although restrictions limit how the bases pair with each other, the number of possible sequences of bases in a strand is virtually unlimited. Because a DNA molecule in a cell can be millions of nucleotides long, it can store enor-

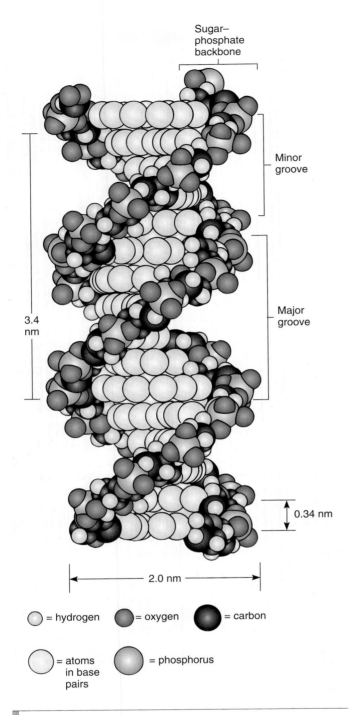

Sugar–
phosphate
backbone

Minor
groove

Major
groove

3.4
nm

0.34 nm

2.0 nm

◯ = hydrogen   ● = oxygen   ⬤ = carbon

◯ = atoms   ◯ = phosphorus
    in base
    pairs

Figure 12-5 *Animated* A three-dimensional model of the DNA double helix

The measurements match those derived from X-ray diffraction images.

mous amounts of information, usually consisting of hundreds of genes.

Review

▮ What types of subunits make up a single strand of DNA? How are the subunits linked?

▮ What is the structure of double-stranded DNA, as determined by Watson and Crick?

▮ How do Chargaff's rules relate to the structure of DNA?

DNA REPLICATION

Learning Objectives

6 Cite evidence from Meselson and Stahl's experiment that enabled scientists to differentiate between semiconservative replication of DNA and alternative models.

7 Summarize how DNA replicates, and identify some unique features of the process.

8 Explain the complexities of DNA replication that make it (a) bidirectional and (b) continuous in one strand and discontinuous in the other.

9 Discuss how enzymes proofread and repair errors in DNA.

10 Define *telomere*, and describe the possible connections between telomerase and cell aging and between telomerase and cancer.

Two immediately apparent and distinctive features of the Watson–Crick model made it seem plausible that DNA is the genetic material. We have already mentioned that the sequence of bases in DNA can carry coded information. The model also suggested a way in which the sequence of nucleotides in DNA could be precisely copied, a process known as **DNA replication.** The connection between DNA replication and the behavior of chromosomes in mitosis was obvious to Watson and Crick. A chromosome becomes duplicated so that it consists of two identical sister chromatids that later separate at anaphase; the genetic material must be precisely duplicated and distributed to the daughter cells. They noted, in a classic and now famous understatement at the end of their first brief paper, "It has not escaped our notice that the specific pairing we have postulated immediately suggests a possible copying mechanism for the genetic material."

The model suggested that because the nucleotides pair with each other in complementary fashion, each strand of the DNA molecule could serve as a template for synthesizing the opposite strand. It would simply be necessary for the hydrogen bonds between the two strands to break (recall that hydrogen bonds are relatively weak) and the two chains to separate. Each strand of the double helix could then pair with new complementary nucleotides to replace its missing partner. The result would be two DNA double helices, each identical to the original one and consisting of one original strand from the parent molecule and one newly synthesized complementary strand. This type of information copying is known as **semiconservative replication** (▮ Fig. 12-7a).

Meselson and Stahl verified the mechanism of semiconservative replication

Although the semiconservative replication mechanism suggested by Watson and Crick was (and is) a simple and compelling model, experimental proof was needed to establish that DNA in fact duplicates in that manner. Researchers first needed to rule out other possibilities. With *conservative replication*, both parent (or old) strands would remain together, and the two newly synthesized strands would form a second double helix (▮ Fig. 12-7b). As a third hypothesis, the parental and newly synthesized strands might become randomly mixed during the replication process; this possibility was known as *dispersive replication* (▮ Fig. 12-7c).

Key Point

Base pairing and the sequence of bases in DNA provide a foundation for understanding both DNA replication and the inheritance of genetic material.

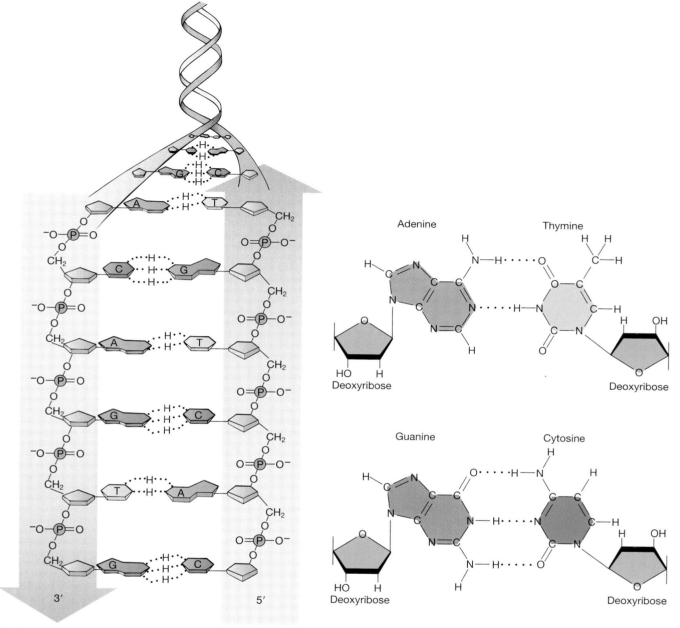

(a) The two sugar–phosphate chains run in opposite directions. This orientation permits the complementary bases to pair.

(b) Hydrogen bonding between base pairs adenine (A) and thymine (T) (*top*) and guanine (G) and cytosine (C) (*bottom*). The AT pair has two hydrogen bonds; the GC pair has three.

Figure 12-6 *Animated* Base pairing and hydrogen bonding

The two strands of a DNA double helix are hydrogen-bonded between the bases.

To discriminate among semiconservative replication and the other models, investigators had to distinguish between old and newly synthesized strands of DNA.

One technique is to use a heavy isotope of nitrogen, ^{15}N (ordinary nitrogen is ^{14}N), to label the bases of the DNA strands, making them more dense. Using **density gradient centrifugation,** scientists can separate large molecules such as DNA on the basis of differences in their density. When DNA is mixed with a solution containing cesium chloride (CsCl) and centrifuged at high speed, the solution forms a density gradient in the centri-

(a) Hypothesis 1: Semiconservative replication

Parental DNA   First generation   Second generation

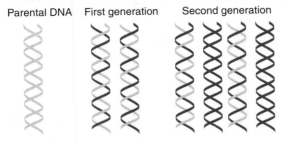

(b) Hypothesis 2: Conservative replication

Parental DNA   First generation   Second generation

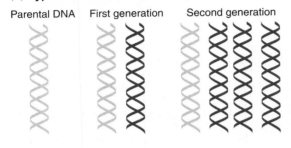

(c) Hypothesis 3: Dispersive replication

Parental DNA   First generation   Second generation

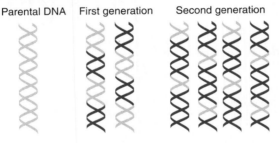

Figure 12-7   Alternative models of DNA replication

The hypothesized arrangement of old (*light blue*) and newly synthesized (*dark blue*) DNA strands after one and two generations, according to **(a)** the semiconservative model, **(b)** the conservative model, and **(c)** the dispersive model.

fuge tube, ranging from a region of lowest density at the top to one of highest density at the bottom. During centrifugation, the DNA molecules migrate to the region of the gradient identical to their own density.

In 1958, Matthew Meselson and Franklin Stahl at the California Institute of Technology grew the bacterium *Escherichia coli* on a medium that contained ^{15}N in the form of ammonium chloride (NH_4Cl). The cells used the ^{15}N to synthesize bases, which then became incorporated into DNA (∎ Fig. 12-8). The resulting DNA molecules, which contained heavy nitrogen, were extracted from some of the cells. When the researchers subjected them to density gradient centrifugation, they accumulated in the high-density region of the gradient. The team transferred the rest of the bacteria (which also contained ^{15}N-labeled DNA) to a different growth

medium in which the NH_4Cl contained the naturally abundant, lighter ^{14}N isotope and then allowed them to undergo additional cell divisions.

Meselson and Stahl expected the newly synthesized DNA strands to be less dense, because they incorporated bases containing the lighter ^{14}N isotope. Indeed, double-stranded DNA from cells isolated after one generation had an intermediate density, indicating they contained half as many ^{15}N atoms as the "parent" DNA. This finding supported the semiconservative replication model, which predicted that each double helix would contain a previously synthesized strand (heavy in this case) and a newly synthesized strand (light in this case). It was also consistent with the dispersive model, which would also yield one class of molecules, all with intermediate density. It was inconsistent with the conservative model, which predicted two classes of double-stranded molecules, those with two heavy strands and those with two light strands.

After another cycle of cell division in the medium with the lighter ^{14}N isotope, two types of DNA appeared in the density gradient, exactly as predicted by the semiconservative replication model. One consisted of DNA with a density intermediate between ^{15}N-labeled DNA and ^{14}N-labeled DNA, whereas the other contained only DNA with a density of ^{14}N-labeled DNA. This finding refuted the dispersive model, which predicted that all strands would have an intermediate density.

Semiconservative replication explains the perpetuation of mutations

The recognition that DNA could be copied by a semiconservative mechanism suggested how DNA could explain a third essential characteristic of genetic material—the ability to mutate. It was long known that **mutations,** or genetic changes, could arise in genes and then be transmitted faithfully to succeeding generations. For example, a mutation in the fruit fly (*Drosophila melanogaster*) produces vestigial wings.

When the double-helix model was proposed, it seemed plausible that mutations could represent a change in the sequence of bases in the DNA. You could predict that if DNA is copied by a mechanism involving complementary base pairing, any change in the sequence of bases on one strand would produce a new sequence of complementary bases during the next replication cycle. The new base sequence would then transfer to daughter molecules by the same mechanism used to copy the original genetic material, as if no change had occurred.

For the example in ∎ Figure 12-9, an adenine base in one of the DNA strands has been changed to guanine. This could occur by a rare error in DNA replication or by one of several other known mechanisms. As discussed later, certain enzymes correct errors when they occur, but not all errors are corrected properly. By one estimate, the rate of uncorrected errors that occur during DNA replication is equal to about one nucleotide in a billion. When the DNA molecule containing an error replicates (left side of Fig. 12-9), one of the strands gives rise to a molecule exactly like its parent strand; the other (mutated) strand gives rise to a

Key Experiment

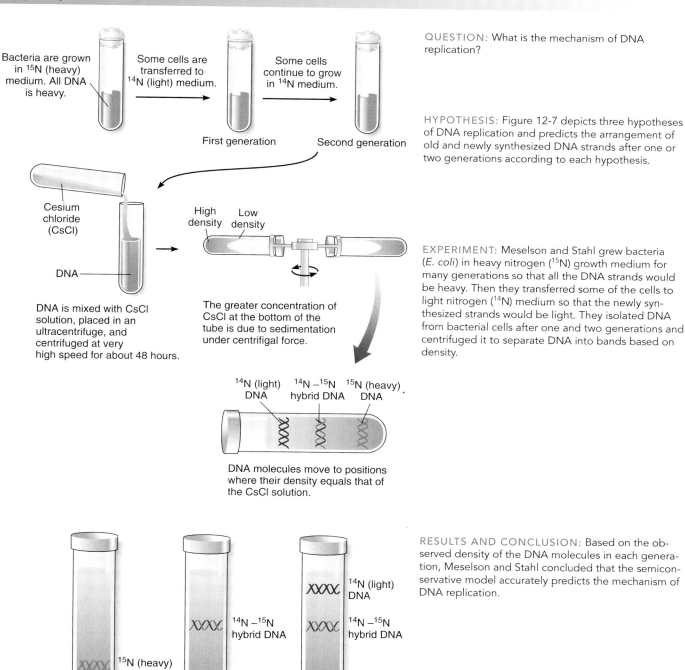

Bacteria are grown in ^{15}N (heavy) medium. All DNA is heavy.

Some cells are transferred to ^{14}N (light) medium.

Some cells continue to grow in ^{14}N medium.

First generation

Second generation

Cesium chloride (CsCl)

DNA

DNA is mixed with CsCl solution, placed in an ultracentrifuge, and centrifuged at very high speed for about 48 hours.

High density   Low density

The greater concentration of CsCl at the bottom of the tube is due to sedimentation under centrifugal force.

^{14}N (light) DNA ^{14}N –^{15}N hybrid DNA ^{15}N (heavy) DNA

DNA molecules move to positions where their density equals that of the CsCl solution.

^{15}N (heavy) DNA

Before transfer to ^{14}N

^{14}N –^{15}N hybrid DNA

One cell generation after transfer to ^{14}N

^{14}N (light) DNA

^{14}N –^{15}N hybrid DNA

Two cell generations after transfer to ^{14}N

The location of DNA molecules within the centrifuge tube can be determined by UV optics. DNA solutions absorb strongly at 260 nm.

QUESTION: What is the mechanism of DNA replication?

HYPOTHESIS: Figure 12-7 depicts three hypotheses of DNA replication and predicts the arrangement of old and newly synthesized DNA strands after one or two generations according to each hypothesis.

EXPERIMENT: Meselson and Stahl grew bacteria (E. coli) in heavy nitrogen (^{15}N) growth medium for many generations so that all the DNA strands would be heavy. Then they transferred some of the cells to light nitrogen (^{14}N) medium so that the newly synthesized strands would be light. They isolated DNA from bacterial cells after one and two generations and centrifuged it to separate DNA into bands based on density.

RESULTS AND CONCLUSION: Based on the observed density of the DNA molecules in each generation, Meselson and Stahl concluded that the semiconservative model accurately predicts the mechanism of DNA replication.

Figure 12-8   The Meselson–Stahl experiment

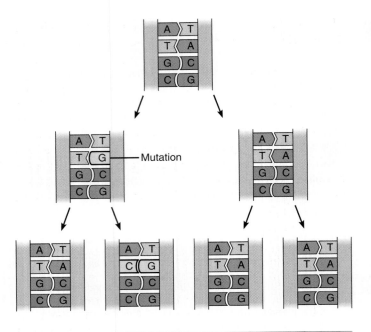

Figure 12-9 The perpetuation of a mutation

The process of DNA replication can stabilize a mutation (*bright yellow*) so that it is transmitted to future generations.

molecule with a new combination of bases that is transmitted generation after generation.

DNA replication requires protein "machinery"

Although semiconservative replication by base pairing appears simple and straightforward, the actual process is highly regulated and requires a "replication machine" containing many types of protein and enzyme molecules. Many essential features of DNA replication are common to all organisms, although prokaryotes and eukaryotes differ somewhat because their DNA is organized differently. In most bacterial cells, such as *E. coli,* most or all the DNA is in the form of a single, *circular,* double-stranded DNA molecule. In contrast, each unreplicated eukaryotic chromosome contains a single, *linear,* double-stranded molecule associated with at least as much protein (by mass) as DNA.

We now present DNA replication, based on the current understanding of the process. Although scientists know much about DNA replication, many aspects of the process remain unclear. For example, in the unicellular yeast *Saccharomyces cerevisiae,* which is considered a relatively "simple" eukaryote, 88 genes are known to be involved in DNA replication! Determining the roles and interactions of all those genes will require the efforts of many scientists over a long period.

DNA strands must be unwound during replication

DNA replication begins at specific sites on the DNA molecule, called **origins of replication,** where small sections of the double helix unwind. **DNA helicases** are helix-destabilizing enzymes

(several have been identified) that bind to DNA at the origin of replication and break hydrogen bonds, thereby separating the two strands (▌Table 12-3). Both DNA strands replicate at the same time at the junction between the separated strands, which is a Y-shaped structure called the **replication fork.** Helicase travels along the helix, opening the double helix like a zipper during movement of the replication fork.

Once DNA helicases separate the strands, **single-strand binding (SSB) proteins** bind to single DNA strands and stabilize them; this prevents the double helix from re-forming until the strands are replicated. SSB proteins also prevent the hydrolysis of the single-strand regions by other enzymes (*nucleases;* as we discuss later in the chapter, nucleases are involved in DNA repair).

Watson and Crick recognized that in their double-helix model, the two DNA strands wrap around each other like the strands of a rope. If you try to pull the strands apart, the rope must either rotate or twist into tighter coils. You could expect similar results when complementary DNA strands are separated for replication. As the DNA strands unwind to open for replication, torsional strain from supercoiling (excessive twisting) occurs in another part of the DNA molecule. Enzymes called **topoisomerases** produce breaks in the DNA molecules and then rejoin the strands, relieving strain and effectively preventing supercoiling and knot formation during replication. There are two ways that topoisomerases reduce supercoiling. Some topoisomerases produce a temporary break in the polynucleotide backbone of a single strand of DNA, pass that strand through the excessively twisted part, and then reseal the break. Other topoisomerases break both DNA strands, pass some of the helix between the cut ends, and then reseal the break. Regardless of their modes of action, topoisomerases give replicating DNA a more relaxed configuration.

DNA synthesis always proceeds in a 5′ ⟶ 3′ direction

The enzymes that catalyze the linking of successive nucleotide subunits are called **DNA polymerases.** They add nucleotides only to the 3′ end of a growing polynucleotide strand, and this strand must be paired with the DNA template strand (❙ Fig. 12-10). Nucleotides with three phosphate groups are substrates for the polymerization reaction. As the nucleotides become linked, two of the phosphates are removed. These reactions are strongly exergonic and do not need additional energy. Because the polynucleotide chain is elongated by the linkage of the 5′ phosphate group of the next nucleotide subunit to the 3′ hydroxyl group of the sugar at the end of the existing strand, the new strand of DNA always grows in the 5′ ⟶ 3′ direction. Some DNA polymerases are very efficient in joining nucleotides to the growing polypeptide chain. **DNA Pol III,** which is one of five DNA polymerases that have been identified in the bacterium *E. coli,* can join 1200 nucleotides per minute.

DNA synthesis requires an RNA primer

As mentioned, DNA polymerases add nucleotides only to the 3′ end of an *existing* polynucleotide strand. Then, how is DNA synthesis initiated once the two strands are separated? The answer is that first a short piece of RNA (5 to 14 nucleotides) called an **RNA primer** is synthesized at the point where replication begins (❙ Fig. 12-11).

RNA, or **ribonucleic acid** (see Chapters 3 and 13), is a nucleic acid polymer consisting of nucleotide subunits that can associate by complementary base pairing with the single-strand DNA template. The RNA primer is synthesized by **DNA primase,** an enzyme that starts a new strand of RNA opposite a short stretch of the DNA template strand. After a few nucleotides have been added, DNA polymerase displaces the primase and subsequently adds subunits to the 3′ end of the short RNA primer. Specific enzymes later degrade the primer (discussed in the next section), and the space fills in with DNA.

DNA replication is discontinuous in one strand and continuous in the other

We mentioned earlier that the complementary DNA strands are antiparallel. DNA synthesis proceeds only in the 5′ ⟶ 3′ direction, which means that the strand being copied is being read in the 3′ ⟶ 5′ direction. Thus, it may seem necessary to copy one of the strands starting at one end of the double helix and the other strand starting at the opposite end. Some viruses replicate their DNA in this way, but this replication method is not

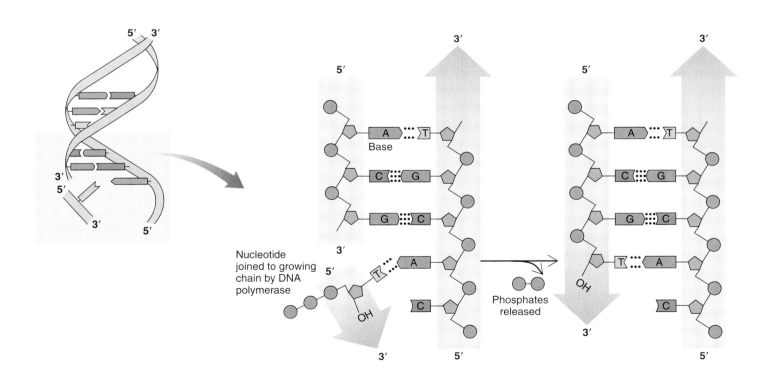

Figure 12-10 *Animated* A simplified view of DNA replication

DNA polymerase adds one nucleotide at a time to the 3′ end of a growing chain.

DNA replication requires several steps and involves several enzymes, proteins, and RNA primers.

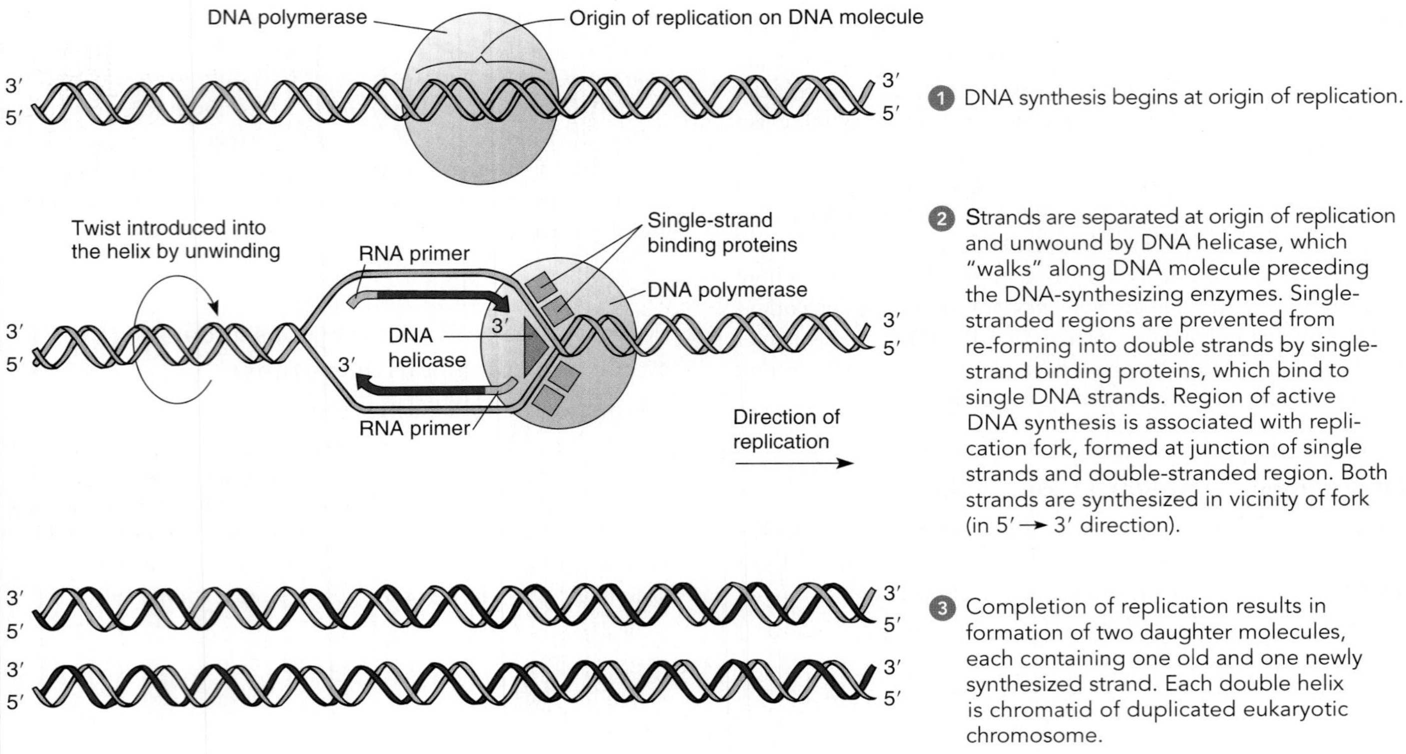

① DNA synthesis begins at origin of replication.

② Strands are separated at origin of replication and unwound by DNA helicase, which "walks" along DNA molecule preceding the DNA-synthesizing enzymes. Single-stranded regions are prevented from re-forming into double strands by single-strand binding proteins, which bind to single DNA strands. Region of active DNA synthesis is associated with replication fork, formed at junction of single strands and double-stranded region. Both strands are synthesized in vicinity of fork (in 5′ → 3′ direction).

③ Completion of replication results in formation of two daughter molecules, each containing one old and one newly synthesized strand. Each double helix is chromatid of duplicated eukaryotic chromosome.

Figure 12-11 An overview of DNA replication

workable in the extremely long DNA molecules in eukaryotic chromosomes.

Instead, as previously mentioned, DNA replication begins at origins of replication, and both strands replicate at the same time at a replication fork (∎ Fig. 12-12; also see chapter introduction photograph). The position of the replication fork is constantly moving as replication proceeds. Two identical DNA polymerase molecules catalyze replication. One of these adds nucleotides to the 3′ end of the new strand that is always growing *toward* the replication fork. Because this strand is synthesized smoothly and continuously, it is called the **leading strand.**

A separate DNA polymerase molecule adds nucleotides to the 3′ end of the other new strand. Called the **lagging strand,** it is always growing *away* from the replication fork. Only short pieces can be synthesized, because if the DNA polymerase were to add continuously to the 3′ end of that strand, it would need to move far from the replication fork. These 100- to 2000-nucleotide pieces are called **Okazaki fragments** after their discoverer, Japanese molecular biologist Reiji Okazaki.

DNA primase periodically catalyzes the synthesis of an RNA primer on the lagging strand. A separate RNA primer initiates each Okazaki fragment, which DNA polymerase then extends toward the 5′ end of the previously synthesized fragment. When the RNA primer of the previously synthesized fragment is reached, DNA polymerase degrades and replaces the primer with DNA. The fragments are then joined by **DNA ligase,** an enzyme that links the 3′ hydroxyl of one Okazaki fragment to the 5′ phosphate of the DNA immediately next to it, forming a phosphodiester linkage. (As you will see later in the chapter, DNA ligase also rejoins broken phosphodiester bonds during DNA repair.)

DNA synthesis is bidirectional

When double-stranded DNA separates, two replication forks form, and the molecule replicates in both directions from the origin of replication. In bacteria, each circular DNA molecule usually has only one origin of replication, so the two replica-

Key Point

Because elongation can proceed only in the 5'⟶3' direction, the two strands at the replication fork are copied in different ways, each by a separate DNA polymerase molecule.

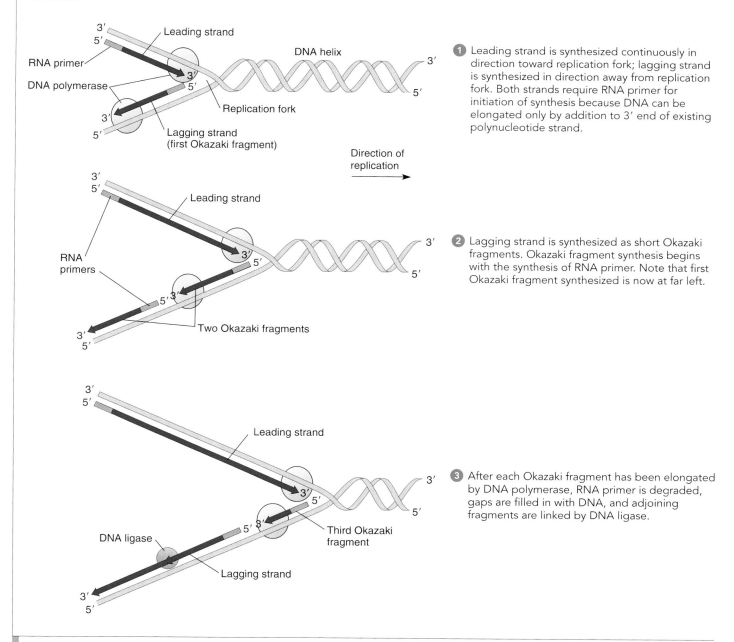

1 Leading strand is synthesized continuously in direction toward replication fork; lagging strand is synthesized in direction away from replication fork. Both strands require RNA primer for initiation of synthesis because DNA can be elongated only by addition to 3' end of existing polynucleotide strand.

2 Lagging strand is synthesized as short Okazaki fragments. Okazaki fragment synthesis begins with the synthesis of RNA primer. Note that first Okazaki fragment synthesized is now at far left.

3 After each Okazaki fragment has been elongated by DNA polymerase, RNA primer is degraded, gaps are filled in with DNA, and adjoining fragments are linked by DNA ligase.

Figure 12-12 Leading and lagging DNA strands

tion forks proceed around the circle and eventually meet at the other side to complete the formation of two new DNA molecules (❙ Fig. 12-13a).

A eukaryotic chromosome consists of one long, linear DNA molecule, so having multiple origins of replication expedites the replication process (❙ Fig. 12-13b and c). Synthesis continues at each replication fork until it meets a newly synthesized strand coming from the opposite direction. This results in a chromosome containing two DNA double helices, each corresponding to a chromatid.

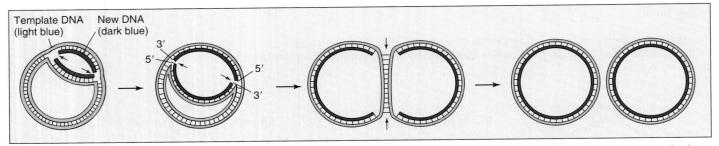

(a) The circular DNA in the prokaryote *E. coli* has only one origin of replication. Because DNA synthesis proceeds from that point in both directions, two replication forks form (*small black arrows*), travel around the circle, and eventually meet.

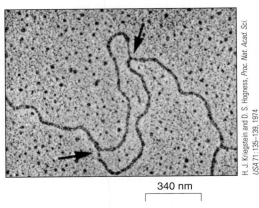

H. J. Kriegstein and D. S. Hogness, *Proc. Nat. Acad. Sci. USA* 71:135–139, 1974

340 nm

(b) This TEM shows two replication forks (*arrows*) in a segment of a eukaryotic chromosome that has partly replicated.

(c) Eukaryotic chromosome DNA contains multiple origins of replication. DNA synthesis proceeds in both directions from each origin until adjacent replication "bubbles" eventually merge.

Figure 12-13   Bidirectional DNA replication in bacteria and eukaryotes
The illustrations do not show leading strands and lagging strands.

Enzymes proofread and repair errors in DNA

DNA replication occurs only once during each cell generation, and it is important that the process be as accurate as possible to avoid harmful, or possibly even lethal, mutations. Although base pairing during DNA replication is very accurate, errors do occur. Mechanisms have evolved that ensure that errors in replication are corrected. During replication, DNA polymerases proofread each newly added nucleotide against its template nucleotide. When an error in base pairing is found, DNA polymerase immediately removes the incorrect nucleotide and inserts the correct one. A few uncorrected mutations still occur, but they are very infrequent—on the order of one error for every 10^9 or 10^{10} base pairs.

When errors have been left uncorrected by the normal repair activities of DNA polymerase during DNA replication, cells make use of other repair mechanisms (although exactly *how* DNA repair enzymes identify these rare, often-subtle errors in the vast amount of normal DNA is not well understood). In **mismatch repair,** special enzymes recognize the incorrectly paired nucleotides and remove them; DNA polymerases then fill in the missing nucleotides. The observation that individuals with a hereditary defect in a mismatch repair enzyme are likely to develop a type of colon cancer demonstrates the crucial role of mismatch repair enzymes in ensuring the fidelity of DNA replication from one generation to the next.

In Chapter 13 you will discover that some types of radiation and chemicals, both in the environment and within the cells themselves, cause mutations in DNA. These mutations are almost always harmful, and cells usually correct mutations by using one or more DNA repair enzymes. About 100 different kinds of repair enzymes in the bacterium *E. coli* and 130 kinds in human cells have been discovered so far.

One type of DNA repair, known as **nucleotide excision repair,** is commonly used to repair DNA lesions (deformed DNA) caused by the sun's ultraviolet radiation or by harmful chemicals (❚ Fig. 12-14). Three enzymes are involved in nucleotide excision repair—a nuclease to cut out the damaged DNA, a DNA polymerase to add the correct nucleotides, and DNA ligase to close the breaks in the sugar–phosphate backbone. Individuals suffering from the disease *xeroderma pigmentosum* have an inherited defect in a nucleotide excision repair enzyme. Affected individuals develop many skin cancers at an early age because DNA lesions caused by ultraviolet radiation are not repaired.

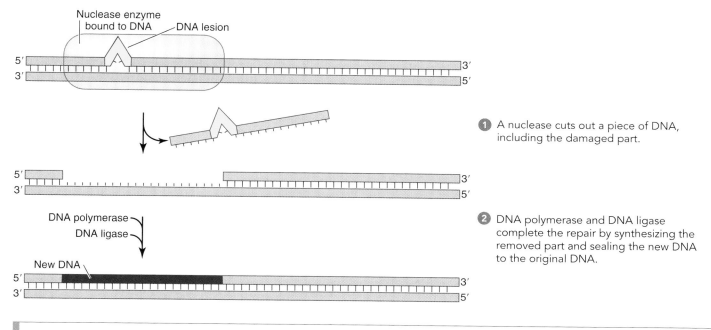

1 A nuclease cuts out a piece of DNA, including the damaged part.

2 DNA polymerase and DNA ligase complete the repair by synthesizing the removed part and sealing the new DNA to the original DNA.

Figure 12-14 Nucleotide excision repair of damaged DNA

Telomeres cap eukaryotic chromosome ends

Unlike bacterial DNA, which is circular, eukaryotic chromosomes have free ends. Because DNA replication is discontinuous in the lagging strand, DNA polymerases do not complete replication of the strand neatly. At the end of the DNA, a small portion is left unreplicated, and a small, single-strand segment of the DNA is lost with each cell cycle (▌Fig. 12-15a).

The important genetic information is retained because chromosomes have protective end caps known as **telomeres** that do not contain protein-coding genes. Telomeres consist of short, simple, noncoding DNA sequences that repeat many times (▌Fig. 12-15b). For example, in human sperm and egg cells, the sequence 5′—TTAGGG—3′ is repeated more than 1000 times at the ends of each chromosome. Therefore, although a small amount of telomeric DNA fails to replicate each time the cell divides, a cell can divide many times before it starts losing essential genetic information.

Telomerase, a special DNA replication enzyme, can lengthen telomeric DNA by adding repetitive nucleotide sequences to the ends of eukaryotic chromosomes. This enzyme, which researchers discovered in 1984, is typically present in cells that divide an unlimited number of times, including protozoa and other unicellular eukaryotes, and most types of **cancer cells.** In humans and other mammals, active telomerase is usually present in germ line cells (which give rise to eggs and sperm) and rapidly dividing cells (such as blood cells, the lining of the intestine, and skin) but not in most somatic cells of adult tissues. When most cells divide for repair or replacement, their chromosome ends shorten. Researchers have demonstrated that in cancer cells of the prostate and pancreas, telomeres are abnormally short: they shorten to a critical point, at which time telomerase is reactivated, which may explain the ability of cancer cells to proliferate in a rapid and uncontrolled manner. Telomere shortening has also been demonstrated in precancerous lesions from the bladder, breast, colon, esophagus, mouth, and cervix.

Research evidence suggests that the shortening of telomeres may contribute to **cell aging** and *apoptosis,* which is programmed cell death. Scientists have analyzed cell aging since the 1960s, following the pioneering studies of American biologist Leonard Hayflick, who showed that when normal somatic cells of the human body are grown in culture, they lose their ability to divide after a limited number of cell divisions. Furthermore, the number of cell divisions is determined by the age of the individual from whom the cells were taken. Cells from a 70-year-old can divide only 20 to 30 times, compared with those from an infant, which can divide 80 to 90 times.

Many investigators have observed correlations between telomerase activity and the ability of cells to undergo unlimited divisions without showing signs of the aging process. However, they could not find evidence of a causal relationship until the early 2000s, when scientists at the Geron Corporation and the University of Texas Southwestern Medical Center conducted a direct test. Using recombinant DNA technology (see Chapter 15), they infected cultured normal human cells with a virus that carried the genetic information coding for the catalytic subunit of telomerase. The infected cells not only produced active telomerase, which elongated the telomeres significantly, but also continued to divide long past the point at which cell divisions would normally cease.

Telomeres and telomerase are an active focus of current research, for both scientific and practical reasons. The ability to give somatic cells telomerase so they can divide many times more than they ordinarily would has many potential therapeutic applications, especially if lost or injured cells must be replaced. However, giving such cells the property of unlimited cell division also has potentially serious consequences. For example, if transplanted into the body, cells with active telomerase might behave like cancer cells.

Cancer cells have the ability to divide hundreds of times in culture; in fact, they are virtually immortal. Most cancer cells, including human cancers of the breast, lung, colon, prostate gland, and pancreas, have telomerase, which allows them to maintain telomere length and possibly to resist apoptosis. Research is underway to develop anticancer drugs that inhibit telomerase activity. An alternative treatment for cancer is also under study—directing the body's immune system to recognize and attack cancer cells that contain telomerase.

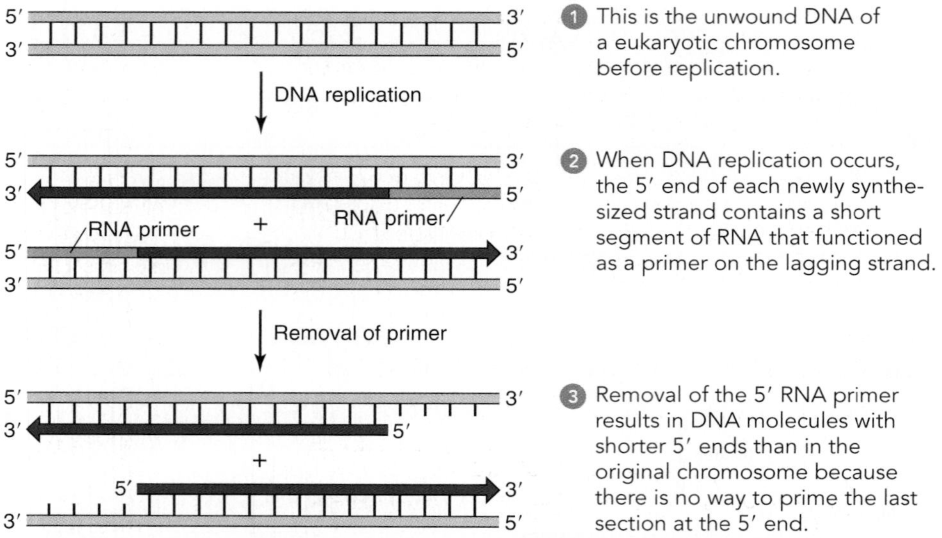

Review

- How did Meselson and Stahl verify that DNA replication is semiconservative?
- Why is DNA replication continuous for one strand but discontinuous for the other?
- How does mismatch repair of DNA compare with nucleotide excision repair?

① This is the unwound DNA of a eukaryotic chromosome before replication.

DNA replication

② When DNA replication occurs, the 5′ end of each newly synthesized strand contains a short segment of RNA that functioned as a primer on the lagging strand.

RNA primer + RNA primer

Removal of primer

③ Removal of the 5′ RNA primer results in DNA molecules with shorter 5′ ends than in the original chromosome because there is no way to prime the last section at the 5′ end.

(a) Why one strand of DNA is shorter than that of the preceding generation.

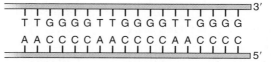

T T G G G G T T G G G G T T G G G G
A A C C C C A A C C C C A A C C C C

(b) The telomeres of chromosomes have short noncoding DNA sequences that repeat many times. Shown is TTGGGG, which is the segment repeated some 20 to 70 times in telomeres of the protozoon *Tetrahymena*, the unicellular organism in which telomerase was first discovered.

Figure 12-15 Replication at chromosome ends

SUMMARY WITH KEY TERMS

Learning Objectives

1 Summarize the evidence that accumulated during the 1940s and early 1950s demonstrating that DNA is the genetic material (page 261).
- Many early geneticists thought genes were made of proteins. They knew proteins are complex and variable, whereas they thought of nucleic acids as simple molecules with a limited ability to store information.
- Several lines of evidence supported the idea that **DNA (deoxyribonucleic acid)** is the genetic material. In **transformation** experiments, the DNA of one strain of bacteria can endow related bacteria with new genetic characteristics.
- When a bacterial cell becomes infected with a **bacteriophage** (virus), only the DNA from the virus enters the cell; this DNA is sufficient for the virus to reproduce and form new viral particles.

2 State the questions that these classic experiments addressed: Griffith's transformation experiment, Avery's contribution

to Griffith's work, and the Hershey–Chase experiments (page 261).
- Griffith's transformation experiment addressed this question: Can a genetic trait be transmitted from one bacterial strain to another? (The answer is yes.)
- Avery's experiments addressed this question: What molecule is responsible for bacterial transformation? (The answer is DNA.)
- The Hershey–Chase experiments addressed this question: Is DNA or protein the genetic material in bacterial viruses (phages)? (The answer is DNA.)

3 Explain how nucleotide subunits link to form a single DNA strand (page 263).
- Watson and Crick's model of the structure of DNA demonstrated how information can be stored in the molecule's structure and how DNA molecules can serve as **templates** for their own replication.

- DNA is a polymer of **nucleotides.** Each nucleotide subunit contains a nitrogenous base, which may be one of the **purines** (**adenine** or **guanine**) or one of the **pyrimidines** (**thymine** or **cytosine**). Each base covalently links to a pentose sugar, **deoxyribose,** which covalently bonds to a phosphate group.
- The backbone of each single DNA chain is formed by alternating sugar and phosphate groups, joined by covalent **phosphodiester linkages.** Each phosphate group attaches to the 5' carbon of one deoxyribose and to the 3' carbon of the neighboring deoxyribose.

4 Describe how the two strands of DNA are oriented with respect to each other (page 263).

- Each DNA molecule consists of two polynucleotide chains that associate as a **double helix.** The two chains are **antiparallel** (running in opposite directions); at each end of the DNA molecule, one chain has a phosphate attached to a 5' deoxyribose carbon, the **5' end,** and the other has a hydroxyl group attached to a 3' deoxyribose carbon, the **3' end.**

ThomsonNOW™ Interact with the structure of the DNA double helix by clicking on the figures in ThomsonNOW.

5 State the base-pairing rules for DNA, and describe how complementary bases bind to each other (page 263).

- Hydrogen bonding between specific base pairs holds together the two chains of the helix. Adenine (A) forms two hydrogen bonds with thymine (T); guanine (G) forms three hydrogen bonds with cytosine (C).
- **Complementary base pairing** between A and T and between G and C is the basis of Chargaff's rules, which state that A equals T and that G equals C.
- Because complementary base pairing holds together the two strands of DNA, it is possible to predict the base sequence of one strand if you know the base sequence of the other strand.

6 Cite evidence from Meselson and Stahl's experiment that enabled scientists to differentiate between semiconservative replication of DNA and alternative models (page 266).

- When *E. coli* cells are grown for many generations in a medium containing heavy nitrogen (^{15}N), they incorporate the ^{15}N into their DNA. When researchers transfer cells from a ^{15}N medium to a ^{14}N medium and isolate them after either one or two generations, the density of the DNA in each group is what would be expected if DNA replication were semiconservative. In **semiconservative replication,** each daughter double helix consists of an original strand from the parent molecule and a newly synthesized complementary strand.

7 Summarize how DNA replicates, and identify some unique features of the process (page 266).

- During **DNA replication,** the two strands of the double helix unwind. Each strand serves as a template for forming a new, complementary strand. Replication is initiated as **DNA primase** synthesizes a short **RNA primer. DNA poly-**

merase then adds new nucleotide subunits to the growing DNA strand.

- Additional enzymes and other proteins are required to unwind and stabilize the separated DNA helix. **DNA helicases** open the double helix, and **topoisomerases** prevent tangling and knotting.

8 Explain the complexities of DNA replication that make it (a) bidirectional and (b) continuous in one strand and discontinuous in the other (page 266).

- DNA replication is bidirectional, starting at the **origin of replication** and proceeding in both directions from that point. A eukaryotic chromosome may have multiple origins of replication and may be replicating at many points along its length at any one time.
- DNA synthesis always proceeds in a 5' $\longrightarrow$ 3' direction. This requires that one DNA strand, the **lagging strand,** be synthesized discontinuously, as short **Okazaki fragments.** DNA primase synthesizes short RNA primers on the lagging strand, and **DNA ligase** links Okazaki fragments of newly synthesized DNA. The opposite strand, the **leading strand,** is synthesized continuously.

ThomsonNOW™ Watch the process of replication by clicking on the figure in ThomsonNOW.

9 Discuss how enzymes proofread and repair errors in DNA (page 266).

- During replication, DNA polymerases proofread each newly added nucleotide against its template nucleotide. When an error in base pairing is found, DNA polymerase immediately removes the incorrect nucleotide and inserts the correct one.
- In **mismatch repair,** enzymes recognize incorrectly paired nucleotides and remove them; DNA polymerases then fill in the missing nucleotides.
- **Nucleotide excision repair** is commonly used to repair DNA lesions caused by the sun's ultraviolet radiation or by harmful chemicals. Three enzymes are involved: a nuclease to cut out the damaged DNA, a DNA polymerase to add the correct nucleotides, and DNA ligase to close the breaks in the sugar–phosphate backbone.

10 Define *telomere,* and describe the possible connections between telomerase and cell aging and between telomerase and cancer (page 266).

- Eukaryotic chromosome ends, known as **telomeres,** are short, noncoding, repetitive DNA sequences. Telomeres shorten slightly with each cell cycle but can be extended by the enzyme **telomerase.**
- The absence of telomerase activity in certain cells may be a cause of **cell aging,** in which cells lose their ability to divide after a limited number of cell divisions.
- Most **cancer cells,** including human cancers of the breast, lung, colon, prostate gland, and pancreas, have telomerase to maintain telomere length and possibly to resist apoptosis.

TEST YOUR UNDERSTANDING

1. When Griffith injected mice with a combination of live rough-strain and heat-killed smooth-strain pneumococci, he discovered that (a) the mice were unharmed (b) the dead mice contained living rough-strain bacteria (c) the dead mice contained living smooth-strain bacteria (d) DNA had been transferred from the smooth-strain bacteria to the mice (e) DNA had been transferred from the rough-strain bacteria to the smooth-strain bacteria

2. Which of the following inspired Avery and his colleagues to perform the experiments demonstrating that the transforming principle in bacteria is DNA? (a) the fact that A is equal to T, and G is equal to C (b) Watson and Crick's model of DNA structure (c) Meselson and Stahl's studies on DNA replication in *E. coli* (d) Griffith's experiments on smooth and rough strains of pneumococci (e) Hershey and Chase's experiments on the reproduction of bacteriophages

3. In the Hershey–Chase experiment with bacteriophages, (a) harmless bacterial cells permanently transformed into virulent cells (b) DNA was shown to be the transforming principle of earlier bacterial transformation experiments (c) the replication of DNA was conclusively shown to be semiconservative (d) viral DNA was shown to enter bacterial cells and cause the production of new viruses within the bacteria (e) viruses injected their proteins, not their DNA, into bacterial cells

4. The two complementary strands of the DNA double helix are held to one another by (a) ionic bonds between deoxyribose molecules (b) ionic bonds between phosphate groups (c) covalent bonds between nucleotide bases (d) covalent bonds between deoxyribose molecules (e) hydrogen bonds between nucleotide bases

5. If a segment of DNA is 5′—CATTAC—3′, the complementary DNA strand is

   (a) 3′—CATTAC—5′   (b) 3′—GTAATG—5′

   (c) 5′—CATTAC—3′   (d) 5′—GTAATG—3′

   (e) 5′—CATTAC—5′

6. Each DNA strand has a backbone that consists of alternating (a) purines and pyrimidines (b) nucleotide bases (c) hydrogen bonds and phosphodiester linkages (d) deoxyribose and phosphate (e) phosphate and phosphodiester linkages

7. The experiments in which Meselson and Stahl grew bacteria in heavy nitrogen conclusively demonstrated that DNA (a) is a double helix (b) replicates semiconservatively (c) consists of repeating nucleotide subunits (d) has complementary base pairing (e) is always synthesized in the 5′ ⟶ 3′ direction

8. The statement "DNA replicates by a semiconservative mechanism" means that (a) only one DNA strand is copied (b) first one DNA strand is copied, and then the other strand is copied (c) the two strands of a double helix have identical base sequences (d) some portions of a single DNA strand are old, and other portions are newly synthesized (e) each double helix consists of one old and one newly synthesized strand

9. What technique did Franklin use to determine many of the physical characteristics of DNA? (a) X-ray diffraction (b) transformation (c) radioisotope labeling (d) density gradient centrifugation (e) transmission electron microscopy

10. Multiple origins of replication (a) speed up replication of eukaryotic chromosomes (b) enable the lagging strands and leading strands to be synthesized at different replication forks (c) help relieve strain as the double helix unwinds (d) prevent mutations (e) are necessary for the replication of a circular DNA molecule in bacteria

11. Topoisomerases (a) synthesize DNA (b) synthesize RNA primers (c) join Okazaki fragments (d) break and rejoin DNA to reduce torsional strain (e) prevent single DNA strands from joining to form a double helix

12. A phosphate in DNA (a) hydrogen-bonds to a base (b) covalently links to two bases (c) covalently links to two deoxyriboses (d) hydrogen-bonds to two additional phosphates (e) covalently links to a base, a deoxyribose, and another phosphate

13. Which of the following depicts the relative arrangement of the complementary strands of a DNA double helix?

    (a) 5′—5′ (b) 3′—5′ (c) 3′—3′ (d) 5′—5′ (e) 3′—5′
        3′—3′   3′—5′   3′—3′   5′—5′   5′—3′

14. A lagging strand forms by (a) joining primers (b) joining Okazaki fragments (c) joining leading strands (d) breaking up a leading strand (e) joining primers, Okazaki fragments, and leading strands

15. The immediate source of energy for DNA replication is (a) the hydrolysis of nucleotides with three phosphate groups (b) the oxidation of NADPH (c) the hydrolysis of ATP (d) electron transport (e) the breaking of hydrogen bonds

16. Which of the following enzymes is *not* involved in nucleotide excision repair? (a) nuclease (b) DNA ligase (c) DNA polymerase (d) telomerase (e) neither a nor d is involved

17. Which of the following statements about eukaryotic chromosomes is *false*? (a) eukaryotic chromosomes have free ends (b) telomeres contain protein-coding genes (c) telomerase lengthens telomeric DNA (d) telomere shortening may contribute to cell aging (e) cells with active telomerase may undergo many cell divisions

CRITICAL THINKING

1. What characteristics must a molecule have to function as genetic material?

2. What important features of the structure of DNA are consistent with its role as the chemical basis of heredity?

3. What do telomeres on the ends of chromosomes protect against?

4. **Evolution Link.** How does the fact that DNA is the universal molecule of inheritance in cells support the theory of evolution?

Additional questions are available in ThomsonNOW at www.thomsonedu.com/login

Gene Expression

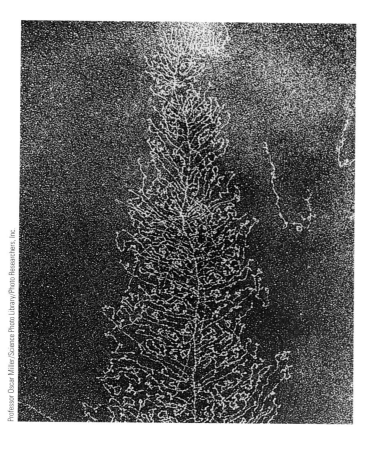

Professor Oscar Miller/Science Photo Library/Photo Researchers, Inc.

Visualizing transcription. In this TEM, RNA molecules (*lateral strands*) are synthesized as complementary copies of a DNA template (*central axis*).

KEY CONCEPTS

Beadle and Tatum demonstrated the relationship between genes and proteins in the 1940s.

The transmission of information in cells is typically from DNA to RNA to polypeptide.

A sequence of DNA base triplets is transcribed into RNA codons.

A sequence of RNA codons is translated into a sequence of amino acids in a polypeptide.

Prokaryotic and eukaryotic cells differ in the details of transcription and translation.

Mutations can cause changes in phenotype.

In Chapter 12, we discussed how the cell accurately replicates the nucleotide sequence in DNA to pass the genetic material unaltered to the next generation. Watson and Crick originally described the basic features of the DNA double helix, which scientists now know to be the same in all cells studied to date—from the simplest bacterial cells to intricate human cells.

By the mid-1950s, researchers had determined that the sequence of bases in DNA contains the information that specifies all the proteins the cell needs. However, more than a decade of intense investigation by many scientists preceded a fundamental understanding of how cells convert DNA information into the amino acid sequences of proteins. Much of that understanding came from studying the functions of bacterial genes. After Watson and Crick discovered the structure of DNA, researchers chose bacterial cells as the organisms for these studies because bacteria grow quickly and easily and because they contain the minimal amount of DNA needed for growth and reproduction. As researchers learned that all organisms have fundamental genetic similarities, the validity, as well as the utility, of this approach was repeatedly confirmed.

In this chapter we examine the evidence that accumulated in the first half of the 20th century indicating that most genes specify the structure of proteins. We then consider how DNA affects the phenotype of the organism at the molecular level through the process of gene expression. **Gene expression** involves a series of steps in which the information in the sequence of bases in DNA specifies the makeup of the cell's proteins. The proteins affect the phenotype in some way; these effects range from readily observable physical traits to subtle changes detectable only at the biochemical level. The first major step of gene expression is **transcription,** the synthesis of RNA molecules complementary to the DNA (see photograph). The second major step is **translation,** in which RNA becomes a coded template to direct polypeptide synthesis. ∎

DISCOVERY OF THE GENE–PROTEIN RELATIONSHIP

Learning Objectives

1 Summarize the early evidence indicating that most genes specify the structure of proteins.
2 Describe Beadle and Tatum's experiments with *Neurospora*.

The idea that genes and proteins are connected originated early in the 20th century, shortly after scientists rediscovered Mendel's principles. In the first edition of his book, *Inborn Errors of Metabolism* (1908), Archibald Garrod, an English physician and biochemist, discussed a rare genetic disease called **alkaptonuria,** which scientists hypothesized had a simple recessive inheritance pattern. The condition involves the metabolic pathway that breaks down the amino acid tyrosine, ultimately converting it to carbon dioxide and water. An intermediate in this pathway, homogentisic acid, accumulates in the urine of affected people, turning it black when exposed to air (❙ Fig. 13-1). Other symptoms of alkaptonuria include the later development of arthritis and, in men, stones in the prostate gland.

In Garrod's time, scientists knew about enzymes but did not recognize they were proteins. Garrod hypothesized that people with alkaptonuria lack the enzyme that normally oxidizes homogentisic acid. Before the second edition of his book had been published, in 1923, researchers found that affected people do indeed lack the enzyme that oxidizes homogentisic acid. Garrod's hypothesis was correct: A mutation in this gene is associated with the absence of a specific enzyme. Shortly thereafter, in 1926, U.S. biochemist James Sumner purified a different enzyme, urease, and showed it to be a protein. This was the first clear identification of an enzyme as a protein. In 1946, Sumner was awarded the Nobel Prize in Chemistry for being the first to crystallize an enzyme.

Beadle and Tatum proposed the one-gene, one-enzyme hypothesis

A major advance in understanding the relationship between genes and enzymes came in the early 1940s, when U.S. geneticists George Beadle and Edward Tatum and their associates developed a new approach to the problem. Most efforts until that time had focused on studying known phenotypes, such as eye color in *Drosophila* or pigments in plants, and seeking to determine what biochemical reactions they affected. Researchers found that a series of biosynthetic reactions controls specific phenotypes, but it was not clear whether the genes themselves were acting as enzymes or whether they controlled the workings of enzymes in more intricate ways.

Beadle and Tatum decided to take the opposite approach. Rather than try to identify the enzymes affected by single genes, they decided to look for mutations interfering with known metabolic reactions that produce essential molecules, such as amino acids and vitamins. They chose as an experimental organism a fungus, the orange mold *Neurospora*, for several important reasons. Wild-type *Neurospora* is easy to grow in culture. (The adjective **wild-type** refers to an individual with the normal phenotype.) *Neurospora* manufactures all its essential biological molecules when grown on a simple growth medium (minimal medium) containing only sugar, salts, and the vitamin biotin. However, a mutant *Neurospora* strain that cannot make a substance such as an amino acid can still grow if that substance is added to the growth medium.

Neurospora is also an ideal experimental organism because it grows primarily as a haploid organism. The haploid condition allows a researcher to immediately identify a recessive mutant allele; there is no homologous chromosome that might carry a dominant allele that would mask its expression.

Beadle and Tatum began by exposing thousands of haploid wild-type *Neurospora* asexual spores to X-rays or ultraviolet radiation to induce mutant strains. They first grew each irradiated strain on a complete growth medium, which contained all the

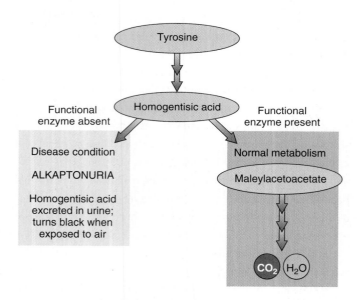

Figure 13-1 An "inborn error of metabolism"

Garrod proposed that alkaptonuria is a genetic disease caused by the absence of the enzyme homogentisic acid oxidase, which normally converts homogentisic acid to maleylacetoacetate. Thus, the acid accumulates in the blood and is excreted in the urine, which turns black on contact with the air.

amino acids and vitamins normally made by *Neurospora.* Then they tested each strain on the minimal medium described previously. About 1% to 2% of the strains that grew on the complete medium failed to grow after transfer to the minimal medium. Beadle and Tatum reasoned that such strains carried a mutation that prevented the fungus from producing a chemical essential for growth. Further testing of the mutant strain on media containing different combinations of amino acids, vitamins, and other nutrients enabled the investigators to determine the exact compound required (▮ Fig. 13-2).

The work on *Neurospora* revealed that each mutant strain had a mutation in only one gene and that each gene affected only one enzyme. Beadle and Tatum stated this one-to-one correspondence between genes and enzymes as the *one-gene, one-enzyme hypothesis.* In 1958, they received the Nobel Prize in Physiology or Medicine for discovering that genes regulate specific chemical events. The work of Beadle, Tatum, and others led to a more precise understanding of what a gene is and to additional predictions about the chemical nature of genes. The idea that a gene encodes the information to produce a single enzyme held for almost a decade, until additional findings required a modification of this definition.

In the late 1940s, researchers began to understand that genes control not only enzymes but other proteins as well. The U.S. chemist Linus Pauling and his colleagues demonstrated in 1949 that a mutation of a single gene alters the structure of hemoglobin. This particular mutant form of hemoglobin is associated with the genetic disease sickle cell anemia (discussed in Chapter 16). British biochemist Vernon Ingram extended Pauling's research in 1957 when he determined that a mutation of a single locus in hemoglobin produced a protein with a single amino acid difference.

Studies by other scientists showed that many proteins are constructed from two or more polypeptide chains, each of which is under the control of a different locus. For example, hemoglobin contains two types of polypeptide chains, the α and β subunits (see Fig. 3-22a). Sickle cell anemia results from a mutation affecting the β subunits.

Scientists therefore extended the definition of a gene to include that one gene is responsible for one polypeptide chain. Even this definition has proved only partially correct, although as you will see later in this chapter, scientists still define a gene in terms of its product.

Although the elegant work of Beadle and Tatum and others demonstrated that genes are expressed in the form of proteins, the mechanism of gene expression was completely unknown. After Watson and Crick's discovery of the structure of DNA, many scientists worked to understand exactly how gene expression takes place. We begin with an overview of gene expression and then consider the various steps in more detail.

Review

▮ What is the one-gene, one-enzyme hypothesis?

▮ What were the contributions of each of the following scientists—Garrod, Beadle and Tatum, and Pauling—to our understanding of the relationship between genes and proteins?

Key Experiment

QUESTION: Can genes code for enzymes?

HYPOTHESIS: Mutations induced in bread mold correspond to the absence of functional enzymes.

EXPERIMENT: Beadle and Tatum irradiated *Neurospora* spores to induce mutations. ❶ They grew cultures derived from these spores on complete growth medium containing all amino acids, vitamins, and other nutrients that *Neurospora* normally makes for itself. They identified nutritional requirements in mutant strains by testing for growth on minimal media supplemented with individual vitamins or amino acids.

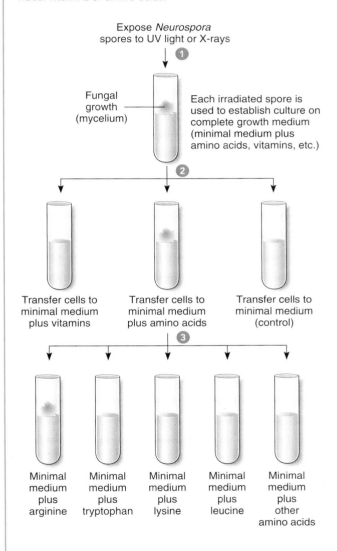

RESULTS AND CONCLUSION: ❷, ❸ In this example, only the medium containing the amino acid arginine supports growth, indicating that the mutation affects some part of arginine's biosynthetic pathway. Beadle and Tatum then identified the enzyme-catalyzed step that was blocked in the synthesis of arginine. They concluded that each gene controls the production of a single enzyme.

Figure 13-2 The Beadle–Tatum experiments

INFORMATION FLOW FROM DNA TO PROTEIN: AN OVERVIEW

Learning Objectives

3 Outline the flow of genetic information in cells, from DNA to RNA to polypeptide.
4 Compare the structures of DNA and RNA.
5 Explain why the genetic code is said to be redundant and virtually universal, and discuss how these features may reflect its evolutionary history.

Although the sequence of bases in DNA determines the sequence of amino acids in polypeptides, cells do not use the information in DNA directly. Instead, a related nucleic acid, **ribonucleic acid (RNA),** is the link between DNA and protein. When a gene that codes for a protein is expressed, first an RNA copy is made of the information in the DNA. It is this RNA copy that provides the information that directs polypeptide synthesis.

Like DNA, RNA is a polymer of nucleotides, but it has some important differences (▌Fig. 13-3). RNA is usually single-stranded, although internal regions of some RNA molecules may have complementary sequences that allow the strand to fold back and pair to form short, double-stranded segments. As shown in Figure 13-3, the sugar in RNA is **ribose,** which is similar to deoxyribose of DNA but has a hydroxyl group at the 2′ position. (Compare ribose with the deoxyribose of DNA, shown in Figure 12-3, which has a hydrogen at the 2′ position.) The base **uracil** substitutes for thymine and, like thymine, is a pyrimidine that can form two hydrogen bonds with adenine. Hence, uracil and adenine are a complementary pair.

DNA is transcribed to form RNA

The process by which RNA is synthesized resembles DNA replication in that the sequence of bases in the RNA strand is determined by complementary base pairing with one of the DNA strands, the **template strand** (▌Fig. 13-4). Because RNA synthesis takes the information in one kind of nucleic acid (DNA) and copies it as another nucleic acid (RNA), this process is called **transcription** ("copying").

Three main kinds of RNA molecules are transcribed: messenger RNA, transfer RNA, and ribosomal RNA. **Messenger RNA (mRNA)** is a single strand of RNA that carries the information for making a protein. Each of the 45 or so kinds of **transfer RNAs (tRNAs)** is a single strand of RNA that folds back on itself to form a specific shape. Each kind of tRNA bonds with only one kind of amino acid and carries it to the *ribosome.* (Because there are more kinds of tRNA molecules than there are amino acids, many amino acids are carried by two or more kinds of tRNA molecules.) **Ribosomal RNA (rRNA),** which is in a globular form, is an important part of the structure of ribosomes and has catalytic functions needed during protein synthesis.

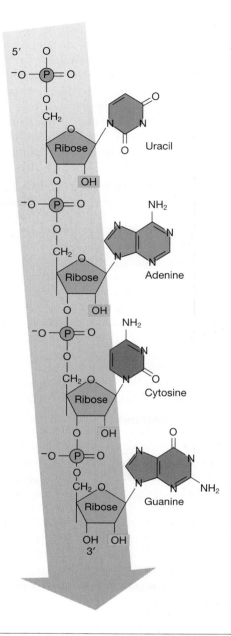

Figure 13-3   The nucleotide structure of RNA

The nucleotide subunits of RNA are joined by 5′ ⟶ 3′ phosphodiester linkages, like those found in DNA. Adenine, guanine, and cytosine are present, as in DNA, but the base uracil replaces thymine. All four nucleotide types contain the five-carbon sugar ribose, which has a hydroxyl group (*blue*) on its 2′ carbon atom.

RNA is translated to form a polypeptide

Following transcription, the transcribed information in the mRNA is used to specify the amino acid sequence of a polypeptide (see Fig. 13-4). This process is called **translation** because it involves conversion of the "nucleic acid language" in the mRNA molecule into the "amino acid language" of protein.

In translation of the genetic instructions to form a polypeptide, a sequence of three consecutive bases in mRNA, called a

Key Point

Protein synthesis requires two major steps:
DNA $\xrightarrow{\text{transcription}}$ RNA $\xrightarrow{\text{translation}}$ protein

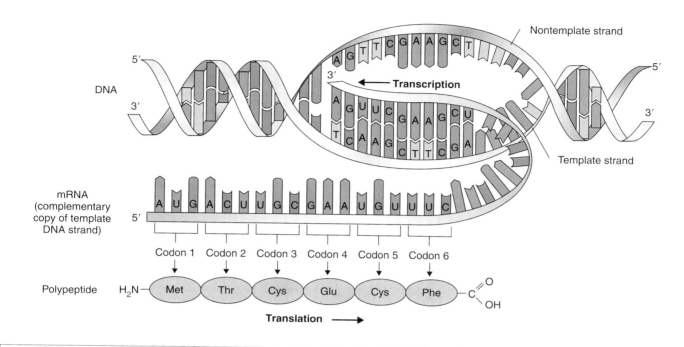

Figure 13-4   An overview of transcription and translation

In transcription, messenger RNA is synthesized as a complementary copy of one of the DNA strands, the template strand. Messenger RNA carries genetic information in the form of sets of three bases, or codons, each of which specifies one amino acid. Codons are translated consecutively, thus specifying the linear sequence of amino acids in the polypeptide chain. Translation requires tRNA and ribosomes (*not shown*). The figure depicts transcription and translation in bacteria. In eukaryotes, transcription takes place in the nucleus and translation occurs in the cytosol.

codon, specifies one amino acid. For example, one codon that corresponds to the amino acid phenylalanine is 5′—UUC—3′. Because each codon consists of three nucleotides, the code is described as a **triplet code.** The assignments of codons for amino acids and for start and stop signals are collectively named the **genetic code** (∎ Fig. 13-5).

Transfer RNAs are crucial parts of the decoding machinery because they act as "adapters" that connect amino acids and nucleic acids. This mechanism is possible because each tRNA can (1) link with a specific amino acid and (2) recognize the appropriate mRNA codon for that particular amino acid (∎ Fig. 13-6). A particular tRNA can recognize a particular codon because it has a sequence of three bases, called the **anticodon,** that hydrogen-bonds with the mRNA codon by complementary base pairing. The exact anticodon that is complementary to the codon for phenylalanine in our example is 3′—AAG—5′.

Translation requires (1) each tRNA anticodon to be hydrogen-bonded to the complementary mRNA codon and (2) the amino acids carried by the tRNAs to be linked in the order specified by the sequence of codons in the mRNA. **Ribosomes,** the site

of translation, are organelles composed of two different subunits, each containing protein and rRNA (see Chapter 4). Ribosomes attach to the 5′ end of the mRNA and travel along it, allowing the tRNAs to attach sequentially to the codons of mRNA. In this way the amino acids carried by the tRNAs take up the proper position to be joined by *peptide bonds* in the correct sequence to form a polypeptide.

Biologists cracked the genetic code in the 1960s

Before the genetic code was deciphered, scientists had become interested in how a genetic code might work. The Watson and Crick model of DNA showed it to be a linear sequence of four different nucleotides. If each nucleotide coded for a single amino acid, the genetic code could specify only 4 amino acids, not the 20 found in the vast variety of proteins in the cell. Scientists saw that the DNA bases could serve as a four-letter "alphabet" and hypothesized that three-letter combinations of the four bases would

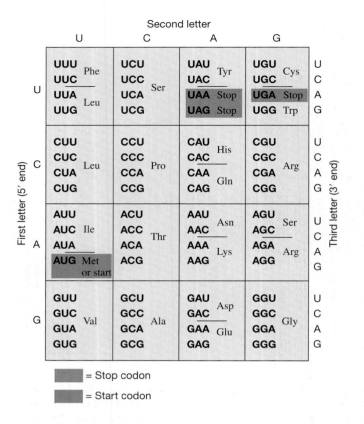

Second letter

	U	C	A	G	
U	UUU Phe UUC UUA Leu UUG	UCU UCC Ser UCA UCG	UAU Tyr UAC UAA Stop UAG Stop	UGU Cys UGC UGA Stop UGG Trp	U C A G
C	CUU CUC Leu CUA CUG	CCU CCC Pro CCA CCG	CAU His CAC CAA Gln CAG	CGU CGC Arg CGA CGG	U C A G
A	AUU AUC Ile AUA AUG Met or start	ACU ACC Thr ACA ACG	AAU Asn AAC AAA Lys AAG	AGU Ser AGC AGA Arg AGG	U C A G
G	GUU GUC Val GUA GUG	GCU GCC Ala GCA GCG	GAU Asp GAC GAA Glu GAG	GGU GGC Gly GGA GGG	U C A G

First letter (5′ end) / Third letter (3′ end)

▮ = Stop codon

▮ = Start codon

Figure 13-5   The genetic code

The genetic code specifies all possible combinations of the three bases that compose codons in mRNA. Of the 64 possible codons, 61 specify amino acids (see Fig. 3-16 for an explanation of abbreviations). The codon AUG specifies the amino acid methionine and also signals the ribosome to initiate translation ("start"). Three codons—UAA, UGA, and UAG—do not specify amino acids; they terminate polypeptide synthesis ("stop").

make it possible to form a total of 64 "words," more than enough to specify all the naturally occurring amino acids. In 1961, Crick and British scientist Sydney Brenner concluded from experimental evidence that the code used nonoverlapping triplets of bases. They predicted that the code is read, one triplet at a time, from a fixed starting point that establishes the **reading frame** for the genetic message. As we will discuss later in the chapter, an alteration in the reading frame that is caused by a mutation results in the incorporation of incorrect amino acids.

Marshall Nirenberg, a U.S. biochemist, and his postdoctoral researcher, Heinrich Matthaei, studied protein synthesis outside living cells in purified *cell-free systems* derived from the bacterium *Escherichia coli*. Nirenberg and Matthaei obtained the first experimental evidence indicating the assignment of triplets to specific amino acids. By constructing artificial mRNA molecules with known base sequences, they determined which amino acids would be incorporated into protein. For example, when they added the synthetic mRNA polyuridylic acid (UUUUUUUUU . . .) to a mixture of purified ribosomes, aminoacyl-tRNAs (amino acids linked to their respective tRNAs), and essential cofactors needed to synthesize polypeptide, only phenylalanine was incorporated

into the resulting polypeptide chain. The inference was inescapable that the UUU triplet codes for phenylalanine. Similar experiments showed that polyadenylic acid (AAAAAAAAA . . .) codes for a polypeptide of lysine and that polycytidylic acid (CCCCCCCCC . . .) codes for a polypeptide of proline.

By using mixed nucleotide polymers (such as a random polymer of A and C) as artificial messengers, researchers such as H. Gobind Khorana, then at the University of Wisconsin, assigned the other codons to specific amino acids. However, three of the codons—UAA, UGA, and UAG—did not specify any amino acids. These codons are now known to be the signals that terminate the coding sequence for a polypeptide chain. By 1967, the genetic code was completely "cracked," and scientists had identified the coding assignments of all 64 possible codons shown in Figure 13-5. In 1968, Nirenberg and Khorana received the Nobel Prize in Physiology or Medicine for their work in deciphering the genetic code.

Remember that the genetic code is an mRNA code. The tRNA anticodon sequences, as well as the DNA sequence from which the message is transcribed, are complementary to the sequences in Figure 13-5. For example, the mRNA codon for the amino acid methionine is 5′—AUG—3′. It is transcribed from the DNA base sequence 3′—TAC—5′, and the corresponding tRNA anticodon is 3′—UAC—5′.

The genetic code is virtually universal

Perhaps the single most remarkable feature of the code is that it is essentially universal. Over the years, biologists have examined the genetic code in a diverse array of species and found it the same in organisms as different as bacteria, redwood trees, jellyfish, and humans. These findings strongly suggest that the code is an ancient legacy that evolved very early in the history of life (discussed in Chapter 21).

Scientists have discovered some very minor exceptions to the universality of the genetic code. In several unicellular protozoa, UAA and UGA code for the amino acid glutamine instead of functioning as stop signals. Other exceptions are found in mitochondria, which contain their own DNA and protein synthesis machinery for some genes (see Chapters 4 and 21). These slight coding differences vary with the organism, but keep in mind that in each case, all the other coding assignments are identical to those of the standard genetic code.

The genetic code is redundant

Given 64 possible codons and only 20 common amino acids, it is not surprising that more than one codon specifies certain amino acids. This redundancy in the genetic code has certain characteristic patterns. The codons CCU, CCC, CCA, and CCG are "synonymous" in that they all code for the amino acid proline. The only difference among the 4 codons involves the nucleotide at the 3′ end of the triplet. Although the code may be read three nucleotides at a time, only the first two nucleotides seem to contain specific information for proline. A similar pattern can be seen for several other amino acids. Only methionine and tryptophan

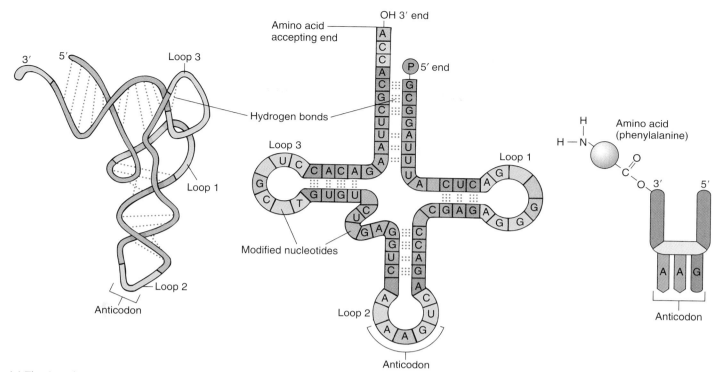

(a) The 3-D shape of a tRNA molecule is determined by hydrogen bonds formed between complementary bases.

(b) One loop contains the anticodon; these unpaired bases pair with a complementary mRNA codon. The amino acid attaches to the terminal nucleotide at the hydroxyl (OH) 3′ end.

(c) This stylized diagram of an aminoacyl-tRNA shows that the amino acid attaches to tRNA by its carboxyl group, leaving its amino group exposed for peptide bond formation.

Figure 13-6 Three representations of a tRNA molecule

are specified by single codons. Each of the other amino acids is specified by 2 to 6 different codons.

Crick first proposed this apparent breach of the base-pairing rules as the **wobble hypothesis.** He reasoned that the third nucleotide of a tRNA anticodon (which is the 5′ base of the anticodon sequence) may sometimes form hydrogen bonds with more than one kind of third nucleotide (the 3′ base) of a codon. Investigators later established this experimentally by determining the anticodon sequences of tRNA molecules and testing their specificities in artificial systems. Some tRNAs bond exclusively to one codon, but other tRNA molecules pair with as many as three separate codons that differ in their third nucleotide but that specify the same amino acid. Thus, the wobble hypothesis accounts for the possible variation in base pairing between the third base of a codon and the corresponding base in its anticodon. "Wobble" results in several acceptable forms of base pairing between mRNA and tRNA.

Review

- Sketch a simple flow diagram that shows the relationships among the following: RNA, translation, DNA, transcription, and polypeptide.
- How are the structures of DNA and RNA similar? How are they different?

TRANSCRIPTION

Learning Objective

6 Compare the processes of transcription and DNA replication, identifying both similarities and differences.

Now that we have presented an overview of information flow from DNA to RNA to polypeptide, let us examine the entire process more closely. The first stage in the flow of information from DNA to polypeptide is the transcription of a DNA nucleotide sequence into an RNA nucleotide sequence.

In eukaryotic transcription, most RNA is synthesized by one of three **RNA polymerases,** enzymes present in all cells. The three RNA polymerases differ in the kinds of RNA they synthesize. *RNA polymerase I* catalyzes the synthesis of several kinds of rRNA molecules that are components of the ribosome; *RNA polymerase II* catalyzes the production of the protein-coding mRNA; and *RNA polymerase III* catalyzes the synthesis of tRNA and one of the rRNA molecules.

RNA polymerases require DNA as a template and have many similarities to the DNA polymerases discussed in Chapter 12. Like DNA polymerases, RNA polymerases carry out synthesis in the 5′ ⟶ 3′ direction; that is, they begin at the 5′ end of the RNA

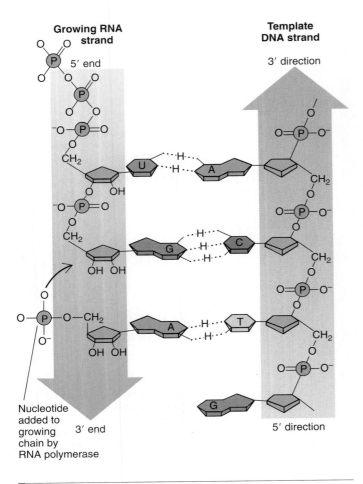

Growing RNA strand
5' end

Template DNA strand
3' direction

Nucleotide added to growing chain by RNA polymerase

3' end

5' direction

Figure 13-7 A molecular view of transcription

Incoming nucleotides with three phosphates pair with complementary bases on the template DNA strand (*right*). RNA polymerase cleaves two phosphates (*not shown*) from each nucleotide and covalently links the remaining phosphate to the 3' end of the growing RNA chain. Thus, RNA, like DNA, is synthesized in the 5' ⟶ 3' direction.

molecule being synthesized and then continue to add nucleotides at the 3' end until the molecule is complete (▌Fig. 13-7). RNA polymerases use nucleotides with three phosphate groups (for example, ATP and GTP) as substrates for the polymerization reaction. As the nucleotides link to the 3' end of the RNA, two of the phosphates are removed. These reactions are strongly exergonic and do not need the input of additional energy.

Recall from Chapter 12 that whenever nucleic acid molecules associate by complementary base pairing, the two strands are **antiparallel.** Just as the two paired strands of DNA are antiparallel, the template strand of the DNA and the complementary RNA strand are also antiparallel. Therefore, when transcription takes place, as RNA is synthesized in its 5' ⟶ 3' direction, the DNA template is read in its 3' ⟶ 5' direction.

Scientists conventionally refer to a sequence of bases in a gene or the mRNA sequence transcribed from it as upstream or downstream from some reference point. **Upstream** is toward the 5' end of the mRNA sequence or the 3' end of the template DNA

strand. **Downstream** is toward the 3' end of the RNA or the 5' end of the template DNA strand.

Upstream		Downstream
5'—A—T—G—A—C—T—3'		Nontemplate DNA strand
3'—T—A—C—T—G—A—5'		Template DNA strand
Direction of transcription ⟶		
5'—A—U—G—A—C—U—3'		RNA

The synthesis of mRNA includes initiation, elongation, and termination

In both bacteria and eukaryotes, the nucleotide sequence in DNA to which RNA polymerase and associated proteins initially bind is called the **promoter.** Because the promoter is not transcribed, RNA polymerase moves past the promoter to begin transcription of the protein-coding sequence of DNA. Different genes may have slightly different promoter sequences, so the cell can direct which genes are transcribed at any one time. Bacterial promoters are usually about 40 bases long and lie in the DNA just upstream of the point at which transcription will begin. Once RNA polymerase has recognized the correct promoter, it unwinds the DNA double helix and *initiates* transcription (▌Fig. 13-8, step **1**). Unlike DNA synthesis, RNA synthesis does not require a primer. However, transcription requires several proteins in addition to RNA polymerase; we discuss these in Chapter 14.

The first nucleotide at the 5' end of a new mRNA chain initially retains its triphosphate group (see Fig. 13-7). However, during the *elongation* stage of transcription (▌Fig. 13-8, step **2**), as each additional nucleotide is incorporated at the 3' end of the growing RNA molecule, two of its phosphates are removed in an exergonic reaction that leaves the remaining phosphate to become part of the sugar–phosphate backbone (as in DNA replication). The last nucleotide to be incorporated has an exposed 3'-hydroxyl group.

Elongation of RNA continues until *termination*, when RNA polymerase recognizes a termination sequence consisting of a specific sequence of bases in the DNA template (▌Fig. 13-8, step **3**). This signal leads to a separation of RNA polymerase from the template DNA and the newly synthesized RNA. Termination of transcription occurs by different mechanisms in bacteria and eukaryotes. In bacteria, transcription stops at the end of the termination sequence. When RNA polymerase comes to the termination sequence, it releases the DNA template and the new RNA strand. In eukaryotic cells, RNA polymerase adds nucleotides to the mRNA molecule after it passes the termination sequence. The 3' end of the mRNA becomes separated from RNA polymerase about 10 to 35 nucleotides past the termination sequence.

Only one strand in a protein-coding region of DNA is used as a template. For example, consider a segment of DNA that contains the following DNA base sequence in the template strand:

3'—TAACGGTCT—5'

The mRNA is synthesized in the 5′ to 3′ direction as the template strand of the DNA molecule is read in the 3′ to 5′ direction.

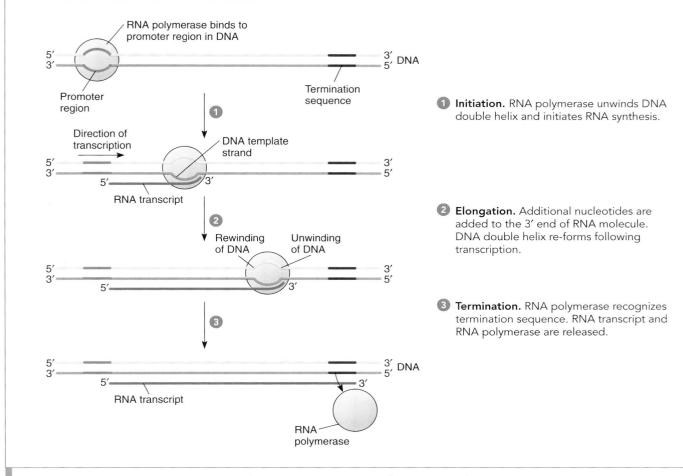

Initiation. RNA polymerase unwinds DNA double helix and initiates RNA synthesis.

Elongation. Additional nucleotides are added to the 3′ end of RNA molecule. DNA double helix re-forms following transcription.

Termination. RNA polymerase recognizes termination sequence. RNA transcript and RNA polymerase are released.

Figure 13-8 *Animated* An overview of transcription: initiation, elongation, and termination

If the complementary DNA strand

5′—ATTGCCAGA—3′

were used as a template, a message would be produced specifying an entirely different (and probably nonfunctional) protein. How-ever, just because one strand is transcribed for a given gene does not mean the same DNA strand is always the template for all genes throughout the length of a chromosome-sized DNA molecule. Instead, a particular strand may serve as the template strand for some genes and the nontemplate strand for others (Fig. 13-9).

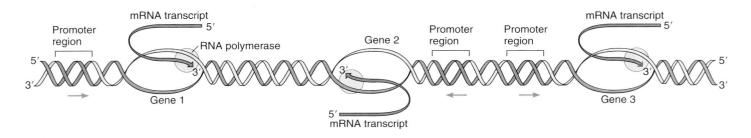

Figure 13-9 The template strand for one gene may be the nontemplate strand for another gene

Only one of the two strands is transcribed for a given gene, but the op-posite strand may be transcribed for a neighboring gene. Each transcript starts at its own promoter region (*orange*). The orange arrow associ-ated with each promoter region indicates the direction of transcription.

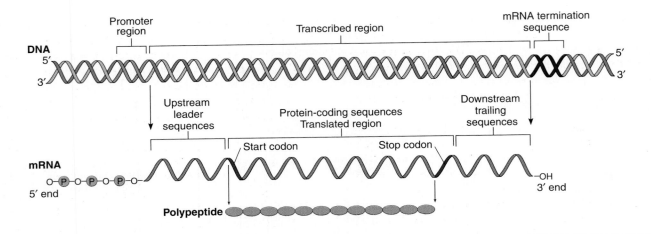

Promoter region | Transcribed region | mRNA termination sequence

DNA 5' 3' ... 5' 3'

Upstream leader sequences | Protein-coding sequences Translated region | Downstream trailing sequences

Start codon | Stop codon

mRNA
5' end

Polypeptide

3' end

Figure 13-10 Bacterial mRNA

This figure compares a bacterial mRNA with the region of DNA from which it was transcribed. RNA polymerase recognizes, but does not transcribe, promoter sequences in the DNA. Initiation of RNA synthesis occurs five to eight bases downstream from the promoter. Ribosome recognition sites are located in the 5' mRNA upstream leader sequences. Protein-coding sequences begin at a start codon, which follows the leader sequences, and end at a downstream stop codon near the 3' end of the molecule. Downstream trailing sequences, which vary in length, follow the polypeptide-coding sequences.

Messenger RNA contains base sequences that do not directly code for protein

A completed mRNA molecule contains more than the nucleotide sequence that codes for a protein. ▌Figure 13-10 shows a typical bacterial mRNA. (The unique features of eukaryotic mRNA are discussed later in the chapter.) In both bacteria and eukaryotes, RNA polymerase starts transcription of a gene upstream of the protein-coding DNA sequence. As a result, the mRNA has a noncoding **leader sequence** at its 5' end. The leader has recognition sites for ribosome binding, which properly position the ribosomes to translate the message. The **start codon** follows the leader sequence and signals the beginning of the **coding sequence** that contains the actual message for the polypeptide. Unlike coding in eukaryotic cells, in bacterial cells it is common for more than one polypeptide to be encoded by a single mRNA molecule (see Chapter 14). At the end of each coding sequence, a **stop codon** signals the end of the protein. The stop codons—UAA, UGA, and UAG—end both bacterial and eukaryotic messages. They are followed by noncoding 3' **trailing sequences,** which vary in length.

Review

- In what ways are DNA polymerase and RNA polymerase similar? How do they differ?
- A certain template DNA strand has the following nucleotide sequence:

  3'—TACTGCATAATGATT—5'

  What would be the sequence of codons in the mRNA transcribed from this strand? What would be the nucleotide sequence of the complementary nontemplate DNA strand?

TRANSLATION

Learning Objectives

7   Identify the features of tRNA that are important in decoding genetic information and converting it into "protein language."
8   Explain how ribosomes function in polypeptide synthesis.
9   Describe the processes of initiation, elongation, and termination in polypeptide synthesis.

Translation adds another level of complexity to the process of information transfer because it involves conversion of the triplet nucleic acid code to the 20–amino acid alphabet of polypeptides. Translation requires the coordinated functioning of more than 100 kinds of macromolecules, including the protein and RNA components of the ribosomes, mRNA, and amino acids linked to tRNAs.

Amino acids are joined by *peptide bonds* to form proteins (see Chapter 3). This joining links the amino and carboxyl groups of adjacent amino acids. The translation process ensures both that peptide bonds form and that the amino acids link in the correct sequence specified by the codons in the mRNA.

An amino acid is attached to tRNA before incorporation into a polypeptide

How do the amino acids align in the proper sequence so they can link? Crick proposed that a molecule was needed to bridge the gap between mRNA and proteins. This molecule was determined to be tRNA. DNA contains genes that are transcribed to form the tRNAs. Each kind of tRNA molecule binds to a specific amino acid. Amino acids are covalently linked to their respective tRNA molecules by enzymes, called **aminoacyl–tRNA synthetases,** which use ATP as an energy source (❙ Fig. 13-11). The resulting complexes, called **aminoacyl-tRNAs,** bind to the mRNA coding sequence to align the amino acids in the correct order to form the polypeptide chain.

The tRNAs are polynucleotide chains 70 to 80 nucleotides long, each with several unique base sequences as well as some sequences that are common to all (see Fig. 13-6). Although considerably smaller than mRNA or rRNA molecules, tRNA molecules have an intricate structure. A tRNA molecule has several properties essential to its function: (1) It has an anticodon, a complementary binding sequence for the correct mRNA codon; (2) it is recognized by an aminoacyl–tRNA synthetase that adds the correct amino acid; (3) it has an attachment site for the specific amino acid specified by the anticodon (the site lies about 180 degrees from the anticodon); and (4) it is recognized by ribosomes.

Complementary base pairing within each tRNA molecule causes it to be doubled back and folded. Three or more loops of unpaired nucleotides are formed, one of which contains the anticodon. The amino acid–binding site is at the 3′ end of the molecule. The carboxyl group of the amino acid is covalently bound to the exposed 3′ hydroxyl group of the terminal nucleotide, leaving the amino group of the amino acid free to participate in peptide bond formation. The pattern of folding in tRNAs keeps a constant distance between the anticodon and amino acid, allowing precise positioning of the amino acids during translation.

The components of the translational machinery come together at the ribosomes

The importance of ribosomes and of protein synthesis in cell metabolism is exemplified by a rapidly growing *E. coli* cell, which contains some 15,000 ribosomes—nearly one third of the total mass of the cell. Although bacterial and eukaryotic ribosomes are not identical, ribosomes from all organisms share many fundamental features and basically work in the same way. Ribosomes consist of two subunits, both made up of protein (about 40% by weight) and rRNA (about 60% by weight). Unlike mRNA and tRNA, rRNA has catalytic functions and does not transfer information. The ribosomal proteins do not appear to be catalytic but instead contribute to the overall structure and stability of the ribosome.

Researchers have isolated each ribosomal subunit intact in the laboratory, then separated each into its RNA and protein constituents. In bacteria, the smaller of these subunits contains 21 proteins and one rRNA molecule, and the larger subunit contains 34 proteins and two rRNA molecules. Under certain conditions researchers can reassemble each subunit into a functional form by adding each component in its correct order. Using this approach, together with sophisticated electron microscopic studies, researchers have determined the 3-D structure of the ribosome (❙ Fig. 13-12a), as well as the way it is assembled in the living cell. The large subunit contains a depression on one surface into which the small subunit fits. During translation, the mRNA fits in a groove between the contact surfaces of the two subunits.

The ribosome has four binding sites, one for mRNA and three for tRNAs. Thus, the ribosome holds the mRNA template, the aminoacyl–tRNA molecules, and the growing polypeptide chain in the correct orientation so the genetic code can be read and the next peptide bond formed. Transfer RNA molecules attach to three depressions on the ribosome, the A, P, and E binding sites (❙ Fig. 13-12b). The **P site,** or peptidyl site, is so named because the tRNA holding the growing polypeptide chain occupies the P site. The **A site** is named the aminoacyl site because the aminoacyl-tRNA delivering the next amino acid in the sequence binds at this location. The **E site** (for *exit*) is where tRNAs that have delivered their amino acids to the growing polypeptide chain exit the ribosome.

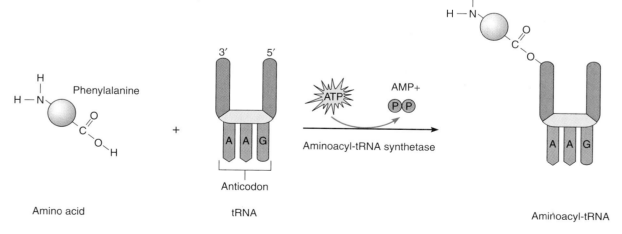

Figure 13-11 Linkage of a specific amino acid to the appropriate tRNA

Amino acids are covalently linked to their respective tRNA molecules by aminoacyl–tRNA synthetases, which use ATP as an energy source.

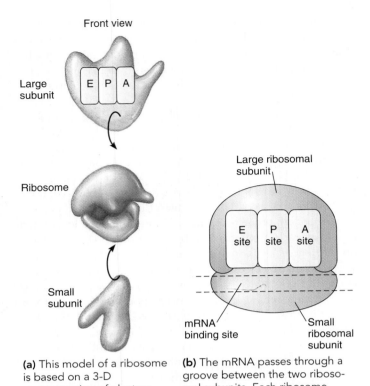

Front view

Large subunit

E P A

Ribosome

Small subunit

Large ribosomal subunit

E site | P site | A site

mRNA binding site

Small ribosomal subunit

(a) This model of a ribosome is based on a 3-D reconstruction of electron microscopic images.

(b) The mRNA passes through a groove between the two ribosomal subunits. Each ribosome contains three binding sites for tRNA molecules.

Figure 13-12 *Animated* Ribosome structure

A ribosome consists of two subunits, one large and one small.

Translation begins with the formation of an initiation complex

The process of protein synthesis has three distinct stages: initiation, repeating cycles of elongation, and termination. Initiation of protein synthesis is essentially the same in all organisms. Here we describe initiation in bacteria and then briefly discuss some differences between bacterial and eukaryotic initiation.

The **initiation** of translation uses proteins called **initiation factors,** which become attached to the small ribosomal subunit. In bacteria, three different initiation factors attach to the small subunit, which then binds to the mRNA molecule in the region of the AUG start codon (❙ Fig. 13-13, step ❶). A leader sequence upstream (toward the 5′ end) of the AUG helps the ribosome identify the AUG sequence that signals the beginning of the mRNA code for a protein.

The tRNA that bears the first amino acid of the polypeptide is the **initiator tRNA.** The amino acid methionine is bound to the initiator tRNA, and as a result, the first amino acid on new polypeptides is methionine. (Quite often, the methionine is removed later.) In bacteria, the initiator methionine is modified by the addition of a one-carbon group derived from formic acid and is designated *fMet* for *N*-formylmethionine. The fMet–initiator tRNA, which has the anticodon 3′—UAC—5′, binds to the AUG

start codon, releasing one of the initiation factors in the process (❙ Fig. 13-13, step ❷). The **initiation complex** is complete when the large ribosomal subunit binds to the small subunit and the remaining initiation factors are released (❙ Fig. 13-13, step ❸).

In eukaryotes, initiation of translation differs in three respects. First, the methionine of the initiator tRNA is unmodified. Second, instead of a leader sequence to help identify the start codon on mRNA, the start codon is embedded within a short sequence (such as ACCAUGG) that indicates the site (the underlined AUG) of translation initiation. Third, the initiation complex, which contains perhaps 10 protein factors, is different from that of bacteria.

During elongation, amino acids are added to the growing polypeptide chain

❙ Figure 13-14 outlines the four steps in a cycle of **elongation,** the stage of translation in which amino acids are added one by one to the growing polypeptide chain. Elongation is essentially the same in bacteria and eukaryotes. In step ❶, the appropriate aminoacyl-tRNA recognizes the codon in the A site and ❷ binds there by base pairing of its anticodon with the complementary mRNA codon. This binding step requires several proteins called **elongation factors.** It also requires energy from **guanosine triphosphate (GTP),** an energy transfer molecule similar to ATP.

The amino group of the amino acid at the A site is now aligned with the carboxyl group of the preceding amino acid attached to the growing polypeptide chain at the P site. In step ❸ of elongation, a peptide bond forms between the amino group of the new amino acid and the carboxyl group of the preceding amino acid. In this way, the polypeptide is released from tRNA at the P site and attaches to the aminoacyl-tRNA at the A site. This reaction is spontaneous (it does not require additional energy), because ATP transferred energy during formation of the aminoacyl-tRNA. Peptide bond formation does, however, require the enzyme **peptidyl transferase.** Remarkably, this enzyme is not a protein but an RNA component of the large ribosomal subunit. Such an RNA catalyst is known as a **ribozyme.** Two biochemists, American Thomas Cech (University of Colorado) and Canadian Sidney Altman (Yale University) received a Nobel Prize in Chemistry in 1989 for their independent discovery of ribozymes.

Recall from Chapter 3 that polypeptide chains have direction, or polarity. The amino acid on one end of the polypeptide chain has a free amino group (the amino end), and the amino acid at the other end has a free carboxyl group (the carboxyl end). Protein synthesis always proceeds from the amino end to the carboxyl end of the growing polypeptide chain:

$$5'\text{———————} 3'\text{ mRNA}$$

Direction of translation →

$$\begin{array}{c} H \\ \diagdown \\ N\text{————————}C \quad \text{polypeptide} \\ \diagup \qquad\qquad \| \\ H \qquad\qquad\quad O \\ \qquad\qquad\qquad OH \end{array}$$

Key Point

Initiation in bacteria involves a ribosome, an mRNA, an initiator tRNA to which formylated methionine (fMet) is bound, and several protein initiation factors.

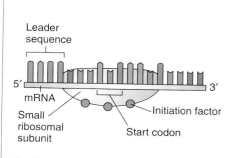

1 The small ribosomal subunit binds to the mRNA at the AUG start codon. The leader sequence upstream of the AUG sequence helps the ribosome identify the AUG sequence.

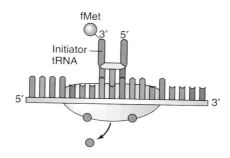

2 The initiator tRNA binds to the start codon, and one of the initiation factors is released.

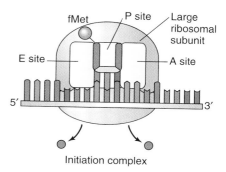

3 When the large ribosomal subunit binds to the small subunit and the remaining initiation factors are released, the initiation complex is complete.

Figure 13-13   The initiation of translation in bacteria

In step **4** of elongation, which is known as **translocation,** the ribosome moves down the mRNA by one codon. As a result, the mRNA codon specifying the next amino acid in the polypeptide chain becomes positioned in the unoccupied A site. The translocation process requires energy, which again GTP supplies. The uncharged tRNA—that is, the tRNA without an attached amino acid—is moved from the P site to the E site. From there it exits the ribosome and joins the cytosolic pool of tRNAs.

Because translocation involves movement of the ribosome in the 3′ direction along the mRNA molecule, translation always proceeds in the 5′ ⟶ 3′ direction. Each peptide bond forms in a fraction of a second; by repeating the elongation cycle, an average-sized protein of about 360 amino acids is assembled by a bacterium in approximately 18 seconds, and by a eukaryotic cell in a little more than 1 minute.

One of three stop codons signals the termination of translation

Termination is the final stage of translation. In termination, the synthesis of the polypeptide chain is terminated by a **release factor,** a protein that recognizes the stop codon at the end of the coding sequence. (Because no tRNA molecule binds to a stop codon, the codon is available for binding by the release factor.) When the release factor binds to the A site, the bond between the tRNA in the P site and the last amino acid of the polypep-

tide chain breaks (❚ Fig. 13-15). This hydrolysis reaction frees the newly synthesized polypeptide and also separates the translation complex into its parts: the mRNA molecule, the release factor, the tRNA molecule in the P site, and the large and small ribosomal subunits. The two ribosomal subunits can be used to form a new initiation complex with another mRNA molecule. Each mRNA is translated only a limited number of times before it is destroyed.

Specialized ribosome-associated proteins, called **molecular chaperones,** assist in the folding of the newly synthesized polypeptide chain into its three-dimensional active shape (see Chapter 3). The sequence of amino acids in the polypeptide chain dictates the shape that it ultimately forms. However, without the molecular chaperone's help, interactions among the various regions of the amino acid chain might prevent the proper folding process from occurring.

Review

❚ What are ribosomes made of? Do ribosomes carry information to specify the amino acid sequence of proteins?

❚ What do the initiation, elongation, and termination stages of protein synthesis each do?

❚ A certain mRNA strand has the following nucleotide sequence:

   5′—AUG—ACG—UAU—AAC—UUU—3′

What is the anticodon for each codon? What is the amino acid sequence of the polypeptide? (Use Fig. 13-5 to help answer this question.)

Each repetition of the elongation process adds one amino acid to the growing polypeptide chain.

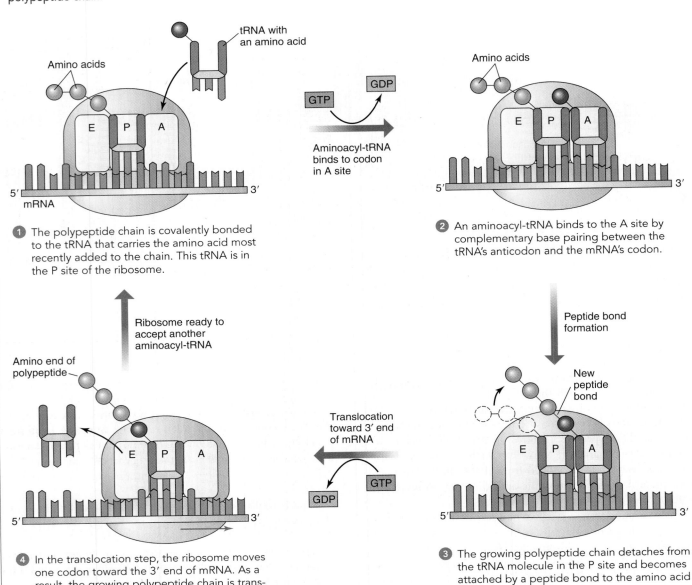

① The polypeptide chain is covalently bonded to the tRNA that carries the amino acid most recently added to the chain. This tRNA is in the P site of the ribosome.

② An aminoacyl-tRNA binds to the A site by complementary base pairing between the tRNA's anticodon and the mRNA's codon.

③ The growing polypeptide chain detaches from the tRNA molecule in the P site and becomes attached by a peptide bond to the amino acid linked to the tRNA at the A site.

④ In the translocation step, the ribosome moves one codon toward the 3′ end of mRNA. As a result, the growing polypeptide chain is transferred to the P site. The uncharged tRNA in the E site exits the ribosome.

Figure 13-14 *Animated* An elongation cycle in translation

This illustration begins after a short chain of amino acids has formed.

VARIATIONS IN GENE EXPRESSION IN DIFFERENT ORGANISMS

Learning Objectives

10 Compare bacterial and eukaryotic mRNAs, and explain the functional significance of their structural differences.

11 Describe the differences in translation in bacterial and eukaryotic cells.

12 Describe retroviruses and the enzyme reverse transcriptase.

Although the basic mechanisms of transcription and translation are quite similar in all organisms, some significant differences exist between bacteria and eukaryotes. Some differences result

A stop signal terminates polypeptide synthesis, because no tRNA molecules exist with an anticodon complementary to a stop codon.

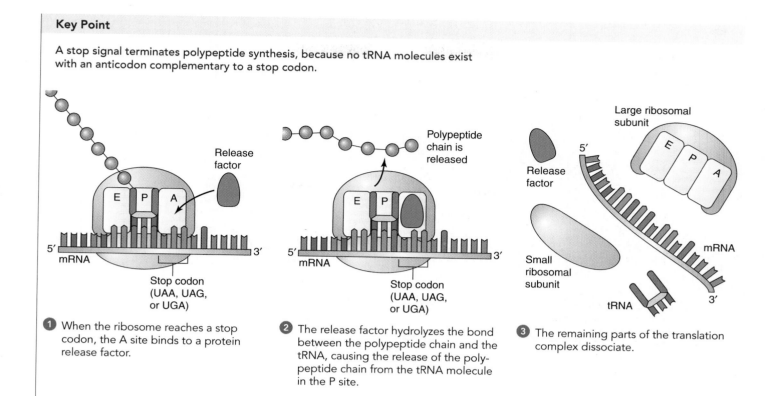

① When the ribosome reaches a stop codon, the A site binds to a protein release factor.

② The release factor hydrolyzes the bond between the polypeptide chain and the tRNA, causing the release of the polypeptide chain from the tRNA molecule in the P site.

③ The remaining parts of the translation complex dissociate.

Figure 13-15 The termination of translation

from prokaryotic versus eukaryotic cell structure. Bacteria lack a nucleus, and bacterial mRNA is translated as it is transcribed from DNA. This cannot occur in eukaryotes, because eukaryotic chromosomes are confined to the cell nucleus and protein synthesis takes place in the cytosol. In eukaryotes, mRNA must move from the nucleus (the site of transcription) to the cytosol (the site of translation, or polypeptide synthesis).

Transcription and translation are coupled in prokaryotes

In *E. coli* and other bacteria, translation begins very soon after transcription, and several ribosomes are attached to the same mRNA (❚ Fig. 13-16). The end of the mRNA molecule synthesized first during transcription is also the first translated to form a polypeptide. Ribosomes bind to the 5′ end of the growing mRNA and initiate translation long before the message is fully synthesized. As many as 15 ribosomes may bind to a single mRNA. An mRNA molecule that is bound to clusters of ribosomes constitutes a **polyribosome**, or *polysome*. Polyribosomes also occur in eukaryotic cells, but not while transcription is still taking place.

Although many polypeptide chains can be actively synthesized on a single mRNA at any one time, the half-life of mRNA molecules in bacterial cells is only about 2 minutes. (Half-life is the time it takes for half the molecules to be degraded.) Usually,

degradation of the 5′ end of the mRNA begins even before the first polypeptide is complete. Once the ribosome recognition sequences at the 5′ end of the mRNA have been degraded, no more ribosomes attach and initiate protein synthesis.

Eukaryotic mRNA is modified after transcription and before translation

Other differences exist between bacterial and eukaryotic mRNA. Bacterial mRNAs are used immediately after transcription, without further processing. In eukaryotes, the original transcript, known as **precursor mRNA,** or **pre-mRNA,** is modified in several ways while it is still in the nucleus. These *posttranscriptional modification and processing* activities prepare mature mRNA for transport and translation.

Modification of the eukaryotic message begins when the growing RNA transcript is about 20 to 30 nucleotides long. At that point, enzymes add a **5′ cap** to the 5′ end of the mRNA chain (❚ Fig. 13-17, step ①). The cap is in the form of an unusual nucleotide, 7-methylguanosine, which is linked to the mRNA transcript by three phosphate groups. Eukaryotic ribosomes cannot bind to an uncapped message.

Capping may also protect the RNA from certain types of degradation and may therefore partially explain why eukaryotic mRNAs are much more stable than bacterial mRNAs. Eukaryotic mRNAs have half-lives ranging from 30 minutes to as long as

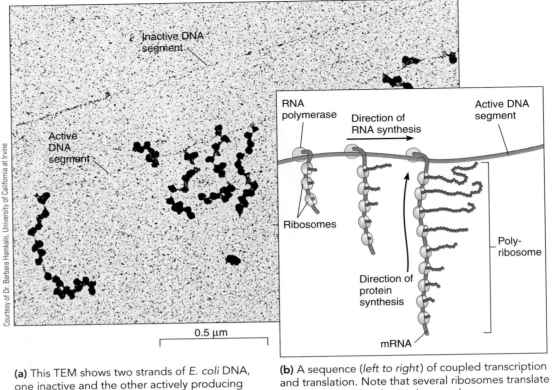

(a) This TEM shows two strands of *E. coli* DNA, one inactive and the other actively producing mRNA. Protein synthesis begins while the mRNA is being completed.

(b) A sequence (*left to right*) of coupled transcription and translation. Note that several ribosomes translate each mRNA molecule simultaneously.

Figure 13-16 Coupled transcription and translation in bacteria

24 hours. The average half-life of an mRNA molecule in a mammalian cell is about 10 hours compared with 2 minutes in a bacterial cell.

A second modification of eukaryotic mRNA, known as **polyadenylation,** may occur at the 3′ end of the molecule (■ Fig. 13-17, step ❷). Near the 3′ end of a completed message usually lies a sequence of bases that serves as a signal for adding many adenine-containing nucleotides, known as a **poly-A tail** (for *polyadenylated*). Enzymes in the nucleus recognize the signal for the poly-A tail and cut the mRNA molecule at that site. This process is followed by the enzymatic addition of a string of 100 to 250 adenine nucleotides to the 3′ end. Polyadenylation appears to have several functions: It helps export the mRNA from the nucleus, stabilizes some mRNAs against degradation in the cytosol (the longer the poly-A tail, the longer the molecule persists in the cytosol), and facilitates initiation of translation by helping ribosomes recognize the mRNA.

Both noncoding and coding sequences are transcribed from eukaryotic genes

One of the greatest surprises in the history of molecular biology was the finding that most eukaryotic genes have **interrupted coding sequences;** that is, long sequences of bases within the protein-coding sequences of the gene do not code for amino acids in the final polypeptide product. The noncoding regions within the gene are called **introns** (*in*tervening sequences), as opposed to **exons** (*ex*pressed sequences), which are parts of the protein-coding sequence.

A typical eukaryotic gene may have multiple exons and introns, although the number varies. For example, the human β-globin gene, which produces one component of hemoglobin, contains 2 introns and 3 exons; the human gene for albumin contains 14 introns and 15 exons; and the human gene for titin, the largest protein known, found in muscle cells, contains 233 introns and 234 exons. In many cases, the combined lengths of the intron sequences are much greater than the combined lengths of the exons. For instance, the ovalbumin gene contains about 7700 base pairs, whereas the total of all the exon sequences is only 1859 base pairs.

When a gene that contains introns is transcribed, the entire gene is copied as a large RNA transcript, the pre-mRNA. A pre-mRNA molecule contains both intron and exon sequences. (Note that *intron* and *exon* refer to corresponding nucleotide sequences in both DNA and RNA.) For the pre-mRNA to be made into a functional message, it must be capped and have a poly-A tail added, and the introns must be removed and the exons spliced together to form a continuous protein-coding message (see Fig. 13-17, step ❷).

Splicing itself occurs by several different mechanisms, depending on which type of RNA is involved. In many instances, splicing involves the association of several **small nuclear ribonucleoprotein complexes** (**snRNPs,** pronounced "snurps") to

Key Point

After transcription, pre-mRNA undergoes extensive modification to produce mature, functional mRNA.

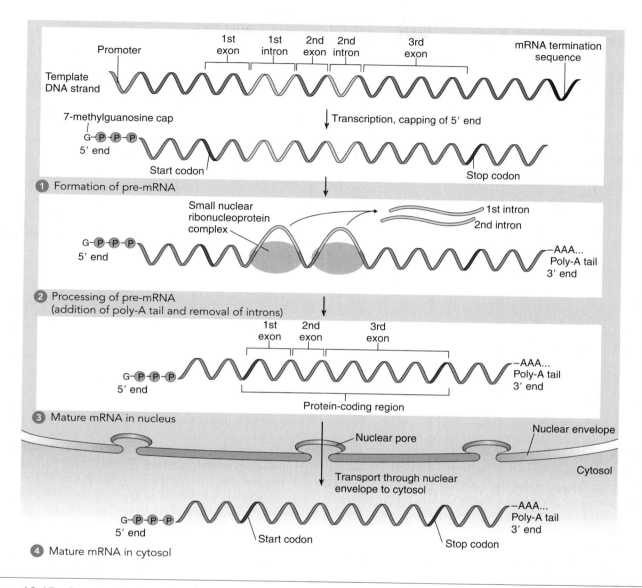

Figure 13-17 Posttranscriptional modification of eukaryotic messenger RNA

form a large nucleoprotein particle called a **spliceosome.** Spliceosomes, which are similar in size to ribosomes, catalyze the reactions that remove introns from pre-mRNA. In other instances the RNA within the intron acts as a ribozyme, splicing itself without the use of a spliceosome or protein enzymes.

Posttranscriptional modification in eukaryotes is summarized as follows:

Pre-mRNA containing introns and exons ⟶ 5′ end of pre-mRNA capped with modified nucleotide ⟶ poly-A tail added to 3′ end ⟶ introns removed and exons spliced together ⟶ mature mRNA transported into cytosol ⟶ translation at ribosome

Following pre-mRNA processing, the mature mRNA (▌ Fig. 13-17, step ❸) is transported through a nuclear pore into the cytosol (▌ Fig. 13-17, step ❹) to be translated by a ribosome.

Biologists debate the evolution of eukaryotic gene structure

The reason for the intricate structure of eukaryotic genes is a matter of ongoing discussion among molecular biologists. Why do introns occur in most eukaryotic nuclear genes but not in the genes of most bacteria (or of mitochondria and chloroplasts)? How did this remarkable genetic system with interrupted coding sequences ("split genes") evolve, and why has it survived? It

TABLE 13-1

Selected Kinds of RNA in Eukaryotic Cells

Kind of RNA	Function
Messenger RNA (mRNA)	Specifies the amino acid sequence of a protein
Transfer RNA (tRNA)	Binds to specific amino acid and serves as adapter molecule when amino acids are incorporated into growing polypeptide chain
Ribosomal RNA (rRNA)	Has both structural and catalytic (ribozyme) roles in ribosome
Small nuclear RNA (snRNA)	Involved in intron removal and regulation of transcription; part of spliceosome particles
Signal-recognition particle RNA (SRP RNA)	Helps direct the ribosome–mRNA–polypeptide complex to the rough ER shortly after translation begins
Small nucleolar RNA (snoRNA)	Processes pre-rRNA in nucleolus during formation of ribosome subunits
Small interfering RNA (siRNA)	Controls gene expression by selective suppression of genes
MicroRNA (miRNA)	Controls expression of genes involved in growth and development by inhibiting the translation of certain mRNAs

seems incredible that as much as 75% of the original transcript of a eukaryotic nuclear gene must be *removed* to make a working message.

In the early 1980s, American Walter Gilbert of Harvard University proposed that exons are nucleotide sequences that code for **protein domains,** regions of protein tertiary structure that may have important functions. For example, the active site of an enzyme may compose one domain. A different domain may enable that enzyme to bind to a particular cell structure, and yet another may be a site involved in allosteric regulation (see Chapter 7). Analyses of the DNA and amino acid sequences of many eukaryotic genes have shown that most exons are too small to code for an entire protein domain, although a block of several exons can code for a domain.

Gilbert further postulated that new proteins with new functions emerge rapidly when genetic recombination produces novel combinations of exons that code for different proteins. This hypothesis has become known as *evolution by exon shuffling.* It has been supported by examples such as the low-density lipoprotein (LDL) receptor protein, a protein found on the surface of human cells that binds to cholesterol transport molecules (see Chapter 5). The LDL receptor protein has domains that are related to parts of other proteins with totally different functions. However, many other genes and their corresponding proteins show no evidence of exon shuffling.

After introns were discovered in a few genes in archaea and bacteria, some biologists hypothesized that introns are relics from the ancient ancestors of all organisms living today. Through billions of years of evolution, the pressure in prokaryotic organisms, which are unicellular, to have a streamlined set of genes may have caused them to lose most of their introns.

Regardless of how split genes originated, intron excision provides one of many ways in which present-day eukaryotes regulate expression of their genes (see Chapter 14). This opportunity for control, together with the fact that eukaryotic RNAs are far more stable than those of bacteria, may balance the energy cost of maintaining a large load of noncoding DNA.

Several kinds of eukaryotic RNA have a role in gene expression

In addition to mRNA, rRNA, and tRNA, eukaryotic cells contain several other kinds of RNA that are an essential part of the protein-synthesizing machinery (Table 13-1). **Small nuclear RNA (snRNA)** molecules bind to specific proteins to form a small nuclear ribonucleoprotein complex (snRNP), which in turn combines with other snRNPs to form a spliceosome. Recall from the discussion in the previous section that a spliceosome catalyzes intron removal.

Some polypeptides enter the endoplasmic reticulum (ER) as they are being synthesized. **Signal-recognition particle RNA (SRP RNA),** in combination with proteins, directs the ribosome–mRNA–polypeptide complex to the rough ER shortly after translation begins. Translation continues after the complex binds to an integral protein in the ER membrane, and the newly synthesized protein is then distributed to its destination within the cell or outside the cell (for example, as a secretion).

A special group of RNA molecules, known as **small nucleolar RNAs (snoRNAs),** processes pre-rRNA molecules in the nucleolus during ribosome subunit formation. Small nucleolar RNAs bind to complementary regions of pre-rRNA and target sites for cleavage or methylation. A pre-rRNA molecule is larger than the combined size of the three rRNA molecules made from it. In mammalian cells, about 48% of the pre-rRNA molecule is removed during the cleavage steps. Pre-rRNA may have methyl groups added to regions that are to be saved during processing.

The discovery that RNA has the ability to regulate gene expression made big news in biology, both because this phenomenon, known as **RNA interference (RNAi),** is found in many organisms and because it has great potential as a research and medical tool (see Chapters 15 and 17). In RNAi, certain small RNA molecules interfere with the expression of genes or their RNA transcripts. RNA interference involves small interfering RNAs, microRNAs, and a few other kinds of short RNA molecules. **Small interfering RNAs (siRNAs)** are double-stranded molecules about 23 nucleotides in length. One way these molecules work is to silence genes at the posttranscriptional level by selectively cleaving mRNA molecules with base sequences complementary to the siRNA, thus inactivating them.

MicroRNAs (miRNAs) are single-stranded RNA molecules about 21 to 22 nucleotides long that inhibit the translation of mRNAs involved in growth and development (Fig. 13-18). MicroRNAs are transcribed from genes and then shortened before they are combined with proteins to form a complex that inhibits the expression of mRNA molecules with sequences complementary to the miRNA. More than 200 distinct miRNA genes have been identified in the human genome so far, but the specific mRNAs that each miRNA targets are not yet known.

Figure 13-18 The effect of an miRNA molecule on plant growth

A normal *Arabidopsis* plant (*top view, leaf, and side view of seedling*) is shown on the left and serves as a control. Biologists modified an miRNA molecule in the plant on the right, and as a result, it exhibits severe developmental abnormalities, such as curly leaves.

The definition of a gene has evolved as biologists have learned more about genes

At the beginning of this chapter, we traced the development of ideas regarding the nature of the gene. For a time scientists found it useful to define a gene as a sequence of nucleotides that codes for one polypeptide chain. However, biologists have also identified many genes that are transcribed to produce the various kinds of RNA molecules (see Table 13-1). Studies have also shown that in eukaryotic cells, a single gene may produce more than one polypeptide chain by modifications in the way the mRNA is processed.

It is perhaps most useful to define a gene in terms of its product: A **gene** is a DNA nucleotide sequence that carries the infor-

mation needed to produce a specific RNA or polypeptide product. As you will find in Chapter 14, genes include both coding regions (that code for RNA) and noncoding regulatory regions.

The usual direction of information flow has exceptions

For several decades, a central premise of molecular biology was that genetic information always flows from DNA to RNA to protein. Through his studies of viruses, U.S. biologist Howard Temin discovered an important exception to this rule in 1964. Although viruses are not cellular organisms, they contain a single type of nucleic acid and reproduce in a host cell. Temin was studying unusual, cancer-causing tumor viruses that have RNA, rather than DNA, as their genetic material. He found that infection of a host cell by one of these viruses is blocked by inhibitors of DNA synthesis and also by inhibitors of transcription. These findings suggested that DNA synthesis and transcription are required for RNA tumor viruses to multiply and that there must be a way for information to flow in the "reverse" direction—from RNA to DNA.

Temin proposed that a **DNA provirus** forms as an intermediary in the replication of RNA tumor viruses. This hypothesis required a new kind of enzyme that would synthesize DNA using RNA as a template. In 1970, Temin and U.S. biologist David Baltimore discovered just such an enzyme, and they shared the Nobel Prize in Physiology or Medicine in 1975 for their discovery. This RNA-directed DNA polymerase, also known as **reverse transcriptase,** is found in all RNA tumor viruses. (Some RNA viruses that do not produce tumors, however, replicate without using a DNA intermediate.) ▌Figure 13-19 shows the steps of RNA tumor virus reproduction. Because they reverse the usual direction

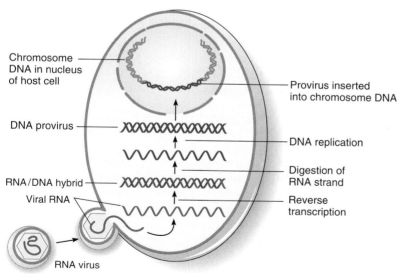

① After an RNA tumor virus enters the host cell, the viral enzyme reverse transcriptase synthesizes a DNA strand that is complementary to the viral RNA. Next, the RNA strand is degraded and a complementary DNA strand is synthesized, thus completing the double-stranded DNA provirus, which is then integrated into the host cell's DNA.

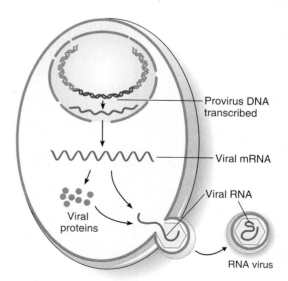

② The provirus DNA is transcribed, and the resulting viral mRNA is translated to form viral proteins. Additional viral RNA molecules are produced and then incorporated into mature viral particles enclosed by protein coats.

Figure 13-19 The infection cycle of an RNA tumor virus

of information flow, viruses that require reverse transcriptase are called **retroviruses.** HIV-1, the virus that causes AIDS, is the most widely known retrovirus. As you will see in Chapter 15, the reverse transcriptase enzyme has become an extremely important research tool for molecular biologists.

Review

- What features do mature eukaryotic mRNA molecules have that bacterial mRNAs lack?
- What is a gene?
- How do retroviruses use the enzyme reverse transcriptase?

∎ MUTATIONS

Learning Objective

13 Give examples of the different classes of mutations that affect the base sequence of DNA, and explain the effects that each has on the polypeptide produced.

One of the first major discoveries about genes was that they undergo **mutations,** changes in the nucleotide sequence of the DNA. However, the overall mutation rate is much higher than the frequency of damage to DNA, because all organisms have systems of enzymes that repair certain kinds of DNA damage.

If a DNA sequence has been changed and not corrected, DNA replication copies the altered sequence just as it would copy a normal sequence, making the mutation stable over an indefinite number of generations. In most cases, the mutant allele has no greater tendency than the original allele to mutate again.

Although most uncorrected mutations are either silent (and have no discernible effect) or harmful (such as those that cause inherited diseases and cancer), a few are useful. Because they can be passed to the next generation (in cell lines that give rise to gametes), mutations are of vital importance in evolution. Mutations provide the variation among individuals that evolutionary forces act on. Mutations are also useful in research (recall Beadle and Tatum's work with mutations) because they provide the diversity of genetic material that enables researchers to study inheritance and the molecular nature of genes.

Mutation alters genes in several ways. Scientists now determine where a particular mutation occurs in a gene by using recombinant DNA methods to isolate the gene and determine its sequence of bases (see Chapter 15).

Base-substitution mutations result from the replacement of one base pair by another

The simplest type of mutation, called a **base substitution,** involves a change in only one pair of nucleotides (∎ Fig. 13-20a). Often these mutations result from errors in base pairing during the replication process. For example, a GC, CG, or TA pair might replace an AT base pair. Such a mutation may cause the altered DNA to be transcribed as an altered mRNA. The altered mRNA may then be translated into a polypeptide chain with only one amino acid different from the normal sequence.

Base substitutions that result in replacement of one amino acid by another are sometimes called **missense mutations.** Missense mutations have a wide range of effects. If the amino acid substitution occurs at or near the active site of an enzyme, the activity of the altered protein may decrease or even be destroyed. Some missense mutations involve a change in an amino acid that is not part of the active site. Others may result in the substitution of a closely related amino acid (one with very similar chemical characteristics). Such silent mutations have no effect on the function of a gene product and may be undetectable. Because silent mutations occur relatively frequently, the true number of mutations in an organism or a species is much greater than is actually observed.

Nonsense mutations are base substitutions that convert an amino acid–specifying codon to a stop codon. A nonsense mutation usually destroys the function of the gene product; in the case of a protein-specifying gene, the part of the polypeptide chain that follows the mutant stop codon is missing.

Frameshift mutations result from the insertion or deletion of base pairs

In **frameshift mutations,** one or two nucleotide pairs are inserted into or deleted from the molecule, altering the reading frame (∎ Fig. 13-20b). As a result, codons downstream of the insertion or deletion site specify an entirely new sequence of amino acids. Depending on where the insertion or deletion occurs in the gene, different effects are generated. In addition to producing an entirely new polypeptide sequence immediately after the change, frameshift mutations may produce a stop codon within a short distance of the mutation. This codon terminates the already altered polypeptide chain. A frame shift in a gene specifying an enzyme usually results in a complete loss of enzyme activity.

Some mutations involve larger DNA segments

Some types of mutations are due to a change in chromosome structure. These changes usually have a wide range of effects because they involve many genes.

One type of mutation is caused by DNA sequences that "jump" into the middle of a gene. These movable sequences of DNA are known as **mobile genetic elements,** or **transposons.** Mobile genetic elements not only disrupt the functions of some genes but also inactivate some previously active genes. Transposons were discovered in maize (corn) by the U.S. geneticist Barbara McClintock in the 1950s. She observed that certain genes appeared to be turned off and on spontaneously. She deduced that the mechanism involved a segment of DNA that moved from one region of a chromosome to another, where it would inactivate a gene. However, biologists did not understand this phenomenon until the development of recombinant DNA methods and the discovery of transposons in a wide variety of organisms. We now know that transposons are segments of DNA that range from a few hundred to several thousand bases. In recognition of

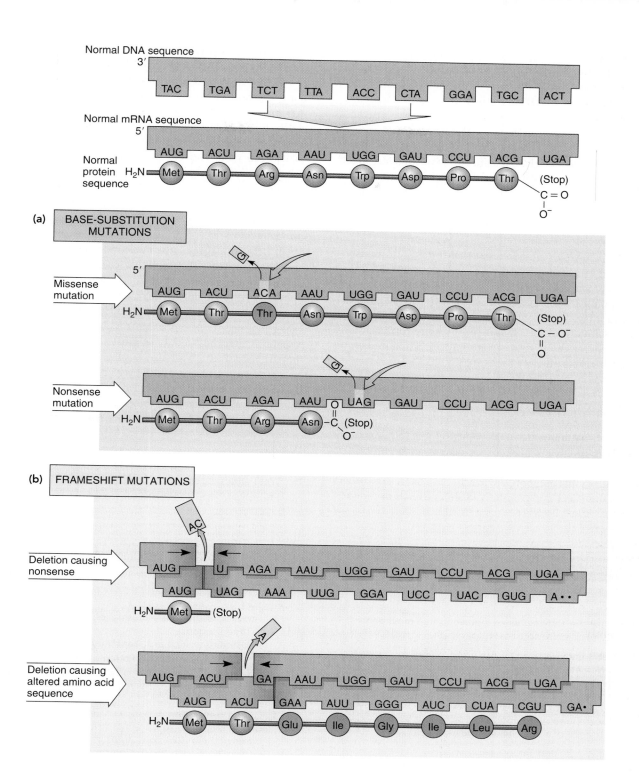

Figure 13-20 *Animated* Mutations

(a) Missense and nonsense mutations are types of base-substitution mutations. A missense mutation results in a polypeptide of normal length, but with an amino acid substitution. A nonsense mutation results in the production of a truncated (shortened) polypeptide, which is usually not functional. **(b)** A frameshift mutation results from the deletion (*shown*) or insertion of one or two bases, causing the base sequence following the mutation to shift to a new reading frame. A frame shift may produce a stop codon downstream of the mutation (which would have the same effect as a nonsense mutation caused by base substitution), or it may produce an entirely new amino acid sequence.

her insightful findings, McClintock was awarded the Nobel Prize in Physiology or Medicine in 1983.

Several kinds of mobile genetic elements have been identified. One of these, called a **DNA transposon,** moves genetic material from one site to another using a "cut-and-paste" method. DNA transposons are common in bacteria and animals, including humans.

Many mobile genetic elements are **retrotransposons,** which replicate by forming an RNA intermediate. Reverse transcriptase converts them to their original DNA sequence before jumping into a gene. Many retrotransposons enlist the host cell's gene expression machinery to produce reverse transcriptase encoded by the retrotransposon. Because retrotransposons use reverse transcriptase, some biologists hypothesize that certain retrotransposons may have evolved from retroviruses, or vice versa. Scientists think that human cells have fewer than 100 retrotransposons actively jumping around. Nevertheless, retrotransposons are extremely important agents of mutation and therefore increase an organism's ability to evolve. By one estimate, retrotransposons are responsible for about 10% of naturally occurring mutations that cause noticeable phenotypic changes.

Mutations have various causes

Most types of mutations occur infrequently but spontaneously from mistakes in DNA replication or defects in the mitotic or meiotic separation of chromosomes. Some regions of DNA, known as mutational *hot spots,* are much more likely than others to undergo mutation. An example is a short stretch of repeated nucleotides, which can cause DNA polymerase to "slip" while reading the template during replication.

Mutations in certain genes increase the overall mutation rate. For example, a mutation in a gene coding for DNA polymerase may make DNA replication less precise, or a mutation in a gene coding for a repair enzyme may allow more mutations to arise as a result of unrepaired DNA damage.

Not all mutations occur spontaneously. **Mutagens,** which cause many of the types of mutations discussed previously, include various types of radiation, such as X-rays, gamma rays, cosmic rays, and ultraviolet rays. Some chemical mutagens react with and modify bases in the DNA, leading to mistakes in complementary base pairing when the DNA molecule is replicated. Other mutagens cause nucleotide pairs to be inserted into or deleted from the DNA molecule, changing the normal reading frame during replication.

Despite the presence of enzymes that repair damage to DNA, some new mutations do arise. In fact, each of us has some mutant alleles that neither of our parents had. Although some of these mutations alter the phenotype, most are not noticeable, because they are recessive.

Mutations that occur in the cells of the body (somatic cells) are not passed on to offspring. However, these mutations are of concern because somatic mutations and cancer are closely related. Many mutagens are also **carcinogens,** agents that cause cancer.

Review

- What are the main types of mutations?
- What effects does each type of mutation have on the polypeptide product produced?

SUMMARY WITH KEY TERMS

Learning Objectives

1 Summarize the early evidence indicating that most genes specify the structure of proteins (page 280).

- Garrod's early work on inborn errors of metabolism provided evidence that genes specify proteins. Garrod studied a rare genetic disease called **alkaptonuria.** He hypothesized that people with alkaptonuria lack the enzyme that normally oxidizes homogentisic acid. Garrod's hypothesis was correct: A mutation in this gene is associated with the absence of a specific enzyme.

2 Describe Beadle and Tatum's experiments with *Neurospora* (page 280).

- Beadle and Tatum looked for mutations that interfere with metabolic reactions that produce essential molecules in the orange mold *Neurospora.* They exposed *Neurospora* spores to X-rays or ultraviolet radiation to induce mutant strains and then identified strains that carried a mutation preventing the fungus from producing a chemical essential for growth. Their work revealed that each mutant strain had a mutation in only one gene and that each gene affected only one enzyme.

3 Outline the flow of genetic information in cells, from DNA to RNA to polypeptide (page 282).

- The process by which information encoded in DNA specifies the sequences of amino acids in proteins involves two steps: transcription and translation.

- During **transcription,** an RNA molecule complementary to the template DNA strand is synthesized. **Messenger RNA (mRNA)** molecules contain information that specifies the amino acid sequences of polypeptide chains.

- During **translation,** a polypeptide chain specified by the mRNA is synthesized. Each sequence of three nucleotide bases in the mRNA constitutes a **codon,** which specifies one amino acid in the polypeptide chain, or a start or stop signal. Translation requires tRNAs and cell machinery, including ribosomes.

4 Compare the structures of DNA and RNA (page 282).

- Like DNA, RNA is formed from nucleotide subunits. However, in RNA each subunit contains the sugar ribose, a base (uracil, adenine, guanine, or cytosine), and three phosphates. Like DNA, RNA subunits are covalently joined by 5'—3' linkages to form an alternating sugar–phosphate backbone.

5 Explain why the genetic code is said to be redundant and virtually universal, and discuss how these features may reflect its evolutionary history (page 282).

- The **genetic code** is read from mRNA as a series of codons that specify a single sequence of amino acids. Of the 64 codons, 61 code for amino acids, and 3 codons serve as stop signals.

- The genetic code is redundant because some amino acids are specified by more than one codon. The genetic code is virtually universal, strongly suggesting that all organisms are descended from a common ancestor. Only a few minor variations are exceptions to the standard code found in all organisms.

6 Compare the processes of transcription and DNA replication, identifying both similarities and differences (page 285).

- **RNA polymerases,** involved in RNA synthesis, have many similarities to DNA polymerases involved in DNA replication. Both enzymes carry out synthesis in the 5' $\longrightarrow$ 3' direction. Both use nucleotides with three phosphate groups as substrates, removing two of the phosphates as the nucleotides are covalently linked to the 3' end of the newly synthesized strand.

- Just as the paired strands of DNA are **antiparallel,** the template strand of the DNA and its complementary RNA strand are also antiparallel. As a result, the DNA template strand is read in its 3' $\longrightarrow$ 5' direction as RNA is synthesized in its 5' $\longrightarrow$ 3' direction.

- The same base-pairing rules are followed as in DNA replication, except that uracil is substituted for thymine.

ThomsonNOW **Watch transcription unfold by clicking on the figure in ThomsonNOW.**

7 Identify the features of tRNA that are important in decoding genetic information and converting it into "protein language" (page 288).

- **Transfer RNAs (tRNAs)** are the "decoding" molecules in the translation process.

- Each tRNA molecule is specific for only one amino acid. One part of the tRNA molecule contains an **anticodon** that is complementary to a codon of mRNA. Attached to one end of the tRNA molecule is the amino acid specified by the complementary mRNA codon.

- Amino acids are covalently bound to tRNA by **aminoacyl–tRNA synthetase** enzymes.

8 Explain how ribosomes function in polypeptide synthesis (page 288).

- **Ribosomes** bring together all the mechanical machinery necessary for translation. They couple the tRNAs to their proper codons on the mRNA, catalyze the formation of peptide bonds between amino acids, and translocate the mRNA so the next codon can be read.

- Each ribosome is made of a large and a small subunit; each subunit contains **ribosomal RNA (rRNA)** and many proteins.

ThomsonNOW **Learn more about the structure of ribosomes by clicking on the figure in ThomsonNOW.**

9 Describe the processes of initiation, elongation, and termination in polypeptide synthesis (page 288).

- **Initiation** is the first stage of translation. **Initiation factors** bind to the small ribosomal subunit, which then binds to mRNA in the region of AUG, the **start codon.** The **initiator tRNA** binds to the start codon, followed by binding of the large ribosomal subunit.

- **Elongation** is a cyclic process in which amino acids are added one by one to the growing polypeptide chain. Elongation proceeds in the 5' $\longrightarrow$ 3' direction along the

mRNA. The polypeptide chain grows from its amino end to its carboxyl end.

- **Termination,** the final stage of translation, occurs when the ribosome reaches one of three **stop codons.** The A site binds to a **release factor,** which triggers the release of the completed polypeptide chain and dissociation of the translation complex.

ThomsonNOW **Learn more about protein synthesis by clicking on the figure in ThomsonNOW.**

10 Compare bacterial and eukaryotic mRNAs, and explain the functional significance of their structural differences (page 292).

- Eukaryotic genes and their mRNA molecules are more complicated than those of bacteria.

- After transcription, a **5' cap** (a modified guanosine triphosphate) is added to the 5' end of a eukaryotic mRNA molecule. The molecule may also have a **poly-A tail** of adenine-containing nucleotides added at the 3' end. These modifications seem to protect eukaryotic mRNA molecules from degradation, giving them a longer life span than bacterial mRNA.

- In many eukaryotic genes the coding regions, called **exons,** are interrupted by noncoding regions, called **introns.** Both exons and introns are transcribed, but the introns are later removed from the original **pre-mRNA,** and the exons are spliced together to produce a continuous polypeptide-coding sequence.

11 Describe the differences in translation in bacterial and eukaryotic cells (page 292).

- Unlike the processes in eukaryotic cells, transcription and translation are coupled in bacterial cells. Bacterial ribosomes bind to the 5' end of the growing mRNA and initiate translation long before the message is fully synthesized. As many as 15 ribosomes may bind to a single mRNA, forming a **polyribosome.**

12 Describe retroviruses and the enzyme reverse transcriptase (page 292).

- **Retroviruses** are viruses that synthesize DNA from an RNA template. In retroviruses the flow of genetic information is reversed by the enzyme **reverse transcriptase.** HIV-1, the virus that causes AIDS, is a retrovirus.

13 Give examples of the different classes of mutations that affect the base sequence of DNA, and explain the effects that each has on the polypeptide produced (page 298).

- Types of **mutations** range from disruption of a chromosome's structure to a change in only a single pair of nucleotide bases. A **base substitution** may alter or destroy the function of a protein if a codon changes so that it specifies a different amino acid (a **missense mutation**) or becomes a stop codon (a **nonsense mutation**). A base substitution has minimal effects if the amino acid is not altered or if the codon is changed to specify a chemically similar amino acid.

- The insertion or deletion of one or two base pairs in a gene invariably destroys the function of that protein because it results in a **frameshift mutation,** which changes the codon sequences downstream from the mutation.

- One type of mutation is caused by movable DNA sequences, known as **transposons,** that "jump" into the middle of a gene. Many transposons are **retrotransposons,** which replicate by forming an RNA intermediate; reverse transcriptase converts them to their original DNA sequence before they jump into a gene.

ThomsonNOW **Learn more about mutations by clicking on the figure in ThomsonNOW.**

1. Beadle and Tatum (a) predicted that tRNA molecules would have anticodons (b) discovered the genetic disease alkaptonuria (c) showed that the genetic disease sickle cell anemia is caused by a change in a single amino acid in a hemoglobin polypeptide chain (d) worked out the genetic code (e) studied the relationship between genes and enzymes in *Neurospora*

2. What is the correct order of information flow in bacterial and eukaryotic cells? (a) DNA $\longrightarrow$ mRNA $\longrightarrow$ protein (b) protein $\longrightarrow$ mRNA $\longrightarrow$ DNA (c) DNA $\longrightarrow$ protein $\longrightarrow$ mRNA (d) protein $\longrightarrow$ DNA $\longrightarrow$ mRNA (e) mRNA $\longrightarrow$ protein $\longrightarrow$ DNA

3. During transcription, how many RNA nucleotide bases would usually be encoded by a sequence of 99 DNA nucleotide bases? (a) 297 (b) 99 (c) 33 (d) 11 (e) answer is impossible to determine with the information given

4. The genetic code is defined as a series of _____ in _____. (a) anticodons; tRNA (b) codons; DNA (c) anticodons; mRNA (d) codons; mRNA (e) codons and anticodons; rRNA

5. Transcription is the process by which _____ is/are synthesized. (a) mRNA only (b) mRNA and tRNA (c) mRNA, tRNA, and rRNA (d) protein (e) mRNA, tRNA, rRNA, and protein

6. RNA differs from DNA in that the base _____ is substituted for _____. (a) adenine; uracil (b) uracil; thymine (c) guanine; uracil (d) cytosine; guanine (e) guanine; adenine

7. RNA grows in the _____ direction, as RNA polymerase moves along the template DNA strand in the _____ direction. (a) 5′ $\longrightarrow$ 3′; 3′ $\longrightarrow$ 5′ (b) 3′ $\longrightarrow$ 5′; 3′ $\longrightarrow$ 5′ (c) 5′ $\longrightarrow$ 3′; 5′ $\longrightarrow$ 3′ (d) 3′ $\longrightarrow$ 3′; 5′ $\longrightarrow$ 5′ (e) 5′ $\longrightarrow$ 5′; 3′ $\longrightarrow$ 3′

8. Which of the following is/are *not* found in a bacterial mRNA molecule? (a) stop codon (b) upstream leader sequences (c) downstream trailing sequences (d) start codon (e) promoter sequences

9. Which of the following is/are typically removed from pre-mRNA during nuclear processing in eukaryotes? (a) upstream leader sequences (b) poly-A tail (c) introns (d) exons (e) all the preceding

10. Which of the following is a spontaneous process, with no direct requirement for ATP or GTP? (a) formation of a peptide bond (b) translocation of a ribosome (c) formation of aminoacyl-tRNA (d) a and b (e) all the preceding

11. The role of tRNA is to transport (a) amino acids to the ribosome (b) amino acids to the nucleus (c) initiation factors to the ribosome (d) mRNA to the ribosome (e) release factors to the ribosome

12. Select the steps of an elongation cycle during protein synthesis from the following list, and place them in the proper sequence:

    1. Peptide bond formation

    2. Binding of the small ribosomal subunit to the 5′ end of the mRNA

    3. Binding of aminoacyl-tRNA to the A site

    4. Translocation of the ribosome

    (a) 1 $\longrightarrow$ 3 $\longrightarrow$ 2 $\longrightarrow$ 4 (b) 2 $\longrightarrow$ 3 $\longrightarrow$ 1 $\longrightarrow$ 4 (c) 3 $\longrightarrow$ 1 $\longrightarrow$ 3 $\longrightarrow$ 2 (d) 2 $\longrightarrow$ 1 $\longrightarrow$ 3 $\longrightarrow$ 4 (e) 4 $\longrightarrow$ 2 $\longrightarrow$ 1 $\longrightarrow$ 3

13. Suppose you mix the following components of protein synthesis in a test tube: amino acids from a rabbit; ribosomes from a dog; tRNAs from a mouse; mRNA from a chimpanzee; and necessary enzymes plus an energy source from a giraffe. If protein synthesis occurs, which animal's protein will be made? (a) rabbit (b) dog (c) mouse (d) chimpanzee (e) giraffe

14. During elongation in translation, the polypeptide chain that is bonded to the tRNA carrying the amino acid most recently added to the chain is in the _____ site, and an incoming aminoacyl-tRNA attaches to the _____ site. (a) E; P (b) P; E (c) E; A (d) A; E (e) P; A

15. What is the minimum number of different tRNA molecules needed to produce a polypeptide chain 50 amino acids long but that has only 16 different kinds of amino acids? (a) 50 (b) 25 (c) 20 (d) 16 (e) 8

16. The statement "the genetic code is redundant" means that (a) some codons specify punctuation (stop and start signals) rather than amino acids (b) some codons specify more than one amino acid (c) certain amino acids are specified by more than one codon (d) the genetic code is read one triplet at a time (e) all organisms have essentially the same genetic code

17. The _____ catalyzes the excision of introns from pre-mRNA. (a) ribosome (b) spliceosome (c) RNA polymerase enzyme (d) aminoacyl–tRNA synthetase enzyme (e) reverse transcriptase enzyme

18. A nonsense mutation (a) causes one amino acid to be substituted for another in a polypeptide chain (b) results from the deletion of one or two bases, leading to a shift in the reading frame (c) results from the insertion of one or two bases, leading to a shift in the reading frame (d) results from the insertion of a transposon (e) usually results in the formation of an abnormally short polypeptide chain

1. Compare and contrast the formation of mRNA in bacterial and eukaryotic cells. How do the differences affect the way in which each type of mRNA is translated? Does one system have any obvious advantage in terms of energy cost? Which system offers greater opportunities for control of gene expression?

2. Biologists hypothesize that transposons eventually lose the ability to replicate and therefore remain embedded in DNA without moving around. Based on what you have learned in this chapter, suggest a possible reason for this loss.

3. How was the genetic code deciphered?

4. Draw an arrow on this figure to show where reverse transcription, catalyzed by the enzyme reverse transcriptase, occurs. Also, show where the enzymes DNA polymerase, RNA polymerase, and peptidyl transferase are involved.

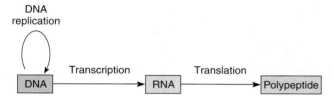

5. **Evolution Link.** Because introns are present in all eukaryotes and a few prokaryotes, biologists hypothesize that introns evolved very early in the history of life. If that is true, suggest why the majority of prokaryotes living today lost most of their introns during the course of evolution.

Additional questions are available in ThomsonNOW at www.thomsonedu.com/ login

Gene Regulation

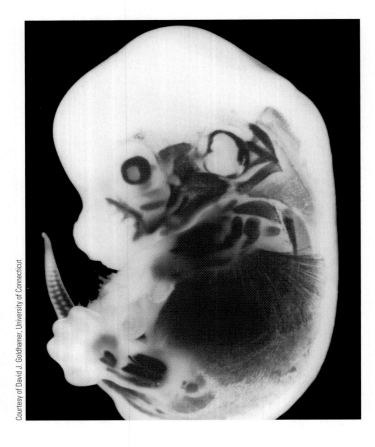

Specific gene expression in a mouse embryo. This 13.5-day-old mouse embryo has been genetically engineered so that cells that transcribe and express the myogenin gene stain blue. The myogenin gene is expressed only in those cells of the embryo that will give rise to muscle tissue. (The gestation period for a mouse is 18 to 20 days.)

KEY CONCEPTS

Each cell has the potential to synthesize several thousand proteins, but not all proteins are required in all cells at all times. Cells regulate which parts of the genome (the complete DNA of a cell) will be expressed and when.

Prokaryotes regulate their gene expression in response to environmental stimuli.

Gene regulation in prokaryotes occurs primarily at the level of transcription.

Gene regulation in eukaryotes occurs at the levels of transcription, posttranscription, translation, and posttranslation.

Each type of cell in a multicellular organism has a characteristic shape, carries out very specific activities, and makes a distinct set of proteins. Yet with few exceptions, all cells in an organism contain the same genetic information. Why, then, aren't they identical in structure, molecular composition, and function? Cells differ because gene expression is regulated, and only certain subsets of the total genetic information are expressed in any given cell (see photograph).

What mechanisms control gene expression? Consider a gene coding for a protein that is an enzyme. Expressing that gene involves three basic steps: transcribing the gene to form messenger RNA (mRNA), translating the mRNA into protein, and activating the protein so it can catalyze a specific reaction in the cell. Thus, gene expression results from a series of processes, each of which is regulated in many ways. The control mechanisms use various signals, some originating within the cell and others coming from other cells or from the environment. These signals interact with DNA, RNA, or protein.

Mechanisms that regulate gene expression include control of the amount of mRNA transcribed, of the rate of translation of mRNA, and of the activity of the protein product. These controls are accomplished in several ways. For example, the rate of transcription and the rate of mRNA degradation both control the amount of available mRNA.

Although bacteria are not multicellular, regulation of gene expression is also essential for their survival. Gene regulation in bacteria often involves controlling the transcription of genes whose products are involved in resource use. In eukaryotes, by contrast, fine-tuning the control systems occurs at *all* levels of gene regulation. Involving all levels of gene regulation is consistent with the greater complexity of eukaryotic cells and the need for developmental controls in multicellular organisms. ∎

GENE REGULATION IN BACTERIA AND EUKARYOTES: AN OVERVIEW

Learning Objective

1   Explain why bacterial and eukaryotic cells have different mechanisms of gene regulation.

Bacterial and eukaryotic cells use distinctly different mechanisms of gene regulation, based on the specific requirements of the organism. Bacterial cells exist independently, and each cell performs all its own essential functions. Because they grow rapidly and have relatively short times between cell division events, bacteria carry fewer chemical components.

The primary requirement of bacterial gene regulation is the production of enzymes and other proteins when they are needed, and **transcriptional-level control** is the most efficient mechanism. The organization of related genes into groups that are rapidly turned on and off as units allows the synthesis of only the gene products needed at any particular time. This type of regulation requires rapid turnover of mRNA molecules to prevent messages from accumulating and continuing to be translated when they are not needed. Bacteria rarely regulate enzyme levels by degrading proteins. Once the synthesis of a protein ends, the previously synthesized protein molecules are diluted so rapidly in subsequent cell divisions that breaking them down is usually not necessary. Only when cells are starved or deprived of essential amino acids do protein-digesting enzymes recycle amino acids by breaking down proteins no longer needed for survival.

Eukaryotic cells have different regulatory requirements. Although transcriptional-level control predominates, control at other levels of gene expression is also very important, especially in multicellular organisms, in which groups of cells cooperate with one another in a division of labor. Because a single gene is regulated in different ways in different types of cells, eukaryotic gene regulation is complex.

Eukaryotic cells usually have a long life span during which they may need to respond repeatedly to many different stimuli. New enzymes are not necessarily synthesized each time the cells respond to a stimulus; instead, in many instances preformed enzymes and other proteins are rapidly converted from an inactive to an active state. Some cells have a large store of inactive messenger RNA; for example, the mRNA of an egg cell becomes activated when the egg is fertilized.

Much gene regulation in multicellular organisms is focused on the differential expression of genes in the cells in various tissues. Each type of cell has certain genes that are active and other genes that may never be used. For example, developing red blood cells produce the oxygen transport protein hemoglobin, whereas muscle cells never produce hemoglobin but instead produce myoglobin, a related protein coded for by a different gene that stores oxygen in muscle tissues. Apparently the selective advantages of cell cooperation in multicellular eukaryotes far outweigh the detrimental effects of carrying a load of inactive genes through many cell divisions.

The domesticated pig provides another example of how gene regulation in multicellular organisms affects expression in different tissues. In 2003, biologists reported details of a genetic mutation in pigs that makes them develop more muscle tissue (∎ Fig. 14-1). The mutation is in a gene, designated **insulin-like growth factor 2 (*IGF2*),** which codes for a protein produced by both muscle and liver tissues. The biologists found that a *base-substitution mutation* (see Chapter 13) in the *IGF2* gene makes the gene 3 times more active in pig muscles, resulting in leaner meat. However, the mutation does not change expression of the gene in liver cells. Interestingly, this mutation is not in the protein-coding portion of the gene but in a nonprotein-coding, and presumably regulatory, region.

Figure 14-1   A lean pig with a mutation in the *IGF2* gene

Years ago, farmers discovered a pig with more muscle and less fat. They selectively bred this pig, and today most commercial pig populations have the mutation that confers this trait. The mutation, which was recently identified as a substitution of a guanine nucleotide for an adenine nucleotide, occurs in a nonprotein-coding region of the *IGF2* gene.

Review

∎   What is the most efficient mechanism of gene regulation in bacteria?

∎   How would you contrast gene regulation in prokaryotes with that in eukaryotes?

GENE REGULATION IN BACTERIA

Learning Objectives

2   Define *operon,* and explain the functions of the operator and promoter regions.

3   Distinguish among inducible, repressible, and constitutive genes.

4   Differentiate between positive and negative control, and show how both types of control operate in regulating the *lac* operon.

5   Describe the types of posttranscriptional control in bacteria.

The *Escherichia coli* bacterium is common in the intestines of humans and other mammals. It has 4288 genes that code for proteins, approximately 60% of which have known functions. Some of these genes encode proteins that are always needed (such as enzymes involved in glycolysis). These genes, which are constantly transcribed, are **constitutive** genes, and biologists describe them as *constitutively expressed.* Other proteins are needed only when the bacterium is growing under certain conditions.

For instance, bacteria living in the colon of an adult cow are not normally exposed to the milk sugar lactose, a disaccharide. If those cells ended up in the colon of a calf, however, they would have lactose available as a source of energy. This situation presents a dilemma. Should a bacterial cell invest energy and materials to produce lactose-metabolizing enzymes just in case it ends up in the digestive system of a calf? Yet if *E. coli* cells could not produce those enzymes, they might starve in the middle of an abundant food supply. Thus, *E. coli* functions by regulating many enzymes to efficiently use available organic molecules.

Cell metabolic activity is controlled in two ways: by regulating the *activity* of certain enzymes (how effectively an enzyme molecule works), and/or by regulating the *number* of enzyme molecules present in each cell. Some enzymes are regulated in both ways.

An *E. coli* cell growing on glucose needs about 800 different enzymes. Some are present in large numbers, whereas only a few molecules of others are required. For the cell to function properly, the quantity of each enzyme must be efficiently controlled.

Bacteria respond to changing environmental conditions. If lactose is added to a culture of *E. coli* cells, they rapidly synthesize the three enzymes needed to metabolize lactose. Bacteria respond so efficiently because functionally related genes—such as the three genes involved in lactose metabolism—are regulated together in gene complexes called *operons.*

Operons in bacteria facilitate the coordinated control of functionally related genes

The French researchers François Jacob and Jacques Monod are credited with the first demonstration, in 1961, of gene regulation at the biochemical level through their studies on the genes that code for the enzymes that metabolize lactose. In 1965, they received the Nobel Prize in Physiology or Medicine for their discoveries relating to genetic control of enzymes.

To use lactose as an energy source, *E. coli* cells first cleave the sugar into the monosaccharides glucose and galactose, using the enzyme β-galactosidase. Another enzyme converts galactose to glucose, and enzymes in the glycolysis pathway further break down the resulting two glucose molecules (see Chapter 8).

E. coli cells growing on glucose produce very little β-galactosidase enzyme. However, each cell grown on lactose as the sole carbon source has several thousand β-galactosidase molecules, accounting for about 3% of the cell's total protein. Amounts of two other enzymes, lactose permease and galactoside transacetylase, also increase when the cells are grown on lactose. The cell needs permease to transport lactose efficiently across the bacterial plasma membrane; without it, only small amounts of lactose enter the cell. The transacetylase may function in a minor aspect of lactose metabolism, although its role is not clear.

Jacob and Monod identified mutant strains of *E. coli* in which a single genetic defect wiped out all three enzymes. This finding, along with other information, led the researchers to conclude that the DNA coding sequences for all three enzymes are linked as a unit on the bacterial DNA and are controlled by a common mechanism. Each enzyme-coding sequence is a **structural gene.** Jacob and Monod coined the term **operon** for a gene complex consisting of a group of structural genes with related functions plus the closely linked DNA sequences responsible for controlling them. The structural genes of the *lac* operon (lactose operon)—*lacZ, lacY,* and *lacA*—code for β-galactosidase, lactose permease, and transacetylase, respectively (❚ Fig. 14-2).

Transcription of the *lac* operon begins as RNA polymerase binds to a single **promoter** region *upstream* from the coding sequences. It then proceeds to transcribe the DNA, forming a single mRNA molecule that contains the coding information for all three enzymes. Each enzyme-coding sequence on this mRNA contains its own start and stop codons; thus, the mRNA is translated to form three separate polypeptides. Because all three enzymes are translated from the same mRNA molecule, their synthesis is coordinated by turning a single molecular "switch" on or off.

The switch that controls mRNA synthesis is the **operator,** which is a sequence of bases upstream from the first structural gene in the operon. In the absence of lactose, a **repressor protein** called the **lactose repressor** binds tightly to the operator. RNA polymerase binds to the promoter but is blocked from transcribing the protein-coding genes of the *lac* operon (see Fig. 14-2a).

The lactose repressor protein is encoded by a **repressor gene,** which in this case is an adjacent structural gene located upstream from the promoter site. Unlike the *lac* operon genes, the repressor gene is constitutively expressed and is therefore constantly transcribed; the cell continuously produces small amounts of the repressor protein.

The repressor protein binds specifically to the *lac* operator sequence. When cells grow in the absence of lactose, a repressor molecule nearly always occupies the operator site. When the operator site is briefly free of the repressor, the cell synthesizes a small amount of mRNA. However, the cell synthesizes very few enzyme molecules, because *E. coli* mRNA is degraded rapidly (it has a half-life of about 2 to 4 minutes).

The structural genes of the *lac* operon are coordinately controlled and transcribed as a single mRNA.

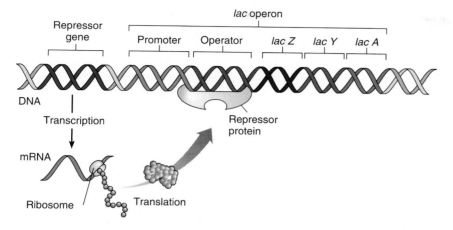

(a) In the absence of lactose, a repressor protein, encoded by an adjacent gene, binds to a region known as the operator, thereby blocking transcription of the structural genes.

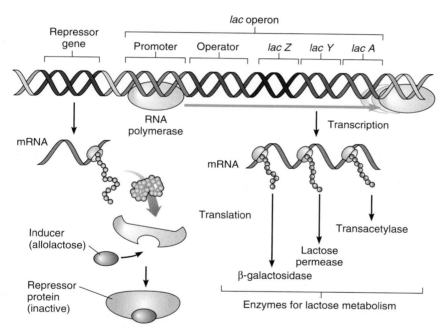

(b) When lactose is present, it is converted to allolactose, which binds to the repressor at an allosteric site, altering the structure of the protein so it no longer binds to the operator. As a result, RNA polymerase is able to transcribe the structural genes.

Figure 14-2 *Animated* The *lac* operon

Lactose "turns on," or *induces*, the transcription of the *lac* operon because the lactose repressor protein contains a second functional region (allosteric site; see Chapter 7) separate from its DNA binding site. This allosteric site binds to allolactose, a structural isomer made from lactose (see Fig. 14-2b). If lactose

is in the growth medium, a few molecules enter the cell and are converted to allolactose by the few β-galactosidase molecules present. The binding of a molecule of allolactose to the repressor alters the shape of the protein so that its DNA-binding site no longer recognizes the operator. When the repressor molecules have allolactose bound to them and are therefore inactivated, RNA polymerase actively transcribes the structural genes of the operon.

The *E. coli* cell continues to produce β-galactosidase and the other *lac* operon enzymes until it uses up virtually all the lactose. When the intracellular level of lactose drops, so too does the concentration of allolactose. Therefore, allolactose dissociates from the repressor protein, which then assumes a shape that lets the repressor bind to the operator region and shut down transcription of the operon.

Jacob and Monod isolated genetic mutants to study the *lac* operon

How did Jacob and Monod elucidate the functioning of the *lac* operon? Their approach involved the use of mutant strains, which even today are an essential tool of researchers trying to unravel the components of various regulatory systems. Genetic crosses of mutant strains allow investigators to determine the *map positions* (linear order) of the genes on the DNA and to infer normal gene functions by studying what happens when they are missing or altered. Researchers usually combine this information with the results of direct biochemical studies.

Jacob and Monod divided their mutant strains into two groups, based on whether a particular mutation affected only one enzyme or all three. In one group (❚ Fig. 14-3 ❶), only one enzyme of the three—β-galactosidase, lactose permease, or galactoside transacetylase—was affected. Subsequent gene-mapping studies showed these were mutations in structural genes located next to each other in a linear sequence.

Jacob and Monod also studied regulatory mutants in which a single mutation affected the expression of all three enzymes (❚ Fig. 14-3 ❷). Some regulatory mutants failed to transcribe the *lac* operon even when lactose was present (❚ Fig. 14-3 ❸). Researchers eventually found that these abnormal genes had an

Jacob and Monod analyzed the properties of various mutant strains of *E. coli* to deduce the structure and function of the *lac* operon.

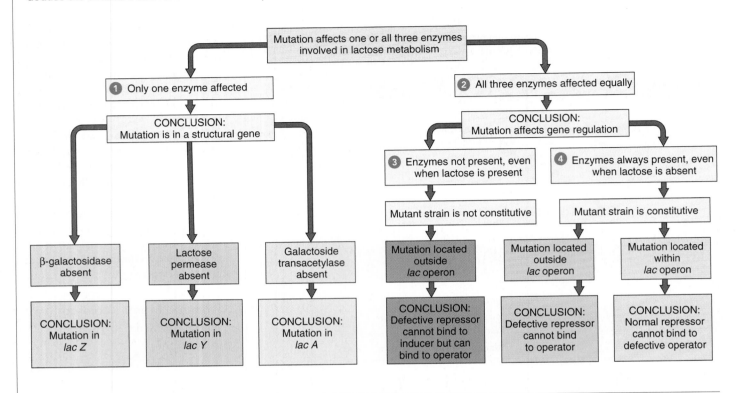

Figure 14-3 Genetic and biochemical characterization of the *lac* operon
The circled numbers are explained in the text.

altered binding site on the repressor protein that prevented al-lolactose from binding, although the repressor could still bind to the operator. Once bound to the operator, such a mutant repressor remained bound, keeping the operon "turned off."

In contrast, some of the regulatory mutants transcribed the structural genes of the *lac* operon at a significant rate, even in the absence of lactose, causing the cell to waste energy in producing unneeded enzymes (Fig. 14-3 ❹). One group of these gene mutations had map positions just outside the *lac* operon. Jacob and Monod hypothesized the existence of a repressor gene that codes for a repressor molecule (later found to be a protein). Although the specific defect may vary, the members of this group of mutants produce defective repressors; hence, no binding to the *lac* operator and promoter takes place, and the *lac* operon is constitutively expressed.

The genes responsible for a second group of mutants had map positions within the *lac* operon but did not directly involve any of the three structural genes. Jacob and Monod hypothesized that the members of this group produced normal repressor molecules but had abnormal operator sequences incapable of binding to the repressor.

An inducible gene is not transcribed unless a specific inducer inactivates its repressor

Geneticists call the *lac* operon an **inducible operon.** A repressor usually controls an inducible gene or operon by keeping it "turned off." The presence of an **inducer** (in this case, allolactose) inactivates the repressor, permitting the gene or operon to be transcribed. Inducible genes or operons usually code for enzymes that are part of catabolic pathways, which break down molecules to provide both energy and components for anabolic reactions. This type of regulatory system lets the cell save the energy cost of making enzymes when no substrates are available on which they can act.

A repressible gene is transcribed unless a specific repressor–corepressor complex is bound to the DNA

Another type of gene regulation system in bacteria is associated mainly with anabolic pathways such as those in which cells synthesize amino acids, nucleotides, and other essential biological molecules from simpler materials. Enzymes coded by repressible genes normally regulate these pathways.

Repressible operons and genes are usually "turned on"; they are turned off only under certain conditions. In most cases the molecular signal for regulating these genes is the end product of the anabolic pathway. When the supply of the end product (such as an amino acid) becomes low, all enzymes in the pathway are actively synthesized. When intracellular levels of the end product rise, enzyme synthesis is repressed. Because the growing cell continuously needs compounds such as amino acids, the most effective mechanism for the cell is to keep the genes that control their production turned on, except when a large supply of the amino acid is available. The ability to turn the genes off prevents cells from overproducing amino acids and other molecules that are essential but expensive to make in terms of energy.

The *trp* operon (tryptophan operon) is an example of a repressible system. In both *E. coli* and a related bacterium, *Salmonella,* the *trp* operon consists of five structural genes that code for the enzymes that the cell needs to synthesize the amino acid tryptophan; these are clustered as a transcriptional unit with a single promoter and a single operator (▌ Fig. 14-4a). A distant repressor gene codes for a diffusible repressor protein, which differs from the lactose repressor in that the cell synthesizes it in an inactive form that cannot bind to the operator region of the *trp* operon.

The DNA-binding site of the repressor becomes effective only when tryptophan, its **corepressor,** binds to an allosteric site on the repressor (▌ Fig. 14-4b).When intracellular tryptophan levels are low, the repressor protein is inactive and cannot bind to the operator region of the DNA. The enzymes required for tryptophan synthesis are produced, and the intracellular concentration of tryptophan increases. Some of it binds to the repressor, altering the repressor's shape so it binds tightly to the operator. This switches the operon off, thereby blocking transcription.

Negative regulators inhibit transcription; positive regulators stimulate transcription

The features of the *lac* and *trp* operons we have described so far are examples of **negative control,** a regulatory mechanism in which the DNA binding regulatory protein is a repressor that turns off transcription of the gene. **Positive control** is regulation by **activator proteins** that bind to DNA and thereby stimulate the transcription of a gene. The *lac* operon is controlled by both a negative regulator (the lactose repressor) and a positively acting activator protein.

Positive control of the *lac* operon requires the cell to recognize the absence of the sugar glucose, which is the initial substrate in the glycolysis pathway. Lactose, like glucose, undergoes stepwise breakdown to yield energy. However, because glucose is a product of the catabolic hydrolysis of lactose, it is most efficient for *E. coli* cells to use the available supply of glucose first, sparing the cell the considerable energy cost of making additional enzymes such as β-galactosidase (▌ Fig. 14-5a).

The *lac* operon has a very inefficient promoter element—that is, it has a low affinity for RNA polymerase, even when the repressor protein is inactivated. However, a DNA sequence adjacent to the promoter site is a binding site for another regulatory protein, the **catabolite activator protein (CAP).** When activated, CAP stimulates transcription of the *lac* operon and several other bacterial operons.

In its active form, CAP is bound to an allosteric site on **cyclic AMP, or cAMP,** an alternative form of adenosine monophosphate (see Fig. 3-25). As the cells become depleted of glucose, cAMP levels increase (▌ Fig. 14-5b). After the cAMP molecules bind to CAP, the resulting active complex then binds to the CAP-binding site near the *lac* operon promoter. This binding of active CAP bends DNA's double helix (▌ Fig. 14-6 on page 312), strengthening the affinity of the promoter region for RNA polymerase so that the rate of transcriptional initiation accelerates in the presence of lactose. Thus, the *lac* operon is fully active only if lactose is available and intracellular glucose levels are low. ▌ Table 14-1 summarizes negative and positive controls in bacteria.

Constitutive genes are transcribed at different rates

Many gene products encoded by *E. coli* DNA are needed only under certain environmental or nutritional conditions. As you have seen, these genes are generally regulated at the level of transcription. They are turned on and off as metabolic and environmental conditions change. By contrast, constitutive genes are continuously transcribed, but they are not necessarily transcribed (or their mRNAs translated) at the same rate. Some enzymes work more effectively or are more stable than others and thus are present in smaller amounts. Constitutive genes that encode proteins required in large amounts generally have greater expression—that is, are transcribed more rapidly—than genes coding for proteins required in smaller amounts. The promoter elements of these genes control their transcriptional rate. Constitutive genes with efficient ("strong") promoters bind RNA polymerase more frequently and consequently transcribe more mRNA molecules than those with inefficient ("weak") promoters.

Genes coding for repressor or activator proteins that regulate metabolic enzymes are usually constitutive and produce their protein products constantly. Because each cell usually needs relatively few molecules of any specific repressor or activator protein, promoters for those genes tend to be relatively weak.

Some posttranscriptional regulation occurs in bacteria

As you have seen, much of the variability in protein levels in *E. coli* is determined by transcriptional-level control. However, regulatory mechanisms after transcription, known as **posttranscriptional controls,** also occur at various levels of gene expression.

Translational controls are posttranscriptional controls that regulate the rate at which an mRNA molecule is translated. Because the life span of an mRNA molecule in a bacterial cell is very short, one that is translated rapidly produces more proteins than one that is translated slowly. Some mRNA molecules in *E. coli* are translated as much as 1000 times faster than other mRNAs. Most of the differences appear to be due to the rate at which ribosomes attach to the mRNA and begin translation.

Structural genes coding for enzymes that synthesize the amino acid tryptophan are organized in a repressible operon.

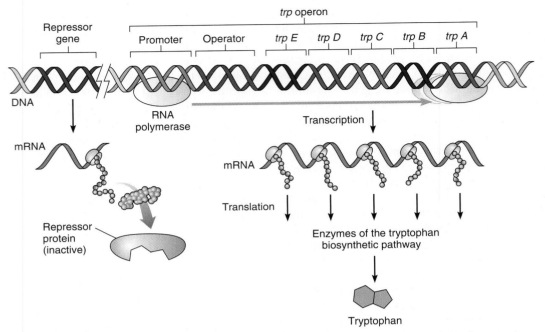

(a) Intracellular tryptophan levels low. Repressor protein is unable to prevent transcription because it cannot bind to the operator.

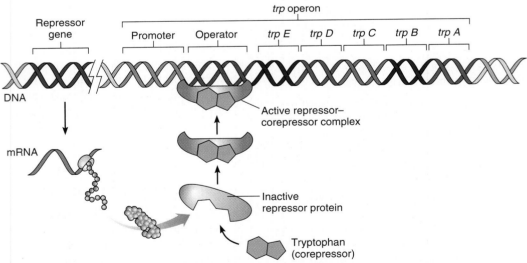

(b) Intracellular tryptophan levels high. The amino acid tryptophan binds to an allosteric site on the repressor protein, changing its conformation. The resulting active form of the repressor binds to the operator region, blocking transcription of the operon until tryptophan is again required by the cell.

Figure 14-4 The *trp* operon

The lactose promoter by itself is weak and binds RNA polymerase inefficiently even when the lactose repressor is inactive.

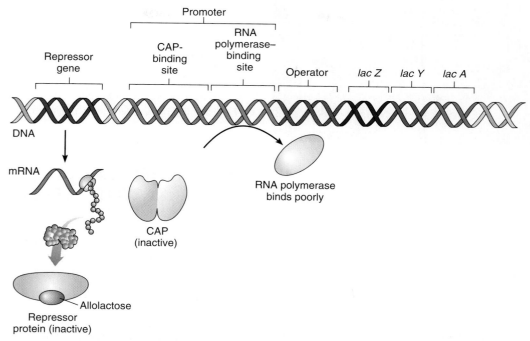

(a) Lactose high, glucose high, cAMP low. When glucose levels are high, cAMP is low. CAP is in an inactive form and cannot stimulate transcription. Transcription occurs at a low level or not at all.

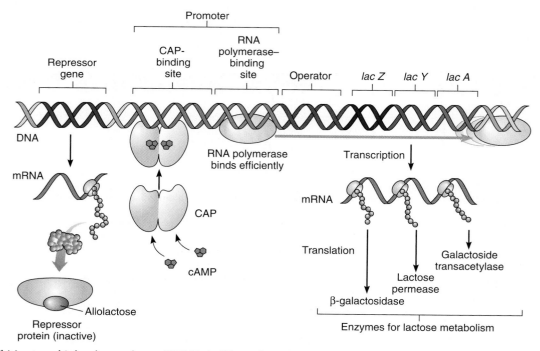

(b) Lactose high, glucose low, cAMP high. When glucose concentrations are low, each CAP polypeptide has cAMP bound to its allosteric site. The active form of CAP binds to the DNA sequence, and transcription becomes activated.

Figure 14-5   Positive control of the *lac* operon

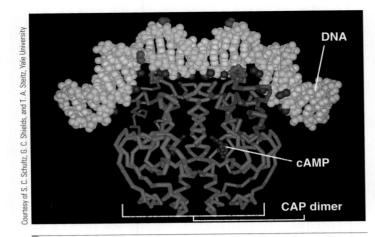

DNA

cAMP

CAP dimer

Courtesy of S. C. Schultz, G. C. Shields, and T. A. Steitz, Yale University

Figure 14-6 The binding of CAP to DNA

This computer-generated image shows the bend formed in the DNA double helix when it binds to CAP. CAP is a dimer consisting of two identical polypeptide chains, each of which binds one molecule of cAMP.

Posttranslational controls act as switches that activate or inactivate one or more existing enzymes, thereby letting the cell respond to changes in the intracellular concentrations of essential molecules, such as amino acids. A common posttranslational control adjusts the rate of synthesis in a metabolic pathway through **feedback inhibition** (see Fig. 7-15). The end product of a metabolic pathway binds to an allosteric site on the first enzyme in the pathway, temporarily inactivating the enzyme. When the first enzyme in the pathway does not function, all the succeeding enzymes are deprived of substrates. Notice that feedback inhibition differs from the repression caused by tryptophan discussed previously. In that case, the end product of the pathway (tryptophan) prevented the formation of *new* enzymes. Feedback inhibition acts as a fine-tuning mechanism that regulates the activity of the *existing* enzymes in a metabolic pathway.

Review

■ What is the function of each of the parts of the *lac* operon (promoter, operator, and structural genes)?

■ What structural features does the *trp* operon share with the *lac* operon?

■ Why do scientists define the *trp* operon as repressible and the *lac* operon as inducible?

■ How is glucose involved in positive control of the *lac* operon?

GENE REGULATION IN EUKARYOTIC CELLS

Learning Objectives

6 Discuss the structure of a typical eukaryotic gene and the DNA sequences involved in regulating that gene.

7 Give examples of some of the ways eukaryotic DNA-binding proteins bind to DNA.

8 Illustrate how a change in chromosome structure may affect the activity of a gene.

9 Explain how a gene in a multicellular organism may produce different products in different types of cells.

10 Identify some of the types of regulatory controls that operate in eukaryotes after mature mRNA is formed.

Like bacterial cells, eukaryotic cells respond to changes in their environment. In addition, multicellular eukaryotes require regulation modes that let individual cells commit to specialized roles

TABLE 14-1	
Transcriptional Control in Bacteria*	

Negative Control	Result
Inducible genes	
Repressor protein alone	Active repressor turns off regulated gene(s)
Lactose repressor alone	*lac* operon not transcribed
Repressor protein + inducer	Inactive repressor–inducer complex fails to turn off regulated gene(s)
Lactose repressor + allolactose	*lac* operon transcribed
Repressible genes	
Repressor protein alone	Inactive repressor fails to turn off regulated gene(s)
Tryptophan repressor alone	*trp* operon transcribed
Repressor protein + corepressor	Active repressor-corepressor complex turns off regulated gene(s)
Tryptophan repressor + tryptophan	*trp* operon not transcribed

Positive Control	Result
Activator protein alone	Activator alone cannot stimulate transcription of regulated gene(s)
CAP alone	Transcription of *lac* operon not stimulated
Activator complex	Functional activator complex stimulates transcription of regulated gene(s)
CAP + cAMP	Transcription of *lac* operon stimulated

*A general description of each type is followed by a specific example.

and let groups of cells organize into tissues and organs. These processes are accomplished mainly by regulating transcription, but posttranscriptional, translational, and posttranslational controls are also important. In Chapters 12 and 13, we observed that in eukaryotes all aspects of information transfer—including replication, transcription, and translation—are more complicated than in prokaryotes. Not surprisingly, this complexity offers more ways to control gene expression.

Unlike many bacterial genes, eukaryotic genes do not normally form operon-like clusters. (A notable exception is the nematodes, or roundworms, in which some genes are organized into operons.) However, each eukaryotic gene has specific regulatory sequences that are essential in controlling transcription.

Many "housekeeping" enzymes (those needed by virtually all cells) appear to be encoded by constitutive genes, which are expressed in all cells at all times. Some inducible genes have also been found; these respond to environmental threats or stimuli, such as heavy-metal ingestion, viral infection, and heat shock. For example, when a cell is exposed to high temperature, many proteins fail to fold properly. These unfolded proteins elicit a survival response in which *heat–shock genes* are transcribed and heat–shock proteins are synthesized. Although the functions of most heat–shock proteins in eukaryotic cells are unknown, some are **molecular chaperones,** which help newly synthesized proteins fold into their proper shape (discussed in Chapters 3 and 13).

Some genes are inducible only during certain periods in the life of the organism; they are controlled by **temporal regulation** mechanisms. Finally, some genes are under **tissue-specific regulation.** For example, a gene involved in production of a particular enzyme may be regulated by one stimulus (such as a hormone) in muscle tissue, by an entirely different stimulus in pancreatic cells, and by a third stimulus in liver cells. We explore these types of regulation in more detail in Chapters 17 and 48.

∎ Table 14-2 summarizes the regulation of gene expression in eukaryotes. You may want to refer to it as you read the following discussion.

Eukaryotic transcription is controlled at many sites and by many different regulatory molecules

Like bacterial genes, most genes of multicellular eukaryotes are controlled at the transcriptional level. As you will see, various base sequences in the DNA are important in transcriptional control. In addition, regulatory proteins and the way the DNA is organized in the chromosome affect the rate of transcription.

Chromosome organization affects the expression of some genes

A chromosome is not only a bearer of genes. Various arrangements of a chromosome's ordered components (see Fig. 10-4) increase or decrease expression of the genes it contains. In multicellular eukaryotes, only a subset of the genes in a cell is active at

TABLE 14-2

Gene Regulation in Eukaryotes

Level of Regulation	Description
Transcriptional control (most common level of gene regulation in eukaryotes)	Chromatin structure regulates transcription; heterochromatin cannot be transcribed
	DNA methylation regulates transcription; methylated DNA is inaccessible to transcription machinery
	Selective transcription: promoter and enhancer elements in DNA interact with protein transcription factors to activate or repress transcription
Posttranscriptional control: mRNA processing and transport	Control mechanisms, such as rate of intron/exon splicing, regulate mRNA processing
	Differential mRNA processing (alternate splicing of exons) produces different proteins from same mRNA
	Controlling access to, or efficiency of, transport through nuclear pores regulates mRNA transport from nucleus to cytosol
Translational control	Translational controls determine how often and how long specific mRNA is translated
	Translational controls determine degree to which mRNA is protected from destruction; proteins that bind to mRNA in cytosol affect stability
Posttranslational control of protein product	Chemical modifications, such as phosphorylation, affect activity of protein after it is produced
	Selective degradation targets specific proteins for destruction by proteasomes

any one time. The inactivated genes differ among cell types and in many cases are irreversibly dormant.

Some of the inactive genes lie in highly compacted chromatin, visible microscopically as densely staining regions of chromosomes during cell division. These regions of chromatin remain tightly coiled and bound to chromosome proteins throughout the cell cycle; even during interphase, they are visible as darkly staining fibers called **heterochromatin** (∎ Fig. 14-7a). Evidence suggests that most of the heterochromatin DNA is not transcribed. Most of the inactive X chromosome of the two X chromosomes in female mammals is heterochromatic and appears as a *Barr body* (see Fig. 11-17).

Active genes are associated with a more loosely packed chromatin structure called **euchromatin** (∎ Fig. 14-7b). The exposure of the DNA in euchromatin lets it interact with *transcription factors* (discussed later in the chapter) and other regulatory proteins.

Cells may have several ways to change chromatin structure from heterochromatin to euchromatin. One way is to chemically modify **histones,** the proteins that associate with DNA to form nucleosomes. Each histone molecule has a so-called tail, a string of amino acids that extends from the DNA-wrapped nucleosome (see Fig. 10-2). Methyl groups, acetyl groups, sugars, and even

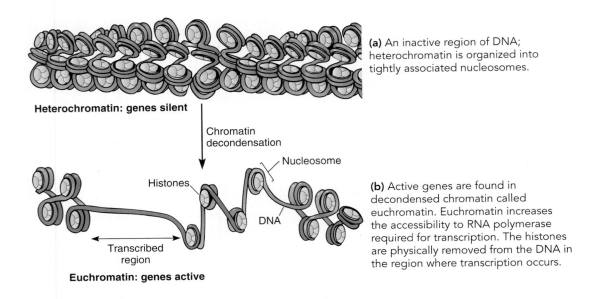

(a) An inactive region of DNA; heterochromatin is organized into tightly associated nucleosomes.

Heterochromatin: genes silent

Chromatin decondensation

Nucleosome

Histones

DNA

Transcribed region

Euchromatin: genes active

(b) Active genes are found in decondensed chromatin called euchromatin. Euchromatin increases the accessibility to RNA polymerase required for transcription. The histones are physically removed from the DNA in the region where transcription occurs.

Figure 14-7   The effect of chromatin structure on transcription

proteins chemically attach to the histone tail and may expose or hide genes, turning them on or off. These chemical modifications of histones are the focus of active research, because they appear to influence gene expression and therefore offer promise of new approaches to treat cancer and other diseases with a genetic component. Biologists have found general correlations between gene activity and certain groups attached to or absent from histone tails. For example, the genes on DNA in nucleosomes where histone tails are flagged with acetyl groups generally tend to be expressed. Biologists are now trying to determine if specific histone modifications are associated with the activity or inactivity of specific genes within a cell.

Epigenetic inheritance: gene inactivation by DNA methylation.

Inactive genes of vertebrates and some other organisms typically show a pattern of **DNA methylation** in which the DNA has been chemically altered by enzymes that add methyl groups to certain cytosine nucleotides in DNA. (The resulting 5-methylcytosine still pairs with guanine in the usual way.) Evidence indicates that certain regulatory proteins selectively bind to methylated DNA and make it inaccessible to the general transcriptional machinery.

DNA methylation, which is found in inactive regions of the cell's genome, probably reinforces gene inactivation rather than serves as the initial mechanism to silence genes. Apparently once a gene has been turned off by some other means, DNA methylation ensures it will remain inactive. For example, the DNA of the inactive X chromosome of a female mammal becomes methylated after the chromosome has become a condensed Barr body. When the DNA replicates in a mitotic cell division, each double strand of DNA has one methylated strand and one unmethylated strand. Methylation enzymes add methyls to the new strand, thereby perpetuating the pre-existing methylation pattern. Hence, the DNA continues to be transcriptionally inactive in all descendants of these cells.

In mammals, DNA methylation maintained in this way accounts for **genomic imprinting** (also called *parental imprinting*), in which the phenotypic expression of certain genes is determined by whether a particular allele is inherited from the female or the male parent. Genomic imprinting is associated with some human genetic diseases, such as *Prader–Willi syndrome* and *Angelman syndrome* (both discussed in Chapter 16).

Genomic imprinting is an example of an epigenetic phenomenon. **Epigenetic inheritance** involves changes in how a gene is expressed. In contrast, *genetic* inheritance involves changes in a gene's nucleotide sequence. Because DNA methylation patterns tend to be passed to successive cell generations, they provide a mechanism for epigenetic inheritance. Evidence that supports the idea that epigenetic inheritance is an important mechanism of gene regulation continues to accumulate. In addition, new phenotypic traits can arise from epigenetic inheritance despite the fact that the nucleotide sequence of the gene itself has not changed. Thus, some inherited characteristics do not depend exclusively on the sequence of nucleotides in DNA.

Multiple copies of genes.

A single gene cannot always provide enough copies of its mRNA to meet the cell's needs. The requirement for high levels of certain products is met if multiple copies of the genes that encode them are present in the chromosome. Genes of this type, whose products are essential for all cells, may occur as multiple copies arranged one after another along the chromosome, in **tandemly repeated gene se-**

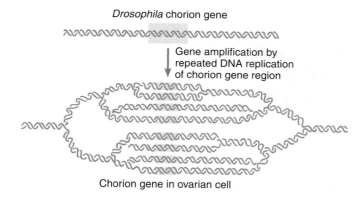

Drosophila chorion gene

Gene amplification by repeated DNA replication of chorion gene region

Chorion gene in ovarian cell

Figure 14-8 Gene amplification

In *Drosophila*, multiple replications of a small region of the chromosome amplify the chorion (eggshell) protein genes. Replication is initiated at a discrete chromosome origin of replication (*top pink box*) for each copy of the gene that is produced. Replication is randomly terminated, resulting in a series of forked structures in the chromosome.

quences. Histone genes are usually multiple copies of 50 to 500 genes in cells of multicellular organisms. Similarly, multiple copies (150 to 450) of genes for rRNA and tRNA are present in cells.

Other genes, which are required only by a small group of cells, are selectively replicated in those cells in a process called **gene amplification.** For example, the *Drosophila* chorion (eggshell) gene product is a protein made specifically in cells of the female insect's reproductive tract. These cells make massive amounts of the protein that envelops and protects the zygote. Amplifying the chorion protein gene by DNA replication meets the demand for chorion mRNA. In other words, the DNA in that small region of the chromosome is copied many times (∎ Fig. 14-8). In other cells of the insect body, however, the gene appears to exist as a single copy in the chromosome.

Eukaryotic promoters vary in efficiency, depending on their upstream promoter elements

In eukaryotic as well as bacterial cells, the transcription of any gene requires a base pair where transcription begins, known as the **transcription initiation site,** plus a sequence of bases, the *promoter,* to which RNA polymerase binds. In multicellular eukaryotes, RNA polymerase binds to a promoter called a **TATA box,** located about 25 to 35 base pairs upstream from the transcription initiation site (∎ Fig. 14-9a). The TATA box is required for transcription to begin. The most commonly found base sequence is shown (either T or A can be present at the positions where they appear together in the figure).

Other eukaryotic promoter elements have a regulatory function and control the expression of the gene. Many eukaryotic promoters contain one or more sequences of 8 to 12 bases within a short distance (about 100 to 200 base pairs) upstream of the

RNA polymerase–binding site. These segments of noncoding DNA are called **upstream promoter elements (UPEs)** or **proximal control elements.** Specific regulatory proteins bind to the UPEs to regulate expression of the gene. The efficient initiation of transcription seems related to the number, location, and types of UPEs, which vary with each gene. A constitutive gene containing only one UPE may be weakly expressed, whereas one with five or six UPEs may be transcribed much more actively (∎ Fig. 14-9b and c).

Enhancers are DNA sequences that facilitate the initiation of transcription

Regulated eukaryotic genes commonly need not only UPEs but also DNA sequences called **enhancers.** Like the upstream promoter elements, which are required for accurate and efficient initiation of mRNA synthesis, enhancers help form an active transcription initiation complex. In this way, enhancers increase the rate of RNA synthesis, often by several orders of magnitude (∎ Fig. 14-9d). Specific regulatory proteins bind to enhancer elements and activate transcription by interacting with the proteins bound to the promoters.

Enhancers are remarkable in many ways. Although they are present in all cells, a particular enhancer is functional only in certain types of cells. An enhancer regulates a gene on the same DNA molecule from very long distances—up to thousands of base pairs away from the promoter—and is either upstream or downstream of the promoter it controls. Furthermore, if an enhancer element is experimentally cut out of the DNA and inverted, it still regulates the gene it normally controls. As you will see, evidence suggests that enhancers work by interacting with proteins that regulate transcription.

Transcription factors are regulatory proteins with several functional domains

We previously discussed some DNA-binding proteins that regulate transcription in bacteria. These included the lactose repressor, the tryptophan repressor, and the catabolite activator protein (CAP). Researchers have identified many more DNA-binding proteins that regulate transcription in eukaryotes than in bacteria; these eukaryotic proteins are collectively known as **transcription factors.** Researchers have identified more than 2000 transcription factors in humans.

It is useful to compare regulatory proteins in bacteria and eukaryotes. Many regulatory proteins are modular molecules; that is, they have more than one **domain,** a region with its own tertiary structure and function. Each eukaryotic transcription factor, like the regulatory proteins of bacteria, has a DNA-binding domain plus at least one other domain that is either an activator or a repressor of transcription for a given gene.

Many transcription factors in eukaryotes (and regulatory proteins in bacteria) have similar DNA-binding domains. One example is the *helix-turn-helix* arrangement, consisting of two α-helical segments. One of these, the recognition helix, inserts

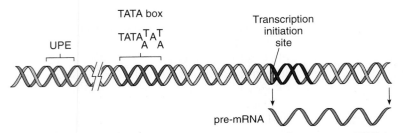

(a) Eukaryotic promoter elements. A eukaryotic promoter usually contains a TATA box located 25 to 35 base pairs upstream from the transcription initiation site. One or more upstream promoter elements (UPEs) are usually present.

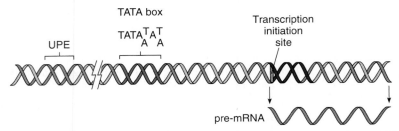

(b) A weak eukaryotic promoter. A weakly expressed gene may contain only one UPE.

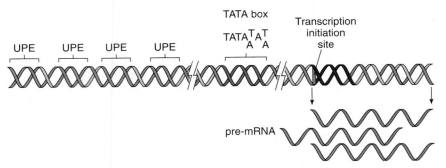

(c) A strong eukaryotic promoter. A strongly expressed gene is likely to contain several UPEs.

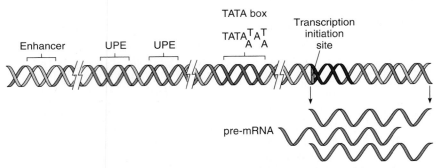

(d) A strong eukaryotic promoter plus an enhancer. Transcription of this eukaryotic gene is stimulated by an enhancer, located several thousand bases from the promoter.

Figure 14-9 The regulation of transcription in eukaryotes

The DNA double helix and other elements are not drawn to scale.

zinc ions. Each loop includes an α-helix that fits into a groove of the DNA (❙ Fig. 14-10b). Certain amino acid functional groups exposed in each finger have been shown to recognize specific DNA sequences.

Many regulatory proteins have DNA-binding domains that function only as pairs, or *dimers*. Many of these transcription factors are known as **leucine zipper proteins** because they are held together by the side chains of leucine and other hydrophobic amino acids (❙ Fig. 14-10c). In some cases the two polypeptides that make up the dimer are identical and form a *homodimer*. In other instances they are different, and the resulting *heterodimer* may have very different regulatory properties from those of a homodimer. For a simple and speculative example, let us assume that three regulatory proteins—A, B, and C—are involved in controlling a particular set of genes. These three proteins might associate as dimers in six different ways: three kinds of homodimers (AA, BB, and CC) and three kinds of heterodimers (AB, AC, and BC). Such multiple combinations of regulatory proteins greatly increase the number of possible ways that transcription is controlled.

Transcription in eukaryotes requires multiple regulatory proteins that are bound to different parts of the promoter. The *general transcriptional machinery* is a protein complex that binds to the TATA box of the promoter near the transcription initiation site. That complex is required for RNA polymerase to bind, thereby initiating transcription.

Both enhancers and UPEs apparently become functional when specific transcription factors bind to them. ❙ Figure 14-11 shows interactions involving an enhancer and a transcription factor that acts as an activator. Each activator has at least two functional domains: a DNA recognition site that usually binds to an enhancer or UPE, and a gene activation site that contacts the target in the general transcriptional machinery. The DNA between the enhancer and promoter elements forms a loop that lets an activator bound to an enhancer come in contact with the target proteins associated with the general transcriptional machinery. When this occurs, the rate of transcription increases.

into the major groove of the DNA without unwinding the double helix. The other helps hold the first one in place (❙ Fig. 14-10a). The "turn" in the helix-turn-helix is a sequence of amino acids that forms a sharp bend in the molecule.

Other regulatory proteins have DNA-binding domains with multiple "zinc fingers," loops of amino acids held together by

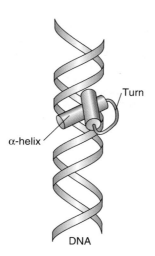

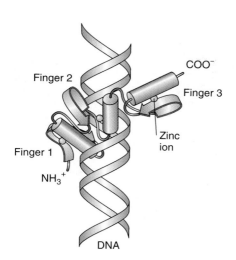

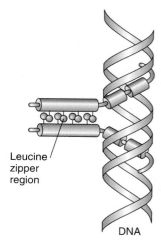

(a) Helix-turn-helix. This portion of a regulatory protein (*purple*) contains the helix-turn-helix arrangement. The recognition helix is inserted into the groove of the DNA and is connected to a second helix that helps hold it in place by a sequence of amino acids that form a sharp bend.

(b) Zinc fingers. Regions of certain transcription factors form projections known as "zinc fingers," which insert into the grooves of the DNA and bind to specific base sequences.

(c) Leucine zipper. This leucine zipper protein is a dimer, held together by hydrophobic interactions involving side chains of leucine and other amino acids.

Figure 14-10 Regulatory proteins

In these illustrations, cylinders represent α-helical regions of regulatory proteins and ribbons represent β-pleated sheets.

The mRNAs of eukaryotes have many types of posttranscriptional control

The half-life of bacterial mRNA is usually minutes long; eukaryotic mRNA, even when it turns over rapidly, is far more stable. Bacterial mRNA is transcribed in a form that is translated immediately. In contrast, eukaryotic mRNA molecules undergo further modification and processing before they are used in protein synthesis. The message is capped, polyadenylated, spliced, and then transported from the nucleus to the cytoplasm to initiate translation (see Fig. 13-17). These events represent potential control points for translation of the message and the production of its encoded protein.

The addition of a **poly-A tail** to eukaryotic mRNA, for example, appears necessary to initiate translation. Researchers have shown that mRNAs with long poly-A tails are efficiently translated, whereas mRNAs with short poly-A tails are essentially dormant. *Polyadenylation,* which commonly occurs in the nucleus, can also take place in the cytosol. When the short poly-A tail of an mRNA is elongated, the mRNA becomes activated and is then translated.

Some pre-mRNAs are processed in more than one way

Investigators have discovered several forms of regulation involving mRNA processing. In some instances, the same gene produces one type of protein in one tissue and a related but somewhat different type of protein in another tissue. This is possible

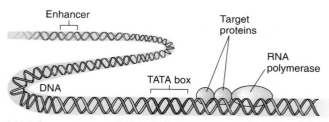

(a) Little or no transcription. This gene is transcribed at a very low rate or not at all, even though the general transcriptional machinery, including RNA polymerase, is bound to the promoter.

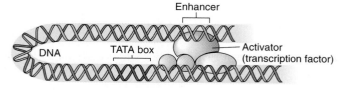

(b) High rate of transcription. A transcription factor that functions as an activator binds to an enhancer. The intervening DNA forms a loop, allowing the transcription factor to contact one or more target proteins in the general transcriptional machinery, thereby increasing the rate of transcription. This diagram is highly simplified, and many more target proteins than the two shown are involved.

Figure 14-11 The stimulation of transcription by an enhancer

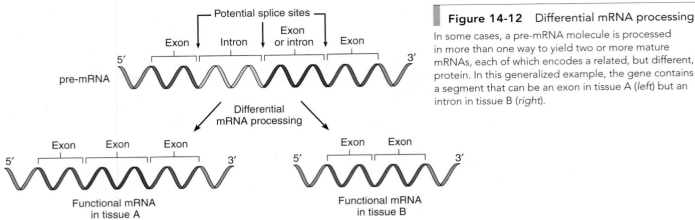

Figure 14-12 Differential mRNA processing

In some cases, a pre-mRNA molecule is processed in more than one way to yield two or more mature mRNAs, each of which encodes a related, but different, protein. In this generalized example, the gene contains a segment that can be an exon in tissue A (*left*) but an intron in tissue B (*right*).

because many genes produce pre-mRNA molecules that have alternative splicing patterns; that is, they are spliced in different ways depending on the tissue.

Typically, such a gene includes at least one segment that can be either an *intron* or an *exon* (see Chapter 13). As an intron, the sequence is removed, but as an exon, it is retained. Through **differential mRNA processing,** the cells in each tissue produce their own version of mRNA corresponding to the particular gene (▌ Fig. 14-12). For example, this mechanism produces different forms of troponin, a protein that regulates muscle contraction, in different muscle tissues.

The stability of mRNA molecules varies

Controlling the life span of a particular kind of mRNA molecule permits control over the number of protein molecules translated from it. In some cases, mRNA stability is under hormonal control. This is true for mRNA that codes for vitellogenin, a protein synthesized in the liver of certain female animals, such as frogs and chickens. Vitellogenin is transported to the oviduct, where it is used in forming egg yolk proteins.

The hormone estradiol regulates vitellogenin synthesis. When estradiol levels are high, the half-life of vitellogenin mRNA in frog liver is about 500 hours. When researchers deprive cells of estradiol, the half-life of the mRNA drops rapidly, to less than 165 hours. This quickly lowers cell vitellogenin mRNA levels and decreases synthesis of the vitellogenin protein. In addition to affecting mRNA stability, the hormone seems to control the rate at which the mRNA is synthesized.

Posttranslational chemical modifications may alter the activity of eukaryotic proteins

Another way to control gene expression is by regulating the activity of the gene product. Many metabolic pathways in eukaryotes, as in bacteria, contain allosteric enzymes regulated through feedback inhibition. In addition, after they are synthesized, many eukaryotic proteins are extensively modified.

In **proteolytic processing,** proteins are synthesized as inactive precursors, which are converted to an active form by the removal of a portion of the polypeptide chain. For example, proinsulin contains 86 amino acids. Removing 35 amino acids yields the hormone insulin, which consists of two polypeptide chains

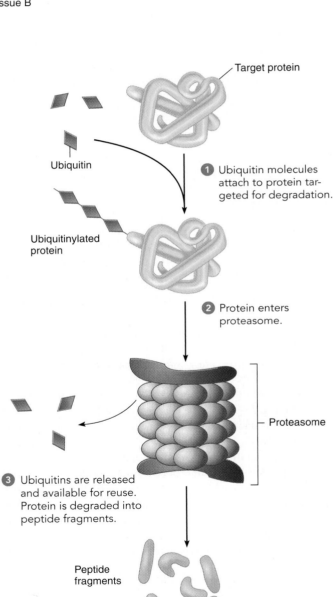

Target protein

Ubiquitin

1 Ubiquitin molecules attach to protein targeted for degradation.

Ubiquitinylated protein

2 Protein enters proteasome.

Proteasome

3 Ubiquitins are released and available for reuse. Protein is degraded into peptide fragments.

Peptide fragments

Figure 14-13 The ubiquitin–proteasome degradation of proteins

containing 30 and 21 amino acids, respectively, linked by disulfide bridges.

Chemical modification, by adding or removing functional groups, reversibly alters the activity of an enzyme. One common way to modify the activity of an enzyme or other protein is to add or remove phosphate groups. Enzymes that add phosphate groups are called **kinases;** those that remove them are **phosphatases.** For example, the cyclin-dependent kinases discussed in Chapter 10 help control the cell cycle by adding phosphate groups to certain key proteins, causing them to become activated or inactivated. Chemical modifications such as protein phosphorylation also let the cell respond rapidly to certain hormones (see Chapter 48), or to fast-changing environmental or nutritional conditions.

Posttranslational control of gene expression can also involve **protein degradation.** The amino acid found at the N-terminal end of a protein correlates with whether the protein is long-lived or short-lived. Some proteins, for example, have a 50% turnover of synthesis and degradation every 3 days. Other proteins, such as proteins in the lens of the human eye, may persist for the entire life of the individual.

Proteins targeted for destruction are covalently bonded to **ubiquitin,** a small polypeptide tag that contains 76 amino acids. Additional molecules of ubiquitin are added, forming short chains attached to the soon-to-be degraded protein (▮ Fig. 14-13). A protein that is tagged by ubiquitin is targeted for degradation in a proteasome. **Proteasomes** are large macromolecular structures that recognize the ubiquitin tags. **Proteases** (protein-degrading enzymes) associated with proteasomes hydrolyze some of the peptide bonds, degrading the protein into short, nonfunctional peptide fragments. As the protein is degraded, the ubiquitin molecules are released intact, which allows them to be used again.

Review

▮ How does regulation of eukaryotic genes differ from regulation of bacterial genes?

▮ Why must certain genes in eukaryotic cells be present in multiple copies?

▮ How does chromosome structure affect the activity of some eukaryotic genes?

▮ How does differential mRNA processing give rise to different proteins?

SUMMARY WITH KEY TERMS

Learning Objectives

1 Explain why bacterial and eukaryotic cells have different mechanisms of gene regulation (page 305).

▮ Bacterial cells grow rapidly and have a relatively short life span. **Transcriptional-level control** is the most common way for the prokaryotic cell to regulate gene expression.

▮ Eukaryotic cells have different regulatory requirements, in part because eukaryotes usually have a relatively long life span, during which they respond to many different stimuli. Also, a single gene may be regulated in different ways in different types of cells. Although transcriptional-level control predominates, control at other levels of gene expression is also important.

2 Define *operon,* and explain the functions of the operator and promoter regions (page 306).

▮ Many regulated genes in bacteria are organized into operons. Each **operon** is a gene complex consisting of a group of structural genes with related functions plus the closely linked DNA sequences responsible for controlling them.

▮ Each operon has a single **promoter** region upstream from the protein-coding regions; the promoter is where RNA polymerase first binds to DNA before transcription begins.

▮ The **operator** serves as the regulatory switch for transcriptional-level control of the operon. The binding of a **repressor protein** to the operator sequence prevents transcription; although RNA polymerase binds to the promoter, it is blocked from transcribing the structural genes. When the repressor is not bound to the operator, transcription proceeds.

3 Distinguish among inducible, repressible, and constitutive genes (page 306).

▮ An **inducible operon,** such as the *lac* operon, is normally turned off. The repressor protein is synthesized in an active form that binds to the operator. If lactose is present, it is converted to allolactose, the **inducer,** which binds to the repressor protein, changing the repressor's shape. The altered repressor cannot bind to the operator, and the operon is transcribed.

ThomsonNOW™ **Watch the lactose operon in action by clicking on the figure in ThomsonNOW.**

▮ A **repressible operon,** such as the *trp* operon, is normally turned on. The repressor protein is synthesized in an inactive form that cannot bind to the operator. A metabolite (usually the end product of a metabolic pathway) acts as a **corepressor.** When intracellular corepressor levels are high, a corepressor molecule binds to the repressor, changing the repressor's shape so that it binds to the operator and thereby turns off transcription of the operon.

▮ **Constitutive genes** are neither inducible nor repressible; they are active at all times. Regulatory proteins such as **catabolite activator protein (CAP)** and the repressor proteins are produced constitutively. These proteins work by recognizing and binding to specific base sequences in the DNA. The activity of constitutive genes is controlled by how efficiently RNA polymerase binds to their promoter regions.

4 Differentiate between positive and negative control, and show how both types of control operate in regulating the *lac* operon (page 306).

▮ Repressible and inducible operons are under **negative control.** When the repressor protein binds to the operator, transcription of the operon is turned off.

▮ Some inducible operons are also under **positive control,** in which an activator protein binds to the DNA and stimulates transcription of the gene. CAP activates the *lac* operon; CAP binds to the promoter region, stimulating transcription by binding RNA polymerase tightly. To bind to the *lac* operon, CAP requires **cyclic AMP (cAMP).** Levels of cAMP increase as levels of glucose decrease.

5 Describe the types of posttranscriptional control in bacteria (page 306).

▮ Some **posttranscriptional controls** operate in bacteria. A **translational control** is a posttranscriptional control that regulates the rate of translation of a particular mRNA. **Posttranslational controls** include **feedback inhibition** of key enzymes in some metabolic pathways.

6 Discuss the structure of a typical eukaryotic gene and the DNA sequences involved in regulating that gene (page 312).

  ▪ Eukaryotic genes are not normally organized into operons. Regulation of eukaryotic genes occurs at the levels of transcription, mRNA processing, translation, and modifications of the protein product.

  ▪ Transcription of a gene requires a **transcription initiation site,** where transcription begins, plus a promoter to which RNA polymerase binds. In multicellular eukaryotes, RNA polymerase binds to a promoter called a **TATA box.**

  ▪ The promoter of a regulated eukaryotic gene consists of an RNA polymerase–binding site and short DNA sequences known as **upstream promoter elements (UPEs)** or **proximal control elements.** The number and types of UPEs within the promoter region determine the efficiency of the promoter.

  ▪ **Enhancers,** which are located thousands of bases away from the promoter, control some eukaryotic genes. Like the upstream promoter elements, enhancers help form an active transcription initiation complex. Specific regulatory proteins bind to enhancer elements and activate transcription by interacting with the proteins bound to the promoters.

7 Give examples of some of the ways eukaryotic DNA-binding proteins bind to DNA (page 312).

  ▪ Eukaryotic genes are controlled by DNA-binding protein regulators called **transcription factors.** Many are transcriptional activators; others are transcriptional repressors.

  ▪ Each transcription factor has a DNA-binding **domain.** Some transcription factors have a helix-turn-helix arrangement and insert one of the helices into the DNA. Other transcription factors have loops of amino acids held together by zinc ions; each loop includes an α-helix that fits into the DNA. Some transcription factors are **leucine zipper proteins** that associate as dimers that insert into the DNA.

8 Illustrate how a change in chromosome structure may affect the activity of a gene (page 312).

  ▪ Genes are inactivated by changes in chromosome structure. Densely packed regions of chromosomes called **heterochromatin** contain inactive genes. Active genes are associated with a loosely packed chromatin structure called **euchromatin.** One way that cells change chroma-

tin structure from heterochromatin to euchromatin is to chemically modify **histones,** the proteins that associate with DNA to form nucleosomes. Methyl groups, acetyl groups, sugars, and proteins may chemically attach to the histone tail, a string of amino acids that extends from the DNA-wrapped nucleosome, and may expose or hide genes, turning them on or off.

  ▪ **DNA methylation** is a mechanism that perpetuates gene inactivation. **Epigenetic inheritance** involves changes in how a gene is expressed. Because DNA methylation patterns tend to be repeated in successive cell generations, they provide a mechanism for epigenetic inheritance. Epigenetic inheritance appears to be an important mechanism of gene regulation.

  ▪ Some genes whose products are required in large amounts exist as multiple copies in the chromosome. In the process of **gene amplification,** some cells selectively amplify genes by DNA replication.

9 Explain how a gene in a multicellular organism may produce different products in different types of cells (page 312).

  ▪ As a result of **differential mRNA processing,** a single gene produces different forms of a protein in different tissues, depending on how the pre-mRNA is spliced. Typically, such a gene contains a segment that can be either an intron or an exon. As an intron, the sequence is removed, and as an exon, the sequence is retained.

10 Identify some of the types of regulatory controls that operate in eukaryotes after mature mRNA is formed (page 312).

  ▪ Certain regulatory mechanisms increase the stability of mRNA, allowing more protein molecules to be synthesized before mRNA degradation. In certain cases mRNA stability is under hormonal control.

  ▪ Posttranslational control of eukaryotic genes occurs by feedback inhibition or by modification of the protein structure. The function of a protein is changed by **kinases** adding phosphate groups or by **phosphatases** removing phosphates.

  ▪ Posttranslational control of gene expression can also involve **protein degradation.** Proteins targeted for destruction are covalently bonded to **ubiquitin.** A protein tagged by ubiquitin is targeted for degradation in a **proteasome,** a large macromolecular structure that recognizes the ubiquitin tags. **Proteases** (protein-degrading enzymes) associated with proteasomes degrade the protein into short peptide fragments.

TEST YOUR UNDERSTANDING

1. The regulation of most bacterial genes occurs at the level of (a) transcription (b) translation (c) replication (d) posttranslation (e) postreplication

2. The operator of an operon (a) encodes information for the repressor protein (b) is the binding site for the inducer (c) is the binding site for the repressor protein (d) is the binding site for RNA polymerase (e) encodes the information for the CAP

3. A mutation that inactivates the repressor gene of the *lac* operon results in (a) the continuous transcription of the structural genes (b) no transcription of the structural genes (c) the binding of the repressor to the operator (d) no production of RNA polymerase (e) no difference in the rate of transcription

4. At a time when the *lac* operon is actively transcribed, (a) the operator is bound to the inducer (b) the lactose repressor protein is bound to the promoter (c) the operator is not bound to the promoter (d) the gene coding for the repressor is not expressed constitutively (e) the lactose repressor protein is bound to the inducer

5. A repressible operon codes for the enzymes of the following pathway. Which component of the pathway is most likely to be the corepressor for that operon?

$$A \xrightarrow{\text{Enzyme 1}} B \xrightarrow{\text{Enzyme 2}} C \xrightarrow{\text{Enzyme 3}} D$$

(a) substance A (b) substance B or C (c) substance D (d) enzyme 1 (e) enzyme 3

6. An mRNA molecule transcribed from the *lac* operon contains nucleotide sequences complementary to (a) structural genes coding for enzymes (b) the operator region (c) the promoter region (d) the repressor gene (e) introns

7. Feedback inhibition is an example of control at the level of _____. (a) transcription (b) translation (c) post-translation (d) replication (e) all the preceding

8. Which of the following control mechanisms is generally the most economical in terms of conserving energy and resources? (a) control by means of operons (b) feedback inhibition (c) selective degradation of mRNA (d) selective degradation of enzymes (e) gene amplification

9. A repressible operon, such as the *trp* operon, is "off" when (a) the gene that codes for the repressor is expressed constitutively (b) the repressor–corepressor complex binds to the operator (c) the repressor binds to the structural genes (d) the corepressor binds to RNA polymerase (e) CAP binds to the promoter

10. Which of the following is an example of positive control? (a) transcription occurs when a repressor binds to an inducer (b) transcription cannot occur when a repressor binds to a corepressor (c) transcription is stimulated when an activator protein binds to DNA (d) a and b (e) a and c

11. Which of the following are typically absent in bacteria? (a) enhancers (b) proteins that regulate transcription (c) repressors (d) promoters (e) operators

12. The "zipper" of a leucine zipper protein attaches (a) specific amino acids to specific DNA base pairs (b) two polypeptide chains to each other (c) one DNA region to another DNA region (d) amino acids to zinc atoms (e) RNA polymerase to the operator

13. Inactive genes tend to be found in (a) highly condensed chromatin, known as euchromatin (b) decondensed chromatin, known as euchromatin (c) highly condensed chromatin, known as heterochromatin (d) decondensed chromatin, known as heterochromatin (e) chromatin that is not organized as nucleosomes

14. Which of the following is characteristic of genes and gene regulation in bacteria, but *not* in eukaryotes? (a) presence of enhancers (b) capping of mRNAs (c) many chromosomes per cell (d) binding of DNA to regulatory proteins (e) no requirement for exon splicing

15. Which of the following is characteristic of genes and gene regulation in *both* bacteria and eukaryotes? (a) promoters (b) noncoding DNA within coding sequences (c) enhancers (d) operons (e) DNA located in a nucleus

16. Which of the following is characteristic of genes and gene regulation in eukaryotes, but *not* in bacteria? (a) enhancers (b) operons (c) promoters (d) a and b (e) a, b, and c

17. Through differential mRNA processing, eukaryotes (a) reinforce gene inactivation (b) prevent transcription of heterochromatin (c) produce related but different proteins in different tissues (d) amplify genes to meet the requirement of high levels of a gene product (e) bind transcription factors to enhancers to activate transcription

CRITICAL THINKING

1. Compare the types of bacterial genes associated with inducible operons, those associated with repressible operons, and those that are constitutive. Predict the category into which each of the following would most likely fit: (a) a gene that codes for RNA polymerase, (b) a gene that codes for an enzyme required to break down maltose, (c) a gene that codes for an enzyme used in the synthesis of adenine.

2. The regulatory gene that codes for the tryptophan repressor is not tightly linked to the *trp* operon. Would it be advantageous if it were? Explain your answer.

3. Under what condition do cells use gene amplification?

4. **Evolution Link.** Hypothesize why evolution resulted in multiple levels of gene regulation in eukaryotes.

5. **Analyzing Data.** Develop a simple hypothesis that would explain the behavior of each of the following types of mutants in *E. coli*:

Mutant a: The map position of this mutation is in the *trp* operon. The mutant cells are constitutive; that is, they produce all the enzymes coded for by the *trp* operon, even if large amounts of tryptophan are present in the growth medium.

Mutant b: The map position of this mutation is in the *trp* operon. The mutant cells do not produce any enzymes coded for by the *trp* operon under any conditions.

Mutant c: The map position of this mutation is some distance from the *trp* operon. The mutant cells are constitutive; that is, they produce all the enzymes coded for by the *trp* operon, even if the growth medium contains large amounts of tryptophan.

Mutant d: The map position of this mutation is some distance from the *trp* operon. The mutant cells do not produce any enzymes coded for by the *trp* operon under any conditions.

Additional questions are available in ThomsonNOW at www.thomsonedu.com/login

DNA Technology and Genomics

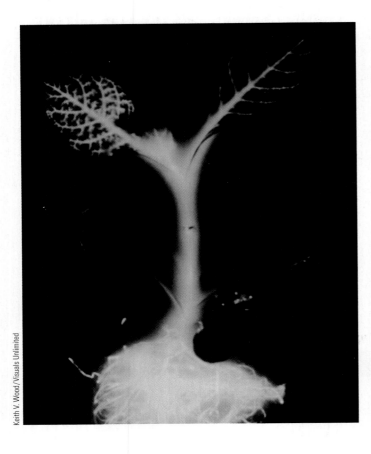

Genetically engineered tobacco plant. This plant glows as it expresses the luciferase gene from a firefly. Luciferase is an enzyme that catalyzes a reaction that produces a flash of light. This classic experiment indicated that animal genes can be expressed in plants.

KEY CONCEPTS

Recombinant DNA techniques allow scientists to clone many copies of specific genes and gene products.

Biologists study DNA using gel electrophoresis, DNA blotting, automated sequencing, and other methods.

Genomics is an emerging field that comprises the structure, function, and evolution of genomes.

DNA technology and genomics have wide applications, from medical to forensic to agricultural.

Beginning in the mid-1970s, the development of new ways to study DNA led to radically new research approaches that have had a major impact in areas from cell biology and evolution to ethical and societal issues. This chapter begins with a consideration of **recombinant DNA technology,** in which researchers splice together DNA from different organisms in the laboratory. One goal of this technology is to enable scientists to obtain many copies of a specific DNA segment for the purpose of studying it. Because new methods to analyze DNA are continually emerging, we do not attempt to explore them all here. Instead, we discuss some of the major approaches that have provided a foundation for the technology.

We then consider how studies of DNA sequences have helped scientists understand the organization of genes and the relationship between genes and their products. In fact, most of our current knowledge of the structure and control of eukaryotic genes, and of the roles of genes in development, comes from applying these methods.

This chapter also explores some of the practical applications of DNA technologies. One of the rapidly advancing areas of study is **genetic engineering**—modifying the DNA of an organism to produce new genes with new traits. Genetic engineering can take many forms, ranging from basic research (see photograph) to the production of strains of bacteria that manufacture useful protein products, to the development of plants and animals that express foreign genes useful in agriculture.

The development of DNA cloning, genetic engineering, and related techniques has transformed people's view of **biotechnology,** the use of organisms to develop useful products. A traditional example of biotechnology is the use of yeast to make alcoholic beverages. Today, biotechnology includes numerous applications in such diverse areas as medicine and the pharmaceutical industry, foods and agriculture, and forensic science. ■

DNA CLONING

Learning Objectives

1 Explain how a typical restriction enzyme cuts DNA molecules, and give examples of the ways in which these enzymes are used in recombinant DNA technology.
2 Distinguish among a genomic DNA library, a chromosome library, and a complementary DNA (cDNA) library; explain why one would clone the same eukaryotic gene from both a genomic DNA library and a cDNA library.
3 Describe the purpose of a genetic probe.
4 Describe how the polymerase chain reaction amplifies DNA in vitro.

Recombinant DNA technology was not developed quickly. It actually had its roots in the 1940s with genetic studies of bacteria and **bacteriophages** ("bacteria eaters"), the viruses that infect them (see Fig. 12-2). After decades of basic research and the accumulation of extensive knowledge, the technology became feasible and available to the many scientists who now use these methods.

In recombinant DNA technology, scientists use enzymes from bacteria, known as **restriction enzymes, to cut DNA molecules only in specific places**. Restriction enzymes enable researchers to cut DNA into manageable segments. Each fragment is then incorporated into a suitable **vector** molecule, a carrier capable of transporting the DNA fragment into a cell. Bacteriophages and DNA molecules called plasmids are two examples of vectors. Bacterial DNA is circular; a **plasmid** is a separate, much smaller, circular DNA molecule that may be present and replicate inside a bacterial cell, such as *Escherichia coli*. Researchers introduce plasmids into bacterial cells by a method called **transformation,** the uptake of foreign DNA by cells (see Chapter 12). For transformation to be efficient, the researcher alters the cells chemically or by *electroporation*—delivering an electric shock—to make the plasma membrane permeable to the plasmid DNA molecules. Once a plasmid enters a cell, it is replicated and distributed to daughter cells during cell division. When a *recombinant plasmid*—one that has foreign DNA spliced into it—replicates in this way, many identical copies of the foreign DNA are made; in other words, the foreign DNA is **cloned.**

Restriction enzymes are "molecular scissors"

Discovering restriction enzymes was a major breakthrough in developing recombinant DNA technology. Today, large numbers of different types of restriction enzymes, each with its own

characteristics, are readily available to researchers. For example, a restriction enzyme known as *Hin*dIII recognizes and cuts a DNA molecule at the restriction site 5′—AAGCTT—3′. (A *restriction site* is a DNA sequence containing the cleavage site that is cut by a particular restriction enzyme.) The sequence 5′—GAATTC—3′ is cut by another restriction enzyme, known as *Eco*RI. The names of restriction enzymes are generally derived from the names of the bacteria from which they were originally isolated. Thus, *Hin*dIII and *Eco*RI are derived from *Hemophilus influenzae* and *E. coli,* respectively.

Why do bacteria produce such enzymes? During infection, a bacteriophage injects its DNA into a bacterial cell. The bacterium can defend itself if it has restriction enzymes that can attack the bacteriophage DNA. The cell protects its own DNA from breakdown by modifying it after replication. An enzyme adds a methyl group to one or more bases in each restriction site so that the restriction enzyme does not recognize and cut the bacterial DNA.

Restriction enzymes enable scientists to cut DNA from chromosomes into shorter fragments in a controlled way. Many of the restriction enzymes used for recombinant DNA studies cut **palindromic sequences,** which means the base sequence of one strand reads the same as its complement when both are read in the 5′ ⟶ 3′ direction. Thus, in our *Hin*dIII example, both strands read 5′—AAGCTT—3′, which as a double-stranded molecule is diagrammed as follows:

5′—AAGCTT—3′
3′—TTCGAA—5′

By cutting both strands of the DNA, but in a staggered fashion, these enzymes produce fragments with identical, complementary, single-stranded ends:

5′—A   AGCTT—3′
3′—TTCGA   A—5′

These ends are called *sticky ends* because they pair by hydrogen bonding with the complementary, single-stranded ends of other DNA molecules that have been cut with the same enzyme (■ Fig. 15-1). Once the sticky ends of two molecules have been joined in this way, they are treated with **DNA ligase,** an enzyme that covalently links the two DNA fragments to form a stable recombinant DNA molecule.

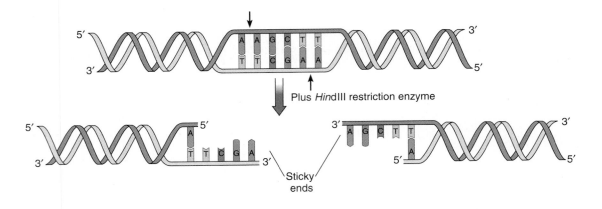

Figure 15-1 *Animated* Cutting DNA with a restriction enzyme

Many restriction enzymes, such as *Hind*III, cut DNA at sequences that are palindromic, producing complementary sticky ends. The small black arrows designate the enzyme's cleavage sites.

Recombinant DNA forms when DNA is spliced into a vector

In recombinant DNA technology, geneticists cut both foreign DNA and plasmid DNA with the same restriction enzyme. The two types of DNA are then mixed under conditions that facilitate hydrogen bonding between the complementary bases of the sticky ends, and the nicks in the resulting recombinant DNA are sealed by DNA ligase (▮ Fig. 15-2).

The plasmids now used in recombinant DNA work have been extensively manipulated in the laboratory to include features helpful in isolating and analyzing cloned DNA (▮ Fig. 15-3). Among these are an origin of replication (see Chapter 12), one or more restriction sites, and genes that let researchers select cells transformed by recombinant plasmids. These genes cause transformed cells to grow under specified conditions that do not allow untransformed cells to grow. In this way, the researchers use features that are also common in naturally occurring plasmids. Typically, plasmids do not contain genes essential to the bacterial cells under normal conditions. However, they often carry genes that are useful under specific environmental conditions, such as genes that confer resistance to particular antibiotics or that let the cells use a specific nutrient. For example, cells transformed with a plasmid that includes a gene for resistance to the antibiotic tetracycline can grow in a medium that contains tetracycline, whereas untransformed cells cannot.

A limiting property of any vector is the size of the DNA fragment it can effectively carry. The size of a DNA segment is often given in kilobases (kb), with 1 kb being equal to 1000 base pairs. Fragments that are smaller than 10 kb are usually inserted into plasmids for use in *E. coli*. However, larger fragments require the use of bacteriophage vectors, which can handle up to 23 kb of DNA. Other vectors are **cosmid cloning vectors,** which are combination vectors with features from both bacteriophages and plasmids, and **bacterial artificial chromosomes (BACs),** which accommodate much larger fragments of DNA. A BAC can include up to about 200 kb of extra DNA, a feature that made

BACs especially useful in the *Human Genome Project*, which is discussed later in the chapter.

Recombinant DNA can also be introduced into cells of eukaryotic organisms. For example, geneticists use engineered viruses as vectors in mammal cells. These viruses are disabled so they do not kill the cells they infect. Instead, the viral DNA, as well as any foreign DNA they carry, becomes incorporated into the cell's chromosomes after infection. As discussed later, other methods do not require a biological vector.

DNA can be cloned inside cells

Because a single gene is only a small part of the total DNA in an organism, isolating the piece of DNA containing that particular gene is like finding a needle in a haystack: a powerful detector is needed. Today, many methods enable biologists to isolate a specific nucleotide sequence from an organism. We start by discussing methods in which DNA is cloned inside bacterial cells. We use the cloning of human DNA as an example, although the procedure is applicable for any organism.

A genomic DNA library contains fragments of all DNA in the genome

The total DNA in a cell is called its **genome.** For example, if DNA is extracted from human cells, we refer to it as human genomic DNA. A **genomic DNA library** is a collection of thousands of DNA fragments that represent all the DNA in the genome. Each fragment is inserted into a plasmid, which is usually incorporated into a bacterial cell. Thus, a human genomic DNA library is stored in a collection of recombinant bacteria, each with a different fragment of human DNA. Scientists use genomic DNA libraries to isolate and study specific genes.

Individual chromosomes can also be isolated to make a **chromosome library** containing all the DNA fragments in that

Recombinant DNA technology involves a series of steps that make use of biochemistry, genetics, and molecular biology.

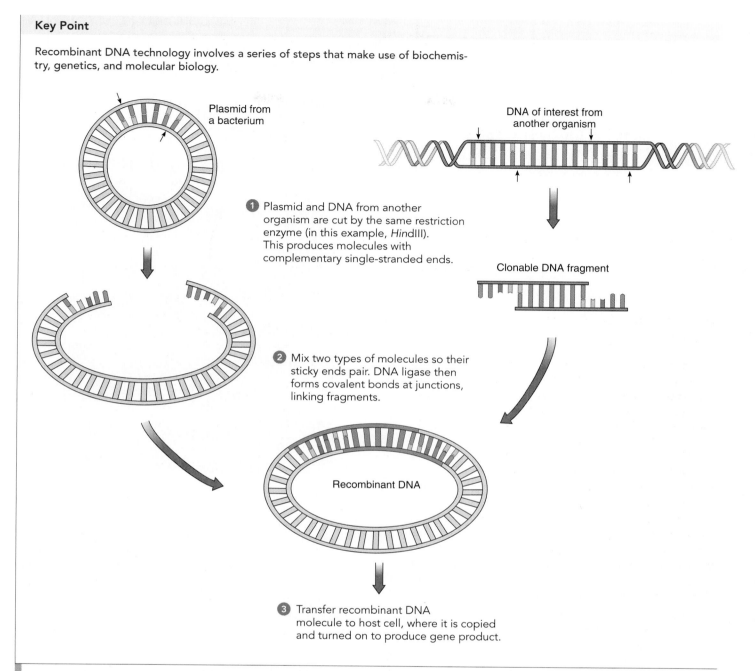

1 Plasmid and DNA from another organism are cut by the same restriction enzyme (in this example, *Hind*III). This produces molecules with complementary single-stranded ends.

Plasmid from a bacterium

DNA of interest from another organism

Clonable DNA fragment

2 Mix two types of molecules so their sticky ends pair. DNA ligase then forms covalent bonds at junctions, linking fragments.

Recombinant DNA

3 Transfer recombinant DNA molecule to host cell, where it is copied and turned on to produce gene product.

Figure 15-2 *Animated* Splicing foreign DNA into a vector

In this example, a bacterial plasmid is the vector.

specific chromosome. If a gene of interest is known to be associated with a particular chromosome, it is easier to isolate that gene from a chromosome library than from a genomic DNA library.

The basic cloning methods just described are used to make genomic DNA or chromosome libraries. First, the DNA is cut with a restriction enzyme, generating a population of DNA fragments (Fig. 15-4 **1**). These fragments vary in size and in the genetic information they carry, but they all have identical sticky ends. Geneticists treat plasmid DNA that will be used as a vector with the same restriction enzyme, which converts the circular

plasmids into linear molecules with sticky ends complementary to those of the human DNA fragments. Recombinant plasmids are produced by mixing the two kinds of DNA (human and plasmid) under conditions that promote hydrogen bonding of complementary bases. Then, DNA ligase is used to covalently bond the paired ends of the plasmid and human DNA, forming recombinant DNA (Fig. 15-4 **2**). Unavoidably, nonrecombinant plasmids also form (*not shown in figure*), because some plasmids revert to their original circular shape without incorporating foreign DNA.

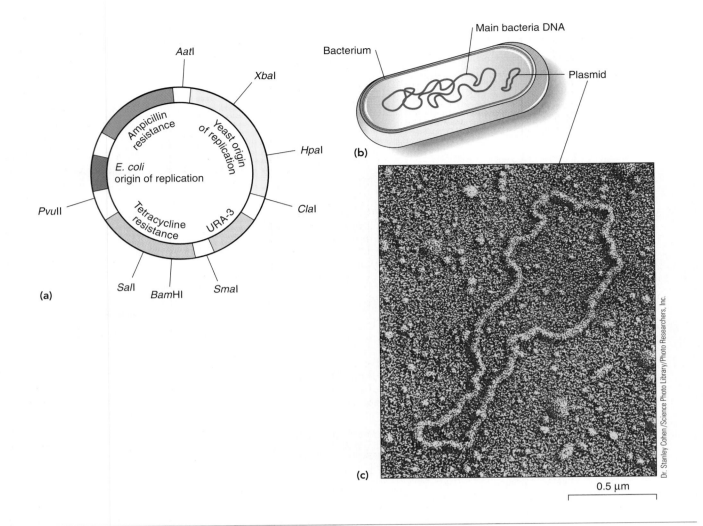

(a) Aatl
Xbal
Ampicillin resistance
Yeast origin of replication
E. coli origin of replication
HpaI
PvuII
Tetracycline resistance
URA-3
ClaI
SalI BamHI SmaI

(a)

Main bacteria DNA
Bacterium
Plasmid

(b)

(c)

0.5 μm

Dr. Stanley Cohen/Science Photo Library/Photo Researchers, Inc.

Figure 15-3 Plasmids

(a) This genetically engineered plasmid vector has many useful features. Researchers constructed it from DNA fragments they had isolated from plasmids, E. coli genes, and yeast genes. The two origins of replication, one for E. coli and one for yeast, Saccharomyces cerevisiae, let it replicate independently in either type of cell. Letters on the outer circle designate sites for restriction enzymes that cut the plasmid only at that position. Resistance genes for the antibiotics ampicillin and tetracycline and the yeast URA-3 gene are also shown. The URA-3 gene is useful when transforming yeast cells lacking an enzyme required for uracil synthesis. Cells that take up the plasmid grow on a uracil-deficient medium. (b) The relative sizes of a plasmid and the main DNA of a bacterium. (c) TEM of a plasmid from E. coli.

The geneticists insert plasmids into antibiotic-sensitive bacterial cells by transformation (❙ Fig. 15-4 ❸). Because the ratio of plasmids to cells is kept very low, it is rare for a cell to receive more than one plasmid molecule, and not all cells receive a plasmid. The researchers incubate normally antibiotic-sensitive cells on a nutrient medium that includes antibiotics, so only those cells grow that have incorporated a plasmid, which contains a gene for antibiotic resistance (❙ Fig. 15-4 ❹). In addition, the plasmid has usually been engineered in ways that enable researchers to identify those cells containing recombinant plasmids.

Genomic DNA and chromosome libraries contain redundancies; that is, certain human DNA sequences have been inserted into plasmids more than once, purely by chance. However, each individual recombinant plasmid—analogous to a book in the library—contains only a single fragment of the total human genome.

To identify a plasmid containing a sequence of interest, the researcher clones each plasmid until there are millions of copies to work with. This process occurs as the bacterial cells grow and divide. A dilute sample of the bacterial culture is spread on solid growth medium, so the cells are widely separated. Each cell divides many times, yielding a visible **colony,** which is a clone of genetically identical cells originating from a single cell. All the cells of a particular colony contain the same recombinant plasmid, so during this process a specific sequence of human DNA is also cloned. The major task is to determine which colony out of

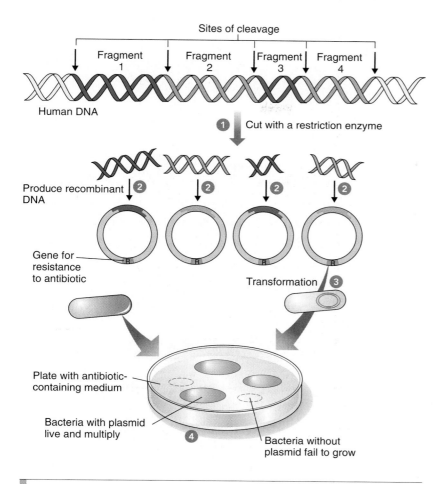

Sites of cleavage

Fragment 1 | Fragment 2 | Fragment 3 | Fragment 4

Human DNA

1 Cut with a restriction enzyme

Produce recombinant DNA 2 2 2 2

Gene for resistance to antibiotic

Transformation 3

Plate with antibiotic-containing medium

Bacteria with plasmid live and multiply 4

Bacteria without plasmid fail to grow

Figure 15-4   Producing a genomic DNA or chromosome library

Only a small part of one chromosome is shown. A great many more DNA fragments would be produced from an entire chromosome or genome. The numbered steps in the figure are explained in the text.

thousands contains a cloned fragment of interest. Specific DNA sequences are identified in various ways.

A complementary genetic probe detects a specific DNA sequence

Suppose a researcher wishes to screen the thousands of recombinant DNA molecules in bacterial cells to find the colony that contains the specific DNA sequence of interest. A common approach to detecting the DNA of interest involves using a **genetic probe,** also called *DNA probe,* which is a segment of DNA that is homologous to part of the sequence of interest. The genetic probe is usually a radioactively labeled segment of single-stranded DNA that can **hybridize**—attach by base pairing—to complementary base sequences in the target DNA.

To find which colony carries the DNA sequence of interest, biologists transfer cells from bacterial colonies containing recombinant plasmids to a nitrocellulose or nylon membrane, which

then becomes a *replica* of the colonies (❚ Fig. 15-5). They treat the cells on the membrane chemically to lyse them, releasing the DNA and making the DNA single stranded. Then they incubate the membrane with the radioactive probe mixture to let the probes hybridize with any complementary strands of DNA that may be present. Each spot on the membrane containing DNA that is complementary to that particular probe becomes radioactive and is detected by *autoradiography* (which was explained in Chapter 2), using X-ray film. Each spot on the film therefore identifies a colony containing a plasmid that includes the DNA of interest. The biologist returns to the original plate and removes cells from the colonies that contain the cloned sequence of interest. These cells are then grown in culture to produce millions of new bacteria, all containing the recombinant DNA of interest.

A cDNA library is complementary to mRNA and does not contain introns

Researchers frequently wish to clone intact genes while avoiding **introns,** which are regions of eukaryotic genes that do not code for proteins. Scientists also may wish to clone only genes that are expressed in a particular cell type. In such cases, they construct libraries of DNA copies of mature mRNA from which introns have been removed. The copies, known as **complementary DNA (cDNA)** because they are complementary to mRNA, also lack introns. The researchers use the enzyme **reverse transcriptase** (see Chapter 13) to synthesize single-stranded cDNA, which they separate from the mRNA and make double stranded with DNA polymerase (❚ Fig. 15-6). A **cDNA library** is formed using mRNA from a single cell type as the starting material. The double-stranded cDNA molecules are inserted into plasmid or virus vectors, which then multiply in bacterial cells.

Analyzing cDNA clones lets investigators determine certain characteristics of the protein encoded by the gene, including its amino acid sequence. They can also study the structure of the mature mRNA. Because the cDNA copy of the mRNA does not contain intron sequences, comparing the DNA base sequences in cDNA and the DNA in genomic DNA or chromosome libraries reveals the locations of intron and exon coding sequences in the gene.

Cloned cDNA sequences are also useful when geneticists want to produce a eukaryotic protein in bacteria. When they introduce an intron-containing human gene, such as the gene for human growth hormone, into a bacterium, it cannot remove the introns from the transcribed RNA to make a functional mRNA for producing its protein product. If they insert a cDNA clone of the gene into the bacterium, however, its transcript contains an uninterrupted coding region. A functional protein is synthesized if the geneticist inserts the gene downstream of an appropriate

DNA Technology and Genomics   **327**

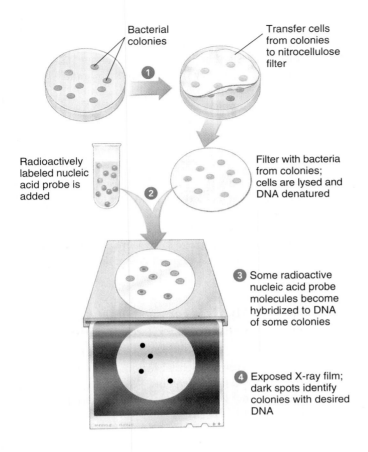

① Bacterial colonies

Transfer cells from colonies to nitrocellulose filter

Radioactively labeled nucleic acid probe is added

② Filter with bacteria from colonies; cells are lysed and DNA denatured

③ Some radioactive nucleic acid probe molecules become hybridized to DNA of some colonies

④ Exposed X-ray film; dark spots identify colonies with desired DNA

Figure 15-5 *Animated* Using a genetic probe to find bacterial cells with a specific recombinant DNA molecule

A radioactive nucleic acid probe, which is usually single-stranded DNA, reveals the presence of complementary sequences of DNA.

bacterial promoter and if he or she places the appropriate bacterial translation initiation sequences in it.

The polymerase chain reaction is a technique for amplifying DNA in vitro

The methods just described for amplifying, or making multiple copies of, a specific DNA sequence involve cloning DNA in cells, usually those of bacteria. These processes are time consuming and require an adequate DNA sample as the starting material. The **polymerase chain reaction (PCR),** which U.S. biochemist Kary Mullis developed in 1985, lets researchers amplify a tiny sample of DNA *without cloning.* The DNA is amplified millions of times in a few hours. In 1993, Mullis received the Nobel Prize in Chemistry for his work.

In PCR, DNA polymerase uses nucleotides and primers to replicate a DNA sequence in vitro (outside a living organism), thereby producing two DNA molecules (█ Fig. 15-7). Then the researchers denature, or separate by heating, the two strands of each molecule and replicate them again, yielding four double-stranded molecules. After the next cycle of heating and replica-

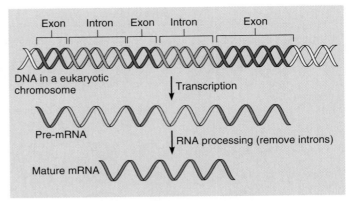

Exon Intron Exon Intron Exon

DNA in a eukaryotic chromosome

Transcription

Pre-mRNA

RNA processing (remove introns)

Mature mRNA

(a) Formation of cDNA relies on RNA processing that occurs in the nucleus to yield mature mRNA.

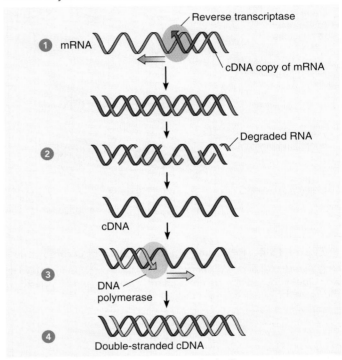

Reverse transcriptase

① mRNA

cDNA copy of mRNA

② Degraded RNA

cDNA

③ DNA polymerase

④ Double-stranded cDNA

(b) Mature mRNA is extracted and purified. ① To make a DNA copy of the mature mRNA, researchers use reverse transcriptase to produce a single-stranded cDNA complementary to the mRNA. ② Specific enzymes then degrade the mRNA, and ③ DNA polymerase is used to synthesize a second strand of DNA. ④ The end result is a double-stranded cDNA molecule.

Figure 15-6 *Animated* The formation of cDNA

tion, there are eight molecules, and so on, with the number of DNA molecules doubling in each cycle. After only 20 heating and cooling cycles are done using automated equipment, this exponential process yields 2^{20}, or more than 1 million, copies of the target sequence!

Because the reaction can be carried out efficiently only if the DNA polymerase can remain stable through many heating cycles, researchers use a heat-resistant DNA polymerase, known as *Taq*

Key Point

PCR amplifies DNA by repeated cycles of denaturation, primer attachment, and strand elongation by complementary base pairing.

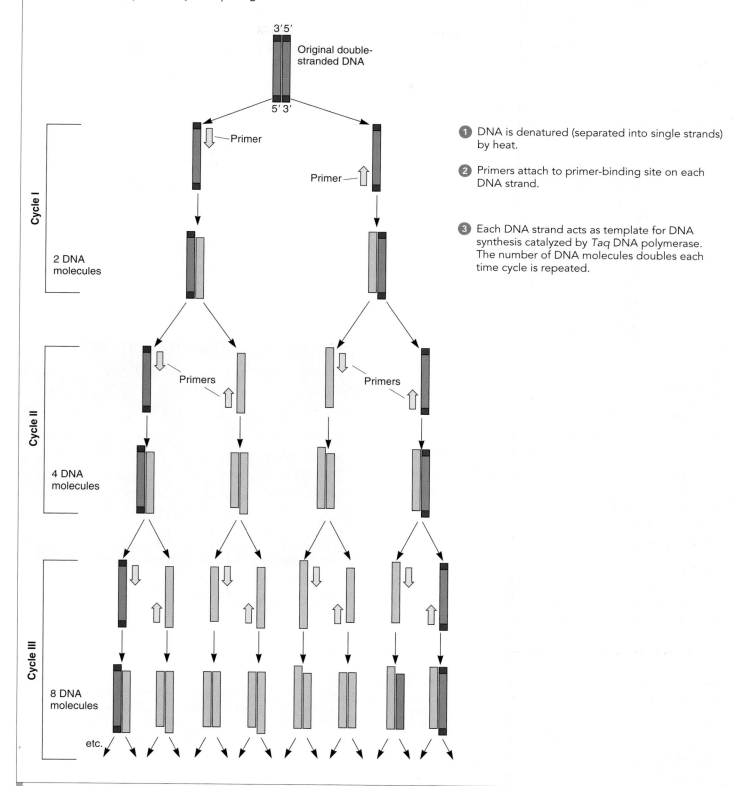

① DNA is denatured (separated into single strands) by heat.

② Primers attach to primer-binding site on each DNA strand.

③ Each DNA strand acts as template for DNA synthesis catalyzed by *Taq* DNA polymerase. The number of DNA molecules doubles each time cycle is repeated.

Figure 15-7 *Animated* Amplification of DNA by PCR

The initial reaction mixture includes a very small amount of double-stranded DNA, DNA precursors (deoxyribonucleotides), specific nucleic acid primers, and heat-resistant *Taq* DNA polymerase.

polymerase. The name of this enzyme reflects its source, *Thermus aquaticus,* a bacterium that lives in hot springs in Yellowstone National Park. Because the water in this environment is close to the boiling point, all enzymes in *T. aquaticus* have evolved to be stable at high temperatures. Bacteria living in deep-sea thermal vents have similar enzymes.

The PCR technique has virtually limitless applications. It enables researchers to amplify and analyze tiny DNA samples from a variety of sources, ranging from crime scenes to archaeological remains. For example, in 1997 investigators reported the first analysis of mitochondrial DNA obtained from the bones of Neandertals (see Chapter 22).

A limitation of the PCR technique is that it is almost too sensitive. Even a tiny amount of contaminant DNA in a sample may become amplified if it includes a DNA sequence complementary to the primers, potentially leading to an erroneous conclusion. Researchers take appropriate precautions to avoid this and other technical pitfalls.

Review

- What are restriction enzymes? How are they used in recombinant DNA research?
- What are the relative merits of genomic DNA libraries, chromosome libraries, and cDNA libraries?
- How are genetic probes used?
- Why is the PCR technique valuable?

DNA ANALYSIS

Learning Objectives

5   Distinguish among DNA, RNA, and protein blotting.
6   Describe the chain termination method of DNA sequencing.

DNA cloning and the PCR technique have had major impacts on genetic analysis, because they provide enough genetic material to address certain fundamental questions about specific genes and their RNA or protein products. For example, when—and how— is a gene expressed? Is a given gene identical from one individual to the next? Is it different from one species to the next? Where is the gene found within the genome? In this and the following section we consider various techniques that help address these questions. We then examine some of the practical applications of DNA technology in the final section of the chapter.

Gel electrophoresis is used for separating macromolecules

Mixtures of certain macromolecules—proteins, polypeptides, DNA fragments, or RNA—are separated by **gel electrophoresis,** a method that exploits the fact that these molecules carry charged groups that cause them to migrate in an electrical field. Figure 15-8 illustrates gel electrophoresis of DNA molecules. Nucleic acids migrate through the gel toward the positive pole

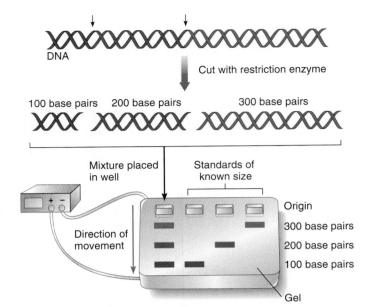

(a) A researcher sets up an electrical field in a gel material, consisting of agarose or polyacrylamide, which is poured as a thin slab on a glass or Plexiglas holder. After the gel has solidified, the researcher loads samples containing a mixture of macromolecules of different sizes in wells formed at one end of the gel, then applies an electrical current. The smallest DNA fragments (*green*) travel the longest distance.

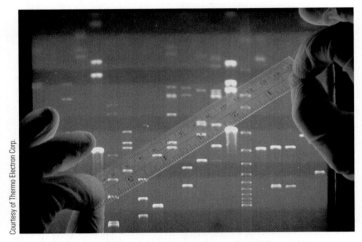

Courtesy of Thermo Electron Corp.

(b) A gel containing separated DNA fragments. The gel is stained with ethidium bromide, a dye that binds to DNA and is fluorescent under UV light.

Figure 15-8   Gel electrophoresis

of the electric field because they are negatively charged due to their phosphate groups. Because the gel slows down the large molecules more than the small molecules, the DNA is separated by size. Including DNA fragments of known size as standards ensures accurate measurements of the molecular weights of the unknown fragments. The gel is stained after the DNA fragments are separated to show their number and location.

DNA, RNA, and protein blots detect specific fragments

Geneticists can identify specific DNA fragments that hybridize with a complementary genetic probe. However, it is impossible to hybridize a probe to DNA fragments contained in a gel. For this reason the DNA is usually denatured and then transferred to a nitrocellulose or nylon membrane, which picks up the DNA like a blotter picks up ink. The resulting blot, which is essentially a replica of the gel, is incubated with a DNA probe, which hybridizes with any complementary DNA fragments. If the probe is radioactive, the blot is subjected to autoradiography. The resulting spots on the X-ray film correspond to the locations of the fragments in the gel that are complementary to the probe. Today, many probes are detected by chemical luminescence, which is analyzed by computer scanners, eliminating the need for autoradiography.

This method of detecting DNA fragments—separating them by gel electrophoresis and then transferring them to a nitrocellulose or nylon membrane—is called a **Southern blot,** named for its inventor, Edward M. Southern of Edinburgh University in Scotland (Fig. 15-9). The procedure has widespread applications. It is often used to diagnose certain types of genetic disorders. For example, in some cases the DNA of a mutant allele is detected because it migrates differently in the gel than the DNA of its normal counterpart does.

Similar blotting techniques are used to study RNA and proteins. When RNA molecules separated by electrophoresis are transferred to a membrane, the result is called, rather in jest, a **Northern blot.** In the same spirit, the term **Western blot** is applied to a blot consisting of proteins or polypeptides previously separated by gel electrophoresis. (So far, no one has invented a type of blot that could be called an Eastern blot.) In the case of Western blotting, scientists recognize the polypeptides of interest by labeled antibody molecules that bind to them specifically. Western blotting is used diagnostically to detect the presence of proteins specific to HIV-1, the AIDS virus.

Restriction fragment length polymorphisms are a measure of genetic relationships

The variability of genes within a population can be studied in several different ways. A traditional approach is based on the fact that random DNA mutations and recombination may result in individuals with different lengths of fragments produced by a given restriction enzyme. Such **restriction fragment length polymorphisms,** commonly known as **RFLPs,** or "riflips," can be used to determine how closely related different members of a population are. (The term *polymorphism* means "many forms" and refers to inherited differences found within a population.) A **genetic polymorphism** is said to exist if individuals of two or more discrete types, or "morphs," are found in a population.

Restriction enzymes are used to cut the DNA from two or more individuals, and the fragments are separated by gel electro-

phoresis, with the DNA from each individual in a separate lane; shorter fragments migrate farther than longer fragments. To visualize the fragments, researchers make a Southern blot of the DNA on the gel, which is then denatured and allowed to hybridize with a genetic probe. The resulting pattern of bands can then be compared.

RFLP analysis has been used when conducting paternity testing and when analyzing evidence at crime scenes. RFLP analysis has also been used to help map the exact location of gene mutations, such as the mutation that causes cystic fibrosis. Although the RFLP technique has provided useful information in many areas of biology, it is rapidly being replaced by newer methods, such as automated DNA sequencing (discussed in the next section).

One way to characterize DNA is to determine its sequence of nucleotides

Investigators use a cloned piece of DNA as a research tool for many different applications. For example, they may clone a gene to obtain the encoded protein for some industrial or pharmaceutical process. Regardless of the particular application, before they engineer a gene, researchers must know a great deal about the gene and how it functions. The usual first step is determining the sequence of nucleotides.

Methods to rapidly sequence DNA exist

In the 1990s, the advent of automated **DNA-sequencing** machines connected to powerful computers let scientists sequence huge amounts of DNA quickly and reliably. Automated DNA sequencing relies on the **chain termination method** that British biochemist Fred Sanger developed in 1974. In 1980, Sanger shared the Nobel Prize in Chemistry for this contribution. This was Sanger's second Nobel Prize; his first was for his work on the structure of the protein insulin. Although DNA sequencing is now fully automated, we briefly consider the essential steps of the chain termination method.

This method of DNA sequencing is based on the fact that a replicating DNA strand that has incorporated a modified synthetic nucleotide, known as a *dideoxynucleotide,* cannot elongate beyond that point. Unlike a "normal" deoxynucleotide, which lacks a hydroxyl group on its 2' carbon, a dideoxynucleotide also lacks a hydroxyl group on its 3' carbon (Fig. 15-10). Recall from Chapter 12 that a 3' hydroxyl group reacts each time a phosphodiester linkage is formed. Thus, dideoxynucleotides terminate elongation during DNA replication.

The researcher prepares four different reaction mixtures. Each contains multiple single-stranded copies of the DNA to be sequenced, DNA polymerase, appropriate radioactively labeled primers, and all four deoxynucleotides needed to synthesize DNA: dATP, dCTP, dGTP, and dTTP. Each mixture also includes a small amount of only one of the four dideoxynucleotides: ddATP, ddCTP, ddGTP, or ddTTP (Fig. 15-11a). The prefix "dd" refers to dideoxynucleotides, to distinguish them from deoxynucleotides, which are designated "d."

The Southern blotting technique allows DNA separated by electrophoresis to be transferred to a filter in the same location as it was on the gel. It can then be hybridized with radioactive complementary probes.

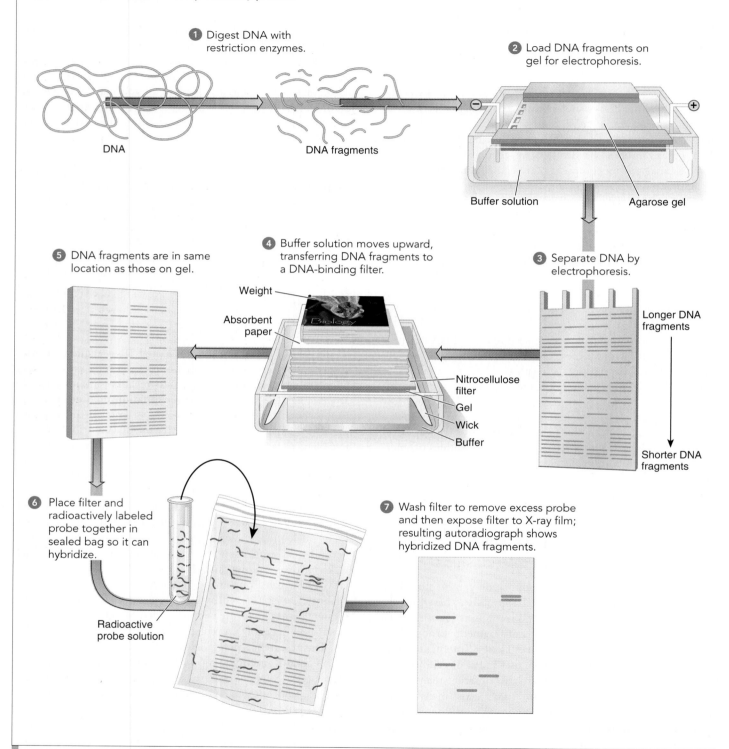

Figure 15-9 The Southern blotting technique

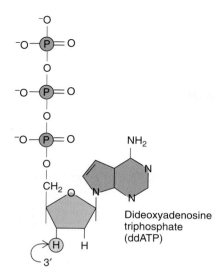

Figure 15-10 Dideoxynucleotide

Dideoxynucleotides are modified nucleotides that lack a 3′ hydroxyl group and thus block further elongation of a new DNA chain. (Note the hydrogen at the 3′ position instead of a hydroxyl group.)

Here's how the reaction proceeds in the mixture that includes ddATP (∥ Fig. 15-11b). At each site where adenine is specified, occasionally a growing strand incorporates a ddATP, halting elongation so a mixture of DNA fragments of varying lengths forms in the reaction mixture. Each fragment that contains a ddATP marks a specific location where adenine would normally be found in the newly synthesized strand. Similarly, in the reaction mixture that includes ddCTP, each fragment that contains ddCTP marks the position of a cytosine in the newly synthesized strand, and so on.

The radioactive fragments from each reaction are denatured and then separated by gel electrophoresis, with each reaction mixture, corresponding to A, T, G, or C, occupying its own lane in the gel. The positions of the newly synthesized fragments in the gel can then be determined by autoradiography (∥ Fig. 15-11c and d). Because the high resolution of the gel makes it possible to distinguish between fragments that differ in length by only a single nucleotide, the researcher can read off the sequence in the newly synthesized DNA one base at a time. For example, consider a film of a DNA-sequencing gel that shows the following:

A	C	G	T
—			
			—
		—	
	—		

Starting at the bottom, the sequence would be read as: 5′—C—G—T—A—3′.

The chain termination method is still used in automated DNA-sequencing machines. However, the method of detection has changed in that the nucleotide sequence is no longer visualized by radioactivity. Instead, researchers label each of the four ddNTPs with a different-colored fluorescent dye. The computer uses a laser to read the fluorescence of the dye markers as the bases emerge from the end of a lane of electrophoresis medium (∥ Fig. 15-12).

Entire genomes have been sequenced using automated DNA sequencing

Sequencing machines today are very powerful and can decode about 1.5 million bases in a 24-hour period. These advances in sequencing technology have made it possible for researchers to study the nucleotide sequences of entire genomes in a wide variety of organisms, both prokaryotic and eukaryotic. Much of this research received its initial impetus from the **Human Genome Project,** which began in 1990. The sequencing of the 3 billion base pairs of the human genome was essentially completed in 2001. The genomes of more than 100 different organisms have been sequenced now, and several hundred more are in various stages of planning or completion. We are in the middle of an extraordinary explosion of gene sequence data, largely due to automated sequencing methods.

DNA sequence information is now kept in large computer databases, many of which are accessed through the Internet. Examples are databases maintained by the National Center for Biotechnology Information and by the Human Genome Organization (HUGO). Geneticists use these databases to compare newly discovered sequences with those already known, to identify genes, and to access many other kinds of information. By searching for DNA or amino acid sequences in a database, researchers can gain a great deal of insight into the function and structure of the gene product, the evolutionary relationships among genes, and the variability among gene sequences within a population.

Review

■ What is a Southern blot?
■ What technique does automated DNA sequencing use? Outline the basic steps of this method.

∥ GENOMICS

Learning Objectives

7 Describe the three main areas of interest in genomics.
8 Explain what a DNA microarray does, and give an example of its research and medical potential.
9 Define *pharmacogenetics* and *proteomics.*

Now that we have sequenced the genomes of humans and many other species, what do we do with that information? **Genomics** is the emerging field of biology that studies the entire DNA sequence of an organism's genome to identify all the genes, determine their RNA or protein products, and ascertain how the genes are regulated. As you will see, genomics has important practical applications in addition to answering scientific questions.

Genomics has three main areas of interest: structural genomics, functional genomics, and comparative genomics. *Structural genomics* is concerned with mapping and sequencing genomes.

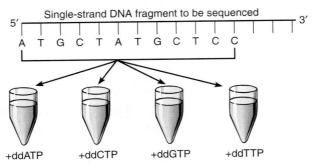

Single-strand DNA fragment to be sequenced

5′ — 3′

A T G C T A T G C T C C

+ddATP   +ddCTP   +ddGTP   +ddTTP

(a) Four different reaction mixtures are used to sequence a DNA fragment; each contains a small amount of a single dideoxynucleotide, such as ddATP. Larger amounts of the four normal deoxynucleotides (dATP, dCTP, dGTP, and dTTP) plus DNA polymerase and radioactively labeled primers are also included.

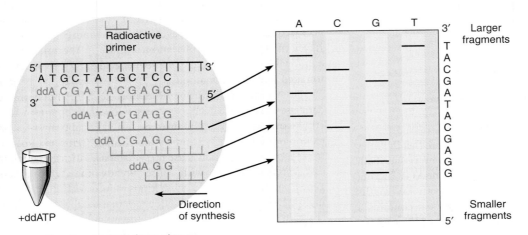

(b) The random incorporation of dideoxyATP (ddATP) into the growing chain generates a series of smaller DNA fragments ending at all the possible positions where adenine is found in the newly synthesized fragments. These correspond to positions where thymine occurs in the original template strand.

(c) The radioactive products of each reaction mixture are separated by gel electrophoresis and located by exposing the gel to X-ray film. The nucleotide sequence of the newly synthesized DNA is read directly from the film (5′ → 3′). The sequence in the original template strand is its complement (3′ → 5′).

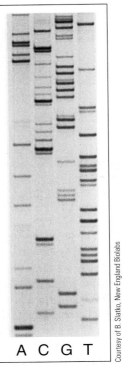

Courtesy of B. Slatko, New England Biolabs

(d) An exposed X-ray film of a DNA sequencing gel. The four lanes represent A, C, G, and T dideoxy reaction mixtures, respectively.

Figure 15-11   The chain termination method of DNA sequencing

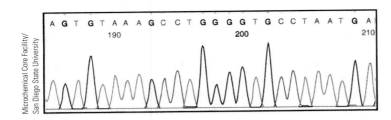

Microchemical Core Facility/ San Diego State University

Figure 15-12   *Animated*   A small segment of the printout of automated DNA-sequencing results

The computer produces the series of peaks shown as it reads the fluorescing bands on the gel. The nucleotide base adenine is shown in green, guanine in black, cytosine in blue, and thymine in red. The DNA sequence assigned by the computer appears at the top of the printout.

Functional genomics deals with the functions of genes and non-gene sequences in genomes. *Comparative genomics* involves comparing the genomes of different species to determine what changes in DNA have occurred during the course of evolution. This knowledge furthers our understanding of evolutionary relationships.

Identifying protein-coding genes is useful for research and for medical applications

The majority of the human genome contains DNA sequences that do not code for proteins. Although the entire genome is of interest to biologists, the protein-coding regions are of particular interest to medical researchers. Scientists have compiled tens of thousands of short (25 to 30 nucleotides) cDNA sequences, known as **expressed sequence tags (ESTs),** that can help identify protein-coding genes.

A large data set containing thousands of human ESTs is available to scientists via the Internet. Using gene-hunting software, biologists working on a particular DNA sequence can access the data set and match their DNA sequence to a known EST sequence.

Many ESTs in the data set have been identified by comparing them to known DNA sequences from other species. For example, the human telomerase gene *TRT* was identified by matching its human EST sequence to that of a known yeast telomerase sequence. (Recall from Chapter 12 that **telomerase** adds repetitive nucleotide sequences to the ends, or telomeres, of eukaryotic chromosomes.)

One way to study gene function is to silence genes one at a time

RNA interference, first discovered in the mid-1990s, can be used to quickly determine the function of a specific gene. You may recall from Chapter 13 that **RNA interference (RNAi)** is caused by small RNA molecules that interfere with the expression of genes or, more precisely, their mature mRNA transcripts. RNA interference involves small interfering RNAs, microRNAs, and a few other kinds of short RNA molecules. Biologists hypothesize that RNAi, which occurs in many kinds of organisms, originally evolved in nature because it protected cells against viruses that have double-stranded RNA molecules (see Chapter 24).

After a protein-coding gene is identified, perhaps by using ESTs, the function of that gene can be studied by using RNA interference to shut the gene off. To do this, biologists synthesize a short stretch of RNA that is complementary to part of the DNA sequence of the gene being examined. After the gene is silenced, biologists observe any changes in the phenotype to help determine the function of the missing protein.

Gene targeting also reveals gene function

Another research tool that reveals gene function is **gene targeting,** a procedure in which the researcher chooses and "knocks out," or inactivates, a single gene in an organism. *Knockout mice* are common models for studying the roles of genes of unknown function in mammals, including humans. The roles of the inactivated gene are determined by observing the phenotype of the mice bearing the knockout gene. If the gene codes for a protein, for example, studying individuals who lack that protein helps the researcher identify the function of the protein. Because 98% to 99% of the loci of mice have human counterparts, information about knockout genes in mice also provides details about human genes.

Unlike RNA interference, gene targeting is a lengthy procedure. It takes about a year for scientists to develop a new strain of knockout mice through gene targeting. A nonfunctional, or knockout, gene is cloned and introduced into mouse **embryonic stem cells (ES cells)** (see Chapter 1). ES cells are particularly easy to handle because, like cancer cells, they grow in culture indefinitely. Most important, if placed in a mouse embryo, ES cells divide and produce all the cell types normally found in the mouse. In a tiny fraction of these ES cells, the introduced knockout gene becomes physically associated with the corresponding gene in a chromosome. If this occurs, the gene on the chromosome and the knockout gene tend to exchange DNA segments in a process called *homologous recombination.* In this way, the knockout allele replaces the normal allele in the mouse chromosome.

Researchers inject into early mouse embryos ES cells they hope are carrying a knockout gene and let the mice develop to maturity. The researchers generally study animals that carry the knockout gene in every cell. However, because many genes are essential to life, researchers have modified the knockout technique to develop strains in which a specific gene is selectively inactivated in only one cell type. Research laboratories around the world have developed thousands of different strains of knockout mice, each displaying its own characteristic phenotype, and the number continues to grow.

Gene targeting in mice is providing answers to basic biological questions relating to the development of embryos, the development of the nervous system, and the normal functioning of the immune system. This technique has great potential for revealing more about various human diseases, especially because many diseases have a genetic component. Geneticists are using gene targeting to study cancer, heart disease, sickle cell anemia, respiratory diseases such as cystic fibrosis, and other disorders.

Mutagenesis screening reveals the genes involved in a particular phenotype

Many large-scale mutagenesis screening projects in mice and other organisms are currently underway. In **mutagenesis screening,** researchers treat male mice with chemical *mutagens* that cause mutations in DNA. They then breed the males and screen their offspring for unusual phenotypes. Unlike gene targeting, mutagenesis does not disable a gene completely; instead, it causes small, random mutations that change the properties of the proteins that the DNA encodes. The scale of mutagenesis screening is enormous; one screening project in Germany has screened some 28,000 different mouse mutants for changes in phenotypes.

DNA microarrays are a powerful tool for studying how genes interact

DNA microarrays (also known as *DNA chips* or *gene chips*) provide a way to study patterns of gene expression. A DNA microarray consists of hundreds of different DNA molecules placed on a glass slide or chip (❚ Figure 15-13). In ❶, a mechanical robot prepares the microarray. It spots each location on the grid with thousands to millions of copies of a specific *complementary DNA (cDNA)* strand. The single-stranded cDNA molecules for each spot, known as a *microdot,* are made using reverse transcriptase from a specific mRNA and then amplified using the polymerase chain reaction.

In ❷, researchers isolate mature mRNA molecules from two cell populations—for example, liver cells treated with a newly

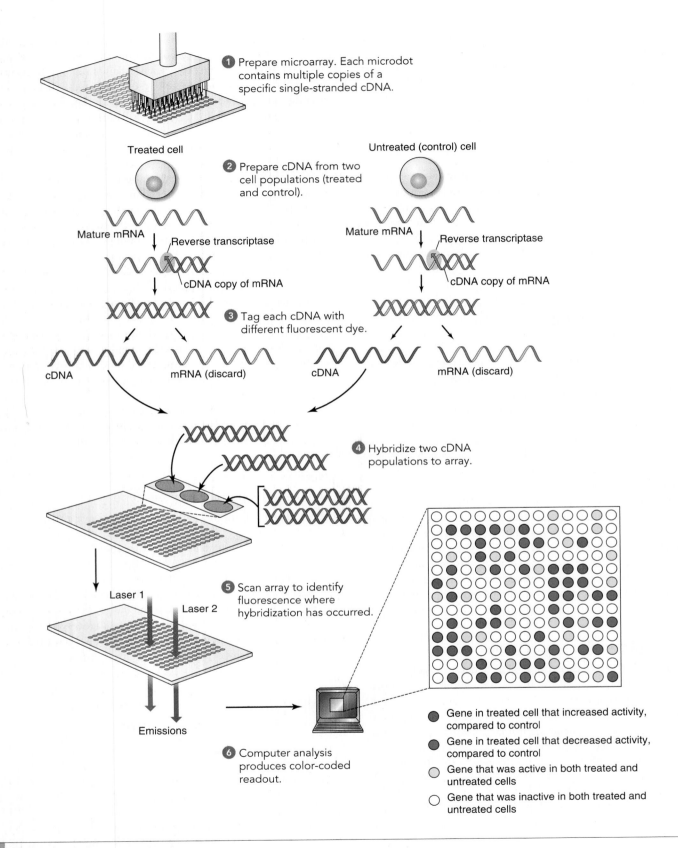

1. Prepare microarray. Each microdot contains multiple copies of a specific single-stranded cDNA.

Treated cell

Untreated (control) cell

2. Prepare cDNA from two cell populations (treated and control).

Mature mRNA

Reverse transcriptase

cDNA copy of mRNA

Mature mRNA

Reverse transcriptase

cDNA copy of mRNA

3. Tag each cDNA with different fluorescent dye.

cDNA

mRNA (discard)

cDNA

mRNA (discard)

4. Hybridize two cDNA populations to array.

5. Scan array to identify fluorescence where hybridization has occurred.

Laser 1

Laser 2

Emissions

6. Computer analysis produces color-coded readout.

- Gene in treated cell that increased activity, compared to control
- Gene in treated cell that decreased activity, compared to control
- Gene that was active in both treated and untreated cells
- Gene that was inactive in both treated and untreated cells

Figure 15-13 A DNA microarray

developed drug and control liver cells not treated with the drug. Liver cells are often tested because the liver produces many enzymes that metabolize foreign molecules, including drug molecules. Drugs the liver cannot metabolize, or metabolizes weakly, may be too toxic to be used.

Researchers use the isolated mRNA molecules to prepare cDNA molecules for each cell population, which ❸ are then tagged with different-colored fluorescent dyes. For example, the treated cells' cDNA molecules might be labeled with red dye and the untreated cells' cDNA molecules with green dye. In ❹, researchers add the two cDNA populations to the array, and some of the cDNA subsequently hybridizes (forms base pairs with) the cDNA on the array.

After washing it to remove any cDNA that has not hybridized to the array, ❺ investigators scan the array with lasers to identify red and green fluorescence where hybridization has occurred. In ❻, computer analysis of the ratio of red to green fluorescence at each spot in the array produces a color-coded readout that researchers can further analyze. For example, a medical researcher might compare the experimental drug's overall pattern of gene activity with that of known drugs and toxins. If the gene activity for the treated cells matches that of a toxin that damages the liver, the drug will probably not go into clinical trials.

DNA microarrays enable researchers to compare the activities of thousands of genes in normal and diseased cells from tissue samples. Because cancer and other diseases exhibit altered patterns of gene expression, DNA microarrays have the potential to identify disease-causing genes or the proteins they code for, which can then be targeted by therapeutic drugs.

Here's an example of the application. In 2002, medical researchers identified 17 genes that are active in different kinds of diffuse large B-cell lymphoma (a cancer of white blood cells). Using DNA microarrays, researchers determined which combination of those 17 genes is active in each subtype of this form of cancer. Armed with the knowledge of which cancer subtype a patient has, physicians can choose the treatment that will probably be most effective for that patient. Examining patterns of gene activity with DNA microarrays also helps medical scientists identify which patients will probably remain free of cancer after treatment and which will probably relapse.

The Human Genome Project stimulated studies on the genome sequences of other species

To aid in analyzing the human genome, investigators carried out comparative genomics by sequencing and mapping studies simultaneously on the mouse and rat genomes. Mice and rats were obvious choices for DNA sequencing because they have been studied for almost a century and much is known about their biology. Mice and rats are sufficiently different from humans that any conserved (that is, identical) sequences found in both rodents and humans are probably functionally important.

In addition, mice are similar enough to humans that both share many physiological traits, including some of the same diseases. To demonstrate the shared similarity between the mouse and human genomes, a consortium of scientists published a study in 2002 that compared the mouse chromosome 16 with the human genome. Only 14 of the 731 known genes on the mouse chromosome had no human counterpart; the remaining 717 genes appeared in one form or another in the human genome.

Comparison of the DNA sequences and chromosome organization of related genes from different species is a powerful tool for identifying the elements essential for their functions. If a human gene has an unknown function, researchers can often deduce clues about its role by studying the equivalent gene in another species, such as a mouse or rat.

Investigators have also sequenced the genomes of other model vertebrate organisms, such as the puffer fish, which has the smallest known genome of all vertebrates, and zebrafish. By comparing the human genome to that of the puffer fish, researchers identified shared sequences that have not changed in several hundred million years. Scientists hypothesize that these sequences may be genes or regulatory elements that are essential in all vertebrates. Further study of the puffer fish and zebrafish genomes may help scientists identify the genes that direct vertebrate development. Biologists have used the zebrafish for many years as a model of vertebrate development because its embryonic stages are transparent and therefore convenient to examine.

Genome analysis of commercially important organisms, such as salmon, chickens, pigs, and rice, is also a priority. Scientists have published the rice genome, the genome of the parasite that causes malaria, and the genome of the parasite that causes African sleeping sickness. One of the most significant genome-sequencing efforts focuses on nonhuman primates, such as the chimpanzee. Comparing the DNA sequences of humans and our closest living relatives is helping biologists understand the genetic changes that occurred during human evolution, including which genes govern mental and linguistic capabilities.

The Human Genome Project has enormous implications for the future

Knowledge of the human genome may revolutionize human health care, and some experts predict that in highly developed countries, the average human life expectancy at birth will be 90 to 95 years by 2050. As a comparison, the current average life expectancy in the United States is 77 years. Nearly every week investigators announce new health-related information about the human genome, such as identifying genes associated with hypertension or with specific cancers.

Several scientific fields have emerged in the wake of the Human Genome Project. These include bioinformatics, pharmacogenetics, and proteomics.

Bioinformatics. We have already mentioned that the amount of genomic data now available requires that it be evaluated using powerful computers. The discipline known as **bioinformatics,** or *biological computing,* includes the storage, retrieval, and comparison of DNA sequences within a given species and among different species. Bioinformatics uses powerful computers and sophisticated software to manage and analyze large amounts of data generated by sequencing and other technologies. For example, as researchers determine new DNA sequences, automated computer programs scan the sequences for patterns typically found in genes. Through comparison of the databases of DNA sequences from different organisms, bioinformatics has already led to insights about gene identification, gene function, and evolutionary relationships.

Pharmacogenetics. The new science of gene-based medicine known as **pharmacogenetics** customizes drugs to match a patient's genetic makeup. Currently, a physician does not know in advance whether a particular medication will benefit a patient or cause severe side effects. An individual's genes, especially those that code for drug-metabolizing enzymes, largely determine that person's response to a specific drug. Pharmacogenetics takes into account the subtle genetic differences among individuals. In as few as 5 to 10 years, patients may take routine genetic screening tests before a physician prescribes a drug.

Pharmacogenetics, like other fields in human genetics, presents difficult ethical questions. The genetic testing that will be an everyday part of pharmacogenetics raises issues of privacy, genetic bias, and potential discrimination. We consider these issues in Chapter 16.

Together, the Human Genome Project and pharmacogenetics will advance knowledge of the role of genes in human health and disease. However, individualized genetic testing may cause people to worry unnecessarily about a genetic disease for which they may never develop symptoms. Most common diseases result from a complex interplay between multiple genes and nongenetic, or environmental, factors. Recall from Chapter 11 that the environment is an important factor influencing gene expression. In humans, healthy environmental factors include proper diet, adequate exercise, and not smoking (see *Focus On: Unwelcome Tissues: Cancers,* in Chapter 38).

Proteomics. The study of all the proteins encoded by the human genome and produced in a person's cells and tissues is **proteomics.** Scientists want to identify all the proteins made by a given kind of cell, but the process is much more complicated than sequencing the human genome. Some genes, for example, encode several different proteins (see Fig. 14-12, differential mRNA processing). Also, every somatic cell in the human body has essentially the same genome, but cells in different tissues vary greatly in the kinds of proteins they produce. Protein expression patterns vary not only in different tissues but also at different stages in the development of a single cell.

Scientists want to understand the role of each protein in a cell, the way the various proteins interact, and the 3-D structure for each protein. While advancing biological knowledge, these goals also promise advances in medicine. If they know the shape of a protein associated with a type of cancer or other disease, pharmacologists may be able to develop drugs that bind to active sites on that protein, turning its activity off. Today, the pharmaceutical industry has drugs that target about 500 proteins in the cell. However, researchers estimate proteomics may yield 10,000 to 20,000 additional protein targets.

Review

- How would you distinguish structural, functional, and comparative genomics?
- What is a DNA microarray?
- What is pharmacogenetics? Proteomics?

APPLICATIONS OF DNA TECHNOLOGIES

Learning Objective

10 Describe at least one important application of recombinant DNA technology in each of the following fields: medicine and pharmacology, DNA fingerprinting, and transgenic organisms.

Recombinant DNA technology has provided both a new and unique set of tools for examining fundamental questions about cells and new approaches to applied problems in many other fields. In some areas, the production of genetically engineered proteins and organisms has begun to have considerable impact on our lives. The most striking of these has been in the fields of pharmacology and medicine.

DNA technology has revolutionized medicine and pharmacology

In increasing numbers of cases, doctors perform **genetic tests** to determine whether an individual has a particular genetic mutation associated with such disorders as Huntington's disease, hemophilia, cystic fibrosis, Tay–Sachs disease, and sickle cell anemia. **Gene therapy,** the use of specific DNA to treat a genetic disease by correcting the genetic problem, is another application of DNA technology that is currently in its infancy. Because genetic testing and gene therapy focus almost exclusively on humans, these applications of DNA technology are discussed in Chapter 16. We focus our discussion here on the use of DNA technology to produce pharmaceutical products.

Human *insulin* produced by *E. coli* became one of the first genetically engineered proteins approved for use by humans. Before the use of recombinant DNA techniques to generate genetically altered bacteria capable of producing the human hormone, insulin was derived exclusively from other animals. Many diabetic patients become allergic to the insulin from animal sources, because its amino acid sequence differs slightly from that of human insulin. The ability to produce the human hormone by recombinant DNA methods has resulted in significant medical benefits to insulin-dependent diabetics.

Genetically engineered human *growth hormone (GH)* is available to children who need it to overcome growth deficiencies, specifically pituitary dwarfism (see Chapter 48). In the past, human GH was obtainable only from cadavers. Only small amounts were available, and evidence suggested that some of the preparations from human cadavers were contaminated with infectious agents similar to those that cause bovine spongiform encephalopathy, or mad cow disease.

The list of products produced by genetic engineering is continually growing. For example, *tissue plasminogen activator (TPA),* a protein that prevents or dissolves blood clots, is used to treat cardiovascular disease. If administered shortly after a heart attack, TPA reduces the risk of a subsequent attack. *Tissue growth factor-beta (TGF-β)* promotes the growth of blood vessels and skin and is used in wound and burn healing. Researchers also use TGF-β in **tissue engineering,** a developing technology to meet the pressing need for human tissues and, eventually, organs for transplantation by growing them from cell cultures. The U.S. Food and Drug Administration (FDA) has approved tissue-engineered skin grafts, and tissue-engineered cartilage is in clinical trials. Hemophilia A is treated with *human clotting factor VIII*. Before the development of recombinant DNA techniques, factor VIII was available only from human- or animal-derived blood, which posed a risk of transmitting infectious agents, such as HIV. *Dornase Alpha (DNase)* improves respiratory function and general well-being in people with cystic fibrosis.

Recombinant DNA technology is increasingly used to produce vaccines that provide safe and effective immunity against infectious diseases in humans and animals. One way to develop a recombinant vaccine is to clone a gene for a surface protein produced by the disease-causing agent, or pathogen; the researcher then introduces the gene into a nondisease-causing vector. The vaccine, when delivered into the human or animal host, stimulates an immune response to the surface-exposed protein. As a result, if the pathogen carrying that specific surface protein is encountered, the immune system targets it for destruction. Human examples of antiviral recombinant vaccines are vaccines for influenza A, hepatitis B, and polio. Recombinant vaccines are also being developed against certain bacterial diseases and human cancers.

DNA fingerprinting has numerous applications

The analysis of DNA extracted from an individual, which is unique to that individual, is known as **DNA fingerprinting,** also called **DNA typing.** DNA fingerprinting has many applications in humans and other organisms, as in the following examples:

1. Analyzing evidence found at crime scenes (forensic analysis) (❚ Fig. 15-14)
2. Identifying mass disaster victims
3. Proving parentage in dogs for pedigree registration purposes
4. Identifying human cancer cell lines
5. Studying endangered species in conservation biology
6. Tracking tainted foods
7. Studying the genetic ancestry of human populations
8. Clarifying disputed parentage

DNA fingerprinting relies on PCR amplification and gel electrophoresis to detect molecular markers. The most useful molecular markers are highly polymorphic within the human population (recall our earlier discussion of *polymorphism*). **Short tandem repeats (STRs)** are molecular markers that are short sequences of repetitive DNA—up to 200 nucleotide bases with a simple repeat pattern such as GTGTGTGTGT or CAGCAGCAGCAG. STRs are highly polymorphic because they vary in length from one individual to another, and this characteristic makes them useful with a high degree of certainty in identifying individuals.

If enough markers are compared, the odds that two people taken at random from the general population would have identical DNA profiles may be as low as one in several billion. The FBI

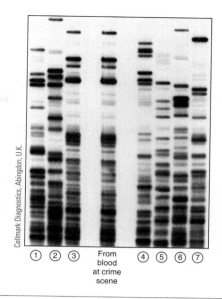

Figure 15-14 *Animated* DNA fingerprinting

These DNA fingerprints show DNA from a crime scene (*in middle*), along with the DNA profiles of seven suspects. Can you pick the suspect whose DNA profile matches blood from the crime scene?

uses a set of STRs from 13 different markers to establish a unique DNA profile for an individual. Such a profile distinguishes that person from every other individual in the United States, except an identical twin.

Recall that DNA is also found in the cell's mitochondria. Whereas nuclear DNA occurs in two copies per cell—one on each of the homologous chromosomes—mitochondrial DNA has as many as 100,000 copies per cell. Therefore, mitochondrial DNA is the molecule of choice for DNA fingerprinting in which biological samples have been damaged—identifying exhumed human remains, for example.

DNA fingerprinting has revolutionized law enforcement. The FBI established the Combined DNA Index System (CODIS) in 1990, consisting of DNA databases from all 50 states. A DNA profile of an unknown suspect can be compared to millions of DNA profiles of convicted offenders in the database, often resulting in identification of the suspect. The DNA from the unknown suspect may come from blood, semen, bones, teeth, hair, saliva, urine, or feces left at the crime scene. Tiny amounts of human DNA have even been extracted from cigarette butts, licked envelopes or postage stamps, dandruff, fingerprints, razor blades, chewing gum, wristwatches, earwax, debris from under fingernails, and toothbrushes.

If applied properly, DNA fingerprinting has the power to identify the guilty and exonerate the innocent. Hundreds of convicted individuals have won new trials and have been subsequently released from incarceration, based on correlating DNA profiles with physical evidence from the crime scene. Such evidence has been ruled admissible in many court cases, including certain trials that have received a great deal of attention in recent years. One limitation arises from the fact that the DNA samples are usually small and may have been degraded. Obviously, great care must be taken to prevent contamination of the samples. This is especially crucial if the PCR technique is used to amplify DNA.

Transgenic organisms have incorporated foreign DNA into their cells

Plants and animals in which foreign genes have been incorporated are referred to as **transgenic organisms.** Researchers use varied approaches to insert foreign genes into plant or animal cells. They often use viruses as vectors, although other methods, such as the direct injection of DNA into cells, have also been applied.

Transgenic animals are valuable in research

Transgenic animals are usually produced by injecting the DNA of a particular gene into the nucleus of a fertilized egg cell or of ES cells. The researcher then implants the eggs into the uterus of a female and lets them develop. Alternatively, the researcher injects genetically modified ES cells into isolated *blastocysts,* an early stage in embryonic development, and then implants them into a foster mother.

Injecting DNA into cells is not the only way to produce transgenic animals. Researchers may use viruses as recombinant DNA vectors. RNA viruses called **retroviruses** make DNA copies of themselves by reverse transcription. Sometimes the DNA copies become integrated into the host chromosomes, where they are replicated along with host DNA.

Transgenic animals provide valuable applications over a wide range of research areas, such as regulation of gene expression, immune system function, genetic diseases, viral diseases, and genes involved in the development of cancer. The laboratory mouse is a particularly important model organism for these studies.

In a classic study of the control of gene expression, University of Pennsylvania geneticist Ralph L. Brinster in 1983 produced transgenic mice carrying a gene for rat growth hormone (see Fig. 17-18). Brinster and his coworkers wanted to understand the controls that let certain genes be expressed in some tissues and not in others. The pituitary gland of a mouse normally produces small amounts of GH, and these researchers reasoned that other tissues might also be capable of producing the hormone. First they isolated the GH gene from a library of genomic rat DNA. They combined it with the promoter region of a mouse gene that normally produces metallothionein, a protein that is active in the liver and whose synthesis is stimulated by the presence of toxic amounts of heavy metals such as zinc. The researchers used metallothionein regulatory sequences as a switch to turn the production of rat GH on and off at will. After injecting the engineered gene into mouse embryo cells, they implanted the embryos into the uterus of a mouse and let them develop.

Because of the difficulty in manipulating the embryos without damaging them, the gene transplant succeeded in only a small percentage of the animals. When exposed to small amounts of zinc, these transgenic mice produced large amounts of growth hormone, because the liver is a much larger organ than the pituitary gland. The mice grew rapidly, and one mouse, which developed from an embryo that had received two copies of the GH gene, grew to more than double the normal size (Fig. 15-15). As might be expected, such mice often transmit their increased growth capability to their offspring.

Figure 15-15 A transgenic mouse

The mouse on the right is normal, whereas the mouse on the left is a transgenic animal that expresses rat growth hormone.

Transgenic animals can produce genetically engineered proteins

Certain transgenic animal strains produce foreign proteins of therapeutic or commercial importance that are secreted into milk. For example, researchers have introduced into sheep the gene for the human protein tissue plasminogen activator (TPA). Producing transgenic livestock, such as pigs, sheep, cows, and goats, that secrete foreign proteins in their milk is known informally as "pharming," a combination of "pharmaceuticals" and "farming" (Fig. 15-16).

In pharming, recombinant genes are fused to the regulatory sequences of the milk protein genes, and such genes are therefore activated only in mammary tissues involved in milk production. The advantages of obtaining the protein from milk are that potentially it can be produced in large quantities and that it can be harvested simply by milking the animal. The protein is then purified from the milk. The introduction of the gene does not harm the animals, and because the offspring of the transgenic animal usually produce the recombinant protein, transgenic strains are established.

Transgenic plants are increasingly important in agriculture

People have selectively bred plants for thousands of years. The success of such efforts depends on desirable traits in the variety of plant selected or in closely related wild or domesticated plants whose traits are transferred by crossbreeding. Local varieties or closely related species of cultivated plants often have traits, such as disease resistance, that agriculturalists could advantageously

Figure 15-16 Transgenic "pharm" cows

These cows contain a human gene that codes for lactoferrin, a protein found in human mothers' milk and in secretions such as tears, saliva, bile, and pancreatic fluids. Lactoferrin is one of the immune system's lines of defense against disease-causing organisms. The cows secrete human lactoferrin in their milk.

introduce into varieties more useful to modern human needs. If genes are introduced into plants from strains or species with which they do not ordinarily interbreed, the possibilities for improvement increase greatly.

The most widely used vector system for introducing recombinant genes into many types of plant cells is the crown gall bacterium, *Agrobacterium tumefaciens.* This bacterium normally produces plant tumors by introducing a plasmid, called the *Ti plasmid,* into the cells of its host; *Ti* stands for "tumor-inducing." The Ti plasmid induces abnormal growth by forcing the plant cells to produce elevated levels of a plant growth hormone called *cytokinin* (discussed in Chapter 37).

Geneticists "disarm" the Ti plasmid so that it does not induce tumor formation and then use it as a vector to insert genes into plant cells. The cells into which the altered plasmid is introduced are essentially normal except for the inserted genes. Genes placed in the plant genome in this fashion may be transmitted sexually, via seeds, to the next generation, but they can also be propagated asexually, for example, by taking cuttings.

Unfortunately, not all plants take up DNA readily, particularly the cereal grains that are a major food source for humans. One useful approach has been the development of a genetic "shotgun." Researchers coat microscopic gold or tungsten fragments with DNA and then shoot them into plant cells, penetrating the cell walls. Some of the cells retain the DNA, which transforms them. Those cells can then be cultured and used to regenerate an entire plant (see Fig. 17-2). Geneticists have successfully used such an approach to transfer a gene for resistance to a bacterial disease into cultivated rice from one of its wild relatives.

An additional complication of plant genetic engineering is that about 120 plant genes lie in the DNA of the chloroplasts;

the other 3000 or so genes that plastids require to function are in the nucleus. Chloroplasts are essential in photosynthesis, which is the basis for plant productivity. Because great agricultural potential hinges on improving the productivity of photosynthesis, developing methods for changing the part of the plant's DNA within the chloroplast is desirable. Dozens of labs are currently studying methods of chloroplast engineering, although progress has been slow. In 2001, Australian plant physiologists reported that they had altered **rubisco,** the key carbon-fixing enzyme of photosynthesis, by changing one of the genes in the chloroplast genome.

The United States is the world's top producer of transgenic crops, also known as **genetically modified (GM) crops.** Currently, 80% of the U.S. soybean crop, 70% of its cotton, and 38% of its corn are GM crops.

Selected applications of transgenic plants. Agricultural geneticists are developing GM plants that are resistant to insect pests, viral diseases, drought, heat, cold, herbicides, and salty or acidic soil. For example, consider the European corn borer, which is the most damaging insect pest in corn in the United States and Canada (❙ Fig. 15-17a). Efforts to control the European corn borer cost farmers more than $1 billion each year. Corn has been genetically modified to contain the *Bt* gene, a bacterial gene that codes for a protein with insecticidal properties; *Bt* stands for the bacterium's scientific name, *Bacillus thuringiensis.* Bt corn, introduced in the United States in 1996, does not need periodic sprays of chemical insecticides to control the European corn borer.

DNA technology also has the potential to develop crops that are more nutritious. For example, in the 1990s geneticists engineered rice to produce high quantities of β-carotene, which the human body uses to make vitamin A (❙ Fig. 15-17b; see also Fig. 3-14). In developing countries, vitamin A deficiency is a leading cause of blindness in children. According to the World Health Organization, 275 million children are vitamin A deficient, and 250,000 to 500,000 become irreversibly blind each year. Vitamin A deficiency also makes children more susceptible to measles and other infectious diseases. Because rice is the staple diet in many countries with vitamin A deficiency, the widespread use of GM rice with β-carotene has the potential to prevent blindness in many of the world's children.

Like some transgenic animals, certain transgenic plants can potentially be "pharmed" to produce large quantities of medically important proteins, such as antibodies against the herpes virus. Methods for developing transgenic plants are well established, but it has been difficult to get plants to produce the desired protein in large enough quantities. To date, most transgenic plants developed for pharming have produced foreign proteins equal to only about 1% of the plant's total protein output. The developers of this technology are working to increase the production of foreign proteins to demonstrate the feasibility of these production methods.

Some people are concerned about the health effects of consuming foods derived from GM crops and think that such foods should be restricted. For example, critics say that some consumers may develop food allergies. Scientists as well recognize this concern and routinely screen new GM crops for allergenicity. There

(a) The European corn borer, shown in the larval form, is the most destructive pest on corn in North America. Genetic engineers have designed *Bt* corn to control the European corn borer without heavy use of chemical insecticides.

(b) "Golden rice," shown here intermixed with regular white rice, contains high concentrations of β-carotene. To make golden rice, scientists insert the gene for carotene from daffodil flowers into the endosperm cells of rice.

Figure 15-17 *Animated* Uses of transgenic plants

is also an ongoing controversy as to whether GM foods should be labeled. The FDA and most scientists think such labeling would be counterproductive, because it would increase public anxiety over a technology that is as safe as traditional breeding methods. In 1996, the U.S. Court of Appeals upheld the FDA position that labeling should not be required.

Review

- Why has the production of human insulin by recombinant DNA methods had significant medical advantages for diabetics?
- What are short tandem repeats (STRs), and why are they so useful in DNA fingerprinting?
- Why do gene targeting and mutagenesis screening in mice have potential benefits for humans?

DNA TECHNOLOGY HAS RAISED SAFETY CONCERNS

Learning Objective

11 Describe at least two safety issues associated with recombinant DNA technology, and explain how these issues are being addressed.

When recombinant DNA technology was introduced in the early 1970s, many scientists considered the potential misuses at least as significant as the possible benefits. The possibility that an organism with undesirable environmental effects might be accidentally produced was of great concern. Scientists feared that new strains of bacteria or other organisms, with which the world has no previous experience, might be difficult to control. The scientists who developed the recombinant DNA methods insisted on stringent guidelines for making the new technology safe.

Experiments in thousands of university and industrial laboratories over more than 30 years have shown that recombinant DNA manipulations can be carried out safely. Geneticists have designed laboratory strains of *E. coli* to die in the outside world. Researchers carry out experiments that might present unusual risks in facilities designed to hold pathogenic organisms; this precaution ensures that researchers can work with them safely. So far, no evidence suggests that researchers have accidentally cloned hazardous genes or have released dangerous organisms into the environment. However, malicious *intentional* manipulations of dangerous genes certainly remain a concern.

As the safety of the experiments has been established, scientists have relaxed many of the restrictive guidelines for using recombinant DNA. Stringent restrictions still exist, however, in certain areas of recombinant DNA research where there are known dangers or where questions about possible effects on the environment are still unanswered.

These restrictions are most evident in research that proposes to introduce transgenic organisms into the wild, such as agricultural strains of plants whose seeds or pollen might spread in an uncontrolled manner. A great deal of research now focuses on determining the effects of introducing transgenic organisms into a natural environment. Carefully conducted tests have shown that transgenic organisms are not dangerous to the environment simply because they are transgenic.

However, it is important to assess the risks of each new recombinant organism. Scientists determine whether the organism has characteristics that might cause it to be environmentally hazardous under certain conditions. For example, if geneticists have engineered a transgenic crop plant to resist an herbicide, could that gene be transferred, via pollen or some other route, to that plant's weedy relatives, generating herbicide-resistant "superweeds"? In 2003, ecologists at the University of Tennessee, Knoxville, announced the crossing of transgenic oilseed rape plants that contained the *Bt* gene with their wild relative, which is a weed. They crossed the resulting hybrids with the wild relative again and then tested its ability to compete with other weeds in a field of wheat. The transgenic weed was a poor competitor and had less effect on wheat production than its wild relatives in a control field. These results, although encouraging, must be interpreted with care. Scientists must evaluate each transgenic crop plant individually to see if there is gene flow to wild relatives, and if so, what effect might result.

Other concerns relate to plants engineered to produce pesticides, such as the *Bt* toxin. The future of the *Bt* toxin in transgenic crops is not secure, because low levels of the insecticide could potentially provide ideal conditions for selection for resistant individuals in the insect population. It appears certain insects may evolve genetic resistance to the *Bt* toxin in transgenic plants in the same way they evolve genetic resistance to chemical insecticides.

Another concern is that nonpest species could be harmed. For example, people paid a great deal of attention to the finding that monarch butterfly larvae raised in the laboratory are harmed if they are fed pollen from *Bt* corn plants. Although more recent studies suggest that monarch larvae living in a natural environment do not consume enough pollen to cause damage, such concerns persist and will have to be addressed individually.

Environmental concerns about transgenic animals also exist. Several countries are in the process of developing fast-growing transgenic fish, usually by inserting a gene that codes for a growth hormone. Transgenic Atlantic salmon, for example, grow up to

6 times as fast as nontransgenic salmon grown for human consumption. The transgenic fish do not grow larger than other fish, just faster. The benefits of such genetically enhanced fish include reduced pressure on wild fisheries and less pollution from fish farms. However, if the transgenic fish escaped from the fish farm, what effect would they have on wild relatives? To address this concern, the researchers are developing only nonreproducing female transgenic salmon.

To summarize, DNA technology in agriculture offers many potential benefits, including higher yields by providing disease resistance, more nutritious foods, and the reduced use of chemical pesticides. However, like other kinds of technology, genetic engineering poses some risks, such as the risk that genetically modified plants and animals could pass their foreign genes to wild relatives, causing unknown environmental problems. The science of **risk assessment,** which uses statistical methods to quantify risks so they can be compared and contrasted, will help society decide whether to ignore, reduce, or eliminate specific risks of genetically engineered organisms.

Review

- What is the potential problem with transgenic salmon? How are scientists addressing this concern?

SUMMARY WITH KEY TERMS

Learning Objectives

1 Explain how a typical restriction enzyme cuts DNA molecules, and give examples of the ways in which these enzymes are used in recombinant DNA technology (page 323).

- **Recombinant DNA technology** isolates and amplifies specific sequences of DNA by incorporating them into **vector** DNA molecules. Researchers then **clone**—propagate and amplify—the resulting recombinant DNA in organisms such as *E. coli*.

- Researchers use **restriction enzymes** to cut DNA into specific fragments. Each type of restriction enzyme recognizes and cuts DNA at a highly specific base sequence. Many restriction enzymes cleave DNA sequences to produce complementary, single-stranded sticky ends.

ThomsonNOW™ **Learn more about restriction enzymes by clicking on the figure in ThomsonNOW.**

- Geneticists may use recombinant DNA vectors from naturally occurring circular bacteria DNA molecules called **plasmids** or from bacterial viruses called **bacteriophages.**

- Geneticists often construct recombinant DNA molecules by allowing the ends of a DNA fragment and a vector, both cut with the same restriction enzyme, to associate by complementary base pairing. Then **DNA ligase** covalently links the DNA strands to form a stable recombinant molecule.

ThomsonNOW™ **Learn more about the formation of recombinant DNA by clicking on the figure in ThomsonNOW.**

2 Distinguish among a genomic DNA library, a chromosome library, and a complementary DNA (cDNA) library; explain

why one would clone the same eukaryotic gene from both a genomic DNA library and a cDNA library (page 323).

- A **genomic DNA library** contains thousands of DNA fragments that represent the total DNA of an organism. A **chromosome library** contains all the DNA fragments from a specific chromosome. Each DNA fragment of a genomic DNA or chromosome library is stored in a specific bacterial strain. Analyzing DNA fragments in genomic DNA and chromosome libraries yields useful information about genes and their encoded proteins.

- A **cDNA library** is produced using **reverse transcriptase** to make DNA copies of mature mRNA isolated from eukaryotic cells. These copies, known as **complementary DNA (cDNA),** are then incorporated into recombinant DNA vectors.

ThomsonNOW™ **Learn more about the formation of cDNA by clicking on the figure in ThomsonNOW.**

- Genes present in genomic DNA and chromosome libraries from eukaryotes contain **introns,** regions that do not code for protein. Those genes can be amplified in bacteria, but the protein is not properly expressed. Because the introns have been removed from mRNA molecules, eukaryotic genes in cDNA libraries can be expressed in bacteria to produce functional protein products.

3 Describe the purpose of a genetic probe (page 323).

- Researchers use a radioactive DNA or RNA sequence as a **genetic probe** to screen thousands of recombinant DNA molecules in bacterial cells to find the colony that contains the DNA of interest.

ThomsonNOW Watch the process for using a genetic probe to find bacterial cells with a specific recombinant DNA molecule by clicking on the figure in ThomsonNOW.

4 Describe how the polymerase chain reaction amplifies DNA in vitro (page 323).

■ The **polymerase chain reaction (PCR)** is a widely used, automated, in vitro technique in which researchers target a particular DNA sequence by specific primers and then clone it using a heat-resistant DNA polymerase.

■ Using PCR, scientists amplify and analyze tiny DNA samples taken from various sites, from crime scenes to archaeological remains.

ThomsonNOW Watch the amplification of DNA by PCR by clicking on the figure in ThomsonNOW.

5 Distinguish among DNA, RNA, and protein blotting (page 330).

■ A **Southern blot** detects DNA fragments by separating them using gel electrophoresis and then transferring them to a nitrocellulose or nylon membrane. A probe is then **hybridized** by complementary base pairing to the DNA bound to the membrane, and the band or bands of DNA are identified by autoradiography or chemical luminescence.

■ When RNA molecules that are separated by electrophoresis are transferred to a membrane, the result is a **Northern blot.**

■ A **Western blot** consists of proteins or polypeptides previously separated by gel electrophoresis.

6 Describe the chain termination method of DNA sequencing (page 330).

■ **DNA sequencing** yields information about the structure of a gene and the probable amino acid sequence of its encoded proteins. Geneticists compare DNA sequences with other sequences stored in massive databases.

■ Automated DNA sequencing is based on the **chain termination method,** which uses dideoxynucleotides, each tagged with a different-colored fluorescent dye, to terminate elongation during DNA replication. Gel electrophoresis separates the resulting fragments, and a laser identifies the nucleotide sequence.

ThomsonNOW Learn more about automated DNA sequencing by clicking on the figure in ThomsonNOW.

7 Describe the three main areas of interest in genomics (page 333).

■ **Genomics** is the emerging field of biology that studies the entire DNA sequence of an organism's **genome.** Structural genomics is concerned with mapping and sequencing genomes. Functional genomics deals with the functions of genes and nongene sequences in genomes. Comparative genomics involves comparing the genomes of different species to further our understanding of evolutionary relationships.

8 Explain what a DNA microarray does, and give an example of its research and medical potential (page 333).

■ Many diagnostic tests involve **DNA microarrays,** in which hundreds of different DNA molecules are placed on a glass slide or chip. DNA microarrays enable researchers to compare the activities of thousands of genes in normal and diseased cells from tissue samples. Because cancer and other diseases exhibit altered patterns of gene expression, DNA microarrays have the potential to identify disease-causing genes or the proteins they code for, which can then be targeted by therapeutic drugs.

9 Define *pharmacogenetics* and *proteomics* (page 333).

■ **Pharmacogenetics,** the new science of gene-based medicine, takes into account the subtle genetic differences among individuals and customizes drugs to match a patient's genetic makeup.

■ The study of all the proteins encoded by the human genome and produced in a person's cells and tissues is **proteomics.** Scientists want to identify all the proteins made by a given kind of cell, but the process is much more complicated than sequencing the human genome.

10 Describe at least one important application of recombinant DNA technology in each of the following fields: medicine and pharmacology, DNA fingerprinting, and transgenic organisms (page 338).

■ Genetically altered bacteria produce many important human protein products, including insulin, growth hormone, tissue plasminogen activator (TPA), tissue growth factor-beta (TGF-β), clotting factor VIII, and Dornase Alpha (DNase).

■ **DNA fingerprinting** is the analysis of an individual's DNA. It is based on a variety of **short tandem repeats (STRs),** molecular markers that are highly **polymorphic** within the human population. DNA fingerprinting has applications in such areas as law enforcement, issues of disputed parentage, and tracking tainted foods.

ThomsonNOW Solve a murder case with DNA fingerprinting by clicking on the figure in ThomsonNOW.

■ **Transgenic organisms** have foreign DNA incorporated into their genetic material. Transgenic livestock produce foreign proteins in their milk. Transgenic plants have great potential in agriculture.

ThomsonNOW Learn more about the development of transgenic rice by clicking on the figure in ThomsonNOW.

11 Describe at least two safety issues associated with recombinant DNA technology, and explain how these issues are being addressed (page 342).

■ Some people are concerned about the safety of genetically engineered organisms. To address these concerns, scientists carry out recombinant DNA technology under specific safety guidelines.

■ The introduction of transgenic plants and animals into the natural environment, where they may spread in an uncontrolled manner, is an ongoing concern that must be evaluated on a case-by-case basis.

TEST YOUR UNDERSTANDING

1. A plasmid (a) is used as a DNA vector (b) is a type of bacteriophage (c) is a type of cDNA (d) is a retrovirus (e) b and c

2. DNA molecules with complementary sticky ends associate by (a) covalent bonds (b) hydrogen bonds (c) ionic bonds (d) disulfide bonds (e) phosphodiester linkages

3. Human DNA and a particular plasmid both have sites that are cut by the restriction enzymes *Hind*III and *Eco*RI. To make recombinant DNA, the scientist should (a) cut the plasmid with *Eco*RI and the human DNA with *Hind*III (b) use *Eco*RI to cut both the plasmid and the human DNA (c) use *Hind*III

to cut both the plasmid and the human DNA (d) a or b (e) b or c

4. Which of the following sequences is *not* palindromic?

    (a) 5′—AAGCTT—3′   (b) 5′—GATC—3′
        3′—TTCGAA—5′   3′—CTAG—5′
    (c) 5′—GAATTC—3′   (d) 5′—CTAA—3′
        3′—CTTAAG—5′   3′—GATT—5′
    (e) b and d

5. Which technique rapidly replicates specific DNA fragments without cloning? (a) gel electrophoresis (b) cDNA libraries (c) genetic probe (d) restriction fragment length polymorphism (e) polymerase chain reaction

6. The PCR technique uses (a) heat-resistant DNA polymerase (b) reverse transcriptase (c) DNA ligase (d) restriction enzymes (e) b and c

7. A cDNA clone contains (a) introns (b) exons (c) anticodons (d) a and b (e) b and c

8. The dideoxynucleotides ddATP, ddTTP, ddGTP, and ddCTP are important in DNA sequencing because they (a) cause premature termination of a growing DNA strand (b) are used as primers (c) cause the DNA fragments that contain them to migrate more slowly through a sequencing gel (d) are not affected by high temperatures (e) have more energy than deoxynucleotides

9. In the Southern blot technique, _____ is/are transferred from a gel to a nitrocellulose or nylon membrane. (a) protein (b) RNA (c) DNA (d) bacterial colonies (e) reverse transcriptase

10. Gel electrophoresis separates nucleic acids on the basis of differences in (a) length (molecular weight) (b) charge (c) nucleotide sequence (d) relative proportions of adenine and guanine (e) relative proportions of thymine and cytosine

11. The Ti plasmid, carried by *Agrobacterium tumefaciens,* is especially useful for introducing genes into (a) bacteria (b) plants (c) animals (d) yeast (e) all eukaryotes

12. A genomic DNA library (a) represents all the DNA in a specific chromosome (b) is made using reverse transcriptase (c) is stored in a collection of recombinant bacteria (d) is a DNA copy of mature mRNAs (e) allows researchers to amplify a tiny sample of DNA

13. Tissue growth factor-beta (a) is a genetic probe for recombinant plasmids (b) is a product of DNA technology used in tissue engineering (c) is necessary to make a cDNA library (d) cannot be synthesized without a heat-resistant DNA polymerase (e) is isolated by the Southern blot technique

14. These highly polymorphic molecular markers are useful in DNA fingerprinting: (a) short tandem repeats (b) cloned DNA sequences (c) palindromic DNA sequences (d) cosmid cloning vectors (e) complementary DNAs

CRITICAL THINKING

1. What are some of the problems that might arise if you were trying to produce a eukaryotic protein in a bacterium? How might using transgenic plants or animals help solve some of these problems?

2. Would genetic engineering be possible if we did not know a great deal about the genetics of bacteria? Explain.

3. What are some of the environmental concerns regarding transgenic organisms? What kinds of information does society need to determine if these concerns are valid?

4. How is proteomics related to functional genomics?

5. **Evolution Link.** DNA technology, such as the production of transgenic animals, is only possible because widely different organisms have essentially identical genetic systems (DNA ⟶ RNA ⟶ protein). What is the evolutionary significance of identical genetic systems in organisms as diverse as bacteria and pigs?

Additional questions are available in ThomsonNOW at www.thomsonedu.com/login

The Human Genome

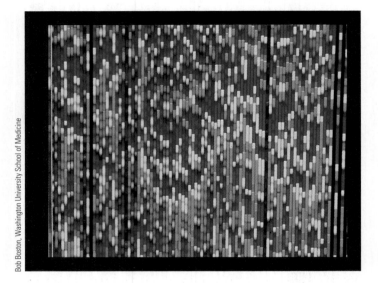

DNA sequencing. This is a display on a computer monitor attached to a DNA-sequencing machine. The machine determines the base sequence of a strand of DNA using the chain termination method and tags each of the four bases with a different-colored fluorescent dye.

Bob Boston, Washington University School of Medicine

KEY CONCEPTS

Pedigree analysis is a basic method used to study human genetics.

Karyotype analysis helps detect chromosome abnormalities.

Mutations in single genes can produce genetic disorders.

Researchers are developing gene therapy methods that could potentially correct many genetic disorders.

Genetic testing and counseling help individuals make reproductive decisions.

Certain advances in human genetics raise ethical questions.

The principles of genetics apply to all organisms, including humans. However, important differences separate genetic research on humans and genetic research on other organisms. To study inheritance in other species, geneticists ideally have stocks of genetically similar individuals—**true-breeding strains** whose traits remain unchanged for many generations because they are homozygous at virtually all loci. Geneticists conduct *controlled matings* between members of different true-breeding strains and raise the offspring under carefully controlled conditions. Of course, experimental matings are not feasible in humans. Also, most human families are small, and 20 to 30 years or more elapse between generations.

Despite the inherent difficulties of studying inheritance in humans, **human genetics,** the science of inherited variation in humans, is progressing rapidly. Researchers conduct population studies of large extended families. Also, the medical attention given to human genetic diseases has expanded our knowledge of human genetics. Genetic studies of other organisms have provided invaluable insights: many human inheritance questions have been explained using model organisms, such as bacteria, yeasts, worms, fruit flies, and mice. More recently, the **human genome,** which represents the totality of genetic information in human cells, has been mapped and sequenced. In **DNA sequencing,** researchers identify the order of nucleotides in DNA to understand the genetic basis of human similarities and differences (see figure).

In this chapter we first examine the methods of human genetics, including new fields that are developing as a result of the Human Genome Project. Then we discuss a variety of human genetic disorders. We explore the use of gene therapy for some of these disorders, as well as the application of genetic testing, screening, and counseling for families at risk. The chapter concludes with a discussion of ethical issues related to human genetics. ■

STUDYING HUMAN GENETICS

Learning Objectives

1 Distinguish between karyotyping and pedigree analysis.
2 Discuss the implications of the Human Genome Project.
3 Discuss the mouse model for studying cystic fibrosis.

Human geneticists use a variety of methods that enable them to make inferences about a trait's mode of inheritance. We consider three of these methods—the identification of chromosomes by karyotyping, the analysis of family inheritance patterns using pedigrees, and DNA sequencing and mapping of genes by genome projects. Investigators often study human inheritance most effectively by combining these and other approaches.

Human chromosomes are studied by karyotyping

Cytogenetics is the study of chromosomes and their role in inheritance. Researchers working with simpler organisms discovered many of the basic principles of genetics. In such organisms, it is often possible to relate genetic data to the number and structure of specific chromosomes. Some organisms used in genetics, such as the fruit fly *Drosophila melanogaster,* have very few chromosomes; the fruit fly has only four pairs. In *Drosophila* larval salivary glands, the chromosomes are large enough that their structural details are readily evident (see Fig. 2-3). This organism, therefore, has provided unique opportunities for correlating certain inherited phenotypic changes with alterations in chromosome structure.

The normal number of chromosomes for the human species is 46: 44 **autosomes** (22 pairs) and 2 sex chromosomes (1 pair). A **karyotype** (from Greek, meaning "nucleus") is an individual's chromosome composition. Until the mid-1950s, when biologists adopted modern methods of karyotyping, the accepted number of chromosomes for the human species was 48, based on a study of human chromosomes published in 1923. The reason researchers counted 48 human chromosomes was the difficulty in separating the chromosomes so they could be accurately counted. In 1952, University of Texas cell biologist T. C. Hsu treated cells with a hypotonic salt solution by mistake; this caused the cells to swell and the chromosomes to spread apart, thereby making them easier to count. Other techniques were also developed, and in 1956 researchers Joe Hin Tjio and Albert Levan, working in Sweden, reported that humans have 46 chromosomes, not 48. Other researchers subsequently verified this report. The story of the human chromosome number is a valuable example of the

"self-correcting" nature of science (although such corrections may take time). The re-evaluation of established facts and ideas, often by using improved techniques or new methods, is an essential part of the scientific process.

Human chromosomes are visible only in dividing cells (see Chapter 10), and it is difficult to obtain dividing cells directly from the human body. Researchers typically use blood, because white blood cells can be induced to divide in a culture medium by treating them with certain chemicals. Other sources of dividing cells include skin and, for prenatal chromosome studies, chorionic villi or fetal cells shed into the amniotic fluid (discussed later in the chapter).

In karyotyping, biologists culture dividing human cells and then treat them with the drug *colchicine,* which arrests the cells at mitotic metaphase or late prophase, when the chromosomes are most highly condensed. Next, the researchers put the cells into a hypotonic solution; the cells swell, and the chromosomes spread out so they are easily observed. The investigators then flatten the cells on microscope slides and stain the chromosomes to reveal the patterns of bands, which are unique for each homologous pair. After the microscopic image has been scanned into a computer, the homologous pairs are electronically matched and placed together (▮ Fig. 16-1a).

By convention, geneticists identify chromosomes by length; position of the centromere; banding patterns, which are produced by staining chromosomes with dyes that produce dark and light crossbands of varying widths; and other features such as *satellites,* tiny knobs of chromosome material at the tips of certain chromosomes. With the exception of the sex chromosomes, all chromosomes but chromosome 21, which is smaller than chromosome 22, are numbered and lined up in order of size. The largest human chromosome (chromosome 1) is about 5 times as long as the smallest one (chromosome 21), but there are only slight size differences among some of the intermediate-sized chromosomes. The X and Y chromosomes of a normal male are homologous only at their tips; normal females have two X chromosomes and no Y chromosome. Differences from the normal karyotype—that is, deviations in chromosome number or structure—are associated with certain disorders such as Down syndrome (discussed later in the chapter).

Another way to distinguish chromosomes in a karyotype is by **fluorescent in situ hybridization (FISH)**. A geneticist tags a DNA strand complementary to DNA in a specific chromosome with a fluorescent dye. The DNA in the chromosome is denatured—that is, the two strands are separated—so the tagged strand can bind to it. A different dye is used for each chromosome, "painting" each with a different color (Fig. 16-1b). A chromosome that is multicolored (*not shown*) indicates breakage and fusion of chromosomes, an abnormality associated with certain genetic diseases and many types of cancer.

Family pedigrees help identify certain inherited conditions

Early studies of human genetics usually dealt with readily identified pairs of contrasting traits and their distribution among members of a family. A "family tree" that shows inheritance patterns, the transmission of genetic traits within a family over several generations, is known as a **pedigree.** Pedigree analysis remains widely used, even in today's world of powerful molecular genetic techniques, because it helps molecular geneticists determine the exact interrelationships of the DNA molecules they analyze from related individuals. Pedigree analysis is also an important tool of genetic counselors and clinicians. However, because human families tend to be small and information about certain family members, particularly deceased relatives, may not be available, pedigree analysis has limitations.

Pedigrees are produced using standardized symbols. Examine Figure 16-2, which shows a pedigree for **albinism,** a lack of the pigment melanin in the skin, hair, and eyes. Each horizontal row represents a separate generation, with the earliest generation (Roman numeral I) at the top and the most recent generation at the bottom. Within a given generation, the individuals are usually numbered consecutively, from left to right, using Arabic numerals. A horizontal line connects two parents, and a vertical line drops from the parents to their children. For example, individuals II-3 and II-4 are parents of four offspring (III-1, III-2, III-3, and III-4). Note that individuals in a given generation can be genetically unrelated. For example, II-1, II-2, and II-3 are unrelated to II-4 and II-5. Within a group of siblings, the oldest is on the left, and the youngest is on the right.

Studying pedigrees enables human geneticists to predict how phenotypic traits that are governed by the genotype at a single locus are inherited. About 10,000 traits have been described in humans. Pedigree analysis most often identifies three modes of single-locus inheritance: autosomal dominant, autosomal recessive, and X-linked recessive. We define and discuss examples of these inheritance modes later in the chapter.

Certain traits that do not show a simple Mendelian inheritance pattern can also be characterized using pedigree analysis. Some of these traits are the result of **genomic imprinting,** in which the expression of a gene in a given tissue or developmental stage is based on its parental origin—that is, whether the individual inherits the gene from the male or female parent. For some imprinted genes, the paternally inherited allele is always

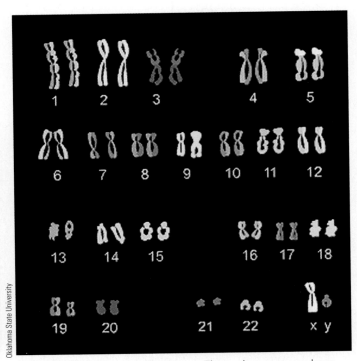

(a) Using an image analysis computer to prepare a karyotype, the biologist matches homologous chromosomes and organizes them by size. Before computers were used, researchers prepared karyotypes by cutting and pasting chromosomes from photographs.

(b) A normal human male karyotype. These chromosomes have been "painted." DNA molecules, to which different-colored fluorescent dyes are attached, hybridize to specific pairs of homologous chromosomes. Painting helps the researcher identify chromosomes under the light microscope.

SIU/Peter Arnold, Inc.

Oklahoma State University

Figure 16-1 *Animated* Karyotyping

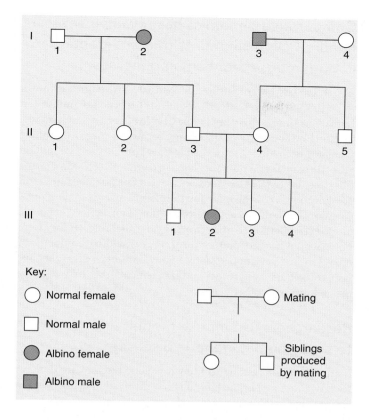

I
1   2   3   4

II
1   2   3   4   5

III
1   2   3   4

Key:

○ Normal female

□ Normal male

● Albino female

■ Albino male

□—○ Mating

Siblings produced by mating

Figure 16-2 *Animated*   A pedigree for albinism

By studying family histories, a researcher can determine the genetic mechanism of an inherited trait. In this example, III-2 represents an albino girl with two phenotypically normal parents, II-3 and II-4. The allele for albinism cannot be dominant because if it were, at least one of III-2's parents would have to be an albino. Also, albinism cannot be an X-linked recessive allele because if it were, her father would have to be an albino (and her mother would have to be a heterozygous carrier). This pedigree is explained if albinism is inherited as an autosomal recessive allele (not carried on a sex chromosome). In such cases, two phenotypically normal parents could produce an albino offspring because they are heterozygotes and could each transmit a recessive allele.

repressed (not expressed); for other imprinted genes, the maternally inherited allele is always repressed. Thus, maternal and paternal genomes have different imprints that result in differential gene expression in the embryo. As discussed in Chapter 14, **epigenetic inheritance,** changes in how a gene is expressed without any change in the coding of the DNA bases, may cause genomic imprinting.

Two rare genetic disorders provide a fascinating demonstration of genomic imprinting. In *Prader–Willi syndrome (PWS),* individuals become compulsive overeaters and obese; they are also short in stature and mildly to moderately retarded. In *Angelman syndrome (AS),* affected individuals are hyperactive, mentally retarded, and unable to speak, and they suffer from seizures. Both PWS and AS are caused by a small *deletion* of several loci

from the same region of chromosome 15. One of these deleted loci is responsible for PWS, another for AS. (Chromosome deletions are discussed later in the chapter.)

Pedigree analysis has shown that when the person inherits the deletion from the father, PWS occurs, whereas when the person inherits the deletion from the mother, AS occurs. This inheritance pattern suggests that the normal PWS gene is expressed only in the paternal chromosome and the normal AS gene is expressed only in the maternal chromosome. PWS occurs because the repressed (imprinted) PWS gene from the mother cannot make up for the absent PWS gene in the paternal chromosome. Similarly, AS occurs because the repressed (imprinted) AS gene from the father cannot make up for the absent AS gene in the maternal chromosome.

The Human Genome Project sequenced the DNA on all human chromosomes

Workers in the **Human Genome Project** sequenced the DNA in the entire nuclear human genome—about 2.9 billion base pairs. (The human mitochondrial genome was sequenced in 1981.) This international undertaking, based on the DNA from 6 to 10 anonymous individuals, was essentially completed in 2001 by hundreds of researchers working in two independent teams, the government-funded International Human Genome Sequencing Consortium and the privately funded company Celera Genomics. The completion of the Human Genome Project in 2003 was a significant milestone in genetics (▌Table 16-1).

Scientists hope to eventually identify where all the genes are located in the sequenced DNA. We do not yet know how many protein-coding genes are in the human genome. In 2001, the National Human Genome Research Institute (NHGRI) estimated there were 35,000 to 45,000 genes in the human genome, but by 2003 the estimate was lowered to about 25,000.

Gene identification represents a formidable challenge. How do you identify a gene in a DNA sequence if you know nothing about it? The human genome is extremely complex. It includes the DNA content of both the nucleus and the mitochondria. However, nuclear DNA accounts for almost all the genetic information in the human genome. Like genomes of other eukaryotic organisms, some (in humans, only about 2%) of the human genome specifies the synthesis of polypeptides, whereas other segments code only for RNA products.

However, most of the genome consists of noncoding regulatory elements, repetitive sequences (multiple copies) of DNA, and gene segments whose functions remain unknown. Sometimes genes overlap one another, and in other cases a single gene codes for multiple proteins. Relating specific genes to the RNAs and proteins they code for and determining the roles these RNAs and proteins play in the body are some of the avenues of human genetic research that scientists are currently pursuing.

In human chromosome 22, for example, computers scanned DNA sequences for identifying markers such as promoter ele-

TABLE 16-1

Major Milestones in Genetics

Year	Scientific Advance
1866	Mendel proposed existence of hereditary factors now known as genes
1871	Nucleic acids discovered
1953	Structure of DNA determined
1960s	Genetic code explained (how proteins are made from DNA)
1977	DNA sequencing began
1986	DNA sequencing automated
1995	Sequencing of first genome (bacterium *Haemophilus influenzae*) completed
1996	Sequencing of first eukaryotic genome (yeast *Saccharomyces cerevisiae*) completed
1998	Sequencing of first multicellular eukaryotic genome (nematode worm *Caenorhabditis elegans*) completed
2001	Draft sequence of entire human genome published
2003	Final completion of DNA sequencing of human genome

ments typically associated with genes; using this approach, researchers identified 545 genes. Many of these are "potential" genes in that their messenger RNAs (mRNAs) and protein products have not yet been isolated. Other genes were identified because they code for proteins similar to those previously identified in humans or other organisms. Despite the difficulties, scientists are making progress, and mapping studies will help scientists understand the physical and functional relationships among genes and groups of genes as revealed by their order on the chromosomes.

Now that the human genome has been sequenced, researchers will be busy for many decades analyzing the growing body of human molecular data. In addition to identifying genes, scientists want to understand what each gene's role is, how each gene interacts with other genes, and how the expression of each gene is regulated in different tissues. Eventually, all the thousands of proteins produced in human cells will be identified, their 3-D structures determined, and their properties and functions evaluated.

Researchers also want to study sequence variations within the human genome to elucidate differences that might be related to susceptibility to illness or disease. The potential medical applications of the Human Genome Project are extremely promising. Genes on human chromosome 22, for example, are associated with at least 27 diseases known to have a genetic component. The causative genes of many of these disorders have not yet been identified. For example, a gene involved in schizophrenia is strongly linked to human chromosome 22, but scientists do not know its exact location or function.

Comparative genomics has revealed several hundred DNA segments that are identical in both mouse and human genomes

About 500 DNA segments longer than 200 base pairs are known to be 100% conserved (that is, identical) between the mouse and human genomes. This remarkable degree of conservation has interesting evolutionary implications, because it means that these segments have not mutated during the approximately 75 million years since mice and humans shared a common ancestor. During this time, other DNA sections underwent considerable mutation and selection, allowing mice and humans to diverge to their present states. Although the functions of these unchanged elements are not yet determined, they clearly have a vital role. (If they were not essential in their present form, they would have undergone mutation and selection.) Many highly conserved segments appear to contain nonprotein-coding elements that may regulate the expression of other genes.

Researchers use mouse models to study human genetic diseases

Many questions relating to human genetic diseases are difficult to answer because of the ethical issues involved in using humans as test subjects. However, research on any disease is greatly facilitated if an animal model is used for experimentation.

A good example is *cystic fibrosis,* a genetic disease caused by a single gene mutation inherited as a recessive allele. In 1994, researchers used **gene targeting** to produce strains of mice that were either homozygous or heterozygous for cystic fibrosis.

The allele that causes cystic fibrosis is a mutant form of a locus involved in controlling the body's water and electrolyte balance. Geneticists have cloned this gene and found that it codes for a protein, the *CFTR protein,* that serves as a chloride ion channel in the plasma membrane. (*CFTR* stands for *cystic fibrosis transmembrane conductance regulator.*) This ion channel transports chloride ions out of the cells lining the digestive tract and the respiratory system. When the chloride ions leave the cells, water follows by osmosis. Thus, the normal secretions of these cells are relatively watery. Because the cells of individuals with cystic fibrosis lack normal chloride ion channels, the individuals' secretions have a low water content and their sweat is very salty. Cells of heterozygous individuals have only half the usual number of functional CFTR ion channels, but these are enough to maintain the normal fluidity of their secretions.

Some researchers are now focusing on understanding the way in which the CFTR channel is activated or inactivated in mice. They hope to use this information to design drugs that enhance chloride transport through the CFTR channels. Such drugs have the potential to treat cystic fibrosis in humans by activating the mutant channels.

Review

■ What kinds of information can a human karyotype provide?

■ What is pedigree analysis?

- What are two possible benefits scientists hope to obtain by further study of the human genome?
- How does using a mouse model for a genetic disease overcome some of the difficulties in studying human inheritance?

ABNORMALITIES IN CHROMOSOME NUMBER AND STRUCTURE

Learning Objectives

4 Explain how nondisjunction in meiosis is responsible for chromosome abnormalities such as Down syndrome, Klinefelter syndrome, and Turner syndrome.
5 Distinguish among the following structural abnormalities in chromosomes: translocations, deletions, and fragile sites.

Polyploidy, the presence of multiple sets of chromosomes, is common in plants but rare in animals. It may arise from the failure of chromosomes to separate during meiosis or from the fertilization of an egg by more than one sperm. When it occurs in all the cells of the body, polyploidy is lethal in humans and many other animals. For example, *triploidy* (3n) is sometimes found in human embryos that have been spontaneously aborted in early pregnancy.

Abnormalities caused by the presence of a single extra chromosome or the absence of a chromosome—called **aneuploidies**—are more common than polyploidy. **Disomy** is the normal state: two of each kind of chromosome. In **trisomy,** a person has an extra chromosome, that is, three of one kind. In **monosomy,** an individual lacks one member of a pair of chromosomes. ▌Table 16-2 summarizes some disorders that aneuploidies produce.

Aneuploidies generally arise as a result of an abnormal meiotic (or, rarely, mitotic) division in which chromosomes fail to separate at anaphase. This phenomenon, called **nondisjunction,** can occur with the autosomes or with the sex chromosomes. In meiosis, chromosome nondisjunction may occur during the first or second meiotic division (or both). For example, two X chromosomes that fail to separate at either the first or the second meiotic division may both enter the egg nucleus. Alternatively, the two joined X chromosomes may go into a *polar body*, leaving the egg with no X chromosome. (Recall from Chapter 10 that a polar body is a nonfunctional haploid cell produced during oogenesis; also see Fig. 49-11.)

Nondisjunction of the XY pair during the first meiotic division in the male may lead to the formation of a sperm with both X and Y chromosomes or a sperm with neither an X nor a Y chromosome (▌Fig. 16-3). Similarly, nondisjunction at the second meiotic division can produce sperm with two Xs or two Ys. When an abnormal gamete unites with a normal one, the resulting zygote has a chromosome abnormality that will be present in every cell of the body.

Meiotic nondisjunction results in an abnormal chromosome number at the zygote stage of development, so all cells in the individual have an abnormal chromosome number. In contrast, nondisjunction during a mitotic division occurs sometime later in development and leads to the establishment of a clone of abnormal cells in an otherwise normal individual. Such a mixture

TABLE 16-2

Chromosome Abnormalities: Disorders Produced by Aneuploidies

Karyotype	Common Name	Clinical Description
Trisomy 13	Patau syndrome	Multiple defects, with death typically by age 3 months.
Trisomy 18	Edwards syndrome	Ear deformities, heart defects, spasticity, and other damage; death typically by age 1 year, but some survive much longer.
Trisomy 21	Down syndrome	Overall frequency is about 1 in 800 live births. Most conceptions involving true trisomy occur in older (age 35+) mothers, but translocation resulting in the equivalent of trisomy is not age-related. Trisomy 21 is characterized by a fold of skin above the eye, varying degrees of mental retardation, short stature, protruding furrowed tongue, transverse palmar crease, cardiac deformities, and increased risk of leukemia and Alzheimer's disease.
X0	Turner syndrome	Short stature, webbed neck, sometimes slight mental retardation; ovaries degenerate in late embryonic life, leading to rudimentary sexual characteristics; gender is female; no Barr bodies.
XXY	Klinefelter syndrome	Male with slowly degenerating testes, enlarged breasts; one Barr body per cell.
XYY	XYY karotype	Many males have no unusual symptoms; others are unusually tall, with heavy acne, and some tendency to mild mental retardation.
XXX	Triplo-X	Despite three X chromosomes, usually fertile females with normal intelligence; two Barr bodies per cell.

Aneuploidy may occur by meiotic nondisjunction, the abnormal segregation of chromosomes during meiosis.

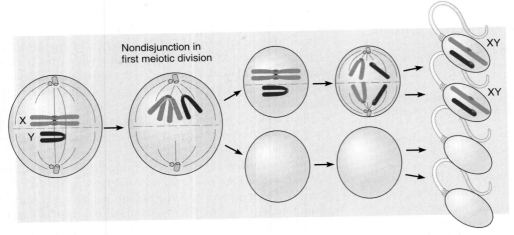

(a) First meiotic division nondisjunction results in two XY sperm and two sperm with neither an X nor a Y.

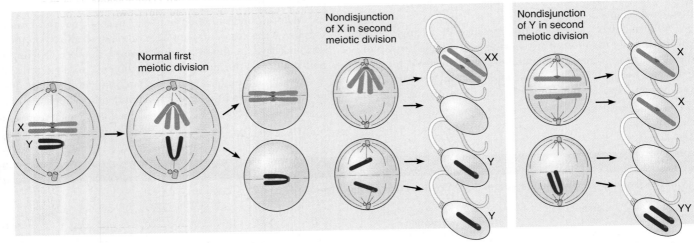

(b) Second meiotic division nondisjunction of the X chromosome results in one sperm with two X chromosomes, two with one Y each, and one with no sex chromosomes. Nondisjunction of the Y chromosome results in one sperm with two Y chromosomes, two with one X each, and one with no sex chromosome (*box on right*).

Figure 16-3 Meiotic nondisjunction

In these examples of nondisjunction of the sex chromosomes in the human male, only the X (*purple*) and Y (*blue*) chromosomes are shown. (The positions of the metaphase chromosomes have been modified to save space.)

of cells with different chromosome numbers may or may not affect somatic (body) or germ line (reproductive) tissues.

Recognizable chromosome abnormalities are seen in less than 1% of all live births, but substantial evidence suggests the rate at conception is much higher. At least 17% of pregnancies recognized at 8 weeks will end in spontaneous abortion (miscarriage). Approximately half of these spontaneously aborted em-

bryos have major chromosome abnormalities, including autosomal trisomies (such as trisomy 21), triploidy, tetraploidy, and Turner syndrome (X0), in which the 0 refers to the absence of a second sex chromosome. Autosomal monosomies are exceedingly rare, possibly because they induce a spontaneous abortion very early in the pregnancy, before a woman is even aware she is pregnant. Some investigators give surprisingly high estimates

(a) This boy with Down syndrome is working on a science experiment in his kindergarten class. Some individuals with Down syndrome learn to read and write.

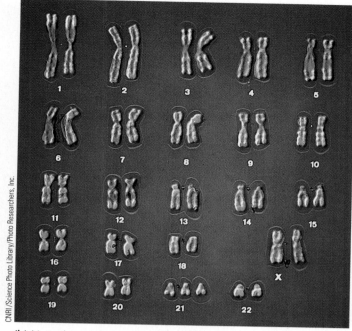

(b) Note the presence of an extra chromosome 21 in this colorized karyotype of a female with Down syndrome.

Figure 16-4 Down syndrome

(50% or more) for the loss rate of very early embryos. Chromosome abnormalities probably induce many of these spontaneous abortions.

Down syndrome is usually caused by trisomy 21

Down syndrome is one of the most common chromosome abnormalities in humans. (The term *syndrome* refers to a set of symptoms that usually occur together in a particular disorder.) It was named after J. Langdon Down, the British physician who in 1866 first described the condition. Affected individuals have abnormalities of the face, eyelids, tongue, hands, and other parts of the body and are often mentally and physically retarded (Fig. 16-4a). They are also unusually susceptible to certain diseases, such as leukemia and Alzheimer's disease.

Cytogenetic studies have revealed that most people with Down syndrome have 47 chromosomes because of *autosomal trisomy:* this condition is known as **trisomy 21** (Fig. 16-4b). Nondisjunction during meiosis is responsible for the presence of the extra chromosome. Although no genetic information is missing in these individuals, the extra copies of chromosome 21 genes bring about some type of genetic imbalance that causes abnormal physical and mental development. Down syndrome is quite variable in expression, with some individuals far more severely

affected than others. Researchers are using DNA technologies to pinpoint genes on chromosome 21 that affect mental development, as well as possible *oncogenes* (cancer-causing genes) and genes that may be involved in Alzheimer's disease. (Like many human conditions, cancer and Alzheimer's disease have both genetic and environmental components.)

Down syndrome occurs in all ethnic groups in about 1 out of 800 live births. Its incidence increases markedly with increasing maternal age. The occurrence of Down syndrome is not affected by the father's age (although other disorders are, including schizophrenia and achondroplasia, the most common form of dwarfism). Down syndrome is 68 times as likely in the offspring of mothers of age 45 than in the offspring of mothers age 20. However, most babies with Down syndrome in the United States are born to mothers younger than 35, in part because these greatly outnumber older mothers, and in part because about 90% of older women who undergo prenatal testing terminate the pregnancy if Down syndrome is diagnosed.

The relationship between increased incidence in Down syndrome and maternal age has been studied for decades, but there is no explanation. Scientists have proposed several hypotheses to explain the maternal age effect, but none are supported unequivocally. One explanation is that older women have held eggs in suspended meiosis too long, leading to a deterioration of the meiotic spindle apparatus. (A woman is born with all the oocytes she will ever have; these oocytes remain in prophase I of meiosis until ovulation.) Another possibility is that an aging uterus is less likely to reject an abnormal fetus.

Most sex chromosome aneuploidies are less severe than autosomal aneuploidies

Sex chromosome aneuploidies are tolerated relatively well (see Table 16-2) This is true, at least in part, because of the mechanism of **dosage compensation:** mammalian cells compensate for extra X chromosome material by rendering all but one X chromosome inactive. The inactive X is seen as a **Barr body,** a region of darkly staining, condensed chromatin next to the nuclear envelope of an interphase nucleus (see Fig. 11-17). Investigators have used the presence of the Barr body in the cells of normal females (but not normal males) as an initial screen to determine whether an individual is genetically female or male. However, as you will see shortly in the context of sex chromosome aneuploidies, the Barr body test has limitations.

Individuals with **Klinefelter syndrome** are males with 47 chromosomes, including two Xs and one Y. They have small testes, produce few or no sperm, and are therefore sterile. The hypothesis that the Y chromosome is the major determinant of the male phenotype has been substantiated by the fact that at least one gene on the Y chromosome acts as a genetic switch, directing male development. Males with Klinefelter syndrome tend to be unusually tall and have female-like breast development. About half show some mental retardation, but many live relatively normal lives. However, each of their cells has one Barr body. On the basis of such a test, they would be erroneously classified as females. About 1 in 1000 live-born males has Klinefelter syndrome.

The sex chromosome composition for **Turner syndrome,** in which an individual has only one sex chromosome, an X chromosome, is designated X0. Because they lack the male-determining effect of the Y chromosome, individuals with Turner syndrome develop as females. However, both their internal and their external genital structures are underdeveloped, and they are sterile. Apparently a second X chromosome is necessary for normal development of ovaries in a female embryo. Examination of the cells of these individuals reveals no Barr bodies, because there is no extra X chromosome to be inactivated. Using the standards of the Barr body test, such an individual would be classified erroneously as a male. About 1 in 2500 live-born females has Turner syndrome.

People with an X chromosome plus two Y chromosomes are phenotypically males, and they are fertile. Other characteristics of these individuals (tall, often with severe acne) hardly qualify as a syndrome; hence the designation **XYY karyotype.** Some years ago, several widely publicized studies suggested that males with this condition are more likely to display criminal tendencies and thus to be imprisoned. However, these studies were flawed because they were based on small numbers of XYY males, without adequate or well-matched control studies of XY males. The prevailing opinion in medical genetics today is that there are many undiagnosed XYY males in the general population who do not have overly aggressive or criminal behaviors and are unlikely to be incarcerated.

Abnormalities in chromosome structure cause certain disorders

Chromosome abnormalities are caused not only by changes in chromosome number but also by distinct changes in the structure of one or more chromosomes. The breakage and rejoining of chromosome parts result in four structural changes within or between chromosomes: duplications, inversions, deletions, and translocations (❚ Fig. 16-5). Breaks in chromosomes are the result of errors in replication or recombination. In a **duplication,** a segment of the chromosome is repeated one or more times; these repeats often appear in tandem with one another. The orientation of a chromosome segment is reversed in an **inversion.** In a **deletion,** breakage causes loss of part of a chromosome, along with the genes on that segment. A deletion can occur at the end of a chromosome or on an internal part of the chromosome. In some cases of **translocation,** a chromosome fragment breaks off and attaches to a nonhomologous chromosome. In a **reciprocal translocation,** two nonhomologous chromosomes exchange segments.

Here we consider three simple examples of structural changes in one or more chromosomes that result in abnormal phenotypes: translocations, deletions, and *fragile sites,* which are chromosome sites that are susceptible to breakage.

Translocation is the attachment of part of one chromosome to another

The consequences of translocations vary considerably. They include deletions, in which some genes are missing, and duplications, in which extra copies of certain genes are present. In about 4% of individuals with Down syndrome, only 46 chromosomes are present, but one is abnormal. The large arm of chromosome 21 has been translocated to the large arm of another chromosome, usually chromosome 14. Individuals with **translocation Down syndrome** have one chromosome 14, one combined 14/21 chromosome, and two normal copies of chromosome 21. All or part of the genetic material from chromosome 21 is thus present in triplicate. When studying the karyotypes of the parents in such cases, geneticists usually find that either the mother or the father has only 45 chromosomes, although she or he is generally phenotypically normal. The parent with 45 chromosomes has one chromosome 14, one combined 14/21 chromosome, and one chromosome 21; although the karyotype is abnormal, there is no extra genetic material. In contrast to trisomy 21, translocation Down syndrome can run in families, and its incidence is not related to maternal age.

A deletion is the loss of part of a chromosome

Sometimes chromosomes break and fail to rejoin. Such breaks result in deletions of as little as a few base pairs to as much as an entire chromosome arm. As you might expect, large deletions are

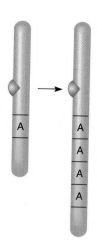

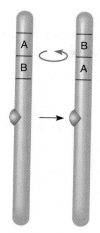

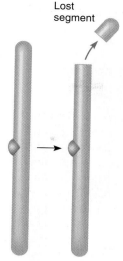

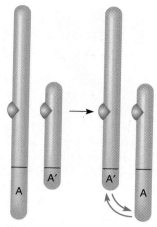

Lost segment

(a) A duplication is a repeated segment of a chromosome. In this example, gene A is repeated.

(b) An inversion is a chromosome segment with a reversed orientation. An inversion does not change the amount of genetic material in the chromosome, only its arrangement.

(c) A deletion is the loss of a chromosome segment. A deletion can occur at the tip (shown) or within the chromosome.

(d) A reciprocal translocation occurs when two non-homologous chromosomes exchange segments.

Figure 16-5 *Animated* Common abnormalities of chromosome structure

generally lethal, whereas small deletions may have no effect or may cause recognizable human disorders.

One deletion disorder (1 in 50,000 live births) is **cri du chat syndrome,** in which part of the short arm of chromosome 5 is deleted. As in most deletions, the exact point of breakage in chromosome 5 varies from one individual to another; some cases of cri du chat involve a small loss, whereas others involve a more substantial deletion of base pairs. Infants born with cri du chat syndrome typically have a small head with altered features described as a "moon face" and a distinctive cry that sounds like a kitten mewing. (The name literally means "cry of the cat" in French.) Affected individuals usually survive beyond childhood but exhibit severe mental retardation.

Fragile sites are weak points at specific locations in chromatids

A **fragile site** is a place where part of a chromatid appears to be attached to the rest of the chromosome by a thin thread of DNA. Fragile sites may occur at a specific location on both chromatids of a chromosome. They have been identified on the X chromosome as well as on certain autosomes. The location of a fragile site is exactly the same in all of an individual's cells, as well as in cells of other family members. Scientists report growing evidence that cancer cells may have breaks at these fragile sites. Whether cancer destabilizes the fragile sites, leading to breakage, or the fragile sites themselves contain genes that contribute to cancer is unknown at this time.

In **fragile X syndrome,** also known as **Martin–Bell syndrome,** a fragile site occurs near the tip of the X chromosome, where the fragile X gene contains a nucleotide triplet CGG that repeats 200 to more than 1000 times (❙ Fig. 16-6a). In a normal chromosome, CGG repeats up to 50 times (❙ Fig. 16-6b).

Fragile X syndrome is the most common cause of inherited mental retardation. The effects of fragile X syndrome, which are more pronounced in males than in females, range from mild learning and attention deficit disorders to severe mental retardation and hyperactivity. According to the National Fragile X Foundation, about 80% of boys and 35% of girls with fragile X syndrome are at least mildly mentally retarded. Females with fragile X syndrome are usually heterozygous (because their other X chromosome is normal) and are therefore more likely to have normal intelligence.

The discovery of the fragile X gene in 1991 and the development of the first fragile X mouse model in 1994 have provided researchers with ways to develop and test potential treatments, including gene therapy. At the microscopic level, the nerve cells of individuals with fragile X syndrome have malformed dendrites (the part of the nerve cell that receives nerve impulses from other nerve cells). At the molecular level, the triplet repeats associated with fragile X syndrome disrupt the functioning of a gene that codes for a certain protein, designated *fragile X mental retardation protein (FMRP).* In normal cells, FMRP binds to dozens of different mRNA molecules (exactly why it binds is not yet understood), but in cells of individuals with fragile X syndrome the mutated allele does not produce functional FMRP.

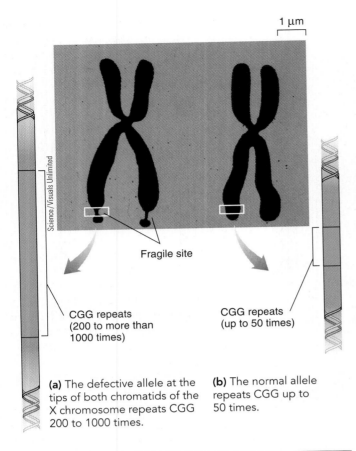

Fragile site

CGG repeats
(200 to more than
1000 times)

CGG repeats
(up to 50 times)

(a) The defective allele at the tips of both chromatids of the X chromosome repeats CGG 200 to 1000 times.

(b) The normal allele repeats CGG up to 50 times.

Figure 16-6 Fragile X syndrome

This colorized SEM shows an X chromosome with a fragile site and a normal X chromosome.

Review

- What are the specific chromosome abnormalities in Down syndrome, Klinefelter syndrome, and Turner syndrome?
- What is the chromosome abnormality in cri du chat syndrome?
- What is the chromosome abnormality in fragile X syndrome?

GENETIC DISEASES CAUSED BY SINGLE-GENE MUTATIONS

Learning Objective

6   State whether each of the following genetic defects is inherited as an autosomal recessive, autosomal dominant, or X-linked recessive: phenylketonuria (PKU), sickle cell anemia, cystic fibrosis, Tay–Sachs disease, Huntington's disease, and hemophilia A.

You have seen that several human disorders involve chromosome abnormalities. Hundreds of human disorders, however, involve enzyme defects caused by mutations of single genes. *Phenylke-*

tonuria (PKU) and *alkaptonuria* (discussed in Chapter 13) are examples of these disorders, which are sometimes referred to as an **inborn error of metabolism,** a metabolic disorder caused by the mutation of a gene that codes for an enzyme needed in a biochemical pathway. Both PKU and alkaptonuria involve blocks in the metabolism of specific amino acids.

Many genetic diseases are inherited as autosomal recessive traits

Many human genetic diseases have a simple autosomal recessive inheritance pattern and therefore appear only in the homozygous state. Why are these traits recessive? Most recessive mutations result in a mutant allele that encodes a product that no longer works (either there is not enough gene product, or it is a defective gene product). In the heterozygous state, there is one functional copy of the gene and one mutated, nonfunctional copy. The normal copy of the gene generally produces enough protein to meet the cell's needs. In homozygous recessive individuals, *both* alleles of the gene are nonfunctional, and the cell's needs are not met. As a result, the person shows symptoms of disease.

Phenylketonuria results from an enzyme deficiency

Phenylketonuria (PKU), which is most common in individuals of western European descent, is an autosomal recessive disease caused by a defect of amino acid metabolism. It affects about 1 in 10,000 live births in North America. Homozygous recessive individuals lack an enzyme that converts the amino acid phenylalanine to another amino acid, tyrosine. These individuals accumulate high levels of phenylalanine, phenylpyruvic acid, and similar compounds.

The accumulating phenylalanine is converted to phenylketones, which damage the central nervous system, including the brain, in children. The ultimate result in untreated cases is severe mental retardation. An infant with PKU is usually healthy at birth because its mother, who is heterozygous, breaks down excess phenylalanine for both herself and her fetus. However, during infancy and early childhood the accumulation of toxic products eventually causes irreversible damage to the central nervous system.

In the 1950s, infants with PKU were identified early and placed on a low-phenylalanine diet, dramatically alleviating their symptoms. The diet is difficult to adhere to because it contains no meat, fish, dairy products, breads, or nuts. Also, individuals with PKU should not consume the sugar substitute aspartame, found in many diet drinks and foods, because it contains phenylalanine. Biochemical tests for PKU have been developed, and screening of newborns through a simple blood test is required in the United States. Because of these screening programs and the availability of effective treatment, thousands of PKU-diagnosed children have not developed severe mental retardation. Most must continue the diet through at least adolescence. Doctors now recommend that patients stay on the diet throughout life, because some adults who have discontinued the low-phenylalanine diet experience certain

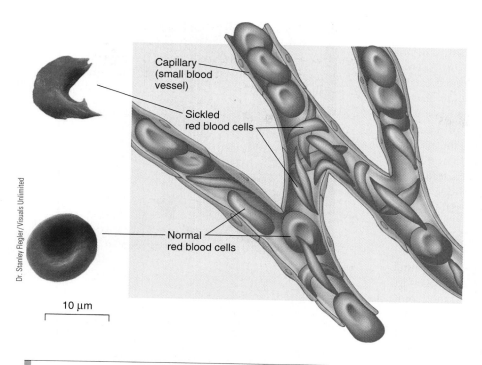

Capillary
(small blood
vessel)

Sickled
red blood cells

Normal
red blood cells

10 μm

Dr. Stanley Fiegler/Visuals Unlimited

Figure 16-7 Sickle cell anemia

Sickled red blood cells do not pass through small blood vessels as easily as unsickled red blood cells do. The sickled cells can cause blockages that prevent oxygen from being delivered to tissues.

mental problems, such as difficulty in concentration and short-term memory loss.

Ironically, the success of PKU treatment in childhood presents a new challenge today. If a homozygous female who has discontinued the special diet becomes pregnant, the high phenylalanine levels in her blood can damage the brain of the fetus she is carrying, even though that fetus is heterozygous. Therefore, she must resume the diet, preferably before becoming pregnant. This procedure is usually (although not always) successful in preventing the effects of **maternal PKU.** It is especially important for women with PKU to be aware of maternal PKU and to obtain appropriate counseling and medical treatment during pregnancy.

Sickle cell anemia results from a hemoglobin defect

Sickle cell anemia is inherited as an autosomal recessive trait. The disease is most common in people of African descent (approximately 1 in 500 African Americans), and about 1 in 12 African Americans is heterozygous. Under low oxygen conditions, the red blood cells of an individual with sickle cell anemia are shaped like sickles, or half-moons, whereas normal red blood cells are biconcave discs.

The mutation that causes sickle cell anemia was first identified more than 50 years ago. The sickled cells contain abnormal hemoglobin molecules, which have the amino acid valine instead of glutamic acid at position 6 (the sixth amino acid from the amino terminal end) in the β-globin chain (see Chapter 3). The substitution of valine for glutamic acid makes the hemoglobin

molecules stick to one another to form fiberlike structures that change the shape of some of the red blood cells. This sickling occurs in the veins after the oxygen has been released from the hemoglobin. The blood cells' abnormal sickled shape slows blood flow and blocks small blood vessels (■ Fig. 16-7), resulting in tissue damage from lack of oxygen and essential nutrients and in episodes of pain. Because sickled red blood cells also have a shorter life span than normal red blood cells, many affected individuals have severe anemia.

Treatments for sickle cell anemia include pain-relief measures, transfusions, and, more recently, medicines such as hydroxyurea, which activates the gene for the production of normal fetal hemoglobin (this gene is generally not expressed after birth). The presence of normal fetal hemoglobin in the red blood cells dilutes the sickle cell hemoglobin, thereby minimizing the painful episodes and reducing the need for blood transfusions. The long-term effects of hydroxyurea are not known at this time, but there are concerns that it may induce tumor formation.

Ongoing research is directed toward providing gene therapy for sickle cell anemia. The development of a mouse model for studying sickle cell anemia has enabled researchers to test gene therapy. The first gene therapy treatments in mice used a mouse retrovirus as a **vector,** a carrier that transfers the genetic information. However, the retrovirus did not effectively transport the normal gene for hemoglobin into the bone marrow, where stem cells produce new blood cells. In 2001, researchers cured sickle cell anemia in mice using a modified HIV as a vector. Before this treatment can be tested in humans, however, researchers must demonstrate that the HIV vector is safe. Bone marrow transplants are also a promising future treatment for seriously ill individuals.

The reason the sickle cell allele occurs at a higher frequency in parts of Africa and Asia is well established. Individuals who are heterozygous ($Hb^A Hb^S$) and carry alleles for both normal hemoglobin (Hb^A) and sickle cell hemoglobin (Hb^S) are more resistant to the malarial parasite, *Plasmodium falciparum*, which causes a severe, often fatal, form of malaria. The malarial parasite, which spends part of its life cycle inside red blood cells, does not thrive when sickle cell hemoglobin is present. (An individual heterozygous for sickle cell anemia produces both normal and sickle cell hemoglobin.) Areas in Africa where falciparum malaria occurs correlate well with areas in which the frequency of the sickle cell allele is more common in the human population. Thus, $Hb^A Hb^S$ individuals, who possess one copy of the mutant sickle cell allele, have a selective advantage over homozygous individuals, both $Hb^A Hb^A$ (who may die of malaria) and $Hb^S Hb^S$ (who may die of sickle cell anemia). This phenomenon, known as **heterozygote advantage,** is discussed further in Chapter 19 (see Fig. 19-7).

Cystic fibrosis results from defective ion transport

Cystic fibrosis is the most common autosomal recessive disorder in children of European descent (1 in 2500 births). About 1 in 25 individuals in the United States is a heterozygous carrier of the mutant cystic fibrosis allele. Abnormal secretions characterize this disorder. Its most severe effect is on the respiratory system, where abnormally viscous mucus clogs the airways. The cilia that line the bronchi cannot easily remove the mucus, and it thus becomes a growth medium for dangerous bacteria. These bacteria or their toxins attack the surrounding tissues, leading to recurring pneumonia and other complications. The heavy mucus also occurs elsewhere in the body, causing digestive difficulties and other effects.

As discussed earlier, the gene responsible for cystic fibrosis codes for CFTR, the protein that regulates the transport of chloride ions across cell membranes. The defective protein, found in plasma membranes of epithelial cells lining the passageways of the lungs, intestines, pancreas, liver, sweat glands, and reproductive organs, results in the production of an unusually thick mucus that eventually leads to tissue damage. Although many forms of cystic fibrosis exist that vary somewhat in the severity of symptoms, the disease is almost always very serious.

Antibiotics are used to control bacterial infections, and daily physical therapy is required to clear mucus from the respiratory system (❙ Fig. 16-8). Treatment with *Dornase Alpha (DNase)*, an enzyme produced by recombinant DNA technology, helps break down the mucus. Without treatment, death would occur in infancy. With treatment, the average life expectancy for individuals with cystic fibrosis is only about 25 years. Because of the serious limitations of available treatments, gene therapy for cystic fibrosis is under development.

The most severe mutant allele for cystic fibrosis predominates in northern Europe, and another, somewhat less serious, mutant allele is more prevalent in southern Europe. Presumably these mutant alleles are independent mutations that have been maintained by natural selection. Some experimental evidence supports the hypothesis that heterozygous individuals are less likely to die from infectious diseases that cause severe diarrhea, such as cholera, another possible example of heterozygote advantage.

Tay–Sachs disease results from abnormal lipid metabolism in the brain

Tay–Sachs disease is an autosomal recessive disease that affects the central nervous system and results in blindness and severe mental retardation. The symptoms begin within the first year of life and result in death before the age of 5 years. Because of the absence of an enzyme, a normal membrane lipid in brain cells fails to break down properly and accumulates in intracellular organelles called *lysosomes* (discussed in Chapter 4). The lysosomes swell and cause the nerve cells to malfunction. Although research is ongoing, no effective treatment for Tay–Sachs disease is available at this time. However, an effective treatment strategy in a mouse model was reported in 1997: oral administration of an inhibitor reduced the synthesis of the lipid that accumu-

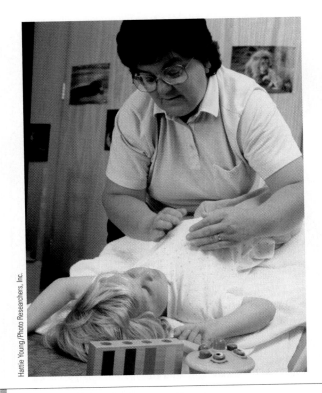

Figure 16-8 Treating cystic fibrosis

One traditional treatment for cystic fibrosis is chest percussion, or gentle pounding on the chest, to clear mucus from clogged airways in the lungs. This technique is typically done after nebulizer therapy, which loosens the mucus.

lates in the lysosomes. This treatment offers the hope of future breakthroughs to deal more effectively with Tay–Sachs disease in humans.

The abnormal allele is especially common in the United States among Jews whose ancestors came from eastern and central Europe (Ashkenazi Jews). In contrast, Jews whose ancestors came from the Mediterranean region (Sephardic Jews) have a very low frequency of the allele.

Some genetic diseases are inherited as autosomal dominant traits

Huntington's disease (HD), named after George Huntington, the U.S. physician who first described it in 1872, is caused by a rare autosomal dominant allele that affects the central nervous system. The disease causes severe mental and physical deterioration, uncontrollable muscle spasms, and personality changes; death ultimately results. No effective treatment has been found. Every child of an affected individual has a 50% chance of also being affected (and, if affected, of passing the abnormal allele to his or her offspring). Ordinarily we would expect a dominant allele with such devastating effects to occur only as a new mutation and not to be transmitted to future generations. Because HD symptoms do not appear until relatively late in life (most

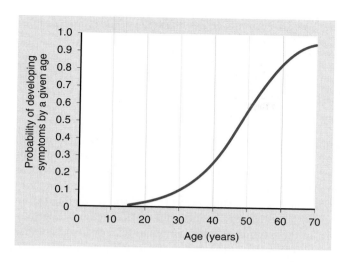

Figure 16-9 The age of onset of Huntington's disease

The graph shows the cumulative probability that an individual affected with a Huntington's disease allele will have developed symptoms at a given age. (Adapted from P. S. Harper, *Genetic Counseling,* 5th ed., Butterworth-Heinemann, Oxford, 1998.)

people do not develop the disease until they are in their thirties or forties), a person may have children before the disease develops (❚ Fig. 16-9). In North America, HD occurs in 1 in 20,000 live births.

The gene responsible for HD is located on the short end of chromosome 4. The mutation is a nucleotide triplet (CAG) that is repeated many times; the normal allele repeats CAG from 6 to 35 times, whereas the mutant allele repeats CAG from 40 to more than 150 times. Because CAG codes for the amino acid glutamine, the resulting protein, called *huntingtin,* has a long strand of glutamines. The number of nucleotide triplet repeats seems to be important in determining the age of onset and the severity of the disease; larger numbers of repeats correlate with an earlier age of onset and greater severity.

Much research now focuses on how the mutation is linked to neurodegeneration in the brain. A mouse model of HD is providing valuable clues about the development of the disease. Using this model, researchers have demonstrated that the defective version of huntingtin binds to enzymes called *acetyltransferases* in brain cells, blocking their action. Acetyltransferases are involved in turning genes on for expression, so in the brain cells of HD individuals, much of normal transcription cannot occur. Once neurologists better understand HD's mechanism of action on nerve cells, it may be possible to develop effective treatments to slow the progression of the disease.

Cloning of the *HD* allele became the basis for tests that allow those at risk to learn presymptomatically if they carry the allele. The decision to be tested for any genetic disease is understandably a highly personal one. The information is, of course, invaluable for those who must decide whether or not to have children.

However, someone who tests positive for the *HD* allele must then live with the virtual certainty of eventually developing this devastating and incurable disease. Researchers hope that information from affected individuals who choose to be identified before the onset of symptoms may ultimately contribute to the development of effective treatments.

Some genetic diseases are inherited as X-linked recessive traits

Hemophilia A was once referred to as a disease of royalty because of its high incidence among male descendants of Queen Victoria, but it is also found in many nonroyal pedigrees. Caused by the absence of a blood-clotting protein called factor VIII, **hemophilia A** is characterized by severe internal bleeding in the head, joints, and other areas from even a slight wound. The mode of inheritance is X-linked recessive. Thus, affected individuals are almost exclusively male, having inherited the abnormal allele on the X chromosome from their heterozygous carrier mothers. (For a female to be affected by an X-linked trait, she would have to inherit the defective allele from both parents, whereas an affected male need only inherit one defective allele from his mother.)

Treatments for hemophilia A consist of blood transfusions and the administration of clotting factor VIII (the missing gene product) by injection. Unfortunately, these treatments are costly. During the 1980s, many clotting factor VIII preparations made from human plasma were unknowingly contaminated with HIV, and many men with hemophilia subsequently died from AIDS. Since 1992, virus-free clotting factor VIII has been available from both human plasma and recombinant DNA technology.

Geneticists are beginning to unravel X-linked genes affecting intelligence

The X chromosome contains a disproportionate number of the more than 200 genes identified so far that affect cognitive abilities, for example, by coding for proteins required for the brain to function normally. Not surprisingly, many kinds of mental impairment are linked to defects in genes on the X chromosome. By one count, the human X chromosome contains less than 4% of the human genome, yet 10% of the genes in which defects are known to cause some form of mental retardation are found on the X chromosome. Because males have only one X chromosome, more boys than girls have some form of mental impairment, a fact that has been observed for more than a century.

Review

❚ Which of the following genetic diseases is/are inherited as an autosomal recessive: phenylketonuria, Huntington's disease, Tay–Sachs disease?

❚ Which of the following genetic diseases is/are inherited as an autosomal dominant: sickle cell anemia, hemophilia A, Huntington's disease?

❚ Which of the following genetic diseases is/are inherited as an X-linked recessive: hemophilia A, cystic fibrosis, Tay–Sachs disease?

GENE THERAPY

Learning Objective

7 Briefly discuss the process of gene therapy, including some of its technical challenges.

Because serious genetic diseases are difficult to treat, scientists have dreamed of developing actual cures. Today, advances in genetics are bringing these dreams closer to reality. One strategy is **gene therapy,** which aims to replace a mutant allele in certain body cells with a normal allele. The rationale is that although a particular allele may be present in all cells, it is expressed only in some. Expression of the normal allele in only the cells that require it may be sufficient to yield a normal phenotype (❙ Fig. 16-10).

This approach presents several technical problems. The solutions to these problems must be tailored to the nature of the gene itself, as well as to its product and the types of cells in which it is expressed. First, the gene is cloned and the DNA introduced into the appropriate cells. One of the most successful techniques is packaging the normal allele in a viral vector, a virus that moves the normal allele into target cells that currently have a mutant allele. Ideally the virus should infect a high percentage of the cells. Most important, the virus should do no harm, especially over the long term.

To date, gene therapy has saved the lives of at least 17 children with *severe combined immunodeficiency (SCID)* by restoring their immune systems. Although many obstacles must be overcome, gene therapies for several other genetic diseases are under development or are being tested on individuals in clinical trials.

Scientists are currently addressing some of the unique problems presented by each disease.

Gene therapy programs are carefully scrutinized

Until recently, major technical advances caused the number of clinical studies involving gene therapy to grow dramatically. However, the death of a young man in a gene therapy trial in 1999, and three more recent cases of cancer (leukemia) in children, one of whom died, led to a shutdown of many trials, pending the outcome of investigations about health risks. The main safety concern in these inquiries is the potential toxicity of viral vectors. The vector used in the young man who died was an adenovirus (see Fig. 24-1b), a virus required in large doses to transfer enough copies of the normal alleles for effective therapy. Unfortunately, the high viral doses triggered a fatally strong immune response in the patient's body. The children who developed leukemia were being treated for SCID. The vector in these cases was a retrovirus that inserted itself into and activated an oncogene that can cause childhood leukemia.

Performing clinical trials on humans always has inherent risks. Researchers carefully select patients and thoroughly explain the potential benefits and risks, as far as they are known, so the patient—or, in the case of children, the parents—can give informed consent for the procedure. However, the problems in gene therapy trials in recent years have researchers busy developing safer alternatives to viral vectors.

Key Point

Mice are a model system for the development of gene therapy, the use of normal genes to correct or alleviate the symptoms of a genetic disease that is caused by defective copies of that gene.

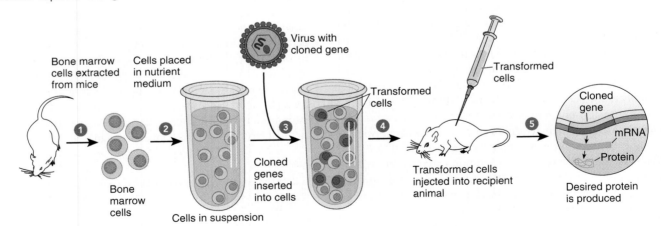

Figure 16-10 Gene therapy in bone marrow cells of a mouse
This procedure is performed on humans for certain types of genetic diseases.

- How are viruses used in human gene therapy?
- Why are viral vectors of potential concern in human gene therapy?

GENETIC TESTING AND COUNSELING

Learning Objectives

8   State the relative advantages and disadvantages of amniocentesis, chorionic villus sampling, and preimplantation genetic diagnosis in the prenatal diagnosis of human genetic abnormalities.

9   Distinguish between genetic screening programs for newborns and adults, and discuss the scope and implications of genetic counseling.

Geneticists have made many advances in detecting genetic disorders in individuals in recent years, including in prenatal diagnosis and genetic screening. With these advances comes increased information for couples at risk of having children with genetic diseases. Helping couples understand and deal with the genetic information now available is part of the rapidly expanding field of genetic counseling.

Prenatal diagnosis detects chromosome abnormalities and gene defects

Health-care professionals are increasingly successful at diagnosing genetic diseases prenatally. In the diagnostic technique called **amniocentesis,** a physician obtains a sample of the *amniotic fluid* surrounding the fetus by inserting a needle through the pregnant woman's abdomen, into the uterus, and then into the amniotic sac surrounding the fetus. Some of the amniotic fluid is withdrawn from the amniotic cavity into a syringe (❙ Fig. 16-11). The fetus is normally safe from needle injuries because **ultrasound imaging** helps determine the positions of the fetus, placenta, and the needle (see Fig. 50-18). However, there is a 0.5%, or 1 in 200, chance that amniocentesis will induce a miscarriage.

Amniotic fluid contains living cells sloughed off the body of the fetus and hence genetically identical to the cells of the fetus. After cells grow for about 2 weeks in culture in the lab, technicians karyotype dividing cells to detect chromosome abnormalities. Other DNA tests have also been developed to identify most chromosome abnormalities. Amniocentesis, which has been performed since the 1960s, is routinely offered for pregnant women older than age 35 because their fetuses have a higher-than-normal risk of Down syndrome.

Researchers have developed other prenatal tests to detect many genetic disorders with a simple inheritance pattern, but these disorders are rare enough that physicians usually order the tests performed only if they suspect a particular problem. Enzyme deficiencies can often be detected by incubating cells recovered

Key Point

In amniocentesis, the fluid surrounding the developing fetus is sampled, usually during the 16th week of pregnancy, to detect genetic and developmental disorders.

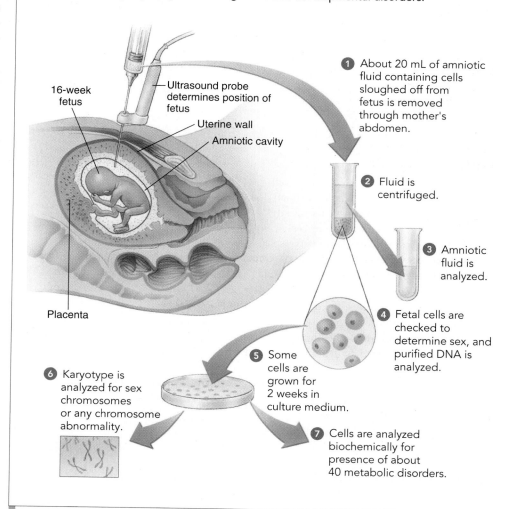

1 About 20 mL of amniotic fluid containing cells sloughed off from fetus is removed through mother's abdomen.

16-week fetus
Ultrasound probe determines position of fetus
Uterine wall
Amniotic cavity

Placenta

2 Fluid is centrifuged.

3 Amniotic fluid is analyzed.

4 Fetal cells are checked to determine sex, and purified DNA is analyzed.

5 Some cells are grown for 2 weeks in culture medium.

6 Karyotype is analyzed for sex chromosomes or any chromosome abnormality.

7 Cells are analyzed biochemically for presence of about 40 metabolic disorders.

Figure 16-11 *Animated* Amniocentesis

Certain genetic diseases and other abnormal conditions are diagnosed prenatally by amniocentesis.

from amniotic fluid with the appropriate substrate and measuring the product; this technique has been useful in prenatal diagnosis of disorders such as Tay–Sachs disease. The tests for several other diseases, including sickle cell anemia, Huntington's disease, and cystic fibrosis, involve directly testing the individual's DNA for the mutant allele.

Amniocentesis is also useful in detecting a condition known as *spina bifida*, in which the spinal cord does not close properly during development. A relatively common malformation (about 1 in 300 births), this birth defect is associated with abnormally high levels of a normally occurring protein, α-*fetoprotein*, in the amniotic fluid. Some of this protein crosses the placenta into the mother's blood, which is tested for *maternal serum* α-*fetoprotein (MSAFP)* as a screen for spinal cord defects. If an elevated level of MSAFP is detected, the physician performs diagnostic tests, such as ultrasound imaging and amniocentesis. (Interestingly, abnormally *low* levels of MSAFP are associated with Down syndrome and other trisomies.)

One problem with amniocentesis is that most of the conditions it detects are unpreventable and incurable, and the results are generally not obtained until well into the second trimester, when terminating the pregnancy is both psychologically and medically more difficult than earlier. Therefore, researchers have developed tests that yield results earlier in the pregnancy. **Chorionic villus sampling (CVS)** involves removing and studying cells that will form the fetal contribution to the placenta (█ Fig. 16-12). CVS, which has been performed in the United States since about 1983, is associated with a slightly greater risk of infection or miscarriage than amniocentesis, but its advantage is that results are obtained earlier in the pregnancy than in amniocentesis, usually within the first trimester.

A relatively new embryo-screening process, known as **preimplantation genetic diagnosis (PGD),** is available for couples who carry alleles for Tay–Sachs disease, hemophilia, sickle cell anemia, and dozens of other inherited genetic conditions. Conception is by **in vitro fertilization (IVF),** in which gametes are collected, eggs are fertilized in a dish in the laboratory, and the resulting embryo is then implanted in the uterus for development (see *Focus On: Novel Origins,* in Chapter 49). Prior to implantation, the physician screens embryos for one or more genetic diseases before placing a healthy embryo into the woman's uterus. PGD differs from amniocentesis and CVS in that the test is performed *before* a woman is pregnant, so it eliminates the decision of whether or not to terminate the pregnancy if an embryo has

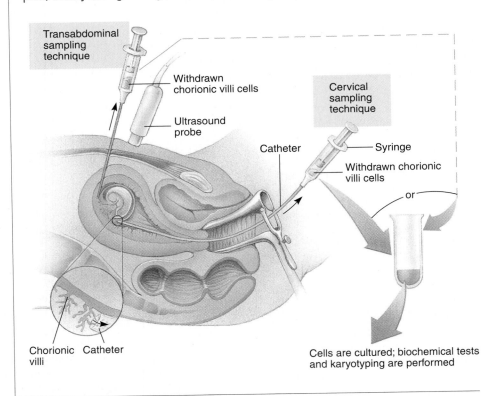

Key Point

In chorionic villus sampling, chorionic cells genetically identical to the embryo are sampled, usually during the eighth or ninth week of pregnancy, to detect genetic disorders.

Transabdominal sampling technique

Withdrawn chorionic villi cells

Ultrasound probe

Cervical sampling technique

Catheter

Syringe

Withdrawn chorionic villi cells

or

Chorionic villi   Catheter

Cells are cultured; biochemical tests and karyotyping are performed

Figure 16-12 Chorionic villus sampling (CVS)

This test allows the early diagnosis of some genetic abnormalities. Samples may be obtained by inserting a needle through the uterine wall or the cervical opening.

a genetic abnormality. However, PGD is not as accurate as amniocentesis or CVS, and it is more expensive. Moreover, PGD is sometimes controversial, because some couples use it to choose the gender of their offspring, not to screen for genetic diseases.

Although using amniocentesis, CVS, and PGD can help physicians diagnose certain genetic disorders with a high degree of accuracy, the tests are not foolproof, and many disorders cannot be diagnosed at all. Therefore, the lack of an abnormal finding is no guarantee of a normal pregnancy.

Genetic screening searches for genotypes or karyotypes

Genetic screening is a systematic search through a population for individuals with a genotype or karyotype that might cause a serious genetic disease in themselves or their offspring. There are two main types of genetic screening, for newborns and for adults, and each serves a different purpose. Newborns are screened primarily as the first step in preventive medicine, and adults are screened to help them make informed reproductive decisions.

Newborns are screened to detect and treat certain genetic diseases before the onset of serious symptoms. The routine screening of infants for PKU began in 1962 in Massachusetts. Laws in all 50 states of the United States, as well as in many other countries, currently require PKU screening. Sickle cell anemia is also more effectively treated with early diagnosis. Screening newborns for sickle cell anemia reduces infant mortality by about 15% because doctors can administer daily doses of antibiotics, thereby preventing bacterial infections common to newborns with the disease. The number of genetic disorders that can be screened in newborns is rapidly increasing, and the March of Dimes currently recommends screening newborns for 29 disorders, most of which are genetic.

Genetic screening of adults identifies carriers (heterozygotes) of recessive genetic disorders. If both prospective parents are heterozygous, the carriers are counseled about the risks involved in having children. Since the 1970s, about 1 million young Jewish adults in the United States, Israel, and other countries have been screened voluntarily for Tay–Sachs disease, and about 1 in 30 has been identified as a carrier. Tay–Sachs screening programs have reduced the incidence of Tay–Sachs disease to almost zero.

Genetic counselors educate people about genetic diseases

Couples who are concerned about the risk of abnormality in their children, because they have either an abnormal child or a relative affected by a hereditary disease, may seek **genetic counseling** for medical and genetic information as well as support and guidance. Genetic clinics, available in most major metropolitan centers, are usually affiliated with medical schools.

Genetic counselors, who have received training in counseling, medicine, and human genetics, provide people with the information they need to make reproductive decisions. They offer advice, tempered with respect and sensitivity, in terms of risk estimates—that is, the *probability* that any given offspring will inherit a particular condition. The counselor uses the complete family histories of both the man and the woman, and a clinical geneticist (a physician who specializes in genetics) may screen for the detection of heterozygous carriers of certain conditions.

When a disease involves only a single gene locus, probabilities can usually be easily calculated. For example, if one prospective parent is affected with a trait that is inherited as an autosomal dominant disorder, such as Huntington's disease, the probability that any given child will have the disease is 0.5, or 50%. The birth to phenotypically normal parents of a child affected with an autosomal recessive trait, such as albinism or PKU, establishes that both parents are heterozygous carriers, and the probability that any subsequent child will be affected is therefore 0.25, or 25%. For a disease inherited through a recessive allele on the X chromosome, such as hemophilia A, a normal woman and an affected man will have daughters who are carriers and sons who are normal. The probability that the son of a carrier mother and a normal father will be affected is 0.5, or 50%; the probability that their daughter will be a carrier is also 0.5.

It is important for identified carriers to receive appropriate genetic counseling. A genetic counselor is trained not only to provide information pertaining to reproductive decisions but also to help individuals understand their situation and avoid feeling stigmatized.

Review

- What are the relative advantages and disadvantages of amniocentesis, chorionic villus sampling, and preimplantation genetic diagnosis?
- What is the purpose of genetic screening for newborns?
- What is the purpose of genetic screening for adults?

HUMAN GENETICS, SOCIETY, AND ETHICS

Learning Objective

10 Discuss the controversies of genetic discrimination.

Many misconceptions exist about genetic diseases and their effects on society. Some people erroneously think of certain individuals or populations as genetically unfit and thus responsible for many of society's ills. They argue, for example, that medical treatment of people affected with genetic diseases, especially those who reproduce, increases the frequency of abnormal alleles in the population. However, such notions are incorrect. Genetic disorders are so rare that modern medical treatments will have only a negligible effect on their incidence.

Recessive mutant alleles are present in *all* individuals and *all* ethnic groups; no one is exempt. According to one estimate, each of us is heterozygous for several (3 to 15) very harmful recessive alleles, any of which could cause debilitating illness or death in the homozygous state. Why, then, are genetic diseases relatively uncommon? Each of us has many thousands of essential genes, any of which can be mutated. It is very unlikely that the abnormal alleles that one individual carries are also carried by that individual's mate. Of course, this possibility is more likely if the harmful allele is a relatively common one, such as the one responsible for cystic fibrosis.

Relatives are more likely than nonrelatives to carry the same harmful alleles, having inherited them from a common ancestor. A greater-than-normal frequency of a particular genetic disease among offspring of **consanguineous matings,** matings between genetically related individuals, is often the first clue that the mode of inheritance is autosomal recessive. The offspring of consanguineous matings have a small but significantly increased risk of genetic disease. In fact, they account for a disproportionately high percentage of those individuals in the population with autosomal recessive disorders. Because of this perceived social cost, marriages of close relatives, including first cousins, are prohibited by about half the states in the United States. However, consanguineous marriages are still relatively common in many other countries.

Genetic discrimination provokes heated debate

One of the fastest-growing areas of medical diagnostics is genetic screening and testing, and the number of new genetic tests that screen for diseases such as cystic fibrosis, sickle cell anemia, HD, colon cancer, and breast cancer increases each year. However, genetic testing raises many social, ethical, and legal issues that we as a society must address.

One of the most difficult issues is whether genetic information should be available to health insurance and life insurance companies. Many people think genetic information should not be given to insurance companies, but others, including employers, insurers, and many organizations representing people affected by genetic disorders, say such a view is unrealistic. If people use genetic tests to help them decide when to buy insurance and how much, then insurers insist they should also have access to this information. Insurers say they need access to genetic data to help calculate equitable premiums (insurance companies average risks over a large population). However, some are concerned that insurers might use the results of genetic tests to discriminate against people with genetic diseases or to deny them coverage.

Physicians argue that people at risk for a particular genetic disease might delay being tested because they fear genetic discrimination from insurers and employers. **Genetic discrimination** is discrimination against an individual or family members because of differences from the "normal" genome in that individual. The perception of genetic discrimination already exists in society. A 1996 study found that 25% of 332 people with family histories of one or more genetic disorders thought they had been refused life insurance, 22% thought they had been refused health insurance, and 13% thought they had been denied employment because of genetic discrimination. In a 1998 survey by the National Center for Genetic Resources, 63% of respondents said they probably or definitely would not take a genetic test if the results could be disclosed to either their employers or insurers.

Complicating the issue even more, genetic tests are sometimes difficult to interpret, in part because of the many complex interactions between genes and the environment. If a woman tests positive for an allele that has been linked to breast cancer, for example, she is at significant risk, but testing positive does not necessarily mean she will develop breast cancer. These uncertainties also make it hard to decide what form of medical intervention, from frequent mammograms to surgical removal of healthy breasts, is appropriate.

The Ethical, Legal, and Social Implications (ELSI) Research Program of the National Human Genome Research Institute has developed principles designed to protect people against genetic discrimination. The Health Insurance Portability and Accountability Act of 1996 provides some safeguards against genetic discrimination, and the Americans with Disabilities Act may also apply to genetic discrimination. As this book goes to press, bills that extend significant protection against workplace discrimination, health discrimination, and invasion of privacy based on genetic information are being considered by both federal and state legislatures. These issues will be debated for years to come.

Many ethical issues related to human genetics must be addressed

Genetic discrimination is only one example of ethical issues arising from our expanding knowledge of human genetics. Consider the following questions, all of which deal with the broad ethical issue of individual rights: What is the youngest age at which genetic testing should be permitted for adult-onset diseases, such as Huntington's disease? What are the emotional and psychological effects on individuals who are told they have tested positive for an incurable genetic disease? Should testing be performed when some family members want testing and others do not? Should parents be able to test their minor children? Should access to genetic test data be permitted in cases of paternity or kinship testing? Should states be able to collect genetic data on their residents? Should school administrators or law enforcement agencies have access to genetic data? These questions are only a sample of the many issues that both ethicists and society must consider now and in the future. As human genetics assumes an increasingly important role in society, issues of genetic privacy and the confidentiality of genetic information must be addressed.

Review

- Why is it incorrect to assume that certain individuals or populations carry most of the abnormal alleles found in humans?
- To be expressed, an autosomal recessive genetic disease must be homozygous. What relationship does this fact have to consanguineous matings?
- Why do health and life insurance companies want genetic information about their clients?

SUMMARY WITH KEY TERMS

Learning Objectives

1 Distinguish between karyotyping and pedigree analysis (page 347).
- Studies of an individual's **karyotype,** the number and kinds of chromosomes present in the nucleus, enable researchers to identify various chromosome abnormalities.
- A **pedigree** is a "family tree" that shows the transmission of genetic traits within a family over several generations. Pedigree analysis is useful in detecting autosomal dominant mutations, autosomal recessive mutations, X-linked

recessive mutations, and defects due to **genomic imprinting,** expression of a gene based on its parental origin.

ThomsonNOW Learn more about pedigrees by clicking on the figure in ThomsonNOW.

2 Discuss the implications of the Human Genome Project (page 347).
- The **human genome** is the total genetic information in human cells. The **Human Genome Project** sequenced the DNA in the nuclear human genome. Now that the human

genome has been sequenced, researchers will be busy for many decades identifying genes and understanding what each gene's role is, how each gene interacts with other genes, and how the expression of each gene is regulated in different tissues.

3   Discuss the mouse model for studying cystic fibrosis (page 347).

- The use of animal models greatly helps researchers investigate human disease. Researchers used **gene targeting** to produce strains of mice that are either homozygous or heterozygous for cystic fibrosis. Results from these studies may yield more effective drugs for treating the disease.

4   Explain how nondisjunction in meiosis is responsible for chromosome abnormalities such as Down syndrome, Klinefelter syndrome, and Turner syndrome (page 351).

- In **aneuploidy,** there are either missing or extra copies of certain chromosomes. Aneuploidies include **trisomy,** in which an individual possesses an extra chromosome, and **monosomy,** in which one member of a pair of chromosomes is missing.

- **Trisomy 21,** the most common form of **Down syndrome,** and **Klinefelter syndrome** (XXY) are examples of trisomy. **Turner syndrome** (X0) is an example of monosomy.

- Trisomy and monosomy are caused by meiotic **nondisjunction,** in which sister chromatids or homologous chromosomes fail to move apart properly during meiosis.

ThomsonNOW™ **Learn more about normal and abnormal karyotypes by clicking on the figure in ThomsonNOW.**

5   Distinguish among the following structural abnormalities in chromosomes: translocations, deletions, and fragile sites (page 351).

- In a **translocation,** part of one chromosome becomes attached to another. About 4% of individuals with Down syndrome have a translocation in which the long arm of chromosome 21 is attached to the long arm of one of the larger chromosomes, such as chromosome 14.

- A **deletion** can result in chromosome breaks that fail to rejoin. The deletion may range in size from a few base pairs to an entire chromosome arm. One deletion disorder in humans is **cri du chat syndrome,** in which part of the short arm of chromosome 5 is deleted.

- **Fragile sites** may occur at specific locations on both chromatids of a chromosome. In **fragile X syndrome,** a fragile site occurs near the tip on the X chromosome, where the nucleotide triplet CGG is repeated many more times than is normal. Fragile X syndrome is the most common cause of inherited mental retardation.

6   State whether each of the following genetic defects is inherited as an autosomal recessive, autosomal dominant, or X-linked recessive: phenylketonuria (PKU), sickle cell anemia, cystic fibrosis, Tay–Sachs disease, Huntington's disease, and hemophilia A (page 356).

- Most human genetic diseases that show a simple inheritance pattern are transmitted as autosomal recessive traits. An **inborn error of metabolism** is a metabolic disorder caused by the mutation of a gene that codes for an enzyme needed for a biochemical pathway.

- **Phenylketonuria (PKU)** is an autosomal recessive disorder in which toxic phenylketones damage the developing nervous system. **Sickle cell anemia** is an autosomal recessive disorder in which abnormal hemoglobin (the protein that transports oxygen in the blood) is produced. **Cystic fibrosis** is an autosomal recessive disorder in which abnormal secretions are produced primarily in organs of the respiratory and digestive systems. **Tay–Sachs disease**

is an autosomal recessive disorder caused by abnormal lipid metabolism in the brain.

- **Huntington's disease** has an autosomal dominant inheritance pattern that results in mental and physical deterioration, usually beginning in adulthood.

- **Hemophilia A** is an X-linked recessive disorder that results in a defect in a blood component required for clotting.

7   Briefly discuss the process of gene therapy, including some of its technical challenges (page 360).

- In **gene therapy,** the normal allele is cloned and the DNA introduced into certain body cells, where its expression may be sufficient to yield a normal phenotype.

- One technical challenge in gene therapy is finding a safe, effective **vector,** usually a virus, to deliver the gene of interest into the cells. Ideally, the virus should infect a high percentage of the cells and do no harm, especially over the long term.

8   State the relative advantages and disadvantages of amniocentesis, chorionic villus sampling, and preimplantation genetic diagnosis in the prenatal diagnosis of human genetic abnormalities (page 361).

- In **amniocentesis,** a physician samples the amniotic fluid surrounding the fetus and cultures and screens the fetal cells suspended in the fluid for genetic defects. Amniocentesis provides results in the second trimester of pregnancy.

ThomsonNOW™ **Learn more about amniocentesis by clicking on the figure in ThomsonNOW.**

- In **chorionic villus sampling (CVS),** a physician removes and studies some of the fetal cells. CVS provides results in the first trimester of pregnancy but is associated with a slightly greater risk of infection and miscarriage than amniocentesis.

- In **preimplantation genetic diagnosis (PGD),** couples conceive by in vitro fertilization. A physician screens the embryos for one or more genetic diseases before placing a healthy embryo into the woman's uterus. PGD is not as accurate as amniocentesis or CVS, and it is more expensive.

9   Distinguish between genetic screening programs for newborns and adults, and discuss the scope and implications of genetic counseling (page 361).

- **Genetic screening** identifies individuals who might carry a serious genetic disease. Screening of newborns is the first step in preventive medicine, and screening of adults helps them make informed reproductive decisions.

- Couples who are concerned about the risk of abnormality in their children may seek **genetic counseling.** A genetic counselor provides medical and genetic information pertaining to reproductive decisions and helps individuals understand their situation and avoid feeling stigmatized.

10 Discuss the controversies of genetic discrimination (page 363).

- **Genetic discrimination** is discrimination against an individual or family member because of differences from the "normal" genome in that individual.

- One of the most difficult issues in avoiding genetic discrimination is whether genetic information should be available to employers and to health and life insurance companies. Physicians are concerned that people at risk for a particular genetic disease might delay being tested because they fear genetic discrimination from insurers and employers.

- As human genetics assumes an increasingly important role in human society, issues of genetic privacy and the confidentiality of genetic information must be addressed.

1. A diagram of a pedigree shows (a) controlled matings between members of different true-breeding strains (b) the total genetic information in human cells (c) a comparison of DNA sequences among genomes of humans and other species (d) the subtle genetic differences among unrelated people (e) the expression of genetic traits in the members of two or more generations of a family

2. Which pattern of inheritance is associated with a trait that (1) is not usually expressed in the parents, (2) is expressed in about one fourth of the children, and (3) is expressed in both male and female children? (a) autosomal recessive (b) autosomal dominant (c) X-linked recessive (d) X-linked dominant (e) Y-linked

3. The Human Genome Project (a) sequenced all the DNA in the human nucleus (b) was exclusively concerned with the comparisons of DNA sequences between human DNA and DNA of other species (c) customized drugs to match an individual's genetic makeup (d) searched for individuals with a genotype that might cause a serious genetic disease in them or their offspring (e) provided risk estimates on human genetic diseases

4. An abnormality in which there is one more or one fewer than the normal number of chromosomes is called a(an) (a) karyotype (b) fragile site (c) aneuploidy (d) trisomy (e) translocation

5. An individual with one extra chromosome (three of one kind) is said to be (a) monosomic (b) triploid (c) trisomic (d) consanguineous (e) true-breeding

6. An individual who is missing one chromosome, having only one member of a given pair, is said to be (a) monosomic (b) haploid (c) trisomic (d) consanguineous (e) true-breeding

7. The failure of chromosomes to separate normally during cell division is called (a) a fragile site (b) an inborn error of metabolism (c) a satellite knob (d) a translocation (e) nondisjunction

8. The transfer of a part of one chromosome to a non-homologous chromosome is called a(an) (a) karyotype (b) inborn error of metabolism (c) pedigree (d) translocation (e) nondisjunction

9. The chromosome composition of an individual or cell is called its (a) karyotype (b) nucleotide triplet repeat (c) pedigree (d) DNA microarray (e) translocation

10. Individuals with trisomy 21, or _____, are often mentally and physically retarded and have abnormalities of the face, tongue, and eyelids. (a) Down syndrome (b) Klinefelter syndrome (c) Turner syndrome (d) Huntington's disease (e) Tay–Sachs disease

11. An inherited disorder caused by a defective or absent enzyme is called a(an) (a) karyotype (b) trisomy (c) reciprocal translocation (d) inborn error of metabolism (e) aneuploidy

12. In _____, a genetic mutation codes for an abnormal hemoglobin molecule that is less soluble than usual and more likely than normal to deform the shape of the red blood cell. (a) Down syndrome (b) Tay–Sachs disease (c) sickle cell anemia (d) PKU (e) hemophilia A

13. In an individual with _____, the mucus is abnormally viscous and tends to plug the ducts of the pancreas and liver and to accumulate in the lungs. (a) Down syndrome (b) Tay–Sachs disease (c) sickle cell anemia (d) PKU (e) cystic fibrosis

14. During this procedure, a sample of the fluid that surrounds the fetus is obtained by inserting a needle through the walls of the abdomen and uterus. (a) DNA marker (b) chorionic villus sampling (c) ultrasound imaging (d) preimplantation genetic diagnosis (e) amniocentesis

15. For which of the following situations would a genetic counselor *not* recommend prenatal diagnosis involving amniocentesis or chorionic villus sampling? (a) an increased risk of a chromosome abnormality (b) an increased risk of a single-locus (Mendelian) disease (c) an increased risk of a spinal cord defect (d) a desire to know the sex of the fetus (e) a pregnant woman older than 35

16–18. Examine the following pedigrees, and decide on the most likely mode of inheritance of each disorder. (a) autosomal recessive (b) autosomal dominant (c) X-linked recessive (d) a, b, or c (e) a or c

16.   17.   18.

19–28. What is the inheritance pattern of each disease? (a) chromosome abnormality (b) autosomal recessive (c) autosomal dominant (d) X-linked recessive (e) unknown

19. Down syndrome

20. Tay–Sachs disease

21. Phenylketonuria

22. Hemophilia A

23. Sickle cell anemia

24. Turner syndrome

25. Huntington's disease

26. Klinefelter syndrome

27. Cri du chat syndrome

28. Fragile X syndrome

1. Imagine that you are a genetic counselor. What advice or suggestions might you give in the following situations?

   a. A couple has come for advice because the woman had a sister who died of Tay–Sachs disease.

   b. A young man and woman who are not related are engaged to be married. However, they have learned that the man's parents are first cousins. They are worried about the possibility of increased risk of genetic defects in their own children.

   c. A young woman's paternal uncle (her father's brother) has hemophilia A. Her father is free of the disease, and there has never been a case of hemophilia A in her mother's family. Should she be concerned about the possibility of hemophilia A in her own children?

   d. A 20-year-old man is seeking counseling because his father was recently diagnosed with Huntington's disease.

   e. A 45-year-old woman has just been diagnosed with Huntington's disease. She says she will not tell her college-age sons because of the burden it will place on them. Given that the woman, not her sons, is your client, do you have a duty to inform the sons? Explain your reasoning.

2. A common belief about human genetics is that an individual's genes alone determine his or her destiny. Explain why this is a misconception.

3. Is a chromosome deletion equivalent to a frameshift mutation (discussed in Chapter 13)? Why or why not?

4. **Evolution Link.** Explain some of the evolutionary implications that one can conclude from the fact that mice and humans have about 500 DNA segments that are completely identical.

5. **Analyzing Data.** Examine Figure 16-9, and estimate the age at which half of individuals carrying a Huntington's disease allele will have developed symptoms. At what age will three fourths of these individuals have symptoms?

Additional questions are available in ThomsonNOW at www.thomsonedu.com/login

Developmental Genetics

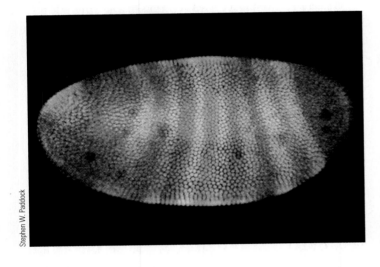

Stephen W. Paddock

Differential gene expression in
the model organism *Drosophila*.

KEY CONCEPTS

Genes control the development of specialized tissues and
organs.

Model organisms with characteristics uniquely suited for
developmental studies include fruit flies, nematodes, mice,
and *Arabidopsis*, a flowering plant.

Many genes that regulate development are quite similar in
a wide range of organisms, from fruit flies to humans.

Mutations in genes that regulate development have provided insights into how those genes function.

Mutations in oncogenes and tumor suppressor genes may
lead to cancer.

Developmental genetics is the study of the genes involved in
cell differentiation and development of an organism. Until
the late 1970s, biologists knew little about how genes interact
to control development. Unraveling the genetic interactions that
take place during development was an intractable problem using
traditional methods. However, rapid progress in recombinant DNA
research led scientists to search for developmental mutants and
to apply sophisticated techniques to study them. (A *mutant* is
an individual with an abnormal phenotype caused by a gene
mutation.)

The organism in the photograph is a developing embryo of
the fruit fly *Drosophila melanogaster*. Geneticists used **immunofluorescence,** in which a fluorescent dye is joined to an antibody
that binds to a specific protein, to localize the protein. In this
case researchers bound three different antibodies—one red, one
blue, and one yellow—to three specific proteins. The patterns of
colored bands indicate that different cells of the embryo have *differential gene expression*—that is, different genes are active at the
same time.

Work with *Drosophila* and other organisms has profound implications for understanding both normal human development,
including aging, and malfunctions that lead to birth defects and
cancer. Striking similarities among genes that govern development
in widely different species suggest that developmentally important
genetic mechanisms are deeply rooted in the evolutionary history
of multicellular organisms. There are also differences in species'

developmental patterns that reflect their separate evolutionary paths. As you will learn in Chapter 18, developmental genes have played a role in reshaping organisms during the course of their evolution.

Biologists are now studying how genes are activated, inactivated, and modified to control development. Eventually, scientists expect to understand how a single cell—a fertilized egg—develops into a multicellular organism as complex as a human. ■

CELL DIFFERENTIATION AND NUCLEAR EQUIVALENCE

Learning Objectives

1 Distinguish between cell determination and cell differentiation, and between nuclear equivalence and totipotency.
2 Describe the classic experiments of Steward, Gurdon, and Wilmut.
3 Define *stem cells*, distinguish between embryonic stem cells and pluripotent stem cells, and describe some of the promising areas of research involving stem cells.

The study of **development,** broadly defined as all the changes that occur in the life of an individual, encompasses some of the most fascinating and difficult problems in biology today. Of particular interest is the process by which cells specialize and organize into a complex organism. During the many cell divisions required for a single cell to develop into a multicellular organism, groups of cells become gradually committed to specific patterns of gene activity through the process of **cell determination.** As cell determination proceeds, it restricts an embryonic cell's developmental pathway so that its fate becomes more and more limited. The final step leading to cell specialization is **cell differentiation.** A differentiated cell, which has a characteristic appearance and characteristic activities, appears to be irreversibly committed to its fate.

Another intriguing part of the developmental puzzle is the building of the body. In **morphogenesis,** the development of form, cells in specific locations differentiate and become spatially organized into recognizable structures. Morphogenesis proceeds through the multistep process of **pattern formation,** the organization of cells into three-dimensional structures. Pattern formation includes signaling between cells, changes in cell shapes, and cell migrations. Depending on their location, cells are exposed to different concentrations of signaling molecules that specify positional information. Thus, *where* a given cell is located often determines *what* it will become when it matures.

The human body, like that of other vertebrates, contains about 250 recognizably different types of cells (❚ Fig. 17-1). Combinations of these specialized cells, known as **differentiated cells,** are organized into diverse and complex structures—such as the eye, hand, and brain—each capable of carrying out many sophisticated activities. Most remarkable of all is the fact that all the structures of the body and the different cells within them descend from a unicellular **zygote,** a fertilized egg.

All multicellular organisms undergo complex patterns of development. The root cells of plants, for example, have structures and functions very different from those of the various types of cells located in leaves. Diversity is also found at the molecular level; most strikingly, each type of plant or animal cell makes a highly specific set of proteins. In some cases, such as the protein hemoglobin in red blood cells, one cell-specific protein may make up more than 90% of the cell's total mass of protein. Other cells may have a complement of cell-specific proteins, each of which is present in small amounts but still plays an essential role. However, because certain proteins are required in every type of cell (all cells, for example, require the same enzymes for glycolysis), cell-specific proteins usually make up only a fraction of the total number of different kinds of proteins.

When researchers first discovered that each type of differentiated cell makes a unique set of proteins, some scientists hypothesized that each group of cells loses the genes it does not need and retains only those required. However, this does not generally seem true. According to the principle of **nuclear equivalence,** the nuclei of essentially all differentiated adult cells of an individual are genetically (though not necessarily metabolically) identical to one another and to the nucleus of the zygote from which they descended. This means that virtually all *somatic cells* in an adult have the same genes. However, different cells express different subsets of these genes.

Somatic cells are all the cells of the body other than **germ line cells,** which ultimately give rise to a new generation. In animals, germ line cells—whose descendants ultimately undergo meiosis and differentiate into gametes—are generally set aside early in development. In plants, the difference between somatic cells and germ line cells is not as distinct, and the determination that certain cells undergo meiosis is made much later in development.

The evidence for nuclear equivalence comes from cases in which differentiated cells or their nuclei have been found to retain the potential of directing the development of the entire organism. Such cells or nuclei are said to exhibit **totipotency.**

Most cell differences are due to differential gene expression

Because genes do not seem to be lost regularly during development (and thus nuclear equivalence is present in different cell types), differences in the molecular composition of cells must be regulated by the activities of different genes. The process of developmental gene regulation is often referred to as **differential gene expression.** As discussed in Chapter 14, the expression of eukaryotic genes is regulated in many ways and at many levels. For example, a particular enzyme may be produced in an inactive form and then be activated later. However, much of the regulation that is important in development occurs at the transcriptional level. The transcription of certain sets of genes is repressed, whereas that of other sets is activated. Even the expression of genes that are *constitutive*—that is, constantly transcribed—is regulated during development so that the *quantity* of each product varies from one tissue type to another.

We can think of differentiation as a series of pathways leading from a single cell to cells in each of the different specialized tissues, arranged in an appropriate pattern. At times a cell makes genetic commitments to the developmental path its descendants will follow. These commitments gradually restrict the development of the descendants to a limited set of final tissue types. Determination, then, is a progressive fixation of the fate of a cell's descendants.

As the development of a cell becomes determined along a differentiation pathway, its physical appearance may or may not change significantly. Nevertheless, when a stage of determination is complete, the changes in the cell usually become self-perpetuating and are not easily reversed. Cell differentiation is usually the last stage in the developmental process. At this stage, a precursor cell becomes structurally and functionally recognizable as a bone cell, for example, and its pattern of gene activity differs from that of a nerve cell, or any other cell type.

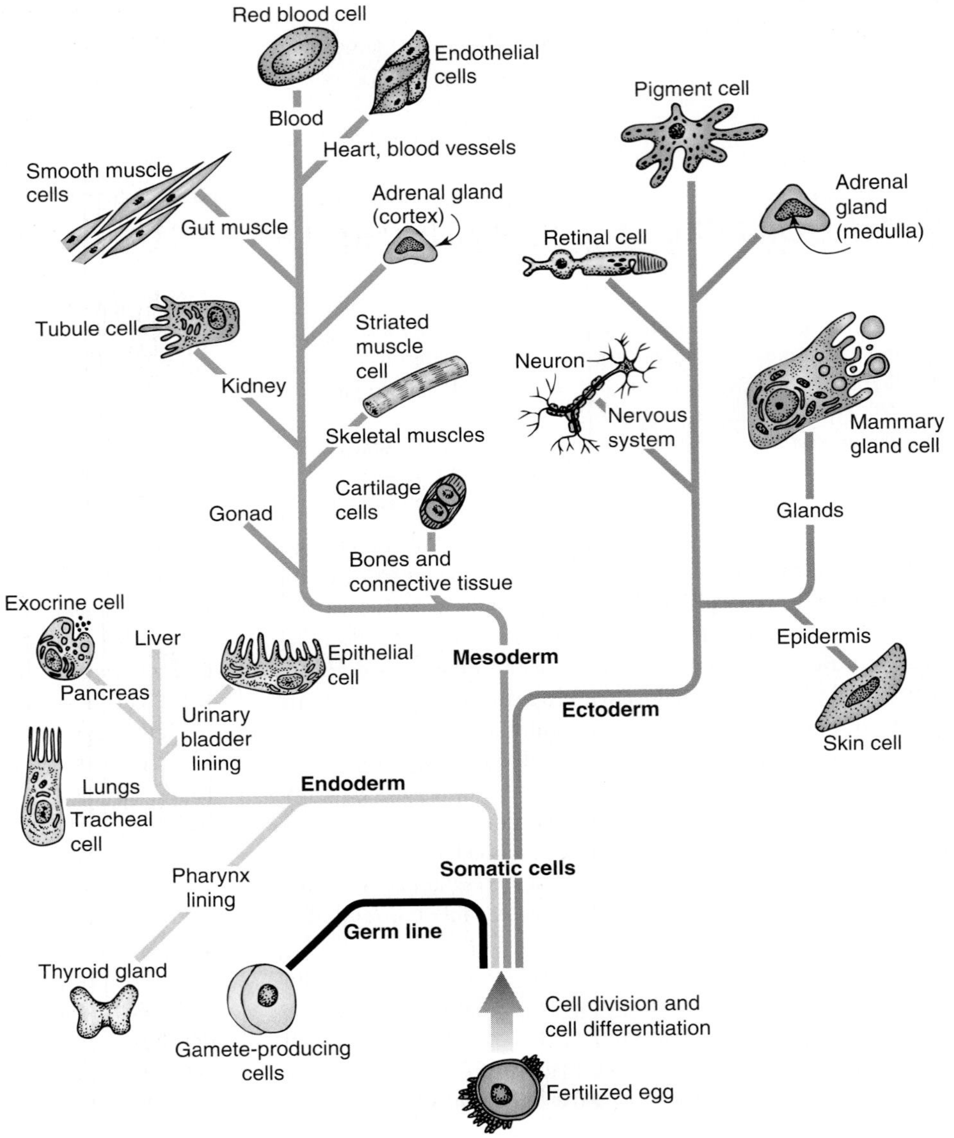

A totipotent nucleus contains all the instructions for development

In plants, some differentiated cells can be induced to become the equivalent of embryonic cells. Biologists use *tissue culture techniques* to isolate individual cells from certain plants and to allow them to grow in a nutrient medium.

In the 1950s, F. C. Steward and his coworkers at Cornell University conducted some of the first experiments investigating cell totipotency in plants (▌ Fig. 17-2). **Totipotent cells** have the potential to give rise to all parts of an organism because they contain a complete set of genetic instructions required to direct the normal development of an entire organism. Steward and his colleagues induced root cells from a carrot to divide in a liquid nutrient medium, forming groups of cells called *embryoid* (embryo-like) *bodies*. These clumps of dividing cells were then transferred to an agar medium, which provided nutrients and a solid supporting structure for the developing plant cells. Some

Figure 17-1 Vertebrate cell lineages

Repeated divisions of the fertilized egg (*bottom*) result in the establishment of tissues containing groups of specialized cells. Germ line cells (cells that produce the gametes) are set aside early in development. Somatic cells progress along various developmental pathways, undergoing a series of commitments that progressively determine their fates.

of the cells of the embryoid bodies gave rise to roots, stems, and leaves. The resulting small plants, called *plantlets* to distinguish them from true seedlings, were then transplanted to soil, where they ultimately developed into adult plants capable of producing flowers and viable seeds.

Because these plants are all derived from the same parent plant, they are genetically alike and therefore constitute a clone. As mentioned in Chapter 15, a **clone** consists of individual organisms, cells, or DNA molecules that are genetically identical to another individual, cell, or DNA molecule, from which it was derived. The methods of plant tissue culture are now extensively

QUESTION: Are differentiated somatic plant cells totipotent?

HYPOTHESIS: Individual carrot cells can be induced to develop into an entire plant.

EXPERIMENT: Carrot root tissues that were cultured in a liquid nutrient medium divided to form clumps of undifferentiated cells. The clumps were then transferred to a solid growth medium.

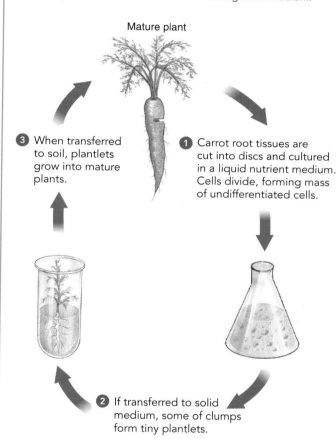

Mature plant

3 When transferred to soil, plantlets grow into mature plants.

1 Carrot root tissues are cut into discs and cultured in a liquid nutrient medium. Cells divide, forming mass of undifferentiated cells.

2 If transferred to solid medium, some of clumps form tiny plantlets.

RESULTS AND CONCLUSION: The development of a complete carrot plant from differentiated somatic cells demonstrated the totipotency of these cells.

Figure 17-2 Steward's experiment on cell totipotency in carrots

used to produce genetically engineered plants, because they enable researchers to regenerate whole plants from individual cells that have incorporated recombinant DNA molecules (see *Focus On: Cell and Tissue Culture,* in Chapter 37).

Similar experiments have been attempted with animal cells, but thus far researchers have not been able to induce a fully differentiated somatic cell to behave like a zygote. Instead, they have tested whether steps in the process of determination are reversible by transplanting the *nucleus* of a cell in a relatively late stage of development into an egg cell that has been *enucleated* (that is, its own nucleus has been destroyed).

In the 1950s, Robert Briggs and Thomas J. King of the Institute for Cancer Research in Pennsylvania pioneered *nuclear transplantation experiments.* They transplanted nuclei from frog cells at different stages of development into egg cells whose nuclei had been removed. Some of the transplants proceeded normally through several developmental stages, and a few even developed into normal tadpoles. As a rule, the nuclei transplanted from cells at earlier stages were most likely to support development to the tadpole stage. As the fate of the cells became more and more determined, the probability quickly declined that a transplanted nucleus could control normal development.

British biologist John B. Gurdon carried out experiments on nuclear transplantation in frogs during the 1960s. In a few cases he demonstrated that nuclei isolated from the intestinal epithelial cells of a tadpole directed development up to the tadpole stage (Fig. 17-3). This result occurred infrequently (about 1.5% of the time); however, in these kinds of experiments success counts more than failure. Therefore, he could safely conclude that at least some nuclei of differentiated animal cells are in fact totipotent.

For many years, because these successes with frogs could not be repeated with mammalian embryos, many developmental biologists concluded that some fundamental feature of mammalian reproductive biology might be an impenetrable barrier to mammalian cloning. This perception changed markedly in 1996 and 1997 with the first reports of the birth of cloned mammals.

The first cloned mammal was a sheep

In 1996, Ian Wilmut and his coworkers at the Roslin Institute in Edinburgh, Scotland, reported that they had succeeded in cloning sheep by using nuclei from an early stage of sheep embryos (the *blastocyst* stage; see Chapter 50). These scientists received worldwide attention in early 1997 when they announced the birth of a lamb, named Dolly (after the singer Dolly Parton). Dolly's genetic material was derived from a cultured sheep mammary gland cell that was fused with an enucleated sheep's egg. The resulting cell divided and developed into an embryo that was then cultured in vitro until it reached a stage at which it could be transferred to a host mother (Fig. 17-4). Not surprisingly, the overall success rate was low: of 277 fused cells, only 29 developed into embryos that could be transferred, and Dolly was the only live lamb produced.

Why did Wilmut's team succeed when so many other researchers had failed? Applying the basic principles of cell biology, they recognized that the **cell cycles** of the egg cytoplasm and the donor nucleus were not synchronous. The egg cell is arrested at metaphase II of meiosis, whereas the actively growing donor somatic cell is usually in the DNA synthesis phase (S), or in G_2. By withholding certain nutrients from the mammary gland cells used as donors, the researchers caused these cells to enter a nondividing state referred to as G_0 (see Chapter 10). This had the effect of synchronizing the cell cycles of the donor nucleus and the

QUESTION: Are nuclei in differentiated animal cells totipotent?

HYPOTHESIS: Nuclei from differentiated cells contain the information required for normal development.

EXPERIMENT: Gurdon injected the nuclei of differentiated cells (tadpole intestinal cells) into eggs whose own nuclei were destroyed by ultraviolet radiation.

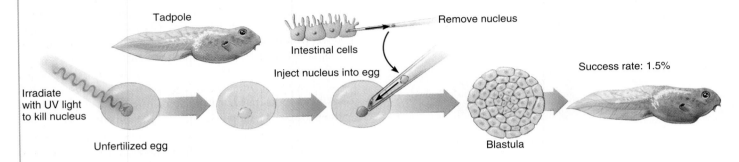

RESULTS AND CONCLUSION: Normal development proceeded to the tadpole stage in about 1.5% of trials, indicating that the genes for programming development up to that point were still present and could be appropriately activated.

Figure 17-3 Gurdon's experiment on nuclear totipotency in frogs

egg. They then used an electric shock to fuse the donor cell with the egg and initiate embryo development.

Although an extremely high level of technical expertise is required, these and other researchers have modified and extended these techniques to produce cloned calves, goats, pigs, mules, rats, and mice, among others. The list of mammals successfully cloned continues to grow. However, the success rate for each set of trials is low, around 1% to 2%, and the incidence of genetic defects is high. Dolly was euthanized at age 6 because she was suffering from a virus-induced lung cancer that infected several sheep where she was housed. However, she developed arthritis at $5\frac{1}{2}$ years, which is relatively young for a sheep to have this degenerative disease. Some biologists speculate that using adult genetic material to produce a clone might produce an animal with prematurely old cells (see discussion of telomeres and cell aging in Chapter 12). Further research may provide some answers to this potential problem.

The main focus of cloning research is the production of **transgenic organisms,** in which foreign genes have been incorporated (see Chapter 15). Researchers are actively pursuing new techniques to improve the efficiency of the cloning process. Only then will it be possible to produce large numbers of cloned transgenic animals for a variety of uses, such as increasing the populations of endangered species. For example, the first healthy clone of an endangered species, a wild relative of cattle known as a *banteng*, was born in 2003. The nucleus for this clone came from a frozen skin cell of a banteng that died in 1980 at the San Diego Zoo.

Stem cells divide and give rise to differentiated cells

Stem cells are undifferentiated cells that can divide to produce differentiated descendants yet retain the ability to divide to maintain the stem cell population. When a stem cell divides, its daughter cells can remain stem cells or differentiate into specialized cells such as muscle cells, nerve cells, or blood cells. What happens depends on the presence or absence of an array of biochemical signals. One of the most challenging areas of stem cell research today involves determining the identity, order, and amounts of the chemical signals that will result in a specific type of cell differentiation.

The most versatile stem cells—zygotes—are totipotent and have the potential to give rise to all tissues of the body and placenta. Stem cells can also be derived from embryos or adult cells. (The term *adult* is somewhat misleading, because adult stem cells can be harvested from umbilical cord tissue, infants, and children as well as adults.) Embryonic and adult stem cells are known as **pluripotent stem cells,** because they can give rise to many, but not all, of the types of cells in an organism. **Embryonic stem cells (ES cells),** formed after a zygote has undergone several rounds of cell division to form a 5- or 6-day-old blastocyst, are more versatile than adult stem cells (❚ Fig. 17-5). For example, ES cells are pluripotent and have the potential to develop into any type of cell in the body; ES cells are not totipotent because they cannot form cells of the placenta.

QUESTION: Are nuclei in differentiated mammalian cells totipotent?

HYPOTHESIS: The nucleus of a differentiated cell from an adult mammal fused with an enucleated egg can provide the genetic information to direct normal development.

EXPERIMENT: Wilmut produced a sheep embryo by fusing a cultured adult sheep mammary cell with an enucleated sheep's egg. He then implanted the embryo into the uterus of a host mother.

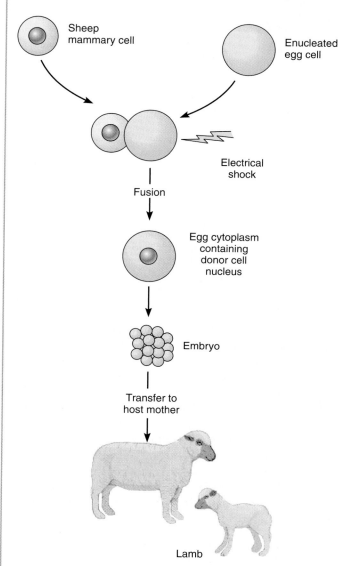

RESULTS AND CONCLUSION: Normal development proceeded, and a female lamb—the world's first cloned mammal—was born. To provide additional evidence that the cloned sheep was fully functioning, when she matured, she was bred and gave birth to a normal offspring.

Figure 17-4 *Animated* Wilmut's experiment on mammalian cloning

Adult stem cells have been found in the human brain, retina, heart, bone marrow, dental pulp, intestines, and other sites. Adult neural stem cells in the brain are pluripotent and differentiate to form neurons and glial cells. Adult stem cells in the bone marrow form both red blood cells and the various types of white blood cells of the immune system. The range of cells that adult stem cells can differentiate into is more limited than that of ES cells. However, recent studies suggest that even specialized stem cells may be more versatile than once thought. For example, neural stem cells form blood cells when transplanted into bone marrow, and bone marrow stem cells can differentiate into muscle cells.

Stem cells are potential sources for cell transplantation into patients to treat serious degenerative conditions. For example, Parkinson's disease results from a progressive loss of cells that produce the neurotransmitter dopamine in a specific region of the brain. Transplantation of stem cells that have been induced to differentiate as dopamine-producing cells holds great promise as an effective long-term treatment. Similarly, stem cells may become a source of insulin-producing cells for transplantation into the pancreas of individuals with diabetes mellitus. Stem cells might also provide replacement nerve cells in people with spinal cord injury or other types of neurological damage.

Researchers ultimately hope to establish lines of human pluripotent stem cells that can grow indefinitely in culture, be induced to differentiate under controlled conditions and stably maintain their differentiated state, and be manipulated genetically. They particularly want to develop ES cell lines from patients with cancer, diabetes, cardiovascular disease, and neurogenerative disorders (such as Parkinson's disease); such cell lines would be invaluable in research on these disorders. Although work on stem cells in mice and other mammals has been conducted for many years, similar studies in humans have progressed slowly, despite the great promise of stem cells as a therapeutic tool.

Thus far, the only known source of ES cells is early human embryos left over from **in vitro fertilization** (see Chapter 49). In the United States and certain other countries, private companies currently fund these studies, largely because of government restrictions on public funding due to ethical considerations related to the origins of ES cells (discussed in the next section). Other countries, notably some Asian countries, have no restrictions on this type of research.

Ethical questions exist relating to human cloning

Cloning research continues to fuel an ongoing debate regarding the potential for human cloning and its ethical implications. In the United States, the National Bioethics Advisory Commission has been established to study this and other questions. In considering these issues, it is important to recognize that *cloning* is a broad term that includes several different processes involved in producing biological cells, tissues, organs, or organisms.

Human reproductive cloning has the goal of making a newborn human that is genetically identical to another, usually adult, human. It would involve placing a human embryo produced by a process other than fertilization into a woman's body. Many countries are opposed to human reproductive cloning.

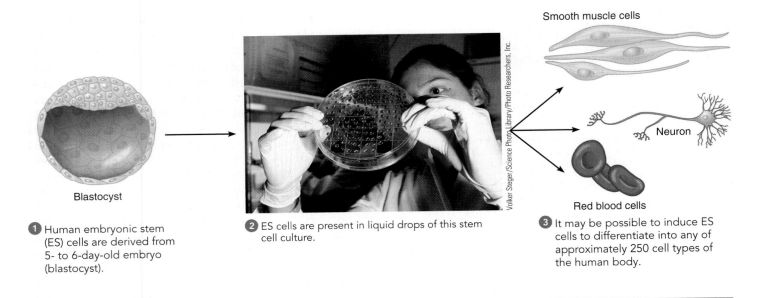

① Human embryonic stem (ES) cells are derived from 5- to 6-day-old embryo (blastocyst).

② ES cells are present in liquid drops of this stem cell culture.

③ It may be possible to induce ES cells to differentiate into any of approximately 250 cell types of the human body.

Blastocyst

Smooth muscle cells

Neuron

Red blood cells

Figure 17-5 Human embryonic stem (ES) cells

In contrast, **human therapeutic cloning** would involve duplication of human cells for scientific study or medical purposes; no newborn human would develop. In human therapeutic cloning, scientists would take the nucleus from—for example, a skin cell of a sick or injured person—and place it in an enucleated egg cell, which would then be treated to develop into an embryo; ES cells would then be extracted from the embryo and grown in culture. If successful, such an accomplishment would be a major advance in using stem cells to provide a supply of replacement tissues, such as heart muscle or nerves, for revolutionary medical procedures. Because the cloned embryo is a genetic match with the patient, transplant rejection problems will likely be avoided.

Many people support research in therapeutic cloning because the potential benefits in curing disease outweigh concerns over destroying week-old embryos. Other people have ethical or religious objections, particularly if the embryos are created specifically for research and then destroyed.

Review

▌ What lines of evidence support the principle of nuclear equivalence?

▌ Why was Wilmut's team successful in mammalian cloning where others had failed?

▌ What are stem cells?

THE GENETIC CONTROL OF DEVELOPMENT

Learning Objectives

4 Indicate the features of *Drosophila melanogaster, Caenorhabditis elegans, Mus musculus,* and *Arabidopsis thaliana* that have made these organisms valuable models in developmental genetics.

5 Distinguish among maternal effect genes, segmentation genes, and homeotic genes in *Drosophila*.

6 Explain the relationship between transcription factors and genes that control development.

7 Define *induction* and *apoptosis,* and give examples of the roles they play in development.

Development has been an important area of research for many years, and biologists have spent considerable time studying the development of invertebrate and vertebrate animals. By investigating patterns of morphogenesis in different species, researchers have identified both similarities and differences in the basic plan of development from a zygote to an adult in organisms ranging from the sea urchin to mammals (see Chapter 50).

In addition to descriptive studies, many classic experiments have demonstrated how groups of cells differentiate and undergo pattern formation. Researchers have developed elaborate screening programs to detect mutations that let them identify many developmental genes in both plants and animals. They then exploit molecular genetic techniques and other sophisticated methodologies to determine how those genes work and how they interact to coordinate developmental processes.

A variety of model organisms provide insights into basic biological processes

In studies of the genetic control of development, the choice of organism to use as an experimental model is important. A **model organism** is a species chosen for biological studies because it has characteristics that allow for the efficient analysis of biological processes. Because most model organisms are small organisms with short generation times, they are easy to grow and study under controlled conditions. For example, mice are a better model organism than kangaroos.

One of the most powerful approaches in developmental genetics involves isolating mutants of a model organism with abnormal development. Not all organisms have useful characteristics that allow researchers to isolate and maintain developmental mutants for future study. Geneticists so thoroughly understand the genetics of the fruit fly, *Drosophila melanogaster*, that this organism has become one of the most important systems for such studies. Other organisms—the yeast *Saccharomyces cerevisiae;* the nematode worm, *Caenorhabditis elegans;* the zebrafish, *Danio rerio;* the laboratory mouse, *Mus musculus;* and certain plants, including *Arabidopsis thaliana*, a tiny weed—have also become important models in developmental genetics. Each of these organisms has attributes that make it particularly useful for examining certain aspects of development (❙ Fig. 17-6).

In the 1990s, developmental geneticists working on *C. elegans* discovered RNA interference, a research tool that can be even more powerful than identifying mutants. As discussed in Chapter 13, in **RNA interference (RNAi)** certain small RNA molecules interfere with the expression of genes or their RNA transcripts. One way these molecules work is to silence genes by selectively cleaving mRNA molecules that have base sequences complementary to the small RNA molecules. The use of RNAi makes it possible to knock out the expression of a specific gene in an organism during its development, thereby deducing the purpose of the gene.

Many examples of genes that control development have been identified in the fruit fly *Drosophila*

The *Drosophila* genome sequence, which was completed in late 1999, includes about 13,600 protein-coding genes. At least 1200 of these are essential for embryonic development.

Two of the traditional advantages of *Drosophila* as a research organism are the abundance of mutant alleles, including those of developmental genes, available for study and the relative ease with which a new mutation is mapped on the chromosomes. U.S. biologist Edward B. Lewis (1918–2004), a pioneering developmental geneticist, began working with *Drosophila* mutants in the 1940s. The work of German researcher Christiane Nüsslein-Volhard and American Eric Wieschaus extended our understanding of development in fruit flies. Lewis, Nüsslein-Volhard, and Wieschaus shared the 1995 Nobel Prize in Physiology or Medicine for their decades of painstaking research on the genetics of *Drosophila* development. Many of the genes they discovered in the fruit fly are now known to be important in the growth and development of all animals.

The *Drosophila* life cycle includes egg, larval, pupal, and adult stages

Development in *Drosophila* consists of several distinct stages (❙ Fig. 17-7). After the egg is fertilized, a period of embryogenesis occurs during which the zygote develops into a sexually immature form known as a **larva** (pl., *larvae*). After hatching from the egg, each larva undergoes several molts (shedding of the external covering, or cuticle). The periods between molts are called *instars*. Each molt results in a size increase until the larva is ready to become a **pupa.** Pupation involves a molt and the hardening of the new external cuticle so that the pupa is completely encased. The insect then undergoes **metamorphosis,** a complete change in form. During that time, most of the larval tissues degenerate and other tissues differentiate to form the body parts of the sexually mature adult fly.

The larvae are wormlike in appearance and look nothing like the adult flies. However, very early in embryogenesis of the developing larvae, precursor cells of many of the adult structures are organized as relatively undifferentiated, paired structures called **imaginal discs.** The name comes from *imago*, the adult form of the insect. Each imaginal disc occupies a definite position in the larva and will form a specific structure, such as a wing or a leg, in the adult body (❙ Fig. 17-8). The discs are formed by the time embryogenesis is complete and the larva is ready to begin feeding. In some respects the larva is a developmental stage that feeds and nurtures the precursor cells that give rise to the adult fly, which is the only form that reproduces.

The organization of the precursors of the adult structures, including the imaginal discs, is under genetic control. Thus far, more than 50 genes have been identified that specify the formation of the imaginal discs, their positions within the larva, and their ultimate functions within the adult fly. Those genes were identified through mutations that either prevent certain discs from forming or alter their structure or ultimate fate.

Drosophila developmental mutations affect the body plan

Scientists have identified many developmental mutations in *Drosophila*. Researchers have examined their effects on development in various combinations and have studied them extensively at the molecular level. In our discussion we pay particular attention to mutations that affect the segmented body plan of the organism, in both the larva and the adult.

Early *Drosophila* development occurs in the following way. The structure of the egg becomes organized as it develops in the ovary of the female. Stores of mRNA, along with yolk proteins and other cytoplasmic molecules, pass from the surrounding maternal cells into the egg. Immediately after fertilization, the zygote nucleus divides, beginning a series of 13 mitotic divisions.

Each division takes only 5 or 10 minutes, which means that the DNA in the nuclei replicates constantly and at a very rapid rate. During that time, the nuclei do not synthesize RNA. Cytokinesis does not take place, and the nuclei produced by the first seven divisions remain at the center of the embryo until the eighth division occurs. At that time, most of the nuclei migrate out from the center and become localized at the periphery of the embryo. This is known as the *syncytial blastoderm* stage, because the nuclei are not surrounded by individual plasma membranes. (A *syncytium* is a structure containing many nuclei residing in a common cytoplasm.) Subsequently, plasma membranes form, and the embryo becomes known as a *cellular blastoderm*.

Maternal effect genes. The genes that organize the structure of the egg cell are called **maternal effect genes.** These genes in the

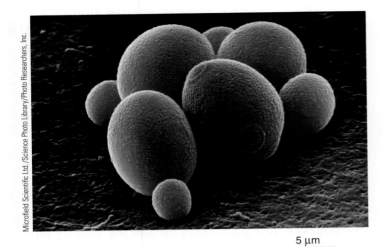

5 µm

(a) Yeast (*Saccharomyces cerevisiae*) is a model organism for studying developmental genetics in haploid, unicellular eukaryotes.

(b) Fruit fly (*Drosophila melanogaster*) development involves an anterior-posterior body plan that is common to many invertebrates and vertebrates.

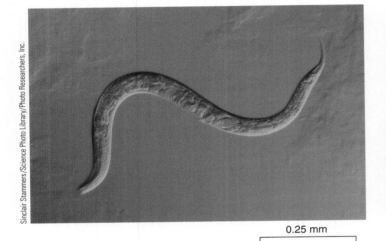

0.25 mm

(c) The nematode (*Caenorhabditis elegans*) adult body consists of only 959 cells, and its transparency allows researchers to trace the development of each cell.

(d) Zebrafish (*Danio rerio*) are small (2 to 4 cm long) and easy to grow, providing an ideal organism for studying the genetic basis of development in vertebrates.

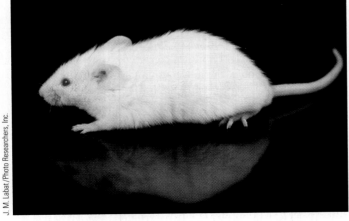

(e) Mouse (*Mus musculus*) is a model organism for studying developmental genetics, including the genetics of cancer, in mammals.

(f) Mouse-ear cress (*Arabidopsis thaliana*) is a small plant with a small genome that is used to study development in flowering plants.

Figure 17-6   Model organisms in developmental genetics

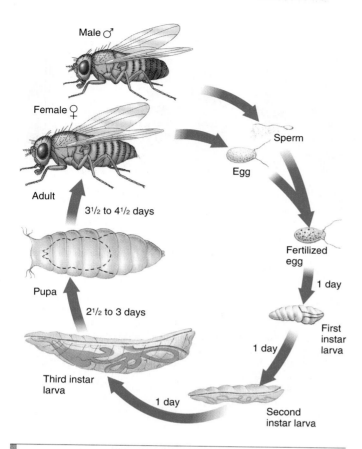

Figure 17-7 The life cycle of *Drosophila*

As it develops from a zygote to a sexually mature adult fly, a fruit fly passes through several stages. It takes about 12 days, at 25°C, to complete the life cycle. The dotted lines within the pupa represent the animal undergoing metamorphosis.

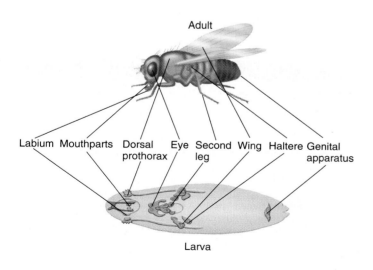

Figure 17-8 The location of imaginal discs

Each pair of discs in a *Drosophila* larva (*bottom*) develops into a specific pair of structures in the adult fly.

surrounding maternal tissues are transcribed to produce mRNA molecules that are transported into the developing egg. Analysis of mutant flies with defective maternal effect genes has revealed that many of these genes are involved in establishing the polarity of the embryo—such as which part of the embryo will become the head and which part will become the tail. Polarity dictates those parts of the egg that are dorsal or ventral and those that are anterior or posterior (see Chapter 29); thus, these genes are known as *egg polarity genes.*

▌Figure 17-9a illustrates concentration gradients for two specific maternal mRNA molecules in the very early embryo. These mRNA transcripts of some of the maternal effect genes are identified by their ability to hybridize with radioactive DNA probes derived from cloned genes (discussed in Chapter 15). Alternatively, researchers use fluorescently tagged antibodies (as in the chapter opening photograph) to bind to specific protein products of the maternal effect genes. These protein gradients organize the early pattern of development in the embryo by determining anterior and posterior regions.

A combination of protein gradients may provide positional information that specifies the fate—that is, the developmental path—of each nucleus within the embryo. For example, muta-

Stage of development   **Gene activity**

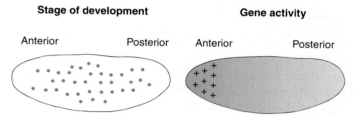

(a) At 1.25 hours after fertilization, the embryo consists of a common cytoplasm with about 128 nuclei (*purple circles*). Maternal effect genes divide the embryo into anterior and posterior sections. The crosses in the anterior region represent mRNA molecules transcribed from one maternal effect gene. The pink shading represents a different maternal mRNA molecule that is more concentrated in the anterior region.

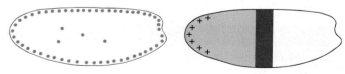

(b) At 2 hours after fertilization, about 1500 nuclei have migrated into the periphery of the embryo and started to make their own mRNA molecules. The gap genes divide the embryo into anterior, middle, and posterior sections. The maternal RNA shown in pink is now transcribed from segmentation genes (*crosses*) in nuclei in the anterior region. The mRNA from another gap gene is transcribed in the middle of the embryo (*black region*).

Figure 17-9 Early development in *Drosophila*

Longitudinal sections on the left show two early stages of development of the *Drosophila* embryo. These are matched with the simplified patterns of gene activity at each stage on the right. (Adapted from M. E. Akam, "The Molecular Basis for Metameric Pattern in the Drosophila Embryo," *Development,* Vol. 101, 1987.)

tions in certain maternal effect genes cause the absence of specific signals, resulting in an embryo with two heads or two posterior ends.

In many cases, injecting normal maternal mRNA into the mutant embryo reverses the phenotype associated with a mutation in a given maternal effect gene. The fly subsequently develops normally, indicating that the gene product is needed only for a short time in the earliest stages of development.

Segmentation genes. As the nuclei start to migrate to the periphery of the embryo, **segmentation genes** in those nuclei begin to produce embryonic mRNA. Thus far, geneticists have identified at least 24 segmentation genes that are responsible for generating a repeating pattern of body segments within the embryo and adult fly. Based on the study of mutant phenotypes, researchers group the segmentation genes into three classes—gap genes, pair–rule genes, and segment polarity genes.

Gap genes are the first set of segmentation genes to act. These genes interpret the maternal anterior-posterior information in the egg and begin organization of the body into anterior, middle, and posterior regions (█ Fig. 17-9b). A mutation in one of the gap genes usually causes the absence of one or more body segments in an embryo (█ Fig.17-10a).

The other two classes of segmentation genes do not act on small groups of body segments; instead, they affect all segments. Mutations in **pair–rule genes** delete every other segment, producing a larva with half the normal number of segments (█ Fig. 17-10b). Mutations in **segment polarity genes** produce segments in which one part is missing and the remaining part is duplicated as a mirror image (█ Fig. 17-10c). █ Table 17-1 summarizes the effects of the different classes of mutants.

Each segmentation gene has distinctive times and places in the embryo in which it is most active (█ Fig. 17-11). The observed pattern of expression of maternal effect genes and segmentation genes indicates that a progressive series of developmental events determines cells destined to form adult structures. First, maternal effect genes that form gradients of morphogens in the egg determine the anterior-posterior (head-to-tail) axis and the dorsal and

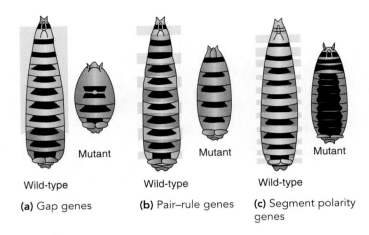

(a) Gap genes   (b) Pair–rule genes   (c) Segment polarity genes

Wild-type   Mutant   Wild-type   Mutant   Wild-type   Mutant

Figure 17-10   A comparison of mutations in *Drosophila* segmentation genes

(a) Gap genes, (b) pair–rule genes, and (c) segment polarity genes control the pattern of body segments in a *Drosophila* embryo. The blue bands mark the regions in which the protein products of these genes are normally expressed in wild-type embryos. These same regions are absent in embryos in which the gene is mutated—that is, the mutant lacks the blue-shaded parts. The resulting phenotype is characteristic of the class to which the gene belongs. (Adapted from C. Nüsslein-Volhard and E. Wieschaus, "Mutations Affecting Segment Number and Polarity in Drosophila," *Nature*, Vol. 287, 1980.)

ventral regions of the embryo. A **morphogen** is a chemical agent that affects the differentiation of cells and the development of form. Morphogen gradients are signals that help cells determine their location within the embryo and promote their eventual differentiation into specialized tissues and organs.

Next, segmentation genes respond to the morphogens at each location to regulate the production of segments from the head to the posterior region. Within each segment, other genes are then activated to specify which body part that segment will become. Every cell's position is further specified with a specific "address"

TABLE 17-1		
Classes of Genes Involved in Pattern Formation of Embryonic Segments in *Drosophila*		
Class of Gene	**Site of Gene Activity**	**Proposed Functions of Genes and Effects of Mutant Alleles**
Maternal effect genes	Maternal tissues surrounding egg	Initiate pattern formation by activating regulatory genes in nuclei in certain locations in embryo; many maternal effect mutations alter polarity of embryo
Segmentation genes		
Gap genes	Embryo	Some may influence activity of pair–rule genes, segment polarity genes, and homeotic genes; mutant alleles cause one or more segments to be missing
Pair–rule genes	Embryo	Some may influence activity of segment polarity genes and homeotic genes; when mutated, cause alternate segments to be missing
Segment polarity genes	Embryo	May influence activity of homeotic genes; mutant alleles delete part of every segment and replace it with mirror image of remaining structure
Homeotic genes	Embryo	Control identities of segments; homeotic mutations cause parts of fly to form structures normally formed in other segments

designated by combinations of the activities of the regulatory genes.

The segmentation genes act in sequence, with the gap genes acting first, then the pair–rule genes, and finally the segment polarity genes. In addition, the genes of the three groups interact. Each time a new group of genes acts, cells for that group become more determined in their development. As the embryo develops, each region is progressively subdivided into smaller regions.

Most segmentation genes code for **transcription factors,** DNA-binding proteins that regulate gene transcription in eukaryotic cells. For example, some segmentation genes code for a "zinc-finger" type of DNA-binding regulatory protein (see Fig. 14-10b). *Homeotic genes,* discussed next, also code for transcription factors. The fact that many of the genes involved in controlling developmental processes code for transcription factors indicates that those proteins act as genetic switches to regulate the expression of other genes.

Once researchers have identified proteins that function as transcription factors, they can use the purified proteins to determine the DNA target sequences to which the proteins bind. This approach has been increasingly useful in identifying additional parts of the regulatory pathway involved in different stages of development. Transcription factors also play a role in cancer (discussed later in the chapter).

Homeotic genes. After segmentation genes have established the basic pattern of segments in the fly body, **homeotic genes** specify the developmental plan for each segment. Mutations in homeotic genes cause one body part to be substituted for another and therefore produce some peculiar changes in the adult. A striking example is the *Antennapedia* mutant fly, which has legs that grow from the head where the antennae would normally be (❚ Fig. 17-12).

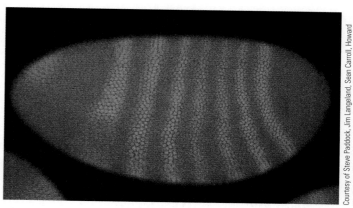

50 µm

Figure 17-11   Segmentation gene activity

The bright bands in this fluorescence LM reveal the presence of mRNA transcribed from one of the pair–rule genes known as *fushi tarazu* (Japanese for "not enough segments"). When this locus is mutated, the segments of the larva that are normally derived from these bands are absent.

Eye   Homeotic leg

250 µm

(b) SEM of the head of a fly with an *Antennapedia* mutation.

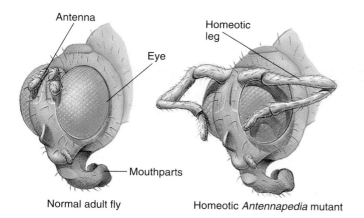

Antenna

Eye

Homeotic leg

Mouthparts

Normal adult fly   Homeotic *Antennapedia* mutant

(a) The head of a normal fly and a fly with an *Antennapedia* mutation.

Figure 17-12   The *Antennapedia* locus

Antennapedia mutations cause homeotic transformations in *Drosophila* in which legs or parts of legs replace the antennae.

Developmental Genetics   **379**

Homeotic genes in *Drosophila* were originally identified by the altered phenotypes produced by mutant alleles. When geneticists analyzed the DNA sequences of several homeotic genes, they discovered a short DNA sequence of approximately 180 base pairs, which characterizes many homeotic genes as well as some other genes that play a role in development. This DNA sequence is called the **homeobox.** Each homeobox codes for a protein functional region called a **homeodomain,** consisting of 60 amino acids that form four α-helices. One of these serves as a recognition helix that binds to specific DNA sequences and thereby affects transcription. Thus, the products of homeotic genes, like those of the earlier-acting segmentation genes, are transcription factors. In fact, some segmentation genes also contain homeoboxes.

Studies of ***Hox* genes,** clusters of homeobox-containing genes that specify the anterior-posterior axis during development, provide insights about evolutionary relationships. *Hox* genes were initially discovered in *Drosophila,* where they are arranged in two adjacent groups on the chromosome: the *Antennapedia complex* and the *bithorax complex.* As *Hox* genes have been identified in other animals, including other arthropods, annelids (segmented worms), roundworms, and vertebrates, researchers have found that these genes are also clustered and that their organization is remarkably similar to that in *Drosophila.*

⏐ Figure 17-13 compares the organization of the *Hox* gene clusters of *Drosophila, M. musculus,* and *C. elegans.* The *Drosophila* and *M. musculus Hox* genes are located in the same order along the chromosome, although the correlation is less exact for *C. elegans.* Moreover, the linear order of the genes on the chromosome reflects the order of the corresponding regions they control, from anterior to posterior, in the animal. This organization apparently reflects the need for these genes to be transcribed in a specific temporal sequence.

Drosophila has only one *Antennapedia–bithorax* complex. However, humans and many other vertebrates have four similar *Hox* gene clusters, each located in a different chromosome. These complexes probably arose through gene duplication. The fact that extra copies of these genes are present helps explain why mutations causing homeotic-like transformations are seldom seen in vertebrate animals. One does not see an extra eye growing where a mouse leg should be, for example. However, one particular type of *Hox* mutation that has been described in both mice and humans causes abnormalities in the limbs and genitalia. The involvement of the genitalia provides a further explanation for the rarity of these mutant alleles, because affected individuals are unlikely to reproduce.

The fact that very similar developmental controls are seen in organisms as diverse as insects, roundworms, and vertebrates (including humans) indicates that the basic mechanism evolved early and has been highly conserved in all animals that have an anterior-posterior axis, even those that are not segmented. Clearly, once a successful way of regulating groups of genes and integrating their activities evolved, it was retained, although it has apparently been modified to provide alterations of the body plan.

The finding of homeobox-like genes in plants suggests these genes originated early during eukaryotic evolution. With further investigations, researchers hope to develop an overall model of how the rudiments of morphogenesis are controlled in all multicellular eukaryotes. These systems of master genes that regulate development are a rich source of "molecular fossils" that are illuminating evolutionary history in new and exciting ways. These kinds of studies have led to an ongoing synthesis of evolution and developmental biology that has come to be known as "Evo Devo."

Caenorhabditis elegans has a relatively rigid developmental pattern

The nematode worm *C. elegans* is an ideal model organism because its system for the genetic control of development is relatively easy to study. Sydney Brenner, a British molecular geneticist, began

Hox genes are arranged in the same order on the chromosome as they are expressed along the anterior-posterior axis of the embryo.

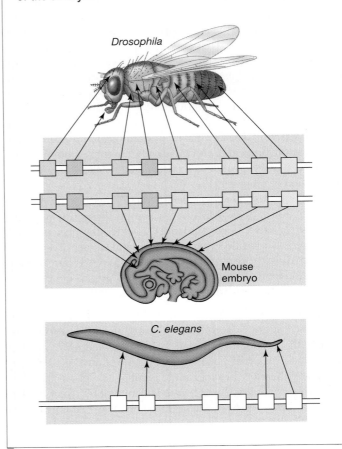

Drosophila

Mouse embryo

C. elegans

Figure 17-13 *Hox* gene clusters

Clusters of *Hox* are found in all animal groups except sponges and cnidarians (jellyfish and their relatives). Note that in each organism, the order of these developmental genes (*squares*) on the chromosome (*white line*) reflects their spatial order of expression in the embryo. (Only one of the four mouse chromosomes with *Hox* gene clusters is shown.) *Caenorhabditis elegans* also has similar *Hox* gene clusters, although the gene order is not identical to that in *Drosophila* and the mouse. (Adapted from C. Kenyon and B. Wang, *Science*, Vol. 253, 1991; and K. Van Auken et al., *Proceedings of the National Academy of Science*, Vol. 97, 2000.)

studying molecular development in this animal at Cambridge University in the 1960s. He selected *C. elegans* because it is small, has a short life cycle (its normal life span is about 3 weeks), and is genetically simple. Consisting of about 19,700 protein-coding genes, it was the first animal genome sequenced. Today, *C. elegans* is an important tool for answering basic questions about the development of individual cells within a multicellular organism.

When an adult, *C. elegans* is only 1.5 mm long and contains only 959 somatic cells (◗ Fig. 17-14). Individuals are either males or **hermaphrodites,** organisms with both sexes in the same individual. Hermaphroditic *C. elegans* are capable of self-fertilization, which makes it easy to obtain offspring that are homozygous for newly induced recessive mutations. The availability of males that can reproduce sexually with the hermaphrodites makes it possible to perform genetic crosses as well.

Because the worm's body is transparent, researchers can follow the development of literally every one of its somatic cells using a Nomarski differential interference microscope, which provides contrast in transparent specimens. As a result of efforts by several research teams, the lineage of each somatic cell in the adult has been determined. Those studies have shown that the nematode has a very rigid, or fixed, developmental pattern. After fertilization, the egg undergoes repeated divisions, producing about 550 cells that make up the small, sexually immature larva. After the larva hatches from the egg case, further cell divisions give rise to the adult worm.

The lineage of each somatic cell in the adult can be traced to a single cell in a small group of **founder cells,** which are formed early in development (◗ Fig. 17-15a). If a particular founder cell is destroyed or removed, the adult structures that would normally develop from that cell are missing. Such a rigid developmental pattern, in which the fates of the cells become restricted early in

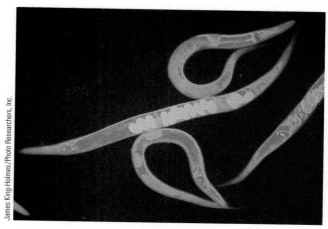

(a) False-color scanning optical micrograph of the adult hermaphrodite nematode. The oval structures are eggs.

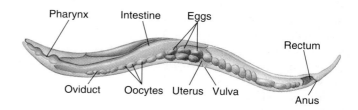

(b) Structures in the adult hermaphrodite. The sperm-producing structures are not shown.

Figure 17-14 *Caenorhabditis elegans*

This transparent organism has a fixed number of somatic cells. [(b) adapted from Fig. 22-6a in W. Walbot and N. Holder, *Developmental Biology.* © 1987. Used with permission from McGraw-Hill companies.]

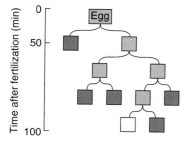

(a) All somatic cells of *C. elegans* are derived from five somatic founder cells (*shown in blue*) produced during the early cell divisions of the embryo. The cell shown in white gives rise to germ line cells.

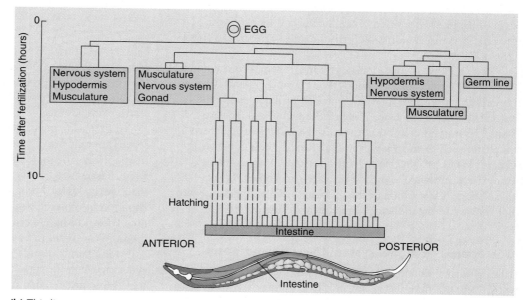

(b) This lineage map traces the development of the cells that form the intestine. The dashed lines represent many cell divisions of a particular lineage.

Figure 17-15 Cell lineages of *Caenorhabditis elegans*

development, is referred to as **mosaic development.** Each cell has a specific fate in the embryo, just as each tile in a mosaic design forms part of the pattern.

Scientists initially hypothesized that every organ system in *C. elegans* might be derived from only one founder cell. Detailed analysis of cell lineages, however, reveals that many of the structures found in the adult, such as the nervous system and the musculature, are in fact derived from more than one founder cell (▌Fig. 17-15b). A few lineages have been identified in which a nerve cell and a muscle cell are derived from the division of a single cell. Researchers have isolated mutations affecting cell lineages, and many of the mutations appear to have properties that would be expected of genes involved in regulating developmental events.

A researcher who uses microscopic laser beams small enough to destroy individual cells can determine the influence one cell has on the development of a neighboring cell. Consistent with the rigid pattern of cell lineages, the destruction of an individual cell in *C. elegans* results, in most cases, in the absence of all of the structures derived from that cell but with the normal differentiation of all the neighboring somatic cells. This observation suggests that development in each cell is regulated through its own internal program.

However, the developmental pattern of *C. elegans* is not entirely mosaic. In some cases, cell differentiation is influenced by interactions with particular neighboring cells, a phenomenon known as **induction.** One example is the formation of the vulva (pl., *vulvae*), the reproductive structure through which the eggs are laid. A single nondividing cell, called the *anchor cell,* is a part of the ovary, the structure in which the germ line cells undergo meiosis to produce the eggs. The anchor cell attaches to the ovary and to a point on the outer surface of the animal, triggering the formation of a passage through which the eggs pass to the outside. When the anchor cell is present, it induces cells on the surface to form the vulva and its opening. If the anchor cell is destroyed by a laser beam, however, the vulva does not form, and the cells that would normally form the vulva remain as surface cells (▌Fig. 17-16).

The analysis of mutations has contributed to our understanding of inductive interactions. For example, several types of mutations cause more than one vulva to form. In such mutant animals, multiple vulvae form even if the anchor cell is destroyed. Thus, the mutant cells do not require an inductive signal from an anchor cell to form a vulva. Evidently in these mutants the gene or genes responsible for vulva formation are constitutive. Conversely, mutants lacking a vulva are also known. In some of these, the cells that would normally form the vulva apparently do not respond to the inducing signal from the anchor cell.

C. elegans is a model system to study apoptosis

During normal development in *C. elegans,* there are instances in which cells are destined to die shortly after they are produced. **Apoptosis,** or genetically programmed cell death, has been observed in a wide variety of organisms, both plant and animal (see Chapter 4). For example, the human hand is formed as a webbed structure, but the fingers become individualized when the cells

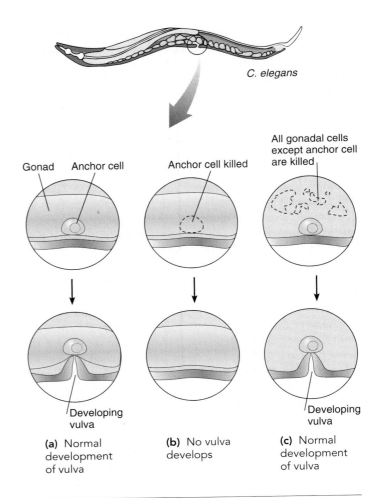

C. elegans

Figure 17-16 Induction

A single anchor cell induces neighboring cells to form the vulva in *C. elegans*. This diagram shows how laser destruction of single cells or groups of cells influences neighboring cells.

between them undergo apoptosis. In *C. elegans,* as in other organisms, apoptosis is under genetic control. The worm embryos undergo mitosis to produce a total of 1090 cells, but 131 of these go through apoptosis during development, resulting in adult worms with 959 cells.

In 1986, U.S. molecular geneticist Robert Horvitz discovered several mutant worms that were defective in some way relating to apoptosis. The mutations and the genes responsible for them were characterized at the molecular level. For example, one of the genes codes for a protein that regulates the release from mitochondria of molecules that trigger apoptosis (mitochondria are now known to have a crucial role in apoptosis in animals ranging from *C. elegans* to humans). Other genes were identified that code for a family of proteins known as **caspases,** proteolytic enzymes active in the early stages of apoptosis. Homologous genes were subsequently identified in other organisms, including humans; some of these loci in mammals also code for caspases. The molecular events that trigger apoptosis remain an area of intense research interest, the results of which will shed considerable light on the general processes of cell aging and cancer.

Robert Horvitz and British scientists Sydney Brenner and John Sulston shared the Nobel Prize in Physiology or Medicine in 2002 for their work on the genetic regulation of organ development and apoptosis in *C. elegans.* Sulston was one of Brenner's students. He traced the lineage of how a single zygote gives rise to the 959 cells in the nematode adult and also observed that some cells die during normal development. Continuing Sulston's work on apoptosis, Horvitz, as mentioned earlier, was the first to discover genes involved in apoptosis.

The mouse is a model for mammalian development

Mammalian embryos develop in markedly different ways from the embryos of *Drosophila* and *C. elegans.* The laboratory mouse, *Mus musculus,* is the best-studied example of mammalian development. Researchers have identified numerous genes affecting development in the mouse. The mouse genome sequence, which was published in 2002, contains 27,000 to 30,000 genes, similar to the number of protein-coding genes in the human genome. Indeed, 99% of the genes in the mouse have counterparts in humans.

The early development of the mouse and other mammals is similar in many ways to human development (see Chapter 50). During the early developmental period, the embryo lives free in the reproductive tract of the female. It then implants in the wall of the uterus, after which the mother meets its nutritional and respiratory needs. Consequently, mammalian eggs are very small and contain little in the way of food reserves. Almost all research on mouse development has concentrated on the stages leading to implantation, because in those stages the embryo is free-living and can be experimentally manipulated. During that period, developmental commitments that have a significant effect on the future organization of the embryo take place.

After fertilization, a series of cell divisions gives rise to a loosely packed group of cells. Research shows that in the very early mouse embryo, all cells are equivalent. For example, at the two-cell stage of mouse embryogenesis, if researchers destroy one cell and implant the remaining cell into the uterus of a surrogate mother, a normal mouse usually develops.

Conversely, if two embryos at the eight-cell stage of development are fused and implanted into a surrogate mother, a normal-sized mouse develops (▌ Fig. 17-17). By using two embryos with different genetic characters, such as coat color, researchers demonstrated that the resulting mouse has four genetic parents. These chimeric mice have fur with patches of different colors derived from clusters of genetically different cells. A **chimera** is an organism containing two or more kinds of genetically dissimilar cells arising from different zygotes. (The term is derived from the name of a mythical beast that was said to have the head of a lion, the body of a goat, and a snakelike tail.) Chimeras let researchers use genetically marked cells to trace the fates of certain cells during development.

The responses of mouse embryos to such manipulations contrast with the mosaic or predetermined nature of early *C. elegans* development, in which destroying one of the founder cells results in the loss of a significant portion of the embryo. For this reason, biologists say that the mouse has highly **regulative development**—the early embryo acts as a self-regulating whole that accommodates missing or extra parts.

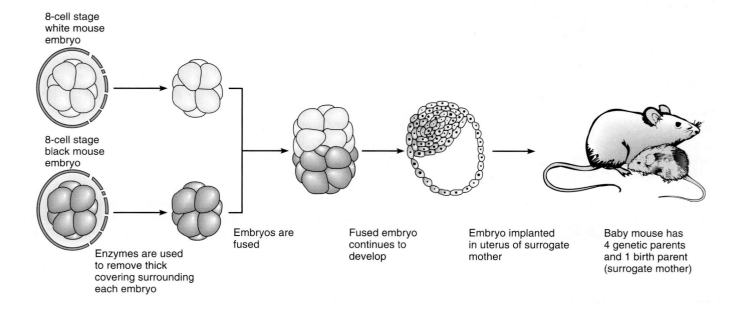

8-cell stage white mouse embryo

8-cell stage black mouse embryo

Enzymes are used to remove thick covering surrounding each embryo

Embryos are fused

Fused embryo continues to develop

Embryo implanted in uterus of surrogate mother

Baby mouse has 4 genetic parents and 1 birth parent (surrogate mother)

Figure 17-17 Chimeric mice

Researchers remove embryos from females of two different strains and combine the cells in vitro. The resulting aggregate embryo continues to develop and is implanted in the uterus of a surrogate mother. The offspring has four different genetic parents. Although the surrogate mother is the birth mother, she is not genetically related.

Transgenic mice are used in studies of developmental regulation

In transformation experiments, foreign DNA injected into fertilized mouse eggs is incorporated into the chromosomes and expressed (Fig. 17-18). The resulting transgenic mice provide insights into how genes are activated during development. Scientists can identify a *transgene* (foreign gene) that has been introduced into a mouse and determine whether it is active by marking the gene in several ways. Sometimes a similar gene from a different species is used; its protein is distinguished from the mouse protein by specific antibodies.

It is also possible to construct a hybrid gene consisting of the regulatory elements of a mouse gene and part of another, nonmouse gene that codes for a "reporter" protein. For example, the reporter protein could be an enzyme not normally found in mice. Such studies have been important in showing which DNA sequences of a mouse homeobox gene determine where the gene is expressed in the embryo.

Many developmentally controlled genes introduced into mice have yielded important information about gene regulation. When researchers introduce developmentally controlled genes from other species, such as humans or rats, into mice, the genes are regulated the same way they normally are in the donor animal. When introduced into the mouse, for example, human genes encoding insulin, globin, and crystallin—which are normally expressed in cells of the pancreas, blood, and eye lens, respectively—are expressed only in those same tissues in the mouse. That these genes are correctly expressed in their appropriate tissues indicates that the signals for tissue-specific gene expression are highly conserved through evolution. Information on the regulation of genes controlling development in one organism can have valuable applications to other organisms, including humans.

Mice and other model organisms are providing insights into the aging process

Aging, which is defined as a progressive decline in the performance of various parts of the body, is an important field of developmental biology. The study of aging has great practical potential because age-related diseases represent one of the biggest challenges of biomedical research today. Scientists have demonstrated that many environmental factors influence aging. For example, severe calorie restriction in rodents and other mammals delays aging. However, how the human genome interacts with the cell environment during the aging process is not well understood.

Researchers working with model organisms, such as *Drosophila* and *C. elegans,* have found that hundreds of protein-coding genes, when mutated, extend the life span of the organism to variable degrees. One of the most potent life span–extending genes, present in the worm *C. elegans,* codes for a protein that is very similar to the membrane receptor that allows cells to respond to the peptide **insulin-like growth factor (IGF)** in humans and other mammals. The binding of IGF to its receptor triggers **signal transduction** within the cell.

Researchers wondered if the IGF receptor influences aging in mammals, including humans. To test this hypothesis, they produced *knockout mice* with an inactivated gene for the insulin-like growth factor receptor (IGF-1R). (Knockout mice were first discussed in Chapter 15.) Because mice are diploid organisms, they have two copies of the gene, designated *igf1r*. When both alleles of *igf1r* are inactivated, the mice develop many abnormalities and die at birth. However, when one allele of *igf1r* is inactivated and the other left in its normal state, the mice thrive and live about 25% longer than control mice. In all other respects, the heterozygous mice are normal. They are virtually indistinguishable from control mice in their rate of development, metabolic rate, ability to reproduce, and body size.

Mice heterozygous at the *igf1r* locus produce fewer IGF receptors, and therefore the cells are not exposed to as much signaling from IGF. As a result, the heterozygous cells may be more

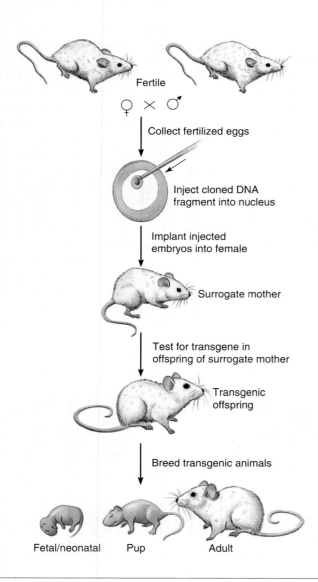

Fertile

♀ × ♂

Collect fertilized eggs

Inject cloned DNA fragment into nucleus

Implant injected embryos into female

Surrogate mother

Test for transgene in offspring of surrogate mother

Transgenic offspring

Breed transgenic animals

Fetal/neonatal   Pup   Adult

Figure 17-18   Producing a transgenic mouse

Researchers inject cloned DNA fragments into the nucleus of a fertilized mouse egg. They then implant the injected eggs into a female, which becomes the surrogate mother. The researchers verify the presence of the foreign gene in the transgenic animal or breed the animal to establish a transgenic line of mice.

resistant to environmental stressors, such as oxygen radicals. (Oxygen radicals and other *free radicals* are atoms, molecules, or ions that have one or more unpaired electrons. Free radicals react readily with biological molecules, often breaking chemical bonds.) When mouse cells that are heterozygous at the *igf1r* locus are grown in culture, they resist attack by oxygen radicals. The same is true at the organism level for mice that are heterozygous at the *igf1r* locus.

Research on the IGF receptor and aging in mice suggests that some of the hundreds of other mutant genes associated with aging in *C. elegans* may also be involved in mammalian aging. Thus, the genetic control of mammalian aging has enough research topics to occupy molecular geneticists for many years.

Arabidopsis is a model for studying plant development, including transcription factors

Botanists use certain well-characterized plants in studying the genetic control of development. Many of these are economically important crop plants such as corn, *Zea mays*. Genes with developmental effects are known in corn, including some that are analogous (but not homologous) to the homeotic genes of *Drosophila*.

Mouse-ear cress (*Arabidopsis thaliana*), the organism most widely used to study the genetic control of development in plants, is a member of the mustard family. Although *Arabidopsis* itself is a weed of no economic importance, it has several advantages for research. The plant completes its life cycle in just a few weeks and is small enough to be grown in a petri dish, yielding thousands of individuals in limited space. Botanists use chemical mutagens to produce mutant strains and have isolated many developmental mutants. When they insert cloned foreign genes into *Arabidopsis* cells, the genes integrate into the chromosomes and are expressed. The researchers then induce these transformed cells to differentiate into transgenic plants.

In 2000, the *Arabidopsis* genome became the first plant genome sequenced. Although its genome is relatively small (the rice genome is about 4 times as large), it includes about 26,000 protein-coding genes. In comparison, *Drosophila* has about 13,600 genes and *C. elegans* about 19,700. Many genes in *Arabidopsis* are functionally equivalent to genes in *Drosophila*, *C. elegans*, and other animal species.

Of particular importance to development are the more than 1500 genes that code for transcription factors in *Arabidopsis* (compared to 635 at last count for *Drosophila*). Not surprisingly, many genes known to specify identities of parts of the *Arabidopsis* flower code for transcription factors. During the development of *Arabidopsis* flowers, four distinct flower parts differentiate: sepals, petals, stamens, and carpels (Fig. 17-19a; also see Fig. 32-2). Sepals cover and protect the flower when it is a bud, petals help attract animal pollinators to the flower, stamens produce pollen grains, and the carpel produces ovules, which develop into seeds following fertilization (discussed in Chapter 28).

The **ABC model** is a working hypothesis that may explain the molecular biology behind how these four organs develop. The *A* gene is needed to specify sepals, both *A* and *B* genes are needed to specify petals, both *B* and *C* genes are needed to specify stamens, and the *C* gene is needed to specify the carpel. Mutations in the *A*, *B*, or *C* organ-identity genes, all of which are homeotic and code for transcription factors, cause one flower part to be substituted for another. For example, class *C* homeotic mutants (which have an inactive *C* gene) have petals in place of stamens and sepals in place of the carpel. Therefore, the entire flower consists of only sepals and petals. Figure 17-19b, c, and d shows three homeotic mutants of *Arabidopsis*.

The ABC model is not the entire explanation for floral development. Another class of genes, designated *SEPALLATA*, interacts with the *B* and *C* genes to specify the development of petals, stamens, and carpels. Remarkably, when *SEPALLATA* genes are turned on permanently in *Arabidopsis*, the resulting plants have white petals growing where the leaves should be. (Because the plants do not photosynthesize, they are grown on a nutrient medium.)

These findings in *Arabidopsis* vastly increase the number of molecular probes available from plants. Researchers use these

Photos courtesy of Jose Luis Reichmann and Elliot Meyerowitz

(a) (b) (c) (d)

Figure 17-19 Homeotic mutants in *Arabidopsis* flowers

(a) A normal *Arabidopsis* flower has four outer leafy green sepals, four white petals, six stamens (the male reproductive structures), and a central carpel (the female reproductive structure). **(b)** This homeotic mutant is missing petals. **(c)** This homeotic mutant has only sepals and carpels. **(d)** This homeotic mutant has sepals and petals but no other floral structures. The homeotic genes in all these plants code for transcription factors.

obes to identify other genes that control development in various plant species and to compare them with genes from a wide range of organisms. The success of the *Arabidopsis* sequencing project led to an international initiative by plant biologists, the 2010 Project. Its goal is to understand the functions of all genes in *Arabidopsis* by the year 2010. This functional genomic information will lead to a far deeper understanding of plant development and evolutionary history.

Review

■ What are the relative merits of *Drosophila*, *C. elegans*, *M. musculus*, and *Arabidopsis* as model organisms for the study of development?

■ What are transcription factors, and how do they influence development?

■ What role does induction play in development?

■ What is apoptosis?

CANCER AND CELL DEVELOPMENT

Learning Objective

8 Discuss the relationship between cancer and mutations that affect cell developmental processes.

Cancer cells lack normal biological inhibitions. Normal cells are tightly regulated by control mechanisms that prompt them to divide when necessary and prevent them from growing and dividing inappropriately. Cells of many tissues in the adult are normally prevented from dividing; they reproduce only to replace a neighboring cell that has died or become damaged. Cancer cells have escaped such controls and can divide continuously.

As a consequence of their abnormal growth pattern, some cancer cells eventually form a mass of tissue called a **tumor,** or *neoplasm.* If the tumor remains at the spot where it originated, it can usually be removed by surgery. One of the major problems with certain forms of cancer is that the cells can escape from the controls that maintain them in one location. **Metastasis** is the spreading of cancer cells to different parts of the body. Cancer cells invade other tissues and form multiple tumors. Lung cancer, for example, is particularly deadly because its cells are highly metastatic; they enter the bloodstream, spread, and form tumors in other parts of the lungs and in other organs, such as the liver and the brain.

Biologists now know that cancer is caused by the altered expression of specific genes critical for cell division. Using recombinant DNA methods, researchers have identified many of the genes that when functioning abnormally, transform normal cells into cancer cells. The traits of each kind of cancer cell come from at least one, and probably several, **oncogenes,** or cancer-causing genes. Oncogenes arise from changes in the expression of certain genes called **proto-oncogenes,** which are *normal* genes found in all cells and are involved in the control of growth and development.

Investigators first discovered oncogenes in viruses that infect mammalian cells and transform them into cancer cells, in a process known as *malignant transformation.* Such viruses have onco-

Growth factors are signaling proteins that stimulate cell division in specific target cells.

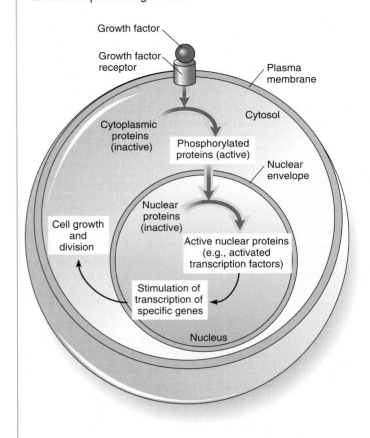

Figure 17-20 A cell growth control cascade

In this example, a growth factor stimulates cell growth. Proto-oncogenes code for the growth factor receptor and some of the other components of the system. When a proto-oncogene mutates, becoming an oncogene, the cell grows and divides even in the absence of the growth factor.

genes as part of their genomes. When these viruses infect a cell, the viral oncogenes are expressed, causing the cell to divide.

A proto-oncogene in a cell that has not been infected by a virus can also mutate and become an oncogene. One of the first oncogenes researchers identified was isolated from a tumor in a human urinary bladder. In the cell that gave rise to the tumor, a proto-oncogene had undergone a single base-pair mutation; the result was that the amino acid valine replaced the amino acid glycine in the protein product of the gene. This subtle change was apparently a critical factor in converting the normal cell into a cancer cell.

Some of the control mechanisms of cell growth are illustrated in greatly simplified form in ▮ Figure 17-20. One or more external signal molecules trigger the growth and division of cells. Some of these substances are **growth factors** that bind to specific **growth factor receptors** associated with the cell surface, initiating a cascade of events inside the cell. Often the growth factor receptor complex acts as a **protein kinase,** an enzyme that phosphory-lates proteins, which then phosphorylates specific amino acids of several cytoplasmic proteins. This posttranslational modification usually results in the activation of previously inactive enzymes. These activated enzymes then catalyze the activation of certain nuclear proteins, many of which are transcription factors. Activated transcription factors bind to their DNA targets and stimulate the transcription of specific sets of genes that initiate growth and cell division (see Chapter 10).

Even in the simplified scenario presented in the figure, it is evident that multiple steps are required to control cell proliferation. Researchers have identified the proto-oncogenes that encode the products responsible for many of these steps. The current list of known proto-oncogenes includes genes that code for various growth factors or growth factor receptors and genes that respond to stimulation by growth factors, including many transcription factors. When a proto-oncogene mutates or is expressed inappropriately (that is, when it becomes an oncogene), the cell may misinterpret the signal and respond by growing and dividing. For example, in some cases a proto-oncogene encoding a growth factor receptor mutates in a way that compromises regulation of the receptor. It is always switched "on," even in the absence of the growth factor that normally controls it.

Not all genes that cause cancer when mutated are proto-oncogenes. About half of all cancers are caused by a mutation in a **tumor suppressor gene.** These genes, also known as *anti-oncogenes,* normally interact with growth-inhibiting factors to block cell division. When mutated, they lose their ability to "put on the brakes," and uncontrolled growth ensues.

Currently, more than 100 oncogenes and 15 tumor suppressor genes have been identified. A change in a single proto-oncogene is usually insufficient to cause a cell to become malig-nant. The development of cancer is usually a multistep process involving both mutations that activate oncogenes and mutations that inactivate tumor suppressor genes.

Additional factors, such as chromosome translocations (see Chapter 16) or the inappropriate activation of the enzyme responsible for the maintenance of telomeres (see Chapter 12), may also play a role in the development of cancer. The Cancer Genome Project, a long-term international initiative based in Great Britain, is currently examining every human gene for cancer-related mutations. As more of these genes are discovered and their interactions unraveled, we will gain a fuller understanding of the control of growth and development. This understanding will improve the diagnosis and treatments of various types of cancer.

Some researchers view cancer as a stem cell disease, because cancer cells are propagated by a small number of cells that have many properties of stem cells. However, the propagation of normal stem cells is tightly regulated, so the question that some biologists are pursuing is how cell mechanisms that regulate stem cell self-renewal are changed to cause cancer cell proliferation. It is possible that the relatively long life of stem cells allows them to accumulate mutations that cause the cells to proliferate in an uncontrolled fashion. Although this line of research is promising, it should be emphasized that the validity of cancer as a stem cell disease remains to be demonstrated for most kinds of cancer.

Review

▌ How are oncogenes and tumor suppressor genes related to genes involved in the control of normal growth and development?

▌ How is a growth factor involved in a typical growth control cascade?

SUMMARY WITH KEY TERMS

Learning Objectives

1 Distinguish between cell determination and cell differentiation, and between nuclear equivalence and totipotency (page 369).

   ▌ An organism contains many types of cells that are specialized both structurally and metabolically to carry out specific functions. These cells are the product of a process of gradual commitment, called **cell determination,** which ultimately leads to the final step in cell specialization, called **cell differentiation.** Differences among various cell types are apparently due to **differential gene expression.**

   ▌ There is no evidence that genes are normally lost during most developmental processes. At least some nuclei from differentiated plant and animal cells contain all the genetic material that would be present in the nucleus of a zygote. **Nuclear equivalence** is the concept that, with a few exceptions, all the nuclei of the differentiated **somatic cells** of an organism are identical to one another and to the nucleus of the single cell from which they descended. **Totipotency** is the capability of cells to direct the development of an entire organism.

2 Describe the classic experiments of Steward, Gurdon, and Wilmut (page 369).

   ▌ Steward induced mature carrot root cells to de-differentiate and express their totipotency, forming an entire plant.

   ▌ Gurdon injected nuclei from tadpole intestinal cells into enucleated eggs; a small fraction of these developed into tadpoles, demonstrating the totipotency of the injected nucleus.

   ▌ Wilmut fused a differentiated cell from an adult sheep cell with an enucleated egg and implanted it into the uterus of a host mother, where it developed into a normal lamb.

ThomsonNOW™ **Learn more about sheep cloning by clicking on the figure in ThomsonNOW.**

3 Define *stem cells,* distinguish between embryonic stem cells and pluripotent stem cells, and describe some of the promising areas of research involving stem cells (page 369).

   ▌ **Stem cells** can divide to produce differentiated descendants yet retain the ability to divide to maintain the stem cell population. Totipotent stem cells give rise to all cell types of the body and placenta, whereas **pluripotent stem cells** give rise to many, but not all, types of cells in an organism.

- **Embryonic stem (ES) cells,** formed after a zygote has undergone several rounds of cell division to form a 5- or 6-day-old blastocyst, are pluripotent and have the potential to develop into any type of cell in the body; ES cells are not totipotent because they cannot form cells of the placenta. **Adult stem cells** have been found in the human brain, retina, heart, bone marrow, dental pulp, intestines, and other sites. Adult stem cells are pluripotent, but the range of cells that adult stem cells can differentiate into is more limited than that of ES cells.

- Stem cells show promise in treating diseases, such as Parkinson's disease and diabetes mellitus.

4 Indicate the features of *Drosophila melanogaster, Caenorhabditis elegans, Mus musculus,* and *Arabidopsis thaliana* that have made these organisms valuable models in developmental genetics (page 374).

- Developmental mutations have been identified in the fruit fly, *D. melanogaster,* many of which affect the organism's segmented body plan. Many developmental genes discovered in the fruit fly are now known to be important in the growth and development of all animals.

- *C. elegans* is a roundworm with **mosaic development,** a rigid developmental pattern in which the fates of cells are restricted early in development. The lineage of every somatic cell in the adult is known, and each can be traced to a single **founder cell** in the early embryo.

- The laboratory mouse, *M. musculus,* is extensively used in studies of mammalian development. In contrast to *C. elegans, M. musculus* shows **regulative development;** the very early embryo is a self-regulating whole and can develop normally even if it has extra or missing cells. **Transgenic** mice, in which foreign genes have been incorporated, have helped researchers determine how genes are activated and regulated during development.

- Genes affecting development have also been identified in certain plants, such as *A. thaliana.* The **ABC model** of interactions among three kinds of genes hypothesizes how floral organs develop in *Arabidopsis.* Mutations in these genes cause one flower part to be substituted for another.

5 Distinguish among maternal effect genes, segmentation genes, and homeotic genes in *Drosophila* (page 374).

- The earliest developmental events to operate in the egg are established by **maternal effect genes** in the surrounding maternal tissues, which are active prior to fertilization. Some produce gradients of **morphogens,** chemical agents that affect the differentiation and development of form. Maternal effect genes establish polarity in the embryo.

- **Segmentation genes** generate a repeating pattern of body segments within the embryo. **Gap genes** are segmentation genes that begin organizing the body into anterior, middle, and posterior regions. **Pair–rule genes** and **segment polarity genes** affect all segments instead of small groups of body segments.

- The later-acting **homeotic genes** are responsible for specifying the identity of each segment.

6 Explain the relationship between transcription factors and genes that control development (page 374).

- **Transcription factors** are DNA-binding proteins that regulate transcription in eukaryotes. Some genes that code for transcription factors contain a DNA sequence called a **homeobox,** which codes for a protein with a DNA-binding region called a **homeodomain.**

- Some homeobox genes are organized into complexes that appear to be systems of master genes specifying an organism's body plan. Parallels exist between the homeobox complexes of *Drosophila* and those of other animals.

7 Define *induction* and *apoptosis,* and give examples of the roles they play in development (page 374).

- **Induction** is developmental interactions with neighboring cells. During development in *C. elegans,* the anchor cell induces cells of the surface to organize to form the vulva, the structure through which the eggs are laid.

- **Apoptosis** is programmed cell death. During human development, the human hand forms as a webbed structure, but the fingers become individualized when the cells between them die.

8 Discuss the relationship between cancer and mutations that affect cell developmental processes (page 386).

- The traits of cancer cells are due to cancer-causing **oncogenes.** Oncogenes arise from changes in the expression of normal genes called **proto-oncogenes,** which exist in all cells and are involved in the control of growth and development.

- Known proto-oncogenes include genes that code for various **growth factors** or **growth factor receptors** and genes that respond to stimulation by growth factors, including many transcription factors. When a proto-oncogene is expressed inappropriately (becomes an oncogene), the cell may misinterpret the signal and respond by growing and dividing.

- **Tumor suppressor genes** normally interact with growth-inhibiting factors to block cell division. A mutation in a tumor suppressor gene may inactivate it and thereby lead to cancer.

TEST YOUR UNDERSTANDING

1. Morphogenesis occurs through the multistep process of (a) differentiation (b) determination (c) pattern formation (d) totipotency (e) selection

2. The cloning experiments carried out on frogs demonstrated that (a) all differentiated frog cells are totipotent (b) some differentiated frog cells are totipotent (c) all nuclei from differentiated frog cells are totipotent (d) some nuclei from differentiated frog cells are totipotent (e) cell differentiation always requires the loss of certain genes

3. *Drosophila* is a particularly good model for developmental studies because (a) a large number of developmental mutants are available (b) it has a fixed number of somatic cells in the adult (c) its embryos are transparent (d) it is a vertebrate (e) all of the preceding

4. The anterior-posterior axis of a *Drosophila* embryo is first established by certain (a) homeotic genes (b) maternal effect genes (c) segmentation genes (d) proto-oncogenes (e) pair–rule genes

5. You discover a new *Drosophila* mutant in which mouthparts are located where the antennae are normally found. You predict that the mutated gene is most likely a (a) homeotic gene (b) gap gene (c) pair—rule gene (d) maternal effect gene (e) segment polarity gene

6. Most segmentation genes code for (a) transfer RNAs (b) enzymes (c) transcription factors (d) histones (e) transport proteins

7. Homeobox genes (a) are found in fruit flies but no other animals (b) tend to be expressed in the order that they appear on a chromosome (c) contain a characteristic DNA sequence (d) b and c (e) a, b, and c

8. The developmental pattern of *C. elegans* is said to be mosaic because (a) development is controlled by gradients of morphogens (b) part of the embryo fails to develop if a founder cell is destroyed (c) some individuals are self-fertilizing hermaphrodites (d) all development is controlled by maternal effect genes (e) apoptosis never occurs

9. The formation of the vulva, the structure through which eggs are laid, in *C. elegans* involves (a) maternal effect genes that organize the egg cytoplasm (b) gradients of morphogens in the eggs (c) groups of *Hox* genes that form the *Antennapedia* complex and *bithorax* complex (d) induction of surface cells by the anchor cell (e) mutations in genes that control developmental timing

10. Which of the following illustrates the regulative nature of early mouse development? (a) the mouse embryo is free-living prior to implantation in the uterus (b) it is possible to produce a transgenic mouse (c) it is possible to produce a mouse in which a specific gene has been knocked out (d) genes related to *Drosophila* homeotic genes have been identified in mice (e) a chimeric mouse can be produced by fusing two mouse embryos

11. When the human gene that codes for insulin is introduced into fertilized mouse eggs that are subsequently allowed to develop, the insulin gene is correctly expressed in the mouse's pancreatic cells. This indicates that (a) the gene that codes for insulin is analogous to the homeotic genes of *Drosophila* (b) the signals for tissue-specific gene expression are highly conserved through evolution (c) like humans, the mouse has an ABC model of organ development (d) unlike the rigid developmental pattern of *C. elegans,* the development of mice and humans is highly regulative (e) imaginal discs are present in the mouse embryo

12. *Arabidopsis* is useful as a model organism for the study of plant development because (a) it is of great economic importance (b) it has a very long generation time (c) many developmental mutants have been isolated (d) it contains a large amount of DNA per cell (e) it has a rigid developmental pattern

13. Pluripotent stem cells (a) lose genetic material during development (b) give rise to many, but not all, types of cells in an organism (c) organize into recognizable structures through pattern formation (d) cannot grow in tissue culture (e) have been used to clone a sheep and several other mammals

14. The genetic material for Dolly, the first cloned sheep, was a nucleus from (a) an early sheep embryo (b) cultured cancer cells (c) intestinal epithelial cells (d) a mouse—sheep chimera (e) a cultured mammary gland cell

15. Which of the following statements about cancer is *false*? (a) oncogenes arise from mutations in proto-oncogenes (b) tumor suppressor genes normally interact with growth-inhibiting factors to block cell division (c) more than 100 oncogenes and 15 tumor suppressor genes have been identified (d) oncogenes were first discovered in mouse models for cancer (e) the development of cancer is usually a multistep process involving both oncogenes and mutated tumor suppressor genes

16. Proto-oncogenes code for (a) morphogens (b) antibodies for immune responses (c) growth factor receptors and other components of the growth control cascade (d) enzymes such as reverse transcriptase (e) ES cells

CRITICAL THINKING

1. Why is an understanding of gene regulation in eukaryotes crucial to an understanding of developmental processes?

2. Why is it necessary for scientists to study development in more than one type of organism?

3. Could a gene be involved in the growth of both stem cells and some kinds of cancer? Explain your answer.

4. Some researchers say that learning how to coax specialized human cells to de-differentiate and become stem cells again is the "Holy Grail" of stem cell research. Explain why.

5. **Evolution Link.** How are the striking similarities among genes that govern development in widely differing species strong evidence for evolution?

6. **Evolution Link.** What is the common ground between evolutionary biologists and developmental biologists who have adopted the perspective known as Evo Devo?

7. **Analyzing Data.** According to the ABC model of floral organ development in *Arabidopsis,* the *A* gene is needed to specify sepals, the *A* and *B* genes to specify petals, the *B* and *C* genes to specify stamens, and the *C* gene to specify the carpel. If a mutation occurs in one of the *B* genes, rendering it inactive, what will the resulting flowers consist of?

Thomson™
NOW!

Additional questions are available in ThomsonNOW at www.thomsonedu.com/login

Introduction to Darwinian Evolution

AKG/Photo Researchers, Inc.

Charles Darwin. This portrait was made shortly after Darwin returned to England from his voyage around the world.

KEY CONCEPTS

Ideas about evolution originated long before Darwin's time.

Darwin's voyage on the *Beagle* provided the basis for his theory of evolution by natural selection.

Natural selection occurs because individuals with traits that make them better adapted to local conditions are more likely to survive and produce offspring than are individuals that are not as well adapted.

The modern synthesis combines Darwin's theory with genetics.

The evidence that evolution has taken place and is still occurring is overwhelming.

A great deal of evidence suggests that the biological diversity represented by the millions of species currently living on our planet evolved from a single ancestor during Earth's long history. Thus, organisms that are radically different from one another are in fact distantly related, linked through numerous intermediate ancestors to a single, common ancestor. The British naturalist Charles Darwin (1809–1882) developed a simple, scientifically testable mechanism to explain the relationship among Earth's diversity of organisms. He argued persuasively that all the species that exist today, as well as the countless extinct species that existed in the past, arose from earlier ones by a process of gradual *divergence* (splitting into separate evolutionary pathways), or *evolution*.

The concept of evolution is the cornerstone of biology, because it links all fields of the life sciences into a unified body of knowledge. As stated by U.S. geneticist Theodosius Dobzhansky, "Nothing in biology makes sense except in the light of evolution."[1] Biologists seek to understand both the remarkable variety and the fundamental similarities of organisms within the context of evolution. The science of evolution allows biologists to compare common threads among organisms as seemingly different as bacteria, whales, lilies, slime molds, and tapeworms. Behavioral evolution, evolutionary developmental biology, evolutionary genetics, evolutionary ecology, evolutionary systematics, and molecular evolution are examples of some of the biological disciplines that focus on evolution.

---

[1] *American Biology Teacher*, Vol. 35, No. 125 (1973).

This chapter discusses Darwin and the scientific development of the theory of evolution by natural selection. It also presents evidence that supports evolution, including fossils, comparative anatomy, biogeography, developmental biology, molecular biology, and experimental studies of ongoing evolutionary change in the laboratory and in nature. ■

WHAT IS EVOLUTION?

Learning Objective

1 Define *evolution*.

In beginning our study of evolution, we define **evolution** as the accumulation of inherited changes within populations over time. A **population** is a group of individuals of one species that live in the same geographic area at the same time. Just as the definition of a gene changed as you studied genetics, you will find that the definition of evolution will change and become more precise in later chapters.

The term *evolution* does not refer to changes that occur in an individual within its lifetime. Instead, it refers to changes in the characteristics of populations over the course of generations. These changes may be so small that they are difficult to detect or so great that the population differs markedly from its ancestral population.

Eventually, two populations may diverge to such a degree that we refer to them as different species. The concept of species is developed extensively in Chapter 20. For now, a simple working definition is that a **species** is a group of organisms, with similar structure, function, and behavior that are capable of interbreeding with one another.

Evolution has two main perspectives—the minor evolutionary changes of populations usually viewed over a few generations (*microevolution,* discussed in Chapter 19); and the major evolutionary events usually viewed over a long period, such as formation of different species from common ancestors (*macroevolution,* discussed in Chapter 20).

Evolution has important practical applications. Agriculture must deal with the evolution of pesticide resistance in insects and other pests. Likewise, medicine must respond to the rapid evolutionary potential of disease-causing organisms such as bacteria and viruses. (Significant evolutionary change occurs in a very short time period in insects, bacteria, and other organisms with short life spans.) Medical researchers use evolutionary principles to predict which flu strains are evolving more quickly, information the scientists need to make the next year's flu vaccine. Also, researchers developing effective treatment strategies for the human immunodeficiency virus (HIV) must understand its evolution, both within and among hosts.

The conservation management of rare and endangered species makes use of the evolutionary principles of population genetics. The rapid evolution of bacteria and fungi in polluted soils is used in the field of **bioremediation,** in which microorganisms are employed to clean up hazardous-waste sites. Evolution even has applications beyond biology. For example, certain computer applications make use of algorithms that mimic natural selection in biological systems.

Review

■ What is evolution?
■ Do individuals evolve? Explain your answer.

PRE-DARWINIAN IDEAS ABOUT EVOLUTION

Learning Objective

2 Discuss the historical development of the theory of evolution.

Although Darwin is universally associated with evolution, ideas of evolution predate Darwin by centuries. Aristotle (384–322 BCE) saw much evidence of natural affinities among organisms. This led him to arrange all the organisms he knew in a "scale of nature" that extended from the exceedingly simple to the most complex. Aristotle visualized organisms as being imperfect but "moving toward a more perfect state." Some scientific historians have interpreted this idea as the forerunner of evolutionary theory, but Aristotle was vague on the nature of this "movement toward perfection" and certainly did not propose that natural processes drove the process of evolution. Furthermore, modern evolutionary theory now recognizes that evolution does not move toward more "perfect" states or even necessarily toward greater complexity.

Long before Darwin's time, fossils had been discovered embedded in rocks. Some of these corresponded to parts of familiar species, but others were strangely unlike any known species. Fossils were often found in unexpected contexts; for example, marine invertebrates (sea animals without backbones) were sometimes discovered in rocks high on mountains. Leonardo da Vinci (1452–1519) was among the first to correctly interpret these unusual finds as the remains of animals that had existed in previous ages but had become extinct.

The French naturalist Jean Baptiste de Lamarck (1744–1829) was the first scientist to propose that organisms undergo change over time as a result of some natural phenomenon rather than divine intervention. According to Lamarck, a changing environment caused an organism to alter its behavior, thereby using some organs or body parts more and others less. Over several generations, a given organ or body part would increase in size if it was used a lot, or shrink and possibly disappear if it was used less. His hypothesis required that organisms pass traits they acquired during their lifetimes to their offspring. For example, Lamarck suggested that the long neck of the giraffe developed when a short-necked ancestor stretched its neck to browse on the leaves of trees. Its offspring inherited the longer neck, which stretched still further as they ate. This process, repeated over many generations, resulted in the long necks of modern giraffes. Lamarck also thought that all organisms were endowed with a vital force that drove them to change toward greater complexity and "perfection" over time.

Lamarck's proposed mechanism of evolution is quite different from the mechanism later proposed by Darwin. However, Lamarck's hypothesis remained a reasonable explanation for evolution until Mendel's basis of heredity was rediscovered at the beginning of the 20th century. At that time, Lamarck's ideas were largely discredited.

Review

▪ Why is Aristotle linked to early evolutionary thought?
▪ What were Jean Baptiste de Lamarck's ideas concerning evolution?

DARWIN AND EVOLUTION

Learning Objectives

3   Explain the four premises of evolution by natural selection as proposed by Charles Darwin.
4   Compare the modern synthesis with Darwin's original theory of evolution.

Darwin, the son of a prominent physician, was sent at the age of 15 to study medicine at the University of Edinburgh. Finding himself unsuited for medicine, he transferred to Cambridge University to study theology. During that time, he became the protégé of the Reverend John Henslow, who was a professor of botany. Henslow encouraged Darwin's interest in the natural world. Shortly after receiving his degree, Darwin embarked on the HMS

Beagle, which was taking a 5-year exploratory cruise around the world to prepare navigation charts for the British navy.

The *Beagle* left Plymouth, England, in 1831 and cruised along the east and west coasts of South America (▪ Fig. 18-1). While other members of the crew mapped the coasts and harbors, Darwin spent many weeks ashore studying the animals, plants, fossils, and geologic formations of both coastal and inland regions, areas that had not been extensively explored. He collected and cataloged thousands of plant and animal specimens and kept notes of his observations, information that became essential in the development of his theory.

The *Beagle* spent almost 2 months at the Galápagos Islands, 965 km (600 mi) west of Ecuador, where Darwin continued his observations and collections. He compared the animals and plants of the Galápagos with those of the South American mainland. He was particularly impressed by their similarities and wondered why the organisms of the Galápagos should resemble those from South America more than those from other islands in different parts of the world. Moreover, although there were similarities between Galápagos and South American species, there were also distinct differences. There were even recognizable differences in the reptiles and birds from one island to the next. Darwin wondered why these remote islands should have such biological diversity. After he returned home, Darwin pondered these observations and attempted to develop a satisfactory explanation for the distribution of species among the islands.

Darwin drew on several lines of evidence when considering how species might have originated. Despite the work of Lamarck, the general notion in the mid-1800s was that Earth was too young for organisms to have changed significantly since they had first appeared. During the early 19th century, however, geologists advanced the idea that mountains, valleys, and other physical features of Earth's surface did not originate in their present forms.

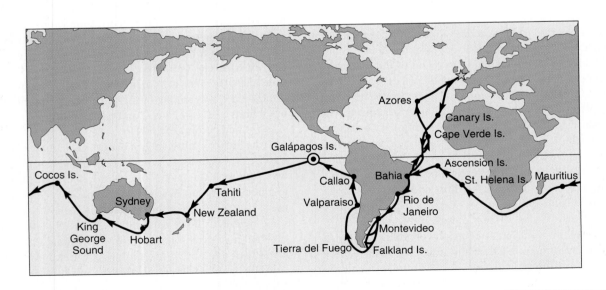

▪ **Figure 18-1** *Animated*   The voyage of HMS *Beagle*

The 5-year voyage began in Plymouth, England (*star*), in 1831. Observations made in the Galápagos Islands (*bull's-eye*) off the western coast of South America helped Darwin develop a satisfactory mechanism to explain how a population of organisms could change over time.

Figure 18-2 Artificial selection in *Brassica oleracea*

An enlarged terminal bud (the "head") was selected in cabbage (*lower left*), flower clusters in broccoli (*upper left*) and cauliflower (*middle right*), axillary buds in brussels sprouts (*bottom middle*), leaves in collards (*upper right*) and kale (*lower right*), and stems in kohlrabi (*middle*).

Instead, these features developed slowly over long periods by the geologic processes of volcanic activity, uplift, erosion, and glaciation. On his voyage Darwin took with him *Principles of Geology*, published by English geologist Charles Lyell in 1830, and studied it carefully. Lyell provided an important concept for Darwin—that the slow pace of geologic processes, which still occur today, indicated that Earth was extremely old.

Other important evidence that influenced Darwin was the fact that breeders and farmers could develop many varieties of domesticated animals in just a few generations. They did so by choosing certain traits and breeding only individuals that exhibited the desired traits, a procedure known as **artificial selection.** Breeders, for example, have produced numerous dog varieties—such as bloodhounds, dalmatians, Airedales, border collies, and Pekinese—by artificial selection.

Many plant varieties were also produced by artificial selection. For example, cabbage, broccoli, brussels sprouts, cauliflower, collard greens, kale, and kohlrabi are distinct vegetable crops that are all members of the same species, *Brassica oleracea* (❚ Fig. 18-2). Selective breeding of the colewort, or wild cabbage, a leafy plant native to Europe and Asia, produced all seven vegetables. Beginning more than 4000 years ago, some farmers artificially selected wild cabbage plants that formed overlapping leaves. Over time, these leaves became so prominent that the plants, which resembled modern cabbages, became recognized as separate and distinct from their wild cabbage ancestor. Other farmers selected different features of the wild cabbage, giving rise to the other modifications. For example, kohlrabi was produced by selection for an enlarged storage stem, and brussels sprouts by selection for enlarged axillary buds. Thus, humans are respon-

sible for the evolution of *B. oleracea* into seven distinct vegetable crops. Darwin was impressed by the changes induced by artificial selection and hypothesized that a similar selective process occurred in nature. Darwin therefore used artificial selection as a model when he developed the concept of natural selection.

The ideas of Thomas Malthus (1766–1834), a British clergyman and economist, were another important influence on Darwin. In *An Essay on the Principle of Population As It Affects the Future Improvement of Society,* published in 1798, Malthus noted that population growth is not always desirable—a view contrary to the beliefs of his day. He observed that populations have the capacity to increase geometrically ($1 \longrightarrow 2 \longrightarrow 4 \longrightarrow 8 \longrightarrow 16$) and thus outstrip the food supply, which only has the capacity to increase arithmetically ($1 \longrightarrow 2 \longrightarrow 3 \longrightarrow 4 \longrightarrow 5$). In the case of humans, Malthus suggested that the conflict between population growth and food supply generates famine, disease, and war, which serve as inevitable brakes on population growth.

Malthus's idea that there is a strong and constant check on human population growth strongly influenced Darwin's explanation of evolution. Darwin's years of observing the habits of animals and plants had introduced him to the struggle for existence described by Malthus. It occurred to Darwin that in this struggle, inherited variations favorable to survival would tend to be preserved, whereas unfavorable ones would be eliminated.

The result would be **adaptation,** an evolutionary modification that improves the chances of survival and reproductive success in a given environment. Eventually, the accumulation of modifications might result in a new species. Time was the only thing required for new species to originate, and the geologists of the era, including Lyell, had supplied evidence that Earth was indeed old enough to provide an adequate period.

Darwin had at last developed a workable explanation of evolution, that of **natural selection,** in which better-adapted organisms are more likely to survive and become the parents of the next generation. As a result of natural selection, the population changes over time; the frequency of favorable traits increases in successive generations, whereas less favorable traits become scarce or disappear. Darwin spent the next 20 years formulating his arguments for natural selection, accumulating an immense body of evidence to support his theory, and corresponding with other scientists.

As Darwin was pondering his ideas, Alfred Russel Wallace (1823–1913), a British naturalist who studied the plants and animals of the Malay Archipelago for 8 years, was similarly struck by the diversity of species and the peculiarities of their distribution. He wrote a brief essay on this subject and sent it to Darwin, by then a world-renowned biologist, asking his opinion. Darwin recognized his own theory and realized Wallace had independently arrived at the same conclusion—that evolution occurs by natural selection. Darwin's colleagues persuaded him to present Wallace's manuscript along with an abstract of his own work, which he had prepared and circulated to a few friends several years earlier. Both papers were presented in July 1858 at a London meeting of the Linnaean Society. Darwin's monumental book, *On the Origin of Species by Natural Selection; or, The Preservation of Favored Races in the Struggle for Life,* was published in 1859. In 1870, Wallace's book, *Contributions to the Theory of Natural Selection,* was published, 8 years after he returned from the Malay Archipelago.

Darwin proposed that evolution occurs by natural selection

Darwin's mechanism of evolution by natural selection consists of observations on four aspects of the natural world: variation, overproduction, limits on population growth, and differential reproductive success.

1. *Variation.* The individuals in a population exhibit variation. Each individual has a unique combination of traits, such as size, color, ability to tolerate harsh environmental conditions, and resistance to certain parasites or infections. Some traits improve an individual's chances of survival and reproductive success, whereas others do not. Remember that the variation necessary for evolution by natural selection must be inherited (Fig. 18-3). Although Darwin recognized the importance to evolution of inherited variation, he did not know the mechanism of inheritance.

2. *Overproduction.* The reproductive ability of each species has the potential to cause its population to geometrically increase over time. A female frog lays about 10,000 eggs, and a female cod produces perhaps 40 million eggs! In each case, however, only about two offspring survive to reproduce. Thus, in every generation each species has the capacity to produce more offspring than can survive.

3. *Limits on population growth, or a struggle for existence.* There is only so much food, water, light, growing space, and other resources available to a population, so organisms compete with one another for these limited resources. Because there are more individuals than the environment can support, not all survive to reproduce. Other limits on population growth include predators, disease organisms, and unfavorable weather conditions.

4. *Differential reproductive success.* Those individuals that have the most favorable combination of characteristics (those that make individuals better adapted to their environment) are more likely to survive and reproduce. Offspring tend to resemble their parents, because the next generation inherits the parents' genetically based traits. Successful reproduction is the key to natural selection: The best-adapted individuals produce the most offspring, whereas individuals that are less well adapted die prematurely or produce fewer or inferior offspring.

Over time, enough changes may accumulate in geographically separated populations (often with slightly different environments) to produce new species. Darwin noted that the Galápagos finches appeared to have evolved in this way. The 14 species are closely related. All descended from a common ancestor—a single species that found its way from the South American mainland 2 million to 3 million years ago.[2] This estimate is based on cal-

culations using a molecular clock (discussed later in the chapter), as fossil evidence is not available. During this 2-million to 3-million year period, the number of islands increased, the climate changed, and the plant life and food supply evolved. The different islands of the Galápagos kept the finches isolated from one another, thereby allowing them to diverge into 14 separate species in response to varying conditions (Fig. 18-4).

Peter Grant, Rosemary Grant, and colleagues have documented natural selection in the Galápagos finches in their natural environment since the early 1970s. As an example of the evolutionary process in action, consider the sharp-beaked ground finch (*Geospiza difficilis*). This species lives on different islands, and each population has evolved different beak shapes and sizes depending on the diet available on the island where it lives.

We revisit the Galápagos finches in later chapters. Some long-term research by Peter Grant, Rosemary Grant, and colleagues on the microevolution of Galápagos finches when droughts affect the food supply is described in Chapter 19; these studies have demonstrated that environmental change can drive natural selection. *Character displacement,* an aspect of evolutionary ecology, is described in Galápagos finches in Chapter 53.

The modern synthesis combines Darwin's theory with genetics

One of the premises on which Darwin based his theory of evolution by natural selection is that individuals transmit traits to the next generation. However, Darwin was unable to explain *how* this occurs or *why* individuals vary within a population. Although he was a contemporary of Gregor Mendel (see Chapter 11), who elucidated the basic patterns of inheritance, Darwin was apparently not acquainted with Mendel's work, which was not recognized by the scientific community until the early part of the 20th century.

 Figure 18-3 Genetic variation in emerald tree boas

These snakes, all of the same species (*Corallus caninus*), were caught in a small section of forest in French Guiana. Many snake species exhibit considerable variation in their coloration and patterns.

---

[2]The closest genetic relatives of the Galápagos finches are small seed-eating birds known as grassquits that live in western South America.

(a) The cactus finch (*Geospiza scandens*) feeds on the fleshy parts of cacti.

(b) The large ground finch (*Geospiza magnirostris*) has an extremely heavy, nutcracker-type bill adapted for eating thick, hard-walled seeds.

(c) The woodpecker finch (*Camarhynchus pallidus*) digs insects out of bark and crevices by using spines, twigs, or even dead leaves.

Figure 18-4 *Animated* Three species of Galápagos finches

Darwin inferred that these birds are derived from a common ancestral population of seed-eating birds from South America. Variation in their beaks is the result of adaptation to different kinds of food.

Beginning in the 1930s and 1940s, biologists experienced a conceptual breakthrough when they combined the principles of Mendelian inheritance with Darwin's theory of natural selection. The result was a unified explanation of evolution known as the **modern synthesis,** or the **synthetic theory of evolution.** In this context, *synthesis* refers to combining parts of several theories to form a unified whole. Some of the founders of the modern synthesis were U.S. geneticist Theodosius Dobzhansky, British geneticist and statistician Ronald Fisher, British geneticist J. B. S. Haldane, British biologist Julian Huxley, U.S. biologist Ernst Mayr, U.S. paleontologist George Gaylord Simpson, U.S. botanist G. Ledyard Stebbins, and U.S. geneticist Sewell Wright.

Today, the modern synthesis incorporates our expanding knowledge in genetics, systematics, paleontology, developmental biology, behavior, and ecology. It explains Darwin's observation of variation among offspring in terms of **mutation,** or changes in DNA, such as nucleotide substitutions. Mutations provide the genetic variability on which natural selection acts during evolution. The modern synthesis, which emphasizes the genetics of populations as the central focus of evolution, has held up well since it was developed. It has dominated the thinking and research of biologists working in many areas and has resulted in an enormous accumulation of new discoveries that validate evolution by natural selection.

Most biologists not only accept the basic principles of the modern synthesis but also try to better understand the causal processes of evolution. For example, what is the role of chance in evolution? How rapidly do new species evolve? These and other questions have arisen in part from a re-evaluation of the fossil record and in part from new discoveries in molecular aspects of inheritance. Such critical analyses are an integral part of the scientific process because they stimulate additional observation and experimentation, along with re-examination of previous evidence. Science is an ongoing process, and information obtained in the future may require modifications to certain parts of the modern synthesis.

We now consider one of the many evolutionary questions currently being addressed by biologists: the relative effects of chance and natural selection on evolution.

Biologists study the effect of chance on evolution

Biologists have wondered whether we would get the same results if we were able to repeat evolution by starting with similar organisms exposed to similar environmental conditions. That is, would the same kinds of changes evolve, as a result of natural selection? Or would the organisms be quite different, as a result of random events? Several recently reported examples of evolution in action suggest that chance may not be as important as natural selection, at least at the population level.

A fruit fly species (*Drosophila subobscura*) native to Europe inhabits areas from Denmark to Spain. Biologists noted that the northern flies have larger wings than southern flies (❙ Fig. 18-5). The same fly species was accidentally introduced to North and South America in the late 1970s, in two separate introductions. Ten years after its introduction to the Americas, biologists determined that no statistically significant changes in wing size had occurred in the different regions of North America. However,

Figure 18-5 Wing size in female fruit flies

In Europe, female fruit flies (*Drosophila subobscura*) in northern countries have larger wings than flies in southern countries. Shown are two flies: one from Denmark (*right*) and the other from Spain (*left*). The same evolutionary pattern emerged in North America after the accidental introduction of *D. subobscura* to the Americas.

20 years after its introduction, the fruit flies in North America exhibited the same type of north–south wing changes as in Europe. (It is not known why larger wings evolve in northern areas and smaller wings in southern climates.)

A study of the evolution of fishes known as *sticklebacks* in three coastal lakes of western Canada yielded intriguingly similar results to the fruit fly study. Molecular evidence indicates that when the lakes first formed several thousand years ago, they were populated with the same ancestral species. (Analysis of the mitochondrial DNA of sticklebacks in the three lakes supports the hypothesis of a common ancestor.) In each lake, the same two species have evolved from the common ancestral fish. One species is large and consumes invertebrates along the bottom of the lake, whereas the other species is smaller and consumes plankton at the lake's surface. Members of the two species within a single lake do not interbreed with one another, but individuals of the larger species from one lake interbreed in captivity with individuals of the larger species from the other lakes. Similarly, smaller individuals from one lake interbreed with smaller individuals from the other lakes.

In these examples, natural selection appears to be a more important agent of evolutionary change than chance. If chance were the most important factor influencing the direction of evolution, then fruit fly evolution would not have proceeded the same way on two different continents, and stickleback evolution would not have proceeded the same way in three different lakes. However, just because we have many examples of the importance of natural selection in evolution, it does not necessarily follow that random events should be discounted as a factor in directing evolutionary change. Proponents of the role of chance think that chance is more important in the evolution of major taxonomic groups (macroevolution) than in the evolution of populations (microevolution). It also may be that random events take place but their

effects on evolution are simply harder to demonstrate than natural selection.

Review

- What is natural selection?
- Why are only inherited variations important in the evolutionary process?
- What part of Darwin's theory was he unable to explain? How does the modern synthesis fill this gap?

EVIDENCE FOR EVOLUTION

Learning Objectives

5 Summarize the evidence for evolution obtained from the fossil record.
6 Relate the evidence for evolution derived from comparative anatomy.
7 Define *biogeography,* and describe how the distribution of organisms supports evolution.
8 Briefly explain how developmental biology and molecular biology provide insights into the evolutionary process.
9 Give an example of how evolutionary hypotheses are tested experimentally.

A vast body of scientific evidence supports evolution, including observations from the fossil record, comparative anatomy, biogeography, developmental biology, and molecular biology. In addition, evolutionary hypotheses are increasingly being tested experimentally. Taken together, this evidence confirms the theory that life unfolded on Earth by the process of evolution.

The fossil record provides strong evidence for evolution

Perhaps the most direct evidence for evolution comes from the discovery, identification, and interpretation of **fossils,** which are the remains or traces typically left in sedimentary rock by previously existing organisms. (The term *fossil* comes from the Latin word *fossilis,* meaning "something dug up.") Sedimentary rock forms by the accumulation and solidification of particles (pebbles, sand, silt, or clay) produced by the weathering of older rocks, such as volcanic rocks. The sediment particles, which are usually deposited on a riverbed, lake bottom, or the ocean floor, accumulate over time and exhibit distinct layers (❙ Fig. 18-6). In an undisturbed rock sequence, the oldest layer is at the bottom, and upper layers are successively younger. The study of sedimentary rock layers, including their composition, arrangement, and correlation (similarity) from one location to another, enables geologists to place events recorded in rocks in their correct sequence.

The fossil record shows a progression from the earliest unicellular organisms to the many unicellular and multicellular organisms living today. The fossil record therefore demonstrates that life has evolved through time. To date, paleontologists (scientists who study extinct species) have described and named about 300,000 fossil species, and others are still being discovered.

Tom Till

Figure 18-6 Exposed layers of sedimentary rock

Shown are weathered rock layers near the Paria River in the Grand Staircase–Escalante National Monument, Utah. The younger layers overlie the older layers. Many layers date to the Mesozoic era, from 248 mya to 65 mya. Characteristic fossils are associated with each layer.

Although most fossils are preserved in sedimentary rock, some more recent remains have been exceptionally well preserved in bogs, tar, amber (ancient tree resin), or ice (❚ Fig. 18-7). For example, the remains of a woolly mammoth deep-frozen in Siberian ice for more than 25,000 years were so well preserved that part of its DNA could be analyzed.

Few organisms that die become fossils. The formation and preservation of a fossil require that an organism be buried under conditions that slow or prevent the decay process. This is most likely to occur if an organism's remains are covered quickly by a sediment of fine soil particles suspended in water. In this way the remains of aquatic organisms may be trapped in bogs, mudflats, sandbars, or deltas. Remains of terrestrial organisms that lived on a floodplain may also be covered by waterborne sediments or, if the organism lived in an arid region, by windblown sand. Over time, the sediments harden to form sedimentary rock, and minerals usually replace the organism's remains so that many details of its structure, even cellular details, are preserved.

The fossil record is not a random sample of past life but instead is biased toward aquatic organisms and those living in the few terrestrial habitats conducive to fossil formation. Relatively few fossils of tropical rainforest organisms have been found, for example, because their remains decay extremely rapidly on the forest floor, before fossils can develop. Another reason for bias

in the fossil record is that organisms with hard body parts such as bones and shells are more likely to form fossils than are those with soft body parts. Also, rocks of different ages are unequally exposed at Earth's surface; some rocks of certain ages are more accessible to paleontologists for fossil study than are rocks of other ages.

Because of the nature of the scientific process, each fossil discovery represents a separate "test" of the theory of evolution. If any of the tests fail, the theory would have to be modified to fit the existing evidence. The verifiable discovery, for example, of fossil remains of modern humans (*Homo sapiens*) in Precambrian rocks, which are more than 570 million years old, would falsify the theory of evolution as currently proposed. However, Precambrian rocks examined to date contain only fossils of simple organisms, such as algae and small, soft-bodied animals, that evolved early in the history of life. The earliest fossils of *H. sapiens* with anatomically modern features do not appear in the fossil record until approximately 195,000 years ago (see Chapter 22).

Fossils provide a record of ancient organisms and some understanding of where and when they lived. Using fossils of organisms from different geologic ages, scientists can sometimes infer the lines of descent (evolutionary relationships) that gave rise to modern-day organisms. In many instances, fossils provide direct evidence of the origin of new species from pre-existing species, including many transitional forms.

Transitional fossils document whale evolution

Over the past century biologists have found evidence suggesting that whales and other cetaceans (an order of marine mammals) evolved from land-dwelling mammals. During the 1980s and 1990s, paleontologists discovered several fossil intermediates in whale evolution that document the whales' transition from land to water. Fossil evidence suggests that one candidate for the ancestor of whales is a now-extinct group of four-legged, land-dwelling mammals called *mesonychians* (❚ Fig. 18-8a). These animals had unusually large heads and teeth that were remarkably similar to those of the earliest whales. About 50 million to 60 million years ago (mya), some descendants of mesonychians had adapted to swimming in shallow seas.

Fossils of *Ambulocetus natans*, a 50-million-year-old transitional form discovered in Pakistan, have many features of modern whales but also possess hind limbs and feet (❚ Fig. 18-8b). (Modern whales do not have hind limbs, although *vestigial* pelvic and hind-limb bones persist. Vestigial structures are discussed later in the chapter.) The vertebrae of *Ambulocetus's* lower back were very flexible, allowing the back to move dorsoventrally (up and down) during swimming and diving, as with modern whales. This ancient whale could swim but also moved about on land, perhaps as sea lions do today.

Rodhocetus is a fossil whale found in more recent rocks in Pakistan (❚ Fig. 18-8c). The vertebrae of *Rodhocetus* were even more flexible than those found in *Ambulocetus*. The flexible vertebrae allowed *Rodhocetus* a more powerful dorsoventral movement during swimming. *Rodhocetus* may have been totally aquatic.

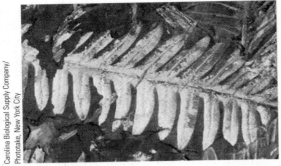

(a) Although some fossils contain traces of organic matter, all that remains in this fossil of a seed fern leaf is an impression, or imprint, in the rock.

(b) Petrified wood from the Petrified Forest National Park in Arizona consists of trees that were buried and infiltrated with minerals. The logs were exposed as the mudstone layers in which they were buried weathered.

(c) A 2-million-year-old insect fossil (a midge) was embedded in amber.

(d) A cast fossil of ancient echinoderms called *crinoids* formed when the crinoids decomposed, leaving a mold that later filled with dissolved minerals that hardened.

(e) Dinosaur footprints, each 75 to 90 cm (2.5 to 3 ft) in length, provide clues about the posture, gait, and behavior of these extinct animals.

Figure 18-7 Fossils develop in different ways

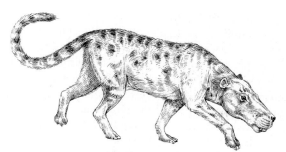

(a) *Mesonychid,* an extinct terrestrial mammal, may have been the ancestor of whales.

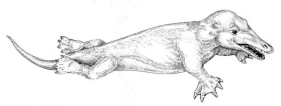

(b) *Ambulocetus natans,* a transitional form between modern whale descendants and their terrestrial ancestors, possessed several recognizable whale features yet retained the hind limbs of its four-legged ancestors.

(c) The more recent *Rodhocetus* had flexible vertebrae that permitted a powerful dorsoventral movement during swimming.

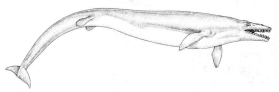

(d) *Basilosaurus* was more streamlined and possessed tiny nonfunctional hind limbs.

(e) *Balaenoptera,* the modern blue whale, contains vestiges of pelvis and leg bones embedded in its body.

Figure 18-8 *Animated* Fossil intermediates in whale evolution

Figures are not drawn to scale. (a–d: Adapted with permission from D. J. Futuyma, *Science on Trial: The Case for Evolution*, Fig. 2, pp. 260–61, Sinauer Associates, Sunderland, MA, 1995.)

By 40 mya, the whale transition from land to ocean was almost complete. Egyptian fossils of *Basilosaurus* show a whale with a streamlined body and front flippers for steering, like those of modern-day whales (❙ Fig. 18-8d). *Basilosaurus* retained vestiges of its land-dwelling ancestors—a pair of reduced hind limbs that were disjointed from the backbone and probably not used in locomotion. Reduction in the hind limbs continued to the present. The modern blue whale has vestigial pelvis and femur bones embedded in its body (❙ Fig. 18-8e).

Various methods determine the age of fossils

Because layers of sedimentary rock occur naturally in the sequence of their deposition, with the more recent layers on top of the older, earlier ones, most fossils are dated by their relative position in sedimentary rock. However, geologic events occurring after the rocks were initially formed have occasionally changed the relationships of some rock layers. Geologists identify specific sedimentary rocks not only by their positions in layers but also by features such as mineral content and by the fossilized remains of certain organisms, known as **index fossils,** that characterize a specific layer over large geographic areas. Index fossils are fossils of organisms that existed for a relatively short geologic time but were preserved as fossils in large numbers. With this information, geologists can arrange rock layers and the fossils they contain in chronological order and identify comparable layers in widely separated locations.

Radioactive isotopes, also called **radioisotopes,** present in a rock provide a means to accurately measure its age (see Chapter 2). Radioisotopes emit invisible radiations. As a radioisotope emits radiation, its nucleus changes into the nucleus of a different element in a process known as **radioactive decay.** The radioactive nucleus of uranium-235, for example, decays into lead-207.

Each radioisotope has its own characteristic rate of decay. The time required for one half of the atoms of a radioisotope to change into a different atom is known as its **half-life** (❙ Fig. 18-9). Radioisotopes differ significantly in their half-lives. For example, the half-life of iodine-132 is only 2.4 hours, whereas the half-life of uranium-235 is 704 million years. The half-life of a particular radioisotope is constant and does not vary with temperature, pressure, or any other environmental factor.

The age of a fossil in sedimentary rock is usually estimated by measuring the relative proportions of the original radioisotope and its decay product in volcanic rock intrusions that penetrate the sediments. For example, the half-life of potassium-40 is 1.3 billion years, meaning that in 1.3 billion years half of the radioactive potassium will have decayed into its decay product, argon-40. The radioactive clock begins ticking when the magma solidifies into volcanic rock. The rock initially contains some potassium but no argon. Because argon is a gas, it escapes from hot rock as soon as it forms, but when potassium decays in rock that has cooled and solidified, the argon accumulates in the crystalline structure of the rock. If the ratio of potassium-40 to argon-40 in the rock being tested is 1:1, the rock is 1.3 billion years old.

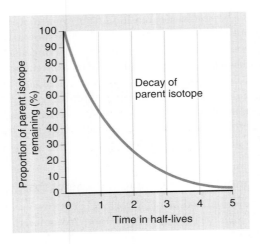

Figure 18-9 *Animated* Radioisotope decay

At time zero, the sample is composed entirely of the radioisotope and the radioactive clock begins ticking. After one half-life, only 50% of the original radioisotope remains. During each succeeding half-life, half of the remaining radioisotope is converted to decay product(s).

Several radioisotopes are commonly used to date fossils. These include potassium-40 (half-life 1.3 billion years), uranium-235 (half-life 704 million years), and carbon-14 (half-life 5730 years). Potassium-40, with its long half-life, is used to date fossils that are many hundreds of millions of years old. Radioisotopes other than carbon-14 are used to date the *rock* in which fossils are found, whereas carbon-14 is used to date the *carbon remains* of anything that was once living, such as wood, bones, and shells. Whenever possible, the age of a fossil is independently verified using two or more different radioisotopes.

Carbon-14, which is continuously produced in the atmosphere from nitrogen-14 (by cosmic radiation), subsequently decays to nitrogen-14. Because the formation and the decay of carbon-14 occur at constant rates, the ratio of carbon-14 to carbon-12 (a more abundant, stable isotope of carbon) is constant in the atmosphere. Organisms absorb carbon from the atmosphere either directly (by photosynthesis) or indirectly (by consuming photosynthetic organisms). Because each organism absorbs carbon from the atmosphere, its ratio of carbon-14 to carbon-12 is the same as that in the atmosphere. When an organism dies, however, it no longer absorbs carbon, and the proportion of carbon-14 in its remains declines as carbon-14 decays to nitrogen-14. Because of its relatively short half-life, carbon-14 is useful for dating fossils that are 50,000 years old or less. It is particularly useful for dating archaeological sites.

Comparative anatomy of related species demonstrates similarities in their structures

Comparing the structural details of features found in different but related organisms reveals a basic similarity. Such features that are derived from the same structure in a common ancestor are termed **homologous features;** the condition is known as **homology.** For example, consider the limb bones of mammals. A human arm, a cat forelimb, a whale front flipper, and a bat wing, although quite different in appearance, have strikingly similar arrangements of bones, muscles, and nerves. ▌Figure 18-10 shows a comparison of their skeletal structures. Each has a single bone (the humerus) in the part of the limb nearest the trunk of the body, followed by the two bones (radius and ulna) of the forearm, a group of bones (carpals) in the wrist, and a variable number of digits (metacarpals and phalanges). This similarity is particularly striking because arms, forelimbs, flippers, and wings are used for different types of locomotion, and there is no overriding mechanical reason for them to be so similar structurally. Similar arrangements of parts of the forelimb are evident in ancestral birds, reptiles, and amphibians and even in the first fishes that came out of water onto land hundreds of millions of years ago.

Leaves are an example of homology in plants. In many plant species, leaves have been modified for functions other than photosynthesis. A cactus spine and a pea tendril, although quite different in appearance, are homologous because both are modified leaves (▌Fig. 18-11). The spine protects the succulent stem tissue of the cactus, whereas the tendril, which winds around a small object once it makes contact, helps support the climbing stem of the pea plant. Such modifications in organs used in different ways are the expected outcome of a common evolutionary origin. The basic structure present in a common ancestor was modified in different ways for different functions as various descendants subsequently evolved.

Not all species with "similar" features have descended from a recent common ancestor, however. Sometimes similar environmental conditions result in the evolution of similar adaptations. Such independent evolution of similar structures in distantly related organisms is known as **convergent evolution.** Aardvarks, anteaters, and pangolins are an excellent example of convergent evolution (▌Fig. 18-12). They resemble one another in lifestyle and certain structural features. All have strong, sharp claws to dig open ant and termite mounds and elongated snouts with long, sticky tongues to catch these insects. Yet aardvarks, anteaters, and pangolins evolved from three distantly related orders of mammals. (See Chapter 23 for further discussion of homology. Also, see Figure 31-28, which shows several examples of convergent evolution in placental and marsupial mammals.)

Structurally similar features that are not homologous but have similar functions that evolved independently in distantly related organisms are said to be **homoplastic features.** Such similarities in different species that are independently acquired by convergent evolution and not by common descent are called **homoplasy.**[3] For example, the wings of various distantly related flying animals, such as insects and birds, resemble one another superficially; they are homoplastic features that evolved over time to meet the common function of flight, although they differ in

---

[3] An older, less precise term that some biologists still use for nonhomologous features with similar functions is *analogy.*

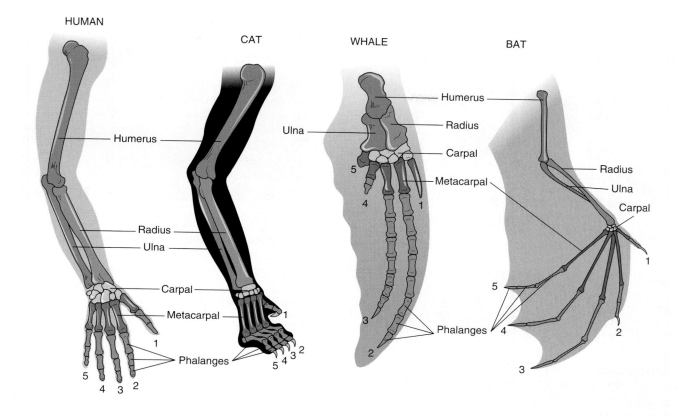

Figure 18-10 Homology in animals

The human arm, cat forelimb, whale flipper, and bat wing have a basic underlying similarity of structure because they are derived from a common ancestor. The five digits are numbered in each drawing.

more fundamental aspects. Bird wings are modified forelimbs supported by bones, whereas insect wings may have evolved from gill-like appendages present in the aquatic ancestors of insects. Spines, which are modified leaves, and thorns, which are modified stems, are an example of homoplasy in plants. Spines and thorns resemble one another superficially but are homoplastic features that evolved independently to solve the common need for protection from herbivores (❚ Fig. 18-13).

Like homology, homoplasy offers crucial evidence of evolution. Homoplastic features are of evolutionary interest because they demonstrate that organisms with separate ancestries may adapt in similar ways to similar environmental demands.

Comparative anatomy reveals the existence of **vestigial structures.** Many organisms contain organs or parts of organs that are seemingly nonfunctional and degenerate, often undersized or lacking some essential part. Vestigial structures are remnants of more developed structures that were present and functional in ancestral organisms. In the human body, more than 100 structures are considered vestigial, including the coccyx (fused tailbones), third molars (wisdom teeth), and the muscles that move our ears. Whales and pythons have vestigial hind-limb bones (❚ Fig. 18-14); pigs have vestigial toes that do not touch the

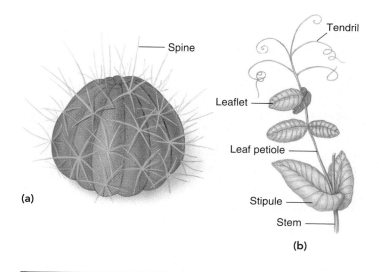

Figure 18-11 Homology in plants

(a) The spines of the fishhook cactus (*Ferocactus wislizenii*) are modified leaves, as are **(b)** the tendrils of the garden pea (*Pisum sativum*). Leaves of the garden pea are compound, and the terminal leaflets are modified into tendrils that are frequently branched.

(a) The aardvark (*Orycteropus afer*) is native to central, southern, and eastern Africa.

(b) A giant anteater (*Myrmecophaga tridactyla*) at a termite mound. The anteater is native to Latin America, from southern Mexico to northern Argentina.

Figure 18-12   Convergent evolution

Three distantly related mammals adapted independently to eat ants and termites in similar grassland/forest environments in different parts of the world.

(c) The pangolin (*Manis crassicaudata*) is native to Africa and to southern and southeastern Asia.

ground; wingless birds such as the kiwi have vestigial wing bones; and many blind, burrowing, or cave-dwelling animals have nonfunctioning, vestigial eyes.

The occasional presence of a vestigial structure is to be expected as a species adapts to a changing mode of life. Some structures become much less important for survival and may end up as vestiges. When a structure no longer confers a selective advantage, it may become smaller and lose much or all of its function with the passage of time. Because the presence of the vestigial structure is usually not harmful to the organism, however, selective pressure for completely eliminating it is weak, and the vestigial structure is found in many subsequent generations.

The distribution of plants and animals supports evolution

The study of the past and present geographic distribution of organisms is called **biogeography.** The geographic distribution of organisms affects their evolution. Darwin was interested in biogeography, and he considered why the species found on ocean islands tend to resemble species of the nearest mainland, even if the environment is different. He also observed that species on ocean islands do not tend to resemble species on islands with similar en-

vironments in other parts of the world. Darwin studied the plants and animals of two sets of arid islands—the Cape Verde Islands, nearly 640 km (400 mi) off western Africa, and the Galápagos Islands, about 960 km (600 mi) west of Ecuador, South America. On each group of islands, the plants and terrestrial animals were indigenous (native), but those of the Cape Verde Islands resembled African species and those of the Galápagos resembled South American species. The similarities of Galápagos species to South American species were particularly striking considering that the Galápagos Islands are dry and rocky and the nearest part of South America is humid and has a lush tropical rain forest. Darwin concluded that species from the neighboring continent migrated or were carried to the islands, where they subsequently adapted to the new environment and, in the process, evolved into new species.

If evolution were not a factor in the distribution of species, we would expect to find a given species everywhere that it could survive. However, the actual geographic distribution of organisms makes sense in the context of evolution. For example, Australia, which has been a separate landmass for millions of years, has distinctive organisms. Australia has populations of egg-laying mammals (monotremes) and pouched mammals (marsupials) not found anywhere else. Two hundred million years ago, Australia and the other continents were joined in a major landmass. Over the course of millions of years, the Australian continent gradually separated from the others. The monotremes and marsupials in Australia continued to thrive and diversify. The isolation of Australia also prevented placental mammals, which arose elsewhere at a later time, from competing with its monotremes and marsupials. In other areas of the world where placental mammals occurred, most monotremes and marsupials became extinct.

We now consider how Earth's dynamic geology has affected biogeography and evolution.

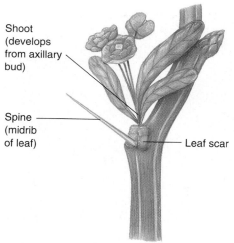

Shoot (develops from axillary bud)

Spine (midrib of leaf)

Leaf scar

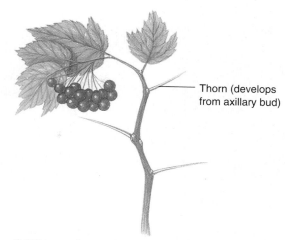

Thorn (develops from axillary bud)

(a) A spine of Japanese barberry (*Berberis thunbergii*) is a modified leaf. (In this example, the spine is actually the midrib of the original leaf, which has been shed.)

(b) Thorns of downy hawthorn (*Crataegus mollis*) are modified stems that develop from axillary buds.

Figure 18-13   Homoplasy in plants

Earth's geologic history is related to biogeography and evolution

In 1915, the German scientist Alfred Wegener, who had noted a correspondence between the geographic shapes of South America and Africa, proposed that all the landmasses had at one time been joined into one huge supercontinent, which he called Pangaea (▌Fig. 18-15a). He further suggested that Pangaea had subsequently broken apart and that the various landmasses had separated in a process known as **continental drift.** Wegener did not know of any mechanism that could have caused continental drift, so his idea, although debated initially, was largely ignored.

In the 1960s, scientific evidence accumulated that provided the explanation for continental drift. Earth's crust is composed of seven large plates (plus a few smaller ones) that float on the mantle, which is the mostly solid layer of Earth lying beneath the crust and above the core.[4] The landmasses are situated on some of these plates. As the plates move, the continents change their relative positions (▌Fig. 18-15b, c, and d). The movement of the crustal plates is termed **plate tectonics.**

Any area where two plates meet is a site of intense geologic activity. Earthquakes and volcanoes are common in such a region. Both San Francisco, noted for its earthquakes, and the Mount Saint Helens volcano are situated where two plates meet.

If landmasses lie on the edges of two adjacent plates, mountains may form. The Himalayas formed when the plate carrying India rammed into the plate carrying Asia. When two plates grind together, one of them is sometimes buried under the other in a process known as *subduction*. When two plates move apart, a ridge of lava forms between them. The Atlantic Ocean is increasing in

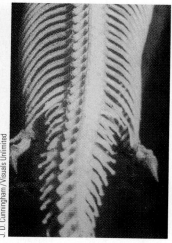

(a) An African rock python (*Python sebae*).

(b) Close-up of part of a python skeleton showing the hind-limb bones.

Figure 18-14   *Animated*   Vestigial structures

All pythons have remnants of hind-limb bones embedded in their bodies.

---

[4]Most of the rock in the upper portion of the mantle is solid, although 1% or 2% is melted. Because of its higher temperature, the solid rock of the mantle is more plastic than the solid rock of the Earth's crust above it.

Pangaea began to break up during the Triassic period.

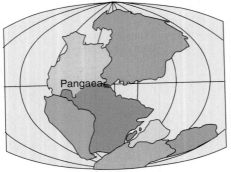

(a) The supercontinent Pangaea, about 240 mya (Triassic period).

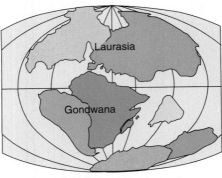

(b) Breakup of Pangaea into Laurasia (Northern Hemisphere) and Gondwana (Southern Hemisphere) began about 180 mya. The landmasses looked this way about 120 mya (Cretaceous period).

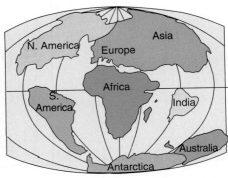

(c) Further separation of landmasses, 60 mya (early Paleogene period). Note that Europe and North America were still joined and that India was a separate landmass.

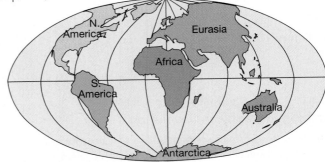

(d) The continents today.

Figure 18-15 *Animated* Continental drift

Geologists hypothesize that the breakup of Pangaea is only the latest in a series of continental breakups and collisions that have taken place since early in Earth's history.

size because of the expanding zone of lava along the Mid-Atlantic Ridge, where two plates are separating.

Knowledge that the continents were at one time connected and have since drifted apart is useful in explaining certain aspects of biogeography (Fig. 18-16). Likewise, continental drift has played a major role in the evolution of different organisms. When Pangaea originally formed during the late Permian period, it brought together terrestrial species that had evolved separately from one another, leading to competition and some extinctions. Marine life was adversely affected, in part because, with the continents joined as one large mass, less coastline existed. (Because coastal areas are shallow, they contain high concentrations of marine species.)

Pangaea separated into several landmasses approximately 180 mya. As the continents began to drift apart, populations became geographically isolated in different environmental conditions and began to diverge along separate evolutionary pathways. As a result, the plants, animals, and other organisms of previously connected continents—South America and Africa, for example—differ. Continental drift also caused gradual changes in ocean and atmospheric currents that have profoundly influenced the biogeography and evolution of organisms. (Biogeography is discussed further in Chapter 55.)

Developmental biology helps unravel evolutionary patterns

How snakes became elongated and lost their limbs has long intrigued evolutionary biologists. Comparative anatomy indicates, for example, that pythons have vestigial hind-limb bones embedded in their bodies (see Fig. 18-14). Other snakes have lost their hind limbs entirely.

Increasingly, developmental biology, particularly at the molecular level, is providing answers to such questions. In many cases, evolutionary changes such as the development of limblessness in snakes occur as a result of changes in genes that regulate the orderly sequence of events that occurs during development. In pythons, for example, the loss of forelimbs and elongation of the body are linked to mutations in several *Hox* genes that affect the expression of body patterns and limb formation in a wide variety of animals (see the discussion of *Hox* gene clusters in Chapter 17). The hind limbs may not develop because python embryonic tissue does not respond to internal signals that trigger leg elongation.

Developmental geneticists at Harvard Medical School and Princeton University are studying the developmental basis for the different beak shapes of the Galápagos finches. They have deter-

(a) *Cynognathus*

(b) *Lystrosaurus*

Africa

India

South America

Australia

Antarctica

(c) *Mesosaurus*

(d) *Glossopteris*

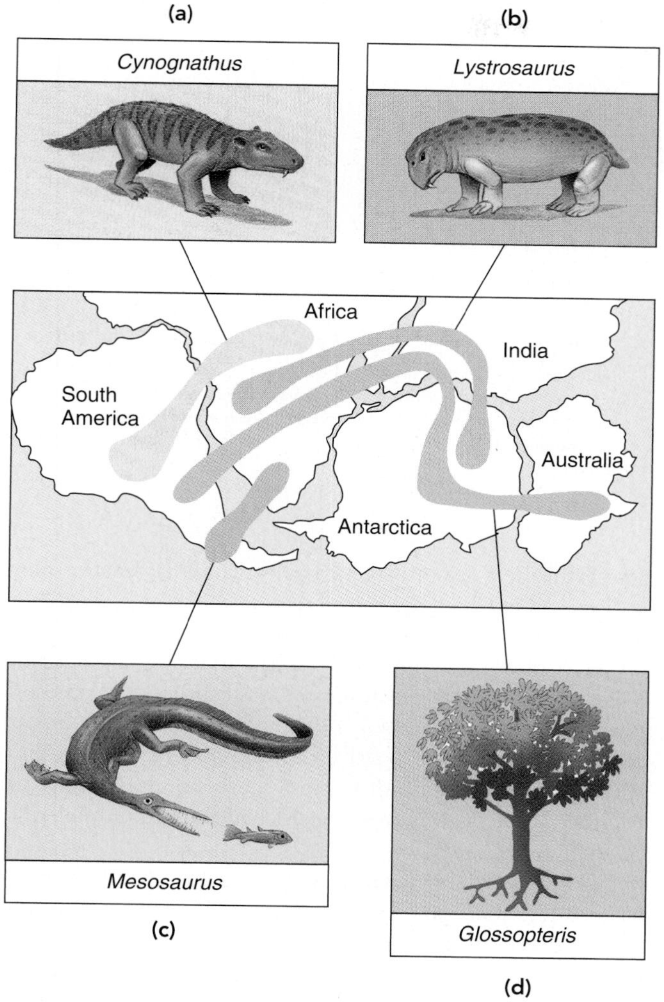

Figure 18-16 Distribution of fossils on continents that were joined during the Permian and Triassic periods (286 mya to 213 mya)

(a) *Cynognathus* was a carnivorous reptile found in Triassic rocks in South America and Africa. **(b)** *Lystrosaurus* was a large, herbivorous reptile with beaklike jaws that lived during the Triassic period. Fossils of *Lystrosaurus* have been found in Africa, India, and Antarctica. **(c)** *Mesosaurus* was a small freshwater reptile found in Permian rocks in South America and Africa. **(d)** *Glossopteris* was a seed-bearing tree dating from the Permian period. *Glossopteris* fossils have been found in South America, Africa, India, Antarctica, and Australia. (Adapted from E. H. Colbert, *Wandering Lands and Animals*, Hutchinson, London, 1973.)

mined that a gene that codes for an important signaling molecule, bone morphogenic protein 4 (BMP4), affects the development of the birds' craniofacial skeletons. The gene for BMP4 is turned on earlier and is expressed more in finch species with larger, thicker beaks than in finches with smaller beaks.

Scientific evidence overwhelmingly demonstrates that development in different animals is controlled by the same kinds of genes; these genetic similarities in a wide variety of organisms reflect a shared evolutionary history. For example, vertebrates have similar patterns of embryological development that indicate they share a common ancestor. All vertebrate embryos have segmented muscles, pharyngeal (throat) pouches, a tubular heart without left and right sides, a system of arteries known as *aortic arches* in the pharyngeal region, and many other shared features. All these structures are necessary and functional in the developing fish. The small, segmented muscles of the fish embryo give rise to the segmented muscles used by the adult fish in swimming. The pharyngeal pouches break through to the surface as gill slits. The adult fish heart remains undivided and pumps blood forward to the gills that develop in association with the aortic arches.

Because none of these embryonic features persists in the adults of reptiles, birds, or mammals, why are these fishlike structures present in their embryos? Evolution is a conservative process, and natural selection builds on what has come before rather than starting from scratch. The evolution of new features often does not require the evolution of new developmental genes but instead depends on a modification in developmental genes that already exist (see discussion of preadaptations in Chapter 20). Terrestrial vertebrates are thought to have evolved from fishlike ancestors; therefore, they share some of the early stages of development still found in fishes today. The accumulation of genetic changes over time in these vertebrates has modified the basic body plan laid out in fish development.

Molecular comparisons among organisms provide evidence for evolution

Similarities and differences in the biochemistry and molecular biology of various organisms provide evidence for evolutionary relationships. Lines of descent based solely on biochemical and molecular characters often resemble lines of descent based on structural and fossil evidence. Molecular evidence for evolution includes the universal genetic code and the conserved sequences of amino acids in proteins and of nucleotides in DNA.

The genetic code is virtually universal

Organisms owe their characteristics to the types of proteins they possess, which in turn are determined by the sequence of nucleotides in their messenger ribonucleic acid (mRNA), as specified by the order of nucleotides in their DNA. Evidence that all life is related comes from the fact that all organisms use a genetic code that is virtually identical.[5] Recall from Chapter 13 that the genetic code specifies a triplet (a sequence of three nucleotides in DNA) that codes for a particular codon (a sequence of three nucleotides in mRNA). The codon then codes for a particular amino acid

---

[5]There is some minor variation in the genetic code. For example, some mitochondria have several deviations from the standard code.

in a polypeptide chain. For example, "AAA" in DNA codes for "UUU" in mRNA, which codes for the amino acid phenylalanine in organisms as diverse as shrimp, humans, bacteria, and tulips. In fact, "AAA" codes for phenylalanine in all organisms examined to date.

The universality of the genetic code—no other code has been found in any organism—is compelling evidence that all organisms arose from a common ancestor. The genetic code has been maintained and transmitted through all branches of the evolutionary tree since its origin in some extremely early (and successful) organism.

Proteins and DNA contain a record of evolutionary change

Researchers have carried out thousands of comparisons of protein and DNA sequences from various species during the past 25 years or so. Sequence-based relationships generally agree with earlier studies, which based evolutionary relationships on similarities in structure among living organisms and on fossil data of extinct organisms.

Investigations of the sequence of amino acids in proteins that play the same roles in many species have revealed both great similarities and certain specific differences. Even organisms that are only remotely related share some proteins, such as cytochrome *c*, which is part of the electron transport chain in aerobic respiration. To survive, all aerobic organisms need a respiratory protein with the same basic structure and function as the cytochrome *c* of their common ancestor. Consequently, not all amino acids that confer the structural and functional features of cytochrome *c* are free to change. Any mutations that changed the amino acid sequence at structurally important sites of the cytochrome *c* molecule would have been harmful, and natural selection would have prevented such mutations from being passed to future generations. However, in the course of the long, independent evolution of different organisms, mutations have resulted in the substitution of many amino acids at less important locations in the cytochrome *c* molecule. The greater the differences in the amino acid sequences of their cytochrome *c* molecules, the longer it has been since two species diverged.

Because a protein's amino acid sequences are coded in DNA, the differences in amino acid sequences indirectly reflect the nature and number of underlying DNA base-pair changes that must have occurred during evolution. Of course, not all DNA codes for proteins (witness introns and transfer RNA genes). **DNA sequencing**—that is, determining the order of nucleotide bases in DNA—of both protein-coding DNA and nonprotein-coding DNA is useful in determining evolutionary relationships.

Generally, the more closely species are considered related on the basis of other scientific evidence, the greater the percentage of nucleotide sequences that their DNA molecules have in common. By using the DNA sequence data in ▌Table 18-1, for example, you can conclude that the closest living relative of humans is the chimpanzee (because its DNA has the lowest percentage of differences in the sequence examined). Which of the primates in Table

TABLE 18-1

Differences in Nucleotide Sequences in DNA as Evidence of Phylogenetic Relationships

Primate Species Pairs	Percent Divergence in a Selected DNA Sequence
Human–chimpanzee	1.7
Human–gorilla	1.8
Human–orangutan	3.3
Human–gibbon	4.3
Human–rhesus monkey (Old World monkey)	7.0
Human–spider monkey (New World monkey)	10.8
Human–tarsier	24.6

Source: From M. Goodman et al., "Primate Evolution at the DNA Level and a Classification of Hominoids," *Journal of Molecular Evolution*, Vol. 30, 1990.

Note: *Percent divergence* refers to how different the base sequences are for the same gene in different organisms. In this example, humans and chimpanzees have a 1.7% difference in their DNA base sequences, which means that 98.3% of the DNA examined is identical. The data shown are for the noncoding sequence of the β-globin gene.

18-1 is the most distantly related to humans?[6] (Primate evolution is discussed in Chapter 22.)

In some cases, molecular evidence challenges traditional evolutionary ideas that were based on structural comparisons among living species and/or on studies of fossil skeletons. Consider artiodactyls, an order of even-toed hoofed mammals such as pigs, camels, deer, antelope, cattle, and hippopotamuses. Traditionally, whales, which do not have toes, are not classified as artiodactyls (although early fossil whales had an even number of toes on their appendages).

▌Figure 18-17 depicts a phylogenetic tree based on molecular data for whales and selected artiodactyls. Such **phylogenetic trees**—diagrams showing lines of descent—can be derived from differences in a given DNA nucleotide sequence. This diagram suggests whales should be classified as artiodactyls and shows that whales are more closely related to hippopotamuses than to any other artiodactyl. The branches representing whales and hippopotamuses probably diverged relatively recently because of the close similarity of DNA sequences in these species. In contrast, camels, which have DNA sequences that are less similar to those of whales, diverged much earlier. The molecular evidence indicates that whales and hippopotamuses share a recent common ancestor, a hippo-like artiodactyl that split from the rest of the artiodactyl line some 55 mya.

However, available fossil evidence does not currently provide support for the molecular hypothesis; a fossil ancestor common to both whales and hippos has not yet been discovered. (Recall from earlier in the chapter that most paleontologists currently suggest that the mesonychians, which are not ancient artiodactyls, may have been the ancestor of whales.) Scientists hope that

---

[6]The tarsier.

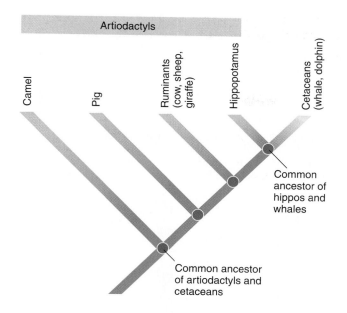

Figure 18-17 Phylogenetic tree of whales and their closest living relatives

This branching diagram, called a *cladogram*, shows hypothetical evolutionary relationships. Based on DNA sequence differences among selected mammals, it suggests that artiodactyls are the close relatives of whales, that the hippopotamus is the closest living artiodactyl relative of whales, and that artiodactyls and whales share a common ancestor in the distant past. The nodes (*circles*) represent branch points where a species splits into two or more lineages. (Ruminants are artiodactyls that have a multichambered stomach and chew regurgitated plant material to make it more digestible.) (Adapted from M. Nikaido et al., "Phylogenetic Relationships among Cetartiodactyls Based on Insertions of Short and Long Interspersed Elements: Hippopotamuses Are the Closest Extant Relatives of Whales," *Proceedings of the National Academy of Sciences*, Vol. 96, Aug. 31, 1999.)

future fossil discoveries will help clarify this discrepancy between molecular and fossil data.

DNA sequencing is used to estimate the time of divergence between two closely related species or taxonomic groups

Imagine that you know the distance from Miami, Florida, to New York City and you also know how long it takes to drive that distance. Now, imagine that you do *not* know the distance from Miami to Chicago but you do know how long it takes to drive that distance. Based on the knowledge you have, you can infer the distance from Miami to Chicago. Similar reasoning is used to estimate the divergence between closely related taxonomic groups. Within a given taxonomic group, mutations are assumed to have occurred at a fairly steady rate over millions of years. Thus, if more differences occur in homologous sequences of DNA of one species compared with another, more time has elapsed since the two species diverged from a common ancestor.

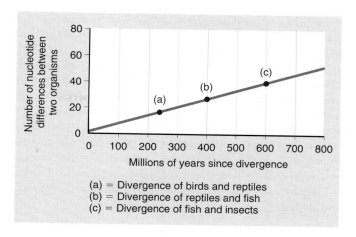

Figure 18-18 Calibration and use of a molecular clock

In this hypothetical example, the DNA of a specific gene is sequenced for birds, reptiles, fishes, and insects. The number of nucleotide differences between birds and reptiles **(a)** is placed on a graph at the time at which birds and reptiles are thought to have diverged as indicated by the fossil evidence. Likewise, the number of nucleotide differences between reptiles and fishes **(b)** is placed on the graph at the time of branching indicated by the fossil record. A line is drawn through points **(a)** and **(b)** and extended, enabling scientists to estimate a much earlier time **(c)** at which the insect line diverged from the vertebrate line.

From the number of alterations in homologous DNA sequences taken from different species, we can develop a **molecular clock** to estimate the time of divergence between two closely related species or higher taxonomic groups. A molecular clock makes use of the average rate at which a particular gene evolves. The clock is calibrated by comparing the number of nucleotide differences between two organisms with the dates of evolutionary branch points that are known from the fossil record. Once a molecular clock is calibrated, past evolutionary events whose timing is not known with certainty are estimated (❚ Fig. 18-18).

Molecular clocks can be used to complement geologic estimates of the divergence of species or to assign tentative dates to evolutionary events that lack fossil evidence. Where there is no fossil record of an evolutionary event, molecular clocks are the only way to estimate the timing of that event. Molecular clocks are also used, along with fossil evidence and structural data, to help reconstruct **phylogeny,** which is the evolutionary history of a group of related species (see Chapter 23). By assigning tentative dates to the divergence of species, molecular clocks show the relative order of branch points in phylogeny.

Molecular clocks must be developed and interpreted with care. Mutation rates vary among different genes and among distantly related taxonomic groups, causing molecular clocks to tick at different rates. Some genes, such as the gene for the respiratory protein cytochrome *c*, code for proteins that lose their function if the amino acid sequence changes slightly; these genes evolve slowly. Other genes, such as genes for blood-clotting proteins, code for proteins that are less constrained by changes in amino acid sequence; these genes evolve rapidly. Scientists consider molecular clocks that are based on several genes to be more accurate than clocks based on a single gene.

Although many dates estimated by molecular clocks agree with fossil evidence, some discrepancies exist. In most of these cases, the molecular clock's estimates of divergence times of particular organisms are dates that are older than the dates at which the groups are first observed in the fossil record. The explanation for this observation may be that earlier fossils have not yet been discovered or that the assumptions on which a particular clock are based need to be re-evaluated. Resolving these differences will require additional research in both molecular biology and paleontology.

Evolutionary hypotheses are tested experimentally

Increasingly, biologists are designing imaginative experiments, often in natural settings, to test evolutionary hypotheses. David Reznick from the University of California at Santa Barbara and John Endler from James Cook University in Australia have studied evolution in guppy populations in Venezuela and in Trinidad, a small island in the southern Caribbean.

Reznick and Endler observed that different streams have different kinds and numbers of fishes that prey on guppies. Predatory fishes that prey on larger guppies are present in all streams at lower elevations; these areas of intense predation pressure are known as *high-predation habitats*. Predators are often excluded from tributaries or upstream areas by rapids and waterfalls. The areas above such barriers are known as *low-predation habitats* because they contain only one species of small predatory fish that occasionally eats smaller guppies.

Differences in predation are correlated with many differences in the guppies, such as male coloration, behavior, and attributes known as *life history traits* (discussed in more detail in Chapter 52). These traits include age and size at sexual maturity, the number of offspring per reproductive event, the size of the offspring, and the frequency of reproduction. For example, guppy adults are larger in streams found at higher elevations and smaller in streams found at lower elevations.

Do predators actually cause these differences to evolve? Reznick and colleagues tested this evolutionary hypothesis by conducting field experiments in Trinidad. Taking advantage of waterfalls that prevent the upstream movement of guppies, guppy predators, or both, they moved either guppies or guppy predators over such barriers. For example, guppies from a high-predation habitat were introduced into a low-predation habitat by moving them over a barrier waterfall into a section of stream that was free of guppies and large predators. The only fish species that lived in this section of stream before the introduction was the predator that occasionally preyed on small guppies.

Eleven years later, the researchers captured adult females from the introduction site (low-predation habitat) and the control site below the barrier waterfall (high-predation habitat). They bred these females in their laboratory and compared the life history traits of succeeding generations. The descendants of

Key Experiment

QUESTION: Can natural selection be observed in a natural population?

HYPOTHESIS: A natural population will respond adaptively to environmental change.

EXPERIMENT: Male and female guppies from a stream in which the predators preferred large adult guppies as prey (*brown bars*) were transferred to a stream in which the predators preferred juveniles and small adults.

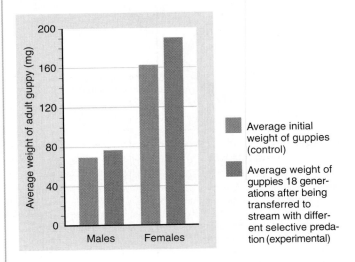

Average initial weight of guppies (control)

Average weight of guppies 18 generations after being transferred to stream with different selective predation (experimental)

RESULTS AND CONCLUSION: After 11 years, the descendants of the transferred guppies (*pink bars*) were measurably larger than their ancestors, indicating that larger guppies had a selective advantage in the new environment.

Figure 18-19 D. N. Reznick's experiment in guppy evolution

(Data used with permission from D. N. Reznick et al., "Evaluation of the Rate of Evolution in Natural Populations of Guppies [*Poecilia reticulata*]," *Science*, Vol. 275, Mar. 28, 1997.)

guppies introduced into the low-predation habitat matured at an older age and larger size than did the descendants of guppies from the control site below the waterfall (Fig. 18-19). They also produced fewer, but larger, offspring. The life histories of the introduced fish had therefore evolved to be similar to those of fishes typically found in such low-predation habitats. Similar studies have demonstrated that predators have played an active role in the evolution of other traits, such as the average number of offspring produced during the lifetime of an individual female (fecundity), male coloration, and behavior.

These and other experiments demonstrate not only that evolution is real but also that it is occurring now, driven by selective environmental forces, such as predation, that can be experimentally manipulated. Darwin incorrectly assumed evolution to be so gradual that humans cannot observe it. As Jonathan Weiner, author of *The Beak of the Finch: A Story of Evolution in Our Time*, notes, "Darwin did not know the strength of his own theory. He vastly underestimated the power of natural selection. Its action is neither rare nor slow. It leads to evolution daily and hourly, all around us, and we can watch."

Review

▋ How do scientists date fossils?

▋ How do homologous and homoplastic features provide evidence of evolution?

▋ How can we explain that fossils of *Mesosaurus*, an extinct reptile that could not swim across open water, are found in the southern parts of both Africa and South America?

▋ How does developmental biology provide evidence of a common ancestry for vertebrates as diverse as reptiles, birds, pigs, and humans?

▋ How do predator preferences drive the evolution of size in guppies?

SUMMARY WITH KEY TERMS

Learning Objectives

1 Define *evolution* (page 391).

▋ **Evolution** is the accumulation of inherited changes within a **population** over time. Evolution is the unifying concept of biology because it links all fields of the life sciences into a unified body of knowledge.

2 Discuss the historical development of the theory of evolution (page 391).

▋ Jean Baptiste de Lamarck was the first scientist to propose that organisms undergo change over time as a result of some natural phenomenon rather than divine intervention. Lamarck thought organisms were endowed with a vital force that drove them to change toward greater complexity over time. He thought organisms could pass traits acquired during their lifetimes to their offspring.

▋ Charles Darwin's observations while voyaging on the HMS *Beagle* were the basis for his theory of evolution. Darwin tried to explain the similarities between animals and plants of the arid Galápagos Islands and the humid South American mainland.

▋ Darwin was influenced by **artificial selection,** in which breeders develop many varieties of domesticated plants and animals in just a few generations. Darwin applied Thomas Malthus's ideas on the natural increase in human populations to natural populations. Darwin was influenced by the idea that Earth was extremely old, an idea promoted by Charles Lyell and other geologists.

3 Explain the four premises of evolution by natural selection as proposed by Charles Darwin (page 392).

▋ Charles Darwin and Alfred Wallace independently proposed the theory of evolution by **natural selection,** which is based on four observations. First, genetic variation exists among the individuals in a population. Second, the reproductive ability of each species causes its populations to geometrically increase in number over time. Third, organisms compete with one another for the resources needed for life, such as food, living space, water, and light. Fourth, offspring with the most favorable combination of characteristics are most likely to survive and reproduce, passing those genetic characteristics to the next generation.

▋ Natural selection results in **adaptations,** evolutionary modifications that improve the chances of survival and reproductive success in a particular environment. Over time, enough changes may accumulate in geographically separated populations to produce new species.

ThomsonNOW™ **Learn more about the Galápagos Islands and the Galápagos Island finches by clicking on the figures in ThomsonNOW.**

4 Compare the modern synthesis with Darwin's original theory of evolution (page 392).

▋ The **modern synthesis,** or **synthetic theory of evolution,** combines Darwin's theory of evolution by natural selection with modern genetics to explain why individuals in a population vary and how species adapt to their environment.

▋ **Mutation** provides the genetic variability that natural selection acts on during evolution.

5 Summarize the evidence for evolution obtained from the fossil record (page 396).

▋ Direct evidence of evolution comes from **fossils,** the remains or traces of ancient organisms.

▋ Layers of sedimentary rock normally occur in their sequence of deposition, with the more recent layers on top of the older, earlier ones. **Index fossils** characterize a specific layer over large geographic areas. **Radioisotopes** present in a rock provide a way to accurately measure the rock's age.

ThomsonNOW™ **Learn more about fossil intermediates by clicking on the figure in ThomsonNOW.**

6 Relate the evidence for evolution derived from comparative anatomy (page 396).

▋ **Homologous features** have basic structural similarities, even though the structures may be used in different ways, because homologous features derive from the same structure in a common ancestor. Homologous features indicate evolutionary affinities among the organisms possessing them.

▋ **Homoplastic features** evolved independently to have similar functions in distantly related organisms. Homoplastic features demonstrate **convergent evolution,** in which organisms with separate ancestries adapt in similar ways to comparable environmental demands.

▋ **Vestigial structures** are nonfunctional or degenerate remnants of structures that were present and functional in ancestral organisms. Structures occasionally become vestigial as species adapt to different modes of life.

ThomsonNOW™ **Learn more about vestigial structures by clicking on the figure in ThomsonNOW.**

7 Define *biogeography,* and describe how the distribution of organisms supports evolution (page 396).

▋ **Biogeography,** the geographic distribution of organisms, affects their evolution. Areas that have been separated from the rest of the world for a long time contain organisms that have evolved in isolation and are therefore unique to those areas.

- At one time the continents were joined to form a super-continent. **Continental drift,** which caused the various landmasses to break apart and separate, has played a major role in evolution.

ThomsonNOW™ **Watch continental drift by clicking on the figure in ThomsonNOW.**

8 Briefly explain how developmental biology and molecular biology provide insights into the evolutionary process (page 396).

- Evolutionary changes are often the result of mutations in genes that affect the orderly sequence of events during development. Development in different animals is controlled by the same kinds of genes, which indicates these animals have a shared evolutionary history.

- The accumulation of genetic changes since organisms diverged, or took separate evolutionary pathways, has modified the pattern of development in more complex vertebrate embryos.

- Molecular evidence for evolution includes the universal genetic code and the conserved sequences of amino acids in proteins and of nucleotides in DNA.

9 Give an example of how evolutionary hypotheses are tested experimentally (page 396).

- Reznick and Endler have studied the effects of predation intensity on the evolution of guppy populations in the laboratory and in nature. Such experiments are a powerful way for investigators to test the underlying processes of natural selection.

TEST YOUR UNDERSTANDING

1. Evolution is based on which of the following concepts? (a) organisms share a common origin (b) over time, organisms have diverged from a common ancestor (c) an animal's body parts can change over its lifetime, and these acquired changes are passed to the next generation (d) a and b (e) a, b, and c

2. Evolution is the accumulation of genetic changes within _____ over time. (a) individuals (b) populations (c) communities (d) a and b (e) a and c

3. Charles Darwin proposed that evolution could be explained by the differential reproductive success of organisms that resulted from their naturally occurring variation. Darwin called this process (a) coevolution (b) convergent evolution (c) natural selection (d) artificial selection (e) homoplasy

4. Which of the following statements is *false*? (a) Darwin was the first to supply convincing evidence for biological evolution (b) Darwin was the first to propose that organisms change over time (c) Wallace independently developed the same theory as Darwin (d) Darwin's theory is based on four observations about the natural world (e) Darwin's studies in the Galápagos Islands strongly influenced his ideas about evolution

5. Which of the following is *not* part of Darwin's mechanism of evolution? (a) differential reproductive success (b) variation in a population (c) inheritance of acquired (nongenetic) traits (d) overproduction of offspring (e) struggle for existence

6. The modern synthesis (a) is based on the sequence of fossils in rock layers (b) uses genetics to explain the source of hereditary variation that is essential to natural selection (c) was first proposed by ancient Greek scholars (d) considers the influence of the geographic distribution of organisms on their evolution (e) is reinforced by homologies that are explained by common descent

7. Jewish and Muslim men have been circumcised for many generations, yet this practice has had no effect on the penile foreskin of their offspring. This observation is inconsistent with evolution as envisioned by (a) Lamarck (b) Darwin (c) Wallace (d) Lyell (e) Malthus

8. Which of the following is *least* likely to have occurred after a small population of finches reached the Galápagos Islands from the South American mainland? (a) after many generations, the finches became increasingly different from the original population (b) over time, the finches adapted to their new environment (c) after many generations, the finches were unchanged and unmodified in any way (d) the finches were unable to survive in their new home and died out (e) the finches survived by breeding with one another

9. The evolution of beak size in the various species of Galápagos finches is associated with their (a) songs (b) diets (c) body sizes (d) predators (e) none of the preceding

10. The fossil record (a) usually occurs in sedimentary rock (b) sometimes appears fragmentary (c) is relatively complete for tropical rainforest organisms but incomplete for aquatic organisms (d) a and b (e) a, b, and c

11. In _____, the selecting agent is the environment, whereas in _____, the selecting agent is humans. (a) natural selection; convergent evolution (b) mutation; artificial selection (c) homoplasy; homology (d) artificial selection; natural selection (e) natural selection; artificial selection

12. Features similar in underlying form in different species because of a common evolutionary origin are called (a) homoplastic (b) homologous (c) vestigial (d) convergent (e) synthetic

13. Similar features in different species that arose through independent evolution instead of descent from a common ancestor are known as (a) homoplasy (b) homology (c) vestigial structures (d) synthetic theory (e) artificial selection

14. Aardvarks, anteaters, and pangolins are only distantly related but are similar in structure and form as a result of (a) homology (b) convergent evolution (c) biogeography (d) vestigial structures (e) artificial selection

15. The species of the Galápagos Islands (a) are similar to those on other islands at the same latitude (b) are similar to those on the South American mainland (c) are identical to those on other islands at the same latitude (d) are identical to those on the South American mainland (e) are similar to those on both the African and South American mainlands

16. Changes in just a few genes that regulate _____ are often responsible for the evolution of new features and form in a population. (a) fossil formation (b) vestigial structures (c) development (d) biogeography (e) adaptation

CRITICAL THINKING

1. **Evolution Link.** The use of model organisms such as the mouse for research and biomedical testing of human diseases is based on the assumption that all organisms share a common ancestor. On what evidence is this assumption based?

2. **Evolution Link.** What adaptations must an animal possess to swim in the ocean? Why are such genetically different organisms as porpoises, which are mammals, and sharks, which are fish, so similar in form?

3. **Evolution Link.** The human fetus grows a coat of fine hair (the lanugo) that is shed before or shortly after birth. Fetuses of chimpanzee and other primates also grow coats of hair, but they are not shed. Explain these observations based on what you have learned in this chapter.

4. **Evolution Link.** Charles Darwin once said, "It is not the strongest of the species that survive, nor the most intelligent, but the one most responsive to change." Explain what he meant.

5. **Evolution Link.** Write short paragraphs explaining each of the following statements:

   a. Natural selection chooses from among the individuals in a population those most suited to *current* environmental conditions. It does not guarantee survival under future conditions.

   b. Individuals do not evolve, but populations do.

   c. The organisms that exist today do so because their ancestors had traits that allowed them and their offspring to thrive.

   d. At the molecular level, evolution can take place by the replacement of one nucleotide by another.

   e. Evolution is said to have occurred within a population when measurable genetic changes are detected.

6. **Analyzing Data.** Examine Figure 18-18. For which two points was fossil evidence used to construct this simple molecular clock? When did the fish and insect lines diverge?

Additional questions are available in ThomsonNOW at www.thomsonedu.com/login

Evolutionary Change in Populations

G. I. Bernard/Animals Animals

Genetic variation in snail shells.
Shown are the shell patterns and colors in a single snail species (*Cepaea nemoralis*), native to Scotland. Variation in shell color may have adaptive value in these snails, because some colors predominate in cooler environments, whereas other colors are more common in warmer habitats.

KEY CONCEPTS

All alleles of all loci of a population constitute its gene pool.

A population's genotype, phenotype, and allele frequencies can be calculated.

The Hardy–Weinberg principle predicts allele and genotype frequencies for a population that is not evolving.

Microevolution is a change in a population's allele or genotype frequencies over successive generations.

Microevolutionary forces include nonrandom mating, mutation, genetic drift, gene flow, and natural selection.

Modes of natural selection include stabilizing selection, directional selection, and disruptive selection.

As you learned in Chapter 18, evolution occurs in populations, not individuals. Although natural selection acts on individuals, causing differential survival and reproduction, individuals themselves do not evolve during their lifetimes. Evolutionary change, which includes modifications in structure, physiology, ecology, and behavior, is inherited from one generation to the next. Although Darwin recognized that evolution occurs in populations, he did not understand how the attributes of organisms are passed to successive generations. One of the most significant advances in biology since Darwin's time has been the demonstration of the genetic basis of evolution.

Recall from Chapter 18 that a **population** consists of all individuals of the same species that live in a particular place at the same time. Individuals within a population vary in many recognizable characters. A population of snails, for example, may vary in shell size, weight, or color (see photograph). Some of this variation is due to heredity, and some is due to environment (nonheritable variation), such as the individual differences observed in the pink color of flamingos, which is partly attributable to differences in diet.

Biologists study variation in a particular character by taking measurements of that character in a population. By comparing the character in parents and offspring, it is possible to estimate the amount of observed variation that is genetic, as represented by the number, frequency, and kinds of alleles in a population. (Recall from Chapter 11 that an **allele** is one of two or more alternate forms of a gene. Alleles occupy corresponding positions, or **loci**, on homologous chromosomes.)

This chapter will help you develop an understanding of the importance of genetic variation as the raw material for evolution and of the basic concepts of **population genetics,** the study of genetic variability within a population and of the evolutionary forces that act on it. Population genetics represents an extension of Gregor Mendel's principles of inheritance (see Chapter 11). You will learn how to distinguish genetic equilibrium from evolutionary change and to assess the roles of the five factors responsible for evolutionary change: nonrandom mating, mutation, genetic drift, gene flow, and natural selection. ■

GENOTYPE, PHENOTYPE, AND ALLELE FREQUENCIES

Learning Objectives

1   Define what is meant by a population's gene pool.
2   Distinguish among genotype, phenotype, and allele frequencies.

Each population possesses a **gene pool,** which includes all the alleles for all the loci present in the population. Because diploid organisms have a maximum of two different alleles at each genetic locus, a single individual typically has only a small fraction of the alleles present in a population's gene pool. The genetic variation that is evident among individuals in a given population indicates that each individual has a different subset of the alleles in the gene pool.

The evolution of populations is best understood in terms of genotype, phenotype, and allele frequencies. Suppose, for example, that all 1000 individuals of a hypothetical population have their genotypes tested, with the following results:

Genotype	Number	Genotype Frequency
AA	490	0.49
Aa	420	0.42
aa	90	0.09
Total	1000	1.00

Each **genotype frequency** is the proportion of a particular genotype in the population. Genotype frequency is usually expressed as a decimal fraction, and the sum of all genotype frequencies is 1.0 (somewhat like probabilities, which were discussed in Chapter 11). For example, the genotype frequency for the Aa genotype is $420 \div 1000 = 0.42$.

A **phenotype frequency** is the proportion of a particular phenotype in the population. If each genotype corresponds to a specific phenotype, then the phenotype and genotype frequencies are the same. If allele A is dominant over allele a, however, the phenotype frequencies in our hypothetical population would be the following:

Phenotype	Number	Phenotype Frequency
Dominant	910	0.91
Recessive	90	0.09
Total	1000	1.00

In this example, the dominant phenotype is the sum of two genotypes, AA and Aa, so the number 910 is obtained by adding $490 + 420$.

An **allele frequency** is the proportion of a specific allele (that is, of A or a) in a particular population. As mentioned earlier, each individual, being diploid, has two alleles at each genetic lo-

cus. Because we started with a population of 1000 individuals, we must account for a total of 2000 alleles. The 490 AA individuals have 980 A alleles, whereas the 420 Aa individuals have 420 A alleles, making a total of 1400 A alleles in the population. The total number of a alleles in the population is $420 + 90 + 90 = 600$. Now it is easy to calculate allele frequencies:

Allele	Number	Allele Frequency
A	1400	0.7
a	600	0.3
Total	2000	1.0

Review

■   Does the term *gene pool* apply to individuals, populations, or both?
■   Can the frequencies of all genotypes in a population be determined directly with respect to a locus that has only two alleles, one dominant and the other recessive?
■   In a human population of 1000, 840 are tongue rollers (TT or Tt), and 160 are not tongue rollers (tt). What is the frequency of the dominant allele (T) in the population?

THE HARDY–WEINBERG PRINCIPLE

Learning Objectives

3   Discuss the significance of the Hardy–Weinberg principle as it relates to evolution, and list the five conditions required for genetic equilibrium.
4   Use the Hardy–Weinberg principle to solve problems involving populations.

In the example just discussed, we observe that only 90 of the 1000 individuals in the population exhibit the recessive phenotype characteristic of the genotype aa. The remaining 910 individuals exhibit the dominant phenotype and are either AA or Aa. You might assume that after many generations, genetic recombination during sexual reproduction would cause the dominant allele to become more common in the population. You might also assume that the recessive allele would eventually disappear altogether. These were common assumptions of many biologists early in the 20th century. However, these assumptions are incorrect, because the frequencies of alleles and genotypes do not change from generation to generation unless influenced by outside factors (discussed later in the chapter).

A population whose allele and genotype frequencies do not change from generation to generation is said to be at **genetic**

equilibrium. Such a population, with no net change in allele or genotype frequencies over time, is not undergoing evolutionary change. A population that is at genetic equilibrium is not evolving with respect to the locus being studied. However, if allele frequencies change over successive generations, evolution is occurring.

The explanation for the stability of successive generations in populations at genetic equilibrium was provided independently by Godfrey Hardy, an English mathematician, and Wilhelm Weinberg, a German physician, in 1908. They pointed out that the expected frequencies of various genotypes in a population can be described mathematically. The resulting **Hardy–Weinberg principle** shows that if the population is large, the process of inheritance does not by itself cause changes in allele frequencies. It also explains why dominant alleles are not necessarily more common than recessive ones. The Hardy–Weinberg principle represents an ideal situation that seldom occurs in the natural world. However, it is useful because it provides a model to help us understand the real world. Knowledge of the Hardy–Weinberg principle is essential to understanding the mechanisms of evolutionary change in sexually reproducing populations.

We now expand our original example to illustrate the Hardy–Weinberg principle. Keep in mind as we go through these calculations that in most cases we know only the phenotype frequencies. When alleles are dominant and recessive, it is usually impossible to visually distinguish heterozygous individuals from homozygous dominant individuals. The Hardy–Weinberg principle lets us use phenotype frequencies to calculate the expected genotype frequencies and allele frequencies, assuming we have a clear understanding of the genetic basis for the character under study.

As mentioned earlier, the frequency of either allele, A or a, is represented by a number that ranges from 0 to 1. An allele that is totally absent from the population has a frequency of zero. If all the alleles of a given locus are the same in the population, then the frequency of that allele is 1.

Because only two alleles, A and a, exist at the locus in our example, the sum of their frequencies must equal 1. If we let p represent the frequency of the dominant (A) allele in the population, and q the frequency of the recessive (a) allele, then we can summarize their relationship with a simple binomial equation, $p + q = 1$. When we know the value of either p or q, we can calculate the value of the other: $p = 1 - q$, and $q = 1 - p$.

Squaring both sides of $p + q = 1$ results in $(p + q)^2 = 1$. This equation can be expanded to describe the relationship of the allele frequencies to the genotypes in the population. When it is expanded, we obtain the frequency of the offspring genotypes:

$$ \underset{\text{Frequency of } AA}{p^2} + \underset{\text{Frequency of } Aa}{2pq} + \underset{\text{Frequency of } aa}{q^2} = \underset{\substack{\text{All individuals} \\ \text{in the population}}}{1} $$

We always begin Hardy–Weinberg calculations by determining the frequency of the homozygous recessive genotype. From the fact that we had 90 homozygous recessive individuals in our population of 1000, we infer that the frequency of the aa genotype, q^2, is 90/1000, or 0.09. Because q^2 equals 0.09, q (the frequency of the recessive a allele) is equal to the square root of 0.09, or 0.3. From the relationship between p and q, we conclude

Genotypes	*AA*	*Aa*	*aa*
Frequency of genotypes in population	0.49	0.42 (0.21 + 0.21)	0.09
Frequency of alleles in gametes	A = 0.49 + 0.21 = 0.7		a = 0.21 + 0.09 = 0.3

(a) Genotype and allele frequencies. The figure illustrates how to calculate frequencies of the alleles *A* and *a* in the gametes produced by each genotype.

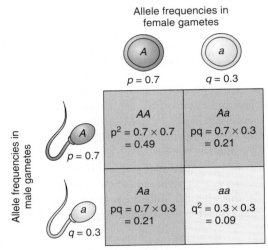

(b) Segregation of alleles and random fertilization. When eggs and sperm containing *A* or *a* alleles unite randomly, the frequency of each of the possible genotypes (*AA*, *Aa*, *aa*) among the offspring is calculated by multiplying the frequencies of the alleles *A* and *a* in eggs and sperm.

Figure 19-1 *Animated* The Hardy–Weinberg principle

that the frequency of the dominant *A* allele, p, equals $1 - q = 1 - 0.3 = 0.7$.

Given this information, we can calculate the frequency of homozygous dominant (AA) individuals: $p^2 = 0.7 \times 0.7 = 0.49$ (Fig. 19-1). The expected frequency of heterozygous individuals (Aa) would be $2pq = 2 \times 0.7 \times 0.3 = 0.42$. Thus, approximately 490 individuals are expected to be homozygous dominant, and 420 are expected to be heterozygous. Note that the sum of homozygous dominant and heterozygous individuals equals 910, the number of individuals showing the dominant phenotype with which we started.

Any population in which the distribution of genotypes conforms to the relation $p^2 + 2pq + q^2 = 1$, whatever the absolute values for p and q may be, is at genetic equilibrium. The Hardy–Weinberg principle allows biologists to calculate allele frequencies in a given population if we know the genotype frequencies, and vice versa. These values provide a basis of comparison with a population's allele or genotype frequencies in succeeding generations. During that time, if the allele or genotype frequencies deviate from the values predicted by the Hardy–Weinberg principle, then the population is evolving.

Genetic equilibrium occurs if certain conditions are met

The Hardy–Weinberg principle of genetic equilibrium tells us what to expect when a sexually reproducing population is not evolving. The relative proportions of alleles and genotypes in successive generations will always be the same, provided the following five conditions are met:

1. *Random mating.* In unrestricted random mating, each individual in a population has an equal chance of mating with any individual of the opposite sex. In our example, the individuals represented by the genotypes *AA, Aa,* and *aa* must mate with one another at random and must not select their mates on the basis of genotype or any other factors that result in nonrandom mating.

2. *No net mutations.* There must be no mutations that convert *A* into *a,* or vice versa. That is, the frequencies of *A* and *a* in the population must not change because of mutations.

3. *Large population size.* Allele frequencies in a small population are more likely to change by random fluctuations (that is, by genetic drift, which is discussed later) than are allele frequencies in a large population.

4. *No migration.* There can be no exchange of alleles with other populations that might have different allele frequencies. In other words, there can be no migration of individuals into or out of a population.[1]

5. *No natural selection.* If natural selection is occurring, certain phenotypes (and their corresponding genotypes) are favored over others. These more successful genotypes are said to have greater **fitness,** which is the relative ability to make a genetic contribution to subsequent generations. Consequently, the allele frequencies will change, and the population will evolve.

Human MN blood groups are a valuable illustration of the Hardy–Weinberg principle

Humans have dozens of antigens on the surfaces of their blood cells. (An *antigen* is a molecule, usually a protein or carbohydrate, that is recognized as foreign by cells of another organism's immune system.) One group of antigens, designated the MN blood group, stimulates the production of antibodies when injected into rabbits or guinea pigs. However, humans do not produce antibodies for M and N, so the MN blood group is not medically

---

[1] Note that evolutionary biologists use the term *migration,* not in its ordinary sense of periodic or seasonal movement of individuals from one location to another but in reference to a movement of individuals that results in a transfer of alleles from one population to another.

important, for example, when giving blood transfusions. (Recall the discussion of the medically important ABO alleles in Chapter 11.) The MN blood group is of interest to population geneticists because the alleles for the MN blood group, usually designated *M* and *N,* are codominant (genotype *MM* produces antigen M only, genotype *NN* produces antigen N only, and the heterozygous genotype *MN* produces both antigens). This allows population geneticists to directly observe all three possible genotype frequencies and compare them with calculated frequencies. The following data are typical of the MN blood group in people in the United States:

Genotype	Observed
MM	320
MN	480
NN	200
Total	1000

Because 1000 diploid individuals are in the sample, there are a total of 2000 alleles. The frequency of *M* alleles in the population $= p = (2 \times 320 + 480) \div 2000 = 0.56$. The frequency of *N* alleles in the population $= q = (2 \times 200 + 480) \div 2000 = 0.44$. As a quick check, the sum of the frequencies should equal 1. Does it?

If this population is in genetic equilibrium, then the expected *MM* genotype frequency $= p^2 = (0.56)^2 = 0.31$. The expected *MN* genotype frequency $= 2pq = 2 \times 0.56 \times 0.44 = 0.49$. The expected *NN* genotype frequency $= q^2 = (0.44)^2 = 0.19$. As a quick check, the sum of the three genotype frequencies should equal 1. Does it? You can use the calculated genotype frequencies to determine how many individuals in a population of 1000 should have the expected genotype frequencies. By comparing the expected numbers with the actual results observed, you see how closely the population is to genetic equilibrium. Simply multiply each genotype frequency by 1000:

Genotype	Observed	Expected
MM	320	313.6
MN	480	492.8
NN	200	193.6
Total	1000	1000.0

The expected numbers closely match the observed numbers, indicating that the *MN* blood groups in the human population are almost at genetic equilibrium. This is not surprising, because the lack of medical significance suggests that the *MN* characteristic is not subject to natural selection and that it does not produce a visible phenotype that might affect random mating.

Review

- In a population at genetic equilibrium, the frequency of the homozygous recessive genotype (*tt*) is 0.16. What are the allele frequencies of *T* and *t,* and what are the expected frequencies of the *TT* and *Tt* genotypes?

- In a population at genetic equilibrium, the frequency of the dominant phenotype is 0.96. What are the frequencies of the dominant (*A*) and recessive (*a*) alleles, and what are the expected frequencies of the *AA, Aa,* and *aa* genotypes?

- The genotype frequencies of a population are determined to be 0.6 *BB*, 0.0 *Bb*, and 0.4 *bb*. Is it likely that this population meets all the conditions required for genetic equilibrium?

MICROEVOLUTION

Learning Objectives

5   Define *microevolution*.

6   Discuss how each of the following microevolutionary forces alters allele frequencies in populations: nonrandom mating, mutation, genetic drift, gene flow, and natural selection.

7   Distinguish among stabilizing selection, directional selection, and disruptive selection, and give an example of each.

Evolution represents a departure from the Hardy–Weinberg principle of genetic equilibrium. The degree of departure between the observed allele or genotype frequencies and those expected by the Hardy–Weinberg principle indicates the amount of evolutionary change. This type of evolution—generation-to-generation changes in allele or genotype frequencies *within* a population—is sometimes referred to as **microevolution,** because it often involves relatively small or minor changes, usually over a few generations.

Changes in the allele frequencies of a population result from five microevolutionary processes: nonrandom mating, mutation, genetic drift, gene flow, and natural selection. These microevolutionary processes are the opposite of the conditions that must be met if a population is in genetic equilibrium. When one or more of these processes acts on a population, allele or genotype frequencies change from one generation to the next.

Nonrandom mating changes genotype frequencies

When individuals select mates on the basis of phenotype (thereby selecting the corresponding genotype), they bring about evolutionary change in the population. Two examples of nonrandom mating are inbreeding and assortative mating.

In many populations, individuals mate more often with close neighbors than with more distant members of the population. As a result, neighbors tend to be more closely related—that is, genetically similar—to one another. The mating of genetically similar individuals that are more closely related than if they had been chosen at random from the entire population is known as **inbreeding.** Although inbreeding does not change the overall allele frequency, the frequency of homozygous genotypes increases with each successive generation of inbreeding. The most extreme example of inbreeding is self-fertilization, which is particularly common in certain plants.

Inbreeding does not appear to be detrimental in some populations, but in others it causes **inbreeding depression,** in which inbred individuals have lower fitness than those not inbred. Fitness is usually measured as the average number of surviving offspring of one genotype compared to the average number of surviving offspring of competing genotypes. Inbreeding depression, as evidenced by fertility declines and high juvenile mortality, is thought to be caused by the expression of harmful recessive alleles as homozygosity increases with inbreeding.

Several studies in the 1990s provided direct evidence of the deleterious consequences of inbreeding in nature. For example,

Key Experiment

QUESTION: Does inbreeding affect survival?

HYPOTHESIS: Non-inbred white-footed mice (*Peromyscus leucopus*) will have a survival advantage over inbred mice in a natural environment.

EXPERIMENT: Field-captured mice were used to establish inbred and non-inbred laboratory populations. The mice were then released, and the populations were sampled six times (each for a 3-day span) during a 10-week period. Values on the *y*-axis are the estimated proportion of mice that survived from one week to the next. Hence, a value of 0.6 means that 60% of the mice alive at the beginning of the week survived through that week.

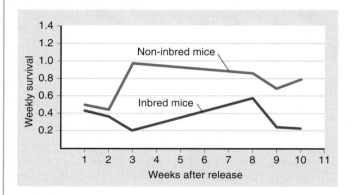

RESULTS AND CONCLUSION: Non-inbred mice (*red*) had a consistently higher survival rate than inbred mice (*blue*). Thus, inbreeding had a negative effect on survival in this species in nature.

Figure 19-2   Survival of inbred and non-inbred mice

(Adapted with permission from J. A. Jiménez et al., "An Experimental Study of Inbreeding Depression in a Natural Habitat," *Science*, Vol. 266, Oct. 14, 1994. Copyright © 1994 American Association for the Advancement of Science.)

white-footed mice (*Peromyscus leucopus*) were taken from a field and used to develop both inbred and non-inbred populations in the laboratory. When these laboratory-bred populations were returned to nature, their survivorship was estimated from release–recapture data. The non-inbred mice had a statistically significant higher rate of survival (❙ Fig. 19-2). It is not known why the inbred mice had a lower survival rate. Some possibilities include higher disease susceptibility, poorer ability to evade predators, less ability to find food, and less ability to win fights with other white-footed mice.

Assortative mating, in which individuals select mates by their phenotypes, is another example of nonrandom mating. For example, biologists selected two phenotypes—high bristle number and low bristle number—in a fruit fly (*Drosophila melanogaster*) population. Although the researchers made no effort to control mating, they observed that the flies preferentially mated with those of similar phenotypes. Females with high bristle number tended to mate with males with high bristle number, and females with low bristle number tended to mate with males with low bristle number. Such selection of mates with the same phenotype is known as *positive assortative mating* (as opposed to the less common phenomenon, *negative assortative mating,* in which mates with opposite phenotypes are selected).

Positive assortative mating is practiced in many human societies, in which men and women tend to marry individuals like themselves in such characteristics as height or intelligence. Like inbreeding, assortative mating usually increases homozygosity at the expense of heterozygosity in the population and does not change the overall allele frequencies in the population. However, assortative mating changes genotype frequencies only at the loci involved in mate choice, whereas inbreeding affects genotype frequencies in the entire genome.

Mutation increases variation within a population

Variation is introduced into a population through **mutation,** which is an unpredictable change in deoxyribonucleic acid (DNA). As discussed in Chapter 13, mutations, which are the source of all new alleles, result from (1) a change in the nucleotide base pairs of a gene, (2) a rearrangement of genes within chromosomes so that their interactions produce different effects, or (3) a change in chromosome structure. Mutations occur unpredictably and spontaneously. A particular locus may have a DNA sequence that causes certain types of mutations to occur more frequently than others. The rate of mutation appears relatively constant for a particular locus but may vary by several orders of magnitude among genes within a single species and among different species.

Not all mutations pass from one generation to the next. Those occurring in somatic (body) cells are not inherited. When an individual with such a mutation dies, the mutation is lost. Some mutations, however, occur in reproductive cells. These mutations may or may not overtly affect the offspring, because most of the DNA in a cell is "silent" and does not code for specific polypeptides or proteins that are responsible for physical characteristics. Even if a mutation occurs in the DNA that codes for a polypeptide, it may still have little effect on the structure or function of that polypeptide (we discuss such *neutral variation* later in the chapter). However, when a polypeptide is sufficiently altered to change its function, the mutation is usually harmful. By acting against seriously abnormal phenotypes, natural selection eliminates or reduces to low frequencies the most harmful mutations. Mutations with small phenotypic effects, even if slightly harmful, have a better chance of being incorporated into the population, where at some later time, under different environmental conditions, they may produce phenotypes that are useful or adaptive.

The "needs" of a population do not determine what mutations will occur. Consider a population living in an increasingly dry environment. A mutation producing a new allele that helps an individual adapt to dry conditions is no more likely to occur than one for adapting to wet conditions or one with no relationship to the changing environment. The production of new mutations simply increases the genetic variability that is acted on by natural selection and, therefore, increases the potential for new adaptations.

Mutation by itself causes small deviations in allele frequencies from those predicted by the Hardy–Weinberg principle. Although allele frequencies may be changed by mutation, these changes are typically several orders of magnitude smaller than changes caused by other evolutionary forces, such as genetic drift. Mutation is usually negligible as an evolutionary force, but it is essential to the evolutionary process because it is the ultimate source of genetic variation.

In genetic drift, random events change allele frequencies

The size of a population has important effects on allele frequencies because random events, or chance, tend to cause changes of relatively greater magnitude in a small population. If a population consists of only a few individuals, an allele present at a low frequency in the population could be completely lost by chance. Such an event would be unlikely in a large population. For example, consider two populations, one with 10,000 individuals and one with 10 individuals. If an uncommon allele occurs at a frequency of 10%, or 0.1, in both populations, then 1900 individuals in the large population have the allele.[2] That same frequency, 0.1, in the smaller population means that only about 2 individuals have the allele.[3] From this exercise, it is easy to see that there is a greater likelihood of losing the rare allele from the smaller population than from the larger one. Predators, for example, might happen to kill one or two individuals possessing the uncommon allele in the smaller population purely by chance, so these individuals would leave no offspring.

The production of random evolutionary changes in small breeding populations is known as **genetic drift.** Genetic drift results in changes in allele frequencies in a population from one generation to another. One allele may be eliminated from the population purely by chance, regardless of whether that allele is beneficial, harmful, or of no particular advantage or disadvantage. Thus, genetic drift decreases genetic variation *within* a population, although it tends to increase genetic differences *among* different populations.

When bottlenecks occur, genetic drift becomes a major evolutionary force

Because of fluctuations in the environment, such as depletion in food supply or an outbreak of disease, a population may rapidly and markedly decrease from time to time. The population is said to go through a **bottleneck** during which genetic drift can occur in the small population of survivors. As the population again increases in size, many allele frequencies may be quite different from those in the population preceding the decline.

---

[2]$2pq + q^2 = 2(0.9)(0.1) + (0.1)^2 = 0.18 + 0.01 = 0.19; 0.19 \times 10,000 = 1900$

[3]$0.19 \times 10 = 1.9$

Figure 19-3 Finns and the founder effect

The Finnish people are thought to have descended from a small founding population that remained separate from the rest of Europe for centuries.

Scientists hypothesize that genetic variation in the cheetah (see Fig. 51-9) was considerably reduced by a bottleneck at the end of the last Ice Age, some 10,000 years ago. Cheetahs nearly became extinct, perhaps from overhunting by humans. The few surviving cheetahs had greatly reduced genetic variability, and as a result, the cheetah population today is so genetically uniform that unrelated cheetahs can accept skin grafts from one another. (Normally, only identical twins accept skin grafts so readily.)

The founder effect occurs when a few "founders" establish a new colony

When one or a few individuals from a large population establish, or found, a colony (as when a few birds separate from the rest of the flock and fly to a new area), they bring with them only a small fraction of the genetic variation present in the original population. As a result, the only alleles among their descendants will be those of the colonizers. Typically, the allele frequencies in the newly founded population are quite different from those of the parent population. The genetic drift that results when a small number of individuals from a large population found a new colony is called the **founder effect.**

The Finnish people may illustrate the founder effect (❙ Fig. 19-3). Geneticists who sampled DNA from Finns and from the European population at large found that Finns exhibit considerably less genetic variation than other Europeans. This evidence supports the hypothesis that Finns are descended from a small group of people who settled about 4000 years ago in the area that is now Finland and, because of the geography, remained separate from other European societies for centuries. The subpopulation in eastern Finland is especially homogeneous because it existed in relative isolation after being established by only several hundred founders in the 1500s.

The founder effect and population bottlenecks have also apparently affected the genetic composition of the population of Iceland. Iceland's 275,000 citizens are descended from a small group of Norse and Celtic people who settled that island in the

ninth century. Isolation and population bottlenecks due to disease and natural disasters have contributed to the relative genetic homogeneity of the Icelandic people.

The founder effect can be of medical importance. For example, by chance one of the approximately 200 founders of the Amish population of Pennsylvania carried a recessive allele that, when homozygous, is responsible for a form of dwarfism, Ellis–van Creveld syndrome. Although this allele is rare in the general population (frequency about 0.001), today it is relatively common in the Amish population (frequency about 0.07).

The Finnish and Icelandic populations, as well as certain Amish populations, are being studied by researchers in the quest to identify genetic contributions to a large number of diseases. The task is simplified in these studies because confounding factors such as variability in the rest of the genome and differences in such environmental factors as nutrition, access to medical care, and exposure to pollutants are relatively controlled.

Gene flow generally increases variation within a population

Individuals of a species tend to be distributed in local populations that are genetically isolated to some degree from other populations. For example, the bullfrogs of one pond form a population separated from those in an adjacent pond. Some exchanges occur by migration between ponds, but the frogs in one pond are much more likely to mate with those in the same pond. Because each population is isolated to some extent from other populations, each has distinct genetic traits and gene pools.

The migration of breeding individuals between populations causes a corresponding movement of alleles, or **gene flow,** that has significant evolutionary consequences. As alleles flow from one population to another, they usually increase the amount of genetic variability within the recipient population. If sufficient gene flow occurs between two populations, they become more similar genetically. Because gene flow reduces the amount of variation between two populations, it tends to counteract the effects of natural selection and genetic drift, both of which often cause populations to become increasingly distinct.

If migration by members of a population is considerable and if populations differ in their allele frequencies, then significant genetic changes occur in local populations. For example, by 10,000 years ago modern humans occupied almost all of Earth's major land areas except a few islands. Because the population density was low in most locations, the small, isolated human populations underwent random genetic drift and natural selection. More recently (during the past 300 years or so), major migrations have increased gene flow, significantly altering allele frequencies within previously isolated human populations.

Natural selection changes allele frequencies in a way that increases adaptation

Natural selection is the mechanism of evolution first proposed by Darwin in which members of a population that are more successfully adapted to the environment have greater fitness—that is, they are more likely to survive and reproduce (see Chapter 18).

Stabilizing, directional, and disruptive selection can change the distribution of phenotypes in a population.

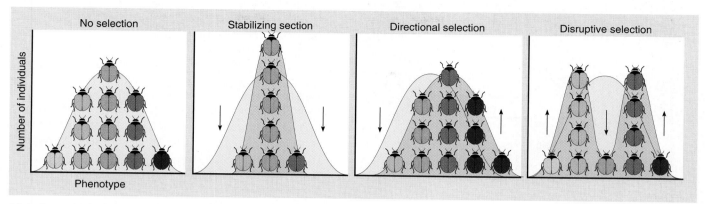

| No selection | Stabilizing section | Directional selection | Disruptive selection |

(a) A character that is under polygenic control (in this example, wing colors in a hypothetical population of beetles) exhibits a normal distribution of phenotypes in the absence of selection.

(b) As a result of stabilizing selection, which trims off extreme phenotypes, variation about the mean is reduced.

(c) Directional selection shifts the curve in one direction, changing the average value of the character.

(d) Disruptive selection, which trims off intermediate phenotypes, results in two or more peaks.

Figure 19-4 *Animated* Modes of selection

The blue screen represents the distribution of individuals by phenotype (in this example, color variation) in the original population. The purple screen represents the distribution by phenotype in the evolved population. The arrows represent the pressure of natural selection on the phenotypes.

Over successive generations, the proportion of favorable alleles increases in the population. In contrast with other microevolutionary processes (nonrandom mating, mutation, genetic drift, and gene flow), natural selection leads to adaptive evolutionary change. Natural selection not only explains why organisms are well adapted to the environments in which they live but also helps account for the remarkable diversity of life. Natural selection enables populations to change, thereby adapting to different environments and different ways of life.

Natural selection results in the differential reproduction of individuals with different traits, or phenotypes (and therefore different genotypes), in response to the environment. Natural selection preserves individuals with favorable phenotypes and eliminates those with unfavorable phenotypes. Individuals that survive and produce fertile offspring have a selective advantage.

The mechanism of natural selection does not develop a "perfect" organism. Rather, it weeds out those individuals whose phenotypes are less adapted to environmental challenges, while allowing better-adapted individuals to survive and pass their alleles to their offspring. By reducing the frequency of alleles that result in the expression of less favorable traits, the probability is increased that favorable alleles responsible for an adaptation will come together in the offspring.

Natural selection operates on an organism's phenotype

Natural selection does not act directly on an organism's genotype. Instead, it acts on the phenotype, which is, at least in part, an expression of the genotype. The phenotype represents an interaction between the environment and all the alleles in the organism's genotype. It is rare that alleles of a single locus determine the phenotype, as Mendel originally observed in garden peas. Much more common is the interaction of alleles of several loci for the expression of a single phenotype (see Chapter 11). Many plant and animal characteristics are under this type of *polygenic* control.

When characters (characteristics) are under polygenic control (as is human height), a range of phenotypes occurs, with most of the population located in the median range and fewer at either extreme. This is a normal distribution or standard bell curve (Fig. 19-4a; see also Fig. 11-22). Three kinds of selection cause changes in the normal distribution of phenotypes in a population: stabilizing, directional, and disruptive selection. Although we consider each process separately, in nature their influences generally overlap.

Stabilizing selection. The process of natural selection associated with a population well adapted to its environment is known as **stabilizing selection.** Most populations are probably influenced by stabilizing forces most of the time. Stabilizing selection selects against phenotypic extremes. In other words, individuals with average, or intermediate, phenotypes are favored.

Because stabilizing selection tends to decrease variation by favoring individuals near the mean of the normal distribution at the expense of those at either extreme, the bell curve narrows (Fig. 19-4b). Although stabilizing selection decreases the amount of variation in a population, variation is rarely elimi-

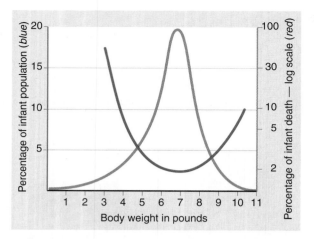

Figure 19-5 Stabilizing selection

The blue curve shows the distribution of birth weights in a sample of 13,730 infants. The red curve shows mortality (death) at each birth weight. Infants with very low or very high birth weights have higher death rates than infants of average weight. The optimum birth weight, that is, the one with the lowest mortality, is close to the average birth weight (about 7 lb). (Adapted from L. L. Cavalli-Sforza and W. F. Bodmer, *The Genetics of Human Populations*, W. H. Freeman and Company, San Francisco, 1971.)

TABLE 19-1

Population Changes in *Geospiza fortis* before and after the 1976–1977 Drought

Character	Average before Drought (634)*	Average after Drought (135)*	Difference
Weight (g)	16.06	17.13	+1.07
Wing length (mm)	67.88	68.87	+0.99
Tarsus (leg, just above the foot) length (mm)	19.08	19.29	+0.21
Bill length (mm)	10.63	10.95	+0.32
Bill depth (mm)	9.21	9.70	+0.49
Bill width (mm)	8.58	8.83	+0.25

From P. R. Grant and B. R. Grant, "Predicting Microevolutionary Responses to Directional Selection on Heritable Variation," *Evolution*, Vol. 49, 1995.

* Number of birds in sample.

Giuliano Gerra and Silvio Sommazzi

nated by this process, because other microevolutionary processes act against a decrease in variation. For example, mutation is slowly but continually adding to the genetic variation within a population.

One of the most widely studied cases of stabilizing selection involves human birth weight, which is under polygenic control and is also influenced by environmental factors. Extensive data from hospitals have shown that infants born with intermediate weights are most likely to survive (❙ Fig. 19-5). Infants at either extreme (too small or too large) have higher rates of mortality. When newborn infants are too small, their body systems are immature, and when they are too large, they have difficult deliveries because they cannot pass as easily through the cervix and vagina. Stabilizing selection operates to reduce variability in birth weight so it is close to the weight with the minimum mortality rate.

Directional selection. If an environment changes over time, **directional selection** may favor phenotypes at one of the extremes of the normal distribution (❙ Fig. 19-4c). Over successive generations, one phenotype gradually replaces another. So, for example, if greater size is advantageous in a new environment, larger individuals will become increasingly common in the population. However, directional selection occurs only if alleles favored under the new circumstances are already present in the population.

Darwin's Galápagos finches provide an excellent example of directional selection. Since 1973, Peter Grant and Rosemary Grant of Princeton University have studied the Galápagos finches. The Grants did a meticulous analysis of finch eating habits and beak sizes on Isla Daphne Major during three extended droughts (1977–1978, 1980, and 1982), one of which was followed by an extremely wet El Niño event (1983). During the droughts, the number of insects and small seeds declined, and large, heavy seeds became the finches' primary food source.

Many finches died during this time, and most of the survivors were larger birds whose beaks were larger and deeper. In a few generations, these larger birds became more common in the population (❙ Table 19-1). After the El Niño event, however, smaller seeds became the primary food source, and smaller finches with average-sized beaks were favored. In this example, natural selection is directional: During the drought, natural selection operated in favor of the larger phenotype; whereas after the wet period, selection occurred in the opposite direction, favoring the smaller phenotype. The guppy populations studied in Venezuela and Trinidad (see Chapter 18) are another example of directional selection.

Directional selection by hunters has been observed in bighorn sheep (*Ovis canadensis*) in Canada. Sport hunters target large rams with rapidly growing horns for their trophy value

Figure 19-6 Bighorn sheep

This ram exhibits the massive horns that characterize this species.

Review

- Which microevolutionary force leads to adaptive changes in allele frequencies?
- Why is mutation important to evolution if it is the microevolutionary force that generally has the smallest effect on allele frequencies?
- Which microevolutionary forces are most associated with an increase in variation within a population? Among populations?
- Which microevolutionary force typically changes genotype frequencies without changing allele frequencies? Explain.

GENETIC VARIATION IN POPULATIONS

Learning Objective

8 Describe the nature and extent of genetic variation, including genetic polymorphism, balanced polymorphism, neutral variation, and geographic variation.

Populations contain abundant genetic variation that was originally introduced by mutation. Sexual reproduction, with its associated crossing-over, independent assortment of chromosomes during meiosis, and random union of gametes, also contributes to genetic variation. The sexual process allows the variability introduced by mutation to be combined in new ways, which may be expressed as new phenotypes.

Genetic polymorphism exists among alleles and the proteins for which they code

One way of evaluating genetic variation in a population is to examine **genetic polymorphism,** which is the presence in a population of two or more alleles for a given locus. Genetic polymorphism is extensive in populations, although many of the alleles are present at low frequencies. Much of genetic polymorphism is not evident, because it does not produce distinct phenotypes.

One way biologists estimate the total amount of genetic polymorphism in populations is by comparing the different forms of a particular protein. Each form consists of a slightly different amino acid sequence that is coded for by a different allele. For example, tissue extracts containing a particular enzyme may be analyzed for different individuals by gel electrophoresis. In *gel electrophoresis,* the enzymes are placed in slots on an agarose gel and an electric current is applied, causing each enzyme to migrate across the gel (see Fig. 15-8). Slight variations in amino acid sequences in the different forms of a particular enzyme cause each to migrate at a different rate, which can be detected using special stains or radioactive labels. ▌ Table 19-2 shows the degree of polymorphism in selected plant and animal groups based on gel electrophoresis of several enzymes. Note that genetic polymorphism tends to be greater in plants than in animals.

Determining the sequence of nucleotides in DNA from individuals in a population provides a *direct* estimate of genetic polymorphism. DNA sequencing is shown in Figures 15-11 and

(▌ Fig. 19-6). Because these rams are typically killed prior to reaching their full reproductive potential, they contribute less to the gene pool; the result has been a 25% decline in average ram weight and horn length over a period of 30 years. Paradoxically, the rams that would normally have the greatest fitness because of their size and ability to use their horns in competition with other males for mates become less fit when these attributes cause them to be singled out by hunters.

Disruptive selection. Sometimes extreme changes in the environment may favor two or more different phenotypes at the expense of the mean. That is, more than one phenotype may be favored in the new environment. **Disruptive selection** is a special type of directional selection in which there is a trend in several directions rather than just one (▌ Fig. 19-4d). It results in a divergence, or splitting apart, of distinct groups of individuals within a population. Disruptive selection, which is relatively rare, selects against the average, or intermediate, phenotype.

Limited food supply during a severe drought caused a population of finches on another island in the Galápagos to experience disruptive selection. The finch population initially exhibited a variety of beak sizes and shapes. Because the only foods available on this island during the drought were wood-boring insects and seeds from cactus fruits, natural selection favored birds with beaks suitable for obtaining these types of food. Finches with longer beaks survived because they could open cactus fruits, and those with wider beaks survived because they could strip off tree bark to expose insects. However, finches with intermediate beaks could not use either food source efficiently and had a lower survival rate.

Natural selection induces change in the types and frequencies of alleles in populations only if there is pre-existing inherited variation. Genetic variation is the raw material for evolutionary change, because it provides the diversity on which natural selection acts. Without genetic variation, evolution cannot occur. In the next section we explore the genetic basis for variation that is acted on by natural selection.

TABLE 19-2

Genetic Polymorphism of Selected Enzymes within Plant and Animal Species

Organism	Number of Species Examined	Percentage of Enzymes Studied That Are Polymorphic
Plants		
Gymnosperms	55	70.9
Flowering plants (monocots)	111	59.2
Flowering plants (eudicots)	329	44.8
Invertebrates		
Marine snails	5	17.5
Land snails	5	43.7
Insects	23	32.9
Vertebrates		
Fishes	51	15.2
Amphibians	13	26.9
Reptiles	17	21.9
Birds	7	15.0
Mammals	46	14.7

Plant data adapted from J. L. Hamrick and M. J. Godt, "Allozyme Diversity in Plant Species," in A. H. D. Brown, M. T. Clegg, A. L. Kahler, and B. J. Weir (eds.), *Plant Population Genetics, Breeding, and Genetic Resources*, Sinauer Associates, Sunderland, MA, 1990. Animal data adapted from D. Hartl, *Principles of Population Genetics*, Sinauer Associates, Sunderland, MA, 1980, and P. W. Hedrick, *Genetics of Populations*, Science Books International, Boston, 1983.

15-12. DNA sequencing of specific alleles in an increasing number of organisms, including humans, indicates that genetic polymorphism is extensive in most populations.

Balanced polymorphism exists for long periods

Balanced polymorphism is a special type of genetic polymorphism in which two or more alleles persist in a population over many generations as a result of natural selection. Heterozygote advantage and frequency-dependent selection are mechanisms that preserve balanced polymorphism.

Genetic variation may be maintained by heterozygote advantage

We have seen that natural selection often eliminates unfavorable alleles from a population, whereas favorable alleles are retained. However, natural selection sometimes helps maintain genetic diversity. In some cases natural selection may even maintain a population's alleles that are unfavorable in the homozygous state. This happens, for example, when the heterozygote *Aa* has a higher degree of fitness than either homozygote *AA* or *aa*. This phenom-

enon, known as **heterozygote advantage**, is demonstrated in humans by the selective advantage of heterozygous carriers of the sickle cell allele.

The mutant allele (Hb^S) for sickle cell anemia produces an altered hemoglobin that deforms or sickles the red blood cells, making them more likely to form dangerous blockages in capillaries and to be destroyed in the liver, spleen, or bone marrow (discussed in Chapter 16). People who are homozygous for the sickle cell allele (Hb^SHb^S) usually die at an early age if medical treatment is not available.

Heterozygous individuals carry alleles for both normal (Hb^A) and sickle cell hemoglobin. The heterozygous condition (Hb^AHb^S) makes a person more resistant to a type of severe malaria (caused by the parasite *Plasmodium falciparum*) than people who are homozygous for the normal hemoglobin allele (Hb^AHb^A). In a heterozygous individual, each allele produces its own specific kind of hemoglobin, and the red blood cells contain the two kinds in roughly equivalent amounts. Such cells sickle much less readily than cells containing only the Hb^S allele. They are more resistant to infection by the malaria-causing parasite, which lives in red blood cells, than are the red blood cells containing only normal hemoglobin.

Where malaria is a problem, each of the two types of homozygous individuals is at a disadvantage. Those homozygous for the sickle cell allele are likely to die of sickle cell anemia, whereas those homozygous for the normal allele may die of malaria. The heterozygote is therefore more fit than either homozygote. In parts of Africa, the Middle East, and southern Asia where falciparum malaria is prevalent, heterozygous individuals survive in greater numbers than either homozygote (❚ Fig. 19-7). The Hb^S allele is maintained at a high frequency in the population, even though the homozygous recessive condition is almost always lethal.

What happens to the frequency of Hb^S alleles in Africans and others who possess it when they migrate to the United States and other countries with few cases of malaria? As might be expected, the frequency of the Hb^S allele gradually declines in such populations, possibly because it confers a selective disadvantage by causing sickle cell anemia in homozygous individuals but no longer confers a selective advantage by preventing malaria in heterozygous individuals. The Hb^S allele never disappears from the population, however, because it is "hidden" from selection in heterozygous individuals and because it is reintroduced into the population by gene flow from the African population.

Genetic variation may be maintained by frequency-dependent selection

Thus far in our discussion of natural selection, we have assumed that the fitness of particular phenotypes (and their corresponding genotypes) is independent of their frequency in the population. However, in cases of **frequency-dependent selection** the fitness of a particular phenotype depends on how frequently it appears in the population. Often a phenotype has a greater selective value when rare than when common in the population. Such phenotypes lose their selective advantage as they become more common.

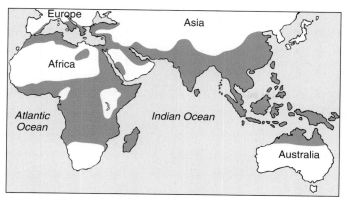

(a) The geographic distribution of falciparum malaria (*green*).

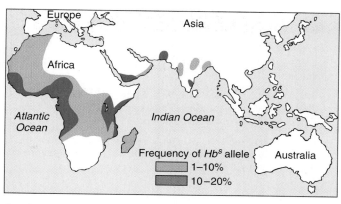

(b) The distribution of sickle cell anemia (*red and orange*).

Frequency of *Hb^s* allele
- 1–10%
- 10–20%

Figure 19-7 Heterozygote advantage

The greater fitness of heterozygous individuals in malarial regions supports the hypothesis of heterozygote advantage, which can be seen by the large area of codistribution. (Adapted from A. C. Allison, "Protection Afforded by Sickle-Cell Traits against Subtertian Malarial Infection," *British Medical Journal*, Vol. 1, 1954.)

Frequency-dependent selection often acts to maintain genetic variation in populations of prey species. In this case, the predator catches and consumes the more common phenotype but may ignore the rarer phenotypes. Consequently, the less common phenotype produces more offspring and therefore makes a greater relative contribution to the next generation.

Frequency-dependent selection is demonstrated in scale-eating fish (cichlids of the species *Perissodus microlepsis*) from Lake Tanganyika in Africa. The scale-eating fish, which obtain food by biting scales off other fish, have either left-pointing or right-pointing mouths. A single locus with two alleles determines this characteristic; the allele for right-pointing mouth is dominant over the allele for left-pointing mouth. These fish attack their prey from behind and from a single direction, depending on mouth morphology. Those with left-pointing mouths always attack the right flanks of their prey, whereas those with right-pointing mouths always attack the left flanks (█ Fig. 19-8a).

The prey species are more successful at evading attacks from the more common form of scale-eating fish. For example, if the cichlids with right-pointing mouths are more common than those with left-pointing mouths, the prey are attacked more often on their left flanks. They therefore become more wary against such attacks, conferring a selective advantage to the less common cichlids with left-pointing mouths. The cichlids with left-pointing mouths would be more successful at obtaining food and would therefore have more offspring. Over time, the frequency of fish with left-pointing mouths would increase in the population, until their abundance causes frequency-dependent selection to work against them and confer an advantage on the now less common fish with right-pointing mouths. Thus, frequency-dependent se-

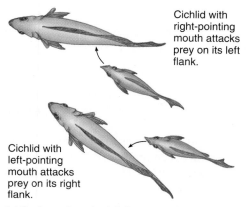

Cichlid with right-pointing mouth attacks prey on its left flank.

Cichlid with left-pointing mouth attacks prey on its right flank.

(a) Scale-eating chiclids have two forms, right-pointing mouths and left-pointing mouths.

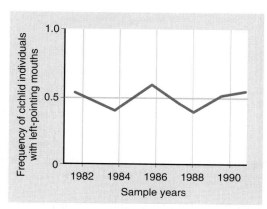

(b) The frequency of fish with left-pointing mouths over a 10-year period. Frequency-dependent selection maintains the frequencies of left-pointing and right-pointing fish in approximately equal numbers, that is, at about 0.5.

Figure 19-8 Frequency-dependent selection in scale-eating cichlids

(Fig. 19-8b adapted with permission from M. Hori, "Frequency-Dependent Natural Selection in the Handedness of Scale-Eating Cichlid Fish," *Science*, Vol. 260, April 9, 1993. Copyright © 1993 American Association for the Advancement of Science.)

Key Experiment

QUESTION: Is clinal variation due to genetic factors, or is it caused by environmental influences?

HYPOTHESIS: The differences in average height exhibited by populations of yarrow (*Achillea millefolium*) growing in their natural habitats at different altitudes (represented in the figure) are due to genetic differences.

EXPERIMENT: Seeds from widely dispersed populations in the Sierra Nevada of California and Nevada were collected and grown for several generations under identical conditions in the same test garden at Stanford, California.

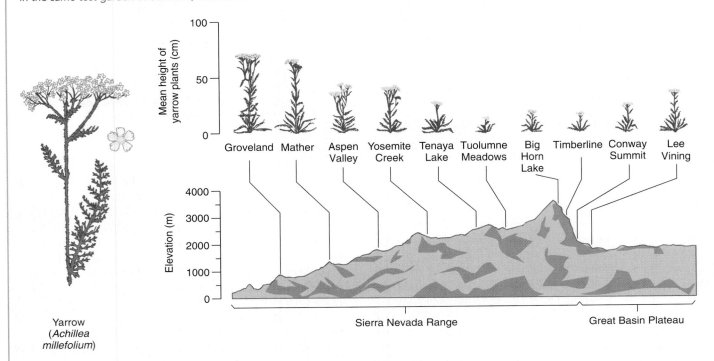

RESULTS AND CONCLUSION: The plants retained their distinctive heights, revealing genetic differences related to the elevation where the seeds were collected.

Figure 19-9   Clinal variation in yarrow (*Achillea millefolium*)

(After J. Clausen, D. D. Keck, and W. M. Hiesey, "Experimental Studies on the Nature of Species: III. Environmental Responses of Climatic Races of *Achillea*," *Carnegie Institute Washington Publication*, Vol. 58, 1948.)

lection maintains both populations of fish at approximately equal numbers (Fig. 19-8b).

Neutral variation may give no selective advantage or disadvantage

Some of the genetic variation observed in a population may confer no apparent selective advantage or disadvantage in a particular environment. For example, random changes in DNA that do not alter protein structure usually do not affect the phenotype.

Variation that does not alter the ability of an individual to survive and reproduce and is, therefore, not adaptive is called **neutral variation.**

The extent of neutral variation in organisms is difficult to determine. It is relatively easy to demonstrate that an allele is beneficial or harmful if its effect is observable. But the variation in alleles that involves only slight differences in the proteins they code for may or may not be neutral. These alleles may be influencing the organism in subtle ways that are difficult to measure or assess. Also, an allele that is neutral in one environment may be beneficial or harmful in another.

Populations in different geographic areas often exhibit genetic adaptations to local environments

In addition to the genetic variation among individuals within a population, genetic differences often exist among different populations within the same species, a phenomenon known as *geographic variation*. One type of geographic variation is a **cline,** which is a gradual change in a species' phenotype and genotype frequencies through a series of geographically separate populations as a result of an environmental gradient. A cline exhibits variation in the expression of such attributes as color, size, shape, physiology, or behavior. Clines are common among species with continuous ranges over large geographic areas. For example, the body sizes of many widely distributed birds and mammals increase gradually as the latitude increases, presumably because larger animals are better able to withstand the colder temperatures of winter.

The common yarrow (*Achillea millefolium*), a wildflower that grows in a variety of North American habitats from low-lands to mountain highlands, exhibits clinal variation in height in response to different climates at different elevations. Although substantial variation exists among individuals within each population, individuals in populations at higher elevations are, on average, shorter than those at lower elevations. The genetic basis of these clinal differences was experimentally demonstrated in a set of classical experiments in which series of populations from different geographic areas were grown in the same environment (❚ Fig. 19-9). Despite being exposed to identical environmental conditions, each experimental population exhibited the traits characteristic of the elevation from which it was collected.

Review

- How does the sickle cell allele illustrate heterozygote advantage?
- How does frequency-dependent selection affect genetic variation within a population over time?
- How can researchers test the hypothesis that clinal variation among populations of a particular species has a genetic basis?

SUMMARY WITH KEY TERMS

Learning Objectives

1 Define what is meant by a population's gene pool (page 413).
- All the individuals that live in a particular place at the same time make up a **population.** Each population has a **gene pool,** which includes all the **alleles** for all the **loci** present in the population.
- **Population genetics** is the study of genetic variability within a population and of the forces that act on it.

2 Distinguish among genotype, phenotype, and allele frequencies (page 413).
- A **genotype frequency** is the proportion of a particular genotype in the population. A **phenotype frequency** is the proportion of a particular phenotype in the population. An **allele frequency** is the proportion of a specific allele of a given genetic locus in the population.

3 Discuss the significance of the Hardy–Weinberg principle as it relates to evolution, and list the five conditions required for genetic equilibrium (page 413).
- The **Hardy–Weinberg principle** states that allele and genotype frequencies do not change from generation to generation (no evolution is occurring) in a population at **genetic equilibrium.**
- The Hardy–Weinberg principle applies only if mating is random in the population, there are no net mutations that change the allele frequencies, the population is large, individuals do not migrate between populations, and natural selection does not occur.

4 Use the Hardy–Weinberg principle to solve problems involving populations (page 413).
- In the Hardy–Weinberg equation, p = the frequency of the dominant allele, q = the frequency of the recessive allele: $p + q = 1$.

- The genotype frequencies of a population are described by the relationship $p^2 + 2pq + q^2 = 1$, where p^2 is the frequency of the homozygous dominant genotype, $2pq$ is the frequency of the heterozygous genotype, and q^2 is the frequency of the homozygous recessive genotype.

ThomsonNOW **Interact with the Hardy–Weinberg principle by clicking on the figure in ThomsonNOW.**

5 Define *microevolution* (page 416).
- **Microevolution** is a change in allele or genotype frequencies within a population over successive generations.

6 Discuss how each of the following microevolutionary forces alters allele frequencies in populations: nonrandom mating, mutation, genetic drift, gene flow, and natural selection (page 416).
- In nonrandom mating, individuals select mates on the basis of phenotype, indirectly selecting the corresponding genotype(s). **Inbreeding** is the mating of genetically similar individuals that are more closely related than if they had been chosen at random from the entire population. Inbreeding in some populations causes **inbreeding depression,** in which inbred individuals have lower **fitness** than non-inbred individuals. In **assortative mating** individuals select mates by their phenotypes. Both inbreeding and assortative mating increase the frequency of homozygous genotypes.
- **Mutations,** unpredictable changes in DNA, are the source of new alleles. Mutations increase the genetic variability acted on by natural selection.
- **Genetic drift** is a random change in the allele frequencies of a small population. Genetic drift decreases genetic variation within a population, and the changes caused by

genetic drift are usually not adaptive. A sudden decrease in population size caused by adverse environmental factors is known as a **bottleneck.** The **founder effect** is genetic drift that occurs when a small population colonizes a new area.

■ **Gene flow,** a movement of alleles caused by the migration of individuals between populations, causes changes in allele frequencies.

■ **Natural selection** causes changes in allele frequencies that lead to adaptation. Natural selection operates on an organism's phenotype, but it changes the genetic composition of a population in a favorable direction for a particular environment.

7 Distinguish among stabilizing selection, directional selection, and disruptive selection, and give an example of each (page 416).

■ **Stabilizing selection** favors the mean at the expense of phenotypic extremes. **Directional selection** favors one phenotypic extreme over another, causing a shift in the phenotypic mean. **Disruptive selection** favors two or more phenotypic extremes.

ThomsonNOW™ **Watch natural selection in action by clicking on the figure in ThomsonNOW.**

8 Describe the nature and extent of genetic variation, including genetic polymorphism, balanced polymorphism, neutral variation, and geographic variation (page 421).

■ **Genetic polymorphism** is the presence in a population of two or more alleles for a given locus. **Balanced polymorphism** is a special type of genetic polymorphism in which two or more alleles persist in a population over many generations as a result of natural selection. **Heterozygote advantage** occurs when the heterozygote exhibits greater fitness than either homozygote. In **frequency-dependent selection,** a genotype's selective value varies with its frequency of occurrence.

■ **Neutral variation** is genetic variation that confers no detectable selective advantage.

■ Geographic variation is genetic variation that exists among different populations within the same species. A **cline** is a gradual change in a species' phenotype and genotype frequencies through a series of geographically separate populations.

TEST YOUR UNDERSTANDING

1. The genetic description of an individual is its genotype, whereas the genetic description of a population is its (a) phenotype (b) gene pool (c) genetic drift (d) founder effect (e) changes in allele frequencies

2. In a diploid species, each individual possesses (a) one allele for each locus (b) two alleles for each locus (c) three or more alleles for each locus (d) all the alleles found in the gene pool (e) half of the alleles found in the gene pool

3. The MN blood group is of interest to population geneticists because (a) people with genotype *MN* cannot receive blood transfusions from either *MM* or *NN* people (b) the *MM*, *MN*, and *NN* genotype frequencies can be observed directly and compared with calculated expected frequencies (c) the *M* allele is dominant to the *N* allele (d) people with the *MN* genotype exhibit frequency-dependent selection (e) people with the *MN* genotype exhibit heterozygote advantage

4. If all copies of a given locus have the same allele throughout the population, then the allele frequency is (a) 0 (b) 0.1 (c) 0.5 (d) 1.0 (e) 10.0

5. If a population's allele and genotype frequencies remain constant from generation to generation, (a) the population is undergoing evolutionary change (b) the population is said to be at genetic equilibrium (c) microevolution has taken place (d) directional selection is occurring, but only for a few generations (e) genetic drift is a significant evolutionary force

6. Comparing the different forms of a particular protein in a population provides biologists with an estimate of (a) genetic drift (b) genetic polymorphism (c) gene flow (d) heterozygote advantage (e) frequency-dependent selection

7. The continued presence of the allele that causes sickle cell anemia in areas where falciparum malaria is prevalent demonstrates which of the following phenomena? (a) inbreeding depression (b) frequency-dependent selection (c) heterozygote advantage (d) genetic drift (e) a genetic bottleneck

8. Frequency-dependent selection maintains _____ in a population. (a) assortative mating (b) genetic drift (c) gene flow (d) genetic variation (e) stabilizing selection

9. According to the Hardy–Weinberg principle, (a) allele frequencies are not dependent on dominance or recessiveness but remain essentially unchanged from generation to generation (b) the sum of allele frequencies for a given locus is always greater than 1 (c) if a locus has only one allele, the frequency of that allele is zero (d) allele frequencies change from generation to generation (e) the process of inheritance, by itself, causes changes in allele frequencies

10. What is the correct equation for the Hardy–Weinberg principle?

   (a) $p^2 + pq + 2q^2 = 1$ (b) $p^2 + 2pq + 2q^2 = 1$

   (c) $2p^2 + 2pq + 2q^2 + 1$ (d) $p^2 + pq + q^2 = 1$

   (e) $p^2 + 2pq + q^2 = 1$

11. The Hardy–Weinberg principle may be applicable if (a) population size is small (b) migration occurs only at the beginning of the breeding season (c) mutations occur at a constant rate (d) matings occur exclusively between individuals of the same genotype (e) natural selection does not occur

12. Which of the following is *not* an evolutionary agent that causes change in allele frequencies? (a) mutation (b) natural

selection (c) genetic drift (d) random mating (e) gene flow due to migration

13. Mutation (a) leads to adaptive evolutionary change (b) adds to the genetic variation of a population (c) is the result of genetic drift (d) almost always benefits the organism (e) a and b

14. Which of the following is *not* true of natural selection? (a) natural selection acts to preserve favorable traits and eliminate unfavorable traits (b) the offspring of individuals that are better adapted to the environment will make up a larger proportion of the next generation (c) natural selec-

tion directs the course of evolution by preserving the traits acquired during an individual's lifetime (d) natural selection acts on a population's genetic variability, which arises through mutation (e) natural selection may result in changes in allele frequencies in a population

15. In _____, individuals with a phenotype near the phenotypic mean of the population are favored over those with phenotypic extremes. (a) microevolution (b) stabilizing selection (c) directional selection (d) disruptive selection (e) genetic equilibrium

CRITICAL THINKING

1. **Evolution Link.** Why are mutations almost always neutral or harmful? If this is true, why are mutations nevertheless essential to evolution?

2. **Evolution Link.** Explain this apparent paradox: Scientists discuss evolution in terms of *genotype* fitness (the selective advantage that a particular genotype confers on an individual), yet natural selection acts on an organism's *phenotype*.

3. **Evolution Link.** Why is it easier for researchers to study the genetic contribution to disease in the populations of Finland or Iceland, as opposed to those of the United States?

4. **Evolution Link.** Evolution is sometimes characterized as "survival of the fittest." Is this consistent with an evolutionary biologist's definitions of *fitness* and *natural selection*? Is this a good way to think about evolution, or a poor one?

5. **Analyzing Data.** Recall that the allele for right-pointing mouth in cichlids (*A*) is dominant over that for left-pointing mouth (*a*). Use the Hardy–Weinberg principle to calculate the allele frequencies that would correspond to equal frequencies of the two phenotypes. Which allele has the higher frequency? Why? How are these scale-eating fish an example of balanced polymorphism?

6. **Analyzing Data.** The recessive allele that causes Ellis–van Creveld syndrome when homozygous has a frequency of about 0.07 in the Amish population of Pennsylvania, although its frequency is only about 0.001 in the general population. How many persons out of 1000 in the Amish population would be expected to have the disease? How many out of a million in the general population?

7. **Analyzing Data.** You study males in populations of a certain species of minnows in a series of lakes at different latitudes. You find that they exhibit clinal variation in average weight at maturity and hypothesize that the weight differences are due to genetic factors. You predict that the average weights at maturity of representatives of each population reared in aquaria will differ in ways that are consistent with the differences you observed among the wild populations. However, when you conduct the experiment, you find no differences among the population averages. What is your conclusion?

Additional questions are available in ThomsonNOW at www.thomsonedu.com/login

Speciation and Macroevolution

Marine iguana basking on a rock. Marine iguanas (*Amblyrhynchus cristatus*) are one of numerous species that evolved on the Galápagos Islands. No other species of iguanas enter the sea.

Michael Lustbader/Photo Researchers, Inc.

KEY CONCEPTS

According to the biological species concept, a species consists of individuals that can successfully interbreed with one another but not with individuals from other species.

The evolution of different species begins with reproductive isolation, in which two populations are no longer able to interbreed successfully.

In allopatric speciation, populations diverge into different species due to geographic isolation, or physical separation.

In sympatric speciation, populations become reproductively isolated from one another despite living in the same geographic area.

Speciation may require millions of years but sometimes occurs much more quickly.

The evolution of species and higher taxa is known as macroevolution.

The various environments on Earth abound in rich assemblages of species. Darwin's studies led him to conclude that the Galápagos Islands were the birthplace of many new species (see photograph), but we now know that some environments are much richer in species. A Brazilian rain forest contains thousands of species of insects, amphibians, reptiles, birds, and mammals. The Great Barrier Reef off the coast of Australia has thousands of species of sponges, corals, mollusks, crustaceans, sea stars, and fishes.

We do not know exactly how many species exist today, but many biologists estimate the number to be on the order of 10 million to 100 million! Often this wondrous diversity has been portrayed as many twig tips on a "tree of life." As one follows the twigs from the tips toward the main body of the tree, they form connections that represent common ancestors that are now extinct. Indeed, more than 99% of all species that ever existed are now extinct. Tracing the branches of the tree of life to larger and larger connections eventually leads to a single trunk that represents the few simple unicellular organisms that evolved early in Earth's history and became the ancestors of all species living today.

Thus, life today is the product of 3 billion to 4 billion years of evolution. How did all these species diversify throughout Earth's history? Our understanding of how species evolve has advanced significantly since Darwin published *On the Origin of Species*. In this chapter we consider the reproductive barriers that isolate species from one another and the possible evolutionary mechanisms that explain how the millions of species that live today or lived in the past originated from ancestral species. We then examine the rate of evolutionary change and the evolution of higher taxonomic categories (species, genus, family, order, class, and phylum), which is the focus of macroevolution. ∎

WHAT IS A SPECIES?

Learning Objective

1   Describe the biological species concept, and list two potential problems with the concept.

The concept of distinct kinds of organisms, known as **species** (from Latin, meaning "kind") is not new. However, there are several different definitions of species, and every definition has some sort of limitation. Linnaeus, the 18th-century biologist who founded modern taxonomy—the science of naming, describing, and classifying organisms—classified plants and other organisms into separate species based on their structural differences (see Chapter 23). This method, known as the *morphological species concept,* is still used to help characterize species, but structure alone is not adequate to explain what constitutes a species.

Population genetics did much to clarify the concept of species. According to the **biological species concept,** first expressed by evolutionary biologist Ernst Mayr in 1942, a species consists of one or more populations whose members interbreed in nature to produce fertile offspring and do not interbreed with—that is, are reproductively isolated from—members of different species. In other words, each species has a *gene pool* separate from that of other species, and reproductive barriers restrict individuals of one species from interbreeding, or exchanging genes, with members of other species. An extension of the biological species concept holds that new species evolve when formerly interbreeding populations become reproductively isolated from one another.

The biological species concept is currently the most widely accepted definition of species, but it has several shortcomings. One problem with the biological species concept is that it applies only to sexually reproducing organisms. Bacteria and other organisms that reproduce asexually do not interbreed, so they cannot be considered in terms of reproductive isolation. These organisms and extinct organisms must be classified based on their structural and biochemical characteristics. Another potential problem with the biological species concept is that although individuals assigned to different species do not normally interbreed, they may *sometimes* successfully interbreed (∎ Fig. 20-1).

Some biologists prefer the **evolutionary species concept,** also called the *phylogenetic species concept,* in which a population is declared a separate species if it has undergone evolution long enough for statistically significant differences in diagnostic traits to emerge. This approach has the advantage of being testable. However, many biologists do not want to abandon the biological species concept, in part because the evolutionary species concept cannot be applied if the evolutionary history of a taxonomic group has not been carefully studied, and most groups have not been rigorously analyzed. Also, if the evolutionary species concept were universally embraced, the number of named species would probably double. This increase would occur because many closely related populations that are classified as subspecies or varieties under the biological species concept would fit the requirements of separate species under the evolutionary species concept.

Thus, the exact definition of a species remains fuzzy. Unless we clearly state otherwise, when we mention *species* in this text, we are referring to the biological species concept.

Review

∎   What is a species, according to the biological species concept? What are some of the limitations of this definition?

∎   What do we mean when we say that the members of a species share a common gene pool?

Figure 20-1   Interbreeding between different species

Shown is a "zebrass," a sterile hybrid formed by a cross between a zebra and a donkey that retains features of both parental species. Although such matings may occur under artificial conditions, such as the wildlife ranch in Texas where this cross took place, zebras and donkeys do not interbreed in the wild.

REPROADUCTIVE ISOLATION

Learning Objective

2 Explain the significance of reproductive isolating mechanisms, and distinguish among the different prezygotic and postzygotic barriers.

Various **reproductive isolating mechanisms** prevent interbreeding between two different species whose *ranges* (areas where each lives) overlap. These mechanisms preserve the genetic integrity of each species, because gene flow between the two species is prevented.

Most species have two or more mechanisms that block a chance occurrence of individuals from two closely related species overcoming a single reproductive isolating mechanism. Most mechanisms occur before mating or fertilization occurs (*prezygotic*), whereas others work after fertilization has taken place (*postzygotic*).

Prezygotic barriers interfere with fertilization

Prezygotic barriers are reproductive isolating mechanisms that prevent fertilization from taking place. Because male and female gametes never come into contact, an *interspecific zygote*—that is, a fertilized egg formed by the union of an egg from one species and a sperm from another species—is never produced. Prezygotic barriers include temporal isolation, habitat isolation, behavioral isolation, mechanical isolation, and gametic isolation.

Sometimes genetic exchange between two groups is prevented because they reproduce at different times of the day, season, or year. Such examples demonstrate **temporal isolation.** For example, two very similar species of fruit flies, *Drosophila pseudoobscura* and *D. persimilis*, have ranges that overlap to a great extent, but they do not interbreed. *Drosophila pseudoobscura* is sexually active only in the afternoon; and *D. persimilis*, only in the morning. Similarly, two frog species have overlapping ranges in eastern Canada and the United States. The wood frog (*Rana sylvatica*) usually mates in late March or early April, when the water temperature is about 7.2°C (45°F), whereas the northern leopard frog (*R. pipiens*) usually mates in mid-April, when the water temperature is 12.8°C (55°F) (▮ Fig. 20-2).

Although two closely related species may be found in the same geographic area, they usually live and breed in different habitats in that area. This provides **habitat isolation** between the two species. For example, the five species of small birds known as flycatchers are nearly identical in appearance and have overlapping ranges in the eastern part of North America. They exhibit habitat isolation because during the breeding season, each species stays in a particular habitat within its range, so potential mates from different species do not meet. The least flycatcher (*Empidonax minimus*) frequents open woods, farms, and orchards; whereas the acadian flycatcher (*E. virescens*) is found in deciduous forests, particularly in beech trees, and swampy woods. The alder flycatcher (*E. alnorum*) prefers wet thickets of alders, the yellow-bellied flycatcher (*E. flaviventris*) nests in conifer woods, and the willow flycatcher (*E. traillii*) frequents brushy pastures and willow thickets.

Many animal species exchange a distinctive series of signals before mating. Such courtship behaviors illustrate **behavioral isolation** (also known as **sexual isolation**). Bowerbirds, for example, exhibit species-specific courtship patterns. The male satin bowerbird of Australia constructs an elaborate bower of twigs, adding decorative blue parrot feathers and white flowers at the entrance (▮ Fig. 20-3).When a female approaches the bower, the male dances around her, holding a particularly eye-catching decoration in his beak. While dancing, he puffs his feathers, extends his wings, and sings a courtship song that consists of a variety of sounds, including loud buzzes and hoots. These specific courtship behaviors keep closely related bird species reproductively isolated from the satin bowerbird. If a male and female of two different species begin courtship, it stops when one member does not recognize or respond to the signals of the other.

(a) The wood frog (*Rana sylvatica*) mates in early spring, often before the ice has completely melted in the ponds.

(b) The leopard frog (*Rana pipiens*) typically mates a few weeks later.

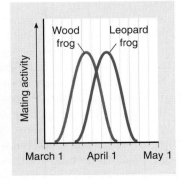

(c) Graph of peak mating activity in wood and leopard frogs. In nature, wood and leopard frogs do not interbreed.

Figure 20-2 *Animated* Temporal isolation in wood and leopard frogs

Figure 20-3 Behavioral isolation in bowerbirds

Each bowerbird species has highly specialized courtship patterns that prevent its mating with another species. The male satin bowerbird (*Ptilonorhynchus violacens*) constructs an enclosed place, or bower, of twigs to attract a female. (The bower is the dark "tunnel" on the left.) Note the white flowers and blue decorations, including human-made objects such as bottle caps, that he has arranged at the entrance to his bower.

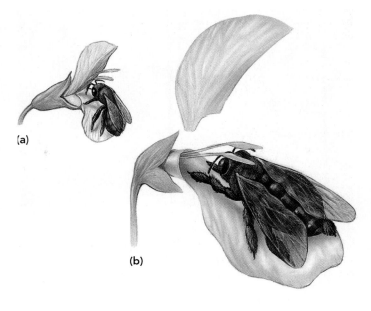

Figure 20-4 Mechanical isolation in black sage and white sage

The differences in their floral structures allow black sage and white sage to be pollinated by different insects. Because the two species exploit different pollinators, they cannot interbreed. **(a)** The petal of the black sage functions as a landing platform for small bees. Larger bees cannot fit on this platform. **(b)** The larger landing platform and longer stamens of white sage allow pollination by larger California carpenter bees (a different species). If smaller bees land on white sage, their bodies do not brush against the stamens. (In the figure, the upper part of the white sage flower has been removed.)

Another example of behavioral isolation involves the wood frogs and northern leopard frogs just discussed as an example of temporal isolation. Males of these two species have very specific vocalizations to attract females of their species for breeding. These vocalizations reinforce each species' reproductive isolation.

Sometimes members of different species court and even attempt copulation, but the incompatible structures of their genital organs prevent successful mating. Structural differences that inhibit mating between species produce **mechanical isolation.** For example, many flowering plant species have physical differences in their flower parts that help them maintain their reproductive isolation from one another. In such plants, the flower parts are adapted for specific insect pollinators. Two species of sage, for example, have overlapping ranges in southern California. Black sage (*Salvia mellifera*), which is pollinated by small bees, has a floral structure different from that of white sage (*S. apiana*), which is pollinated by large carpenter bees (❚ Fig. 20-4). Interestingly, black sage and white sage are also prevented from mating by a temporal barrier: black sage flowers in early spring, and white sage flowers in late spring and early summer. Presumably, mechanical isolation prevents insects from cross-pollinating the two species should they happen to flower at the same time.

If mating takes place between two species, their gametes may still not combine. Molecular and chemical differences between species cause **gametic isolation,** in which the egg and sperm of different species are incompatible. In aquatic animals that release their eggs and sperm into the surrounding water simultaneously, interspecific fertilization is extremely rare. The surface of the egg contains specific proteins that bind only to complementary molecules on the surface of sperm cells of the same species (see Chapter 50). A similar type of molecular recognition often occurs between pollen grains and the stigma (receptive surface of the female part of the flower) so that pollen does not germinate on the stigma of a different plant species.

Postzygotic barriers prevent gene flow when fertilization occurs

Fertilization sometimes occurs between gametes of two closely related species despite the existence of prezygotic barriers. When this happens, **postzygotic barriers** that increase the likelihood of reproductive failure come into play. Generally, the embryo of an interspecific hybrid spontaneously aborts. **Development,** the orderly sequence of events that occurs as an organism grows and matures, is a complex process requiring the precise interaction and coordination of many genes. Apparently the genes from parents belonging to different species do not interact properly in regulating the mechanisms for normal development. In this case, reproductive isolation occurs by **hybrid inviability.** For example, nearly all the hybrids die in the embryonic stage when the eggs of a bullfrog are fertilized artificially with sperm from a leopard frog. Similarly, in crosses between different species of irises the embryos die before reaching maturity.

If an interspecific hybrid does live, it may not reproduce, for several reasons. Hybrid animals may exhibit courtship behaviors incompatible with those of either parental species, and as a re-

Figure 20-5 Hybrid sterility in mules

Mules are interspecific hybrids formed by mating a female horse (*left*) with a male donkey (*right*). Although the mule (*center*) exhibits valuable characteristics of each of its parents, it is sterile.

TABLE 20-1

Reproductive Isolating Mechanisms

Mechanism	How It Works
Prezygotic Barriers	Prevent fertilization
Temporal isolation	Similar species reproduce at different times
Habitat isolation	Similar species reproduce in different habitats
Behavioral isolation	Similar species have distinctive courtship behaviors
Mechanical isolation	Similar species have structural differences in their reproductive organs
Gametic isolation	Gametes of similar species are chemically incompatible
Postzygotic Barriers	Reduce viability or fertility of hybrid
Hybrid inviability	Interspecific hybrid dies at early stage of embryonic development
Hybrid sterility	Interspecific hybrid survives to adulthood but is unable to reproduce successfully
Hybrid breakdown	Offspring of interspecific hybrid are unable to reproduce successfully

sult they will not mate. More often, **hybrid sterility** occurs when problems during meiosis cause the gametes of an interspecific hybrid to be abnormal. Hybrid sterility is particularly common if the two parental species have different chromosome numbers. For example, a mule is the offspring of a female horse ($2n = 64$) and a male donkey ($2n = 62$) (▌ Fig. 20-5). This type of union almost always results in sterile offspring ($2n = 63$), because synapsis (the pairing of homologous chromosomes during meiosis) and segregation cannot occur properly. The zebrass shown in Figure 20-1 also exhibits hybrid sterility because of different chromosome numbers. Donkeys have 62 chromosomes, and zebras, depending on the species, have 32, 44, or 46 chromosomes.

Occasionally, a mating between two F_1 hybrids produces a second hybrid generation (F_2). The F_2 hybrid may exhibit **hybrid breakdown,** the inability of a hybrid to reproduce because of some defect. For example, hybrid breakdown in the F_2 generation of a cross between two sunflower species was 80%. In other words, 80% of the F_2 generation were defective in some way and could not reproduce successfully. Hybrid breakdown also occurs in the F_3 and later generations.

▌ Table 20-1 summarizes the various prezygotic and postzygotic barriers that prevent interbreeding between two species.

Biologists are discovering the genetic basis of isolating mechanisms

Progress has been made in identifying some of the genes involved in reproductive isolation. For example, scientists have determined the genetic basis for prezygotic isolation in species of abalone, large mollusks found along the Pacific coast of North America. In abalone, the fertilization of eggs by sperm requires *lysin,* a sperm protein that attaches to a lysin receptor protein located on the egg envelope. After attachment, the lysin produces a hole in the egg envelope that permits the sperm to penetrate the egg. Scientists cloned the lysin receptor gene and demonstrated that this gene varies among abalone species. Differences in the lysin receptor

protein in various abalone species determine sperm compatibility with the egg. Sperm of one abalone species do not attach to a lysin receptor protein of an egg of a different abalone species.

Review

▌ What barriers prevent wood frogs and leopard frogs from interbreeding in nature?

▌ How is temporal isolation different from behavioral isolation?

▌ How is mechanical isolation different from gametic isolation?

▌ Of which type of postzygotic barrier is the mule an example?

SPECIATION

Learning Objectives

3   Explain the mechanism of allopatric speciation, and give an example.

4   Explain the mechanisms of sympatric speciation, and give both plant and animal examples.

We are now ready to consider how entirely new species may arise from previously existing ones. The evolution of a new species is called **speciation.** The formation of two species from a single species occurs when a population becomes reproductively isolated from other populations of the species and the gene pools of the two separated populations begin to diverge in genetic composition. When a population is sufficiently different from its ancestral species that there is no genetic exchange between them, we consider speciation to have occurred. Speciation happens in two ways: allopatric speciation and sympatric speciation.

Figure 20-6 Allopatric speciation of pupfishes (*Cyprinodon*)

Shown is one of the 20 or so pupfish species that apparently evolved when larger lakes in southern Nevada dried up about 10,000 years ago, leaving behind small, isolated desert pools fed by springs. The pupfish's short, stubby body is characteristic of fish that live in springs; fish that live in larger bodies of water are more streamlined.

Long physical isolation and different selective pressures result in allopatric speciation

Speciation that occurs when one population becomes geographically separated from the rest of the species and subsequently evolves by natural selection and/or genetic drift is known as **allopatric speciation** (from the Greek *allo*, "different," and *patri*, "native land"). Recall from Chapter 18 that *natural selection* occurs as individuals that possess favorable adaptations to the environment survive and become parents of the next generation. *Genetic drift* is a random change in allele frequency resulting from the effects of chance on the survival and reproductive success of individuals in a small breeding population (see Chapter 19). Both natural selection and genetic drift result in changes in allele frequencies in a population, but only in natural selection is the change in allele frequency adaptive.

Allopatric speciation is the most common method of speciation and accounts for almost all evolution of new animal species. The geographic isolation required for allopatric speciation may occur in several ways. Earth's surface is in a constant state of change. Such change includes rivers shifting their courses; glaciers migrating; mountain ranges forming; land bridges forming that separate previously united aquatic populations; and large lakes diminishing into several smaller, geographically separated pools.

What may be an imposing geographic barrier to one species may be of no consequence to another. Birds and cattails, for example, do not become isolated when a lake subsides into smaller pools; birds easily fly from one pool to another, and cattails disperse their pollen and fruits by air currents. Fishes, in contrast, usually cannot cross the land barriers between the pools and so become reproductively isolated.

In the Death Valley region of California and Nevada, large interconnected lakes formed during wetter climates of the last Ice Age. These lakes were populated by one or several species of pupfishes. Over time, the climate became drier, and the large lakes dried up, leaving isolated pools. Presumably, each pool contained a small population of pupfish that gradually diverged from the common ancestral species by genetic drift and natural selection in response to habitat differences—the high temperatures, high salt concentrations, and low oxygen levels characteristic of desert springs. Today, there are 20 or so distinct species, subspecies, and populations in the area of Death Valley, California, and Ash Meadows, Nevada. Many, such as the Devil's Hole pupfish (*Cyprinodon diabolis*) and the Owens pupfish (*C. radiosus*), are restricted to one or two isolated springs (❚ Fig. 20-6).

Allopatric speciation also occurs when a small population migrates or is dispersed (such as by a chance storm) and colonizes a new area away from the range of the original species. This colony is geographically isolated from its parental species, and the small *microevolutionary* changes that accumulate in the isolated gene pool over many generations may eventually be sufficient to form a new species. Because islands provide the geographic isolation required for allopatric speciation, they offer excellent opportunities to study this evolutionary mechanism. A few individuals of a few species probably colonized the Galápagos Islands and the Hawaiian Islands, for example. The hundreds of unique species presently found on each island presumably descended from these original colonizers (❚ Fig. 20-7).

Speciation is more likely to occur if the original isolated population is small. Recall that genetic drift, including the founder effect, is more consequential in small populations (see Chapter 19). Genetic drift tends to result in rapid changes in allele frequencies in the small, isolated population. The different selective pressures

Figure 20-7 Allopatric speciation of the Hawaiian goose (the nene)

Nene (pronounced "nay'-nay"; *Branta sandvicensis*) are geese originally found only on volcanic mountains on the geographically isolated islands of Hawaii and Maui, which are some 4200 km (2600 mi) from the nearest continent. Compared with those of other geese, the feet of Hawaiian geese are not completely webbed, their toenails are longer and stronger, and their foot pads are thicker; these adaptations enable Hawaiian geese to walk easily on lava flows. Nene are thought to have evolved from a small population of geese that originated in North America. Photographed in Hawaii Volcanoes National Park, Hawaii.

of the new environment to which the population is exposed further accentuate the divergence caused by genetic drift.

The Kaibab squirrel may be an example of allopatric speciation in progress

About 10,000 years ago, when the American Southwest was less arid, the forests in the area supported a tree squirrel with conspicuous tufts of hair sprouting from its ears. A small tree squirrel population living on the Kaibab Plateau of the Grand Canyon became geographically isolated when the climate changed, thus causing areas to the north, west, and east to become desert. Just a few miles to the south lived the rest of the squirrels, which we know as Abert squirrels, but the two groups were separated by the Grand Canyon. With changes over time in both its appearance and its ecology, the Kaibab squirrel is on its way to becoming a new species.

During its many years of geographic isolation, the small population of Kaibab squirrels has diverged from the widely distributed Abert squirrels in several ways. Perhaps most evident are changes in fur color. The Kaibab squirrel now has a white tail and a gray belly, in contrast to the gray tail and white belly of the Abert squirrel (Fig. 20-8). Biologists think these striking changes arose in Kaibab squirrels as a result of genetic drift.

Some scientists consider the Kaibab squirrel and the Abert squirrel to be distinct populations of the same species (*Sciurus aberti*). Because the Kaibab and Abert squirrels are reproductively isolated from one another, however, some scientists have classified the Kaibab squirrel as a different species (*S. kaibabensis*).

Porto Santo rabbits may be an example of extremely rapid allopatric speciation

Allopatric speciation has the potential to occur quite rapidly. Early in the 15th century, a small population of rabbits was released on Porto Santo, a small island off the coast of Portugal. Because there were no other rabbits or competitors and no predators on the island, the rabbits thrived. By the 19th century, these rabbits were markedly different from their European ancestors. They were only half as large (weighing slightly more than 500 g, or 1.1 lb), with a different color pattern and a more nocturnal lifestyle. Most significant, attempts to mate Porto Santo rabbits with mainland European rabbits failed. Many biologists concluded that within 400 years, an extremely brief period in evolutionary history, a new species of rabbit had evolved.

Not all biologists agree that the Porto Santo rabbit is a new species. The objection stems from a more recent breeding experiment and is based on biologists' lack of a consensus about the definition of a species. In the experiment, foster mothers of the wild Mediterranean rabbit raised newborn Porto Santo rabbits. When they reached adulthood, these Porto Santo rabbits mated successfully with Mediterranean rabbits to produce healthy, fertile offspring. To some biologists, this experiment clearly demonstrated that Porto Santo rabbits are not a separate species but a **subspecies,** which is a taxonomic subdivision of a species. These biologists cite the Porto Santo rabbits as an example of speciation in progress, much like the Kaibab squirrels just discussed.

Other biologists think the Porto Santo rabbit is a separate species, because it does not interbreed with the other rabbits under natural conditions. They point out that the breeding experiment was successful only after the baby Porto Santo rabbits were raised under artificial conditions that probably modified their natural behavior.

Two populations diverge in the same physical location by sympatric speciation

Although geographic isolation is an important factor in many cases of evolution, it is not an absolute requirement. In **sympatric speciation** (from the Greek *sym*, "together," and *patri*, "native land"), a new species evolves within the same geographic region as the parental species. The divergence of two populations in the same geographic range occurs when reproductive isolating mechanisms evolve at the *start* of the speciation process. Sympatric speciation is especially common in plants. The role of sympatric speciation in animal evolution is probably much less important than allopatric speciation; until recently, sympatric speciation in animals has been difficult to demonstrate in nature.

Sympatric speciation occurs in at least two ways: a change in **ploidy** (the number of chromosome sets making up an organism's genome) and a change in ecology. We now examine each of these mechanisms.

(a) The Kaibab squirrel, with its white tail and gray belly, is found north of the Grand Canyon.

(b) The Abert squirrel, with its gray tail and white belly, is found south of the Grand Canyon.

Figure 20-8 *Animated* Allopatric speciation in progress

Allopolyploidy is an important mechanism of sympatric speciation in plants

As a result of reproductive isolating mechanisms discussed earlier, the union of two gametes from different species rarely forms viable offspring; if offspring are produced, they are usually sterile. Before gametes form, meiosis reduces the chromosome number (see Chapter 10). For the chromosomes to be parceled correctly into the gametes, homologous chromosome pairs must come together (a process called *synapsis*) during prophase I. This cannot usually occur in interspecific hybrid offspring, because not all the chromosomes are homologous. However, if the chromosome number doubles *before* meiosis, then homologous chromosomes can undergo synapsis. Although not common, this spontaneous doubling of chromosomes has been documented in a variety of plants and a few animals. It produces nuclei with multiple sets of chromosomes.

Polyploidy, the possession of more than two sets of chromosomes, is a major factor in plant evolution. Reproductive isolation occurs in a single generation when a polyploid species with multiple sets of chromosomes arises from diploid parents. There are two kinds of polyploidy: autopolyploidy and allopolyploidy. An **autopolyploid** contains multiple sets of chromosomes from a single species, and an **allopolyploid** contains multiple sets of chromosomes from two or more species. We discuss only allopolyploidy, because it is much more common in nature.

Allopolyploidy occurs in conjunction with **hybridization,** which is sexual reproduction between individuals from closely related species. Allopolyploidy produces a fertile interspecific hybrid because the polyploid condition provides the homologous chromosome pairs necessary for synapsis during meiosis. As a result, gametes may be viable (❚ Fig. 20-9). An allopolyploid, that is, an interspecific hybrid produced by allopolyploidy, reproduces with itself (self-fertilization) or with a similar individual. However, allopolyploids are reproductively isolated from both parents, because their gametes have a different number of chromosomes from those of either parent.

If a population of allopolyploids (that is, a new species) becomes established, selective pressures cause one of three outcomes. First, the new species may not compete successfully against species that are already established, so it becomes extinct. Second, the allopolyploid individuals may assume a new role in the environment and so coexist with both parental species. Third, the new species may successfully compete with either or both of its parental species. If it has a combination of traits that confers greater fitness than one or both parental species for all or part of the original range of the parent(s), the hybrid species may replace the parent(s).

Although allopolyploidy is extremely rare in animals, it is significant in the evolution of flowering plant species. As many as 80% of all flowering plant species are polyploids, and most of these are allopolyploids. Moreover, allopolyploidy provides a mechanism for extremely rapid speciation. A single generation is all that is needed to form a new, reproductively isolated species. Allopolyploidy may explain the rapid appearance of many flowering plant species in the fossil record and their remarkable diversity today (about 235,000 species).

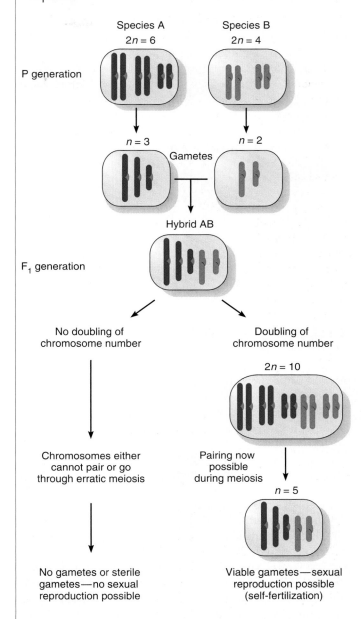

Key Point

Changes in chromosome number have often led to speciation in plants.

Figure 20-9   Sympatric speciation by allopolyploidy in plants

When two species (designated the P generation) successfully interbreed, the interspecific hybrid offspring (the F₁ generation) are almost always sterile (*bottom left*). If the chromosomes double, proper synapsis and segregation of the chromosomes can occur and viable gametes may be produced (*bottom right*). (Unduplicated chromosomes are shown for clarity.)

The kew primrose (*Primula kewensis*) is an example of sympatric speciation that was documented at the Royal Botanic Gardens at Kew, England, in 1898 (❚ Fig. 20-10). Plant breeders developed *P. kewensis* as an interspecific hybrid of two primrose

| Primula floribunda | Primula kewensis | Primula verticillata |

Figure 20-10 *Animated* Sympatric speciation of a primrose

A new species of primrose, *Primula kewensis,* arose in 1898 as an allo-polyploid derived from the interspecific hybridization of *P. floribunda* and *P. verticillata.* Today, *P. kewensis* is a popular houseplant.

species, *P. floribunda* ($2n = 18$) and *P. verticillata* ($2n = 18$). *Primula kewensis* had a chromosome number of 18 but was sterile. Then, at three different times, it was reported to have spontaneously formed a fertile branch, which was an allopolyploid ($2n = 36$) that produced viable seeds of *P. kewensis.*

The mechanism of sympatric speciation has been experimentally verified for many plant species. One example is a group of species, collectively called *hemp nettles,* that occurs in temperate parts of Europe and Asia. One hemp nettle, *Galeopsis tetrahit* ($2n = 32$), is a naturally occurring allopolyploid hypothesized to have formed by the hybridization of two species, *G. pubescens* ($2n = 16$) and *G. speciosa* ($2n = 16$). This process occurred in nature but was experimentally reproduced. *Galeopsis pubescens* and *G. speciosa* were crossed to produce F_1 hybrids, most of which were sterile. Nevertheless, both F_2 and F_3 generations were produced. The F_3 generation included a polyploid plant with $2n = 32$ that self-fertilized to yield fertile F_4 offspring that could not mate with either of the parental species. These allopolyploid plants had the same appearance and chromosome number as the naturally occurring *G. tetrahit.* When the experimentally produced plants were crossed with the naturally occurring *G. tetrahit,* a fertile F_1 generation was formed. Thus, the experiment duplicated the speciation process that occurred in nature.

Changing ecology causes sympatric speciation in animals

Biologists have observed the occurrence of sympatric speciation in animals, but its significance—how often it happens and under what conditions—is still debated. Many examples of sympatric speciation in animals involve parasitic insects and rely on genetic mechanisms other than polyploidy. For example, in the 1860s in the Hudson River Valley of New York, a population of fruit maggot flies (*Rhagoletis pomonella*) parasitic on the small red fruits of native hawthorn trees was documented to have switched to a new host, domestic apples, which had been introduced from Europe. Although the sister populations (hawthorn maggot flies and apple maggot flies) continue to occupy the same geographic area, no gene flow occurs between them because they eat, mate, and lay their eggs on different hosts (❚ Fig. 20-11). In other words, because the hawthorn and apple maggot flies have diverged and are reproductively isolated from one another, they have effectively become separate species. Most entomologists, however, still recognize hawthorn and apple maggot flies as a single species, because their appearance is virtually identical.

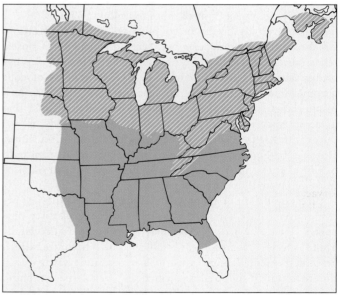

◼ Hawthorn maggot fly range

▨ Apple maggot fly and hawthorn maggot fly range

Figure 20-11 Ranges of apple and hawthorn maggot flies

Apple and hawthorn maggot flies are sympatric throughout the northern half of the hawthorn maggot fly's range. (Adapted from G. L. Bush, "Sympatric Host Race Formation and Speciation in Frugivorous Flies of the Genus *Rhagoletis* [Diptera, Tephritidae]," *Evolution,* Vol. 23, No. 2, June 1969.)

(a) *Pundamilia pundamilia* males have bluish-silver bodies.

(b) *Pundamilia nyererei* males have red backs.

(c) *Pundamilia* "red head" males have a red "chest." (The "red head" species has not yet been scientifically named.)

Figure 20-12 Color variation in Lake Victoria cichlids

Some evidence suggests that changes in male coloration may be the first step in speciation of Lake Victoria cichlids. Later, other traits, including ecological characteristics, diverge. Female cichlids generally have cryptic coloration; their drab colors help them blend into their surroundings.

In situations such as that of the fruit maggot flies, a mutation arises in an individual and spreads through a small group of insects by sexual reproduction. The particular mutation leads to *disruptive selection* (see Chapter 19), in which both the old and new phenotypes are favored. Selection is disruptive because the original population is still favored when it parasitizes the original host, and the mutants are favored by a new ecological opportunity—in this case, to parasitize a different host species. Additional mutations may occur that cause the sister populations to diverge even further.

Biologists have studied the speciation of colorful fishes known as *cichlids* (pronounced "sick'-lids") in several East African lakes. The different species of cichlids in a given lake have remarkably different eating habits, which are reflected in the shapes of their jaws and teeth. Some graze on algae; some consume dead organic material at the bottom of the lake; and others are predatory and eat plankton (microscopic aquatic organisms), insect larvae, the scales off fish, or even other cichlid species. In some cichlids, food preferences are related to size (smaller cichlids consume plankton), which in turn is related to mating preference (small, plankton-eating cichlids mate only with other small, plankton-eating cichlids).

In other cichlids, related species do not differ in size but only in color and mating habits. Each species has a distinct male coloration, which matches the preferences of females of that species in choosing mates (Fig. 20-12). Charles Darwin called choosing a mate based on its color or some other characteristic **sexual selection.** He recognized that female preference for showy colors or elaborate ornaments on males, if it was based on *inherited* variation, could result in these male sexual traits becoming more common and more pronounced over time. Female preference for a particular male trait confers a selective advantage to the males exhibiting that trait, because they have an improved chance to mate and pass on their genes. Darwin suggested that sexual selection, like natural selection, could eventually result in the evolution of new species. (Sexual selection is discussed in greater detail in Chapter 51.)

DNA sequence data indicate that the cichlid species within a single lake are more closely related to one another than to fishes in nearby lakes or rivers. These molecular data suggest that cichlid species evolved sympatrically, or at least within the confines of a lake, rather than by repeated colonizations by fish populations in nearby rivers.

How rapidly did sympatric speciation occur in cichlids? Seismic and drill core data provide evidence that Lake Victoria dried up completely during the Late Pleistocene, when much of north and equatorial Africa was arid. The diverse cichlid species may have arisen after the climate became wetter and Lake Victoria refilled; that is, the more than 500 **endemic** (found nowhere else) cichlid species in Lake Victoria may have evolved from one or a few common ancestors in a remarkably short period—as little as 12,000 years. Other biologists have offered competing hypotheses about the timing and geographic origin of Lake Victoria's cichlids. Regardless of the uncertainties of certain details, Lake Victoria's cichlids are widely recognized as exemplifying the fastest rate of evolution known for such a large group of vertebrate species.

Reproductive isolation breaks down in hybrid zones

When two populations have significantly diverged as a result of geographic separation, there is no easy way to determine if the speciation process is complete (recall the disagreement about whether Porto Santo rabbits are a separate species or a subspecies). If such populations, subspecies, or species come into contact, they may hybridize where they meet, forming a **hybrid zone,** or area of overlap in which interbreeding can take place. Hybrid zones are typically narrow, presumably because the hybrids are not well adapted for either parental environment, and the hybrid population is typically very small compared with the parental populations.

On the Great Plains of North America, red-shafted and yellow-shafted flickers (types of woodpeckers) meet and interbreed. The red-shafted flicker, named for the male's red under-wings and tail, is found in the western part of North America, from the Great Plains to the Pacific Ocean. Yellow-shafted flicker males, which have yellow underwings and tails, range east of the Rockies. Hybrid flickers, which form a stable hybrid zone from Texas to southern Alaska, are varied in appearance, although many have orange underwings and tails.

Biologists do not agree about whether the red-shafted and yellow-shafted flickers are separate species or geographic subspecies. According to the biological species concept, if red-shafted and yellow-shafted flickers are two species, they should maintain their reproductive isolation. However, the flicker hybrid zone has not expanded, that is, the two types of flickers have maintained their distinctiveness and have not rejoined into a single, freely interbreeding population.

The study of hybrid zones has made important contributions to what is known about speciation. As in other fields of science, disagreements and differences of opinion are an important part of the scientific process because they stimulate new ideas, hypotheses, and experimental tests that expand our base of scientific knowledge.

Review

- What are five geographic barriers that might lead to allopatric speciation?
- How do hybridization and polyploidy cause a new plant species to form in as little as one generation?

- What is the likely mechanism of speciation for pupfishes? For cichlids?

THE RATE OF EVOLUTIONARY CHANGE

Learning Objective

5 Take either side in a debate on the pace of evolution by representing the views of either punctuated equilibrium or gradualism.

We have seen that speciation is hard for us to directly observe as it occurs. Does the fossil record provide clues about how rapidly new species arise? Biologists have long recognized that the fossil record lacks many transitional forms; the starting points (ancestral species) and the end points (new species) are present, but the intermediate stages in the evolution from one species to another are often lacking. This observation has traditionally been explained by the incompleteness of the fossil record. Biologists have attempted to fill in the missing parts with new fossil discoveries.

Two different models—punctuated equilibrium and gradualism—have been developed to explain evolution as observed in the fossil record (Fig. 20-13). The **punctuated equilibrium** model was proposed by paleontologists who questioned whether the fossil record really is as incomplete as it initially appeared. In the history of a species, long periods of **stasis** (little or no evolu-

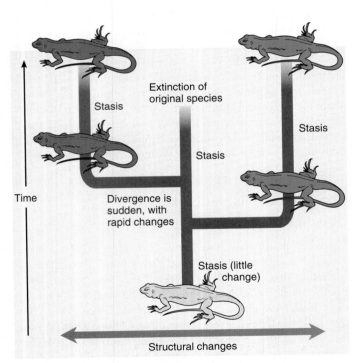

(a) In punctuated equilibrium, long periods of stasis are interrupted by short periods of rapid speciation.

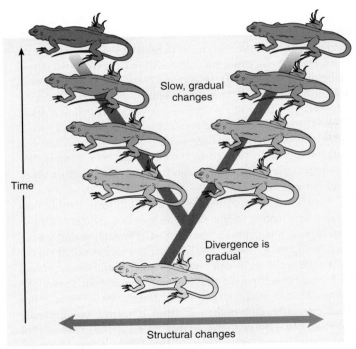

(b) In gradualism, a slow, steady change in species occurs over time.

Figure 20-13 Punctuated equilibrium and gradualism

In this figure, structural changes in the lizards are represented by changes in skin color.

tionary change) are punctuated, or interrupted, by short periods of rapid speciation that are perhaps triggered by changes in the environment—that is, periods of great evolutionary stress. Thus, speciation normally proceeds in "spurts." These relatively short periods of active evolution (perhaps 100,000 years) are followed by long periods (perhaps 2 million years) of stability.

In the punctuated equilibrium model, speciation occurs in a relatively short period. Keep in mind, however, that a "short" amount of time for speciation may be thousands of years. Such a span is short when compared with the several million years of a species' existence. Biologists who support the idea of punctuated equilibrium emphasize that sympatric speciation and even allopatric speciation occur in such relatively short periods. Punctuated equilibrium accounts for the abrupt appearance of a new species in the fossil record, with little or no evidence of intermediate forms. Proponents hypothesize that few transitional forms appear in the fossil record, because few transitional forms occurred during speciation.

In contrast, the traditional view of evolution espouses the **gradualism** model, in which evolution proceeds continuously over long periods. Gradualism is rarely observed in the fossil record, because the record is incomplete. (Recall from Chapter 18 that precise conditions are required for fossil formation. Most organisms leave no trace of their existence because they decompose when they die.) Occasionally a complete fossil record of transitional forms is discovered and cited as a strong case for gradualism. The gradualism model maintains that populations slowly diverge from one another by the gradual accumulation of adaptive characteristics within each population. These adaptive characteristics accumulate as a result of different selective pressures encountered in different environments.

The abundant fossil evidence of long periods with no change in a species has been used to argue against the gradualism model. Gradualists, however, maintain that any periods of stasis evident in the fossil record are the result of *stabilizing selection* (see Chapter 19). They also emphasize that stasis in fossils is deceptive, because fossils do not reveal all aspects of evolution. Although fossils display changes in external structure and skeletal structure, other genetic changes—in physiology, internal structure, resistance to infection, and behavior—all of which also represent evolution, are not evident. Gradualists recognize rapid evolution only when strong *directional selection* occurs.

Many biologists embrace both models to explain the fossil record; they also contend that the pace of evolution may be abrupt in certain instances and gradual in others and that neither punctuated equilibrium nor gradualism exclusively characterizes the changes associated with evolution. Other biologists do not view the distinction between punctuated equilibrium and gradualism as real. They suggest that genetic changes occur gradually and at a roughly constant pace and that the majority of these mutations do not cause speciation. When the mutations that do cause speciation occur, they are dramatic and produce a pattern consistent with the punctuated equilibrium model.

Review

■ Are the gradualism and punctuated equilibrium models mutually exclusive? Explain your answer.

MACROEVOLUTION

Learning Objectives

6   Define *macroevolution*.
7   Discuss macroevolution in the context of novel features, including preadaptations, allometric growth, and paedomorphosis.
8   Discuss the macroevolutionary significance of adaptive radiation and extinction.

Macroevolution is large-scale phenotypic changes in populations that warrant their placement in taxonomic groups at the species level and higher—that is, new species, genera, families, orders, classes, and even phyla, kingdoms, and domains. One concern of macroevolution is to explain evolutionary novelties, which are large phenotypic changes such as the appearance of jointed limbs during the evolution of arthropods (crustaceans, insects, and spiders). These phenotypic changes are so great that the new species possessing them are assigned to different genera or higher taxonomic categories. Studies of macroevolution also seek to discover and explain major changes in species diversity through time such as occur during *adaptive radiation,* when many species appear, and *mass extinction,* when many species disappear. Thus, evolutionary novelties, adaptive radiation, and mass extinction are important aspects of macroevolution.

Evolutionary novelties originate through modifications of pre-existing structures

New designs arise from structures already in existence. A change in the basic pattern of an organism produces something unique, such as wings on insects, flowers on plants, and feathered wings on birds. Usually these evolutionary novelties are variations of some pre-existing structures, called **preadaptations,** that originally fulfilled one role but subsequently changed in a way that was adaptive for a different role. Feathers, which evolved from reptilian scales and may have originally provided thermal insulation in primitive birds and some dinosaurs, represent a preadaptation for flight. With gradual modification, feathers evolved to function in flight as well as to fulfill their original thermoregulatory role. (Interestingly, a few feather-footed bird species exist; this phenotype is the result of a change in gene regulation that alters scales, normally found on bird feet, into feathers.)

How do such evolutionary novelties originate? Many are probably due to changes during development. *Regulatory genes* exert control over hundreds of other genes during development, and very slight genetic changes in regulatory genes could ultimately cause major structural changes in the organism (see Chapters 17 and 18).

For example, during development most organisms exhibit varied rates of growth for different parts of the body, known as **allometric growth** (from the Greek *allo,* "different," and *metr,* "measure"). The size of the head in human newborns is large in proportion to the rest of the body. As a human grows and matures, the torso, hands, and legs grow more rapidly than the head. Allometric growth is found in many organisms, including

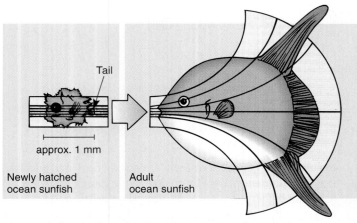

Tail

approx. 1 mm

Newly hatched
ocean sunfish

Adult
ocean sunfish

(a) A newly hatched
ocean sunfish, only
1 mm long, has an
extremely small tail.

(b) This allometric transformation
is visualized by drawing rectangular
coordinate lines through a picture of the
juvenile fish and then changing the
coordinate lines mathematically.

(c) An ocean sunfish swims off the coast of southern California.
The adult ocean sunfish may reach 4 m (13 ft) and weigh about
1500 kg (3300 lb).

Richard Herrmann

Figure 20-14 Allometric growth in the ocean sunfish

The tail end of an ocean sunfish (*Mola mola*) grows faster than the head end, resulting in the
unique shape of the adult.

the male fiddler crab with its single, oversized claw, and the ocean
sunfish with its enlarged tail (Fig. 20-14). If growth rates are
altered even slightly, drastic changes in the shape of an organism may result, changes that may or may not be adaptive. For
example, allometric growth may help explain the extremely small
and relatively useless forelegs of the dinosaur *Tyrannosaurus rex*,
compared with those of its ancestors.

Sometimes novel evolutionary changes occur when a species
undergoes changes in the *timing* of development. Consider, for
example, the changes that would occur if juvenile characteristics
were retained in the adult stage, a phenomenon known as **paedo-**

Stephen Dalton/Photo Researchers, Inc.

Figure 20-15 Paedomorphosis in a salamander

An adult axolotl salamander (*Ambystoma mexicanum*) retains the juvenile characteristics of external gills (feathery structures protruding from
the neck) and a tail fin (*not visible*). Paedomorphosis allows the axolotl
to remain permanently aquatic and to reproduce without developing
typical adult characteristics.

morphosis (from the Greek *paed,* "child," and *morph,* "form").
Adults of some salamander species have external gills and tail
fins, features found only in the larval (immature) stages of other
salamanders.

Retention of external gills and tail fins throughout life obviously alters the salamander's behavioral and ecological characteristics (Fig. 20-15). Perhaps such salamanders succeeded because
they had a selective advantage over "normal" adult salamanders;
that is, by remaining aquatic, they did not have to compete for
food with the terrestrial adult forms of related species. The paedomorphic forms also escaped the typical predators of terrestrial
salamanders (although they had other predators in their aquatic
environment). Studies suggest that paedomorphosis in salamanders is probably the result of mutations in genes that block the
production of hormones that stimulate metamorphic changes.
When paedomorphic salamanders receive hormone injections,
they develop into adults lacking external gills and tail fins.

Adaptive radiation is the diversification of an ancestral species into many species

Some ancestral species have given rise to far more species than
have other evolutionary lineages. **Adaptive radiation** is the evolutionary diversification of many related species from one or a
few ancestral species in a relatively short period.

Evolutionary innovations such as those we considered in the
last section may trigger an adaptive radiation. Alternatively, an
adaptive radiation may occur because the ancestral species happens to have been in the right place at the right time to exploit
numerous ecological opportunities. Biologists developed the
concept of adaptive zones to help explain this type of adaptive
radiation. **Adaptive zones** are new ecological opportunities that

Key Point

Adaptive radiation occurs when a single ancestral species diversifies into a variety of species, each adapted to a different ecological niche.

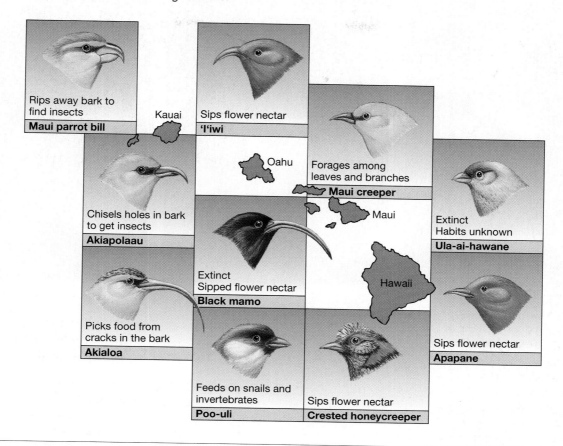

Figure 20-16 Adaptive radiation in Hawaiian honeycreepers

Compare the various beak shapes and methods of obtaining food. Many honeycreeper species are now extinct or nearing extinction as a result of human activities, including the destruction of habitat and the introduction of predators such as rats, dogs, and pigs.

were not exploited by an ancestral organism. At the species level, an adaptive zone is essentially identical to one or more similar *ecological niches* (the functional roles of species within a community; see Chapter 53). Examples of adaptive zones include nocturnal flying to catch small insects, grazing on grass while migrating across a savanna, and swimming at the ocean's surface to filter out plankton. When many adaptive zones are empty, as they were in Lake Victoria when it refilled some 12,000 years ago (discussed earlier in the chapter), colonizing species such as the cichlids may rapidly diversify and exploit them.

Because islands have fewer species than do mainland areas of similar size, latitude, and topography, vacant adaptive zones are more common on islands than on continents. Consider the Hawaiian honeycreepers, a group of related birds found in the Hawaiian Islands. When the honeycreeper ancestors reached Hawaii, few other birds were present. The succeeding generations of honeycreepers quickly diversified into many new species and, in the process, occupied the many available adaptive zones that on the mainland are occupied by finches, honeyeaters, treecreep-

ers, and woodpeckers. The diversity of their bills is a particularly good illustration of adaptive radiation (▌Fig. 20-16). Some honeycreeper bills are curved to extract nectar out of tubular flowers, whereas others are short and thick for ripping away bark in search of insects.

Another example of vacant adaptive zones involves the Hawaiian silverswords, 28 species of closely related plants found only in the Hawaiian Islands. When the silversword ancestor, a California plant related to daisies, reached the Hawaiian Islands, many diverse environments, such as cool, arid mountains; exposed lava flows; dry woodlands; shady, moist forests; and wet bogs were present and more or less unoccupied. The succeeding generations of silverswords quickly diversified in structure and physiology to occupy the many adaptive zones available to them. The diversity in their leaves, which changed during the course of natural selection as different populations adapted to various levels of light and moisture, is a particularly good illustration of adaptive radiation (▌Fig. 20-17). Leaves of silverswords that are adapted to shady, moist forests are large, for example, whereas

(a) The Haleakala silversword (*Argyroxyphium sandwicense ssp. macrocephalum*) is found only in the cinders on the upper slope of Haleakala Crater on the island of Maui. This plant is adapted to low precipitation and high levels of ultraviolet radiation.

(b) This silversword species (*Wilkesia gymnoxiphium*), which superficially resembles a yucca, is found only along the slopes of Waimea Canyon on the island of Kauai.

(c) *Daubautia scabra* is a small, herbaceous silversword found in moist to wet environments on several Hawaiian islands. (The fern fronds in the background give an idea of the small size of *D. scabra.*)

Figure 20-17 Adaptive radiation in Hawaiian silverswords

The 28 silversword species, which are classified in three closely related genera, live in a variety of habitats.

those of silverswords living in arid areas are small. The leaves of silverswords living on exposed volcanic slopes are covered with dense, silvery hairs that may reflect some of the intense ultraviolet radiation off the plant.

Adaptive radiation appears more common during periods of major environmental change, but it is difficult to determine how these changes are related to adaptive radiation. Possibly major environmental change indirectly affects adaptive radiation by increasing the extinction rate. Extinction produces empty adaptive zones, which provide new opportunities for species that remain. Mammals, for example, had existed mainly as small nocturnal insectivores (insect eaters) for millions of years before undergoing adaptive radiation that led to the modern mammalian orders. This radiation was presumably triggered by the extinction of the dinosaurs. Mammals diversified and exploited a variety of adaptive zones relatively soon after the dinosaurs' demise. Flying bats, running gazelles, burrowing moles, and swimming whales all originated from the small, insect-eating, ancestral mammals.

Extinction is an important aspect of evolution

Extinction, the end of a lineage, occurs when the last individual of a species dies. The loss is permanent, for once a species is extinct, it never reappears. Extinctions have occurred continually since the origin of life; by one estimate, only 1 species is alive today for every 2000 that have become extinct. Extinction is the eventual fate of all species, in the same way that death is the eventual fate of all individual organisms.

Although extinction has a negative short-term impact on the number of species, it facilitates evolution over a period of thousands to millions of years. As mentioned previously, when species become extinct, their adaptive zones become vacant. Consequently, surviving species are presented with new evolutionary opportunities and may diverge, filling in some of the unoccupied zones. In other words, the extinct species may eventually be replaced by new species.

During the long history of life, extinction appears to have occurred at two different rates. The continuous, low-level extinction of species is sometimes called **background extinction.** In contrast, five or possibly six times during Earth's history, **mass extinctions** of numerous species and higher taxonomic groups have taken place in both terrestrial and marine environments. The most recent mass extinction, which occurred about 65 million years ago (mya), killed off many marine organisms, terrestrial plants, and vertebrates, including the last of the dinosaurs (Fig. 20-18). The time span over which a mass extinction occurs may be several million years, but that is relatively short compared with the 3.5 billion years or so of Earth's history of life. Each pe-

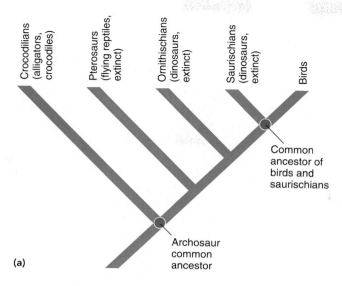

(a)

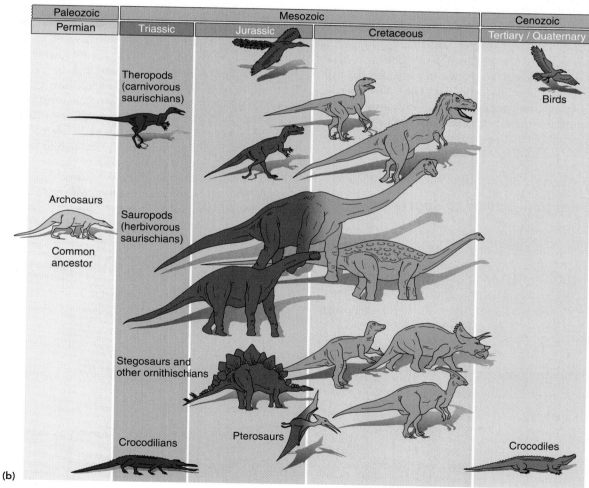

(b)

Figure 20-18 Mass extinction of the archosaurs

(a) Cladogram of the descendants of archosaurs, reptiles that were the ancestors of the dinosaurs. (b) At the end of the Cretaceous period, approximately 65 mya, a mass extinction of many organisms, including the remaining dinosaurs, occurred. (Dinosaurs had already been declining in diversity throughout the latter part of the Cretaceous period.) The only archosauran lines to survive were crocodiles and birds.

riod of mass extinction has been followed by a period of adaptive radiation of some of the surviving groups.

The causes of past episodes of mass extinction are not well understood. Both environmental and biological factors seem to have been involved. Major changes in climate could have adversely affected those plants and animals that lacked the genetic flexibility to adapt. Marine organisms, in particular, are adapted to a steady, unchanging climate. If Earth's temperature were to increase or decrease by just a few degrees overall, many marine species would probably perish.

It is also possible that past mass extinctions were due to changes in the environment induced by catastrophes. If a large comet or small asteroid collided with Earth, the dust ejected into the atmosphere on impact could have blocked much of the sunlight. In addition to disrupting the food chain by killing many plants (and therefore terrestrial animals), this event would have lowered Earth's temperature, leading to the death of many marine organisms. Evidence that the extinction of dinosaurs was caused by an extraterrestrial object's collision with Earth continues to accumulate (see Chapter 21).

Biological factors also trigger extinction. Competition among species may lead to the extinction of species that cannot compete effectively. The human species, in particular, has had a profound impact on the rate of extinction. The habitats of many animal and plant species have been altered or destroyed by humans, and habitat destruction can result in a species' extinction. Some biologists think we have entered the greatest mass extinction episode in Earth's history. (Extinction is discussed further in Chapter 56.)

Is microevolution related to speciation and macroevolution?

The concepts presented in Chapters 18 and 19 represent the **modern synthesis,** or *synthetic theory of evolution,* in which mutation provides the genetic variation on which natural selection acts. The modern synthesis combines Darwin's theory with important aspects of genetics. Many aspects of the modern synthesis have been tested and verified at the population and subspecies levels.

Many biologists contend that microevolutionary processes (natural selection, mutation, genetic drift, and gene flow) account for the genetic variation within species and for the origin of new species. Biologists also hypothesize that microevolutionary processes explain macroevolutionary patterns.

A considerable body of data from many fields supports the modern synthesis as it relates to speciation and macroevolution. Consider, for example, the *fish–tetrapod transition* (the evolution of four-legged amphibians from fishes), which was a major macroevolutionary event in the history of vertebrates. The fish–tetrapod transition is well documented in the fossil record. Studies of fossil intermediates show that the transition from aquatic fishes to terrestrial amphibians occurred as evolutionary novelties, such as changes in the limbs and skull roof, were added. These novelties accumulated as a succession of small changes over a period of 9 million to 14 million years. This time scale is sufficient to have allowed natural selection and other microevolutionary processes to produce the novel characters.

Although few biologists doubt the role of natural selection and microevolution in generating specific adaptations, some question the *extent* of microevolution's role in the overall pattern of life's history. These biologists ask whether speciation and macroevolution have been dominated by microevolutionary processes or by external, chance events, such as an impact by an asteroid. Chance events do not "care" about adaptive superiority but instead lead to the random extinction or survival of species. In the case of an asteroid impact, for example, certain species may survive because they were "lucky" enough to be in a protected environment at the time of impact. If chance events have been the overriding factor during life's history, then microevolution cannot be the exclusive explanation for the biological diversity present today (see additional discussion of the role of chance in evolution in Chapter 18).

Review

- Why are evolutionary novelties important to scientists studying macroevolution?
- What is a preadaptation? What is paedomorphosis?
- What roles do extinction and adaptive radiation play in macroevolution?

SUMMARY WITH KEY TERMS

Learning Objectives

1 Describe the biological species concept, and list two potential problems with the concept (page 429).

- According to the **biological species concept,** a species consists of one or more populations whose members interbreed in nature to produce fertile offspring and do not interbreed with members of different species.
- One problem with the biological species concept is that it applies only to sexually reproducing organisms. Another potential problem is that although individuals assigned to different species do not normally interbreed, they may occasionally successfully interbreed.

2 Explain the significance of reproductive isolating mechanisms, and distinguish among the different prezygotic and postzygotic barriers (page 430).

- **Reproductive isolating mechanisms** restrict gene flow between species.
- **Prezygotic barriers** are reproductive isolating mechanisms that prevent fertilization from taking place. **Temporal isolation** occurs when two species reproduce at different times of the day, season, or year. In **habitat isolation,** two closely related species live and breed in different habitats in the same geographic area. In **behavioral isolation,** distinctive courtship behaviors prevent mating between species. **Mechanical isolation** is due to incompatible structural differences in the reproductive organs of similar species. In **gametic isolation,** gametes from different species are incompatible because of molecular and chemical differences.

- **Postzygotic barriers** are reproductive isolating mechanisms that prevent gene flow after fertilization has taken place. **Hybrid inviability** is the death of interspecific embryos during development. **Hybrid sterility** prevents interspecific hybrids that survive to adulthood from reproducing successfully. **Hybrid breakdown** prevents the offspring of hybrids that survive to adulthood and successfully reproduce from reproducing beyond one or a few generations.

ThomsonNOW™ **Explore reproductive isolation by clicking on the figures in ThomsonNOW.**

3   Explain the mechanism of allopatric speciation, and give an example (page 432).
- **Speciation** is the evolution of a new species from an ancestral population. **Allopatric speciation** occurs when one population becomes geographically isolated from the rest of the species and subsequently diverges. Speciation is more likely to occur if the original isolated population is small, because genetic drift is more significant in small populations. Examples of allopatric speciation include Death Valley pupfishes, Kaibab squirrels, and Porto Santo rabbits.

ThomsonNOW™ **Explore allopatric speciation by clicking on the figures in ThomsonNOW.**

4   Explain the mechanisms of sympatric speciation, and give both plant and animal examples (page 432).
- **Sympatric speciation** does not require geographic isolation.
- Sympatric speciation in plants results almost exclusively from **allopolyploidy,** in which a **polyploid** individual (one with more than two sets of chromosomes) is a hybrid derived from two species. Two examples of sympatric speciation by allopolyploidy are the kew primroses and hemp nettles.
- Sympatric speciation occurs in animals, such as fruit maggot flies and cichlids, but how often it occurs and under what conditions remain to be determined.

ThomsonNOW™ **Explore sympatric speciation by clicking on the figure in ThomsonNOW.**

5   Take either side in a debate on the pace of evolution by representing the views of either punctuated equilibrium or gradualism (page 438).
- Biologists debate the interpretation of evolution, as observed in the fossil record. According to the **punctuated equilibrium** model, evolution of species proceeds in spurts. Short periods of active speciation are interspersed with long periods of **stasis.** According to the **gradualism** model, populations slowly diverge from one another by the accumulation of adaptive characteristics within a population.

6   Define *macroevolution* (page 439).
- **Macroevolution** is large-scale phenotypic changes in populations that warrant the placement of the populations in taxonomic groups at the species level and higher—that is, new species, genera, families, orders, classes, and even phyla, kingdoms, and domains.

7   Discuss macroevolution in the context of novel features, including preadaptations, allometric growth, and paedomorphosis (page 439).
- Macroevolution includes the appearance of evolutionary novelties, which may be due to changes during **development,** the orderly sequence of events that occurs as an organism grows and matures. Slight genetic changes in regulatory genes could cause major structural changes in the organism.
- Evolutionary novelties may originate from **preadaptations,** structures that originally fulfilled one role but changed in a way that was adaptive for a different role. Feathers are an example of a preadaptation.
- **Allometric growth,** varied rates of growth for different parts of the body, results in overall changes in the shape of an organism. Examples include the ocean sunfish and the male fiddler crab.
- **Paedomorphosis,** the retention of juvenile characteristics in the adult, occurs because of changes in the timing of development. Adult axolotl salamanders, with external gills and tail fins, are an example of paedomorphosis.

8   Discuss the macroevolutionary significance of adaptive radiation and extinction (page 439).
- **Adaptive radiation** is the process of diversification of an ancestral species into many new species. **Adaptive zones** are new ecological opportunities that were not exploited by an ancestral organism. When many adaptive zones are empty, colonizing species rapidly diversify and exploit them. Hawaiian honeycreepers and silverswords both underwent adaptive radiation after their ancestors colonized the Hawaiian Islands.
- **Extinction** is the death of a species. When species become extinct, the adaptive zones that they occupied become vacant, allowing other species to evolve and fill those zones. **Background extinction** is the continuous, low-level extinction of species. **Mass extinction** is the extinction of numerous species and higher taxonomic groups in both terrestrial and marine environments.

TEST YOUR UNDERSTANDING

1. According to the biological species concept, two populations belong to the same species if (a) their members freely interbreed (b) individuals from the two populations produce fertile offspring (c) their members do not interbreed with individuals of different species (d) a and c (e) a, b, and c

2. The zebrass is an example of (a) a fertile hybrid (b) a sterile hybrid (c) prezygotic barriers (d) a biological species (e) allopolyploidy

3. A prezygotic barrier prevents (a) the union of egg and sperm (b) reproductive success by an interspecific hybrid (c) the development of the zygote into an embryo (d) allopolyploidy from occurring (e) changes in allometric growth

4. The reproductive isolating mechanism in which two closely related species live in the same geographic area but reproduce at different times is (a) temporal isolation (b) behavioral isolation (c) mechanical isolation (d) gametic isolation (e) hybrid inviability

5. Interspecific hybrids, if they survive, are (a) always sterile (b) always fertile (c) usually sterile (d) usually fertile (e) never sterile

6. The process by which populations of one species evolve into separate species is known as (a) the evolutionary species concept (b) speciation (c) behavioral isolation (d) ploidy (e) hybridization

7. The first step leading to allopatric speciation is (a) hybrid inviability (b) hybrid breakdown (c) adaptive radiation (d) geographic isolation (e) paedomorphosis

8. The pupfishes in the Death Valley region are an example of which evolutionary process? (a) background extinction (b) allopatric speciation (c) sympatric speciation (d) allopolyploidy (e) paedomorphosis

9. Sympatric speciation (a) is most common in animals (b) does not require geographic isolation (c) accounts for the evolution of the Hawaiian goose (the nene) (d) involves the accumulation of gradual genetic changes (e) usually takes millions of years

10. Which of the following evolutionary processes is associated with allopolyploidy? (a) gradualism (b) allometric growth (c) sympatric speciation (d) mass extinction (e) preadaptation

11. According to the punctuated equilibrium model, (a) populations slowly diverge from one another (b) the evolution of species occurs in spurts interspersed with long periods of stasis (c) evolutionary novelties originate from preadaptations (d) reproductive isolating mechanisms restrict gene flow between species (e) the fossil record, being incomplete, does not accurately reflect evolution as it actually occurred

12. The evolutionary conversion of reptilian scales into feathers is an example of (a) allometric growth (b) paedomorphosis (c) gradualism (d) hybrid breakdown (e) a preadaptation

13. Adaptive radiation is common following a period of mass extinction, probably because (a) the survivors of a mass extinction are remarkably well adapted to their environment (b) the unchanging environment following a mass extinction drives the evolutionary process (c) many adaptive zones are empty (d) many ecological niches are filled (e) the environment induces changes in the timing of development for many species

14. Adaptive radiations do not appear to have ever occurred (a) on isolated islands (b) in birds such as honeycreepers (c) in environments colonized by few species (d) in plants such as silverswords (e) in environments with many existing species

15. The Hawaiian silverswords are an excellent example of which evolutionary process? (a) allometry (b) preadaptation (c) microevolution (d) adaptive radiation (e) extinction

CRITICAL THINKING

1. **Evolution Link.** Why is allopatric speciation more likely to occur if the original isolated population is small?

2. **Evolution Link.** Based on what you have learned about prezygotic and postzygotic isolating mechanisms, which reproductive isolating mechanism(s) would you say is/are probably at work between the Porto Santo rabbit and its mainland relative?

3. **Evolution Link.** Could hawthorn and apple maggot flies be considered an example of assortative mating, which was discussed in Chapter 19? Explain your answer.

4. **Evolution Link.** Because mass extinction is a natural process that may facilitate evolution during the period of thousands to millions of years that follow it, should humans be concerned about the current mass extinction that we are causing? Why or why not?

5. **Analyzing Data.** Examine Figure 20-2c, and predict which date is likeliest for researchers to have collected wood and leopard frogs and interbred them in the lab: March 10, April 1, or April 15. Explain your answer.

6. **Analyzing Data.** Examine Figure 20-18a, which shows the relationships among archosaur descendants. Which line was the earliest to diverge? Which line is most closely related to birds?

 Additional questions are available in ThomsonNOW at www.thomsonedu.com/login

21

The Origin and Evolutionary History of Life

William E. Ferguson

Fossil trilobites. These extinct arthropods (*Phacops rana*), which were about 3 cm (1.2 in) long, flourished in the ocean during the Paleozoic era. Trilobites ranged from 1 mm to 1 m in length, depending on the species. Note the large, well-developed eyes visible on either side of the head region.

KEY CONCEPTS

Although there is no direct fossil evidence of the origin of life, biochemical experiments have demonstrated how the complex organic molecules that are found in all living organisms may have formed.

Photosynthesis, aerobic respiration, and eukaryotic cell structure represent several major advances that occurred during the early history of life.

The fossil record tells us much of what we know about the history of life, such as what kinds of organisms existed and where and when they lived.

Certain organisms appear in the fossil record, then disappear and are replaced by others.

Scientists identify and demonstrate relationships among fossils in rock layers from different periods of geologic time.

The preceding three chapters were concerned with biological evolution. However, we have not dealt with what many regard as a fundamental question: How did life begin? Although biologists generally accept the hypothesis that life developed from nonliving matter, exactly how this process, called **chemical evolution,** occurred is not certain. Current models suggest that small organic molecules first formed spontaneously and accumulated over time. Large organic macromolecules such as proteins and nucleic acids could have then assembled from the smaller molecules. The macromolecules interacted, combining into more complicated structures that could eventually metabolize and replicate. Their descendants eventually became the first true cells. After the first cells originated, they diverged over several billion years into the rich biological diversity that characterizes our planet today.

Initially unicellular prokaryotes predominated, followed by unicellular eukaryotes. The first multicellular eukaryotes were soft-bodied marine animals that did not leave many fossils. Shelled animals and other marine invertebrates (animals without backbones) appeared next, as exemplified by trilobites, primitive aquatic arthropods (see photograph). Marine invertebrates were followed by the first vertebrates (animals with backbones). The first fishes with jaws appeared and diversified. Some of these gave rise to amphibians, the first vertebrates with limbs that were capable of moving about on land. Amphibians gave rise to reptiles, which diversified and populated the land. Reptiles, in turn, gave rise independently to birds and to mammals. Plants underwent a comparable evolutionary history and diversification.

In this chapter we survey life over a vast span of time, starting when our planet was relatively young. We examine proposed models about how life began and trace life's long evolutionary history. ■

CHEMICAL EVOLUTION ON EARLY EARTH

Learning Objectives

1 Describe the conditions that geologists think existed on early Earth.
2 Contrast the prebiotic soup hypothesis and the iron–sulfur world hypothesis.

Many biologists speculate that life originated only once and that life's beginnings occurred under environmental conditions quite different from those of today. We must therefore examine the conditions of early Earth to understand the origin of life. Although we will never be certain about the exact conditions that existed when life arose, scientific evidence from many sources provides us with valuable clues that help us formulate plausible scenarios. Study of the origin of life is an active area of scientific research today, and many important contributions are adding to our understanding of how life began.

Astrophysicists and geologists estimate that Earth is about 4.6 billion years old. The atmosphere of early Earth apparently included carbon dioxide (CO_2), water vapor (H_2O), carbon monoxide (CO), hydrogen (H_2), and nitrogen (N_2). It may also have contained some ammonia (NH_3), hydrogen sulfide (H_2S), and methane (CH_4), although ultraviolet radiation from the sun most likely broke these reduced molecules down rapidly. The early atmosphere probably contained little or no free oxygen (O_2).

Four requirements must have existed for the chemical evolution of life: little or no free oxygen, a source of energy, the availability of chemical building blocks, and time. First, life could have begun only in the absence of free oxygen. Oxygen is quite reactive and would have oxidized the organic molecules that are necessary building blocks in the origin of life. Earth's early atmosphere was probably strongly reducing, which means that any free oxygen would have reacted with other elements to form oxides. Thus, oxygen would have been tied up in compounds.

The origin of life would also have required energy to do the work of building biological molecules from simple inorganic chemicals. Early Earth was a place of high energy with violent thunderstorms; widespread volcanic activity; bombardment from meteorites and other extraterrestrial objects; and intense radiation, including ultraviolet radiation from the sun (❚ Fig. 21-1). The young sun probably produced more ultraviolet radiation than it does today, and ancient Earth had no protective ozone layer to filter it.

A third requirement would have been the presence of the chemical building blocks needed for chemical evolution. These included water, dissolved inorganic minerals (present as ions), and the gases present in the early atmosphere. A final requirement for the origin of life was time for molecules to accumulate and react with one another. Earth is approximately 4.6 billion years old, and the earliest traces of life are approximately 3.5 billion years old; therefore, life had several hundred million years to get started.

Courtesy of Reader's Digest Books. Drawing by H. K. Wimmer.

❚ **Figure 21-1** *Animated* An artist's interpretation of conditions on early Earth

Volcanoes erupted, spewing gases that contributed to the atmosphere, and violent thunderstorms triggered torrential rainfall that eroded the land. Meteorites and other extraterrestrial objects continu-ally bombarded Earth, cataclysmically changing the crust, ocean, and atmosphere.

Organic molecules formed on primitive Earth

Because organic molecules are the building materials for organisms, it is reasonable to first consider how they may have originated. Two main models seek to explain how the organic precursors of life originated: The **prebiotic soup hypothesis** proposes that these molecules formed near Earth's surface, whereas the **iron–sulfur world hypothesis** proposes that organic precursors formed at cracks in the ocean's floor.

Organic molecules may have been produced at Earth's surface

The concept that simple organic molecules such as sugars, nucleotide bases, and amino acids could form spontaneously from simpler raw materials was first advanced in the 1920s by two scientists working independently: A. I. Oparin, a Russian biochemist, and J. B. S. Haldane, a Scottish physiologist and geneticist.

Their hypothesis was tested in the 1950s by U.S. biochemists Stanley Miller and Harold Urey, who designed a closed apparatus that simulated conditions that presumably existed on early Earth (▮ Fig. 21-2). They exposed an atmosphere rich in H_2, CH_4, H_2O, and NH_3 to an electric discharge that simulated lightning. Their analysis of the chemicals produced during one week revealed that amino acids and other organic molecules had formed. Although more recent data suggest that Earth's early atmosphere was not rich in methane or ammonia, similar experiments using different combinations of gases have produced a wide variety of organic molecules that are important in contemporary organisms. These include all 20 amino acids, several sugars, lipids, the nucleotide bases of RNA and DNA, and ATP (when phosphate is present). Thus, before life began, its chemical building blocks were probably accumulating as a necessary step in chemical evolution.

Oparin envisioned that the organic molecules would, over vast spans of time, accumulate in the shallow seas to form a "sea of organic soup." He envisioned smaller organic molecules (monomers) combining to form larger ones (polymers) under such conditions. Evidence gathered since Oparin's time indicates that organic polymers may have formed and accumulated on rock or clay surfaces rather than in the primordial seas. Clay, which consists of microscopic particles of weathered rock, is particularly intriguing as a possible site for early polymerizations because it binds organic monomers and contains zinc and iron ions that may have served as catalysts. Laboratory experiments have confirmed that organic polymers form spontaneously from monomers on hot rock or clay surfaces.

Organic molecules may have formed at hydrothermal vents

Some biologists hypothesize a different scenario of chemical evolution: that early polymerizations leading to the origin of life may have occurred in cracks in the deep-ocean floor where hot water, carbon monoxide, and minerals such as sulfides of iron and nickel spew forth—the iron–sulfur world hypothesis. Such **hydrother-**

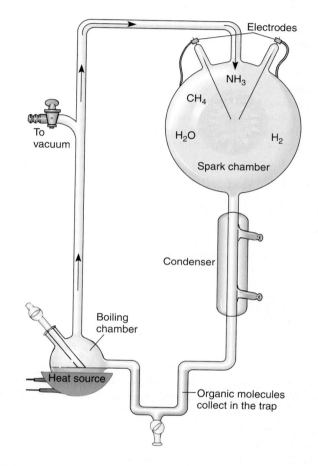

Key Experiment

QUESTION: Could organic molecules have formed in the conditions of early Earth?

HYPOTHESIS: Organic molecules can form in a reducing atmosphere similar to that thought to be present on early Earth.

EXPERIMENT: The apparatus that Miller and Urey used to simulate the reducing atmosphere of early Earth contained nitrogen, hydrogen, methane, ammonia, and water. An electrical spark was produced in the upper right flask to simulate lightning.

Electrodes

NH_3

CH_4

To vacuum

H_2O

H_2

Spark chamber

Condenser

Boiling chamber

Heat source

Organic molecules collect in the trap

RESULTS AND CONCLUSION: The gases present in the flask reacted together, and in one week a variety of simple organic compounds, such as amino acids, accumulated in the trap at the bottom. Thus, the formation of organic molecules—the first step in the origin of life—could have occurred in the conditions present on early Earth.

Figure 21-2 *Animated* Miller and Urey's experiment in chemical evolution

mal vents would have been better protected than Earth's surface from the catastrophic effects of meteorite bombardment.

Today these hot springs produce precursors of biological molecules and of energy-rich "food," including the highly reduced compounds hydrogen sulfide and methane. These chemicals support a diverse community of microorganisms, clams, crabs, tube worms, and other animals (see *Focus On: Life without the Sun,* in Chapter 54).

Testing the iron–sulfur world hypothesis at hydrothermal vents is difficult, but laboratory experiments simulating the high pressures and temperatures at the vents have yielded intriguing results. For example, iron and nickel sulfides catalyze reactions between carbon monoxide and hydrogen sulfide, producing acetic acid and other simple organic compounds. Also, experiments show that ammonia, one of the precursors of proteins and nucleic acids, is produced in abundance, suggesting that vents were ammonia-rich environments in the prebiotic world and that the earliest organisms formed in this seemingly inhospitable environment.

Review

■ What are the four requirements for chemical evolution, and why is each essential?

■ What is the prebiotic soup hypothesis?

■ How does the iron–sulfur world hypothesis differ from the prebiotic soup hypothesis?

▮ THE FIRST CELLS

Learning Objectives

3 Outline the major steps hypothesized to have occurred in the origin of cells.

4 Explain how the evolution of photosynthetic autotrophs affected both the atmosphere and other organisms.

5 Describe the hypothesis of serial endosymbiosis.

After the first polymers formed, could they have assembled spontaneously into more complex structures? Scientists have synthesized several different **protobionts,** which are assemblages of abiotically produced (that is, not produced by organisms) organic polymers. They have recovered protobionts that resemble living cells in several ways, thus providing clues as to how aggregations of complex nonliving molecules took that "giant leap" and became living cells. These protobionts exhibit many functional and structural attributes of living cells. They often divide in half (binary fission) after they have sufficiently "grown." Protobionts maintain an internal chemical environment that is different from the external environment (homeostasis), and some of them show the beginnings of metabolism (catalytic activity). They are highly organized, considering their relatively simple composition.

Microspheres are a type of protobiont formed by adding water to abiotically formed polypeptides (▮ Fig. 21-3). Some microspheres are excitable: they produce an electrical potential across their surfaces, reminiscent of electrochemical gradients in cells. Microspheres can also absorb materials from their surroundings (selective permeability) and respond to changes in osmotic pressure as though membranes enveloped them, even though they contain no lipid.

The study of protobionts shows that relatively simple "precells" have some of the properties of contemporary life. However, it is a major step (or several steps) to go from simple mo-

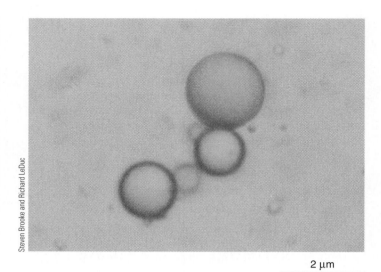

2 μm

▮ Figure 21-3 Microspheres

These tiny protobionts exhibit some of the properties of life.

lecular aggregates such as protobionts to living cells. Although we have learned many things about how organic molecules may have formed on primitive Earth, the problem of how precells evolved into living cells remains to be solved. One of the most significant parts of that process was the evolution of molecular reproduction.

Molecular reproduction was a crucial step in the origin of cells

In living cells, genetic information is stored in the nucleic acid DNA, which is transcribed into messenger RNA (mRNA), which in turn is translated into the proper amino acid sequence in proteins. All three macromolecules in the DNA ⟶ RNA ⟶ protein sequence contain precise information, but only DNA and RNA are capable of self-replication, although only in the presence of the proper enzymes. Because both RNA and DNA can form spontaneously on clay in much the same way that other organic polymers do, the question becomes, which molecule, DNA or RNA, first appeared in the prebiotic world?

Many scientists have suggested that RNA was the first informational molecule to evolve in the progression toward a selfsustaining, self-reproducing cell and that proteins and DNA came along later. According to a model known as the **RNA world,** the chemistry of prebiotic Earth gave rise to self-replicating RNA molecules that functioned as both enzymes and substrates for their own replication. We represent the replication of RNA in the RNA world scenario as a circular arrow:

RNA often has catalytic properties; such enzymatic RNAs are called **ribozymes.** Before the evolution of true cells, ribozymes may have catalyzed their own replication in the clays, shallow rock pools, or hydrothermal vents where life originated. When

RNA strands are added to a test tube containing RNA nucleotides but no enzymes, the nucleotides combine to form short RNA molecules.

The occurrence of an RNA world early in the history of life can never be proven, but experiments with **in vitro evolution,** also called **directed evolution,** have shown it is feasible. These experiments address an important question about the RNA world, namely, could RNA molecules have catalyzed the many different chemical reactions needed for life? In directed evolution, a large pool of RNA molecules with different sequences is mixed, and molecules are selected for their ability to catalyze a single biologically important reaction (❚ Fig. 21-4). Those molecules that have at least some catalytic ability are then amplified into many copies and mutated (for example, by exposure to certain chemicals) before being exposed to another round of selection. After this cycle is repeated several times, the RNA molecules at the end of the selection process function efficiently as catalysts for the reaction. In vitro evolution studies have shown that RNA has a large functional repertoire—that is, RNA can catalyze a variety of biologically important reactions.

Biologists hypothesize that in the RNA world, ribozymes initially catalyzed protein synthesis and other important biological reactions; only later did protein enzymes catalyze these reactions.

$$\circlearrowleft \text{RNA} \longrightarrow \text{protein}$$

Interestingly, in test tube experiments, RNA molecules have evolved that direct protein synthesis by catalyzing peptide bond formation. Some single-stranded RNA molecules fold back on themselves as a result of interactions among the nucleotides composing the RNA strand. Sometimes the conformation (shape) of the folded RNA molecule is such that it weakly binds to an amino acid. Amino acids held close to one another by RNA molecules may bond, forming a polypeptide.

We have considered how the evolution of informational molecules may have given rise to RNA and later to proteins. If a self-replicating RNA capable of coding for proteins appeared prior to DNA, how did DNA, the universal molecule of heredity in cells, become involved? Perhaps RNA made double-stranded copies of itself that eventually evolved into DNA.

$$\text{DNA} \longleftarrow \text{RNA} \longrightarrow \text{protein}$$

The incorporation of DNA into the information transfer system was advantageous because the double helix of DNA is more stable (less reactive) than the single strand of RNA. Such stability in a molecule that stores genetic information would have provided a decided advantage in the prebiotic world (as it does today).

Thus, in the DNA/RNA/protein world, DNA became the information storage molecule, RNA remained involved in protein synthesis, and protein enzymes catalyzed most cell reactions, including DNA replication and RNA synthesis.

$$\circlearrowleft \text{DNA} \longrightarrow \text{RNA} \longrightarrow \text{protein}$$

RNA is still a necessary component of the information transfer system, because DNA is not catalytic. Thus, natural selection at

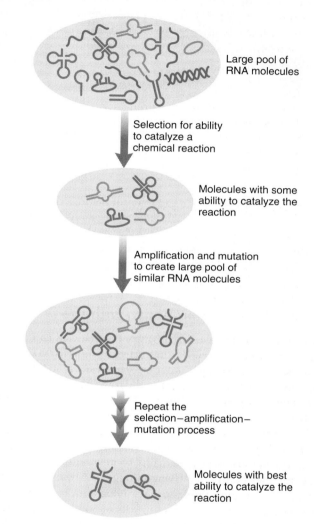

Key Experiment

QUESTION: Can in vitro natural selection result in chemical evolution of catalysts?

HYPOTHESIS: Self-replicating RNA molecules can evolve and become more efficient catalysts.

EXPERIMENT: RNA molecules are selected from a large pool, based on their ability to catalyze a specific reaction, then amplified and mutated (*see figure*). This process is repeated 7 to 20 additional times.

Large pool of RNA molecules

Selection for ability to catalyze a chemical reaction

Molecules with some ability to catalyze the reaction

Amplification and mutation to create large pool of similar RNA molecules

Repeat the selection–amplification–mutation process

Molecules with best ability to catalyze the reaction

RESULTS AND CONCLUSION: The final group of RNA molecules is the most efficient at catalyzing the chemical reaction selected for. Scientists have developed more than two dozen synthetic RNA catalysts by in vitro evolution.

Figure 21-4 In vitro evolution of RNA molecules

the molecular level favored the DNA ⟶ RNA ⟶ protein information sequence. Once DNA was incorporated into this sequence, RNA molecules assumed their present role as an intermediary in the transfer of genetic information.

(b) Cutaway view of a fossil stromatolite showing the layers of microorganisms and sediments that accumulated over time. This stromatolite, also from Western Australia, is about 3.5 billion years old.

(a) These living stromatolites at Hamlin Pool in Western Australia consist of mats of cyanobacteria and minerals such as calcium carbonate. They are several thousand years old.

Figure 21-5 Stromatolites

Several additional steps had to occur before a true living cell could evolve from macromolecular aggregations. For example, the self-replicating genetic code must have arisen extremely early in the prebiotic world because all organisms possess it—but how did it originate? Also, how did an encapsulating membrane of lipid and protein evolve?

Biological evolution began with the first cells

No one knows exactly when or where the first cells appeared on Earth. Did they live at hydrothermal vents or at Earth's surface? Were they bacteria? We cannot answer these questions, in part because no fossils exist that trace the transition from nonlife to life. Nonfossil evidence—isotopic "fingerprints" of organic carbon in ancient rocks in Greenland—suggest to some researchers that life existed as early as 3.8 billion years ago (bya). However, other scientists vigorously debate this conclusion.

Microfossils, ancient remains of microscopic life, suggest that cells may have been thriving as long as 3.5 bya. Rich deposits of microfossils appear to exist in the Pilbara rocks of northwestern Australia and the Barberton rocks in South Africa; both of these deposits are about 3.3 billion to 3.5 billion years old. However, some scientists have challenged the interpretation of the oldest carbon-rich "squiggles" in ancient rocks as microfossils of prokaryotic organisms. They suggest that microfossils are not fossils at all but are formed by natural geologic processes.

Evidence suggests that the earliest cells were prokaryotic. **Stromatolites,** another type of fossil evidence of the earliest cells, are rocklike columns composed of many minute layers of prokaryotic cells (■ Fig. 21-5). Over time, sediment collects around the cells and mineralizes. Meanwhile, a new layer of living cells grows over the older, dead cells. Fossil stromatolite reefs are found in several places in the world, including Great Slave Lake in Canada and the Gunflint Iron Formations along Lake Superior in the United States. Some fossil stromatolites are extremely ancient. One group in Western Australia, for example, is several billion years old. Stromatolite reefs are still living in hot springs and in warm, shallow pools of fresh and salt water.

The first cells were probably heterotrophic

The earliest cells probably did not synthesize the organic molecules they needed but instead obtained them from the environment. These primitive **heterotrophs** may have consumed many types of organic molecules that had formed spontaneously—for example, sugars, nucleotides, and amino acids. By fermenting these organic compounds, they obtained the energy they needed to support life. *Fermentation* is, of course, an anaerobic process (performed in the absence of oxygen), and the first cells were almost certainly **anaerobes.**

When the supply of spontaneously generated organic molecules gradually declined, only certain organisms could survive. Mutations had probably already occurred that permitted some cells to obtain energy directly from sunlight, perhaps by using sunlight to make ATP. These cells, which did not require the energy-rich organic compounds that were now in short supply in the environment, had a distinct selective advantage.

Photosynthesis requires both light energy and a source of electrons, which are used to reduce CO_2 to form organic molecules such as glucose (see Chapter 9). Most likely, the first photosynthetic **autotrophs**—organisms that produce their own food from simple raw materials—used the energy of sunlight to split hydrogen-rich molecules such as H_2S, releasing elemental sulfur (not oxygen) in the process. Indeed, the green sulfur bacteria and the purple sulfur bacteria still use H_2S as a hydrogen (electron) source for photosynthesis.[1]

---

[1]Members of a third group of bacteria, the purple nonsulfur bacteria, use other organic molecules or hydrogen gas as a hydrogen source.

The first photosynthetic autotrophs to obtain hydrogen electrons by splitting *water* were the **cyanobacteria.** Water was quite abundant on early Earth, as it is today, and the selective advantage that splitting water gave cyanobacteria allowed them to thrive. The process of splitting water released oxygen as a gas (O_2). Initially, the oxygen released during photosynthesis oxidized minerals in the ocean and in Earth's crust, and oxygen did not begin to accumulate in the atmosphere for a long time. Eventually, however, oxygen levels increased in the ocean and the atmosphere.

Scientists estimate the timing of the events just described on the basis of geologic and fossil evidence. Fossils from that period, which include rocks containing traces of chlorophyll, as well as the fossil stromatolites discussed earlier, indicate that the first photosynthetic organisms appeared approximately 3.1 bya to 3.5 bya. This evidence suggests that heterotrophic forms may have existed even earlier.

Aerobes appeared after oxygen increased in the atmosphere

Based on sulfur isotope data from ancient rocks in South Africa, it appears that cyanobacteria had produced enough oxygen to begin significantly changing the composition of the atmosphere by 2.3 bya. The increase in atmospheric oxygen affected life profoundly. The oxygen poisoned *obligate anaerobes* (organisms that cannot use oxygen for cellular respiration), and many species undoubtedly perished. Some anaerobes, however, survived in environments where oxygen did not penetrate; others evolved adaptations to neutralize the oxygen so it could not harm them. In some organisms, called **aerobes,** a respiratory pathway evolved that *used* oxygen to extract more energy from food. Aerobic respiration was joined with the existing anaerobic process of glycolysis.

The evolution of organisms that could use oxygen in their metabolism had several consequences. Organisms that respire aerobically gain much more energy from a single molecule of glucose than anaerobes gain by fermentation (see Chapter 8). As a result, the newly evolved aerobic organisms were more efficient and more competitive than anaerobes. Coupled with the poisonous nature of oxygen to many anaerobes, the efficiency of aerobes forced anaerobes into relatively minor roles. Today the vast majority of organisms, including plants, animals, and most fungi, protists, and prokaryotes, use aerobic respiration; only a few bacteria and even fewer protists and fungi are anaerobic.

The evolution of aerobic respiration stabilized both oxygen and carbon dioxide levels in the biosphere. Photosynthetic organisms used carbon dioxide as a source of carbon for synthesizing organic compounds. This raw material would have been depleted from the atmosphere in a relatively brief period without the advent of aerobic respiration, which released carbon dioxide as a waste product from the complete breakdown of organic molecules. Carbon thus started cycling in the biosphere, moving from the nonliving physical environment, to photosynthetic organisms, to heterotrophs that ate the photosynthetic organisms (see Chapter 54). Aerobic respiration released carbon back into the physical environment as carbon dioxide, and the carbon cycle continued. In a similar manner, molecular oxygen was produced by photosynthesis and used during aerobic respiration.

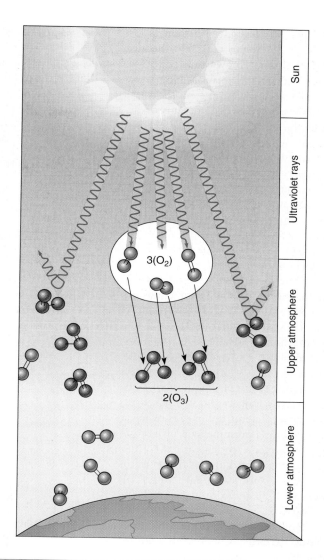

Figure 21-6 The formation of ozone

Ozone (O_3) forms in the upper atmosphere when ultraviolet radiation from the sun breaks the double bonds of oxygen molecules.

Another significant consequence of photosynthesis occurred in the upper atmosphere, where molecular oxygen reacted to form **ozone** (O_3) (Fig. 21-6). A layer of ozone eventually blanketed Earth, preventing much of the sun's ultraviolet radiation from penetrating to the surface. With the ozone layer's protection from the mutagenic effect of ultraviolet radiation, organisms could live closer to the surface in aquatic environments and eventually move onto land. Because the energy in ultraviolet radiation may have been necessary to form organic molecules, however, their abiotic synthesis decreased.

Eukaryotic cells descended from prokaryotic cells

Eukaryotes may have appeared in the fossil record as early as 2.2 bya, and geochemical evidence suggests that eukaryotes were present much earlier. *Steranes*, molecules derived from steroids, have been discovered in Australian rocks dated at 2.7 billion years old. Because bacteria are not known to produce steroids, the steranes may be biomarkers for eukaryotes. These ancient rocks lack fossil traces of ancient organisms because the rocks have since been exposed to heat and pressure that would have destroyed any fossilized cells. Steranes, however, are very stable in the presence of heat and pressure.

Eukaryotes arose from prokaryotes. Recall from Chapter 4 that prokaryotic cells lack nuclear envelopes as well as other membranous organelles such as mitochondria and chloroplasts. How did these organelles arise? According to the hypothesis of **serial endosymbiosis,** organelles such as mitochondria and chloroplasts may have originated from mutually advantageous symbiotic relationships between two prokaryotic organisms (Fig. 21-7). Chloroplasts apparently evolved from photosynthetic bacteria (cyanobacteria) that lived inside larger heterotrophic cells, whereas mitochondria presumably evolved from aerobic bacteria (perhaps ancient purple bacteria) that lived inside larger anaerobic cells. Thus, early eukaryotic cells were assemblages of formerly free-living prokaryotes.

How did these bacteria come to be **endosymbionts,** which are organisms that live symbiotically inside a host cell? They may originally have been ingested, but not digested, by a host cell. Once incorporated, they could have survived and reproduced along with the host cell so that future generations of the host also contained endosymbionts. The two organisms developed a mutualistic relationship in which each contributed something to the other. Eventually the endosymbiont lost the ability to exist outside its host, and the host cell lost the ability to survive without its endosymbionts. This hypothesis stipulates that each of these partners brought to the relationship something the other lacked. For example, mitochondria provided the ability to carry out the aerobic respiration lacking in the original anaerobic host cell. Chloroplasts provided the ability to use a simple carbon source (CO_2) to produce needed organic molecules. The host cell provided endosymbionts with a safe environment and raw materials or nutrients.

The principal evidence in favor of serial endosymbiosis is that mitochondria and chloroplasts possess some (although not all) of their own genetic material and translational components. They have their own DNA (as a circular molecule similar to that of prokaryotes; discussed in Chapter 24) and their own ribosomes (which resemble prokaryotic rather than eukaryotic

Chloroplasts and mitochondria of eukaryotic cells are hypothesized to have originated from various bacteria that lived as endosymbionts inside other cells.

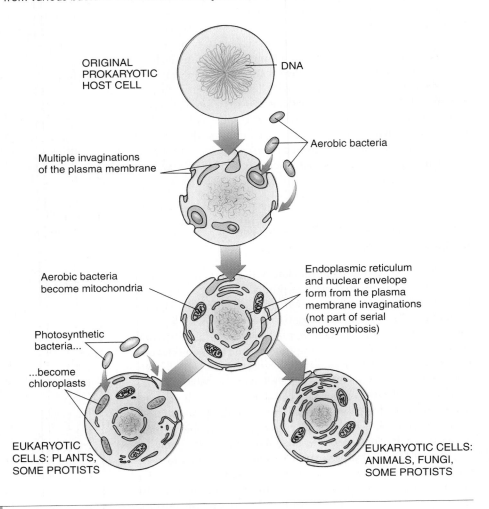

ORIGINAL PROKARYOTIC HOST CELL

DNA

Aerobic bacteria

Multiple invaginations of the plasma membrane

Aerobic bacteria become mitochondria

Endoplasmic reticulum and nuclear envelope form from the plasma membrane invaginations (not part of serial endosymbiosis)

Photosynthetic bacteria...

...become chloroplasts

EUKARYOTIC CELLS: PLANTS, SOME PROTISTS

EUKARYOTIC CELLS: ANIMALS, FUNGI, SOME PROTISTS

Figure 21-7 *Animated* Serial endosymbiosis

ribosomes). Mitochondria and chloroplasts also possess some of the machinery for protein synthesis, including tRNA molecules, and conduct protein synthesis on a limited scale independent of the nucleus. Furthermore, it is possible to poison mitochondria and chloroplasts with an antibiotic that affects prokaryotic but not eukaryotic cells. As discussed in Chapter 4, double membranes envelope mitochondria and chloroplasts. The outer membrane apparently developed from the invagination (infolding) of the host cell's plasma membrane, whereas the inner membrane is derived from the endosymbiont's plasma membrane. (Serial endosymbiosis is discussed in greater detail in Chapter 25.)

Many endosymbiotic relationships exist today. Many corals have algae living as endosymbionts within their cells (see Fig. 53-11). Protozoa (*Myxotricha paradoxa*) live symbiotically in the gut of termites; in turn, several different endosymbionts, including spirochete bacteria, attach to the protozoa and function as whiplike flagella, allowing the protozoa to move.

Although much evidence supports serial endosymbiosis, it does not completely explain the evolution of eukaryotic cells from prokaryotes. For example, serial endosymbiosis does not explain how a double membranous envelope came to surround the genetic material in the nucleus.

Review

- What major steps probably occurred in the origin of cells?
- How did the presence of molecular oxygen in the atmosphere affect early life?
- What kinds of evidence support serial endosymbiosis?

THE HISTORY OF LIFE

Learning Objectives

6   Briefly describe the distinguishing organisms and major biological events of the Ediacaran period and the Paleozoic, Mesozoic, and Cenozoic eras.

The sequence of biological, climate, and geologic events that make up the history of life is recorded in rocks and fossils. The sediments of Earth's crust consist of five major rock strata (layers), each subdivided into minor strata, lying one on top of the other. Very few places on Earth show all layers, but the strata present typically occur in the correct order, with younger rocks on top of older ones. These sheets of rock formed from the accumulation of mud and sand at the bottoms of the ocean, seas, and lakes. Each layer contains certain characteristic **index fossils** that identify deposits made at about the same time in different parts of the world.

Geologists divide Earth's history into units of time based on major geologic, climate, and biological events. **Eons** are the largest divisions of the geologic time scale. Eons are divided into **eras;** where fossil evidence exists, these divisions are based primarily on organisms that characterized each era. Eras are subdivided into **periods,** which in turn are composed of **epochs.**

Relatively little is known about Earth from its beginnings approximately 4.6 bya up to 2.5 bya, a time known as the **Archaean eon** (**Table 21-1**). Life originated on Earth during the Archaean eon. Signs of life date as early as 3.5 bya. Not much physical evidence is available, because the rocks of the Archaean eon, being extremely ancient, are deeply buried in most parts of the world. Ancient rocks are exposed in a few places, including the bottom of the Grand Canyon and along the shores of Lake Superior. Many Archaean rock formations have revealed what appear to be microfossils of prokaryotic cells such as cyanobacteria and other bacteria.

Rocks from the Ediacaran period contain fossils of cells and simple animals

That part of time from 2500 mya to 542 mya is known as the **Proterozoic eon.** This enormous span of time is easier to study than the preceding Archaean eon, because the rocks are less altered by heat and pressure. Life at the beginning of the Proterozoic eon

consisted of prokaryotes such as cyanobacteria. Stromatolites were abundant. About 2.2 bya the first eukaryotic cells appeared. By the end of the Proterozoic eon, multicellularity was evident in the abundant fossils of small, soft-bodied, invertebrate animals.

The **Ediacaran period** (pronounced "ee-dee-ack'-uh-run"), from 600 mya to 542 mya, is the last (most recent) period of the Proterozoic eon. It is named for the fossil deposits in the Ediacara Hills in South Australia, although the oldest, simplest Ediacaran fossils are from 580-million- to 600-million-year-old rocks in Newfoundland, the Mackenzie Mountains of northwestern Canada, and the Doushantuo Formation of China.

Ediacaran fossils are the oldest known fossils of multicellular animals (**Fig. 21-8**). Experts have not yet identified all the simple, soft-bodied animals found in the Ediacara Hills and at other sites. Some paleontologists interpret many of these fossils as early sponges, jellyfish, corals, and comb jellies; however, other scientists think that the Ediacaran animals were not ancestral to modern-day species but instead became extinct at the end of the Ediacaran period.

A diversity of organisms evolved during the Paleozoic era

The **Paleozoic era** began approximately 542 mya and lasted about 291 million years. It is divided into six periods: Cambrian, Ordovician, Silurian, Devonian, Carboniferous, and Permian.

Rocks rich in fossils represent the oldest subdivision of the Paleozoic era, the **Cambrian period.** For about 40 million years, evolution was in such high gear, with the sudden appearance of many new animal body plans, that this period is called the **Cambrian Radiation,** or more informally, *Cambrian explosion.* Fossils of all contemporary animal phyla are present, along with many bizarre, extinct phyla, in marine sediments. The seafloor was covered with sponges, corals, sea lilies, sea stars, snails, clamlike bivalves, primitive squidlike cephalopods, lamp shells (brachiopods), and trilobites (see chapter opening photograph; also see Fig. 30-19). In addition, small vertebrates—cartilaginous fishes—became established in the marine environment.

Scientists have not determined the factors responsible for the Cambrian Radiation, a period unmatched in the evolutionary history of life. There is some evidence that oxygen concentrations, which had continued to gradually increase in the atmosphere, passed some critical threshold late in the Proterozoic eon. Scientists who advocate the *oxygen enrichment hypothesis* note that until late in the Proterozoic eon, Earth did not have enough oxygen to support larger animals. The most important fossil sites that document the Cambrian Radiation are the **Chengjiang site** in China (for Early Cambrian fossils) and the **Burgess Shale** in British Columbia (for Middle Cambrian fossils) (**Fig. 21-9**).

According to geologists, seas gradually flooded the continents during the Cambrian period. In the **Ordovician period,** much land was covered by shallow seas, in which there was another burst of evolutionary diversification, although not as dramatic as the Cambrian Radiation. The Ordovician seas were inhabited by giant cephalopods, squidlike animals with straight shells 5 to 7 m (16 to 23 ft) long and 30 cm (12 in) in diameter. Coral reefs first appeared during this period, as did small, jawless, bony-armored

TABLE 21-1

Some Important Biological Events in Geologic Time

Eon	Era	Period	Epoch	Time	Some Important Biological Events
Phanerozoic	Cenozoic	Neogene	Holocene	0.01 (10,000 years ago)	Decline of some woody plants; rise of herbaceous plants; age of *Homo sapiens*
			Pleistocene	2	Extinction of some plant species; extinction of many large mammals at end
			Pliocene	5	Expansion of grasslands and deserts; many grazing animals
			Miocene	23	Flowering plants continue to diversify; diversity of songbirds and grazing mammals
		Paleogene	Oligocene	34	Spread of forests; apes appear; present mammalian families are represented
			Eocene	56	Flowering plants dominant; modern mammalian orders appear and diversify; modern bird orders appear
			Paleocene	66	Semitropical vegetation (flowering plants and conifers) widespread; primitive mammals diversify
	Mesozoic	Cretaceous		146	Rise of flowering plants; dinosaurs reach peak, then become extinct at end; toothed birds become extinct
		Jurassic		200	Gymnosperms common; large dinosaurs; first toothed birds
		Triassic		251	Gymnosperms dominant; ferns common; first dinosaurs; first mammals
	Paleozoic	Permian		299	Conifers diversify; cycads appear; modern insects appear; mammal-like reptiles; extinction of many invertebrates and vertebrates at end of Permian
		Carboniferous		359	Forests of ferns, club mosses, horsetails, and gymnosperms; many insect forms; spread of ancient amphibians; first reptiles
		Devonian		416	First forests; gymnosperms appear; many trilobites; wingless insects appear; fishes with jaws appear and diversify; amphibians appear
		Silurian		444	Vascular plants appear; coral reefs common; jawless fishes diversify; terrestrial arthropods
		Ordovician		488	Fossil spores of terrestrial plants (bryophytes?); invertebrates dominant; coral reefs appear; first fishes appear
		Cambrian		542	Bacteria and cyanobacteria; algae; fungi; age of marine invertebrates; first chordates
Proterozoic		Ediacaran		600 mya	Algae and soft-bodied invertebrates diversify
		Early Proterozoic		2500 mya	Eukaryotes evolve
Archaean				4600 mya to 2500 mya	Oldest known rocks; prokaryotes evolve; atmospheric oxygen begins to increase

Figure 21-8 Reconstruction of life in an Ediacaran sea

The organisms shown here are based on fossils from the Ediacara Hills of South Australia, although similar associations have been found in Ediacaran rocks from every continent except Antarctica. Shown are organisms that some scientists have interpreted as jellyfish, flatworms, algae, and paddle-shaped organisms similar in appearance to soft corals.

fishes called *ostracoderms* (█ Fig. 21-10). Lacking jaws, these fishes typically had round or slitlike mouth openings that may have sucked in small food particles from the water or scooped up organic debris from the bottom. Ordovician deposits also contain fossil spores of terrestrial (land-dwelling) plants, suggesting the colonization of land had begun.

During the **Silurian period,** jawless fishes diversified considerably, and jawed fishes first appeared. Definitive evidence of two life-forms of great biological significance appeared in the Silurian period: terrestrial plants and air-breathing animals. The evolution of plants allowed animals to colonize the land because plants provided the first terrestrial animals with food and shelter. All air-breathing land animals discovered in Silurian rocks were

arthropods—millipedes, spiderlike arthropods, and possibly centipedes. From an ecological perspective, the energy flow from plants to animals probably occurred via detritus, which is organic debris from decomposing organisms, rather than directly from living plant material. Millipedes eat plant detritus today, and spiders and centipedes prey on millipedes and other animals.

The **Devonian period** is frequently called the Age of Fishes. This period witnessed the explosive radiation of fishes with jaws, an adaptation that lets a vertebrate chew and bite. Armored *placoderms,* an extinct group of jawed fishes, diversified to exploit varied lifestyles (see Fig. 31-10b). Appearing in Devonian deposits are sharks and the two predominant types of bony fishes: lobe-finned fishes and ray-finned fishes, which gave rise to the major orders of modern fishes (see Chapter 31). Upper (more recent) Devonian sediments contain fossil remains of salamander-like amphibians (*labyrinthodonts*) that were often quite large, with short necks and heavy, muscular tails. Wingless insects also originated in the Late Devonian period.

The early vascular plants (plants with specialized tissue to conduct water and nutrients) diversified during the Devonian period in a burst of evolution that rivaled that of animals during the Cambrian Radiation. With the exception of flowering plants, all major plant groups appeared during the Devonian period. Forests of ferns, club mosses, horsetails, and seed ferns (an extinct group of ancient plants that had fernlike foliage but reproduced by forming seeds) flourished.

The **Carboniferous period** is named for the great swamp forests whose remains persist today as major coal deposits. Much of the land during this time was covered with low swamps filled with horsetails, club mosses, ferns, seed ferns, and gymnosperms (seed-bearing plants such as conifers) (█ Fig. 21-11). Amphibians, which underwent an **adaptive radiation** and exploited both aquatic and terrestrial ecosystems, were the dominant terrestrial carnivores of the Carboniferous period. Reptiles first appeared and diverged to form two major lines at this time. One line con-

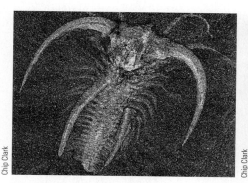

(a) *Marrella splendens* was a small arthropod.

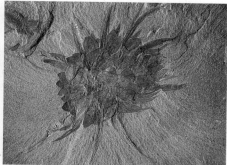

(b) *Wiwaxia* was a bristle-covered marine worm that was distantly related to earthworms. It had scaly armor and needlelike spines for protection.

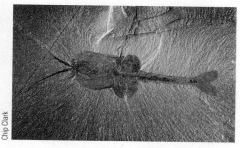

(c) *Waptia fieldensis* was a crustacean that may have been an ancestor of modern crustaceans such as shrimp.

Figure 21-9 Fossils from the Cambrian Radiation

All three of these fossils were discovered in the Burgess Shale in the Canadian Rockies of British Columbia.

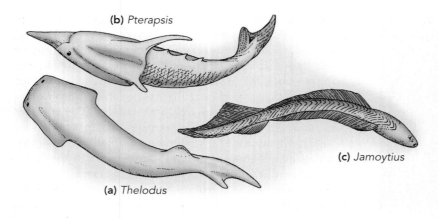

(b) *Pterapsis*

(c) *Jamoytius*

(a) *Thelodus*

Figure 21-10   Ostracoderms

Ostracoderms, primitive jawless fishes that lived in the Devonian period, ranged from 10 to 50 cm (4 to 20 in) in length.

sisted of mostly small and midsize insectivorous (insect-eating) lizards; this line later led to lizards, snakes, crocodiles, dinosaurs, and birds. The other reptilian line led to a diverse group of Permian and Early Mesozoic mammal-like reptiles. Two groups of winged insects, cockroaches and dragonflies, appeared in the Carboniferous period.

Amphibians continued in importance during the **Permian period,** but they were no longer the dominant carnivores in terrestrial ecosystems. During the Permian period, reptiles diversified and dominated both carnivorous and herbivorous terrestrial lifestyles. One important group of mammal-like reptiles, originating in the Permian and extending into the Mesozoic era, were the *therapsids,* a group that included the ancestor of mammals (see Fig. 31-24). During the Permian period, seed plants diversified and dominated most plant communities. Cone-bearing conifers were widespread, and cycads (plants with crowns of fernlike leaves and large, seed-containing cones) and ginkgoes (trees with fan-shaped leaves and exposed, fleshy seeds) appeared.

The greatest **mass extinction** of all time occurred at the end of the Paleozoic era, between the Permian and Triassic periods, 251 mya. More than 90% of all existing marine species became extinct at this time, as did more than 70% of the vertebrate genera living on land. There is also evidence of a major extinction of plants. Many causes for the Late Permian mass extinction have been suggested, from meteor impacts to global warming to changes in ocean chemistry. Regardless of cause, evidence suggests that the extinction occurred globally in a very compressed period, within a few hundred thousand years. This is an extremely short period in the geologic time scale and suggests that some sort of catastrophic event caused the mass extinction.

Field Museum of Natural History, Chicago, No. GE086638c

Figure 21-11   Reconstruction of a Carboniferous forest

Plants of this period included giant ferns, horsetails, club mosses, seed ferns, and early gymnosperms.

Figure 21-12 Ichthyosaurs

Grendelius was an ichthyosaur that superficially resembled a shark or porpoise. It was about 4 m (13 ft) long.

Dinosaurs and other reptiles dominated the Mesozoic era

The **Mesozoic era** began about 251 mya and lasted some 185 million years. It is divided into the Triassic, Jurassic, and Cretaceous periods. Fossil deposits from the Mesozoic era occur worldwide. Notable sites include the **Yixian Formation** in northeastern China, the Solnhofen Limestone in Germany, northwestern Patagonia in Argentina, the Sahara in central Niger, the Badlands in South Dakota, and other sites in western North America.

The outstanding feature of the Mesozoic era was the origin, differentiation, and ultimately the extinction of a large variety of reptiles. For this reason, the Mesozoic era is commonly called the Age of Reptiles. Most of the modern orders of insects appeared during the Mesozoic era. Snails and bivalves (clams and their relatives) increased in number and diversity, and sea urchins reached their peak diversity. From a botanical viewpoint, the Mesozoic era was dominated by gymnosperms until the Mid-Cretaceous period, when the flowering plants first diversified.

During the **Triassic period,** reptiles underwent an adaptive radiation leading to the formation of many groups. On land, the dominant Triassic groups were the mammal-like therapsids, which ranged from small-sized insectivores (insect-eating reptiles) to moderately large herbivores (plant-eating reptiles), and a diverse group of *thecodonts,* early "ruling reptiles," that were primarily carnivores. Thecodonts are the ancestral archosaurans that gave rise to crocodilians, flying reptiles, dinosaurs, and birds (see Fig. 20-18a).

In the ocean, several important marine reptile groups, the plesiosaurs and ichthyosaurs, appeared in the Triassic. *Plesiosaurs* had bodies up to 15 m (about 49 ft) long and paddlelike fins. *Ichthyosaurs* had body forms superficially resembling those of sharks or porpoises, with short necks, large dorsal fins, and shark-type tails (▌Fig. 21-12). Ichthyosaurs had very large eyes, which may have helped them see at diving depths of 500 m (1650 ft) or more.

Pterosaurs, the first flying reptiles, appeared and underwent considerable diversification during the Mesozoic era (▌Fig. 21-13). This group produced some quite spectacular forms, most notably the giant *Quetzalcoatlus,* known from fragmentary Cretaceous fossils in Texas to have had a wingspan of 11 to 15 m (36 to 49 ft). Pterosaur wings were leathery membranes of skin that were supported by an elongated fourth finger bone. Claws extended from the other finger bones.

The first mammals to appear in the Triassic period were small insectivores that evolved from the mammal-like therapsids. Mammals diversified into a variety of mostly small, nocturnal insectivores during the remainder of the Mesozoic, with marsupial and placental mammals appearing later in the Mesozoic era.

During the **Jurassic** and **Cretaceous periods,** crocodiles, lizards, snakes, and birds appeared, and the dinosaurs diversified dramatically (▌Figure 21-14). One group of lizards, the *mosasaurs,* were large, voracious marine predators during the Late Cretaceous period. The mosasaurs, which are now extinct, attained lengths of 10 m (33 ft) or more.

There were two main groups of dinosaurs: the *saurischians,* with pelvic bones similar to those of modern-day lizards, and the *ornithischians,* with pelvic bones similar to those of birds (▌Fig. 21-15). Some saurischians were fast, bipedal (walking on two feet) forms ranging from those the size of a dog to the ultimate representatives of this group, the gigantic carnivores of the Cretaceous period—*Tyrannosaurus, Giganotosaurus,* and *Carcharodontosaurus.* Other saurischians were huge, quadrupedal (walking on four feet) dinosaurs that ate plants. Some of these were the largest terrestrial animals that have ever lived, including *Argentinosaurus,* with an estimated length of 30 m (98 ft) and an estimated weight of 72 to 90 metric tons (80 to 100 tons).

The other group of dinosaurs, the ornithischians, was entirely herbivorous. Although some ornithischians were bipedal, most

Figure 21-13 Pterosaurs

Shown are *Peteinosaurus* (*left*), with a wingspan of 60 cm (2 ft), and *Eudimorphodon* (*right foreground*), with a wingspan of 75 cm (2.5 ft). Both species had long, sharp teeth for catching insects or fishes while flying.

Figure 21-14 Dinosaurs

Three *Deinonychus* dinosaurs attack a larger *Tenontosaurus*. The name *Deinonychus* means "terrible claw" and refers to the enlarged, sharp claw on the second digit of its hind feet. *Deinonychus* dinosaurs were small (3 m, or 10 ft in length) but fearsome predators that hunted in packs. *Tenontosaurus* adults were as long as 7.5 m (24 ft).

were quadrupedal. Some had no front teeth and possessed stout, horny, birdlike beaks. In some species these beaks were broad and ducklike, hence the common name *duck-billed dinosaurs.* Other ornithischians had great armor plates, possibly as protection against carnivorous saurischians. *Ankylosaurus,* for example, had a broad, flat body covered with armor plates (actually bony scales embedded in the skin) and large, laterally projecting spikes.

Over the past few decades scientists have reconsidered many traditional ideas about dinosaurs and no longer think they were all cold-blooded, slow-moving monsters living in swamps. Recent evidence suggests that at least some dinosaurs were warm-blooded, agile, and able to move extremely fast. Many dinosaurs appear to have had complex social behaviors, including courtship rituals and parental nurturing of their young. Some species lived in social groups and hunted in packs.

Birds appeared by the late Jurassic period, and fossil evidence indicates they evolved directly from saurischian dinosaurs (see *Focus On: The Origin of Flight in Birds*). *Archaeopteryx,* the oldest known bird in the fossil record, lived about 150 mya (see Fig. 31-22a). It was about the size of a pigeon and had rather feeble wings that it used to glide rather than actively fly. Although *Archaeopteryx* is considered a bird, it had many reptilian features, including a mouthful of teeth and a long, bony tail.

Thousands of well-preserved bird fossils have been found in Early Cretaceous deposits in China. These include *Sinornis,* a 135-million-year-old sparrow-sized bird capable of perching, and the magpie-sized *Confuciusornis,* the earliest known bird

with a toothless beak. *Confuciusornis* may date back as far as 142 mya.

At the end of the Cretaceous period, 66 mya, dinosaurs, pterosaurs, and many other animals abruptly became extinct (see Fig. 20-18b). Many gymnosperms, with the exception of conifers, also perished. Evidence suggests that a catastrophic collision of a large extraterrestrial body with Earth dramatically changed the climate at the end of the Cretaceous period. Part of the evidence is a thin band of dark clay, with a high concentration of iridium, located between Mesozoic and Cenozoic sediments at more than 200 sites around the world. Iridium is rare on Earth but abundant in meteorites. (The force of the impact would have driven the iridium into the atmosphere, to be deposited later on the land by precipitation.)

The Chicxulub crater, buried under the Yucatán Peninsula in Mexico, is the apparent site of this collision. The impact produced giant tsunamis (tidal waves) that deposited materials from the extraterrestrial body around the perimeter of the Gulf of Mexico, from Alabama to Guatemala. It may have caused global forest fires and giant smoke and dust clouds that lowered global temperatures for many years.

Although scientists widely accept that a collision with an extraterrestrial body occurred 66 mya, they have reached no consensus about the effects of such an impact on organisms. The extinction of many marine organisms at or immediately after the time of the impact was probably the result of the environmental upheaval that the collision produced. However, many clam spe-

The two orders of dinsosaurs are distinguished primarily by differences in their pelvic bones.

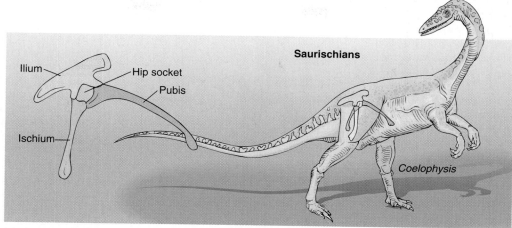

(a) The saurischian pelvis. Note the opening (hip socket), a trait possessed by no quadrupedal vertebrates other than dinosaurs.

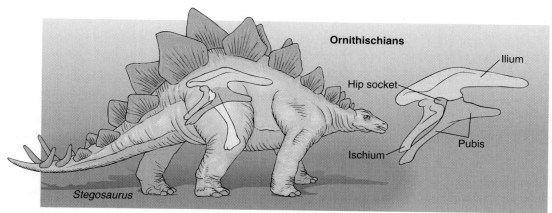

(b) The ornithischian pelvis also has the hole in the hip socket but differs from the saurischian pelvis in that it has a backward-directed extension of the pubis.

Figure 21-15 Saurischian and ornithischian dinosaurs

In each dinosaur figure, the pale yellow femur is shown relative to the pelvic bone.

cies associated with the mass extinction at the end of the Cretaceous period seem to have gone extinct *before* the impact, suggesting that other factors caused some of the massive extinctions occurring then.

The Cenozoic era is the Age of Mammals

With equal justice the **Cenozoic era** could be called the Age of Mammals, the Age of Birds, the Age of Insects, or the Age of Flowering Plants. This era is marked by the appearance of all these forms in great variety and numbers of species. The Cenozoic era extends from 66 mya to the present. It is subdivided into two periods: the **Paleogene period,** encompassing some 43 million years, and the **Neogene period,** which covers the last 23 million years. The Paleogene period is subdivided into three epochs, named from earliest to latest: Paleocene, Eocene, and Oligocene. The Neogene period is subdivided into four epochs: Miocene, Pliocene, Pleistocene, and Holocene.

Flowering plants, which arose during the Cretaceous period, continued to diversify during the Cenozoic era. During the Paleocene and Eocene epochs, fossils indicate that tropical and semitropical plant communities extended to relatively high latitudes. Palms, for example, are found in Eocene deposits in Wyoming. Later in the Cenozoic era, there is evidence of more

THE ORIGIN OF FLIGHT IN BIRDS

The evolution of birds is arguably one of the most interesting chapters in Earth's history of life. Given the substantial fossil evidence, most paleontologists have concluded that the ancestors of birds were dinosaurs, specifically the *dromaeosaurs*, a group of ground-dwelling, bipedal theropods (carnivorous saurischians). Beginning in 1997, paleontologists made several discoveries of fossil dinosaurs with feathers, indicating that feathers appeared before birds. Scientists think that feathers evolved as one or a series of evolutionary novelties, or **preadaptations.** The first feathers may have provided thermal insulation but were subsequently modified for flight (see Chapter 20).

Once dinosaurs and early birds had feathers, how did they fly? Did tree-dwelling animals glide as an intermediate step in bird flight, or did ground-dwelling animals flap their wings and run to provide thrust and lift for a takeoff? The question about how flight originated in birds has intrigued biologists for more than a century. In 1915, U.S. zoologist William Beebe hypothesized that the ancestors of birds were probably tree-dwelling gliders that had feathers on all four limbs.* Because no fossil evidence supported his suggestion, scientists did not consider it seriously at the time.

Almost a century later, in 2003, a group of Chinese paleontologists announced the discovery of two nearly complete fossils

*W. H. Beebe, "A Tetrapteryx Stage in the Ancestry of Birds," *Zoologica*, Vol. 2, 1915.

An artist's interpretation of *Microraptor gui.*

Portia Sloan

of the organism that Beebe had hypothesized.† The fossils of a small, feathered dromaeosaur dinosaur were found in Liaoning Province in northeastern China. The dinosaur, *Microraptor gui*, had feathers on both forelimbs and hind limbs as well as on its long tail (see figure). It was small— 77 cm in length, including the tail—and appeared to be adapted to life in the trees.

The feathers on *M. gui* were similar to those of modern-day birds. Downy feathers covered the body, and each limb had about 12 "primary" flight feathers and about 18 shorter, "secondary" feathers. *Microraptor gui*'s flight feathers were asymmetrical, a characteristic associated

†X. Xu, Z. Zhou, X. Wang, X. Kuang, F. Zhang, and X. Du, "Four-Winged Dinosaurs from China," *Nature*, Vol. 421, Jan. 23, 2003.

with flight or gliding in modern birds. The primary and secondary feathers followed a similar pattern on both the forelimbs and hind limbs, and this pattern resembles that on modern birds. In part because *M. gui*'s breastbone was not structured to attach large flight muscles, the dinosaur probably glided rather than flapped its wings.

Microraptor gui, which is about 126 million years old, is not a direct ancestor of birds. Birds had already evolved when *M. gui* existed. The earliest known bird, *Archaeopteryx*, lived about 145 mya and therefore predates *M. gui* by about 20 million years. *Microraptor gui* is considered to have evolved from a *basal member*—that is, an earlier evolutionary branch—of the most recent ancestor of *Archaeopteryx* and other birds. Like *M. gui*, earlier feathered dinosaurs may have been four-winged organisms that lived and glided in the trees.

During the course of bird evolution, the feathered hind limbs may have become reduced and eventually lost. However, an alternative hypothesis is that feathered hind limbs may have been a failed evolutionary experiment restricted to dromaeosaurs and not important as an intermediate step in bird evolution. Further analysis of *M. gui* and future discoveries of older dromaeosaur fossils may shed some light on the importance of feathered hind limbs.

Microraptor gui has given paleontologists and biologists much to consider in the evolutionary transition from dinosaurs to birds. Scientists will continue to study and debate the evolution of flight in birds for many years.

---

open habitats. Grasslands and savannas spread throughout much of North America during the Miocene epoch, with deserts developing later in the Pliocene and Pleistocene epochs. During the Pleistocene epoch, plant communities changed dynamically in response to the fluctuating climates associated with the multiple advances and retreats of continental glaciers.

During the Eocene epoch, there was an explosive radiation of birds, which acquired adaptations for different habitats. Paleontologists hypothesize that the jaws and beak of the flightless giant bird *Diatryma*, for example, may have been adapted primarily for crushing and slicing vegetation in Eocene forests, marshes, and grasslands. Other paleontologists hypothesize that these giant birds were carnivores that killed or scavenged mammals and other vertebrates (▌Fig. 21-16).

During the Paleocene epoch, an explosive radiation of primitive mammals occurred. Most of these were small forest dwellers that are not closely related to modern mammals. During the Eocene epoch, mammals continued to diverge, and all the modern orders first appeared. Again, many of the mammals were small, but there were also some larger herbivores.

During the Oligocene epoch, many modern families of mammals evolved, including the first apes in Africa. Many lineages showed adaptations that suggest a more open type of habitat, such as grassland or savanna. Many mammals were larger and had longer legs for running, specialized teeth for chewing coarse vegetation or for preying on animals, and increases in their relative brain sizes. The *indricotheres*, for example, are extinct relatives of the rhinoceros. These mammals, which lived on the

Figure 21-16 A bird from the Eocene epoch

The flightless bird *Diatryma,* which stood 2.1 m (7 ft) tall and weighed about 175 kg (385 lb), may have been a herbivore or a formidable predator. In this picture, *Diatryma* has captured a small, horselike perissodactyl.

Figure 21-17 A mammal from the Oligocene epoch

Paraceratherium was an indricothere, a hornless relative of the rhinoceros. This huge land mammal was about 8 m (26 ft) long and weighed about 15 to 20 tons. It probably browsed on leaves and branches of deciduous trees, much as a modern-day giraffe does.

grassless plains of Eurasia, became progressively larger during the Oligocene epoch (Fig. 21-17).

Human ancestors appeared in Africa during the Late Miocene and Early Pliocene epochs. *Homo,* the genus to which humans belong, appeared approximately 2.3 mya. (Primate evolution, including human evolution, is discussed in Chapter 22.)

The Pliocene and Pleistocene epochs witnessed a spectacular North and South American large-mammal fauna, including mastodons, saber-toothed cats, camels, giant ground sloths, and giant armadillos. However, many of the large mammals became extinct at the end of the Pleistocene epoch. This extinction was possibly due to climate change—the Pleistocene epoch was marked by several ice ages—and/or to the influence of humans, which had spread from Africa to Europe and Asia, and later to

North and South America by crossing a land bridge between Siberia and Alaska. Archaeological evidence indicates that this mass extinction event was concurrent with the appearance of human hunters.

Review

- What is the correct order of appearance in the fossil record, starting with the earliest: eukaryotic cells, multicellular organisms, prokaryotic cells?
- What is the correct order of appearance in the fossil record, starting with the earliest: reptiles, mammals, amphibians, fishes?
- What is the correct order of appearance in the fossil record, starting with the earliest: flowering plants, ferns, gymnosperms?

SUMMARY WITH KEY TERMS

Learning Objectives

1   Describe the conditions that geologists think existed on early Earth (page 448).

- Biologists generally agree that life originated from nonliving matter by **chemical evolution.** Although the origin of life is very difficult to test experimentally, hypotheses about chemical evolution are testable.
- Four requirements for chemical evolution are (1) the absence of oxygen, which would have reacted with and oxidized abiotically produced organic molecules; (2) energy to form organic molecules; (3) chemical building

blocks, including water, minerals, and gases present in the atmosphere, to form organic molecules; and (4) sufficient time for molecules to accumulate and react.

ThomsonNOW™   **Learn more about conditions on early Earth by clicking on the figure in ThomsonNOW.**

2   Contrast the prebiotic soup hypothesis and the iron–sulfur world hypothesis (page 448).

- During chemical evolution, small organic molecules formed spontaneously and accumulated. The **prebiotic** soup hypothesis proposes that organic molecules formed near Earth's surface in a "sea of organic soup" or on rock or clay surfaces. The **iron–sulfur world hypothesis** suggests that organic molecules were produced at **hydrothermal vents,** cracks in the deep ocean floor.

ThomsonNOW™ **See the Miller–Urey experiment unfold by clicking on the figure in ThomsonNOW.**

3 Outline the major steps hypothesized to have occurred in the origin of cells (page 450).
  ▮ After small organic molecules formed and accumulated, macromolecules assembled from the small organic molecules. Macromolecular assemblages called **protobionts** formed from macromolecules. Cells arose from the protobionts.
  ▮ According to a model known as the **RNA world,** RNA was the first informational molecule to evolve in the progression toward a self-sustaining, self-reproducing cell. Natural selection at the molecular level eventually resulted in the information sequence DNA ⟶ RNA ⟶ protein.

4 Explain how the evolution of photosynthetic autotrophs affected both the atmosphere and other organisms (page 450).
  ▮ The first cells were prokaryotic **heterotrophs** that obtained organic molecules from the environment. They were almost certainly **anaerobes.** Later, **autotrophs**—organisms that produce their own organic molecules by photosynthesis—evolved.
  ▮ The evolution of oxygen-generating photosynthesis ultimately changed early life. The accumulation of molecular oxygen in the atmosphere permitted the evolution of **aerobes,** organisms that could use oxygen for a more efficient type of cellular respiration.

5 Describe the hypothesis of serial endosymbiosis (page 450).
  ▮ Eukaryotic cells arose from prokaryotic cells. According to the hypothesis of **serial endosymbiosis,** certain eukaryotic organelles (mitochondria and chloroplasts) evolved from

prokaryotic **endosymbionts** incorporated within larger prokaryotic hosts.

ThomsonNOW™ **Learn more about endosymbiosis by clicking on the figure in ThomsonNOW.**

6 Briefly describe the distinguishing organisms and major biological events of the Ediacaran period and the Paleozoic, Mesozoic, and Cenozoic eras (page 455).
  ▮ Life at the beginning of the **Proterozoic eon** (2500 mya to 542 mya) consisted of prokaryotes. About 2.2 bya the first eukaryotic cells appeared. The **Ediacaran period,** from 600 mya to 542 mya, is the last period of the Proterozoic eon. Ediacaran fossils are the oldest known fossils of multicellular animals. Ediacaran fauna were small, soft-bodied invertebrates.
  ▮ During the **Paleozoic era,** which began about 542 mya and lasted approximately 291 million years, all major groups of plants, except flowering plants, and all animal phyla appeared. Fishes and amphibians flourished, and reptiles appeared. The greatest mass extinction of all time occurred at the end of the Paleozoic era, 251 mya. More than 90% of marine species and 70% of land-dwelling vertebrate genera became extinct, as well as many plant species.
  ▮ The **Mesozoic era** began about 251 mya and lasted some 185 million years. Flowering plants appeared, and reptiles diversified. Dinosaurs, which descended from early reptiles, dominated. Insects flourished, and birds and early mammals appeared. At the end of the Cretaceous period, 66 mya, many species abruptly became extinct. A collision of a large extraterrestrial body with Earth may have resulted in dramatic climate changes that played a role in this mass extinction.
  ▮ In the **Cenozoic era,** which extends from 66 mya to the present, flowering plants, birds, insects, and mammals diversified greatly. Human ancestors appeared in Africa during the Late Miocene and Early Pliocene epochs.

TEST YOUR UNDERSTANDING

1. Energy, the absence of oxygen, chemical building blocks, and time were the requirements for (a) chemical evolution (b) biological evolution (c) the Cambrian Radiation (d) the mass extinction episode at the end of the Cretaceous period (e) directed evolution

2. Protobionts are (a) spontaneously formed in hydrothermal vents in the ocean floor (b) heterotrophs that obtain the organic molecules they need from the environment (c) assemblages of abiotically produced organic polymers that resemble living cells in several ways (d) autotrophs that use sunlight to split hydrogen sulfide (e) fossilized mats of cyanobacteria

3. Many scientists think that _____ was the first information molecule to evolve. (a) DNA (b) RNA (c) a protein (d) an amino acid (e) a lipid

4. The first cells were probably (a) heterotrophs (b) autotrophs (c) anaerobes (d) a and c (e) b and c

5. According to the hypothesis of serial endosymbiosis, (a) life originated from nonliving matter (b) the pace of evolution quickened at the start of the Cambrian period (c) chloroplasts, mitochondria, and possibly other organelles origi-

nated from intimate relationships among prokaryotic cells (d) banded iron formations reflect the buildup of sufficient oxygen in the atmosphere to oxidize iron at Earth's surface (e) the first photosynthetic organisms appeared 3.1 bya to 3.5 bya

6. During the Early _____, life consisted of prokaryotic cells, but by the end of this geologic time span, multicellular eukaryotic organisms had evolved. (a) Cenozoic era (b) Paleozoic era (c) Mesozoic era (d) Archaean eon (e) Proterozoic eon

7. Geologists divide the eons into (a) periods (b) epochs (c) eras (d) millennia (e) none of the preceding

8. Ediacaran fossils (a) are the oldest known fossils of multicellular animals (b) come from the Burgess Shale in British Columbia (c) contain remains of large salamander-like organisms (d) are the oldest fossils of early vascular plants (e) contain a high concentration of iridium

9. The correct chronological order of geologic eras, starting with the oldest, is (a) Paleozoic, Cenozoic, Mesozoic (b) Mesozoic, Cenozoic, Paleozoic (c) Mesozoic, Paleozoic, Cenozoic

(d) Paleozoic, Mesozoic, Cenozoic (e) Cenozoic, Paleozoic, Mesozoic

10. The time of greatest evolutionary diversification in the history of life occurred during the (a) Cambrian period (b) Ordovician period (c) Silurian period (d) Carboniferous period (e) Permian period

11. The greatest mass extinction episode in the history of life occurred at what boundary? (a) Pliocene–Pleistocene (b) Permian–Triassic (c) Mesozoic–Cenozoic (d) Cambrian–Ordovician (e) Triassic–Jurassic

12. The Mesozoic era is divided into which three periods? (a) Cambrian, Ordovician, and Silurian (b) Devonian, Carboniferous, and Permian (c) Triassic, Jurassic, and Cretaceous (d) Cretaceous, Paleogene, and Neogene (e) Pliocene, Pleistocene, and Holocene

13. The Age of Reptiles corresponds to the (a) Paleozoic era (b) Mesozoic era (c) Cenozoic era (d) Pleistocene epoch (e) Permian period

14. Evidence exists that a catastrophic collision between Earth and a large extraterrestrial body occurred 66 mya, resulting in the extinction of (a) worms, mollusks, and soft-bodied arthropods (b) jawless ostracoderms and jawed placoderms (c) dinosaurs, pterosaurs, and many gymnosperm species (d) mastodons, saber-toothed cats, and giant ground sloths (e) ferns, horsetails, and club mosses

15. Flowering plants and mammals diversified and became dominant during the (a) Paleozoic era (b) Mesozoic era (c) Cenozoic era (d) Devonian period (e) Cambrian period

CRITICAL THINKING

1. **Evolution Link.** If you were studying how protobionts evolved into cells and you developed a protobiont that was capable of self-replication, would you consider it a living cell? Why or why not?

2. **Evolution Link.** If living cells were produced in a test tube from nonbiological components by chemical processes, would this accomplishment prove that life evolved in a similar manner billions of years ago? Why or why not?

3. **Evolution Link.** Why did the evolution of multicellular organisms such as plants and animals have to be preceded by the evolution of oxygen-producing photosynthesis?

Additional questions are available in ThomsonNOW at www.thomsonedu.com/login

The Evolution of Primates

Des Bartlett/Photo Researchers, Inc.

Studying human evolution. The late Drs. Mary and Louis Leakey studied fossil teeth from *Australopithecus boisei,* an early hominid (human) that lived in Africa. This photo was taken in Olduvai Gorge in Tanzania. The Leakey family has contributed much of what we know about human evolution. Son Richard, daughter-in-law Meave, and granddaughter Louise have followed in the footsteps of Mary and Louis.

KEY CONCEPTS

Humans are classified in the order Primates, along with lemurs, tarsiers, monkeys, and apes. This classification is based on close evolutionary ties.

The study of living primates provides clues to help scientists reconstruct the adaptations and lifestyles of early primates, some of which were ancestors of humans.

Fossil evidence indicates that the earliest hominids (human ancestors) evolved in Africa and shared many features with their apelike ancestors.

The human brain did not begin to enlarge to its present size and complexity until long after human ancestors had evolved bipedal locomotion.

Human culture began when human ancestors started making stone tools.

Twelve years after Darwin wrote *On the Origin of Species by Natural Selection,* he published another controversial book, *The Descent of Man,* which addressed human evolution. In this book, Darwin hypothesized that humans and apes share a common ancestry. For nearly a century after Darwin's studies, fossil evidence of human ancestry remained fairly incomplete. However, research during the last several decades, especially in Africa, has yielded a rapidly accumulating set of fossils that gives an increasingly clear answer to the question, "Where did we come from?" (see photograph).

Fossil evidence from **paleoanthropology,** the study of human evolution, has allowed scientists to infer not only the structure but also the habits of early humans and other primates. Teeth and bones are the main fossil evidence that paleoanthropologists study. Much information can be obtained by studying teeth, which have changed dramatically during the course of primate and human evolution. Because tooth enamel is more mineralized (harder) than bone, teeth are more likely to fossilize. The teeth of each primate species, living or extinct, are distinctive enough to identify the species, approximate age, diet, and even sex of the individual.

Paleontologists hypothesize, based on fossil evidence, that the first primates descended from small, shrewlike **placental mammals** that lived in trees and ate insects. (Placental mammals are the largest and most successful group of mammals. They possess a placenta, an organ that exchanges materials between the mother and the embryo/fetus developing in the uterus.) Many traits of the 233 living primate species are related to their **arboreal** (tree-dwelling) past. This chapter focuses on humans and their ancestors, who differ from most other primates because they did not remain in the trees but instead adapted to a terrestrial way of life. ■

PRIMATE ADAPTATIONS

Learning Objective

1 Describe the structural adaptations that primates have for life in treetops.

Humans and other **primates**—including lemurs, tarsiers, monkeys, and apes—are mammals that share such traits as flexible hands and feet with five digits, a strong social organization, and front-facing eyes, which permit depth perception. Mammals (class Mammalia) evolved from mammal-like reptiles known as *therapsids* more than 200 million years ago (mya), during the Mesozoic era (see Fig. 31-24). These early mammals remained a minor component of life on Earth for almost 150 million years and then rapidly diversified during the Cenozoic era (from 66 mya to the present).

Fossil evidence indicates that the first primates with traits characteristic of modern primates appeared by the Early Eocene epoch about 56 mya. These early primates had digits with nails, and their eyes were directed somewhat forward on the head.

Several novel adaptations evolved in early primates that allowed them to live in trees. One of the most significant primate features is each limb has five highly flexible digits: four lateral digits (fingers) plus a partially or fully opposable first digit (thumb and, in many primates, big toe; ▌Fig. 22-1). An **opposable thumb** positions the fingers opposite the thumb, enabling primates to grasp objects such as tree branches with precision. Nails (instead of claws) provide a protective covering for the tips of the digits, and the fleshy pads at the ends of the digits are sensitive to touch. Another arboreal feature is long, slender limbs that rotate freely at the hips and shoulders, giving primates full mobility to climb and search for food in the treetops.

Having eyes located in the front of the head lets primates integrate visual information from both eyes simultaneously; they have *stereoscopic* (three-dimensional) vision important in judging distance and in depth perception. Stereoscopic vision is vital in an arboreal environment, especially for species that leap from branch to branch, because an error in depth perception may cause a fatal fall. In addition to having sharp sight, primates hear acutely.

Primates share several other characteristics, including a relatively large brain size. Biologists have suggested that the increased sensory input associated with primates' sharp vision and greater agility favored the evolution of larger brains. Primates are generally very social and intelligent animals that reach sexual maturity

Key Point

Primates have five grasping digits at the end of each limb, and the thumb or first toe is often partially or fully opposable.

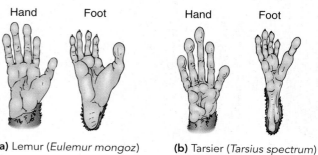

(a) Lemur (*Eulemur mongoz*)

(b) Tarsier (*Tarsius spectrum*)

(c) Woolly spider monkey (*Brachyteles arachnoides*)

(d) Gorilla (*Gorilla gorilla*)

Figure 22-1 Right hands and feet of selected primates

(*Figures not drawn to scale.*) (Adapted from A. H. Schultz, *The Life of Primates*, Weidenfeld & Nicholson, London, 1969.)

relatively late in life. They typically have long life spans. Females usually bear one offspring at a time; the baby is helpless and requires a long period of nurturing and protection.

Review

■ How are primate hands and feet adapted to an arboreal existence?

■ Why is the location of primate eyes in the front of the head an important adaptation for an arboreal existence?

Several kinds of data support the hypothesis that chimpanzees are the closest living relative of humans.

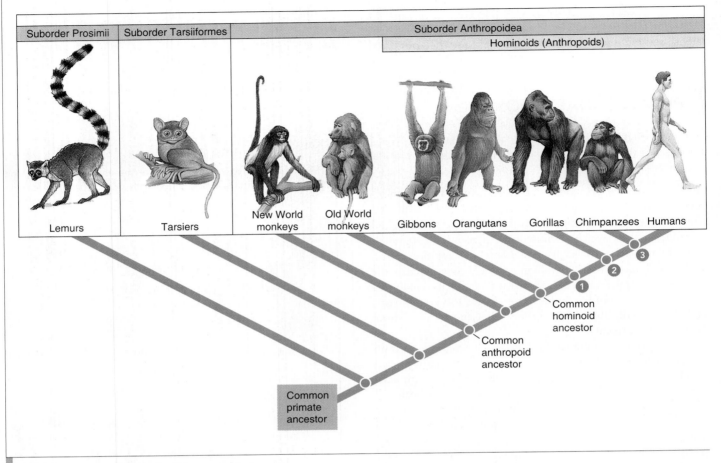

Figure 22-2 *Animated* Primate evolution

This branching diagram, called a cladogram, shows evolutionary relationships among living primates, based on current scientific evidence. The nodes (*circles*) represent branch points where a species splits into two or more lineages. ❶ The divergence of orangutans from the ape/ hominid line occurred some 12 mya to 16 mya. ❷ Gorillas separated from the chimpanzee/hominid line an estimated 8 mya, and ❸ the hominid (human) lineage diverged from that of chimpanzees about 6 mya. (*Figures not drawn to scale.*)

PRIMATE CLASSIFICATION

Learning Objectives

2 List the three suborders of primates, and give representative examples of each.

3 Distinguish among anthropoids, hominoids, and hominids.

Now that we have surveyed the general characteristics of primates, let us look at how they are classified. Many biologists currently divide the order Primates into three groups, or suborders (❙ Fig. 22-2). The suborder Prosimii includes lemurs, galagos, and lorises; the suborder Tarsiiformes includes tarsiers; and the suborder Anthropoidea includes **anthropoids** (monkeys, apes, and humans).

All lemurs are restricted to the island of Madagascar off the coast of Africa. Because of extensive habitat destruction and hunting, they are very endangered. Lorises, which live in tropical areas of Southeast Asia and Africa, resemble lemurs in many respects, as do galagos, which live in sub-Saharan Africa. Lemurs, lorises, and galagos have retained several early mammalian features, such as elongated, pointed faces.

Tarsiers are found in rain forests of Indonesia and the Philippines (❙ Fig. 22-3). They are small primates (about the size of a small rat) and are very adept leapers. These nocturnal primates resemble anthropoids in a number of ways, including their shortened snouts and forward-pointing eyes.

Suborder Anthropoidea includes monkeys, apes, and humans

Anthropoid primates arose during the Middle Eocene epoch, at least 45 mya. Several different fossil anthropoids have been identified, from Asia and North Africa, and there is a growing consensus about the relationships of these fossil groups to one another and to living anthropoids. Evidence indicates that anthropoids originated in Africa or Asia. The oldest known anthropoid fossils, such as 42-million-year-old *Eosimias,* have been found in China and Myanmar. Given details about their dentition and the few bones that have been discovered, scientists infer that *Eosimias* and other ancestral anthropoids were small, insect-eating arboreal primates that were active during the day. Once they evolved, anthropoids quickly spread throughout Europe, Asia, and Africa and arrived in South America much later. (Paleoanthropologists date the oldest known South American primate, *Branisella,* from Bolivia, at 26 million years old.)

One significant difference between anthropoids and other primates is the size of their brains. The cerebrum, in particular, is more developed in monkeys, apes, and humans, where it functions as a highly complex center for learning, voluntary movement, and interpretation of sensation.

Monkeys are generally diurnal (active during the day) tree dwellers. They tend to eat fruit and leaves, with nuts, seeds, buds, insects, spiders, birds' eggs, and even small vertebrates playing a smaller part in their diets. The two main groups of monkeys, New World monkeys and Old World monkeys, are named for the hemispheres where they diversified. Monkeys in South and Central America are called New World monkeys, whereas monkeys in Africa, Asia, and Europe are called Old World monkeys. New and Old World monkeys have been evolving separately for tens of millions of years.

One of the most important unanswered questions in anthropoid evolution concerns *how* monkeys arrived in South America. Africa and South America had already drifted apart (see Fig. 18-15), so the ancestors of New World monkeys may have rafted from Africa to South America on floating masses of vegetation. The South Atlantic Ocean would have been about half as wide as it is today, and islands could have provided "stepping stones." Alternatively, the ancestors of New World monkeys may have dispersed from Asia to North America to South America. Once established in the New World, these monkeys rapidly diversified.

New World monkeys are restricted to Central and South America and include marmosets, capuchins, howler monkeys, squirrel monkeys, and spider monkeys. New World monkeys are arboreal, and some have long, slender limbs that permit easy movement in the trees (Fig. 22-4a). A few have **prehensile tails** capable of wrapping around branches and serving as fifth limbs. Some New World monkeys have shorter thumbs, and in certain cases the thumbs are totally absent. Their facial anatomy differs from that of the Old World monkeys; New World monkeys have flattened noses with the nostrils opening to the side. They live in groups and engage in complex social behaviors.

Old World monkeys are distributed in tropical parts of Africa and Asia. In addition to baboons and macaques (pronounced

Figure 22-3 A tarsier

The huge eyes of the tarsier (*Tarsius bancanus*) help it find insects, lizards, and other prey when it hunts at night. When a tarsier sees an insect, it pounces on it and grasps the prey with its hands. Tarsiers live in the rain forests of Indonesia and the Philippines.

"muh-kacks'"), the Old World group includes guenons, mangabeys, langurs, and colobus monkeys. Most Old World monkeys are arboreal, although some, such as baboons and macaques, spend much of their time on the ground (Fig. 22-4b). The ground dwellers, which are **quadrupedal** (four footed; they walk on all fours), arose from arboreal monkeys. None of the Old World monkeys has a prehensile tail, and some have extremely short tails. They have a fully opposable thumb, and unlike the New World monkeys, their nostrils are closer together and directed downward. Old World monkeys are intensely social animals.

Apes are our closest living relatives

Old World monkeys shared a common ancestor with the **hominoids,** a group composed of apes and **hominids** (humans and their ancestors). A fairly primitive anthropoid, discovered in Egypt, was named *Aegyptopithecus* (Fig. 22-5a). A cat-sized, forest-dwelling arboreal monkey with a few apelike characteristics, *Aegyptopithecus* lived during the Oligocene epoch, approximately 34 mya.

Fossil evidence indicates that apes and Old World monkeys had diverged between 23 mya and 25 mya, although a 2004 genetic analysis of living apes and monkeys suggests that this divergence may have occurred even earlier. During the Miocene epoch as many as 100 species of apes lived in Africa, Asia, and Europe.

Paleoanthropologists have discovered the oldest fossils with hominoid features in East Africa, mostly in Kenya. *Proconsul*, for example, appeared early in the Miocene epoch, about 20 mya (❚ Fig. 22-5b). It had a larger brain than that of monkeys, apelike teeth and diet (fruits), but a monkeylike body. At least 30 other early hominoid species lived during the Miocene epoch, but most of them became extinct and were not the common ancestor of modern apes and humans.

Miocene fossils of forest-dwelling, chimpanzee-sized apes called **dryopithecines**, which lived about 15 mya, are of special interest because this hominoid lineage may have given rise to modern apes as well as to humans (❚ Fig. 22-5c). The dryopithecines, such as *Dryopithecus, Kenyapithecus,* and *Morotopithecus,* were distributed widely across Europe, Africa, and Asia. As the climate gradually cooled and became drier, their range became more limited. These apes had highly modified bodies for swinging through the branches of trees, although there is also evidence that some of them may have left the treetops for the ground as dense forest gradually changed into open woodland. Many

(a) New World monkey. The white-faced monkey (*Cebus capucinus*) has a prehensile tail and a flattened nose with nostrils directed to the side.

(b) Old World monkey. The Anubis baboon (*Papio anubis*) is native to Africa. Note that its nostrils are directed downward.

❚ **Figure 22-4**   New World and Old World monkeys

questions about the relationships among the various early apes have been generated by the discovery of these and other Miocene hominoids. As future fossil finds are evaluated, they may lead to a rearrangement of ancestors in the hominoid family tree.

Five genera of hominoids exist today: gibbons (*Hylobates*), orangutans (*Pongo*), gorillas (*Gorilla*), chimpanzees (*Pan*), and humans (*Homo*). Gibbons are known informally as small apes; orangutans, gorillas, and chimpanzees are known as great apes. Molecular evidence indicates a close relationship between

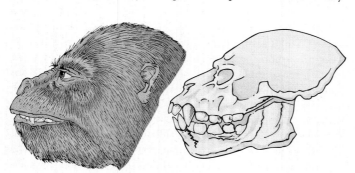

(a) Fossils of *Aegyptopithecus*, a fairly primitive anthropoid, were discovered in Egypt.

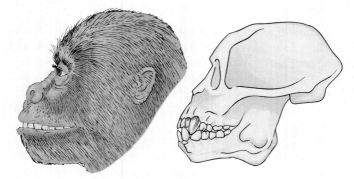

(c) *Dryopithecus*, a more advanced ape, may have been ancestral to modern hominoids.

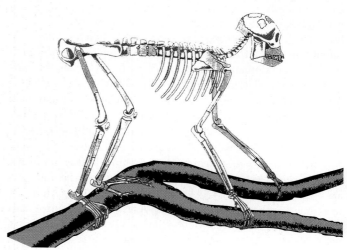

(b) Skeletal reconstruction of *Proconsul*. (The reconstructed parts are white.) This anthropoid had the limbs and body proportions of a monkey but lacked a tail, which is characteristic of apes.

❚ **Figure 22-5**   Ape evolution
(*Figures not drawn to scale.*)

humans and the great apes, particularly chimpanzees.

Gibbons are natural acrobats that can **brachiate,** or arm-swing, with their weight supported by one arm at a time (❚ Fig. 22-6a). Orangutans are also tree dwellers, but chimpanzees and especially gorillas have adapted to life on the ground (❚ Fig. 22-6b–d). Gorillas and chimpanzees have retained long arms typical of brachiating primates but use these to assist in quadrupedal walking, sometimes known as **knuckle walking** because of the way they fold (flex) their digits when moving. Like humans, apes lack tails, a characteristic that makes them easy to distinguish from monkeys. They are generally much larger than monkeys, although gibbons are a notable exception.

Evidence of the close relatedness of orangutans, gorillas, chimpanzees, and humans is abundant at the molecular level. The amino acid sequence of the chimpanzee's hemoglobin is identical to that of the human; hemoglobin molecules of the gorilla and rhesus monkey differ from the human's by 2 and 15 amino acids, respectively. DNA sequence analyses indicate that chimpanzees are likely to be our nearest living relatives among the apes (see Table 18-1). Molecular evidence suggests that orangutans may have diverged from the gorilla, chimpanzee, and hominid lines about 12 mya to 16 mya. Gorillas may have diverged from the chimpanzee and hominid lines some 8 mya, whereas chimpanzee and hominid lines probably separated about 6 mya.

Review

❚ What are the three suborders of primates?

❚ How can you distinguish between anthropoids and hominoids?

HOMINID EVOLUTION

Learning Objectives

4 Describe skeletal and skull differences between apes and hominids.

5 Briefly describe the following early hominids: *Sahelanthropus, Orrorin, Ardipithecus ramidus,* and *Australopithecus anamensis, A. afarensis,* and *A. africanus.*

6 Distinguish among the following members of genus *Homo: H. habilis, H. ergaster, H. erectus, H. neanderthalensis,* and *H. sapiens.*

7 Discuss the origin of modern humans.

(a) A mother white-handed gibbon (*Hylobates lar*) nurses her baby. Gibbons are extremely acrobatic and often move through the trees by brachiation.

(b) An orangutan (*Pongo pygmaeus*) mother and baby. Orangutan anatomy is adapted to living in trees.

(c) A young lowland gorilla (*Gorilla gorilla*) in knuckle-walking stance. Gorillas spend most of the day eating plants.

(d) A mother bonobo chimpanzee (*Pan paniscus*) holds her sleeping baby. Bonobos are endemic to a single country, the Democratic Republic of Congo.

Figure 22-6 Apes

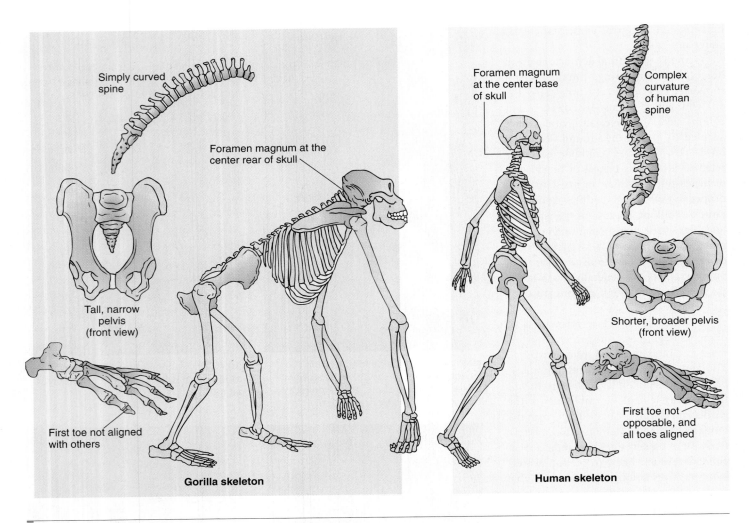

Simply curved spine

Foramen magnum at the center rear of skull

Tall, narrow pelvis (front view)

First toe not aligned with others

Gorilla skeleton

Foramen magnum at the center base of skull

Complex curvature of human spine

Shorter, broader pelvis (front view)

First toe not opposable, and all toes aligned

Human skeleton

Figure 22-7 *Animated* Gorilla and human skeletons

When gorilla and human skeletons are compared, the skeletal adaptations for bipedalism in humans become apparent.

Scientists have a growing storehouse of hundreds of hominid fossils, which provide useful data about general trends in the body design, appearance, and behavior of ancestral humans. For example, before their brains enlarged, early hominids clearly adopted a **bipedal** (two-footed) posture. Despite the wealth of fossil evidence, scientists continue to vigorously debate interpretations of hominid characteristics, classification, and evolution, and new discoveries raise new questions. As in other scientific fields, ideas about hominid evolution are influenced by the different perspectives of the various workers studying it. The lack of scientific consensus regarding certain aspects of hominid evolution is, therefore, an expected part of the scientific process.

Evolutionary changes from the earliest hominids to modern humans are evident in some of the characteristics of the skeleton and skull. Compared with the ape skeleton, the human skeleton shows distinct differences that reflect humans' ability to stand erect and walk on two feet (❙ Fig. 22-7). These differences also reflect the habitat change for early hominids, from an arboreal existence in the forest to a life spent at least partly on the ground.

The curvature of the human spine provides better balance and weight distribution for bipedal locomotion. The human

pelvis is shorter and broader than the ape pelvis, allowing better attachment of muscles used for upright walking. In apes the **foramen magnum,** the hole in the base of the skull for the spinal cord, is located in the rear of the skull. In contrast, the human foramen magnum is centered in the skull base, positioning the head for erect walking. An increase in the length of the legs relative to the arms, and alignment of the first toe with the rest of the toes, further adapted the early hominids for bipedalism.

Another major trend in hominid evolution was an increase in brain size relative to body size (❙ Fig. 22-8). The ape skull has prominent **supraorbital ridges** above the eye sockets, whereas modern human skulls lack these ridges. Human faces are flatter than those of apes, and the jaws are different. The arrangement of teeth in the ape jaw is somewhat rectangular, compared with a rounded, or U-shaped, arrangement in humans. Apes have larger front teeth (canines and incisors) than do humans, and their canines are especially large. Gorillas and orangutans also have larger back teeth (premolars and molars) than do humans.

We now examine some of the increasing number of fossil hominids in the human lineage. As you read the following descriptions of human evolution, keep in mind that much of what

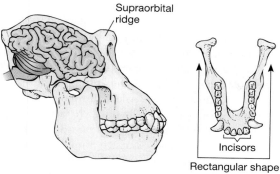

(a) The ape skull has a pronounced supraorbital ridge.

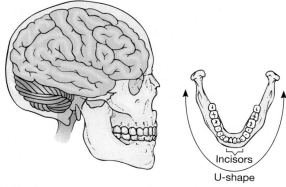

(b) The human skull is flatter in the front and has a pronounced chin. The human brain, particularly the cerebrum (*purple*), is larger than that of an ape, and the human jaw is structured so that the teeth are arranged in a U shape. Human canines and incisors are also smaller than those of apes.

Figure 22-8 Gorilla and human heads

is discussed is still open to interpretation and major revision as additional discoveries are made. It is also important to remember that although we present human evolution in a somewhat linear fashion, from ancient hominids to anatomically and behaviorally modern humans, the human family tree is not a single trunk but has several branches (Fig. 22-9). Perhaps it is most useful to think of the known hominid fossils as a sampling of human evolution instead of a continuous sequence.

Homo sapiens is the only species of hominid in existence today, but more than one hominid species coexisted at any given time for most of the past 4 million years. In addition, do not make the mistake of thinking that your smaller-brained ancestors were inferior to yourself. Ancestral hominids were evolutionarily successful in that they were well adapted to their environment and survived for millions of years.

The earliest hominids may have lived 6 mya to 7 mya

Hominid evolution began in Africa. Although most hominid fossils have been discovered in Ethiopia and Kenya, in 2002 French paleontologist Michel Brunet and an international team made a stunning discovery in a dry lake bed in Chad, which is in central

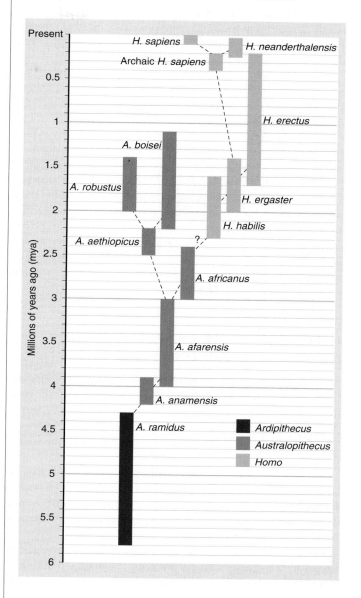

Figure 22-9 One interpretation of human evolution

This figure does not include hominid fossils earlier than *Ardipithecus*. Dashed lines show possible evolutionary relationships.

Africa. The fossil, which has been reliably dated at 6 million to 7 million years old, may be the earliest known hominid. Viewed from the back, the skull of **Sahelanthropus,** with its small braincase, resembles that of a chimpanzee. However, viewed from the front, the face and teeth have many characteristics of larger-brained human ancestors. Most paleoanthropologists place *Sahelanthropus* close to the base of the human family tree—that is, close to when genetic analyses suggest the last common ancestor of hominids and chimpanzees existed. The discovery of *Sahel-*

anthropus is important in its own right, but it is also significant because it shows that early hominids exhibited more variation and lived over a larger area of Africa than had been originally hypothesized.

Orrorin, discovered in 2001 in Kenya, may also represent one of the earliest known hominids. The sediments in which fossils of *Orrorin* were found have been reliably dated at 6 million years old. Researchers studying the fossil leg bones of *Orrorin* think that it walked upright and was bipedal.

As with many aspects of human evolution, scientists have widely differing interpretations of *Sahelanthropus* and *Orrorin*. For example, some paleoanthropologists hypothesize that *Sahelanthropus* was a forerunner of modern apes, specifically the gorilla, and not one of the earliest humans. Other researchers think the similar features of *Sahelanthropus*, *Orrorin*, and *Ardipithecus* (discussed in the next section) fossils mean that they are all members of the same genus, *Ardipithecus*. This point remains controversial, largely because there are currently no skeletal bones (leg, pelvis, and foot bones) to indicate if *Sahelanthropus* walked upright, a hallmark characteristic of hominids. (A 2005 reconstruction of the skull has features that strongly suggest *Sahelanthropus* was bipedal.) Future fossil discoveries and additional analyses of existing fossils should help clarify the evolutionary relationships among *Sahelanthropus*, *Orrorin*, and *Ardipithecus*.

Australopithecines are the immediate ancestors of genus *Homo*

After *Sahelanthropus* and *Orrorin*, the next-oldest hominid belongs to the genus **Ardipithecus,** which lived in eastern Africa more than 5 mya. Although not as primitive as *Sahelanthropus*, *Ardipithecus* is close to the "root" of the human family tree. The shape of the toe bones indicates that *Ardipithecus* walked upright.

Ardipithecus gave rise to **Australopithecus,** a genus that includes several species that lived between 4 mya and 1 mya. *Ardipithecus* and *Australopithecus* are referred to as **australopithecines,** or "southern man apes." Both had longer arms, shorter legs, and smaller brains relative to those of modern humans.

Hominids that existed between 4.2 mya and 3.9 mya are assigned to the species **Australopithecus anamensis,** first named by paleoanthropologist Meave Leakey and her coworkers in 1995 from fossils discovered in East Africa. This hominid species, which has a mixture of apelike and humanlike features, presumably evolved from *Ardipithecus ramidus*. A comparison of male and female *Australopithecus anamensis* body sizes and canine teeth reveals **sexual dimorphism,** marked phenotypic differences between the two sexes of the same species. (The modern-day gorilla is sexually dimorphic.) The back teeth and jaws of *A. anamensis* are larger than those of modern chimpanzees, whereas the front teeth are smaller and more like those of later hominids. A fossil leg bone, the tibia, indicates that *A. anamensis* had an upright posture and was bipedal, although it also may have foraged in the trees. Thus, bipedalism occurred early in human evolution and appears to have been the first human adaptation.

Australopithecus afarensis, another primitive hominid, probably evolved directly from *A. anamensis*. Many fossils of *A. afarensis* skeletal remains have been discovered in Africa, including a remarkably complete 3.2-million-year-old skeleton, nicknamed Lucy, found in Ethiopia in 1974 by a team led by U.S. paleoanthropologist Donald Johanson. In 1978, British paleoanthropologist Mary Leakey and coworkers discovered beautifully preserved fossil footprints of three *A. afarensis* individuals who walked more than 3.6 mya. In 1994, paleoanthropologists found the first adult skull of *A. afarensis*. The skull, characterized by a relatively small brain, pronounced supraorbital ridges, a jutting jaw, and large canine teeth, is an estimated 3 million years old. It is probable that *A. afarensis* did not construct tools or make fires, because no evidence of tools or fire has been found at fossil sites.

Many paleoanthropologists think that *A. afarensis* gave rise to several australopithecine species, including **Australopithecus africanus,** which may have appeared as early as 3 mya. The first *A. africanus* fossil was discovered in South Africa in 1924, and since then hundreds have been found. This hominid walked erect and possessed hands and teeth that were distinctly humanlike. Given the characteristics of the teeth, paleoanthropologists think that *A. africanus* ate both plants and animals. Like *A. afarensis*, it had a small brain, more like that of its primate ancestors than of present-day humans.

Three australopithecine species (*Australopithecus robustus* from South Africa, and *A. aethiopicus* and *A. boisei*, both from East Africa) were larger than *A. africanus* and had extremely large molars, very powerful jaws, relatively small brains, and bony skull crests. Most females lacked the skull crests and had substantially smaller jaws, another example of sexual dimorphism in early hominids. The teeth and jaws suggest a diet, perhaps of tough roots and tubers, that would require powerful grinding. These so-called *robust australopithecines* may or may not be closely related but are generally thought to represent evolutionary offshoots, or side branches, of human evolution. The first robust australopithecine, *A. aethiopicus*, appeared about 2.5 mya. Some researchers classify robust australopithecines in a separate genus, *Paranthropus*.

The actual number of australopithecine species for which fossil evidence has been found is under debate. In some cases, differences in the relatively few skeletal fragments could indicate either variation among individuals within a species or evidence of separate species. Most paleoanthropologists recognize at least six species of australopithecines.

Homo habilis is the oldest member of genus *Homo*

The first hominid to have enough uniquely human features to be placed in the same genus as modern humans is **Homo habilis.** It was first discovered in the early 1960s at Olduvai Gorge in Tanzania. Since then, paleoanthropologists have discovered other fossils of *H. habilis* in East and South Africa. *Homo habilis* was a small hominid with a larger brain and smaller premolars and molars than those of the australopithecines. This hominid appeared approximately 2.3 mya and persisted for about 0.75 million years. Fossils of *H. habilis* have been found in numerous areas in Africa. These sites contain primitive tools, stones that had been chipped, cracked, or hammered to make sharp edges for cutting or scraping. *Oldowan* pebble choppers and flakes, for example, were probably used to cut through animal hides to obtain meat and to break bones for their nutritious marrow.

The relationship between the australopithecines and *H. habilis* is not clear. Using physical characteristics of their fossilized skeletons as evidence, many paleoanthropologists have inferred that the australopithecines were ancestors of *H. habilis*. Some researchers do not think that *H. habilis* belongs in the genus *Homo*, and they suggest it should be reclassified as *Australopithecus habilis*. Discoveries of additional fossils may help clarify these relationships.

Homo erectus apparently evolved from *Homo habilis*

Investigators found the first fossil evidence of **Homo erectus** in Indonesia in the 1890s. Since then, searchers have found numerous fossils of *H. erectus* throughout Africa and Asia. Paleoanthropologists think that *H. erectus* originated in Africa about 1.7 mya and then spread to Europe and Asia. Peking man and Java man, discovered in Asia, were later examples of *H. erectus*, which existed until at least 200,000 years ago; some populations of *H. erectus* may have persisted more recently.

Homo erectus was taller than *H. habilis*. Its brain, which was larger than that of *H. habilis*, got progressively larger during the course of its evolution. Its skull, although larger, did not possess totally modern features, retaining the heavy supraorbital ridge and projecting face that are more characteristic of its ape ancestors (❙ Fig. 22-10). *Homo erectus* is the first hominid to have fewer differences between the sexes.

The increased intelligence associated with an increased brain size enabled these early humans to make more advanced stone tools, known as *Acheulean* tools, including hand axes and other implements that scientists have interpreted as choppers, borers, and scrapers. Their intelligence also allowed these humans to survive in cold areas. *Homo erectus* obtained food by hunting or scavenging and may have worn clothing, built fires, and lived in caves or shelters. Evidence of weapons (spears) has been unearthed at *H. erectus* sites in Europe.

Ideas regarding *H. erectus*, like many other aspects of human evolution, are changing with each new fossil discovery. Some scientists hypothesize that the fossils classified as *H. erectus* really represent two species, **Homo ergaster,** an earlier African species, and *H. erectus*, a later East Asian offshoot. The best-known fossils of *H. ergaster* come from the Lake Turkana region in Kenya. Researchers who support this split speculate that *H. ergaster* may be the direct ancestor of later humans, whereas *H. erectus* may be an evolutionary dead end. (For the remarkable story of an evolutionary offshoot from *H. erectus*, see *Focus On: The Smallest Humans*.)

Other paleoanthropologists do not think that *H. erectus* should be split into two or more species. They cite evidence of a 1-million-year-old *H. erectus* fossil that was discovered in Africa in 2002. This fossil shares striking similarities with Asian *H. erectus* fossils and enables paleoanthropologists to compare *H. erectus* fossils from the same period but from two different

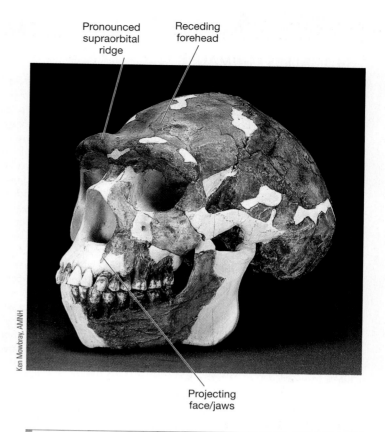

Pronounced supraorbital ridge

Receding forehead

Projecting face/jaws

Ken Mowbray, AMNH

❙ **Figure 22-10** *Animated* *Homo erectus* skull from China

The reconstructed parts are white. Note the receding forehead, pronounced supraorbital ridge, and projecting face and jaws.

continents. Scientists hope that further analysis of this fossil and of future fossil discoveries will clarify the status of *H. erectus*.

Archaic *Homo sapiens* appeared between 400,000 and 200,000 years ago

Archaic Homo sapiens are regionally diverse descendants of *H. erectus* or *H. ergaster* that lived in Africa, Asia, and Europe from about 400,000 to 200,000 years ago. The brains of archaic *H. sapiens* were essentially the same size as our brains, although their skulls retained some ancestral characters. Populations of archaic *H. sapiens* had rich and varied cultures (learned traditions) that included making many kinds of tools and objects with symbolic and ceremonial meaning. Some researchers classify European fossils of archaic *H. sapiens* as a separate species, **Homo heidelbergensis,** that existed until the appearance of the Neandertals.

Neandertals appeared approximately 230,000 years ago

Fossils of **Neandertals** were first discovered in the Neander Valley in Germany.[1] They lived throughout Europe and western Asia

---

[1] *Neandertal* was formerly spelled *Neanderthal*. The silent *h* has been dropped in modern German but not in the scientific name.

THE SMALLEST HUMANS

In 2004, paleoanthropologists reported a startling discovery—fossils of seven humans in a cave on the Indonesian island of Flores. The finding was completely unexpected, because the fossil bones and teeth were from adult humans that were about 1 m (about 3 ft) tall. The tiny humans, which represent a new species named *Homo floresiensis*, also had small, ape-sized brains. Yet sophisticated stone tools, fireplaces, and charred bones found with the fossils indicate that *H. floresiensis* was capable of complex thinking and activities. The charred bones were primarily from pygmy elephants (*Stegodon*) and Flores giant rats (*Papagomys*) and provide clues about the food that *H. floresiensis* hunted and killed (see figure). Based on several dating methods, *H. floresiensis* is thought to have existed from about 38,000 years ago to as recently as 13,000 years ago, when it went extinct.

Detailed studies of the *H. floresiensis* braincase, published in 2005, revealed that it was similar in many ways to the larger brain of *H. erectus,* which is known to have lived on nearby islands. Many researchers have concluded from this and other evidence that *H. floresiensis* was an evolutionary offshoot of *H. erectus.*

Why was *H. floresiensis* so small? Biologists have often observed two evolutionary trends of mammals living on remote islands: large mammals tend to evolve into much smaller species, and small mammals tend to evolve into much larger species. It is not unreasonable to assume that the small population of *H. erectus* ancestors that colonized Flores, perhaps by rafting to the island on a log, underwent natural selection and dwindled in size over the thousands of years they were isolated on Flores. (If *H. erectus* had been a frequent visitor to Flores, gene flow between the indigenous population and the visitors would have prevented the indigenous population from becoming measurably smaller.)

Not everyone agrees with the hypothesis that *H. floresiensis* is a separate species of tiny humans. Some researchers think the fossils have not been identified properly. In a study published in 2006, these researchers suggest that the fossils are from *H. sapiens* individuals suffering from a rare genetic defect (microcephaly) that causes small brains and bodies.

The research team that unearthed the fossils has returned to the cave and will continue excavations. It is anticipated that future discoveries will help scientists answer the many questions we have about Earth's smallest humans.

Peter Schouten/National Geographic Society/University of Wollongong

Homo floresiensis—or pygmy *H. sapiens*—carrying a giant rat

---

from about 230,000 to 30,000 years ago. These early humans had short, sturdy builds. Their faces projected slightly, their chins and foreheads receded, they had heavy supraorbital ridges and jawbones, and their brains and front teeth were larger than those of modern humans. They had large nasal cavities and receding cheekbones. Scientists have suggested that the large noses provided larger surface areas in Neandertal sinuses, enabling them to better warm the cold air of Ice Age Eurasia as inhaled air traveled through the head to the lungs.

Scientists have not reached a consensus about whether the Neandertals are a separate species from modern humans. Many think the anatomical differences between Neandertals and modern humans mean that they were separate species, ***Homo neanderthalensis*** and ***Homo sapiens.*** Other scientists disagree and think that Neandertals were a group of *H. sapiens.*

Neandertal tools, known as *Mousterian* tools, were more sophisticated than those of *H. erectus* (❚ Fig. 22-11). Studies of Neandertal sites indicate that Neandertals hunted large animals. The existence of skeletons of elderly Neandertals and of some with healed fractures may demonstrate that Neandertals cared for the aged and the sick, an indication of advanced social cooperation.

They apparently had rituals, possibly of religious significance, and sometimes buried their dead.

The disappearance of the Neandertals about 28,000 years ago is a mystery that has sparked debate among paleoanthropologists. Other groups of *H. sapiens* with more modern features coexisted with the Neandertals for several thousand years. Perhaps the other humans outcompeted or exterminated the Neandertals, leading to their extinction. It is also possible that the Neandertals interbred with these humans, diluting their features beyond recognition.

Analysis of **mitochondrial DNA (mtDNA)** contributes useful data for such controversies. Each of the several hundred mitochondria within a cell has about 10 copies of a small loop of DNA that codes for transfer RNAs, ribosomal RNAs, and certain respiratory enzymes. The mtDNA is transmitted only through the maternal line because eggs, not sperm, contribute mitochondria. Because mtDNA mutates more rapidly than nuclear DNA, mtDNA is a sensitive indicator of evolution. Researchers have extracted and evaluated the mtDNA from seven separate Neandertals. Its sequence differs significantly from all modern human mtDNA sequences, although it is more similar to human than

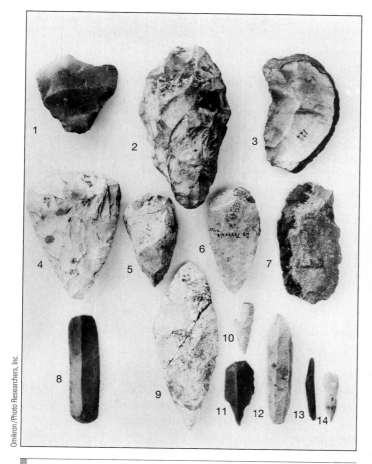

Figure 22-11 Mousterian tools

Paleoanthropologists named Mousterian tools after a Neandertal site in Le Moustier, France. Mousterian tools included a variety of skillfully made stone tools, such as hand axes, flakes, scrapers, borers, and spear points (*1* to *7* are earlier tools, and *8* to *14* are later tools).

Figure 22-12 Cro-Magnon man painting an outline of his hand

to chimpanzee mtDNA. This finding suggests that Neandertals are an evolutionary dead end that did not interbreed appreciably with more modern humans. However, the amount of interbreeding that may have occurred between Neandertals and anatomically modern humans remains controversial.

Biologists debate the origin of modern *Homo sapiens*

Homo sapiens with anatomically modern features existed in Africa about 195,000 years ago. The *H. sapiens* skull lacked a heavy brow ridge and had a distinct chin. By about 30,000 years ago, anatomically modern humans were the only members of genus *Homo* remaining, excluding small, isolated populations.

Paleoanthropologists refer to European populations of these ancient people as **Cro-Magnons.** Their weapons and tools were complex and often made of materials other than stone, including bone, ivory, and wood. They made stone blades that were extremely sharp. Cro-Magnons developed art, including cave paintings, engravings, and sculpture, possibly for ritual purposes (❙ Fig. 22-12). Their sophisticated tools and art indicate they may

have possessed language, which they would have used to transmit their culture to younger generations.

Two opposing hypotheses have been suggested to explain the origin of these modern humans: the *out-of-Africa hypothesis* and the *multiregional hypothesis.* The out-of-Africa hypothesis holds that modern *H. sapiens* evolved from *H. erectus* in Africa sometime after 200,000 years ago and then migrated to Europe and Asia, displacing the Neandertals and other *Homo* species living there. The out-of-Africa hypothesis predicts that fossils of the earliest humans will be found in Africa.

According to the multiregional hypothesis, modern humans originated as separately evolving *H. erectus* populations living in several parts of Africa, Asia, and Europe. Each of these populations evolved in its own distinctive way but occasionally met and interbred with other populations, preventing complete reproductive isolation. The variation found today in different human populations therefore represents a continuation of this multiregional process. The multiregional hypothesis predicts that fossils of the earliest humans will be found scattered throughout Africa, Europe, and Asia.

Recent fossil finds and molecular analyses have promoted the out-of-Africa hypothesis as the main explanation for the origin of modern humans. New data include the discovery of the earliest fossilized remains of modern *H. sapiens* in Africa and analyses of DNA from mitochondria and the Y chromosome. In 2005, fossils of the earliest known modern *H. sapiens* were reliably dated at 195,000 years old; these fossils were found in southern Ethiopia. No *H. sapiens* fossils of comparable age have been found in Europe or Asia. The earliest fossils of anatomically modern *H. sapiens* in Europe and most parts of Asia date to 45,000 to 40,000 years ago. One exception is in Israel, where *H. sapiens* fossils have been dated at 100,000 years before present.

Molecular anthropology, the comparison of biological molecules from present-day individuals of regional human populations, provides clues that help scientists unravel the origin of modern humans and trace human migrations. A series of recent genetic studies of both mtDNA and Y-chromosome DNA has strengthened the case for Africa as the birthplace of modern humans. These studies provide detailed comparisons of DNA in present-day populations around the world as well as ancient DNA extracted from Neandertal and early *H. sapiens* remains.

Based on the data, the ancestors common to all humans were probably ancient Khoisans, an indigenous group in southern Africa. This conclusion is based on separate analyses of DNA from both mitochondria and the Y chromosome. Both studies indicated that the Khoisan people are the most ancient of all human groups.

Review

- How do the skulls of apes and humans differ?
- How does an ape skeleton differ from a human skeleton?
- How do australopithecines and *Homo* differ?
- How can you distinguish between *H. habilis* and *H. erectus*?
- How do *H. neanderthalensis* and *H. sapiens* differ?
- What two hypotheses have been proposed to explain where modern humans originated? Which hypothesis is strongly supported by fossil and molecular evidence?

CULTURAL CHANGE

Learning Objective

8 Describe the impact of human culture on the biosphere.

Genetically speaking, humans are not very different from other primates. At the level of our DNA sequences, we are roughly 98% identical to gorillas and 99% identical to chimpanzees. Our relatively few genetic differences, however, give rise to several important distinguishing features, such as greater intelligence and the ability to capitalize on it by transmitting knowledge from one generation to the next. Human culture is not inherited in the biological sense but instead is learned, largely through the medium of language.[2] Human culture is dynamic and is modified as people obtain new knowledge. Human culture is generally divided into three stages: (1) the development of hunter–gatherer societies, (2) the development of agriculture, and (3) the Industrial Revolution.

Early humans were hunters and gatherers who relied on what was available in their immediate environment. They were nomadic, and as the resources in a given area were exhausted or as the population increased, they migrated to a different area. These societies required a division of labor and the ability to make tools and weapons, which were needed to kill game, scrape hides, dig up roots and tubers, and cook food. Although scientists are not certain when hunting was incorporated into human society, they do know it declined in importance approximately 15,000 years ago. This may have been due to a decrease in the abundance of large animals, triggered in part by overhunting. A few isolated groups of hunter–gatherer societies, such as the Inuit of the northern polar region and the Mbuti of Africa, survived into the 21st century.

Development of agriculture resulted in a more dependable food supply

Evidence that humans had begun to cultivate crops approximately 10,000 years ago includes the presence of agricultural tools and plant material at archaeological sites. Agriculture resulted in a more dependable food supply. Archaeological evidence suggests that agriculture arose in several steps. Although there is variation from one site to another, plant cultivation, in combination with hunting, usually occurred first. Animal domestication generally followed later, although in some areas, such as Australia, early humans did not domesticate animals.

Agriculture, in turn, often led to more permanent dwellings, because considerable time was invested in growing crops in one area. Villages and cities often grew up around the farmlands, but the connection between agriculture and the establishment of villages and towns is complicated by certain discoveries. For example, Abu Hureyra in Syria was a village founded *before* agriculture arose. The villagers subsisted on the rich plant life of the area and the migrating herds of gazelle. Once people turned to agriculture, however, they seldom went back to hunting and gathering to obtain food.

Other advances in agriculture include the domestication of animals, which people kept to supply food, milk, and hides. Archaeological evidence indicates that wild goats and sheep were probably the first animals to be domesticated in southwestern Turkey, northern Iraq, and Iran. In the Old World, people also used animals to prepare fields for planting. Another major advance in agriculture was irrigation, which began more than 5000 years ago in Egypt.

Producing food agriculturally was more time-consuming than hunting and gathering, but it was also more productive. In hunter–gatherer societies, everyone shares the responsibility for obtaining food. In agricultural societies, fewer people are needed to provide food for everyone. Thus, agriculture freed some people to pursue other endeavors, including religion, art, and various crafts.

---

[2]Humans are not the only animals to have culture. Chimpanzees have primitive cultures that include tool-using techniques, hunting methods, and social behaviors that vary from one population to another. These cultural traditions are passed to the next generation by teaching and imitation; see Chapter 51.

Cultural evolution has had a profound impact on the biosphere

Cultural evolution has had far-reaching effects on both human society and on other organisms. The Industrial Revolution, which began in the 18th century, drew populations to concentrate in urban areas near centers of manufacturing. Advances in agriculture encouraged urbanization, because fewer and fewer people were needed in rural areas to produce food for everyone. The spread of industrialization increased the demand for natural resources to supply the raw materials for industry.

Cultural evolution has permitted the human population, which reached 6.5 billion in 2006, to expand so dramatically that there are serious questions about Earth's ability to support so many people indefinitely (see Chapter 52). According to the United Nations (UN) Food and Agricultural Organization, more than 800 million people lack access to the food needed to be healthy and lead productive lives. To further compound the problem, the UN projects that 2.8 billion *additional* people will be added to the world population by the year 2050.

Cultural evolution has resulted in large-scale disruption and degradation of the environment. Tropical rain forests and other natural environments are rapidly being eliminated. Soil, water, and air pollution occur in many places. Since World War II, soil degradation caused by poor agricultural practices, overgrazing, and deforestation has occurred in an area equal to 17% of Earth's total vegetated surface area. Many species cannot adapt to the rapid environmental changes caused by humans and thus are becoming extinct. The decrease in biological diversity from species extinction is alarming (see Chapter 56).

On a positive note, people are aware of the damage we are causing and have the intelligence to modify behavior to improve these conditions. Education, including the study of biology, may help future generations develop environmental sensitivity, making cultural evolution our salvation rather than our destruction.

Review

▌ What is cultural evolution?

▌ How has cultural evolution affected planet Earth?

SUMMARY WITH KEY TERMS

Learning Objectives

1   Describe the structural adaptations that primates have for life in treetops (page 467).

   ▌ **Primates** are **placental mammals** that arose from small, **arboreal** (tree-dwelling), shrewlike mammals. Primates possess five grasping digits, including an **opposable thumb** or toe; long, slender limbs that move freely at the hips and shoulders; and eyes located in front of the head.

2   List the three suborders of primates, and give representative examples of each (page 468).

   ▌ Primates are divided into three suborders. The suborder Prosimii includes lemurs, galagos, and lorises. The suborder Tarsiiformes includes tarsiers. The suborder Anthropoidea includes anthropoids—monkeys, apes, and humans.

ThomsonNOW™   **Learn more about primate evolution by clicking on the figure in ThomsonNOW.**

3   Distinguish among anthropoids, hominoids, and hominids (page 468).

   ▌ Anthropoids include monkeys, apes, and humans. Early anthropoids branched into two groups: New World monkeys and Old World monkeys.

   ▌ **Hominoids** include apes and humans. Hominoids arose from the Old World monkey lineage. There are four modern genera of apes: gibbons, orangutans, gorillas, and chimpanzees.

   ▌ The **hominid** line consists of humans and their ancestors.

4   Describe skeletal and skull differences between apes and hominids (page 471).

   ▌ Unlike ape skeletons, hominid skeletons have adaptations that reflect the ability to stand erect and walk on two feet. These adaptations include a complex curvature of the spine; a shorter, broader pelvis; repositioning of the **foramen magnum** to the base of the skull; and a first toe that is aligned with the other toes.

   ▌ The human skull lacks a pronounced **supraorbital ridge,** is flatter than ape skulls in the front, and has a pronounced chin. The human brain is larger than ape brains, and the jaw is structured so that the teeth are arranged in a U shape.

ThomsonNOW™   **Learn more about monkey, gorilla, and human skeletons by clicking on the figure in ThomsonNOW.**

5   Briefly describe the following early hominids: *Sahelanthropus, Orrorin, Ardipithecus ramidus,* and *Australopithecus anamensis, A. afarensis,* and *A. africanus* (page 471).

   ▌ Hominid evolution began in Africa. The earliest known hominids, dated at 6 to 7 million years old, belong to *Sahelanthropus.* Although *Sahelanthropus* had a small brain, its face and teeth had many characteristics of larger-brained human ancestors.

   ▌ *Orrorin* is an early hominid that arose about 6 mya. Researchers studying the fossil leg bones of *Orrorin* think that it walked upright and was bipedal.

   ▌ *Ardipithecus* and *Australopithecus* species are often referred to as **australopithecines.** *Australopithecus* species were **bipedal** (walked on two feet), a hominid feature. *Ardipithecus ramidus,* which appeared about 5.8 mya to 5.2 mya, presumably gave rise to *Australopithecus anamensis,* which in turn may have given rise to another primitive hominid, *Australopithecus afarensis.* Many paleoanthropologists think that *A. afarensis* gave rise to several australopithecine species, including *Australopithecus africanus.*

   ▌ The genus *Australopithecus* contains the immediate ancestors of the genus *Homo.*

6   Distinguish among the following members of genus *Homo*: *H. habilis, H. ergaster, H. erectus, H. neanderthalensis,* and *H. sapiens* (page 471).

   ▌ *Homo habilis* was the earliest known hominid with some of the human features lacking in the australopithecines, including a slightly larger brain. *H. habilis* fashioned crude tools from stone.

- **Homo erectus** had a larger brain than *H. habilis*; made more sophisticated tools; and may have worn clothing, built fires, and lived in caves or shelters. Some scientists hypothesize that fossils identified as *H. erectus* represent two different species, **Homo ergaster**, an earlier African species that gave rise to archaic *H. sapiens*, and *H. erectus*, a later Asian offshoot that may be an evolutionary dead end.

- **Archaic Homo sapiens** are regionally diverse descendants of *H. erectus* or *H. ergaster* that lived in Africa, Asia, and Europe from about 400,000 to 200,000 years ago. The brains of archaic *H. sapiens* were essentially the same size as our brains, although their skulls retained some ancestral characters, and they had rich and varied cultures.

- **Neandertals** lived from about 230,000 to 30,000 years ago. Neandertals had short, sturdy builds; receding chins and foreheads; heavy supraorbital ridges and jawbones; larger front teeth; and nasal cavities with unusual triangular bony projections. Many scientists think that Neandertals were a separate species, **Homo neanderthalensis**, whereas some scientists think that Neandertals were a type of modern human.

- **Homo sapiens**, anatomically modern humans, existed in Africa about 195,000 years ago. By about 30,000 years ago, anatomically modern humans were the only members of genus *Homo* remaining, excluding small, isolated populations. European remains of these ancient people are referred to as **Cro-Magnons.**

ThomsonNOW **Learn more about *Homo* skulls by clicking on the figure in ThomsonNOW.**

7 Discuss the origin of modern humans (page 471).
- Two hypotheses purport to explain the origin of modern humans. The out-of-Africa hypothesis holds that modern *H. sapiens* arose in Africa and migrated to Europe and Asia, displacing the more primitive humans living there. According to the multiregional hypothesis, modern humans originated as separately evolving populations of *H. erectus* living in several parts of Africa, Asia, and Europe. Each of these populations occasionally met and interbred with other populations, thereby preventing complete reproductive isolation. Recent fossil discoveries and molecular data support the out-of-Africa hypothesis as the main explanation for the origin of modern humans.

8 Describe the impact of human culture on the biosphere (page 478).
- Large human brain size makes possible the transmission of knowledge from one generation to the next. Two significant advances in human culture were the development of agriculture and the Industrial Revolution.

TEST YOUR UNDERSTANDING

1. The first primates evolved about 56 mya from (a) shrewlike monotremes (b) therapsids (c) shrewlike placental mammals (d) tarsiers (e) shrewlike marsupials

2. The anthropoids are more closely related to _____ than to _____. (a) tarsiers; lemurs (b) lemurs; monkeys (c) tree shrews; tarsiers (d) lemurs; tarsiers (e) tree shrews; monkeys

3. Unlike Old World monkeys, some New World monkeys have (a) body hair (b) five grasping digits (c) a well-developed cerebrum (d) a bipedal walk (e) a prehensile tail

4. Apes and humans are collectively called (a) mammals (b) primates (c) anthropoids (d) hominoids (e) hominids

5. With what group do hominoids share the most recent common ancestor? (a) Old World monkeys (b) New World monkeys (c) tarsiers (d) lemurs (e) lorises and galagos

6. The _____ in humans is centered at the base of the skull, positioning the head for erect walking. (a) supraorbital ridge (b) foramen magnum (c) pelvis (d) bony skull crest (e) femur

7. Scientists collectively call humans and their *immediate* ancestors (a) mammals (b) primates (c) anthropoids (d) hominoids (e) hominids

8. The earliest hominid fossil found so far belongs to the genus (a) *Aegyptopithecus* (b) *Dryopithecus* (c) *Sahelanthropus* (d) *Homo* (e) *Australopithecus*

9. The earliest hominid that scientists placed in the genus *Homo* is (a) *H. habilis* (b) *H. ergaster* (c) *H. erectus* (d) *H. heidelbergensis* (e) *H. neanderthalensis*

10. Some scientists now think that fossils identified as *Homo erectus* represent which two different species? (a) *H. habilis* and *H. erectus* (b) *H. ergaster* and *H. erectus* (c) *H. heidelbergensis* and *H. ergaster* (d) *H. neanderthalensis* and *H. erectus* (e) *H. neanderthalensis* and *H. sapiens*

11. Archaic *Homo sapiens* appeared as early as _____ years ago. (a) 5 million (b) 400,000 (c) 230,000 (d) 100,000 (e) 5000

12. _____ were an early group of humans with short, sturdy builds and heavy supraorbital ridges that lived throughout Europe and western Asia from about 230,000 to 30,000 years ago. (a) Australopithecines (b) Dryopithecines (c) Archaic *Homo sapiens* (d) Neandertals (e) Cro-Magnons

13. The modern human skull *lacks* (a) small canines (b) a foramen magnum centered in the base of the skull (c) pronounced supraorbital ridges (d) a U-shaped arrangement of teeth on the jaw (e) a large cranium (braincase)

14. The comparison of genetic material from individuals of regional populations of humans, used to help unravel the origin and migration of modern humans, is known as (a) paleoarchaeology (b) cultural anthropology (c) molecular anthropology (d) cytogenetics (e) genetic dimorphism

15. Place the following hominids in chronological order of appearance in the fossil record, beginning with the earliest:

1. *H. habilis* 2. modern *H. sapiens* 3. *H. neanderthalensis* 4. *H. erectus*

(a) 1, 4, 2, 3 (b) 4, 1, 2, 3 (c) 4, 1, 3, 2 (d) 3, 1, 4, 2 (e) 1, 4, 3, 2

1. What types of as-yet-undiscovered scientific evidence might help explain how monkeys got to South America from the Old World?

2. If you were evaluating whether other early humans exterminated the Neandertals, what kinds of archaeological evidence might you look for?

3. The remains of Cro-Magnons have been found in southern Europe alongside reindeer bones, but reindeer currently exist only in northern Europe and Asia. Explain the apparent discrepancy.

4. **Evolution Link.** What was the common ancestor of chimpanzees and humans—a chimpanzee, a human, or neither? Explain your answer.

5. **Evolution Link.** Some scientists say that modern medicine and better sanitation are slowing down or altering the course of human evolution in highly developed countries today. As a result, evolution in humans today has gone from survival of the fittest to survival of almost everyone. Do you think this is a valid idea? Why or why not?

6. **Analyzing Data.** If you were to add *H. floresiensis* to Figure 22-9, where would you put it? Justify your answer.

7. **Analyzing Data.** Using the information in Figure 22-9, draw a simple cladogram that represents a reasonable hypothesis of human evolution at the genus level. (*Hint:* Your diagram will have three branches.)

Additional questions are available in ThomsonNOW at www.thomsonedu.com/login

Appendix A

Periodic Table of the Elements

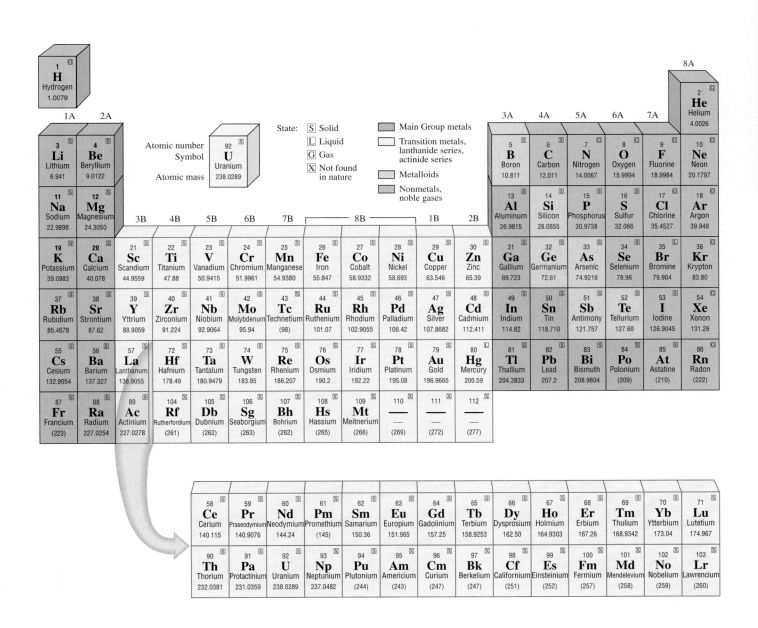

Classification of Organisms

The system of cataloguing organisms used in this book is described in Chapter 1 and in Chapters 23 through 31. In this eighth edition of *Biology*, we use the three-domain and six-kingdom classification. The three domains are **Bacteria, Archaea,** and **Eukarya** (eukaryotes). The six kingdoms are **Bacteria** (which corresponds to domain Bacteria); **Archaea** (which corresponds to domain Archaea); **Protista, Plantae, Fungi,** and **Animalia** (these four kingdoms are assigned to domain Eukarya). In this classification overview, we have included select groups. (We have also omitted many groups, especially extinct ones.) We have omitted viruses from this survey, because they do not fit into the three domains or six kingdoms.

PROKARYOTES

Domains Bacteria and Archaea are made up of prokaryotic organisms. They are distinguished from eukaryotic organisms by their smaller ribosomes and absence of membranous organelles, including the absence of a discrete nucleus surrounded by nuclear envelope. Prokaryotes reproduce mainly asexually by binary fission. When present, flagella are simple and solid; they do not have the 9 + 2 microfilament structure typical of eukaryotes.

DOMAIN BACTERIA, KINGDOM BACTERIA

Very large, diverse group of prokaryotic organisms. Typically unicellular, but some form colonies or filaments. Mainly heterotrophic, but some groups are photosynthetic or chemosynthetic. Bacteria are nonmotile or move by rotating flagella. Typically have peptidoglycan in their cell walls. Estimated 100,000 to 200,000 species.

Bacterial nomenclature and taxonomic practices are controversial and changing. See Table 24-4.

Proteobacteria. Large, diverse group of gram-negative bacteria. Five subgroups are designated alpha, beta, gamma, delta, and epsilon. Proteobacteria include rickettsias, enterobacteria, purple sulfur bacteria, myxobacteria.

Cyanobacteria. Gram-negative, photosynthetic.

Gram-positive bacteria. Diverse group; includes actinomycetes, lactic acid bacteria, mycobacteria, streptococci, staphylococci, clostridia. Thick cell wall of peptidoglycan; many produce spores.

Mycoplasmas. Lack cell walls. Extremely small bacteria bounded by plasma membrane.

Chlamydias. Lack peptidoglycan in their cell walls. Energy parasites dependent on host for ATP.

Spirochetes. Spiral-shaped bacteria with flexible cell walls.

DOMAIN ARCHAEA, KINGDOM ARCHAEA

Prokaryotes with unique cell membrane structure and cell walls lacking peptidoglycan. Also distinguished by their ribosomal RNA, lipid structure, and specific enzymes. Archaea are found in extreme environments—hot springs, sea vents, dry and salty seashores, boiling mud, and near ash-ejecting volcanoes. Three main types based on their metabolism and ecology are *methanogens, extreme halophiles,* and *extreme thermophiles.* Fewer than 225 named species. See Table 24-3 for a classification based on molecular data (as outlined in *Bergey's Manual*).

Methanogens. Anaerobes that produce methane gas from simple carbon compounds.

Extreme halophiles. Inhabit saturated salt solutions.

Extreme thermophiles. Grow at 70°C or higher; some thrive above boiling point.

DOMAIN EUKARYA, KINGDOM PROTISTA

Primarily unicellular or simple multicellular eukaryotic organisms that do not form tissues and that exhibit relatively little division of labor. Most modes of nutrition occur in this kingdom. Life cycles may include both sexually and asexually reproducing phases and may be extremely complex, especially in parasitic forms. Locomotion is by cilia, flagella, amoeboid movement, or by other means. Flagella and cilia have 9 + 2 structure.

Excavates. Anaerobic zooflagellates that have deep, or excavated, oral groove. Have atypical mitochondria or lack them. Most are endosymbionts. Include diplomonads and trichonymphs.

Diplomonads. Excavates with one or two nuclei, no mitochondria, no Golgi complex, and up to eight flagella.

Trichonymphs. Excavates with hundreds of flagella; live in guts of termites and wood-eating cockroaches.

Discicristates. Zooflagellates with disc-shaped cristae in their mitochondria. Include euglenoids and trypanosomes.

Euglenoids and trypanosomes. Unicellular flagellates; some free-living and some pathogenic. Many are heterotrophic, but some (about one third) are photosynthetic. At least 900 species.

Alveolates. Unicellular protists with alveoli, flattened vesicles located inside plasma membrane. Include ciliates, dinoflagellates, and apicomplexans.

Ciliates. Unicellular protists that move by means of cilia. Reproduction is asexual by binary fission or sexual by conjugation. About 7200 species.

Dinoflagellates. Unicellular (some colonial), photosynthetic, biflagellate. Cell walls composed of overlapping cell plates; contain cellulose. Contain chlorophylls *a* and *c* and carotenoids, including fucoxanthin. About 2000 to 4000 species.

Apicomplexans. Parasitic unicellular protists that lack specific structures for locomotion. At some stage in life cycle, they develop spores (small infective agents). Some pathogenic. About 3900 species.

Heterokonts. Diverse group; all have motile cells with two flagella, one with tiny, hairlike projections off shaft. Include water molds, diatoms, golden algae, and brown algae.

Water molds. Consist of branched, coenocytic mycelia. Cellulose and/or chitin in cell walls. Produce biflagellate asexual spores. Sexual stage involves production of oospores. Some parasitic. About 700 species.

Diatoms. Unicellular (some colonial), photosynthetic. Most nonmotile, but some move by gliding. Cell walls composed of silica rather than cellulose. Contain chlorophylls *a* and *c* and carotenoids, including fucoxanthin. At least 100,000 species estimated.

Golden algae. Unicellular (some colonial), photosynthetic, biflagellate (some lack flagella). Cells covered by tiny scales of either silica or calcium carbonate. Contain chlorophylls *a* and *c* and carotenoids, including fucoxanthin. About 1000 species.

Brown algae. Multicellular, often quite large (kelps). Photosynthetic; contain chlorophylls *a* and *c* and carotenoids, including fucoxanthin. Biflagellate reproductive cells. About 1500 species.

"Plants." Photosynthetic organisms with chloroplasts bounded by outer and inner membranes. This monophyletic group includes organisms currently placed in two kingdoms: Plantae (land plants) and Protista (red algae and green algae).

Red algae. Most multicellular (some unicellular), mainly marine. Some (coralline algae) have bodies impregnated with calcium carbonate. No motile cells. Photosynthetic; contain chlorophyll *a*, carotenoids, phycocyanin, and phycoerythrin. About 5000 species.

Green algae. Unicellular, colonial, siphonous, and multicellular forms. Some motile and flagellate. Photosynthetic; contain chlorophylls *a* and *b* and carotenoids. About 17,000 species.

Cercozoa. Amoeboid cells, often with hard outer shells through which cytoplasmic projections extend. Include foraminiferans and actinopods.

Foraminiferans. Unicellular protists that produce calcareous tests (shells) with pores through which cytoplasmic projections extend, forming sticky net to entangle prey.

Actinopods. Unicellular protists that produce axopods (long, filamentous cytoplasmic projections) that protrude through pores in their siliceous shells.

Amoebozoa. Amoeboid protists that lack tests and move by means of lobose pseudopodia. Include amoebas, plasmodial slime molds, and cellular slime molds.

Amoebas. Free-living and parasitic unicellular protists whose movement and capture of food are associated with pseudopodia.

Plasmodial slime molds. Spend part of life cycle as thin, streaming, multinucleate plasmodium that creeps along on decaying leaves or wood. Flagellate or amoeboid reproductive cells; form spores in sporangia. About 700 species.

Cellular slime molds. Vegetative (nonreproductive) form is unicellular; move by pseudopods. Amoeba-like cells aggregate to form multicellular pseudoplasmodium that eventually develops into fruiting body that bears spores. About 50 species.

Opisthokonts. Monophyletic group that includes members of three kingdoms: Fungi, Animalia, and Protista (choanoflagellates). Opisthokonts (Greek *opistho,* "rear," and *kontos,* "pole") have single posterior flagellum in flagellate cells.

Choanoflagellates. Zooflagellates with single flagellum surrounded at base by collar of microvilli. May be related to common ancestor of animals.

DOMAIN EUKARYA, KINGDOM PLANTAE

Multicellular eukaryotic organisms with differentiated tissues and organs. Cell walls contain cellulose. Cells frequently contain large vacuoles; photosynthetic pigments in plastids. Photosynthetic pigments are chlorophylls *a* and *b* and carotenoids. Nonmotile. Reproduce both asexually and sexually, with alternation of gametophyte (n) and sporophyte ($2n$) generations.

Phylum Bryophyta. *Mosses.* Nonvascular plants that lack xylem and phloem. Marked alternation of generations with dominant gametophyte generation. Motile sperm. Gametophytes generally form dense green mat consisting of individual plants. At least 9900 species.

Phylum Hepatophyta. *Liverworts.* Nonvascular plants that lack xylem and phloem. Marked alternation of generations with dominant gametophyte generation. Motile sperm. Gametophytes of certain species have flat, liverlike thallus; other species more mosslike in appearance. About 6000 species.

Phylum Anthocerophyta. *Hornworts.* Nonvascular plants that lack xylem and phloem. Marked alternation of generations with dominant gametophyte generation. Motile sperm. Gametophyte is small, flat, green thallus with scalloped edges. Spores produced on erect, hornlike stalk. About 100 species.

Phylum Pteridophyta. Vascular plants with dominant sporophyte generation. Reproduce by spores. Motile sperm.

Ferns. Generally homosporous. Gametophyte is free-living and photosynthetic. About 11,000 species.

Whisk ferns. Homosporous. Sporophyte stem branches dichotomously; lacks true roots and leaves. Gametophyte is subterranean and nonphotosynthetic and forms mycorrhizal relationship with fungus. About 12 species.

Horsetails. Homosporous. Sporophyte has hollow, jointed stems and reduced, scalelike leaves. Gametophyte is tiny photosynthetic plant. About 15 species.

Phylum Lycopodiophyta. *Club mosses.* Sporophyte plants are vascular with branching rhizomes and upright stems that bear microphylls. Although modern representatives are small, some extinct species were treelike. Some homosporous; others heterosporous. Motile sperm. About 1200 species.

Phylum Coniferophyta. *Conifers.* Heterosporous vascular plants with woody tissues (trees and shrubs) and needle-shaped or scalelike leaves. Most are evergreen. Seeds usually borne naked on surface of cone scales. Nutritive tissue in seed is haploid female gametophyte tissue. Nonmotile sperm. About 630 species.

Phylum Cycadophyta. *Cycads.* Heterosporous, vascular, dioecious plants that are small and shrubby or larger and palmlike. Produce naked seeds in conspicuous cones. Flagellate sperm. About 140 species.

Phylum Ginkgophyta. *Ginkgo.* Broadleaf deciduous trees that bear naked seeds directly on branches. Dioecious. Contain vascular tissues. Flagellate sperm. Ginkgo tree is only living representative. One species.

Phylum Gnetophyta. *Gnetophytes.* Woody shrubs, vines, or small trees that bear naked seeds in cones. Contain vascular tissues. Possess many features similar to flowering plants. About 70 species.

Phylum Anthophyta. *Flowering plants* or *angiosperms.* Largest, most successful group of plants. Heterosporous; dominant sporophytes with extremely reduced gametophytes. Contain vascular tissues. Bear flowers, fruits, and seeds (enclosed in fruit; seeds contain endosperm as nutritive tissue). Double fertilization. More than 300,000 species.

DOMAIN EUKARYA, KINGDOM FUNGI

Eukaryotic, mainly multicellular organisms with cell walls containing chitin. Body form is often mycelium. Cells usually haploid or dikaryotic, with brief diploid period following fertilization. Heterotrophs that secrete digestive enzymes onto food source and then absorb predigested food. Most decomposers, but some parasites. Reproduce by means of spores, which may be produced sexually or asexually. No flagellate stages except in chytrids. Classified as opisthokonts (along with choanoflagellates and animals) because flagellate cells, where present, have a single posterior flagellum (Greek *opistho,* "rear," and *kontos,* "pole").

Phylum Chytridiomycota. *Chytridiomycetes* or *chytrids.* Parasites and decomposers found mainly in fresh water. Motile cells (gametes and zoospores) contain single, posterior flagellum. Reproduce both sexually and asexually. About 790 species.

Phylum Zygomycota. *Zygomycetes (molds).* Important decomposers; some are insect parasites. Produce sexual resting spores called *zygospores;* and nonmotile, haploid, asexual spores in sporangium. Hyphae are coenocytic. Many are heterothallic (two mating types). About 1060 species.

Phylum Glomeromycota. *Glomeromycetes.* Symbionts that form intracellular mycorrhizal associations within roots of most trees and herbaceous plants. Reproduce asexually with large, multinucleate spores called *blastospores.* About 160 species.

Phylum Ascomycota. *Ascomycetes* or *sac fungi (yeasts, powdery mildews, molds, morels, truffles).* Important symbionts; 98% of lichen-forming fungi are ascomycetes; some form mycorrhizae. Sexual reproduction: form ascospores in sacs called *asci.* Asexual reproduction: produce spores called *conidia,* which pinch off from conidiophores. Hyphae usually have perforated septa. Dikaryotic stage. About 32,300 species.

Phylum Basidiomycota. *Basidiomycetes* or *club fungi (mushrooms, bracket fungi, puffballs).* Many form mycorrhizae with tree roots. Sexual reproduction: form basidiospores on basidium. Asexual reproduction uncommon. Heterothallic. Hyphae usually have perforated septa. Dikaryotic stage. About 22,300 species.

DOMAIN EUKARYA, KINGDOM ANIMALIA

Eukaryotic, multicellular heterotrophs with differentiated cells. In most animals, cells are organized to form tissues, tissues form organs, and tissues and organs form specialized organ systems that carry on specific functions. Most have well-developed nervous system and respond adaptively to changes in their environment. Most are capable of locomotion during some time in their life cycle. Most diploid and reproduce sexually; flagellate haploid sperm unites with large, non-motile, haploid egg, forming diploid zygote that undergoes cleavage. Classified as opisthokonts (along with choanoflagellates and fungi) because flagellate cells, where present, have a single posterior flagellum (Greek *opistho,* "rear," and *kontos,* "pole").

Phylum Porifera. *Sponges.* Mainly marine; solitary or colonial. Body bears many pores through which water circulates. Food is filtered from water by collar cells (choanocytes). Asexual reproduction by budding; external sexual reproduction in which sperm are released and swim to internal egg. Larva is motile. About 10,000 species.

Phylum Cnidaria. *Hydras, jellyfish, sea anemones, corals.* Marine, with a few freshwater species; solitary or colonial; polyp and medusa forms. Radial symmetry. Tentacles surrounding mouth. Stinging cells (cnidocytes) contain stinging structures called *nematocysts.* Planula larva. About 10,000 species.

Phylum Ctenophora. *Comb jellies.* Marine; free-swimming. Biradial symmetry. Two tentacles and eight longitudinal rows of cilia resembling combs; animal moves by means of these bands of cilia. About 100 species.

Protostomes

Coelomates (characterized by true coelom, that is, body cavity completely lined with mesoderm). Spiral, determinate cleavage, and mouth typically develops from blastopore. Two branches of protostomes are Lophotrochozoa and Ecdysozoa.

Lophotrochozoa

Include the platyhelminths, nemerteans (ribbon worms), mollusks, annelids, the lophophorate phyla, and the rotifers. The name *Lophotrochozoa* comes from the names of the two major animal groups included: the Lophophorata and the Trochozoa. The Lophophorata include the lophophorate phyla, characterized by a lophophore, a ciliated ring of tentacles surrounding the mouth. The name *Trochozoa* is derived from the trochophore larva that characterizes its two major groups—the mollusks and annelids.

Phylum Platyhelminthes. *Flatworms.* Acoelomate (no body cavity); region between body wall and internal organs filled with tissue. Planarians are free-living; flukes and tapeworms are parasitic. Body dorsoventrally flattened; cephalization; three tissue layers. Simple nervous system with ganglia in head region. Excretory organs are protonephridia with flame cells. About 20,000 species.

Phylum Nemertea. *Proboscis worms* (also called *ribbon worms*). Long, dorsoventrally flattened body with complex proboscis used for defense and for capturing prey. Functionally acoelomate but have small true coelom in proboscis. Definite organ systems. Complete digestive tract. Circulatory system with blood. About 1000 species.

Phylum Mollusca. *Snails, clams, squids, octopods.* Unsegmented, soft-bodied animals usually covered by dorsal shell. Have ventral, muscular foot. Most organs located above foot in visceral mass. Shell-secreting mantle covers visceral mass and forms mantle cavity, which contains gills. Trochophore and/or veliger larva. About 50,000 species.

Phylum Annelida. *Segmented worms: polychaetes, earthworms, leeches.* Both body wall and internal organs are segmented. Body segments separated by septa. Some have nonjointed appendages. Setae used in locomotion. Closed circulatory system; metanephridia; specialized regions of digestive tract. Trochophore larva. About 15,000 species.

Phylum Brachiopoda. *Lamp shells.* One of lophophorate phyla. Marine; body enclosed between two shells. About 350 species.

Phylum Phoronida. One of lophophorate phyla. Tube-dwelling marine worms. About 12 species.

Phylum Bryozoa. One of lophophorate phyla. Mainly marine; sessile colonies produced by asexual budding. About 5000 species.

Phylum Rotifera. *Wheel animals.* Aquatic, microscopic, wormlike animals. Anterior end has ciliated crown that looks like wheel when cilia beat. Posterior end tapers to foot. Characterized by pseudocoelom (body cavity not completely lined with mesoderm). Constant number of cells in adult. About 1800 species.

Ecdysozoa

Animals in this group characterized by ecdysis (molting).

Phylum Nematoda. *Roundworms: ascaris, hookworms, pinworms.* Slender, elongated, cylindrical worms; covered with cuticle. Characterized by pseudocoelom (body cavity not completely lined with mesoderm). Free-living and parasitic forms. About 15,000 species.

Phylum Arthropoda. *Arachnids (spiders, mites, ticks), crustaceans (lobsters, crabs, shrimp), insects, centipedes, millipedes.* Segmented animals with paired, jointed appendages and hard exoskeleton made of chitin. Open circulatory system with dorsal heart. Hemocoel occupies most of body cavity, and coelom is reduced. More than 1 million species.

Deuterostomes

Coelomates with radial, indeterminate cleavage. Blastopore develops into anus, and mouth forms from second opening.

Phylum Echinodermata. *Sea stars, sea urchins, sand dollars, sea cucumbers.* Marine animals. Pentaradial symmetry as adults; bilateral symmetry as larvae. Endoskeleton of small, calcareous plates. Water vascular system; tube feet for locomotion. About 7000 species.

Phylum Hemichordata. *Acorn worms.* Marine animals with ring of cilia around mouth. Anterior muscular proboscis is connected by collar region to long, wormlike body. Larval form resembles echinoderm larva. About 85 species.

Phylum Chordata. *Subphylum Urochordata (tunicates), subphylum Cephalochordata (lancelets), subphylum Vertebrata (fishes, amphibians, reptiles, birds, mammals).* Notochord; pharyngeal slits; dorsal, tubular nerve cord; and postanal tail present at some time in life cycle. About 48,000 species.

Appendix C

Understanding Biological Terms

Your task of mastering new terms will be greatly simplified if you learn to dissect each new word. Many terms can be divided into a prefix, the part of the word that precedes the main root; the word root itself; and often a suffix, a word ending that may add to or modify the meaning of the root. As you progress in your study of biology, you will learn to recognize the more common prefixes, word roots, and suffixes. Such recognition will help you analyze new terms so that you can more readily determine their meaning and will help you remember them.

Prefixes

a-, ab- from, away, apart (*abduct*, move away from the midline of the body)

a-, an-, un- less, lack, not (*asymmetrical*, not symmetrical)

ad- (also **af-, ag-, an-, ap-**) to, toward (*adduct*, move toward the midline of the body)

allo- different (*allometric growth*, different rates of growth for different parts of the body during development)

ambi- both sides (*ambidextrous*, able to use either hand)

andro- a man (*androecium*, the male portion of a flower)

anis- unequal (*anisogamy*, sexual reproduction in which the gametes are of unequal sizes)

ante- forward, before (*anteflexion*, bending forward)

anti- against (*antibody*, proteins that have the capacity to react against foreign substances in the body)

auto- self (*autotroph*, organism that manufactures its own food)

bi- two (*biennial*, a plant that takes two years to complete its life cycle)

bio- life (*biology*, the study of life)

circum-, circ- around (*circumcision*, a cutting around)

co-, con- with, together (*congenital*, existing with or before birth)

contra- against (*contraception*, against conception)

cyt- cell (*cytology*, the study of cells)

di- two (*disaccharide*, a compound made of two sugar molecules chemically combined)

dis- apart (*dissect*, cut apart)

ecto- outside (*ectoderm*, outer layer of cells)

end-, endo- within, inner (*endoplasmic reticulum*, a network of membranes found within the cytoplasm)

epi- on, upon (*epidermis*, upon the dermis)

ex-, e-, ef- out from, out of (*extension*, a straightening out)

extra- outside, beyond (*extraembryonic membrane*, a membrane that encircles and protects the embryo)

gravi- heavy (*gravitropism*, growth of a plant in response to gravity)

hemi- half (*cerebral hemisphere*, lateral half of the cerebrum)

hetero- other, different (*heterozygous*, having unlike members of a gene pair)

homeo- unchanging, steady (*homeostasis*, reaching a steady state)

homo-, hom- same (*homologous*, corresponding in structure; *homozygous*, having identical members of a gene pair)

hyper- excessive, above normal (*hypersecretion*, excessive secretion)

hypo- under, below, deficient (*hypotonic*, a solution whose osmotic pressure is less than that of a solution with which it is compared)

in-, im- not (*incomplete flower*, a flower that does not have one or more of the four main parts)

inter- between, among (*interstitial*, situated between parts)

intra- within (*intracellular*, within the cell)

iso- equal, like (*isotonic*, equal osmotic pressure)

macro- large (*macronucleus*, a large, polyploid nucleus found in ciliates)

mal- bad, abnormal (*malnutrition*, poor nutrition)

mega- large, great (*megakaryocyte*, giant cell of bone marrow)

meso- middle (*mesoderm*, middle tissue layer of the animal embryo)

meta- after, beyond (*metaphase*, the stage of mitosis after prophase)

micro- small (*microscope*, instrument for viewing small objects)

mono- one (*monocot*, a group of flowering plants with one cotyledon, or seed leaf, in the seed)

oligo- small, few, scant (*oligotrophic lake*, a lake deficient in nutrients and organisms)

oo- egg (*oocyte*, cell that gives rise to an egg cell)

paedo- a child (*paedomorphosis*, the preservation of a juvenile characteristic in an adult)

para- near, beside, beyond (*paracentral*, near the center)

peri- around (*pericardial membrane*, membrane that surrounds the heart)

photo- light (*phototropism*, growth of a plant in response to the direction of light)

poly- many, much, multiple, complex (*polysaccharide*, a carbohydrate composed of many simple sugars)

post- after, behind (*postnatal*, after birth)

pre- before (*prenatal*, before birth)

pseudo- false (*pseudopod*, a temporary protrusion of a cell, i.e., "false foot")

retro- backward (*retroperitoneal*, located behind the peritoneum)

semi- half (*semilunar*, half-moon)

sub- under (*subcutaneous tissue*, tissue immediately under the skin)

super, supra- above (*suprarenal*, above the kidney)

sym- with, together (*sympatric speciation*, evolution of a new species within the same geographic region as the parent species)

syn- with, together (*syndrome*, a group of symptoms that occur together and characterize a disease)

trans- across, beyond (*transport*, carry across)

Suffixes

-able, -ible able (*viable*, able to live)

-ad used in anatomy to form adverbs of direction (*cephalad*, toward the head)

-asis, -asia, -esis condition or state of (*euthanasia*, state of "good death")

-cide kill, destroy (*biocide*, substance that kills living things)

-emia condition of blood (*anemia*, a blood condition in which there is a lack of red blood cells)

-gen something produced or generated or something that produces or generates (*pathogen*, an organism that produces disease)

-gram record, write (*electrocardiogram*, a record of the electrical activity of the heart)

-graph record, write (*electrocardiograph*, an instrument for recording the electrical activity of the heart)

-ic adjective-forming suffix that means *of* or *pertaining to* (*ophthalmic*, of or pertaining to the eye)

-itis inflammation of (*appendicitis*, inflammation of the appendix)

-logy study or science of (*cytology*, study of cells)

-oid like, in the form of (*thyroid*, in the form of a shield, referring to the shape of the thyroid gland)

-oma tumor (*carcinoma*, a malignant tumor)

-osis indicates disease (*psychosis*, a mental disease)

-pathy disease (*dermopathy*, disease of the skin)

-phyll leaf (*mesophyll*, the middle tissue of the leaf)

-scope instrument for viewing or observing (*microscope*, instrument for viewing small objects)

Some Common Word Roots

abscis cut off (*abscission*, the falling off of leaves or other plant parts)

angi, angio vessel (*angiosperm*, a plant that produces seeds enclosed within a fruit or "vessel")

apic tip, apex (*apical meristem*, area of cell division located at the tips of plant stems and roots)

arthr joint (*arthropods*, invertebrate animals with jointed legs and segmented bodies)

aux grow, enlarge (*auxin*, a plant hormone involved in growth and development)

blast a formative cell, germ layer (*osteoblast*, cell that gives rise to bone cells)

brachi arm (*brachial artery*, blood vessel that supplies the arm)

bry grow, swell (*embryo*, an organism in the early stages of development)

cardi heart (*cardiac*, pertaining to the heart)

carot carrot (*carotene*, a yellow, orange, or red pigment in plants)

cephal head (*cephalad*, toward the head)

cerebr brain (*cerebral*, pertaining to the brain)

cervic, cervix neck (*cervical*, pertaining to the neck)

chlor green (*chlorophyll*, a green pigment found in plants)

chondr cartilage (*chondrocyte*, a cartilage cell)

chrom color (*chromosome*, deeply staining body in nucleus)

cili small hair (*cilium*, a short, fine cytoplasmic hair projecting from the surface of a cell)

coleo a sheath (*coleoptile*, a protective sheath that encircles the stem in grass seedlings)

conjug joined (*conjugation*, a sexual phenomenon in certain protists)

cran skull (*cranial*, pertaining to the skull)

decid falling off (*deciduous*, a plant that sheds its leaves at the end of the growing season)

dehis split (*dehiscent fruit*, a fruit that splits open at maturity)

derm skin (*dermatology*, study of the skin)

ecol dwelling, house (*ecology*, the study of organisms in relation to their environment, i.e., "their house")

enter intestine (*enterobacteria*, a group of bacteria that includes species that inhabit the intestines of humans and other animals)

evol to unroll (*evolution*, descent with modification, gradual directional change)

fil a thread (*filament*, the thin stalk of the stamen in flowers)

gamet a wife or husband (*gametangium*, the part of a plant, protist, or fungus that produces reproductive cells)

gastr stomach (*gastrointestinal tract*, the digestive tract)

glyc, glyco sweet, sugar (*glycogen*, storage form of glucose)

gon seed (*gonad*, an organ that produces gametes)

gutt a drop (*guttation*, loss of water as liquid "drops" from plants)

gymn naked (*gymnosperm*, a plant that produces seeds that are not enclosed with a fruit, i.e., "naked")

hem blood (*hemoglobin*, the pigment of red blood cells)

hepat liver (*hepatic*, of or pertaining to the liver)

hist tissue (*histology*, study of tissues)

hydr water (*hydrolysis*, a breakdown reaction involving water)

leuk white (*leukocyte*, white blood cell)

menin membrane (*meninges*, the three membranes that envelop the brain and spinal cord)

morph form (*morphogenesis*, development of body form)

my, myo muscle (*myocardium*, muscle layer of the heart)

myc a fungus (*mycelium*, the vegetative body of a fungus)

nephr kidney (*nephron*, microscopic unit of the kidney)

neur, nerv nerve (*neuromuscular*, involving both the nerves and muscles)

occiput back part of the head (*occipital*, back region of the head)

ost bone (*osteology*, study of bones)

path disease (*pathologist*, one who studies disease processes)

ped, pod foot (*bipedal*, walking on two feet)

pell skin (*pellicle*, a flexible covering over the body of certain protists)

phag eat (*phagocytosis,* process by which certain cells ingest particles and foreign matter)

phil love (*hydrophilic,* a substance that attracts, i.e., "loves," water)

phloe bark of a tree (*phloem,* food-conducting tissue in plants that corresponds to bark in woody plants)

phyt plant (*xerophyte,* a plant adapted to xeric, or dry, conditions)

plankt wandering (*plankton,* microscopic aquatic protists that float or drift passively)

rhiz root (*rhizome,* a horizontal, underground stem that superficially resembles a root)

scler hard (*sclerenchyma,* cells that provide strength and support in the plant body)

sipho a tube (*siphonous,* a type of tubular body form found in certain algae)

som body (*chromosome,* deeply staining body in the nucleus)

sor heap (*sorus,* a cluster or "heap" of sporangia in a fern)

spor seed (*spore,* a reproductive cell that gives rise to individual offspring in plants, protists, and fungi)

stom a mouth (*stoma,* a small pore, i.e., "mouth," in the epidermis of plants)

thigm a touch (*thigmotropism,* plant growth in response to touch)

thromb clot (*thrombus,* a clot within a blood vessel)

troph nourishment (*heterotroph,* an organism that must depend on other organisms for its nourishment)

tropi turn (*thigmotropism,* growth of a plant in response to contact with a solid object, such as a tendril "turning" or wrapping around a wire fence)

visc pertaining to an internal organ or body cavity (*viscera,* internal organs)

xanth yellow (*xanthophyll,* a yellowish pigment found in plants)

xyl wood (*xylem,* water-conducting tissue in plant, the "wood" of woody plants)

zoo an animal (*zoology,* the science of animals)

Appendix D

Abbreviations

The biological sciences use a great many abbreviations and with good reason. Many technical terms in biology and biological chemistry are both long and difficult to pronounce. Yet it can be difficult for beginners, when confronted with something like NADPH or EPSP, to understand the reference. Here, for your ready reference, are some of the common abbreviations used in biology.

A adenine
ABA abscisic acid
ABC transporters ATP-binding cassette transporters
ACTH adrenocorticotropic hormone
AD Alzheimer's disease
ADA adenosine deaminase
ADH antidiuretic hormone
ADP adenosine diphosphate
AIDS acquired immunodeficiency syndrome
AMP adenosine monophosphate
amu atomic mass unit (dalton)
APC anaphase-promoting complex *or* antigen-presenting cell
ATP adenosine triphosphate
AV node or valve atrioventricular node or valve (of heart)
B lymphocyte or B cell lymphocyte responsible for antibody-mediated immunity
BAC bacterial artificial chromosome
BH brain hormone (of insects)
BMR basal metabolic rate
bya billion years ago
C cytosine
C_3 three-carbon pathway for carbon fixation (Calvin cycle)
C_4 four-carbon pathway for carbon fixation
CAM crassulacean acid metabolism
cAMP cyclic adenosine monophosphate
CAP catabolite activator protein
CD4 T cell T helper cell (TH); has a surface marker designated CD4
CD8 T cell T cell with a surface marker designated CD8; includes T cytotoxic cells
Cdk cyclin-dependent protein kinase
cDNA complementary deoxyribonucleic acid
CFCs chlorofluorocarbons
CFTR cystic fibrosis transmembrane conductance regulator
CITES Convention on International Trade in Endangered Species of Wild Flora and Fauna
CNS central nervous system
CO cardiac output
CoA coenzyme A
COPD chronic obstructive pulmonary disease
CP creatine phosphate
CPR cardiopulmonary resuscitation
CREB cyclic AMP response element binding protein

CSF cerebrospinal fluid
CVS cardiovascular system
DAG diacylglycerol
DNA deoxyribonucleic acid
DOC dissolved organic carbon
E_A activation energy (of an enzyme)
ECG electrocardiogram
ECM extracellular matrix
EEG electroencephalogram
EKG electrocardiogram
EM electron microscope or micrograph
ENSO El Niño–Southern Oscillation
EPSP excitatory postsynaptic potential (of a neuron)
ER endoplasmic reticulum
ES cells embryonic stem cells
EST expressed sequence tag
F_1 first filial generation
F_2 second filial generation
Fab portion the part of an antibody that binds to an antigen
Factor VIII blood-clotting factor (absent in hemophiliacs)
FAD/FADH$_2$ flavin adenine dinucleotide (oxidized and reduced forms, respectively)
FAP fixed action pattern
Fc portion the part of an antibody that interacts with cells of the immune system
FISH fluorescent in situ hybridization
FSH follicle-stimulating hormone
G guanine
G_1 phase first gap phase (of the cell cycle)
G_2 phase second gap phase (of the cell cycle)
G3P glyceraldehyde-3-phosphate
G protein cell-signaling molecule that requires GTP
GA_3 gibberellin
GAA Global Amphibian Assessment (by World Conservation Union)
GABA gamma-aminobutyric acid
GH growth hormone (somatotropin)
GnRH gonadotropin-releasing hormone
GTP guanosine triphosphate
HB hemoglobin
HBEF Hubbard Brook Experimental Forest
HBO$_2$ oxyhemoglobin
HCFCs hydrochlorofluorocarbons
hCG human chorionic gonadotropin
HD Huntington disease
HDL high-density lipoprotein
HFCs hydrofluorocarbons
hGH human growth hormone
HIV human immunodeficiency virus

HLA human leukocyte antigen
HUGO Human Genome Organization
IAA indole acetic acid (natural auxin)
Ig immunoglobulin, as in IgA, IgG, etc.
IGF insulin-like growth factor
IP$_3$ inositol trisphosphate
IPCC United Nations Intergovernmental Panel on Climate Change
IPSP inhibitory postsynaptic potential (of a neuron)
IUCN World Conservation Union
IUD intrauterine device
JH juvenile hormone (of insects)
kb kilobase
LDH lactate dehydrogenase enzyme
LDL low-density lipoprotein
LH luteinizing hormone
LM light microscope or micrograph
LSD lysergic acid diethylamide
LTP long-term potentiation
MAO monoamine oxidase
MAPs microtubule-associated proteins
MHC major histocompatibility complex
MI myocardial infarction
miRNA microribonucleic acid
MPF mitosis-promoting factor
MRI magnetic resonance imaging
mRNA messenger RNA
MSAFP maternal serum α-fetoprotein
MSH melanocyte-stimulating hormone
mtDNA mitochondrial DNA
MTOC microtubule organizing center
MVP minimum viable population
mya million years ago
9 + 2 structure cilium or flagellum (of a eukaryote)
9 × 3 structure centriole or basal body (of a eukaryote)
n, 2n the chromosome number of a gamete and of a zygote, respectively
NAD$^+$/NADH nicotinamide adenine dinucleotide (oxidized and reduced forms, respectively)
NADP$^+$/NADPH nicotinamide adenine dinucleotide phosphate (oxidized and reduced forms, respectively)
NAG *N*-acetyl glucosamine
NK cell natural killer cell
NMDA *N*-methyl-D aspartate (an artificial ligand)
NO nitric oxide
NSF National Science Foundation
P generation parental generation
P53 a tumor suppressor gene
P680 reaction center of photosystem II
P700 reaction center of photosystem I
PABA para-aminobenzoic acid

PAMPs pathogen-associated molecular patterns
PCR polymerase chain reaction
PEP phosphoenolpyruvate
Pfr phytochrome (form that absorbs far-red light)
PGA phosphoglycerate
PID pelvic inflammatory disease
PIF3 phytochrome-interacting factor-3
PKU phenylketonuria
PNS peripheral nervous system
pre-mRNA precursor messenger RNA (in eukaryotes)
Pr phytochrome (form that absorbs red light)
PTH parathyroid hormone
RAS reticular activating system
RBC red blood cell (erythrocyte)
REM sleep rapid-eye-movement sleep
RFLP restriction fragment length polymorphism
RNA ribonucleic acid
RNAi ribonucleic acid interference
rRNA ribosomal RNA
rubisco ribulose bisphosphate carboxylase/oxygenase
RuBP ribulose bisphosphate
S phase DNA synthetic phase (of the cell cycle)
SA node sinoatrial node (of heart)
SCID severe combined immunodeficiency
SEM scanning electron microscope or micrograph
siRNA small interfering ribonucleic acid
snRNA small nuclear ribonucleic acid
snoRNA small nucleolar ribonucleic acid
snRNP small nuclear ribonucleoprotein complex
SRP RNA signal recognition particle ribonucleic acid
SSB protein single-strand binding protein
ssp subspecies
STD sexually transmitted disease
STR short tandem repeat
T thymine
T lymphocyte or T cell lymphocyte responsible for cell-mediated immunity
T$_C$ lymphocyte T cytotoxic cell
T$_H$ lymphocyte T helper cell (CD4 T cell)
TATA box base sequence in eukaryotic promoter
TCA cycle tricarboxylic acid cycle (synonym for citric acid cycle)
TCR T-cell receptor
TEM transmission electron microscope or micrograph
Tm tubular transport maximum
TNF tumor necrosis factor
tRNA transfer RNA
U uracil
UPE upstream promoter element
UV light ultraviolet light
WBC white blood cell (leukocyte)

Appendix E

Answers to Test Your Understanding Questions

Chapter 1: 1. a 2. a 3. e 4. c 5. a 6. b 7. a 8. c 9. c 10. b 11. a 12. b 13. b 14. d 15. c 16. b 17. b 18. e

Chapter 2: 1. c 2. d 3. e 4. a 5. b 6. b 7. c 8. d 9. a 10. b 11. a 12. a 13. d 14. e 15. e 16. d 17. e

Chapter 3: 1. a 2. d 3. b 4. e 5. e 6. d 7. c 8. c 9. a 10. c 11. d 12. a 13. b 14. e 15. c 16. e 17. b

Chapter 4: 1. b 2. a 3. d 4. d 5. d 6. a 7. e 8. d 9. a 10. a 11. b 12. a 13. e 14. a 15. d

Chapter 5: 1. d 2. c 3. c 4. c 5. a 6. b 7. a 8. c 9. b 10. b 11. b 12. e 13. d 14. a 15. c 16. e 17. a

Chapter 6: 1. b 2. d 3. a 4. e 5. e 6. c 7. a 8. b 9. d 10. c 11. a 12. e 13. b 14. d 15. c

Chapter 7: 1. a 2. c 3. c 4. d 5. b 6. d 7. b 8. d 9. e 10. e 11. a 12. e 13. c 14. b 15. a

Chapter 8: 1. c 2. a 3. d 4. e 5. a 6. a 7. b 8. c 9. c 10. a 11. c 12. d 13. e 14. b 15. c

Chapter 9: 1. a 2. d 3. c 4. a 5. a 6. c 7. a 8. e 9. e 10. a 11. c 12. d 13. c 14. a 15. c 16. b

Chapter 10: 1. d 2. c 3. a 4. d 5. e 6. d 7. c 8. a 9. a 10. d 11. c 12. e 13. d 14. b 15. b

Chapter 11: 1. c 2. d 3. c 4. c 5. d 6. b 7. e 8. c 9. e 10. c 11. b 12. a

13. a. all yellow b. yellow green c. all yellow d. yellow green

14. 150

15. The short-winged condition is recessive. Both parents are heterozygous.

16. Repeated matings of the roan bull and the white cow will yield an approximate 1:1 ratio of roan-to-white offspring. Repeated matings among roan offspring will yield red, roan, and white offspring in an approximate 1:2:1 ratio. The mating of two red individuals will yield only red offspring.

17. There are 36 possible outcomes when a pair of dice is rolled. There are six ways of obtaining a 7: 1,6; 6,1; 2,5; 5,2; 3,4; and 4,3. There are five ways of rolling a 6: 1,5; 5,1; 2,4; 4,2; and 3,3. There are also five ways of rolling an 8: 2,6; 6,2; 3,5; 5,3; and 4,4.

18. The genotype of the brown, spotted rabbit is *bbSS*. The genotype of the black, solid rabbit is *BBss*. An F₁ rabbit would be black, spotted (*BbSs*). The F₂ is expected to be black, spotted (*B_S_*), black, solid (*B_ss*), brown, spotted (*bbS_*), and brown, solid (*bbss*).

19. The F₁ offspring are expected to be black, short-haired. There is a chance of brown and tan, long-haired offspring in the F₂ generation.

20. yes.

21. pleiotropic

22. The rooster is *PpRr;* hen A is *PpRR;* hen B is *Pprr;* and hen C is *PPRR.*

23. a. Both parents are heterozygous for a single locus.   b. One parent is heterozygous, and the other is homozygous recessive (for a single locus).   c. Both parents are heterozygous (for two loci).   d. One parent is heterozygous, and the other is homozygous recessive (for two loci).

24. These are the expected types of F₂ plants: of the plants produce 2-kg fruits; produce 1.75-kg fruits; produce 1.5-kg fruits; produce 1.25-kg fruits; and produce 1-kg fruits.

25. All male offspring of this cross are barred, and all females are nonbarred, thus allowing the sex of the chicks to be determined by their phenotypes.

26. This is a two-point test cross involving linked loci. The parental class of offspring are *Aabb* and *aaBb;* the recombinant classes are *AaBb* and *aabb.* There is 4.6% recombination, which corresponds to 4.6 map units between the loci.

27. If genes B and C are 10 map units apart, gene A is in the middle. If genes B and C are 2 map units apart, gene C is in the middle.

Chapter 12: 1. c 2. d 3. d 4. e 5. b 6. d 7. b 8. e 9. a 10. a 11. d 12. c 13. e 14. b 15. a 16. d 17. b

Chapter 13: 1. e 2. a 3. b 4. d 5. c 6. b 7. a 8. e 9. c 10. a 11. a 12. b 13. d 14. e 15. d 16. c 17. b 18. e

Chapter 14: 1. a 2. c 3. a 4. e 5. c 6. a 7. c 8. a 9. b 10. c 11. a 12. b 13. c 14. e 15. a 16. a 17. c

Chapter 15: 1. a 2. b 3. e 4. d 5. e 6. a 7. b 8. a 9. c 10. a 11. b 12. c 13. b 14. a

Chapter 16: 1. e 2. a 3. a 4. c 5. c 6. a 7. e 8. d 9. a 10. a 11. d 12. c 13. e 14. e 15. d 16. d 17. e 18. a 19. a 20. b 21. b 22. d 23. b 24. a 25. c 26. a 27. a 28. a

Chapter 17: 1. c 2. d 3. a 4. b 5. a 6. c 7. d 8. b 9. d 10. e 11. b 12. c 13. b 14. e 15. d 16. c

Chapter 18: 1. d 2. b 3. c 4. b 5. c 6. b 7. a 8. c 9. b 10. c 11. e 12. b 13. a 14. b 15. b 16. c

Chapter 19: 1. b 2. b 3. b 4. d 5. b 6. b 7. c 8. d 9. a 10. e 11. e 12. d 13. b 14. c 15. b

Chapter 20: 1. e 2. b 3. a 4. a 5. c 6. b 7. d 8. b 9. b 10. c 11. b 12. e 13. c 14. e 15. d

Chapter 21: 1. a 2. c 3. b 4. d 5. c 6. e 7. c 8. a 9. d 10. a 11. b 12. c 13. b 14. c 15. c

Chapter 22: 1. c 2. a 3. d 4. d 5. a 6. b 7. e 8. c 9. a 10. b 11. b 12. d 13. c 14. c 15. e

Chapter 23: 1. b 2. c 3. a 4. d 5. b 6. b 7. e 8. c 9. a 10. a 11. a 12. e 13. d 14. a 15. b 16. e 17. a

Chapter 24: 1. e 2. a 3. c 4. b 5. c 6. a 7. d 8. b 9. b 10. a 11. b 12. a 13. d 14. c 15. e 16. c 17. b 18. a 19. d 20. a

Chapter 25: 1. c 2. a 3. c 4. d 5. b 6. d 7. d 8. e 9. b 10. b 11. e 12. d 13. a 14. e 15. c

Chapter 26: 1. a 2. b 3. e 4. d 5. d 6. a 7. b 8. c 9. a 10. c 11. e 12. c 13. a 14. b 15. d 16. d 17. c 18. a

Chapter 27: 1. d 2. a 3. b 4. c 5. d 6. b 7. a 8. e 9. c 10. d 11. c 12. c 13. d 14. b 15. d

Chapter 28: 1. e 2. b 3. c 4. d 5. a 6. e 7. d 8. d 9. c 10. b 11. d 12. a 13. c 14. a 15. c

Chapter 29: 1. d 2. e 3. d 4. a 5. b 6. d 7. e 8. a 9. b 10. c 11. a 12. c 13. c 14. e 15. b 16. d

Chapter 30: 1. b 2. a 3. c 4. d 5. c 6. a 7. c 8. a 9. b 10. e 11. c 12. d 13. e 14. b 15. d 16. a 17. d 18. b 19. b 20. e

Chapter 31: 1. d 2. b 3. a 4. a 5. b 6. e 7. c 8. b 9. a 10. d 11. c 12. c 13. a 14. d 15. c 16. e

Chapter 32: 1. c 2. a 3. a 4. c 5. d 6. e 7. b 8. c 9. c 10. b 11. e 12. c 13. b 14. a 15. c 16. b 17. d

Chapter 33: 1. c 2. b 3. d 4. d 5. b 6. e 7. c 8. a 9. d 10. d 11. a 12. e 13. c 14. c 15. a 16. c 17. a

Chapter 34: 1. e 2. b 3. d 4. d 5. e 6. b 7. b 8. a 9. c 10. d 11. b 12. e 13. c 14. b 15. a 16. b

Chapter 35: 1. c 2. b 3. b 4. b 5. e 6. a 7. c 8. c 9. d 10. d 11. a 12. d 13. b 14. e 15. c 16. a

Chapter 36: 1. e 2. c 3. d 4. e 5. c 6. a 7. c 8. b 9. c 10. b 11. d 12. c 13. a 14. d 15. a

Chapter 37: 1. c 2. a 3. a 4. b 5. c 6. c 7. e 8. e 9. b 10. d 11. d 12. b 13. e 14. c 15. e 16. d

Chapter 38: 1. b 2. b 3. d 4. b 5. a 6. c 7. b 8. b 9. a 10. e 11. c 12. e 13. a 14. d 15. b 16. d 17. a 18. a 19. e

Chapter 39: 1. d 2. a 3. c 4. c 5. a 6. b 7. d 8. b 9. a 10. e 11. b 12. a 13. c 14. e 15. b 16. d

Chapter 40: 1. c 2. e 3. a 4. b 5. e 6. d 7. e 8. e 9. a 10. c 11. c 12. c 13. c 14. b 15. d 16. e

Chapter 41: 1. b 2. d 3. c 4. b 5. d 6. a 7. a 8. d 9. a 10. c 11. e 12. d 13. b 14. b 15. d 16. a

Chapter 42: 1. e 2. c 3. a 4. d 5. e 6. c 7. b 8. a 9. e 10. d 11. b 12. a 13. c 14. b 15. c 16. b

Chapter 43: 1. b 2. d 3. b 4. a 5. b 6. c 7. a 8. e 9. a 10. c 11. e 12. d 13. a 14. c 15. e 16. d 17. d

Chapter 44: 1. b 2. a 3. d 4. c 5. e 6. c 7. a 8. b 9. c 10. a 11. d 12. e 13. e 14. e 15. d 16. b 17. a

Chapter 45: 1. c 2. a 3. b 4. a 5. c 6. b 7. b 8. a 9. d 10. e 11. b 12. a 13. e 14. c 15. d 16. e

Chapter 46: 1. a 2. a 3. c 4. d 5. d 6. e 7. e 8. b 9. b 10. d 11. a 12. b 13. c 14. c 15. a 16. e

Chapter 47: 1. c 2. b 3. a 4. c 5. d 6. e 7. b 8. e 9. d 10. d 11. c 12. a 13. b 14. d 15. e

Chapter 48: 1. b 2. c 3. a 4. d 5. c 6. a 7. b 8. a 9. d 10. e 11. b 12. d 13. a 14. c 15. e 16. b

Chapter 49: 1. b 2. c 3. a 4. e 5. b 6. d 7. a 8. c 9. e 10. d 11. c 12. a 13. e 14. b 15. e

Chapter 50: 1. b 2. d 3. a 4. a 5. d 6. b 7. e 8. b 9. b 10. c 11. b 12. c 13. c 14. d

Chapter 51: 1. c 2. b 3. a 4. e 5. c 6. e 7. a 8. a 9. b 10. a 11. b 12. a 13. e 14. a 15. b 16. b 17. a 18. e

Chapter 52: 1. b 2. b 3. a 4. e 5. b 6. c 7. b 8. a 9. d 10. e 11. a 12. b 13. b 14. e 15. c

Chapter 53: 1. d 2. b 3. d 4. d 5. e 6. c 7. c 8. d 9. a 10. d 11. e 12. d 13. e 14. d

Chapter 54: 1. c 2. d 3. e 4. d 5. a 6. d 7. e 8. e 9. a 10. b 11. c 12. d 13. e 14. d 15. b 16. a

Chapter 55: 1. c 2. b 3. a 4. d 5. e 6. c 7. b 8. d 9. b 10. a 11. b 12. c 13. a 14. e 15. e

Chapter 56: 1. d 2. e 3. c 4. c 5. c 6. b 7. e 8. c 9. e 10. a 11. d 12. d 13. c 14. a 15. b

Glossary

G-1

abiotic factors Elements of the nonliving, physical environment that affect a particular organism. Compare with *biotic factors*.

abscisic acid (ab-sis′ik) A plant hormone involved in dormancy and responses to stress.

abscission (ab-sizh′en) The normal (usually seasonal) fall of leaves or other plant parts, such as fruits or flowers.

abscission layer The area at the base of the petiole where the leaf will break away from the stem. Also known as *abscission zone*.

absorption (ab-sorp′shun) (1) The movement of nutrients and other substances through the wall of the digestive tract and into the blood or lymph. (2) The process by which chlorophyll takes up light for photosynthesis.

absorption spectrum A graph of the amount of light at specific wavelengths that is absorbed as light passes through a substance. Each type of molecule has a characteristic absorption spectrum. Compare with *action spectrum*.

accessory fruit A fruit consisting primarily of tissue other than ovary tissue, e.g., apple, pear. Compare with *aggregate, simple, and multiple fruits*.

acclimatization Adjustment to seasonal changes.

acetyl coenzyme A (acetyl CoA) (as′uh-teel) A key intermediate compound in metabolism; consists of a two-carbon acetyl group covalently bonded to coenzyme A.

acetyl group A two-carbon group derived from acetic acid (acetate).

acetylcholine (ah′see-til-koh′leen) A common neurotransmitter released by cholinergic neurons, including motor neurons.

achene (a-keen′) A simple, dry fruit with one seed in which the fruit wall is separate from the seed coat, e.g., sunflower fruit.

acid A substance that is a hydrogen ion (proton) donor; acids unite with bases to form salts. Compare with *base*.

acid precipitation Precipitation that is acidic as a result of both sulfur and nitrogen oxides forming acids when they react with water in the atmosphere.

acidic solution A solution in which the concentration of hydrogen ions $[H^+]$ exceeds the concentration of hydroxide ions $[OH^-]$. An acidic solution has a pH less than 7. Compare with *basic solution* and *neutral solution*.

acoelomate (a-seel′oh-mate) An animal lacking a body cavity (coelom). Compare with *coelomate* and *pseudocoelomate*.

acquired immune responses See *specific immune responses*.

acquired immunodeficiency syndrome (AIDS) A serious, potentially fatal disease caused by the human immunodeficiency virus (HIV).

acromegaly (ak″roh-meg′ah-lee) A condition characterized by overgrowth of the extremities of the skeleton, fingers, toes, jaws, and nose. It may be produced by excessive secretion of growth hormone by the anterior pituitary gland.

acrosome reaction (ak′roh-sohm) A series of events in which the acrosome, a caplike structure covering the head of a sperm cell, releases proteolytic (protein-digesting) enzymes and undergoes other changes that permit the sperm to penetrate the outer covering of the egg.

actin (ak′tin) The protein of which microfilaments consist. Actin, together with the protein myosin, is responsible for muscle contraction.

actin filaments Thin filaments consisting mainly of the protein actin; actin and myosin filaments make up the myofibrils of muscle fibers.

actinopods (ak-tin′o-podz) Protozoa characterized by axopods that protrude through pores in their shells. See *radiolarians*.

action potential An electrical signal resulting from depolarization of the plasma membrane in a neuron or muscle cell. Compare with *resting potential*.

action spectrum A graph of the effectiveness of light at specific wavelengths in promoting a light-requiring reaction. Compare with *absorption spectrum*.

activation energy (E_A) The kinetic energy required to initiate a chemical reaction.

activator protein A positive regulatory protein that stimulates transcription when bound to DNA. Compare with *repressor protein*.

active immunity Immunity that develops as a result of exposure to antigens; it can occur naturally after recovery from a disease or can be artificially induced by immunization with a vaccine. Compare with *passive immunity*.

active site A specific region of an enzyme (generally near the surface) that accepts one or more substrates and catalyzes a chemical reaction. Compare with *allosteric site*.

active transport Transport of a substance across a membrane that does not rely on the potential energy of a concentration gradient for the substance being transported and therefore requires an additional energy source (often ATP); includes carrier-mediated active transport, endocytosis, and exocytosis. Compare with *diffusion* and *facilitated diffusion*.

adaptation (1) An evolutionary modification that improves an organism's chances of survival and reproductive success. (2) A decline in the response of a receptor subjected to repeated or prolonged stimulation.

adaptive immune responses See *specific immune responses*.

adaptive radiation The evolution of a large number of related species from an unspecialized ancestral organism.

adaptive zone A new ecological opportunity that was not exploited by an ancestral organism; used by evolutionary biologists to explain the ecological paths along which different taxa evolve.

addiction Physical dependence on a drug, generally based on physiological changes that take place in response to the drug; when the drug is withheld, the addict may suffer characteristic withdrawal symptoms.

adenine (ad′eh-neen) A nitrogenous purine base that is a component of nucleic acids and ATP.

adenosine triphosphate (ATP) (a-den′oh-seen) An organic compound containing adenine, ribose, and three phosphate groups; of prime importance for energy transfers in cells.

adhering junction A type of anchoring junction between cells; connects epithelial cells.

adhesion The property of sticking to some other substance. Compare with *cohesion*.

adipose tissue (ad′i-pohs) Tissue in which fat is stored.

adrenal cortex (ah-dree′nul kor′teks) The outer region of each adrenal gland; secretes steroid hormones, including mineralocorticoids and glucocorticoids.

adrenal glands (ah-dree′nul) Paired endocrine glands, one located just superior to each kidney; secrete hormones that help regulate metabolism and help the body cope with stress.

adrenal medulla (ah-dree′nul meh-dull′uh) The inner region of each adrenal gland; secretes epinephrine and norepinephrine.

adrenergic neuron (ad-ren-er′jik) A neuron that releases norepinephrine or epinephrine as a neurotransmitter. Compare with *cholinergic neuron*.

adventitious (ad″ven-tish′us) Of plant organs, such as roots or buds, that arise in an unusual position on a plant.

aerobe Organism that grows or metabolizes only in the presence of molecular oxygen. Compare with *anaerobe*.

aerobic (air-oh′bik) Growing or metabolizing only in the presence of molecular oxygen. Compare with *anaerobic*.

aerobic respiration See *respiration*.

afferent (af′fer-ent) Leading toward some point of reference. Compare with *efferent*.

afferent neurons Neurons that transmit action potentials from sensory receptors to the brain or spinal cord. Compare with *efferent neurons*.

age structure The number and proportion of people at each age in a population. Age structure diagrams represent the number of males and females at each age, from birth to death, in the population.

aggregate fruit A fruit that develops from a single flower with many separate carpels, e.g., raspberry. Compare with *simple, accessory,* and *multiple fruits*.

aggregated distribution See *clumped dispersion*.

aging Progressive changes in development in an adult organism.

agnathans (ag-na′thanz) Jawless fishes; historical class of vertebrates, including lampreys, hagfishes, and many extinct forms.

AIDS See *acquired immunodeficiency syndrome*.

albinism (al′bih-niz-em) A hereditary inability to form melanin pigment, resulting in light coloration.

albumin (al-bew′min) A class of protein found in most animal tissues; a fraction of plasma proteins.

aldehyde An organic molecule containing a carbonyl group bonded to at least one hydrogen atom. Compare with *ketone*.

aldosterone (al-dos′tur-ohn) A steroid hormone produced by the vertebrate adrenal cortex; stimulates sodium reabsorption. See *mineralocorticoids*.

algae (al′gee) (sing., *alga*) An informal group of unicellular, or simple multicellular, photosynthetic protists that are important producers in aquatic ecosystems.

allantois (a-lan′toe-iss) An extraembryonic membrane of reptiles, birds, and mammals that stores the embryo's nitrogenous wastes; most of the allantois is detached at hatching or birth.

allele frequency The proportion of a specific allele in the population.

alleles (al-leelz′) Genes governing variation of the same character that occupy corresponding positions (loci) on homologous chromosomes; alternative forms of a gene.

allelopathy (uh-leel′uh-path″ee) An adaptation in which toxic substances secreted by roots or shed leaves inhibit the establishment of competing plants nearby.

allergen A substance that stimulates an allergic reaction.

allergy A hypersensitivity to some substance in the environment, manifested as hay fever, skin rash, asthma, food allergies, etc.

allometric growth Variation in the relative rates of growth for different parts of the body during development.

allopatric speciation (al-oh-pa′trik) Speciation that occurs when one population becomes geographically separated from the rest of the species and subsequently evolves. Compare with *sympatric speciation*.

allopolyploid (al″oh-pol′ee-ployd) A polyploid whose chromosomes are derived from two species. Compare with *autopolyploid*.

all-or-none law The principle that neurons transmit an impulse in a similar way no matter how weak or strong the stimulus; the neuron either transmits an action potential (all) or does not (none).

allosteric regulators Substances that affect protein function by binding to allosteric sites.

allosteric site (al-oh-steer′ik) A site on an enzyme other than the active site, to which a specific substance binds, thereby changing the shape and activity of the enzyme. Compare with *active site*.

alpha (α) helix A regular, coiled type of secondary structure of a polypeptide chain, maintained by hydrogen bonds. Compare with *beta (β)-pleated sheet*.

alpine tundra An ecosystem located in the higher elevations of mountains, above the tree line and below the snow line. Compare with *tundra*.

alternation of generations A type of life cycle characteristic of plants and a few algae and fungi in which they spend part of their life in a multicellular *n* gametophyte stage and part in a multicellular *2n* sporophyte stage.

altruistic behavior Behavior in which one individual helps another, seemingly at its own risk or expense.

alveolus (al-vee′o-lus) (pl., *alveoli*) (1) An air sac of the lung through which gas exchange with the blood takes place. (2) Saclike unit of some glands, e.g., mammary glands. (3) One of several flattened vesicles located just inside the plasma membrane in certain protists.

Alzheimer's disease (AD) A progressive, degenerative brain disorder characterized by amyloid plaques and neurofibrillary tangles.

amino acid (uh-mee′no) An organic compound containing an amino group (—NH₂) and a carboxyl group (—COOH); may be joined by peptide bonds to form a polypeptide chain.

amino group A weakly basic functional group; abbreviated —NH₂.

aminoacyl-tRNA (uh-mee″no-ace′seel) Molecule consisting of an amino acid covalently linked to a transfer RNA.

aminoacyl-tRNA synthetase One of a family of enzymes, each responsible for covalently linking an amino acid to its specific transfer RNA.

ammonification (uh-moe″nuh-fah-kay′shun) The conversion of nitrogen-containing organic compounds to ammonia (NH_3) by certain soil bacteria (ammonifying bacteria); part of the nitrogen cycle.

amniocentesis (am″nee-oh-sen-tee′sis) Sampling of the amniotic fluid surrounding a fetus to obtain information about its development and genetic makeup. Compare with *chorionic villus sampling*.

amnion (am′nee-on) In terrestrial vertebrates, an extraembryonic membrane that forms a fluid-filled sac for the protection of the developing embryo.

amniotes Terrestrial vertebrates: reptiles, birds, and mammals; animals whose embryos are enclosed by an amnion.

amoeba (a-mee′ba) (pl., *amoebas*) A unicellular protozoon that moves by means of pseudopodia.

amphibians Members of vertebrate class that includes salamanders, frogs, and caecilians.

amphipathic molecule (am″fih-pa′thik) A molecule containing both hydrophobic and hydrophilic regions.

ampulla Any small, saclike extension, e.g., the expanded structure at the end of each semicircular canal of the ear.

amylase (am′-uh-laze) Starch-digesting enzyme, e.g., human salivary amylase or pancreatic amylase.

amyloplasts See *leukoplasts*.

anabolic steroids Synthetic androgens that increase muscle mass, physical strength, endurance, and aggressiveness but cause serious side effects; these drugs are often abused.

anabolism (an-ab′oh-lizm) The aspect of metabolism in which simpler substances are combined to form more complex substances, resulting in the storage of energy, the production of new cell materials, and growth. Compare with *catabolism*.

anaerobe Organism that grows or metabolizes only in the absence of molecular oxygen. See *facultative anaerobe* and *obligate anaerobe*. Compare with *aerobe*.

anaerobic (an″air-oh′bik) Growing or metabolizing only in the absence of molecular oxygen. Compare with *aerobic*.

anaerobic respiration See *respiration*.

anaphase (an′uh-faze) Stage of mitosis in which the chromosomes move to opposite poles of the cell; anaphase occurs after metaphase and before telophase.

anaphylaxis (an″uh-fih-lak′sis) An acute allergic reaction following sensitization to a foreign substance or other substance.

ancestral characters See *shared ancestral characters*.

androgen (an′dro-jen) Any substance that has masculinizing properties, such as a sex hormone. See *testosterone*.

androgen-binding protein (ABP) A protein produced by Sertoli cells in the testes; binds and concentrates testosterone.

anemia (uh-nee′mee-uh) A deficiency of hemoglobin or red blood cells.

aneuploidy (an′you-ploy-dee) Any chromosomal aberration in which there are either extra or missing copies of certain chromosomes.

angiosperms (an′jee-oh-spermz″) The traditional name for flowering plants, a very large (more than 300,000 species), diverse phylum of plants that form flowers for sexual reproduction and produce seeds enclosed in fruits; include monocots and eudicots.

angiotensin I (an-jee-o-ten′sin) A polypeptide produced by the action of renin on a plasma protein (angiotensinogen).

angiotensin II A peptide hormone formed by the action of angiotensin-converting enzyme on angiotensin I; stimulates aldosterone secretion by the adrenal cortex.

animal pole The nonyolky, metabolically active pole of a vertebrate or echinoderm egg. Compare with *vegetal pole*.

anion (an′eye-on) A particle with one or more units of negative charge, such as a chloride ion (Cl⁻) or hydroxide ion (OH⁻). Compare with *cation*.

anisogamy (an″eye-sog′uh-me) Sexual reproduction involving motile gametes of similar form but dissimilar size. Compare with *isogamy* and *oogamy*.

annelid (an′eh-lid) A member of phylum Annelida; segmented worm such as earthworm.

annual plant A plant that completes its entire life cycle in 1 year or less. Compare with *perennial* and *biennial*.

antenna complex The arrangement of chlorophyll, accessory pigments, and pigment-binding proteins into light-gathering units in the thylakoid membranes of photoautotrophic eukaryotes. See *reaction center* and *photosystem*.

antennae (sing., *antenna*) Sensory structures characteristic of some arthropod groups.

anterior Toward the head end of a bilaterally symmetrical animal. Compare with *posterior*.

anther (an′thur) The part of the stamen in flowers that produces microspores and, ultimately, pollen grains.

antheridium (an″thur-id′ee-im) (pl., *antheridia*) In plants, the multicellular male gametangium (sex organ) that produces sperm cells. Compare with *archegonium*.

anthropoid (an′thra-poid) A member of a suborder of primates that includes monkeys, apes, and humans.

antibody (an′tee-bod″ee) A specific protein (immunoglobulin) that recognizes and binds to specific antigens; produced by plasma cells.

antibody-mediated immunity A type of specific immune response in which B cells differentiate into plasma cells and produce antibodies that bind with foreign antigens, leading to the destruction of pathogens. Compare with *cell-mediated immunity*.

anticodon (an″tee-koh′don) A sequence of three nucleotides in transfer RNA that is complementary to, and combines with, the three-nucleotide codon.

antidiuretic hormone (ADH) (an″ty-dy-uh-ret′ik) A hormone secreted by the posterior lobe of the pituitary that controls the rate of water reabsorption by the kidney.

antigen (an′tih-jen) Any molecule, usually a protein or large carbohydrate, that is specifically recognized as foreign by cells of the immune system.

antigen–antibody complex The combination of antigen and antibody molecules.

antigen-presenting cell (APC) A cell that displays foreign antigens as well as its own surface proteins. Dendritic cells, macrophages, and B cells are APCs.

antimicrobial peptides Soluble molecules that destroy pathogens.

anti-oncogene See *tumor suppressor gene*.

antioxidants Certain enzymes (e.g., catalase and peroxidase), vitamins, and other substances that destroy free radicals and other reactive molecules. Compare with *oxidants*.

antiparallel Said of a double-stranded nucleic acid in which the 5′ to 3′ direction of the sugar–phosphate backbone of one strand is reversed in the other strand.

anus (ay′nus) The distal end and outlet of the digestive tract.

aorta (ay-or′tah) The largest and main systemic artery of the vertebrate body; arises from the left ventricle and branches to distribute blood to all parts of the body except the lungs.

aphotic region (ay-fote′ik) The lower layer of the ocean (deeper than 100 m or so) where light does not penetrate.

apical dominance (ape′ih-kl) The inhibition of lateral buds by a shoot tip.

apical meristem (mehr′ih-stem) An area of dividing tissue, located at the tip of a shoot or root, that gives rise to primary tissues; apical meristems cause an increase in the length of the plant body. Compare with *lateral meristems*.

apicomplexans A group of parasitic protozoa that lack structures for locomotion and that produce sporozoites as infective agents; malaria is caused by an apicomplexan. Also called *sporozoa*.

apoenzyme (ap″oh-en′zime) Protein portion of an enzyme; requires the presence of a specific coenzyme to become a complete functional enzyme.

apomixis (ap″uh-mix′us) A type of reproduction in which fruits and seeds are formed asexually.

apoplast A continuum consisting of the interconnected, porous plant cell walls, along which water moves freely. Compare with *symplast*.

apoptosis (ap-uh-toe′sis) Programmed cell death; apoptosis is a normal part of an organism's development and maintenance. Compare with *necrosis*.

aposematic coloration The conspicuous coloring of a poisonous or distasteful organism that enables potential predators to easily see and recognize it. Also called *warning coloration*. Compare with *cryptic coloration*.

aquaporin One of a family of transport proteins located in the plasma membrane that facilitate the rapid movement of water molecules into or out of cells.

arachnids (ah-rack′nidz) Eight-legged arthropods, such as spiders, scorpions, ticks, and mites.

arachnoid The middle of the three meningeal layers that cover and protect the brain and spinal cord; see *pia mater* and *dura mater*.

archaea (ar′key-ah) Prokaryotic organisms with a number of features, such as the absence of peptidoglycan in their cell walls, that set them apart from the bacteria. Archaea is the name of one of the two prokaryotic domains. Compare with *bacteria*.

Archaean eon The period of Earth's history from its beginnings approximately 4.6 billion up to 2.5 billion years ago; life originated during the Archaean.

archaic *Homo sapiens* Regionally diverse descendants of *H. erectus* or *H. ergaster* that lived in Africa, Asia, and Europe from about 400,000 to 200,000 years ago.

archegonium (ar′ke-go′nee-um) (pl., *archegonia*) In plants, the multicellular female gametangium (sex organ) that contains an egg. Compare with *antheridium*.

archenteron (ark-en′ter-on) The central cavity of the gastrula stage of embryonic development that is lined with endoderm; primitive digestive system.

arctic tundra See *tundra*.

arterial pulse See *pulse, arterial*.

arteriole (ar-teer′ee-ole) A very small artery. Vasoconstriction and vasodilation of arterioles help regulate blood pressure.

artery A thick-walled blood vessel that carries blood away from a heart chamber and toward the body organs. Compare with *vein*.

arthropod (ar′throh-pod) An invertebrate that belongs to phylum Arthropoda; characterized by a hard exoskeleton; a segmented body; and paired, jointed appendages.

artificial insemination The impregnation of a female by artificially introducing sperm from a male.

artificial selection The selection by humans of traits that are desirable in plants or animals and breeding only those individuals that have the desired traits.

ascocarp (ass′koh-karp) The fruiting body of an ascomycete.

ascomycete (ass″koh-my′seat) Member of a phylum of fungi characterized by the production of nonmotile asexual conidia and sexual ascospores.

ascospore (ass′koh-spor) One of a set of sexual spores, usually eight, contained in a special spore case (an ascus) of an ascomycete.

ascus (ass′kus) A saclike spore case in ascomycetes that contains sexual spores called *ascospores*.

asexual reproduction Reproduction in which there is no fusion of gametes and in which the genetic makeup of parent and of offspring is usually identical. Compare with *sexual reproduction*.

assimilation (of nitrogen) The conversion of inorganic nitrogen (nitrate, NO_3^-, or ammonia, NH_3) to the organic molecules of living things; part of the nitrogen cycle.

association areas Areas of the brain that link sensory and motor areas; responsible for thought, learning, memory, language abilities, judgment, and personality.

association neuron See *interneuron*.

assortative mating Sexual reproduction in which individuals pair nonrandomly, i.e., select mates on the basis of phenotype.

asters Clusters of microtubules radiating out from the poles in dividing cells that have centrioles.

astrocyte A type of glial cell; some are phagocytic; others regulate the composition of the extracellular fluid in the central nervous system.

atherosclerosis (ath″ur-oh-skle-row′sis) A progressive disease in which lipid deposits accumulate in the inner lining of arteries, leading eventually to impaired circulation and heart disease.

atom The smallest quantity of an element that retains the chemical properties of that element.

atomic mass The total number of protons and neutrons in an atom; expressed in atomic mass units or daltons.

atomic mass unit (amu) The approximate mass of a proton or neutron; also called a *dalton*.

atomic number The number of protons in the atomic nucleus of an atom, which uniquely identifies the element to which the atom corresponds.

ATP See *adenosine triphosphate*.

ATP synthase Large enzyme complex that catalyzes the formation of ATP from ADP and inorganic phosphate by chemiosmosis; located in the inner mitochondrial membrane, the thylakoid membrane of chloroplasts, and the plasma membrane of bacteria.

atrial natriuretic peptide (ANP) A hormone released by the atrium of the heart; helps regulate sodium excretion and lowers blood pressure.

atrioventricular (AV) node (ay″tree-oh-ven-trik′you-lur) Mass of specialized cardiac tissue that receives an impulse from the sinoatrial node (pacemaker) and conducts it to the ventricles.

atrioventricular (AV) valve (of the heart) A valve between each atrium and its ventricle that prevents backflow of blood. The right AV valve is the tricuspid valve; the left AV valve is the mitral valve.

atrium (of the heart) (ay′tree-um) A heart chamber that receives blood from the veins.

australopithecines Early hominids that lived between about 4 mya and 1 mya, based on fossil evidence. Includes several species in two genera, *Ardipithecus* and *Australopithecus*.

autocrine regulation A type of regulation in which a signaling molecule (e.g., a hormone) is secreted into interstitial fluid and then acts on the cells that produce it. Compare with *paracrine regulation*.

autoimmune disease (aw″toh-ih-mune′) A disease in which the body produces antibodies against its own cells or tissues. Also called *autoimmunity*.

autonomic nervous system (aw-tuh-nom′ik) The portion of the peripheral nervous system that controls the visceral functions of the body, e.g., regulates smooth muscle, cardiac muscle, and glands. Its divisions are the sympathetic and parasympathetic nervous systems. Compare with *somatic nervous system*.

autopolyploid A polyploid whose chromosomes are derived from a single species. Compare with *allopolyploid*.

autoradiography Method for detecting radioactive decay; radiation causes the appearance of dark silver grains in special X-ray film.

autosome (aw′toh-sohm) A chromosome other than the sex (X and Y) chromosomes.

autotroph (aw′toh-trof) An organism that synthesizes complex organic compounds from simple inorganic raw materials; also called *producer* or *primary producer*. Compare with *heterotroph*. See *chemoautotroph* and *photoautotroph*.

auxin (awk′sin) A plant hormone involved in various aspects of growth and development, such as stem elongation, apical dominance, and root formation on cuttings, e.g., indole acetic acid (IAA).

avirulence Properties that render an infectious agent nonlethal, i.e., unable to cause disease in its host. Compare with *virulence*.

Avogadro's number The number of units (6.02×10^{23}) present in one mole of any substance.

axillary bud A bud in the axil of a leaf. Compare with *terminal bud*.

axon (aks′on) The long extension of the neuron that transmits nerve impulses away from the cell body. Compare with *dendrite*.

axopods (aks′o-podz) Long, filamentous, cytoplasmic projections characteristic of actinopods.

B cell (B lymphocyte) The type of white blood cell responsible for antibody-mediated immunity. When stimulated, B cells differentiate to become plasma cells that produce antibodies. Compare with *T cell*.

bacillus (bah-sill′us) (pl., *bacilli*) A rod-shaped bacterium. Compare with *coccus, spirillum, vibrio*, and *spirochete*.

background extinction The continuous, low-level extinction of species that has occurred throughout much of the history of life. Compare with *mass extinction*.

bacteria (bak-teer′ee-ah) Prokaryotic organisms that have peptidoglycan in their cell walls; most are decomposers, but some are parasites and others are autotrophs. Bacteria is the name of one of the two prokaryotic domains. Compare with *archaea*.

bacterial artificial chromosome (BAC) Genetically engineered segments of chromosomal DNA with the ability to carry large segments of foreign DNA.

bacteriophage (bak-teer′ee-oh-fayj) A virus that infects a bacterium (literally, "bacteria eater"). Also called *phage*.

balanced polymorphism (pol″ee-mor′fizm) The presence in a population of two or more genetic variants that are maintained in a stable frequency over several generations.

bark The outermost covering over woody stems and roots; consists of all plant tissues located outside the vascular cambium.

baroreceptors (bare″oh-ree-sep′torz) Receptors within certain blood vessels that are stimulated by changes in blood pressure.

Barr body A condensed and inactivated X chromosome appearing as a distinctive dense spot in the interphase nucleus of certain cells of female mammals.

basal angiosperms (bay′sl) Three clades of angiosperms that are thought to be ancestral to all other flowering plants. Compare with *core angiosperms*.

basal body Structure involved in the organization and anchorage of a cilium or flagellum. Structurally similar to a centriole; each is in the form of a cylinder composed of nine triplets of microtubules (9×3 structure).

basal metabolic rate (BMR) The amount of energy expended by the body at resting conditions, when no food is being digested and no voluntary muscular work is being performed.

base (1) A substance that is a hydrogen ion (proton) acceptor; bases unite with acids to form salts. Compare with *acid*. (2) A nitrogenous base in a nucleotide or nucleic acid. See *purines* and *pyrimidines*.

basement membrane The thin, noncell layer of an epithelial membrane that attaches to the underlying tissues; consists of tiny fibers and polysaccharides produced by the epithelial cells.

base-substitution mutation A change in one base pair in DNA. See *missense mutation* and *nonsense mutation*.

basic solution A solution in which the concentration of hydroxide ions [OH^-] exceeds the concentration of hydrogen ions [H^+]. A basic solution has pH greater than 7. Compare with *acidic solution* and *neutral solution*.

basidiocarp (ba-sid′ee-o-karp) The fruiting body of a basidiomycete, e.g., a mushroom.

basidiomycete (ba-sid″ee-o-my′seat) Member of a phylum of fungi characterized by the production of sexual basidiospores.

basidiospore (ba-sid′ee-o-spor) One of a set of sexual spores, usually four, borne on a basidium of a basidiomycete.

basidium (ba-sid′ee-um) The clublike spore-producing organ of basidiomycetes that bears sexual spores called *basidiospores*.

basilar membrane The multicellular tissue in the inner ear that separates the cochlear duct from the tympanic canal; the sensory cells of the organ of Corti rest on this membrane.

Batesian mimicry (bate′see-un mim′ih-kree) The resemblance of a harmless or palatable species to one that is dangerous, unpalatable, or poisonous so that predators are more likely to avoid them. Compare with *Müllerian mimicry*.

behavioral ecology The scientific study of behavior in natural environments from the evolutionary perspective.

behavioral isolation A prezygotic reproductive isolating mechanism in which reproduction between similar species is prevented because each group exhibits its own characteristic courtship behavior; also called *sexual isolation*.

bellwether species An organism that provides an early warning of environmental damage. Examples include lichens, which are very sensitive to air pollution, and amphibians, which are sensitive to a wide variety of environmental stressors.

benthos (ben′thos) Bottom-dwelling sea organisms that fix themselves to one spot, burrow into the sediment, or simply walk about on the ocean floor.

berry A simple, fleshy fruit in which the fruit wall is soft throughout, e.g., tomato, banana, grape.

beta (β) oxidation Process by which fatty acids are converted to acetyl CoA before entry into the citric acid cycle.

beta (β)-pleated sheet A regular, folded, sheetlike type of protein secondary structure, resulting from hydrogen bonding between two different polypeptide chains or two regions of the same polypeptide chain. Compare with *alpha (α) helix*.

biennial plant (by-en′ee-ul) A plant that takes 2 years to complete its life cycle. Compare with *annual* and *perennial*.

bilateral symmetry A body shape with right and left halves that are approximately mirror images of each other. Compare with *radial symmetry*.

bile The fluid secreted by the liver; emulsifies fats.

binary fission (by′nare-ee fish′un) Equal division of a prokaryotic cell into two; a type of asexual reproduction.

binomial system of nomenclature (by-nome′ee-ul) System of naming a species by the combination of the genus name and a specific epithet.

bioaccumulation The buildup of a persistent toxic substance, such as certain pesticides, in an organism's body.

biodiversity See *biological diversity*.

biofilm An irregular layer of microorganisms embedded in the slime they secrete and concentrated at a solid or liquid surface.

biogenic amines A class of neurotransmitters that includes norepinephrine, serotonin, and dopamine.

biogeochemical cycle (bye″o-jee″o-kem′ee-kl) Process by which matter cycles from the living world to the nonliving, physical environment and back again, e.g., the carbon cycle, the nitrogen cycle, and the phosphorus cycle.

biogeography The study of the past and present geographic distributions of organisms.

bioinformatics The storage, retrieval, and comparison of biological information, particularly DNA or protein sequences within a given species and among different species.

biological clocks Mechanisms by which activities of organisms are adapted to regularly recurring changes in the environment. See *circadian rhythm.*

biological diversity The variety of living organisms considered at three levels: genetic diversity, species diversity, and ecosystem diversity. Also called *biodiversity.*

biological magnification The increased concentration of toxic chemicals, such as PCBs, heavy metals, and certain pesticides, in the tissues of organisms at higher trophic levels in food webs.

biological species concept See *species.*

biomass (bye′o-mas) A quantitative estimate of the total mass, or amount, of living material in a particular ecosystem.

biome (by′ohm) A large, relatively distinct terrestrial region characterized by a similar climate, soil, plants, and animals, regardless of where it occurs on Earth.

bioremediation A method to clean up a hazardous waste site that uses microorganisms to break down toxic pollutants, or plants to selectively accumulate toxins.

biosphere All of Earth's living organisms, collectively.

biotic factors Elements of the living world that affect a particular organism, i.e., its relationships with other organisms. Compare with *abiotic factors.*

biotic potential See *intrinsic rate of increase.*

bipedal Walking on two feet.

bipolar cell A type of neuron in the retina of the eye; receives input from the photoreceptors (rods and cones) and synapses on ganglion cells.

biramous appendages Appendages with two jointed branches at their ends; characteristic of crustaceans.

bivalent (by-vale′ent *or* biv′ah-lent) See *tetrad.*

blade (1) The thin, expanded part of a leaf. (2) The flat, leaflike structure of certain multicellular algae.

blastocoel (blas′toh-seel) The fluid-filled cavity of a blastula.

blastocyst The mammalian blastula. See *blastula.*

blastodisc A small disc of cytoplasm at the animal pole of a reptile or bird egg; cleavage is restricted to the blastodisc (meroblastic cleavage).

blastomere A cell of an early embryo.

blastopore (blas′toh-pore) The primitive opening into the body cavity of an early embryo that may become the mouth (in protostomes) or anus (in deuterostomes) of the adult organism.

blastula (blas′tew-lah) In animal development, a hollow ball of cells produced by cleavage of a fertilized ovum. In mammalian development, known as a *blastocyst.*

blood A fluid, circulating connective tissue that transports nutrients and other materials through the bodies of many types of animals.

blood pressure The force exerted by blood against the inner walls of the blood vessels.

bloom The sporadic occurrence of huge numbers of algae in freshwater and marine ecosystems.

body mass index (BMI) An index of weight in relation to height; calculated by dividing the square of the weight (square kilograms) by height (meters).

Bohr effect Increased oxyhemoglobin dissociation due to lowered pH; occurs as carbon dioxide concentration increases.

bolting The production of a tall flower stalk by a plant that grows vegetatively as a rosette (growth habit with a short stem and a circular cluster of leaves).

bond energy The energy required to break a particular chemical bond.

bone tissue Principal vertebrate skeletal tissue; a type of connective tissue.

boreal forest (bor′ee-uhl) The northern coniferous forest biome found primarily in Canada, northern Europe, and Siberia; also called *taiga.*

bottleneck A sudden decrease in a population size caused by adverse environmental factors; may result in genetic drift; also called *genetic bottleneck* or *population bottleneck.*

bottom-up processes Control of ecosystem function by nutrient cycles and other parts of the abiotic environment. Compare with *top-down processes.*

Bowman's capsule A double-walled sac of cells that surrounds the glomerulus of each nephron.

brachiopods (bray′kee-oh-podz) The phylum of solitary marine invertebrates having a pair of shells and, internally, a pair of coiled arms with ciliated tentacles; one of the lophophorate phyla.

brain A concentration of nervous tissue that controls neural function; in vertebrates, the anterior, enlarged portion of the central nervous system.

brain stem The part of the vertebrate brain that includes the medulla, pons, and midbrain.

branchial Pertaining to the gills or gill region.

brassinosteroid (BR) One of a group of steroids that function as plant hormones and are involved in several aspects of growth and development.

bronchiole (bronk′ee-ole) Air duct in the lung that branches from a bronchus; divides to form air sacs (alveoli).

bronchus (bronk′us) (pl., *bronchi*) One branch of the trachea and its immediate branches within the lung.

brown alga One of a phylum of predominantly marine algae that are multicellular and contain the pigments chlorophyll *a* and *c*, and carotenoids, including fucoxanthin.

bryophytes (bry′oh-fites) Nonvascular plants including mosses, liverworts, and hornworts.

bryozoans Animals belonging to phylum Bryozoa, one of the three lophophorate phyla; form sessile colonies by asexual budding.

bud An undeveloped shoot that develops into flowers, stems, or leaves. Buds are enclosed in bud scales.

bud scale A modified leaf that covers and protects a dormant bud.

bud scale scar Scar on a twig left when a bud scale abscises from the terminal bud.

budding Asexual reproduction in which a small part of the parent's body separates from the rest and develops into a new individual; characteristic of yeasts and certain other organisms.

buffer A substance in a solution that tends to lessen the change in hydrogen ion concentration (pH) that otherwise would be produced by adding an acid or base.

bulb A globose, fleshy, underground bud that consists of a short stem with fleshy leaves, e.g., onion.

bundle scar Marks on a leaf scar left when vascular bundles of the petiole break during leaf abscission.

bundle sheath cells Tightly packed cells that form a sheath around the veins of a leaf.

bundle sheath extension Support cells that extend from the bundle sheath of a leaf vein toward the upper and/or lower epidermis.

buttress root A bracelike root at the base of certain trees that provides upright support.

C₃ plant Plant that carries out carbon fixation solely by the Calvin cycle. Compare with *C₄ plant* and *CAM plant.*

C₄ plant Plant that fixes carbon initially by a pathway, in which the reaction of CO_2 with phosphoenolpyruvate is catalyzed by PEP carboxylase in leaf mesophyll cells; the products are transferred to the bundle sheath cells, where the Calvin cycle takes place. Compare with *C₃ plant* and *CAM plant.*

calcitonin (kal-sih-toh′nin) A hormone secreted by the thyroid gland that rapidly lowers the calcium content in the blood.

callus (kal′us) Undifferentiated tissue formed on an explant (excised tissue or organ) in plant tissue culture.

calmodulin A calcium-binding protein; when bound, it alters the activity of certain enzymes or transport proteins.

calorie The amount of heat energy required to raise the temperature of 1g of water 1°C; equivalent to 4.184 joules. Compare with *kilocalorie.*

Calvin cycle Cyclic series of reactions in the chloroplast stroma in photosynthesis; fixes carbon dioxide and produces carbohydrate. See C_3 *plant.*

calyx (kay′liks) The collective term for the sepals of a flower.

CAM plant Plant that carries out crassulacean acid metabolism; carbon is initially fixed into organic acids at night in the reaction of CO_2 and phosphoenolpyruvate, catalyzed by PEP carboxylase; during the day the acids break down to yield CO_2, which enters the Calvin cycle. Compare with C_3 *plant* and C_4 *plant.*

cambium See *lateral meristems.*

Cambrian Radiation A span of 40 million years, beginning about 542 mya, during which many new animal groups appeared in the fossil record.

cAMP See *cyclic AMP.*

cancer cells See *malignant cells.*

CAP See *catabolite activator protein.*

capillaries (kap′i-lare-eez) Microscopic blood vessels in the tissues that permit exchange of materials between cells and blood.

capillary action The ability of water to move in small-diameter tubes as a consequence of its cohesive and adhesive properties.

capping See *mRNA cap.*

capsid Protein coat surrounding the nucleic acid of a virus.

capsule (1) The portion of the moss sporophyte that contains spores. (2) A simple, dry, dehiscent fruit that develops from two or more fused carpels and opens along many sutures or pores to release seeds. (3) A gelatinous coat that surrounds some bacteria.

carbohydrate Compound containing carbon, hydrogen, and oxygen, in the approximate ratio of C:2H:O, e.g., sugars, starch, and cellulose.

carbon cycle The worldwide circulation of carbon from the abiotic environment into living things and back into the abiotic environment.

carbon fixation reactions Reduction reactions of photosynthesis in which carbon from carbon dioxide becomes incorporated into organic molecules, leading to the production of carbohydrate. See *Calvin cycle.*

carbonyl group A polar functional group consisting of a carbon attached to an oxygen by a double bond; found in aldehydes and ketones.

carboxyl group A weakly acidic functional group; abbreviated —COOH.

carcinogen (kar-sin′oh-jen) An agent that causes cancer or accelerates its development.

cardiac cycle One complete heartbeat.

cardiac muscle Involuntary, striated type of muscle found in the vertebrate heart. Compare with *smooth muscle* and *skeletal muscle.*

cardiac output The volume of blood pumped by the left ventricle into the aorta in 1 minute.

cardiovascular disease Disease of the heart or blood vessels; the leading cause of death in most industrial societies.

carnivore (kar′ni-vor) An animal that feeds on other animals; flesh eater; also called *secondary* or *tertiary consumer.* Secondary consumers eat primary consumers (herbivores), whereas tertiary consumers eat secondary consumers.

carotenoids (ka-rot′n-oidz) A group of yellow to orange plant pigments synthesized from isoprene subunits; include carotenes and xanthophylls.

carpel (kar′pul) The female reproductive unit of a flower; carpels bear ovules. Compare with *pistil.*

carrier-mediated active transport Transport across a membrane of a substance from a region of low concentration to a region of high concentration; requires both a transport protein with a binding site for the specific substance and an energy source (often ATP).

carrier-mediated transport Any form of transport across a membrane that uses a membrane-bound transport protein with a binding site for a specific substance; includes both facilitated diffusion and carrier-mediated active transport.

carrying capacity The largest population that a particular habitat can support and sustain for an indefinite period, assuming there are no changes in the environment.

cartilage A flexible skeletal tissue of vertebrates; a type of connective tissue.

caryopsis See *grain.*

Casparian strip (kas-pare′ee-un) A band of waterproof material around the radial and transverse walls of endodermal root cells.

caspase Any of a group of proteolytic enzymes that are active in the early stages of apoptosis.

catabolism The aspect of metabolism in which complex substances are broken down to form simpler substances; catabolic reactions are particularly important in releasing chemical energy stored by the cell. Compare with *anabolism.*

catabolite activator protein (CAP) A positively acting regulator that becomes active when bound to cAMP; active CAP stimulates transcription of the *lac* operon and other operons that code for enzymes used in catabolic pathways.

catalyst (kat′ah-list) A substance that increases the speed at which a chemical reaction occurs without being used up in the reaction. Enzymes are biological catalysts.

catecholamine (cat″eh-kole′-ah-meen) A class of compounds including dopamine, epinephrine, and norepinephrine; these compounds serve as neurotransmitters and hormones.

cation A particle with one or more units of positive charge, such as a hydrogen ion (H^+) or calcium ion (Ca^{2+}). Compare with *anion.*

CD4 T cell See *T helper cell.*

CD8 T cell See *T cytotoxic cell.*

cDNA library A collection of recombinant plasmids that contain complementary DNA (cDNA) copies of mRNA templates. The cDNA, which lacks introns, is synthesized by reverse transcriptase. Compare with *genomic DNA library.*

cell The basic structural and functional unit of life, which consists of living material enclosed by a membrane.

cell cycle Cyclic series of events in the life of a dividing eukaryotic cell; consists of mitosis, cytokinesis, and the stages of interphase.

cell-cycle control system Regulatory molecules that control key events in the cell cycle; common to all eukaryotes.

cell determination See *determination.*

cell differentiation See *differentiation.*

cell fractionation The technique used to separate the components of cells by subjecting them to centrifugal force. See *differential centrifugation* and *density gradient centrifugation.*

cell-mediated immunity A type of specific immune response carried out by T cells. Compare with *antibody-mediated immunity.*

cell plate The structure that forms during cytokinesis in plants, separating the two daughter cells produced by mitosis.

cell signaling Mechanisms of communication between cells. Cells signal one another with secreted signaling molecules, or a signaling molecule on one cell combines with a receptor on another cell. See *signal transduction*.

cell theory The theory that the cell is the basic unit of life, of which all living things are composed, and that all cells are derived from pre-existing cells.

cell wall The structure outside the plasma membrane of certain cells; may contain cellulose (plant cells), chitin (most fungal cells), peptidoglycan and/or lipopolysaccharide (most bacterial cells), or other material.

cellular respiration See *respiration*.

cellular slime mold A phylum of funguslike protists whose feeding stage consists of unicellular, amoeboid organisms that aggregate to form a pseudoplasmodium during reproduction.

cellulose (sel′yoo-lohs) A structural polysaccharide consisting of beta glucose subunits; the main constituent of plant primary cell walls.

Cenozoic era A geologic era that began about 66 million years ago and extends to the present.

center of origin The geographic area where a given species originated.

central nervous system (CNS) In vertebrates, the brain and spinal cord. Compare with *peripheral nervous system*.

centrifuge A device used to separate cells or their components by subjecting them to centrifugal force.

centriole (sen′tree-ohl) One of a pair of small, cylindrical organelles lying at right angles to each other near the nucleus in the cytoplasm of animal cells and certain protist and plant cells; each centriole is in the form of a cylinder composed of nine triplets of microtubules (9×3 structure).

centromere (sen′tro-meer) A specialized constricted region of a chromatid; contains the kinetochore. In cells at prophase and metaphase, sister chromatids are joined in the vicinity of their centromeres.

centrosome (sen′tro-sowm) An organelle in animal cells that is the main microtubule-organizing center; typically contains a pair of centrioles and is important in cell division.

cephalization The evolution of a head; the concentration of nervous tissue and sense organs at the front end of the animal.

cephalochordates Members of the chordate subphylum that includes the lancelets.

cerebellum (ser-eh-bel′um) A convoluted subdivision of the vertebrate brain concerned with coordination of muscular movements, muscle tone, and balance.

cerebral cortex (ser-ee′brul kor′teks) The outer layer of the cerebrum composed of gray matter and consisting mainly of nerve cell bodies.

cerebrospinal fluid (CSF) The fluid that bathes the central nervous system of vertebrates.

cerebrum (ser-ee′brum) A large, convoluted subdivision of the vertebrate brain; in humans, it functions as the center for learning, voluntary movement, and interpretation of sensation.

chaos The tendency of a simple system to exhibit complex, erratic dynamics; used by some ecologists to model the state of flux displayed by some populations.

chaparral (shap″uh-ral′) A biome with a Mediterranean climate (mild, moist winters and hot, dry summers). Chaparral vegetation is characterized by drought-resistant, small-leaved evergreen shrubs and small trees.

chaperones See *molecular chaperones*.

character displacement The tendency for two similar species to diverge (become more different) in areas where their ranges overlap; reduces interspecific competition.

Chargaff's rules A relationship in DNA molecules based on nucleotide composition data; the number of adenines equals the number of thymines, and the number of guanines equals the number of cytosines.

chelicerae (keh-lis′er-ee) The first pair of appendages in certain arthropods; clawlike appendages located immediately anterior to the mouth and used to manipulate food into the mouth.

chemical bond A force of attraction between atoms in a compound. See *covalent bond, hydrogen bond,* and *ionic bond*.

chemical compound Two or more elements combined in a fixed ratio.

chemical evolution The origin of life from nonliving matter.

chemical formula A representation of the composition of a compound; the elements are indicated by chemical symbols with subscripts to indicate their ratios. See *molecular formula, structural formula,* and *simplest formula*.

chemical symbol The abbreviation for an element; usually the first letter (or first and second letters) of the English or Latin name.

chemiosmosis Process by which phosphorylation of ADP to form ATP is coupled to the transfer of electrons down an electron transport chain; the electron transport chain powers proton pumps that produce a proton gradient across the membrane; ATP is formed as protons diffuse through transmembrane channels in ATP synthase.

chemoautotroph (kee″moh-aw′toh-trof) Organism that obtains energy from inorganic compounds and synthesizes organic compounds from inorganic raw materials; includes some bacteria. Compare with *photoautotroph, photoheterotroph,* and *chemoheterotroph*.

chemoheterotroph (kee″moh-het′ur-oh-trof) Organism that uses organic compounds as a source of energy and carbon; includes animals, fungi, and many bacteria. Compare with *photoautotroph, photoheterotroph,* and *chemoautotroph*.

chemoreceptor (kee″moh-ree-sep′tor) A sensory receptor that responds to chemical stimuli.

chemotroph (kee′moh-trof) Organism that uses organic compounds or inorganic substances, such as iron, nitrate, ammonia, or sulfur, as sources of energy. Compare with *phototroph*. See *chemoautotroph* and *chemoheterotroph*.

chiasma (ky-az′muh) (pl., *chiasmata*) An X-shaped site in a tetrad (bivalent) usually marking the location where homologous (nonsister) chromatids previously crossed over.

chimera (ky-meer′uh) An organism consisting of two or more kinds of genetically dissimilar cells.

chitin (ky′tin) A nitrogen-containing structural polysaccharide that forms the exoskeleton of insects and the cell walls of many fungi.

chlorophyll (klor′oh-fil) A group of light-trapping green pigments found in most photosynthetic organisms.

chlorophyll-binding proteins About 15 different proteins associated with chlorophyll molecules in the thylakoid membrane.

chloroplasts (klor′oh-plastz) Membranous organelles that are the sites of photosynthesis in eukaryotes; occur in some plant and algal cells.

cholinergic neuron (kohl″in-air′jik) A neuron that releases acetylcholine as a neurotransmitter. Compare with *adrenergic neuron*.

chondrichthyes (kon-drik′-thees) The class of cartilaginous fishes that includes the sharks, rays, and skates.

chondrocytes Cartilage cells.

chordates (kor′dates) Deuterostome animals that, at some time in their lives, have a cartilaginous, dorsal skeletal structure called a *notochord*; a dorsal, tubular, nerve cord; pharyngeal gill grooves; and a postanal tail.

chorion (kor′ee-on) An extraembryonic membrane in reptiles, birds, and mammals that forms an outer cover around the embryo, and in mammals contributes to the formation of the placenta.

chorionic villus sampling (CVS) (kor″ee-on′ik) Study of extra-embryonic cells that are genetically identical to the cells of an embryo, making it possible to assess its genetic makeup. Compare with *amniocentesis*.

choroid layer A layer of cells filled with black pigment that absorbs light and prevents reflected light from blurring the image that falls on the retina; the layer of the eyeball outside the retina.

chromatid (kroh′mah-tid) One of the two identical halves of a duplicated chromosome; the two chromatids that make up a chromosome are referred to as *sister chromatids*.

chromatin (kro′mah-tin) The complex of DNA and protein that makes up eukaryotic chromosomes.

chromoplasts Pigment-containing plastids; usually found in flowers and fruits.

chromosome theory of inheritance A basic principle in biology that states that inheritance can be explained by assuming that genes are linearly arranged in specific locations along the chromosomes.

chromosomes Structures in the cell nucleus that consist of chromatin and contain the genes. The chromosomes become visible under the microscope as distinct structures during cell division.

chylomicrons (kie-low-my′kronz) Protein-covered fat droplets produced in the intestinal cells; they enter the lymphatic system and are transported to the blood.

chytrid See *chytridiomycete*.

chytridiomycete (ki-trid″ee-o-my′seat) A member of a phylum of fungi characterized by the production of flagellate cells at some stage in their life history. Also called *chytrid*.

ciliate (sil′e-ate) A unicellular protozoon covered by many short cilia.

cilium (sil′ee-um) (pl., *cilia*) One of many short, hairlike structures that project from the surface of some eukaryotic cells and are used for locomotion or movement of materials across the cell surface.

circadian rhythm (sir-kay′dee-un) An internal rhythm that approximates the 24-hour day. See *biological clocks*.

circulatory system The body system that functions in internal transport and protects the body from disease.

cisternae (sing., *cisterna*) Stacks of flattened membranous sacs that make up the Golgi complex.

citrate (citric acid) A six-carbon organic acid.

citric acid cycle Series of chemical reactions in aerobic respiration in which acetyl coenzyme A is completely degraded to carbon dioxide and water with the release of metabolic energy that is used to produce ATP; also known as the *Krebs cycle* and the *tricarboxylic acid (TCA) cycle*.

clade A group of organisms containing a common ancestor and all its descendants; a monophyletic group.

cladistics An approach to classification based on recency of common ancestry rather than degree of structural similarity. Also called *phylogenetic systematics*. Compare with *phenetics* and *evolutionary systematics*.

cladogram A branching diagram that illustrates taxonomic relationships based on the principles of cladistics.

class A taxonomic category made up of related orders.

classical conditioning A type of learning in which an association is formed between some normal response to a stimulus and a new stimulus, after which the new stimulus elicits the response.

cleavage Series of mitotic cell divisions, without growth, that converts the zygote to a multicellular blastula.

cleavage furrow A constricted region of the cytoplasm that forms and progressively deepens during cytokinesis of animal cells, thereby separating the two daughter cells.

cline Gradual change in phenotype and genotype frequencies among contiguous populations that is the result of an environmental gradient.

clitoris (klit′o-ris) A small, erectile structure at the anterior part of the vulva in female mammals; homologous to the male penis.

cloaca (klow-a′ka) An exit chamber in some animals that receives digestive wastes and urine; may also serve as an exit for gametes.

clonal selection Lymphocyte activation in which a specific antigen causes activation, cell division, and differentiation only in cells that express receptors with which the antigen binds.

clone (1) A population of cells descended by mitotic division from a single ancestral cell. (2) A population of genetically identical organisms asexually propagated from a single individual. Also see *DNA cloning*.

cloning The process of forming a clone.

closed circulatory system A type of circulatory system in which the blood flows through a continuous circuit of blood vessels; characteristic of annelids, cephalopods, and vertebrates. Compare with *open circulatory system*.

closed system An entity that does not exchange energy with its surroundings. Compare with *open system*.

club mosses A phylum of seedless vascular plants with a life cycle similar to that of ferns.

clumped dispersion The spatial distribution pattern of a population in which individuals are more concentrated in specific parts of the habitat. Also called *aggregated distribution* and *patchiness*. Compare with *random dispersion* and *uniform dispersion*.

cnidarians (ni-dah′ree-anz) Phylum of animals that have stinging cells called *cnidocytes,* two tissue layers, and radial symmetry; include hydras and jellyfish.

cnidocytes Stinging cells characteristic of cnidarians.

coated pit A depression in the plasma membrane, the cytosolic side of which is coated with the protein clathrin; important in receptor-mediated endocytosis.

cochlea (koke′lee-ah) The structure of the inner ear of mammals that contains the auditory receptors (organ of Corti).

coccus (kok′us) (pl., *cocci*) A bacterium with a spherical shape. Compare with *bacillus, spirillum, vibrio,* and *spirochete*.

codominance (koh″dom′in-ants) Condition in which two alleles of a locus are expressed in a heterozygote.

codon (koh′don) A triplet of mRNA nucleotides. The 64 possible codons collectively constitute a universal genetic code in which each codon specifies an amino acid in a polypeptide, or a signal to either start or terminate polypeptide synthesis.

coelacanths A genus of lobe-finned fishes that have survived to the present day.

coelom (see′lum) The main body cavity of most animals; a true coelom is lined with mesoderm. Compare with *pseudocoelom*.

coelomate (seel′oh-mate) Animal that has a true coelom. Compare with *acoelomate* and *pseudocoelomate*.

coenocyte (see′no-site) An organism consisting of a multinucleate cell, i.e., the nuclei are not separated from one another by septa.

coenzyme (koh-en′zime) An organic cofactor for an enzyme; generally participates in the reaction by transferring some component, such as electrons or part of a substrate molecule.

coenzyme A (CoA) Organic cofactor responsible for transferring groups derived from organic acids.

coevolution The reciprocal adaptation of two or more species that occurs as a result of their close interactions over a long period.

cofactor A nonprotein substance needed by an enzyme for normal activity; some cofactors are inorganic (usually metal ions); others are organic (coenzymes).

cohesion The property of sticking together. Compare with *adhesion*.

cohort A group of individuals of the same age.

colchicine A drug that blocks the division of eukaryotic cells by binding to tubulin subunits, which make up the microtubules, the major component of the mitotic spindle.

coleoptile (kol-ee-op′tile) A protective sheath that encloses the young stem in certain monocots.

collagens (kol′ah-gen) Proteins found in the collagen fibers of connective tissues.

collecting duct A tube in the kidney that receives filtrate from several nephrons and conducts it to the renal pelvis.

collenchyma (kol-en′kih-mah) Living cells with moderately but unevenly thickened primary cell walls; collenchyma cells help support the herbaceous plant body.

colony An association of loosely connected cells or individuals of the same species.

commensalism (kuh-men′sul-izm) A type of symbiosis in which one organism benefits and the other one is neither harmed nor helped. Compare with *mutualism* and *parasitism*.

commercial harvest The collection of commercially important organisms from the wild. Examples include the commercial harvest of parrots (for the pet trade) and cacti (for houseplants).

community An association of populations of different species living together in a defined habitat with some degree of interdependence. Compare with *ecosystem*.

community ecology The description and analysis of patterns and processes within the community.

compact bone Dense, hard bone tissue found mainly near the surfaces of a bone.

companion cell A cell in the phloem of flowering plants that governs loading and unloading sugar into the sieve tube element for translocation.

compass sense The sense of direction an animal requires to travel in a straight line toward a destination.

competition The interaction among two or more individuals that attempt to use the same essential resource, such as food, water, sunlight, or living space. See *interspecific* and *intraspecific competition*. See *interference* and *exploitation competition*.

competitive exclusion principle The concept that no two species with identical living requirements can occupy the same ecological niche indefinitely.

competitive inhibitor A substance that binds to the active site of an enzyme, thus lowering the rate of the reaction catalyzed by the enzyme. Compare with *noncompetitive inhibitor*.

complement A group of proteins in blood and other body fluids that are activated by an antigen–antibody complex and then destroy pathogens.

complementary DNA (cDNA) DNA synthesized by reverse transcriptase, using RNA as a template.

complete flower A flower that has all four parts: sepals, petals, stamens, and carpels. Compare with *incomplete flower*.

compound eye An eye, such as that of an insect, consisting of many light-sensitive units called *ommatidia*.

concentration gradient A difference in the concentration of a substance from one point to another, as for example, across a cell membrane.

condensation synthesis A reaction in which two monomers are combined covalently through the removal of the equivalent of a water molecule. Compare with *hydrolysis*.

cone (1) In botany, a reproductive structure in many gymnosperms that produces either microspores or megaspores. (2) In zoology, one of the conical photoreceptive cells of the retina that is particularly sensitive to bright light and, by distinguishing light of various wavelengths, mediates color vision. Compare with *rod*.

conidiophore (kah-nid′e-o-for″) A specialized hypha that bears conidia.

conidium (kah-nid′e-um) (pl., *conidia*) An asexual spore that is usually formed at the tip of a specialized hypha called a *conidiophore*.

conifer (kon′ih-fur) Any of a large phylum of gymnosperms that are woody trees and shrubs with needlelike, mostly evergreen leaves and with seeds in cones.

conjugation (kon″jew-gay′shun) (1) A sexual process in certain protists that involves exchange or fusion of a cell with another cell. (2) A mechanism for DNA exchange in bacteria that involves cell-to-cell contact.

connective tissue Animal tissue consisting mostly of intercellular substance (fibers scattered through a matrix) in which the cells are embedded, e.g., bone.

conservation biology A multidisciplinary science that focuses on the study of how humans impact organisms and on the development of ways to protect biological diversity.

constitutive gene A gene that is constantly transcribed.

consumer See *heterotroph*.

consumption overpopulation A situation in which each individual in a human population consumes too large a share of resources; results in pollution, environmental degradation, and resource depletion. Compare with *people overpopulation*.

contest competition See *interference competition*.

continental drift The theory that continents were once joined and later split and drifted apart.

contraception Any method used to intentionally prevent pregnancy.

contractile root (kun-trak′til) A specialized type of root that contracts and pulls a bulb or corm deeper into the soil.

contractile vacuole A membrane-enclosed organelle found in certain freshwater protists, such as *Paramecium*; appears to have an osmoregulatory function.

control group In a scientific experiment, a group in which the experimental variable is kept constant. The control provides a standard of comparison used to verify the results of the experiment.

controlled mating A mating in which the genotypes of the parents are known.

convergent circuit (kun-vur′jent) A neural pathway in which a postsynaptic neuron is controlled by signals coming from two or more presynaptic neurons. Compare with *divergent circuit*.

convergent evolution (kun-vur′jent) The independent evolution of structural or functional similarity in two or more distantly related species, usually as a result of adaptations to similar environments.

core angiosperms The clade to which most angiosperm species belong. Core angiosperms are divided into three subclades: magnoliids, monocots, and eudicots. Compare with *basal angiosperms*.

corepressor Substance that binds to a repressor protein, converting it to its active form, which is capable of preventing transcription.

Coriolis effect (kor″e-o′lis) The tendency of moving air or water to be deflected from its path to the right in the Northern Hemisphere and to the left in the Southern Hemisphere. Caused by the direction of Earth's rotation.

cork cambium (kam'bee-um) A lateral meristem that produces cork cells and cork parenchyma; cork cambium and the tissues it produces make up the outer bark of a woody plant. Compare with *vascular cambium*.

cork cell A cell in the bark that is produced outwardly by the cork cambium; cork cells are dead at maturity and function for protection and reduction of water loss.

cork parenchyma (par-en'kih-mah) One or more layers of parenchyma cells produced inwardly by the cork cambium.

corm A short, thickened underground stem specialized for food storage and asexual reproduction, e.g., crocus, gladiolus.

cornea (kor'nee-ah) The transparent covering of an eye.

corolla (kor-ohl'ah) Collectively, the petals of a flower.

corpus callosum (kah-loh'sum) In mammals, a large bundle of nerve fibers interconnecting the two cerebral hemispheres.

corpus luteum (loo'tee"um) The temporary endocrine tissue in the ovary that develops from the ruptured follicle after ovulation; secretes progesterone and estrogen.

cortex (kor'teks) (1) The outer part of an organ, such as the cortex of the kidney. Compare with *medulla*. (2) The tissue between the epidermis and vascular tissue in the stems and roots of many herbaceous plants.

cortical reaction Process occurring after fertilization that prevents additional sperm from entering the egg; also known as the "slow block to polyspermy."

cortisol A steroid hormone, secreted by the adrenal cortex, that helps the body adjust to long-term stress; stimulates conversion of other nutrients to glucose in the liver, resulting in increased blood glucose concentration.

cosmid cloning vector A cloning vector with features of both bacteriophages and plasmids and with the ability to carry large segments of foreign DNA.

cosmopolitan species Species that have a nearly worldwide distribution and occur on more than one continent or throughout much of the ocean. Compare with *endemic species*.

cotransport The active transport of a substance from a region of low concentration to a region of high concentration by coupling its transport to the transport of a substance down its concentration gradient.

cotyledon (kot"uh-lee'dun) The seed leaf of a plant embryo, which may contain food stored for germination.

cotylosaurs The first reptiles; also known as *stem reptiles*.

countercurrent exchange system A biological mechanism that enables maximum exchange between two fluids. The two fluids must be flowing in opposite directions and have a concentration gradient between them.

coupled reactions A set of reactions in which an exergonic reaction provides the free energy required to drive an endergonic reaction; energy coupling generally occurs through a common intermediate.

covalent bond The chemical bond involving shared pairs of electrons; may be single, double, or triple (with one, two, or three shared pairs of electrons, respectively). Compare with *ionic bond* and *hydrogen bond*.

covalent compound A compound in which atoms are held together by covalent bonds; covalent compounds consist of molecules. Compare with *ionic compound*.

cranial nerves The 10 to 12 pairs of nerves in vertebrates that emerge directly from the brain.

cranium The bony framework that protects the brain in vertebrates.

crassulacean acid metabolism See *CAM plant*.

creatine phosphate An energy-storing compound in muscle cells.

cretinism (kree'tin-izm) A chronic condition caused by lack of thyroid secretion during fetal development and early childhood; results in retarded physical and mental development if untreated.

cri du chat syndrome A human genetic disease caused by losing part of the short arm of chromosome 5 and characterized by mental retardation, a cry that sounds like a kitten mewing, and death in infancy or childhood.

cristae (kris'tee) (sing., *crista*) Shelflike or fingerlike inward projections of the inner membrane of a mitochondrion.

Cro-Magnons Prehistoric humans (*Homo sapiens*) with modern features (tall, erect, lacking a heavy brow) who lived in Europe some 30,000 years ago.

cross bridges The connections between myosin and actin filaments in muscle fibers; formed by the binding of myosin heads to active sites on actin filaments.

crossing-over A process in which genetic material (DNA) is exchanged between paired, homologous chromosomes.

cryptic coloration Colors or markings that help some organisms hide from predators by blending into their physical surroundings. Compare with *aposematic coloration*.

cryptochrome A proteinaceous pigment that strongly absorbs blue light; implicated in resetting the biological clock in plants, fruit flies, and mice.

ctenophores (ten'oh-forz) Phylum of marine animals (comb jellies) whose bodies consist of two layers of cells enclosing a gelatinous mass. The outer surface is covered with comblike rows of cilia, by which the animal moves.

cuticle (kew'tih-kl) (1) A noncell, waxy covering over the epidermis of the aerial parts of plants that reduces water loss. (2) The outer covering of some animals, such as roundworms.

cyanobacteria (sy-an"oh-bak-teer'ee-uh) Prokaryotic photosynthetic microorganisms that possess chlorophyll and produce oxygen during photosynthesis. Formerly known as blue-green algae.

cycad (sih'kad) Any of a phylum of gymnosperms that live mainly in tropical and semitropical regions and have stout stems (to 20 m in height) and fernlike leaves.

cyclic AMP (cAMP) A form of adenosine monophosphate in which the phosphate is part of a ring-shaped structure; acts as a regulatory molecule and second messenger in organisms ranging from bacteria to humans.

cyclic electron transport In photosynthesis, the cyclic flow of electrons through photosystem I; ATP is formed by chemiosmosis, but no photolysis of water occurs, and O_2 and NADPH are not produced. Compare with *noncyclic electron transport*.

cyclin-dependent kinases (Cdks) Protein kinases involved in controlling the cell cycle.

cyclins Regulatory proteins whose levels oscillate during the cell cycle; activate cyclin-dependent kinases.

cystic fibrosis A genetic disease with an autosomal recessive inheritance pattern; characterized by secretion of abnormally thick mucus, particularly in the respiratory and digestive systems.

cytochromes (sy'toh-krohmz) Iron-containing heme proteins of an electron transport system.

cytokines Signaling proteins that regulate interactions between cells in the immune system. Important groups include interferons, interleukins, tumor necrosis factors, and chemokines.

cytokinesis (sy"toh-kih-nee'sis) Stage of cell division in which the cytoplasm divides to form two daughter cells.

cytokinin (sy"toh-ky'nin) A plant hormone involved in various aspects of plant growth and development, such as cell division and delay of senescence.

cytoplasm The plasma membrane and cell contents with the exception of the nucleus.

cytosine A nitrogenous pyrimidine base that is a component of nucleic acids.

cytoskeleton The dynamic internal network of protein fibers that includes microfilaments, intermediate filaments, and microtubules.

cytosol The fluid component of the cytoplasm in which the organelles are suspended.

cytotoxic T cell See *T cytotoxic cell*.

dalton See *atomic mass unit (amu)*.

day-neutral plant A plant whose flowering is not controlled by variations in day length that occur with changing seasons. Compare with *long-day, short-day,* and *intermediate-day plants*.

deamination (dee-am-ih-nay′shun) The removal of an amino group (—NH₂) from an amino acid or other organic compound.

decarboxylation A reaction in which a molecule of CO₂ is removed from a carboxyl group of an organic acid.

deciduous A term describing a plant that sheds leaves or other structures at regular intervals, e.g., during autumn. Compare with *evergreen*.

decomposers Microbial heterotrophs that break down dead organic material and use the decomposition products as a source of energy. Also called *saprotrophs* or *saprobes*.

deductive reasoning The reasoning that operates from generalities to specifics and can make relationships among data more apparent. Compare with *inductive reasoning*. See *hypothetico-deductive approach*.

deforestation The temporary or permanent removal of forest for agriculture or other uses.

dehydrogenation (dee-hy″dro-jen-ay′shun) A form of oxidation in which hydrogen atoms are removed from a molecule.

deletion (1) A chromosome abnormality in which part of a chromosome is missing, e.g., cri du chat syndrome. (2) The loss of one or more base pairs from DNA, which can result in a frameshift mutation.

demographics The science that deals with human population statistics, such as size, density, and distribution.

denature (dee-nay′ture) To alter the physical properties and three-dimensional structure of a protein, nucleic acid, or other macromolecule by treating it with excess heat, strong acids, or strong bases.

dendrite (den′drite) A branch of a neuron that receives and conducts nerve impulses toward the cell body. Compare with *axon*.

dendritic cells A set of immune cells present in many tissues that capture antigens and present them to T cells.

dendrochronology (den″dro-kruh-naal′uh-gee) A method of dating using the annual rings of trees.

denitrification (dee-nie″tra-fuh-kay′shun) The conversion of nitrate (NO₃⁻) to nitrogen gas (N₂) by certain bacteria (denitrifying bacteria) in the soil; part of the nitrogen cycle.

dense connective tissue A type of tissue that may be irregular, as in the dermis of the skin, or regular, as in tendons.

density-dependent factor An environmental factor whose effects on a population change as population density changes; tends to retard population growth as population density increases and enhance population growth as population density decreases. Compare with *density-independent factor*.

density gradient centrifugation Procedure in which cell components are placed in a layer on top of a density gradient, usually a sucrose solution and water. Cell structures migrate during centrifugation, forming a band at the position in the gradient where their own density equals that of the sucrose solution.

density-independent factor An environmental factor that affects the size of a population but is not influenced by changes in population density. Compare with *density-dependent factor*.

deoxyribonucleic acid (DNA) Double-stranded nucleic acid; contains genetic information coded in specific sequences of its constituent nucleotides.

deoxyribose Pentose sugar lacking a hydroxyl (—OH) group on carbon-2′; a constituent of DNA.

depolarization (dee-pol″ar-ih-zay′shun) A decrease in the charge difference across a plasma membrane; may result in an action potential in a neuron or muscle cell.

derived characters See *shared derived characters*.

dermal tissue system The tissue that forms the outer covering over a plant; the epidermis or periderm.

dermis (dur′mis) The layer of dense connective tissue beneath the epidermis in the skin of vertebrates.

desert A temperate or tropical biome in which lack of precipitation limits plant growth.

desertification The degradation of once-fertile land into nonproductive desert; caused partly by soil erosion, deforestation, and overgrazing by domestic animals.

desmosomes (dez′moh-sohmz) Buttonlike plaques, present on two opposing cell surfaces, that hold the cells together by means of protein filaments that span the intercellular space.

determinate growth Growth of limited duration, as for example, in flowers and leaves. Compare with *indeterminate growth*.

determination The developmental process by which one or more cells become progressively committed to a particular fate. Determination is a series of molecular events usually leading to differentiation. Also called *cell determination*.

detritivore (duh-try′tuh-vore) An organism, such as an earthworm or crab, that consumes fragments of freshly dead or decomposing organisms; also called *detritus feeder*.

detritus (duh-try′tus) Organic debris from decomposing organisms.

detritus feeder See *detritivore*.

deuteromycetes (doo″ter-o-my′seats) An artificial grouping of fungi characterized by the absence of sexual reproduction but usually having other traits similar to ascomycetes; also called *imperfect fungi*.

deuterostome (doo′ter-oh-stome) Major division of the animal kingdom in which the anus develops from the blastopore; includes the echinoderms and chordates. Compare with *protostome*.

development All the progressive changes that take place throughout the life of an organism.

diabetes mellitus (mel′i-tus) The most common endocrine disorder. In type 1 diabetes, there is a marked decrease in the number of beta cells in the pancreas, resulting in insulin deficiency. In the more common type 2 diabetes, insulin receptors on target cells do not bind with insulin (insulin resistance).

diacylglycerol (DAG) (di″as-il-glis′er-ol) A lipid consisting of glycerol combined chemically with two fatty acids; also called *diglyceride*. Compare with *monoacylglycerol* and *triacylglycerol*.

dialysis The diffusion of certain solutes across a selectively permeable membrane.

diaphragm In mammals, the muscular floor of the chest cavity; contracts during inhalation, expanding the chest cavity.

diastole (di-ass′toh-lee) Phase of the cardiac cycle in which the heart is relaxed. Compare with *systole*.

diatom (die′eh-tom″) A usually unicellular alga that is covered by an ornate, siliceous shell consisting of two overlapping halves; an important component of plankton in both marine and fresh waters.

dichotomous branching (di-kaut′uh-mus) In botany, a type of branching in which one part always divides into two more or less equal parts.

diencephalon See *forebrain*.

differential centrifugation Separation of cell particles according to their mass, size, or density. In differential centrifugation, the supernatant is spun at successively higher revolutions per minute.

differential gene expression The expression of different subsets of genes at different times and in different cells during development.

differentiated cell A specialized cell; carries out unique activities, expresses a specific set of proteins, and usually has a recognizable appearance.

differentiation (dif″ah-ren-she-ay′shun) Development toward a more mature state; a process changing a young, relatively unspecialized cell to a more specialized cell. Also called *cell differentiation*.

diffusion The net movement of particles (atoms, molecules, or ions) from a region of higher concentration to a region of lower concentration (i.e., down a concentration gradient), resulting from random motion. Compare with *facilitated diffusion* and *active transport*.

digestion The breakdown of food to smaller molecules.

diglyceride See *diacylglycerol*.

dihybrid cross (dy-hy′brid) A genetic cross that takes into account the behavior of alleles of two loci. Compare with *monohybrid cross*.

dikaryotic (dy-kare-ee-ot′ik) Condition of having two nuclei per cell (i.e., *n* + *n*), characteristic of certain fungal hyphae. Compare with *monokaryotic*.

dimer An association of two monomers (e.g., a disaccharide or a dipeptide).

dinoflagellate (dy″noh-flaj′eh-late) A unicellular, biflagellate, typically marine alga that is an important component of plankton; usually photosynthetic.

dioecious (dy-ee′shus) Having male and female reproductive structures on separate plants; compare with *monoecious*.

dipeptide See *peptide*.

diploid (dip′loyd) The condition of having two sets of chromosomes per nucleus. Compare with *haploid* and *polyploid*.

diplomonads Small, mostly parasitic zooflagellates with one or two nuclei, no mitochondria, and one to four flagella.

direct fitness An individual's reproductive success, measured by the number of viable offspring it produces. Compare with *inclusive fitness*.

directed evolution See *in vitro evolution*.

directional selection The gradual replacement of one phenotype with another because of environmental change that favors phenotypes at one of the extremes of the normal distribution. Compare with *stabilizing selection* and *disruptive selection*.

disaccharide (dy-sak′ah-ride) A sugar produced by covalently linking two monosaccharides (e.g., maltose or sucrose).

disomy The normal condition in which both members of a chromosome pair are present in a diploid cell or organism. Compare with *monosomy* and *trisomy*.

dispersal The movement of individuals among populations. See *immigration and emigration*.

dispersion The pattern of distribution in space of the individuals of a population relative to their neighbors; may be clumped, random, or uniform.

disruptive selection A special type of directional selection in which changes in the environment favor two or more variant phenotypes at the expense of the mean. Compare with *stabilizing selection* and *directional selection*.

distal Remote; farther from the point of reference. Compare with *proximal*.

distal convoluted tubule The part of the renal tubule that extends from the loop of Henle to the collecting duct. Compare with *proximal convoluted tubule*.

disturbance In ecology, any event that disrupts community or population structure.

divergent circuit A neural pathway in which a presynaptic neuron stimulates many postsynaptic neurons. Compare with *convergent circuit*.

diving reflex A group of physiological mechanisms, such as decrease in metabolic rate, that are activated when a mammal dives to its limit.

dizygotic twins Twins that arise from the separate fertilization of two eggs; commonly known as *fraternal twins*. Compare with *monozygotic twins*.

DNA See *deoxyribonucleic acid*.

DNA cloning The process of selectively amplifying DNA sequences so their structure and function can be studied.

DNA fingerprinting The analysis of DNA extracted from an individual, which is unique to that individual; also called *DNA typing*.

DNA ligase Enzyme that catalyzes the joining of the 5′ and 3′ ends of two DNA fragments; essential in DNA replication and used in recombinant DNA technology.

DNA methylation A process in which gene inactivation is perpetuated by enzymes that add methyl groups to DNA.

DNA microarray A diagnostic test involving thousands of DNA molecules placed on a glass slide or chip.

DNA polymerases Family of enzymes that catalyze the synthesis of DNA from a DNA template by adding nucleotides to a growing 3′ end.

DNA probe See *genetic probe*.

DNA provirus Double-stranded DNA molecule that is an intermediate in the life cycle of an RNA tumor virus (retrovirus).

DNA replication The process by which DNA is duplicated; ordinarily a semiconservative process in which a double helix gives rise to two double helices, each with an "old" strand and a newly synthesized strand.

DNA sequencing Procedure by which the sequence of nucleotides in DNA is determined.

DNA typing See *DNA fingerprinting*.

domain (1) A structural and functional region of a protein. (2) The broadest taxonomic category; each domain includes one or more kingdoms.

dominance hierarchy A linear "pecking order" into which animals in a population may organize according to status; regulates aggressive behavior within the population.

dominant allele (al-leel′) An allele that is always expressed when it is present, regardless of whether it is homozygous or heterozygous. Compare with *recessive allele*.

dominant species In a community, a species that as a result of its large biomass or abundance exerts a major influence on the distribution of populations of other species.

dopamine A neurotransmitter of the biogenic amine group.

dormancy A temporary period of arrested growth in plants or plant parts such as spores, seeds, bulbs, and buds.

dorsal (dor′sl) Toward the uppermost surface or back of an animal. Compare with *ventral*.

dosage compensation Genetic mechanism by which the expression of X-linked genes in mammals is made equivalent in XX females and XY males by rendering all but one X chromosome inactive.

double fertilization A process in the flowering plant life cycle in which there are two fertilizations; one fertilization results in formation of a zygote, whereas the second results in formation of endosperm

double helix The structure of DNA, which consists of two antiparallel polynucleotide chains twisted around each other.

doubling time The amount of time it takes for a population to double in size, assuming that its current rate of increase does not change.

Down syndrome An inherited condition in which individuals have abnormalities of the face, eyelids, tongue, and other parts of the body and are physically and mentally retarded; usually results from trisomy of chromosome 21.

drupe (droop) A simple, fleshy fruit in which the inner wall of the fruit is a hard stone, e.g., peach, cherry.

duodenum (doo″o-dee′num) The portion of the small intestine into which the contents of the stomach first enter.

duplication An abnormality in which a set of chromosomes contains more than one copy of a particular chromosomal segment; the translocation form of Down syndrome is an example.

dura mater The tough, outer meningeal layer that covers and protects the brain and spinal cord. Also see *arachnoid* and *pia mater*.

dynamic equilibrium The condition of a chemical reaction when the rate of change in one direction is exactly the same as the rate of change in the opposite direction, i.e., the concentrations of the reactants and products are not changing, and the difference in free energy between reactants and products is zero.

ecdysis Molting; shedding outer skin; common process in insects, crustaceans, snakes.

ecdysone (ek′dih-sone) See *molting hormone*.

Ecdysozoa A branch of the protostomes that includes animals that molt, such as the rotifers, nematodes, and arthropods.

echinoderms (eh-kine′oh-derms) Phylum of spiny-skinned marine deuterostome invertebrates characterized by a water vascular system and tube feet; include sea stars, sea urchins, and sea cucumbers.

echolocation Determination of the position of objects by detecting echoes of high-pitched sounds emitted by an animal; a type of sensory system used by bats and dolphins.

ecological niche See *niche*.

ecological pyramid A graphical representation of the relative energy value at each trophic level. See *pyramid of biomass* and *pyramid of energy*.

ecological succession See *succession*.

ecology (ee-kol′uh-jee) A discipline of biology that studies the interrelations among living things and their environments.

ecosystem (ee′koh-sis-tem) The interacting system that encompasses a community and its nonliving, physical environment. Compare with *community*.

ecosystem management A conservation focus that emphasizes restoring and maintaining ecosystem quality rather than the conservation of individual species.

ecosystem services Important environmental services, such as clean air to breathe, clean water to drink, and fertile soil in which to grow crops, that ecosystems provide.

ecotone The transition zone where two communities meet and intergrade.

ectoderm (ek′toh-derm) The outer germ layer of the early embryo; gives rise to the skin and nervous system. Compare with *mesoderm* and *endoderm*.

ectoparasite A tick or other parasite that lives outside its host's body. Compare with *endoparasite*.

ectotherm An animal whose temperature fluctuates with that of the environment; may use behavioral adaptations to regulate temperature; sometimes referred to as *cold-blooded*. Compare with *endotherm*.

edge effect The ecological phenomenon in which ecotones between adjacent communities often contain a greater number of species or greater population densities of certain species than either adjacent community.

Ediacaran period (ee-dee-ack′uh-ran″) The last (most recent) period of the Proterozoic eon, from 600 million to 542 million years ago; named for early animal fossils found in the Ediacara Hills in South Australia.

effector A muscle or gland that contracts or secretes in direct response to nerve impulses.

efferent (ef′fur-ent) Leading away from some point of reference. Compare with *afferent*.

efferent neurons Neurons that transmit action potentials from the brain or spinal cord to muscles or glands. Compare with *afferent neurons*.

ejaculation (ee-jak″yoo-lay′shun) A sudden expulsion, as in the ejection of semen from the penis.

electrolyte A substance that dissociates into ions when dissolved in water; the resulting solution can conduct an electric current.

electron A particle with one unit of negative charge and negligible mass, located outside the atomic nucleus. Compare with *neutron* and *proton*.

electron configuration The arrangement of electrons around the atom. In a Bohr model, the electron configuration is depicted as a series of concentric circles.

electron microscope A microscope capable of producing high-resolution, highly magnified images through the use of an electron beam (rather than light). Transmission electron microscopes (TEMs) produce images of thin sections; scanning electron microscopes (SEMs) produce images of surfaces.

electron shell Group of orbitals of electrons with similar energies.

electron transport system A series of chemical reactions during which hydrogens or their electrons are passed along an electron transport chain from one acceptor molecule to another, with the release of energy.

electronegativity A measure of an atom's attraction for electrons.

electrophoresis, gel See *gel electrophoresis*.

electroreceptor A receptor that responds to electrical stimuli.

element A substance that cannot be changed to a simpler substance by a normal chemical reaction.

elimination Ejection of undigested food from the body. Compare with *excretion*.

El Niño–Southern Oscillation (ENSO) (el nee′nyo) A recurring climatic phenomenon that involves a surge of warm water in the Pacific Ocean and unusual weather patterns elsewhere in the world.

elongation (in protein synthesis) Cyclic process by which amino acids are added one by one to a growing polypeptide chain. See *initiation* and *termination*.

embryo (em′bree-oh) (1) A young organism before it emerges from the egg, seed, or body of its mother. (2) Developing human until the end of the second month, after which it is referred to as a fetus. (3) In plants, the young sporophyte produced following fertilization and subsequent development of the zygote.

embryo sac The female gametophyte generation in flowering plants.

embryo transfer See *host mothering*.

emergent properties Characteristics of an object, process, or behavior that could not be predicted from its component parts; emergent properties can be identified at each level as we move up the hierarchy of biological organization.

emigration The movement of individuals out of a population. Compare with *immigration*.

enantiomers (en-an′tee-oh-merz) Two isomeric chemical compounds that are mirror images.

end product inhibition See *feedback inhibition*.

endangered species A species whose numbers are so severely reduced that it is in imminent danger of extinction throughout all or part of its range. Compare with *threatened species*.

endemic species Localized, native species that are not found anywhere else in the world. Compare with *cosmopolitan species*.

endergonic reaction (end′er-gon″ik) A nonspontaneous reaction; a reaction requiring a net input of free energy. Compare with *exergonic reaction*.

endocrine gland (en′doh-crin) A gland that secretes hormones directly into the blood or tissue fluid instead of into ducts. Compare with *exocrine gland*.

endocrine system The body system that helps regulate metabolic activities; consists of ductless glands and tissues that secrete hormones.

endocytosis (en″doh-sy-toh′sis) The active transport of substances into the cell by the formation of invaginated regions of the plasma membrane that pinch off and become cytoplasmic vesicles. Compare with *exocytosis*.

endoderm (en′doh-derm) The inner germ layer of the early embryo; becomes the lining of the digestive tract and the structures that develop from the digestive tract—liver, lungs, and pancreas. Compare with *ectoderm* and *mesoderm*.

endodermis (en″doh-der′mis) The innermost layer of the plant root cortex. Endodermal cells have a waterproof Casparian strip around their radial and transverse walls that ensures that water and minerals enter the xylem only by passing through the endoderm cells.

endolymph (en′doh-limf) The fluid of the membranous labyrinth and cochlear duct of the ear.

endomembrane system See *internal membrane system*.

endometrium (en″doh-mee′tree-um) The uterine lining.

endoparasite A parasite such as a tapeworm that lives within the host. Compare with *ectoparasite*.

endoplasmic reticulum (ER) (en′doh-plaz″mik reh-tik′yoo-lum) An interconnected network of internal membranes in eukaryotic cells enclosing a compartment, the ER lumen. Rough ER has ribosomes attached to the cytosolic surface; smooth ER, a site of lipid biosynthesis, lacks ribosomes.

endorphins (en-dor′finz) Neuropeptides released by certain brain neurons; block pain signals.

endoskeleton (en″doh-skel′eh-ton) Bony and/or cartilaginous structures within the body that provide support. Compare with *exoskeleton*.

endosperm (en′doh-sperm) The 3*n* nutritive tissue that is formed at some point in the development of all angiosperm seeds.

endospore A resting cell formed by certain bacteria; highly resistant to heat, radiation, and disinfectants.

endostyle In nonvertebrate chordates, a groove in the floor of the pharynx that secretes mucus and traps food particles in sea water passing through the pharynx. In vertebrates, the thyroid gland is derived from the endostyle.

endosymbiont (en″doe-sim′bee-ont) An organism that lives inside the body of another kind of organism. Endosymbionts may benefit their host (mutualism) or harm their host (parasitism).

endothelium (en-doh-theel′ee-um) The tissue that lines the cavities of the heart, blood vessels, and lymph vessels.

endotherm (en′doh-therm) An animal that uses metabolic energy to maintain a constant body temperature despite variations in environmental temperature; e.g., birds and mammals. Compare with *ectotherm*.

endotoxin A poisonous substance in the cell walls of gram-negative bacteria. Compare with *exotoxin*.

energy The capacity to do work; expressed in kilojoules or kilocalories.

energy of activation See *activation energy*.

enhanced greenhouse effect See *greenhouse effect*.

enhancers Regulatory DNA sequences that can be located long distances away from the actual coding regions of a gene.

enkephalins (en-kef′ah-linz) Neuropeptides released by certain brain neurons that block pain signals.

enterocoely (en′ter-oh-seely) The process by which the coelom forms as a cavity within mesoderm produced by outpocketings of the primitive gut (archenteron); characteristic of many deuterostomes. Compare with *schizocoely*.

enthalpy The total potential energy of a system; sometimes referred to as the "heat content of the system."

entropy (en′trop-ee) Disorderliness; a quantitative measure of the amount of the random, disordered energy that is unavailable to do work.

environmental resistance Unfavorable environmental conditions, such as crowding, that prevent organisms from reproducing indefinitely at their intrinsic rate of increase.

environmental sustainability The ability to meet humanity's current needs without compromising the ability of future generations to meet their needs.

enzyme (en′zime) An organic catalyst (usually a protein) that accelerates a specific chemical reaction by lowering the activation energy required for that reaction.

enzyme–substrate complex The temporary association between enzyme and substrate that forms during the course of a catalyzed reaction; also called *ES complex*.

eon The largest division of the geologic time scale; eons are divided into eras.

eosinophil (ee-oh-sin′oh-fil) A type of white blood cell whose cytoplasmic granules absorb acidic stains; functions in parasitic infestations and allergic reactions.

epidermis (ep-ih-dur′mis) (1) An outer layer of cells that covers the body of plants and functions primarily for protection. (2) The outer layer of vertebrate skin.

epididymis (ep-ih-did′ih-mis) (pl., *epididymides*) A coiled tube that receives sperm from the testis and conveys it to the vas deferens.

epigenetic inheritance Inheritance that involves changes in how a gene is expressed without any change in that gene's nucleotide sequence.

epiglottis A thin, flexible structure that guards the entrance to the larynx, preventing food from entering the airway during swallowing.

epinephrine (ep-ih-nef′rin) Hormone produced by the adrenal medulla; stimulates the sympathetic nervous system.

epistasis (ep-ih-sta′-sis) Condition in which certain alleles of one locus alter the expression of alleles of a different locus.

epithelial tissue (ep-ih-theel′ee-al) The type of animal tissue that covers body surfaces, lines body cavities, and forms glands; also called *epithelium*.

epoch The smallest unit of geologic time; a subdivision of a period.

equilibrium See *dynamic equilibrium, genetic equilibrium*, and *punctuated equilibrium*.

era An interval of geologic time that is a subdivision of an eon; eras are divided into periods.

erythroblastosis fetalis (eh-rith″row-blas-toe′sis fi-tal′is) Serious condition in which Rh$^+$ red blood cells (which bear antigen D) of a fetus are destroyed by maternal anti-D antibodies.

erythrocyte (eh-rith′row-site) A vertebrate red blood cell; contains hemoglobin, which transports oxygen.

erythropoietin (eh-rith″row-poy′ih-tin) A peptide hormone secreted mainly by kidney cells; stimulates red blood cell production.

ES complex See *enzyme–substrate complex.*

esophagus (e-sof′ah-gus) The part of the digestive tract that conducts food from the pharynx to the stomach.

essential nutrient A nutrient that must be provided in the diet because the body cannot make it or cannot make it in sufficient quantities to meet nutritional needs, e.g., essential amino acids and essential fatty acids.

ester linkage Covalent linkage formed by the reaction of a carboxyl group and a hydroxyl group, with the removal of the equivalent of a water molecule; the linkage includes an oxygen atom bonded to a carbonyl group.

estivation A state of torpor caused by lack of food or water during periods of high temperature. Compare with *hibernation.*

estrogens (es′troh-jenz) Female sex hormones produced by the ovary; promote the development and maintenance of female reproductive structures and of secondary sex characteristics.

estuary (es′choo-wear-ee) A coastal body of water that connects to an ocean, in which fresh water from the land mixes with salt water.

ethology (ee-thol′oh-jee) The study of animal behavior under natural conditions from the point of view of adaptation.

ethyl alcohol A two-carbon alcohol.

ethylene (eth′ih-leen) A gaseous plant hormone involved in various aspects of plant growth and development, such as leaf abscission and fruit ripening.

euchromatin (yoo-croh′mah-tin) A loosely coiled chromatin that is generally capable of transcription. Compare with *heterochromatin.*

eudicot (yoo-dy′kot) One of the two clades of flowering plants; eudicot seeds contain two cotyledons, or seed leaves. Compare with *monocot.*

euglenoids (yoo-glee′noidz) A group of mostly freshwater, flagellate, unicellular algae that move by means of an anterior flagellum and are usually photosynthetic.

eukaryote (yoo″kar′ee-ote) An organism whose cells have nuclei and other membrane-enclosed organelles. Includes protists, fungi, plants, and animals. Compare with *prokaryote.*

euphotic zone The upper reaches of the ocean, in which enough light penetrates to support photosynthesis.

eustachian tube (yoo-stay′shee-un) The auditory tube passing between the middle-ear cavity and the pharynx in vertebrates; permits the equalization of pressure on the tympanic membrane.

eutrophic lake A lake enriched with nutrients such as nitrate and phosphate and consequently overgrown with plants or algae.

evergreen A plant that sheds leaves over a long period, so some leaves are always present. Compare with *deciduous.*

Evo Devo The study of the evolution of the genetic control of development.

evolution Any cumulative genetic changes in a population from generation to generation. Evolution leads to differences in populations and explains the origin of all the organisms that exist today or have ever existed.

evolutionary species concept An alternative to the biological species concept in which for a population to be declared a separate species, it must have undergone evolution long enough for statistically significant differences to emerge. Compare with *species.*

evolutionary systematics An approach to classification that considers both evolutionary relationships and the extent of divergence that has occurred since a group branched from an ancestral group. Compare with *cladistics* and *phenetics.*

excitatory postsynaptic potential (EPSP) A change in membrane potential that brings a neuron closer to the firing level. Compare with *inhibitory postsynaptic potential (IPSP).*

excretion (ek-skree′shun) The discharge from the body of a waste product of metabolism (not to be confused with the elimination of undigested food materials). Compare with *elimination.*

excretory system The body system in animals that functions in osmoregulation and in the discharge of metabolic wastes.

exergonic reaction (ex′er-gon″ik) A reaction characterized by a release of free energy. Also called *spontaneous reaction.* Compare with *endergonic reaction.*

exocrine gland (ex′oh-crin) A gland that excretes its products through a duct that opens onto a free surface, such as the skin (e.g., sweat glands). Compare with *endocrine gland.*

exocytosis (ex″oh-sy-toh′sis) The active transport of materials out of the cell by fusion of cytoplasmic vesicles with the plasma membrane. Compare with *endocytosis.*

exon (1) A protein-coding region of a eukaryotic gene. (2) The mRNA transcribed from such a region. Compare with *intron.*

exoskeleton (ex″oh-skel′eh-ton) An external skeleton, such as the shell of mollusks or outer covering of arthropods; provides protection and sites of attachment for muscles. Compare with *endoskeleton.*

exotoxin A poisonous substance released by certain bacteria. Compare with *endotoxin.*

explicit memory Factual knowledge of people, places, or objects; requires conscious recall of the information.

exploitation competition An intraspecific competition in which all the individuals in a population "share" the limited resource equally, so at high population densities, none of them obtains an adequate amount. Also called *scramble competition.* Compare with *interference competition.*

exponential population growth The accelerating population growth rate that occurs when optimal conditions allow a constant per capita growth rate. Compare with *logistic population growth.*

ex situ conservation Conservation efforts that involve conserving individual species in human-controlled settings, such as zoos. Compare with *in situ conservation.*

exteroceptor (ex′tur-oh-sep″tor) One of the sense organs that receives sensory stimuli from the outside world, such as the eyes or touch receptors. Compare with *interoceptor.*

extinction The elimination of a species; occurs when the last individual member of a species dies.

extracellular matrix (ECM) A network of proteins and carbohydrates that surrounds many animal cells.

extraembryonic membranes Multicellular membranous structures that develop from the germ layers of a terrestrial vertebrate embryo but are not part of the embryo itself. See *chorion, amnion, allantois,* and *yolk sac.*

F$_1$ generation (first filial generation) The first generation of hybrid offspring resulting from a cross between parents from two different true-breeding lines.

F$_2$ generation (second filial generation) The offspring of the F$_1$ generation.

facilitated diffusion The passive transport of ions or molecules by a specific carrier protein in a membrane. As in simple diffusion, net transport is down a concentration gradient, and no additional energy has to be supplied. Compare with *diffusion* and *active transport.*

facilitation (1) In neurology, a process in which a neuron is brought close to its threshold level by stimulation from various presynaptic neurons. (2) In ecology, a situation in which one species has a positive effect on other species in the community, for example, by enhancing the local environment.

facultative anaerobe An organism capable of carrying out aerobic respiration but able to switch to fermentation when oxygen is unavailable, e.g., yeast. Compare with *obligate anaerobe*.

FAD/FADH$_2$ Oxidized and reduced forms, respectively, of flavin adenine dinucleotide, a coenzyme that transfers electrons (as hydrogen) in metabolism, including cellular respiration.

fallopian tube See *oviduct*.

family A taxonomic category made up of related genera.

fatty acid A lipid that is an organic acid containing a long hydrocarbon chain, with no double bonds (saturated fatty acid), one double bond (monounsaturated fatty acid), or two or more double bonds (polyunsaturated fatty acid); components of triacylglycerols and phospholipids, as well as monoacylglycerols and diacylglycerols.

fecundity The potential capacity of an individual to produce offspring.

feedback inhibition A type of enzyme regulation in which the accumulation of the product of a reaction inhibits an earlier reaction in the sequence; also known as *end product inhibition*.

fermentation An anaerobic process by which ATP is produced by a series of redox reactions in which organic compounds serve both as electron donors and terminal electron acceptors.

fern One of a phylum of seedless vascular plants that reproduce by spores produced in sporangia; ferns undergo an alternation of generations between the dominant sporophyte and the gametophyte (prothallus).

fertilization The fusion of two *n* gametes; results in the formation of a 2*n* zygote. Compare with *double fertilization*.

fetus The unborn human offspring from the third month of pregnancy to birth.

fiber (1) In plants, a type of sclerenchyma cell; fibers are long, tapered cells with thick walls. Compare with *sclereid*. (2) In animals, an elongated cell such as a muscle or nerve cell. (3) In animals, the microscopic, threadlike protein and carbohydrate complexes scattered through the matrix of connective tissues.

fibrin An insoluble protein formed from the plasma protein fibrinogen during blood clotting.

fibroblasts Connective tissue cells that produce the fibers and the protein and carbohydrate complexes of the matrix of connective tissues.

fibronectins Glycoproteins of the extracellular matrix that bind to integrins (receptor proteins in the plasma membrane).

fibrous root system A root system consisting of several adventitious roots of approximately equal size that arise from the base of the stem. Compare with *taproot system*.

Fick's law of diffusion A physical law governing rates of gas exchange in animal respiratory systems; states that the rate of diffusion of a substance across a membrane is directly proportional to the surface area and to the difference in pressure between the two sides.

filament In flowering plants, the thin stalk of a stamen; the filament bears an anther at its tip.

first law of thermodynamics The law of conservation of energy, which states that the total energy of any closed system (any object plus its surroundings, i.e., the universe) remains constant. Compare with *second law of thermodynamics*.

fitness See *direct fitness*.

fixed action pattern (FAP) An innate behavior triggered by a sign stimulus.

flagellum (flah-jel′um) (pl., *flagella*) A long, whiplike structure extending from certain cells and used in locomotion. (1) Eukaryote flagella consist of two central, single microtubules surrounded by nine double microtubules (9 + 2 structure), all covered by a plasma membrane. (2) Prokaryote flagella are filaments rotated by special structures located in the plasma membrane and cell wall.

flame cells Collecting cells that have cilia; part of the osmoregulatory system of flatworms.

flavin adenine dinucleotide See *FAD/FADH$_2$*.

flowering plants See *angiosperms*.

flowing-water ecosystem A river or stream ecosystem.

fluid-mosaic model The currently accepted model of the plasma membrane and other cell membranes, in which protein molecules "float" in a fluid phospholipid bilayer.

fluorescence The emission of light of a longer wavelength (lower energy) than the light originally absorbed.

fluorescent in situ hybridization (FISH) A technique to detect specific DNA segments by hybridization directly to chromosomes; visualized microscopically by using a fluorescent dye.

follicle (fol′i-kl) (1) A simple, dry, dehiscent fruit that develops from a single carpel and splits open at maturity along one suture to liberate the seeds. (2) A small sac of cells in the mammalian ovary that contains a maturing egg. (3) The pocket in the skin from which a hair grows.

follicle-stimulating hormone (FSH) A gonadotropic hormone secreted by the anterior lobe of the pituitary gland; stimulates follicle development in the ovaries of females and sperm production in the testes of males.

food chain The series of organisms through which energy flows in an ecosystem. Each organism in the series eats or decomposes the preceding organism in the chain. See *food web*.

food web A complex interconnection of all the food chains in an ecosystem.

foramen magnum The opening in the vertebrate skull through which the spinal cord passes.

foraminiferan (for″am-in-if′er-an) A marine protozoon that produces a shell, or test, that encloses an amoeboid body.

forebrain In the early embryo, one of the three divisions of the developing vertebrate brain; subdivides to form the telencephalon, which gives rise to the cerebrum, and the diencephalon, which gives rise to the thalamus and hypothalamus. Compare with *midbrain* and *hindbrain*.

forest decline A gradual deterioration (and often death) of many trees in a forest; can be caused by a combination of factors, such as acid precipitation, toxic heavy metals, and surface-level ozone.

fossil Parts or traces of an ancient organism usually preserved in rock.

fossil fuel Combustible deposits in Earth's crust that are composed of the remnants of prehistoric organisms that existed millions of years ago, e.g., oil, natural gas, and coal.

founder cell A cell from which a particular cell lineage is derived.

founder effect Genetic drift that results from a small population colonizing a new area.

fovea (foe′vee-ah) The area of sharpest vision in the retina; cone cells are concentrated here.

fragile site A weak point at a specific location on a chromosome where part of a chromatid appears attached to the rest of the chromosome by a thin thread of DNA.

fragile X syndrome A human genetic disorder caused by a fragile site that occurs near the tip on the X chromosome; effects range from mild learning disabilities to severe mental retardation and hyperactivity.

frameshift mutation A mutation that results when one or two nucleotide pairs are inserted into or deleted from the DNA. The change causes the mRNA transcribed from the mutated DNA to have an altered reading frame such that all codons downstream from the mutation are changed.

fraternal twins See *dizygotic twins.*

free energy The maximum amount of energy available to do work under the conditions of a biochemical reaction.

free radicals Toxic, highly reactive compounds with unpaired electrons that bond with other compounds in the cell and interfere with normal function.

frequency-dependent selection Selection in which the relative fitness of different genotypes is related to how frequently they occur in the population.

freshwater wetlands Land that is transitional between freshwater and terrestrial ecosystems and is covered with water for at least part of the year, e.g., marshes and swamps.

frontal lobes In mammals, the anterior part of the cerebrum.

fruit In flowering plants, a mature, ripened ovary. Fruits contain seeds and usually provide seed protection and dispersal.

fruiting body A multicellular structure that contains the sexual spores of certain fungi; refers to the ascocarp of an ascomycete and the basidiocarp of a basidiomycete.

fucoxanthin (few"koh-zan'thin) The brown carotenoid pigment found in brown algae, golden algae, diatoms, and dinoflagellates.

functional genomics The study of the roles of genes in cells.

functional group A group of atoms that confers distinctive properties on an organic molecule (or region of a molecule) to which it is attached, e.g., hydroxyl, carbonyl, carboxyl, amino, phosphate, and sulfhydryl groups.

fundamental niche The potential ecological niche that an organism could occupy if there were no competition from other species. Compare with *realized niche.*

fungus (pl., *fungi*) A heterotrophic eukaryote with chitinous cell walls and a body usually in the form of a mycelium of branched, threadlike hyphae. Most fungi are decomposers; some are parasitic.

G protein One of a group of proteins that bind GTP and are involved in the transfer of signals across the plasma membrane.

G$_1$ phase The first gap phase within the interphase stage of the cell cycle; G$_1$ occurs before DNA synthesis (S phase) begins. Compare with *S* and *G$_2$ phases.*

G$_2$ phase Second gap phase within the interphase stage of the cell cycle; G$_2$ occurs after DNA synthesis (S phase) and before mitosis. Compare with *S* and *G$_1$ phases.*

gallbladder A small sac that stores bile.

gametangium (gam"uh-tan'gee-um) Special multicellular or unicellular structure of plants, protists, and fungi in which gametes are formed.

gamete (gam'eet) A sex cell; in plants and animals, an egg or sperm. In sexual reproduction, the union of gametes results in the formation of a zygote. The chromosome number of a gamete is designated *n.*

gametic isolation (gam-ee'tik) A prezygotic reproductive isolating mechanism in which sexual reproduction between two closely related species cannot occur because of chemical differences in the gametes.

gametogenesis The process of gamete formation. See *spermatogenesis* and *oogenesis.*

gametophyte generation (gam-ee'toh-fite) The *n,* gamete-producing stage in the life cycle of a plant. Compare with *sporophyte generation.*

gamma-aminobutyric acid (GABA) A neurotransmitter that has an inhibitory effect.

ganglion (gang'glee-on) (pl., *ganglia*) A mass of neuron cell bodies.

ganglion cell A type of neuron in the retina of the eye; receives input from bipolar cells.

gap junction Structure consisting of specialized regions of the plasma membrane of two adjacent cells; contains numerous pores that allow the passage of certain small molecules and ions between them.

gastrin (gas'trin) A hormone released by the stomach mucosa; stimulates the gastric glands to secrete pepsinogen.

gastrovascular cavity A central digestive cavity with a single opening that functions as both mouth and anus; characteristic of cnidarians and flatworms.

gastrula (gas'troo-lah) A three-layered embryo formed by the process of gastrulation.

gastrulation (gas-troo-lay'shun) Process in embryonic development during which the three germ layers (ectoderm, mesoderm, and endoderm) form.

gel electrophoresis Procedure by which proteins or nucleic acids are separated on the basis of size and charge as they migrate through a gel in an electric field.

gene A segment of DNA that serves as a unit of hereditary information; includes a transcribable DNA sequence (plus associated sequences regulating its transcription) that yields a protein or RNA product with a specific function.

gene amplification The developmental process in which certain cells produce multiple copies of a gene by selective replication, thus allowing for increased synthesis of the gene product. Compare with *nuclear equivalence* and *genomic rearrangement.*

gene flow The movement of alleles between local populations due to the migration of individuals; can have significant evolutionary consequences.

gene locus See *locus.*

gene pool All the alleles of all the genes present in a freely interbreeding population.

gene therapy Any of a variety of methods designed to correct a disease or alleviate its symptoms through the introduction of genes into the affected person's cells.

genetic bottleneck See *bottleneck.*

genetic code See *codon.*

genetic counseling Medical and genetic information provided to couples who are concerned about the risk of abnormality in their children.

genetic drift A random change in allele frequency in a small breeding population.

genetic engineering Manipulation of genes, often through recombinant DNA technology.

genetic equilibrium The condition of a population that is not undergoing evolutionary change, i.e., in which allele and genotype frequencies do not change from one generation to the next. See *Hardy–Weinberg principle.*

genetic polymorphism (pol"ee-mor'fizm) The presence in a population of two or more alleles for a given gene locus.

genetic probe A single-stranded nucleic acid (either DNA or RNA) used to identify a complementary sequence by hydrogen-bonding to it.

genetic recombination See *recombination, genetic.*

genetic screening A systematic search through a population for individuals with a genotype or karyotype that might cause a serious genetic disease in them or their offspring.

genetics The science of heredity; includes genetic similarities and genetic variation between parents and offspring or among individuals of a population.

genome (jee′nome) Originally, all the genetic material in a cell or individual organism. The term is used in more than one way depending on context, e.g., an organism's haploid genome is all the DNA contained in one haploid set of its chromosomes, and its mitochondrial genome is all the DNA in a mitochondrion. See *human genome.*

genomic DNA library A collection of recombinant plasmids in which all the DNA in the genome is represented. Compare with *cDNA library.*

genomic imprinting See *imprinting,* first definition.

genomic rearrangement A physical change in the structure of one or more genes that occurs during the development of an organism and leads to an alteration in gene expression; compare with *nuclear equivalence* and *gene amplification.*

genomics The emerging field of biology that studies the entire DNA sequence of an organism's genome to identify all the genes, determine their RNA or protein products, and ascertain how the genes are regulated.

genotype (jeen′oh-type) The genetic makeup of an individual. Compare with *phenotype.*

genotype frequency The proportion of a particular genotype in the population.

genus (jee′nus) A taxonomic category made up of related species.

geometric isomer One of two or more chemical compounds having the same arrangement of covalent bonds but differing in the spatial arrangement of their atoms or groups of atoms.

germ layers In animals, three embryonic tissue layers: endoderm, mesoderm, and ectoderm.

germ line cell In animals, a cell that is part of the line of cells that will ultimately undergo meiosis to form gametes. Compare with *somatic cell.*

germination Resumption of growth of an embryo or spore; occurs when a seed or spore sprouts.

germplasm Any plant or animal material that may be used in breeding; includes seeds, plants, and plant tissues of traditional crop varieties and the sperm and eggs of traditional livestock breeds.

gibberellin (jib″ur-el′lin) A plant hormone involved in many aspects of plant growth and development, such as stem elongation, flowering, and seed germination.

gills (1) The respiratory organs characteristic of many aquatic animals, usually thin-walled projections from the body surface or from some part of the digestive tract. (2) The spore-bearing, platelike structures under the caps of mushrooms.

ginkgo (ging′ko) A member of an ancient gymnosperm group that consists of a single living representative (*Ginkgo biloba*), a hardy, deciduous tree with broad, fan-shaped leaves and naked, fleshy seeds (on female trees).

gland See *endocrine gland* and *exocrine gland.*

glial cells (glee′ul) In nervous tissue, cells that support and nourish neurons; they also communicate with neurons and have several other functions; also see *astrocyte, microglia,* and *oligodendrocyte.*

globulin (glob′yoo-lin) One of a class of proteins in blood plasma, some of which (gamma globulins) function as antibodies.

glomeromycetes A group of fungal symbionts that form arbuscular mycorrhizae with the roots of many plants; belong to phylum Glomeromycota.

glomerulus (glom-air′yoo-lus) The cluster of capillaries at the proximal end of a nephron; the glomerulus is surrounded by Bowman's capsule.

glucagon (gloo′kah-gahn) A hormone secreted by the pancreas that stimulates glycogen breakdown, thereby increasing the concentration of glucose in the blood. Compare with *insulin.*

glucose A hexose aldehyde sugar that is central to many metabolic processes.

glutamate An amino acid that functions as the major excitatory neurotransmitter in the vertebrate brain.

glyceraldehyde-3-phosphate (G3P) Phosphorylated three-carbon compound that is an important intermediate in glycolysis and in the Calvin cycle.

glycerol A three-carbon alcohol with a hydroxyl group on each carbon; a component of triacylglycerols and phospholipids, as well as monoacylglycerols and diacylglycerols.

glycocalyx (gly″koh-kay′lix) A coating on the outside of an animal cell, formed by the polysaccharide portions of glycoproteins and glycolipids associated with the plasma membrane.

glycogen (gly′koh-jen) The principal storage polysaccharide in animal cells; formed from glucose and stored primarily in the liver and, to a lesser extent, in muscle cells.

glycolipid A lipid with covalently attached carbohydrates.

glycolysis (gly-kol′ih-sis) The first stage of cellular respiration, literally the "splitting of sugar." The metabolic conversion of glucose into pyruvate, accompanied by the production of ATP.

glycoprotein (gly′koh-pro-teen) A protein with covalently attached carbohydrates.

glycosidic linkage Covalent linkage joining two sugars; includes an oxygen atom bonded to a carbon of each sugar.

glyoxysomes (gly-ox′ih-sohmz) Membrane-enclosed structures in cells of certain plant seeds; contain a large array of enzymes that convert stored fat to sugar.

gnetophyte (nee′toe-fite) One of a small phylum of unusual gymnosperms that have some features similar to those of flowering plants.

goblet cells Unicellular glands that secrete mucus.

goiter (goy′ter) An enlargement of the thyroid gland.

golden alga A member of a phylum of algae, most of which are biflagellate, are unicellular, and contain pigments, including chlorophylls *a* and *c* and carotenoids, including fucoxanthin.

Golgi complex (goal′jee) Organelle composed of stacks of flattened, membranous sacs. Mainly responsible for modifying, packaging, and sorting proteins that will be secreted or targeted to other organelles of the internal membrane system or to the plasma membrane; also called *Golgi body* or *Golgi apparatus.*

gonad (goh′nad) A gamete-producing gland; an ovary or a testis.

gonadotropic hormones (go-nad-oh-troh′pic) Hormones produced by the anterior pituitary gland that stimulate the testes and ovaries; include follicle-stimulating hormone (FSH) and luteinizing hormone (LH).

gonadotropin-releasing hormone (GnRH) A hormone secreted by the hypothalamus that stimulates the anterior pituitary to secrete the gonadotropic hormones: follicle-stimulating hormone (FSH) and luteinizing hormone (LH).

graded potential A local change in electrical potential that varies in magnitude depending on the strength of the applied stimulus.

gradualism The idea that evolution occurs by a slow, steady accumulation of genetic changes over time. Compare with *punctuated equilibrium.*

graft rejection An immune response directed against a transplanted tissue or organ.

grain A simple, dry, one-seeded fruit in which the fruit wall is fused to the seed coat, e.g., corn and wheat kernels. Also called *caryopsis*.

granulosa cells In mammals, cells that surround the developing oocyte and are part of the follicle; produce estrogens and inhibin.

granum (pl., *grana*) A stack of thylakoids within a chloroplast.

gravitropism (grav″ih-troh′pizm) Growth of a plant in response to gravity.

gray crescent The grayish area of cytoplasm that marks the region where gastrulation begins in an amphibian embryo.

gray matter Nervous tissue in the brain and spinal cord that contains cell bodies, dendrites, and unmyelinated axons. Compare with *white matter*.

green alga A member of a diverse phylum of algae that contain the same pigments as plants (chlorophylls *a* and *b* and carotenoids).

greenhouse effect The natural global warming of Earth's atmosphere caused by the presence of carbon dioxide and other gases that trap the sun's radiation. The additional warming produced when increased levels of greenhouse gases absorb infrared radiation is known as the *enhanced greenhouse effect*.

greenhouse gases Trace gases in the atmosphere that allow the sun's energy to penetrate to Earth's surface but do not allow as much of it to escape as heat.

gross primary productivity The rate at which energy accumulates (is assimilated) in an ecosystem during photosynthesis. Compare with *net primary productivity*.

ground state The lowest energy state of an atom.

ground tissue system All tissues in the plant body other than the dermal tissue system and vascular tissue system; consists of parenchyma, collenchyma, and sclerenchyma.

growth factors A group of more than 50 extracellular peptides that signal certain cells to grow and divide.

growth hormone (GH) A hormone secreted by the anterior lobe of the pituitary gland; stimulates growth of body tissues; also called *somatotropin*.

growth rate The rate of change of a population's size on a per capita basis.

guanine (gwan′een) A nitrogenous purine base that is a component of nucleic acids and GTP.

guanosine triphosphate (GTP) An energy transfer molecule similar to ATP that releases free energy with the hydrolysis of its terminal phosphate group.

guard cell One of a pair of epidermal cells that adjust their shape to form a stomatal pore for gas exchange.

guttation (gut-tay′shun) The appearance of water droplets on leaves, forced out through leaf pores by root pressure.

gymnosperm (jim′noh-sperm) Any of a group of seed plants in which the seeds are not enclosed in an ovary; gymnosperms frequently bear their seeds in cones. Includes four phyla: conifers, cycads, ginkgoes, and gnetophytes.

habitat The natural environment or place where an organism, population, or species lives.

habitat fragmentation The division of habitats that formerly occupied large, unbroken areas into smaller pieces by roads, fields, cities, and other human land-transforming activities.

habitat isolation A prezygotic reproductive isolating mechanism in which reproduction between similar species is prevented because they live and breed in different habitats.

habituation (hab-it″yoo-ay′shun) A type of learning in which an animal becomes accustomed to a repeated, irrelevant stimulus and no longer responds to it.

hair cell A vertebrate mechanoreceptor found in the lateral line of fishes, the vestibular apparatus, semicircular canals, and cochlea.

half-life The period of time required for a radioisotope to change into a different material.

haploid (hap′loyd) The condition of having one set of chromosomes per nucleus. Compare with *diploid* and *polyploid*.

"hard-wiring" Refers to how neurons signal one another, how they connect, and how they carry out basic functions such as regulating heart rate, blood pressure, and sleep–wake cycles.

Hardy–Weinberg principle The mathematical prediction that allele frequencies do not change from generation to generation in a large population in the absence of microevolutionary processes (mutation, genetic drift, gene flow, natural selection).

haustorium (hah-stor′ee-um) (pl., *haustoria*) In parasitic fungi, a specialized hypha that penetrates a host cell and obtains nourishment from the cytoplasm.

Haversian canals (ha-vur′zee-un) Channels extending through the matrix of bone; contain blood vessels and nerves.

heat The total amount of kinetic energy in a sample of a substance.

heat energy The thermal energy that flows from an object with a higher temperature to an object with a lower temperature.

heat of vaporization The amount of heat energy that must be supplied to change one gram of a substance from the liquid phase to the vapor phase.

helicases Enzymes that unwind the two strands of a DNA double helix.

helper T cell See *T helper cell*.

hemichordates A phylum of sedentary, wormlike deuterostomes.

hemizygous (hem″ih-zy′gus) Possessing only one allele for a particular locus; a human male is hemizygous for all X-linked genes. Compare with *homozygous* and *heterozygous*.

hemocoel Blood cavity characteristic of animals with an open circulatory system.

hemocyanin A hemolymph pigment that transports oxygen in some mollusks and arthropods.

hemoglobin (hee′moh-gloh″bin) The red, iron-containing protein pigment in blood that transports oxygen and carbon dioxide and aids in regulation of pH.

hemolymph (hee′moh-limf) The fluid that bathes the tissues in animals with an open circulatory system, e.g., arthropods and most mollusks.

hemophilia (hee″moh-feel′ee-ah) A hereditary disease in which blood does not clot properly; the form known as *hemophilia A* has an X-linked, recessive inheritance pattern.

Hensen's node See *primitive streak*.

hepatic (heh-pat′ik) Pertaining to the liver.

hepatic portal system The portion of the circulatory system that carries blood from the intestine through the liver.

herbivore (erb′uh-vore) An animal that feeds on plants or algae. Also called *primary consumer*.

heredity The transmission of genetic information from parent to offspring.

hermaphrodite (her-maf′roh-dite) An organism that has both male and female sex organs.

heterochromatin (het″ur-oh-kroh′mah-tin) Highly coiled and compacted chromatin in an inactive state. Compare with *euchromatin*.

heterocyst (het′ur-oh-sist″) An oxygen-excluding cell of cyanobacteria that is the site of nitrogen fixation.

heterogametic A term describing an individual that produces two classes of gametes with respect to their sex chromosome constitutions. Human males (XY) are heterogametic, producing X and Y sperm. Compare with *homogametic*.

heterospory (het″ur-os′pur-ee) Production of two types of *n* spores, microspores (male) and megaspores (female). Compare with *homospory*.

heterothallic (het″ur-oh-thal′ik) Pertaining to certain algae and fungi that have two mating types; only by combining a plus strain and a minus strain can sexual reproduction occur. Compare with *homothallic*.

heterotroph (het′ur-oh-trof) An organism that cannot synthesize its own food from inorganic raw materials and therefore must obtain energy and body-building materials from other organisms. Also called *consumer*. Compare with *autotroph*. See *chemoheterotroph* and *photoheterotroph*.

heterozygote advantage A phenomenon in which the heterozygous condition confers some special advantage on an individual that either homozygous condition does not (i.e., *Aa* has a higher degree of fitness than does *AA* or *aa*).

heterozygous (het-ur″oh-zye′gus) Having a pair of unlike alleles for a particular locus. Compare with *homozygous*.

hexose A monosaccharide containing six carbon atoms.

hibernation Long-term torpor in response to winter cold and scarcity of food. Compare with *estivation*.

high-density lipoprotein (HDL) See *lipoprotein*.

hindbrain In the early embryo, one of the three divisions of the developing vertebrate brain; subdivides to form the metencephalon, which gives rise to the cerebellum and pons, and the myelencephalon, which gives rise to the medulla. Compare with *forebrain* and *midbrain*.

histamine (his′tah-meen) Substance released from mast cells that is involved in allergic and inflammatory reactions.

histones (his′tohnz) Small, positively charged (basic) proteins in the cell nucleus that bind to the negatively charged DNA. See *nucleosomes*.

holdfast The basal structure for attachment to solid surfaces found in multicellular algae.

holoblastic cleavage A cleavage pattern in which the entire embryo cleaves; characteristic of eggs with little or moderate yolk (isolecithal or moderately telolecithal), e.g., the eggs of echinoderms, amphioxus, and mammals. Compare with *meroblastic cleavage*.

home range A geographic area that an individual animal seldom or never leaves. Compare with *range*.

homeobox A short (180-nucleotide) DNA sequence that characterizes many homeotic genes as well as some other genes that play a role in development.

homeodomain A functional region of certain transcription factors; consists of approximately 60 amino acids specified by a homeobox DNA sequence and includes a recognition alpha helix, which binds to specific DNA sequences and affects their transcription.

homeostasis (home″ee-oh-stay′sis) The balanced internal environment of the body; the automatic tendency of an organism to maintain such a steady state.

homeotic gene (home″ee-ah′tik) A gene that controls the formation of specific structures during development. Such genes were originally identified through insect mutants in which one body part is substituted for another.

hominid (hah′min-id) Any of a group of extinct and living humans.

hominoid (hah′min-oid) The apes and hominids.

homogametic Term describing an individual that produces gametes with identical sex chromosome constitutions. Human females (XX) are homogametic, producing all X eggs. Compare with *heterogametic*.

homologous chromosomes (hom-ol′ah-gus) Chromosomes that are similar in morphology and genetic constitution. In humans there are 23 pairs of homologous chromosomes; one member of each pair is inherited from the mother, and the other from the father.

homologous features See *homology*.

homology Similarity in different species that results from their derivation from a common ancestor. The features that exhibit such similarity are called *homologous features*. Compare with *homoplasy*.

homoplastic features See *homoplasy*.

homoplasy Similarity in the characters in different species that is due to convergent evolution, not common descent. Characters that exhibit such similarity are called *homoplastic features*. Compare with *homology*.

homospory (hoh″mos′pur-ee) Production of one type of *n* spore that gives rise to a bisexual gametophyte. Compare with *heterospory*.

homothallic (hoh″moh-thal′ik) Pertaining to certain algae and fungi that are self-fertile. Compare with *heterothallic*.

homozygous (hoh″moh-zy′gous) Having a pair of identical alleles for a particular locus. Compare with *heterozygous*.

hormone An organic chemical messenger in multicellular organisms that is produced in one part of the body and often transported to another part where it signals cells to alter some aspect of metabolism.

hornwort A phylum of spore-producing, nonvascular, thallose plants with a life cycle similar to that of mosses.

host mothering The introduction of an embryo from one species into the uterus of another species, where it implants and develops; the host mother subsequently gives birth and may raise the offspring as her own.

***Hox* genes** Clusters of homeobox-containing genes that specify the anterior-posterior axis of various animals during development.

human chorionic gonadotropin (hCG) A hormone secreted by cells surrounding the early embryo; signals the mother's corpus luteum to continue to function.

human genetics The science of inherited variation in humans.

human genome The totality of genetic information in human cells; includes the DNA content of both the nucleus and mitochondria. See *genome*.

human immunodeficiency virus (HIV) The retrovirus that causes AIDS (acquired immunodeficiency syndrome).

human leukocyte antigen (HLA) See *major histocompatibility complex*.

humus (hew′mus) Organic matter in various stages of decomposition in the soil; gives soil a dark brown or black color.

Huntington disease A genetic disease that has an autosomal dominant inheritance pattern and causes mental and physical deterioration.

hybrid The offspring of two genetically dissimilar parents.

hybrid breakdown A postzygotic reproductive isolating mechanism in which, although an interspecific hybrid is fertile and produces a second (F_2) generation, the F_2 has defects that prevent it from successfully reproducing.

hybrid inviability A postzygotic reproductive isolating mechanism in which the embryonic development of an interspecific hybrid is aborted.

hybrid sterility A postzygotic reproductive isolating mechanism in which an interspecific hybrid cannot reproduce successfully.

hybrid vigor The genetic superiority of an F_1 hybrid over either parent, caused by the presence of heterozygosity for a number of different loci.

hybrid zone An area of overlap between two closely related populations, subspecies, or species in which interbreeding occurs.

hybridization (1) Interbreeding between members of two different taxa. (2) Interbreeding between genetically dissimilar parents. (3) In molecular biology, complementary base pairing between nucleic acid (DNA or RNA) strands from different sources.

hydration Process of association of a substance with the partial positive and/or negative charges of water molecules.

hydrocarbon An organic compound composed solely of hydrogen and carbon atoms.

hydrogen bond A weak attractive force existing between a hydrogen atom with a partial positive charge and an electronegative atom (usually oxygen or nitrogen) with a partial negative charge. Compare with *covalent bond* and *ionic bond*.

hydrologic cycle The water cycle, which includes evaporation, precipitation, and flow to the ocean; supplies terrestrial organisms with a continual supply of fresh water.

hydrolysis Reaction in which a covalent bond between two subunits is broken through the addition of the equivalent of a water molecule; a hydrogen atom is added to one subunit and a hydroxyl group to the other. Compare with *condensation synthesis*.

hydrophilic Interacting readily with water; having a greater affinity for water molecules than they have for each other. Compare with *hydrophobic*.

hydrophobic Not readily interacting with water; having less affinity for water molecules than they have for each other. Compare with *hydrophilic*.

hydroponics (hy″dra-paun′iks) Growing plants in an aerated solution of dissolved inorganic minerals, i.e., without soil.

hydrostatic skeleton A type of skeleton found in some invertebrates in which contracting muscles push against a tube of fluid.

hydroxide ion An anion (negatively charged particle) consisting of oxygen and hydrogen; usually written OH^-.

hydroxyl group (hy-drok′sil) Polar functional group; abbreviated —OH.

hyperpolarize To change the membrane potential so that the inside of the cell becomes more negative than its resting potential.

hypertonic A term referring to a solution having an osmotic pressure (or solute concentration) greater than that of the solution with which it is compared. Compare with *hypotonic* and *isotonic*.

hypha (hy′fah) (pl., *hyphae*) One of the threadlike filaments composing the mycelium of a water mold or fungus.

hypocotyl (hy′poh-kah″tl) The part of the axis of a plant embryo or seedling below the point of attachment of the cotyledons.

hypothalamus (hy-poh-thal′uh-mus) Part of the vertebrate brain; in mammals it regulates the pituitary gland, the autonomic system, emotional responses, body temperature, water balance, and appetite; located below the thalamus.

hypothesis A testable statement about the nature of an observation or relationship. Compare with *theory*.

hypothetico-deductive approach Emphasizes the use of deductive reasoning to test hypotheses. Compare with *hypothetico-inductive approach*. See *deductive reasoning*.

hypothetico-inductive approach Emphasizes the use of inductive reasoning to discover new general principles. Compare with *hypothetico-deductive approach*. See *inductive reasoning*.

hypotonic A term referring to a solution having an osmotic pressure (or solute concentration) less than that of the solution with which it is compared. Compare with *hypertonic* and *isotonic*.

hypotrichs A group of dorsoventrally flattened ciliates that exhibit an unusual creeping–darting locomotion.

identical twins See *monozygotic twins*.

illuviation The deposition of material leached from the upper layers of soil into the lower layers.

imaginal discs Paired structures in an insect larva that develop into specific adult structures during complete metamorphosis.

imago (ih-may′go) The adult form of an insect.

imbibition (im″bi-bish′en) The absorption of water by a seed prior to germination.

immigration The movement of individuals into a population. Compare with *emigration*.

immune response Process of recognizing foreign macromolecules and mounting a response aimed at eliminating them. See *specific* and *nonspecific immune responses; primary* and *secondary immune responses*.

immunoglobulin (im-yoon″oh-glob′yoo-lin) See *antibody*.

imperfect flower A flower that lacks either stamens or carpels. Compare with *perfect flower*.

imperfect fungi See *deuteromycetes*.

implantation The embedding of a developing embryo in the inner lining (endometrium) of the uterus.

implicit memory The unconscious memory for perceptual and motor skills, e.g., riding a bicycle.

imprinting (1) The expression of a gene based on its parental origin; also called *genomic imprinting*. (2) A type of learning by which a young bird or mammal forms a strong social attachment to an individual (usually a parent) or object within a few hours after hatching or birth.

in situ conservation Conservation efforts that concentrate on preserving biological diversity in the wild. Compare with *ex situ conservation*.

in vitro Occurring outside a living organism (literally "in glass"). Compare with *in vivo*.

in vitro evolution Test tube experiments that demonstrate that RNA molecules in the RNA world could have catalyzed the many different chemical reactions needed for life. Also called *directed evolution*.

in vitro fertilization The fertilization of eggs in the laboratory prior to implantation in the uterus for development..

in vivo Occurring in a living organism. Compare with *in vitro*.

inborn error of metabolism A metabolic disorder caused by the mutation of a gene that codes for an enzyme needed for a biochemical pathway.

inbreeding The mating of genetically similar individuals. Homozygosity increases with each successive generation of inbreeding. Compare with *outbreeding*.

inbreeding depression The phenomenon in which inbred offspring of genetically similar individuals have lower fitness (e.g., decline in fertility and high juvenile mortality) than do noninbred individuals.

inclusive fitness The total of an individual's direct and indirect fitness; includes the genes contributed directly to offspring and those contributed indirectly by kin selection. Compare with *direct fitness*. See *kin selection*.

incomplete dominance A condition in which neither member of a pair of contrasting alleles is completely expressed when the other is present.

incomplete flower A flower that lacks one or more of the four parts: sepals, petals, stamens, and/or carpels. Compare with *complete flower*.

independent assortment, principle of The genetic principle, first noted by Gregor Mendel, that states that the alleles of unlinked loci are randomly distributed to gametes.

indeterminate growth Unrestricted growth, as for example, in stems and roots. Compare with *determinate growth*.

index fossils Fossils restricted to a narrow period of geologic time and found in the same sedimentary layers in different geographic areas.

indoleacetic acid See *auxin*.

induced fit Conformational change in the active site of an enzyme that occurs when it binds to its substrate.

inducer A molecule that binds to a repressor protein, converting it to its inactive form, which is unable to prevent transcription.

inducible operon An operon that is normally inactive because a repressor molecule is attached to its operator; transcription is activated when an inducer binds to the repressor, making it incapable of binding to the operator. Compare with *repressible operon*.

induction The process by which the differentiation of a cell or group of cells is influenced by interactions with neighboring cells.

inductive reasoning The reasoning that uses specific examples to draw a general conclusion or discover a general principle. Compare with *deductive reasoning*. See *hypothetico-inductive approach*.

infant mortality rate The number of infant deaths per 1000 live births. (A child is an infant during its first 2 years of life.)

inflammatory response The response of body tissues to injury or infection, characterized clinically by heat, swelling, redness, and pain, and physiologically by increased dilation of blood vessels and increased phagocytosis.

inflorescence A cluster of flowers on a common floral stalk.

ingestion The process of taking food (or other material) into the body.

ingroup See *outgroup*.

inhibin A hormone that inhibits FSH secretion; produced by Sertoli cells in the testes and by granulosa cells in the ovaries.

inhibitory postsynaptic potential (IPSP) A change in membrane potential that takes a neuron farther from the firing level. Compare with *excitatory postsynaptic potential (EPSP)*.

initiation (of protein synthesis) The first steps of protein synthesis, in which the large and small ribosomal subunits and other components of the translation machinery bind to the 5' end of mRNA. See *elongation* and *termination*.

initiation codon See *start codon*.

innate behavior Behavior that is inherited and typical of the species; also called *instinct*.

innate immune responses See *nonspecific immune responses*.

inner cell mass The cluster of cells in the early mammalian embryo that gives rise to the embryo proper.

inorganic compound A simple substance that does not contain a carbon backbone. Compare with *organic compound*.

inositol trisphosphate (IP$_3$) A second messenger that increases intracellular calcium concentration and activates enzymes.

insight learning A complex learning process in which an animal adapts past experience to solve a new problem that may involve different stimuli.

instinct See *innate behavior*.

insulin (in'suh-lin) A hormone secreted by the pancreas that lowers blood glucose concentration. Compare with *glucagon*.

insulin-like growth factors (IGFs) Somatomedins; proteins that mediate responses to growth hormone.

insulin resistance See *diabetes mellitus*.

insulin shock A condition in which the blood glucose concentration is so low that the individual may appear intoxicated or may become unconscious and even die; caused by the injection of too much insulin or by certain metabolic malfunctions.

integral membrane protein A protein that is tightly associated with the lipid bilayer of a biological membrane; a transmembrane integral protein spans the bilayer. Compare with *peripheral membrane protein*.

integration The process of summing (adding and subtracting) incoming neural signals.

integrins Receptor proteins that bind to specific proteins in the extracellular matrix and to membrane proteins on adjacent cells; transmit signals into the cell from the extracellular matrix.

integumentary system (in-teg"yoo-men'tar-ee) The body's covering, including the skin and its nails, glands, hair, and other associated structures.

integuments The outer cell layers that surround the megasporangium of an ovule; develop into the seed coat.

intercellular substance In connective tissues, the combination of matrix and fibers in which the cells are embedded.

interference competition Intraspecific competition in which certain dominant individuals obtain an adequate supply of the limited resource at the expense of other individuals in the population. Also called *contest competition*. Compare with *exploitation competition*.

interferons (in"tur-feer'onz) Cytokines produced by animal cells when challenged by a virus; prevent viral reproduction and enable cells to resist a variety of viruses.

interkinesis The stage between meiosis I and meiosis II. Interkinesis is usually brief; the chromosomes may decondense, reverting at least partially to an interphase-like state, but DNA synthesis and chromosome duplication do not occur.

interleukins A diverse group of cytokines produced mainly by macrophages and lymphocytes.

intermediate-day plant A plant that flowers when it is exposed to days and nights of intermediate length but does not flower when the day length is too long or too short. Compare with *long-day*, *short-day*, and *day-neutral plants*.

intermediate disturbance hypothesis In community ecology, the idea that species richness is greatest at moderate levels of disturbance, which create a mosaic of habitat patches at different stages of succession.

intermediate filaments Cytoplasmic fibers that are part of the cytoskeletal network and are intermediate in size between microtubules and microfilaments.

internal membrane system The group of membranous structures in eukaryotic cells that interact through direct connections by vesicles; includes the endoplasmic reticulum, outer membrane of the nuclear envelope, Golgi complex, lysosomes, and the plasma membrane; also called *endomembrane system*.

interneuron (in"tur-noor'on) A nerve cell that carries impulses from one nerve cell to another and is not directly associated with either an effector or a sensory receptor. Also known as an *association neuron*.

internode The region on a stem between two successive nodes. Compare with *node*.

interoceptor (in'tur-oh-sep"tor) A sense organ within a body organ that transmits information regarding chemical composition, pH, osmotic pressure, or temperature. Compare with *exteroceptor*.

interphase The stage of the cell cycle between successive mitotic divisions; its subdivisions are the G$_1$ (first gap), S (DNA synthesis), and G$_2$ (second gap) phases.

interspecific competition The interaction between members of different species that vie for the same resource in an ecosystem (e.g., food or living space). Compare with *intraspecific competition*.

interstitial cells (of testis) The cells between the seminiferous tubules that secrete testosterone.

interstitial fluid The fluid that bathes the tissues of the body; also called *tissue fluid*.

intertidal zone The marine shoreline area between the high-tide mark and the low-tide mark.

intraspecific competition The interaction between members of the same species that vie for the same resource in an ecosystem (e.g., food or living space). Compare with *interspecific competition*.

intrinsic rate of increase The theoretical maximum rate of increase in population size occurring under optimal environmental conditions. Also called *biotic potential*.

intron A non-protein-coding region of a eukaryotic gene and also of the pre-mRNA transcribed from such a region. Introns do not appear in mRNA. Compare with *exon*.

invasive species A foreign species that, when introduced into an area where it is not native, upsets the balance among the organisms living there and causes economic or environmental harm.

inversion A chromosome abnormality in which the breakage and rejoining of chromosome parts results in a chromosome segment that is oriented in the opposite (reverse) direction.

invertebrate An animal without a backbone (vertebral column); invertebrates account for about 95% of animal species.

ion An atom or group of atoms bearing one or more units of electric charge, either positive (cation) or negative (anion).

ion channels Channels for the passage of ions through a membrane; formed by specific membrane proteins.

ionic bond The chemical attraction between a cation and an anion. Compare with *covalent bond* and *hydrogen bond*.

ionic compound A substance consisting of cations and anions, which are attracted by their opposite charges; ionic compounds do not consist of molecules. Compare with *covalent compound*.

ionization The dissociation of a substance to yield ions, e.g., the ionization of water yields H^+ and OH^-.

iris The pigmented portion of the vertebrate eye.

iron–sulfur world hypothesis The hypothesis that simple organic molecules that are the precursors of life originated at hydrothermal vents in the deep-ocean floor. Compare with *prebiotic soup hypothesis*.

irreversible inhibitor A substance that permanently inactivates an enzyme. Compare with *reversible inhibitor*.

islets of Langerhans (eye′lets of lahng′er-hanz) The endocrine portion of the pancreas that secretes glucagon and insulin, hormones that regulate the concentration of glucose in the blood.

isogamy (eye-sog′uh-me) Sexual reproduction involving motile gametes of similar form and size. Compare with *anisogamy* and *oogamy*.

isolecithal egg An egg containing a relatively small amount of uniformly distributed yolk. Compare with *telolecithal egg*.

isomer (eye′soh-mer) One of two or more chemical compounds having the same chemical formula but different structural formulas, e.g., structural and geometrical isomers and enantiomers.

isoprene units Five-carbon hydrocarbon monomers that make up certain lipids such as carotenoids and steroids.

isotonic (eye″soh-ton′ik) A term applied to solutions that have identical concentrations of solute molecules and hence the same osmotic pressure. Compare with *hypertonic* and *hypotonic*.

isotope (eye′suh-tope) An alternative form of an element with a different number of neutrons but the same number of protons and electrons. See *radioisotopes*.

iteroparity The condition of having repeated reproductive cycles throughout a lifetime. Compare with *semelparity*.

jasmonate One of a group of lipid-derived plant hormones that affect several processes, such as pollen development, root growth, fruit ripening, and senescence; also involved in defense against insect pests and disease-causing organisms.

jelly coat One of the acellular coverings of the eggs of certain animals, such as echinoderms.

joint The junction between two or more bones of the skeleton.

joule A unit of energy, equivalent to 0.239 calorie.

juvenile hormone (JH) An arthropod hormone that preserves juvenile structure during a molt. Without it, metamorphosis toward the adult form takes place.

juxtaglomerular apparatus (juks″tah-glo-mer′yoo-lar) A structure in the kidney that secretes renin in response to a decrease in blood pressure.

K selection A reproductive strategy recognized by some ecologists in which a species typically has a large body size, slow development, and long life span and does not devote a large proportion of its metabolic energy to the production of offspring. Compare with *r selection*.

karyogamy (kar-e-og′uh-me) The fusion of two haploid nuclei; follows fusion (plasmogamy) of cells from two sexually compatible mating types.

karyotype (kare′ee-oh-type) The chromosomal composition of an individual.

keratin (kare′ah-tin) A horny, water-insoluble protein found in the epidermis of vertebrates and in nails, feathers, hair, and horns.

ketone An organic molecule containing a carbonyl group bonded to two carbon atoms. Compare with *aldehyde*.

keystone species A species whose presence in an ecosystem largely determines the species composition and functioning of that ecosystem.

kidney The paired vertebrate organ important in excretion of metabolic wastes and in osmoregulation.

killer T cell See *T cytotoxic cell*.

kilobase (kb) 1000 bases or base pairs of a nucleic acid.

kilocalorie The amount of heat required to raise the temperature of 1 kg of water 1°C; also called *Calorie*, which is equivalent to 1000 calories.

kilojoule 1000 joules. See *joule*.

kinases Enzymes that catalyze the transfer of phosphate groups from ATP to acceptor molecules. See *protein kinases*.

kinetic energy Energy of motion. Compare with *potential energy*.

kinetochore (kin-eh′toh-kore) The portion of the chromosome centromere to which the mitotic spindle fibers attach.

kingdom A broad taxonomic category made up of related phyla; many biologists currently recognize six kingdoms of living organisms.

kin selection A type of natural selection that favors altruistic behavior toward relatives (kin), thereby ensuring that although the chances of an individual's survival are lessened, some of its genes will survive through successful reproduction of close relatives; increases inclusive fitness.

Klinefelter syndrome Inherited condition in which the affected individual is a sterile male with an XXY karyotype.

Koch's postulates A set of guidelines used to demonstrate that a specific pathogen causes specific disease symptoms.

Krebs cycle See *citric acid cycle*.

krummholz The gnarled, shrublike growth habit found in trees at high elevations, near their upper limit of distribution.

labyrinth The system of interconnecting canals of the inner ear of vertebrates.

labyrinthodonts The first successful group of tetrapods.

lactate (lactic acid) A three-carbon organic acid.

lactation (lak-tay′shun) The production or release of milk from the breast.

lacteal (lak′tee-al) One of the many lymphatic vessels in the intestinal villi that absorb fat.

lagging strand A strand of DNA that is synthesized as a series of short segments, called *Okazaki fragments,* which are then covalently joined by DNA ligase. Compare with *leading strand*.

lamins Polypeptides attached to the inner surface of the nuclear envelope that provide a type of skeletal framework.

landscape A large land area (several to many square kilometers) composed of interacting ecosystems.

landscape ecology The subdiscipline in ecology that studies the connections in a heterogeneous landscape.

large intestine The portion of the digestive tract of humans (and other vertebrates) consisting of the cecum, colon, rectum, and anus.

larva (pl., *larvae*) An immature form in the life history of some animals; may be unlike the parent.

larynx (lare′inks) The organ at the upper end of the trachea that contains the vocal cords.

lateral meristems Areas of localized cell division on the side of a plant that give rise to secondary tissues. Lateral meristems, including the vascular cambium and the cork cambium, cause an increase in the girth of the plant body. Compare with *apical meristem*.

leaching The process by which dissolved materials are washed away or carried with water down through the various layers of the soil.

leader sequence Noncoding sequence of nucleotides in mRNA that is transcribed from the region that precedes (is upstream to) the coding region.

leading strand Strand of DNA that is synthesized continuously. Compare with *lagging strand*.

learning A change in the behavior of an animal that results from experience.

legume (leg′yoom) (1) A simple, dry fruit that develops from a single carpel and splits open at maturity along two sutures to release seeds. (2) Any member of the pea family, e.g., pea, bean, peanut, alfalfa.

lek A small territory in which males compete for females.

lens The oval, transparent structure located behind the iris of the vertebrate eye; bends incoming light rays and brings them to a focus on the retina.

lenticels (len′tih-sels) Porous swellings of cork cells in the stems of woody plants; facilitate the exchange of gases.

leptin A hormone produced by adipose tissue that signals brain centers about the status of energy stores.

leukocytes (loo′koh-sites) White blood cells; colorless amoeboid cells that defend the body against disease-causing organisms.

leukoplasts Colorless plastids; include amyloplasts, which are used for starch storage in cells of roots and tubers.

lichen (ly′ken) A compound organism consisting of a symbiotic fungus and an alga or cyanobacterium.

life history traits Significant features of a species' life cycle, particularly traits that influence survival and reproduction.

life span The maximum duration of life for an individual of a species.

life table A table showing mortality and survival data by age of a population or cohort.

ligament (lig′uh-ment) A connective tissue cable or strap that connects bones to each other or holds other organs in place.

ligand A molecule that binds to a specific site in a receptor or other protein.

light-dependent reactions Reactions of photosynthesis in which light energy absorbed by chlorophyll is used to synthesize ATP and usually NADPH. Include *cyclic electron transport* and *noncyclic electron transport*.

lignin (lig′nin) A substance found in many plant cell walls that confers rigidity and strength, particularly in woody tissues.

limbic system In vertebrates, an action system of the brain. In humans, plays a role in emotional responses, motivation, autonomic function, and sexual response.

limiting resource An environmental resource that because it is scarce or unfavorable tends to restrict the ecological niche of an organism.

limnetic zone (lim-net′ik) The open water away from the shore of a lake or pond extending down as far as sunlight penetrates. Compare with *littoral zone* and *profundal zone*.

linkage The tendency for a group of genes located on the same chromosome to be inherited together in successive generations.

lipase (lip′ase) A fat-digesting enzyme.

lipid Any of a group of organic compounds that are insoluble in water but soluble in nonpolar solvents; lipids serve as energy storage and are important components of cell membranes.

lipoprotein (lip-oh-proh′teen) A large molecular complex consisting of lipids and protein; transports lipids in the blood. High-density lipoproteins (HDLs) transport cholesterol to the liver; low-density lipoproteins (LDLs) deliver cholesterol to many cells of the body.

littoral zone (lit′or-ul) The region of shallow water along the shore of a lake or pond. Compare with *limnetic zone* and *profundal zone*.

liver A large, complex organ that secretes bile, helps maintain homeostasis by removing or adding nutrients to the blood, and performs many other metabolic functions.

liverworts A phylum of spore-producing, nonvascular, thallose or leafy plants with a life cycle similar to that of mosses.

local hormones See *local regulators*.

local regulators Prostaglandins (a group of local hormones), growth factors, cytokines, and other soluble molecules that act on nearby cells by paracrine regulation or act on the cells that produce them (autocrine regulation).

locus The place on the chromosome at which the gene for a given trait occurs, i.e., a segment of the chromosomal DNA containing information that controls some feature of the organism; also called *gene locus*.

logistic population growth Population growth that initially occurs at a constant rate of increase over time (i.e., exponential) but then levels out as the carrying capacity of the environment is approached. Compare with *exponential population growth*.

long-day plant A plant that flowers in response to shortening nights; also called *short-night plant*. Compare with *short-day, intermediate-day,* and *day-neutral plants*.

long-night plant See *short-day plant*.

long-term potentiation (LTP) Long-lasting increase in the strength of synaptic connections that occurs in response to a series of high-frequency electrical stimuli. Compare with *long-term synaptic depression (LTD)*.

long-term synaptic depression (LTD) Long-lasting decrease in the strength of synaptic connections that occurs in response to low-frequency stimulation of neurons. Compare with *long-term potentiation (LTP)*.

loop of Henle (hen′lee) The U-shaped loop of a mammalian kidney tubule that extends down into the renal medulla.

loose connective tissue A type of connective tissue that is widely distributed in the body; consists of fibers strewn through a semifluid matrix.

lophophorate phyla Three related invertebrate protostome phyla, characterized by a ciliated ring of tentacles that surrounds the mouth.

Lophotrochozoa A branch of the protostomes that includes the flatworms, nemerteans (proboscis worms), mollusks, annelids, and the lophophorate phyla.

low-density lipoprotein (LDL) See *lipoprotein*.

lumen (loo′men) (1) The space enclosed by a membrane, such as the lumen of the endoplasmic reticulum or the thylakoid lumen. (2) The cavity or channel within a tube or tubular organ, such as a blood vessel or the digestive tract. (3) The space left within a plant cell after the cell's living material dies, as in tracheids.

lung An internal respiratory organ that functions in gas exchange; enables an animal to breathe air.

luteinizing hormone (LH) (loot'eh-ny-zing) Gonadotropic hormone secreted by the anterior pituitary; stimulates ovulation and maintains the corpus luteum in the ovaries of females; stimulates testosterone production in the testes of males.

lymph (limf) The colorless fluid within the lymphatic vessels that is derived from blood plasma; contains white blood cells; ultimately lymph is returned to the blood.

lymph node A mass of lymph tissue surrounded by a connective tissue capsule; manufactures lymphocytes and filters lymph.

lymphatic system A subsystem of the cardiovascular system; returns excess interstitial fluid (lymph) to the circulation; defends the body against disease organisms.

lymphocyte (lim'foh-site) White blood cell with nongranular cytoplasm that governs immune responses. See *B cell* and *T cell*.

lysis (ly'sis) The process of disintegration of a cell or some other structure.

lysogenic conversion The change in properties of bacteria that results from the presence of a prophage.

lysosomes (ly'soh-sohmz) Intracellular organelles present in many animal cells; contain a variety of hydrolytic enzymes.

lysozyme An enzyme found in many tissues and in tears and other body fluids; attacks the cell wall of many gram-positive bacteria.

macroevolution Large-scale evolutionary events over long time spans. Macroevolution results in phenotypic changes in populations that are significant enough to warrant their placement in taxonomic groups at the species level and higher. Compare with *microevolution*.

macromolecule A very large organic molecule, such as a protein or nucleic acid.

macronucleus A large nucleus found, along with one or several micronuclei, in ciliates. The macronucleus regulates metabolism and growth. Compare with *micronucleus*.

macronutrient An essential element required in fairly large amounts for normal growth. Compare with *micronutrient*.

macrophage (mak'roh-faje) A large phagocytic cell capable of ingesting and digesting bacteria and cell debris. Macrophages are also antigen-presenting cells.

magnoliid One of the clades of flowering plants; magnoliids are core angiosperms that were traditionally classified as "dicots," but molecular evidence indicates they are neither eudicots or monocots.

major histocompatibility complex (MHC) A group of membrane proteins, present on the surface of most cells, that are slightly different in each individual. In humans, the MHC is called the *HLA (human leukocyte antigen) group*.

malignant cells Cancer cells; tumor cells that are able to invade tissue and metastasize.

malignant transformation See *transformation*.

malnutrition Poor nutritional status; results from dietary intake that is either below or above required needs.

Malpighian tubules (mal-pig'ee-an) The excretory organs of many arthropods.

mammals The class of vertebrates characterized by hair, mammary glands, a diaphragm, and differentiation of teeth.

mandible (man'dih-bl) (1) The lower jaw of vertebrates. (2) Jaw-like, external mouthparts of insects.

mangrove forest A tidal wetland dominated by mangrove trees in which the salinity fluctuates between that of sea water and fresh water.

mantle In the mollusk, a fold of tissue that covers the visceral mass and that usually produces a shell.

marine snow The organic debris (plankton, dead organisms, fecal material, etc.) that "rains" into the dark area of the oceanic province from the lighted region above; the primary food of most organisms that live in the ocean's depths.

marsupials (mar-soo'pee-ulz) A subclass of mammals, characterized by the presence of an abdominal pouch in which the young, which are born in a very undeveloped condition, are carried for some time after birth.

mass extinction The extinction of numerous species during a relatively short period of geologic time. Compare with *background extinction*.

mast cell A type of cell found in connective tissue; contains histamine and is important in an inflammatory response and in allergic reactions.

maternal effect genes Genes of the mother that are transcribed during oogenesis and subsequently affect the development of the embryo. Compare with *zygotic genes*.

matrix (may'triks) (1) In cell biology, the interior of the compartment enclosed by the inner mitochondrial membrane. (2) In zoology, nonliving material secreted by and surrounding connective tissue cells; contains a network of microscopic fibers.

matter Anything that has mass and takes up space.

maxillae Appendages used for manipulating food; characteristic of crustaceans.

mechanical isolation A prezygotic reproductive isolating mechanism in which fusion of the gametes of two species is prevented by morphological or anatomical differences.

mechanoreceptor (meh-kan'oh-ree-sep"tor) A sensory cell or organ that perceives mechanical stimuli, e.g., touch, pressure, gravity, stretching, or movement.

medulla (meh-dul'uh) (1) The inner part of an organ, such as the medulla of the kidney. Compare with *cortex*. (2) The most posterior part of the vertebrate brain, lying next to the spinal cord.

medusa A jellyfish-like animal; a free-swimming, umbrella-shaped stage in the life cycle of certain cnidarians. Compare with *polyp*.

megaphyll (meg'uh-fil) Type of leaf found in horsetails, ferns, gymnosperms, and angiosperms; contains multiple vascular strands (i.e., complex venation). Compare with *microphyll*.

megaspore (meg'uh-spor) The *n* spore in heterosporous plants that gives rise to a female gametophyte. Compare with *microspore*.

meiosis (my-oh'sis) Process in which a 2*n* cell undergoes two successive nuclear divisions (meiosis I and meiosis II), potentially producing four *n* nuclei; leads to the formation of gametes in animals and spores in plants.

melanin A dark pigment present in many animals; contributes to the color of the skin.

melanocortins A group of peptides that appear to decrease appetite in response to increased fat stores.

melatonin (mel-ah-toh'nin) A hormone secreted by the pineal gland that plays a role in setting circadian rhythms.

memory cell B or T cell (lymphocyte) that permits rapid mobilization of immune response on second or subsequent exposure to a particular antigen. Memory B cells continue to produce antibodies after the immune system overcomes an infection.

meninges (meh-nin'jeez) (sing., *meninx*) The three membranes that protect the brain and spinal cord: the dura mater, arachnoid, and pia mater.

menopause The period (usually occurring between 45 and 55 years of age) in women when the recurring menstrual cycle ceases.

menstrual cycle (men'stroo-ul) In the human female, the monthly sequence of events that prepares the body for pregnancy.

menstruation (men-stroo-ay'shun) The monthly discharge of blood and degenerated uterine lining in the human female; marks the beginning of each menstrual cycle.

meristem (mer'ih-stem) A localized area of mitotic cell division in the plant body. See *apical meristem* and *lateral meristems*.

meroblastic cleavage Cleavage pattern observed in the telolecithal eggs of reptiles and birds, in which cleavage is restricted to a small disc of cytoplasm at the animal pole. Compare with *holoblastic cleavage*.

mesencephalon See *midbrain*.

mesenchyme (mes'en-kime) A loose, often jellylike connective tissue containing undifferentiated cells; found in the embryos of vertebrates and the adults of some invertebrates.

mesoderm (mez'oh-derm) The middle germ layer of the early embryo; gives rise to connective tissue, muscle, bone, blood vessels, kidneys, and many other structures. Compare with *ectoderm* and *endoderm*.

mesophyll (mez'oh-fil) Photosynthetic tissue in the interior of a leaf; sometimes differentiated into palisade mesophyll and spongy mesophyll.

Mesozoic era That part of geologic time extending from roughly 251 million to 66 million years ago.

messenger RNA (mRNA) RNA that specifies the amino acid sequence of a protein; transcribed from DNA.

metabolic pathway A series of chemical reactions in which the product of one reaction becomes the substrate of the next reaction.

metabolic rate Energy use by an organism per unit time. See *basal metabolic rate (BMR)*.

metabolism The sum of all the chemical processes that occur within a cell or organism; the transformations by which energy and matter are made available for use by the organism. See *anabolism* and *catabolism*.

metamorphosis (met"ah-mor'fuh-sis) Transition from one developmental stage to another, such as from a larva to an adult.

metanephridia (sing., *metanephridium*) The excretory organs of annelids and mollusks; each consists of a tubule open at both ends; at one end a ciliated funnel opens into the coelom, and the other end opens to the outside of the body.

metaphase (met'ah-faze) The stage of mitosis in which the chromosomes line up on the midplane of the cell. Occurs after prometaphase and before anaphase.

metapopulation A population that is divided into several local populations among which individuals occasionally disperse.

metastasis (met-tas'tuh-sis) The spreading of cancer cells from one organ or part of the body to another.

metencephalon See *hindbrain*.

methyl group A nonpolar functional group; abbreviated —CH₃.

micro RNAs (miRNAs) Single-stranded RNA molecules about 21 to 22 nucleotides long that inhibit the translation of mRNAs involved in growth and development.

microclimate Local variations in climate produced by differences in elevation, in the steepness and direction of slopes, and in exposure to prevailing winds.

microevolution Small-scale evolutionary change caused by changes in allele or genotype frequencies that occur within a population over a few generations. Compare with *macroevolution*.

microfilaments Thin fibers consisting of actin protein subunits; form part of the cytoskeleton.

microfossils Ancient traces (fossils) of microscopic life.

microglia Phagocytic glial cells found in the CNS.

micronucleus One or more smaller nuclei found, along with the macronucleus, in ciliates. The micronucleus is involved in sexual reproduction. Compare with *macronucleus*.

micronutrient An essential element that is required in trace amounts for normal growth. Compare with *macronutrient*.

microphyll (mi'kro-fil) Type of leaf found in club mosses; contains one vascular strand (i.e., simple venation). Compare with *megaphyll*.

microsphere A protobiont produced by adding water to abiotically formed polypeptides.

microspore (mi'kro-spor) The *n* spore in heterosporous plants that gives rise to a male gametophyte. Compare with *megaspore*.

microsporidia Small, unicellular, fungal parasites that infect eukaryotic cells; classified with the zygomycetes.

microtubule-associated proteins (MAPs) Include structural proteins that help regulate microtubule assembly and cross-link microtubules to other cytoskeletal polymers; and motors, such as kinesin and dynein, that use ATP to produce movement.

microtubule-organizing center (MTOC) The region of the cell from which microtubules are anchored and possibly assembled. The MTOCs of many organisms (including animals, but not flowering plants or most gymnosperms) contain a pair of centrioles.

microtubules (my-kroh-too'bewls) Hollow, cylindrical fibers consisting of tubulin protein subunits; major components of the cytoskeleton and found in mitotic spindles, cilia, flagella, centrioles, and basal bodies.

microvilli (sing., *microvillus*) Minute projections of the plasma membrane that increase the surface area of the cell; found mainly in cells concerned with absorption or secretion, such as those lining the intestine or the kidney tubules.

midbrain In vertebrate embryos, one of the three divisions of the developing brain. Also called *mesencephalon*. Compare with *forebrain* and *hindbrain*.

middle lamella The layer composed of pectin polysaccharides that serves to cement together the primary cell walls of adjacent plant cells.

midvein The main, or central, vein of a leaf.

migration (1) The periodic or seasonal movement of an organism (individual or population) from one place to another, usually over a long distance. See *dispersal*. (2) In evolutionary biology, a movement of individuals that results in a transfer of alleles from one population to another. See *gene flow*.

mineralocorticoids (min"ur-al-oh-kor'tih-koidz) Hormones produced by the adrenal cortex that regulate mineral metabolism and, indirectly, fluid balance. The principal mineralocorticoid is aldosterone.

minerals Inorganic nutrients ingested as salts dissolved in food and water.

minimum viable population (MVP) The smallest population size at which a species has a high chance of sustaining its numbers and surviving into the future.

mismatch repair A DNA repair mechanism in which special enzymes recognize the incorrectly paired nucleotides and remove them. DNA polymerases then fill in the missing nucleotides.

missense mutation A type of base-substitution mutation that causes one amino acid to be substituted for another in the resulting protein product. Compare with *nonsense mutation*.

mitochondria (my"toh-kon'dree-ah) (sing., *mitochondrion*) Intracellular organelles that are the sites of oxidative phosphorylation in eukaryotes; include an outer membrane and an inner membrane.

mitochondrial DNA (mtDNA) DNA present in mitochondria that is transmitted maternally, from mothers to their offspring. Mitochondrial DNA mutates more rapidly than nuclear DNA.

mitosis (my-toh′sis) The division of the cell nucleus resulting in two daughter nuclei, each with the same number of chromosomes as the parent nucleus; mitosis consists of prophase, prometaphase, metaphase, anaphase, and telophase. Cytokinesis usually overlaps the telophase stage.

mitotic spindle Structure consisting mainly of microtubules that provides the framework for chromosome movement during cell division.

mitral valve See *atrioventricular valve.*

mobile genetic element See *transposon.*

model organism A species chosen for biological studies because it has characteristics that allow for the efficient analysis of biological processes. Most model organisms are small, have short generation times, and are easy to grow and study under controlled conditions.

modern synthesis A comprehensive, unified explanation of evolution based on combining previous theories, especially of Mendelian genetics, with Darwin's theory of evolution by natural selection; also called the *synthetic theory of evolution.*

mole The atomic mass of an element or the molecular mass of a compound, expressed in grams; one mole of any substance has 6.02×10^{23} units (Avogadro's number).

molecular anthropology The branch of science that compares genetic material from individuals of regional human populations to help unravel the origin and migrations of modern humans.

molecular chaperones Proteins that help other proteins fold properly. Although not dictating the folding pattern, chaperones make the process more efficient.

molecular clock analysis A comparison of the DNA nucleotide sequences of related organisms to estimate when they diverged from one another during the course of evolution.

molecular formula The type of chemical formula that gives the actual numbers of each type of atom in a molecule. Compare with *simplest formula* and *structural formula.*

molecular mass The sum of the atomic masses of the atoms that make up a single molecule of a compound; expressed in atomic mass units (amu) or daltons.

molecule The smallest particle of a covalently bonded element or compound; two or more atoms joined by covalent bonds.

mollusks A phylum of coelomate protostome animals characterized by a soft body, visceral mass, mantle, and foot.

molting The shedding and replacement of an outer covering such as an exoskeleton.

molting hormone A steroid hormone that stimulates growth and molting in insects. Also called *ecdysone.*

monoacylglycerol (mon″o-as″il-glis′er-ol) Lipid consisting of glycerol combined chemically with a single fatty acid. Also called *monoglyceride.* Compare with *diacylglycerol* and *triacylglycerol.*

monocot (mon′oh-kot) One of two classes of flowering plants; monocot seeds contain a single cotyledon, or seed leaf. Compare with *eudicot.*

monoclonal antibodies Identical antibody molecules produced by cells cloned from a single cell.

monocyte (mon′oh-site) A type of white blood cell; a large, phagocytic, nongranular leukocyte that enters the tissues and differentiates into a macrophage.

monoecious (mon-ee′shus) Having male and female reproductive parts in separate flowers or cones on the same plant; compare with *dioecious.*

monogamy A mating system in which a male animal mates with a single female during a breeding season.

monoglyceride See *monoacylglycerol.*

monohybrid cross A genetic cross that takes into account the behavior of alleles of a single locus. Compare with *dihybrid cross.*

monokaryotic (mon″o-kare-ee-ot′ik) The condition of having a single *n* nucleus per cell, characteristic of certain fungal hyphae. Compare with *dikaryotic.*

monomer (mon′oh-mer) A molecule that can link with other similar molecules; two monomers join to form a dimer, whereas many form a polymer. Monomers are small (e.g., sugars or amino acids) or large (e.g., tubulin or actin proteins).

monophyletic group (mon″oh-fye-let′ik) A group of organisms that evolved from a common ancestor. Compare with *polyphyletic group* and *paraphyletic group.*

monosaccharide (mon-oh-sak′ah-ride) A sugar that cannot be degraded by hydrolysis to a simpler sugar (e.g., glucose or fructose).

monosomy The condition in which only one member of a chromosome pair is present and the other is missing. Compare with *trisomy* and *disomy.*

monotremes (mon′oh-treemz) Egg-laying mammals such as the duck-billed platypus of Australia.

monounsaturated fatty acid See *fatty acid.*

monozygotic twins Genetically identical twins that arise from the division of a single fertilized egg; commonly known as *identical twins.* Compare with *dizygotic twins.*

morphogen Any chemical agent thought to govern the processes of cell differentiation and pattern formation that lead to morphogenesis.

morphogenesis (mor-foh-jen′eh-sis) The development of the form and structures of an organism and its parts; proceeds by a series of steps known as *pattern formation.*

mortality The rate at which individuals die; the average per capita death rate.

morula (mor′yoo-lah) An early embryo consisting of a solid ball of cells.

mosaic development A rigid developmental pattern in which the fates of cells become restricted early in development. Compare with *regulative development.*

mosses A phylum of spore-producing nonvascular plants with an alternation of generations in which the dominant *n* gametophyte alternates with a 2*n* sporophyte that remains attached to the gametophyte.

motor neuron An efferent neuron that transmits impulses away from the central nervous system to skeletal muscle.

motor program A coordinated sequences of muscle actions responsible for many behaviors we think of as automatic.

motor unit All the skeletal muscle fibers that are stimulated by a single motor neuron.

mRNA cap An unusual nucleotide, 7-methylguanylate, that is added to the 5′ end of a eukaryotic messenger RNA. Capping enables eukaryotic ribosomes to bind to mRNA.

mucosa (mew-koh′suh) See *mucous membrane.*

mucous membrane A type of epithelial membrane that lines a body cavity that opens to the outside of the body, e.g., the digestive and respiratory tracts; also called *mucosa.*

mucus (mew′cus) A sticky secretion composed of covalently linked protein and carbohydrate; serves to lubricate body parts and trap particles of dirt and other contaminants. (The adjectival form is spelled *mucous.*)

Müllerian mimicry (mul-ler′ee-un mim′ih-kree) The resemblance of dangerous, unpalatable, or poisonous species to one another so that potential predators recognize them more easily. Compare with *Batesian mimicry*.

multiple alleles (al-leelz') Three or more alleles of a single locus (in a population), such as the alleles governing the ABO series of blood types.

multiple fruit A fruit that develops from many ovaries of many separate flowers, e.g., pineapple. Compare with *simple, aggregate,* and *accessory fruits.*

muscle (1) A tissue specialized for contraction. (2) An organ that produces movement by contraction.

mutagen (mew'tah-jen) Any agent capable of entering the cell and producing mutations.

mutation Any change in DNA; may include a change in the nucleotide base pairs of a gene, a rearrangement of genes within the chromosomes so that their interactions produce different effects, or a change in the chromosomes themselves.

mutualism (1) In ecology, a symbiotic relationship in which both partners benefit from the association. Compare with *parasitism* and *commensalism.* (2) In animal behavior, cooperative behavior in which each animal in the group benefits.

mycelium (my-seel'ee-um) (pl., *mycelia*) The vegetative body of most fungi and certain protists (water molds); consists of a branched network of hyphae.

mycorrhizae (my"kor-rye'zee) Mutualistic associations of fungi and plant roots that aid in the plant's absorption of essential minerals from the soil.

mycotoxins Poisonous chemical compounds produced by fungi, e.g., aflatoxins that harm the liver and are known carcinogens.

myelencephalon See *hindbrain.*

myelin sheath (my'eh-lin) The white, fatty material that forms a sheath around the axons of certain nerve cells, which are then called *myelinated fibers.*

myocardial infarction (MI) Heart attack; serious consequence occurring when the heart muscle receives insufficient oxygen.

myofibrils (my-oh-fy'brilz) Tiny threadlike structures in the cytoplasm of striated and cardiac muscle that are composed of myosin filaments and actin filaments; these filaments are responsible for muscle contraction; see *myosin filaments* and *actin filaments.*

myoglobin (my'oh-glo"bin) A hemoglobin-like, oxygen-transferring protein found in muscle.

myosin (my'oh-sin) A protein that together with actin is responsible for muscle contraction.

myosin filaments Thick filaments consisting mainly of the protein myosin; actin and myosin filaments make up the myofibrils of muscle fibers.

n The chromosome number of a gamete. The chromosome number of a zygote is 2*n*. If an organism is not polyploid, the *n* gametes are haploid and the 2*n* zygotes are diploid.

NAD$^+$/NADH Oxidized and reduced forms, respectively, of nicotinamide adenine dinucleotide, a coenzyme that transfers electrons (as hydrogen), particularly in catabolic pathways, including cellular respiration.

NADP$^+$/NADPH Oxidized and reduced forms, respectively, of nicotinamide adenine dinucleotide phosphate, a coenzyme that acts as an electron (hydrogen) transfer agent, particularly in anabolic pathways, including photosynthesis.

nanoplankton Extremely minute (<20 μm in length) algae that are major producers in the ocean because of their great abundance; part of phytoplankton.

natality The rate at which individuals produce offspring; the average per capita birth rate.

natural killer cell (NK cell) A large, granular lymphocyte that functions in both nonspecific and specific immune responses; releases cytokines and proteolytic enzymes that target tumor cells and cells infected with viruses and other pathogens.

natural selection The mechanism of evolution proposed by Charles Darwin; the tendency of organisms that have favorable adaptations to their environment to survive and become the parents of the next generation. Evolution occurs when natural selection results in changes in allele frequencies in a population.

necrosis Uncontrolled cell death that causes inflammation and damages other cells. Compare with *apoptosis.*

nectary (nek'ter-ee) In plants, a gland or other structure that secretes nectar.

negative feedback mechanism A homeostatic mechanism in which a change in some condition triggers a response that counteracts, or reverses, the changed condition, restoring homeostasis, e.g., how mammals maintain body temperature. Compare with *positive feedback mechanism.*

nekton (nek'ton) Free-swimming aquatic organisms such as fish and turtles. Compare with *plankton.*

nematocyst (nem-at'oh-sist) A stinging structure found within cnidocytes (stinging cells) in cnidarians; used for anchorage, defense, and capturing prey.

nematodes The phylum of animals commonly known as *roundworms.*

nemerteans The phylum of animals commonly known as *ribbon worms;* each has a proboscis (tubular feeding organ) for capturing prey.

neonate Newborn individual.

neoplasm See *tumor.*

nephridial organ (neh-frid'ee-al) The excretory organ of many invertebrates; consists of simple or branching tubes that usually open to the outside of the body through pores; also called *nephridium.*

nephron (nef'ron) The functional, microscopic unit of the vertebrate kidney.

neritic province (ner-ih'tik) Ocean water that extends from the shoreline to where the bottom reaches a depth of 200 m. Compare with *oceanic province.*

nerve A bundle of axons (or dendrites) wrapped in connective tissue that conveys impulses between the central nervous system and some other part of the body.

nerve net A system of interconnecting nerve cells found in cnidarians and echinoderms.

nervous tissue A type of animal tissue specialized for transmitting electrical and chemical signals.

net primary productivity The energy that remains in an ecosystem (as biomass) after cellular respiration has occurred; net primary productivity equals gross primary productivity minus respiration. Compare with *gross primary productivity.*

neural crest (noor'ul) A group of cells along the neural tube that migrate and form various parts of the embryo, including parts of the peripheral nervous system.

neural plasticity The ability of the nervous system to change in response to experience.

neural plate See *neural tube.*

neural transmission See *transmission, neural.*

neural tube The hollow, longitudinal structure in the early vertebrate embryo that gives rise to the brain and spinal cord. The neural tube forms from the neural plate, a flattened, thickened region of the ectoderm that rolls up and sinks below the surface.

neuroendocrine cells Neurons that produce neurohormones.

neurohormones Hormones produced by neuroendocrine cells; transported down axons and released into interstitial fluid; common in invertebrates; in vertebrates, hypothalamus produces neurohormones.

neuron (noor'on) A nerve cell; a conducting cell of the nervous system that typically consists of a cell body, dendrites, and an axon.

neuropeptide One of a group of peptides produced in neural tissue that function as signaling molecules; many are neurotransmitters.

neuropeptide Y A signaling molecule produced by the hypothalamus that increases appetite and slows metabolism; helps restore energy homeostasis when leptin levels and food intake are low.

neurotransmitter A chemical signal used by neurons to transmit impulses across a synapse.

neutral solution A solution of pH 7; there are equal concentrations of hydrogen ions [H^+] and hydroxide ions [OH^-]. Compare with *acidic solution* and *basic solution*.

neutral variation Variation that does not appear to confer any selective advantage or disadvantage to the organism.

neutron (noo'tron) An electrically neutral particle with a mass of 1 atomic mass unit (amu) found in the atomic nucleus. Compare with *proton* and *electron*.

neutrophil (new'truh-fil) A type of granular leukocyte important in immune responses; a type of phagocyte that engulfs and destroys bacteria and foreign matter.

niche (nich) The totality of an organism's adaptations, its use of resources, and the lifestyle to which it is fitted in its community; how an organism uses materials in its environment as well as how it interacts with other organisms; also called *ecological niche*. See *fundamental niche* and *realized niche*.

nicotinamide adenine dinucleotide See *NAD$^+$/NADH*.

nicotinamide adenine dinucleotide phosphate See *NADP$^+$/NADPH*.

nitric oxide (NO) A gaseous signaling molecule; a neurotransmitter.

nitrification (nie"tra-fuh-kay'shun) The conversion of ammonia (NH_3) to nitrate (NO_3^-) by certain bacteria (nitrifying bacteria) in the soil; part of the nitrogen cycle.

nitrogen cycle The worldwide circulation of nitrogen from the abiotic environment into living things and back into the abiotic environment.

nitrogen fixation The conversion of atmospheric nitrogen (N_2) to ammonia (NH_3) by certain bacteria; part of the nitrogen cycle.

nitrogenase (nie-traa'jen-ase) The enzyme responsible for nitrogen fixation under anaerobic conditions.

nociceptors (no'sih-sep-torz) Pain receptors; free endings of certain sensory neurons whose stimulation is perceived as pain.

node The area on a stem where each leaf is attached. Compare with *internode*.

nodules Swellings on the roots of plants, such as legumes, in which symbiotic nitrogen-fixing bacteria (*Rhizobium*) live.

noncompetitive inhibitor A substance that lowers the rate at which an enzyme catalyzes a reaction but does not bind to the active site. Compare with *competitive inhibitor*.

noncyclic electron transport In photosynthesis, the linear flow of electrons, produced by photolysis of water, through photosystems II and I; results in the formation of ATP (by chemiosmosis), NADPH, and O_2. Compare with *cyclic electron transport*.

nondisjunction Abnormal separation of sister chromatids or of homologous chromosomes caused by their failure to disjoin (move apart) properly during mitosis or meiosis.

nonpolar covalent bond Chemical bond formed by the equal sharing of electrons between atoms of approximately equal electronegativity. Compare with *polar covalent bond*.

nonpolar molecule Molecule that does not have a positively charged end and a negatively charged end; nonpolar molecules are generally insoluble in water. Compare with *polar molecule*.

nonsense mutation A base-substitution mutation that results in an amino acid–specifying codon being changed to a termination (stop) codon; when the abnormal mRNA is translated, the resulting protein is usually truncated and nonfunctional. Compare with *missense mutation*.

nonspecific immune responses Mechanisms such as physical barriers (e.g., the skin) and phagocytosis that provide immediate and general protection against pathogens. Also called *innate immunity*. Compare with *specific immune responses*.

norepinephrine (nor-ep-ih-nef'rin) A neurotransmitter that is also a hormone secreted by the adrenal medulla.

Northern blot A technique in which RNA fragments, previously separated by gel electrophoresis, are transferred to a nitrocellulose membrane and detected by autoradiography or chemical luminescence. Compare with *Southern blot* and *Western blot*.

notochord (no'toe-kord) The flexible, longitudinal rod in the anterior-posterior axis that serves as an internal skeleton in the embryos of all chordates and in the adults of some.

nuclear area Region of a prokaryotic cell that contains DNA; not enclosed by a membrane. Also called *nucleoid*.

nuclear envelope The double membrane system that encloses the cell nucleus of eukaryotes.

nuclear equivalence The concept that the nuclei of all differentiated cells of an adult organism are genetically identical to one another and to the nucleus of the zygote from which they were derived. Compare with *genomic rearrangement* and *gene amplification*.

nuclear pores Structures in the nuclear envelope that allow passage of certain materials between the cell nucleus and the cytoplasm.

nucleoid See *nuclear area*.

nucleolus (new-klee'oh-lus) (pl., *nucleoli*) Specialized structure in the cell nucleus formed from regions of several chromosomes; site of assembly of the ribosomal subunits.

nucleoplasm The contents of the cell nucleus.

nucleoside triphosphate Molecule consisting of a nitrogenous base, a pentose sugar, and three phosphate groups, e.g., adenosine triphosphate (ATP).

nucleosomes (new'klee-oh-sohmz) Repeating units of chromatin structure, each consisting of a length of DNA wound around a complex of eight histone molecules. Adjacent nucleosomes are connected by a DNA linker region associated with another histone protein.

nucleotide (noo'klee-oh-tide) A molecule consisting of one or more phosphate groups, a five-carbon sugar (ribose or deoxyribose), and a nitrogenous base (purine or pyrimidine).

nucleotide excision repair A DNA repair mechanism commonly used to repair a damaged segment of DNA caused by the sun's ultraviolet radiation or by harmful chemicals.

nucleus (new'klee-us) (pl., *nuclei*) (1) The central region of an atom that contains the protons and neutrons. (2) A cell organelle in eukaryotes that contains the DNA and serves as the control center of the cell. (3) A mass of nerve cell bodies in the central nervous system. Compare with *ganglion*.

nut A simple, dry fruit that contains a single seed and is surrounded by a hard fruit wall.

nutrients The chemical substances in food that are used as components for synthesizing needed materials and/or as energy sources.

nutrition The process of taking in and using food (nutrients).

obesity Excess accumulation of body fat; a person is considered obese if the body mass index (BMI) is 30 or higher.

obligate anaerobe An organism that grows only in the absence of oxygen. Compare with *facultative anaerobe*.

occipital lobes Posterior areas of the mammalian cerebrum; interpret visual stimuli from the retina of the eye.

oceanic province That part of the open ocean that overlies an ocean bottom deeper than 200 m. Compare with *neritic province*.

Okazaki fragment One of many short segments of DNA, each 100 to 1000 nucleotides long, that must be joined by DNA ligase to form the lagging strand in DNA replication.

olfactory epithelium Tissue containing odor-sensing neurons.

oligodendrocyte A type of glial cell that forms myelin sheaths around neurons in the CNS.

oligosaccharin One of several signaling molecules in plants that trigger the production of phytoalexins and affect aspects of growth and development.

ommatidium (om″ah-tid′ee-um) (pl., *ommatidia*) One of the light-detecting units of a compound eye, consisting of a lens and a crystalline cone that focus light onto photoreceptors called *retinular cells*.

omnivore (om′nih-vore) An animal that eats a variety of plant and animal materials.

oncogene (on′koh-jeen) An abnormally functioning gene implicated in causing cancer. Compare with *proto-oncogene* and *tumor suppressor gene*.

oocytes (oh′oh-sites) Meiotic cells that give rise to egg cells (ova).

oogamy (oh-og′uh-me) The fertilization of a large, nonmotile female gamete by a small, motile male gamete. Compare with *isogamy* and *anisogamy*.

oogenesis (oh″oh-jen′eh-sis) Production of female gametes (eggs) by meiosis. Compare with *spermatogenesis*.

oospore A thick-walled, resistant spore formed from a zygote during sexual reproduction in water molds.

open circulatory system A type of circulatory system in which the blood bathes the tissues directly; characteristic of arthropods and many mollusks. Compare with *closed circulatory system*.

open system An entity that exchanges energy with its surroundings. Compare with *closed system*.

operant conditioning A type of learning in which an animal is rewarded or punished for performing a behavior it discovers by chance.

operator site One of the control regions of an operon; the DNA segment to which a repressor binds, thereby inhibiting the transcription of the adjacent structural genes of the operon.

operculum In bony fishes, a protective flap of the body wall that covers the gills.

operon (op′er-on) In prokaryotes, a group of structural genes that are coordinately controlled and transcribed as a single message, plus their adjacent regulatory elements.

opisthokont A member of a clade of eukaryotes, including certain protists (choanoflagellates), fungi, and animals; flagellate cells in this group have a single, posterior flagellum.

opposable thumb The arrangement of the fingers so that they are positioned opposite the thumb, enabling the organism to grasp objects.

optimal foraging The process of obtaining food in a manner that maximizes benefits and/or minimizes costs.

orbital Region in which electrons occur in an atom or molecule

order A taxonomic category made up of related families.

organ A specialized structure, such as the heart or liver, made up of tissues and adapted to perform a specific function or group of functions.

organ of Corti The structure within the inner ear of vertebrates that contains receptor cells that sense sound vibrations.

organ system An organized group of tissues and organs that work together to perform a specialized set of functions, e.g., the digestive system or circulatory system.

organelle One of the specialized structures within the cell, such as the mitochondria, Golgi complex, ribosomes, or contractile vacuole; many organelles are membrane-enclosed.

organic compound A compound consisting of a backbone made up of carbon atoms. Compare with *inorganic compound*.

organism Any living system consisting of one or more cells.

organismic respiration See *respiration*.

organogenesis The process of organ formation.

orgasm (or′gazm) The climax of sexual excitement.

origin of replication A specific site on the DNA where replication begins.

osmoconformer An animal in which the salt concentration of body fluids varies along with changes in the surrounding sea water so that it stays in osmotic equilibrium with its surroundings. Compare with *osmoregulator*.

osmoregulation (oz″moh-reg-yoo-lay′shun) The active regulation of the osmotic pressure of body fluids so that they do not become excessively dilute or excessively concentrated.

osmoregulator An animal that maintains an optimal salt concentration in its body fluids despite changes in salinity of its surroundings. Compare with *osmoconformer*.

osmosis (oz-moh′sis) The net movement of water (the principal solvent in biological systems) by diffusion through a selectively permeable membrane from a region of higher concentration of water (a hypotonic solution) to a region of lower concentration of water (a hypertonic solution).

osmotic pressure The pressure that must be exerted on the hypertonic side of a selectively permeable membrane to prevent diffusion of water (by osmosis) from the side containing pure water.

osteichthyes (os″tee-ick′thees) Historically, the vertebrate class of bony fishes. Biologists now divide bony fishes into three classes: Actinopterygii, the ray-finned fishes; Actinistia, the lobe-finned fishes; and Dipnoi, the lungfishes.

osteoblast (os′tee-oh-blast) A type of bone cell that secretes the protein matrix of bone. Also see *osteocyte*.

osteoclast (os′tee-oh-clast) Large, multinucleate cell that helps sculpt and remodel bones by dissolving and removing part of the bony substance.

osteocyte (os′tee-oh-site) A mature bone cell; an osteoblast that has become embedded within the bone matrix and occupies a lacuna.

osteon (os′tee-on) The spindle-shaped unit of bone composed of concentric layers of osteocytes organized around a central Haversian canal containing blood vessels and nerves.

otoliths (oh′toe-liths) Small calcium carbonate crystals in the saccule and utricle of the inner ear; sense gravity and are important in static equilibrium.

outbreeding The mating of individuals of unrelated strains, also called *outcrossing*. Compare with *inbreeding*.

outcrossing See *outbreeding*.

outgroup In cladistics, a taxon that represents an approximation of the ancestral condition; the outgroup is related to the *ingroup* (the members of the group under study) but separated from the ingroup lineage before they diversified.

ovary (oh′var-ee) (1) In animals, one of the paired female gonads responsible for producing eggs and sex hormones. (2) In flowering plants, the base of the carpel that contains ovules; ovaries develop into fruits after fertilization.

oviduct (oh′vih-dukt) The tube that carries ova from the ovary to the uterus, cloaca, or body exterior. Also called *fallopian tube* or *uterine tube*.

oviparous (oh-vip′ur-us) Bearing young in the egg stage of development; egg laying. Compare with *viviparous* and *ovoviviparous*.

ovoviviparous (oh″voh-vih-vip′ur-us) A type of development in which the young hatch from eggs incubated inside the mother's body. Compare with *viviparous* and *oviparous*.

ovulation (ov-u-lay′shun) The release of an egg from the ovary.

ovule (ov′yool) The structure (i.e., megasporangium) in the plant ovary that develops into the seed following fertilization.

ovum (pl., *ova*) Female gamete of an animal.

oxaloacetate Four-carbon compound; important intermediate in the citric acid cycle and in the C_4 and CAM pathways of carbon fixation in photosynthesis.

oxidants Highly reactive molecules such as free radicals, peroxides, and superoxides that are produced during normal cell processes that require oxygen; can damage DNA and other molecules by snatching electrons. Compare with *antioxidants*.

oxidation The loss of one or more electrons (or hydrogen atoms) by an atom, ion, or molecule. Compare with *reduction*.

oxidative phosphorylation (fos″for-ih-lay′shun) The production of ATP using energy derived from the transfer of electrons in the electron transport system of mitochondria; occurs by chemiosmosis.

oxygen-carrying capacity The maximum amount of oxygen transported by hemoglobin.

oxygen debt The oxygen necessary to metabolize the lactic acid produced during strenuous exercise.

oxygen–hemoglobin dissociation curve A curve depicting the percentage saturation of hemoglobin with oxygen, as a function of certain variables such as oxygen concentration, carbon dioxide concentration, or pH.

oxyhemoglobin Hemoglobin that has combined with oxygen.

oxytocin (ok″see-tow′sin) Hormone secreted by the hypothalamus and released by the posterior lobe of the pituitary gland; stimulates contraction of the pregnant uterus and the ducts of mammary glands.

ozone A blue gas, O_3, with a distinctive odor that is a human-made pollutant near Earth's surface (in the troposphere) but a natural and essential component of the stratosphere.

P generation (parental generation) Members of two different true-breeding lines that are crossed to produce the F_1 generation.

P680 Chlorophyll *a* molecules that serve as the reaction center of photosystem II, transferring photoexcited electrons to a primary acceptor; named by their absorption peak at 680 nm.

P700 Chlorophyll *a* molecules that serve as the reaction center of photosystem I, transferring photoexcited electrons to a primary acceptor; named by their absorption peak at 700 nm.

pacemaker (of the heart) See *sinoatrial (SA) node*.

Pacinian corpuscle (pah-sin′ee-an kor′pus-el) A receptor located in the dermis of the skin that responds to pressure.

paedomorphosis Retention of juvenile or larval features in a sexually mature animal.

pair bond A stable relationship between animals of opposite sex that ensures cooperative behavior in mating and rearing the young.

paleoanthropology (pay″lee-o-an-thro-pol′uh-gee) The study of human evolution.

Paleozoic era That part of geologic time extending from roughly 542 million to 251 million years ago.

palindromic Reading the same forward and backward; DNA sequences are palindromic when the base sequence of one strand reads the same as its complement when both are read in the 5′ to 3′ direction.

palisade mesophyll (mez′oh-fil) The vertically stacked, columnar mesophyll cells near the upper epidermis in certain leaves. Compare with *spongy mesophyll*.

pancreas (pan′kree-us) Large gland located in the vertebrate abdominal cavity. The pancreas produces pancreatic juice containing digestive enzymes; also serves as an endocrine gland, secreting the hormones insulin and glucagon.

panspermia The idea that life did not originate on Earth but began elsewhere in the galaxy and drifted through space to Earth.

parabronchi (sing., *parabronchus*) Thin-walled ducts in the lungs of birds; gases are exchanged across their walls.

paracrine regulation A type of regulation in which a signal molecule (e.g., certain hormones) diffuses through interstitial fluid and acts on nearby target cells. Compare with *autocrine regulation*.

paraphyletic group A group of organisms made up of a common ancestor and some, but not all, of its descendants. Compare with *monophyletic group* and *polyphyletic group*

parapodia (par″uh-poh′dee-ah) (sing., *parapodium*) Paired, thickly bristled paddlelike appendages extending laterally from each segment of polychaete worms.

parasite A heterotrophic organism that obtains nourishment from the living tissue of another organism (the host).

parasitism (par′uh-si-tiz″m) A symbiotic relationship in which one member (the parasite) benefits and the other (the host) is adversely affected. Compare with *commensalism* and *mutualism*.

parasympathetic nervous system A division of the autonomic nervous system concerned with the control of the internal organs; functions to conserve or restore energy. Compare with *sympathetic nervous system*.

parathyroid glands Small, pea-sized glands closely adjacent to the thyroid gland; they secrete parathyroid hormone, which regulates calcium and phosphate metabolism.

parathyroid hormone (PTH) A hormone secreted by the parathyroid glands; regulates calcium and phosphate metabolism.

parenchyma (par-en′kih-mah) Highly variable living plant cells that have thin primary walls; function in photosynthesis, the storage of nutrients, and/or secretion.

parsimony The principle based on the experience that the simplest explanation is most probably the correct one.

parthenogenesis (par″theh-noh-jen′eh-sis) The development of an unfertilized egg into an adult organism; common among honeybees, wasps, and certain other arthropods.

partial pressure (of a gas) The pressure exerted by a gas in a mixture, which is the same pressure it would exert if alone. For example, the partial pressure of atmospheric oxygen (P_{O_2}) is 160 mm Hg at sea level.

parturition (par″to-rish′un) The birth process.

passive immunity Temporary immunity that depends on the presence of immunoglobulins produced by another organism. Compare with *active immunity*.

passive ion channel A channel in the plasma membrane that permits the passage of specific ions such as Na^+, K^+, or Cl^-.

patch–clamp technique A method that allows researchers to study the ion channels of a tiny patch of membrane by tightly sealing a micropipette to the patch and measuring the flow of ions through the channels.

patchiness See *clumped dispersion*.

pathogen (path′oh-gen) An organism, usually a microorganism, capable of producing disease.

pathogen-associated molecular patterns (PAMPs) Molecules on bacteria and other pathogens that combine with Toll-like receptors on macrophages, stimulating them to produce cytokines.

pattern formation See *morphogenesis.*

pedigree A chart constructed to show an inheritance pattern within a family through multiple generations.

peduncle The stalk of a flower or inflorescence.

pellicle A flexible outer covering consisting of protein; characteristic of certain protists, e.g., ciliates and euglenoids.

penis The male sexual organ of copulation in reptiles, mammals, and a few birds.

pentose A sugar molecule containing five carbons.

people overpopulation A situation in which there are too many people in a given geographic area; results in pollution, environmental degradation, and resource depletion. Compare with *consumption overpopulation.*

pepsin (pep′sin) An enzyme produced in the stomach that initiates digestion of protein.

pepsinogen The precursor of pepsin; secreted by chief cells in the gastric glands of the stomach.

peptide (pep′tide) A compound consisting of a chain of amino acid groups linked by peptide bonds. A dipeptide consists of two amino acids, a polypeptide of many.

peptide bond A distinctive covalent carbon-to-nitrogen bond that links amino acids in peptides and proteins.

peptidoglycan (pep″tid-oh-gly′kan) A modified protein or peptide having an attached carbohydrate; component of the bacterial cell wall.

peptidyl transferase The ribosomal enzyme that catalyzes the formation of a peptide bond.

perennial plant (purr-en′ee-ul) A woody or herbaceous plant that grows year after year, i.e., lives more than 2 years. Compare with *annual* and *biennial.*

perfect flower A flower that has both stamens and carpels. Compare with *imperfect flower.*

pericentriolar material Fibrils surrounding the centrioles in the microtubule-organizing centers in cells of animals and other organisms having centrioles.

pericycle (pehr′eh-sy″kl) A layer of meristematic cells typically found between the endodermis and phloem in roots.

periderm (pehr′ih-durm) The outer bark of woody stems and roots; composed of cork cells, cork cambium, and cork parenchyma, along with traces of primary tissues.

period An interval of geologic time that is a subdivision of an era; each period is divided into epochs.

peripheral membrane protein A protein associated with one of the surfaces of a biological membrane. Compare with *integral membrane protein.*

peripheral nervous system (PNS) In vertebrates, the nerves and receptors that lie outside the central nervous system. Compare with *central nervous system (CNS).*

peristalsis (pehr″ih-stal′sis) Rhythmic waves of muscular contraction and relaxation in the walls of hollow tubular organs, such as the ureter or parts of the digestive tract, that serve to move the contents through the tube.

permafrost Permanently frozen subsoil characteristic of frigid areas such as the tundra.

peroxisomes (pehr-ox′ih-sohmz) In eukaryotic cells, membrane-enclosed organelles containing enzymes that produce or degrade hydrogen peroxide.

persistence A characteristic of certain chemicals that are extremely stable and may take many years to be broken down into simpler forms by natural processes.

petal One of the parts of the flower attached inside the whorl of sepals; petals are usually colored.

petiole (pet′ee-ohl) The part of a leaf that attaches to a stem.

pH The negative logarithm of the hydrogen ion concentration of a solution (expressed as moles per liter). Neutral pH is 7, values less than 7 are acidic, and those greater than 7 are basic.

phage See *bacteriophage.*

phagocytosis (fag″oh-sy-toh′sis) Literally, "cell eating"; a type of endocytosis by which certain cells engulf food particles, microorganisms, foreign matter, or other cells.

pharmacogenetics A new field of gene-based medicine in which drugs are personalized to match a patient's genetic makeup.

pharyngeal slits (fair-in′jel) Openings that lead from the pharyngeal cavity to the outside; evolved as part of a filter-feeding system in chordates and later became modified for other functions, including gill slits in many aquatic vertebrates.

pharynx (fair′inks) Part of the digestive tract. In complex vertebrates, it is bounded anteriorly by the mouth and nasal cavities and posteriorly by the esophagus and larynx; the throat region in humans.

phenetics (feh-neh′tiks) An approach to classification based on measurable similarities in phenotypic characters without consideration of homology or other evolutionary relationships. Compare with *cladistics* and *evolutionary systematics.*

phenotype (fee′noh-type) The physical or chemical expression of an organism's genes. Compare with *genotype.*

phenotype frequency The proportion of a particular phenotype in the population.

phenylketonuria (PKU) (fee″nl-kee″toh-noor′ee-ah) An inherited disease in which there is a deficiency of the enzyme that normally converts phenylalanine to tyrosine; results in mental retardation if untreated.

pheromone (fer′oh-mone) A substance secreted by an organism to the external environment that influences the development or behavior of other members of the same species.

phloem (flo′em) The vascular tissue that conducts dissolved sugar and other organic compounds in plants.

phosphate group A weakly acidic functional group that can release one or two hydrogen ions.

phosphodiester linkage Covalent linkage between two nucleotides in a strand of DNA or RNA; includes a phosphate group bonded to the sugars of two adjacent nucleotides.

phosphoenolpyruvate (PEP) Three-carbon phosphorylated compound that is an important intermediate in glycolysis and is a reactant in the initial carbon fixation step in C_4 and CAM photosynthesis.

phosphoglycerate (PGA) Phosphorylated three-carbon compound that is an important metabolic intermediate.

phospholipids (fos″foh-lip′idz) Lipids in which two fatty acids and a phosphorus-containing group are attached to glycerol; major components of cell membranes.

phosphorus cycle The worldwide circulation of phosphorus from the abiotic environment into living things and back into the abiotic environment.

phosphorylation (fos″for-ih-lay′shun) The introduction of a phosphate group into an organic molecule. See *kinases.*

photoautotroph An organism that obtains energy from light and synthesizes organic compounds from inorganic raw materials; includes plants, algae, and some bacteria. Compare with *photoheterotroph, chemoautotroph,* and *chemoheterotroph.*

photoheterotroph An organism that can carry out photosynthesis to obtain energy but cannot fix carbon dioxide and therefore requires organic compounds as a carbon source; includes some bacteria. Compare with *photoautotroph, chemoautotroph,* and *chemoheterotroph.*

photolysis (foh-tol'uh-sis) The photochemical splitting of water in the light-dependent reactions of photosynthesis; a specific enzyme is needed to catalyze this reaction.

photon (foh'ton) A particle of electromagnetic radiation; one quantum of radiant energy.

photoperiodism (foh"teh-peer'ee-o-dizm) The physiological response (such as flowering) of plants to variations in the length of daylight and darkness.

photophosphorylation (foh"toh-fos-for-ih-lay'shun) The production of ATP in photosynthesis.

photoreceptor (foh"toh-ree-sep'tor) (1) A sense organ specialized to detect light. (2) A pigment that absorbs light before triggering a physiological response.

photorespiration (foh"toh-res-pur-ay'shun) The process that reduces the efficiency of photosynthesis in C_3 plants during hot spells in summer; consumes oxygen and produces carbon dioxide through the degradation of Calvin cycle intermediates.

photosynthesis The biological process that captures light energy and transforms it into the chemical energy of organic molecules (e.g., carbohydrates), which are manufactured from carbon dioxide and water.

photosystem One of two photosynthetic units responsible for capturing light energy and transferring excited electrons; photosystem I strongly absorbs light of about 700 nm, whereas photosystem II strongly absorbs light of about 680 nm. See *antenna complex* and *reaction center*.

phototroph (foh'toh-trof) Organism that uses light as a source of energy. Compare with *chemotroph*. See *photoautotroph* and *photoheterotroph*.

phototropism (foh"toh-troh'pizm) The growth of a plant in response to the direction of light.

phycocyanin (fy"koh-sy-ah'nin) A blue pigment found in cyanobacteria and red algae.

phycoerythrin (fy"koh-ee-rih'thrin) A red pigment found in cyanobacteria and red algae.

phylogenetic systematics See *cladistics*.

phylogenetic tree A branching diagram that shows lines of descent among a group of related species.

phylogeny (fy-loj'en-ee) The complete evolutionary history of a group of organisms.

phylum (fy'lum) A taxonomic grouping of related, similar classes; a category beneath the kingdom and above the class.

phytoalexins Antimicrobial compounds produced by plants that limit the spread of pathogens such as fungi.

phytochemicals Compounds found in plants that play important roles in preventing certain diseases; some function as antioxidants.

phytochrome (fy'toh-krome) A blue-green, proteinaceous pigment involved in a wide variety of physiological responses to light; occurs in two interchangeable forms depending on the ratio of red to far-red light.

phytoplankton (fy"toh-plank'tun) Microscopic floating algae and cyanobacteria that are the base of most aquatic food webs. Compare with *zooplankton*. See *plankton* and *nanoplankton*.

pia mater (pee'a may'ter) The inner membrane covering the brain and spinal cord; the innermost of the meninges; also see *dura mater* and *arachnoid*.

pigment A substance that selectively absorbs light of specific wavelengths.

pili (pie'lie) (sing., *pilus*) Hairlike structures on the surface of many bacteria; function in conjugation or attachment.

pineal gland (pie-nee'al) Endocrine gland located in the brain.

pinocytosis (pin"oh-sy-toh'sis) Cell drinking; a type of endocytosis by which cells engulf and absorb droplets of liquids.

pioneer The first organism to colonize an area and begin the first stage of succession.

pistil The female reproductive organ of a flower; consists of either a single carpel or two or more fused carpels. See *carpel*.

pith The innermost tissue in the stems and roots of many herbaceous plants; primarily a storage tissue.

pituitary gland (pi-too'ih-tehr"ee) An endocrine gland located below the hypothalamus; secretes several hormones that influence a wide range of physiological processes.

placenta (plah-sen'tah) The partly fetal and partly maternal organ whereby materials are exchanged between fetus and mother in the uterus of placental mammals.

placoderms (plak'oh-durmz) A group of extinct jawed fishes.

plankton Free-floating, mainly microscopic aquatic organisms found in the upper layers of the water; consisting of phytoplankton and zooplankton. Compare with *nekton*.

planula larva (plan'yoo-lah) A ciliated larval form found in cnidarians.

plasma The fluid portion of blood in which red blood cells, white blood cells, and platelets are suspended.

plasma cell Cell that secretes antibodies; a differentiated B lymphocyte (B cell).

plasma membrane The selectively permeable surface membrane that encloses the cell contents and through which all materials entering or leaving the cell must pass.

plasma proteins Proteins such as albumins, globulins, and fibrinogen that circulate in the blood plasma.

plasmid (plaz'mid) Small, circular, double-stranded DNA molecule that carries genes separate from those in the main DNA of a cell.

plasmodesmata (sing., *plasmodesma*) Cytoplasmic channels connecting adjacent plant cells and allowing for the movement of molecules and ions between cells.

plasmodial slime mold (plaz-moh'dee-uhl) A funguslike protist whose feeding stage consists of a plasmodium.

plasmodium (plaz-moh'dee-um) A multinucleate mass of living matter that moves and feeds in an amoeboid fashion.

plasmogamy (1) Fusion of the cytoplasm of two cells without fusion of nuclei. (2) A stage in the asexual reproduction of some fungi; hyphae of two compatible mating types come together, and their cytoplasm fuses.

plasmolysis (plaz-mol'ih-sis) The shrinkage of cytoplasm and the pulling away of the plasma membrane from the cell wall when a plant cell (or other walled cell) loses water, usually in a hypertonic environment.

plastids (plas'tidz) A family of membrane-enclosed organelles occurring in photosynthetic eukaryotic cells; include chloroplasts, chromoplasts, and amyloplasts and other leukoplasts.

platelets (playt'lets) Cell fragments in vertebrate blood that function in clotting; also called *thrombocytes*.

platyhelminths The phylum of acoelomate animals commonly known as *flatworms*.

pleiotropy The ability of a single gene to have multiple effects.

plesiomorphic characters See *shared ancestral characters*.

pleural membrane (ploor'ul) The membrane that lines the thoracic cavity and envelops each lung.

ploidy The number of chromosome sets in a nucleus or cell. See *haploid, diploid,* and *polyploid*.

plumule (ploom'yool) The embryonic shoot apex, or terminal bud, located above the point of attachment of the cotyledon(s).

pluripotent (ploor-i-poh′tent) A term describing a stem cell that can divide to give rise to many, but not all, types of cells in an organism. Compare with *totipotent*.

pneumatophore (noo-mat′uh-for″) Roots that extend up out of the water in swampy areas and are thought to provide aeration between the atmosphere and submerged roots.

polar body A small *n* cell produced during oogenesis in female animals that does not develop into a functional ovum.

polar covalent bond Chemical bond formed by the sharing of electrons between atoms that differ in electronegativity; the end of the bond near the more electronegative atom has a partial negative charge, and the other end has a partial positive charge. Compare with *nonpolar covalent bond*.

polar molecule Molecule that has one end with a partial positive charge and the other with a partial negative charge; polar molecules are generally soluble in water. Compare with *nonpolar molecule*.

polar nucleus In flowering plants, one of two *n* cells in the embryo sac that fuse with a sperm during double fertilization to form the 3*n* endosperm.

pollen grain The immature male gametophyte of seed plants (gymnosperms and angiosperms) that produces sperm capable of fertilization.

pollen tube In gymnosperms and flowering plants, a tube or extension that forms after germination of the pollen grain and through which male gametes (sperm cells) pass into the ovule.

pollination (pol″uh-nay′shen) In seed plants, the transfer of pollen from the male to the female part of the plant.

poly-A tail See *polyadenylation*.

polyadenylation (pol″ee-a-den-uh-lay′shun) That part of eukaryotic mRNA processing in which multiple adenine-containing nucleotides (a poly-A tail) are added to the 3′ end of the molecule.

polyandry A mating system in which a female mates with several males during a breeding season. Compare with *polygyny*.

polygenic inheritance (pol″ee-jen′ik) Inheritance in which several independently assorting or loosely linked nonallelic genes modify the intensity of a trait or contribute to the phenotype in additive fashion.

polygyny A mating system in which a male animal mates with many females during a breeding season. Compare with *polyandry*.

polymer (pol′ih-mer) A molecule built up from repeating subunits of the same general type (monomers); examples include proteins, nucleic acids, or polysaccharides.

polymerase chain reaction (PCR) A method by which a targeted DNA fragment is amplified in vitro to produce millions of copies.

polymorphism (pol″ee-mor′fizm) (1) The existence of two or more phenotypically different individuals within a population. (2) The presence of detectable variation in the genomes of different individuals in a population.

polyp (pol′ip) A hydralike animal; the sessile stage of the life cycle of certain cnidarians. Compare with *medusa*.

polypeptide See *peptide*.

polyphyletic group (pol″ee-fye-let′ik) A group made up of organisms that evolved from two or more different ancestors. Compare with *monophyletic group* and *paraphyletic group*.

polyploid (pol′ee-ployd) The condition of having more than two sets of chromosomes per nucleus. Compare with *diploid* and *haploid*.

polyribosome A complex consisting of a number of ribosomes attached to an mRNA during translation; also known as a *polysome*.

polysaccharide (pol-ee-sak′ah-ride) A carbohydrate consisting of many monosaccharide subunits (e.g., starch, glycogen, and cellulose).

polysome See *polyribosome*.

polyspermy The fertilization of an egg by more than one sperm.

polytene A term describing a giant chromosome consisting of many (usually >1000) parallel DNA double helices. Polytene chromosomes are typically found in cells of the salivary glands and some other tissues of certain insects, such as the fruit fly, *Drosophila*.

polyunsaturated fatty acid See *fatty acid*.

pons (ponz) The white bulge that is the part of the brain stem between the medulla and the midbrain; connects various parts of the brain.

population A group of organisms of the same species that live in a defined geographic area at the same time.

population bottleneck See *bottleneck*.

population crash An abrupt decline in the size of a population.

population density The number of individuals of a species per unit of area or volume at a given time.

population dynamics The study of changes in populations, such as how and why population numbers change over time.

population ecology That branch of biology that deals with the numbers of a particular species that are found in an area and how and why those numbers change (or remain fixed) over time.

population genetics The study of genetic variability in a population and of the forces that act on it.

population growth momentum The continued growth of a population after fertility rates have declined, as a result of a population's young age structure.

poriferans Sponges; members of phylum Porifera.

positive feedback mechanism A homeostatic mechanism in which a change in some condition triggers a response that intensifies the changing condition. Compare with *negative feedback mechanism*.

posterior Toward the tail end of a bilaterally symmetrical animal. Compare with *anterior*.

postsynaptic neuron A neuron that transmits an impulse away from a synapse. Compare with *presynaptic neuron*.

postzygotic barrier One of several reproductive isolating mechanisms that prevent gene flow between species after fertilization has taken place, e.g., hybrid inviability, hybrid sterility, and hybrid breakdown. Compare with *prezygotic barrier*.

potential energy Stored energy; energy that can do work as a consequence of its position or state. Compare with *kinetic energy*.

potentiation A form of synaptic enhancement (increase in neurotransmitter release) that can last for several minutes; occurs when a presynaptic neuron continues to transmit action potentials at a high rate for a minute or longer.

preadaptation A novel evolutionary change in a pre-existing biological structure that enables it to have a different function; feathers, which evolved from reptilian scales, represent a preadaptation for flight.

prebiotic soup hypothesis The hypothesis that simple organic molecules that are the precursors of life originated and accumulated at Earth's surface, in shallow seas or on rock or clay surfaces. Compare with *iron–sulfur world hypothesis*.

predation Relationship in which one organism (the predator) kills and devours another organism (the prey).

pre-mRNA RNA precursor to mRNA in eukaryotes; contains both introns and exons.

prenatal Pertaining to the time before birth.

pressure-flow hypothesis The mechanism by which dissolved sugar is thought to be transported in phloem; caused by a pressure gradient between the source (where sugar is loaded into the phloem) and the sink (where sugar is removed from phloem).

presynaptic neuron A neuron that transmits an impulse to a synapse. Compare with *postsynaptic neuron*.

prezygotic barrier One of several reproductive isolating mechanisms that interfere with fertilization between male and female gametes of different species, e.g., temporal isolation, habitat isolation, behavioral isolation, mechanical isolation, and gametic isolation. Compare with *postzygotic barrier*.

primary consumer See *herbivore*.

primary growth An increase in the length of a plant that occurs at the tips of the shoots and roots due to the activity of apical meristems. Compare with *secondary growth*.

primary immune response The response of the immune system to first exposure to an antigen. Compare with *secondary immune response*.

primary mycelium A mycelium in which the cells are monokaryotic and haploid; a mycelium that grows from either an ascospore or a basidiospore. Compare with *secondary mycelium*.

primary producer See *autotroph*.

primary productivity The amount of light energy converted to organic compounds by autotrophs in an ecosystem over a given period. Compare with *secondary productivity*. See *gross primary productivity* and *net primary productivity*.

primary structure (of a protein) The complete sequence of amino acids in a polypeptide chain, beginning at the amino end and ending at the carboxyl end. Compare with *secondary, tertiary,* and *quaternary protein structure*.

primary succession An ecological succession that occurs on land that has not previously been inhabited by plants; no soil is present initially. See *succession*. Compare with *secondary succession*.

primates Mammals that share such traits as flexible hands and feet with five digits; a strong social organization; and front-facing eyes; includes lemurs, tarsiers, monkeys, apes, and humans.

primer See *RNA primer*.

primitive streak Dynamic, constantly changing structure that forms at the midline of the blastodisc in birds, mammals, and some other vertebrates and is active in gastrulation. The anterior end of the primitive streak is Hensen's node.

primosome A complex of proteins responsible for synthesizing the RNA primers required in DNA synthesis.

principle In science, a statement of a rule that explains how something works. A scientific principle has withstood repeated testing and has the highest level of scientific confidence.

prion (pri'on) An infectious agent that consists only of protein.

producer See *autotroph*.

product Substance formed by a chemical reaction. Compare with *reactant*.

product rule The rule for combining the probabilities of independent events by multiplying their individual probabilities. Compare with *sum rule*.

profundal zone (pro-fun'dl) The deepest zone of a large lake, located below the level of penetration by sunlight. Compare with *littoral zone* and *limnetic zone*.

progesterone (pro-jes'ter-own) A steroid hormone secreted by the ovary (mainly by the corpus luteum) and placenta; stimulates the uterus (to prepare the endometrium for implantation) and breasts (for milk secretion).

progymnosperm (pro-jim'noh-sperm) An extinct group of plants that may have been the ancestors of gymnosperms.

prokaryote (pro-kar'ee-ote) A cell that lacks a nucleus and other membrane-enclosed organelles; includes the bacteria and archaea (kingdoms Eubacteria and Archaea). Compare with *eukaryote*.

prometaphase Stage of mitosis during which spindle microtubules attach to kinetochores of chromosomes, which begin to move toward the cell's midplane; occurs after prophase and before metaphase.

promoter The nucleotide sequence in DNA to which RNA polymerase attaches to begin transcription.

prop root An adventitious root that arises from the stem and provides additional support for a plant such as corn.

prophage (pro'faj) Bacteriophage nucleic acid that is inserted into the bacterial DNA.

prophase The first stage of mitosis. During prophase the chromosomes become visible as distinct structures, the nuclear envelope breaks down, and a spindle forms.

proplastids Organelles that are plastid precursors; may mature into various specialized plastids, including chloroplasts, chromoplasts, or leukoplasts.

proprioceptors (pro″pree-oh-sep'torz) Receptors in muscles, tendons, and joints that respond to changes in movement, tension, and position; enable an animal to perceive the position of its body.

prostaglandins (pros″tah-glan'dinz) A group of local regulators derived from fatty acids; synthesized by most cells of the body and produce a wide variety of effects; sometimes called local *hormones*.

prostate gland A gland in male animals that produces an alkaline secretion that is part of the semen.

proteasome A large multiprotein structure that recognizes and degrades protein molecules tagged with ubiquitin into short, nonfunctional peptide fragments.

protein A large, complex organic compound composed of covalently linked amino acid subunits; contains carbon, hydrogen, oxygen, nitrogen, and sulfur.

protein kinase One of a group of enzymes that activate or inactivate other proteins by phosphorylating (adding phosphate groups to) them.

proteomics The study of all the proteins encoded by the human genome and produced in a person's cells and tissues.

Proterozoic eon The period of Earth's history that began approximately 2.5 billion years ago and ended 542 million years ago; marked by the accumulation of oxygen and the appearance of the first multicellular eukaryotic life-forms.

prothallus (pro-thal'us) (pl., *prothalli*) The free-living, *n* gametophyte in ferns and other seedless vascular plants.

protist (pro'tist) One of a vast kingdom of eukaryotic organisms, primarily unicellular or simple multicellular; mostly aquatic.

protobionts (pro″toh-by'ontz) Assemblages of organic polymers that spontaneously form under certain conditions. Protobionts may have been involved in chemical evolution.

proton A particle present in the nuclei of all atoms that has one unit of positive charge and a mass of 1 atomic mass unit (amu). Compare with *electron* and *neutron*.

protonema (pro″toh-nee'mah) (pl., *protonemata*) In mosses, a filament of *n* cells that grows from a spore and develops into leafy moss gametophytes.

protonephridia (pro″toh-nef-rid'ee-ah) (sing., *protonephridium*) The flame-cell excretory organs of flatworms and some other simple invertebrates.

proto-oncogene A gene that normally promotes cell division in response to the presence of certain growth factors; when mutated, it may become an oncogene, possibly leading to the formation of a cancer cell. Compare with *oncogene*.

protostome (pro'toh-stome) A major division of the animal kingdom in which the blastopore develops into the mouth, and the anus forms secondarily; includes the annelids, arthropods, and mollusks. Compare with *deuterostome*.

protozoa (proh″toh-zoh′a) (sing., *protozoon*) An informal group of unicellular, animal-like protists, including amoebas, foraminiferans, actinopods, ciliates, flagellates, and apicomplexans. (The adjectival form is *protozoan*.)

provirus (pro-vy′rus) A part of a virus, consisting of nucleic acid only, that was inserted into a host genome. See *DNA provirus*.

proximal Closer to the point of reference. Compare with *distal*.

proximal convoluted tubule The part of the renal tubule that extends from Bowman's capsule to the loop of Henle. Compare with *distal convoluted tubule*.

proximate causes (of behavior) The immediate causes of behavior, such as genetic, developmental, and physiological processes that permit the animal to carry out a specific behavior. Compare with *ultimate causes of behavior*.

pseudocoelom (sue″doh-see′lom) A body cavity between the mesoderm and endoderm; derived from the blastocoel. Compare with *coelom*.

pseudocoelomate (sue″doh-seel′oh-mate) An animal having a pseudocoelom. Compare with *coelomate* and *acoelomate*.

pseudoplasmodium (sue″doe-plaz-moh′dee-um) In cellular slime molds, an aggregation of amoeboid cells that forms a spore-producing fruiting body during reproduction.

pseudopodium (sue″doe-poe′dee-um) (pl., *pseudopodia*) A temporary extension of an amoeboid cell that is used for feeding and locomotion.

puff In a polytene chromosome, a decondensed region that is a site of intense RNA synthesis.

pulmonary circulation The part of the circulatory system that delivers blood to and from the lungs for oxygenation. Compare with *systemic circulation*.

pulse, arterial The alternate expansion and recoil of an artery.

punctuated equilibrium The idea that evolution proceeds with periods of little or no genetic change, followed by very active phases, so that major adaptations or clusters of adaptations appear suddenly in the fossil record. Compare with *gradualism*.

Punnett square The grid structure, first developed by Reginald Punnett, that allows direct calculation of the probabilities of occurrence of all possible offspring of a genetic cross.

pupa (pew′pah) (pl., *pupae*) A stage in the development of an insect, between the larva and the imago (adult); a form that neither moves nor feeds and may be in a cocoon.

purines (pure′eenz) Nitrogenous bases with carbon and nitrogen atoms in two attached rings, e.g., adenine and guanine; components of nucleic acids, ATP, GTP, NAD$^+$, and certain other biologically active substances. Compare with *pyrimidines*.

pyramid of biomass An ecological pyramid that illustrates the total biomass, as, for example, the total dry weight, of all organisms at each trophic level in an ecosystem.

pyramid of energy An ecological pyramid that shows the energy flow through each trophic level of an ecosystem.

pyrimidines (pyr-im′ih-deenz) Nitrogenous bases, each composed of a single ring of carbon and nitrogen atoms, e.g., thymine, cytosine, and uracil; components of nucleic acids. Compare with *purines*.

pyruvate (pyruvic acid) A three-carbon compound; the end product of glycolysis.

quadrupedal (kwad′roo-ped″ul) Walking on all fours.

quantitative trait A trait that shows continuous variation in a population (e.g., human height) and typically has a polygenic inheritance pattern.

quaternary structure (of a protein) The overall conformation of a protein produced by the interaction of two or more polypeptide chains. Compare with *primary, secondary*, and *tertiary protein structure*.

r selection A reproductive strategy recognized by some ecologists, in which a species typically has a small body size, rapid development, and short life span and devotes a large proportion of its metabolic energy to the production of offspring. Compare with *K selection*.

radial cleavage The pattern of blastomere production in which the cells are located directly above or below one another; characteristic of early deuterostome embryos. Compare with *spiral cleavage*.

radial symmetry A body plan in which any section through the mouth and down the length of the body divides the body into similar halves. Jellyfish and other cnidarians have radial symmetry. Compare with *bilateral symmetry*.

radicle (rad′ih-kl) The embryonic root of a seed plant.

radioactive decay The process in which a radioactive element emits radiation, and as a result, its nucleus changes into the nucleus of a different element.

radioisotopes Unstable isotopes that spontaneously emit radiation; also called *radioactive isotopes*.

radiolarians Those actinopods that secrete elaborate shells of silica (glass).

radula (rad′yoo-lah) A rasplike structure in the digestive tract of chitons, snails, squids, and certain other mollusks.

rain shadow An area that has very little precipitation, found on the downwind side of a mountain range. Deserts often occur in rain shadows.

random dispersion The spatial distribution pattern of a population in which the presence of one individual has no effect on the distribution of other individuals. Compare with *clumped dispersion* and *uniform dispersion*.

range The area where a particular species occurs. Compare with *home range*.

ray A chain of parenchyma cells (one to many cells thick) that functions for lateral transport in stems and roots of woody plants.

ray-finned fishes A class (Actinopterygii) of modern bony fishes; contains about 95% of living fish species.

reabsorption The selective removal of certain substances from the glomerular filtrate by the renal tubules and collecting ducts of the kidney, and their return into the blood.

reactant Substance that participates in a chemical reaction. Compare with *product*.

reaction center The portion of a photosystem that includes chlorophyll *a* molecules capable of transferring electrons to a primary electron acceptor, which is the first of several electron acceptors in a series. See *antenna complex* and *photosystem*.

realized niche The lifestyle that an organism actually pursues, including the resources that it actually uses. An organism's realized niche is narrower than its fundamental niche because of interspecific competition. Compare with *fundamental niche*.

receptacle The end of a flower stalk where the flower parts (sepals, petals, stamens, and carpels) are attached.

reception Process of detecting a stimulus.

receptor (1) In cell biology, a molecule on the surface of a cell, or inside a cell, that serves as a recognition or binding site for signaling molecules such as hormones, antibodies, or neurotransmitters. (2) A sensory receptor. See *sensory receptor*.

receptor down-regulation The process by which some hormone receptors decrease in number, thereby suppressing the sensitivity of target cells to the hormone. Compare with *receptor up-regulation*.

receptor up-regulation The process by which some hormone receptors increase in number, thereby increasing the sensitivity of the target cells to the hormone. Compare with *receptor down-regulation*.

receptor-mediated endocytosis A type of endocytosis in which extracellular molecules become bound to specific receptors on the cell surface and then enter the cytoplasm enclosed in vesicles.

recessive allele (al-leel′) An allele that is not expressed in the heterozygous state. Compare with *dominant allele.*

recombinant DNA Any DNA molecule made by combining genes from different organisms.

recombination, genetic The appearance of new gene combinations. Recombination in eukaryotes generally results from meiotic events, either crossing-over or shuffling of chromosomes.

red alga A member of a diverse phylum of algae that contain the pigments chlorophyll *a,* carotenoids, phycocyanin, and phycoerythrin.

red blood cell (RBC) See *erythrocyte.*

red tide A red or brown coloration of ocean water caused by a population explosion, or bloom, of dinoflagellates.

redox reaction (ree′dox) The chemical reaction in which one or more electrons are transferred from one substance (the substance that becomes oxidized) to another (the substance that becomes reduced). See *oxidation* and *reduction.*

reduction The gain of one or more electrons (or hydrogen atoms) by an atom, ion, or molecule. Compare with *oxidation.*

reflex action An automatic, involuntary response to a given stimulus that generally functions to restore homeostasis.

refractory period The brief period that elapses after the response of a neuron or muscle fiber, during which it cannot respond to another stimulus.

regulative development The very plastic developmental pattern in which each individual cell of an early embryo retains totipotency. Compare with *mosaic development.*

regulatory gene Gene that turns the transcription of other genes on or off.

releasing hormone A hormone secreted by the hypothalamus that stimulates secretion of a specific hormone by the anterior lobe of the pituitary gland.

renal (ree′nl) Pertaining to the kidney.

renal pelvis The funnel-shaped chamber of the kidney that receives urine from the collecting ducts; urine then moves into the ureters.

renin (reh′nin) An enzyme released by the kidney in response to a decrease in blood pressure; activates a pathway leading to production of angiotensin II, a hormone that increases aldosterone release; aldosterone increases blood pressure.

replacement-level fertility The number of children a couple must produce to "replace" themselves. The average number is greater than two, because some children die before reaching reproductive age.

replication See *DNA replication.*

replication fork Y-shaped structure produced during the semiconservative replication of DNA.

repolarization The process of returning membrane potential to its resting level.

repressible operon An operon that is normally active, but can be controlled by a repressor protein, which becomes active when it binds to a corepressor; the active repressor binds to the operator, making the operon transcriptionally inactive. Compare with *inducible operon.*

repressor protein A negative regulatory protein that inhibits transcription when bound to DNA; some repressors require a corepressor to be active; some other repressors become inactive when bound to an inducer molecule. Compare with *activator protein.*

reproduction The process by which new individuals are produced. See *asexual reproduction* and *sexual reproduction.*

reproductive isolating mechanisms The reproductive barriers that prevent a species from interbreeding with another species; as a result, each species' gene pool is isolated from those of other species. See *prezygotic barrier* and *postzygotic barrier.*

reptiles A class of vertebrates characterized by dry skin with horny scales and adaptations for terrestrial reproduction; include turtles, snakes, and alligators; reptiles are a paraphyletic group.

residual capacity The volume of air that remains in the lungs at the end of a normal exhalation.

resin A viscous organic material that certain plants produce and secrete into specialized ducts; may play a role in deterring disease organisms or plant-eating insects.

resolution See *resolving power.*

resolving power The ability of a microscope to show fine detail, defined as the minimum distance between two points at which they are seen as separate images; also called *resolution.*

resource partitioning The reduction of competition for environmental resources such as food that occurs among coexisting species as a result of each species' niche differing from the others in one or more ways.

respiration (1) Cellular respiration is the process by which cells generate ATP through a series of redox reactions. In aerobic respiration the terminal electron acceptor is molecular oxygen; in anaerobic respiration the terminal acceptor is an inorganic molecule other than oxygen. (2) Organismic respiration is the process of gas exchange between a complex animal and its environment, generally through a specialized respiratory surface, such as a lung or gill.

respiratory centers Centers in the medulla and pons that regulate breathing.

resting potential The membrane potential (difference in electric charge between the two sides of the plasma membrane) of a neuron in which no action potential is occurring. The typical resting potential is about −70 millivolts. Compare with *action potential.*

restoration ecology The scientific field that uses the principles of ecology to help return a degraded environment as closely as possible to its former undisturbed state.

restriction enzyme One of a class of enzymes that cleave DNA at specific base sequences; produced by bacteria to degrade foreign DNA; used in recombinant DNA technology.

restriction map A physical map of DNA in which sites cut by specific restriction enzymes serve as landmarks.

reticular activating system (RAS) (reh-tik′yoo-lur) A diffuse network of neurons in the brain stem; responsible for maintaining consciousness.

retina (ret′ih-nah) The innermost of the three layers (retina, choroid layer, and sclera) of the eyeball, which is continuous with the optic nerve and contains the light-sensitive rod and cone cells.

retinular cells See *ommatidium.*

retrovirus (ret′roh-vy″rus) An RNA virus that uses reverse transcriptase to produce a DNA intermediate, known as a *DNA provirus,* in the host cell. See *DNA provirus.*

reverse transcriptase An enzyme produced by retroviruses that catalyzes the production of DNA using RNA as a template.

reversible inhibitor A substance that forms weak bonds with an enzyme, temporarily interfering with its function; a reversible inhibitor is either competitive or noncompetitive. Compare with *irreversible inhibitor.*

Rh factors Red blood cell antigens, known as *D antigens,* first identified in *Rhesus* monkeys. People who have these antigens are Rh⁺; people lacking them are Rh⁻. See *erythroblastosis fetalis.*

rhizome (ry′zome) A horizontal underground stem that bears leaves and buds and often serves as a storage organ and a means of asexual reproduction, e.g., iris.

rhodopsin (rho-dop′sin) Visual purple; a light-sensitive pigment found in the rod cells of the vertebrate eye; a similar molecule is employed by certain bacteria in the capture of light energy to make ATP.

ribonucleic acid (RNA) A family of single-stranded nucleic acids that function mainly in protein synthesis

ribose The five-carbon sugar present in RNA and in important nucleoside triphosphates such as ATP.

ribosomal RNA (rRNA) See *ribosomes.*

ribosomes (ry′boh-sohmz) Organelles that are part of the protein synthesis machinery of both prokaryotic and eukaryotic cells; consist of a larger and smaller subunit, each composed of ribosomal RNA (rRNA) and ribosomal proteins.

ribozyme (ry′boh-zime) A molecule of RNA that has catalytic properties.

ribulose bisphosphate (RuBP) A five-carbon phosphorylated compound with a high energy potential that reacts with carbon dioxide in the initial step of the Calvin cycle.

ribulose bisphosphate carboxylase/oxygenase See *rubisco.*

RNA interference (RNAi) Phenomenon in which certain small RNA molecules interfere with the expression of genes or their RNA transcripts; RNA interference involves small interfering RNAs, microRNAs, and a few other kinds of short RNA molecules.

RNA polymerase An enzyme that catalyzes the synthesis of RNA from a DNA template.

RNA primer The sequence of about five RNA nucleotides that are synthesized during DNA replication to provide a 3′ end to which DNA polymerase adds nucleotides. The RNA primer is later degraded and replaced with DNA.

RNA world A model that proposes that during the evolution of cells, RNA was the first informational molecule to evolve, followed at a later time by proteins and DNA.

rod One of the rod-shaped, light-sensitive cells of the retina that are particularly sensitive to dim light and mediate black-and-white vision. Compare with *cone.*

root cap A covering of cells over the root tip that protects the delicate meristematic tissue directly behind it.

root graft The process of roots from two different plants growing together and becoming permanently attached to each other.

root hair An extension, or outgrowth, of a root epidermal cell. Root hairs increase the absorptive capacity of roots.

root pressure The pressure in xylem sap that occurs as a result of the active absorption of mineral ions followed by the osmotic uptake of water into roots from the soil.

root system The underground portion of a plant that anchors it in the soil and absorbs water and dissolved minerals.

rough ER See *endoplasmic reticulum.*

rubisco The common name of ribulose bisphosphate carboxylase/oxygenase, the enzyme that catalyzes the fixation of carbon dioxide in the Calvin cycle.

rugae (roo′jee) Folds, such as those in the lining of the stomach.

runner See *stolon.*

S phase Stage in interphase of the cell cycle during which DNA and other chromosomal constituents are synthesized. Compare with G_1 and G_2 phases.

saccule The structure within the vestibule of the inner vertebrate ear that along with the utricle houses the receptors of static equilibrium.

salicylic acid A signaling molecule that helps plants defend against insect pests and pathogens such as viruses by helping activate systemic acquired resistance.

salinity The concentration of dissolved salts (e.g., sodium chloride) in a body of water.

salivary glands Accessory digestive glands found in vertebrates and some invertebrates; in humans there are three pairs.

salt An ionic compound consisting of an anion other than a hydroxide ion and a cation other than a hydrogen ion. A salt is formed by the reaction between an acid and a base.

salt marsh A wetland dominated by grasses in which the salinity fluctuates between that of sea water and fresh water; salt marshes are usually located in estuaries.

saltatory conduction The transmission of a neural impulse along a myelinated neuron; ion activity at one node depolarizes the next node along the axon.

saprobe See *decomposer.*

saprotroph (sap′roh-trof) See *decomposer.*

sarcolemma (sar″koh-lem′mah) The muscle cell plasma membrane.

sarcomere (sar′koh-meer) A segment of a striated muscle cell located between adjacent Z lines that serves as a unit of contraction.

sarcoplasmic reticulum The system of vesicles in a muscle cell that surrounds the myofibrils and releases calcium in muscle contraction; a modified endoplasmic reticulum.

saturated fatty acid See *fatty acid.*

savanna (suh-van′uh) A tropical grassland containing scattered trees; found in areas of low rainfall or seasonal rainfall with prolonged dry periods.

scaffolding proteins (1) Proteins that organize groups of intracellular signaling molecules into signaling complexes. (2) Nonhistone proteins that help maintain the structure of a chromosome.

schizocoely (skiz′oh-seely) The process of coelom formation in which the mesoderm splits into two layers, forming a cavity between them; characteristic of protostomes. Compare with *enterocoely.*

Schwann cells Supporting cells found in nervous tissue outside the central nervous system; produce the myelin sheath around peripheral neurons.

sclera (skler′ah) The outer coat of the eyeball; a tough, opaque sheet of connective tissue that protects the inner structures and helps maintain the rigidity of the eyeball.

sclereid (skler′id) In plants, a sclerenchyma cell that is variable in shape but typically not long and tapered. Compare with *fiber.*

sclerenchyma (skler-en′kim-uh) Cells that provide strength and support in the plant body, are often dead at maturity, and have extremely thick walls; includes fibers and sclereids.

scramble competition See *exploitation competition.*

scrotum (skroh′tum) The external sac of skin found in most male mammals that contains the testes and their accessory organs.

second law of thermodynamics The physical law stating that the total amount of entropy in the universe continually increases. Compare with *first law of thermodynamics.*

second messenger A substance, e.g., cyclic AMP or calcium ions, that relays a message from a hormone bound to a cell-surface receptor; leads to some change in the cell.

secondary consumer See *carnivore.*

secondary growth An increase in the girth of a plant due to the activity of the vascular cambium and cork cambium; secondary growth results in the production of secondary tissues, i.e., wood and bark. Compare with *primary growth.*

secondary immune response The rapid production of antibodies induced by a second exposure to an antigen several days, weeks, or even months after the initial exposure. Compare with *primary immune response.*

secondary mycelium A dikaryotic mycelium formed by the fusion of two primary hyphae. Compare with *primary mycelium.*

secondary productivity The amount of food molecules converted to biomass by consumers in an ecosystem over a given period. Compare with *primary productivity.*

secondary structure (of a protein) A regular geometric shape produced by hydrogen bonding between the atoms of the uniform polypeptide backbone; includes the alpha helix and the beta-pleated sheet. Compare with *primary, tertiary,* and *quaternary protein structure.*

secondary succession An ecological succession that takes place after some disturbance destroys the existing vegetation; soil is already present. See *succession.* Compare with *primary succession.*

secretory vesicles Small cytoplasmic vesicles that move substances from an internal membrane system to the plasma membrane.

seed A plant reproductive body consisting of a young, multicellular plant and nutritive tissue (food reserves), enclosed by a seed coat.

seed coat The outer protective covering of a seed.

seed fern An extinct group of seed-bearing woody plants with fern-like leaves; seed ferns probably descended from progymnosperms and gave rise to cycads and possibly ginkgoes.

segmentation genes In *Drosophila,* genes transcribed in the embryo that are responsible for generating a repeating pattern of body segments within the embryo and adult fly.

segregation, principle of The genetic principle, first noted by Gregor Mendel, that states that two alleles of a locus become separated into different gametes.

selectively permeable membrane A membrane that allows some substances to cross it more easily than others. Biological membranes are generally permeable to water but restrict the passage of many solutes.

self-incompatibility A genetic condition in which the pollen cannot fertilize the same flower or flowers on the same plant.

semelparity The condition of having a single reproductive effort in a lifetime. Compare with *iteroparity.*

semen The fluid consisting of sperm suspended in various glandular secretions that is ejaculated from the penis during orgasm.

semicircular canals The passages in the vertebrate inner ear containing structures that control the sense of equilibrium (balance).

semiconservative replication See *DNA replication.*

semilunar valves Valves between the ventricles of the heart and the arteries that carry blood away from the heart; aortic and pulmonary valves.

seminal vesicles (1) In mammals, glandular sacs that secrete a component of semen (seminal fluid). (2) In some invertebrates, structures that store sperm.

seminiferous tubules (sem-ih-nif′er-ous) Coiled tubules in the testes in which spermatogenesis takes place in male vertebrates.

senescence (se-nes′cents) The aging process.

sensory neuron A neuron that transmits an impulse from a receptor to the central nervous system.

sensory receptor A cell (or part of a cell) specialized to detect specific energy stimuli in the environment.

sepal (see′pul) One of the outermost parts of a flower, usually leaf-like in appearance, that protect the flower as a bud.

septum (pl., *septa*) A cross wall or partition, e.g., the walls that divide a hypha into cells.

sequencing See *DNA sequencing.*

serial endosymbiosis The hypothesis that certain organelles such as mitochondria and chloroplasts originated as symbiotic prokaryotes that lived inside other, free-living prokaryotic cells.

serotonin A neurotransmitter of the biogenic amine group.

Sertoli cells (sur-tole′ee) Supporting cells of the tubules of the testis.

sessile (ses′sile) Permanently attached to one location, e.g., coral animals.

setae (sing., *seta*) Bristlelike structures that aid in annelid locomotion.

set point A normal condition maintained by homeostatic mechanisms.

sex chromosome Chromosome that plays a role in sex determination.

sex-influenced trait A genetic trait that is expressed differently in males and females.

sex-linked gene A gene carried on a sex chromosome. In mammals almost all sex-linked genes are borne on the X chromosome, i.e., are X-linked.

sexual dimorphism Marked phenotypic differences between the two sexes of the same species.

sexual isolation See *behavioral isolation.*

sexual reproduction A type of reproduction in which two gametes (usually, but not necessarily, contributed by two different parents) fuse to form a zygote. Compare with *asexual reproduction.*

sexual selection A type of natural selection that occurs when individuals of a species vary in their ability to compete for mates; individuals with reproductive advantages are selected over others of the same sex.

shade avoidance The tendency of plants that are adapted to high light intensities to grow taller when they are closely surrounded by other plants.

shared ancestral characters Traits that were present in an ancestral species that have remained essentially unchanged; suggest a distant common ancestor. Also called *plesiomorphic characters.* Compare with *shared derived characters.*

shared derived characters Homologous traits found in two or more taxa that are present in their most recent common ancestor but not in earlier common ancestors. Also called *synapomorphic characters.* Compare with *shared ancestral characters.*

shoot system The aboveground portion of a plant, such as the stem and leaves.

short tandem repeats (STRs) Molecular markers that are short sequences of repetitive DNA; because STRs vary in length from one individual to another, they are useful in identifying individuals with a high degree of certainty.

short-day plant A plant that flowers in response to lengthening nights; also called *long-night plant.* Compare with *long-day, intermediate-day,* and *day-neutral plants.*

short-night plant See *long-day plant.*

sickle cell anemia An inherited form of anemia in which there is abnormality in the hemoglobin beta chains; the inheritance pattern is autosomal recessive.

sieve tube elements Cells that conduct dissolved sugar in the phloem of flowering plants.

sign stimulus Any stimulus that elicits a fixed action pattern in an animal.

signal amplification The process by which a few signaling molecules can effect major responses in the cell; the strength of each signaling molecule is magnified.

signal transduction A process in which a cell converts and amplifies an extracellular signal into an intracellular signal that affects some function in the cell. Also see *cell signaling.*

signaling molecule A molecule such as a hormone, local regulator, or neurotransmitter that transmits information when it binds to a receptor on the cell surface or within the cell.

signal-recognition particle (SRP) A protein–RNA complex that directs the ribosome–mRNA–polypeptide complex to the surface of the endoplasmic reticulum.

simple fruit A fruit that develops from a single ovary. Compare with *aggregate, accessory,* and *multiple fruits.*

simplest formula A type of chemical formula that gives the smallest whole-number ratio of the component atoms. Compare with *molecular formula* and *structural formula.*

single-strand binding proteins (SSBs) Proteins involved in DNA replication that bind to single DNA strands and prevent the double helix from re-forming until the strands are copied.

sink habitat A lower-quality habitat in which local reproductive success is less than local mortality. Compare with *source habitat.*

sinoatrial (SA) node The mass of specialized cardiac muscle in which the impulse triggering the heartbeat originates; the pacemaker of the heart.

sister chromatids See *chromatid.*

sister taxa Groups of organisms that share a more recent common ancestor with one another than either taxon does with any other group shown on a cladogram.

skeletal muscle The voluntary striated muscle of vertebrates, so called because it usually is directly or indirectly attached to some part of the skeleton. Compare with *cardiac muscle* and *smooth muscle.*

slash-and-burn agriculture A type of agriculture in which tropical rain forest is cut down, allowed to dry, and burned. The crops that are planted immediately afterward thrive because the ashes provide nutrients; in a few years, however, the soil is depleted and the land must be abandoned.

slow block to polyspermy See *cortical reaction.*

small interfering RNAs (siRNAs) Double-stranded RNA molecules about 23 nucleotides in length that silence genes at the post-transcriptional level by selectively cleaving mRNA molecules with base sequences complementary to the siRNA.

small intestine Portion of the vertebrate digestive tract that extends from the stomach to the large intestine.

small nuclear ribonucleoprotein complexes (snRNP) Aggregations of RNA and protein responsible for binding to pre-mRNA in eukaryotes; catalyze the excision of introns and the splicing of exons.

smooth ER See *endoplasmic reticulum.*

smooth muscle Involuntary muscle tissue that lacks transverse striations; found mainly in sheets surrounding hollow organs, such as the intestine. Compare with *cardiac muscle* and *skeletal muscle.*

social behavior Interaction of two or more animals, usually of the same species.

sociobiology The branch of biology that focuses on the evolution of social behavior through natural selection.

sodium–potassium pump Active transport system that transports sodium ions out of, and potassium ions into, cells.

soil erosion The wearing away or removal of soil from the land; although soil erosion occurs naturally from precipitation and run-off, human activities (such as clearing the land) accelerate it.

solute A dissolved substance. Compare with *solvent.*

solvent Substance capable of dissolving other substances. Compare with *solute.*

somatic cell In animals, a cell of the body not involved in formation of gametes. Compare with *germ line cell.*

somatic nervous system That part of the vertebrate peripheral nervous system that keeps the body in adjustment with the external environment; includes sensory receptors on the body surface and within the muscles, and the nerves that link them with the central nervous system. Compare with *autonomic nervous system.*

somatomedins See *insulin-like growth factors.*

somatotropin See *growth hormone.*

somites A series of paired blocks of mesoderm that develop on each side of the notochord in cephalochordates and vertebrates. Somites define the segmentation of the embryo, and in vertebrates, they give rise to the vertebrae, ribs, and certain skeletal muscles.

sonogram See *ultrasound imaging.*

soredium (sor-id′e-um) (pl., *soredia*) In lichens, a type of asexual reproductive structure that consists of a cluster of algal cells surrounded by fungal hyphae.

sorus (soh′rus) (pl., *sori*) In ferns, a cluster of spore-producing sporangia.

source habitat A good habitat in which local reproductive success is greater than local mortality. Surplus individuals in a source habitat may disperse to other habitats. Compare with *sink habitat.*

Southern blot A technique in which DNA fragments, previously separated by gel electrophoresis, are transferred to a nitrocellulose or nylon membrane and detected by autoradiography or chemical luminescence. Compare with *Northern blot* and *Western blot.*

speciation Evolution of a new species.

species According to the biological species concept, one or more populations whose members are capable of interbreeding in nature to produce fertile offspring and do not interbreed with members of other species. Compare with *evolutionary species concept.*

species diversity A measure of the relative importance of each species within a community; represents a combination of species richness and species evenness.

species richness The number of species in a community.

specific epithet The second part of the name of a species; designates a specific species belonging to that genus.

specific heat The amount of heat energy that must be supplied to raise the temperature of 1 g of a substance 1°C.

specific immune responses Defense mechanisms that target specific macromolecules associated with a pathogen. Includes cell-mediated immunity and antibody-mediated immunity. Also known as *acquired* or *adaptive immune responses.*

sperm The motile male gamete of animals and some plants and protists; also called a *spermatozoan.*

spermatid (spur′ma-tid) An immature sperm cell.

spermatocyte (spur-mah′toh-site) A meiotic cell that gives rise to spermatids and ultimately to mature sperm cells.

spermatogenesis (spur″mah-toh-jen′eh-sis) The production of male gametes (sperm) by meiosis and subsequent cell differentiation. Compare with *oogenesis.*

spermatozoan (spur-mah-toh-zoh′un) See *sperm.*

sphincter (sfink′tur) A group of circularly arranged muscle fibers, the contractions of which close an opening, e.g., the pyloric sphincter at the exit of the stomach.

spinal cord In vertebrates, the dorsal, tubular nerve cord.

spinal nerves In vertebrates, the nerves that emerge from the spinal cord.

spindle See *mitotic spindle.*

spine A leaf that is modified for protection, such as a cactus spine.

spiracle (speer′ih-kl) An opening for gas exchange, such as the opening of a trachea on the body surface of an insect.

spiral cleavage A distinctive spiral pattern of blastomere production in an early protostome embryo. Compare with *radial cleavage.*

spirillum (pl., *spirilla*) A long, rigid, helical bacterium. Compare with *spirochete, vibrio, bacillus,* and *coccus.*

spirochete A long, flexible, helical bacterium. Compare with *spirillum, vibrio, bacillus,* and *coccus.*

spleen An abdominal organ located just below the diaphragm that removes worn-out blood cells and bacteria from the blood and plays a role in immunity.

spliceosome A large nucleoprotein particle that catalyzes the reactions that remove introns from pre-mRNA.

spongy mesophyll (mez′oh-fil) The loosely arranged mesophyll cells near the lower epidermis in certain leaves. Compare with *palisade mesophyll*.

spontaneous reaction See *exergonic reaction*.

sporangium (spor-an′jee-um) (pl., *sporangia*) A spore case, found in plants, certain protists, and fungi.

spore A reproductive cell that gives rise to individual offspring in plants, fungi, and certain algae and protozoa.

sporophyll (spor′oh-fil) A leaflike structure that bears spores.

sporophyte generation (spor′oh-fite) The 2*n*, spore-producing stage in the life cycle of a plant. Compare with *gametophyte generation*.

sporozoa See *apicomplexans*.

sporozoite The infective sporelike state in apicomplexans.

stabilizing selection Natural selection that acts against extreme phenotypes and favors intermediate variants; associated with a population well adapted to its environment. Compare with *directional selection* and *disruptive selection*.

stamen (stay′men) The male part of a flower; consists of a filament and anther.

standing-water ecosystem A lake or pond ecosystem.

starch A polysaccharide composed of alpha glucose subunits; made by plants for energy storage.

start codon The codon AUG, which signals the beginning of translation of messenger RNA. Compare with *stop codon*.

stasis Long periods in the fossil record in which there is little or no evolutionary change.

statocyst (stat′oh-sist) An invertebrate sense organ containing one or more granules (statoliths); senses gravity and motion.

statoliths (stat′uh-liths) Granules of loose sand or calcium carbonate found in statocysts.

stele The cylinder in the center of roots and stems that contains the vascular tissue.

stem cell A relatively undifferentiated cell capable of repeated cell division. At each division at least one of the daughter cells usually remains a stem cell, whereas the other may differentiate as a specific cell type.

stereocilia Hairlike projections of hair cells; microvilli that contain actin filaments.

sterilization A procedure that renders an individual incapable of producing offspring; the most common surgical procedures are vasectomy in the male and tubal ligation in the female.

steroids (steer′oids) Complex molecules containing carbon atoms arranged in four attached rings, three of which contain six carbon atoms each and the fourth of which contains five, e.g., cholesterol and certain hormones, including the male and female sex hormones of vertebrates.

stigma The portion of the carpel where pollen grains land during pollination (and before fertilization).

stipe A short stalk or stemlike structure that is a part of the body of certain multicellular algae.

stipule (stip′yule) One of a pair of scalelike or leaflike structures found at the base of certain leaves.

stolon (stow′lon) An aboveground, horizontal stem with long internodes; stolons often form buds that develop into separate plants, e.g., strawberry; also called *runner*.

stomach Muscular region of the vertebrate digestive tract, extending from the esophagus to the small intestine.

stomata (sing., *stoma*) Small pores located in the epidermis of plants that provide for gas exchange for photosynthesis; each stoma is flanked by two guard cells, which are responsible for its opening and closing.

stop codon Any of the three codons in mRNA that do not code for an amino acid (UAA, UAG, or UGA) but signal the termination of translation. Compare with *start codon*.

stratosphere The layer of the atmosphere between the troposphere and the mesosphere. It contains a thin ozone layer that protects life by filtering out much of the sun's ultraviolet radiation.

stratum basale (strat′um bah-say′lee) The deepest sublayer of the human epidermis, consisting of cells that continuously divide. Compare with *stratum corneum*.

stratum corneum The most superficial sublayer of the human epidermis. Compare with *stratum basale*.

strobilus (stroh′bil-us) (pl., *strobili*) In certain plants, a conelike structure that bears spore-producing sporangia.

stroke volume The volume of blood pumped by one ventricle during one contraction.

stroma A fluid space of the chloroplast, enclosed by the chloroplast inner membrane and surrounding the thylakoids; site of the reactions of the Calvin cycle.

stromatolite (stroh-mat′oh-lite) A columnlike rock that consists of many minute layers of prokaryotic cells, usually cyanobacteria.

structural formula A type of chemical formula that shows the spatial arrangement of the atoms in a molecule. Compare with *simplest formula* and *molecular formula*.

structural isomer One of two or more chemical compounds having the same chemical formula but differing in the covalent arrangement of their atoms, e.g., glucose and fructose.

style The neck connecting the stigma to the ovary of a carpel.

subsidiary cell In plants, a structurally distinct epidermal cell associated with a guard cell.

substance P A peptide neurotransmitter released by certain sensory neurons in pain pathways; signals the brain regarding painful stimuli; also stimulates other structures, including smooth muscle in the digestive tract.

substrate A substance on which an enzyme acts; a reactant in an enzymatically catalyzed reaction.

succession The sequence of changes in the species composition of a community over time. See *primary succession* and *secondary succession*.

sucker A shoot that develops adventitiously from a root; a type of asexual reproduction.

sulcus (sul′kus) (pl., *sulci*) A groove, trench, or depression, especially one occurring on the surface of the brain, separating the convolutions.

sulfhydryl group Functional group abbreviated —SH; found in organic compounds called thiols.

sum rule The rule for combining the probabilities of mutually exclusive events by adding their individual probabilities. Compare with *product rule*.

summation The process of adding together excitatory postsynaptic potentials (EPSPs).

suppressor T cell T lymphocyte that suppresses the immune response.

supraorbital ridge (soop″rah-or′bit-ul) The prominent bony ridge above the eye socket; ape skulls have prominent supraorbital ridges.

surface tension The attraction that the molecules at the surface of a liquid may have for one another.

survivorship The probability that a given individual in a population or cohort will survive to a particular age; usually presented as a survivorship curve.

survivorship curve A graph of the number of surviving individuals of a cohort, from birth to the maximum age attained by any individual.

suspensor (suh-spen′sur) In plant embryo development, a multicellular structure that anchors the embryo and aids in nutrient absorption from the endosperm.

sustainability See *environmental sustainability.*

swim bladder The hydrostatic organ in bony fishes that permits the fish to hover at a given depth.

symbiosis (sim-bee-oh′sis) An intimate relationship between two or more organisms of different species. See *commensalism, mutualism,* and *parasitism.*

sympathetic nervous system A division of the autonomic nervous system; its general effect is to mobilize energy, especially during stress situations; prepares the body for fight-or-flight response. Compare with *parasympathetic nervous system.*

sympatric speciation (sim-pa′trik) The evolution of a new species within the same geographic region as the parental species. Compare with *allopatric speciation.*

symplast A continuum consisting of the cytoplasm of many plant cells, connected from one cell to the next by plasmodesmata. Compare with *apoplast.*

synapomorphic characters See *shared derived characters.*

synapse (sin′aps) The junction between two neurons or between a neuron and an effector (muscle or gland).

synapsis (sin-ap′sis) The process of physical association of homologous chromosomes during prophase I of meiosis.

synaptic enhancement An increase in neurotransmitter release thought to occur as a result of calcium ion accumulation inside the presynaptic neuron.

synaptic plasticity The ability of synapses to change in response to certain types of stimuli. Synaptic changes occur during learning and memory storage.

synaptonemal complex The structure, visible with the electron microscope, produced when homologous chromosomes undergo synapsis.

synthetic theory of evolution See *modern synthesis.*

systematics The scientific study of the diversity of organisms and their evolutionary relationships. Taxonomy is an aspect of systematics. See *taxonomy.*

systemic acquired resistance A defensive response in infected plants that helps fight infection and promote wound healing.

systemic anaphylaxis A rapid, widespread allergic reaction that can lead to death.

systemic circulation The part of the circulatory system that delivers blood to and from the tissues and organs of the body. Compare with *pulmonary circulation.*

systems biology A field of biology that synthesizes knowledge of many small parts to understand the whole. Also referred to as *integrative biology* or *integrative systems biology.*

systole (sis′tuh-lee) The phase of the cardiac cycle when the heart is contracting. Compare with *diastole.*

T cell (T lymphocyte) The type of white blood cell responsible for a wide variety of immune functions, particularly cell-mediated immunity. T cells are processed in the thymus. Compare with *B cell.*

T cytotoxic cell (T_C) T lymphocyte that destroys cancer cells and other pathogenic cells on contact. Also known as *CD8 T cell* and *killer T cell.*

T helper cell (T_H) T lymphocyte that activates B cells (B lymphocytes) and stimulates T cytotoxic cell production. Also known as *CD4 T cell.*

T tubules Transverse tubules; system of inward extensions of the muscle fiber plasma membrane.

taiga (tie′gah) See *boreal forest.*

taproot system A root system consisting of a prominent main root with smaller lateral roots branching off it; a taproot develops directly from the embryonic radicle. Compare with *fibrous root system.*

target cell or tissue A cell or tissue with receptors that bind a hormone.

TATA box A component of a eukaryotic promoter region; consists of a sequence of bases located about 30 base pairs upstream from the transcription initiation site.

taxon A formal taxonomic group at any level, e.g., phylum or genus.

taxonomy (tax-on′ah-mee) The science of naming, describing, and classifying organisms; see *systematics.*

Tay–Sachs disease A serious genetic disease in which abnormal lipid metabolism in the brain causes mental deterioration in affected infants and young children; inheritance pattern is autosomal recessive.

tectorial membrane (tek-tor′ee-ul) The roof membrane of the organ of Corti in the cochlea of the ear.

telencephalon See *forebrain.*

telolecithal egg An egg with a large amount of yolk, concentrated at the vegetal pole. Compare with *isolecithal egg.*

telomerase A special DNA replication enzyme that can lengthen telomeric DNA by adding repetitive nucleotide sequences to the ends of eukaryotic chromosomes; typically present in cells that divide an unlimited number of times.

telomeres The protective end caps of chromosomes that consist of short, simple, noncoding DNA sequences that repeat many times.

telophase (teel′oh-faze or tel′oh-faze) The last stage of mitosis and of meiosis I and II when, having reached the poles, chromosomes become decondensed, and a nuclear envelope forms around each group.

temperate deciduous forest A forest biome that occurs in temperate areas where annual precipitation ranges from about 75 cm to 125 cm.

temperate grassland A grassland characterized by hot summers, cold winters, and less rainfall than is found in a temperate deciduous forest biome.

temperate rain forest A coniferous biome characterized by cool weather, dense fog, and high precipitation, e.g., the north Pacific coast of North America.

temperate virus A virus that integrates into the host DNA as a prophage.

temperature The average kinetic energy of the particles in a sample of a substance.

temporal isolation A prezygotic reproductive isolating mechanism in which genetic exchange is prevented between similar species because they reproduce at different times of the day, season, or year.

tendon A connective tissue structure that joins a muscle to another muscle, or a muscle to a bone. Tendons transmit the force generated by a muscle.

tendril A leaf or stem that is modified for holding or attaching onto objects.

tension–cohesion model The mechanism by which water and dissolved inorganic minerals are thought to be transported in xylem; water is pulled upward under tension because of transpiration while maintaining an unbroken column in xylem because of cohesion; also called *transpiration–cohesion model.*

teratogen Any agent capable of interfering with normal morphogenesis in an embryo, thereby causing malformations; examples include radiation, certain chemicals, and certain infectious agents.

terminal bud A bud at the tip of a stem. Compare with *axillary bud*.

termination (of protein synthesis) The final stage of protein synthesis, which occurs when a termination (stop) codon is reached, causing the completed polypeptide chain to be released from the ribosome. See *initiation* and *elongation*.

termination codon See *stop codon*.

territoriality Behavior pattern in which one organism (usually a male) stakes out a territory of its own and defends it against intrusion by other members of the same species and sex.

tertiary consumer See *carnivore*.

tertiary structure (of a protein) (tur′she-air″ee) The overall three-dimensional shape of a polypeptide that is determined by interactions involving the amino acid side chains. Compare with *primary, secondary,* and *quaternary protein structure*.

test A shell.

test cross The genetic cross in which either an F_1 individual, or an individual of unknown genotype, is mated to a homozygous recessive individual.

testis (tes′tis) (pl., *testes*) The male gonad that produces sperm and the male hormone testosterone; in humans and certain other mammals, the testes are located in the scrotum.

testosterone (tes-tos′ter-own) The principal male sex hormone (androgen); a steroid hormone produced by the interstitial cells of the testes; stimulates spermatogenesis and is responsible for primary and secondary sex characteristics in the male.

tetrad The chromosome complex formed by the synapsis of a pair of homologous chromosomes (i.e., four chromatids) during meiotic prophase I; also known as a *bivalent*.

tetrapods (tet′rah-podz) Four-limbed vertebrates: the amphibians, reptiles, birds, and mammals.

thalamus (thal′uh-mus) The part of the vertebrate brain that serves as a main relay center, transmitting information between the spinal cord and the cerebrum.

thallus (thal′us) (pl., *thalli*) The simple body of an alga, fungus, or nonvascular plant that lacks root, stems, or leaves, e.g., a liverwort thallus or a lichen thallus.

theca cells The layer of connective tissue cells that surrounds the granulosa cells in an ovarian follicle; stimulated by luteinizing hormone (LH) to produce androgens, which are converted to estrogen in the granulosa cells.

theory A widely accepted explanation supported by a large body of observations and experiments. A good theory relates facts that appear unrelated; it predicts new facts and suggests new relationships. Compare with *hypothesis*.

therapsids (ther-ap′sidz) A group of mammal-like reptiles of the Permian period; gave rise to the mammals.

thermal stratification The marked layering (separation into warm and cold layers) of temperate lakes during the summer. See *thermocline*.

thermocline (thur′moh-kline) A marked and abrupt temperature transition in temperate lakes between warm surface water and cold deeper water. See *thermal stratification*.

thermodynamics Principles governing energy transfer (often expressed in terms of heat transfer). See *first law of thermodynamics* and *second law of thermodynamics*.

thermoreceptor A sensory receptor that responds to heat.

thigmomorphogenesis (thig″moh-mor-foh-jen′uh-sis) An alteration of plant growth in response to mechanical stimuli, such as wind, rain, hail, and contact with passing animals.

thigmotropism (thig′moh-troh′pizm) Plant growth in response to contact with a solid object, such as the twining of plant tendrils.

threatened species A species in which the population is small enough for it to be at risk of becoming extinct throughout all or part of its range but not so small that it is in imminent danger of extinction. Compare with *endangered species*.

threshold level The potential that a neuron or other excitable cell must reach for an action potential to be initiated.

thrombocytes See *platelets*.

thylakoid lumen See *thylakoids*.

thylakoids (thy′lah-koidz) An interconnected system of flattened, saclike, membranous structures inside the chloroplast.

thymine (thy′meen) A nitrogenous pyrimidine base found in DNA.

thymus gland (thy′mus) An endocrine gland that functions as part of the lymphatic system; processes T cells; important in cell-mediated immunity.

thyroid gland An endocrine gland that lies anterior to the trachea and releases hormones that regulate the rate of metabolism.

thyroid hormones Hormones, including thyroxin, secreted by the thyroid gland; stimulate rate of metabolism.

tidal volume The volume of air moved into and out of the lungs with each normal resting breath.

tight junctions Specialized structures that form between some animal cells, producing a tight seal that prevents materials from passing through the spaces between the cells.

tissue A group of closely associated, similar cells that work together to carry out specific functions.

tissue culture The growth of tissue or cells in a synthetic growth medium under sterile conditions.

tissue engineering A developing technology that is striving to grow human tissues and organs (for transplantation) in cell cultures.

tissue fluid See *interstitial fluid*.

tolerance A decreased response to a drug over time.

Toll-like receptors Cell-surface receptors on phagocytes that recognize certain common features of classes of pathogens. See also *pathogen-associated molecular patterns (PAMPs)*.

tonoplast The membrane surrounding a vacuole.

top-down processes Control of ecosystem function by trophic interactions, particularly from the highest trophic level. Compare *bottom-up processes*.

topoisomerases (toe-poe-eye-sahm′er-ases) Enzymes that relieve twists and kinks in a DNA molecule by breaking and rejoining the strands.

torpor An energy-conserving state of low metabolic rate and inactivity. See *estivation* and *hibernation*.

torsion The twisting of the visceral mass characteristic of gastropod mollusks.

total fertility rate The average number of children born to a woman during her lifetime.

totipotent (toh-ti-poh′tent) A term describing a cell or nucleus that contains the complete set of genetic instructions required to direct the normal development of an entire organism. Compare with *pluripotent*.

trace element An element required by an organism in very small amounts.

trachea (tray′kee-uh) (pl., *tracheae*) (1) Principal thoracic air duct of terrestrial vertebrates; windpipe. (2) One of the microscopic air ducts (or tracheal tubes) branching throughout the body of most terrestrial arthropods and some terrestrial mollusks.

tracheal tubes See *trachea*.

tracheid (tray′kee-id) A type of water-conducting and supporting cell in the xylem of vascular plants.

tract A bundle of nerve fibers within the central nervous system.

transcription The synthesis of RNA from a DNA template.

transcription factors DNA-binding proteins that regulate transcription in eukaryotes; include positively acting activators and negatively acting repressors.

transduction (1) The transfer of a genetic fragment from one cell to another, e.g., from one bacterium to another, by a virus. (2) In the nervous system, the conversion of energy of a stimulus to electrical signals.

transfer RNA (tRNA) RNA molecules that bind to specific amino acids and serve as adapter molecules in protein synthesis. The tRNA anticodons bind to complementary mRNA codons.

transformation (1) The incorporation of genetic material into a cell, thereby changing its phenotype. (2) The conversion of a normal cell to a cancer cell (called a malignant transformation).

transgenic organism A plant or animal that has foreign DNA incorporated into its genome.

translation The conversion of information provided by mRNA into a specific sequence of amino acids in a polypeptide chain; process also requires transfer RNA and ribosomes.

translocation (1) The movement of organic materials (dissolved food) in the phloem of a plant. (2) Chromosome abnormality in which part of one chromosome has become attached to another. (3) Part of the elongation cycle of protein synthesis in which a transfer RNA attached to the growing polypeptide chain is transferred from the A site to the P site.

transmembrane protein An integral membrane protein that spans the lipid bilayer.

transmission, neural The conduction of a neural impulse along a neuron or from one neuron to another.

transpiration The loss of water vapor from the aerial surfaces of a plant (i.e., leaves and stems).

transpiration–cohesion model See *tension–cohesion model.*

transport vesicles Small cytoplasmic vesicles that move substances from one membrane system to another.

transposon (tranz-poze'on) A DNA segment that is capable of moving from one chromosome to another or to different sites within the same chromosome; also called a *mobile genetic element.*

transverse tubules See *T tubules.*

triacylglycerol (try-ace"il-glis'er-ol) The main storage lipid of organisms, consisting of a glycerol combined chemically with three fatty acids; also called *triglyceride.* Compare with *monoacylglycerol* and *diacylglycerol.*

tricarboxylic acid (TCA) cycle See *citric acid cycle.*

trichome (try'kohm) A hair or other appendage growing out from the epidermis of a plant.

tricuspid valve See *atrioventricular valve.*

triglyceride See *triacylglycerol.*

triose A sugar molecule containing three carbons.

triplet A sequence of three nucleotides that serves as the basic unit of genetic information.

triplet code The sequences of three nucleotides that compose the codons, the units of genetic information in mRNA that specify the order of amino acids in a polypeptide chain.

trisomy (try'sohm-ee) The condition in which each chromosome has two copies, except one, which is present in triplicate; the cell contains one more chromosome than the diploid number. Compare with *monosomy* and *disomy.*

trochophore larva (troh'koh-for) A larval form found in mollusks and many polychaetes.

trophic level (troh'fik) Each sequential step of matter and energy in a food web, from producers to primary, secondary, or tertiary consumers; each organism is assigned to a trophic level based on its primary source of nourishment.

trophoblast (troh'foh-blast) The outer cell layer of a late blastocyst, which in placental mammals gives rise to the chorion and to the fetal contribution to the placenta.

tropic hormone (trow'pic) A hormone, produced by one endocrine gland, that targets another endocrine gland.

tropical dry forest A tropical forest where enough precipitation falls to support trees but not enough to support the lush vegetation of a tropical rain forest; often occurs in areas with pronounced rainy and dry seasons.

tropical rain forest A lush, species-rich forest biome that occurs in tropical areas where the climate is very moist throughout the year. Tropical rain forests are also characterized by old, infertile soils.

tropism (troh'pizm) In plants, a directional growth response that is elicited by an environmental stimulus.

tropomyosin (troh-poh-my'oh-sin) A muscle protein involved in regulation of contraction.

true-breeding strain A genetic strain of an organism in which all individuals are homozygous at the loci under consideration.

tube feet Structures characteristic of echinoderms; function in locomotion and feeding.

tuber A thickened end of a rhizome that is fleshy and enlarged for food storage, e.g., white potato.

tubular transport maximum (Tm) The maximum rate at which a substance is reabsorbed from the renal tubules of the kidney.

tumor A mass of tissue that grows in an uncontrolled manner; a neoplasm.

tumor necrosis factors (TNFs) Cytokines that kill tumor cells and stimulate immune cells to initiate an inflammatory response.

tumor suppressor gene A gene (also known as an *anti-oncogene*) whose normal role is to block cell division in response to certain growth-inhibiting factors; when mutated, may contribute to the formation of a cancer cell. Compare with *oncogene.*

tundra (tun'dra) A treeless biome between the boreal forest in the south and the polar ice cap in the north that consists of boggy plains covered by lichens and small plants. Also called *arctic tundra.* Compare with *alpine tundra.*

tunicates Chordates belonging to subphylum Urochordata; sea squirts.

turgor pressure (tur'gor) Hydrostatic pressure that develops within a walled cell, such as a plant cell, when the osmotic pressure of the cell's contents is greater than the osmotic pressure of the surrounding fluid.

Turner syndrome An inherited condition in which only one sex chromosome (an X chromosome) is present in cells; karyotype is designated XO; affected individuals are sterile females.

tyrosine kinase An enzyme that phosphorylates the tyrosine part of proteins.

tyrosine kinase receptor A plasma membrane receptor that phosphorylates the tyrosine part of proteins; when a ligand binds to the receptor, the conformation of the receptor changes and it may phosphorylate itself as well as other molecules; important in immune function and serves as a receptor for insulin.

ultimate causes (of behavior) Evolutionary explanations for why a certain behavior occurs. Compare with *proximate causes of behavior.*

ultrasound imaging A technique in which high-frequency sound waves (ultrasound) are used to provide an image (sonogram) of an internal structure.

ultrastructure The fine detail of a cell, generally only observable by use of an electron microscope.

umbilical cord In placental mammals, the organ that connects the embryo to the placenta.

uniform dispersion The spatial distribution pattern of a population in which individuals are regularly spaced. Compare with *random dispersion* and *clumped dispersion.*

unsaturated fatty acid See *fatty acid.*

upstream promoter elements (UPEs) Components of a eukaryotic promoter, found upstream of the RNA polymerase-binding site; the strength of a promoter is affected by the number and type of UPEs present.

upwelling An upward movement of water that brings nutrients from the ocean depths to the surface. Where upwelling occurs, the ocean is very productive.

uracil (yur′ah-sil) A nitrogenous pyrimidine base found in RNA.

urea (yur-ee′ah) The principal nitrogenous excretory product of mammals; one of the water-soluble end products of protein metabolism.

ureter (yur′ih-tur) One of the paired tubular structures that conducts urine from the kidney to the bladder.

urethra (yoo-ree′thruh) The tube that conducts urine from the bladder to the outside of the body.

uric acid (yoor′ik) The principal nitrogenous excretory product of insects, birds, and reptiles; a relatively insoluble end product of protein metabolism; also occurs in mammals as an end product of purine metabolism.

urinary bladder An organ that receives urine from the ureters and temporarily stores it.

urinary system The body system in vertebrates that consists of kidneys, urinary bladder, and associated ducts.

urochordates A subphylum of chordates; includes the tunicates.

uterine tube (yoo′tur-in) See *oviduct.*

uterus (yoo′tur-us) The hollow, muscular organ of the female reproductive tract in which the fetus undergoes development.

utricle The structure within the vestibule of the vertebrate inner ear that, along with the saccule, houses the receptors of static equilibrium.

vaccine (vak-seen′) A commercially produced, weakened or killed antigen associated with a particular disease that stimulates the body to make antibodies.

vacuole (vak′yoo-ole) A fluid-filled, membrane-enclosed sac found within the cytoplasm; may function in storage, digestion, or water elimination.

vagina The elastic, muscular tube, extending from the cervix to its external opening, that receives the penis during sexual intercourse and serves as the birth canal.

valence electrons The electrons in the outer electron shell, known as the *valence shell,* of an atom; in the formation of a chemical bond, an atom can accept electrons into its valence shell or donate or share valence electrons.

van der Waals interactions Weak attractive forces between atoms; caused by interactions among fluctuating charges.

vas deferens (vas def′ur-enz) (pl., *vasa deferentia*) One of the paired sperm ducts that connects the epididymis of the testis to the ejaculatory duct.

vascular cambium A lateral meristem that produces secondary xylem (wood) and secondary phloem (inner bark). Compare with *cork cambium.*

vascular tissue system The tissues specialized for translocation of materials throughout the plant body, i.e., the xylem and phloem.

vasoconstriction Narrowing of the diameter of blood vessels.

vasodilation Expansion of the diameter of blood vessels.

vector (1) Any carrier or means of transfer. (2) Agent, e.g., a plasmid or virus, that transfers genetic information. (3) Agent that transfers a parasite from one host to another.

vegetal pole The yolky pole of a vertebrate or echinoderm egg. Compare with *animal pole.*

vein (1) A blood vessel that carries blood from the tissues toward a chamber of the heart (compare with *artery*). (2) A strand of vascular tissue that is part of the network of conducting tissue in a leaf.

veliger larva The larval stage of many marine gastropods (snails) and bivalves (e.g., clams); often is a second larval stage that develops after the trochophore larva.

ventilation The process of actively moving air or water over a respiratory surface.

ventral Toward the lowermost surface or belly of an animal. Compare with *dorsal.*

ventricle (1) A cavity in an organ. (2) One of the several cavities of the brain. (3) One of the chambers of the heart that receives blood from an atrium.

vernalization (vur″nul-uh-zay′shun) The induction of flowering by a low-temperature treatment.

vertebrates A subphylum of chordates that possess a bony vertebral column; includes fishes, amphibians, reptiles, birds, and mammals.

vesicle (ves′ih-kl) Any small sac, especially a small, spherical, membrane-enclosed compartment, within the cytoplasm.

vessel element A type of water-conducting cell in the xylem of vascular plants.

vestibular apparatus Collectively, the saccule, utricle, and semicircular canals of the inner ear.

vestigial (ves-tij′ee-ul) Rudimentary; an evolutionary remnant of a formerly functional structure.

vibrio A spirillum (spiral-shaped bacterium) that is shaped like a comma. Compare with *spirillum, spirochete, bacillus,* and *coccus.*

villus (pl., *villi*) A multicellular, minute, elongated projection from the surface of an epithelial membrane, e.g., villi of the mucosa of the small intestine.

viroid (vy′roid) A tiny, naked, infectious particle consisting only of nucleic acid.

virulence Properties that render an infectious agent pathogenic (and often lethal) to its host. Compare with *avirulence.*

virus A tiny pathogen consisting of a core of nucleic acid usually encased in protein and capable of infecting living cells; a virus is characterized by total dependence on a living host.

viscera (vis′ur-uh) The internal body organs, especially those located in the abdominal or thoracic cavities.

visceral mass The concentration of body organs (viscera) located above the foot in mollusks.

vital capacity The maximum volume of air a person exhales after filling the lungs to the maximum extent.

vitamin A complex organic molecule required in very small amounts for normal metabolic functioning.

vitelline envelope An acellular covering of the eggs of certain animals (e.g., echinoderms), located just outside the plasma membrane.

viviparous (vih-vip′er-us) Bearing living young that develop within the body of the mother. Compare with *oviparous* and *ovoviviparous.*

voltage-activated ion channels Ion channels in the plasma membrane of neurons that are regulated by changes in voltage. Also called *voltage-gated channels.*

vomeronasal organ In mammals, an organ in the epithelium of the nose, made up of specialized chemoreceptor cells that detect pheromones.

vulva The external genital structures of the female.

warning coloration See *aposematic coloration*.

water mold A funguslike protist with a body consisting of a coenocytic mycelium that reproduces asexually by forming motile zoospores and sexually by forming oospores.

water potential Free energy of water; the water potential of pure water is zero and that of solutions is a negative value. Differences in water potential are used to predict the direction of water movement (always from a region of less negative water potential to a region of more negative water potential).

water vascular system Unique hydraulic system of echinoderms; functions in locomotion and feeding.

wavelength The distance from one wave peak to the next; the energy of electromagnetic radiation is inversely proportional to its wavelength.

weathering processes Chemical or physical processes that help form soil from rock; during weathering processes, the rock is gradually broken into smaller and smaller pieces.

Western blot A technique in which proteins, previously separated by gel electrophoresis, are transferred to paper. A specific labeled antibody is generally used to mark the location of a particular protein. Compare with *Southern blot* and *Northern blot*.

white matter Nervous tissue in the brain and spinal cord that contains myelinated axons. Compare with *gray matter*.

wild type The phenotypically normal (naturally occurring) form of a gene or organism.

wobble The ability of some tRNA anticodons to associate with more than one mRNA codon; in these cases the 5′ base of the anticodon is capable of forming hydrogen bonds with more than one kind of base in the 3′ position of the codon.

work Any change in the state or motion of matter.

X-linked gene A gene carried on an X chromosome.

X-ray diffraction A technique for determining the spatial arrangement of the components of a crystal.

xylem (zy′lem) The vascular tissue that conducts water and dissolved minerals in plants.

XYY karyotype Chromosome constitution that causes affected individuals (who are fertile males) to be unusually tall, with severe acne.

yeast A unicellular fungus (ascomycete) that reproduces asexually by budding or fission and sexually by ascospores.

yolk sac One of the extraembryonic membranes; a pouchlike outgrowth of the digestive tract of embryos of certain vertebrates (e.g., birds) that grows around the yolk and digests it. Embryonic blood cells are formed in the mammalian yolk sac, which lacks yolk.

zero population growth Point at which the birth rate equals the death rate. A population with zero population growth does not change in size.

zona pellucida (pel-loo′sih-duh) The thick, transparent covering that surrounds the plasma membrane of a mammalian ovum.

zooflagellate A unicellular, nonphotosynthetic protozoon that has one or more long, whiplike flagella.

zooplankton (zoh″oh-plank′tun) The nonphotosynthetic organisms present in plankton, e.g., protozoa, tiny crustaceans, and the larval stages of many animals. See *plankton*. Compare with *phytoplankton*.

zoospore (zoh′oh-spore) A flagellated motile spore produced asexually by certain algae, chytrids, and water molds and other protists.

zooxanthellae (zoh″oh-zan-thel′ee) (sing., *zooxanthella*) Endosymbiotic, photosynthetic dinoflagellates found in certain marine invertebrates; their mutualistic relationship with corals enhances the corals' reef-building ability.

zygomycetes (zy″gah-my′seats) Fungi characterized by the production of nonmotile asexual spores and sexual zygospores.

zygosporangium (zy″gah-spor-an′gee-um) A thick-walled sporangium containing a zygospore.

zygospore (zy′gah-spor) A sexual spore produced by a zygomycete.

zygote The 2n cell that results from the union of n gametes in sexual reproduction. Species that are not polyploid have haploid gametes and diploid zygotes.

zygotic genes Genes that are transcribed after fertilization, either in the zygote or in the embryo. Compare with *maternal effect genes*.

Index

Archosaurs, *443*
Ardipithecus (genus), *473, 474*
Ardipithecus ramidus, 473, 474
Argentinosaurus, 459
Arginine, *61, 62, 281*
Argon-40, in fossil dating, 399
Argyroxyphium sandwicense ssp. macrocephalum (Haleakala silversword), *442*
Aristotle, 21, 391
Armadillos, 463
Arthropods/Arthropoda (phylum), trilobites, *447*, 455
Artificial selection, 393, *393*
Artiodactyls/Artiodactyla (order), 406, *407*
Arum maculatum (lords-and-ladies), 182
Asexual reproduction, 5, *5*, 223. *See also* Mitosis; Reproduction
Ash Meadows, Nevada, allopatric speciation, 433
Asparagine, *60*
Aspartame, 356
Aspartate/aspartic acid, *60*, 186
Assortative mating, genotype frequency and, 416
Asteroid impact, 444
Asters, 218, *219*
Astrophysicists, 448
Asymmetry, membrane protein, 112
Atlantic Ocean, 403–404, 469
Atlantic salmon, transgenic, 343
Atom(s), 6, *7*, 26–29
    compounds/molecules formed by, 29 (*see also* Chemical bonds)
    light interactions with, 192, 193, *193*
Atomic mass, 28, *28*
Atomic mass unit (amu), 28
Atomic nucleus, 26
Atomic number, 27, *27*
Atomic orbitals, 29, *30*, 192
Atomic weight, *28*
ATP, 74
    cellular ratio of, to ADP, 159
    as coenzyme, 163
    energy currency of cells, 68, 156, 157–159
    exergonic and endergonic reactions linked by, 158–159, *159*
    hydrolysis of, *158*, 158–159, *159*
    mitochondria in synthesis of, 94–96, *95*
    synthesis of, *158*, 158–159, *159*, 173–188, *187* (*see also* Cellular respiration)
        active transport and, 115, 120–123, *121, 122*
        in aerobic respiration, 173–185 (*see also* Aerobic respiration)
        in anaerobic respiration, 186–188, *187*
        in fermentation, 186–188, *187*
        in photosynthesis, 197, *197*, 198–202, *199, 201*
    yield from aerobic respiration and, 183, *185*
ATP synthase, 182, 183, *184*, 200, 201, *201*
ATP-driven plasma membrane pumps, 120–122, *121, 122*. *See also* Active transport
Australia, 402
    fossil discoveries, 452, *452*, 454, 455
Australopithecines ("southern man apes"), 474
Australopithecus (genus), *473, 474*
Australopithecus aethiopicus, 473, 474
Australopithecus afarensis, 473, 474
Australopithecus africanus, 473, 474
Australopithecus anamensis, 473, 474
Australopithecus boisei, 466, 473, 474

Australopithecus robustus, 473, 474
Automated DNA-sequencing machines, 331, 333, *334*
Autopolyploidy, 435
Autoradiography, 28, *28*, 216, 327, 331, 333
Autosomal dominant trait, genetic disease and, 358–359
Autosomal monosomies, 352
Autosomal recessive trait, genetic disease and, 356–358, 363
Autosomal trisomy, in Down syndrome, 353
Autosomes, 248, 249, 347, *348*
Autotrophs (producers), *14*, 14–15
    defined, 207
    early cells as, 452–453
Avery, Oswald T., 262, *263*
Aves (class). *See* Bird(s)
Avirulence, 261
Avogadro, Amadeo, 31
Avogadro's number, 31
Axolotl salamander (*Ambystoma mexicanum*), *440*
Axon(s), 97–98

B cells/B lymphocytes. *See* Antibodies
Baboon(s), Anubis (*Papio anubis*), 469, *470*
Bacillus anthracis, 12
Bacillus thuringiensis (Bt) gene, 341, *342*, 343
Background extinction, 442
Bacteria
    aerobic respiration and, 174
    antigens and, 114
    bidirectional DNA replication in, 272–273, *274*
    cell communication across prokaryote–eukaryote boundary, 134, *134*, 135
    in cell signaling, 146, 148
    coupled transcription and translation in, 293, *294*
    DNA and transformation in, *261*, 261–262
    drug effects on enzymes, 167
    fermentation and, 187
    first cells, 452–453, 454
    gene regulation in, 305, 306–312
    heat tolerant, 163–164, *164*, 330
    isoleucine synthesis, *166*
    motile, 195
    mRNA of, 288, *288*
    nutrition in, 207
    operons in, 306–309, *307, 308, 310, 311*
    posttranscriptional regulation in, 309, 312
    proton gradients, 122
    transcription in, 288, *288*, 294, 305, 306–309, *312* (*see also* Transcription)
    transformation in, 323
    translation in, 293, *294* (*see also* Translation)
        elongation and, 290–291, *292*
        initiation of, 290, *291*
    viruses attacking (bacteriophages), 263, 323, 324
        phagocytosis, 123–124, *125*
    See also Prokaryotes/prokaryotic cells; *specific type*
Bacteria (domain), 11, *12*
Bacteria (kingdom), 11, *12*
Bacteria (prokaryotic cells), 3, *4*
    as decomposers, *14*, 15
    digestion of, in cells, 92–93, *125*
    in discovery of penicillin, 16
Bacterial artificial chromosomes (BACs), 324

Bacteriophages (phages), *262*, 263, 323, 324
Badlands, South Dakota, 459
Balaenoptera (blue whale), 399, *399*
Balanced polymorphism, 422–424, *423*
Baltimore, David, 297
Banteng, cloning, 372
Barberry, Japanese (*Berberis thunbergii*), *403*
Barberton rocks, South Africa, 452
Barbiturates, 90, 140
Barr body, 251, *251*, 313, 314, 354
Basal body, 99
Basal member, 462
Base(s), 39–42, *40, 41*
Base composition, DNA, 264, *264*
Base pairing, 274
    DNA, 265–266, *267*
Base sequences, 286, 288, *288*
Base-pair mutations, 386
Base-substitution mutations, 298, *299, 305, 305*
Basic amino acids, *61, 62*
Basic solution, 40, *41*
Basilosaurus, 399, 399
Bats, homology in, 400, *401*
Battery acid, *41*
Beadle, George, 280–281, *281*
Beagle, Darwin's voyage on, 392, *392*
Beak of the Finch: A Story of Evolution in Our Time, The (Weiner), 408
Bear, grizzly (*Ursus arctos*), *171*
Beebe, William, 462
Beer, *41*, 187
Bees, mechanical isolation, 431
Beetle, bombardier (*Stenaptinus insignis*), 161, *161*
Begonia plant, *119*
Behavioral isolation (sexual isolation), *430*, 430–431, *431, 432*
Benson, Andrew, 202
Benzene, 46
Benzodiazepine drugs, 140
Berberis thunbergii (Japanese barberry), *403*
Beta (β)-carotene, 57, *59*, 341
Beta (β)-galactosidase, 306, 307
Beta (β)-glucose, 52, *52*, 54
Beta (β)-oxidation, 186
Beta (β) particles, 28
Beta (β)-pleated sheet, 63–65, *64*, 111, *112*
Beta (β)-tubulin, in microtubules, 97, *98*
Beta chains, hemoglobin, 66, *66*
Bias, in scientific study, 18
Bicarbonate ions, 41
Bidirectional DNA replication, 272–273, *274*
Big Bang theory, 20
Bighorn sheep (*Ovis canadensis*), directional selection and, 420–421, *421*
Bile salts, 58
Binary fission, 221, *221*, 450
Binding, in facilitated diffusion, 119–120, *120*
Binomial system of nomenclature, 9
Biochemical techniques to study cells, 78–80
Biochemists, 25, 166, 260
Biodiversity. *See* Biological diversity
Biofilm, 135
Biogeography, 402–404, *404, 405*
Bioinformatics, 333, 337–338, *339*, 349–350
Biological computing, 333, 337–338, *339*, 349–350
Biological diversity (biodiversity)
    cultural evolution and, 479
    Darwin's observations on, 392

Carbonyl group, 48, *49*, 51, 193
Carboxyl group, 48, *49, 62*
Carcharodontosaurus, 459
Carcinogens (cancer-causing agents), 90, 297, 300
Cardiac muscle, *1*
  cells and gap junctions, 129
Cardiovascular disease, cholesterol and, 124
Carnivora (order), 9
Carotenoids, 57, *59, 69,* 96
  in chloroplasts, 194, 195
Carpals (bones in wrist), 400, *401*
Carpels, plant development studies, 385, *385*
Carrier (genetic), 250, 363
Carrier proteins, 115, 118–123
  in active transport, 120–122, *121, 122*
  in cotransport, 122–123, *123*
  in facilitated diffusion, 119–120, *120*
Carrier-mediated transport, 115, 120–123
  active transport and, 120–123, *121, 122, 123*
  cotransport systems in, 122–123, *123*
  facilitated diffusion and, 118–120, *119, 120*
Carrots (*Daucus carota*), 370, *371*
Cartilage, 54
Caspases, in apoptosis, 96, 382
Cat(s)
  autosomes in, 248
  domestic (*Felis catus*), 9, *9*
  homology in, 400, *401*
  saber-toothed (*Smilodon*), 463
  variegation in, 251, *251*
Catabolism/catabolic pathways, 155, 158, 171, 172, *202*
Catabolite activator protein (CAP), 309, *311, 312*
Catalase, 94, 161, *161*
Catalysts, 161, 450, 451
  biological, 50
  See also Enzyme(s)
Catastrophic events on Earth, 458, 460
Catecholamines. *See* Epinephrine
Cation(s), 33–34, 42
Cattle, 253
Cauliflower, *393*
Cave paintings, 477
Cdks (cyclin-dependent kinases), 222, *222*
cDNA, 265, 327–328, *328,* 335
cDNA library, 327–328
Cebus capucinus (white-faced monkey), *470*
Cech, Thomas, 290
Cell(s), 2–3, 18, 73–102, 106–130, 134–148
  aging of, 275
  cancer, *386,* 386–387
  diversity of, 369
  energy use and, 14, *14*
  energy use and enzymes, 160–161
  enzymatic activity regulated by, *165,* 165–166, *166*
  eukaryotic, 3, 11, 80–102
  in levels of biological organization, 6, *7*
  malignant transformation of, 386 (*see also* Cancer)
  origin of, *450–454, 450*–455
  origin of name, 76
  prokaryotic, 3, 11, 80–81
  signaling, 166
  structure, 3, 73–102, 106–130
  surface area–to-volume ratio of, 75, *76*
  totipotent, 370–371, *371*
  See also Animal cells; Eukaryotes/eukaryotic cells; Plant cells; Prokaryotes/prokaryotic cells

Cell, structure, 73–105, *89*
  coverings, 102, *102*
  cytoskeleton, 97–101, *97–101*
  eukaryotic, 3, 11, 80–102
  membranes, 57, 58, 81, 84, *84, 85* (*see also* Plasma membrane (cell))
  methods of study of, 76–80, *77, 79, 80*
  nucleus, 3, 18, *82–85,* 84–88, *88* (*see also* Nucleus (cell))
  organelles, 3, *7,* 88–97, *89–96* (*see also* Organelles)
  plasma membrane, 106–130
  prokaryotic, 3, 11, 80–81, *81*
  size, 74–76, *75*
  theory, 2, 74
  See also Cell walls
Cell adhesion molecules, 106, 127
Cell biologists, 74, 87, 88, 108–109, 111–112, 135, 214, 219, 223
Cell coat (glycocalyx), 102
Cell communication, 21, 134–151
  evolution of, 147–148
  overview, 135–136, *136*
  reception, 137–140, *138*
  responses to signals, 136, *136,* 145–147, *147, 148*
  sending signals, 136–137, *137*
  signal transduction, 136, *136,* 140–145, *141– 144, 148*
Cell cortex, 99–100
Cell coverings, 102, *102*
Cell cycle, *215,* 215–221, 371–372
  regulation of, 221–223, *222*
Cell cycle checkpoints, 222
Cell cycle control system, 222
Cell death
  programmed (apoptosis), 21, 95–96, 102, 382– 383
    telomere shortening and, 275–276
  uncontrolled (necrosis), 95
Cell determination, 369, 370
Cell differentiation, 368, 369–374, *370*
  differential gene expression, 369–370
  first cloned mammal, 371–372
  stem cells, 372–373, *374*
  totipotent nucleus, 370–371, *371*
Cell division
  in eukaryotic cells (*see also* Cell cycle)
    cancer and, 222, 223
    by meiosis, 212–215
    microtubule assembly/disassembly in, 99
    by mitosis, 215–221, *220*
    regulation of, 221–223
  signaling cascade, 144
Cell extract, 79
Cell fractionation, 78–80, *80,* 109
Cell junctions, 127–130, *129, 130*
  membrane proteins and, *127,* 127–130, *128*
  See also specific type
Cell lineages
  of *Caenorhabditis elegans, 381,* 381–382
  of vertebrates, 369, *370*
Cell membrane. *See* Cell(s); Plasma membrane (cell)
Cell plate, 220, *220*
Cell proteins, functions of, 113–114, *114*
Cell recognition, membrane proteins in, 127
Cell signaling, 8, 114, 115, 134–151
  area of research, 8

  definition of, 135
  integrins in, 102
  nucleotides in, 68
  overview of, 135–136, *136, 137*
  reception, 135–136, *136,* 137–140
  responses to signals, 136, *136,* 145–147, *147, 148*
  sending signals, 136–137
  termination, 136, 147
Cell surface proteins, 8
Cell theory, 2, 74
Cell totipotency, 370–371, *371*
Cell walls, 81, 102
  eukaryotic, 81, *82, 84*
  plant, *82, 84,* 102, *102*
  prokaryotic/bacterial, 80, *81*
  turgor pressure and, 118, *118*
Cell-free systems, 284
Cell-specific proteins, 369
Cellular blastoderm, 375
Cellular respiration, 14, *14, 95,* 153, 156, 172, *187*
  aerobic, 172, *173,* 173–185, *185,* 188
    electron transport and chemiosmosis, *201*
  anaerobic, 172, 186–188
  in mitochondria, 94–96, *95*
  See also ATP
Cellulose, 50, 52, 53–54, *55,* 102
Cenozoic era, 461–463, *463*
  mammal evolution in, 467
Central nervous system (CNS), 358. *See also* Brain
Centrifuge/centrifugation, 79, *80,* 267–268
Centrioles, *83, 89,* 98–99, *99*
  in meiosis, 223
  in mitosis, 217–218
Centromere(s), 217
  in meiosis, 223
  in mitosis, 217, 218, *218*
Centrosome, 98–99
Cepaea nemoralis (garden snail), *412*
Cephalopods, 455
Cesium chloride, 267
Cetaceans/Cetacea (order), 397–399, *399. See also specific type*
CFTR (cystic fibrosis transmembrane conductance regulator) protein, 350, 358
Chain, Ernst Boris, 16
Chain termination method, for automated DNA sequencing, 331, 333, *334*
Chains, of hydrocarbons, *46, 47*
Challenger (space shuttle) explosion, 161, *161*
Chance
  random events in evolution, 395–396
  in scientific discovery, 16
Channel proteins, 115, 118–120
Channels. *See* Ion channels
Chaperones, molecular, 66, 91, 291, 313
Character displacement, 394
Characteristics, polygenic control, 419
Characters
  heritable, 235, *236,* 255
  See also Trait(s)
Chargaff, Erwin, *263,* 264
Chargaff's rules, 264, *264,* 265
Charges
  in ionic bonds, 33–34
  partial, in covalent bonds, 33
  in van der Waals interactions, 36
  in water molecules, 37

Coenzyme Q (CoQ/ubiquinone), 179
Cofactors of enzymes, 163
Coffee, *41*
Cognitive abilities, 359
Cohesin, 217, *218*
Cohesion, of water molecules, 37
Colchicine, for karyotyping, 347
Colewort, *393*
Collagen(s), 62, 66, *66*, 102, *102*
Collards, *393*
Colobus monkeys, 469
Colonies
    cloned, 326
    replica, 327
Colorblindness, 250, *250*, 251
Colorectal cancer, 274
Combined DNA Index System (CODIS), 339
Combs, inheritance of, 254, *254*
Communication, information transfer, 6–8
Community biodiversity. *See* Biological diversity
Community (communities), 6, *7*. *See also*
    Ecosystem(s)
Comparative anatomy, evolution and, 400–402,
    401, 402, 403
Comparative genomics, 334, 350
Competitive inhibition, of enzymes, 166–167,
    167
Complementary base pairing
    anticodon and mRNA, 283
    antiparallel strands, 286
    DNA, 265–266, *267*, 268, 274 (*see also*
        Complementary DNA)
    tRNA, 289
Complementary DNA (cDNA), 265, 327–328,
    328, 335
Complementary DNA (cDNA) library, 327–328
Complexes I–IV, in electron transport chain, 179,
    181, *181*
Compound microscope, 76
Compounds. *See* Chemical compounds
Computers
    searching databases of polypeptides, 67
    simulations on, 17
    use in bioinformatics, 333, 337–338, 339, 349–
        350
Concentration gradient
    in active transport, 120–123, *121, 122, 123*
    in facilitated diffusion, 118–120, *120*
    free energy changes and, *156,* 156–157
    in simple diffusion, 116, *116*
Conclusions, in scientific method, 16, 18
Condensation reactions, 50, *50*, 56, 62–63
Condensin, 214
Confocal fluorescence microscope, 73, 78
    images, 98, 101
Conformation, of proteins, 63, 66–67, 163, 164
Confuciusornis, 460
Conifers, 458
Connexin, 129, *129*
Consanguineous matings, 363
Conservative replication, 266, 268, *268*
Conserved sequences, in rodents and humans,
    337
Constitutive genes, 306, 309, 313, 370
Consumers (heterotrophs), *14,* 15, 207
    first cells as, 452–453, 454
Continental drift, 402–404, *404, 405*
Contractile vacuoles, 93

Contributions to the Theory of Natural Selection
    (Wallace), 393–394
Control group, in controlled experiment, 18, *18,*
    19, *19*
Controlled experiments, in scientific method,
    17–18
Controlled matings, 346
Convergent evolution, 400, *402*
Cooling, evaporative, 38
Copper, 26
    as cofactor, 163
Coral(s)/coral reefs
    endosymbiotic relationships, 454
    in Ordovician period, 455
Corallus caninus, 394
Corepressor, 309, *310*
Corn (maize/*Zea mays*)
    DNA of, *264*
    genetic control of development and, 385
    genetically engineered, 341, *342*
    transposons discovered in, 298
Correns, Carl, 242
Cortisol (hydrocortisone), 58, *59*
    reception in cell signaling, 140
Cosmid cloning vectors, 324
Cotransport systems, 122–123, *123*
Cotton, 224
    cellulose and, 53
Counseling, genetic, 363
Coupled reactions, 157, 164–165
Coupled transcription and translation in bacteria,
    293, *294*
Courtship patterns/rituals, 8
    behavioral isolation and, *431,* 431–432
    See also Mating
Covalent bonds, 31–33, *32, 34*, 35
    of carbon, 32, *33,* 46, 46–47
    nucleotides and polymers, 263–264, *264*
Covalent compound, 32
Cow milk, *41*
Cows
    and cellulose, 54
    transgenic, 340, *341*
Crabgrass, 205
Crabs, 215
Crassulaceae, carbon dioxide fixation by, 205–206
Crassulacean acid metabolism (CAM pathway),
    205–206
Crataegus mollis (downy hawthorn), *403*
Cretaceous period, *404, 456,* 459–461
Cri du chat syndrome, 355
Crick, Francis
    DNA chemical composition, 265
    DNA structure, *263,* 263–264
    double-helix model, 270
    gene expression, 279, 283, 284, 285, 288
    work with Watson, 8
Crime scenes, DNA analysis at, 331, 339
Crinoids, *398*
Crista(e), *82, 83,* 95, *95*
Crocodiles, 459
    mass extinction of archosaurs and, *443*
Cro-Magnons, 477, *477*
Crop plants, 204
Crops
    cultivating, and development of agriculture,
        478
    genetically modified, 340–342

Crossing-over, 224–225, 247, *247*
    double, 248, *249*
    frequency of, gene linkage and, *247,* 247–248
Cross-pollination, 431
Crown gall bacterium (*Agrobacterium tumefa-
    ciens*), 341
Crustaceans/Crustacea (subphylum), fossils from
    Cambrian period, *457*
Cryptochromes, 138
Cultural change, of humans, 478–479
Cuticle, plant, 11
Cyanide
    effect on aerobic respiration, 181
    effect on enzymes, 167
Cyanobacteria, *452,* 453, 455
    evolution of aerobes and, 453
    evolution of chloroplasts and, 454
Cycads, 458
Cyclic adenosine monophosphate (cyclic AMP/
    cAMP), 68, *70*
    CAP regulation of *lac* operon and, 309, *311, 312*
    cellular slime mold aggregation, 135, *135*
    in enzyme regulation, 166, *166*
    signal responses, 146, 147, 148
    signal transduction, *141,* 141–143, *142, 143*
Cyclic AMP–dependent protein kinase, 166, *166*
Cyclic electron transport, 200, *200*
Cyclic guanosine monophosphate (cGMP), 68,
    145
Cyclin(s), 222, *222*
Cyclin–Cdk complex, *222,* 222–223
Cyclin-dependent kinases, 222, *222*
Cyclopentane, *46*
Cynognathus, 405
Cyprinodon (pupfish), *433, 433*
Cysteine, *49, 61*
Cystic fibrosis, 115, 253–254, 331, 338, 339, 358,
    358
    defective ion transport, caused by, 358
    mouse model, 350
    prenatal genetic testing, 362
Cystic fibrosis transmembrane conductance
    regulator (CFTR) protein, 350, 358
Cytochrome a_3, 181
Cytochrome *c*, 96, 179, 406, *407*
Cytochrome *c* oxidase, 179, *181*
Cytochrome oxidase, 167
Cytochromes, 160, 179, 181, *181*
Cytogenetics, 347
Cytokinesis, 215, 220, *220,* 225, *227*
Cytokinin(s), 223
    Ti plasmid affecting production of, 341
Cytoplasm, 81, 88, 120, 229. *See also* Organelles
Cytosine, 67, *68,* 263, *264*
    bonding of, to guanine, 265–266, *267*
Cytoskeletal elements, 114
Cytoskeleton, *73, 74, 89, 97,* 97–101
Cytosol, 81, *108,* 117
    aerobic respiration in, 174, 175

da Vinci, Leonardo. *See* Leonardo da Vinci
Daffodil flowers, transgenic, *342*
Dalton (atomic mass unit/amu), 28
Dalton, John, 28
Danielli, James, 108–109
Danio rerio, 375, *376*
Dark reactions, 197. *See also* Carbon fixation
    reactions

Dark-field microscopy, 77, *77*
Darwin, Charles, 11–13, 390, *390*, 392–396, 408–409, 466
   natural selection, 393, 396, 409, 418
   sexual selection, 437
   See also On the Origin of Species
Data, in process of science, 15
Databases of DNA sequences, 333, 337–338, 339, 349–350
Dating fossils, 399–400
Daubautia scabra, 442
Daughter cells
   from cytokinesis/mitosis, *219*, 220, 221, *229*
   DNA replication, 266
   from meiosis (haploid cells), 224, *227*, 228, 229, *229*, 230
Davson, Hugh, 108–109
Davson–Danielli "sandwich" model, *108*, 108–109
ddATP (dideoxy adenosine triphosphate), 333, *333*
Deamination, 185–186
Death Valley region, allopatric speciation, 433
Decarboxylations, 174, 175, *178*
Decay of isotopes, 28, *28*
Deception in science, 21
Decomposers, *14*, 15
Deductive reasoning, 16
Deep-sea vents, 163, 330
Defense, internal. *See* Immune response/immunity
Dehydration synthesis (condensation reaction), 50, *50*
Dehydrogenations, 174
Deinonychus, 460
Deletions, 349, 354–355, *355*
   Angelman and Prader–Willi syndromes caused by, 349
   mutations caused by, 298, *299*
Delta (Δ), 155
Denaturation, of protein, 67
Denaturation of DNA, 328, *329*
Density, of ice and water, 38
Density gradient centrifugation, 79, 267–268, *269*
Deoxyribonucleic acid (DNA). *See* DNA
Deoxyribose, 51, *51*, 67, 263, *264*
Depolymerization of microtubules, 219
Descent of Man, The (Darwin), 466
Deserts, 206
Desmosomes, 127, *127*
Desmotubule, 130, *130*
Detergents, amphipathic properties of, *107*, 107–108
Detoxification, 90
Deuterium, 28
Development, 369
   as characteristic of life, 3
   definition of, 431
   evolutionary novelties and, 439–440
   genetic regulation of, evolution and, 404–405, 439
   mammalian, 383–385
   mosaic, 381–382
   plant (*see* Plant(s))
   regulative, 383
   See also Developmental genetics
Developmental biology, evolution and, *403*, 404–405

Developmental genetics, 368–389
   cancer and cell development, 386–387
   cell differentiation and nuclear equivalence in, 369–374
   genetic control of, 374–386
   use of model organisms (*see* Model organisms)
Devil's Hole pupfish (*Cyprinodon diabolis*), 433
Devonian period, *456*, 457
DeVries, Hugo, 242
Diabetes mellitus
   insulin-dependent, 338
   stem cells and, 373
Diacylglycerol (DAG/diglyceride), 56, 143, *144*
Diatryma, 462, 463
Dictyostelium discoideum, 135, 135
Dideoxynucleotide, DNA sequencing and, 331, *333*
Differential centrifugation, 79, *80*
Differential gene expression, 368, *368*, 369–370, 374–386
   evolution and, 404–405
Differential mRNA processing, 318, *318*
Differential reproductive success, natural selection and, 394
Differential-interference-contrast (Nomarski) microscopy, 77, *77*
Differentiated cells, 369, *370*
   from stem cells, 372–373
   See also Cell differentiation
Difflugia, 5
Diffuse large B-cell lymphoma, 337
Diffusion, 115–118, *116, 117, 118*
   as exergonic process, 156, *156*
   facilitated, 118–120, *119, 120*, 122
Digestive enzymes, 92–93, *93*
Dihybrid cross, 240–241, *242, 245*
Dihydroxyacetone, 51, *51*
Dihydroxyacetone phosphate, 174, *176*
Dimers
   microtubule, 97, *98*
   regulatory protein, 316
Dimethyl ether, *47*
Dimorphism, sexual, 474
Dinosaurs, 459–461, *459–461*
   allometric growth, 440
   bird evolution and, 462, *462, 463*
   extinction of, 442, *443*, 460
   footprints of, *398*
Dinucleotides, 68
Dionaea muscipula (Venus flytrap), 4, *4*
Dipeptide, 62–63, *63*
Diploid cells, 224, 225
   zygote as, 224, 230, *230*
Diploid parents, 435
Diploid sporophytes, 230
Directed evolution (in vitro evolution), 451, *451*
Directional selection, *419*, 420–421, 439
   bighorn sheep (*Ovis canadensis*), 420–421, *421*
Disaccharides, 50, 52, *53, 69*, 207
Discrimination, genetic, 364
Disease
   genetic contributions to, 418
Disomy, 351
Dispersive replication, 266, 268, *268*
Disruptive selection, *419*, 421, 437
Dissociation of compounds, 34, 40
Dissolved substance, 34, 37, 42
Disulfide bonds/bridges, 65, 66

Divergence, 390, *407*, 407–408. *See also* Evolution
Diversification of species, 440–442
Diversity
   biological (biodiversity), 11, *12, 13* (*see also* Biological diversity)
   cell differentiation, 369
   of cells, *75*
   early evolution of organisms, 455–463
   evidence of, 390
   metabolic, and photosynthesis, 206–207
Division, meiotic, 351, *352*
DNA, 3, 25, 67, 260–278
   bacterial transformation and, *261*, 261–262, 323
   base compositions of, 264, *264*
   base pairing in, 265–266, *267*, 268, 274
   in cell organization, 85
   complementary (cDNA), 265, 327–328, *328*, 335
   discoveries related to, *263*
   helicases, 270, *270*
   as hereditary material, 261–263
   in information transfer, 6–8, *8*, 282–285, *282–285* (*see also* Gene expression)
   mismatch repair, 274
   mitochondrial (*see* Mitochondrial DNA)
   mutations, 417 (*see also* Mutation(s))
   as nucleic acid class, 67–68, *69*
   nucleotide subunits of, 263–264, *264*, 265
   origin of cells and, 450–451
   packaging of, 212–214, *213, 214*
   radioactive decay and, *28*
   recombinant (*see* Recombinant DNA technology)
   repair enzymes in, 274
   replication of, 211 (*see also* DNA replication)
   segments, 350
   sequencing (*see* DNA sequencing)
   strands, 270–273
   structure of, *8*, 263–266, *265–267*, 270
   supercoiling, 270
   synthesis, 271
   transcription of, *279*, 279–280, 282, *283*, 285–288, *286, 287* (*see also* Transcription)
   translation of, 282–283, *283, 284*
   virus, *262*, 263
   X-ray diffraction of, 264, *265*
DNA analysis, 330–333
   at crime scenes, 331, 339
   fragment detection, 331
   gel electrophoresis in, 330, *330*, 331, 333, 339
   nucleotide sequencing, 331–333, *333, 334*
   restriction fragment length polymorphisms, 331
DNA chips, 335
DNA cloning, 323–330
   cloned DNA sequences, 324–328
   polymerase chain reaction (PCR), 328–330
   restriction enzymes in, 323, *324, 325*, 331
   vector(s), in recombinant DNA technology, 324
DNA databases, 333, 337–338, 339, 349–350
DNA fingerprinting, 339, *339*
DNA helicases, 270, *270*
DNA ligase, *270, 272, 273*, 323
DNA methylation, gene inactivation and, 314
DNA microarray(s), 335–337, *336*
DNA Pol III, 271
DNA polymerase(s)
   in DNA replication, *270*, 271, *271, 272, 273*, 274
   heat resistance, 328

Gamma rays, 192, *192*
Gamma-aminobutyric acid (GABA), 140
Gap genes, 378, *378*, 379
Gap junctions, 128–129, *129*
Gap phases, in cell cycle, 215, *215*
Garden pea, 419
Garden snail (*Cepaea nemoralis*), *412*
Garrod, Archibald, 280
Gas particles, 116
Gasoline, energy from, 154
Gated channels, in plasma membrane, 115, 119, *119*
Gazelle, 478
Geese, Hawaiian (nene/*Branta sandvicensis*), *433*
Gel electrophoresis, 421
   in recombinant DNA technology, 330, *330*, 331, 333, 339
Gender, determination of, 248–252, *249*
Gene(s), 8, 85, 211, 212
   alleles, 412
   constitutive, 306, 309, 313, 370
   development and, 368–387
      cancer and, *386*, 386–387
      differentiation and, 369–374
      evolution and, 404–405
      regulation and, 368, 369–370, 374–386 (*see also* Differential gene expression)
   environmental interaction, 256
   enzyme synthesis controlled by, 165–166
   evolution of definition, *296*, *297*
   expression of (*see* Gene expression)
   function, 335
   homologous, 382
   identification of, 349–350
   inactivation of, in knockout mice, 335
   linked
      independent assortment and, 246, *246*
      linear order of, 247–248, *248*
   mapping, 247–248, *248*
   as Mendel's hereditary factors, 236
   multiple copies of, 314–315
   mutation of, 298–300, *299* (*see also* Mutation(s))
   one-gene, one-enzyme hypothesis and, 280–281, *281*
   regulation of (*see* Gene regulation)
   structural, 306
   X-linked, 250–252
      abnormalities involving, 249–251, *250*
      dosage compensation and, 250–252, *251*
   See also DNA
Gene amplification, 315, *315*
Gene chips, 335
Gene expression, 279–303
   cancer and, *386*, 386–387
   differential, 368, 369–370, 374–386 (*see also* Differential gene expression)
   DNA-to-protein information flow, 282–285, *282–285*
   environment affecting, 256
   gene–protein relationship, 279, *280*, 280–281, *281*
   mutations and, 298–300, *299*
   regulation of, 304–305 (*see also* Gene regulation)
   transcription, 280, 285–288, *286–288*, 294, 295 (*see also* Transcription)
   translation, 288–291, *289–294* (*see also* Translation)

variations in organisms, 292–298, *297*
   See also Gene regulation
Gene flow, 418
Gene interaction, *254*, 254–255, *255*
   with environment, 256
Gene linkage, *246*, 246–248
Gene loci, 239, *239*
   linked (*see also* Linkage/linked genes)
      independent assortment and, 246, *246*
   multiple alleles for, 253, *254*
Gene pool, 9, 13, 413, 429, 432
Gene regulation, 304–321
   in bacteria, 305, 306–312
   developmental, evolution and, 404–405
   in eukaryotic cells, 305, 312–319, *313*
   See also Gene expression
Gene silencing, 335
Gene targeting, 335, 350
Gene therapy, 338, 360, *360*
General transcriptional machinery, 316
Generation(s)
   alternation of (*see* Alternation of generations)
   designations for, *245*
   filial (F₁/F₂), *245*
   parental (P), *245*
Generation time, of eukaryotic cells, 221–222
Generic name. *See* Genus (genera), in taxonomic classification system
Genetic code, 8, *8*, 283–285, *284*, 405–408
Genetic constitution of organisms, 240
Genetic control of development, 374–386
   Arabidopsis, 375, *376*, *385*, 385–386
   Caenorhabditis elegans, 375, *376*, 380–382, 380–383
   Drosophila, 368, 368–369, 375–380, *376*, *380*
   model organisms to study, 374–375
   mouse, 375, *376*, *380*, *383*, 383–385, *384*
Genetic counseling, 363
Genetic counselors, 363
Genetic discrimination, 364
Genetic disorders
   chromosome abnormalities and, *351*, 351–355, *355*
   family pedigrees in identification of, 348–349, *349*
   gene therapy for, 360, *360*
   inheritance of
      autosomal dominant, 358–359
      autosomal recessive, 356–358
      X-linked recessive, 359
   mouse model for, 350, 355, 357, 358, 359
   pleiotropy in, 253–254
   single-gene mutations and, 356–359
   testing and counseling and, 361–363
Genetic diversity. *See* Biological diversity
Genetic drift, 417–418, *418*, 433, 434
Genetic engineering, 322
   animals, 340–342
   plant, 341–342, *342*
   protein conformation study and, 67
   safety concerns about, 342–343
Genetic equilibrium
   conditions for, 415
   Hardy–Weinberg principle of, 413–415, *414*
Genetic inheritance, compared to epigenetic inheritance, 314
Genetic material
   DNA as, 261–263, 266

protein, thought to be, 261
   storage of, 260, 265–266
   in viruses, 263
Genetic mutation, 5
   in transport proteins, 115
Genetic polymorphism(s), 331, 421–422, *422*
   balanced, 422–424, *423*
   in DNA typing, 339
   plants, invertebrates, and vertebrates, 421, *421*
Genetic probes, 327, *328*
Genetic recombination, 225. *See also* Recombination
Genetic screening, 356, 362–363, 364
Genetic "shotgun," in transgenic plants, 341
Genetic testing, 361–363
   for cystic fibrosis, 362
   DNA technology and, 338
   ethical issues and, 363–364
   for Huntington's disease, 359
   individualized, 338
   for phenylketonuria, 356, 363
   prenatal, *361*, 361–362
   reproductive decisions, 363
Genetic variation, 234, *412*, 413, 421–425
   frequency-dependent selection and, 422–424, *423*
   gene flow and, 418
   geographic, 422, *423*
   heterozygote advantage and, 422, *423*
   mutation and, 395, 417
   natural selection and, 394, *394*, 412
   neutral, 424
   polymorphisms and, 421–422, *422*
   in populations, 421–425
   sexual reproduction and, 5, 223–224, 225
Genetically modified (GM) crops, 340–342
Geneticists, 238, 327, 335, 341, 348, 361
Genetics, 234
   chromosome abnormalities and, *351*, 351–355, *355*
   and diseases, 356–359, 418
   DNA, carrier of genetic information, 260–278 (*see also* DNA)
   DNA technology, 322–345 (*see also* DNA technology)
   ethical issues in, 21, 363–364
   evolution and, 404–408, *406*, *407*, 412 (*see also* Population genetics)
   gene therapy, 360, *360*
   human, 346, 347–350
   human genome, 346–367 (*see also* Human genome)
   milestones in study of, *350*
   problem-solving exercises in, *245*
   testing and counseling, 361–363
   See also DNA; Heredity
Genetics problems, solving, *245*
Genome(s), 211, 212, 324–327
   comparison of mouse and human, 350
   DNA methylation and gene inactivation, 314
   mitochondrial DNA (mtDNA), 349
   sequencing, 338, *346*, 349–350
   See also Gene(s); Human genome
Genome analysis, 337
Genomic imprinting, 314, 348–349
Genomic library, 324–327, *327*
Genomics, 333–338
   areas of, 333–334

Hydrogen, 26, *26*, *27*, 29
  covalent bonding, 31–32, *32*, *33*
  isotopes of, 28, 29
  reactions with oxygen, 161, *161*
  in redox reactions, 159–160, *160*
Hydrogen acceptors, in redox reactions, 159–160, *160*
Hydrogen bonds, *35*, 35–36
  base pairing in DNA and, 265–266, *267*
  structure of protein and, *64*, 64–66, *65*
  water and, *37*, 37–38, *38*
Hydrogen ion concentration, 40, *40*
Hydrogen peroxide, 94, 161, *161*
Hydrogen sulfide, 449–450
Hydrogenation, 57
Hydrolases, 163, *163*
Hydrolysis, 50, *50*
  ATP, 158–159, *159*
  of disaccharides, 52, *52*
Hydronium ions, 40
Hydrophilic molecules/interactions, 38, 48
  amino acids, *60*, 62
  carbohydrates, 51
  lipids, 57, *58*
  phospholipid, 107, *107*
Hydrophobic molecules/interactions, 38, 48
  amino acids, *60*, 62
  lipids, 56, 57, *58*
  phospholipid, 107, *107*
  structure of protein and, 65, 66
Hydrosphere, 6
Hydrothermal vents, 449–450
Hydroxide ions, 40
Hydroxyl group, 48, *49*, 51
  in phospholipids, 110
5-Hydroxytryptamine/5-HT (seratonin), 146
Hydroxyurea, as treatment for sickle cell anemia, 357
Hylobates (gibbons), *468*, 470, 471, *471*
Hypertonic solution, *117*, 117–118, *118*, *119*
Hypothalamus, 142
Hypothesis, in scientific method, 15, 17, *18*, *19*, 20–21
Hypotonic solution, *117*, 117–118, *118*, *119*

Ice, and hydrogen bonding, 38, *39*
Ice ages, 463
Iceland, population bottlenecks, 418
Ichthyosaurs, 459, *459*
igf1r locus, 384–385
IGF2 gene mutation, 305, *305*
'I'iwi (*Vestiaria cocciniea*), *13*, 441
Imaginal discs, 375, *377*
Imago, 375
Immune response/immunity, 253
Immunofluorescence, of *Drosophila*, 368, *368*, *379*
Imprinting, genomic, 348–349
In situ hybridization, fluorescent (FISH), 347–348, *348*
In vitro, protein study, 66
In vitro evolution (directed evolution), of RNA, 451, *451*
In vitro fertilization (IVF)
  embryos for, stem cells from, 373
  preimplantation genetic diagnosis and, 362
In vivo, protein study, 66
Inborn errors of metabolism, 280, 356
Inborn Errors of Metabolism (Garrod), 280

Inbreeding, and genotype frequency, *416*, 416–417
Inbreeding depression, 416
Incomplete dominance, *252*, 252–253
Independent assortment, Mendel's principle of, *237*, 241–242, *243*, 246–247
  linked genes as exception to, *246*, 246–247
Independent events, 244
Index fossils, 399, 455
Indirect active transport, 122–123, *123*
Individualized genetic testing, 338
Indricotheres, 462, *463*
Induced fit, in enzymatic binding, 162, *162*
Inducer, 308
Inducible operon, 308, *312*
Induction, 382
  development of *Caenorhabditis elegans* and, 382, *382*
  of *lac* operon transcription, 307, *307*
Inductive reasoning, 16
Industrial Revolution, 479
Infection cycle of RNA tumor virus, *297*, 297–298
Inflammation, 58
Influenza A, 339
Information transfer, 6–8, 282–285
  origin of cells and, 450
  reverse transcriptase, 297–298, *298*
  as theme of life, 2
  transcription, 280, 282, 285–288, *286–288*, 294, 295 (see also Transcription)
  translation, 282, 288–291, *289–294* (see also Translation)
Ingestion, by phagocytes, 123, *125*
Ingram, Vernon, 281
Inheritance, 234–259
  blending, 235
  chromosomal basis of, 246–252
  chromosome theory of, 242, 246
  dominant, 236, *237*, 358–359
  epigenetic, 349
  evolution and, 394–395
  Mendel's principles of, 234–242, *236*, *237*
  in monohybrid cross, 238–240, *244*
  patterns of, 260–261, 348–349, *349*, 394–395
  pedigree analysis and, 348–349, *349*
  polygenic, 255–256, *256*
  probability and, 234–244, *244*
  problem-solving exercises in, *245*
  recessive, 236, *237*, 356–358
  See also Heredity
Inherited variation, 437
Inhibition, feedback, 312
Inhibition of enzymes, 165–167
  antibiotic mechanism of action and, 167, *167*
  feedback, 165–166, *166*
Initiation
  of transcription, 286, *287*, 315
  of translation, 290, *291*
Initiation complex, 290, *291*
Initiation factors, 290, *291*
Initiator tRNA, 290, *291*
Inner mitochondrial membrane, 95, *95*
Inorganic compounds, 26, 45. See also *specific type*
Inositol trisphosphate (IP$_3$), 143, 144, *144*
Insect(s)
  in Mesozoic era, 459
  in Paleozoic era, 458

Insecticide(s), and transgenic plants, 343
Insectivorous lizards, 458, 459
Insertions, mutations caused by, 298
Inside-out signaling, 145
Instars, 375
Institute for Cancer Research, 371
Insulin, *62*
  in cell signaling, 135–136, 138
  genetically engineered, 338
  See also Diabetes mellitus
Insulin-like growth factors (IGFs), 305, 384
Insurance, genetic discrimination and, 364
Integral membrane proteins, 110–111, *111*, 112, *112*, 129, *129*
Integrative biology, 21
Integrative systems biology, 21
Integrins, 102, *102*, 114, *114*, 145
Intelligence, 256, 359
Interbreeding, 429, *429*, 431
  in hybrid zones, 437–438
  plants, *435*
  Porto Santo rabbits, 434
  reproductive isolating mechanisms preventing, *430*, 430–432, *431*
Intercellular junctions
  membrane proteins and, *114*
  See also Cell junctions
Interkinesis, 225
Intermediate filaments, *89*, 101, *101*
Intermembrane space, 94–95, *96*, *194*
  mitochondria, 182, *183*
International Human Genome Sequencing Consortium, 349
Interphase, 251
  meiosis and, 225, *226–227*
  mitosis and, 215–216, *216*, 221
Interrupted coding sequences, 294, 295
Interspecific hybrid, 431–432, 435
Interspecific zygote, 430
Intestinal lining cell, 128, *128*
Intestine(s), of *Caenorhabditis elegans*, 381
Intracellular receptors, 140, 144–145, *145*
Intracellular signaling agents, 141–143
Intron(s), 294, 295, *295*, *296*, 318, 327–328, *329*
Intron excision, 296
Invagination (infolding) of plasma membrane, 454
Inversions, chromosomes, 354, *355*
Invertebrates, genetic polymorphism, 421
Iodine, 26
Iodine-132, in fossil dating, 399
Ion(s), 33
  membrane permeability and, 115
Ion channel–linked receptors, 138, *139*, 140–141
Ion channels, 119
Ion pumps, 122, *122*
Ionic bonds, 33–35, *34*, *35*
  in protein structures, 65, *65*, 66
Ionic compound, 34, *35*
Ionic functional groups, 48, *49*
Ionization, of water, 40
Iridium, in rock sediments, mass extinction and, 460
Irises, hybrid inviability, 431
Iron, *26*, 36, 450
  as cofactor, 163
Iron–sulfur proteins, 179
Iron–sulfur world hypothesis, 449

Irreversible inhibition, of enzymes, 167
Irrigation, 478
Isla Daphne Major, droughts, 420
Islands
    allopatric speciation and, 434–435
    evolutionary trends on, 476
    vacant adaptive zones on, 441
Isobutane, 46, 48
Isocitrate dehydrogenase, 180
Isolation
    allopatric speciation and, 433–434
    reproductive, 430, 430–432, 431, 432, 436
Isoleucine, 61, 62
    bacterial synthesis, 166
Isomerases, 163
Isomers, 47, 47–48
Isopentane, 46
Isoprene units, 57, 58, 59, 69
Isotonic solution, 117, 117, 118
Isotopes, 28, 28–29

Jacob, François, 306, 307–308, 308
Jaguar (Panthera onca), 25
Jamoytius ostracoderm, 458
Japanese barberry (Berberis thunbergii), 403
Java man, 475
Johanson, Donald, 474
Jurassic period, 456, 459–460

Kaibab Plateau, Grand Canyon, 434
Kaibab squirrel (Sciurus kaibabensis), 434, 434
Karyotype, 219, 347–348, 348
    genetic screening and, 362–363, 364
    preparation of during metaphase, 225, 347
    X0 karyotype in Turner syndrome, 249, 351, 354
    XXX karyotype, 351
    XXY karyotype, 249, 351, 354
    XY karyotype, 248, 249, 249
    XYY karyotype, 351
Keating, Mark T., 1
Kelvin units, 155
Kenyapithecus, 470
Ketones, 48, 49, 51, 52
Kew primrose (Primula kewensis), 435–436, 436
Khoisans, 478
Khorana, H. Gobind, 284
Kidney
    cells, 94
    mitochondria in, 185
Kilocalorie (kcal), 153
Kilojoule (kJ), 153
Kinases, 163
    in posttranslational control, 319
Kinesin, 98, 98, 225
Kinetic energy, 38, 153, 153
Kinetochore(s)
    in meiosis, 228
    in mitosis, 217, 218, 218–219, 219
Kinetochore microtubules (chromosomal spindle
        fibers), 219, 219, 220
King, Thomas J., 371
Kingdoms, in taxonomic classification system, 9,
        9, 10, 11, 12, 21
Kiwi bird, 16, 16, 402
Klinefelter syndrome, 249, 351, 354
Knockout mice, 335, 384
Knuckle-walking, 471, 471
Kohlrabi, 393, 393

Krebs, Hans, 178
Krebs cycle. See Citric acid cycle

Labrador retriever, epistasis in, 254–255, 255
Labyrinthodonts/labyrinthodont fossils, 457
lac operon (lactose operon), 306–308
    mutant strains in study of, 307–308, 308
    negative control of, 309
    positive control of, 309, 311, 312
Lactate (lactic acid) fermentation, 187, 187–188
    efficiency of, 187, 187–188
Lactoferin, human, transgenic cows producing,
        341
Lactose, 52
    lac operon transcription induced by, 306–308,
        307, 309, 311
Lactose operon. See lac operon
Lactose permease, 306, 307
Lactose repressor, 306
Lady's slipper (Phragmipedium caricinum), 12
Lagging strand, 272, 273, 275
Lake Superior
    Gunflint Iron Formations, 452
    rock strata along, 455
Lake Tanganyika, 423
Lake Turkana, Kenya, 475
Lake Victoria cichlids, 437, 437, 441
Lamarck, Jean Baptiste de, 391–392
Land bridge between Siberia and Alaska, 463
Landmasses, Earth's, 403–404
Language, in early humans, 477, 478
Langurs, 469
Larva (larvae), 375
    Drosophila, 375, 377
Laurasia, 404
Law enforcement, use of DNA, 339
Laws of thermodynamics, 154, 156, 160
Le Moustier, France, 477
Lead, 167
Leader sequences, 288, 288
Leading strand, 272, 273
Leaf/leaves
    evolution of, 400, 401
Leakey, Mary, 466, 474
Leakey, Meave, 466, 474
Leakey family, 466
Learning, in chimpanzee populations, 18–19, 19
Leaves. See Leaf/leaves
Lecithin, 58
Leeuwenhoek, Anton, 76
Lemur (Eulemur mongoz), 467, 467, 468, 468
Leonardo da Vinci, 391
Leopard frog (Rana pipiens), 430, 430, 431
Leucine, 60, 62
Leucine zipper proteins, 316, 317
Leukemia, 353, 360
Leukoplasts, 97
Levan, Albert, 347
Lewis, Edward B., 375
Lewis, G. N., 32
Lewis structure, 32
Liaoning Province, China, 462
Life, 1–24
    characteristics of, 2–5
    energy of, 2, 14, 14–15
    evolution as unifying concept of, 9–13
    evolution/origin of, 447–465, 456
    information transfer, 6–8

levels of biological organization, 6, 7, 10
    scientific method in study of, 15–21
    See also Chemical basis of life; Evolution
Life cycles, and timing of meiosis, 229–230, 230
Life history traits, natural selection and, 408
Ligand, 124–125, 137
Ligand-gated channels, 138
Ligases, 163
    DNA, 270, 272, 273, 323
Light, 76, 192, 192, 193
Light compensation point, 207
Light energy, 198, 199
Light microscopes, 76–78, 77, 212, 215
Light-dependent reactions, in photosynthesis, 196,
        197, 197–201, 198–202, 204
Light-independent reactions. See Carbon fixation
        reactions
Lilies (Liliaceae), and carbon dioxide fixation, 206
Limb bones, 397, 400
Linkage maps, 248
Linkage/linked genes
    independent assortment and, 246–247
    linear order of, 246–247
Linnaean Society, 393
Linnaeus, Carolus, 9, 429
Linoleic acid, 56, 57
Lions (Panthera leo), 12
Lipase, 163
Lipid(s), 56–59, 69
    accumulation of, in Tay–Sachs disease, 358
    carotenoids, 57, 59, 96
    in cell membranes, 107, 107, 108, 108–109, 109
    as chemical mediators, 58–59
    energy yield of, 186
    phospholipids, 57, 58 (see also Phospholipids)
    saturated/unsaturated fatty acids, 56, 56–57
    steroids, 57–58, 59
    triacylglycerol, 56, 56
Lipid bilayer. See Phospholipid bilayer
Lipoproteins, low-density (LDLs)
    cholesterol transport and, 124–125, 126
    recycling of, 124–125, 126
Liposomes, glucose diffusion and, 119
Liquid(s), 116
Liquid crystals, phospholipid bilayer as, 109, 109–
        110
Lithosphere, 6
Liver, 264
    glycogen storage, 53, 90
    human, 90, 94
    lactate and, 188
    mitochondria in, 185
    in transgenic animals, 340
Liver cells, 337
Lizards, 459
Local regulator, 136
Loci, 412
Locomotion
    cell crawling, 100
    dorsoventral movement, 397, 399
    knuckle-walking, 471, 471
    in prokaryotic cells, 80
    See also Movement
Locus (loci), gene, 238, 239
    linked (see also Linkage/linked genes)
        independent assortment and, 246, 246
    multiple alleles for, 253–254, 254
Lords-and-ladies (Arum maculatum), 182

Lorises, 468
Lou Gehrig's disease (amyotrophic lateral sclerosis/ALS), 101
Low-density lipoprotein (LDL)
    cholesterol transport and, 124–125, *126*
    receptor protein, 296
    receptors, recycling of, 124–125, *126*
Lowland gorilla, *471*
Low-predation habitats, 408
Loxiodes bailleui (palila), *13*
Luciferase gene, *322*
Lucy (skeleton), 474
Lumen, in endoplasmic reticulum, 90, 91
Lung cancer, 386
Lupines, blue (*Lupinus hirsutus*), *191*
Lyases, *163*
Lye, *41*
Lyell, Charles, 393
Lyon, Mary, 251
Lysin, 432
Lysine, *61*, 62, 284
Lysosomal storage diseases, 93
Lysosomes, *83*, *89*, 91, 92–93, *93*
    in phagocytosis, 107, 124, *125*
    in Tay–Sachs disease, 358
Lysozyme, 163
Lystrosaurus, 405

M phase, 215, *215*. See also Cytokinesis; Mitosis
Macaques, 469
Mackenzie Mountains, 455
MacKinnon, Roderick, 115
MacLeod, Colin M., 262, *263*
Macroevolution, 391, 396, 439–444
    adaptive radiation and, 439, 440–442, *441, 442,* 444
    extinction and, 442–444, *443*
    microevolution and, 444
    modification of pre-existing structures, 439–440, *440*
    See also Evolution; Speciation
Macromolecules, *7*, 46, 50, 59, 63, 447
Macromolecules, separation, 330
Macrophages, 125
Mad cow disease, 338
Magnesium, 26, *26, 27,* 193, *195*
    as cofactor, 163
Magnification, of microscope, 76
Maize. See Corn
Malaria, 337
    resistance to, sickle cell anemia and, 357, 422, *423*
Malate, *180*
    in C$_4$ pathway, 205, *206*
    in CAM pathway, 206
    decarboxylation of, 206
Malate dehydrogenase, *180*
Male sex, determination of, 248–249, *249*
Males, X-linked genetic disorders expressed in, 250–252
Malignant transformation, 386
Malignant tumors, 386
Malthus, Thomas, 393
Maltose (malt sugar), 52, *53,* 207
Mammal(s)
    Age of, 461–463, *463*
    cloning, 371–372, *373*
    diversification of, 442, 461–463
    egg-laying (monotremes), 402

evolution of, 462, 467
    mouse model of development in, *383,* 383–385, *384*
    placental, 402, 467
    pouched (marsupials), 402
    in Triassic period, 459
Mammalia (class), *10,* 11
Mammary gland cell, *124*
Mangabeys, 469
Manganese, *26*
    as cofactor, 163
Manis crassicaudata (pangolin), 400, *402*
Mantle, Earth's, 403
MAP kinases, 144, 145
Map positions of genes, 307, 308
Map units, 247–248, *248*
Maples, 135, *220*
March of Dimes, 363
Marine iguanas (*Amblyrhynchus cristatus*), *428*
Marine organisms, 442, 444, 457
Marmosets, 469
Marrella splendens fossil, *457*
Marriage, consanguineous, 363
Marsupials, 402
Martin–Bell syndrome, 355
Mass, *28*
    atomic, 28, *28*
    molecular, 31
Mass extinction, 439, 442, *443,* 444
    at end of Cretaceous period, 442, *443,* 444, 460
    at end of Paleozoic era, 458
    at end of Pleistocene epoch, 463
    See also Extinction
Mastodons, 463
Maternal age, Down syndrome and, 353–354
Maternal effect genes, 375–378, *377, 378*
Maternal homologue, 224, 225
Maternal phenylketonuria (maternal PKU), 357
Maternal serum α-fetoprotein (MSAFP), 362
Mating
    consanguineous, 363
    controlled, 346
    nonrandom, genotype frequency and, *416,* 416–417
    random, genetic equilibrium and, *416,* 416–417
    reproductive isolating mechanisms and, *430,* 430–432, *431*
    See also Reproduction
Matrix
    extracellular (ECM), 102, *102*
    mitochondrial, 94–95, *95*
Matter, 26, 153
    states of, 116
Matthaei, Heinrich, 284
Mayr, Ernst, 395, 429
McCarty, Maclyn, 262, *263*
McClintock, Barbara, 298, 300
M-Cdk, in cell cycle regulation, 222–223
Mechanical energy, 153, 154
Mechanical isolation, 430, 431, *431, 432*
Medial region, Golgi stack, 91
Medical researchers, 335, 337, 391
Medicine
    applications in protein-coding genes, 334–335
    DNA technology affecting, 338–339
    ethics in biological research, 21
    radioisotopes, use of, 29
    See also Drug(s)

Mediterranean rabbits, 434
Meiosis, 212, 223–230, *226–227*
    crossing-over during, 224–225, *247,* 247–248, *249*
    haploid cell production, 224, *227,* 228, 229, *229,* 230
    independent assortment and, *243,* 243–244
    Mendel's findings and, 234, 236, 242
    mitosis compared with, 225, 228, *229*
    species variation in life cycle, 228–230, *230*
Meiosis I, 224, 225, *226,* 228
Meiosis II, 224, 225, *226,* 228
Meiotic division, 351, *352*
Meiotic nondisjunction, 351, *352,* 353
Meiotic spindle, 353
Meiotic tetrads, *228*
Membrane(s), 81–84, 106–126
    as barriers, 115
    cell junctions and, *127,* 127–130, *128, 129, 130*
    cell signaling and, 114, *114*
    fluidity of, 109–110, *110*
    mitochondrial, 94–95, *95*
    plasma, 3, 74, 75, 80, *82–84,* 107 (*see also* Plasma membrane (cell))
    polarization of, 107
    proteins, 182
    selectively permeable, 116–117, *117*
    structure of, 107–114, *108, 111, 112, 113*
    transport across, *114,* 114–115 (*see also* Transport)
        active transport, 120–123, *121, 122, 123*
        fluidity and, *109,* 109–110, *110*
        passive transport, *114,* 115–120, *116, 117, 118, 119, 120*
    See also specific type
Membrane fusion, vesicle formation and, 110, 113, *113,* 125, *125*
Membrane potential, 122
Membrane proteins, *106,* 106–115, *111, 112, 113, 114*
    in active transport, 120–123, *121, 122, 123*
    asymmetric orientation of, 111–113, *112, 113*
    enzymatic activities of, 114, *114*
    formation/transport of, 114, *114,* 115
        membrane fluidity and, *109,* 109–110, *110*
    functions of, 113–114, *114*
    in passive transport, *118,* 118–120, *119, 120*
Mendel, Gregor, 212, 234, *234,* 235–236, 240, 242, 394
    garden peas experiments, 419
    principles of inheritance, 413
Mendelian genetics/inheritance, 234, 235–242, *237, 242*
    chromosomes and, 246–252, *247*
    evolution and, 394–395
    extensions of, 252–256
    probability and, 243–244
    problem-solving exercises in, *245*
    recognition of work, 242
Mendel's principle of independent assortment, *237,* 241–242, *243,* 246–247
    linked genes as exception to, *246,* 246–247
Mendel's principle of segregation, 236–238, *237, 238*
Mental illness, 350, 353
Mental retardation, 94
Mercury, 167
Meselson, Matthew, *263,* 268, *269*

Mouse (*continued*)
    DNA sequencing, use in, 337
    gene expression in embryo, *304*
    gene therapy developed in, *360*
    genome, 383
        compared to human genome, 350
    heterozygous, 384–385
    homeobox genes in, 384
    Hox genes of, 380, *380*
    human genetic diseases studied in, 350, 355, 357, 358, 359
    inbreeding studied in, 416, *416*
    knockout, 335, 384
    transgenic, 340, *340*, 384, *384*
    white-footed (*Peromyscus leucopus*), 416, *416*
    See also Mouse model
Mouse model, 350
    in cystic fibrosis, 350
    in fragile X syndrome, 355
    in Huntington's disease, 359
    in sickle cell anemia, 357
    in Tay–Sachs disease, 358
Mouse-ear cress. *See Arabidopsis thaliana*
Mouse–human cell fusion, 109–110, *110*
Mousterian tools, 476, *477*
Movement
    amoeboid, 4
    biological, 4, *4*
    See also Locomotion
Mucus, 55, *92*
Mules, hybrid sterility in, 432, *432*
Mullis, Kary, 328
Multicellular organisms, 2–3, *3*, 6, 11
Multienzyme complex, 165
Multinucleated cells, 215
Multiple alleles, 253–254, *254*
Multiregional hypothesis, 477
Mus musculus
    developmental genetics studied in, 375, *376*, *380*, 383, 383–385, *384*
    See also Mouse
Muscle fibers (cells)
    glycogen storage, 53
    lactate produced by, 188
    mitochondria in, 185
Muscle(s)/muscle tissue
    contraction of, 58, 115, 138
    skeletal, 215
Mushroom(s), fly agaric (*Amanita muscaria*), *12*
Mustard plant, 146, 385
Mutagen(s), 300, 335
Mutagenesis screening, 335
Mutant(s), 368
    developmental, 368, 375
Mutation(s), 5, 13, 260, 298–300, *299*
    allele frequency, 417
    base substitution, 298
    in *Caenorhabditis elegans*, 382
    cancer-related, *386*, 386–387
    caused by mutagenesis, 335
    causes of, 274, 300
    developmental, *Drosophila* body plan affected by, 375–380, *378, 379*
    disorders associated with, 338–339
    DNA mutations, 298–300, 417
    in evolution, 395, 404, 406, 407–408, 417–418
    in first cells, 452
    frameshift, 298, *299*

    genetic diseases caused by, 338–339, 356–359, *357, 358*
    genetic equilibrium and, 415
    genetic isolation to study *lac* operon, 307–308, *308*
    genetic variation and, 395, 417
    mitochondrial DNA mutations, 96, 476–477
    one-gene, one-enzyme hypothesis and, 280–281, *281*
    in peroxisome-membrane synthesis, 94
    perpetuation of, 268, 270, *270*
    in pigs, with growth factor, 305, *305*
    in proteins, 85
    recessive, 356
    single-gene, 356–359
    somatic (body) cells, 417
    transposons and, 298–300
    variation within populations, 417
Myocardial infarction, 1–2
Myoglobin, 66, 305
Myosin, 62, *62*, 100, 146
Myosin filaments, *220*
Myrmecophaga tridactyla (anteater), 400, *402*
Myxotricha paradoxa, 454

N-acetyl glucosamine, in chitin, 54, *55*
NAD^+ (nicotinamide adenine dinucleotide), 68, 159–160, *160*
    in citric acid cycle, *173*, 174n2, 178, *179, 180*
    in fermentation, 186, *187*
    in glycololysis, 174–175, *175, 177, 178*
NADH, 68, 159–160, *160*, 163
    in anaerobic respiration, 186–187, *187*
    in citric acid cycle, *173*, 174, 178–179, *179, 180*, 183, *185*
    in electron transport chain, 179, *181, 181*
    in fermentation, 186–187, *187*
    in glycolysis, *173*, 174–175, *175, 177, 178*, 183, *185*
    transfer of electrons of, 179, *181*, 181–183, *184, 185*, 186–187, *187*
NADH–ubiquinone oxidoreductase, in electron transport chain, 179, *181*
$NADP^+$ (nicotinamide adenine dinucleotide phosphate), 160, *160*, 197, *197*, 198, 199, *201*
NADPH, 160, *160*, 163
    in C_4 pathway, 205, *206*
    light-dependent reaction product, 197, *197*, 198, *199*
Nails, 64
Nanometer (nm), 75, *75*
National Bioethics Advisory Commission, 373
National Center for Biotechnology Information, 333
National Center for Genetic Resources, 364
National Fragile X Foundation, 355
National Human Genome Research Institute (NHGRI), 349, 364
National Institute of Environmental Health Sciences, 141
Natural resources, cultural evolution affecting, 479
Natural selection, 11–13, *13*, 393–394, 412, 418–421
    adaptation and, 418–421, *419*
    allopatric speciation and, 433
    experimental evidence of, *408*, 408–409
    extinction and, 444

    genetic equilibrium and, 415
    heterozygote advantage, 422
    at molecular level, 451
    predation and, 408, *408*
    vs. chance in evolution, 395–396
Nature–nurture question, 256
Neander Valley, Germany, 475
Neandertals, *473*, 475–477
Necrosis, 95
Negative assortative mating, 417
Negative control, bacteria regulators, 309
Nelumbo nucifera (sacred lotus), 182
Nematodes, developmental genetics studied in, 375, *376*, 380–383, *381*
Nene (*Branta sandvicensis*/Hawaiian goose), *433*
Neogene period, *456, 461, 463*
Neon, 27, 29
Neoplasm (tumor), 386. *See also* Cancer
Nerve cells, 75, 76
    gap junctions and, 129
Nerve gases, effect on enzymes, 167
Nervous systems, to transmit information, 8
Net movement of water, 116–119, *117, 118, 119*
Nettles, hemp, 436
Neurologic disease, and membrane proteins, 115
Neurological disorders, 94
Neurons, 135, 137, *137*
Neurospora, 248, 280–281, *281*
Neurotransmitters, 8, 129
    in cell signaling, 136, 137, 140
Neutral solution, 40, *41*
Neutral variation, 417, 424
Neutrons, 26, 28, *28*
Neutrophils (WBCs/white blood cells), 146
    phagocytosis, *125*
New World monkeys, 468, 469, *470*
Newborns, screening for PKU, 356, 363
Newfoundland, fossil discoveries, 455
Newts, *211*
Ngorongoro Crater, Tanzania, *5*
Nickel sulfides, 450
Nicolson, Garth, 108–109, 112
Nicotiana tabacum (tobacco), *94*
Nicotinamide adenine dinucleotide (NAD^+). *See* NAD^+
Nicotinamide adenine dinucleotide phosphate ($NADP^+$). *See* $NADP^+$
9 + 2 arrangement, 99, *100*
9 × 3 structures, 98, 99, *99*
Nirenberg, Marshall, 284
Nitrate, 42, 186, *187*
Nitric oxide (NO), cell signaling, 136–137, 140, 144–145
Nitrocellulose membrane, in DNA hybridization, 331
Nitrogen, 26, *26, 27*, 28
    covalent bonding, 32, 33, *33*
    heavy isotope, 267, 268
Nitrogen cycle, 186
Nitrogen-14, in fossil dating, 400
Noble gases, 29
Nomarski (differential-interference-contrast) microscopy, 77, *77*, 381
Nomenclature, binomial, 9
Noncoding sequences, 294
Noncompetitive inhibition, of enzymes, 167, *167*
Noncyclic electron transport, in photosynthesis, 198–200, *199, 200*

Nondisjunction, 351, *352,* 353
Nonelectrolytes, 42
Nonhomologous chromosomes, 240–242
Nonpolar amino acids, *60,* 62
Nonpolar covalent bonds, 33
Nonpolar molecules, and van der Waals interactions, 36
Nonrandom mating, microevolution and, *416,* 416–417
Nonsense mutations, 298, *299*
Nonsulfur purple bacteria, 207
Nonvascular plants, 11
Norm of reaction, 256
Normal distribution curve, 255, *256,* 418, *418,* 419
Northern blot, 331
Nuclear area/nucleoid, of prokaryotic cell, 80, *81*
Nuclear envelope, *82, 83,* 85, *88,* 107, 218, 225
Nuclear equivalence, 369–374
Nuclear lamina, 85
Nuclear pore proteins, *88*
Nuclear pores, *82–83,* 85, *88*
Nuclear totipotency, 370–371, *371, 372, 373*
Nuclear transplantation experiments, 371, *372*
Nuclei. *See* Nucleus (cell)
Nucleic acids, 67–68, *68, 69*
    component of, *26*
    See also DNA; RNA
Nucleoid/nuclear area, of prokaryotic cell, 80, *81*
Nucleolar organizer, 85
Nucleolus/nucleoli, *81, 82–84,* 85, *85, 88,* 218
Nucleoplasm, *81, 88*
Nucleosomes, *213,* 213–214, *214*
Nucleotide excision repair, 274, *275*
Nucleotide sequencing, 331–333, *333, 334*
    evolution and, 405–408, *406, 407*
Nucleotide triplet repeats, in Huntington's disease, 359
Nucleotides, 8, *8,* 67–68, *70*
    ATP, 158, *158*
    origin of cells, 451
    RNA structure, 282, *282*
    structure of DNA, 263–265, *264*
Nucleus (atomic), 26
Nucleus (cell), 3, *82–85, 84–88, 88, 89*
    as control center of cell, *17–18, 18,* 84, *86–87*
    formation of, in telophase, 225
    organization in cell, *82–85, 84–88, 88, 89*
    removing from amoeba, *17–18, 18*
    transplantation of, 371, *372*
Nurse, Paul, 222
Nüsslein-Volhard, Christiane, 375
Nutrients
    transport of, 115
    See also specific type
Nylon membrane, in DNA hybridization, 331

Oak, white
    American (*Quercus alba*), 9, *9*
    European (*Quercus robur*), 9
Obligate anaerobes, 453
Observations, in scientific method, 16
Ocean sunfish (*Mola mola*), *440*
Offspring
    differential reproductive success, 394
    Punnett square prediction, 239
    of sexual reproduction, 223
Oil, as hydrophobic substance, 38
Oilseed rape plant, transgenic, 342–343

Okazaki, Reiji, 272
Okazaki fragments, *270,* 272, *273*
Old World monkeys, *468,* 469, *470*
Oldowan pebble choppers, 475
Olduvai Gorge, Tanzania, *466,* 474
Oleic acid, *56,* 57
Oligocene epoch, *456,* 462–463, *463*
    Aegyptopithecus from, 469, *470*
On the Origin of Species by Natural Selection (Darwin), 11, 393, 429, 466
Oncogene(s), 353, *386,* 386–387
One-gene, one-enzyme hypothesis, 280–281, *281*
Oocytes, 353–354
Oogenesis, 229
Oparin, A. I., 449
Open system, 154, *154*
Operator
    for *lac* operon, 306, *307*
    for *trp* operon, 309, *310*
Operons, 306–309
    inducible, 308
    repressible, 308–309, *310*
    trp, 309, *310*
    See also lac operon
Opposable thumb, primate evolution and, 467, *467*
Opuntia (prickly pear cactus), *206*
Orangutans (*Pongo*), *468,* 470, 471, *471*
Orbital hybridization, 33, *33*
Orbitals, atomic, 29, *30*
Orchids (*Orchidaceae*), carbon dioxide fixation by, 206
Orders, in taxonomic classification system, 9, *9, 10*
Ordovician period, 455, *456,* 457, *458*
Organ systems, 6, *7*
Organelles, 3, *7,* 107, 130
    nonmitotic division of, 221
    organization in cell, 88–97, *89–96*
    study of, 77, 78, 79
Organic chemists, 77
Organic compounds, 26, 45–72, *46, 69*
    carbohydrates, 50–55, *69*
    carbon atoms and molecules, *46,* 46–50, *49*
    definition of, 45
    diversity of, 45–46
    lipids, 56–59, *69*
    nucleic acids, 67–68, *69*
    origination of, on primitive Earth, *449,* 449–450
    proteins, 59–67, *69*
    See also specific type
Organism(s), 2–5
    common ancestry of, 406
    energy use and, 14, *14*
    functions of elements in, 26
    growth and development of, 3
    levels of organization and, 6, *7*
    metabolic regulation and, 3
    model, 374–375, *376,* 380 (*see also* Model organisms)
    multicellular, 2, *3,* 6
    in populations, 5, *5*
    reproduction by, 5, *5*
    response to stimuli, 4, *4*
    unicellular, 2, *3*
Organismic respiration, 172. *See also* Respiration
Organs, 6, *7. See also specific type*
Origins of replication, 270, 272, *272, 274,* 324, 326

Ornithischians, 459–460, *461*
Ornithorhynchus anatinus (duck-billed platypus), 17, *17*
Orrorin, 474
Orycteropus afer (aardvark), 400, *402*
Osmosis, 116–118, *117, 118*
Osmotic pressure, 116–118, *117, 118*
Osmotic stress, 146, *147*
Ostracoderms, 457, *458*
Outer mitochondrial membrane, 95, *95*
Out-of-Africa hypothesis, 477–478
Ova. *See* Ovum (egg)
Ovalbumin, *62,* 294
Oven cleaner, *41*
Overproduction, natural selection and, 394
Ovis canadensis (bighorn sheep), directional selection and, 420–421, *421*
Ovum (egg), 5
    formation of, 229
Owens pupfish (*Cyprinodon radiosus*), 433
Oxaloacetate, 174, 186
    in C$_4$ pathway, 204–205
    in CAM pathway, 206
    in citric acid cycle, 178, *179, 180*
Oxidation, 36, 159, 172
Oxidation–reduction (redox) reactions, 36, 172, *172*
    energy transfer in, 159–160, *160,* 172, *172*
Oxidative decarboxylation, 175, *178*
Oxidative phosphorylation, 179, 181, 183, 185, *187*
Oxidized state, cytochromes in, 160
Oxidizing agent, 36
Oxidoreductases, *163*
Oxygen, 26, *26, 27*
    in atmosphere of early Earth, 448, 453
    combining with iron (rusting), *26,* 36
    covalent bonding, 32, *32,* 33, *33*
    electron transport chain affected by, 179, 181, *181*
    photosynthesis in production of, *199, 199*
    reactions with hydrogen, 161, *161*
Oxygen enrichment hypothesis, 455
Oxygen radicals (free radicals), 95, 384–385
Ozone (O$_3$), 453
    stratospheric, 453, *453*

P (parental) generation, 235, *237,* 435
P site, 289, *290*
p27, in cell cycle regulation, 222
p38 protein, 1
P680 (photosystem II), 198, *199, 199,* 200, 207
P700 (photosystem I), 198, *199,* 200
Paedomorphosis, 440, *440*
Pair–rule genes, *378,* 378–379, *379*
Paleoanthropologists, 469, 470, 473–474, 475, 476, 477
Paleoanthropology, 466
Paleocene epoch, *456,* 461
Paleogene period, *404, 456,* 461–463
Paleontologists, 396, 397, 462, 467
Paleozoic era, 455, *456, 457,* 457–458, *458*
Palila (*Loxiodes bailleui*), 13
Palindromic sequences, 323, *324*
Palisade mesophyll, *194*
Palmitic acid, *56,* 57
Palms, 461
Pan. See Chimpanzees

Primary lysosomes, 92, *93*
Primary structure of protein, 63, *64*
Primase, DNA, *270, 271,* 272
Primates, 466–481
  classification of, *468,* 468–471, *470, 471*
  cultural change, 478–479
  definition of, 467
  hominid evolution, 471–478, *472, 473, 475–477*
  primate adaptations, 467, *467*
Primates (order), 9, *10*
Primrose, 435–436, *436*
Primula, 435–436, *436*
Principal energy level, 29, *30*
Principle of independent assortment, *237,* 241–
    242, *243,* 246–247
  linked genes as exception to, *246,* 246–247
Principle of segregation, 236–238, *237, 238,* 239,
    240
Principles of Geology (Lyell), 393
Principles of inheritance (Mendel), 413
Privacy, and ethics, 21
Privacy issues, in genetic testing, 338
Probability
  genetic counseling and, 363
  rules of, 243–244, *244, 245*
  statistical, 20, *20*
Probe(s), genetic, 327, *328*
Procariotique. *See* Prokaryotes/prokaryotic cells
Process of science. *See* Science, process of
Proconsul, 470
Producers (autotrophs), *14,* 14–15, 207
  early cells as, 452–453
Product rule, 244
Products, in chemical reactions, 31, 36, 156–157
Programmed cell death (apoptosis), 95–96, 102,
    382–383
  telomere shortening and, 275–276
Prokaryotes/prokaryotic cells
  aerobic respiration, 174, 175
  bidirectional DNA replication in, 270, *274*
  defined, 3, 11
  division of, by binary fission, 221, *221*
  DNA content of, 212
  electron transport chain, 179
  in evolutionary history, 148, 453–455, *454*
  internal organization of, 80–81, *81*
  size of, *75,* 80
  transcription and translation coupling, 293, *294*
  See also Bacteria
Proline, *60,* 62, 284
Prometaphase, in mitosis, 218
Promoter(s), 286, *287*
  eukaryotic, 315, *316*
  for *lac* operon, 306, *307*
  for *trp* operon, 309, *310*
Promoter elements, 309, *316*
Promoter elements, upstream (UPEs), 306, 315, *316*
Propane, *46*
Prophase
  in meiosis, 225, *226, 229*
  in mitosis, *216,* 216–218, *229*
Prophase I, 224–225
Prophase II, 225, *229*
Proplastids, 96–97
Prosimii (suborder), *468, 468*
Prostaglandins, 58, 136
Prostate cancer, 275
Proteases, 319

Proteasomes, 91, *318,* 319
Protective proteins, *62*
Protein(s), 8, 50, 59–67, *62,* 69
  amino acids, 46, 49, 50, *60–61,* 62–63, *63*
  carrier, 115, 118–123, *120, 121, 122, 123*
  cell surface, 8
  channel, 115
  conformation of, 66–67, 163, 164
  energy yield of, 185–186, *186*
  enzymes, 161, 163
  gene–protein relationship, 279, *280,* 280–281,
      281
  genetically engineered, 340
  inhibitors, 1
  integral membrane, 110–111, *111*
  molecular conformation (3-D shape), 63, 66–
      67, 162
  motor, 218, 219
  origin of cells, 450, 454
  peripheral membrane, 110–111, *111*
  in plasma membranes, *108,* 108–114, *111, 112,
      113, 114*
  porins/aquaporins, 115
  as receptors, 8
  scaffolding, 213, 214
  sequence of, 66–67, 405–408, *406, 407*
  structure/organization of, 63–66, *64, 65, 66*
  synthesis of (*see also* Transcription; Translation)
      Golgi complex in, 91–92, *92*
      rough endoplasmic reticulum in, 90–91
  thought to be genetic material, 261
  transport, 91–92, *92,* 115, 118, 120–122, *121*
  See also specific type
Protein degradation, 319
Protein domains, 296
Protein expression patterns, 338
Protein growth factors, 223
Protein kinase(s)
  activation of by cAMP, 166, *166*
  in cell cycle regulation, 222, *222*
  growth factor receptor complex as, 386–387
Protein kinase A, 142
Protein kinase C, 143
Protein-coding genes, 334–335
Proteolytic processing, 318
Proteomics, 338
Proterozoic eon, 455, *456*
Protista (kingdom), 11, *12*
Protobionts, 450
Protofilaments, *101*
Proton(s), 26–28, *28*
  in covalent bonding, 31–32, *32*
Proton acceptor, 40
Proton concentration gradients, 122, *122*
Proton donor, 40
Proton gradient, 182–183, *183, 200,* 200–202
Proton pumps, in ATP synthesis, 122, *122*
Proto-oncogenes, *386,* 386–387
Protoplasm, 81
Protoplasmic face (P-face), *112, 129*
Protozoa, 93
  Myxotricha paradoxa, 454
Proximal control elements, 315
Pseudopodia, 100
Pterapsis ostracoderm, *458*
Pterosaurs, 459, *459*
Ptilonorhynchus violacens (satin bowerbird), 430,
    431

Puffer fish, 337
Punctuated equilibrium, *438,* 438–439
Pundamilia (cichlids), *437*
Punnett, Reginald, 239
Punnett square, 239, 241, 244, *245*
Pupa, 375, *377*
Pupfish (*Cyprinodon*), 433, *433*
Purines, 67, *68,* 263, 265
Purple bacteria, 452, 454
  nonsulfur, 207
Pygmy elephants (*Stegodon*), 476
Pyrimidines, 67, *68,* 263, 265
Pyruvate
  in C₄ pathway, 205, *206*
  in cellular respiration, *173,* 174–175, *175, 177–
      180,* 178, 183, *184–186,* 186
  in fermentation, *187, 187*
Pyruvate dehydrogenase, 165, 178
Pyruvate kinase, *177*
Pyruvic acid, 174n1
Python sebae, vestigial structures in, 401, *403*
Pythons, 401, *403,* 404

Quadrupedal posture, 469, 471, *471*
Quantitative experimental methods, 242
Quaternary structure of protein, 66, *66,* 163, 164
Quercus (oak)
  American white (*Q. alba*), 9, *9*
  European white (*Q. robur*), 9
Quetzalcoatlus, 459
Quorum sensing, 134–135

R groups, 48, 62, 63, *63, 64, 65, 65*
Rabbit(s), 253, *254,* 256, 434
Radiant energy, 153
Radiation, as mutagen, 300
Radio waves, 192, *192*
Radioactive decay, 28, *28*
  fossil age and, 399–400, *400*
Radioactive thymidine, 216
Radioisotopes, 28–29, 399–400
  decay product of, 399
  rate of decay of, 399–400, *400*
Radius, 400, *401*
Rain forests, tropical
  fossils of organisms in, 397
  organisms in, 25, 468, *469*
Rana pipiens (leopard frog), 430, *430,* 431
Rana sylvatica (wood frog), *13,* 430, *430,* 431
Random events
  in evolution, 395–396
  and genetic drift, 417–418
Random mating, genetic equilibrium and, *414,*
    414–415
Range, reproductive isolating mechanisms and,
    430, *431*
Ras proteins, in signal transduction, 144
Rate of decay, in fossil dating, 399–400, *400*
Rats
  DNA sequencing, use in, 337
  giant (*Papagomys*), 476
  growth hormone, 340, *340*
Reactants, in chemical reaction, 31, 36, 156–157
Reaction center, of photosystem, 198, *198*
Reaction coupling, 157, 164–165
Reading frame, 284, 298, 300
Reasoning, deductive/inductive, in scientific
    method, 16

Species, 391, 429
   common ancestry of, 406
   distribution of, evolution and, 402–404, *404*, *405*
   divergence of, 407–408
   diversification of, 440–442
   extinction of, 442, *443*, 444 (*see also* Extinction)
   habitat isolation, *430*, 430–432, *431*
   interbreeding between, *430*, 430–432, *431*
   parasitizing, 437
   reproductive isolation, *430*, 430–432, *431*, 436
   stasis, 438–439
   in taxonomic classification system, 9, *9*, *10*
   See also Biological diversity; Speciation
Specific epithet, in taxonomic classification system, 9
Specific heat, 38
Spectrophotometer, 194
Sperm, 5
   sea urchin, *264*
Sperm cells, 76, 99, 223
Spermatogenesis, 229
Spider monkeys, 469
Spiders, proteins in silk, 65
Spina bifida, 362
Spindle
   in meiosis, 225, 353
   in mitosis, 99, *216*, *217*, *217*
Spines (leaf modification), 401
Spiny anteater, 17
Spirochete bacteria, 454
Spirogyra, 194–195, *196*
Spliceosome, 295, *296*
Splicing, RNA, 294–295, *296*
Split genes, 295, *296*
Splitleaf philodendron (*Philodendron selloum*), *182*
Spongy mesophyll, *194*
Spontaneous abortion, 352–353, 431
Spontaneous reactions, 155–156
Spores, *227*, 228
Sporophytes/sporophyte generation, 229–230, *230*
Squirrel(s), 434, *434*
Squirrel monkeys, 469
Stabilizing selection, *419*, 419–420, 439
Stahl, Franklin, *263*, 268, *269*
Stains, fluorescent, 77–78
Stamens, plant development studies, 385, *385*
Standard bell curve (normal distribution curve), 255–256, *256*, 419, *419*
Staphylococcus (genus), 16
Starch, 50, 52, 53, *54*
   animal, 53 (*see also* Glycogen)
Start codon, in transcription, 288, *288*
Stasis, rate of evolutionary change/speciation and, *438*, 438–439
Statistical probability, 20, *20*
Stebbins, G. Ledyard, 395
Stegodon (pygmy elephants), 476
Stegosaurus, 461
Steinhoff, Gustav, 1–2
Stem cell research, 21
Stem cells, 372–373, *374*
   adult, 372–373
   cancer cells and, 387
   embryonic, 372–373, *374*
   human therapeutic cloning and, 374
   pluripotent, 373

transgenic animal production and, 335
transplantation of, 373
Stenaptinus insignis (bombardier beetle), 161, *161*
Steranes, as early eukaryote markers, 453–454
Stereoscopic vision, primate evolution and, 467
Sterility, hybrid, 431–432, *432*
Steroid(s), 57–59, *59*, 69
   in evolution of cells, 453–454
   membrane fluidity and, 110
Steroid hormones, 58, 90
   brassinolides/brassinosteroids, 140
   in cell signaling, 146
Steward, F. C., 370, *371*
Stickleback fish, 396
Sticky ends, 323, *324*, *325*
Stigma, 235, *235*
Stimulatory G proteins, 142
Stimuli, 4, *4*
Stoma (stomata)
   and photosynthesis, 193, *194*, 204
   plant feature, 11
Stomach acid, *41*
Stomach ulcers, 4
Stonecrop plant family (Crassulaceae), 205–206
Stop codon
   mutations and, 298, *299*
   in transcription, 288, *288*
   in translation, 291, *293*
Storage proteins, *62*
Stratosphere, ozone in, 453, *453*
Stroma, chloroplast, *82*, 96, *96*, 193, *194*, *197*, 200, 202
Stromatolites, 452, *452*, 453, 455
Structural formula(s), 31
   of amino acids, *60–61*
   of cAMP, *70*
   of disaccharides, *53*
   of functional groups, *49*
   of glucose, *52*
   of monosaccharides, *51*
   of nucleic acids, *68*
   of organic molecules, *46*
   of peptide bonds, *63*
   of phospholipid, *58*
   of polysaccharides, *54*, *55*
   of proteins, *64–66*
   of steroids, *59*
   of triacylglycerol, *56*
Structural gene, 306
Structural genomics, 333
Structural isomers, *47*, 47–48, 51–52
Structural MAPs, 97
Structural proteins, *62*
Structure, in hierarchy of biological organization, 6
Subatomic particles, 26
Subduction, 403
Subphyla (subphylum), in taxonomic classification system, *9*
Subspecies, 434
Substrate
   concentration of, reaction rate and, 165, *165*
   enzyme–substrate complex, *162*, 162–163, *165*, 165–166, *166*
Substrate-level phosphorylation, 175, *175–177*, 178–179, *180*, 183, 186, *187*
Succinate, *180*

Succinate dehydrogenase, *180*
Succinate–ubiquinone reductase, in electron transport chain, 179, *181*
Succinyl CoA synthetase, *180*
Succinyl coenzyme A, *180*
Sucrase, 162–163
Sucrose, 158, 162
   common table sugar, 52, *53*
   hydrophilic substance, 38
   molecular mass, 31
   synthesis of, in Calvin cycle, 203–204, 207
Suffixes (biological terms), A-7
Sugar(s), simple (monosaccharides), 50, *51*, 51–52, *52*
Sulfa drugs, bacterial enzyme inhibition by, 167, *167*
Sulfate, in anaerobic respiration, 186, *187*
Sulfhydryl group, 49, *49*, 66
Sulfur, *26*, *27*, 263
   and bacteria, 452
   covalent bonding, *33*
Sulston, John, 383
Sum rule, 244
Sumner, James, 280
Sunfish, ocean (*Mola mola*), *440*
Sunflowers, 432
Supercoiling, DNA, 270
Supernatant (centrifuge), in cell fractionation, 79
Supernatural, 16
"Superweeds," 342
Supraorbital ridges, 472, *473*, 475, *475*, 476
Surface area–to-volume ratio, 75, *76*
Surface tension, of water, 37, *38*
Sutherland, Earl, 141
Sutton, Walter, 212, 242
Sutton–Boveri theory, 242
Swamp forests, 457
Sympatric speciation, 434–437, *435*, *436*, 439
Symplocarpus foetidus (skunk cabbage), 182, *182*
Symporter carrier proteins, 122
Synapsis, 137, 224, 432, 435
Synaptonemal complex, 224, *228*
Syncerus caffer (African buffalo), 3
Syncytial blastoderm stage, in *Drosophila* development, 375
Syncytium, 375
Syndrome(s), definition of, 353
Synthesis of theories, 395
Synthesis (S) phase, of cell cycle, 215, *215*
Synthetases, tRNA, 289
Synthetic theory of evolution (modern synthesis), 394–395, 444
Systematics, 9–11
Systems biologists, 21
Systems biology, 21

Tail, prehensile, 469
Tandemly repeated gene sequences, 314–315
Taq polymerase, for PCR, 328–330
Tardigrades (water bears), 37
Target cells, 135
Tarsiers (*Tarsius*), 467, *467*, 468, *468*, 469
Tarsiiformes (suborder), 468, *468*
Tarsius bancanus, 469
Tarsius spectrum, 467
TATA box, 315, *316*
Tatum, Edward, 280–281, *281*
Taxon (taxa), 9